Douglas A. Granger • Marcus K. Taylor
Editors

# Salivary Bioscience

## Foundations of Interdisciplinary Saliva Research and Applications



Springer

*Editors*
Douglas A. Granger
Institute for Interdisciplinary Salivary
Bioscience Research
University of California at Irvine
Irvine, CA, USA

Johns Hopkins University
Baltimore, MD, USA

Marcus K. Taylor
Institute for Interdisciplinary Salivary
Bioscience Research
University of California at Irvine
Irvine, CA, USA

ISBN 978-3-030-35783-2 ISBN 978-3-030-35784-9 (eBook)
https://doi.org/10.1007/978-3-030-35784-9



This Springer imprint is published by the registered company Springer Nature Switzerland AG.
The registered company address is: Gewerbestrasse 11, 6330 Cham, Switzerland

# Preface

Thousands of investigators worldwide are actively engaged in basic, applied, and clinical research involving salivary bioscience. Our literature search reveals that across the past two decades the number of empirical papers published annually has increased substantially; in 2019 more than 1500 empirical salivary bioscience papers were published. Investigators engaged in salivary bioscience span many academic disciplines including medicine, public health, psychology, sociology, education, neuroscience, biological science, animal behavior and welfare, infectious disease epidemiology, social neuroscience of human–animal interaction, drugs/drug abuse, social networks, nursing, psychoneuroendocrinology, anthropology, cognitive science, bioengineering, dentistry, oncology, oral health, and pediatrics.

To date, the foundational information in salivary bioscience (i.e., 25 years of literature) is not easy or efficient to find. It is scattered across many different journals, over at least two decades, and some of the early work has been subsequently shown to be in error. This makes it challenging for new investigators interested in the topic to find the "right" information on their own. Unfortunately, and to the best of our knowledge, there is no definitive state-of-the-art guide to interdisciplinary salivary bioscience research. Over the years there have been edited volumes from the proceedings of highly specific conferences. Understandably, the nature of these presentations is highly technical and narrow in scope, and the content chapters in those texts are written for an audience of highly trained experts. By contrast, this edited volume is written by leaders in multiple fields and fulfills a demand for a *broad understanding of salivary bioscience* across a range of disciplines.

Douglas A. Granger

# Acknowledgements

The history of the emergence of interdisciplinary salivary bioscience has been influenced by many mentors, advisors, early adopters and visionaries over many years. Here we call attention to some of these key individuals for the significant roles they played in the development of the foundation of knowledge that scripted the "big picture"—Daniel Malamud, Lawrence Tabak, Harold Slavkin, John R. Weisz, James T. McCracken, John L. Fahey, Barbara Henker, Herbert Weiner, Dirk Hellhammer, Ben Weigand, Margaret Kemeny, Lynn Kozlowski, Elizabeth Susman, Alan Booth, James Dabbs, Jr., Peter Ellison, Lynn Vernon-Feagans, Ann Crouter, Megan Gunnar, Dante Cicchetti, Dan Leri, Clancy Blair, Martha N. Hill, Gayle Page, Deborah Gross, Robert Blum, Janet Dipietro, Tina Chang, Keith Crnic, Cary Savage, Karen Rook, Dele Ogunseitan, and Nancy Guerra. We also call attention to some of the many technical and operational experts who have made significant contributions—Najib Aziz, Eve Schwartz, Mary Curran, Skip Nelson, Laurie O'Brien, Tracy Hand, Jon Peterson, Rebecca Zavacky, Jessica Acevedo, Lillian Buitenhuys, Kelly Henning, Greg Reinhard, John Stebbins, Kaitlin Smith, Hillary Piccerillo, and Anthony Tette.

# Contents

# Editors and Section Editors

## About the Editors

**Douglas A. Granger** Ph.D., is engaged in multi-institution research focused on the discovery, measurement, and application of analytes (e.g., enzymes, hormones, antibodies, chemicals, elements, and cytokines) in saliva. He is a Chancellor's Professor and Director of the Institute for Interdisciplinary salivary bioscience research (IISBR) at the University of California at Irvine and holds adjunct faculty positions at Johns Hopkins University. His studies have been instrumental in the conceptualization and analysis of biosocial relationships involving child well-being, parent–child and family relationships, as well as how these biosocial links moderate and mediate the effects of adversity and stress on health and development.

**Marcus K. Taylor** Ph.D., is an Institute for Interdisciplinary salivary bioscience research (IISBR) Faculty Affilitate at the University of California at Irvine and a Fellow of the American College of Sports Medicine. He previously held faculty positions at Duke University and San Diego State University. His program of research focuses on stress physiology in extreme environments, blast exposure, and gene–environment interactions. His recent work appears in *Journal of Special Operations Medicine, Psychoneuroendocrinology, and Psychiatry Research.*

## Section Editors

**Elizabeth A. Thomas** Ph.D., is an Adjunct Associate Professor in the Department of Neuroscience at the Scripps Research Institute and researcher in the Department of Epidemiology and IISBR faculty affiliate at the University of California at Irvine. Her research program seeks to investigate epigenetic, biochemical, and translational biomarkers in clinical populations with neurodegenerative and/or neuropsychiatric disorders. Overall, the goals of these discovery and translational research studies are to improve the diagnosis, prevention, and treatment of patients with these devastating disorders.

**Jenna L. Riis** Ph.D., is an Assistant Professor of Psychological Science at the University of California at Irvine and member of the Institute for Interdisciplinary Salivary Bioscience Research. Her program of research focuses on the etiology of health disparities and the processes by which early-life adversity affects child development and lifelong health. Dr. Riis' research examines the biopsychosocial mechanisms underlying health and development from an integrated, multisystem perspective and includes studies of novel salivary markers of the immune, metabolic, and purinergic systems.

**Kate R. Kuhlman** Ph.D., conducts research on the biobehavioral processes through which adversity in childhood leads to health disparities across the life span, with a particular interest in stress-related disorders such as depression. She is currently an Assistant Professor in the Department of Psychological Science and the Institute for Interdisciplinary Salivary Bioscience at the University of California Irvine, as well as a member of the faculty in the UCLA Cousins Center for Psychoneuroimmunology.

# Part I
# What Is Salivary Bioscience, Why Is It Important, and How Do We Study It?

**Douglas A. Granger and Marcus K. Taylor**

# Part I
# What Is Salivary Bioscience, Why Is It Important and How Do We Study It?

Douglas A. Granger and Marcus K. Taylor

# Chapter 1
# Foundations of Interdisciplinary Salivary Bioscience: An Introduction

**Douglas A. Granger and Marcus K. Taylor**

History reveals that advances in our scientific understanding often accelerate after technological innovations improve upon our ability to observe and measure phenomenon more precisely (e.g., Kuhn, 1962). Also, more often than not, big leaps in knowledge are the result of the collective effort of teams of collaborating investigators rather than by individual scientists working in relative isolation independently. At least part of the argument in favor of team science is that the nature of most key phenomenon under study involves factors operating at multiple levels of analysis. Indeed, contemporary theorists assume that complex phenomenon, such as human development, disease, poverty, and public health, are determined by a confluence of effects involving interacting intrinsic individual differences, behavior, biological, and contextual factors. A major advantage of team science is that individual team members represent a deeper level of knowledge in a particular field or level of analysis than is efficient for any particular investigator to achieve and maintain. That, in theory, enables problems to be approached from dynamic inter- and trans-disciplinary

In the interest of full disclosure, Douglas A. Granger is the founder and chief scientific and strategy advisor of Salimetrics LLC and Salivabio LLC (Carlsbad, CA) and the nature of these relationships is managed by the policies of the committees on conflict of interest at Johns Hopkins University School of Medicine and the University of California at Irvine.

D. A. Granger (✉)
Institute for Interdisciplinary Salivary Bioscience Research, University of California at Irvine, Irvine, CA, USA

Johns Hopkins University, Baltimore, MD, USA
e-mail: dagrange@uci.edu

M. K. Taylor
Institute for Interdisciplinary Salivary Bioscience Research, University of California at Irvine, Irvine, CA, USA

 
D. A. Granger, M. K. Taylor (eds.), *Salivary Bioscience*,
https://doi.org/10.1007/978-3-030-35784-9_1

perspectives with the intellectual capacity of the scientific collective being greater than the sum of the individual's knowledge (Stokols, Hall, Taylor, & Moser, 2008).

As a partial consequence of these macro-level scientific trends, there has been widespread integration of biological processes into contemporary conceptual and measurement models in traditionally nonbiologically focused areas of inquiry. The last 2–3 cohorts of PhD trainees would most likely accept this as the norm, given the adoption of this approach prior to the period in which they have been trained. Yet, on the scientific evolutionary time scale, this has been just recently realized. Until advances in technology enabled the minimally invasive measurement of biological variables this was not commonplace. What is interesting is that before 1950, discoveries enabled the minimally invasive measurement of biomarkers and analytes in oral fluids. Yet, the impact of these advances was largely limited to sub-specialties within oral biology generally, and clinical dental research specifically, until perhaps the 1980s.

In the 1980s, observations by investigators interested in the psychobiology of stress, such as Dirk Hellhammer and his students at Trier University, explored the possibility that circulating levels of cortisol could be accurately estimated using salivary measurements. Advances like these sharply focused empirical attention of many investigators and the results of that effort ushered in more than 4 decades of research involving the integration of salivary analytes in behavioral and social sciences. In parallel, within the field of infectious disease surveillance, the early stages of the HIV epidemic in the 1980–1990s focused empirical and commercialization (e.g., Saliva Diagnostics Systems, Orasure) efforts to enable a minimally invasive means of screening HIV antibody sero status (e.g., without risk of accidental needle stick). Pioneers of this early era of salivary bioscience included dental researchers such as Irwin Mandel, Lawrence Tabak, and Daniel Malamud. The need for a safe and effective screening method in the context of this widespread epidemic began the effort to chart the potential of oral fluid as an alternative to traditional diagnostic specimens. As a result of their individual scientific achievements and the efforts of their extensive network of international colleagues, a series of scientific meetings was organized and sponsored by the New York Academy of Sciences. The proceedings of these conferences resulted in at least two seminal publications describing the current state of knowledge. The first was a 342-page edited volume titled "*Saliva as a Diagnostic Fluid*" (Malamud & Tabak, 1994), which was followed up years later by the 514-page edited volume titled "*Oral-based Diagnostics*" (Malamud & Niebala, 2007). Concurrent efforts in the 1980–1990s by Peter Ellison and James Dabbs expanded the range of small molecule salivary measurements to include progesterone, estrogen, and testosterone—adding new dimensions to the study of hormones and human behavior, health, and development. The widespread enthusiasm surrounding the potential of oral fluids as a diagnostic specimen lead to several new research initiatives at the US National Institute for Craniofacial and Dental Research (NICDR). Noteworthy is the fact that this effort, in addition to funding an extensive portfolio of individual investigator initiated projects, supported a groundbreaking multisite collaborative to characterize the Human Salivary Proteome. The library produced by the project opened the windows of opportunity widely—hundreds if not thousands of analytes were identified in oral fluids (e.g., Yan et al., 2009). Among the many indications of the high impact of this effort is that salivary bioscience was included by Harold Slavkin as one of the key pillars in the first US Surgeon General's report on oral health (US Department of Health and Human Services, 2000). Moreover, as a consequence of the human genome project, a strong pulse of research activity demonstrated the capability to extract high quality and quantity DNA from oral fluids (e.g., Nemoda et al., 2011). Another major research emphasis was focused on the assessment of drugs of abuse and/or their metabolites in oral fluids. These capabilities, in particular, lead to the realization of niche commercial enterprises with specialized applications in consumer-based testing, health care, insurance, and law enforcement.

Quite surprisingly, given the obvious advantages, the scientific and commercial enthusiasm for *saliva as the diagnostic fluid of the future*, with a few notable exceptions (i.e., DNA, HIV screening, and some drug testing) has waxed and waned over the years. Many saliva-based opportunities have attracted considerable public interest and also substantial financial investment, but more often than not these ventures have failed to deliver commercially viable devices or clinically meaningful saliva-based assessments. There are of course isolated examples of success (e.g., Orasure, Salimetrics, 23-and-me, Tecan-IBL), but generally speaking, saliva as a *diagnostic* specimen has yet to live up to the initial enthusiasm expressed by the clinical and business communities about its potential (see for instance Malamud & Tabak, 1994). Explanations range widely from legal, regulatory, and economic barriers to entry, to concerns about precision and reproducibility, to the interpretation of measurements made from oral fluid samples.

By stark contrast, in the last two decades, this effort has led to a renaissance in the traditionally behaviorally oriented basic and applied sciences. In fact, the number of peer-reviewed scientific publications involving salivary analytes has increased from a handful to multiple thousands annually within the past 20 years. For instance, Fig. 1.1 displays the results of a pubmed.gov search using only the key words "salivary cortisol"—revealing the number of annual publications increased from a handful in 1985 to more than 400 in 2019. Similar "accelerating trends" are evident in the publication rates for many other salivary analytes (e.g., sIgA, testosterone, C-reactive protein, and alpha-amylase). The scientific research often involves a focus on testing innovative theoretical models of individual differences in health, behavior, and cognition as a function of multilevel biosocial processes in the context of everyday life.

The ease of use and minimally invasive nature of saliva collection is especially valued in this endeavor because complex multilevel models of individual difference can be studied in the laboratory, in quasi-naturalistic settings, or in response to the trials and tribulations of people's (and animals') everyday social worlds. Also, multiple time point sampling can be undertaken, in many circumstances, without adding significant participant burden or interrupting the natural flow of activity. Significant progress has been made, especially during the 15-year period between 1995 and 2010, to develop and refine saliva collection and measurement methods,

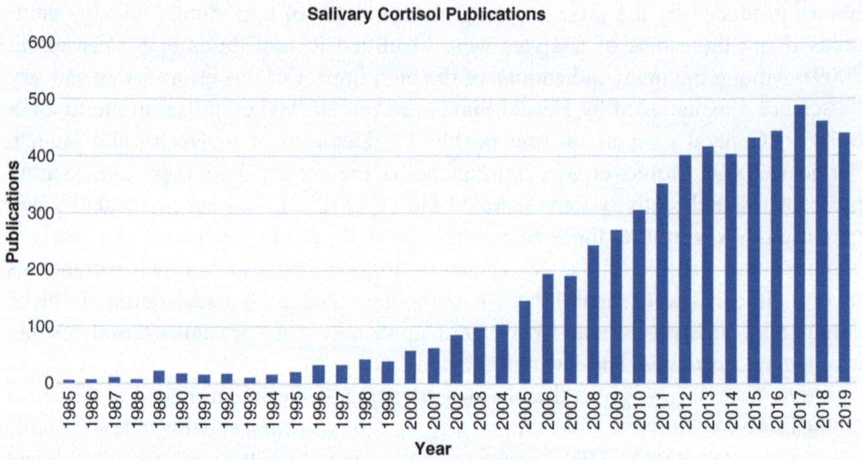


**Fig. 1.1** An illustration of the increasing number of salivary bioscience publications. The search period was from 1985 to 2019 (the last year of complete data as of the publication date). The search engine was PubMed, and keywords for this illustrative search were "salivary cortisol." As can be seen in these historical data, the level of research activity in salivary bioscience has substantially increased over the last 20 years. Similar findings emerge when searches are conducted with many other salivary analytes

many of which are now commercially viable. It is reasonable to conclude that since mid-2000, the scientific community has had a very advanced understanding of how to collect, handle, transport, store, assay, and interpret saliva-derived biological data.

Simple database searches (e.g., PubMed, Psychinfo, Scopus, and Web of Science) are sufficient to reveal that not only is the number of publications incorporating salivary measures increasing, the number of scientific fields in which salivary analytes has been employed is also expanding. The scan of PubMed search results between 1980 and 2019 for keywords "salivary cortisol," noted above, for example, yields more than 5500 total publications. Across the years, the range of scientific fields (represented by the journal titles) is also increasing. In 2019, publications involving salivary bioscience appeared in journals linked to dentistry, psychiatry, psychology, sports science, endocrinology, sleep, neuroscience, nursing, circadian biology, obesity, anthropology, public health, child development, psychosomatic medicine, and aging. The number of different salivary analytes involved in this research effort is also expanding, and now includes, for instance, measures of hormones, cytokines, immunoglobulins, enzymes, DNA, environmental chemicals, and elements and metals.

The significance, importance, and impact of this scientific integration was marked on a scientific evolutionary time line in 2010, when Johns Hopkins University created the Center for Interdisciplinary Salivary Bioscience Research, then again in 2016 when the University of California executed a campus wide strategic initiative to create the Institute for Interdisciplinary Salivary Bioscience Research. In parallel, on the international scientific stage, similar events mark this developmental milestone with the creation of salivary bioscience laboratories and centers at the University of New South Wales in Sydney Australia, Angelia Ruskin University in Cambridge United Kingdom, Charite University in Berlin Germany, and the Institute for Experimental Medicine of the Hungarian Academy of Science in Budapest Hungary. Noteworthy is the absence of the term "diagnostics" and the inclusion of the terms "interdisciplinary" and "bioscience" in the labeling of many of these academic units.

A core assumption in the "saliva diagnostic era" was that a salivary measure is primarily of interest because of the extent to which that measurement is highly associated with the corresponding measurement in the general circulation. The core idea has been *saliva as a surrogate of blood*. Today, this possibility reflects only a subcomponent of a much broader set of assumptions. That is, consistent with the hypothesis forwarded by Harold Slavkin, *Salivary Bioscience* assumes that the oral cavity serves as a window to the body, measurements made in oral fluids (even if they are not directly correlated with parallel measures in circulation) have the potential to be important indicators of health, disease, and physiological states in their own right, and in addition to the measurements made in traditional biological specimens (US Department of Health and Human Services, 2000). On the other hand, it has also been realized that not every measure derived from a blood sample can be determined in an oral fluid sample. Therefore, saliva is unlikely to be a replacement for traditional diagnostic specimens in many circumstances. For instance, in a clinical setting it is often the case that a panel of measurements is determined from a blood sample. If only a subset of those measures is possible in an oral fluid sample, then a blood draw is necessary and the determinations from an oral fluid sample are redundant and add little value.

The contemporary interdisciplinary perspective—*salivary bioscience* in contrast to *saliva diagnostics*—allows investigators to consider that the conceptual, methodological, and empirical advances to date will enable us to determine for whom, under which circumstances, which measurements made in oral fluid might add value. Beyond the minimally invasive nature of sample collection is the opportunity saliva affords for investigators to collect multiple measurements from individuals and explore intraindividual variation. Multiple sampling time points, without adding participant burden, and multiple measurements per time point affords researchers and clinicians many advantages. For instance, multiple time point salivary samples enabling intraindividual assessments of therapeutic drug metabolites might provide clinicians higher resolution in titrating drug dosage to the individual, and might increase the probability of detecting exposure to an environmental contaminant with a short half-life. Furthermore, multiple measurements enable aggregation that may improve the reliability of the estimates of a highly variable analytes concentration, and multiple measures across the day enables estimation of an analytes diurnal pattern of production. Not surprisingly, these advantages of saliva as a research specimen have somewhat outpaced our statistical strategies and tactics—raising new challenges and opportunities.

As our basic knowledge has developed so too have measurement tactics been influenced by changing technologies. The first generation of salivary measurements were made possible by in-house research use only modifications of radioimmunoassays designed for use with serum/plasma. The modifications were focused on minimizing matrix effects and improving upon the lower limits of sensitivity. These assays were commonplace in research prior to the late 1990s. The next generation of measures were enzyme-based immunoassays specifically designed for use with saliva. These assays are largely still in use and focused on minimizing sample test volumes, optimizing precision, and reducing lower limits of sensitivity to the pg/mL range. The most recent applications of these tactics involve multiplex-coated bead or plate technologies (i.e., Luminex and Mesoscale Discovery). The third-generation measurement strategies involve the use of a variety of point-of-care lateral flow, layered paper, or microfluidic technologies paired in some circumstances with mobile phone applications. POC applications have been proven technically feasible, and qualitative assays have performed well. On the other hand, at the time of this writing, quantitative measurements in saliva using POC technologies struggle with reliability and precision below the low ng/mL range.

The purpose of this edited volume is to serve as a foundational reference text in salivary research methods, data collection, analysis, and interpretation; as well as applications to medicine, surveillance, and monitoring. More specifically, to highlight, document, and benchmark the current state of the art and breath of salivary bioscience from an interdisciplinary perspective. To this end, the editorial team has tasked multiple writing teams to summarize progress in the application of salivary bioscience. Chapters summarize the integration of salivary bioscience in the context of research on neuroscience, stress, genetics, microbiome, immunity, circadian biology, pain, infectious disease epidemiology, behavioral medicine, psychiatry, oncology, periodontal medicine, pain, environmental exposure, drugs of abuse and therapeutic drug monitoring, social neuroscience of human–animal interaction, animal well-being, precision medicine, health policy, aging, and military, and space research. The charge to the writing teams was to summarize progress, to speak to the history, the promises and the prospects, and estimate the "impact" salivary bioscience has on these specific fields, but also identifying the problems and the pitfalls with an eye on the gaps in knowledge, and also best practices and worthwhile future directions. As a collective, the editors and authors anticipate this volume will be particularly inspiring to those scientists, practitioners, and students who wish to make significant contributions to the evolution and eventual maturity of this exciting emerging field.

## References

Kuhn, T. S. (1962). *The structure of scientific revolutions*. Chicago: University of Chicago press.

Malamud, D., & Niebala, R. S. (2007). *Oral-based diagnostics* (Vol. 1098). New York: Annals of the New York Academy of Sciences.

Malamud, D., & Tabak, L. (1994). *Saliva as a diagnostic fluid* (Vol. 694). New York: Annals of the New York Academy of Sciences.

Nemoda, Z., Horvat-Gordon, M., Fortunato, C. K., Beltzer, E. K., Scholl, J. L., & Granger, D. A. (2011). Assessing genetic polymorphisms using DNA extracted from cells present in saliva samples. *BMC Medical Research Methodology, 11*, 170.

Stokols, D., Hall, K. L., Taylor, B. K., & Moser, R. P. (2008). The science of team science: Overview of the field and introduction to the supplement. *American Journal of Preventive Medicine, 35*, S77–S289.

US Department of Health and Human Services. (2000). *Oral health in America: A report of the surgeon general-executive summary*. Rockville, MD: National Institute of Dental and Craniofacial Research.

Yan, W., Apweiler, R., Balgley, B. M., Boontheung, P., Bundy, J. L., Cargile, B. J., . . . Wong, D. T. (2009). Systematic comparison of the human saliva and plasma proteomes. *Proteomics Clinical Applications, 3*, 116–134.

# Chapter 2
# Salivary Gland Anatomy and Physiology

**Lisa M. Hernández and Marcus K. Taylor**

**Abstract** In mammals, saliva is a mildly acidic secretion made mostly of water (99.5–99.8%). In a healthy state, humans produce between 500 mL and 1.5 L of saliva per day. Saliva has numerous functions including lubrication, digestion, and immunity. Salivary glands are classified as exocrine, and as such, they produce secretions (i.e., saliva) onto an epithelial surface via a system of ducts. Saliva secretion and production are mediated by the autonomic nervous system (ANS) and thus; salivary glands have both parasympathetic and sympathetic innervation.

Within the oral cavity, there are three major salivary glands; parotid, submandibular, and sublingual, as well as hundreds of minor glands. These glands produce serous, mucous, or seromucous secretions that contain proteins and compounds, which are significant to salivary bioscience studies. Saliva composition depends upon health status and overall physiologic need.

This chapter will delve further into the macro- and microanatomy of the normal salivary gland and will detail the physiologic and neural regulation of saliva production, composition, and secretion.

**Keywords** Oral cavity · Duct · Submandibular · Parotid · Sublingual

## 2.1 Basic Anatomy of the Oral Cavity

As the entrance to the digestive system, the oral cavity senses, mechanically processes, lubricates, and initiates the digestion of food (Martini, Timmons, & Tallitsch, 2009). The mouth is lined with epithelial cells and is structurally supported by fat (buccal fat pads) and muscle (buccinator muscles) in the cheeks. The teeth are

L. M. Hernández (✉)
Leidos Inc., San Diego, CA, USA

M. K. Taylor
Institute for Interdisciplinary Salivary Bioscience Research, University of California at Irvine, Irvine, CA, USA



D. A. Granger, M. K. Taylor (eds.), *Salivary Bioscience*,
https://doi.org/10.1007/978-3-030-35784-9_2

anchored to the maxilla (upper jaw) and mandible (lower jaw) and surrounded by gum tissue. The space where a tooth meets the gum tissue is known as the gingival sulcus where gingival crevicular fluid (GCF) is found. The tongue muscle provides mechanical processing and manipulation of food for chewing and swallowing. It also plays a role in sensory analysis (i.e., temperature and taste) and produces secretions for digestion (i.e., mucins and lipase).

## 2.2 Salivary Gland Anatomy

Salivary glands are classified as exocrine glands just like sweat, mammary, and sebaceous glands. By definition, they produce and secrete substances via a duct onto an epithelial surface (e.g., the skin or oral cavity). All exocrine glands are categorized into three types, based on the type of secretions that they deliver; (1) Serous glands produce a watery liquid, which usually contains enzymes; (2) Mucous glands yield mucins, which combine with water to form mucus; and (3) Seromucous (mixed) exocrine glands yield both serous and mucous secretions. Additionally, salivary glands are considered merocrine glands as they secrete their products via exocytosis.

Anchored deep to epithelial lining of the mouth, salivary glands are histologically classified according to their structure and secretion, but the basic components are the acinar/alveolar cells, a duct system, and myoepithelial cells. Acinar cells form clusters called acini, which act as the secretory unit of the gland. The epithelium-derived duct system delivers products (Anatomy & Physiology, 2018) to the oral cavity. The length and diameter of the duct system depends upon the type of gland and secretion that is produced (Pedersen, Sørensen, Proctor, Carpenter, & Ekström, 2018). Saliva is first secreted by the acini and thus, the type of acinar cell of the gland dictates the type of secretion to be produced (e.g., serous, mucus, or mixed secretion). Even if a salivary gland has a combination of acinar cells, the individual gland will still predominantly produce serous or mucous secretions.

The fluid that initially constitutes saliva is isotonic, but as it reaches the duct system, it eventually becomes hypotonic (a liquid with more water and less solute than that of blood serum). The tonicity of saliva can be indicative of a basal (unstimulated, more hypotonic) or stimulated state (less hypotonic). The duct system is composed of various cell types; intercalated, striated, and excretory duct cells. Intercalated cells make up the first segment of a duct. The second and third duct portions are made up of striated and excretory cells, respectively. The composition of saliva is altered in these last duct sections where striated cells regulate electrolytes by resorbing sodium. This sodium resorption continues in the excretory duct cells, which also secrete potassium. Finally, the saliva reaches the oral cavity with the help of myoepithelial cells. Located at the base of the acini, and sometimes the intercalated duct cells, myoepithelial cells contract to facilitate salivary secretion. The contraction/relaxation of myoepithelial cells is regulated by the sympathetic or parasympathetic nervous systems, which act upon brain salivary centers (Garrett,

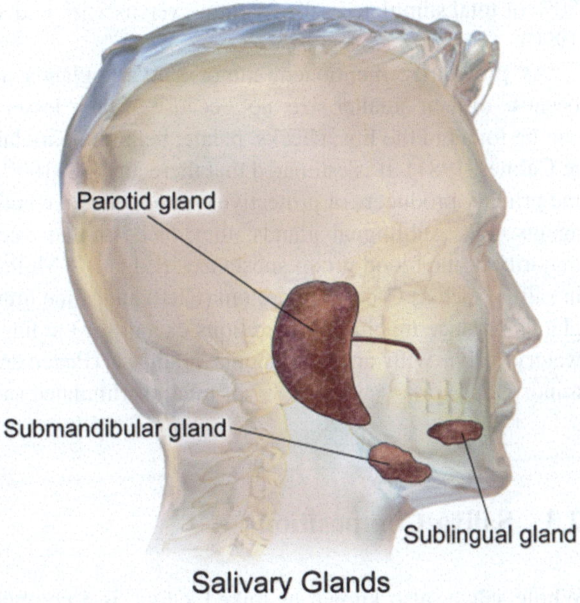


**Fig. 2.1** The three major human salivary glands. From https://en.wikipedia.org/wiki/Salivary_gland

1987). Although they are contractile in nature, the myoepithelial cells are not necessary for saliva secretion (Pedersen et al., 2018).

There are three major pairs of salivary glands; the submandibular, parotid, and sublingual glands (Fig. 2.1; Blausen.com staff, 2014). These are classified as major glands based on their anatomical size, and they all have long, branched duct systems as describe above. However, the ducts of the sublingual glands lack striated cells, which means that they resorb sodium to a lesser extent than the other glands. While the major glands produce greater quantities of saliva, they do not necessarily add more to the overall quality of saliva. In fact, major salivary glands contribute the most to volume and electrolyte content, but little to other significant proteins of interest. The submandibular glands produce about 60% of total unstimulated saliva. Inferred by their name, they are located on the floor of the mouth, medial to the lower jaw (mandible). As a mixed gland, it produces a viscous fluid, rich in enzymes and mucins. Mucins combine with water to form mucus, which protects the epithelial lining of the oral cavity by coating food as it makes its way to the esophagus. The parotid glands are the largest glands, but they produce only about 20–25% of total unstimulated saliva. These glands are located just inferior to the cheekbones (zygomatic arches) and are anterior to the ears. Saliva from these glands is serous and abundant in enzymes (e.g., amylase). The sublingual glands are located under the tongue (Fig. 2.1; Blausen.com staff, 2014) and primarily produce viscous saliva that is rich with mucins. In combination with a multitude of other minor salivary glands, they contribute to the remaining 5–10% of total unstimulated saliva. The percent contribution of each gland toward saliva production changes when saliva flow rate is highly stimulated. For instance, the parotid glands increase their yield to account for

50% of total stimulated saliva volume versus 25% under resting conditions (Edgar, 1990).

As previously mentioned, minor salivary glands are only classified as such because of their smaller size, not because of their lesser significance. Minor glands can be found in the lips, cheeks, palate, behind the molars, and on the tongue (Roth & Calmes, 1981). It is estimated that there are 600–1000 minor glands and some are the primary producers of protective components like antibacterial and antimicrobial agents. Like sublingual glands, they lack striated duct cells and they supply a majority of the blood group substances (Edgar, O'Mullane, & Dawes, 2004) found in saliva, such as C-reactive protein (CRP) and some immunoglobulins. Most minor glands produce mucin-rich secretions except for the lingual glands, which generate watery saliva with ample amounts of lipase (Pedersen et al., 2018). Altogether, minor glands produce about 8% of total unstimulated saliva.

## 2.3 Saliva Composition

Whole saliva, also known as mixed saliva, is a combination of unstimulated and stimulated saliva, microorganisms, gingival crevicular fluid, food debris, and medication metabolites, if any (Fig. 2.2). Habits (e.g., oral hygiene), behavior (e.g., physical activity), and nutritional intake drive the relative contributions of these

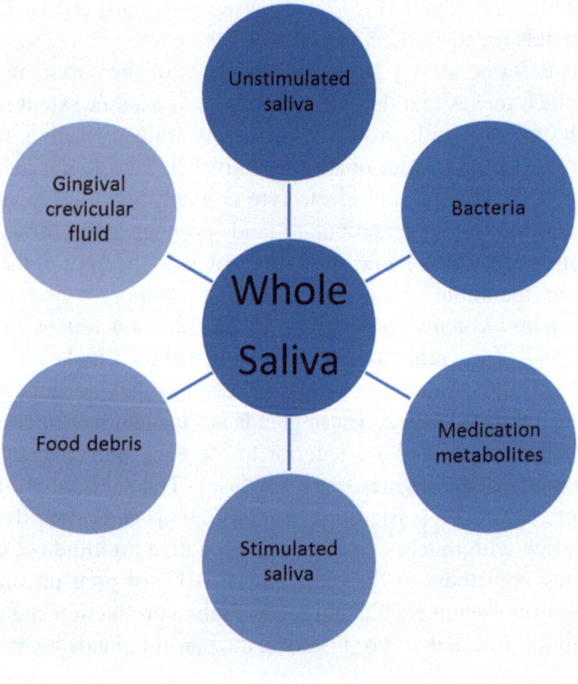


**Fig. 2.2** Major components of whole saliva

components. Thus, the composition of whole saliva is dynamic and influenced by an exact combination of stimuli. The electrolyte and protein concentration of whole saliva are regulated by circadian rhythms and salivary protein values generally peak in the late afternoon (Rudney, 1995). Electrolyte concentration is largely dependent upon saliva flow rate, which is influenced by health status (e.g., hydration) and overall physiologic needs. In the long-term, whole saliva composition is stable, however; short-term changes in proteins can occur due to daily emotions (Jemmott et al., 1983), respiratory infections (Cockle & Harkness, 1983), inflammation (Henskens, Veerman, Mantel, Van der Velden, & Nieuw Amerongen, 1994), and reproductive status (Cockle & Harkness, 1983; Tenovuo, Laine, Söderling, & Irjala, 1981; Widerström & Bratthall, 1984). These temporary changes may be limited by genetic factors (Rudney, 1995). Systemic disease such as metabolic or immunologic disease also impact saliva protein composition over time. It is therefore important to record an individual's initial health status and then any significant changes thereafter.

Unstimulated and stimulated saliva are the two basic components of whole saliva. Unstimulated saliva is the basal level of saliva production as opposed to stimulated saliva, which is produced in response to chewing (mastication). The submandibular glands are the primary generators of unstimulated saliva. During sleep, this basal production is almost absent. Most stimulated saliva comes from the parotid glands. The sublingual and minor glands contribute equally to both unstimulated and stimulated saliva production. Unstimulated saliva is very hypotonic and has a pH that is neutral or slightly acidic. Stimulated saliva is less hypotonic and has an alkaline pH.

The protein concentration of saliva is inversely proportional to the flow; if there is a high rate of flow, there is less time for the acinar and duct cells to modify saliva and protein concentration is lower. At highest flow rates, saliva is the most isotonic to plasma. Conversely, if flow is at a low rate, protein concentration is increased and saliva is more hypotonic than plasma. The protein content of saliva is highly specific to each person as it is influenced by the individual's genetics, environment, and habits (Rudney, 1995). Most salivary proteins originate solely from the salivary glandular cells and not the blood. Salivary proteins represent only about 10% of the 2500 proteins found in the whole saliva. The remaining 90% is from microorganisms and epithelial cells that are shed from the lining of the mouth (Ekström, Khosravani, Castagnola, & Messana, 2011).

Another innate component of whole saliva is gingival crevicular fluid (GCF), which is found in the gingival sulcus (the space between a tooth and gum tissue). It has a varied composition that is similar to whole saliva, but is produced only in small amounts. It is thought that the main role of GCF is to help to clear food debris and impart antimicrobial/immune protection. Alternatively, GCF has been linked to inflammatory processes, which cause an increase in vessel permeability. Cytokines such as IL-1$\beta$, TNF-$\alpha$, and immunoglobulins G and M, can be found in GCF (Gupta, 2012). Similar to unstimulated saliva, GCF production is regulated by a circadian rhythm with an increase during the typical waking hours and a decline in the late evening. Although it is quite stable during waking hours (Suppipat, Johanasen, &

Gjermo, 1977), GCF production is higher after periodontal interventions, while eating (chewing), if a person smokes, and during hormonal fluctuations in females (e.g., menstruation, oral contraceptives, and pregnancy).

## 2.4 Functions of Saliva

The constituents of saliva (Fig. 2.3) help to maintain oral health and also facilitate systemic health. Saliva contains electrolytes, immunoglobulins, proteins, enzymes, mucins, and nitrogen products. These entities are multifunctional and work in concert to perform the primary functions of saliva, which are: (1) immunity and antibacterial activity, (2) buffering action, (3) lubrication and tissue protection, (4) taste and predigestion, and (5) tooth integrity. As a primary gateway to the external environment, the oral cavity bears a strong capacity for protection and immune function. Secretory immunoglobulin A (sIgA), the largest immunologic component of saliva, is produced in connective tissue and translocated through the duct cells of major and minor salivary glands (Humphrey & Williamson, 2001). Antibacterial activities are provided by immunoglobulins, proteins, and enzymes. Glycoproteins (proteins attached to oligosaccharide chains) and mucins help to rid the mouth of microorganisms and reduce dental plaque. The pH of saliva in a healthy state ranges from 6.6 to 7.6 (Choi, Lyons, Kieser, & Waddell, 2017) and its buffering capacity is imparted by bicarbonates, phosphates, and urea. Saliva pH is influenced by consumption of sugary or acidic foods (e.g., cherries) and drinks (e.g., soft

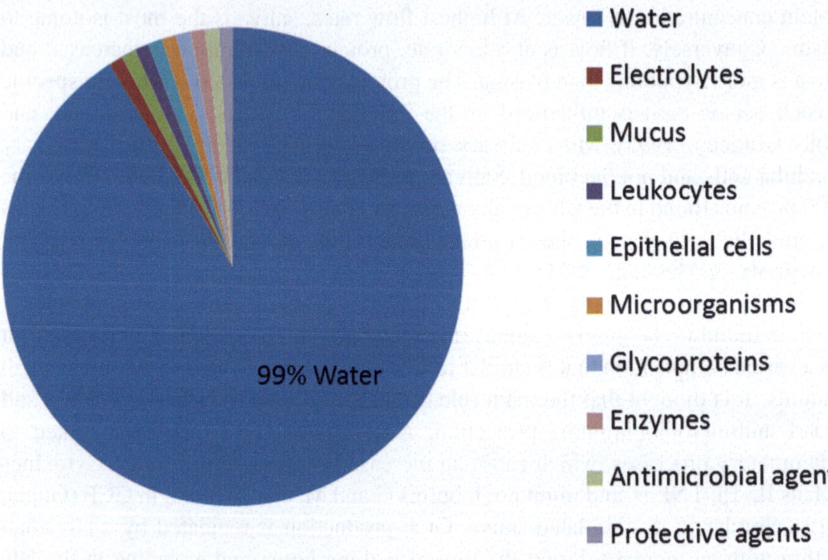


**Fig. 2.3** Minor components of whole saliva

drinks), which can temporarily reduce the pH to about 5.5. Mucins help to form mucus, which provides a physical barrier to protect tissues, to soften food for chewing and swallowing, and to facilitate speech. Another major saliva substance is epidermal growth factor, which promotes healing by stimulating DNA synthesis and cell growth/differentiation. Enzymes, such as salivary alpha amylase and lingual lipase, initiate the breakdown of carbohydrates and fats, respectively. Finally, one of the most important functions of saliva is to support tooth integrity. It has been written that "Saliva is to tooth enamel what blood is to the cells of the body." (Moss, 1995). Together, calcium, phosphate, and other proteins combine to form an "antisolubility factor," which modulates the formation of tooth enamel (Humphrey & Williamson, 2001).

## 2.5 Salivary Flow

Unstimulated saliva production is mediated by the parasympathetic nervous system, which is colloquially known as the "rest and digest" portion of the autonomic nervous system (ANS). Saliva production or volume can also be influenced by environmental and pharmacologic factors, however; the average amount of saliva produced by a healthy adult is estimated to be about 1–1.5 liters/day. Salivary flow is unique to an individual and their responses to internal and/ or external stimuli, but the general range for unstimulated flow rate is about 0.3–0.4 mL/min. Stimulated saliva is the major contributor to changes in waking saliva production. For example, the flow rate in response to food can increase up to a maximum of 7 mL/min (Humphrey & Williamson, 2001). As previously mentioned, unstimulated salivary flow is almost completely absent during sleep. Salivary flow is an important consideration in both clinical and research contexts. Since saliva affords a great deal of immune function and protection, low flow, or hypofunction, can be detrimental. It is important to identify low saliva production, or hyposalivation, under stimulated circumstances, which is defined as a rate $<0.1$ mL/min. Gland hypofunction can result in an increase in cavities, soft tissue ulcerations, infections, and altered taste (dysguesia) as well as reduced healing from aesthetic dental surgeries and the loss of prosthetic dental restorations (Moss, 1995). Overall, salivary gland function has been reported to be quite robust to the aging process (Pedersen et al., 2018). But, illness, disease, prescription medications (Humphrey & Williamson, 2001; Rudney, 1995), chemotherapy, and radiation (to the head and neck), can cause hypofunction. The anticholinergic side effects of antihistamines or antidepressants may reduce saliva flow and cause xerostomia (dry mouth). Any medications that act upon the beta-adrenergic receptors (e.g., asthma or heart medications) may affect acinar cell protein production. Some antipsychotics, blood pressure, and Parkinson's disease medications may also act on these receptors, resulting in lower flow. Furthermore, hydration status will affect saliva secretion and it has been demonstrated that flow is reduced following restriction of liquid and food (Pedersen et al., 2018). With respect to research, recording hydration status, stress level, presence of respiratory infection, signs of inflammation, and hormone changes (e.g., ovulation and pregnancy) is prudent since it has been shown that these factors can profoundly affect short-term saliva composition (Rudney, 1995).

Salivary flow is not uniform throughout the oral cavity and there are specific intraoral areas known as the "salivary highways and byways" (Moss, 1995) where flow is either larger or smaller. For example, areas of the lower mouth produce a high volume of saliva while the upper front of the mouth produces very little. Intraoral flow influences the composition of whole saliva as well as composition within different areas of the mouth. This is especially important when instructing a patient or study participant on saliva collection. The analyte(s) being detected will dictate the ideal intraoral area to be sampled (e.g., under the tongue, in the cheek pocket, etc.). Finally, salivary flow is known to change predictably throughout the day (day versus night) and also between seasons (e.g., summer versus winter). Clock genes, which are implicated in circadian rhythm function, have been identified within the salivary glands of mice (Zheng, Seon, McHugh, Papagerakis, & Papagerakis, 2012) and this suggests that flow is regulated by circadian rhythms. To date, the protein expression and characterization of the periodicity of clock genes in human salivary glands have not been reported.

## 2.6 Neural Regulation of Salivary Glands

Salivary glands are primarily controlled by salivatory nuclei, which are called the "salivary centers." This cluster of nuclei is located in the brainstem (medulla); specifically, in the dorsal pons. Salivary glands also receive input from other brain centers and are influenced by gastrointestinal hormones (Pedersen et al., 2018). The superior salivatory nucleus innervates the submandibular and the sublingual glands. The inferior salivatory nucleus innervates the parotid gland. Both nuclei are components of the main cranial nerves; the superior salivatory nucleus is part of the facial nerve (cranial nerve VII) and the inferior salivatory nucleus belongs to the glossopharyngeal nerve (cranial nerve IX). These nuclei confer parasympathetic input to the glands to produce vasodilation and saliva secretion. In a normal state, there are different types of sensory stimuli for secretion: (1) mechanical, (2) gustatory, and (3) olfactory (Humphrey & Williamson, 2001). In altered states, secretion can be stimulated by pain and pharmacological agents. Other conditions such as depression, fatigue, and fear, can reduce saliva flow (Feher, 2017). In these instances, the common misconception is that salivary flow is reduced by sympathetic inhibitory fibers. In fact, flow is decreased due to "supranuclear control," the influence of higher brain regions like the hypothalamus. Nevertheless, knowledge about the neural regulation of salivary glands is mostly derived from animal studies and the precise connections between the salivary centers and higher brain centers remain unidentified in humans (Ekström et al., 2011).

Under control of the ANS, salivary glands are innervated predominantly by parasympathetic fibers, but they also receive sympathetic input. Binding of autonomic neurotransmitters with their respective receptors in salivary glands produces a myriad of outcomes and effects. When cholinergic parasympathetic nerves release acetylcholine (ACh), which binds to muscarinic receptors, saliva is secreted from the acini. Adrenergic (adrenaline and noradrenaline) and cholinergic neurotransmitters are the first messengers of a sympathetic secretory response (Garrett, Ekström, & Anderson, 1999). Sympathetic nerves release noradrenaline to activate adrenergic receptors, which induce smaller volumes of saliva, but with larger amounts of protein, to be expressed from the acini and duct cells (Proctor & Carpenter, 2007). Other neuropeptides released from autonomic nerves can also increase saliva production and alter membrane permeability.

Neural actions on salivary glands include water mobilization, protein secretion, stimulation of cell synthesis, and maintenance of cell function and size. Activation of the parasympathetic and sympathetic systems results in saliva secretions, which interact synergistically to secrete fluid and proteins (Ekström et al., 2011) to meet physiologic demand. Additionally, capillary vessels, which are adjacent to the salivary ducts, indirectly influence saliva secretion. As mentioned earlier, the cranial nerves confer parasympathetic innervation, which causes vasodilation of these vessels. This type of parasympathetic stimulation favors abundant volumes of serous or watery secretions. Conversely, sympathetic innervation is conferred directly by the spinal nerves (i.e., thoracic and cervical), and indirectly by the capillary plexus, which supplies the glands (Garrett, 1987). Indirect input by the capillary plexus is exerted more so by vascular control and not by reflexive sympathetic pathway (Garrett, 1987).

This described model of neuronal control becomes more complicated when considering input from second messenger systems, like cyclic adenosine monophosphate, nitric oxide, and calcium or, neuropeptides such as vasoactive intestinal peptide. Intracellular signaling and co-transmitter receptor activation allow for enhanced coupling between the ANS and the current salivary protein content to create a "real-time" response that ensures the maintenance of optimal saliva composition. Irrespective of the source of stimulation (either parasympathetic or sympathetic), if saliva production is increased, there is a concomitant rise in other salivary ingredients like water, electrolytes, proteins, and other organic molecules (Proctor & Carpenter, 2007).

## 2.7 Summary

An understanding of the salivary gland anatomy and the regulation of saliva production in a normal state is the first step toward maximizing the power of salivary bioscience. Precise knowledge of gland location can optimize the sampling of analytes. Familiarity with the nuances of saliva production can help to identify sampling confounds (i.e., controlling for flow rates and circadian rhythm) and can also enhance data interpretation. Altogether, this foundational information supports pristine research results and may significantly augment patient-centered care in a clinical setting.

## References

Anatomy & Physiology. A learning initiative. Retrieved July 15, 2018, from http://anatomyandphysiologyi.com/glandular-epithelium/

Blausen.com staff. (2014). Medical gallery of Blausen medical 2014. *WikiJournal of Medicine, 1* (2). doi:https://doi.org/10.15347/wjm/2014.010. ISSN 2002-4436.

Choi, J. E., Lyons, K. M., Kieser, J. A., & Waddell, N. J. (2017). Diurnal variation of intraoral pH and temperature. *BDJ Open, 3*, 17015.

Cockle, S. M., & Harkness, R. A. (1983). Changes in the electrophoretic pattern of salivary peroxidase at the middle of the menstrual cycle. *Biochemical Medicine, 30*(2), 188–195.

Edgar, W. M. (1990). Saliva and dental health. Clinical implications of saliva: Report of a consensus meeting. *British Dental Journal, 169*(4), 96–98.

Edgar, W. M., O'Mullane, D. M., & Dawes, C. (Eds.). (2004). *Saliva and oral health* (pp. 14–50). London: British Dental Association.

Ekström, J., Khosravani, N., Castagnola, M., & Messana, I. (2011). Saliva and the control of its secretion. In *Dysphagia* (pp. 19–47). Berlin, Heidelberg: Springer.

Feher, J. J. (2017). *Quantitative human physiology: An introduction*. Amsterdam: Academic Press.

Garrett, J. R. (1987). The proper role of nerves in salivary secretion: A review. *Journal of Dental Research, 66*(2), 387–397.

Garrett, J. R., Ekström, J., & Anderson, L. C. (Eds.). (1999). *Neural mechanisms of salivary gland secretion*. Basel: Karger Medical and Scientific Publishers.

Gupta, G. (2012). Gingival crevicular fluid as a periodontal diagnostic indicator-I: Host derived enzymes and tissue breakdown products. *Journal of Medicine and Life, 5*(4), 390.

Henskens, Y. M. C., Veerman, E. C. I., Mantel, M. S., Van der Velden, U., & Nieuw Amerongen, A. V. (1994). Cystatins S and C in human whole saliva and in glandular salivas in periodontal health and disease. *Journal of Dental Research, 73*(10), 1606–1614.

Humphrey, S. P., & Williamson, R. T. (2001). A review of saliva: Normal composition, flow, and function. *The Journal of Prosthetic Dentistry, 85*(2), 162–169.

Jemmott, J., III, Borysenko, M., Chapman, R., Borysenko, J., Mcclelland, D., Meyer, D., … Benson, H. (1983). Academic stress, power motivation, and decrease in secretion rate of salivary secretory immunoglobulin A. *The Lancet, 321*(8339), 1400–1402.

Martini, F. H., Timmons, M. J., & Tallitsch, R. B. (2009). *Human anatomy* (6th ed.). San Francisco: Pearson.

Moss, S. J. (1995). Clinical implications of recent advances in salivary research. *Journal of Esthetic and Restorative Dentistry, 7*(5), 197–203.

Pedersen, A. M. L., Sørensen, C. E., Proctor, G. B., Carpenter, G. H., & Ekström, J. (2018). Salivary secretion in health and disease. *Journal of Oral Rehabilitation*. https://doi.org/10.1111/joor.12664

Proctor, G. B., & Carpenter, G. H. (2007). Regulation of salivary gland function by autonomic nerves. *Autonomic Neuroscience, 133*(1), 3–18.

Roth, G., & Calmes, R. (Eds.). (1981). Salivary glands and saliva. In *Oral biology* (pp. 196–236). St Louis: C.V. Mosby.

Rudney, J. D. (1995). Does variability in salivary protein concentrations influence oral microbial ecology and oral health? *Critical Reviews in Oral Biology & Medicine, 6*(4), 343–367.

Suppipat, N., Johanasen, J. R., & Gjermo, P. (1977). Influence of "time of day", pocket depth and scaling on gingivai fluid flow. *Journal of Clinical Periodontology, 4*, 48–55.

Tenovuo, J., Laine, M., Söderling, E., & Irjala, K. (1981). Evaluation of salivary markers during the menstrual cycle: Peroxidase, protein, and electrolytes. *Biochemical Medicine, 25*(3), 337–345.

Widerström, L., & Bratthall, D. (1984). Increased IgA levels in saliva during pregnancy. *Scandinavian Journal of Dental Research, 92*(1), 33.

Zheng, L., Seon, Y. J., McHugh, J., Papagerakis, S., & Papagerakis, P. (2012). Clock genes show circadian rhythms in salivary glands. *Journal of Dental Research, 91*(8), 783–788.

# Chapter 3
# Saliva Collection, Handling, Transport, and Storage: Special Considerations and Best Practices for Interdisciplinary Salivary Bioscience Research

**Genieleah A. Padilla, Jessica L. Calvi, Marcus K. Taylor, and Douglas A. Granger**

**Abstract** This chapter reviews common practices, identifies special considerations, and recommends best practices for saliva specimen collection methods employed in interdisciplinary salivary bioscience research. We discuss the practical aspects of collecting saliva samples across a variety of research applications, considerations for target populations, environmental factors that affect salivary biomarkers, collection methods, cold chain management and sample handling, transport, and archiving. Best practices are recommended based on the current state of the art.

**Keywords** Saliva collection · Cold chain management · Best practices · Research design

In the interest of full disclosure, Douglas A. Granger is the founder and chief scientific and strategy advisor of Salimetrics LLC and Salivabio LLC (Carlsbad, CA) and the nature of these relationships is managed by the policies of the committees on conflict of interest at Johns Hopkins University School of Medicine and the University of California at Irvine.

G. A. Padilla (✉)
Leidos, Inc., Reston, VA, USA

J. L. Calvi
Salivary Bioscience Laboratory, University of Nebraska-Lincoln, Lincoln, NE, USA

M. K. Taylor
Institute for Interdisciplinary Salivary Bioscience Research, University of California at Irvine, Irvine, CA, USA

D. A. Granger
Institute for Interdisciplinary Salivary Bioscience Research, University of California at Irvine, Irvine, CA, USA

Johns Hopkins University, Baltimore, MD, USA



D. A. Granger, M. K. Taylor (eds.), *Salivary Bioscience*,
https://doi.org/10.1007/978-3-030-35784-9_3

In comparison to traditional biospecimens, such as blood serum/plasma or urine, advantages of oral fluid include that its collection is minimally invasive, convenient, and non-painful (Granger et al., 2012; Hodgson & Granger, 2013; Wilde, Out, Johnson, & Granger, 2013); multiple samples can be collected quickly; samples can be obtained in a variety of settings; in many circumstances samples can be self-collected; and samples can also be obtained from multiple individuals simultaneously (Granger et al., 2012; Wilde et al., 2013). These advantages are largely derived from the fact that saliva collection is relatively uncomplicated and rarely requires skilled professionals or special equipment (Ramirez et al., 2017; Wilde et al., 2013). The literature reviewed in this volume (e.g., Chaps. 15 and 24) illustrates how well sampling can be accomplished while patients or research participants continue their normal activity in a laboratory, at home, or in work settings. The purposes of this chapter are to review common practices, identify special considerations, and recommend best practices for saliva specimen collection methods employed in interdisciplinary salivary bioscience research.

## 3.1 Saliva Collection Considerations Related to Research Design

Collection of oral fluid allows for the monitoring of biological reactivity and regulation in everyday life. Saliva sampling is especially advantageous when collecting synchronized samples in real time from individuals, pairs (Ha et al., 2016; Rankin, Swearingen-Stanborough, Granger, & Byrd-Craven, 2018), or large groups of individuals such as sports teams or military populations (Gaviglio & Cook, 2014; Taylor et al., 2008), which expands the possibilities for examining contextually dependent, real-time questions regarding rapid reactivity to external/environmental stimuli of interest to researchers. Salivary measurement enables investigators to collect normative and/or longitudinal data, allowing for the comparison of changes between different treatment groups or for monitoring the effectiveness of interventions (Hodgson & Granger, 2013). Complex, multidisciplinary studies can benefit greatly from the collection of oral fluid by enabling investigators to collect valuable physiological measurements.

Investigators should bear in mind that participant compliance can be significantly impacted by increasing sample volume requirements. An expert panel from the International Society of Psychoneuroendocrinology (ISPNE) wrote a consensus statement (Stalder et al., 2016), sharing strategies and recommendations, for maximizing participant compliance for saliva sampling. Some of those strategies include: (1) explain the importance of the research and adherence to saliva collection timing and sample handling; (2) outline the protocol in detail and/or practice with the participant before samples will be taken, allowing participants to ask questions; (3) provide clear, detailed instructions with contact information for the research team members; (4) provide all supplies in "kits" or "packages" for participants for ease of collection; and (5) contact participants at or before sample collections with reminders of saliva collection times in order to increase compliance. The method of collection must balance participant burden, developmentally appropriate expectations, and availability of resources to manage the cold chain. Several demographic factors may influence the use of a specific type of saliva collection method, some of which we outline below.

**Age/Developmental Considerations** Age can have dramatic effects on different analytes found in saliva. Some changes are gradual, while others occur rapidly at certain times in life (e.g., neonatal period, puberty, and menopause); therefore, different reference ranges should be applied to different age groups. Additionally, levels of sex hormones, such as dehydroepiandrosterone (DHEA), testosterone, and progesterone change at adrenarche and puberty (Marceau, Ruttle, Shirtcliff, Essex, & Susman, 2015; Romeo, 2005).

**Race/Ethnicity** Racial and ethnic differences have been found in salivary biomarkers (e.g., salivary amylase gene; Perry et al., 2007). Additionally, some ethnic/racial minority populations display lower levels of osteocalcin and 25-hydroxy vitamin D. African and West Indian men have higher levels of prostate-specific antigen (PSA), whereas Asian-Pacific populations show no age-related increases in PSA (Nielson et al., 2016). Ramirez et al. (2017) found that in working with ethnically diverse and disadvantaged populations, self-collection of saliva was especially helpful in overcoming cultural and logistical barriers associated with the collection of blood.

**Socioeconomic Status** In light of extensive research on the influences of early life environment, including socioeconomic status (Hertzman & Boyce, 2010; Koss & Gunnar, 2018), we encourage researchers to consider socioeconomic status (SES) in preparing research designs. For example, if SES limits access to a consistent electrical power source for sample refrigeration or freezing, then researchers should provide a cooler or other cold storage option to research participants in free-living or field research environments. Likewise, access to technology (e.g., ability to send reminders via text message, e-mail, or phone call) for those conducting field studies in remote areas (e.g., anthropological studies), and lack of access to these technologies may limit the types of biomarkers examined.

**Physiological/Medical Status** Contextually dependent salivary biomarkers dictate specific research collection schemes, and the physiological status of the participant should be taken into account when the research protocol begins. For instance, when examining the effect of a laboratory-based stress induction task on human stress response, there is a requirement to control or account for physiological status during screening, pre-participation instructions, procedural design, and/or pre/post-task questions asked of the participant.

**Exercise** It is advisable to document vigorous exercise (Henskens et al., 1996; Kivlighan & Granger, 2006) because physical activity can stimulate the production and secretion of hormones such as cortisol and testosterone (Hayes, Grace, Baker, & Sculthorpe, 2015). An individual's fitness level and the amount of exercise performed may dictate the extent of these hormone increases. When resting-state levels of analytes are studied, investigators should instruct participants to refrain from vigorous exercise prior to sample collection (if possible). However, hormone production in response to controlled exercise may be of interest to an investigator, in which case salivary sampling should occur before and after physical exertion (e.g., Beaven, Gill, & Cook, 2008).

**Menstrual Cycles** Menstrual cycle's effects on reproductive and immune biomarkers are well-documented. It is also important, when working with women of child-bearing age, to document the use of contraceptives among female participants, which affects salivary hormone concentrations (Wiegratz et al., 2003; Zimmerman, Eijkemans, Coelingh Bennink, Blankenstein, & Fauser, 2014).

**Medication and Illicit Drug Use** Medication and illicit drugs may create difficulties when interpreting immunoassay results due to their effect on physiological mechanisms or by affecting the analytical process. It is important to document all drug use at the time of biospecimen sampling so that atypical results may be cross-referenced with drug-test interaction data banks (Wilde et al., 2013). Disrupting drug therapy before saliva sampling is typically impractical, unfeasible, and potentially unethical. However, for some assays, it is necessary to cease medication intake momentarily in order to interpret results. Residual substances left in the oral cavity by medications that are inhaled, taken intranasally, or applied as oral topicals can dramatically alter saliva composition. Medications of these categories transport various molecules directly into oral fluids that may potentially interfere with immunoassays. In some cases, the effects on salivary biomarkers such as cortisol are well-documented (Granger, Hibel, Fortunato, & Kapelewski, 2009). Various medications reduce salivary flow (e.g., diuretics, hypotensives, antipsychotics, antihistamines, barbiturates, hallucinogens, cannabis, and alcohol) and can cause alterations in saliva pH and viscosity, thus affecting the movement of molecules from serum into oral fluid. A single medication can potentially affect salivary analytes via numerous pathways or mechanisms; multiple medications taken simultaneously may compound this phenomenon. Wilde et al. (2013) suggest that investigators monitor the use of prescription and over-the-counter medications that: (1) influence the subjective experience of "stress," emotion, novelty threat, or pain; (2) have agonistic or antagonistic effects within the physiological system of interest; (3) interfere with the biosynthesis of the analyte; (4) affect physiological systems networked with the subsystem of interest; (5) alter levels of binding globulins and the fraction of biologically active analyte; and (6) have active ingredients that cross-react or cause nonspecific interference in immunoassays for the salivary analyte in question. If possible, it is highly recommended to document the name, dosage, and schedule of all medications taken within the last 48 h prior to sampling. Investigators should utilize this information to rule out the possibility that the medication is driving the analyte–outcome relationships of interest.

**Acute Illness and Disease States** Illness can influence salivary biomarkers, especially if mucous increases in response to an illness, and/or certain salivary biomarkers are suppressed during acute bouts of illness (e.g., upper respiratory tract infection). Secretory Immunoglobulin A (SigA; see also Chap. 9) has been studied extensively in the context of exercise-induced changes in infection risk and immune function (Karacabey et al., 2005; Keaney, Kilding, Merien, & Dulson, 2018; Orysiak, Malczewska-Lenczowska, Szyguła, & Pokrywka, 2012). It is critical to take into consideration potential medical conditions or disease states that can affect biomarkers of interest.

### *3.1.1 Environmental Factors*

For all settings and regardless of the specific scenario, the key to quality saliva collection is following procedures that decrease the possibility of contaminating saliva and documenting any factors that may affect results. When designing a study, different collection settings such as a laboratory, at home, or in "the field," have various advantages and disadvantages. While laboratory-based tasks typically offer more control over environmental factors, some free-living and field-based studies offer greater ecological validity.

**Lab** Conducting saliva collection in a laboratory presents many inherent benefits: Investigators have direct oversight of the collection process with the ability to control sampling times in a sterile environment; collection at a laboratory often provides immediate analysis or storage and freezing capabilities; and unwanted freeze–thaw cycles can either be avoided, easily monitored, or recorded. Laboratory tasks are especially useful to address both individual differences in reactivity to standardized tasks and commonalities in stress-based tasks (e.g., see Dickerson and Kemeny 2004). However, when examining the biological reactivity to a laboratory-based task, it may be important to allow participants to habituate to the laboratory space, for example, by performing innocuous tasks for 30–40 min before completing the task designed to elicit a stress response. Additionally, participants may visit the laboratory in advance to practice saliva collection and prevent anticipatory stress.

**Free-Living** Although choosing a laboratory as the sampling site may appear to be the best possible choice, it may not always be an ecologically valid or practical option. Certain analytes, such as cortisol, may be anchored to the awakening response, rendering at-home self-collection as the ideal protocol. In free-living environments, the use of "participant adherence" recommendations is especially important, as participants will need to collect with little to no assistance. When the self-collection paradigm is selected, compliance must always be taken into consideration. If particular sampling times should be followed, participants can record the times in handwritten logs (provided by the research team in a user-friendly format) or take time-stamped photographs when they self administer each sample. These studies offer advantages in examining biological processes in real time and can offer insights into the nature of those underlying biological process (e.g., shift workers; Curtis, Bellet, Sassone-Corsi, & O'Neill, 2014; Leproult, Holmbäck, & Van Cauter, 2014).

**Field** For studies conducted in "the field," maintaining the integrity of samples is paramount. Preventing, or at least mitigating, uncontrolled environmental factors or debris from spoiling biospecimens should be a main concern. Weather elements such as wind and rain should be given careful thought prior to data collection. If possible, planning for and allowing participants a sheltered environment for sample collection will increase the likelihood of preserving sample integrity. Keeping tubes properly sealed prior to and after sampling will aid in the prevention of sample contamination. Additionally, careful consideration must be given for sample storage, transport, and cold chain management. Dry ice and travel coolers have proven effective when conducting research for long periods of time with limited freezer resources. When possible, it is practical to pilot saliva collection methods in the field to confirm that they are not creating measurement error.

### *3.1.2 Biologically Relevant Factors*

Despite its ease of use, salivary biomarkers are often dependent on several biologically relevant factors, of which we discuss three major, broad categories: (1) biological rhythms (e.g., circadian, ultradian); (2) salivary biomarker mechanism of entry into the saliva; and (3) the anatomical location of the various salivary glands and subsequent influences on salivary flow rate. With these factors in mind, it is important to examine previous multidisciplinary research in order to determine when and how to collect saliva samples to target specific biomarkers.

**Biological Rhythms** Timing and biological rhythms affect certain analyte levels and must be controlled to enable investigators meaningful interpretation of test results. Various biological systems follow and change according to well-established diurnal, circadian, infradian, and ultradian rhythms. Melatonin and cortisol are examples of analytes that follow diurnal or circadian rhythms, with a periodicity of approximately 24 h. An example of a common infradian rhythm influencing the fluctuation of progesterone and estradiol within human females is the menstrual cycle with a periodicity of approximately 28 days. Therefore, when measuring biological substances by immunoassay, it is critical to understand their rhythmic patterns and to time sample collection accordingly. For analytes affected by biological rhythmicity, maintaining consistency in sampling times along the rhythm is important when comparing within or between individuals in order to make accurate conclusions about the nature of the biomarker of interest.

Most hormones and certain salivary proteins demonstrate a diurnal rhythm of expression, which should be understood in order to optimize research design. Investigators should understand the response and recovery characteristics of their analytes of interest in order to plan the timing of the saliva sampling scheme (e.g., salivary cortisol; Dickerson & Kemeny, 2004). Biospecimen collection should be conducted at specific, standardized times or within a well-defined window of time for effective measurements. For intervention-based studies or studies measuring stress reactions to lab-based stimuli, saliva sampling should be appropriately timed to capture accurate responses to acute stimuli. Furthermore, it is important to note that diurnal patterns can confound reactivity measurements, so there is a need to control or account for this influence when attempting to isolate a stress response. Along these lines, there is a distinct difference between adjusting for individualized patterns (using, for example, wake times) and standardized sampling (e.g., all subjects provide a 9:00 sampling). The analyte that is most well-known for its sensitivity to such methodological differences is cortisol, but many other analytes (e.g., DHEA, testosterone) may also vary as a function of the chosen method.

**Mechanisms of Entry into Saliva** For most steroid hormones and other small, neutral molecules, the major conduit into saliva is provided by passive diffusion. Dense beds of capillaries surround the salivary glands, with many blood constituents passing easily through the capillary walls. These blood components then wash over the salivary glands, which harbor secretory cells. Neutral steroids readily and rapidly diffuse through the lipoprotein membranes of these secretory cells into saliva. Stimulation of saliva flow does not appear to influence these concentrations of these neutral steroids in the saliva (Hofman, 2001).

However, molecules such as DHEA-sulfate (DHEA-S) and other charged or conjugated steroids filter into saliva from blood; thus, assay results for DHEA-S may be difficult to interpret without reference to flow rate. The challenge to interpret meaningful results increases since saliva flow may vary from person to person or for a single individual at different times.

The protein SIgA, which is synthesized in the mouth and not derived from blood, as well as sAA, which is produced locally, are also flow rate dependent. B-lymphocytes secrete polymeric Immunoglobulin A (IgA) in the area of the salivary glands. Polymeric Immunoglobulin receptors then actively transport the polymeric IgA through the cell membranes to be released into saliva. Nervous stimulation of the salivary glands increases secretion of SIgA. However, as flow rate increases, levels of SIgA in saliva decrease. Therefore, there must be an upper limit to the speed of SIgA's transport (Brandtzaeg, 2013). The release of sAA is controlled by nervous signals in the sympathetic and/or parasympathetic systems that also manipulate the flow rates of saliva.

A correction must be made in order to express the measure of flow rate-dependent analytes, such as SIgA, DHEA-S, and sAA, as "output as a function of time." It is important to note that absorbent devices are susceptible to ceiling effects. That is, estimates of flow rate may be inaccurate unless the swab is removed from the mouth before it reaches absorbance capacity (Beltzer et al., 2010). This is especially of concern with smaller swabs, since they may reach their saturation limit quickly. That said, there may be additional analytes that are influenced to varying degrees by flow rate, implying that it is best practice to measure flow rate when feasible.

**Anatomical Collection Site and Salivary Flow Rate** Whole saliva is a mixture of oral fluids secreted from various glands within the mouth (see Chap. 2). The main sources of saliva are the parotid gland region (upper posterior of the oral cavity), the submandibular gland region (lower area of the mouth between the cheek and jaw), and the sublingual gland region (under the tongue). Some ancillary secretory glands are found in the cheek, lip, palate, and tongue. Crevicular fluid from serum leakage in the cleft area between each tooth, and its surrounding gums composes a small percentage of oral fluid. Additionally, mucosal injury or inflammation from oral disease can cause serum leakage. Lastly, nasal and bronchial secretions, bacteria and its derivative products, and tears are all components of saliva. In this state, whole saliva is rich in glycoproteins, known as mucins, which come from the mucous cells of the sublingual, submandibular, and minor glands.

The composition of whole saliva is complex. Each source gland's contribution to the oral fluid pool varies due to each secretory gland generating fluids of different makeup and volume. The most active glands are those of the submandibular area, generating about 65% of saliva in the mouth when in a state of rest and under minimal stimulation. The parotid gland does not produce oral fluid that contains mucins, which make saliva viscous and sticky. However, once the parotid glands are stimulated via taste, smell, or chewing motions of the jaw, watery saliva production increases, which contain enzymes like sAA and lipase (Edgar, 1990). The increased flow assists with chewing and commences the digestive process. When utilizing the absorbent device method, it is possible to collect different types of oral fluid depending on where the swab is placed in the mouth; this of course introduces variability in terms of measured levels or activity of some analytes if methods vary between and/or within participants' collections. For analytes collected optimally via parotid saliva, swabs may be placed next to the second upper molar, between the cheek and upper gum. This area, where the parotid duct opens into the mouth, has a lower flow rate of unstimulated parotid saliva; therefore, absorbent devices should be left in place for a longer duration to collect a sufficient amount of oral fluid. As parotid flow increases, mucin concentration decreases in whole saliva, potentially altering the balance of other components found in saliva (Proctor, 2016).

If utilizing a swab, it is recommended that while sampling, swabs are not moved around in the mouth. This is especially important when measuring for analytes that are known to be influenced by mouth location. Swabs may collect localized saliva rather than whole saliva, affecting some analyte results like levels of sAA and SIgA that vary by mouth location (Beltzer et al., 2010; Crawford, Taubman, & Smith, 1975; Veerman, van den Keybus, Vissink, & Nieuw Amerongen, 1996).

DNA can be collected from all areas of the mouth and can be extracted from samples collected for other immunoassay testing. Collecting separate and additional samples for DNA testing is often unnecessary. Once samples have been centrifuged, the pellet that is formed at the bottom of the sample contains cells; DNA is found within the nuclei of these cells (see Chap. 6). Therefore, if genetic testing is desired, it is recommended to keep the pellet or any absorbent devices used for collection since the fibers of the collection device capture much of the cellular material (Nemoda et al., 2011).

**Sample Volume and Salivary Stimulants** Inadequate volumes of saliva for analysis are a genuine concern, making xerostomia (dry mouth) a threat to testing validity. Technology in the field of salivary bioscience has improved considerably over the past decade; thus, modern immunoassays have the capacity to utilize extremely small amounts of saliva for successful processing. Often, less than 100 μL of oral fluid is necessary for analyses. Stimulants are typically not necessary and, in some cases, will interfere with assay functioning. Introducing oral stimulants while collecting saliva samples may potentially cause assay interference or alter the levels of some analytes, rendering artificially high or low levels (Granger et al., 2007; Schwartz, Granger, Susman, Gunnar, & Laird, 1998). Since oral stimulants are a potential source of variation, it is wise to avoid them altogether. The use and act of chewing on seemingly innocuous stimulants like unflavored paraffin or wax to encourage saliva production may possibly affect flow-dependent analytes.

If oral fluid cannot be collected successfully and the use of stimulants is absolutely necessary, a few recommendations should be noted. Talge, Donzella, Kryzer, Gierens, and Gunnar (2005) advise sparing use of stimulants in a consistent manner throughout the entirety of the study. Before investigators introduce oral stimulants into their collection methods, they should explore the option of olfactory or visual stimulants; the inhalation of lemon or orange oil extracts or the use of visual aids like pictures of food are acceptable methods to stimulate oral fluid production (Schwartz et al., 1998). Even audio cues like simply mentioning the word "lemons" to invoke mental imagery has personally and anecdotally been a successful catalyst for initiating salivary flow. Prior to actual study sampling, it may be wise to practice the saliva stimulation methods with participants to verify effectiveness and ensure a participant's comfort level. If any oral stimulants are employed in a study, investigators should conduct a pilot study to confirm that it does not cause assay or analyte interference (Granger, Kivlighan, Fortunato, et al., 2007). Investigators should consult with testing services, the kit manufacturer's website, or kit inserts for minimum collection volumes needed for testing various analytes. In general, it is important to collect enough saliva to repeat the test (if needed), with an additional 25% to account for mucus and liquid loss during assaying. For example, salivary cortisol enzyme immunoassay testing requires 50 μL to assay in duplicate; therefore, it is recommended that researchers collect approximately 125 μL, i.e., $(50\ \mu L \times 2) = 100\ \mu L$, then $(100\ \mu L \times 1.25) = 125\ \mu L$.

For best results, investigators should standardize instructions given to research staff and participants about oral swab placement and monitor compliance. Another way to combat measurement variability related to swab placement is to use the passive drool method and collect whole saliva that has pooled on the floor of the mouth. This is especially useful for investigators wishing to biobank samples for future testing.

In order to measure multiple analytes from the same sample, investigators can sum the recommended sample volumes of all target analytes, including a supplemental 300 μL (0.3 mL) to determine total volume. The additional 300 μL of oral fluid is to account for any loss of sample during handling and if repeat analyses are desired and/or mucus that cannot be assayed for salivary biomarkers. A minimum volume of 500 μL (0.5 mL) of whole saliva obtained via the passive drool method is adequate to ensure the extraction of a sufficient amount of DNA for multiple polymorphism assays. However, collecting 1.5–2.0 mL of saliva will give the flexibility to analyze for DNA polymorphisms and for other hormones or biomarkers of interest (Pulford, Mosteller, Briley, Johansson, & Nelsen, 2013).

**Blood Contamination** Blood contamination within saliva can pose a legitimate concern since analytes found in serum have much higher levels than those same analytes present in oral fluid. In other words, levels of most analytes are higher in the general circulation (10- to 100-fold) than in saliva, making any blood leakage into the mouth a serious confound. Even visually undetectable amounts of blood contamination can elevate analyte levels of oral fluid (Kivlighan et al., 2004; Kivlighan, Granger, & Schwartz, 2005; Schwartz & Granger, 2004).

It is possible for blood and blood components to leak into saliva in numerous ways, such as abrasions, burns, or cuts to the cheek, gums, or tongue. Populations more prone to having blood-contaminated saliva are children with loose deciduous teeth, individuals with poor oral health (e.g., gingivitis, open sores, and periodontal disease), those suffering from certain infectious diseases (e.g., human immunodeficiency), and individuals who engage in behavior that negatively affects oral health (e.g., tobacco usage). Based on the amount of contamination, saliva samples visibly contaminated with whole blood appear yellow-brownish in color, to varying degrees. There are several mechanisms by which to address potential blood contamination in saliva of participants: (1) Screening participants by asking about recent oral history (e.g., "Have you recently lost any teeth?") before the study; (2) Pre-study instructions to participants to avoid dental work and teeth brushing; (3) Systematically inspect saliva samples and recollect if possible; (4) Include questions during the study for post-assay screening of results (if prescreening is not practical or possible); and/or (5) Quantify blood contamination. For example, Kivlighan et al. (2004) created the Blood Contamination in Saliva Scale (BCSS), which is a 5-point scale to assess blood contamination, or researchers can utilize a blood transferrin assay to test samples suspected of blood leakage.

## 3.2 Data Collection Methods

Before choosing an appropriate saliva collection method, several criteria should be considered, including: the participant age and species, the target analytes and whether DNA will also be analyzed, the number of required samples, the required sample volume, whether collection will be self-administered or assisted, and whether samples will be archived or discarded.

When collecting biospecimen for DNA analysis, participants should abstain from eating a meal at least 1 h prior to collection. Additionally, participants should rinse their mouths with water thoroughly to remove food particles or contaminants and allow 10 min to expire before saliva collection. Sampling too soon after rinsing the mouth may decrease the amount of extractable DNA and may affect any hormone or biomarker analyses. In order to prevent degradation of the DNA by enzymes found in the saliva and skin, participants and investigators handling samples should wash their hands prior to saliva collection and wear disposable gloves during actual sampling.

It is not recommended to use cotton as an absorbent device because cotton is a biological material; it increases variability within results that differ significantly from "true passive drool values" (Shirtcliff, Granger, Schwartz, & Curran, 2001). For DNA sampling, buccal swabs may also be acceptable. To collect buccal cells, an investigator or participant providing the biospecimen should rub the inside of the cheeks with the swab for 30–60 s, applying firm pressure.

Regardless of collection technique utilized, the following supplies should be used:

1. Only high-quality polypropylene cryovials and swab tubes for sample storage to maintain analyte integrity. (Measured values of analytes can be unfavorably affected by using polystyrene or other non-validated plastic tubes.)
2. Disposable gloves for researcher (following universal precautions).
3. Cryolabels and/or freezer-proof markers to label samples.
4. Storage boxes for samples after collection.
5. Napkins or tissue (in the event of a biohazard spill).
6. Access to trash can or biohazard bag (in the event of blood leakage).

Other optional items include:

1. −20 °C freezer (laboratory-based research).
2. Small containers to transport samples to freezer.
3. Dry ice or ice packs for storage.

When investigators have identified their population of choice, whether it is adults, children, infants, individuals needing assistance, or nonhuman animals, certain factors should be taken into account. Various analytes may be influenced heavily by specific biological processes that occur during certain age ranges. See Fig. 3.1 for a brief summary list of demographic and biomarker considerations.

**Fig. 3.1** Special considerations for selecting a salivary collection technique

| Special Considerations for Selecting an Appropriate Collection Technique |
|---|
| • Participant age and species<br>• Volume of sample available and needed<br>• Single or multi-analyte measurement<br>• Participant burden<br>• Number of samples to be collected<br>• Self- or assisted collection<br>• Test and discard or archive samples for future<br>• Setting (e.g., home, lab, and field) |

### 3.2.1 Adults and Older Children

Adults and older children are often, with few exceptions (e.g., populations needing assistance), able to follow instructions for passive drool. As the "gold standard" of saliva samples (i.e., "whole saliva" discussed above), instructions are slightly more complex than other methods. However, in some instances, collecting saliva via passive drool is not feasible or significantly interferes with the research protocol. An alternative option that limits salivary biomarker options, but can streamline the saliva collection process, is to collect via swab method.

**Passive Drool** Because some analytes can only be tested with saliva collected via passive drool, selecting the collection method that is compatible with the analyte of interest is vital. In order to sustain the consistency in the type of sample collected, the majority of investigators prefer to use unstimulated, whole saliva acquired by the passive drool method. There are additional advantages to using the passive drool technique, such as its cost effectiveness and its ability to maintain sample integrity. By drooling into a cryovial using a plastic straw device, a large sample of volume can be collected within a relatively short duration (3–5 min), depending on timing and volume needs. Furthermore, target collection volumes can be instantly and visually confirmed. When using the passive drool procedure, the acquired oral fluid is a collective mixture of the output from all salivary glands and this method does not present interference related to stimulating or absorbing saliva. If storing samples long term, vials should always seal tightly and withstand temperatures as low as −80 °C. Because concentrations of SIgA, DHEA-S, and sAA are affected by saliva flow rate, investigators interested in these analytes should typically choose passive drool as their collection method (Beltzer et al., 2010; Kugler, Hess, & Haake, 1992; Vining, McGinley, & Symons, 1983), measure sample collection time in the field, and request sample weights upon intake to the processing laboratory.

Passive drool materials: (1) Polypropylene cryovials, (2) Short straw or Saliva Collection Aid (e.g., SalivaBio LLC, Baltimore, MD). Collection instructions: (1) Participants should first swallow all saliva in the mouth to prevent unnecessary interference from any leftover saliva. At the same time, researcher or participant should start a timer if measuring salivary flow rate; (2) Allow saliva to pool in mouth by refraining from swallowing saliva; (3) With head tilted forward, participants must then drool down a short straw-like device or collection aid, gently guiding the saliva down the straw and into the cryovial; (4) Participants should repeat steps 2–3 of this process until the desired volume of saliva is collected; (5) Participants can then remove/discard the straw or collection aid and tightly attach the cap of the cryovial; (6) Keep sample cold on ice and as soon as possible, freeze at a minimum of −20 °C; and (7) Participants should record the time it takes them to provide the sample and label sample with freezer-proof label or marker.

Passive drool tips: If the participant is having difficulty salivating, asking them to imagine eating lemons or their favorite foods may help. However, see section above on salivary stimulants for more specific guidance on the use of stimulants. When collecting whole saliva, it is normal for the saliva to foam. It is recommended to use a larger vial with extra capacity to account for foam and for the expansion of the oral fluid once the sample is frozen. It is often helpful to mark vials for the specified volume needed in order to give the participants guidance on how much saliva to collect.

**Oral Swab** For participants who are unable or unwilling to passively drool into a cryovial, the absorbent device method may be utilized. Sliding a swab out of a collection tube and into the mouth is rather simple; therefore, participants sometimes find the absorbent device method less burdensome to complete than the passive drool method. Swabs come in different sizes in order to collect oral fluid from different age groups and species. Due to the risk of choking, the adult swabs are not recommended for children under the age of 6 years old. There are long swabs that can be held by an adult for young children. It is important to recognize that not all swabs are created equal and some manufactured brands are limited to certain analytes. For example, some brands of swabs are specifically manufactured and validated for the use of collecting and measuring salivary cortisol only. Using this particular brand of swab to measure other hormones may produce erroneous results. Investigators should use nontoxic, inert polymer swabs that have been validated for the measurement of the analyte of interest and that are consistent across all lots. Another advantage to using swabs for saliva collection is that they tend to improve immunoassay results, as they assist in filtering large macro molecules, mucus, and other particulate matter from the sample. When centrifugation is unavailable, it is possible to express saliva from the swab into a cryovial using a needle-less 5 cc plastic syringe; this method will result in smaller sample volumes recovered. Investigators interested in genetic analysis may find the swab method beneficial because genetic material adheres to the swab.

### 3.2.2 *Adult and Older Children*

Oral swab materials: (1) Oral swab (inert polymer), (2) Swab storage tube (polypropylene); and (3) Optional: Disposable gloves, 5 cc syringe, and cryovials (polypropylene). Collection instructions: (1) Participants should first swallow all saliva in the mouth to prevent unnecessary interference from any leftover saliva. At the same time, the researcher or participant should start a timer if measuring salivary flow rate; (2) Participant should place the swab into the proper mouth location, as directed; (3) To ensure saturation, participants should keep the swab in place for 1–2 min; (4) Without touching the swab with their hands, participants should place the swab back into the storage tube using their teeth. If participants are unable to do so, using disposable gloves to remove the swab from their mouth is acceptable; (5) The cap should be replaced and snapped securely onto the tube; (6) Keep sample cold on ice and as soon as possible, freeze at a minimum of −20 °C; (7) Participants should record the time it takes them to provide the sample and label the sample with a freezer-proof label or a marker.

If utilizing a syringe to confirm sample volume, follow these additional instructions: (8) The investigator/participant must remove the plunger from the syringe and insert the swab into the syringe barrel; (9) The plunger can then be replaced into the syringe and the syringe can be squeezed to express saliva into a cryovial; (10) Collection procedures should be repeated if additional volume is needed and collection is not time dependent; (11) The tube should be capped tightly and swabs may be discarded (unless DNA analysis is required).

Additional oral swab tips: Placing the absorbent devices in the location where saliva is pooling is recommended. For an adult or child sitting upright, this is typically the floor of the mouth under the tongue. Investigators should note that the swab may cause temporary dryness of the oral cavity or of the mucosal membranes; therefore, it may be beneficial to provide participants with access to water for post-testing consumption.

### *3.2.3 Small Children and Infants*

Some older preschoolers are able to provide saliva using the passive drool method. However, when collecting from small children it is more conventional to use the absorbent device technique. To eliminate the risk of choking, oral swabs come in smaller diameters to fit the mouths of children and in longer lengths so a parent or technician can hold onto the other end. Adult assistance and supervision is always required when sampling from small children and infants. In contrast to unfamiliar staff or technicians, parents may have more success in getting the child to accept the absorbent device into his or her mouth.

**Children's Oral Swab** Materials: (1) Children's oral swab (inert polymer, length: 125 mm, diameter: 8 mm); (2) Swab storage tube (polypropylene); (3) Disposable gloves; and (4) Optional: scissors, 5 cc syringe, and cryovials (polypropylene). Collection instructions: (1) Researcher/parent should first instruct the child to swallow all saliva in the mouth to prevent unnecessary interference from any leftover saliva. At the same time, researcher/parent should start a timer if measuring salivary flow rate; (2) Using disposable gloves, parents or a technician should securely hold one end of the swab and place the other end under the child's tongue; (3) To ensure saturation, parents/technicians should keep the swab in place for 60–90 s (or collect in intervals by reintroducing the swab into the mouth as necessary until a third of the swab is saturated); (4) Parents/technicians should place the swab into the storage tube, cutting off or folding over the dry end of the swab to fit the tube; (5) The cap should be replaced and snapped securely onto tube; (6) Participants should record the time it takes them to provide the sample and label the sample with a freezer-proof label or a marker, (7) Keep sample cold on ice and as soon as possible, freeze at a minimum of −20 °C.

If sample volume must immediately be assessed, follow the same instructions as above in the adult swab section. Children's swab tips: Investigators and parents should be sure to collect enough sample volume. Typically, the volume of sample recovered from children's swabs falls in the range of 200–1000 μL (0.2–1.0 mL). It may be possible to collect pooling saliva at the corners of the child's mouth. Lastly, all swabs and collection materials should be stored out of the reach of children.

**Infant Oral Swab** Special considerations must be taken when collecting saliva from infants less than 6 months old because of the increased potential for choking when collection devices are placed in the infant's mouth. Again, adult assistance and supervision is critical when using infant oral swabs. Like the children's swabs, infant swabs are longer in length to allow one end to be held by a parent or technician while the other end is placed in the infant's mouth. When sampling from children, the Sorbette (BD Opthalmic Systems Visipear, prod. No. 581089), which is a small, spear-shaped device, composed of a hydrocellulose sponge head and a plastic shaft, has been used in past studies. However, the Sorbette presented challenges of low recovery volume, limited absorbent capacity, and assay interference with analytes other than cortisol, a-amylase, cotinine, and SIgA. It is important to note that similar "eyespear" devices from other manufacturers are made for the collection of tears only and should not be used for saliva collection (Harmon, Hibel, Rumyantseva, & Granger, 2007).

Infant swab materials: (1) Infant oral swab (inert polymer, length: 90 mm, diameter: 6.3 mm); (2) Swab storage tube (polypropylene); (3) Disposable gloves; and (4) Optional: scissors, 5 cc syringe, and cryovials (polypropylene). Collection instructions: (1) Participants should record the time it takes them to provide the sample if the analytes of interest are flow rate dependent. (2) Using disposable gloves, parents or a technician should securely hold one end of the swab and place the other end under the infant's tongue; (3) To ensure saturation, parents/technicians should keep the swab in place for 60–90 s (or collect in intervals by reintroducing the swab into the mouth as necessary until a third of the swab is saturated); (4) Parents/ technicians should place the swab into the storage tube, cutting off or folding over the dry end of the swab to fit the tube; (5) The cap should be replaced and snapped securely onto tube; (6) Participants should label the sample with freezer-proof label or marker; (7) Keep sample cold on ice and as soon as possible, freeze at a minimum of −20 °C.

If sample volume must immediately be assessed, follow the same instructions as the adult swab instructions listed above. Infant swab tips: For infants (and even children and adults) in a reclined position, it is recommended to turn the head to the side. Parents or investigators should then collect biospecimen from the side of the cheek where saliva is pooling.

### 3.2.4 *Geriatric and Special Populations*

It is important to note that most individuals are able to utilize the passive drool or adult swab methods. However, when sampling from subsets of the geriatric population and special populations of participants who are not able to collect samples themselves, investigators and caretakers may use the same collection protocols employed for young children. For geriatric patients, it is possible that the use of multiple medications may cause dry mouth, so collection times may need to be extended.

### 3.2.5 *Nonhuman Animals*

Salivary bioscience has proven to be incredibly functional in monitoring the health and well-being of animals. Testing salivary analytes in animals has provided investigators of biobehavioral studies with valid, consistent, and repeatable results. Early techniques involving saliva collection from deer (Millspaugh et al., 2002), dogs (Dreschel & Granger, 2009, 2016; Horváth, Igyártó, Magyar, & Miklósi, 2007), guinea pigs (Emack, Kostaki, Walker, & Matthews, 2008), pigs (Gutiérrez, Martínez-Subiela, Eckersall, & Cerón, 2009), nonhuman primates (Lutz, Tiefenbacher, Jorgensen, Meyer, & Novak, 2000; Newman, Perry, & Carroll, 2007), and other animal species (Gómez, Jewell, Walker, & Brown, 2004) have used cotton ropes, hydrocellulose sponges, plain or flavored pads, Salivettes (Sarstedt, Inc.), swabs, and other devices. When collecting saliva from small animals such as mice and rats, other devices such as capillary tubes, filter paper strips, plastic pipettes, and more sophisticated suction devices have been used in prior research.

Newer absorbent devices have significantly reduced saliva collection burden. For comfort and ease of collection, infant and children's swab devices are highly recommended for use with animals that have mouths 8 mm and larger in diameter. There are several benefits to using infant and children's swabs for sampling with animals. Swabs are nontoxic, comfortable, durable, and resistant to rips and tears. Additionally, swabs provide quick, capillary action collection. Depending on the device, a significant volume of approximately 1.0–2.0 mL of saliva can be collected at one time. Essentially, investigators can apply the same protocol for collecting oral fluid from children or infants when working with animals.

Additional nonhuman animal collection tips for larger animals include: (1) Introduce swabs slowly and train animals prior to sampling; (2) Dab the swab in areas of pooling saliva; (3) If needed, swabs can be flavored with simple solutions to increase animal acceptance of the swab. If this technique is used, a pilot study is recommended; (4) When collecting saliva from non-domesticated animals, "extended reach" devices like poles and sticks, are often used.

### 3.2.6 *Practical (Everyday) Sources of Variability*

When preparing for a study, participant state and preparation must be taken into consideration. Stress levels should be documented before saliva samples are collected. Depending on the nature of an investigator's study and the analytes of interest, it is essential to consider other factors such as: drug interactions with assays and effects of food and drinks.

**Medical/Dental Procedures** Documenting recent medical procedures offers insights for the interpretation of assay results within the context of the research design of interest. Levels of circulating analytes may be influenced by short-term effects of some medical procedures and postsurgical stress. As a result, sufficient recovery time should be given to a participant before saliva sample collection. For example, levels of certain thyroid hormones are often depressed post surgery. Medical interventions such as radiation influence the secretion of oral fluids (Wilde et al., 2013). The presence of oral diseases should be recorded since blood contamination may confound results (Henskens et al., 1996; Kivlighan et al., 2005).

**Food and Drink** Due to their potent influence, food and drink play a critical role in the integrity of saliva samples. After eating or drinking, particulate matter may be left in the oral cavity having the potential to influence salivary pH levels or viscosity. It may contain substances that cross-react in immune or kinetic reaction assays, such as bovine hormones, active ingredients in medications, or enzymes. For this reason, it is recommended that research participants abstain from food or drink consumption 10–20 min prior to saliva sampling.

**pH of Saliva** Avoid foods that alter saliva pH and increase bacterial growth, such as those with alcohol content, high sugar or acidity contents, or high caffeine levels, immediately before sample donation; these may compromise assay results (Klein, Bennett, Whetzel, Granger, & Ritter, 2010; Schwartz et al., 1998). Document the consumption of caffeine, nicotine, prescription, and over-the-counter medications within 12 h prior to saliva sampling (Granger et al., 2007, 2009; Hibel, Granger, Cicchetti, & Rogosch, 2007; Sheth, López-Pedrajas, González-Martínez, & Veses, 2018). Alcohol ingestion varyingly alters body fluid composition depending on whether an individual is a casual drinker or an abuser; therefore, alcohol consumption and time of consumption should be documented (Sheth et al., 2018). Saliva flow is stimulated by alcohol, which is respirated out of the body. After consumption, this increased ethyl alcohol (ethanol) will be present in saliva samples. Enzyme immunoassays are compromised by high levels of ethanol in saliva biospecimens because it interferes with the binding of antibodies. This only presents an issue if the participant has consumed high quantities of alcohol within 12 h of sampling or is still experiencing the effects of intoxication (i.e., "hung over").

**Dilution of Salivary Biomarkers** If anything is consumed before saliva sampling, participants should rinse their mouth with water and allow 10 min to pass prior to sample donation. This will allow saliva to equilibrate within the mouth and the dilution effect should wear off. Not allowing enough time to pass after rinsing could lead to artificially lowered concentration/volume (μg/dL, ng/mL, pg/mL) or activity/volume (U/mL) estimates of salivary analytes (Beltzer et al., 2010; Fenoll-Palomares et al., 2004). Rinsing the mouth for awakening samples is not recommended or necessary since food consumption theoretically would not have occurred during slumber.

**Context-Dependent Stressors** There are several basic contextual stressors that can interfere with the interpretation of the research design. Some of these contextual stressors are: sitting versus standing when participants collect saliva samples (e.g., Deacon & Arendt, 1994), anticipatory stress for a stressful task or event (Pulopulos, Vanderhasselt, & De Raedt, 2018; Van Paridon, Timmis, Nevison, & Bristow, 2017), and novel environments such as a laboratory space (Federenko, Nagamine, Hellhammer, Wadhwa, & Wüst, 2004; Zmyj, Schneider, & Seehagen, 2017).

## 3.3 Sample Handling, Transport, and Storage

Once samples are collected, it is important to manage handling, transport, and short- or long-term storage of samples in order to maintain the integrity of the salivary biomarkers of interest. Below we outline a few of the methods and handling maxims that should be followed when handling saliva samples.

### *3.3.1 Cold Chain Management and Additives*

It is crucial to protect the integrity of saliva samples by enforcing proper handling, storage, and transport methods. Although some analytes are stable, numerous salivary hormone, peptide, and protein levels endure significant declines at extended exposure to room temperatures. Refrigeration of samples prevents degradation of some analytes and limits the activity of proteolytic enzymes and bacterial growth (Whembolua, Granger, Singer, Kivlighan, & Marguin, 2006). Therefore, saliva samples should typically be kept cold or frozen as soon as possible once they have been collected. Ideally immediately after collection, samples should be frozen at or below −20 °C; many household freezers can accommodate these conditions. If freezing is not feasible, then at a minimum samples should be kept cold on ice or refrigerated at 4 °C and maintained at this temperature for no longer than necessary. Samples should then be frozen as soon as possible in a non-cycling freezer (Pramanik et al., 2012).

If stored long term, it is recommended that saliva be expressed from the swab into cryovials for storage at −80 °C. This can be achieved most effectively by centrifuging the swab or less effectively by squeezing the swab in the chamber of a 5 cc syringe, and then placing the saliva into the cryovials. At −80 °C, samples can be stored for several years. However, the precise time has yet to be determined and could differ by analyte. Some investigators have found that samples, which were properly stored for over 4 years, have displayed little or no degradation (Garde & Hansen, 2005; Gröschl, 2008; Janardhanam, Zunt, & Srinivasan, 2012).

When considering DNA analyses, it is still ideal for investigators to freeze saliva samples as soon as possible. However, the quality of DNA can remain uncompromised for up to 5 days when stored at room temperature for some genetic testing. DNA is hardy against multiple freeze–thaw cycles without its quality being significantly affected (Nemoda et al., 2011). If hormone or biomarker analysis is to be combined with DNA testing, investigators should always follow storage instructions for the more delicate analytes; therefore, biomarkers take precedence over DNA in regards to storage procedures. One significant obstacle to using saliva as a biospecimen of choice is its requirement of transporting samples through refrigerated packing methods to maintain sample integrity. This delicate process is otherwise known as cold chain management. For at-home collections, individuals can ship saliva samples via regular mail in Styrofoam bioshipment containers and packed on an adequate amount of dry ice. Packing samples on enough dry ice to last the duration of transport should sufficiently safeguard the integrity of the samples and maintain the cold chain for the entire transport period. If shipping on dry ice is not feasible or practical, participants should freeze their samples and package them along with ice packs in an insulated cooler. Cold chain management can be logistically complex and cost-prohibitive for studies completed at home, for large-scale national surveys, or for studies conducted in remote locations. For global health studies, cold chain management may be even impossible. Investigators must note that sample integrity can be significantly compromised when the cold chain is not properly managed, whereas other biomarkers may remain stable for long periods of time at varying temperatures (Nalla, Thomsen, Knudsen, & Frokjaer, 2015).

Since it is approved for all salivary analytes, the passive drool technique is the recommended, gold standard method for saliva collection. To protect some analytes from rapid degradation, like neuropeptides, sampling directly into chilled storage vials or treating biospecimens with inhibitors (such as EDTA or aprotinin) may be required (Ng, Koh, Choo, & Chia, 2006). When collecting oral fluid for genotyping analysis alone, the swab method is a convenient method since DNA can be extracted directly from the fibers of the swab; therefore, it is important to evaluate the type of swab material utilized for DNA analysis (Bruijns, Tiggelaar, & Gardeniers, 2018). This remains true unless an investigator desires to extract high yields of DNA for microarray genomic analysis studies; then passive drool is the optimal method of collection (Pramanik et al., 2012).

### *3.3.2 Prioritizing Analyte Testing*

When prioritizing testing and before freezing samples, investigators should know whether their analytes of choice are stable or more sensitive to repeated freeze–thaw cycles. Testing multiple analytes on the same day but with fewer samples can aid in minimizing freeze–thaw cycles. Additionally, this strategy of assaying smaller batches of samples for multiple analytes may reduce the burden for the investigator conducting the assays. Alternatively, before freezing, investigators can aliquot fresh saliva into smaller volumes, with the understanding that the following analyte concentrations may differ slightly between each aliquot.

For large sample volumes, investigators should prioritize the order of assays by first testing unstable analytes and applying this order of operations consistently across all samples. When testing for various freeze–thaw sensitive analytes, investigators may find it beneficial to aliquot samples prior to the first freeze–thaw. Alternatively, aliquoting the supernatant into smaller test volumes subsequent to the first freeze–thaw is another option (Pramanik et al., 2012; Slavish, Graham-Engeland, Smyth, & Engeland, 2015).

**Freeze–Thaw Cycles** Regardless of the type of biospecimen, freezing then thawing samples may have a dramatic effect on the quality of the sample and on analyte levels. Therefore, as a general rule, investigators should avoid repeated freeze–thaw cycles. Some analytes such as DHEA, progesterone, estradiol, and CRP are extremely sensitive to freeze–thaw (Wilde et al., 2013). Other analytes are more robust and can withstand up to at least three cycles, like DNA, cortisol, testosterone, and sAA. This general rule of avoiding multiple freeze–thaw cycles remains consistent when aliquoting and archiving frozen samples to be assayed for future analyses (Pramanik et al., 2012).

**Aliquoting Samples for Future Analysis** When testing for multiple or temperature sensitive analytes, it is often more practical to collect whole saliva in a single cryovial and aliquot the sample into several tubes immediately after collection. Samples are then better able to avoid multiple freeze–thaw impacts. Albeit, this may be somewhat difficult due to the high viscosity of saliva. Additionally, analyte values within the aliquots of a given sample might vary slightly. However, aliquoting samples is still more beneficial than subjecting the samples to repeated freeze–thaw cycles. When pipetting highly viscous whole saliva, investigators should aspirate slowly to achieve greater accuracy in sample volume and to avoid spillage or the formation of bubbles. To decrease sample viscosity and ensure more uniform concentrations of a particular analyte across all aliquots, creating the aliquots after one freeze–thaw cycle and centrifugation is recommended. This allows the mucins to precipitate and fall out of the solution. The resulting solution or supernatant will be clear, water-like in consistency, and more easily pipettable, enabling investigators to create more reproducible aliquots (Slavish et al., 2015). Note though that this process adds an additional freeze–thaw that may have detrimental effects on certain analytes (e.g., see Gröschl, 2008). Collecting multiple samples sequentially, or back-to-back, is not recommended since substantial variability in analyte concentrations could be present from one sample to the next; instead, researchers should collect a single sample and aliquot after collection.

If sampling opportunities or collection materials are limited, researchers may find biobanking (archiving biospecimens for future investigations) an attractive option to mitigate time constraints and maximize cost effectiveness. In the USA, biorepositories have been in existence for over 150 years starting with the pathology specimens of the Civil War era and organized by the Armed Forces Institute of Pathology (Eiseman & Haga, 1999). Since then, biorepositories of blood, urine, tissue, and saliva have existed in various forms from small clinics to modern-day, automated facilities responsible for managing millions of samples (Vaught, Henderson, & Compton, 2012).

Investigators interested in developing a saliva biorepository should note that storing samples at −40 °C or below is the most optimal and economic option for ensuring consistent quality of the samples. The stability of analytes in saliva that have been frozen at −40 °C or below does not differ from the stability of those found in other biospecimens such as blood or urine (Barranco et al., 2019; Karched, Bhardwaj, Pauline, George, & Asikainen, 2017). The number of freeze–thaw cycles the samples endure should always be documented.

## 3.4 Concluding Comments

Saliva collection offers myriad benefits to researchers in contextually rich environments, which expands the possibilities in understanding interactions between individuals and their environments. The various target populations, environmental factors, common sources of variability, collection, and storage methods can affect the interpretation of salivary biomarkers within the context of the research program. Therefore, we recommend addressing these factors in a variety of ways: (1) prescreening participants for participation, (2) providing pre-study instructions to participants, (3) creating standardized procedures that apply between and within participants across time, and (4) documenting sources of variability by asking participants questions that assess the likelihood of interference from extraneous variables. Figure 3.2 outlines the variables of interest and how to address them within a research design of interest.

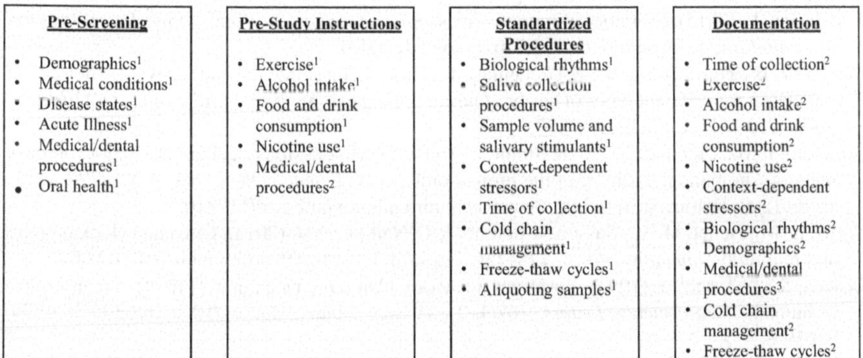


**Fig. 3.2** Prioritizing sources of error in salivary sampling. Note: Superscript numbers indicate the preferred order of prioritizing the sources of error

If not feasible to prescreen participants or participants do not comply with pre-study requests, research design can account for the possibility of noncompliance. For example, pre-task procedures can require the research participant to sit quietly, answering innocuous questionnaires or performing tasks that require little effort in order to habituate participants to the laboratory space and saliva collection procedures. However, this may not be possible in free-living or field-based research. Therefore, it is important to utilize objective methods to verify participant compliance in order to account for those factors; for example, in order to verify cortisol awakening response, researchers may use technologies such as polysomnography in order to verify waking time (Adam & Kumari, 2009; Stalder et al., 2016). By utilizing the collection guidelines in this chapter, it is possible to address contextually rich questions in a variety of research settings.

## References

Adam, E. K., & Kumari, M. (2009). Assessing salivary cortisol in large-scale, epidemiological research. *Psychoneuroendocrinology, 34*(10), 1423–1436. https://doi.org/10.1016/j.psyneuen.2009.06.011

Barranco, T., Rubio, C. P., Tvarijonaviciute, A., Rubio, M., Damia, E., Lamy, E., … Escribano, D. (2019). Changes of salivary biomarkers under different storage conditions: Effects of temperature and length of storage. *Biochemia Medica, 29*(1), 94–111. https://doi.org/10.11613/BM.2019.010706

Beaven, C. M., Gill, N. D., & Cook, C. J. (2008). Salivary testosterone and cortisol responses in professional rugby players after four resistance exercise protocols. *Journal of Strength and Conditioning Research, 22*(2), 426–432. https://doi.org/10.1519/JSC.0b013e3181635843

Beltzer, E. K., Fortunato, C. K., Guaderrama, M. M., Peckins, M. K., Garramone, B. M., & Granger, D. A. (2010). Salivary flow and alpha-amylase: Collection technique, duration, and oral fluid type. *Physiology and Behavior, 101*(2), 289–296. https://doi.org/10.1016/j.physbeh.2010.05.016

Brandtzaeg, P. (2013). Secretory immunity with special reference to the oral cavity. *Journal of Oral Microbiology, 5*. https://doi.org/10.3402/jom.v5i0.20401

Bruijns, B. B., Tiggelaar, R. M., & Gardeniers, H. (2018). The extraction and recovery efficiency of pure DNA for different types of swabs. *Journal of Forensic Sciences, 63*(5), 1492–1499. https://doi.org/10.1111/1556-4029.13837

Crawford, J. M., Taubman, M. A., & Smith, D. J. (1975). Minor salivary glands as a major source of secretory immunoglobulin A in the human oral cavity. *Science (New York, N.Y.), 190*(4220), 1206–1209. Retrieved from http://www.ncbi.nlm.nih.gov/pubmed/1198107

Curtis, A. M., Bellet, M. M., Sassone-Corsi, P., & O'Neill, L. A. J. (2014). Circadian clock proteins and immunity. *Immunity, 40*(2), 178–186. https://doi.org/10.1016/j.immuni.2014.02.002

Deacon, S., & Arendt, J. (1994). Posture influences melatonin concentrations in plasma and saliva in humans. *Neuroscience Letters, 167*(1–2), 191–194. https://doi.org/10.1016/0304-3940(94)91059-6

Dickerson, S. S., & Kemeny, M. E. (2004). Acute stressors and cortisol responses: A theoretical integration and synthesis of laboratory research. *Psychological Bulletin, 130*(3), 355–391. https://doi.org/10.1037/0033-2909.130.3.355

Dreschel, N. A., & Granger, D. A. (2009). Methods of collection for salivary cortisol measurement in dogs. *Hormones and Behavior, 55*(1), 163–168. https://doi.org/10.1016/j.yhbeh.2008.09.010

Dreschel, N. A., & Granger, D. A. (2016). Advancing the social neuroscience of human-animal interaction: The role of salivary bioscience. In L. S. Freund, S. McCune, L. Esposito, N. R. Gee, & P. McCardle (Eds.), *The social neuroscience of human-animal interaction* (pp. 195–216). Washington, DC: American Psychological Association. https://doi.org/10.1037/14856-012

Edgar, W. M. (1990). Saliva and dental health. Clinical implications of saliva: Report of a consensus meeting. *British Dental Journal, 169*, 96–98. https://doi.org/10.1038/sj.bdj.4807284

Eiseman, E., & Haga, S. B. (1999). *Large tissue banks, repositories, and core facilities. Handbook of human tissue sources: A national resource of human tissue samples*. Santa Monica, CA: RAND Corporation.

Emack, J., Kostaki, A., Walker, C. D., & Matthews, S. G. (2008). Chronic maternal stress affects growth, behaviour and hypothalamo-pituitary-adrenal function in juvenile offspring. *Hormones and Behavior, 54*(4), 514–520. https://doi.org/10.1016/j.yhbeh.2008.02.025

Federenko, I. S., Nagamine, M., Hellhammer, D. H., Wadhwa, P. D., & Wüst, S. (2004). The heritability of hypothalamus pituitary adrenal axis responses to psychosocial stress is context dependent. *Journal of Clinical Endocrinology and Metabolism, 89*(12), 6244–6250. https://doi.org/10.1210/jc.2004-0981

Fenoll-Palomares, C., Muñoz Montagud, J. V., Sanchiz, V., Herreros, B., Hernández, V., Mínguez, M., & Benages, A. (2004). Unstimulated salivary flow rate, pH and buffer capacity of saliva in healthy volunteers. *Revista Espanola de Enfermedades Digestivas: Organo Oficial de La Sociedad Espanola de Patologia Digestiva, 96*(11), 773–783. Retrieved from http://www.ncbi.nlm.nih.gov/pubmed/15584851

Garde, A. H., & Hansen, Å. M. (2005). Long-term stability of salivary cortisol. *Scandinavian Journal of Clinical and Laboratory Investigation, 65*(5), 433–436. https://doi.org/10.1080/00365510510025773

Gaviglio, C. M., & Cook, C. J. (2014). Relationship between midweek training measures of testosterone and cortisol concentrations and game outcome in professional rugby union matches. *Journal of Strength and Conditioning Research, 28*(12), 3447–3452. https://doi.org/10.1519/JSC.0000000000000574

Gómez, A., Jewell, E., Walker, S. L., & Brown, J. L. (2004). Use of salivary steroid analyses to assess ovarian cycles in an Indian rhinoceros at the National Zoological Park. *Zoo Biology, 23*(6), 501–512. https://doi.org/10.1002/zoo.20028

Granger, D. A., Blair, C., Willoughby, M., Kivlighan, K. T., Hibel, L. C., Fortunato, C. K., & Wiegand, L. E. (2007). Individual differences in salivary cortisol and alpha-amylase in mothers and their infants: Relation to tobacco smoke exposure. *Developmental Psychobiology, 49*(7), 692–701. https://doi.org/10.1002/dev.20247

Granger, D. A., Fortunato, C. K., Beltzer, E. K., Virag, M., Bright, M. A., & Out, D. (2012). Focus on methodology: Salivary bioscience and research on adolescence: An integrated perspective. *Journal of Adolescence, 35*(4), 1081–1095. https://doi.org/10.1016/j.adolescence.2012.01.005

Granger, D. A., Hibel, L. C., Fortunato, C. K., & Kapelewski, C. H. (2009). Medication effects on salivary cortisol: Tactics and strategy to minimize impact in behavioral and developmental science. *Psychoneuroendocrinology, 34*(10), 1437–1448. https://doi.org/10.1016/j.psyneuen.2009.06.017

Granger, D. A., Kivlighan, K. T., Fortunato, C., Harmon, A. G., Hibel, L. C., Schwartz, E. B., & Whembolua, G. L. (2007). Integration of salivary biomarkers into developmental and behaviorally-oriented research: Problems and solutions for collecting specimens. *Physiology and Behavior, 92*(4), 583–590. https://doi.org/10.1016/j.physbeh.2007.05.004

Gröschl, M. (2008). Current status of salivary hormone analysis. *Clinical Chemistry, 54*(11), 1759–1769. https://doi.org/10.1373/clinchem.2008.108910

Gutiérrez, A. M., Martínez-Subiela, S., Eckersall, P. D., & Cerón, J. J. (2009). C-reactive protein quantification in porcine saliva: A minimally invasive test for pig health monitoring. *Veterinary Journal, 181*(3), 261–265. https://doi.org/10.1016/j.tvjl.2008.03.021

Ha, T., Yeung, E. W., Rogers, A. A., Poulsen, F. O., Kornienko, O., & Granger, D. A. (2016). Supportive behaviors in adolescent romantic relationships moderate adrenocortical attunement. *Psychoneuroendocrinology, 74*, 189–196. https://doi.org/10.1016/j.psyneuen.2016.09.007

Harmon, A. G., Hibel, L. C., Rumyantseva, O., & Granger, D. A. (2007). Measuring salivary cortisol in studies of child development: Watch out—What goes in may not come out of saliva collection devices. *Developmental Psychobiology, 49*(5), 495–500. https://doi.org/10.1002/dev.20231

Hayes, L. D., Grace, F. M., Baker, J. S., & Sculthorpe, N. (2015). Exercise-induced responses in salivary testosterone, cortisol, and their ratios in men: A meta-analysis. *Sports Medicine, 45*(5), 713–726. https://doi.org/10.1007/s40279-015-0306-y

Henskens, Y. M. C., Van Den Keijbus, P. A. M., Veerman, E. C. I., Van Der Weijden, G. A., Timmerman, M. F., Snoek, C. M., . . . Nieuw Amerongen, A. V. (1996). Protein composition of whole and parotid saliva in healthy and periodontitis subjects: Determination of cystatins, albumin, amylase and IgA. *Journal of Periodontal Research, 31*(1), 57–65. https://doi.org/10.1111/j.1600-0765.1996.tb00464.x

Hertzman, C., & Boyce, T. (2010). How experience gets under the skin to create gradients in developmental health. *Annual Review of Public Health, 31*(1), 329–347. https://doi.org/10.1146/annurev.publhealth.012809.103538

Hibel, L. C., Granger, D. A., Cicchetti, D., & Rogosch, F. (2007). Salivary biomarker levels and diurnal variation: Associations with medications prescribed to control children's problem behavior. *Child Development, 78*(3), 927–937. https://doi.org/10.1111/j.1467-8624.2007.01041.x

Hodgson, N. A., & Granger, D. A. (2013). Collecting saliva and measuring salivary cortisol and alpha-amylase in frail community residing older adults via family caregivers. *Journal of Visualized Experiments, 82*, e50815. https://doi.org/10.3791/50815

Hofman, L. F. (2001). Human saliva as a diagnostic specimen. *The Journal of Nutrition, 131*(5), 1621S–1625S. Retrieved from http://www.ncbi.nlm.nih.gov/pubmed/11340128

Horváth, Z., Igyártó, B. Z., Magyar, A., & Miklósi, Á. (2007). Three different coping styles in police dogs exposed to a short-term challenge. *Hormones and Behavior, 52*(5), 621–630. https://doi.org/10.1016/j.yhbeh.2007.08.001

Janardhanam, S. B., Zunt, S. L., & Srinivasan, M. (2012). Quality assessment of saliva bank samples. *Biopreservation and Biobanking, 10*(3), 282–287. https://doi.org/10.1089/bio.2011.0039

Karacabey, K., Saygin, O., Ozmerdivenli, R., Zorba, E., Godekmerdan, A., & Bulut, V. (2005). The effects of exercise on the immune system and stress hormones in sportswomen. *Neuroendocrinology Letters, 26*(4), 361–366.

Karched, M., Bhardwaj, R. G., Pauline, E. M., George, S., & Asikainen, S. (2017). Effect of preparation method and storage period on the stability of saliva DNA. *Archives of Oral Biology, 81*, 21–25. https://doi.org/10.1016/j.archoralbio.2017.04.011

Keaney, L. C., Kilding, A. E., Merien, F., & Dulson, D. K. (2018). The impact of sport related stressors on immunity and illness risk in team-sport athletes. *Journal of Science and Medicine in Sport, 21*, 1192–1199. https://doi.org/10.1016/j.jsams.2018.05.014

Kivlighan, K. T., & Granger, D. A. (2006). Salivary α-amylase response to competition: Relation to gender, previous experience, and attitudes. *Psychoneuroendocrinology, 31*(6), 703–714. https://doi.org/10.1016/j.psyneuen.2006.01.007

Kivlighan, K. T., Granger, D. A., & Schwartz, E. B. (2005). Blood contamination and the measurement of salivary progesterone and estradiol. *Hormones and Behavior, 47*(3), 367–370. https://doi.org/10.1016/j.yhbeh.2004.12.001

Kivlighan, K. T., Granger, D. A., Schwartz, E. B., Nelson, V., Curran, M. J., & Shirtcliff, E. A. (2004). Quantifying blood leakage into the oral mucosa and its effects on the measurement of cortisol, dehydroepiandrosterone, and testosterone in saliva. *Hormones and Behavior, 46*(1), 39–46. https://doi.org/10.1016/j.yhbeh.2004.01.006

Klein, L. C., Bennett, J. M., Whetzel, C. A., Granger, D. A., & Ritter, F. E. (2010). Caffeine and stress alter salivary α-amylase activity in young men. *Human Psychopharmacology, 25*(5), 359–367. https://doi.org/10.1002/hup.1126

Koss, K. J., & Gunnar, M. R. (2018). Annual research review: Early adversity, the hypothalamic–pituitary–adrenocortical axis, and child psychopathology. *Journal of Child Psychology and Psychiatry and Allied Disciplines, 59*(4), 327–346. https://doi.org/10.1111/jcpp.12784

Kugler, J., Hess, M., & Haake, D. (1992). Secretion of salivary immunoglobulin a in relation to age, saliva flow, mood states, secretion of albumin, cortisol, and catecholamines in saliva. *Journal of Clinical Immunology, 12*(1), 45–49. https://doi.org/10.1007/BF00918272

Leproult, R., Holmbäck, U., & Van Cauter, E. (2014). Circadian misalignment augments markers of insulin resistance and inflammation, independently of sleep loss. *Diabetes, 63*(6), 1860–1869. https://doi.org/10.2337/db13-1546

Lutz, C. K., Tiefenbacher, S., Jorgensen, M. J., Meyer, J. S., & Novak, M. A. (2000). Techniques for collecting saliva from awake, unrestrained, adult monkeys for cortisol assay. *American Journal of Primatology, 52*(2), 93–99. https://doi.org/10.1002/1098-2345(200010)52:2<93::aid-ajp3>3.0.co;2-b

Marceau, K., Ruttle, P. L., Shirtcliff, E. A., Essex, M. J., & Susman, E. J. (2015). Developmental and contextual considerations for adrenal and gonadal hormone functioning during adolescence: Implications for adolescent mental health. *Developmental Psychobiology, 57*(6), 742–768. https://doi.org/10.1002/dev.21214

Millspaugh, J. J., Washburn, B. E., Milanick, M. A., Beringer, J., Hansen, L. P., & Meyer, T. M. (2002). Non-invasive techniques for stress assessment in white-tailed deer. *Wildlife Society Bulletin, 30*(3), 899–907. Retrieved from http://www.jstor.org/stable/3784245

Nalla, A. A., Thomsen, G., Knudsen, G. M., & Frokjaer, V. G. (2015). The effect of storage conditions on salivary cortisol concentrations using an enzyme immunoassay. *Scandinavian Journal of Clinical and Laboratory Investigation, 75*(1), 92–95. https://doi.org/10.3109/00365513.2014.985252

Nemoda, Z., Horvat-Gordon, M., Fortunato, C. K., Beltzer, E. K., Scholl, J. L., & Granger, D. A. (2011). Assessing genetic polymorphisms using DNA extracted from cells present in saliva samples. *BMC Medical Research Methodology, 11*(1), 170. https://doi.org/10.1186/1471-2288-11-170

Newman, J. L., Perry, J. L., & Carroll, M. E. (2007). Social stimuli enhance phencyclidine (PCP) self-administration in rhesus monkeys. *Pharmacology Biochemistry and Behavior, 87*(2), 280–288. https://doi.org/10.1016/j.pbb.2007.05.004

Ng, D. P. K., Koh, D., Choo, S., & Chia, K. S. (2006). Saliva as a viable alternative source of human genomic DNA in genetic epidemiology. *Clinica Chimica Acta, 367*(1–2), 81–85. https://doi.org/10.1016/j.cca.2005.11.024

Nielson, C. M., Jones, K. S., Chun, R. F., Jacobs, J. M., Wang, Y., Hewison, M., . . . Orwoll, E. S. (2016). Free 25-hydroxyvitamin D: Impact of vitamin D binding protein assays on racial-genotypic associations. *Journal of Clinical Endocrinology and Metabolism, 101*(5), 2226–2234. https://doi.org/10.1210/jc.2016-1104

Orysiak, J. O., Malczewska-Lenczowska, J., Szyguła, Z., & Pokrywka, A. (2012). The role of salivary immunoglobulin a in the prevention of the upper respiratory tract infections in athletes—An overview. *Biology of Sport, 29*(4), 311–315. https://doi.org/10.5604/20831862.1022653

Perry, G. H., Dominy, N. J., Claw, K. G., Lee, A. S., Fiegler, H., Redon, R., . . . Stone, A. C. (2007). Diet and the evolution of human amylase gene copy number variation. *Nature Genetics, 39*(10), 1256–1260. https://doi.org/10.1038/ng2123

Pramanik, R., Thompson, H., Kistler, J. O., Wade, W. G., Galloway, J., Peakman, T., & Proctor, G. B. (2012). Effects of the UK biobank collection protocol on potential biomarkers in saliva. *International Journal of Epidemiology, 41*(6), 1786–1797. https://doi.org/10.1093/ije/dys166

Proctor, G. B. (2016). The physiology of salivary secretion. *Periodontology 2000, 70*(1), 11–25. https://doi.org/10.1111/prd.12116

Pulford, D. J., Mosteller, M., Briley, J. D., Johansson, K. W., & Nelsen, A. J. (2013). Saliva sampling in global clinical studies: The impact of low sampling volume on performance of DNA in downstream genotyping experiments. *BMC Medical Genomics, 6*(1), 20. https://doi.org/10.1186/1755-8794-6-20

Pulopulos, M. M., Vanderhasselt, M. A., & De Raedt, R. (2018). Association between changes in heart rate variability during the anticipation of a stressful situation and the stress-induced cortisol response. *Psychoneuroendocrinology, 94*, 63–71. https://doi.org/10.1016/j.psyneuen.2018.05.004

Ramirez, J., Elmofty, M., Castillo, E., DeRouen, M., Shariff-Marco, S., Allen, L., . . . Márquez-Magaña, L. (2017). Evaluation of cortisol and telomere length measurements in ethnically diverse women with breast cancer using culturally sensitive methods. *Journal of Community Genetics, 8*(2), 75–86. https://doi.org/10.1007/s12687-016-0288-y

Rankin, A., Swearingen-Stanborough, C. S., Granger, D. A., & Byrd-Craven, J. (2018). The role of co-rumination and adrenocortical attunement in young women's close friendships. *Psychoneuroendocrinology, 98*, 61–66. https://doi.org/10.1016/j.psyneuen.2018.07.027

Romeo, R. D. (2005). Neuroendocrine and behavioral development during puberty: A tale of two axes. *Vitamins and Hormones, 71*, 1–25. https://doi.org/10.1016/S0083-6729(05)71001-3

Schwartz, E. B., & Granger, D. A. (2004). Transferrin enzyme immunoassay for quantitative monitoring of blood contamination in saliva. *Clinical Chemistry, 50*(3), 654–656. https://doi.org/10.1373/clinchem.2003.028266

Schwartz, E. B., Granger, D. A., Susman, E. J., Gunnar, M. R., & Laird, B. (1998). Assessing salivary cortisol in studies of child development. *Child Development, 69*, 1503–1513. https://doi.org/10.1111/j.1467-8624.1998.tb06173.x

Sheth, C. C., López-Pedrajas, R. M., Jovani-Sancho, M. D. M., González-Martínez, R., & Veses, V. (2018). Modulation of salivary cytokines in response to alcohol, tobacco and caffeine consumption: A pilot study. *Scientific Reports, 8*(1), 16687. https://doi.org/10.1038/s41598-018-35094-z

Shirtcliff, E. A., Granger, D. A., Schwartz, E. B., & Curran, M. J. (2001). Use of salivary biomarkers in biobehavioral research: Cotton-based sample collection methods can interfere with salivary immunoassay results. *Psychoneuroendocrinology, 26*(2), 165–173. Retrieved from http://www.ncbi.nlm.nih.gov/pubmed/11087962

Slavish, D. C., Graham-Engeland, J. E., Smyth, J. M., & Engeland, C. G. (2015). Salivary markers of inflammation in response to acute stress. *Brain, Behavior, and Immunity, 44*, 253–269. https://doi.org/10.1016/j.bbi.2014.08.008

Stalder, T., Kirschbaum, C., Kudielka, B. M., Adam, E. K., Pruessner, J. C., Wüst, S., . . . Clow, A. (2016). Assessment of the cortisol awakening response: Expert consensus guidelines. *Psychoneuroendocrinology, 63*, 414–432. https://doi.org/10.1016/j.psyneuen.2015.10.010

Talge, N. M., Donzella, B., Kryzer, E. M., Gierens, A., & Gunnar, M. R. (2005). It's not that bad: Error introduced by oral stimulants in salivary cortisol research. *Developmental Psychobiology, 47*(4), 369–376. https://doi.org/10.1002/dev.20097

Taylor, M. K., Reis, J. P., Sausen, K. P., Padilla, G. A., Markham, A. E., Potterat, E. G., & Drummond, S. P. A. (2008). Trait anxiety and salivary cortisol during free living and military stress. *Aviation Space and Environmental Medicine, 79*(2), 129–135. https://doi.org/10.3357/ASEM.2131.2008

Van Paridon, K. N., Timmis, M. A., Nevison, C. M., & Bristow, M. (2017). The anticipatory stress response to sport competition; a systematic review with meta-analysis of cortisol reactivity. *BMJ Open Sport and Exercise Medicine, 3*(1), e000261. https://doi.org/10.1136/bmjsem-2017-000261

Vaught, J. B., Henderson, M. K., & Compton, C. C. (2012). Biospecimens and biorepositories: From afterthought to science. *Cancer Epidemiology Biomarkers and Prevention, 21*, 253–255. https://doi.org/10.1158/1055-9965.EPI-11-1179

Veerman, E. C., van den Keybus, P. A., Vissink, A., & Nieuw Amerongen, A. V. (1996). Human glandular salivas: Their separate collection and analysis. *European Journal of Oral Sciences, 104*(4 (Pt 1)), 346–352. Retrieved from http://www.ncbi.nlm.nih.gov/pubmed/8930581

Vining, R. F., McGinley, R. A., & Symons, R. G. (1983). Hormones in saliva: Mode of entry and consequent implications for clinical interpretation. *Clinical Chemistry, 29*(10), 1752–1756. Retrieved from http://clinchem.aaccjnls.org/content/29/10/1752.long

Whembolua, G. L. S., Granger, D. A., Singer, S., Kivlighan, K. T., & Marguin, J. A. (2006). Bacteria in the oral mucosa and its effects on the measurement of cortisol, dehydroepiandrosterone, and testosterone in saliva. *Hormones and Behavior, 49*(4), 478–483. https://doi.org/10.1016/j.yhbeh.2005.10.005

Wiegratz, I., Kutschera, E., Lee, J. H., Moore, C., Mellinger, U., Winkler, U. H., & Kuhl, H. (2003). Effect of four different oral contraceptives on various sex hormones and serum-binding globulins. *Contraception, 67*(1), 25–32. https://doi.org/10.1016/S0010-7824(02)00436-5

Wilde, C., Out, D., Johnson, S., & Granger, D. A. (2013). Sample collection, including participant preparation and sample handling. In D. Wild (Ed.), *The immunoassay handbook* (pp. 427–440). Oxford: Elsevier. https://doi.org/10.1016/B978-0-08-097037-0.00029-4

Zimmerman, Y., Eijkemans, M. J. C., Coelingh Bennink, H. J. T., Blankenstein, M. A., & Fauser, B. C. J. M. (2014). The effect of combined oral contraception on testosterone levels in healthy women: A systematic review and meta-analysis. *Human Reproduction Update, 20*(1), 76–105. https://doi.org/10.1093/humupd/dmt038

Zmyj, N., Schneider, S., & Seehagen, S. (2017). Fifteen-month-old infants' cortisol levels decrease after a 30-min-warm-up procedure. *Psychoneuroendocrinology, 76*, 11–13. https://doi.org/10.1016/j.psyneuen.2016.11.010

# Chapter 4
# Analytical Strategies and Tactics in Salivary Bioscience

**Jenna L. Riis, Frances R. Chen, Amy L. Dent, Heidemarie K. Laurent, and Crystal I. Bryce**

**Abstract** This chapter provides an overview of many of the methodological obstacles and opportunities of salivary analyte data analysis. Salivary data processing procedures essential for data validity and reliability, but rarely described in the literature (e.g., flow-rate adjustments, the handling of censored data points, and data transformations), are discussed. Statistical modeling approaches appropriate for salivary biomeasure data (e.g., multilevel modeling and latent state-trait analysis) are also described. Recommendations for data processing and statistical modeling are proposed to improve validity, reliability, precision, and cross-study consistency for the next generation of studies.

## 4.1 Introduction

The use of saliva as a biospecimen has greatly expanded the integration of biologic data into research studies conducted across a wide range of scientific disciplines. Alongside the expansion of salivary bioscience, saliva collection devices and assay

J. L. Riis (✉)
Department of Psychological Science, School of Social Ecology, University of California, Irvine, Irvine, CA, USA

Institute for Interdisciplinary Salivary Bioscience Research, University of California, Irvine, CA, USA
e-mail: jriis@uci.edu

F. R. Chen
Department of Criminal Justice and Criminology, Georgia State University, Altanta, GA, USA

A. L. Dent
Department of Psychological Science, School of Social Ecology, University of California, Irvine, Irvine, CA, USA

H. K. Laurent
Department of Psychology, The University of Illinois at Urbana-Champaign, Champaign, IL, USA

C. I. Bryce
The Sanford School, Arizona State University, Tempe, AZ, USA



D. A. Granger, M. K. Taylor (eds.), *Salivary Bioscience*,
https://doi.org/10.1007/978-3-030-35784-9_4

reagents have become commercially available and widely accessible for an extensive range of biologic analytes and markers of environmental exposure. Using saliva, researchers from all disciplines can collect repeated biologic samples and examine multiple physiologic analytes per observation without high participant burden. While these practical advantages enable vast opportunities to examine the interplay of physiology, psychology, and sociocultural and environmental factors, they also introduce methodological and statistical challenges. This chapter discusses many of these challenges and presents strategies to address them, so researchers can take advantage of the opportunities provided by rich salivary biomeasure datasets.

## 4.2 Power Analysis with Salivary Biomeasure Data

Statistical challenges may arise even in the very early stages of the research process for scientists planning to use salivary biomeasures in their investigations. Estimations of statistical power are essential to the planning of any research study, and the nature of salivary analyte data and the state of the science regarding specific analytes complicate power calculations for salivary biomeasure studies. Here we provide a brief review of statistical power concepts, highlight areas of potential challenges for salivary bioscience researchers, and provide recommendations for addressing these issues.

Statistical power is the likelihood of a statistical test uncovering a phenomenon (e.g., an effect or relation of interest) in the population if it exists. Statistical power can be evaluated either before or after a study is conducted, typically to explore target sample sizes beforehand and to explain nonsignificant findings afterward. Whether evaluated prospectively or retrospectively, sample size is an important determinant of statistical power and often the only one under a researcher's control (apart from altering the conventional alpha level of 0.05 or power level of 0.80). The remaining determinants of statistical power arise from our data, or its underlying population, and can only be estimated in a priori power analyses (Cohen, 1988, 1992). For complex statistical models, the determinants of power likewise become more complex [e.g., see Snijders (2005) for a discussion of power estimations in multilevel modeling].

The uncertainty and complexity of statistical power calculations present important challenges in salivary bioscience research. As an accelerating area of study characterized by complex data and modeling approaches, it can be difficult to accurately project unknown parameters that influence the power of a statistical test. For example, estimating regression coefficients, variance components, and even population means for many analytes, especially newer ones such as salivary cytokines, is particularly challenging given the dearth of previous research examining them and the lack of accepted population norms or expected ranges. Yet these estimates are critical when conducting a priori power analyses, as the target sample size of a study hinges on them and can dramatically differ based on them as a result. With resource-intensive data collection and sample processing procedures, the stakes are high for salivary bioscience researchers to identify an appropriate sample size. Overestimating unknown parameters may decrease the target sample size but doing so risks nonsignificant results due to inadequate power. Underestimating unknown parameters may protect against this possibility but requires potentially considerable upfront costs of additional participants.

So, how should researchers balance these tradeoffs? Thoughtfully considering and conducting an a priori power analysis that mirrors eventual analytic specifications as closely as possible is a valuable first step. Doing so should involve empirically projecting the unknown components of statistical power based on previous research [see Anderson, Kelley, and Maxwell (2017) for important controversies and caveats] or commonly used benchmarks. For example, Granger et al. (2012) suggested that a meaningful within-individual change in analyte concentration should be at least two times the lower limit of the assay and a difference of at least two times the average intra-assay coefficient of variation (CV). While this threshold is a benchmark for change that is meaningful based on analyte measurement limitations and error rates, it does not imply a biologically or clinically meaningful change. The benchmarks for changes in analyte levels that correlate with physiological changes and clinical significance are analyte specific and largely unknown, thereby complicating their inclusion in a priori power analyses.

As our field conducts more studies and creates more knowledge about the components of power analyses (e.g., the typical or moderated magnitude of regression coefficients), developing a public repository of these components for a priori power analyses would promote optimal sample sizes and accurate results. At a critical juncture of our field's acceleration and innovation, this repository could enhance the feasibility, validity, and transparency of salivary bioscience research.

## 4.3 Early Processing of Salivary Biomeasure Data: Unique Challenges and Solutions

The next phase of statistical considerations occurs after the saliva samples are collected, processed, and assayed when investigators receive raw data reports from the saliva processing laboratory. As with all raw data, salivary biomeasure data received from the laboratory require data processing, including careful examination and "cleaning" before use in statistical models. This section reviews these data processing steps, highlights specific challenges, and presents recommendations for salivary analyte data processing. Decisions made by laboratory technicians during the saliva assay process that have the potential to impact data validity are briefly discussed. Details of in-laboratory sample processing can be found in Wild (2013). Researchers are encouraged to review this text, as a basic understanding of in-laboratory sample assay procedures helps inform data processing decisions, as well as evaluations of data quality, validity, and reliability.

This section focuses on the choices researchers must make after receiving analyte data from the laboratory. Details of these early data processing steps, such as handling duplicate, extreme, and censored values, addressing kurtotic and skewed distributions, and computing indicators of individual differences, are often omitted from scientific reports. Basic statistical methods that are particularly useful for salivary analyte data analysis, such as transforming data and modeling group parameters, are also discussed in this section. While guidance about these salivary analyte data processing decision points is remarkably thin in the literature, researchers' choices in this data processing phase can impact the reliability and validity of the study data and findings.

### 4.3.1 Examining Data Parameters Related to In-Laboratory Saliva Processing Procedures

Saliva assay protocols include strict criteria to ensure the consistency of liquid handling, the standard curve, and data reliability and precision. Raw data generated by this process typically include determinations for the analyte of interest (in singlet or replicate), indices of inter- and intra-assay precision (i.e., CVs), and metrics to indicate salivary flow rate (i.e., volume, or volume estimated by weight, and time). In the earliest steps of data processing, researchers should review these indices to assess the quality and nature of the data produced by the laboratory.

***Intra-assay Precision*** Intra-assay precision refers to the reliability of individual measurements of a sample. It represents how well the assay was performed in the hands of the laboratory operator. The intra-assay CV indexes the reliability of the assay for individual samples as well as for all the study samples as a whole. To calculate the intra-assay CV, all, or a subset of, samples are run in replicate (e.g., duplicate or triplicate). On an individual sample basis, the CV is used to determine whether that sample's assay results are reliable or whether the test should be repeated. The criterion by which researchers judge replicate values as reliable is rarely reported. Adequate reliability across an entire study is indexed by an intra-assay CV that is typically less than 5% on average across all samples. Intra-assay CVs are generally provided by the saliva processing laboratory, however, they can be calculated by researchers using replicate values [Chard (1990); see Box 4.1 for an example].

When samples are assayed in replicate, the average replicate concentration is typically used in data analyses to represent the analyte concentration with assay and operator-specific error "averaged out" across the replicate determinations. Researchers using complex modeling approaches, such as latent modeling, may instead choose to use individual determinations of analytes rather than the average (see Sect. 4.4.5). Researchers may also choose to include singlet determinations of an analyte for specific cases or under certain conditions (see Sect. 4.3.2).

***Inter-assay Precision*** Inter-assay precision refers to variation in salivary analyte data between assay plates or batches. Inter-assay precision is typically calculated using the high and low controls that are tested on each plate, as a measure of reliability. When assays are processed without controls (e.g., in some multiplex testing), a control (or pooled) sample can be added to each plate and used to calculate the inter-assay CV. Immunodiagnostic industry standards expect the inter-assay CV to be less than 15% (Chard, 1990). With up to 15% variation in assay results accepted as normative between plates, it might be important for all samples collected from the same participant to be assayed on the same plate. Researchers should consider requesting plate numbers and values for high and low, and/or pooled control samples in order to examine the impact of plate or batch assignment on analyte determinations. See Box 4.1 for an example calculation of inter-assay CV.

***Salivary Flow Rate*** Levels of analytes that move from the circulation into the oral fluid by ultrafiltration (i.e., through the tight junctions between salivary gland acinar cells) or are secreted by the salivary glands have the potential to be influenced by salivary flow rate [e.g., alpha-amylase (Beltzer et al., 2010); secretory immunoglobulin A (SIgA; (Brandtzaeg, 2007)); and dehydroepiandrosterone-sulfate (DHEA-S; (Vining & McGinley, 1987))]. For serum constituents passing into oral fluid via ultrafiltration, increased saliva flow may be associated with lower analyte concentration/volume measurements because increased saliva flow dilutes the concentration/volume measurement. In contrast, for analytes released via salivary glands, increased saliva flow may be associated with higher levels of the analyte as more analyte is delivered into oral fluid as a function of higher salivary flow.

To measure flow rate, researchers typically allow participants a standard time (e.g., 2 min) in the field to donate saliva, either by passive drool or by mouthing an absorbent swab, then have the laboratory estimate the volume generated during that period by weight. These metrics are used to compute flow rate (flow rate = sample volume/sample collection duration), which is expressed in units of volume/time. If flow rate is associated with analyte levels (two tailed, $p < 0.05$), the main analyses should be adjusted for flow rate. As an alternative to statistical covariation, a correction can be computed by multiplying concentration/volume by volume/time. This corrected index is expressed in units of concentration/time or units of activity/time and is referred to as "output." While statistical covariation allows for clearer interpretation of the analyte parameters, using a corrected index may be more appropriate in models with limited power. Failing to account for salivary flow rate has the potential to considerably restrict the utility and interpretation of salivary analyte data.

If saliva collection duration was not recorded or no starting weight of the sample was obtained, salivary analyte concentrations can instead be adjusted for the total protein concentration in the saliva sample. Associations between analyte concentrations and the concentration of total protein can be assessed statistically. If total protein concentrations are related to concentrations of the analyte of interest, the main analyses can then be adjusted for total protein concentration in the sample. The utility of total protein as corrective index that approximates salivary flow rate adjustment is debated in the field as changes in total protein have been reported in response to stress, exercise, and oral health problems, and salivary protein secretions likely vary by gland and fluid type (Bishop & Gleeson, 2009; Brandtzaeg, 2007; Burgener et al., 2010; Lee, Chung, Kim, Chung, & Kho, 2007; Trueba, Mizrachi, Auchus, Vogel, & Ritz, 2012).

### 4.3.2 Examining and Addressing Missing, Unreliable, or Invalid Values

Unreliable or invalid analyte determinations may be related to several laboratory, assay, analyte, and participant factors, including laboratory technician and mechanical errors, singlet testing, complications related to low analyte concentrations and the limits of assay technology, and saliva contamination by blood or food. For each of the factors, decisions regarding the inclusion of potentially unreliable or invalid determinations should be carefully considered and reported in the Methods section of the research report. If potentially unreliable or invalid determinations are included in the analytic dataset, sensitivity analyses that exclude and/or replace unreliable and invalid data should test the impact these determinations have on the final study results, and these findings should be included in all research reports.

***Singlet Testing*** Many analytes are assayed in duplicate (or triplicate) to improve precision of the analyte determination. Yet researchers may choose, due to financial or saliva sample volume constraints, to perform assays in singlet. Singlet determinations are generally less precise than determinations derived from the mean of replicate tests. Researchers must decide if and how singlet test determinations will be used in statistical analyses. Intra-assay CVs provide valuable information about the reliability of analyte determinations across all samples tested, and this information can help researchers decide if singlet determinations should be used in the final dataset or treated as missing.

***Saliva Contamination*** Saliva contamination can occur at any point during the collection, storage, and assay process. During collection, common saliva contaminants include cosmetics (e.g., lipstick), food particles, and blood. Specific guidelines are available for assessing blood contamination (Kivlighan et al., 2004) as blood leakage into saliva can atypically elevate concentrations of salivary analytes. Researchers should record all known or visible contaminations during the collection of each sample as this information may be helpful for technicians involved in assay procedures. Errors during sample and assay processing in the laboratory can also result in sample contamination. Determinations with known contamination events should be flagged by research and laboratory staff and assessed as potentially unreliable or invalid values.

***Limits of Assay Technology*** Advancements in assay technology have increased the precision with which salivary analytes can be measured. While the range of measurable concentrations has expanded for many analytes, every assay has thresholds beyond which analyte concentrations cannot be reliably assessed. In general, analyte determinations that are outside the assay range are neither reliable nor valid.

However, modern microplate and reader technology may extrapolate and report values for determinations outside the assay range. Researchers should discuss issues related to assay sensitivity with their laboratory staff and assay manufacturers and ensure extrapolated values are noted in the data report, so that these values can be excluded from analysis if appropriate. If the laboratory does not report values outside the assay range or if the reliability and validity of these estimations is inadequate, researchers can choose from several methods for handling these missing data (see *Censored Data and Data Missing Not at Random* in Sect. 4.3.3).

### 4.3.3 Approaches for Addressing Unreliable, Invalid, and Missing Data

Researchers should consider the nature and amount of missing data, as well as the mechanism(s) of missingness (e.g., missing at random, missing not at random) when deciding their approach to addressing missing salivary analyte data. Data missing due to high CVs, inadequate sample volume, and/or sample contamination are likely missing at random, so missing data approaches that assume a random missingness mechanism (e.g., multiple imputation, maximum likelihood estimation; discussed below) can appropriately model these data. In contrast, out-of-range determinations that are missing due to the limits of assay technology are censored data that are not missing at random, so other approaches (described below) should be considered.

***Data Missing at Random*** Drawing on advanced approaches to missing data, multiple imputation (MI) and maximum likelihood estimation (ML, often in the form of full information maximum likelihood or FIML) are appropriate for data missing at random and data missing completely at random [e.g., data missing due to insufficient, contaminated, or missing saliva samples, or high CV determinations; see Enders (2010) for a detailed explanation of missing data and approaches to handling missing data]. These approaches assume the missing data are not related to the true values underlying the missing scores themselves, but the likelihood of missingness may be related to other variables available in the dataset. For example, younger participants may be less compliant with saliva collection protocols resulting in more missingness in analyte data for younger children compared to adolescents or adults. When analyte data are missing at random (e.g., due to small saliva volume) or missing completely at random (e.g., due to unsystematic laboratory processing errors) MI and/or ML should be explored as approaches to maximizing statistical power and minimizing bias. Importantly, as sample size increase, the results from MI and ML will converge (Enders, 2010).

Multiple imputation estimates missing data through researcher-specified regression models and substitutes each missing score with an imputed estimate that is derived through multiple iterations of the regression model for the missing data. The result is multiple (often more than 100) datasets in which the missing scores are replaced with imputed ones (Lubin et al., 2004; Uh, Hartgers, Yazdanbakhsh, & Houwing-Duistermaat, 2008). Statistical analyses can then be performed with the imputed data, and results are pooled across imputations (Uh et al., 2008). It is important to note that multiple imputation assumes multivariate normality of the outcome variable residuals within an ordinary least squares (OLS) single-level regression framework. This assumption may be violated with salivary analyte data, which typically display skewed distributions. If this assumption is violated, the accuracy of the imputed data is compromised (Baccarelli et al., 2005). In addition, the accuracy of the imputed data depends on the specification of the regression model for the missing data. Yet for many analytes, particularly novel and experimental biomeasures, appropriate model specification and inclusion of explanatory (also called "predictor" or "independent") variables are not known. For this reason, it is important that researchers create many imputed datasets and carefully consider their specification of the imputation model and accuracy of its imputed scores when employing this approach with salivary biomeasure data.

In general, the ML (commonly referred to as FIML) approach is preferred when data are missing at random, because it provides unbiased parameter estimates (Allison, 2001; Enders, 2010). By utilizing a ML estimator, missing scores are estimated for the specified model with computational steps conducted "behind the scenes." In other words, ML works through many iterations of imputation options until the best suited scores are determined [see Enders (2010)]. ML does not produce a dataset where those scores have been imputed (like in MI), but instead produces the results of the specified model. Similar to MI, ML assumes multivariate normality of the outcome variable residuals. However, there are other types of ML estimators, such as maximum likelihood estimation with robust standard errors (MLR), that can estimate models fitted to both missing and non-normal data. Often, researchers view ML as a more advantageous, because it is less taxing on statistical programs than MI, does not rely on the creation of multiple new imputed datasets, and can be implemented for many different analyses.

When determining the appropriate approach to missing data in salivary bioscience studies, it is important to consider the missingness mechanism, as well as the variables included in the final statistical model. For example, if composite measures such as area under the curve (AUC) or diurnal slopes are of interest, researchers should consider if missingness is best handled at the level of the composite variable or at the level of the indicators contributing to it.

***Censored Data and Data Missing Not at Random*** Determinations that are missing due to the limits of the assay technology are censored and missing not at random. As a result, the MI and ML methods discussed above are generally not appropriate approaches for handling these unavailable or untrustworthy analyte determinations. There are several more suitable options for handing these missing data.

Traditionally, one of the most common approaches to handling data outside the assay range (i.e., "non-detects" or "censored data") involves dropping or substituting censored data points. Dropping censored data points is not generally recommended, because it introduces bias, sacrifices information, and decreases power (Baccarelli et al., 2005; Cole, Chu, Nie, & Schisterman, 2009; Helsel, 2006). Substituting out-of-range concentrations with a constant (e.g., the lower or upper limit of the assay or a percent of these values) also introduces bias and reduces variance (Cole et al., 2009; Helsel, 2006; Lubin et al., 2004). The amount of bias associated with substitution methods depends on many factors including the distribution of the data, the level of measurement error, the percent of censored data, and the assay limits relative to the observed concentrations (Hewett & Ganser, 2007; Richardson & Ciampi, 2003; Schisterman, Vexler, Whitcomb, & Liu, 2006; Uh et al., 2008).

In addition to these traditional approaches, parametric and nonparametric statistical modeling methods for censored data and data missing not at random are available. For example, Kaplan–Meier, Cox, quantile, and tobit regression approaches can be employed when analyte data include missing or unreliable out-of-range concentrations on both ends of the assay range (in some approaches). There are also "not at random" adjustments of commonly used missing at random approaches, such as MI, that may be useful for handling non-detect data (e.g., Galimard, Chevret, Curis, & Resche-Rigon, 2018). There is currently no single approach that is preferred in the salivary bioscience field, and modeling approaches should be tailored to the specific characteristics of the data and nature of the research question. Researchers should also examine the assumptions of the approach employed to ensure it is appropriate for the data. While a comprehensive discussion of these, and other advanced modeling, approaches is beyond the scope of this chapter, the reader is referred to Antweiler and Taylor (2008), Austin (2002), Austin, Escobar, and Kopec (2000), Dinse et al. (2014), Eilers, Röder, Savelkoul, and van Wijk (2012), Epstein, Lin, and Boehnke (2003), Fu et al. (2012), Gillespie et al. (2010), Helsel (2005, 2010), Helsel and Lopaka (2006), Hewett and Ganser (2007), Jin, Hein, Deddens, and Hines (2011), and Lubin et al. (2004) for more information about them.

### *4.3.4 Examining the Distribution of the Data*

After carefully examining indices related to in-laboratory saliva processing and potentially unreliable and invalid values, the final steps in the early processing of salivary data involve examining the distribution and spread of salivary determinations within the study sample.

***Outliers and Extreme Values*** The first step when examining the distribution of raw analyte data is to assess outlying and extreme data points. For some analytes, such as salivary cortisol, investigators can refer to a wealth of research from different populations and settings to decide the level at which a determination should be considered an outlier or biologically implausible. Biologically implausible determinations should be deleted from the analytic dataset, as they are likely the result of laboratory error or contamination. However, identifying true outliers is complicated when examining new biomeasures for which there are no established norms and when using assays not optimized for use with saliva, because these kits will not provide researchers with the expected range of analyte concentrations. For these analytes, careful examination of the distribution of the data is critical. Often in the salivary bioscience literature, researchers have treated values more than three or four standard deviations from the study sample mean as "outliers." However, the threshold at which a determination becomes biologically implausible and/or a true outlier is unsettled. Researchers are encouraged to consider the characteristics of their sample and draw upon previous study findings and expert opinions from laboratory staff and assay manufacturers when deciding how to handle potential outliers and extreme data points in their sample.

Conventional approaches to handling extreme data points include dropping these determinations from the analytic dataset or Winsorizing them (Kruskal, Ferguson, Tukey, & Gumbel, 1960). Winsorizing the data, or replacing extreme data points with values approximate or equal to the next most extreme score, will improve the distribution of the data (Kruskal et al., 1960). If Winsorizing is applied, it is important to consider which groups of data (e.g., participant groups and time points) should be Winsorized together and to set an acceptable range of variation from the mean. When Winsorizing longitudinal data by time point, individual trajectories should be preserved. Both these approaches (dropping and Winsorizing extreme data points) artificially reduce variance in the analyte data, and dropping extreme data points also restricts sample size.

These conventional approaches are likely inappropriate given recent advances in statistical modeling. Many modern statistical approaches and estimators are robust to violations of normality. This means that biologically plausible extreme values can be retained in the dataset. After model estimation, however, the impact of extreme analyte determinations on model fit and parameter estimates should be assessed through diagnostic tests of influence before proceeding to interpret the results. When doing so, it is important to remember that just one influential case can impact parameter estimates in OLS regression (Cohen, Cohen, West, & Aiken, 2003). Regardless of the approach employed, investigators should report their procedures for addressing extreme scores, along with the impact of these decisions on study results, in the Methods and Results sections of research reports.

***Non-normal Distributions*** Salivary analyte data typically display a strong positive skew with a disproportionate amount of low concentrations and a long tail of high concentrations. During statistical planning, researchers should identify the assumptions of their chosen statistical model(s) and address analyte distribution issues accordingly. For example, if researchers plan to use parametric statistical models and the analyte as their outcome variable, normality of the outcome residuals is often assumed. Under this assumption, many data transformation techniques can be employed to adjust the distribution of the outcome variable (e.g., log transformation, inverse square root, and square root). The impact of these approaches should be tested for each analyte and for each analysis conducted with a transformed outcome variable. Rather than transform the analyte data, researchers may choose to use a statistical estimator, such as MLR, that accounts for non-normality in the data when conducting analyses. It is important to note that there is typically no assumption of normality for explanatory variables. As a result, the biomeasure data do not need to be transformed if the biomeasure is intended to be an explanatory variable in a regression model.

### 4.3.5 *Computing Composites and Indicators of Individual Differences*

The goal of early data processing is to generate a dataset that best represents the underlying variables and can be appropriately included in statistical models to produce sound inferences. Once this is achieved, salivary bioscience researchers are often interested in exploring commonly used composite variables that represent biologic function and changes within individuals. Some of these common approaches for modeling analyte change over time are presented below.

#### 4.3.5.1 Within-Individual Analyte Change

***Comparing "Responders" to "Non-responders"*** Often studies aim to examine the psychological or health correlates of a specific analyte's response to a stimulus. A simple method for assessing these relations is to divide the study sample into stimulus "responders" and "non-responders" then compare these groups on a range of selected factors. For example, researchers may examine the personality characteristics of participants who display increases in cortisol in response to a cold-pressor task ("responders") compared to those who do not ("non-responders"). While statistically simple, this approach is fraught with conceptual complexities. There are no established standards for distinguishing these groups, and differences in the number of groups and thresholds for group membership make it difficult to compare results across studies. While Granger et al. (2012) suggested a threshold for meaningful change based on assay limitations and laboratory error (see Sect. 4.2), other studies have used difference scores, proportion change percentiles, cluster analysis, and mean and standard deviation parameters to determine group cutoffs (e.g., Izawa et al., 2013; Kunz-Ebrecht, Mohamed-Ali, Feldman, Kirschbaum, & Steptoe, 2003; Miller, Plessow, Kirschbaum, & Stalder, 2013). It is important to consider thresholds used by previous researchers, the distribution of the data, and the biological processes being examined when deciding how to distinguish responder groups.

Although methodologically simple, comparing responders and nonresponders sacrifices information as continuous data are categorized and participants may be dropped from analyses if their responses do not meet group thresholds. More robust methods for examining the correlates of changes in biomeasures across time are discussed below (see Sects. 4.4.3 and 4.4.4 for example approaches).

***Examining Composite Measures*** A common and simple approach to examining change in analytes over time is to use composite measures of change, such as proportion change, difference scores, diurnal slopes, and AUC measures [i.e., AUC with respect to ground (AUCg) and with respect to increase (AUCi); Pruessner, Kirschbaum, Meinlschmid, and Hellhammer (2003)]. Proportion change and difference scores summarize analyte change across two saliva samples.

Slope measures further incorporate the elapsed time between samples. AUCg and AUCi are composite measures that can express change in concentration over many samples and account for the time between samples (Pruessner et al., 2003). AUCg is a measure of total analyte output, while AUCi measures output across time with respect to the initial concentration [i.e., reactivity; Pruessner et al. (2003)]. Pruessner et al. (2003) standardized the formula for AUCg and AUCi to improve across-study comparisons.

In general, all composite measures (e.g., AUCs, slopes, and difference scores) should be calculated using non-transformed analyte data. If composite measures are included as outcome variables in parametric statistical models that assume multivariate normality of the outcome variable residuals, the distribution of composite measures should be examined and data transformations may be employed to improve skewed outcome distributions. Composite measures express concentration changes with a single value, thereby avoiding the complexities of modeling repeated measures and correlated data points. However, more robust approaches to addressing these research questions are available and discussed below.

## 4.4 Advanced Statistical Analysis of Salivary Biomeasure Data

Once the preliminary processing of analyte data is completed, the data are ready for inclusion in statistical models to address the study aims and research questions. The statistical approach should be selected based on the research question, the hypothesized biopsychosocial and environmental processes underlying the hypothesis, and the available data (e.g., number of analyte determinations per person and number of analytes). A critical step in selecting a statistical approach is a clear conceptual and operational definition of the target biologic processes. For example, stress study paradigms can be used to examine questions regarding changes in analytes related to anticipatory, reactivity, and recovery processes in a single biologic system [e.g., the hypothalamic–pituitary–adrenal (HPA) axis] or across multiple systems (e.g., the neuroendocrine-immune system). Investigators should design specific research questions that focus on the biologic processes of interest and tailor their statistical approach best match the nature of these specific processes.

In addition to appropriately identifying and conceptualizing the biologic processes of interest, it is important to consider the nature of the relation(s) between the analyte and the other study variables when choosing a statistical approach. Many statistical approaches (e.g., Pearson's correlation and linear regression), assume a linear relation between explanatory and outcome variables. However, this assumption of linearity may not be an appropriate conceptualization of physiologic relations. For example, there may be thresholds beyond which concentrations of a specific analyte no longer have an effect on physiologic processes (i.e., a saturation point). In contrast, a certain concentration may be needed for some analytes to have a physiologic effect (i.e., a threshold effect). For many salivary biomeasures, these relations have not yet been examined, making it essential that researchers anticipate and test the nature of relations between their analytes and other study variables.

### 4.4.1 *Conceptualizing Biomeasures Within a Mediation Framework*

Salivary biomeasure data provide opportunities to examine the complex web of biological, psychological, and social relations underlying behavioral and health outcomes. Addressing these questions requires statistical approaches that go beyond examining simple relations between two variables to include the hypothesized mechanisms linking them. For example, SIgA (a measure of the mucosal immune system) may be examined as a potential mechanism linking stress (the explanatory variable) and respiratory infections (the outcome variable). In this example, and in other studies examining the mechanisms linking two phenomena, a mediation model could be conducted.

Mediation modeling allows researchers to enhance the complexity of their research questions, yet the actual statistical approach is relatively straightforward. With a traditional approach, attributed to Baron and Kenny (1982), mediation can be assessed through a series of regression models that are examined in combination to determine if there is an indirect effect of the explanatory variable on the outcome variable via the intervening (or mediating) variable (Fig. 4.1). From this Baron and Kenny (1982) approach, mediation can be tested with separate regression models or in a structural equation modeling framework where both the *a* path (the explanatory variable to the mediator) and the *b* path (the mediator to the outcome variable) are estimated. Rather than conducting a series of regression models, other mediation approaches focus on the variance and significance of the mediation pathway (also called the *indirect effect*, represented by *ab* in Fig. 4.1) (Hayes, 2009). For example, a Sobel test assesses the indirect effect relative to its standard error, while a bootstrapping approach obtains bias-corrected confidence intervals around the indirect effect (Hayes, 2009). A review of statistical approaches to mediation can be found in MacKinnon, Lockwood, Hoffman, West, and Sheets (2002).

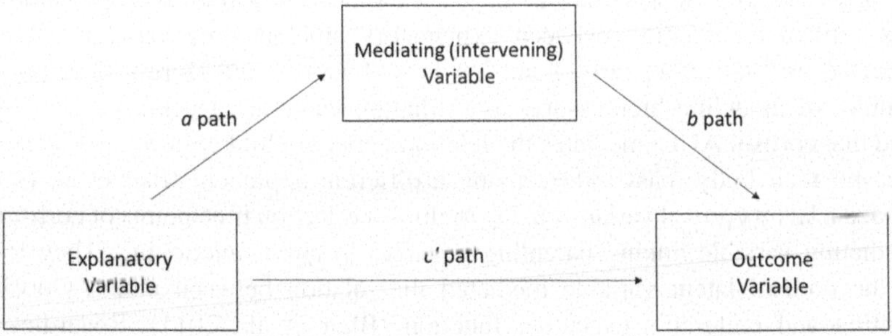


**Fig. 4.1** Example of a mediation model. Mediation is assessed by examining the product of the *a* and *b* paths. In addition, some approaches require that the $c'$ path be statistically significant to establish mediation (MacKinnon et al., 2002), while other approaches do not include this requirement (Hayes, 2009)

To examine mediation, it is ideal that the explanatory, mediating, and outcome variables were collected at three different time points (Little, 2013). If this timing sequence is not possible, the findings can be discussed in terms of indirect effects, rather than mediation.

Researchers interested in salivary biomeasures often have small sample sizes due to resource constraints (e.g., time, saliva sample volumes, and budget). A small sample size does not preclude researchers from examining mediation or indirect effects, yet they should take some precaution when conducting these models for other reasons. For example, many mediation models assume normality. In a parametric regression framework, both the mediating and the outcome variable must be normally distributed, or researchers will need to use an estimator that accounts for non-normality of their distribution(s). Research has shown that in small samples, confidence intervals around the indirect or mediating effect (*ab*) are susceptible to bias. Therefore, researchers should use bootstrap methods to obtain bias-corrected confidence intervals, which, especially for non-normal data, yield more accurate parameter estimates and minimize the risk of Type I error (MacKinnon et al., 2002; MacKinnon, Lockwood, & Williams, 2004).

Many studies have used biomeasures as either the explanatory or the mediator variable, but ultimately it is up to the researchers to decide the question of interest. If researchers are interested in how a biomeasure is associated with the outcome variable via a mediator, then the biomeasure would be included as the explanatory variable. For example, Shoal, Giancola, and Kirillova (2003) used the Baron and Kenny (1982) approach to examine if self-control mediates the relation between cortisol and aggressive behavior. That is, they were interested in knowing if cortisol levels were associated with self-control, and, in turn, if self-control was associated aggressive behaviors. In another study, researchers found that the relation between cortisol and depressive symptoms was mediated by social functioning (Tse & Bond, 2004). Interested readers should review these studies by Shoal et al. (2003) and Tse and Bond (2004) for additional details regarding the application of modeling strategies that conceptualize salivary biomeasures as primary agents (i.e., explanatory variables) in mechanistic pathways that include mediating and outcome variables.

Conversely, researchers may focus on a biomeasure's role as a mediator that helps explain the relation between explanatory and outcome variables. Studies conducted by Blair et al. (2011) and Dockray, Susman, and Dorn (2009) provide examples of modeling biomeasures as mediating variables. Dockray et al. (2009) found that cortisol AUCi mediates the relation between children's depressive symptoms and their body mass index. Using a different approach, Blair et al. (2011) included a latent cortisol variable, using multiple collection time points of cortisol, as a mediating variable linking parenting practices to child functioning. They found that the cortisol latent variable mediated the relation between parent's positive parenting and children's executive function (Blair et al., 2011). Regardless of which biomeasure is studied or how the variable was created, the goal when utilizing a biomeasure as a mediating variable is to be able to state that the specified analyte acts as a mechanism through which the explanatory variable is related to the outcome variable. Mediation modeling thus enables research questions that extend beyond description toward explanation and promotes further progress in the field as mechanisms are identified and conceptual models are specified to reflect them.

### 4.4.2 Modeling Coordination Among Multiple Analytes in Cross-Sectional Studies[1]

Recent developments in salivary bioscience assay technology have greatly expanded the number of different analytes that can be assayed from a single saliva sample. In the case of multiple biomeasures within a person, there is an opportunity to examine *cross-system coordination*, or associations of functioning of two or more different physiologic systems represented by the biomeasures. While the ability to model multiple physiologic systems within the same saliva sample presents new opportunities for understanding biologic functioning and health, it also presents statistical challenges in the modeling and interpretation of relations between the analytes and with other antecedents and/or outcomes. Analytes from multiple components of a response system (e.g., stress system) work in coordination, and they could do so in an additive or nonadditive way. Additive effects can be modeled using simple, main effects models. In this discussion, we focus on strategies for modeling the more complex, nonadditive, combined effects of different biomeasures in relation to an outcome (e.g., behavior problems or anxiety). Our discussion in this section is limited to approaches for analyzing multiple analytes that are collected cross-sectionally. See Sect. 4.4.3 for a discussion of modeling multiple analytes within a longitudinal design.

The two most common methods for examining nonadditive combined effects of analytes on behavioral and health outcomes are the ratio and interaction approaches. Interested readers are encouraged to consult other resources to learn more about the interaction approach (Hayes, 2013; Jaccard & Turrisi, 2003) and the rationale and procedures for probing significant interactions using simple slopes effects (Aiken, West, & Reno, 1991; Cohen, Cohen, West, & Aiken, 2002) and regions of significance with the Johnson–Neyman technique (Hayes & Matthes, 2009; Johnson & Fay, 1950). Both the ratio and interaction approaches allow the strength of the relation between one analyte ($X_1$) and the outcome ($Y$) to vary according to the level of the other analyte ($X_2$) [see Eqs. (4.1) and (4.2)]. While both approaches help us examine the joint effects of the biomeasures on the outcome, neither provides specific information about coordination between the two biomeasures (i.e., how

[1]This section is based on *Chen, F. R., Raine, A., & Granger, D. A. (2015). Tactics for modeling multiple salivary analyte data in relation to behavior problems: Additive, ratio, and interaction effects. Psychoneuroendocrinology, 51, 188–200*. The reader is referred to this paper for additional discussion of these concepts.

strongly they are associated with each other), and the approaches are not interchangeable. There are two important statistical and conceptual differences that distinguish them.

First, the ratio and interaction approaches assume different functions of how the $X_2$ analyte impacts the relation between the $X_1$ analyte and outcome variable ($Y$). These effects can be in either a nonlinear (the ratio approach; Eq. 4.1) or linear (the interaction approach; Eq. 4.2) function. In these models, we are most interested in testing whether the coefficients $d_0$ and $d_1$ are significant, because these two coefficients capture the coordinated effects in the ratio and the interaction approach, respectively.

$$\begin{aligned}\text{Ratio approach}: \quad Y &= a_0 + d_0 * X_1/X_2 + \varepsilon_0 \\ &= a_0 + (d_0 * 1/X_2) \ * \ X_1 + \varepsilon_0 \end{aligned} \tag{4.1}$$

$$\begin{aligned}\text{Interaction approach}: \quad Y &= a_1 + b * X_1 + c * X_2 + d_1 * X_1 * X_2 + \varepsilon_1 \\ &= a_1 + (b + d_1 * X_2) * X_1 + c * X_2 + \varepsilon_1 \end{aligned} \tag{4.2}$$

The ratio approach assumes that the relation between the analyte $X_1$ and the outcome $Y$ (i.e., the slope of $X_1$ on $Y$) varies as a nonlinear function of the analyte $X_2$. As shown in Eq. (4.1), a one-degree increase in $X_1$ corresponds to $d_0 * 1/X_2$ degree change in $Y$. Thus, the slope of $X_1$ (i.e., $d_0 * 1/X_2$) is a nonlinear function of $X_2$. So, for example, when $X_2$ changes from 2 to 3, the slope of $X_1$ changes from 1/2 $d_0$ to 1/3 $d_0$, which is a 1/6 $d_0$ decrease; but if $X_2$ changes from 3 to 4, the slope of $X_1$ changes from 1/3 $d_0$ to 1/4 $d_0$, which is a 1/12 $d_0$ decrease. Although both illustrate a one-degree change in $X_2$, the slope of $X_1$ on $Y$ does not change in a fixed manner. As seen in the left panel of Fig. 4.2, at the lower levels of $X_2$, the linear relation between $X_1$ and $Y$ (i.e., the slope) is much larger in magnitude for either a positive or a negative relation; whereas at higher levels of $X_2$, the linear relation between $X_1$ and $Y$ is smaller. The magnitude in the change of the slopes ($Y$ on $X_1$) drops considerably as $X_2$ increases.

In contrast to the ratio approach, the interaction approach assumes that the slope of analyte $X_1$ on $Y$ is a linear function of analyte $X_2$. A one-degree increase in $X_1$ results in $b + d_1 * X_2$ degree increase in $Y$. Thus, the slope of $X_1$ (i.e., $b + d_1 * X_2$) is a linear function of $X_2$. That is, regardless of the level of $X_2$, a one-degree increase in $X_2$ always corresponds to a fixed degree of change in the slope of $X_1$ (right panel of Fig. 4.2).

The second major difference between the ratio and interaction approaches is whether the approach allows for a bidirectional (positive and negative) relation between $X_1$ and $Y$. This distinction is the byproduct of the coefficients modeled in each approach. The ratio approach allows only a unidirectional relation between $X_1$ and $Y$ (the relation can only be positive or negative), but the interaction approach allows both bidirectional and unidirectional relations. Because the coefficient that captures the coordinated effects of analytes $X_1$ and $X_2$ is $d_0$ in the ratio approach, the sign (positive or negative) for the slope of $X_1$ on $Y$ will remain the same (see the left panel of Fig. 4.2). In contrast, the interaction approach allows the sign for the slope of analyte $X_1$ on $Y$ to be always negative, always positive, or switch from positive to

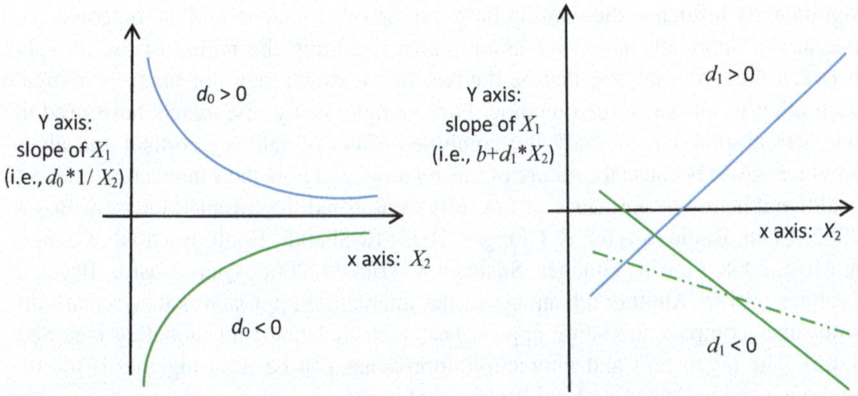


**Fig. 4.2** Illustration of the nonlinear (left) and linear (right) slopes of analyte $X_1$ on outcome $Y$ that are modeled using the ratio (left) and interaction (right) approaches. With the ratio approach (left), a one-degree increase in $X_2$ corresponds to a much bigger change in the slope of $X_1$ at lower levels of $X_2$ than at higher levels of $X_2$ (nonlinear change in the $X_1$ slope). With the interaction approach (right), a one-degree increase in $X_2$ corresponds to a fixed degree of change in the slope of $X_1$ at all levels of $X_2$ (linear change in the slope)

negative (see the solid green line in the right panel of Fig. 4.2 for an illustration of the classic crossover interaction pattern and the dashed green line for an illustration of a relation between $X_1$ and $Y$ that is always negative).

Given these distinctions between the ratio and interaction approaches, they address different research questions. The ratio approach is best suited for modeling relations between two analytes that are thought to have opposite effects on the same outcome or when one of the biomeasures cancels out or reduces the effect of the other. For example, modeling the cortisol–DHEA-S relation as a ratio term may be appropriate given the anti-glucocorticoid and neuroprotective effect of DHEA-S. Previous studies have found higher cortisol-to-DHEA-S ratios are linked with depressive symptoms among adolescents and adults (Goodyer, Herbert, & Tamplin, 2003; Goodyer, Park, Netherton, & Herbert, 2001; Markopoulou et al., 2009), anxiety in men aged 62–76 years old (van Niekerk, Huppert, & Herbert, 2001), and internalizing problems in children (Chen, Raine, & Granger, 2015). A large cortisol-to-DHEA-S ratio could result from many combinations of the two analytes as long as cortisol overwhelms the effect of DHEA-S, including, for example, (a) low DHEA-S and high cortisol, (b) high DHEA-S and even higher cortisol, and (c) low DHEA-S and normal levels of cortisol. One way to interpret the ratio effects is, as DHEA-S increases, the magnitude of the positive relationship between cortisol and depression is reduced. Another example of the ratio approach involves testosterone and cortisol and their relation to psychopathy (Glenn, Raine, Schug, Gao, & Granger, 2011).

The interaction approach is most appropriate when researchers are interested in whether one analyte has a different relation with the outcome at different levels of the other analyte. This can be conceptualized as modeling a contextual effect—where the level of one biomeasure represents a “context” that may (or may not) significantly influence the relation between the other analyte and the outcome. The interaction approach has fewer assumptions regarding the nature of the interplay between the two analytes; that is, the two biomeasures may potentiate *or* mitigate each other to influence the outcome. For example, many researchers have used the interaction approach to study the combined effect of salivary cortisol and alpha-amylase (sAA) because the nature of this relation, and how their interaction relates to health and behavior outcomes, is not fully established (e.g., Bauer, Quas, & Boyce, 2002; Chen, Raine, Soyfer, & Granger, 2015; El-Sheikh, Erath, Buckhalt, Granger, & Mize, 2008; Gordis, Granger, Susman, & Trickett, 2006; Quas, Castro, Bryce, & Granger, 2018). Another advantage of the interaction approach is its compatibility with other complex modeling approaches, such as latent trait modeling (see Sect. 4.4.5). The latent trait and interaction approaches can be used together if the two analytes were collected with multiple repeated measures, because these approaches are complementary.

While this section focuses on examining two analytes as interacting variables, it is important to note that the approaches discussed can also be used to examine relations between a single analyte and a non-biomeasure variable (e.g., a personality or behavioral characteristic) or between more than two biomeasures. Researchers should carefully consider their conceptualization of hypothesized cross-system, physiologic relations and how they interact with other health or behavioral variables, so that they can appropriately specify their statistical models.

### 4.4.3 Modeling Individual Biomeasure Response Trajectories Using Multilevel Modeling

When there are more than two measures of a particular analyte within a person over time, it is possible to examine the *trajectory* of that biomeasure using growth curve modeling in either a multilevel or structural equation modeling framework. Here, we focus on the former approach, which divides the variance in outcome data nested within units (i.e., observations nested within a person over time and/or from people nested within groups) into within-unit and between-unit components [see Raudenbush and Bryk (2002), Snijders and Bosker (2002) for further background on the rationale for multilevel modeling]. The examples that follow deal primarily with one common occasion for modeling biomeasure trajectories—the response to an acute stressor—but it should be noted that these approaches can address varying time scales of biomeasure change such as diurnal rhythms and growth across days, weeks, or years. For example, we can separately model each individual's trajectory of salivary cortisol across a given stress episode (level 1) and between-person differences in these trajectories (level 2). Below, we describe some of the main approaches to modeling biomeasure trajectories in response to an event such as a stressor and how to connect model estimates to substantive interpretations about an individual's responses.

One way to model stress responding is with a curvilinear function, usually a quadratic curve that captures an expected rise in the biomeasure in response to the stressor followed by a decline during the recovery period. This means that each person's analyte concentrations over time are modeled with an intercept, linear slope (time) term, and quadratic slope (time squared) term. In certain cases, a cubic or other more complex function may best capture the shape of response trajectories, which can be verified by comparing model fit with and without further growth terms. Each of the typical quadratic growth parameters carries meaning. Whereas the intercept represents a person's biomeasure level at the time point where the model is centered (i.e., where time = 0), linear and quadratic slopes represent the person's response dynamics: the instantaneous slope of the curve at the centering point and overall rate of analyte deceleration (or acceleration) across the session, respectively. More concretely, the intercept points to a person's level of biomeasure activity, and the slopes point to when/how quickly they respond and recover. Figure 4.3 illustrates how each of these terms translates into potential stress response patterns.

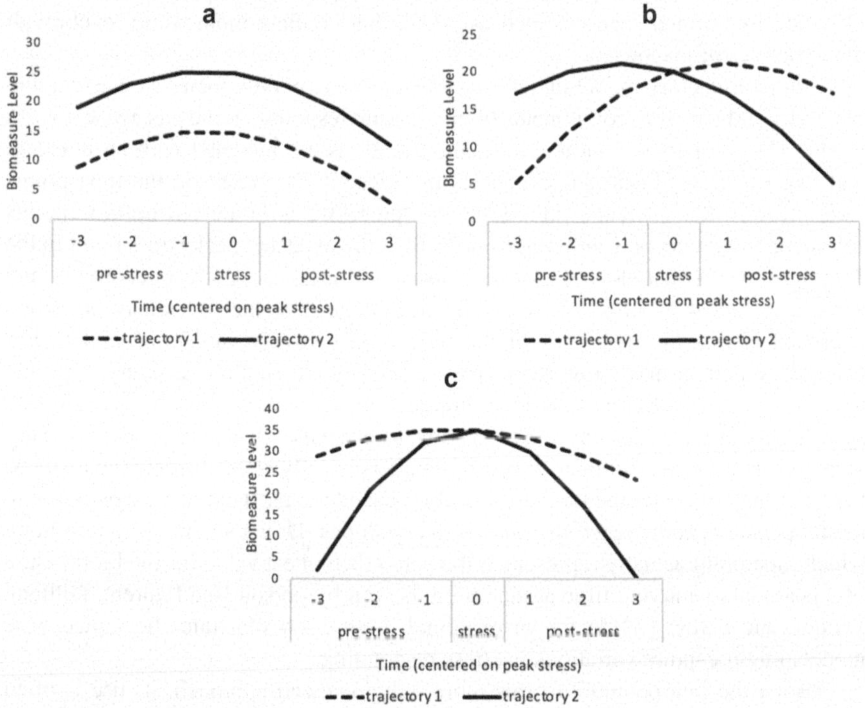


**Fig. 4.3** Curvilinear (quadratic) modeling of stress-related biomeasure trajectories. Panel (**a**) depicts differences in trajectory intercepts: trajectory 1 involves lower levels of activation than trajectory 2. Panel (**b**) depicts differences in linear slopes: trajectory 1 peaks later than trajectory 2. Panel (**c**) depicts differences in quadratic slopes: trajectory 2 is steeper (shorter response duration, with more marked reactivity/recovery) than trajectory 1

While much of the focus on adaptive stress responding deals with response levels (intercept) and steepness versus flatness of response curves (quadratic slope), the linear slope offers potentially useful information about the timing of a person's response and whether they are reacting (typically a positive slope) or recovering (typically a negative slope) as expected at a given time. In fact, a little calculus can be used to determine when a given person would be expected to peak in their stress response based on their quadratic growth parameters. Knowing that the peak of the curve is the point where the instantaneous slope is zero, and that the slope of the curve is the first derivative of the growth function, we could use parameters from the following within-person quadratic function $biomeasure\ y = a*time^2 + b*time + c$ to calculate $peak\ time = -b/2a$. Because the sign of $a$ in the usual case of a rising and falling curve is negative, we can see that a more positive linear slope ($b$) translates into a later peak and a more negative linear slope into an earlier peak. Why would this matter? In the case of stress responding, we might be interested in the (mal) adaptive implications of early peaking anticipatory stress reactivity that has recovered to lower levels by the time the person is actually confronted with the stressor, as opposed to more gradually building stress reactivity that peaks after the stressor is over. This potentially important dimension of a person's response tendencies is obscured by simpler metrics, such as AUC, that conflate biomeasure levels with the dynamics of response.

Another trajectory modeling approach that more explicitly focuses on estimating reactivity and recovery components of biomeasure responses is the piecewise growth model. Here, a person's biomeasure concentrations are modeled with an intercept and two (or more) linear slopes; slope (or "piece") 1 is based on the time points leading up to the intercept, and slope (or "piece") 2 is based on the time points following the intercept. Centering (time = 0) at the expected peak stress point in the paradigm means that slope 1 can be considered an index of stress reactivity, and slope 2 an index of stress recovery. An advantage over the previously presented approach is that the dynamics of reactivity and recovery can be separated. For example, a person may show steep stress reactivity but shallow recovery, a profile that likely carries different functional implications from those of either a uniformly steep (marked reactivity and recovery slopes) or shallow (reactivity and recovery slopes close to zero) responder. While researchers usually select a single (post-stress) time at which to center the models, it is also possible to use landmark registration to model person-specific reactivity and recovery slopes. Doing so entails determining which time point actually represented the person's highest value for the biomeasure (the peak) and centering time at that point, person by person [see Laurent, Gilliam, Wright, and Fisher (2015) for an example]. Figure 4.4 illustrates how piecewise models might capture variations in stress responding.

Despite the rich potential of this more individualized approach, its use is often limited in practice by the number of available biomeasure time points; there must be at least two observations on either side of the intercept (preceding and following it) in order to model separate slopes. This means a minimum of five repeated measures are needed to model two slopes, with additional observations needed to model different person-specific peaks. For researchers with longitudinal biomeasure data, the choice of a trajectory modeling approach should involve both theory-driven and

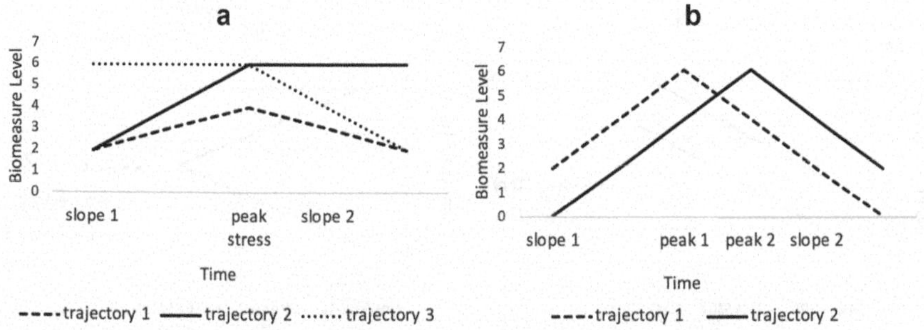


**Fig. 4.4** Piecewise modeling of biomeasure trajectories. Panel (**a**) depicts differences in both intercepts and piecewise slopes—trajectory 1 shows relatively shallow reactivity and recovery (and lower activation overall); trajectory 2 shows steep reactivity but no recovery; trajectory 3 shows no reactivity but does show steep recovery. Panel (**b**) depicts different peaks at which the intercept is centered—trajectory 1 peaks earlier than trajectory 2 (with comparable reactivity/recovery slopes). See also Lopez-Duran, Mayer, and Abelson (2014) for an illustration of individual landmark registration modeling of cortisol trajectories

practical considerations. When choosing an approach, researchers should consider the conceptual model guiding expectations about how the biomeasure should behave; the aspects of reactivity and/or recovery that are of interest and should be dissociated; and the number of time points available per person.

## 4.4.4 Additional Applications of Multilevel Modeling

So far, we have considered how to model multiple observations of an analyte within a person using a growth trajectory. What happens when we expand the scope to look at multiple growth trajectories across biomeasures and/or people? A multilevel modeling framework can accommodate this added layer of complexity.

### 4.4.4.1 Modeling Multiple Biomeasure Trajectories with Multilevel Modeling

Multilevel modeling allows researchers to move beyond cross-sectional approaches to examine multi-biomeasure data and assess cross-system coordination with multiple analytes, representing function in different physiologic systems from the same person across time. Figure 4.5 illustrates the different possibilities for associations among biomeasures across response trajectories. For example, some researchers have proposed that stress-related HPA axis activation (indexed by salivary cortisol) may be more or less adaptive depending on whether the response is coordinated with a concurrent autonomic nervous system response (indexed by sAA) [see Laurent, Powers, and Granger (2013)]. With multilevel modeling, we can investigate such coordination at both a between-person and a within-person level of analysis.

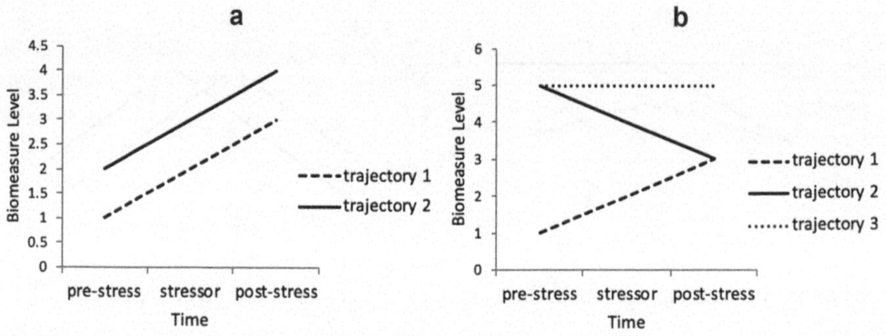


**Fig. 4.5** Associations between biomeasure trajectories within people (cross-system coordination) or across people (dyadic attunement). In panel (**a**), trajectories 1 and 2 show matched-phase association, but not average-level association. In panel (**b**), trajectories 1 and 2 show average-level association, but inverse matched-phase association; trajectory 3 is independent of trajectories 1 and 2 (not associated in either a matched-phase or average-level manner)

On one hand, we might distinguish whether people who show higher absolute mean levels of sAA also tend to show higher levels of cortisol during stress, or a between-person effect. Beyond this average-level difference, we can also determine whether, for a given person, a relative increase in cortisol from one sample to the next (representing a particular phase of the stress response) tends to be accompanied by a parallel increase in sAA, or a within-person effect we have called "matched phase coordination." Disambiguating these forms of coordination can help to shed further light on both what is normative and what can be considered an optimal multisystem stress response. We have found that adults responding to psychosocial stressors may (but do not necessarily) show average-level coordination of cortisol and sAA, but that the degree of within-person matched-phase coordination varies and is associated with markers of well-being including stress appraisals and depressive symptoms (Laurent, Ablow, & Measelle, 2011; Laurent, Lucas, Pierce, Goetz, & Granger, 2016).

Statistically, cross-system coordination can be modeled by selecting one biomeasure as the outcome in a multilevel model and adding measures of the other analytes at both level 1 (within-person) and level 2 (between-person). The within-person effect, a time-varying covariate comprising the person's repeated samples, is centered around the group mean to represent increases and decreases relative to the person's own average. The between-person effect, a single score based on the person's average concentration across sampling times, is centered around the grand mean to represent higher or lower levels relative to the sample as a whole. For such statistical tests, the decision of which biomeasure to model as the outcome variable should be guided by what makes sense theoretically and methodologically. In the case of concurrently sampled cortisol and sAA, modeling sAA as the outcome makes for a more straightforward temporal interpretation (i.e., given the longer time lag for HPA axis response to appear in salivary cortisol than for the autonomic response to appear in sAA, the cortisol value from a given saliva sample represents an earlier stress system response than the sAA value). However, this is not the only reasonable approach to modeling coordination of these systems. Given the reality that, over the course of a stress response, the HPA axis and sympathetic nervous system exert bidirectional permissive, stimulatory, and suppressive effects (Sapolsky, Romero, & Munck, 2000), one might model either cortisol or sAA as the outcome while accounting for the differential time lag in the matching of samples and interpretation of matched phase effects. More broadly, researchers should aim to designate explanatory and outcome biomeasures in such a way that the direction of effects makes sense from both a methodological standpoint (i.e., the sampling structure for each biomeasure allows earlier activation of one system to predict concurrent or later activation of the other) and a biological standpoint (i.e., the activation of one physiological system is known to influence the activation of the other). Finally, practicalities of the data may also play a role in this decision, with researchers selecting the biomeasure with a better known response course and/or distribution of scores as the outcome in a given analysis.

#### 4.4.4.2 Modeling Biomeasure Trajectories from Multiple Individuals with Multilevel Modeling

Extending beyond an individual person's trajectories, we might ask whether a person's biomeasure response tends to be associated with that of another person, such as a romantic partner or parent. We have called this type of association *cross-partner attunement*. Here, the two systems being modeled are not different biomeasures from the same person, but the same analyte from different dyad partners. The trajectories shown in Fig. 4.5 could also represent different possibilities for associations between individuals' biomeasures response trajectories.

Dyadic relationship research examines whether physiologic attunement across individuals—also called synchrony or co-regulation—is present and/or adaptive. Much of this research in salivary bioscience has examined the degree of co-regulation in individuals' stress responses, and how coordination of HPA and sympathetic nervous system activation across individuals is related to health and relational outcomes. Attunement can be examined using an approach similar to that described above (i.e., with one biomeasure modeled as both a level 1 time-varying covariate and a level 2 mean score explanatory variable of the second biomeasure trajectory). For example, we investigated whether mother–infant HPA axis attunement was normative and/or adaptive by modeling maternal cortisol as a level 1 explanatory variable of concurrent infant cortisol outcomes across a dyadic stress task, then testing maternal depressive symptom scores during pregnancy and postpartum as level 2 explanatory variables of between-dyad variation in this level 1 attunement term (Laurent et al., 2011). We demonstrated that greater matched-phase coordination of mothers' and infants' cortisol responses within dyads was associated with increasing maternal depressive symptoms (Laurent et al., 2011). This finding adds an important counterpoint to earlier suggestions that stronger mother–child attunement marks better adjusted dyads, which would have been impossible using more traditional measures of synchrony such as bivariate correlations.

There are other ways of approaching dyadic biomeasure data that may be appropriate for the questions under investigation. So far, we have considered the case where one partner's trajectory is modeled at level 1 as the outcome, with the other partner's scores added as explanatory variables at both level 1 and level 2. This allows for a focus on cross-partner attunement, but not a concurrent examination of each partner's response trajectory and explanatory variables of each. When researchers wish to explain both dyad partners' response trajectories and compare model parameters across partners, the partners' trajectories can be modeled simultaneously as within-dyad outcomes. For example, a model of romantic partners' cortisol trajectories could include an intercept, linear slope, and quadratic slope for each partner as level 1 explanatory variables (six terms total, three of which are dummy-coded to designate Partner 1 and three to designate Partner 2). Direct comparisons across partners in terms of average trajectory characteristics and the predictive strength of level 2 explanatory variables can then be made by comparing the size of Partner 1 and Partner 2's model coefficients.

As an illustration of this approach, Laurent et al. (2013) compared heterosexual partners' cortisol responses to conflict and associations with behavior during the conflict. They found that while male partners tended to exhibit higher cortisol levels (comparison of intercepts) during conflict compared to their female counterparts, males recovered more quickly following the conflict (comparison of linear and quadratic slopes) and appeared to benefit more from positive behaviors and suffer less from negative behaviors during the discussion (comparison of coefficients for positive/negative behavioral explanatory variables) (Laurent, Powers, Laws, et al., 2013). Such a direct comparison across partners offers a more solid basis for conclusions about dyad partner differences than simply modeling each partner's outcomes separately and tallying the number of significant effects for each.

Again, researchers with access to multiple analytes within a person or dyad must decide which approach is most appropriate based on their particular aims and available data. One consideration is whether it is plausible and/or theoretically important to distinguish within-person/dyad, between-person/dyad, or both types of effects. Another consideration is whether the most important dyad characteristic under investigation involves associations between partners or differences across partners. While the above offers some possible ways to proceed, this is not meant to be an exhaustive list, and the reader is encouraged to consult other resources [e.g., Lyons and Sayer (2010), which outlines further approaches to modeling longitudinal dyadic data] to develop a plan for addressing their unique research questions.

### *4.4.5 Isolating Trait, State, and Error Components of Salivary Biomeasure Data Using Latent Variable Modeling*

Multilevel modeling allows for a detailed examination of multiple trajectories and responses in analytes within and between individuals. However, for some research questions the primary aim may be to model stable, rather than changing, aspects of a specific analyte. For these research questions, there is interest and value in isolating variance in biomeasures levels related to stable trait-like factors, changing state-like factors, and error-related factors.

Variance in biomeasures at any given moment can be related to a variety of different sources, including stable individual differences, momentary situational influences (e.g., time of day, mood, food and drug intake, exercise, and environmental exposures), and measurement error (Kenny & Zautra, 2001; Shirtcliff, Granger, Booth, & Johnson, 2005). For these reasons, obtaining reliable measures for stable individual differences in analyte levels necessitates strict saliva sample collection, handling, and assay protocols be established and implemented by all study participants and staff (Granger et al., 2012). Given feasibility constraints and considerations of participant burden, a much more common practice is to collect multiple saliva samples and use the average analyte level across all samples to represent an individual's analyte level (e.g., El-Sheikh et al., 2008). However, this approach fails to account for the mixed nature of the average analyte concentration, which represents a combination of the true analyte concentration, measurement error, and other sources of momentary variability. For example previous studies have found that daily variation in salivary cortisol output accounts for over 70% of the variance in the awakening response (CAR) and at least 50% of the variance in the diurnal slope (Doane, Chen, Sladek, Van Lenten, & Granger, 2015; Ross, Murphy, Adam, Chen, & Miller, 2014). These findings demonstrate the fluctuations within an individual's salivary cortisol and illustrate the difficulty of assessing stable individual differences in cortisol using conventional statistical methods. Researchers can adopt latent modeling alternatives to mean or composite variable approaches when attempting to model both stable and non-stable aspects of salivary biomeasures.

Statistical advances have supported latent variables as an appropriate approach to improve the estimates of trait-like biomeasure concentrations. Within a structural equation modeling (SEM) framework, measurement models can be used to create latent variables that represent a more "true score" of salivary analyte concentrations. Specifically, two approaches to measuring trait and state levels of biomeasures are outlined below. Presented first is the latent state trait approach (LST; Steyer, Ferring, & Schmitt, 1992; Steyer, Geiser, & Fiege, 2012), which allows for the isolation and measurement of varying "state" and stable "trait" aspects of analyte concentrations. Statistically, this approach may be more difficult to model as the model requirements are more stringent. The second approach, which is less statistically stringent, is a latent variable approach to model the "trait-like" variance in the biomeasures utilizing a confirmatory factor analysis model.

**Latent State Trait Models** By parsing sources of variance in biomeasure concentrations, LST can be used to address research questions that focus on both stable, trait analyte levels and their associations with behavior, health, and physiology, as well as changing, state analyte levels (e.g., related to circadian rhythm or stress reactivity). When conducting an LST model, one approach is to use the replicate values for each analyte determination in the lowest level of the model (Fig. 4.6). Including replicate

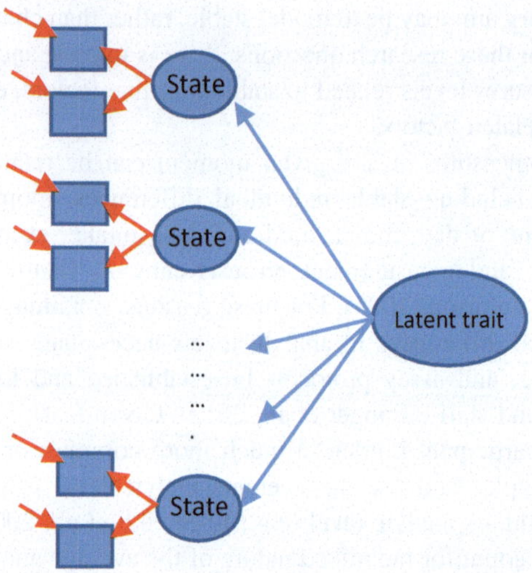


**Fig. 4.6** Illustration of a Latent State Trait (LST) model of salivary analyte data. Replicate determinations of the salivary analyte are used as indicators in the LST model. The latent trait analyte concentration is modeled as a second-order factor. Circles represent latent variables, and rectangles represent indicators. Red arrows refer to error variance and factor loadings that are fixed based on the reliability of the two replicate determinations. Blue arrows are estimated factor loadings that should be fixed to be equal when measurement invariance is upheld

values as indicators in LST models, rather than using the mean of the replicates as the indicators, improves statistical estimates of the relationships between the analytes and other antecedents/outcomes because measurement error inherent in the salivary analyte assay is explicitly modeled. The LST approach requires that a series of models be examined to ensure measurement invariance is upheld, including constraints across all the residual variances of individual analyte determinations, across all state factor loadings, across all state residual variances, and across all trait factor loadings [see Riis et al. (2018) and Shirtcliff et al. (2005) for examples of LST models with salivary biomeasures]. In a "true" LST model, measurement invariance must be upheld at each level of measurement without allowing for correlations/covariances among any variables or residual variances (Steyer et al., 2012). Yet some assays are extremely precise, resulting in small residual variances, which can cause problems with model convergence even when the factors are accurately modeled (Browne, MacCallum, Kim, Andersen, & Glaser, 2002). Should this occur, researchers must make an informed decision based on the accuracy of the assay, trends in previous research, and model parameter estimates when faced with small residual variances and inadequate model fit.

Depending on the research question, LST modeling can also be used to isolate the situational or state variance from the measurement error variance and calculate the proportion of variance attributed to the state and trait factors [see Kirschbaum et al. (1990) for calculations]. An example of the early adoption of this method comes from Kirschbaum et al. (1990) who used an LST approach to examine diurnal cortisol. The authors found a high proportion of the variance in morning cortisol concentrations was accounted for by a latent trait factor (60% variance) while afternoon cortisol concentrations were largely accounted for by latent state factors (75% variance; Kirschbaum et al., 1990). LST modeling has since been used to examine if the state and/or trait components of an analyte, such as salivary cortisol and uric acid (Riis et al., 2018; Shirtcliff et al., 2005; Yeung et al., 2016), are related to other constructs of interest. For more information on the statistical requirements for LST models review Kirschbaum et al. (1990), Steyer et al. (1992), and Steyer et al. (2012).

**Latent "Trait-Like" Modeling** Often researchers are not be able to use LST modeling because they do not have access to replicate analyte data, they do not have enough samples to create multiple state variables, or because measurement invariance is not upheld. When this occurs, trait-like variance can be modeled via an alternative latent variable approach. In this framework, a latent trait-like variable is modeled using the mean analyte concentrations of each saliva sample as indicators of the latent variable (Fig. 4.7). Given the typical quality control measures implemented when assaying biomeasures (e.g., duplicates with low CVs), assay measurement error is dramatically reduced and may be regarded as minimal. Thus, we can use the mean of the two replicate determinations as indicators when modeling the trait factor. With this approach, the error terms in the measurement model will effectively capture both the measurement error and the situational influences. To estimate a latent trait-like variable, three indicators (or saliva samples) are needed to allow for a just identified model that does not produce any model fit indices. Thus, to estimate how well the proposed latent model fits the data, at least four saliva

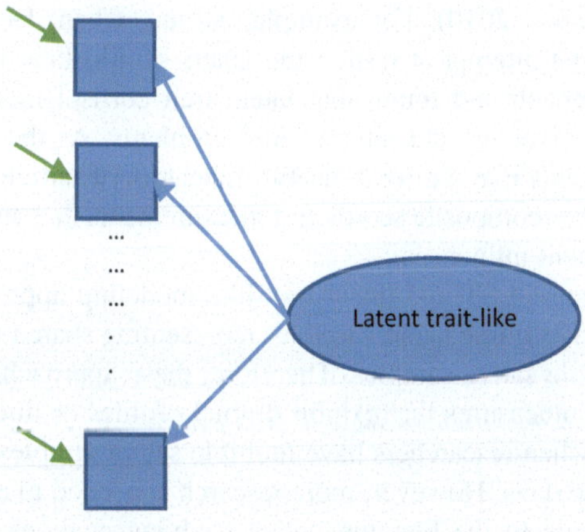


**Fig. 4.7** Illustration of a latent trait-like model of salivary analyte data. The means of salivary analyte replicate determinations are used as indicators in the trait-like model. Circles represent latent variables, and rectangles represent indicators. Green arrows refer to error variances that are freely estimated, which capture both measurement errors and moment-to-moment fluctuations in analyte concentrations. Blue arrows are freely estimated factor loadings

samples, without correlated residual variances, are needed. For more information on modeling latent variables, review Brown (2014). It is recommended that saliva samples are collected over multiple days to derive a latent trait-like factor. For example, previous studies showed that two saliva samples in the morning collected over 2 (Giesbrecht et al., 2015) or 3 days appear to be sufficient to capture a latent trait-like cortisol measure (Doane et al., 2015; Stroud, Chen, Doane, & Granger, 2016). Interested readers are encouraged to review the aforementioned studies for example applications of the latent trait-like approach with salivary bioscience data.

Isolating the variance in salivary biomeasures attributable to stable, intrinsic "trait-like" sources has been applied primarily in analyzing cortisol. A study conducted by Doane et al. (2015) found latent trait cortisol in a sample of young adults was distinct from other diurnal indicators of cortisol such as the awakening response (CAR) and diurnal slope (Doane et al., 2015). Latent trait cortisol in this study was also stable within and across three assessments spanning 9 months and was largely reflective of between-person variability in HPA activity (Doane et al., 2015).

A latent trait modeling approach may be advantageous over other approaches to modeling multiple analyte concentrations, such as means across saliva samples or AUCg, if the research question of interest is related to between-person differences. For example, retrospectively reported childhood trauma has been found to correlate with latent trait cortisol but not with the CAR or cortisol diurnal slope (Doane et al., 2015). Trait-like cortisol, but not mean cortisol concentrations, has also been linked to behavior problems in other studies (Chen, Raine, Soyfer, & Granger, 2015). Among pregnant women, daytime trait-like latent cortisol was shown to be relatively stable across trimesters, and third trimester trait-like cortisol predicted infant birthweight (Giesbrecht et al., 2015). Investigating latent trait factors in biomeasures may allow researchers to better tap into a "set point" of the biological system, which has advantages in testing some theoretical propositions such as allostatic load (McEwen & Gianaros, 2010). For example, Stroud, Chen, Doane, and Granger (2018) made the first attempt at testing the tenets of allostatic load theory with a latent variable approach and found that latent trait cortisol mediated the relation between early-life adversity and internalizing problems. As these studies suggest, latent trait components may be more useful indicators of individual differences in physiology than other composite scores and measurements that are more susceptible to moment-to-moment influences.

A key advantage of LST and latent trait-like modeling approaches is that they allow researchers to examine latent variables representing shared variance in analyte levels across multiple saliva samples. Therefore, these approaches are most useful when examining biomeasures that exhibit diurnal profiles or fluctuate dynamically across the day, or when researchers have multiple saliva samples taken at the same time across multiple days. However, more research is needed to examine how these approaches contribute to the literature when studying analytes that are relatively stable, and how these latent variables may be related to common composite measures such as slopes and AUC [see Dariotis, Chen, and Granger (2016) for an example with testosterone]. In addition, research is needed that compares latent trait analyte measures from LST models to latent trait analyte measures derived from the trait-like approach. Both areas of research will provide valuable information regarding the utility of this approach across biomeasures and samples.

## 4.5 Conclusion

Salivary biomeasure data hold great promise for advancing our understanding of the complex relations underlying health and development. However, there are inherent challenges with these data associated with their measurement, sensitivity to external and internal factors, and the nature of the questions salivary biomeasures are used to address. In addition, the relative ease and efficiency of collecting and assaying saliva samples for multiple biomeasures makes it easy for researchers to become overwhelmed by the complexity of multisystem biologic data. This chapter is meant as an overview of the important considerations to bear in mind when working with salivary biomeasure data, including the limitations of the data as well as the great opportunities and approaches for answering multifaceted and interdisciplinary research questions.

When generating and testing their hypotheses, researchers working with salivary biomeasure data should have a strong theoretical and conceptual foundation for their work, as well as a nuanced understanding of the specific analyte(s) of interest and the associated methodological considerations. We discussed some validity and reliability challenges of salivary biomeasure data in Sect. 4.3 of this chapter, with researchers encouraged to learn more about the laboratory and assay procedures through close collaborations with their laboratory staff and assay developers. Beyond these considerations, researchers should strive for a complete understanding of how their specific salivary analyte(s) enter the oral compartment and the important confounders influencing salivary levels. This information can be found in other sections of our handbook and the extant literature that serves as its inspiration. In addition, laboratory staff and assay developers can provide important insight into issues of analyte biology and measurement. Understanding these salivary bioscience basics is foundational to developing theoretically sound research questions and appropriate statistical strategies. When reporting their findings, researchers should include details about the assay methods along with early data processing and cleaning procedures. Making these methodological details standard in salivary bioscience manuscripts will support standardization of early data processing methods, help improve cross-study comparisons, and enhance both the reliability and replicability of salivary analyte study results.

Using the advanced statistical modeling approaches presented in this chapter, salivary bioscientists can answer complex questions about how our experiences and environments interact with how we feel, act, grow, and relate to each other. The examples discussed in this chapter are provided as both scaffolding for salivary bioscientists examining similar research questions and inspiration for those interested in building on these existing methods to develop innovative approaches to modeling the biopsychosocial processes that shape health and well-being. Researchers are encouraged to be both deliberate and creative in their statistical modeling approaches. A well-defined research question and clear conceptualization of both the salivary analyte measures and their relations with other variables are essential to appropriately selecting and specifying a statistical model. As illustrated above with the LST approach—which grew out of a common interest in stable, person-specific analyte levels—a clear conceptualization of biomeasure concentrations can spark innovative uses of statistical approaches from other disciplines. In this chapter, we aimed to highlight opportunities to use the approaches presented, as well as encourage the adaptation of other approaches yet to be applied to the field of salivary bioscience.

As salivary bioscience research progresses, the practices and approaches discussed in this chapter will continue to develop and with them new standards will emerge. With advancements in salivary bioscience assay technology, the need for more complex modeling of multiple physiologic systems will push the evolution of advanced statistical methods suited to answering questions about how these systems work together. Likewise, advances in statistical approaches can encourage us to think more broadly and creatively about study design and scale, as well as the range of questions that salivary bioscience data can help us answer. This synergistic scientific momentum will continue to push salivary bioscience research forward and introduce new opportunities to integrate biologic data into studies across several fields. Salivary bioscience researchers with statistical skills and interest will play a critical role in these advancements.

**Box 4.1 Illustration of Calculations for Inter- and Intra-Assay Coefficients of Variation**

| Control | Plate | Concentration | Plate mean | SD of the high control means | Mean of the high control means | Inter-assay CV for the high controls | Average inter-assay CV for the study |
|---|---|---|---|---|---|---|---|
| High | 1 | 1.005 | 0.994 | 0.050 | 1.040 | 4.78% | 5.92% |
| | | 0.981 | | | | | |
| | | 0.987 | *(1.005 + 0.981 + 0.987 + 1.002)/4* | *SD of: 0.994, 1.093, 1.033* | *(0.994 + 1.093 + 1.033)/3* | *0.050/1.040* | *(4.78% + 7.06%)/2* |
| | | 1.002 | | | | | |
| High | 2 | 1.111 | 1.093 | | | | |
| | | 1.036 | | | | | |
| | | 1.121 | | | | | |
| | | 1.102 | | | | | |
| High | 3 | 1.026 | 1.033 | | | | |
| | | 0.997 | | | | | |
| | | 1.103 | | | | | |
| | | 1.005 | | | | | |
| **Control** | **Plate** | **Concentration** | **Plate mean** | **SD of the low control means** | **Mean of the low control means** | **Inter-assay CV for the low controls** | |
| Low | 1 | 0.079 | 0.085 | 0.006 | 0.087 | 7.06% | |
| | | 0.083 | | | | | |
| | | 0.081 | | | | | |
| | | 0.095 | | | | | |
| Low | 2 | 0.101 | 0.094 | | | | |
| | | 0.095 | | | | | |
| | | 0.089 | | | | | |
| | | 0.091 | | | | | |

(continued)

**Box 4.1** (continued)

| | | | | | | | |
|---|---|---|---|---|---|---|---|
| Low | 3 | 0.079 | 0.083 | | | | |
| | | 0.075 | | | | | |
| | | 0.085 | | | | | |
| | | 0.091 | | | | | |

| ID | Analyte Rep. 1 concentration | Analyte Rep. 2 concentration | Mean of Reps. 1 and 2 *(Rep. 1 + Rep. 2)/2* | Standard deviation of Reps. 1 and 2 | Intra-assay CV (%) *(SD of Reps. 1 and 2/mean of Reps. 1 and 2)∗100* | Ave. intra-assay CV across the study (%) |
|---|---|---|---|---|---|---|
| 1 | 0.084 | 0.084 | 0.084 | 0.000 | 0.00 | 3.36 |
| 2 | 0.071 | 0.065 | 0.068 | 0.004 | 5.88 | |
| 3 | 0.079 | 0.076 | 0.078 | 0.002 | 2.56 | |
| 4 | 0.086 | 0.085 | 0.086 | 0.001 | 1.16 | |
| 5 | 0.073 | 0.076 | 0.075 | 0.002 | 2.67 | |
| 6 | 0.040 | 0.047 | 0.044 | 0.005 | 11.36 | |
| 7 | 0.063 | 0.067 | 0.065 | 0.003 | 4.62 | |
| 8 | 0.104 | 0.097 | 0.101 | 0.005 | 4.95 | |
| 9 | 0.044 | 0.044 | 0.044 | 0.000 | 0.00 | |
| 10 | 0.236 | 0.238 | 0.237 | 0.001 | 0.42 | |

Notes: Example calculations and formulas are shown in italics
*CV* = coefficient of variation, *SD* = standard deviation, *Ave.* = average, *Rep.* = replicate

## References

Aiken, L., West, S., & Reno, R. (1991). *Multiple regression: Testing and interpreting interactions*. Thousand Oaks, CA: Sage.

Allison, P. (2001). Missing data. In R. E. Millsap & A. Maydeu-Olivares (Eds.), *The Sage handbook of quantitative methods in psychology* (136th ed.). Thousand Oaks, CA: Sage.

Anderson, S., Kelley, K., & Maxwell, S. (2017). Sample-size planning for more accurate statistical power: A method adjusting sample effect sizes for publication bias and uncertainty. *Psychological Science, 28*, 1547–1562.

Antweiler, R. C., & Taylor, H. E. (2008). Evaluation of statistical treatments of left-censored environmental data using coincident uncensored data sets: I. Summary statistics. *Environmental Science & Technology, 42*, 3732–3738.

Austin, P. (2002). Bayesian extensions of the Tobit model for analyzing measures of health status. *Medical Decision Making, 22*, 152–162. https://doi.org/10.1177/02729890222063035

Austin, P. C., Escobar, M., & Kopec, J. A. (2000). The use of the Tobit model for analysing measures of health status. *Quality of Life Research, 9*, 901–910.

Baccarelli, A., Pfeiffer, R., Consonni, D., Pesatori, A. C., Bonzini, M., Patterson, D. G., . . . Landi, M. T. (2005). Handling of dioxin measurement data in the presence of non-detectable values: Overview of available methods and their application in the Seveso chloracne study. *Chemosphere, 60*, 898–906. https://doi.org/10.1016/j.chemosphere.2005.01.055

Baron, R. M., Kenny, D. A. (1982). The moderator-mediator variable distinction in social the moderator-mediator variable distinction in social psychological research: Conceptual, strategic, and statistical considerations. *Sociological Methodology, 13*, 1173–1182. https://doi.org/10.1037/0022-3514.51.6.1173

Bauer, A. M., Quas, J. A., & Boyce, W. T. (2002). Associations between physiological reactivity and children's behavior: Advantages of a multisystem approach. *Journal of Developmental and Behavioral Pediatrics, 23*, 102–113. https://doi.org/10.1097/00004703-200204000-00007

Beltzer, E. K., Fortunato, C. K., Guaderrama, M. M., Peckins, M. K., Garramone, B. M., & Granger, D. A. (2010). Salivary flow and alpha-amylase: Collection technique, duration, and oral fluid type. *Physiology & Behavior, 101*, 289–296. https://doi.org/10.1016/j.physbeh.2010.05.016

Bishop, N. C., & Gleeson, M. (2009). Acute and chronic effects of exercise on markers of mucosal immunity. *Frontiers in Bioscience (Landmark Edition), 14*, 4444–4456.

Blair, C., Granger, D. A., Willoughby, M., Mills-Koonce, R., Cox, M., Greenberg, M. T., . . . Fortunato, C. K. (2011). Salivary cortisol mediates effects of poverty and parenting on executive functions in early childhood. *Child Development, 82*, 1970–1984. https://doi.org/10.1111/j.1467-8624.2011.01643.x

Brandtzaeg, P. (2007). Do salivary antibodies reliably reflect both mucosal and systemic immunity? *Annals of the New York Academy of Sciences, 1098*, 288–311. https://doi.org/10.1196/annals.1384.012

Brown, T. (2014). *Confirmatory factor analysis for applied research*. New York: Guilford.

Browne, M., MacCallum, R., Kim, C., Andersen, B., & Glaser, R. (2002). When fit indices and residuals are incompatible. *Psychological Methods, 7*, 403–421.

Burgener, B., Ford, A., Situ, H., Fayad, M., Hao, J., Wenckus, C., . . . George, A. (2010). Biologic markers for odontogenic periradicular periodontitis. *Journal of Endodontia, 36*, 1307–1310. https://doi.org/10.1158/0008-5472.CAN-10-4002.BONE

Chard, T. (1990). *An introduction to radioimmunoassay and related techniques* (4th ed.). Amsterdam: Elsevier.

Chen, F. R., Raine, A., & Granger, D. A. (2015). Tactics for modeling multiple salivary analyte data in relation to behavior problems: Additive, ratio, and interaction effects. *Psychoneuroendocrinology, 51*, 188–200. https://doi.org/10.1016/j.psyneuen.2014.09.027

Chen, F. R., Raine, A., Soyfer, L., & Granger, D. A. (2015). Interaction of adrenocortical activity and autonomic arousal on children's externalizing and internalizing behavior problems. *Journal of Abnormal Child Psychology, 43*, 189–202. https://doi.org/10.1007/s10802-014-9900-y

Cohen, J. (1988). *Statistical power analysis for the behavioral sciences* (2nd ed.). Hillsdale, NJ: Lawrence Erlbaum Associates.

Cohen, J. (1992). A power primer. *Psychological Bulletin, 112*, 155–159.

Cohen, J., Cohen, P., West, S., & Aiken, L. (2002). *Applied multiple regression/correlation analysis for the behavioral sciences* (3rd ed.). Mahwah, NJ: Routledge.

Cohen, J., Cohen, P., West, S., & Aiken, L. (2003). *Applied multiple regression/correlation analysis for the behavioral sciences* (3rd ed.). Mahwah, NJ: Lawrence Erlbaum Associates.

Cole, S. R., Chu, H., Nie, L., & Schisterman, E. F. (2009). Estimating the odds ratio when exposure has a limit of detection. *International Journal of Epidemiology, 38*, 1674–1680. https://doi.org/10.1093/ije/dyp269

Dariotis, J. K., Chen, F. R., & Granger, D. A. (2016). Latent trait testosterone among 18–24 year olds: Methodological considerations and risk associations. *Psychoneuroendocrinology, 67*, 1–9. https://doi.org/10.1016/j.psyneuen.2016.01.019

Dinse, G. E., Jusko, T. A., Ho, L. A., Annam, K., Graubard, B. I., Hertz-Picciotto, I., . . . Weinberg, C. R. (2014). Accommodating measurements below a limit of detection: A novel application of cox regression. *American Journal of Epidemiology, 179*, 1018–1024. https://doi.org/10.1093/aje/kwu017

Doane, L. D., Chen, F. R., Sladek, M. R., Van Lenten, S. A., & Granger, D. A. (2015). Latent trait cortisol (LTC) levels: Reliability, validity, and stability. *Psychoneuroendocrinology, 55*, 21–35. https://doi.org/10.1016/j.psyneuen.2015.01.017

Dockray, S., Susman, E., & Dorn, L. (2009). Depression, cortisol reactivity and obesity in childhood and adolescence. *Journal of Adolescent Health, 45*, 344–350. https://doi.org/10.1016/j.freeradbiomed.2008.10.025.The

Eilers, P. H., Röder, E., Savelkoul, H. F., & van Wijk, R. (2012). Quantile regression for the statistical analysis of immunological data with many non-detects. *BMC Immunology, 13*, 37. https://doi.org/10.1186/1471-2172-13-37

El-Sheikh, M., Erath, S. A., Buckhalt, J. A., Granger, D. A., & Mize, J. (2008). Cortisol and children's adjustment: The moderating role of sympathetic nervous system activity. *Journal of Abnormal Child Psychology, 36*, 601–611. https://doi.org/10.1007/s10802-007-9204-6

Enders, C. (2010). *Applied missing data analysis*. New York: Guilford.

Epstein, M. P., Lin, X., & Boehnke, M. (2003). A tobit variance-component method for linkage analysis of censored trait data. *American Journal of Human Genetics, 72*, 611–620. https://doi.org/10.1086/367924

Fu, P., Hughes, J., Zeng, G., Hanook, S., Orem, J., Mwanda, O. W., & Remick, S. C. (2012). A comparative investigation of methods for longitudinal data with limits of detection through a case study. *Statistical Methods in Medical Research, 25*(1), 153–166. https://doi.org/10.1177/0962280212444800

Galimard, J. E., Chevret, S., Curis, E., & Resche-Rigon, M. (2018). Heckman imputation models for binary or continuous MNAR outcomes and MAR predictors. *BMC Medical Research Methodology, 18*, 1–13. https://doi.org/10.1186/s12874-018-0547-1

Giesbrecht, G. F. G., Bryce, C. C. I., Letourneau, N., Granger, D. A., The APrON Study Team, Kaplan, B. J., . . . Sing-hal, N. (2015). Latent trait cortisol (LTC) during pregnancy: Composition, continuity, change and concomitants. *Psychoneuroendocrinology, 62*, 149–158. https://doi.org/10.1016/j.psyneuen.2015.08.009

Gillespie, B. W., Chen, Q., Reichert, H., Franzblau, A., Hedgeman, E., Lepkowski, J., . . . Garabrant, D. H. (2010). Estimating population distributions when some data are below a limit of detection by using a reverse Kaplan-Meier estimator. *Epidemiology, 21*, S64–S70. https://doi.org/10.1097/EDE.0b013e3181ce9f08

Glenn, A. L., Raine, A., Schug, R. A., Gao, Y., & Granger, D. A. (2011). Increased testosterone-to-cortisol ratio in psychopathy. *Journal of Abnormal Psychology, 120*, 389–399.

Goodyer, I., Herbert, J., & Tamplin, A. (2003). Psychoendocrine antecedents of persistent first-episode major depression in adolescents: A community-based longitudinal enquiry. *Psychological Medicine, 33*, 601–610.
Goodyer, I., Park, R., Netherton, C., & Herbert, J. (2001). Possible role of cortisol and dehydroepiandrosterone in human development and psychopathology. *The British Journal of Psychiatry, 179*, 243–249.
Gordis, E. B., Granger, D. A., Susman, E. J., & Trickett, P. K. (2006). Asymmetry between salivary cortisol and alpha-amylase reactivity to stress: Relation to aggressive behavior in adolescents. *Psychoneuroendocrinology, 31*, 976–987. https://doi.org/10.1016/j.psyneuen.2006.05.010
Granger, D. A., Fortunato, C. K., Beltzer, E. K., Virag, M., Bright, M. A., & Out, D. (2012). Focus on methodology: Salivary bioscience and research on adolescence: An integrated perspective. *Journal of Adolescence, 35*, 1081–1095. https://doi.org/10.1016/j.adolescence.2012.01.005
Hayes, A. F. (2009). Beyond Baron and Kenny: Statistical mediation analysis in the new millennium. *Communication Monographs, 76*, 408–420. https://doi.org/10.1080/03637750903310360
Hayes, A. (2013). *Introduction to mediation, moderation, and conditional process analysis: A regression-based approach*. New York, NY: Guilford.
Hayes, A., & Matthes, J. (2009). Computational procedures for probing interactions in OLS and logistic regression: SPSS and SAS implementations. *Behavior Research Methods, 41*, 924–936. https://doi.org/10.3758/BRM.41.3.924
Helsel, D. R. (2005). More than obvious: Better methods for interpreting nondetect data. *Environmental Science & Technology, 39*, 419A–423A. https://doi.org/10.1021/es053368a
Helsel, D. R. (2006). Fabricating data: How substituting values for nondetects can ruin results, and what can be done about it. *Chemosphere, 65*, 2434–2439. https://doi.org/10.1016/j.chemosphere.2006.04.051
Helsel, D. (2010). Much ado about next to nothing: Incorporating nondetects in science. *The Annals of Occupational Hygiene, 54*, 257–262. https://doi.org/10.1093/annhyg/mep092
Helsel, D., & Lopaka, R. L. (2006). *Analysis of environmental data with nondetects: Statistical methods for censored environmental data.*
Hewett, P., & Ganser, G. H. (2007). A comparison of several methods for analyzing censored data. *The Annals of Occupational Hygiene, 51*, 611–632. https://doi.org/10.1093/annhyg/mem045
Izawa, S., Sugaya, N., Kimura, K., Ogawa, N., Yamada, K. C., Shirotsuki, K., … Nomura, S. (2013). An increase in salivary interleukin-6 level following acute psychosocial stress and its biological correlates in healthy young adults. *Biological Psychology, 94*, 249–254. https://doi.org/10.1016/j.biopsycho.2013.06.006
Jaccard, J., & Turrisi, R. (2003). *Interaction effects in multiple regression* (2nd ed.). Thousand Oaks, CA: Sage.
Jin, Y., Hein, M. J., Deddens, J. A., & Hines, C. J. (2011). Analysis of lognormally distributed exposure data with repeated measures and values below the limit of detection using SAS. *The Annals of Occupational Hygiene, 55*, 97–112. https://doi.org/10.1093/annhyg/meq061
Johnson, P., & Fay, L. (1950). The Johnson-Neyman technique, its theory and application. *Psychometrika, 15*, 349–367. https://doi.org/10.1007/BF02288864
Kenny, D., & Zautra, A. (2001). Trait–state models for longitudinal data. In L. Collins & A. Sayer (Eds.), *Decade of behavior. New methods for the analysis of change* (pp. 243–263). Washington, DC: American Psychological Association.
Kirschbaum, C., Steyer, R., Eid, M., Patalla, U., Schwenkmezger, P., & Hellhammer, D. H. (1990). Cortisol and behavior: 2. Application of a latent state-trait model to salivary cortisol. *Psychoneuroendocrinology, 15*, 297–307.
Kivlighan, K., Granger, D. A., Schwartz, E., Nelson, V., Curran, M., & Shirtcliff, E. (2004). Quantifying blood leakage into the oral mucosa and its effects on the measurement of cortisol, dehydroepiandrosterone, and testosterone in saliva Katie. *Hormones and Behavior, 46*, 39–46. https://doi.org/10.1016/j.yhbeh.2005.10.005

Kruskal, W., Ferguson, T. S., Tukey, J. W., & Gumbel, E. J. (1960). Discussion of the papers of Messrs. Anscombe and Daniel. *Technometrics, 2*, 157–166. https://doi.org/10.1080/00401706.1960.10489890

Kunz-Ebrecht, S. R., Mohamed-Ali, V., Feldman, P. J., Kirschbaum, C., & Steptoe, A. (2003). Cortisol responses to mild psychological stress are inversely associated with proinflammatory cytokines. *Brain, Behavior, and Immunity, 17*, 373–383. https://doi.org/10.1016/S0889-1591(03)00029-1

Laurent, H. K., Ablow, J. C., & Measelle, J. (2011). Risky shifts: How the timing and course of mothers' depressive symptoms across the perinatal period shape their own and infant's stress response profiles. *Development and Psychopathology, 23*, 521–538. https://doi.org/10.1017/S0954579411000083

Laurent, H., Gilliam, K., Wright, D., & Fisher, P. (2015). Child anxiety symptoms related to longitudinal cortisol trajectories and acute stress responses: Evidence of developmental stress sensitization Heidemarie. *Journal of Abnormal Psychology, 124*, 68–79. https://doi.org/10.1037/abn0000009.Child

Laurent, H., Lucas, T., Pierce, J., Goetz, S., & Granger, D. A. (2016). Coordination of cortisol response to social evaluative threat with autonomic and inflammatory responses is moderated by stress appraisals and affect. *Biological Psychology, 118*, 17–24. https://doi.org/10.1016/j.biopsycho.2016.04.066.Coordination

Laurent, H. K., Powers, S. I., & Granger, D. A. (2013). Refining the multisystem view of the stress response: Coordination among cortisol, alpha-amylase, and subjective stress in response to relationship conflict. *Physiology & Behavior, 119*, 52–60. https://doi.org/10.1016/j.physbeh.2013.05.019

Laurent, H., Powers, S., Laws, H., Gunlicks-Stoessel, M., Bent, E., & Balaban, S. (2013). HPA regulation and dating couples' behaviors during conflict: Gender-specific associations and cross-partner interactions. *Physiology & Behavior, 13*, 218–226. https://doi.org/10.1016/j.physbeh.2013.05.037.HPA

Lee, J. Y., Chung, J. W., Kim, Y. K., Chung, S. C., & Kho, H. S. (2007). Comparison of the composition of oral mucosal residual saliva with whole saliva. *Oral Diseases, 13*, 550–554. https://doi.org/10.1111/j.1601-0825.2006.01332.x

Little, T. (2013). *Longitudinal structural equation modeling*. New York: Guilford.

Lopez-Duran, N., Mayer, S., & Abelson, J. (2014). Modeling neuroendocrine stress reactivity in salivary cortisol: Adjusting for peak latency variability. *Stress, 17*, 285–295.

Lubin, J. H., Colt, J. S., Camann, D., Davis, S., Cerhan, J. R., Severson, R. K., . . . Hartge, P. (2004). Epidemiologic evaluation of measurement data in the presence of detection limits. *Environmental Health Perspectives, 112*, 1691–1696. https://doi.org/10.1289/ehp.7199

Lyons, K. S., & Sayer, A. G. (2010). Using multilevel modeling in caregiving research. *Aging & Mental Health, 9*, 189–195. https://doi.org/10.1080/13607860500089831

MacKinnon, D., Lockwood, C., Hoffman, J., West, S., & Sheets, V. (2002). A comparison of methods to test mediation and other intervening variable effects. *Psychological Methods, 7*, 83. https://doi.org/10.3816/CLM.2009.n.003.Novel

MacKinnon, D., Lockwood, C., & Williams, J. (2004). Confidence limits for the indirect effect: Distribution of the product and resampling methods. *Multivariate Behavioral Research, 39*(1), 99. https://doi.org/10.1207/s15327906mbr3901

Markopoulou, K., Papadopoulos, A., Juruena, M., Poon, L., Pariante, C., & Cleare, A. (2009). The ratio of cortisol/DHEA in treatment resistant depression. *Psychoneuroendocrinology, 2*(34), 19–26.

McEwen, B. S., & Gianaros, P. J. (2010). Central role of the brain in stress and adaptation: Links to socioeconomic status, health, and disease. *Annals of the New York Academy of Sciences, 1186*, 190–222. https://doi.org/10.1111/j.1749-6632.2009.05331.x

Miller, R., Plessow, F., Kirschbaum, C., & Stalder, T. (2013). Classification criteria for distinguishing cortisol responders from nonresponders to psychosocial stress. *Psychosomatic Medicine, 75*, 832–840. https://doi.org/10.1097/PSY.0000000000000002

Pruessner, J. C., Kirschbaum, C., Meinlschmid, G., & Hellhammer, D. H. (2003). Two formulas for computation of the area under the curve represent measures of total hormone concentration versus time-dependent change. *Psychoneuroendocrinology, 28*, 916–931. https://doi.org/10.1016/S0306-4530(02)00108-7

Quas, J. A., Castro, A., Bryce, C. I., & Granger, D. A. (2018). Stress physiology and memory for emotional information: Moderation by individual differences in pubertal hormones. *Developmental Psychology, 54*, 1606–1620.

Raudenbush, S., & Bryk, A. (2002). *Hierarchical linear models: Applications and data analysis methods*. Thousand Oaks, CA: Sage.

Richardson, D. B., & Ciampi, A. (2003). Effects of exposure measurement error when an exposure variable is constrained by a lower limit. *American Journal of Epidemiology, 157*, 355–363. https://doi.org/10.1093/aje/kwf217

Riis, J., Bryce, C., Matin, M., Stebbins, J., Kornienko, O., van Huisstede, L., & Granger, D. A. (2018). The validity, stability, and utility of measuring uric acid in saliva. *Biomarkers in Medicine, 12*(6), 583–596. https://doi.org/10.2217/bmm-2017-0336

Ross, K., Murphy, M., Adam, E., Chen, E., & Miller, G. (2014). How stable are diurnal cortisol activity indices in healthy individuals? Evidence from three multi-wave studies. *Psychoneuroendocrinology, 39*, 184–193. https://doi.org/10.1016/j.psyneuen.2013.09.016. How

Sapolsky, R., Romero, M., & Munck, A. (2000). How do glucocorticoids influence stress responses? Preparative actions. *Endocrine Reviews, 21*, 55–89.

Schisterman, E. F., Vexler, A., Whitcomb, B. W., & Liu, A. (2006). The limitations due to exposure detection limits for regression models. *American Journal of Epidemiology, 163*, 374–383. https://doi.org/10.1093/aje/kwj039

Shirtcliff, E. A., Granger, D. A., Booth, A., & Johnson, D. (2005). Low salivary cortisol levels and externalizing behavior problems in youth. *Development and Psychopathology, 17*, 167–184. https://doi.org/10.1017/S0954579405050091

Shoal, G. D., Giancola, P. R., & Kirillova, G. P. (2003). Salivary cortisol, personality, and aggressive behavior in adolescent boys: A 5-year longitudinal study. *Journal of the American Academy of Child and Adolescent Psychiatry, 42*, 1101–1107. https://doi.org/10.1097/01.CHI.0000070246.24125.6D

Snijders, T. (2005). Power and sample size in multilevel linear models. In B. Everitt & D. Howell (Eds.), *Encyclopedia of statistics in behavioral science* (pp. 1570–1573). Chichester: Wiley.

Snijders, T., & Bosker, R. (2002). *Multilevel analysis: An introduction to basic and advanced multilevel modeling*. London: Sage.

Steyer, R., Ferring, D., & Schmitt, M. J. (1992). States and traits in psychological assessment. *European Journal of Psychological Assessment, 8*, 79–98.

Steyer, R., Geiser, C., & Fiege, C. (2012). Latent state-trait models. In H. Cooper (Ed.), *APA handbook of research methods in psychology: Vol. 3. Data analysis and research publication* (pp. 291–308). Washington, DC: American Psychological Association.

Stroud, C. B., Chen, F. R., Doane, L. D., & Granger, D. A. (2016). Individual differences in early adolescents' latent trait cortisol (LTC): Relation to early adversity. *Developmental Psychobiology, 58*, 700–713. https://doi.org/10.1002/dev.21410

Stroud, C. B., Chen, F. R., Doane, L. D., & Granger, D. A. (2018). Early adversity and internalizing symptoms in adolescence: Mediation by individual differences in latent trait cortisol. *Development and Psychopathology, 21*, 1–16. https://doi.org/10.1017/S0954579418000044

Trueba, A. F., Mizrachi, D., Auchus, R. J., Vogel, P. D., & Ritz, T. (2012). Effects of psychosocial stress on the pattern of salivary protein release. *Physiology & Behavior, 105*, 841–849. https://doi.org/10.1016/j.physbeh.2011.10.014

Tse, W. S., & Bond, A. J. (2004). Relationship between baseline cortisol, social functioning and depression: A mediation analysis. *Psychiatry Research, 126*, 197–201. https://doi.org/10.1016/j.psychres.2004.02.002

Uh, H.-W., Hartgers, F. C., Yazdanbakhsh, M., & Houwing-Duistermaat, J. J. (2008). Evaluation of regression methods when immunological measurements are constrained by detection limits. *BMC Immunology, 9*, 59. https://doi.org/10.1186/1471-2172-9-59

van Niekerk, J., Huppert, F., & Herbert, J. (2001). Salivary cortisol and DHEA: Association with measures of cognition and well-being in normal older men, and effects of three months of DHEA supplementation. *Psychoneuroendocrinology, 26*, 591–612.

Vining, R. F., & McGinley, R. A. (1987). The measurement of hormones in saliva: Possibilities and pitfalls. *Journal of Steroid Biochemistry, 27*, 81–94. https://doi.org/10.1016/0022-4731(87)90297-4

Wild, D. (Ed.). (2013). *The immunoassay handbook: Theory and applications of ligand binding, ELISA and related techniques* (4th ed.). Oxford: Elsevier.

Yeung, E. W., Place, R., Gordish-Dressman, H., Visich, P., Hoffman, E., Walker, S. O., & Granger, D. A. (2016). Salivary latent trait cortisol (LTC): Relation to lipids, blood pressure, and body composition in middle childhood. *Psychoneuroendocrinology, 71*, 110–118. https://doi.org/10.1016/j.psyneuen.2016.05.013

# Part II
# Physiology and Development Research and Applications

Jenna L. Riis, Section Editor

# Chapter 5
# Salivary Bioscience, Human Stress and the Hypothalamic–Pituitary–Adrenal Axis

**Nina Smyth and Angela Clow**

**Abstract** Since validation in the late 1990s measurement of cortisol in saliva has revolutionised the study of human stress. Cortisol secretion is controlled by the brain and in addition to being a principal regulator of circadian rhythms, it is the body's main stress hormone. The use of salivary measures has spawned an exponential growth in studies exploring individual differences in reactions to stress and consequent impact on health and circadian rhythms. Several key features have facilitated this work: saliva samples can be self-collected both within and outside of laboratory settings; stress-free repeated sampling enables examination of short-term changes in concentration. Researchers have utilised these advantages to measure relationships between stress and health and have explored an astonishingly wide array of research questions that have shone a torch on how the way we feel affects our physical and mental health and well-being. In this chapter, we explore the main approaches to using salivary cortisol in human stress research both within the laboratory and in the domestic setting. We highlight crucial methodological considerations when measuring basal diurnal cortisol patterns as well as stress reactivity. Key findings from this body of work will be summarised as well as exciting potential future directions discussed.

**Keywords** Cortisol · HPA axis · Stress reactivity · Diurnal patterns · Health

---

N. Smyth · A. Clow (✉)
Psychophysiology and Stress Research Group, Psychology, School of Social Sciences, University of Westminster, London, UK
e-mail: clowa@westminster.ac.uk



D. A. Granger, M. K. Taylor (eds.), *Salivary Bioscience*,
https://doi.org/10.1007/978-3-030-35784-9_5

## 5.1 History of Salivary Bioscience in Stress Psychoneuroendocrinology

Human stress psychophysiology is the science that links mood and cognition with physiology and eventually health and ageing outcomes. This connection is orchestrated via physiological pathways linking the brain (thoughts and emotions) with the main organs and systems of the body, for example, the heart, metabolic, and immune systems. There are two main outflow pathways from the brain that can be studied in psychophysiology research. One is the autonomic nervous system, including activation of the sympathetic nervous system and sympathetic adrenal medullary system with epinephrine and α-amylase as the main biomarkers. The other is the neuroendocrine hypothalamic–pituitary–adrenal (HPA) axis, which releases the potent steroid hormone cortisol. It is this stress response system that is addressed here as cortisol is the most widely studied biomarker in psychoneuroendocrine research.

Cortisol, synthesised from cholesterol, is secreted from the zona fasciculata of the adrenal cortex in response to circulating adrenocorticotrophic hormone (ACTH). ACTH is secreted into the blood from the anterior pituitary in response to corticotrophic hormone (CRH) delivered via the hypophyseal portal system following activation of the periventricular nucleus (PVN) of the hypothalamus. This cascade is known as the hypothalamic–pituitary–adrenal (HPA) axis (see Fig. 5.1).

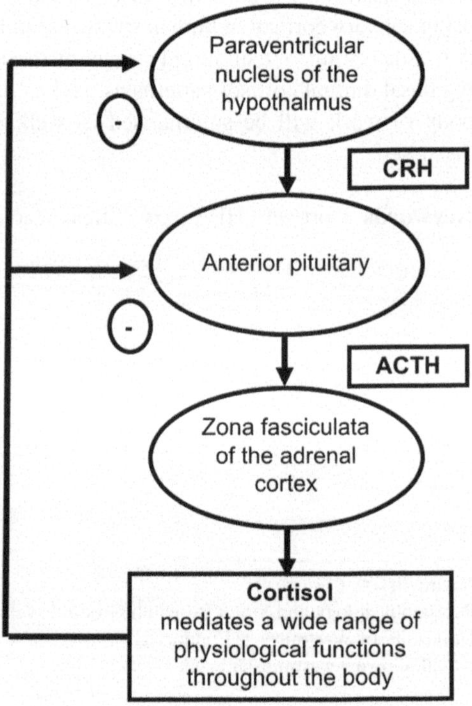


**Fig. 5.1** The hypothalamic–pituitary–adrenal (HPA) axis. *CRH* corticotropin releasing hormone; *ACTH* adrenocorticotrophic hormone. Cortisol exerts negative feedback (indicated by the minus symbol)

The PVN responds to negative moods and thoughts detected as stressors as well as internal biological cues such as the 24-h rhythm of biological clock located in the suprachiasmatic nucleus (SCN) of the hypothalamus. The role of cortisol is very widespread as virtually every bodily cell has cortisol receptors. For example, it helps regulate blood sugar levels, blood pressure and metabolism as well as assists with memory function and help reduce inflammation. In short, the hormone cortisol is essential for life as it helps maintain daily homeostasis as well as providing a potent response to stress that prepares us to fight or flight. As cortisol secretion responds to brain activity it can be viewed as a 'window on the brain' and has become a key focus of psychoneuroendocrinology, especially with respect to the antecedents of physical and mental ill-health.

For many years, cortisol was measured either in blood samples (to provide single time point estimates of concentration) or in 24-h urine samples (an index of daily output when expressed per unit of creatinine). The work of Katz and Shannon (1964) pioneered measurement in saliva during the early 1960s, but adoption of this practise was not widespread due to methodological problems. However, the development of sensitive radioimmunoassay techniques in the late 1970s kindled growing interest in the potential of this new technique as well as being more reliable it has many advantages over both blood and urine sampling (see Table 5.1 for a list of these advantages). However, there were several validation queries that needed resolution prior to full acceptance of salivary cortisol as a measure in psychoneuroendocrine research. The most important of these was whether it accurately represented circulating levels. On a more practical basis determination of whether salivary cortisol concentration was dependent on salivary flow rate as well an index of normative daily and stress-related values measured with various analytical techniques were needed. Such validation studies were spearheaded and summarised by Kischbaum and Hellhammer in a seminal methodological review in 1989 (Kirschbaum & Hellhammer, 1989).

Most of the circulating cortisol is bound to blood-borne carriers (corticosteroid-binding globulin and albumin) with only 5–10% available in a biologically active form. Crucially, salivary cortisol was shown to accurately reflect this biologically active 'free' component as bound cortisol is unable to pass into saliva. This fact makes saliva a superior and ideal medium in which to study changing levels of the active hormone. Once this, along with other practical matters, was resolved there was a consequential burgeoning growth in the popularity of salivary cortisol as a marker in human psychoneuroendocrinology. It became the ideal medium in which to study the biology of human stress; it provides an accurate index of circulating free cortisol sampling that is not inherently stressful and proved acceptable to the very young, old and those with a wide range of ill-health conditions. Furthermore, use of saliva enabled repeated sampling (as frequently as every 5–15 min) that enabled the study of the dynamic time course of changing concentrations. In addition, saliva sampling meant studies could be taken out of the laboratory with participants able to self-collect samples (in specialised saliva collection devices, e.g. salivettes) within any chosen setting, e.g. the home, work and school. This new approach provided the advantage of ecological validity as participants could go about their usual daily

**Table 5.1** Salivary cortisol: advantages

| |
|---|
| *A: Collection* |
| • Saliva is not considered a biological hazard (if it is not contaminated with blood). |
| • Self-collected in adolescents and adults, and there is no requirement for a medically trained person. |
| • Sampling in other populations (e.g. babies, children and vulnerable adults) is possible, with assistance from parents or researcher (special collection devices can be used for babies < 6 years old and infants, e.g. devices from Salimertics). |
| • Non-invasive collection process. Cortisol concentrations are not affected by collection (like other methods, e.g. venepuncture can significantly increase cortisol levels for blood sampling) or salivary flow rate (thus it is not necessary to record the time taken to collect the sample). Quick process; collection takes approximately 1–2 min. |
| • Can be collected repeatedly over the day (e.g. every 15–30 min) enabling measurement of the dynamic cortisol profile (e.g. over the day or in response to a stressor). |
| *B: Storage* |
| • Refrigeration or freezing after collection is advisory. |
| • Samples can be posted to and from participants however, postage of saliva samples should be via safe postage with postage services (e.g. Royal Mail Safebox®). |
| • Samples should be frozen at −20 °C; they can be stored for several years, and thawing or refreezing does not affect them. |
| *C: Assay* |
| • Relatively easy, only a small volume is required, using a range of commercial assay providers. Equipment needed is typically available in biological or medical laboratories. Samples can be assayed by commercial laboratories that offer labelled salivettes, packaging materials and cortisol values after assay (typically costs less than £10 per sample, but prices vary depending on the number of samples). |

routine on sampling days, without interference by trained personnel or laboratory visits. However, as discussed later in this chapter this advantage has a downside: reliance on participants to self-collect in a reliable manner is not always advisable, especially when timing is critical to the interpretation of results (see Table 5.1 for summary of advantages of using salivary measures of cortisol). To date there have been 3317 original peer-reviewed articles and 131 reviews on salivary cortisol (Web of Science; Sep 2018). The average number of citations per article is over 27, making a total of nearly 100,000 citations. Cleary the area of salivary cortisol as a research tool in psychiatry, neuroscience, and the behavioural and health sciences is very substantial.

Initial laboratory-based investigations explored individual differences in salivary cortisol responses to psychosocial stress. Whilst laboratory stressors such as public speaking, mirror drawing, and the Stroop test had been in common use they were found not to reliably induce a cortisol response in the majority of participants (see Wetherell, Craw, & Smith, 2018). Again, Kischbaum and Hellhammer were at the forefront of the development of a 'gold standard' laboratory stress induction protocol, which reliability activated the HPA axis in an average of over 70% of participants, enabling meaningful evaluation of individual differences. The Trier Social

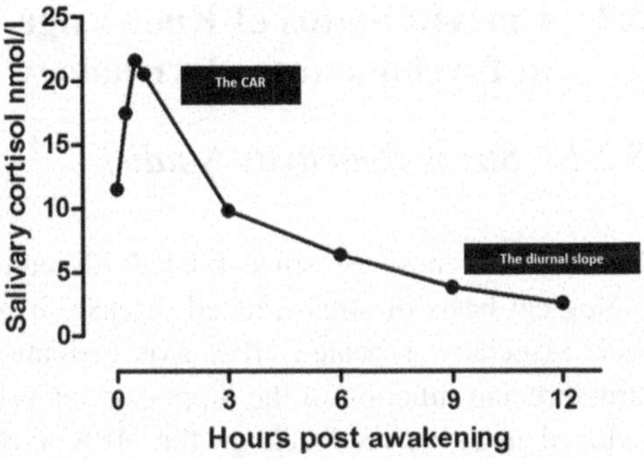


**Fig. 5.2** Typical diurnal profile of salivary cortisol in healthy participants from the moment of awakening till 12 h after awakening. *CAR* cortisol awakening response

Stress Test (TSST), a widely used and effective laboratory method for inducing psychosocial stress. It was first published in 1993 (Kirschbaum, Pirke, & Hellhammer, 1993) and to date has received 2381 citations (Web of Science, Sep 2018). The test comprises a very specific set of conditions to induce social evaluative threat (task performance that could be negatively judged by others) and lack of control, key ingredients for HPA axis activation (see Dickerson & Kemeny, 2004). The TSST (and now various adaptations, e.g. TSST for children and groups: TSST-C and TSST-G) remains widely used and has helped illuminate a range of individual difference predictors of stress responding (see details in a later section).

More recently there has been intense research interest in the basal (unstimulated) pattern of cortisol secretion. As mentioned earlier the PVN has strong anatomical links with the central body clock (suprachiasmatic nucleus: SCN). The resultant basal pattern of cortisol secretion has a very distinct rhythmic circadian (24-h) pattern. Originally these changing daily levels were merely considered a nuisance to be controlled for by sampling at specific clock times. However, salivary cortisol research has revealed the importance of synchronising sampling to awakening time—not clock time (for more details, see the methodological considerations section). Such research has identified distinct elements of the daily cortisol rhythm: the diurnal slope and the cortisol awakening response: CAR (see Fig. 5.2).

Typically, in healthy people the diurnal slope is the decline in cortisol concentrations from about 2–3 h after awakening till bedtime. The CAR, on the other hand, is the marked (typically doubling) post-awakening rise in cortisol concentrations peaking 30–45 min post-awakening. Both the diurnal slope and CAR have emerged as constructs of great interest. Researchers have examined the wide range of individual differences in, as well as the potential health consequences of, variation in these two elements of the diurnal cortisol rhythm.

## 5.2 Current Status of Knowledge in Psychoneuroendocrinology

### *5.2.1 Stress Reactivity Studies*

Greater awareness of stress-related ill-health has led to investigations into the biological basis of stress-related disease. It has been proposed that chronic stress, with associated repeated HPA axis activation, is associated with changes in the structure and function of the hippocampus (via changes in neurogenesis) leading to reduced negative feedback on the HPA axis. This generates a damaging vicious circle of dysregulated HPA axis function thought influential in the genesis of a range of mental and physical health conditions (Mahar, Bambico, Mechawar, & Nobrega, 2014). Such theories make it valuable to investigate individual differences in HPA axis reactivity.

An obvious way to explore these pathways is to examine the determinants of HPA axis reactivity to a standardised stressor. This can easily be assessed by examination of changing levels of salivary cortisol in response to a test such as the TSST (and its adaptations). There are marked inter-individual differences in stress reactivity and some of these differences can be attributed to socio-demographic variables. For example, healthy males reliably show larger cortisol responses (e.g. 200–400% increases) compared with healthy females (50–150% increases) (e.g. Kirschbaum et al., 1993). Larger responses are observed in pre-menopausal females during the luteal menstrual phase compared to the follicular phase or in women taking oral contraceptives (Kudielka, Hellhammer, & Wust, 2009). Not surprisingly, given the association with sex hormones, pregnancy and lactation are also associated with changes (attenuation) in stress reactivity. Age-related changes are also observed with older adults showing higher basal cortisol levels and an enhanced response to psychosocial stress (Kudielka, Schmidt-Reinwald, Hellhammer, & Kirschbaum, 1999; Nicolson, Storms, Ponds, & Sulon, 1997) but findings are mixed (Kudielka, Buske-Kirschbaum, Hellhammer, & Kirschbaum, 2004; Rohleder, Kudielka, Hellhammer, Wolf, & Kirschbaum, 2002). Blunted stress responses (defined as a smaller increase from base to peak) are associated with chronic alcohol consumption (Lovallo, Dickensheets, Myers, Thomas, & Nixon, 2000) whereas results for social drinkers and those with a family history are more mixed (Kudielka et al., 2009). Habitual smokers also have chronically elevated cortisol levels (Kirschbaum, Wust, & Strasburger, 1992) but a blunted response to a standardised stressor (Kudielka et al., 2009).

It is beyond the scope of this chapter to give a full review of the stress reactivity literature including genetic and epigenetic antecedents (see Kudielka et al., 2009 for an overview, Allen, Kennedy, Cryan, Dinan, & Clarke, 2014, Allen et al., 2017). However, in summary, larger cortisol responses to psychosocial stress are typically associated with disadvantages such as low birth weight, low self-esteem and anxious attachment style. However, more recently the opposite, i.e. lower (i.e. blunted) stress reactivity, is also acknowledged to be implicated in a range of adverse health and behavioural outcomes (Carroll, Ginty, Whittaker, Lovallo, & de Rooij, 2017). For example, lower stress reactivity is associated with low socioeconomic status, anxiety disorders, smaller hippocampal volume and poorer cognitive performance (Almela et al., 2014; Domes, Heinrichs, Reichwald, & Hautzinger, 2002; Raffington et al., 2018). The realisation that both heightened and blunted cortisol reactivity to stress is an index of poor functioning is somewhat challenging. For example, depression in children and adolescents is associated with increased cortisol stress reactivity (Dockray, Susman, & Dorn, 2009), whilst an attenuated stress reactivity is reported in older depressed patients. It remains unclear precisely what level of cortisol reactivity is optimal, especially as this appears to change over the life course and be different for males and females.

Relatively fewer studies have examined the power of stress reactivity to predict long-term health outcomes. However, a recent systematic review has highlighted that exaggerated reactivity (in both ANS and HPA axis systems) is associated with an increased prevalence of cardiovascular disorder whereas blunted reactivity predicted an increased likelihood of obesity, depression, reduced cognitive ability and greater physical decline at follow-up at least 3 years later (Turner et al., 2019). So, it seems that both exaggerated and blunted stress reactivity are related to poorer health outcomes at follow-up. Future research should examine biological pathways by which both exaggerated and blunted stress reactivity may impact health.

Another approach has been to examine the short-term impact of stress-induced salivary cortisol increases on behavior, e.g. cognitive function. Such studies have provided extensive evidence that short-term rises in cortisol are critically involved in promoting consolidation of emotionally arousing experiences whilst inhibiting memory retrieval (de Quervain, Aerni, Schelling, & Roozendaal, 2009). In this and other ways (see below with respect to the diurnal rhythm of cortisol secretion) salivary cortisol studies have been able to enlighten cognitive processes as well as health outcomes.

### 5.2.2 *Unstimulated Diurnal Cortisol Secretion*

#### 5.2.2.1 Cortisol Secretion and the Body Clock

This changing daily pattern of cortisol secretion is regulated by the hypothalamic SCN, which in turn is synchronised to day/night cycles by light, a process called photoentrainment. Ambient light is detected by the photosensitive melanopsin retinal ganglion cells that project via the retinohypothalamic tract to the SCN, which is the master pacemaker of circadian rhythms. Patients suffering from a range of physical and mental ill health often exhibit a loss of regulation of their circadian rhythms as indicated by altered sleep/wake cycle and body temperature. Historically these symptoms were considered a result from the underlying pathology, however, increasing evidence now indicates that dysregulation of the circadian system may be more directly involved in the aetiology of ill health. This view originated with the discovery that clock genes are not restricted to the SCN master clock but are expressed widely across body tissues, including in many brain structures. This raised the possibility that these remote peripheral (or 'slave') clocks might malfunction and generate ill health (see Menet & Rosbash, 2011). The changing daily pattern of cortisol secretion permeates all body tissues and is a major pathway by which the SCN regulates its peripheral clocks, synchronising them to time of day and each other to ensure optimal functioning (Nader, Chrousos, & Kino, 2010). The intriguing possibility arose that stress-induced aberrant circadian patterns of cortisol secretion might be a root cause of peripheral clock misalignment and be instrumental in stress-induced ill health.

This possibility has been widely investigated using salivary measures of cortisol secretion collected at timed sampling points (synchronised to awakening time) across the day. In this way, salivary cortisol could justly be described as providing a 'window on the brain' as the cortisol rhythm reflects brain activation of the HPA axis cascade, although it should be borne in mind that examples of dissociation between circulating ACTH and cortisol are documented (see Hellhammer, Wust, & Kudielka, 2009 for a discussion). There are two distinct aspects of the diurnal cortisol rhythm that need to be studied separately (Edwards, Evans, Hucklebridge, & Clow, 2001): the diurnal slope (decline of cortisol over the day) and the post-awakening rise known as the cortisol awakening response (CAR).

#### 5.2.2.2 The Cortisol Diurnal Slope

Early research on salivary cortisol diurnal rhythms typically focused on average concentration across the day, providing an equivalent index to 24-h urinary measures. However, this way of quantifying the daily rhythm failed to maximise on a key advantage of multiple salivary sampling, i.e. information on the rate of change over hours and minutes, within the same day. It soon became evident that it was this daily dynamic aspect of changing cortisol secretion that held the key to understanding the HPA axis function. Changes in the level of cortisol secretion from mid-morning to evening are referred to as the diurnal cortisol slope (or diurnal decline). Over the years flattened diurnal cortisol slopes were proposed as a mediator linking chronic stress and ill health and this has been supported by a recent systematic review and meta-analysis (Adam et al., 2017). This cross-sectional correlational analysis demonstrated that flatter cortisol slopes were associated with conditions including cancer, depression, fatigue and externalising symptoms, with the biggest effect size for immune/inflammatory conditions. In addition, flatter diurnal cortisol slopes predicted increased risk of death up to 10 years follow-up. As cross-sectional studies should be viewed with caution as they cannot demonstrate causation, it is interesting that results from prospective studies showed similar average effect sizes. It is argued that attenuated diurnal cortisol slopes result from chronic stress and contribute to stress-related ill health via dysregulation of central and peripheral circadian mechanisms (i.e. the SCN and the peripheral clocks located within body organs) with consequent downstream effects on multiple aspects of biology and health.

#### 5.2.2.3 The Cortisol Awakening Response

The cortisol awakening response (CAR) has distinct regulatory processes to the cortisol diurnal slope (see Clow, Hucklebridge, Stalder, Evans, & Thorn, 2010 for a full review of CAR regulation). These distinctive features mean that both aspects of the cortisol diurnal pattern (the diurnal decline and CAR) need to be examined as separate entities. In healthy people the CAR generates the highest levels (i.e. the acrophase) of daily cortisol secretion, peaking between 30 and 45 min following morning awakening. There can be marked intra-individual day differences in the size of the CAR as it is sensitive to transient state variables such as ambient light and daily psychological state (Law, Hucklebridge, Thorn, Evans, & Clow, 2013). Consequently, to determine a representative measure of individual trait CAR it is necessary to sample for more than one day (see methodological issues below). The decision about which of the two indices of basal diurnal cortisol secretion to measure will depend on the specific research question and participant demand. It is more onerous for participants (and researchers) to accurately assess the CAR compared to the diurnal slope as the timing of sampling immediately post-awakening is more critical than during the past awakening period. If in doubt about capacity to follow the expert consensus guidelines on CAR (Stalder et al., 2016) assessment it is advised to study the diurnal decline alone.

There are many studies exploring the CAR in relation to health and illness (e.g. Desantis, Kuzawa, & Adam, 2015; Gardner et al., 2013; Kudielka & Kirschbaum, 2003; O'Connor et al., 2009). However, the literature is inconsistent, which has been attributed partly to poor accuracy in estimating sampling times from awakening (Smyth, Thorn, Hucklebridge, Clow, & Evans, 2016; Stalder et al., 2016). The most prominent theory for the function of the CAR is that it provides an allostatic 'boost' upon awakening (Adam, Hawkley, Kudielka, & Cacioppo, 2006; Clow, Huckleldridge, & Thorn, 2010; Fries, Dettenborn, & Kirschbaum, 2009; Law et al., 2013). Accumulating evidence suggests that the CAR is associated with cognitive functions dependent upon brain regions with high densities of glucocorticoid receptors, such as the hippocampus (HC) and frontal cortex (FC). CAR magnitude is positively associated with hippocampal integrity and associated functions (Bäumler et al., 2014; Buchanan, Kern, Allen, Tranel, & Kirschbaum, 2004; Pruessner, Pruessner, Hellhammer, Pike, & Lupien, 2007; Rimmele, Meier, Lange, & Born, 2010; Wolf, Fujiwara, Luwinski, Kirschbaum, & Markowitsch, 2005). It has been demonstrated that pharmacologic suppression of the CAR inhibits HC-dependent declarative memory (Rimmele et al., 2010), and that CAR magnitude is positively associated with prospective memory performance in children (Bäumler et al., 2014) and episodic memory in healthy adults (Ennis, Moffat, & Hertzog, 2016). It has also been suggested that the CAR may serve as a powerful time-of-day marker, synchronising circadian rhythms in peripheral CLOCKs within the body and brain under the regulatory influence of the suprachiasmatic nucleus of the hypothalamus (Clow et al., 2010; Law et al. 2013).

#### 5.2.2.4 Methods of Measuring Salivary Cortisol in the Laboratory and Domestic Setting; Methodological Recommendations, Issues, Challenges and Considerations

In all studies of salivary cortisol secretion (both within and outside of the laboratory), it is necessary to avoid activities know to affect cortisol secretion or measurement for at least 30 min prior to saliva sampling. These include eating, drinking (except water), smoking and strenuous exercising. If participants do engage in these activities, it is important that researchers ask participants to be honest in reporting if they have engaged in any of the activities. Potential exclusions, unless it is the specific focus of the study, include individuals who are suffering from jet-lag, those working night shifts, pregnant and/or breastfeeding women, those diagnosed with a neuroendocrine disorder (e.g. Cushing's and Addison's disease) and those taking steroid-based medication (e.g. to alleviate asthma, Crohn's disease and rheumatoid arthritis).

### 5.2.3 *Stress Reactivity Studies*

The TSST has repeatedly been shown to reliably activate the HPA axis (Dickerson & Kemeny, 2004; Kirschbaum et al., 1993; McRae et al., 2006). It includes elements of socio-evaluative threat and uncontrollability. After a 10-min anticipation period, individual participants engage in a 10-min public speaking and a mental arithmetic task in front of an authoritarian and unsympathetic audience (see Kirschbaum et al., 1993 for an overview, Smyth, Hucklebridge, Thorn, Evans, & Clow, 2013). See Fig. 5.3 for an overview of the protocol.

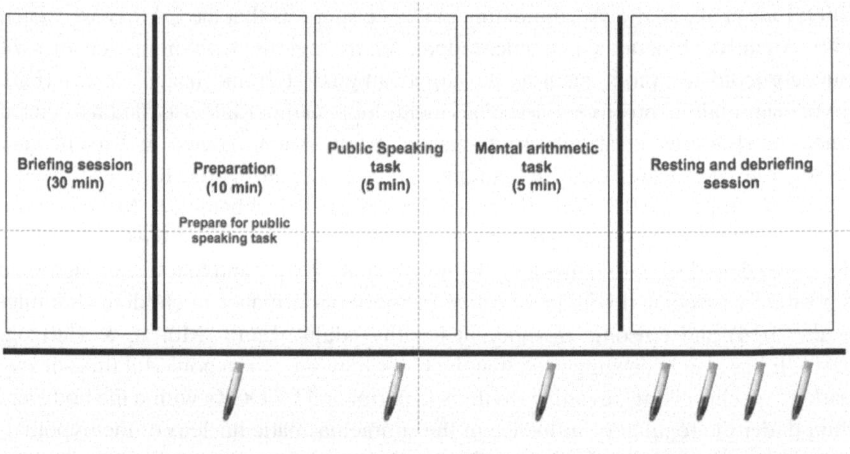


**Fig. 5.3** The TSST; includes a period of preparation followed by a speech and mental arithmetic task (Kirschbaum, Pirke, & Hellhammer, 1993). Samples are typically collected pre- (baseline), immediately after TSST tasks and again every 10 min for 40 min

Different versions of the original individual adult TSST have been developed to accommodate varying research needs. The TSST for groups (TSST-G von Dawans, Kirschbaum, & Heinrichs, 2011) involves testing up to six participants at a time; individually performing the same, but abbreviated tasks, in the presence of other group members. The public speaking task lasts for 2-min per participant and the mental arithmetic task lasts for 80-s each. The group version makes it quick and relatively easy to collect data. More recently, a virtual reality of the TSST (TSST-VR) has been developed. The same protocol to the original TSST is followed but with a virtual panel and pre-recorded responses. Preliminary research shows HPA axis activation (e.g. Jönsson et al., 2010), although it was not shown to induce cortisol response with certain VR technology (Kotlyar et al., 2008). Such a protocol is cost effective, and the protocol is more standardised (e.g. panel and participant interaction), which potentially will increase reliability. The version for use with children (TSST-C: Buske-Kirschbaum et al., 1997) is recommended for ages between 7 and 16 years. This version of the TSST uses age and ability appropriate mental arithmetic tasks, and the panel provides children encouragement to complete the task. The public speaking task involves the children being given the beginning of a story and they are asked to continue the story in an exciting and interesting way. They are told that the ending of the story should be more interesting than the stories of other children. Unlike for the adult version the panel provides positive feedback during the task. A control version of the TSST has also been developed and instead of the social-evaluative component participants are exposed to a friendly and supportive panel (Wiemers, Schoofs, & Wolf, 2013). This version does not result in cortisol response as the presence of others is not sufficient to induce a cortisol response (Dickerson, Mycek, & Zaldivar, 2008), although there may be individual differences in this.

The HPA axis is activated immediately in response to the stressor, peak cortisol concentrations are reached approximately 20–30 min following stressor onset and return to baseline levels 1-h later. To measure the stress response and obtain a measure of the peak repeated saliva samples should be collected. Typically sampling points are immediately prior to the stressor and every 10–15 min up to 40–45 min following the end of the stressor (e.g. S1 −10, +10, +20, +30, +40, +50, and +60). It is advisable for testing to commence in the afternoon (at least 1 h after lunch) due to changes in basal cortisol secretion in the morning and postprandial period. If testing is to occur in the morning it should be synchronised to awakening time. Time of testing should be consistent throughout the study; testing completed in the morning period or afternoon period should remain consistent throughout the study (Dickerson & Kemeny, 2004; Kirschbaum et al., 1993; Kudielka et al., 2004).

Stress induced in the laboratory has the advantage of researcher control and standardisation across testing sessions. This means that confounding factors can be controlled or monitored. These advantages enable comparisons within and between studies, as well as other physiological domains (e.g. heart rate and blood pressure) and immune measures can be obtained. The TSST is limited by its habituation effects (Foley & Kirschbaum, 2010), unless habituation rate itself is the variable of interest. Although some individuals do not habituate to the TSST; people with lower self-esteem, feeling unattractive and higher feelings of depression and reports of poor health continue to respond to the TSST with repeated testing (e.g. Kirschbaum et al., 1995). Another disadvantage is the lack of ecological validity, however, protocols like the TSST do resemble tasks that are representative of their natural environments.

An important methodological factor, typically ignored in cortisol stress reactivity research, is the issue of non-responding. Although the TSST has been shown to reliably induce a two-to-three fold increase in cortisol levels in approximately 70–80% of study participants (Dickerson & Kemeny, 2004), non-responding is apparent and it is important to distinguish blunted cortisol stress responding from non-responding as endocrine stress reactivity can only by validly assessed in responders (Bellingrath & Kudielka, 2008). Non-responding is variously defined as responses less than 1.0–2.5 nmol/l but more recently, Miller, Plessow, Kirschbaum, and Stalder (2013), recommended a <1.5 nmol/l or <15.5% cortisol increase from baseline to peak as a 'non-responding' criteria. Calculation of stress reactivity has recently been recommended as the individual peak minus baseline and recovery of the stressor as the minimum sample (Miller et al., 2018) for overall cortisol calculate area under the curve with respect to ground (AUCg Pruessner, Kirschbaum, Meinlschmidt, & Hellhammer, 2003). In our TSST-G data (Smyth et al., 2015, 2019; Wood, Clow, Hucklebridge, Law, & Smyth, 2018), we have investigated if non-responding to the TSST-G was associated with trait or state factors. In healthy females, demographics (e.g. age, socio-economic status, ethnicity) or health variables (e.g. smoking status, body mass index, use of medication and/or oral contraceptives, days since menstruation or self-reported health) did not differ between responders or non-responders. Nor were there any differences in perceived stress between the two groups. However, there were differences in stress reported prior to the stressor. Findings are limited to healthy young females and other research suggests that males are more likely to respond to the TSST compared to females, as well as exhibit larger responses (Herbison et al., 2016). As well more trait-like factors such as being mindful have been associated with cortisol responding to the TSST (Manigault, Woody, Zoccola, & Dickerson, 2018). Non-responding might not be due to the participants not perceiving the tasks as stressful but due to the timing of stress induction relative to the underlying rise or fall of ultradian bursts or individual differences in exposure to chronic stress across the life course (Elzinga et al., 2008; Petrowski, Herold, Joraschky, Wittchen, & Kirschbaum, 2010). Additionally, anticipatory stress (i.e. elevated cortisol during the preparatory phase) might be a potential reason for cortisol non-responding. Further research is needed on correlates of non-responding. With increased interest in blunted stress reactivity it is important to distinguish non-responding and to report whether non-responders are different from responders in any measured variables.

### 5.2.4 Unstimulated Diurnal Studies

To measure the diurnal cortisol profile samples should be synchronised to morning awakening. For measurement of the CAR samples should be collected on awakening and 15, 30 and 45 min post-awakening. Costs and participant demand increases with the more samples in the protocol, thus larger epidemiological studies might typically employ a two-sample protocol to measure the CAR (i.e. 0 min and 30 or 45 min post-awakening). However, it is not advisable to employ a two-sample protocol as the peak of cortisol could be missed, and the timing of the peak cannot be obtained, which has been associated with sex, cognitive function and menstrual phase adolescent development (Evans, Clow, Hucklebridge, & Loveday, 2012; Oskis, Loveday, Hucklebridge, Thorn, & Clow, 2009; Schlotz, Hellhammer, Schulz, & Stone, 2004; Wolfram, Bellingrath, & Kudielka, 2011; Wust et al., 2000). If samples need to be limited it is advisable to employ a 3-sample protocol of 0, 30 and 45 min post-awakening in adult populations. Such a protocol is likely to capture peak concentrations of both male (typically 30 min) and female (typically 45 min) adult populations. However, for children and prepubertal adolescents who do not exhibit sex-specific CAR patterns a two-sample protocol (0 and 30 min) is acceptable. It has been estimated that to reliably capture trait CAR characteristics requires sampling on six consecutive days (Hellhammer et al., 2007). However, this is costly and time consuming and so it is recommended that measurement of the CAR should be sampled on at the very least two consecutive days. Nevertheless, intervention studies using the CAR as a measure should employ repeated sampling days (between 4 and 6 days), particularly for pre-intervention assessments since the first-time of sampling is novel and usually results in a bigger CAR (which might confound predicted pre-post CAR effects unless accounted for). Sampling days should commence on typical weekdays since the CAR can be bigger on weekdays and attenuated on weekend days (Kunz-Ebrecht, Kirschbaum, Marmot, & Steptoe, 2004; Schlotz et al., 2004). Typical weekdays should be discussed with the study population as typical weekdays are likely to differ for student/college population and those who do shift work.

The CAR is influenced by several state factors and when these factors are not the focus of the study researchers should record them for use as covariates in analyses. A bigger CAR is associated with earlier awakening, greater ambient light and anticipated busyness and demands/challenges of the next day (see Stalder et al., 2016). Awakening time should be measured electronically as discussed below. Photosensor devices can be used to measure ambient light but this is costly and where not practical season should be reported. People reporting seasonal variation in mood have bigger CARs in the summer months (Thorn, Evans, Cannon, Hucklebridge, & Clow, 2011) and thus the season of assessment should be kept constant or season of assessment should be used as a covariate in analyses when data is collected over an extended time period. Keeping collection of sampling times constant for intervention studies is particularly important to ensure that any differences in the CAR are not due to seasonal changes. If seasons are different at time of testing it is recommended that a measure of seasonality is obtained and participants that are seasonal should not be included, unless they are the focus of the study. There is some evidence that alcohol consumed the evening prior to sampling days is associated with a flatter CAR (Stalder, Hucklebridge, Evans, & Clow, 2009) whilst chronic consumption of alcohol and smoking is associated with increased cortisol levels (e.g. Badrick et al., 2008; Badrick, Kirschbaum, & Kumari, 2007), recording alcohol consumption is recommended.

Trait sociodemographic and health-related variables have been suggested to be potential confounds of the CAR. A reasonably consistent finding is that women exhibit a larger and later peaking CAR compared to males. Higher cortisol levels, flatter CARs and decreased diurnal decline are associated with increasing age (Evans et al., 2011; Ice, 2005). The CAR is also sensitive to specific developmental stages. For example, the timing of the peak of cortisol during CAR period is observed in premenarche women (Oskis et al., 2009) increased basal cortisol profiles over the day and steeper diurnal decline is associated with pubertal developmental changes (Adam, 2006). Generally, phase of the menstrual cycle has not been shown to influence the size of the CAR (Kudielka & Kirschbaum, 2003) but one study showed the CAR was elevated during ovulation (Wolfram et al., 2011). Lower mean cortisol in the CAR period has been observed in women taking oral contraceptives (Pruessner et al., 1997). Medication use can affect cortisol levels (for an overview see Granger et al., 2009). Cortisol levels are influenced by ethnicity and/or socio-economic status (see Hajat et al., 2010) and higher body fat (see Steptoe, Kunz-Ebrecht, Brydon, & Wardle, 2004; Therrien et al., 2007).

Samples in the immediate post-awakening period (i.e. up to 60 min) should not be included in measurement of the diurnal decline (Edwards, Clow, Evans, & Hucklebridge, 2001), but sampling times should be synchronised to morning awakening at 3, 6, 9 and 12 h post-awakening. To capture the full diurnal profile it is ideal to collect multiple samples across the day (e.g. 4-sample protocol). To minimise the number of samples due to costs and/or to reduce participant demand it is advisable to include measurement at 3 and 12 h post-awakening. This 2-sample protocol provides the best estimate of the slope of cortisol over the day and the 12-h sample will provide pre-night time cortisol levels. Sampling over two consecutive weekdays (at least) is required to avoid influence of random state factors (Thorn et al., 2011). Absolute levels and slope of cortisol should be reported in order to differentiate between the overall levels of cortisol (i.e. high or low) and the rate of decline (flat or steep slope). Researchers should choose the aspect of the diurnal cycle that best meets the need of question and capacity of participants (slope or CAR, or both). Measurement of the CAR is very demanding so should only be attempted if the protocol is designed in line with the consensus guidelines that were set out by internationally recognised CAR researchers (Stalder et al., 2016).

#### 5.2.4.1 Saliva Sampling: Adherence to Protocol

Cortisol rises rapidly in the post-awakening period, thus collecting samples according to the protocol is crucial to ensure reliable measurement. The CAR measured in a domestic setting requires participant self-collection of repeated saliva samples. Numerous studies have shown that participants are inaccurate in collecting saliva samples at the desired sampling times and that their self-reports of sampling times are often inaccurate, when verified against electronic devices, such as actigraphy and track caps.

Wrist-worn actigraphy devices, that have been validated against polysomnography (Lichstein et al., 2006), can be used to monitor awakening times (indicated by increased movement on awakening) in the domestic setting. Wake times determined by actigraphy are usually earlier than participant self-reported wake times (Smyth, Clow, Thorn, Hucklebridge, & Evans, 2013), and this might be due to sleep inertia and not inaccuracy in reporting wake times (see Smyth, Clow, et al., 2013; Smyth, Hucklebridge, et al., 2013). Track caps (e.g. Medication Event Monitoring: MEM Caps) are electronic caps attached to bottles. The bottles contain the cotton swabs for saliva sampling and participants should be instructed to open the bottle at the time of sampling only. Each time the bottle is opened, the date and time are recorded, and this is used as a proxy for timing of sample collection. Comparison of track caps and self-reports shows that participants are typically inaccurate when reporting their sampling times but are more accurate when they are informed that their sampling times will be verified by the track caps (Broderick et al., 2004; Kudielka et al., 2003).

Track caps can be costly; one caps cost approx. £80 plus the required software and caps expire after 2–3 years. Furthermore, the date and time of the cap opening are not always recorded (due to the cap needing to be open for several seconds). A cost-effective alternative to track caps are time-stamped photographs; using a smartphone, with location turned on, participants take a photo of themselves collecting each saliva sample. This is a relatively new method that needs further investigation, but preliminary research suggests that participants reported little difficulty collecting the time-stamped photos, including taking and sending photographs. Low levels of feelings of intrusiveness or discomfort were reported, although participants reported slightly more difficulty for the awakening sample. The time-stamped photos were found to be successful in identifying sampling inaccuracy when compared with self-reports, track caps or actigraphy (Dockray et al., 2017).

Sampling inaccuracy leads to erroneous CAR measurement (e.g. AUCi or MnInc, see Table 5.2). For example, an overestimated CAR is obtained for short delays and an underestimated CAR for longer delays: the delay between awakening and collection of sample 1 between 3 and 15 min and >15 min, respectively (see Smyth et al., 2016; Stalder et al., 2016). The delay between awakening and collection of S1 is the most problematic in producing erroneous CARs but in line with best practise it is advisable to omit further delays for collection of samples after the awakening point. If samples are collected ±7.5 min from the desired interval, it is recommended

**Table 5.2** Treatment of data and statistical analysis for salivary cortisol

| |
|---|
| *Outliers* |
| • Potentially artifactual influences (e.g. blood contamination or rare problems from assaying) should be considered as outliers and excluded from analysis. |
| • Genuine extreme scores should be winsorised (i.e. data above or below a suitable cut-off point are held at the cut-off value) to acknowledge highness or lowness of value but not to influence the analyses. |
| *Data transformation/removal* |
| • Normalise data with a transformation (e.g. log or root). Transformed values should be used for inferential analyses (raw cortisol values can be presented to illustrate results in tables and figures). |
| • Remove extreme outliers if they contribute as a minority of the data and judged to be veridical. |
| *Calculation of cortisol composites:* |
| *CAR:* |
| • *CAR magnitude*: Cortisol rise post-awakening |
| – Mean Increase (MnInc) calculated as (s2 + s3 + s4)/3−s1 or Area Under the Curve relative to increase (AUCi) calculated as s2 + s3 + [(s4_s1)/2]−2s1. Both provide an estimate of CAR magnitude. |
| – Area Under the Curve relative to ground (AUCg) provides an estimate of total post-awakening cortisol secretion and is calculated as s1 + s2 + s3 + [(s4_s1)/2]; or AUCi=AUCG-3s1. Formulae for samples collected at 0 min (sample 1), 15-min (sample 2); 30 min (sample 3) and 45 min (sample 4) post-awakening and equal time intervals between all samples. |
| • CAR Salience: The shape of the cortisol curve in the post-awakening period |
| – Calculated as the difference between an individual's mean secretion rate before |
| – 30-min, [s30 – s0]/30) and after 30-min, [s45 – s30]/15 (see Evans, Smyth, Thorn, Hucklebridge, & Clow, 2019). |
| *Diurnal Decline:* |
| • 2-sample protocol: Calculate the difference between 3- and 12-h post-awakening. |
| • 4-sample protocol: Use linear regression of cortisol levels against collection time across each sampling day. |

to check analyses omitting these data (see Smyth et al., 2013, 2016). Samples collected in the post-awakening period that are within 15 min, as verified by electronic devices, can be used with growth curve multilevel modelling. This is where cortisol values are mapped onto the real saliva sampling times (obtained by electronic verification of awakening and sampling times). We have shown that sampling inaccuracy does not affect the cortisol curve in this way, but sampling inaccuracy is problematic when using cortisol composites in analyses (Smyth et al., 2016).

Sampling inaccuracy is less problematic for the measurement of the diurnal decline. Some studies show that deviations from protocol timings (±60 min) do not impact the slope of cortisol over the day or total amount of cortisol (e.g. Jacobs et al., 2005), whilst other studies report flatter decline for samples collected ±60 min from desired sampling time (Broderick et al., 2004). However, these studies measured spot samples that were not anchored to awakening time. In our datasets (unpublished) we have not observed significant impact of sampling inaccuracy (±60 min) on the diurnal decline anchored to awakening. It is recommended that participants are encouraged to report sampling times accurately and track caps are used to verify timings.

#### 5.2.4.2 Maximizing Participant Adherence

Measurement of diurnal pattern of cortisol in the domestic setting confers ecological validity but can be costly. Sampling inaccuracy, particularly for the CAR, typically results in the inclusion of inaccurate data that contributes to the inconsistency of the CAR literature or more recently in the last decade exclusion of inaccurate data. But recently, we have shown that sampling inaccuracy up to 15 min can be adjusted for in statistical models, only if real-timing of sample collection relative to awakening is obtained (see section above). Figure 5.4 and Table 5.3 provide an overview of the strategies for maximising sampling accuracy during the saliva sampling period.

#### 5.2.4.3 CAR Non-Responders

A CAR is not always observed in all healthy individuals on all days, for some, cortisol levels decrease following awakening. It was typically thought that this 'non-responding' was observed because participants delayed collecting saliva samples in the post-awakening period and since sampling inaccuracy (>15 min) is associated with flattened CARs the peak of cortisol was simply missed. As such it was accepted practise to remove any 'non-responders' from CAR analyses (Thorn, Hucklebridge, Evans, & Clow, 2006). However, more recently several studies have shown that non-responding has been observed in those fully accurate to the sampling protocol in healthy and clinical samples (e.g. Smyth et al., 2013). Thus, removing non-responders would result in excluding people with genuine negative CARs and thus introduce bias. The issue of genuine non-responding is unclear—it is not clear if this is a trait or state characteristic, and is an area that requires further investigation.

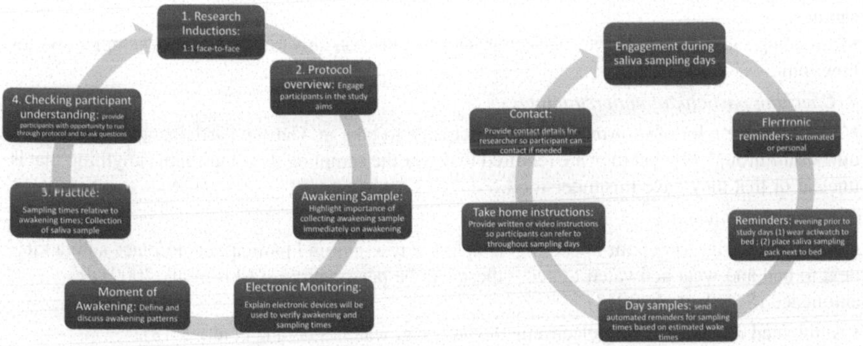


**Fig. 5.4** Strategies for maximising sampling accuracy during the saliva sampling period

**Table 5.3** Recommendations for research inductions for saliva sampling in the domestic period

| |
|---|
| *1. Type of research induction* |
| • One-to-one inductions: Researcher can fully explain the research process; practise key components of the protocol (e.g. collecting saliva samples); time for questions; check participant understanding of protocol by asking participants to talk through the protocol. |
| • Face-to-face inductions are recommended as this enables rapport between the researcher and participant which in turn may motivate participants in the protocol. However, recruiting participants from different locations is difficult. |
| • Telephone/skype inductions: Enable recruitment of participants from different locations. Study packs should be sent prior to induction so they can be referred to during the induction. |
| • Video instructions: Our work has shown that an instructional video was useful, and instructions were clear, however, to ensure participant understood instructions when participants collected sampling kit a brief overview of the sampling protocol was provided and they were given the opportunity to ask questions. We recommend that where video instructions are used a researcher follows up with an email or telephone/skype call to check that participants understand the protocol and to give them the opportunity to ask any questions. |
| *2. Protocol overview* |
| • Sample accuracy: Highlighting the importance of sampling accuracy, particularly, obtaining the first awakening sample, makes participants aware of the importance of their accuracy to the results. Without such knowledge, participants may not realise the impact delayed sampling could have on the results, and thus not strive to collect the samples at the desired times. |
| • Electronic devices: Researcher should inform participants that sampling times will be used to verify sampling accuracy to the protocol to encourage adherence and accuracy. |
| • Instructions on waking behaviours: |
| – Define the moment of 'awakening': "When you are awake, i.e. you are conscious: you know who and where you are; you are in a state that is clearly different from when you were sleeping even though you may still feel tired." (Stalder et al., 2016). |
| – What to do after waking during the night or earlier than expected: Sampling should not be initiated if woken during the night or if woken up much earlier than expected. Participants should sample when they are awake for the final time, e.g. waking up expectantly is classed as waking at 4 am with plans to wake at 6 am. |
| – Dozing has not been systematically examined, however, it is recommended that participants refrain from 'dozing' following waking but are able to stay in bed whilst collecting the samples. |
| *3. Practise:* |
| • Providing participants with the opportunity to practise their saliva sampling times relative to awakening is recommended to ensure that participant understands the time intervals between samples.<br>• Collecting saliva samples gives participants reassurance on how to collect saliva samples and for how long. |
| *4. Checking participant understanding* |
| • There is quite a lot of information for participants to take in. Getting participants to provide a quick run-through of what they are required to do for the sampling days highlights anything that is unclear or that they have misunderstood. |
| *Electronic Reminders* |
| • Email/text reminders about upcoming sampling days—remind participants to place saliva kit next to bed and wear actiwatch to bed—these can be personalised (Oskis et al., 2009) or automated (Smyth et al., 2013). |
| • Automated reminders using electronic devices, e.g. watches (Franz et al., 2013). |

(continued)

**Table 5.3** (continued)

| |
|---|
| • Timers activated by participants to signal collection of the first sample, and thereafter a beep/flash reminds participants to collect later samples (see Doane & Adam, 2010; Griefahn & Robens, 2011). |
| • Telephone calls are useful for those with well-defined wake times only. |
| *Additional instructions:* |
| • Highlight take-home instructions in the form of written or video. |
| • Clearly set out dates of sampling days and sampling times. |
| *Contact:* |
| • It is reassuring for participants to have contact details of the researcher(s) should they need to ask any questions during the sampling protocol. |
| *Return of sampling kits:* |
| • Until the return of saliva samples ask participants to store in home freezers or in a cool place. |
| • Arrange a convenient time for return of saliva samples. |
| • If samples are returned via post, ensure that prepaid package is provided and check with mail system for safe delivery of postage in accord with the human tissue act. |

#### 5.2.4.4 Treatment of Data and Statistical Analysis

Cortisol data is typically positively skewed and requires transformation. Where possible mixed regression modelling should be used for cortisol data as it enables multilevel analysis of repeated measures data and modelling of dynamic aspects of the diurnal cortisol profile. As well as such analyses can manage missing data when data is 'randomly' missing (Blackwell, de Leon, & Miller, 2006). Use of growth curve modelling (Llabre, Spitzer, Siegel, Saab, & Schneiderman, 2004) accounts for continuous dynamics of time that enables inaccuracies in the timing of sample collection to be modelled when sampling times are known (see Smyth et al., 2016). See Table 5.2 for an overview of cortisol composites.

## 5.3 Future Directions and Opportunities

Salivary cortisol, as a measure in psychoneuroendocrine research, has been very useful for investigation of the antecedents and outcomes of HPA axis dysregulation. However, despite the large array of published work it is still early days in its utility: there is much to do and there are many opportunities for important new work using salivary cortisol as the measure. Some of the early work has been affected by methodological flaws, but despite these the area shows enormous promise as an avenue to understanding mind–body links in human development, health and disease.

The area of HPA reactivity to stress has attracted much work and following the initial discovery of cortisol hyper-reactivity being associated with poor prognosis, it is now apparent that blunted reactivity is also an indicator of homeostatic dysregulation. The area needs clarity as to the biological processes by which life events lead to these opposite outcomes as well as pathways by which both blunted and hyper-reactivity impact physiological systems and health.

Much evidence shows that flatter diurnal cortisol slopes are related to ill-health and vulnerability. However, the research on the CAR has suffered from methodological issues contributing to inconsistent findings. Accumulating evidence, where the CAR is measured accurately, indicates that a bigger CAR is associated with better health and cognition. More recently, the shape of the CAR (salience) in addition to the size (magnitude) has been shown to be relevant for health but research on this is in its infancy. More research is needed to clarify the best way to optimally measure the CAR characteristics relevant to health and health outcomes. This is especially important given the costs associated with salivary cortisol research. Another avenue for CAR research is examination of daily CAR variability, i.e. the capacity of the CAR to be matched to the demands of the day. It is proposed that in healthy people the CAR is larger on days with more demands; however, this has not been fully explored.

The salivary cortisol literature is largely cross-sectional. Little research has explored the long-term consequences of aberrant cortisol patterns and stress reactivity on health outcomes and cognition. Longitudinal studies will require large-scale collaborative working to shed light on the pathways by which cortisol impacts health, illness and cognition. This knowledge is needed in order to justify and devise effective interventions to restore HPA axis function, with consequent long-term benefits for health.

An additional unresolved issue for both stress reactivity and diurnal cortisol pattern research is non-responding, i.e. failure to mount a cortisol response to stress and failure to generate a CAR. This is frequently not reported and may affect findings reported as non-responding is not fully understood and may be unrelated to the variables under investigation. It is recommended that researchers report non-responding rates and characteristics so that we can untangle whether this is a trait or state characteristics and if it is marker of ill-health in its own right or simply random.

In conclusion the development of salivary cortisol as an accurate measure of human HPA axis activity had provided a very valuable tool with which to investigate brain–body interactions. Despite years of illuminating research there is much more important work to do. This work has the capacity to transform our understanding of human health and provide opportunities for the development of life-changing interventions that targets the HPA axis.

## References

Adam, E. K. (2006). Transactions among adolescent trait and state emotion and diurnal and momentary cortisol activity in naturalistic settings. *Psychoneuroendocrinology, 31*, 664–679.

Adam, E. K., Hawkley, L. C., Kudielka, B. M., & Cacioppo, J. T. (2006). Day-to-day dynamics of experience-cortisol associations in a population-based sample of older adults. *Proceedings of the National Academy of Sciences of the United States of America, 103*, 17058–17063.

Adam, E. K., Quinn, M. E., Tavernier, R., McQuillan, M. T., Dahlke, K. A., & Gilbert, K. E. (2017). Diurnal cortisol slopes and mental and physical health outcomes: A systematic review and meta-analysis. *Psychoneuroendocrinology, 83*, 25–41.

Allen, A. P., Kennedy, P. J., Cryan, J. F., Dinan, T. G., & Clarke, G. (2014). Biological and psychological markers of stress in humans: Focus on the Trier Social Stress Test. *Neuroscience and Biobehavioral Reviews, 38*, 94–124.

Allen, A. P., Kennedy, P. J., Dockray, S., Cryan, J. F., Dinan, T. G., & Clarke, G. (2017). The Trier Social Stress Test: Principles and practice. *Neurobiology of Stress, 6*, 113–126.

Almela, M., Hidalgo, V., van der Meij, L., Pulopulos, M. M., Villada, C., & Salvador, A. (2014). A low cortisol response to acute stress is related to worse basal memory performance in older people. *Frontiers in Aging Neuroscience, 6*, 157.

Badrick, E., Bobak, M., Britton, A., Kirschbaum, C., Marmot, M., & Kumari, M. (2008). The relationship between alcohol consumption and cortisol secretion in an aging cohort. *Journal of Clinical Endocrinology and Metabolism, 93*, 750–757.

Badrick, E., Kirschbaum, C., & Kumari, M. (2007). The relationship between smoking status and cortisol secretion. *Journal of Clinical Endocrinology and Metabolism, 92*, 819–824.

Bäumler, D., Voigt, B., Miller, R., Stalder, T., Kirschbaum, C., & Kliegel, M. (2014). The relation of the cortisol awakening response and prospective memory functioning in young children. *Biological Psychology, 99*, 41–46.

Bellingrath, S., & Kudielka, B. M. (2008). Effort-reward-imbalance and overcommitment are associated with hypothalamus–pituitary–adrenal (HPA) axis responses to acute psychosocial stress in healthy working schoolteachers. *Psychoneuroendocrinology, 33*, 1335–1343.

Blackwell, E., de Leon, C. F. M., & Miller, G. E. (2006). Applying mixed regression models to the analysis of repeated-measures data in psychosomatic medicine. *Psychosomatic Medicine, 68*, 870–878.

Broderick, J. E., Arnold, D., Kudielka, B. M., & Kirschbaum, C. (2004). Salivary cortisol sampling compliance: Comparison of patients and healthy volunteers. *Psychoneuroendocrinology, 29*, 636–650.

Buchanan, T. W., Kern, S., Allen, J. S., Tranel, D., & Kirschbaum, C. (2004). Circadian regulation of cortisol after hippocampal damage in humans. *Biological Psychiatry, 56*, 651–656.

Buske-Kirschbaum, A., Jobst, S., Wustmans, A., Kirschbaum, C., Rauh, W., & Hellhammer, D. (1997). Attenuated free cortisol response to psychosocial stress in children with atopic dermatitis. *Psychosomatic Medicine, 59*, 419–426.

Carroll, D., Ginty, A. T., Whittaker, A. C., Lovallo, W. R., & de Rooij, S. R. (2017). The behavioural, cognitive, and neural corollaries of blunted cardiovascular and cortisol reactions to acute psychological stress. *Neuroscience and Biobehavioral Reviews, 77*, 74–86.

Clow, A., Hucklebridge, F., Stalder, T., Evans, P., & Thorn, L. (2010). The cortisol awakening response: More than a measure of HPA axis function. *Neuroscience and Biobehavioral Reviews, 35*, 97–103.

Clow, A., Huckleldridge, F., & Thorn, L. (2010). The cortisol awakening response in context. *Science of Awakening.*

de Quervain, D. J., Aerni, A., Schelling, G., & Roozendaal, B. (2009). Glucocorticoids and the regulation of memory in health and disease. *Frontiers in Neuroendocrinology, 30*, 358–370.

Desantis, A. S., Kuzawa, C. W., & Adam, E. K. (2015). Developmental origins of flatter cortisol rhythms: Socioeconomic status and adult cortisol activity. *American Journal of Human Biology, 27*, 458–467.

Dickerson, S. S., & Kemeny, M. E. (2004). Acute stressors and cortisol responses: A theoretical integration and synthesis of laboratory research. *Psychological Bulletin, 130*, 355–391.

Dickerson, S. S., Mycek, P. J., & Zaldivar, F. (2008). Negative social evaluation, but not mere social presence, elicits cortisol responses to a laboratory stressor task. *Health Psychology, 27*, 116.

Doane, L. D., & Adam, E. K. (2010). Loneliness and cortisol: Momentary, day-to-day, and trait associations. *Psychoneuroendocrinology, 35*, 430–441.

Dockray, S., Smyth, N., Flynn, M., Kearney, I., Cusack, K., Mulally, A., & Collins, M. R. (2017). Picture yourself here: A novel method for determine sampling protocol adherence in ecological settings. *Psychoneuroendocrinology, 83*, 16–17.

Dockray, S., Susman, E. J., & Dorn, L. D. (2009). Depression, cortisol reactivity, and obesity in childhood and adolescence. *Journal of Adolescent Health, 45*, 344–350.

Domes, G., Heinrichs, A., Reichwald, U., & Hautzinger, M. (2002). Hypothalamic-pituitary-adrenal axis reactivity to psychological stress and memory in middle-aged women: High responders exhibit enhanced declarative memory performance. *Psychoneuroendocrinology, 27*, 843–853.

Edwards, S., Clow, A., Evans, P., & Hucklebridge, F. (2001). Exploration of the awakening cortisol response in relation to diurnal cortisol secretory activity. *Life Sciences, 68*, 2093–2103.

Edwards, S., Evans, P., Hucklebridge, F., & Clow, A. (2001). Association between time of awakening and diurnal cortisol secretory activity. *Psychoneuroendocrinology, 26*, 613–622.

Elzinga, B. M., Roelofs, K., Tollenaar, M. S., Bakvis, P., van Pelt, J., & Spinhoven, P. (2008). Diminished cortisol responses to psychosocial stress associated with lifetime adverse events: A study among healthy young subjects. *Psychoneuroendocrinology, 33*, 227–237.

Ennis, G. E., Moffat, S. D., & Hertzog, C. (2016). The cortisol awakening response and cognition across the adult lifespan. *Brain and Cognition, 105*, 66–77.

Evans, P. D., Clow, A., Hucklebridge, F., & Loveday, C. (2012). The cortisol awakening response is related to executive function in older age. *International Journal of Psychophysiology, 84*, 201–204.

Evans, P. D., Fredhoi, C., Loveday, C., Hucklebridge, F., Aitchison, E., Forte, D., & Clow, A. (2011). The diurnal cortisol cycle and cognitive performance in the healthy old. *International Journal of Psychophysiology, 79*, 371–377.

Evans, P., Smyth, N., Thorn, L., Hucklebridge, F., & Clow, A. (2019). Salience versus magnitude in the measurement of the cortisol awakening response. *Psychoneuroendocrinology.*

Foley, P., & Kirschbaum, C. (2010). Human hypothalamus-pituitary-adrenal axis responses to acute psychosocial stress in laboratory settings. *Neuroscience and Biobehavioral Reviews, 35*, 91–96.

Franz, C. E., Spoon, K., Thompson, W., Hauger, R. L., Hellhammer, D. H., Jacobson, K. C., … Kremen, W. S. (2013). Adult cognitive ability and socioeconomic status as mediators of the effects of childhood disadvantage on salivary cortisol in aging adults. *Psychoneuroendocrinology, 38*, 2127–2139.

Fries, E., Dettenborn, L., & Kirschbaum, C. (2009). The cortisol awakening response (CAR): Facts and future directions. *International Journal of Psychophysiology, 72*, 67–73.

Gardner, M. P., Lightman, S., Sayer, A. A., Cooper, C., Cooper, R., Deeg, D., … Halcyon Study, T. (2013). Dysregulation of the hypothalamic pituitary adrenal (HPA) axis and physical performance at older ages: An individual participant meta-analysis. *Psychoneuroendocrinology, 38*, 40–49.

Granger, D. A., Hibel, L. C., Fortunato, C. K., & Kapelewski, C. H. (2009). Medication effects on salivary cortisol: Tacticsand strategy to minimize impact in behavioral and developmental science. *Psychoneuroendocrinology, 34*, 1437–1448.

Griefahn, B., & Robens, S. (2011). Cortisol awakening response—Sampling delays of 15 minutes are not acceptable. *International Journal of Psychophysiology, 82*, 202–205.

Hajat, A., Diez-Roux, A., Franklin, T. G., Seeman, T., Shrager, S., Ranjit, N., … Kirschbaum, C. (2010). Socioeconomic and race/ethnic differences in daily salivary cortisol profiles: The multi-ethnic study of atherosclerosis. *Psychoneuroendocrinology, 35*, 932–943.

Hellhammer, J., Fries, E., Schweisthal, O. W., Schlotz, W., Stone, A. A., & Hagemann, D. (2007). Several daily measurements are necessary to reliably assess the cortisol rise after awakening: State and trait components. *Psychoneuroendocrinology, 32*, 80–86.

Hellhammer, D. H., Wust, S., & Kudielka, B. M. (2009). Salivary cortisol as a biomarker in stress research. *Psychoneuroendocrinology, 34*, 163–171.

Herbison, C. E., Henley, D., Marsh, J., Atkinson, H., Newnham, J. P., Matthews, S. G., . . . Pennell, C. E. (2016). Characterization and novel analyses of acute stress response patterns in a population-based cohort of young adults: Influence of gender, smoking, and BMI. *Stress, 19*, 139–150.

Ice, G. H. (2005). Factors influencing cortisol level and slope among community dwelling older adults in Minnesota. *Journal of Cross-Cultural Gerontology, 20*, 91–108.

Jacobs, N., Nicolson, N. A., Derom, C., Delespaul, P., van Os, J., & Myin-Germeys, I. (2005). Electronic monitoring of salivary cortisol sampling compliance in daily life. *Life Sciences, 76*, 2431–2443.

Jönsson, P., Wallergård, M., Österberg, K., Hansen, Å. M., Johansson, G., & Karlson, B. (2010). Cardiovascular and cortisol reactivity and habituation to a virtual reality version of the Trier Social Stress Test: A pilot study. *Psychoneuroendocrinology, 35*, 1397–1403.

Katz, F. H., & Shannon, I. L. (1964). Identification and significance of parotid fluid corticosteroids. *Acta endocrinologica, 46*, 393–404.

Kirschbaum, C., & Hellhammer, D. H. (1989). Salivary cortisol in psychobiological research: An overview. *Neuropsychobiology, 22*, 150–169.

Kirschbaum, C., Pirke, K. M., & Hellhammer, D. H. (1993). The 'Trier Social Stress Test' – A tool for investigating psychobiological stress responses in a laboratory setting. *Neuropsychobiology, 28*, 76–81.

Kirschbaum, C., Prussner, J. C., Stone, A. A., Federenko, I., Gaab, J., Lintz, D., . . . Hellhammer, D. H. (1995). Persistent high cortisol responses to repeated psychological stress in a subpopulation of healthy men. *Psychosomatic Medicine, 57*, 468–474.

Kirschbaum, C., Wust, S., & Strasburger, C. J. (1992). Normal cigarette-smoking increases free cortisol in habitual smokers. *Life Sciences, 50*, 435–442.

Kotlyar, M., Donahue, C., Thuras, P., Kushner, M. G., O'Gorman, N., Smith, E. A., & Adson, D. E. (2008). Physiological response to a speech stressor presented in a virtual reality environment. *Psychophysiology, 45*, 1034–1037.

Kudielka, B. M., Broderick, J. E., & Kirschbaum, C. (2003). Compliance with saliva sampling protocols: Electronic monitoring reveals invalid cortisol daytime profiles in noncompliant subjects. *Psychosomatic Medicine, 65*, 313–319.

Kudielka, B. M., Buske-Kirschbaum, A., Hellhammer, D. H., & Kirschbaum, C. (2004). HPA axis responses to laboratory psychosocial stress in healthy elderly adults, younger adults, and children: Impact of age and gender. *Psychoneuroendocrinology, 29*, 83–98.

Kudielka, B. M., Hellhammer, D. H., & Wust, S. (2009). Why do we respond so differently? Reviewing determinants of human salivary cortisol responses to challenge. *Psychoneuroendocrinology, 34*, 2–18.

Kudielka, B. M., & Kirschbaum, C. (2003). Awakening cortisol responses are influenced by health status and awakening time but not by menstrual cycle phase. *Psychoneuroendocrinology, 28*, 35–47.

Kudielka, B. M., Schmidt-Reinwald, A. K., Hellhammer, D. H., & Kirschbaum, C. (1999). Psychological and endocrine responses to psychosocial stress and dexamethasone/corticotropin-releasing hormone in healthy postmenopausal women and young controls: The impact of age and a two-week estradiol treatment. *Neuroendocrinology, 70*, 422–430.

Kunz-Ebrecht, S. R., Kirschbaum, C., Marmot, M., & Steptoe, A. (2004). Differences in cortisol awakening response on work days and weekends in women and men from the Whitehall II cohort. *Psychoneuroendocrinology, 29*, 516–528.

Law, R., Hucklebridge, F., Thorn, L., Evans, P., & Clow, A. (2013). State variation in the cortisol awakening response. *Stress, 16*, 483–492.

Lichstein, K. L., Stone, K. C., Donaldson, J., Nau, S. D., Soeffing, J. P., Murray, D., . . . Aguillard, N. (2006). Actigraphy validation with insomnia. *Sleep, 29*, 232–239.

Llabre, M. M., Spitzer, S., Siegel, S., Saab, P. G., & Schneiderman, N. (2004). Applying latent growth curve modeling to the investigation of individual differences in cardiovascular recovery from stress. *Psychosomatic Medicine, 66*, 29–41.

Lovallo, W. R., Dickensheets, S. L., Myers, D. A., Thomas, T. L., & Nixon, S. J. (2000). Blunted stress cortisol response in abstinent alcoholic and polysubstance-abusing men. *Alcoholism-Clinical and Experimental Research, 24*, 651–658.

Mahar, I., Bambico, F. R., Mechawar, N., & Nobrega, J. N. (2014). Stress, serotonin, and hippocampal neurogenesis in relation to depression and antidepressant effects. *Neuroscience & Biobehavioral Reviews, 38*, 173–192.

Manigault, A. W., Woody, A., Zoccola, P. M., & Dickerson, S. S. (2018). Trait mindfulness predicts the presence but not the magnitude of cortisol responses to acute stress. *Psychoneuroendocrinology, 90*, 29–34.

McRae, A. L., Saladin, M. E., Brady, K. T., Upadhyaya, H., Back, S. E., & Timmerman, M. A. (2006). Stress reactivity: Biological and subjective responses to the cold pressor and Trier Social stressors. *Human Psychopharmacology-Clinical and Experimental, 21*, 377–385.

Menet, J. S., & Rosbash, M. (2011). When brain clocks lose track of time: Cause or consequence of neuropsychiatric disorders. *Current Opinion in Neurobiology, 21*, 849–857.

Miller, R., Plessow, F., Kirschbaum, C., & Stalder, T. (2013). Classification criteria for distinguishing cortisol responders from nonresponders to psychosocial stress: Evaluation of salivary cortisol pulse detection in panel designs. *Psychosomatic Medicine, 75*, 832–840.

Miller, R., Wojtyniak, J.-G., Weckesser, L. J., Alexander, N. C., Engert, V., & Lehr, T. (2018). How to disentangle psychobiological stress reactivity and recovery: A comparison of model-based and non-compartmental analyses of cortisol concentrations. *Psychoneuroendocrinology, 90*, 194–210.

Nader, N., Chrousos, G. P., & Kino, T. (2010). Interactions of the circadian CLOCK system and the HPA axis. *Trends in Endocrinology & Metabolism, 21*, 277–286.

Nicolson, N., Storms, C., Ponds, R., & Sulon, J. (1997). Salivary cortisol levels and stress reactivity in human aging. *Journals of Gerontology Series a-Biological Sciences and Medical Sciences, 52*, M68–M75.

O'Connor, D. B., Hendrickx, H., Dadd, T., Elliman, T. D., Willis, T. A., Talbot, D., . . . Dye, L. (2009). Cortisol awakening rise in middle-aged women in relation to psychological stress. *Psychoneuroendocrinology, 34*, 1486–1494.

Oskis, A., Loveday, C., Hucklebridge, F., Thorn, L., & Clow, A. (2009). Diurnal patterns of salivary cortisol across the adolescent period in healthy females. *Psychoneuroendocrinology, 34*, 307–316.

Petrowski, K., Herold, U., Joraschky, P., Wittchen, H.-U., & Kirschbaum, C. (2010). A striking pattern of cortisol non-responsiveness to psychosocial stress in patients with panic disorder with concurrent normal cortisol awakening responses. *Psychoneuroendocrinology, 35*, 414–421.

Pruessner, J. C., Kirschbaum, C., Meinlschmidt, G., & Hellhammer, D. H. (2003). Two formulas for computation of the area under the curve represent measures of total hormone concentration versus time-dependent change. *Psychoneuroendocrinology, 28*, 916–931.

Pruessner, M., Pruessner, J. C., Hellhammer, D. H., Pike, G. B., & Lupien, S. J. (2007). The associations among hippocampal volume, cortisol reactivity, and memory performance in healthy young men. *Psychiatry Research-Neuroimaging, 155*, 1–10.

Pruessner, J. C., Wolf, O. T., Hellhammer, D. H., Buske-Kirschbaum, A., von Auer, K., Jobst, S., . . . Kirschbaum, C. (1997). Free cortisol levels after awakening: A reliable biological marker for the assessment of adrenocortical activity. *Life Sciences, 61*, 2539–2549.

Raffington, L., Prindle, J., Keresztes, A., Binder, J., Heim, C., & Shing, Y. L. (2018). Blunted cortisol stress reactivity in low-income children relates to lower memory function. *Psychoneuroendocrinology, 90*, 110–121.

Rimmele, U., Meier, F., Lange, T., & Born, J. (2010). Suppressing the morning rise in cortisol impairs free recall. *Learning & Memory, 17*, 186–190.
Rohleder, N., Kudielka, B. M., Hellhammer, D. H., Wolf, J. M., & Kirschbaum, C. (2002). Age and sex steroid-related changes in glucocorticoid sensitivity of pro-inflammatory cytokine production after psychosocial stress. *Journal of Neuroimmunology, 126*, 69–77.
Schlotz, W., Hellhammer, J., Schulz, P., & Stone, A. A. (2004). Perceived work overload and chronic worrying predict weekend-weekday differences in the cortisol awakening response. *Psychosomatic Medicine, 66*, 207–214.
Smyth, N., Clow, A., Thorn, L., Hucklebridge, F., & Evans, P. (2013). Delays of 5-15 min between awakening and the start of saliva sampling matter in assessment of the cortisol awakening response. *Psychoneuroendocrinology, 38*, 1476–1483.
Smyth, N., Flynn, M., Rajcani, J., Hucklebridge, M. F., Thorn, L., Wood, C., . . . Clow, A. (2019). Attenuated cortisol reactivity to psychosocial stress is associated with greater visual dependency in postural control. *Psychoneuroendocrinology*.
Smyth, N., Hucklebridge, F., Thorn, L., Evans, P., & Clow, A. (2013). Salivary cortisol as a biomarker in social science research. *Social and Personality Psychology Compass, 7*, 605–625.
Smyth, N., Thorn, L., Hucklebridge, F., Clow, A., & Evans, P. (2016). Assessment of the cortisol awakening response: Real-time analysis and curvilinear effects of sample timing inaccuracy. *Psychoneuroendocrinology, 74*, 380–386.
Smyth, N., Thorn, L., Oskis, A., Hucklebridge, F., Evans, P., & Clow, A. (2015). Anxious attachment style predicts an enhanced cortisol response to group psychosocial stress. *Stress-The International Journal on the Biology of Stress, 18*, 143–148.
Stalder, T., Hucklebridge, F., Evans, P., & Clow, A. (2009). Use of a single case study design to examine state variation in the cortisol awakening response: Relationship with time of awakening. *Psychoneuroendocrinology, 34*, 607–614.
Stalder, T., Kirschbaum, C., Kudielka, B. M., Adam, E. K., Pruessner, J. C., Wuest, S., . . . Clow, A. (2016). Assessment of the cortisol awakening response: Expert consensus guidelines. *Psychoneuroendocrinology, 63*, 414–432.
Steptoe, A., Kunz-Ebrecht, S. R., Brydon, L., & Wardle, J. (2004). Central adiposity and cortisol responses to waking in middle-aged men and women. *International Journal Of Obesity, 28*, 1168.
Therrien, F., Drapeau, V., Lalonde, J., Lupien, S. J., Beaulieu, S., Tremblay, A., & Richard, D. (2007). Awakening cortisol response in lean, obese, and reduced obese individuals: Effect of gender and fat distribution. *Obesity, 15*, 377–385.
Thorn, L., Evans, P., Cannon, A., Hucklebridge, F., & Clow, A. (2011). Seasonal differences in the diurnal pattern of cortisol secretion in healthy participants and those with self-assessed seasonal affective disorder. *Psychoneuroendocrinology, 36*, 816–823.
Thorn, L., Hucklebridge, F., Evans, P., & Clow, A. (2006). Suspected non-adherence and weekend versus week day differences in the awakening cortisol response. *Psychoneuroendocrinology, 31*, 1009–1018.
Turner, A., Smyth, N., Hall, S., Torres, S., Hussein, M., Jayasinghe, S., . . . Clow, A. (2019). How strong is the evidence linking stress reactivity with future health and disease outcomes?—A systematic review of prospective evidence. *Psychoneuroendocrinology, 100*, S52–S53.
von Dawans, B., Kirschbaum, C., & Heinrichs, M. (2011). The Trier Social Stress Test for Groups (TSST-G): A new research tool for controlled simultaneous social stress exposure in a group format. *Psychoneuroendocrinology, 36*, 514–522.
Wetherell, M. A., Craw, O., & Smith, M. A. (2018). Techniques for inducing stress: Problems and opportunities. In *The Routledge international handbook of psychobiology*. London: Routledge.
Wiemers, U. S., Schoofs, D., & Wolf, O. T. (2013). A friendly version of the Trier Social Stress Test does not activate the HPA axis in healthy men and women. *Stress, 16*, 254–260.
Wolf, O. T., Fujiwara, E., Luwinski, G., Kirschbaum, C., & Markowitsch, H. J. (2005). No morning cortisol response in patients with severe global amnesia. *Psychoneuroendocrinology, 30*, 101–105.

Wolfram, M., Bellingrath, S., & Kudielka, B. M. (2011). The cortisol awakening response (CAR) across the female menstrual cycle. *Psychoneuroendocrinology, 36*, 905–912.
Wood, C. J., Clow, A., Hucklebridge, F., Law, R., & Smyth, N. (2018). Physical fitness and prior physical activity are both associated with less cortisol secretion during psychosocial stress. *Anxiety, Stress, & Coping, 31*, 135–145.
Wust, S., Wolf, J., Hellhammer, D. H., Federenko, I., Schommer, N., & Kirschbaum, C. (2000). The cortisol awakening response—Normal values and confounds. *Noise and Health, 2*, 79–88.

# Chapter 6
# The Use of Saliva for Genetic and Epigenetic Research

**Zsofia Nemoda**

**Abstract** Minimally invasive sampling methods for collecting oral fluids or cells are preferred in genetic research involving children and in large-scale studies where the biological sample could be mailed to the laboratory at room temperature. Saliva is an easily accessible source of cells, containing both epithelial cells exfoliating from the oral mucosa, and leukocytes filtrating from blood vessels. While it does not matter which type of somatic cells are used for genetic analyses, extreme caution and proper correction are necessary in DNA methylation analyses to address the technical heterogeneity in samples with different cell composition, since the majority of the epigenetic marks are tissue specific.

High-quantity and high-quality DNA can be obtained from whole saliva samples (collected by passive drooling) or from collection devices which usually absorb cells when collecting saliva. Although salivary DNA is a mix of human and bacterial DNA, due to the species-specific amplification step in the genotyping procedures, it can be used both in candidate gene analyses and genome-wide association studies. With a careful design and appropriate additional analytical steps, salivary DNA samples can be also successfully applied in epigenetic association studies. The technical recommendations for these studies are highlighted in this chapter.

**Keywords** Salivary DNA sample · Genetic polymorphism · DNA methylation · Cell composition · Tissue-specificity

## 6.1 Introduction

Over the past decades, the number of genetic studies using saliva samples has been steadily increasing (for a comprehensive review see Sun & Reichenberger, 2014). Since all normal somatic cells contain the same genetic sequence in our body,

Z. Nemoda (✉)
Department of Medical Chemistry, Molecular Biology and Pathobiochemistry, Semmelweis University, Budapest, Hungary
e-mail: nemoda.zsofia@med.semmelweis-univ.hu



D. A. Granger, M. K. Taylor (eds.), *Salivary Bioscience*,
https://doi.org/10.1007/978-3-030-35784-9_6

analyzing genetic markers from any kind of cell other than gametes should provide the same information. There are some exceptions to this general rule, for example in genetic mosaicism or in the case of foreign human cells present in some individuals (such as the donor cells after transplantation or in twins exposed to fused circulation), and in precancerous cells where the original genomic DNA (gDNA) sequence is already altered. Epigeneticists also started to use salivary specimen, based on the assumption that saliva can be as good alternative source of DNA as blood in epigenetic association studies of non-blood-based diseases. However, there are still a few technical issues (due to tissue specificity and reversibility of epigenetic signals) to be considered in this area of research (see details at the last sections).

## 6.2 Using Saliva as a Minimally Invasive Genetic Sample

In hospital-based genetic laboratories, blood samples are the preferred DNA source for a couple of reasons: (1) reliable DNA yield with high quantity and quality, (2) more experience has been accumulated, (3) blood sampling can be easily combined with other (routine clinical) blood tests, and automatic sample processing is often available at hospital centers (reducing workload). However, in field studies where sample storage and processing cannot be easily achieved, or at participants' homes—where medical assistance for blood drawing is difficult to manage—saliva can serve as an ideal surrogate tissue for genetic analyses. Noninvasive sampling methods of collecting cells from the mouth are also preferred in research involving children, the elderly, and healthy nonclinical participants, especially in large-scale studies where biological samples could be mailed to the laboratory at ambient temperature. Table 6.1 summarizes the fields where saliva sampling is preferred. Since 1 ml of blood or saliva contains similar number of cells (~half a million), comparable amounts of DNA can be extracted from both (Sun & Reichenberger, 2014). However, blood samples have greater volume (typically ~8 ml) than saliva samples (maximum 4 ml, usually 1–2 ml, see Chap. 3). Therefore, the total DNA yield from blood samples is mostly higher.

Saliva is an easily accessible source of cells, containing both epithelial cells exfoliating from the oral mucosa and leukocytes filtrating from blood vessels, as it has been shown via microscopic observations (see schematic representation at Fig. 6.1 upper part). The approximate ratio of the different cells was demonstrated by genetic analyses of oral samples from patients after allogeneic blood stem cell transplantation using informative DNA markers to differentiate between the host's epithelial cells and donor's blood cells (Thiede, Prange-Krex, Freiberg-Richter, Bornhauser, & Ehninger, 2000). Nevertheless, a major drawback of oral biospecimens is that bacterial and fungal or viral contamination cannot be avoided, which can cause problems if bacterial deoxyribonuclease (DNase) enzymes degrade the human DNA. However, this issue can be resolved simply with proper design. For example, rinsing the mouth and avoid eating for at least half an hour prior to saliva collection reduces contamination. If the sample has to remain at room temperature

**Table 6.1** Types of genetic analyses when saliva samples are preferred

| Number of subjects | Type of analyses |
|---|---|
| Individual | ✓ At-home genetic tests (e.g., ancestry tests, familial genetic risk tests)<br>✓ Detection of genetic mutations in oral precancerous cells (sputum is used for detection of respiratory tract diseases)<br>~ Forensic tests (depends on the available tissue)<br>∅ Genetic tests of blood cells (e.g., leukemia)<br>∅ Clinical diagnosis of genetic disorders (lower failure rate is required) |
| Group-level | ✓ Field studies (when resources are limited), home visits (blood drawing is not easily manageable)<br>~ Genetic research conducted with children or the elderly in a clinical setting (additional blood drawing can be avoided by saliva sampling)<br>∅ Genetic research with adult patients in hospital settings (when blood is drawn anyway for other laboratory tests) |
| Population-based | ✓ Epidemiological studies, especially when samples are mailed<br>~ Biobanking samples of the general population (when higher consent rate is important, saliva can substitute blood samples)<br>∅ Genetic studies of blood-based (e.g., immunological) diseases |

✓ Saliva is preferred (because of easier access or lower cost)
~ Saliva and blood-based samples are equally accepted
∅ Blood-based samples are preferred

for a longer time, a preservative can be added. There are several commercially available kits providing convenient protocols for saliva collection when the stabilizing solution is released into the sample after closing the device. This solution is a lysis buffer containing the ionic detergent sodium dodecyl sulfate at a proper pH and salt concentration that is optimal for DNA storage (i.e., removing divalent cations, which are required for DNase activity). Since all kinds of cells, either human or microbial, are disrupted by the lysis buffer, it also renders the sample nonhazardous, which might be important for complying with biosafety regulations.

## 6.3 Quantity and Quality of Salivary DNA Samples

If saliva is collected in a laboratory setting, the passive drooling method is advised (see Chap. 3), and bacterial growth can be avoided without adding any chemical by storing the sample at a temperature −20 °C or less. In this way, the salivary sample can be used for multiple assays (e.g., using the supernatant for hormone or other saliva-biomarker analyses, while using the cell pellet fraction for genetic analyses). Since DNA is a stable nucleic acid, repeated freeze-thaw cycles have a negligible effect on the DNA yield (Nemoda et al., 2011). Although several freeze-thaw cycles can damage the gDNA, this fragmentation rarely results in non-amplifiable sample (Digestion by bacterial DNase enzymes imposes a bigger risk if the sample is not stored properly). In addition, storage for extended time periods reduces the DNA yield but does not affect substantially the usability of gDNA in genotyping

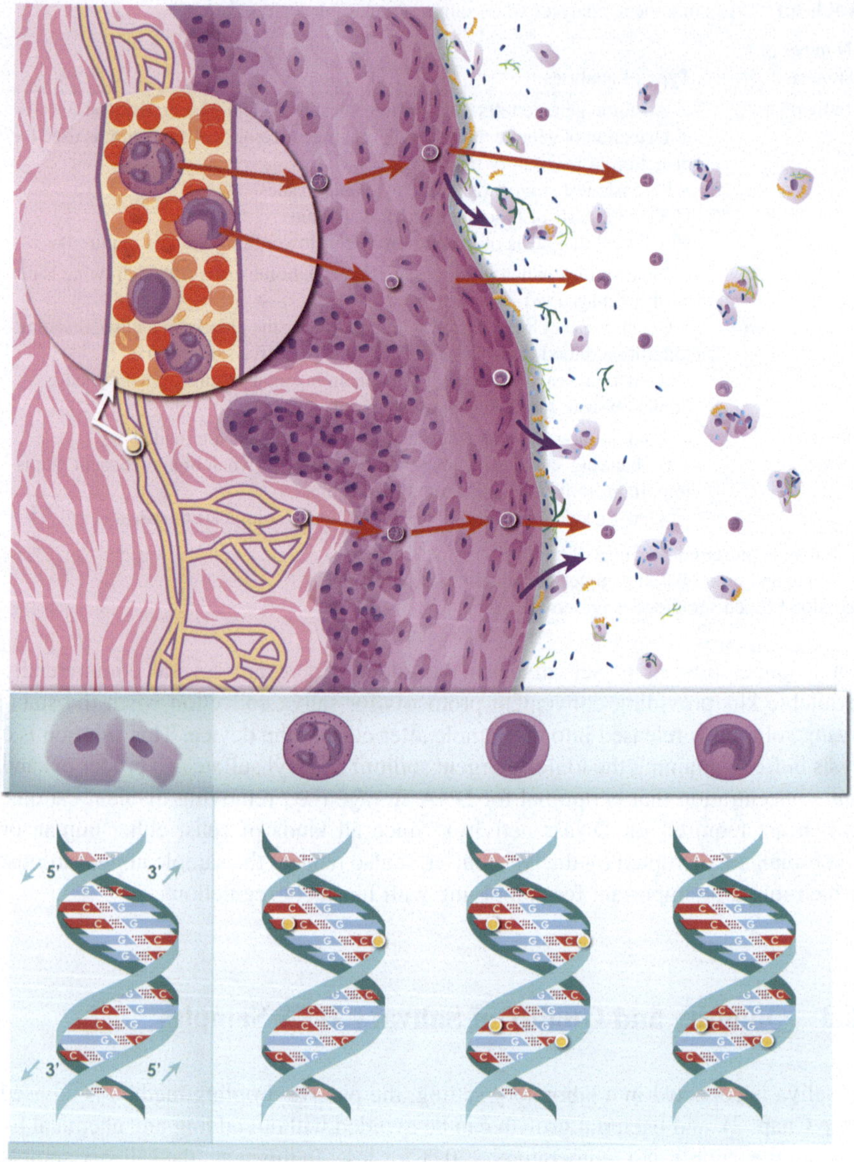


**Fig. 6.1** Cell types and their possible epigenetic modifications in saliva. Top: Structure of the nonkeratinized buccal epithelium. Cells are shown in purple as they appear in the histological hematoxylin & eosin staining. The underlying collagen fibers are indicated by the pink stripes, and the capillary branches are represented in purple with yellow symbolizing plasma. A magnified section of a blood vessel demonstrates the different cell types in blood (beside the most abundant red blood cells and tiny thrombocytes, nucleus containing leukocytes are shown). To achieve their immunological functions, leukocytes can exit blood vessels (red arrows) and transverse through the epithelial cell layers. Hence the cell composition of saliva is quite heterogenous, which is shown on the right side. Based on microscopic observations and DNA methylation data of salivary samples, the two major cell types are buccal epithelial cells (large, pink cells) and granulocytes (smaller, purple cells with segmented nuclei). The different types of bacteria are represented by light blue

(Durdiaková, Kamodyová, Ostatníková, Vlková, & Celec, 2012). It is to note that routine DNA concentration measurements (using UV absorbance at 260 nm) are based only on the chemical properties of the nucleic acids. Thus, the integrity of high-molecular weight gDNA should be checked by other methods (e.g., gel electrophoresis, see Rethmeyer, Tan, Manzardo, Schroeder, & Butler, 2013) before the costly and time-consuming genetic analyses.

High-quantity and high-quality DNA can be obtained not only from whole saliva samples but also from the collection devices, that usually absorb cells present in saliva. Therefore, it is advisable to keep the collection device after the centrifugation of saliva if genetic analyses are also planned beside hormone measurements. In this way, the biological sample collection procedure can be simplified, which can be a crucial point in studies involving children. In terms of collection medium, if saliva is obtained via cotton or hydrocellulose absorbent device, most of the nucleic acid content can be recovered from the device after incubating it in cell lysis buffer (similar to DNA isolation starting from a cell pellet). However, if a synthetic swab is used, equal amounts of DNA could be isolated from the saliva filtrate and the collection device (Nemoda et al., 2011). Thus, preliminary analyses are advised to check the approximate DNA yield when a new protocol is planned with a collection device, especially when collecting saliva from infants, since much lower saliva volume can be obtained from them.

It is important to emphasize that the total amount of DNA obtained from saliva samples shows huge variability. In a review paper by Sun and Reichenberger (2014), the cited research groups report on average 20–40 μg DNA yield per ml saliva (ranging from 1 to 160 μg/ml). Larger DNA yield was observed when rubbing the tongue against the inside of the mouth before saliva collection (Nunes et al., 2012). Therefore, even though 0.1 ml of saliva is sufficient for genetic analyses (yielding approximately 1–2 μg DNA), researchers are advised to collect at least 1 ml saliva from children and adults, so that a larger number of analyses can be conducted (Usually one genotyping assay requires 10–20 ng of DNA). Better quality and higher amount of DNA (hence larger volume of saliva) is required for genome-wide association studies (GWAS), where thousands of gene variants are measured in parallel on microarrays (for a detailed review on GWAS see Stranger, Stahl, & Raj, 2011). Although the current methodologies can work with as low as 0.5 μg gDNA at

**Fig. 6.1** (continued) dots, dark blue rods, and orange conglomerates, whereas fungi are illustrated by green filaments. These microorganisms live in the mucinous layer (shown in gray) and are often attached to the exfoliating epithelial cells. Therefore, genetic samples obtained from saliva contain both human and foreign DNA. Bottom: Epithelial cells (first panel) and the three main white blood cell types (granulocyte, lymphocyte, and monocyte), which can be present in saliva. The nuclei (shown in dark purple) are the source of gDNA. Epigenetic variations are shown at the bottom: The two parallel green ribbons represent the double stranded DNA with the sense sequence in 5′–3′ direction. Bases are: A, adenine; C, cytosine; G, guanine; and T, thymine. Hydrogen bonds are denoted as dashed lines. Although the genetic sequence is the same in every somatic cell type of a healthy individual, epigenetic marks such as DNA methylation can show cell type-specific patterns, as illustrated by the yellow circles on the C bases

high-throughput, large-scale measurements, using DNA samples with concentration below 30–50 ng/μl is not recommended. Thus, the optimal protocol for saliva collection has to be selected according to the aims of each study so that enough good quality gDNA can be obtained for the genetic analyses while considering the lower end of the DNA yield range.

The quality of salivary DNA can also vary substantially depending on the DNA isolation technique (e.g., classical phenol-chloroform extraction versus silica membrane-based purification kit, Durdiaková et al., 2012). Interestingly, the ratio of human and microbial DNA can be also affected by the extraction method (Vesty, Biswas, Taylor, Gear, & Douglas, 2017). The pros and cons of the most frequently used DNA isolation methods are presented in Table 6.2. It has to be noted, that the preparatory step of cell lysis (to release DNA molecules in the sample) can also affect the quality of the DNA specimen. For example, using preloaded lysis buffer at saliva collection can be disadvantageous if the expected volume of saliva is not achieved (e.g., only 0.5 ml saliva is provided instead of the recommended 2 ml at an Oragene self-collection kit), because it can affect the efficiency of the molecular analyses (Pulford, Mosteller, Briley, Johansson, & Nelsen, 2013). Remaining chemicals, such as sodium dodecyl sulfate, phenol, or ethanol can inhibit or degrade the enzyme amplifying the DNA template in the polymerase chain reaction (PCR) (Rossen et al., 1992). While this problem could be easily detected in samples where the reaction is completely inhibited (i.e., not yielding sequence-specific amplicons), more subtle differences in amplification efficiencies could result in biases of quantitative measurements, like at DNA methylation analyses (Soriano-Tárraga et al., 2013).

The remaining organic compounds can also affect DNA concentration measurements using UV absorbance on a spectrophotometer, potentially resulting in overestimated (or confounded) DNA quantity and quality. Therefore, companies recommend using concentration measurements based on colorimetric reactions that can estimate the double stranded gDNA in salivary samples. However, this technique still does not provide precise information about the amount of useful human gDNA in a sample, since saliva always has a portion of foreign DNA due to its microbial content (even from healthy individuals). Importantly, the human/microbial DNA ratio can be estimated by real-time PCR technique using human-specific primers (for more details see methods by Nishita et al., 2009). Remarkably, varying portions of human DNA were reported by different research groups: Mean percentages of amplifiable human DNA varied between 40 and 80% in saliva samples, ranging from about 10 to 100% in most of the studies (see references by Sun & Reichenberger, 2014). Still, the human DNA yield can be kept on the higher end of this range by thoroughly rinsing the mouth with water 5–30 min before saliva collection. Notably, Hu et al. (2012) showed that salivary samples with at least 31% human-specific amplifiable DNA performed as well as blood-derived DNA samples.

In summary, oral cells can be used for various genetic analyses (see Table 6.1). Saliva can be easily collected in a broad age range, increasing the consent rate for providing biological sample, especially among healthy participants, which is a

**Table 6.2** Techniques for genomic DNA isolation from saliva

| Extraction method | Contamination problems[a] | Laboratory requirement | Cost | Most frequent applications Pros & Cons |
|---|---|---|---|---|
| Phenol-chloroform and isoamyl alcohol | Organic solvents (230 nm abs.): phenol, chloroform, alcohol | Equipment: centrifuge, fume hood Time: hours | $ | Classical laboratory experiments Pros: good for small DNA fragments & single stranded DNA Cons: working with hazardous solutions, higher chance of solvent contamination at low DNA yield samples |
| High-salt | Proteins (280 nm abs.), alcohol (230 nm abs.) | Equipment: centrifuge Time: hours | $ | Routine laboratory experiments Pros: cheap, relatively quick Cons: higher chance of impurity at low DNA yield samples |
| Silica membrane-based (solid phase) | Alcohol (230 nm abs.) | Equipment: centrifuge or vacuum manifold Time: <hour | $$ | High-throughput, automatic processing (using plates); routine laboratory experiments (with columns in separate tubes) Pros: separation of high-molecular weight gDNA from small fragments Cons: different kits are required for the various DNA fragment sizes |
| Anion exchange resin based | | Equipment: none (gravity) or centrifuge Time: <hour | $$ | High-throughput, automatic processing & clinical or forensic diagnostics, when there is no need for longer DNA storage Pros: quick, efficient |
| Magnetic particles/ beads based | Alcohol (230 nm abs.) | Equipment: magnetic rack (it can be included) Time: <hour | $$$ | High-throughput, automatic processing (in plates); molecular laboratory experiments (in tubes) Pros: good for small amount of DNA, as the recovery rate is high |

The laboratory time and costs are presented only in relative amounts, as they can vary according to the preparatory steps (e.g., mechanical or enzymatic cell lysis procedure), number of samples, and other specific features of the genetic study. Note that alcohol contamination can affect almost every type of extraction method, since DNA is washed with 70% ethanol before the final dissolving step. Contamination can also arise from the preparatory steps, such as sodium dodecyl sulfate from the lysis buffer, which are not shown in the table

[a]The routine concentration measurement of DNA is based on the absorbance of the sample at 260 nm in a spectrophotometer, the absorbance at 230 nm and 280 nm shows the presence of organic compounds and proteins, respectively

crucial point in large-scale epidemiological studies (e.g., in follow-up analyses of specific birth cohorts). The processing of this noninvasively obtained biological specimen is similar to that of blood-based samples, yielding comparatively good quality DNA samples for a wide range of genetic analyses. Until recently, the majority of genetic and epigenetic studies used blood as a source of gDNA, hence saliva is often referred to as surrogate tissue in these studies. Although salivary DNA is a mix of human, bacterial, and fungal DNA (Fig. 6.1), due to the species-specific PCR amplification step in the genotyping procedures (for a visualized experiment see Lorenz, 2012), it can be readily used for genetic analyses. In the following sections, applications of salivary DNA samples in human genetic studies are discussed. These studies assess the sequence of human DNA which is present in every normal somatic cell (hence the source of cells does not matter in these genetic analyses). Measurements of malignant cells and DNA adducts used in oral cancer diagnostics are presented in Chap. 19, whereas Chap. 8 describes studies assessing salivary cell-free DNA (called as liquid biopsy in cancer diagnostics, see review by Siravegna, Marsoni, Siena, & Bardelli, 2017). For the usage of microbial DNA in salivary samples, see Chaps. 7 and 13.

## 6.4 Applications of Saliva in Genetic Analyses

There are two main types of genetic analyses where saliva samples are used: genetic tests on the individual level and genetic association studies that compare groups (see Table 6.1). Genetic tests aim to detect inherited risk factors for specific diseases (e.g., sickle cell disease, cystic fibrosis) helping diagnosis, whereas the current association studies try to reveal genetic susceptibility for developing common diseases (e.g., diabetes mellitus, hypertension, and Alzheimer's disease) by linking certain genetic variants to disease state or associated medical, physical, and psychological characteristics (e.g., blood sugar level, blood pressure, memory functions, respectively). In these studies, common gene variants—the so-called polymorphisms with allele frequencies higher than 5%—are the most often analyzed (for more information on human genetic topics, check NIH website: https://ghr.nlm.nih.gov/primer). Since individual genetic factors usually explain only a small portion of the heritability in complex (multifactorial) diseases, large numbers of study participants are needed to detect their modest effect (for an educational review, see Craig, 2008). Especially, GWAS require exceptionally large sample sizes, as they analyze thousands of polymorphisms in order to identify new genes which could be linked to the phenotypes of interest, without a priori hypotheses. In these large-scale epidemiological studies salivary samples became popular, since the consent rate for providing saliva is higher than for blood (Hansen, Simonsen, Nielsen, & Hundrup, 2007; Randell et al., 2016).

Genetic studies targeting children and the elderly, or with nonpersonal recruitment procedures (i.e., via mail or Internet) particularly benefit from the use of saliva. However, there are a few technical issues to consider when planning to recruit participants providing this noninvasive biospecimen for a genetic study. For example, higher consent and return rates were reported from patients with pediatric Crohn's disease compared to controls in a pilot study by Kappelman et al. (2018): 75% of the contacted adolescent patients gave consent to their participation in the genetic study and returned saliva sample by mail, while only 44% of the sex and age matched controls gave their consent and saliva sample (expecting a gift card after the successful study enrollment). Another study investigated the effect of monetary incentive in donating biospecimen for genetic study: 43% of adult patients with inflammatory bowel disease participating in an internet-based survey gave salivary samples when 20$ was offered, and only 26% mailed back the saliva collection kit when no compensation was offered (Randell et al., 2016). The age of the targeted population also matters, because older adults are more willing to donate saliva and send the home-collected kit by mail. In a UK study, 84% consent rate was reported among older individuals with a chronic disease compared to 59% of the contacted families with a sick child (Bhutta et al., 2013). Importantly, the prospective or retrospective nature of the recruitment procedure (i.e., calling families before or after the doctor's visit) can also impact consent rate. Where parental consent is necessary for the genetic study, contacting families before the doctor's visit in order to provide detailed information about the aims of the study is advised, and taking the saliva at the clinic personally would result in higher rate of sample donation (Bhutta et al., 2013). In sum, adult patients can be easily recruited for a genetic study via mail, especially if a telephone call is made by a specialist physician providing detailed information before the sample collection at home. In the recruitment procedure of healthy adults, a follow-up telephone call is also advised after sending the information via mail.

Nowadays, biobanks all over the world store various kinds of biological samples from patients with specific diseases, and often DNA samples (derived either from blood or saliva) of participants from the general population (see https://biobanking.org/). Besides the classical clinical case-control studies, geneticists investigate population-based cohorts at an increasing rate, since well-characterized subjects with data on thousands of genetic markers are a valuable research resource for association studies, such as the UK Biobank (see publications at http://www.ukbiobank.ac.uk/genetic-publications/). Large registries of patients and healthy individuals are building up in almost every country, where—for the sake of non-biased inclusion—saliva samples are also accepted for genetic analyses (see the All of Us Program at https://allofus.nih.gov/).

## 6.5 Comparison of Salivary and Blood Samples in Genetic Analyses

Following the spread of easily accessible and affordable genotyping methods, many research groups could try out saliva collection methods and compare the resulting DNA samples to the "gold standard" blood DNA. The simplest type of genetic variation is the Single Nucleotide Polymorphism (SNP) with only two types of alleles. It is also the most common type of genetic polymorphism (Genomes Project Consortium et al., 2015). For more details on the biological background of genetic variations see the NIH Biological Sciences Curriculum Study (2007). Since usually small fragments (100–200 base pairs) are amplified from the human gDNA during SNP genotyping (see Table 6.3), even degraded DNA samples can give reliable

**Table 6.3** Types of genetic variants in the human genome and their analytic methods

| Genetic variant | Variation length (bp) | Genotyping method | Amplicon length (bp) |
|---|---|---|---|
| SNP (Single Nucleotide Polymorphism), point-mutation | 1 | *Classical PCR & electrophoresis* | |
| | | • Allele-specific amplification with two primer pairs | 70–500 |
| Small indel (insertion/deletion) | 1–15 | *Enzyme-based methods* | |
| | | • Restriction fragment length polymorphism (RFLP) | 100–500 |
| | | • Primer extension | 60–150 |
| | | *Hybridization-based methods* | |
| | | • Quantitative real-time-PCR (e.g., molecular beacon, Taqman assay) | 100–150 |
| | | • SNP microarray (chip) | 100–200 |
| | | *Sequencing methods* | |
| | | • Sanger-sequencing | 200–500 |
| | | • Pyrosequencing | 150–250 |
| | | • Next generation sequencing | 100–200 |
| STR (Short Tandem Repeats), microsatellites | 2–6 | • PCR & capillary electrophoresis | 100–500 |
| | | • Sequencing | 100–500 |
| VNTR (Variable Number of Tandem Repeats) | 10–100 | • PCR & electrophoresis | 100–1000 |
| | | • Sequencing | 100–500 |
| Larger indel (insertion/deletion) | 100–1000 | • PCR & electrophoresis | 200–1000 |
| | | • Sequencing | 100–500 |
| CNP (Copy Number Polymorphism) | 1000–10,000 | • Multiplex ligation-dependent probe amplification (MLPA) | 70–500 |
| Chromosomal microdeletion/microduplication | 10,000–1,000,000 | • Quantitative real-time-PCR | 100–150 |
| | | • Multiple probes from SNP-chip | 100–200 |

Note that chromosomal abnormalities (typically larger than a million base pairs) are not part of the list, as they are not diagnosed from saliva. Generally, these large genomic variants are determined with microscopic karyotyping or with fluorescence in situ hybridization (FISH) without PCR amplification, although current techniques apply amplification of cell-free DNA and next generation sequencing. For details on human genetic variations and their genotyping method see book chapter by Gonzalez-Bosquet and Chanock (2011)

*bp* base pair, *PCR* polymerase chain reaction

results, which is an important issue in forensic applications (reviewed by Sobrino, Brion, & Carracedo, 2005). As most of the genotyping methods include an amplification step with sequence-specific primer pair, mixed origin, low-concentration salivary DNA samples with a tiny amount (picograms) of human DNA can be used. Of course, the DNA input requirements are higher at high-throughput, multiplex PCR methods, where unbiased amplification should be achieved for multiple primer pairs.

With recent genome-wide analyses, precise estimates were gained for the accuracy of genetic analyses performed with salivary DNA samples. Approximately 99% (or higher) concordance rates have been reported with matched saliva and blood-derived DNA samples on high-density SNP-microarrays (Abraham et al., 2012; Bahlo et al., 2010; Gudiseva et al., 2016; Hu et al., 2012). Using similar, hybridization-based genotyping assays for the detection of larger Copy Number Variation (CNV), paired blood and saliva specimens were compared on chromosomal microarrays obtained from three different companies. Importantly, the bacterial content (ranging from 3 to 21%) of salivary DNA did not affect the genotyping quality of any platform used (Reiner et al., 2017), proving that saliva is a reliable alternative DNA source for genetic testing.

As for the genetic methodology, applying two types of SNP genotyping (SNP microarray and Taqman assays), Abraham et al. (2012) showed a high concordance rate (>99%) between paired blood and saliva samples in the genotype results. In addition, high-quality Sanger-sequencing could be produced from most saliva samples (Gudiseva et al., 2016). Failed genotyping was reported only for samples with DNA concentration below 10 ng/μl, highlighting the need for concentrating samples with low DNA yield. Fewer studies were published on saliva collection issues in connection with Short Tandem Repeats (STR) or Variable Number of Tandem Repeats (VNTR), where the length of the targeted genomic region can range from 100 to 1000 base pairs (Table 6.3). Genotyping performance of VNTRs was not influenced by either saliva or DNA sample characteristics (Nemoda et al., 2011; Nishita et al., 2009), but degradation of DNA samples can affect long-range PCR amplification (reviewed by Alaeddini, Walsh, & Abbas, 2010). In conclusion, saliva is as good a source of human cells and gDNA as blood, performing similarly in a wide range of genetic analyses. Although the human-specific portion is lower, and the risk of impurity and DNA degradation are higher in salivary samples, with a careful quality checkup step most of the problematic samples can be recognized and excluded from the analyses.

## 6.6 Applications of Saliva in Epigenetic Analyses

Following the disappointing results of the first wave of GWAS, the pursuit for the "missing heritability" prompted researchers to measure epigenetic variants in order to study the underlying biological mechanisms of gene–environment interactions (Manolio et al., 2009). Until recently, epigenetic analyses have been restricted to the affected, disease-relevant tissues, since a substantial portion of the epigenetic marks is tissue specific. However, using the appropriate tissue for epigenetic association studies is often not feasible (e.g., having liver samples for metabolic diseases or brain samples for neurological disorders). Therefore, researchers started to use surrogate, easily accessible peripheral tissues, such as blood, saliva, or buccal cells. Previously, most of the epigenome-wide association studies (EWAS) used blood-derived DNA samples, although buccal cells could potentially serve as better surrogate tissue in non-blood-based diseases (Lowe et al., 2013). Importantly, saliva contains both buccal and blood cells (see Fig. 6.1); hence, it can serve as a good alternative source of gDNA in several EWAS. Although it is still questionable which peripheral tissue is more relevant for studying certain non-blood-based diseases or traits, the answers would be hopefully revealed by current bioinformatic analyses of epigenomic and transcriptomic datasets of various tissues, which are publicly available for researchers (e.g., Gene Expression Omnibus, GEO, https://www.ncbi.nlm.nih.gov/geo/).

Epigenetic marks create important information above the genetic sequence (*epi*— in Greek means on, above, over) which govern gene expression in multiple ways. These mechanisms are responsible for long-term regulation, switching on exclusively those genes which an individual cell requires (Almouzni & Cedar, 2016). Once the cell (and tissue) identity is established, epigenetic marks are transferred from the mother cell to the daughter cells during somatic cell divisions (contributing to the cellular memory). These marks include covalent modifications of the gDNA and the chromatin-associated histone proteins (such as acetylation, methylation, phosphorylation, and ubiquitination), controlling the accessibility of the chromatin structure. Importantly, there exists a reciprocal cross talk between these processes, therefore many studies measure only one type of epigenetic marks. Due to the stability of DNA, studying the chemical modifications of gDNA is one of the most popular epigenetic analyses, which will be discussed in this section. Analyses of other types of epigenetic mechanisms (such as chromatin structure and histone modifications) are more sensitive (requiring freshly frozen or processed samples), technically laborious, and expensive. Hence, mostly disease-specific tissues are studied by these detailed methods in cancer and infectious disease research; for these types of epigenetic analyses using saliva and oral tissues see Chaps. 9 and 19.

## 6.7 An Overview of Epigenetic Modifications on the DNA Molecule

The most frequent and most widely studied covalent modification in the human genome is the methylation of the cytosine base of a CpG dinucleotide which makes up about 1% of the genome (The letter "p" represents the phosphodiester bond between cytosine and guanine). The 5-methylcytosine (5mC) is often called as the fifth base of the DNA, present in 0.6–0.8% of all bases in the human genome, depending on the developmental stage and tissue type. The majority of methylated CpG sites is located in repetitive sequences and confers repression on transposable elements (the so-called "junk" DNA). When DNA methylation occurs at crucial regulatory regions of protein-coding genes, such as promoters and enhancers, it usually correlates with gene silencing, especially at the so-called CpG islands (CpG-rich regions with high G & C base content). This transcription repression can be achieved by either directly inhibiting transcription factor binding or recruiting chromatin-modifying proteins (see review by Deaton & Bird, 2011). However, methylation at CpG islands located in gene bodies might result in the opposite effect by preventing aberrant transcription initiation events in order to guarantee the correct mRNA transcription (Neri et al., 2017). Therefore, it is important to note that an increase in DNA methylation level can either repress or enhance gene expression depending on the genetic position of the methylation change. Since the majority of previous analyses focused on promoter regions, the repressive feature of DNA methylation would be applied in the subsequent sections when discussing the role of this epigenetic process. The level of DNA methylation is usually expressed in percentages, although it is a binary signal (i.e., a certain CpG-site can be either methylated or non-methylated on the chromosome). It is due to the fact that the DNA content of a biological sample comes from a pool of cells whose DNA methylation patterns might differ substantially (see examples in Fig.6.1 lower part). Therefore, the percentage of DNA methylation is more representative of the proportion of cells in which certain CpG-sites have been methylated in order to shut down the transcription of the respective gene (according to the simplified model using the repressive feature of DNA methylation).

The 5mC can be further modified by hydroxylation catalyzed by a family of oxidases, the Ten Eleven Translocation (TET) enzymes (for a review, see Huang & Rao, 2014). Relatively high 5-hydroxymethylcytosine (5hmC) levels are reported in the brain, while other organs have various amount of this epigenetic modification (on average ~0.1% of all bases in the human genome, for details see Nestor et al., 2012). Growing evidence supports the hypothesis that 5hmC is an intermediate in active DNA demethylation processes, and it might also be involved in gene expression regulation (reviewed by Wu & Zhang, 2017). Additionally, 5hmC can be further oxidized to formyl- and carboxylcytosine, but these covalent cytosine modifications are observed at much lower rates in the genome of mature cells compared to 5mC and 5hmC (1000 and 10,000 times less, respectively). Hence, the formyl and carboxyl modifications are not measured routinely. It is noteworthy that stem cells can have a substantial level of these oxidized forms during organogenesis, as they are involved in DNA demethylation, a process yielding activation of genes, which is needed for acquiring new cell type-specific features.

It is important to mention that, until recently, the most commonly used techniques in epigenetic analyses were based on the classical bisulfite conversion of the gDNA, which cannot distinguish between 5mC and 5hmC (see Fig. 6.2c). Therefore, the more accurate terminology for 5mC and its oxidized forms together is "modified cytosine" when referring to previous DNA methylation array or pyrosequencing data

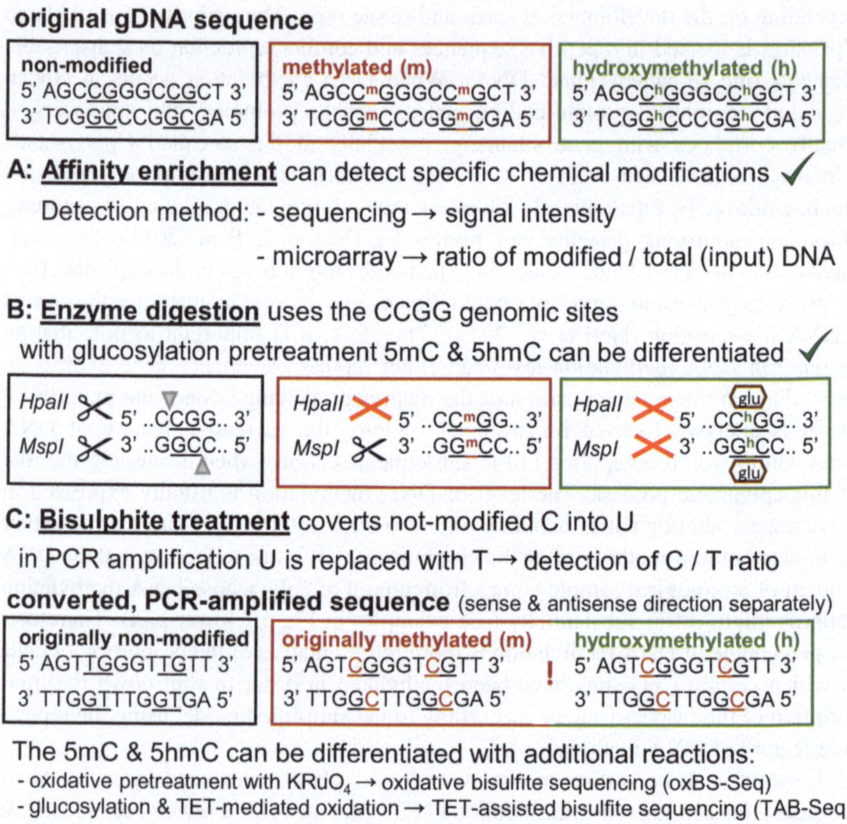


**Fig. 6.2** Techniques for detection of epigenetic modifications in human DNA. The non-modified cytosine base (C) can be methylated (m) in a CpG dinucleotide sequence (underlined), which can be further modified (hydroxymethylation: h). For details on the biochemical processes creating 5-methylcytosine (5mC) and 5-hydroxymethylcytosine (5hmC) see the overview of epigenetic modifications. (**a**) In the affinity enrichment methods, specific antibodies are applied in order to differentiate 5mC and 5hmC and also distinguish them from formyl- and carboxylcytosine. (**b**) Specific restriction enzymes can also differentiate covalently modified C from non-modified C. The scissors show when the CCGG sequence can be cut (∇ indicates cleavage site), whereas the red X indicates when an enzyme cannot cut the DNA. Both *HpaII* and *MspI* enzymes cleave the CCGG sequence with a non-modified CpG-site (left panel), but *HpaII* is blocked by any kind of chemical modification (other two panels). *MspI* can cut the CCGG sequence with 5mC or 5hmC in the middle, but its cleavage is blocked by glycosylated 5hmC (right panel). Therefore, the classical enzyme pair *HpaII* and *MspI* cleavage ratio gives information about the modified C (5mC & 5hmC together)/non-modified C level, but with a prior glycosylation (catalyzed by β-glucosyltransferase using UDP-glucose) the 5hmC can be differentiated from 5mC. (**c**) With the classical bisulfite treatment—used as a first step in bisulfite cloning, pyrosequencing, and at Infinium Human Methylation BeadChip arrays—the non-methylated C is converted to uracil (U is replaced by T in the PCR), whereas both 5mC and 5hmC stay as C. This way, the C/T ratio readout shows the proportion of modified C (5mC & 5hmC together)/non-modified C, which is an inaccurate measure for DNA methylation in certain tissues (indicated by a red exclamation mark). Note that after bisulfite conversion, the two parallel DNA strands are not complementary anymore, therefore one strand has to be selected for amplification and subsequent bisulfite sequencing (BS-seq). Additional chemical reactions can separate 5mC and 5hmC signals. For example, the oxidative pretreatment with potassium perruthenate ($KRuO_4$) changes 5hmC to 5-formylcytosine, which would be

where bisulfite converted samples were used. In newly developed methods, 5mC and 5hmC signals can be differentiated with additional chemical reactions (reviewed by Nestor, Reddington, Benson, & Meehan, 2014). On Fig. 6.2, green check marks indicate techniques which are specific enough to be used for accurate DNA methylation and hydroxymethylation analyses. However, most of the accumulated EWAS results obtained from blood samples can be still regarded relevant for DNA methylation data, because normal white blood cells have negligible 5hmC levels (around 0.02% of all bases, less than 5% of modified cytosines), hence one can argue that there is no need to distinguish it from the 5mC mark. There is less data on the different DNA modifications in oral mucosa, which can affect salivary 5mC and 5hmC levels (see the different cell types in saliva at Fig. 6.1). One study reported higher global 5hmC level in saliva compared to blood (0.036 vs. 0.027%, Godderis et al., 2015), but this is still in the range of <5% of DNA modifications. Although further studies with larger sample sizes are needed for more accurate estimates of salivary 5hmC levels, it seems that the proportion of 5hmC in saliva is similar to that in blood. In conclusion, the distinction between 5mC and 5hmC can be biologically important for certain cell types such as neurons and stem cells, but marginally important for others like leukocytes and buccal cells. Therefore, in the following section, only DNA methylation studies of saliva compared to blood samples would be discussed in detail.

## 6.8 Epigenetic Studies Measuring Environmental Effects in Salivary Samples

Besides its crucial role in embryogenesis, where innate developmental signals elicit epigenetic changes in a highly predictable pattern, DNA methylation is also responsive to the external environmental cues (Szyf & Bick, 2013). Growing evidence shows interindividual variation in every stage of life, even after birth in monozygotic twins. These differences between monozygotic twins potentially reflect the cumulative effects of environmental exposure (Tan, Christiansen, von Bornemann Hjelmborg, & Christensen, 2015). Analyzing monozygotic twin pairs is an important model in epigenetic studies, because it enables us to rule out the effects of

**Fig. 6.2** (continued) changed to U at the subsequent bisulfite conversion step, then to T in the PCR amplification (similarly as non-methylated C is converted to U, then to T). This technique is called as oxidative bisulfite sequencing (oxBS-Seq, developed by Booth et al., 2013) where the readout of 5mC is achieved, whereas for 5hmC detection the comparison of the classical BS-seq and oxBS-Seq information is needed. Another alternative method is the TET-assisted bisulfite sequencing (TAB-seq, Yu et al., 2012), which involves a β-glucosyltransferase-mediated protection of 5hmC (The big glucose moiety protects this modified C from further chemical reactions), and subsequent oxidation with recombinant TET enzyme, which turns 5mC to 5-carboxylcytosine (5caC). The following bisulfite treatment and PCR amplification would change both originally non-methylated C and 5caC (derived from 5mC) into T, whereas 5hmC is read as C

genetic polymorphisms. While DNA methylation is a dynamic and reversible process, it is the most stable epigenetic mark as it is part of the covalent structure of the DNA itself; therefore, it has the potential to serve as a biomarker. The overall stability of DNA methylation patterns has been shown in longitudinal study samples (Forest et al., 2018), although at certain CpG-sites there could be significant age-related changes (Horvath et al., 2016). Consequently, the use of DNA methylation analyses in association studies has been increasing over the last decade (for brief summaries of recent epigenetic publications visit https://www.whatisepigenetics.com/). These epigenetic studies are often conducted on surrogate peripheral tissues, such as blood or saliva. Since saliva is easily accessible, and sampling is less invasive than blood, a relevant direction in epigenetic research is to establish if the changes in DNA methylation patterns of saliva samples are comparable to that of the target tissue (e.g., liver or brain).

Although many studies reported high correlations in methylation levels between blood-derived and salivary DNA samples (reviewed by Langie et al., 2017), one should not forget that the majority of CpG-sites are located within repetitive sequences with high methylation levels and at CpG islands of housekeeping genes with low methylation levels, resulting in little variability within or between individuals. Hence, these sequences are not ideal for association studies measuring correlations between DNA methylation level and environmental exposure. A recent epigenome sequencing analysis demonstrated that only about 10% of the human CpG-sites showed interindividual variability, representing 2 million out of the 26.8 million autosomal CpG-sites (Hachiya et al., 2017). However, less than 2% of the total CpG-sites have been analyzed with previous array-based techniques, which measured 27–450,000 sites. The presently available DNA methylation array (EPIC BeadChip analyzing more than 850,000 sites) has increased number of CpG-sites but it is still cancer research oriented, meaning that it covers most of the human genes; it does not focus on regions which are variable between healthy individuals, and could be informative for association studies.

When using surrogate tissue for their analyses, researchers are warned that only a minority of the variable CpG-sites show correlations between the DNA methylation levels of different tissues, as it was shown for example in paired blood and brain samples (Hannon, Lunnon, Schalkwyk, & Mill, 2015). Interestingly, DNA methylation profiles of saliva samples were more similar to publicly available data of brain samples, compared to that of whole blood samples (Smith et al., 2015). Comparative DNA methylation analyses of matched brain, blood, saliva, and buccal samples showed high overall correlation between brain and peripheral tissue ($r = 0.90$ for saliva-brain, $r = 0.86$ for blood-brain, $r = 0.85$ for buccal-brain) when assessing the average methylation level at each CpG-site in a group of 21 patients undergoing brain resection (Braun et al., 2019). However, the proportion of CpG-sites showing significantly similar DNA methylation levels between the target and surrogate tissue was the highest in blood samples (20.8% compared to 17.4% in buccal and 15.1% in saliva samples). The main conclusion of this study is that the similarity of DNA methylation patterns of different tissues highly depends on the actual chromosomal region. Researchers can check the degree of cross-tissue correlation of the analyzed

CpG-sites on the study website (Iowa Methylation Array Graphing for Experimental Comparison of Peripheral tissue & Gray matter, IMAGE-CpG, at https://han-lab.org/methylation/default/imageCpG#). As for blood-saliva comparisons, epigenome-wide array data of paired samples showed that 2–4% of the assayed CpG-sites were differentially methylated (Langie et al., 2017). Thus, selection of informative CpG-sites is highly recommended for EWAS to: (1) reduce the number of analyzed loci, which would be crucial for the detection of moderate-small effects given the available statistical methods; (2) focus only on those gene regions which are responsive to environmental stimuli (i.e., the variable CpG-sites); (3) select CpG-sites with good reported correlations of DNA methylation level between the surrogate and target tissue.

Based on animal and human epigenetic studies, DNA methylation mechanisms are proposed to be involved in recording early life experiences, thus influencing gene expression in order to fine-tune the activity of physiological systems. In particular, the prenatal environment, where the majority of epigenetic modifications are established, can have long-lasting effects on DNA methylation patterns. This has been shown in relation to both physical and psychosocial environmental exposure (see reviews by Marsit, 2015; Nemoda & Szyf, 2017). Based on previous epigenome-wide and targeted DNA methylation analyses, it is hypothesized that epigenetic changes involved in life-long responses to the intrauterine and early life environment are system-wide; hence, potentially detectable in multiple tissues. For example, after the pioneering animal studies, psychosocial stress evoked DNA methylation changes have been reported at the glucocorticoid receptor gene promoter in human studies using different tissues (reviewed by Turecki & Meaney, 2016). Increased methylation at the 1F promoter region of the glucocorticoid receptor gene was associated with childhood adversity in brain hippocampal samples of deceased adults (McGowan et al., 2009). It was also associated with prenatal exposure to maternal stress in newborns' cord blood, and in infants' salivary samples, although the affected CpG-sites varied (see meta-analysis by Palma-Gudiel, Cordova-Palomera, Eixarch, Deuschle, & Fananas, 2015). Therefore, saliva could be a suitable surrogate tissue in DNA methylation analyses, enabling measurement from an early age, even from early infancy.

However, other studies using blood or saliva to assess epigenetic changes caused by different intrauterine environment (e.g., birth weight discordant monozygotic twins) did not show significant differences in DNA methylation patterns of adult twin pairs (Souren et al., 2013; Tan et al., 2014). It has to be emphasized that these studies analyzed approximately 450,000 sites (using Illumina's Infinium HumanMethylation450 BeadChip array), without reducing the informative CpG-sites in their statistical analyses, as it was later suggested by Edgar, Jones, Robinson, and Kobor (2017) in their data reduction method, which lists more than 100,000 non-variable CpG-sites in both blood and buccal epithelial cells. Considering the limitations of current genome-wide studies assessing thousands of sites with potentially small individual effects, it is not surprising that none of the associations reached statistical significance. In addition, epigenetic changes triggered by early life adversity could be overshadowed by later environmental exposure (As of note, 34-

and 63-year-old adults were analyzed in the mentioned twin EWAS yielding no significant associations). Studying a younger age group and using a reduced number of CpG-sites, birth weight discordance was associated with within-pair differences of salivary DNA methylation at genes involved in neurodevelopment, as well as with differences in brain shape and size of the adolescent MZ twins (Casey et al., 2017). Using another approach to reveal biological processes, Zaghlool et al. (2018) analyzed intermediate molecular phenotypes, including blood, urinary, and salivary metabolite levels, and reported associations with DNA methylation levels at selected CpG-sites previously linked to diabetes mellitus, obesity, and smoking. Salivary tyramine metabolite, for instance, was associated with CpG-sites linked to smoking.

Finally, there are still a lot of technical issues that must be considered when using surrogate peripheral tissues for DNA methylation analyses. For example, although the type of somatic cells used for genetic analysis is irrelevant, tissue type variation and intraindividual differences in cell composition of non-sorted biological samples can hide authentic epigenetic differences. In addition, DNA methylation levels in blood and saliva samples can be affected by age, sex, and ethnicity (Horvath et al., 2016), and also by genetic variants (i.e., methylation quantitative trait loci, for details, see Do et al., 2017). Therefore, proper data processing is necessary to control for heterogeneity in samples—even when a cohort is homogenous and the biological sample type is the same throughout an epigenetic study—because different cell composition ratios can still substantially affect DNA methylation patterns. When using epigenome-wide arrays, the different proportion of leukocytes (mostly granulocytes) and buccal epithelial cells in salivary samples can be adjusted by reference-based or reference-free statistical methods (Langie et al., 2017). Cell ratios can also be assessed by measuring specific markers selected from cell type-specific CpG-lists in candidate gene analyses (Eipel et al., 2016). Lastly, caution should be taken when interpreting differences in DNA methylation levels, since current laboratory methods measuring epigenetic marks can be biased. This can be a common issue when using bisulfite-converted templates due to the different chemical properties of C- and T-rich DNA strands. However, this technical problem can be easily detected with internal controls and solved by suitable correction methods (see Moskalev et al., 2011). In conclusion, with careful design and appropriate additional analytical steps, salivary DNA samples can be successfully applied in epigenetic association studies.

## 6.9 Future Directions and Opportunities

Based on recent (often negative) findings of GWAS, it seems that individual genetic factors linked to complex diseases or traits explain only a small proportion of the inherited component of phenotypic variance. To improve the ability to detect moderate effects, researchers in medical genetic fields are aiming at: (1) increasing sample size in specific GWAS cohorts and pool samples in international consortia which would allow for conducting meta- and mega-analyses to identify genetic variants with small effects; (2) applying more precise intermediate or endophenotypes, which are influenced by fewer genetic variants (Blanco-Gómez et al., 2016); (3) studying gene–gene and gene–environment interactions in order to reveal the "missing heritability" (Manolio et al., 2009). The use of saliva as a biospecimen seems valuable to these goals. Genetic studies over the last decade have shown that saliva is a reliable source to study inherited genetic variants present in every somatic cell of an individual. Moreover, because sequencing the coding gene regions (i.e., exome) or the whole genome in large patient cohorts is now a reality, studying rare genetic variants in the pathomechanisms of complex inheritance diseases became possible, thus supplementing present GWAS that measure common genetic polymorphisms. Importantly, the bioinformatic and statistical tools dealing with this enormous data have to be constantly updated in research laboratories. Fortunately, the research community is providing free program packages which can be easily applied (e.g., the R Project for Statistical Computing at https://www.r-project.org/).

In the wake of technical developments, the number of studies using saliva is likely to increase exponentially both in genetic and epigenetic analyses. However, caution is needed in DNA methylation studies due to numerous technical issues (Langie et al., 2017). In order to overcome the various biological and statistical challenges, improvement of bioinformatic analyses is continuously needed in this area of research. Fortunately, open access to publications has been increasing. These include databases of analytical procedures (e.g., European Bioinformatics Institute, https://www.ebi.ac.uk/) and genome-wide datasets (such as dbGAP, https://www.ncbi.nlm.nih.gov/gap), which are helping the research community to achieve scientific goals (see Complex Disease Epigenetics Group at https://www.epigenomicslab.com/). Similarly to genetic studies, the spread of sequencing methods could also widen the repertoire of analyzed CpG-sites in epigenetic association studies. Likewise, pooling different samples with the help of international consortia could facilitate the generalization of EWAS findings (Flanagan, 2015). Once the technical issues are controlled, epigenetic studies should offer great possibilities in disease prevention and management, as proposed by the developmental origins of health and disease (DOHaD) concept (Rosenfeld, 2015). This could be achieved, for instance, via longitudinal studies using epigenetic analyses at multiple time points starting with the in utero environment (assessed at birth), then in infancy and childhood, thus focusing on the most sensitive periods to adverse environmental effects. The potential outcomes of these studies would help intervention programs concentrate on specific time points. Furthermore, epigenetic changes are dynamic and could be modified by the reverse enzymatic processes later in life, potentially even in adulthood, hypothetically allowing epigenetic treatments. According to present theories of disease development, the early life environment can alter the genetically determined program to prepare the individual for the anticipated environment later in life (e.g., poor nutrition during early life would predict life-long undernutrition), prompting functional epigenetic changes. However, the adaptive responses may become maladaptive when there is an inconsistency between the anticipated and the real environments later in life, resulting in metabolic, cardiovascular, or mental health problems (Gluckman, Hanson, Cooper, & Thornburg, 2008). Finally, although there are still lots of technical obstacles in clinical epigenetics (Aslibekyan, Claas, & Arnett, 2015), linking specific epigenetic alterations to disease-specific gene expression changes in the background of common diseases would pave the way for the development of targeted epigenetic treatments (Szyf, 2015).

**Acknowledgments** I thank Zsanett Botos-Hofbauer graphic designer for drawing Fig. 6.1, and my colleagues and collaborators—especially Melissa Nicolosi—for their helpful comments. I would also like to acknowledge my Marie Curie International Outgoing Fellowship within the 7th European Community Framework Programme (FP7-PEOPLE-2010-IOF N° 276107), which gave me the possibility to explore epigenetic techniques.

## References

Abraham, J. E., Maranian, M. J., Spiteri, I., Russell, R., Ingle, S., Luccarini, C., … Caldas, C. (2012). Saliva samples are a viable alternative to blood samples as a source of DNA for high throughput genotyping. *BMC Medical Genomics, 5*, 19. https://doi.org/10.1186/1755-8794-5-19

Alaeddini, R., Walsh, S. J., & Abbas, A. (2010). Forensic implications of genetic analyses from degraded DNA—A review. *Forensic Science International Genetics, 4*(3), 148–157. https://doi.org/10.1016/j.fsigen.2009.09.007

Almouzni, G., & Cedar, H. (2016). Maintenance of epigenetic information. *Cold Spring Harbor Perspectives in Biology, 8*(5), a019372. https://doi.org/10.1101/cshperspect.a019372

Aslibekyan, S., Claas, S. A., & Arnett, D. K. (2015). Clinical applications of epigenetics in cardiovascular disease: The long road ahead. *Translational Research: The Journal of Laboratory and Clinical Medicine, 165*(1), 143–153. https://doi.org/10.1016/j.trsl.2014.04.004

Bahlo, M., Stankovich, J., Danoy, P., Hickey, P. F., Taylor, B. V., Browning, S. R., … Rubio, J. P. (2010). Saliva-derived DNA performs well in large-scale, high-density single-nucleotide polymorphism microarray studies. *Cancer Epidemiology, Biomarkers & Prevention, 19*(3), 794–798. https://doi.org/10.1158/1055-9965.EPI-09-0812

Bhutta, M. F., Hobson, L., Lambie, J., Scaman, E. S., Burton, M. J., Giele, H., … Furniss, D. (2013). Alternative recruitment strategies influence saliva sample return rates in community-based genetic association studies. *Annals of Human Genetics, 77*(3), 244–250. https://doi.org/10.1111/ahg.12009

Blanco-Gómez, A., Castillo-Lluva, S., Del Mar Saez-Freire, M., Hontecillas-Prieto, L., Mao, J. H., Castellanos-Martin, A., & Perez-Losada, J. (2016). Missing heritability of complex diseases: Enlightenment by genetic variants from intermediate phenotypes. *BioEssays: News and Reviews in Molecular, Cellular and Developmental Biology, 38*(7), 664–673. https://doi.org/10.1002/bies.201600084

Booth, M. J., Ost, T. W., Beraldi, D., Bell, N. M., Branco, M. R., Reik, W., & Balasubramanian, S. (2013). Oxidative bisulfite sequencing of 5-methylcytosine and 5-hydroxymethylcytosine. *Nature Protocols, 8*(10), 1841–1851. https://doi.org/10.1038/nprot.2013.115

Braun, P. R., Han, S., Hing, B., Nagahama, Y., Gaul, L. N., Heinzman, J. T., … Shinozaki, G. (2019). Genome-wide DNA methylation comparison between live human brain and peripheral tissues within individuals. *Translational Psychiatry, 9*, 47.

Casey, K. F., Levesque, M. L., Szyf, M., Ismaylova, E., Verner, M. P., Suderman, M., … Booij, L. (2017). Birth weight discordance, DNA methylation, and cortical morphology of adolescent monozygotic twins. *Human Brain Mapping, 38*(4), 2037–2050. https://doi.org/10.1002/hbm.23503

Craig, J. (2008). Complex diseases: Research and applications. *Nature Education, 1*(1), 184.
Deaton, A. M., & Bird, A. (2011). CpG islands and the regulation of transcription. *Genes & Development, 25*(10), 1010–1022. https://doi.org/10.1101/gad.2037511
Do, C., Shearer, A., Suzuki, M., Terry, M. B., Gelernter, J., Greally, J. M., & Tycko, B. (2017). Genetic-epigenetic interactions in cis: A major focus in the post-GWAS era. *Genome Biology, 18*(1), 120. https://doi.org/10.1186/s13059-017-1250-y
Durdiaková, J., Kamodyová, N., Ostatníková, D., Vlková, B., & Celec, P. (2012). Comparison of different collection procedures and two methods for DNA isolation from saliva. *Clinical Chemistry and Laboratory Medicine, 50*(4), 643–647. https://doi.org/10.1515/CCLM.2011.814
Edgar, R. D., Jones, M. J., Robinson, W. P., & Kobor, M. S. (2017). An empirically driven data reduction method on the human 450K methylation array to remove tissue specific non-variable CpGs. *Clinical Epigenetics, 9*, 11. https://doi.org/10.1186/s13148-017-0320-z
Eipel, M., Mayer, F., Arent, T., Ferreira, M. R., Birkhofer, C., Gerstenmaier, U., ... Wagner, W. (2016). Epigenetic age predictions based on buccal swabs are more precise in combination with cell type-specific DNA methylation signatures. *Aging, 8*(5), 1034–1048. https://doi.org/10.18632/aging.100972
Flanagan, J. M. (2015). Epigenome-wide association studies (EWAS): Past, present, and future. *Methods in Molecular Biology, 1238*, 51–63. https://doi.org/10.1007/978-1-4939-1804-1_3
Forest, M., O'Donnell, K. J., Voisin, G., Gaudreau, H., MacIsaac, J. L., McEwen, L. M., ... Greenwood, C. M. T. (2018). Agreement in DNA methylation levels from the Illumina 450K array across batches, tissues, and time. *Epigenetics, 13*(1), 19–32. https://doi.org/10.1080/15592294.2017.1411443
Genomes Project Consortium, Auton, A., Brooks, L. D., Durbin, R. M., Garrison, E. P., Kang, H. M., ... Abecasis, G. R. (2015). A global reference for human genetic variation. *Nature, 526*(7571), 68–74. https://doi.org/10.1038/nature15393
Gluckman, P. D., Hanson, M. A., Cooper, C., & Thornburg, K. L. (2008). Effect of in utero and early-life conditions on adult health and disease. *The New England Journal of Medicine, 359*(1), 61–73. https://doi.org/10.1056/NEJMra0708473
Godderis, L., Schouteden, C., Tabish, A., Poels, K., Hoet, P., Baccarelli, A. A., & Van Landuyt, K. (2015). Global methylation and hydroxymethylation in DNA from blood and saliva in healthy volunteers. *BioMed Research International, 2015*, 845041. https://doi.org/10.1155/2015/845041
Gonzalez-Bosquet, J., & Chanock, S. J. (2011). Basic principles and laboratory analysis of genetic variation. *IARC Scientific Publications, 163*, 99–120.
Gudiseva, H. V., Hansen, M., Gutierrez, L., Collins, D. W., He, J., Verkuil, L. D., ... O'Brien, J. M. (2016). Saliva DNA quality and genotyping efficiency in a predominantly elderly population. *BMC Medical Genomics, 9*, 17. https://doi.org/10.1186/s12920-016-0172-y
Hachiya, T., Furukawa, R., Shiwa, Y., Ohmomo, H., Ono, K., Katsuoka, F., ... Shimizu, A. (2017). Genome-wide identification of inter individually variable DNA methylation sites improves the efficacy of epigenetic association studies. *NPJ Genomic Medicine, 2*, 11. https://doi.org/10.1038/s41525-017-0016-5
Hannon, E., Lunnon, K., Schalkwyk, L., & Mill, J. (2015). Interindividual methylomic variation across blood, cortex, and cerebellum: Implications for epigenetic studies of neurological and neuropsychiatric phenotypes. *Epigenetics, 10*(11), 1024–1032. https://doi.org/10.1080/15592294.2015.1100786
Hansen, T. V., Simonsen, M. K., Nielsen, F. C., & Hundrup, Y. A. (2007). Collection of blood, saliva, and buccal cell samples in a pilot study on the Danish nurse cohort: Comparison of the response rate and quality of genomic DNA. *Cancer Epidemiology, Biomarkers & Prevention, 16*(10), 2072–2076. https://doi.org/10.1158/1055-9965.EPI-07-0611
Horvath, S., Gurven, M., Levine, M. E., Trumble, B. C., Kaplan, H., Allayee, H., ... Assimes, T. L. (2016). An epigenetic clock analysis of race/ethnicity, sex, and coronary heart disease. *Genome Biology, 17*(1), 171. https://doi.org/10.1186/s13059-016-1030-0

Hu, Y., Ehli, E. A., Nelson, K., Bohlen, K., Lynch, C., Huizenga, P., … Davies, G. E. (2012). Genotyping performance between saliva and blood-derived genomic DNAs on the DMET array: A comparison. *PLoS One, 7*(3), e33968. https://doi.org/10.1371/journal.pone.0033968

Huang, Y., & Rao, A. (2014). Connections between TET proteins and aberrant DNA modification in cancer. *Trends in Genetics, 30*(10), 464–474. https://doi.org/10.1016/j.tig.2014.07.005

Kappelman, M. D., Lange, A., Randell, R. L., Basta, P. V., Sandler, R. S., Laugesen, K., … Erichsen, R. (2018). Feasibility of salivary DNA collection in a population-based case-control study: A pilot study of pediatric Crohn's disease. *Clinical Epidemiology, 10*, 215–222. https://doi.org/10.2147/CLEP.S143322

Langie, S. A. S., Moisse, M., Declerck, K., Koppen, G., Godderis, L., Vanden Berghe, W., … De Boever, P. (2017). Salivary DNA methylation profiling: Aspects to consider for biomarker identification. *Basic & Clinical Pharmacology & Toxicology, 121*(Suppl 3), 93–101. https://doi.org/10.1111/bcpt.12721

Lorenz, T. C. (2012). Polymerase chain reaction: Basic protocol plus troubleshooting and optimization strategies. *Journal of Visualized Experiments, 63*, e3998. https://doi.org/10.3791/3998

Lowe, R., Gemma, C., Beyan, H., Hawa, M. I., Bazeos, A., Leslie, R. D., … Ramagopalan, S. V. (2013). Buccals are likely to be a more informative surrogate tissue than blood for epigenome-wide association studies. *Epigenetics, 8*(4), 445–454. https://doi.org/10.4161/epi.24362

Manolio, T. A., Collins, F. S., Cox, N. J., Goldstein, D. B., Hindorff, L. A., Hunter, D. J., … Visscher, P. M. (2009). Finding the missing heritability of complex diseases. *Nature, 461* (7265), 747–753. https://doi.org/10.1038/nature08494

Marsit, C. J. (2015). Influence of environmental exposure on human epigenetic regulation. *The Journal of Experimental Biology, 218*(Pt 1), 71–79. https://doi.org/10.1242/jeb.106971

McGowan, P. O., Sasaki, A., D'Alessio, A. C., Dymov, S., Labonte, B., Szyf, M., … Meaney, M. J. (2009). Epigenetic regulation of the glucocorticoid receptor in human brain associates with childhood abuse. *Nature Neuroscience, 12*(3), 342–348. https://doi.org/10.1038/nn.2270

Moskalev, E. A., Zavgorodnij, M. G., Majorova, S. P., Vorobjev, I. A., Jandaghi, P., Bure, I. V., & Hoheisel, J. D. (2011). Correction of PCR-bias in quantitative DNA methylation studies by means of cubic polynomial regression. *Nucleic Acids Research, 39*(11), e77. https://doi.org/10.1093/nar/gkr213

National Institutes of Health (Bethesda, MD, US) Biological Sciences Curriculum Study. (2007). NIH Curriculum Supplement Series [Internet] Understanding Human Genetic Variation. Retrieved from https://www.ncbi.nlm.nih.gov/books/NBK20363/

Nemoda, Z., Horvat-Gordon, M., Fortunato, C. K., Beltzer, E. K., Scholl, J. L., & Granger, D. A. (2011). Assessing genetic polymorphisms using DNA extracted from cells present in saliva samples. *BMC Medical Research Methodology, 11*, 170. https://doi.org/10.1186/1471-2288-11-170

Nemoda, Z., & Szyf, M. (2017). Epigenetic alterations and prenatal maternal depression. *Birth Defects Research, 109*(12), 888–897. https://doi.org/10.1002/bdr2.1081

Neri, F., Rapelli, S., Krepelova, A., Incarnato, D., Parlato, C., Basile, G., … Oliviero, S. (2017). Intragenic DNA methylation prevents spurious transcription initiation. *Nature, 543*(7643), 72–77. https://doi.org/10.1038/nature21373

Nestor, C. E., Ottaviano, R., Reddington, J., Sproul, D., Reinhardt, D., Dunican, D., … Meehan, R. R. (2012). Tissue type is a major modifier of the 5-hydroxymethylcytosine content of human genes. *Genome Research, 22*(3), 467–477. https://doi.org/10.1101/gr.126417.111

Nestor, C. E., Reddington, J. P., Benson, M., & Meehan, R. R. (2014). Investigating 5-hydroxymethylcytosine (5hmC): The state of the art. *Methods in Molecular Biology, 1094*, 243–258. https://doi.org/10.1007/978-1-62703-706-8_19

Nishita, D. M., Jack, L. M., McElroy, M., McClure, J. B., Richards, J., Swan, G. E., & Bergen, A. W. (2009). Clinical trial participant characteristics and saliva and DNA metrics. *BMC Medical Research Methodology, 9*, 71. https://doi.org/10.1186/1471-2288-9-71

Nunes, A. P., Oliveira, I. O., Santos, B. R., Millech, C., Silva, L. P., Gonzalez, D. A., … Barros, F. C. (2012). Quality of DNA extracted from saliva samples collected with the Oragene DNA self-collection kit. *BMC Medical Research Methodology, 12*, 65. https://doi.org/10.1186/1471-2288-12-65

Palma-Gudiel, H., Cordova-Palomera, A., Eixarch, E., Deuschle, M., & Fananas, L. (2015). Maternal psychosocial stress during pregnancy alters the epigenetic signature of the glucocorticoid receptor gene promoter in their offspring: A meta-analysis. *Epigenetics, 10*(10), 893–902. https://doi.org/10.1080/15592294.2015.1088630

Pulford, D. J., Mosteller, M., Briley, J. D., Johansson, K. W., & Nelsen, A. J. (2013). Saliva sampling in global clinical studies: The impact of low sampling volume on performance of DNA in downstream genotyping experiments. *BMC Medical Genomics, 6*, 20. https://doi.org/10.1186/1755-8794-6-20

Randell, R. L., Gulati, A. S., Cook, S. F., Martin, C. F., Chen, W., Jaeger, E. L., . . . Kappelman, M. D. (2016). Collecting biospecimens from an internet-based prospective cohort study of inflammatory bowel disease (CCFA partners): A feasibility study. *JMIR Research Protocols, 5* (1), e3. https://doi.org/10.2196/resprot.5171

Reiner, J., Karger, L., Cohen, N., Mehta, L., Edelmann, L., & Scott, S. A. (2017). Chromosomal microarray detection of constitutional copy number variation using saliva DNA. *The Journal of Molecular Diagnostics, 19*(3), 397–403. https://doi.org/10.1016/j.jmoldx.2016.11.006

Rethmeyer, J. A., Tan, X., Manzardo, A., Schroeder, S. R., & Butler, M. G. (2013). Comparison of biological specimens and DNA collection methods for PCR amplification and microarray analysis. *Clinical Chemistry and Laboratory Medicine, 51*(5), e79–e83. https://doi.org/10.1515/cclm-2012-0429

Rosenfeld, C. S. (Ed.). (2015). *The epigenome and developmental origins of health and disease*. London: Academic. ISBN: 9780128016725.

Rossen, L., Norskov, P., Holmstrom, K., & Rasmussen OF. (1992). Inhibition of PCR by components of food samples, microbial diagnostic assays and DNA-extraction solutions. *International Journal of Food Microbiology, 17*(1), 37–45.

Siravegna, G., Marsoni, S., Siena, S., & Bardelli, A. (2017). Integrating liquid biopsies into the management of cancer. *Nature Reviews Clinical Oncology, 14*(9), 531–548. https://doi.org/10.1038/nrclinonc.2017.14

Smith, A. K., Kilaru, V., Klengel, T., Mercer, K. B., Bradley, B., Conneely, K. N., . . . Binder, E. B. (2015). DNA extracted from saliva for methylation studies of psychiatric traits: Evidence tissue specificity and relatedness to brain. *American Journal of Medical Genetics Part B, Neuropsychiatric Genetics, 168B*(1), 36–44. https://doi.org/10.1002/ajmg.b.32278

Sobrino, B., Brion, M., & Carracedo, A. (2005). SNPs in forensic genetics: A review on SNP typing methodologies. *Forensic Science International, 154*(2–3), 181–194. https://doi.org/10.1016/j.forsciint.2004.10.020

Soriano-Tárraga, C., Jimenez-Conde, J., Giralt-Steinhauer, E., Ois, A., Rodriguez-Campello, A., Cuadrado-Godia, E., . . . GeneStroke "The Spanish Stroke Genetics Consortium". (2013). DNA isolation method is a source of global DNA methylation variability measured with LUMA. Experimental analysis and a systematic review. *PLoS One, 8*(4), e60750. https://doi.org/10.1371/journal.pone.0060750

Souren, N. Y., Lutsik, P., Gasparoni, G., Tierling, S., Gries, J., Riemenschneider, M., . . . Walter, J. (2013). Adult monozygotic twins discordant for intra-uterine growth have indistinguishable genome-wide DNA methylation profiles. *Genome Biology, 14*(5), R44. https://doi.org/10.1186/gb-2013-14-5-r44

Stranger, B. E., Stahl, E. A., & Raj, T. (2011). Progress and promise of genome wide association studies for human complex trait genetics. *Genetics, 187*(2), 367–383. https://doi.org/10.1534/genetics.110.120907

Sun, F., & Reichenberger, E. J. (2014). Saliva as a source of genomic DNA for genetic studies: Review of current methods and applications. *Oral Health and Dental Management, 13*(2), 217–222.

Szyf, M. (2015). Epigenetics, a key for unlocking complex CNS disorders? Therapeutic implications. *European Neuropsychopharmacology, 25*(5), 682–702. https://doi.org/10.1016/j.euroneuro.2014.01.009

Szyf, M., & Bick, J. (2013). DNA methylation: A mechanism for embedding early life experiences in the genome. *Child Development, 84*(1), 49–57. https://doi.org/10.1111/j.1467-8624.2012.01793.x

Tan, Q., Christiansen, L., von Bornemann Hjelmborg, J., & Christensen, K. (2015). Twin methodology in epigenetic studies. *The Journal of Experimental Biology, 218*(Pt 1), 134–139. https://doi.org/10.1242/jeb.107151

Tan, Q., Frost, M., Heijmans, B. T., von Bornemann Hjelmborg, J., Tobi, E. W., Christensen, K., & Christiansen, L. (2014). Epigenetic signature of birth weight discordance in adult twins. *BMC Genomics, 15*, 1062. https://doi.org/10.1186/1471-2164-15-1062

Thiede, C., Prange-Krex, G., Freiberg-Richter, J., Bornhauser, M., & Ehninger, G. (2000). Buccal swabs but not mouthwash samples can be used to obtain pretransplant DNA fingerprints from recipients of allogeneic bone marrow transplants. *Bone Marrow Transplantation, 25*(5), 575–577. https://doi.org/10.1038/sj.bmt.1702170

Turecki, G., & Meaney, M. J. (2016). Effects of the social environment and stress on glucocorticoid receptor gene methylation: A systematic review. *Biological Psychiatry, 79*(2), 87–96. https://doi.org/10.1016/j.biopsych.2014.11.022

Vesty, A., Biswas, K., Taylor, M. W., Gear, K., & Douglas, R. G. (2017). Evaluating the impact of DNA extraction method on the representation of human oral bacterial and fungal communities. *PLoS One, 12*(1), e0169877. https://doi.org/10.1371/journal.pone.0169877

Wu, X., & Zhang, Y. (2017). TET-mediated active DNA demethylation: Mechanism, function and beyond. *Nature Reviews Genetics, 18*(9), 517–534. https://doi.org/10.1038/nrg.2017.33

Yu, M., Hon, G. C., Szulwach, K. E., Song, C. X., Jin, P., Ren, B., & He, C. (2012). Tet-assisted bisulfite sequencing of 5-hydroxymethylcytosine. *Nature Protocols, 7*(12), 2159–2170. https://doi.org/10.1038/nprot.2012.137

Zaghlool, S. B., Mook-Kanamori, D. O., Kader, S., Stephan, N., Halama, A., Engelke, R., … Suhre, K. (2018). Deep molecular phenotypes link complex disorders and physiological insult to CpG methylation. *Human Molecular Genetics, 27*(6), 1106–1121. https://doi.org/10.1093/hmg/ddy006

# Chapter 7
# Saliva as a Window into the Human Oral Microbiome and Metabolome

**Heather Maughan and Katrine Whiteson**

**Abstract** The human oral cavity is colonized by one of the densest bacterial communities on the planet, with consistent microbial composition throughout life and in response to perturbations such as tooth cleaning. The oral microbiome has important associations with health in the context of gingivitis and even heart disease and preterm birth. In this chapter, we will provide an overview of what is known about the human oral microbiome, starting from the first glimpse of microbes when van Leeuwenhoek looked at his own oral microbes with the first microscope, to recent expansion of knowledge with both high-throughput sequencing and microscopy. We will describe what is known about healthy oral microbial communities, including how they develop in childhood and become shared within families and others who cohabitate, and how the biogeographic and biophysical forces drive construction of largely anaerobic microbial communities unique to each tooth. Finally, we touch on what is known about health-and disease-associated human oral microbial communities and briefly discuss the potential for using saliva samples to monitor health.

**Keywords** Oral microbiome · Metabolomics · Microbial communities

## 7.1 Introduction

Saliva comprises millions or billions of human and microbial cells along with molecules and proteins of various sizes. Salivary microbes are contributed from a variety of intraoral surfaces; there are about a thousand different kinds of bacteria in

H. Maughan
Ronin Institute, Montclair, NJ, USA

K. Whiteson (✉)
Department of Molecular Biology and Biochemistry, University of California at Irvine, Irvine, CA, USA
e-mail: katrine@uci.edu



D. A. Granger, M. K. Taylor (eds.), *Salivary Bioscience*,
https://doi.org/10.1007/978-3-030-35784-9_7

an average healthy mouth (Lazarevic, Whiteson, Francois, & Schrenzel, 2011), and more bacteria in one person's mouth than people on the Earth. The study of microbes living in our mouths began with the birth of microbiology, when in the late 1600s Antonie van Leeuwenhoek first observed a microbe in his dental plaque scrapings with a home-made microscope. Since that time, microbiologists have developed numerous methods for isolating and characterizing the microbes inhabiting the oral cavity. These centuries of research have taught us that oral microbial communities are dominated by Bacteria, along with their viruses, known as bacteriophages, and with some Archaea and Fungi. Oral microbial communities include a surprising proportion of anaerobic bacteria, are some of the most densely situated and most diverse communities on Earth, do not typically support pathogenic microbes, and sustain themselves through an incredible diversity of metabolisms.

Characterizing the interactions of oral microbes with each other and the human host is essential for improving oral and overall health, and offers a window into the functioning of one of the most diverse ecosystems on Earth. Fortunately, oral microbes can usually be studied with noninvasive sampling by swab or saliva collection, which provides aggregate material from the microhabitats within the oral cavity and is easily accessible and obtainable.

In this chapter, we will discuss the microbes and molecules found in the human oral cavity, along with the historical and modern approaches used to study them. We then look at the structure of oral microbial communities at several scales, starting from a single microbial cell on a tooth and then zooming out to view the oral microbial communities in individual people, families, and within a geographic region. Finally, we touch on the ways in which oral microbes influence our general health, and whether manipulating oral communities is feasible.

## 7.2 Methods for Sampling and Studying the Microbiome and Metabolome in Saliva

Since van Leeuwenhoek's first glimpse at oral microbes in the 1600s, sophisticated methods have been developed to sample and examine both live and dead microbes that reside in the oral cavity. Of particular importance is the development of culture techniques that enable isolation and study of anaerobic bacteria, and the development of high-throughput DNA sequencing methods for culture-independent identification of microbes (Fig. 7.1).

***Sampling Microbes in Saliva and Plaque*** Saliva and dental plaque are the most common samples obtained for studies of oral microbes. Saliva sampling is easier to carry out and is preferred in situations where teeth are not present (edentulous infants or adults) or when training someone to take a dental plaque sample consistently is not feasible. Saliva includes a mix of microbes contributed from several distinct niches in the oral cavity, along with others that arrive with food or drink. For this reason, saliva is sometimes considered a less stable sample. [However, in

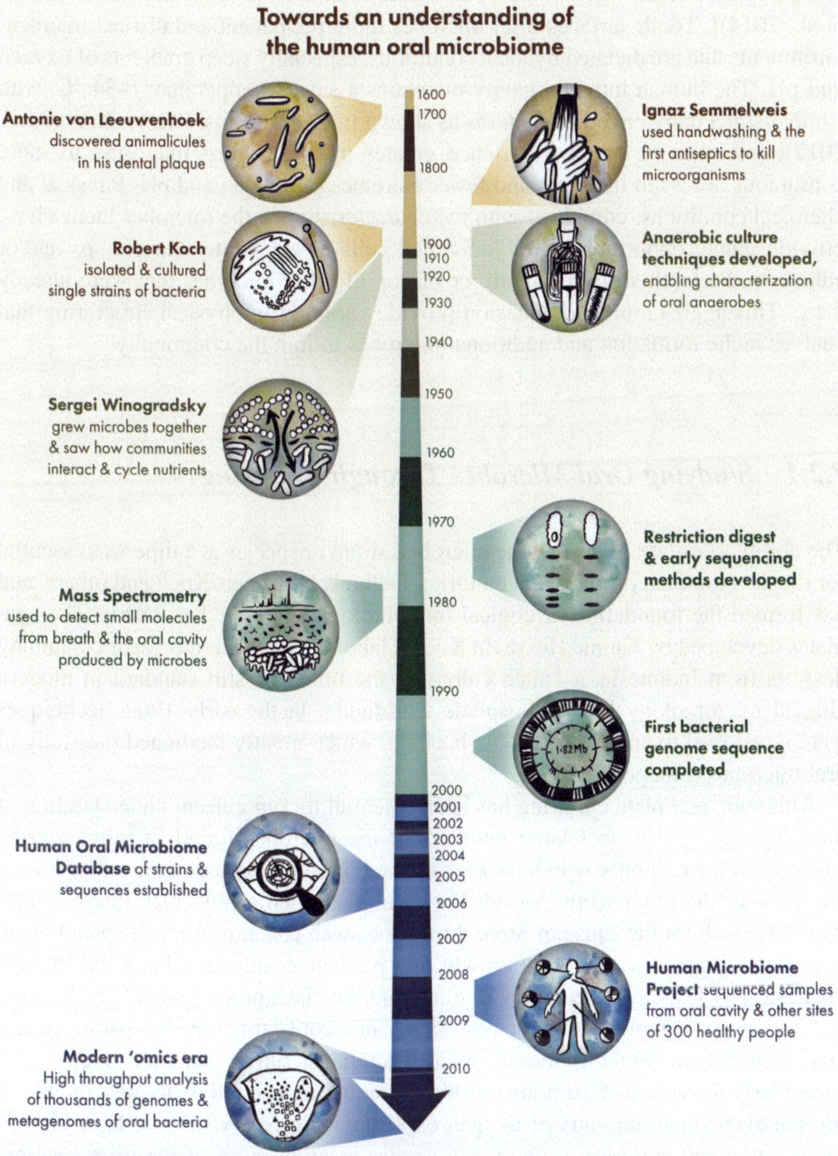


**Fig. 7.1** Timeline of milestones in oral microbiology. Beginning at the top in the 1600s and finishing in the present, major advances in the detection and characterization of microbes are shown. The top of the timeline focuses on the first microbiologists who invented some of the techniques still used today. Following in their footsteps were several major technological advances that enabled detection of microbes that had previously been elusive (e.g., anaerobes or other difficult to culture bacteria). Finally, the bottom of the timeline shows multinational collaborations formed to produce and house unprecedented levels of -omics data. Bars in the timeline are not to scale. Illustration credit: Dr. Eliza Wolfson (https://lizawolfson.co.uk)

comparison to fecal microbiomes, saliva microbiomes are relatively stable (David et al., 2014)]. Tooth surfaces offer microbes more permanent and distinct microenvironments that are dictated by local conditions, especially steep gradients of oxygen and pH. The human intraoral cavity maintains a steady temperature (~34 °C, with some distinct daily circadian patterns as shown in Choi, Lyons, Kieser, & Waddell, 2017); microbes in saliva experience greater mixing, which may lead to more continuous access to nutrients and fewer extremes of oxygen and pH. Physical and chemical conditions, combined with the characteristics of the microbes themselves, provide ample opportunities for individual cells to aggregate into clumps and/or adhere to the tooth surface directly or on top of other microbes that were already there. This aggregation and adhesion provide variation in physical structuring that enables niche formation and additional microbes to join the community.

### *7.2.1 Studying Oral Microbes Through Culturing*

The ability to isolate and grow one microbial strain or species at a time was essential for the development of microbial culturing methods by Robert Koch and others, and has formed the foundation of clinical microbiology since the late 1800s. The agar plates developed by Fannie Hesse (in Koch's laboratory) inspired by agar containing desserts from Indonesia, a Dutch colony at the time, are still standard in modern clinical microbiology labs in hospitals worldwide. In the early 1900s techniques were developed to culture anaerobic bacteria, which greatly facilitated the study of oral microbes (Finegold, 1993).

Although agar plate culturing has been essential for our current understanding of microbiology, and in fact large numbers of the microbes found in saliva can be cultured on them, it only works for a small fraction of microbes, and has thus biased our view of the microscopic world. For example, when a physician takes an oral swab to search for the cause of Strep throat, the swab contents are then spread on an agar plate designed to favor the growth of a particular pathogen. For these clinical microbiology tests, it is necessary to eliminate the "background signal" of bacteria that are nonpathogenic so that the physician can identify the infecting pathogen and treat their patient. For this reason, many bacteria in saliva had been overlooked, particularly those with fastidious nutritional requirements and anaerobes unable to survive even small amounts of oxygen exposure. In recent years, we have significantly improved culturing techniques in order to grow more of the microbes that inhabit the human body (Browne et al., 2016; Sibley et al., 2011).

Instead of the iconic culture flask or petri dish used to isolate a single strain of microbe, many investigators are improving culture models to closely mimic the structure of the oral cavity. One important component is to offer the microbes naturally occurring gradients of pH, oxygen, and nutrients, which have been shown to affect transcription and production of metabolites in oral microbes (Edlund et al., 2015; McLean et al., 2012). These gradients have been recreated in a fermentation model that uses rotating disks in rich media, which were then inoculated with pooled saliva samples; the microbes eventually formed stable and diverse populations (Hope & Wilson, 2006).

Another important aspect of culture is to offer conditions that promote the formation of biofilms, which are polysaccharide and protein-rich matrices that offer bacteria a structured environment; biofilms are a natural state for many oral microbes, e.g., plaque is biofilm. Classic biofilm culturing has relied on bacteria adhering to the walls in plastic 96-well plates. This method works for some bacterial types, such as *Pseudomonas aeruginosa* (O'toole & Kolter, 1998). However, some strains do not attach to a surface, but rather form small aggregates, which are hard to study in a surface attachment assay. New efforts to allow biofilms and aggregates of cells to form have been shown to recreate physiologically relevant gradients of oxygen and pH (Sonderholm et al., 2018).

Finally, efficiency is a key to making progress with culturing. Anna Edlund at the Craig Venter Institute has a higher throughput and well-characterized oral microbiome model culture system with saliva pooled from 10 people and growing in 96-well plates (Edlund et al., 2013). This system produced highly replicable data and enabled culturing of stable oral communities in biofilms, including many taxa that were previously considered unculturable.

### *7.2.2 Studying Oral Microbes Through Culture-Independent Approaches*

The discrepancy between what we routinely grow in the lab and the diversity of microbes observed in the native environment has been referred to as the "great plate anomaly" by environmental microbiologists since the mid-twentieth century. Scientists would observe enormous numbers of microbes in a natural sample, but only a few would grow on an agar plate in the lab. Attempting multiple culturing techniques was sometimes successful, but it was the revolution in DNA sequencing capacity that truly enabled our current high-tech inventories of the thousands of microbes living in saliva. In this section, we describe what we have learned from such high-throughput approaches that provide a relatively unbiased view of the microbes.

***Omics*** While sequencing the first human genome in the late twentieth century took multiple governments and nearly a billion dollars (National Human Genome Institute, 2016), today it requires only a desktop machine and hundreds of dollars. Indeed, new high-throughput sequencing methods have reduced the cost and time investment requirements at rates that surpass even the pace of improvements in computing power (Muir et al., 2016; Wetterstrand, 2019). Some of our most important glimpses at oral microbial communities came from DNA sequencing. Most of the data have come from amplicon sequencing, where a segment of a universal bacterial gene (the 16S rRNA gene) is targeted and sequenced from each bacterial genome in a sample. The resultant hundreds to millions of sequences are then analyzed to identify, and infer abundances for, the bacteria present, and to infer characteristics of the community's ecology using statistics. Amplicon-based approaches have also been used to identify Fungi and Archaea in oral samples, but do not work for viruses, as there are no universal viral genes. Instead, viruses are usually physically separated from the rest of the sample using filtration or density gradient ultracentrifugation, and then all the viral DNA is sequenced (Reyes, Semenkovich, Whiteson, Rohwer, & Gordon, 2012). Note that sequencing all of the DNA in microbial communities is also a commonly used approach for identifying all microorganisms present and inferring their functional or metabolic abilities (Belda-Ferre et al., 2012).

16S rRNA sequencing has shown us that saliva in most people includes *Streptococcus*, among other facultative anaerobes such as *Rothia mucilaginosa*, and also a diversity of anaerobic bacteria, including *Prevotella*, *Veillonella*, and *Gemella* (Human Microbiome Project, 2012; Lazarevic, Whiteson, Hernandez, Francois, & Schrenzel, 2010; Lim, Totsika, Morrison, & Punyadeera, 2017; Stahringer et al., 2012). Large comparative studies have shown that the resilience of oral communities after antibiotic treatment is high, and that saliva communities are less affected by antibiotics than gut communities (Zaura et al., 2015). In one study of more than 2000 Japanese adults, oral microbiomes were found to cluster into two main types that are associated with overall health, and confirm the widespread nature of the observations made to date with regard to the typical composition of bacteria in saliva, dominated by *Streptococcus* spp. and other facultative anaerobes (Takeshita et al., 2016).

Fungi frequently found (through ITS amplicon sequencing) include members of the genera *Candida* and *Aspergillus* (Ghannoum et al., 2010; Peters, Wu, Hayes, & Ahn, 2017), though a diversity of fungi were present in all sampled individuals. Fungal diversity or phyla present do not seem to vary with oral health (Peters et al., 2017). Archaea are also present, although they may be more common in cases of reduced immune health (Lepp et al., 2004).

***Microscopy*** Although microscopy was the first method used for examining bacteria and other microorganisms, it is normally hindered by the limited morphological differences between bacterial taxa. However, new microscopy methods that combine nucleic acid-binding fluorescent probes with confocal microscopy have quickly propelled microscopy to the forefront of microbiome research. Microscopy allows us to examine the physical structure, or biogeography, of microbial communities to see which cells are neighbors and whether cells are organized or random (discussed more in the next section). For example, Kolenbrander (Kolenbrander, 2011) used confocal microscopy to follow colonization dynamics of bacteria growing in flow cells with saliva as their only nutrient source.

***Metabolomics*** Saliva comprises small molecules/metabolites produced by the microbes and human host. Over a thousand such molecules have been identified with a variety of methods, including NMR, GC–MS and LC–MS, and they are profiled in the Saliva Metabolome Database (Dame et al., 2015). The saliva metabolome is a promising tool for disease diagnosis and health monitoring; biomarkers in saliva have been identified for many diseases or conditions, including diabetes, dementia, and cancer (Barnes et al., 2014; Figueira et al., 2016; Ishikawa et al., 2016). The importance and promise of using saliva metabolomes to inform clinical research is evidenced by the NIH Common Fund Metabolomics Program, which supports a metabolome database and multiple institutes for generating metabolomics data.

In summary, culture-dependent and culture-independent methods have provided the identities of microbes that live in the oral cavity, resulting in a catalog of species that are commonly found and maintained throughout an individual's life. Some of the metabolites produced by these microbes are also being investigated. These studies have resulted in the creation of several public repositories of data related to oral microbes. The expanded HOMD includes genomic and taxonomic information for over 700 microbial species detected in the oral cavity, pharynx, nasal passages, sinuses, and esophagus (Chen et al., 2010; Escapa et al., 2018). The Human Microbiome Project includes amplicon sequence data, shotgun metagenomic sequence data, and a repository of bacterial cultures that can be requested (Human Microbiome Project, 2012). The Oral Microbe Bank of China (OMBC) (Xian et al., 2018) houses hundreds of bacterial strains and human samples, both available for research purposes.

**Box 7.1 Sampling and Processing Saliva: Tips and Best Practices**
Standardization of sampling and methods is an important priority in the microbiome and metabolome fields (Knight et al., 2018). This is because DNA, RNA, metabolites, or even whole cells, are often altered upon sample storage. Decisions about the strategy for sample collection, storage, processing, and analysis can all alter the downstream results. During the early days of the human microbiome project, samples were processed and sequenced at four facilities, and the biggest signal in the early datasets emerged as the core facility and processing methods, rather than anything biological. Batch effects are still important, as are positive and negative controls. Comparisons of each step in the process of saliva analysis suggest that the earlier experimental steps, such as DNA extraction from saliva samples, have a bigger impact on the resulting data than later data processing steps (Lazarevic, Gaia, Girard, Francois, & Schrenzel, 2013). Regular meetings have been held at the National Institute of Standards and Technology, where an International Metagenomics and Microbiome Standards Alliance (IMMSA, 2019) has been established (Standards).

The top considerations when characterizing the salivary microbiome and metabolome include:

1. Keep everything consistent within a batch of samples that will be compared to one another. This includes sample collection timing, method, storage, and processing. Several studies suggest abstaining from food or drink for 1 h before collecting saliva, and collecting at the same time of day, usually in the morning (Dame et al., 2015; Lazarevic et al., 2010, 2011, 2013; Lim et al., 2017). Lim et al. (2017), show that food or drink does not affect DNA quality and quantity. Although optimal microbiome and metabolome

(continued)

**Box 7.1** (continued)

focused saliva collection times have not been established, microbiome profiles are quite consistent for an individual over the course of months to years (David et al., 2014; Human Microbiome Project Consortium, 2012; Lazarevic et al., 2010). Standardized collection kits for passive drool or swabs are available—it is important to save the entire swab and not just the flow through, because several common swabbing materials trap microbes.

2. Saliva collected with the intention of sequencing microbial DNA can be stored in a commercially available nucleic acid preservation buffer that will be frozen, preferably within 1 h of collection. For metabolomics, there are not commercial buffers to preserve metabolome composition at room temperature, and freezing the samples within an hour is the gold standard. Limited storage studies with similar samples, such as sputum from Cystic Fibrosis patients, have shown that metabolome composition shifts significantly within a few hours of storage at 4 °C (Wandro, Carmody, Gallagher, Lipuma, & Whiteson, 2017).
3. Including positive and negative controls, along with batch controls. Running blank samples using the same collection tubes and reagents to understand the background, especially for microbiome samples that will include DNA amplification steps, is more important for very low abundance samples. Saliva is densely colonized by microbes, but negative controls are still an important way to understand the background. Positive controls with known mixtures of microbes are commercially available and can be included with each batch. In a big study with many batches, making many aliquots of a large pooled representative sample that can be stored and included in every batch will allow for some comparison across batches.

## 7.3 Biogeography of Microbial Communities of the Oral Cavity

While most studies of the oral microbiome have relied on saliva samples, there are a few studies using 16S rRNA amplicon sequencing or microscopy that have highlighted the variability in microbial communities at several scales. Examining where particular species or genera of bacteria reside in comparison to others, i.e., their biogeography, is the first step to using these oral communities as signatures of health or markers of disease.

***Structured microbial communities form around a single tooth*** Mark Welch, Rossetti, Rieken, Dewhirst, and Borisy (2016) used fluorescently tagged bacteria and confocal microscopy to determine the structure of bacteria living in dental plaque. They found that filamentous cells of Corynebacteria, which formed central pillars, and nine other bacterial species radially emanated from the tooth surface

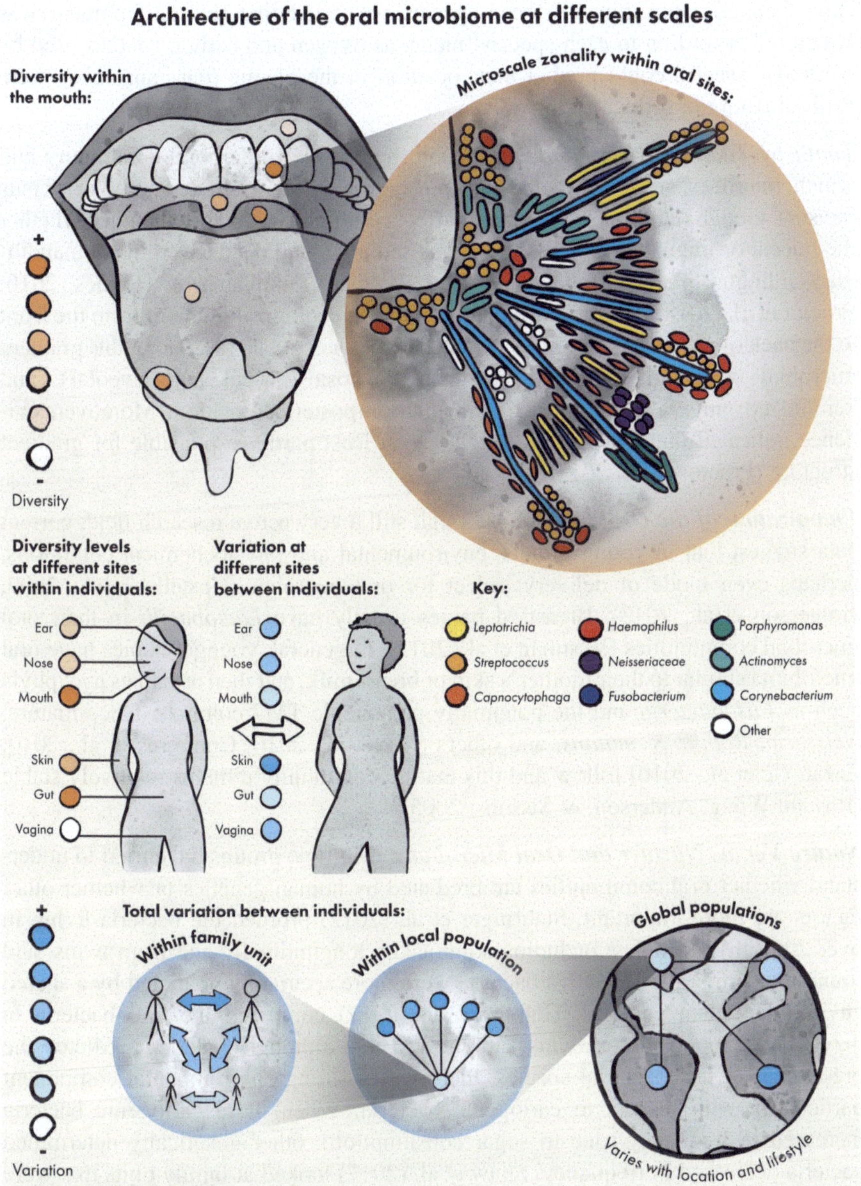


**Fig. 7.2** Architecture of the oral microbiome at different scales. Variation in microbial communities at differing scales. Beginning at the top in the oral cavity, levels of diversity are represented by orange circles. The tooth inset shows the organization of bacterial cells of different species based on microscopy data from Mark Welch et al. (2016). In the middle, diversity levels within an individual and variation in community diversity between individuals are shown with orange and blue circles, respectively. At the bottom, variation in community diversity between individuals of a family unit, a local population, and global populations are indicated by blue circles. Illustration credit: Dr. Eliza Wolfson (https://lizawolfson.co.uk)

(Fig. 7.2). The positions of these species suggested that this environment was structured according to each species' niche, as oxygen and carbon sources used by particular species could predict their position in the plaque (e.g., anaerobes were furthest from the surface).

***Tooth by Tooth Environment*** Each tooth has its own microbial community and which microbes are present depends most on type of tooth (e.g., molar versus incisor), though other things that could affect community structure included whether the buccal or lingual side of the tooth was sampled, and distance from submandibular/sublingual glands (Callahan, Proctor, Relman, Fukuyama, & Holmes, 2016; Proctor et al., 2018; Proctor & Relman, 2017). An ecological gradient from the front to the back of mouth has also been revealed (Proctor et al., 2018). Along this gradient microbial communities varied on teeth, mucosa (buccal and alveolar), and keratinized gingiva according to their anterior–posterior position. Moreover, evidence indicated that the flow of saliva was at least partly responsible for gradient structure (Proctor et al., 2018).

***Colonization of the Oral Cavity*** Although still a very active research field, current data suggest that in young infants, environmental and physiochemical conditions, perhaps even mode of delivery, select for oral microbes (Costello et al., 2009; Holgerson et al., 2013). Breastfed babies usually have *Lactobacilli* in their oral microbial communities (Vestman et al., 2013). In general, younger babies have oral microbiota similar to their mother's skin or breast milk, and then as babies age, phyla such as *Fusobacteria* and the potentially pathogenic TM7 colonize. Once mature, *Neisseria, Rothia, S. mutans*, and others (Bik et al., 2010; Contreras et al., 2010; Lazarevic et al., 2010) follow and this mature community remains relatively stable (Rasiah, Wong, Anderson, & Sissons, 2005).

***Nature Versus Nurture and Oral Microbiota*** Multiple groups have tried to understand whether oral communities are predicted by human genetics or whether other factors are more important. Stahringer et al. (2012) profiled the bacteria living in over 200 saliva samples, including some taken longitudinally and from twins, and found that similarities in saliva bacteria were more accurately predicted by a shared environment than genetics. Gomez et al. (2017) compared the oral bacteria in dizygotic or monozygotic twins and showed that although genetics predicted the presence of a few bacterial species, there was a strong environmental component particularly with respect to cariogenic bacteria. When these cariogenic bacteria increased in frequency (due to sugar consumption), other genetically determined bacteria decreased in frequency. Shaw et al. (2017) looked at family units that were cohabiting and found that cohabitation was the strongest predictor of oral microbial communities. This is consistent with results from other studies that found that sharing households was associated with very similar oral microbiota (e.g., Abeles et al., 2016; Song et al., 2013). Cohabitation was also associated with similar bacteriophage communities (Ly et al., 2016). Not surprisingly, diet plays a strong role in community composition (Hansen et al., 2018).

***Global and Ethnic Differences in the Oral Microbiota*** Four ethnic groups in the USA were shown to have specific bacterial signatures in their oral microbiomes; the distinct saliva microbiomes of Caucasian and African-Americans who shared a longer history of US diet and lifestyle heritage (in comparison to Chinese and Latino saliva donors who more recently immigrated to the USA) suggested that this was not due to environmental or dietary differences alone, although this is difficult to disentangle (Mason, Nagaraja, Camerlengo, Joshi, & Kumar, 2013). Thus, although within an ethnic group there is a weak influence of genetics on oral microbiota (e.g., twin study discussed above), comparisons between ethnic groups suggest that genetics may play some broader role. For example, Gupta, Paul, and Dutta (2017) suggested that differences in root and tooth morphologies and innate immunity between ethnic groups, and hunter-gatherer versus urban lifestyles, may influence which microbes can colonize and persist in the oral cavity.

***Individuality*** However, despite these comparisons at many levels, i.e., between twins, cohabiting relatives or strangers, and ethnic groups across the globe, the most important source of variation in any microbiome or metabolome study comes from which individual the sample was taken from. For example, Mukherjee, Beall, Griffen, and Leys (2018) showed that oral microbiomes are highly personalized. Although each person's oral microbiome did change over the year, they resulted in differences that were minimal compared to the differences between individuals. Thus, for studies that involve oral bacteria, carrying out cross-sectional studies is difficult because each individual is quite unique. This point reiterates the importance of having baseline samples as part of a longitudinal study.

## 7.4 The Oral Microbiota's Influence on Overall Health

Two major conceptual advances regarding oral microbes have occurred in the last few decades. First, doctors are beginning to understand that disease may be caused by a change in community structure, rather than the presence of an individual pathogen. In other words, Koch's postulates are not applicable to all infectious diseases. Second, we have recently begun to understand how oral health affects our overall health. There is reasonable evidence that oral microbes can enter the bloodstream via lesions in the oral cavity (associated with periodontitis) and then travel to various regions in the body (e.g., blood vessels, placenta, and colon). Oral microbes may also travel through inhalation (to the lungs) and swallowing (to the stomach/gut). Once these oral microbes arrive in their new home, they can significantly affect health.

*Streptococcus* species are a major cause of dental caries, along with a collection of other microbes (Belda-Ferre et al., 2012). These bacteria cause caries by consuming dietary sugar and producing acids that degrade tooth enamel. Inflammation of the gums (i.e., gingivitis, which can progress to periodontitis) is also typically a response to the bacterial biofilms lining the teeth. The cause of periodontitis was originally attributed to shifts in all microbes or due to infection by individual pathogens. However, investigations into oral communities of microbes have resulted in an ecological model of caries and periodontitis (reviewed in Kilian et al., 2016) where complex microbial communities interact with the immune system and dietary factors to eventually determine oral health.

In addition to the obvious influence of oral microbes on caries and periodontitis, these microbes and/or their associated oral diseases have been shown to affect overall health. For example, periodontitis is associated with progression to cardiovascular disease (Leishman, Do, & Ford, 2010), preterm and low birth weight (Puertas et al., 2018). Abundances of *Veillonella* and *Streptococcus* in atherosclerotic plaque correlated with their oral abundances (Koren et al., 2011). The presence of specific oral microbes may eventually be a useful predictor of atherosclerosis (Chhibber-Goel et al., 2016); however, at this point, more data are needed.

Oral bacteria have also been shown to colonize the intestines, which can cause issues in individuals with IBD (Atarashi et al., 2017) or colon cancer (Flynn, Baxter, & Schloss, 2016). In individuals with colon cancer, biofilms described as "oral-like" were identified on the colonic mucosa (Flemer et al., 2018). Finally, respiratory infections are also affected by oral microbes. Half of all cases of pneumonia have oral etiology (Yamasaki et al., 2013), and the proximity of the densely colonized oral cavity to the airways has important implications, especially when immune health and airway clearance are compromised. Part of health is maintaining distinct microbial communities in different parts of the oropharynx and the airways. Aging and alcoholism have both been shown to lead to blurred microbial communities (Samuelson et al., 2018; Thevaranjan et al., 2018), where the oropharynx and the nasopharynx, for example, are no longer distinct.

The diversity of microbes present in and on the human body may be important for health, although defining the composition and diversity of a microbiome that promotes health is still an underexplored frontier. In some regions of the body, for example, in the gut, greater microbial diversity is associated with health. This also appears to be the case for the distinct communities of microbes colonizing dental plaque in the oral cavity, where increased diversity was found in healthy individuals compared to those with caries (Espinoza et al., 2018). However, in saliva, the opposite has been found to be true, where greater diversity is associated with gingivitis (Takeshita et al., 2016).

## 7.5 Summary of Current Knowledge and a Look to the Future

In summary, the oral microbiome is a diverse population of microbes that live in various niches within the oral cavity. Often the community of microbes is specific for each individual, and cohabiting individuals tend to share many oral microbes.

Saliva samples are accessible, and have contributed to our knowledge of oral microbes, in addition to samples of dental plaque and other oral specimens.

Although we have learned a lot since Antonie van Leeuwenhoek first gazed at his oral microbes under the microscope, there is still much to learn. First, we have a good idea of which bacteria are common oral inhabitants, but we know less about what these diverse organisms are doing in the mouth and whether they are beneficial, detrimental, or neutral, and if their roles change as health improves or declines. Second, associations between the presence of particular microbes and health need to be better understood. Improved culturing, high-throughput sequencing, metabolic analyses, and microscopy will no doubt enable a deeper look into these associations to tease apart correlation and causation. Third, as we identify particular microbes that have a significant role in oral or overall health, we can begin to manipulate their abundances to improve health. Saliva will no doubt play a crucial role in all of these future endeavors.

Definitions of key terms:

- *Microbiome*: The microbes living in a given environment, including viruses, bacteria, archaea, and small eukaryotes.
- *Metabolome*: The metabolites produced by all organisms living in a given environment.
- *16S rRNA gene*: A gene that is highly conserved in all organisms and is often amplified and sequenced to identify microorganisms. It is the gene most often used for culture-independent study of microbial communities.
- *Metagenome*: All of the genes encoded by the organisms living in a given environment.
- *High-throughput sequencing*: DNA sequencing technologies that produce large amounts of sequence data in short amounts of time. These technologies are typically much lower in cost than earlier methods.

## References


Abeles, S. R., Jones, M. B., Santiago-Rodriguez, T. M., Ly, M., Klitgord, N., Yooseph, S., … Pride, D. T. (2016). Microbial diversity in individuals and their household contacts following typical antibiotic courses. *Microbiome, 4*, 39.

Atarashi, K., Suda, W., Luo, C., Kawaguchi, T., Motoo, I., Narushima, S., … Honda, K. (2017). Ectopic colonization of oral bacteria in the intestine drives TH1 cell induction and inflammation. *Science, 358*, 359–365.

Barnes, V. M., Kennedy, A. D., Panagakos, F., Devizio, W., Trivedi, H. M., Jonsson, T., … Scannapieco, F. A. (2014). Global metabolomic analysis of human saliva and plasma from healthy and diabetic subjects, with and without periodontal disease. *PLoS One, 9*, e105181.

Belda-Ferre, P., Alcaraz, L. D., Cabrera-RUBIO, R., Romero, H., Simon-Soro, A., Pignatelli, M., & Mira, A. (2012). The oral metagenome in health and disease. *The ISME Journal, 6*, 46–56.

Bik, E. M., Long, C. D., Armitage, G. C., Loomer, P., Emerson, J., Mongodin, E. F., … Relman, D. A. (2010). Bacterial diversity in the oral cavity of 10 healthy individuals. *The ISME Journal, 4*, 962–974.

Browne, H. P., Forster, S. C., Anonye, B. O., Kumar, N., Neville, B. A., Stares, M. D., … Lawley, T. D. (2016). Culturing of 'unculturable' human microbiota reveals novel taxa and extensive sporulation. *Nature, 533*, 543–546.

Callahan, B., Proctor, D., Relman, D., Fukuyama, J., & Holmes, S. (2016). Reproducible research workflow in R for the analysis of personalized human microbiome data. *Pacific Symposium on Biocomputing, 21*, 183–194.

Chen, T., Yu, W. H., Izard, J., Baranova, O. V., Lakshmanan, A., & Dewhirst, F. E. (2010). The Human Oral Microbiome Database: a web accessible resource for investigating oral microbe taxonomic and genomic information. *Database (Oxford), 2010*, baq013.

Chhibber-Goel, J., Singhal, V., Bhowmik, D., Vivek, R., Parakh, N., Bhargava, B., & Sharma, A. (2016). Linkages between oral commensal bacteria and atherosclerotic plaques in coronary artery disease patients. *NPJ Biofilms Microbiomes, 2*, 7.

Choi, J. E., Lyons, K. M., Kieser, J. A., & Waddell, N. J. (2017). Diurnal variation of intraoral pH and temperature. *BDJ Open, 3*, 17015.

Contreras, M., Costello, E. K., Hidalgo, G., Magris, M., Knight, R., & Dominguez-Bello, M. G. (2010). The bacterial microbiota in the oral mucosa of rural Amerindians. *Microbiology, 156*, 3282–3287.

Costello, E. K., Lauber, C. L., Hamady, M., Fierer, N., Gordon, J. I., & Knight, R. (2009). Bacterial community variation in human body habitats across space and time. *Science, 326*, 1694–1697.

Dame, Z. T., Aziat, F., Mandal, R., Krishnamurthy, R., Bouatra, S., Borzouie, S., … Wishart, D. S. (2015). The human saliva metabolome. *Metabolomics, 11*, 1864–1883.

David, L. A., Materna, A. C., Friedman, J., Campos-Baptista, M. I., Blackburn, M. C., Perrotta, A., … Alm, E. J. (2014). Host lifestyle affects human microbiota on daily timescales. *Genome Biology, 15*, R89.

Edlund, A., Yang, Y., Hall, A. P., Guo, L., Lux, R., He, X., … Mclean, J. S. (2013). An in vitro biofilm model system maintaining a highly reproducible species and metabolic diversity approaching that of the human oral microbiome. *Microbiome, 1*, 25.

Edlund, A., Yang, Y., Yooseph, S., Hall, A. P., Nguyen, D. D., Dorrestein, P. C., … Mclean, J. S. (2015). Meta-omics uncover temporal regulation of pathways across oral microbiome genera during in vitro sugar metabolism. *The ISME Journal, 9*, 2605–2619.

Escapa, I. F., Chen, T., Huang, Y., Gajare, P., Dewhirst, F. E., & Lemon, K. P. (2018). New insights into human nostril microbiome from the expanded Human Oral Microbiome Database (eHOMD): A resource for the microbiome of the human aerodigestive tract. *mSystems, 3*, e00187.

Espinoza, J. L., Harkins, D. M., Torralba, M., Gomez, A., Highlander, S. K., Jones, M. B., … Dupont, C. L. (2018). Supragingival plaque microbiome ecology and functional potential in the context of health and disease. *MBio, 9*, e01631.

Figueira, J., Jonsson, P., Nordin Adolfsson, A., Adolfsson, R., Nyberg, L., & Ohman, A. (2016). NMR analysis of the human saliva metabolome distinguishes dementia patients from matched controls. *Molecular BioSystems, 12*, 2562–2571.

Finegold, S. M. (1993). A century of anaerobes: A look backward and a call to arms. *Clinical Infectious Diseases, 16*(Suppl 4), S453–S457.

Flemer, B., Warren, R. D., Barrett, M. P., Cisek, K., Das, A., Jeffery, I. B., … O'toole, P. W. (2018). The oral microbiota in colorectal cancer is distinctive and predictive. *Gut, 67*, 1454–1463.

Flynn, K. J., Baxter, N. T., & Schloss, P. D. (2016). Metabolic and community synergy of oral bacteria in colorectal cancer. *mSphere, 1*, e00102.

Ghannoum, M. A., Jurevic, R. J., Mukherjee, P. K., Cui, F., Sikaroodi, M., Naqvi, A., & Gillevet, P. M. (2010). Characterization of the oral fungal microbiome (mycobiome) in healthy individuals. *PLoS Pathogens, 6*, e1000713.

Gomez, A., Espinoza, J. L., Harkins, D. M., Leong, P., Saffery, R., Bockmann, M., … Nelson, K. E. (2017). Host genetic control of the oral microbiome in health and disease. *Cell Host and Microbe, 22*(269–278), e3.

Gupta, V. K., Paul, S., & Dutta, C. (2017). Geography, ethnicity or subsistence-specific variations in human microbiome composition and diversity. *Frontiers in Microbiology, 8*, 1162.

Hansen, T. H., Kern, T., Bak, E. G., Kashani, A., Allin, K. H., Nielsen, T., . . . Pedersen, O. (2018). Impact of a vegan diet on the human salivary microbiota. *Scientific Reports, 8*, 5847.

Holgerson, P. L., Vestman, N. R., Claesson, R., Ohman, C., Domellof, M., Tanner, A. C., . . . Johansson, I. (2013). Oral microbial profile discriminates breast-fed from formula-fed infants. *Journal of Pediatric Gastroenterology and Nutrition, 56*, 127–136.

Hope, C. K., & Wilson, M. (2006). Biofilm structure and cell vitality in a laboratory model of subgingival plaque. *Journal of Microbiological Methods, 66*, 390–398.

Human Microbiome Project Consortium. (2012). Structure, function and diversity of the healthy human microbiome. *Nature, 486*, 207–214.

International Metagenomics and Microbiome Standards. (2019). Accessed January 2019, from https://microbialstandards.org/

Ishikawa, S., Sugimoto, M., Kitabatake, K., Sugano, A., Nakamura, M., Kaneko, M., . . . Iino, M. (2016). Identification of salivary metabolomic biomarkers for oral cancer screening. *Scientific Reports, 6*, 31520.

Kilian, M., Chapple, I. L., Hannig, M., Marsh, P. D., Meuric, V., Pedersen, A. M., . . . Zaura, E. (2016). The oral microbiome – An update for oral healthcare professionals. *British Dental Journal, 221*, 657–666.

Knight, R., Vrbanac, A., Taylor, B. C., Aksenov, A., Callewaert, C., Debelius, J., . . . Dorrestein, P. C. (2018). Best practices for analysing microbiomes. *Nature Reviews. Microbiology, 16*, 410–422.

Kolenbrander, P. E. (2011). Multispecies communities: Interspecies interactions influence growth on saliva as sole nutritional source. *International Journal of Oral Science, 3*, 49–54.

Koren, O., Spor, A., Felin, J., Fak, F., Stombaugh, J., Tremaroli, V., . . . Backhed, F. (2011). Human oral, gut, and plaque microbiota in patients with atherosclerosis. *Proceedings of the National Academy of Sciences of the United States of America, 108*(Suppl 1), 4592–4598.

Lazarevic, V., Gaia, N., Girard, M., Francois, P., & Schrenzel, J. (2013). Comparison of DNA extraction methods in analysis of salivary bacterial communities. *PLoS One, 8*, e67699.

Lazarevic, V., Whiteson, K., Francois, P., & Schrenzel, J. (2011). The salivary microbiome assessed by a high-throughput and culture-independent approach. *Journal of Integrated Omics, 1*, 28–35.

Lazarevic, V., Whiteson, K., Hernandez, D., Francois, P., & Schrenzel, J. (2010). Study of inter- and intra-individual variations in the salivary microbiota. *BMC Genomics, 11*, 523.

Leishman, S. J., Do, H. L., & Ford, P. J. (2010). Cardiovascular disease and the role of oral bacteria. *Journal of Oral Microbiology, 2*.

Lepp, P. W., Brinig, M. M., Ouverney, C. C., Palm, K., Armitage, G. C., & Relman, D. A. (2004). Methanogenic archaea and human periodontal disease. *Proceedings of the National Academy of Sciences of the United States of America, 101*, 6176–6181.

Lim, Y., Totsika, M., Morrison, M., & Punyadeera, C. (2017). The saliva microbiome profiles are minimally affected by collection method or DNA extraction protocols. *Scientific Reports, 7*, 8523.

Ly, M., Jones, M. B., Abeles, S. R., Santiago-Rodriguez, T. M., Gao, J., Chan, I. C., . . . Pride, D. T. (2016). Transmission of viruses via our microbiomes. *Microbiome, 4*, 64.

Mark Welch, J. L., Rossetti, B. J., Rieken, C. W., Dewhirst, F. E., & Borisy, G. G. (2016). Biogeography of a human oral microbiome at the micron scale. *Proceedings of the National Academy of Sciences of the United States of America, 113*, E791–E800.

Mason, M. R., Nagaraja, H. N., Camerlengo, T., Joshi, V., & Kumar, P. S. (2013). Deep sequencing identifies ethnicity-specific bacterial signatures in the oral microbiome. *PLoS One, 8*, e77287.

Mclean, J. S., Fansler, S. J., Majors, P. D., Mcateer, K., Allen, L. Z., Shirtliff, M. E., . . . Shi, W. (2012). Identifying low pH active and lactate-utilizing taxa within oral microbiome communities from healthy children using stable isotope probing techniques. *PLoS One, 7*, e32219.

Muir, P., Li, S., Lou, S., Wang, D., Spakowicz, D. J., Salichos, L., . . . Gerstein, M. (2016). The real cost of sequencing: Scaling computation to keep pace with data generation. *Genome Biology, 17*, 53.

Mukherjee, C., Beall, C. J., Griffen, A. L., & Leys, E. J. (2018). High-resolution ISR amplicon sequencing reveals personalized oral microbiome. *Microbiome, 6*, 153.

National Human Genome Research Institute. (2016). *The cost of sequencing a human genome*. Accessed January 2019, from https://www.genome.gov/27565109/the-cost-of-sequencing-a-human-genome/

O'toole, G. A., & Kolter, R. (1998). Initiation of biofilm formation in *Pseudomonas fluorescens* WCS365 proceeds via multiple, convergent signalling pathways: A genetic analysis. *Molecular Microbiology, 28*, 449–461.

Peters, B. A., Wu, J., Hayes, R. B., & Ahn, J. (2017). The oral fungal mycobiome: Characteristics and relation to periodontitis in a pilot study. *BMC Microbiology, 17*, 157.

Proctor, D. M., Fukuyama, J. A., Loomer, P. M., Armitage, G. C., Lee, S. A., Davis, N. M., . . . Relman, D. A. (2018). A spatial gradient of bacterial diversity in the human oral cavity shaped by salivary flow. *Nature Communications, 9*, 681.

Proctor, D. M., & Relman, D. A. (2017). The landscape ecology and microbiota of the human nose, mouth, and Throat. *Cell Host Microbe, 21*, 421–432.

Puertas, A., Magan-Fernandez, A., Blanc, V., Revelles, L., O'valle, F., Pozo, E., . . . Mesa, F. (2018). Association of periodontitis with preterm birth and low birth weight: A comprehensive review. *The Journal of Maternal-Fetal & Neonatal Medicine, 31*, 597–602.

Rasiah, I. A., Wong, L., Anderson, S. A., & Sissons, C. H. (2005). Variation in bacterial DGGE patterns from human saliva: Over time, between individuals and in corresponding dental plaque microcosms. *Archives of Oral Biology, 50*, 779–787.

Reyes, A., Semenkovich, N. P., whiteson, K., Rohwer, F., & Gordon, J. I. (2012). Going viral: Next-generation sequencing applied to phage populations in the human gut. *Nature Reviews. Microbiology, 10*, 607–617.

Samuelson, D. R., Burnham, E. L., Maffei, V. J., Vandivier, R. W., Blanchard, E. E., Shellito, J. E., . . . Welsh, D. A. (2018). The respiratory tract microbial biogeography in alcohol use disorder. *American Journal of Physiology. Lung Cellular and Molecular Physiology, 314*, L107–L117.

Shaw, L., Ribeiro, A. L. R., Levine, A. P., Pontikos, N., Balloux, F., Segal, A. W., . . . Smith, A. M. (2017). The human salivary microbiome is shaped by shared environment rather than genetics: Evidence from a large family of closely related individuals. *MBio, 8*, e01237.

Sibley, C. D., Grinwis, M. E., Field, T. R., Eshaghurshan, C. S., Faria, M. M., Dowd, S. E., . . . Surette, M. G. (2011). Culture enriched molecular profiling of the cystic fibrosis airway microbiome. *PLoS One, 6*, e22702.

Sonderholm, M., Koren, K., Wangpraseurt, D., Jensen, P. O., Kolpen, M., Kragh, K. N., . . . Kuhl, M. (2018). Tools for studying growth patterns and chemical dynamics of aggregated *Pseudomonas aeruginosa* exposed to different electron acceptors in an alginate bead model. *NPJ Biofilms Microbiomes, 4*, 3.

Song, S. J., Lauber, C., Costello, E. K., Lozupone, C. A., Humphrey, G., Berg-Lyons, D., . . . Knight, R. (2013). Cohabiting family members share microbiota with one another and with their dogs. *eLife, 2*, e00458.

Stahringer, S. S., Clemente, J. C., Corley, R. P., Hewitt, J., Knights, D., Walters, W. A., . . . KRAUTER, K. S. (2012). Nurture trumps nature in a longitudinal survey of salivary bacterial communities in twins from early adolescence to early adulthood. *Genome Research, 22*, 2146–2152.

Takeshita, T., Kageyama, S., Furuta, M., Tsuboi, H., Takeuchi, K., Shibata, Y., . . . Yamashita, Y. (2016). Bacterial diversity in saliva and oral health-related conditions: The Hisayama study. *Scientific Reports, 6*, 22164.

Thevaranjan, N., Puchta, A., Schulz, C., Naidoo, A., Szamosi, J. C., Verschoor, C. P., . . . Bowdish, D. M. E. (2018). Age-associated microbial dysbiosis promotes intestinal permeability, systemic inflammation, and macrophage dysfunction. *Cell Host Microbe, 23*, 570.

Vestman, N. R., timby, N., Holgerson, P. L., Kressirer, C. A., Claesson, R., Domellof, M., ... Johansson, I. (2013). Characterization and in vitro properties of oral lactobacilli in breastfed infants. *BMC Microbiology, 13*, 193.

Wandro, S., Carmody, L., Gallagher, T., Lipuma, J. J., & Whiteson, K. (2017). Making it last: Storage time and temperature have differential impacts on metabolite profiles of airway samples from cystic fibrosis patients. *mSystems, 2*, e00100.

Wetterstrand, K. A. (2019) *Data from the NHGRI Genome Sequencing Program (GSP)*. Accessed January 2019, from *www.genome.gov/sequencingcostsdata*.

Xian, P., Xuedong, Z., Xin, X., Yuqing, L., Yan, L., Jiyao, L., ... Ga, L. (2018). The Oral Microbiome Bank of China. *International Journal of Oral Science, 10*, 16.

Yamasaki, K., Kawanami, T., Yatera, K., Fukuda, K., Noguchi, S., Nagata, S., ... Mukae, H. (2013). Significance of anaerobes and oral bacteria in community-acquired pneumonia. *PLoS One, 8*, e63103.

Zaura, E., Brandt, B. W., Teixeira DE Mattos, M. J., Buijs, M. J., Caspers, M. P., Rashid, M. U., ... Crielaard, W. (2015). Same exposure but two radically different responses to antibiotics: Resilience of the salivary microbiome versus long-term microbial shifts in feces. *MBio, 6*, e01693–e01615.

# Chapter 8
# Salivaomics, Saliva-Exosomics, and Saliva Liquid Biopsy

**Jordan Cheng, Taichiro Nonaka, Qianlin Ye, Fang Wei, and David T. W. Wong**

**Abstract** The concept of liquid biopsy has emerged into the cancer lexicon to describe detected alterations of tumor biomarkers within body fluids which reflect the presence and the biology of cancer. This is typically performed by assessing circulating tumor cells (CTCs), circulating tumor DNAs (ctDNAs), tumor-derived extracellular vesicles (EVs), microRNAs (miRNAs), and proteins. Although plasma, urine, and cerebral spinal fluid (CSF) are all viable biofluids, growing attention has recently been cast on saliva. Saliva is readily available, can be obtained noninvasively, is easily collected and stored, and also demonstrates compelling pathophysiological association with systemic diseases. Importantly, saliva liquid biopsy delivers the best clinical performance to detect ctDNAs in lung cancer patients. In conjunction with validated biomarkers, and reliable and robust analytical detection tools, saliva has the potential to pioneer a new landscape of real-time point-of-care testing in personalized medicine. This chapter will review the salivaomics, the disease-related biomarker properties of saliva, and the scientific advances in the pathophysiological foundations which make saliva an ideal candidate for noninvasive liquid biopsy. Moreover, a novel liquid biopsy technology termed electric field-induced release and measurement (EFIRM), which works in tandem with the technological advances to extract crucial disease information in saliva, is reviewed.

**Keywords** Liquid biopsy · Salivaomics · Saliva-exosomics · Biomarker · Cancer · Point-of-care

J. Cheng · T. Nonaka (✉) · Q. Ye · F. Wei · D. T. W. Wong (✉)
Center for Oral/Head and Neck Oncology Research, School of Dentistry, University of California, Los Angeles, Los Angeles, CA, USA

Division of Oral Biology and Medicine, School of Dentistry, University of California, Los Angeles, Los Angeles, CA, USA
e-mail: tnonaka@ucla.edu; dtww@ucla.edu



D. A. Granger, M. K. Taylor (eds.), *Salivary Bioscience*,
https://doi.org/10.1007/978-3-030-35784-9_8

**Abbreviations**

| | |
|---|---|
| AUC | area under the curve |
| CSF | cerebral spinal fluid |
| CSW | cyclic square wave |
| CTC | circulating tumor cell |
| ctDNA | circulating tumor DNA |
| ddPCR | droplet digital PCR |
| EFIRM | electric field-induced release and measurement |
| EV | extracellular vesicle |
| exRNA | extracellular RNA |
| HNSCC | head and neck squamous cell carcinoma |
| HPV | human papiloma virus |
| HRP | horseradish peroxidase |
| mRNA | messenger RNA |
| miRNA | microRNA |
| MVB | multivesicular body |
| NGS | next generation sequencing |
| NSCLC | non-small cell lung carcinoma |
| OSCC | oral squamous cell carcinoma |
| PCR | polymerase chain reaction |
| piRNA | piwi-interacting RNA |
| TMB | tetramethylbenzidine |

## 8.1 Introduction

Globally, cancer persists as a tremendous burden. It has been projected that 1.7 million new cancer cases will arise in the USA and 609,640 will result in death (Siegel, Miller, & Jemal, 2018). Accordingly, there is an impetus to advance the understanding of cancer biology, diagnosis, and therapeutic strategies. Optimizing the methodologies for screening and early detection of developing lesions before they reach an advanced stage is highly sought after by cancer therapy researchers. For many cancer types, such as non-small cell lung carcinoma (NSCLC), the major subtype of lung cancer, the current gold standard of the traditional tissue biopsy is invasive, and requires that the lesion reaches a critical mass before it is detectable by imaging modalities to warrant a surgical biopsy procedure (Ilié & Hofman, 2016).

An emerging concept, liquid biopsy, holds promise as an alternative to solid tissue biopsy by identifying and detecting alterations in biofluids that may reflect the presence of primary cancerous lesions. As the number of publications involving liquid biopsy increases, the definition continues to evolve. Initially, it referred to the diagnosis and characterization of a solid cancerous lesion by collecting and analyzing CTCs from blood (Chatterjee, 2016). More recently, the definition of liquid biopsy has been extended beyond CTCs in the blood to include other biomarkers that reflect the presence of a tumor. Additionally, liquid biopsy now includes the analysis of other biofluids besides blood. Urine, cerebral spinal fluid (CSF), and saliva are all potential mediums where cancer-associated molecular targets can be found. The

**Table 8.1** Considerations of utilizing saliva as a sample type for liquid biopsy

| Advantages | Disadvantages |
|---|---|
| • Saliva contains biomarkers that can be used for clinical liquid biopsy (e.g. ctDNA *EGFR* mutation).<br>• Collection method is easy and noninvasive.<br>• Patients can self-collect repeatedly.<br>• Biomarkers are stable when stored properly (e.g. −80 °C for RNA, protein, ctDNA) (Henson & Wong, 2010). | • Inherent salivary proteins (e.g., amylase) may compete with selected biomarker proteins during analysis (Henson & Wong, 2010).<br>• Production and composition of saliva are influenced by circadian rhythm (Henson & Wong, 2010). Note: inherent to all biofluids.<br>• Patient's unstimulated (basal) and stimulated (during mastication) salivary state affect the tonicity and pH balance of saliva (Chap. 2. By Hernández and Taylor).<br>• Complicated sample processing (e.g., addition of RNase inhibitor and serin protease inhibitor (SUPERase· In™ is a trademark. Aprotinin and PMFS are serine protease inhibitor) are required to protect salivary constituents) (Henson & Wong, 2010). |

National Cancer Institute now defines liquid biopsy as "A test done on a sample of blood to look for cancer cells from a tumor that are circulating in the blood or for pieces of DNA from tumor cells that are in the blood" (National Cancer Institute, 2011). While plasma, urine, and CSF are all viable biofluid candidates, more recent attention has been given to saliva. Saliva is readily available (0.5–1.5 l per day), can be obtained noninvasively, is easily collected and stored, and demonstrates compelling pathophysiological association with systemic diseases (Table 8.1).

Saliva is composed of secretions from three major glands (parotid, submandibular, and sublingual) and numerous minor salivary glands located throughout the oral cavity. Secretions from minor glands, gingival cervical fluid, mucosal exudates, microflora, and dislodgements form the oral-esophageal pathway also contributes to the composition (Nonaka & Wong, 2017). The composition of saliva is described in greater detail in Chap. 2 by Hernández and Taylor.

Salivary glands are highly vascularized and are composed of epithelial cells enriched with transporters and channels which provide a conducive environment for molecule exchange between blood and saliva (Gröschl, 2008). Trans blood to salivary pathways of hormones and cytokines have been described using passive and active transport mechanisms through transporters or passage through lipophilic layers of capillaries and glandular epithelial cells (Gröschl, 2008). Due to this proximity to the blood circulation and the predilection for molecule exchange, pathological biomarkers can appear in saliva. Saliva is therefore a practical sample type for liquid biopsy as that reflects systemic health.

## 8.2 Salivaomics

*Salivaomics* is the study of saliva and all its -omics constituents, contents, functions, and related techniques. Historically, in 2004, the transcriptome in saliva of stage T1/T2 oral squamous cell carcinoma (OSCC) patients was first examined and compared with equivalent healthy controls, showing that 1679 extracellular RNAs (exRNAs) exhibited significantly differential expression levels (Li, St. John et al., 2004). Subsequently, a proteomic study showed there was a 20–30% similarity between salivary and plasma proteomes, suggesting that saliva predominantly contains proteins synthesized in the salivary gland along with proteins originating in the blood or lymph sources (Yan et al., 2009). Furthermore, genomic analysis revealed that 30% of salivary exRNAs originate from the host and 70% are derived from oral microbiota (Bonne & Wong, 2012; Looi, Zakaria, Osman, & Jamal, 2012). Analysis of microgram amounts of protein from oral swabs has resulted in over 3700 quantified human proteins (Grassl et al., 2016). These findings provided initial evidence that saliva could be used to differentiate between health and disease and solidified the study of salivaomics as an independent field.

Since saliva is not a homogenous substance, the study of all its components; genomics, epigenomics, transcriptomics, proteomics, metabolomics, and microbiomics are all key components that make up salivaomics. This section will cover the three major—omic groups of salivaomics that are associated with cancer liquid biopsy, RNA (transcriptomics), circulating tumor DNA (genomics), and protein (proteomics).

### 8.2.1 Salivary Extracellular RNA (exRNA)

Since many different RNA types are present in saliva, the transcriptomic profile of saliva is complex. Saliva contains mRNAs, microRNAs (miRNAs), and other small noncoding RNAs (e.g., piwi-interacting RNAs). Thus far, a major focus in salivary transcriptomic work has been on describing the mRNA and miRNA alterations that differentiate the patients from the healthy subjects (Han et al., 2018; Park, Li, Yu, Brinkman, & Wong, 2006). The salivary transcriptome was first described using high-density microarray technology. This revealed a transcriptomic profile that was highly fragmented with coding and noncoding gene transcripts originating from both the host and oral microbiota (Li, Zhou, St. John, & Wong, 2004; Park et al., 2006, 2007; Spielmann & Wong, 2011).

One of the primary goals of liquid biopsy is to use it as a screening tool for early detection of cancer. Thus, there is merit in discovering viable biomarkers in saliva that signal disease. Transcriptome analysis of saliva from OSCC patients revealed that mRNA biomarkers (*DUSP1*, *H3F3A*, *IL1B*, *IL8*, *OAZ1*, *S100P*, and *SAT*) exhibited at least a 3.5-fold elevation in saliva from OSCC patients (Li, St John, et al., 2004). Upon using four of the candidate RNA biomarkers (*IL1B*, *OAZ1*, *SAT*, and *IL8*) and a cutoff of 50% with a logistic regression model, 91% sensitivity and 91% specificity was determined with an area under the curve (AUC) of 0.95.

In the saliva of pancreatic cancer patients, a similar approach was used, and salivary candidate exRNA biomarkers *KRAS*, *MBD3L2*, *ACRV1*, and *DPM1* were deemed associated with the early stage resectable pancreatic ductal adenocarcinomas. Using these markers, investigators could differentiate pancreatic cancer patients from non-cancer subjects (chronic pancreatitis and healthy control), yielding an AUC of 0.971 with 90.0% sensitivity and 95.0% specificity (Zhang, Farrell et al., 2010).

Similarly, in breast cancer a combination of eight (*CSTA*, *TPT1*, *IGF2BP1*, *GRM1*, *GRIK1*, *H6PD*, *MDM4*, and *S100A8*) salivary exRNAs and one proteomic marker (CA6) could discriminate between saliva samples of patients and control group with 83% sensitivity and 97% specificity (Zhang, Xiao, et al., 2010).

Ovarian cancer salivary transcriptome analysis identified four upregulated and sixteen downregulated exRNAs (Lee, Kim, Zhou, Kim, & Wong, 2012). Using five salivary exRNA biomarkers from the discovery phase (*AGPAT1*, *B2M*, *IER3*, *IL1B*, and *BASP1*) could differentiate healthy controls from patients with 85.7% sensitivity and 91.4% specificity. Using a similar approach, salivary exRNA markers (*CCNI*, *FGF19*, *GREB1*, *FRS2*, and *EGFR*) could be used to separate lung cancer and control with 93.75% sensitivity and 82.81% specificity with an AUC of 0.925.

Another discriminatory salivary exRNA target in salivary liquid biopsy is miRNAs. miRNAs are a class of 21–25 nucleotide long noncoding RNAs that play major roles in the regulation of gene expression and other processes (Ha & Kim, 2014). Recently, miRNAs have been identified in exosomes, membrane-bound vesicles 40–100 nm in diameter, released directly from the plasma membrane (Zhang et al., 2015). Exosomal miRNAs have been shown to regulate oncogenic and tumor suppressor genes (Chen, Liang, Zhang, Zen, & Zhang, 2012). There are vast differences in the miRNA profile between normal cells and cancer cells (Lin & Gregory, 2015).

In an analysis of miRNAs in plasma of head and neck squamous cell carcinoma (HNSCC) patients, elevated levels of miR-21 and miR-24 were detected in plasma from HNSCC patients (Hsu et al. 2012; Lin et al. 2010). Moreover, amplified miR-31 was detected in the plasma of HNSCC patients and was observed to have reduced after tumor resection, suggesting its tumor origin (Liu et al. 2010). Overexpression of miR-106b cluster and underexpression of miR-375 have been associated with HNSCC, and miR-451 appears to resurface during HNSCC recurrence (Hui et al., 2010). Similarly, a high expression of circulating miR-142, miR-186, miR-195, miR-374b, and miR-574 was reported (Summerer et al., 2015).

In the examination of saliva, lower levels of miR-125a and miR-200a were seen in saliva samples of OSCC patients (Park et al., 2009). In one study examining miRNAs for salivary gland tumors, miR-132, miR-15b, miR-140, and miR-223 had a distinguishing ability of 69% sensitivity and 95% specificity with an AUC of 0.90 (Matse et al., 2013). In another study, saliva from patients with esophageal cancer identified miR-144, miR-451, miR-98, miR-10b, and miR-363 as differentiable markers (Du & Zhang, 2017). The novelty in their study was they proposed a regulation network pathway in which the miRNA targets were involved in oncogenesis of the esophageal cancer. miRNAs appear to demonstrate greater stability in saliva compared to mRNAs and thus may be a viable alternative target for salivary liquid biopsy (Gallo, Tandon, Alevizos, & Illei, 2012).

Furthermore, piwi-interacting RNAs (piRNAs), the largest class of small long noncoding RNA molecules (26–31 nucleotides) are present in saliva (Bahn et al., 2015). Their activity has been linked with epigenetic and posttranscriptional gene silencing. Due to their size, piRNAs are able to pass through cell membranes easily and avoid degradation compared to longer RNAs (Han, Li et al., 2017). Salivary piRNAs present another exRNA biomarker source for salivary liquid biopsy.

## 8.2.2 *Salivary Genomics*

### 8.2.2.1 Circulating Tumor DNA

It was reported that blood from cancer patients have higher concentrations of circulating DNA as compared to healthy individuals (Stroun, Anker, Maurice, & Gahan, 1977). Further work from Stroun et al. revealed that DNA could be extracted from the plasma of cancer patients and their source from cancer cells could be determined by strand stability (Stroun et al., 1989). This work shaped and defined the concept of circulating tumor DNA (ctDNA)—the subset of circulating cell-free DNA that can potentially be used as a genetic "fingerprint" for a primary tumor.

ctDNA is believed to originate from tumor cells undergoing apoptosis or necrosis thereby releasing DNA strands into the systemic circulation (Jahr et al., 2001). Pathogenic and physiological processes such as phagocytosis and exocytosis may also contribute (Thierry, El Messaoudi, Gahan, Anker, & Stroun, 2016). Typically, remnant DNA in the blood is degraded by nucleases and eliminated by the liver, spleen, and kidneys (Barra et al., 2015). ctDNA is often described to be roughly 180–200 base pair in length which is characteristic of the apoptotic process and corresponds to the inter-nucleosomal length of DNA that is wrapped around the nucleosome including the linker segment (Diaz & Bardelli, 2014). Recently, it has been shown that ultrashort single-stranded cell-free DNA is present in plasma, which alludes to the possibility that a similar nucleic acid demographic may appear in saliva (Burnham et al., 2016).

A large-scale study on multiple cancer types including HNSCC demonstrated that increasing concentration of ctDNA is associated with advancing stage of disease (Bettegowda et al., 2014). It is not clear if ctDNA promotes carcinogenesis or is merely a by-product of cellular waste disposal as a result of apoptosis. However, there is evidence that ctDNA can promote cancer by transfecting healthy cells (García-Olmo et al., 2010). Laboratory methods for the assessment of ctDNA include allele-specific polymerase chain reaction (PCR), droplet digital PCR (ddPCR), and next generation sequencing (NGS) (Han, Wang, & Sun, 2017). These methods are predominantly PCR-based and have varying strengths and limitations and are mainly used in probing plasma ctDNA. At our laboratory, an emerging electrochemical platform of liquid biopsy named electric field-induced release and measurement (EFIRM) has been developed. This procedure can detect and quantify the ctDNA in saliva of NSCLC patients with superior performance as compared to the current technologies of ddPCR and NGS (Pu et al., 2016; Wei et al., 2014).

In an intriguing study, the saliva from 93 HNSCC patients was analyzed for human papilloma virus (HPV) DNA and/or somatic mutations related to HNSCC. In patients with tumors in the oral cavity, ctDNAs were detected in saliva with 100% concordance to tissue biopsy as compared to saliva-ctDNA concordance in patients from the oropharynx (47%), larynx (70%), and hypopharynx (67%) (Wang et al., 2015).

## 8.3 Salivary Proteomics

The saliva proteome is the first salivaomics constituent advocated for salivary biomarker development. An NIDCR/NIH-funded collaborative effort between The Scripps Research Institutes in San Diego, the University of California San Francisco, and the University of California Los Angeles resulting in an annotated catalog of 1166 proteins in salivary proteome of healthy individuals (Denny et al., 2008). This was then deposited into an open access Saliva Proteome Knowledge Base (Ai, Smith, & Wong, 2010). Additionally, a three-dimensional peptide fractionation technique was used to generate a deeper data set including 2340 proteins involved in integral functions in the oral cavity (Bandhakavi, Stone, Onsongo, Van Riper, & Griffin, 2009). Functional analysis shows salivary proteins occupy a higher proportion of metabolic and catabolic processes compared with plasma (Loo, Yan, Ramachandran, & Wong, 2010). This may be a clinical advantage of probing saliva if this particular subtype of proteins is oncogenically relevant (Schulz, Cooper-White, & Punyadeera, 2013).

Saliva proteins such as histatins, statherin, acidic proline rich proteins (PRPs), basic non-glycosylated PRPs are prone to degradation (Helmerhorst & Oppenheim, 2007). Careful considerations must therefore be taken to prevent skewing of downstream analysis of the potential protein biomarkers including other extracellular RNAs and possibly ctDNA that may be present. Preemptive strategies have been developed, published, and curated in order to stabilize the salivary proteins with protease inhibitors and preserve their integrity (Xiao & Wong, 2012).

Presently, the majority of proteomics in cancer biomarkers discovery utilize high-throughput mass spectrometry to identify salivary proteins associated with specific cancers. Breast cancer is a well-studied cancer in terms of salivary proteins. Elevations in EGF (Navarro et al., 1997), c-erbB-2 (Streckfus, Bigler, Dellinger et al., 2000), and CA15 3 (Streckfus, Bigler, Tucci, & Thigpen, 2000) were demonstrated in the saliva of breast cancer patients compared to healthy subjects. Salivary detection of CA15-3 is of particular interest because the same elevation of CA15-3 occurs in serum of patients (Duffy, Shering, Sherry, McDermott, & O'Higgins, 2000). CA15-3 is a transmembrane glycoprotein, frequently overexpressed and glycosylated in cancer. It appears to contribute to cell adhesion involved in metastasis (Duffy et al., 2000). In fact, it has been approved as an

FDA-approved biomarker to track the metastatic progression of breast cancer in serum (Füzéry, Levin, Chan, & Chan, 2013).

Another set of breast cancer salivary biomarker, VEGF, EGF, and CEA was analyzed in saliva of breast cancer patients by testing their predictive power individually or in combination revealing that VEGF and EGF leading to a combined 83% sensitivity and 74% specificity with an AUC of 0.84 (Brooks et al., 2008). CA6, another salivary candidate protein, demonstrated the ability to differentiate between cancer and healthy controls in saliva (Zhang, Xiao, et al., 2010). Lung resistant protein in saliva was observed to present at higher concentrations in Stage 1 breast cancer patients exhibiting the ability to detect the disease at its early stages (Wood & Streckfus, 2015). Liu et al. using 9 candidate salivary lectins with alterations in salivary glycopatterns can predict the presence of Stage 1 breast cancer with an accuracy of 0.902 in a double-blind cohort (Liu et al., 2018).

Salivary proteomic biomarker development for OSCC revealed that M2BP, MRP14, CD59, catalase, and profilin were associated and with a regression model which achieved 90% sensitivity and 83% specificity with a predication rate of 85% (AUC = 0.93) (Hu et al., 2008). In other studies IL-8, M2BP, and IL-1B were discovered as viable biomarkers. IL-8 and M2BP, however, showed better statistical distinction compared to control groups (Elashoff et al., 2012; Hu et al., 2008; Li, St John, et al., 2004). Analysis of the whole saliva from three HNSCC patients with LC–MS/MS revealed alpha-1-B-glycoprotein and complement factor B as unique proteins in cancer (Ohshiro et al., 2007). Additionally, using 2D DIGE analysis and subsequent mass spectrometry showed that beta-fibrin, S100 calcium-binding, transferrin, immunoglobulin heavy chain constant region gamma, and cofilin-1 were increased in HNSCC patients (Dowling et al., 2008). Studies in tongue cancer patients revealed that salivary adenosine deaminase activity may be a good differentiator between healthy and control subjects (Rai, Kaur, Jacobs, & Anand, 2011). Another group utilized selected reaction monitoring (SRM) tandem mass spectrometry to observe that C1R, LCN2, SLPI, FAM49B, TAGLN2, CFB, C3, C4B, LRG1, and SERPINA1 proteins are elevated in the saliva of OSCC patients and that, in particular, CFB, C3, C4B, SERPINA1, and LRG1 were associated an increased risk for developing OSCC (Kawahara et al., 2016).

Analysis of lung cancer saliva samples revealed that three proteins (HP, AZGP1, and CALPR) were higher in lung cancer patients compared to healthy controls. Logistic regression analysis of the diagnostic screening potential for theses biomarkers elicited a result of 88.5% sensitivity and 92.3% specificity (AUC = 0.90) (Xiao et al., 2012). Recently, using an exosome and microvesicle isolation approach coupled with LC–MS/MS-based label-free quantification showcased that four salivary exosome/microvesicle associated proteins (BPIFA1, CRNN, MUC5B, and IQGAP) were dysregulated in lung cancer patients (Sun et al., 2018).

In gastric cancer, four salivary proteins (*mass-to-charge ratio (m/z)* 1472.78, 2936.49, 6556.81, and 7081.17) identified by mass spectrometry appeared differ between gastric cancer and control groups (Wu, Wang, & Zhang, 2009). Xiao et al. identified 519 proteins in the saliva of gastric cancer patients and the data suggest that 48 proteins demonstrated a significantly different gastric profile in gastric cancer patients (Xiao et al., 2016). From these 48, six proteins (CSTB, TPI1, DMBT1, CALML3, IGH, and IL1RA) were selected as a gastric cancer screening verified with ELISA demonstrating downregulating. Regression model analysis depicted that by using three proteins as a screening protocol (CTSB, TPI1, and DMBT1) would have an 85% sensitivity and 80% specificity (AUC = 0.930).

For ovarian cancer, CA125 is elevated in the saliva discriminating between malignant and benign groups with an 81.3% sensitivity (Chen, Schwartz, & Li, 1990). More recently, 25 overexpressed and 19 underexpressed ($p < 0.05$) proteins between healthy controls and cancer patients were uncovered by fluorescence-based 2D-DIGE coupled with matrix-assisted laser desorption/ionization-time of flight Mass Spectrometry. Discriminatory candidate salivary proteins Lipocalin-2, indoleamine-2, 3-dioxygenase1 (IDO1), and S100A8 were identified and validated using western blotting and ELISA and when validated with 40 ovarian cancer patients and 40 control patients revealing a combined sensitivity of 87.5% and specificity of 86.7% with an ROC of 0.93 (Tajmul et al., 2018).

These proteomic association studies show the vast potential for cancer prediction and that salivary protein can provide for translational applications. Further work must be performed to validate these candidate proteomic biomarkers in prospective clinical trials adherent to the PRoBE (prospective-specimen-collection and retrospective-blinded-evaluation) design in a specific clinical context of use (Pepe et al., 2001).

## 8.4 Biological Mechanisms of Salivary Exosomes

Recent exploration of salivary genomics, transcriptomics, and proteomics have uncovered associated respective biomarkers source with disease phenotypes. Yet, little is known of the pathophysiological relationship with the distal cancer. How these biomarkers biologically arrive in the saliva is largely unknown. For diseases like pancreatic, gastric, and ovarian cancers, since the neoplastic lesion is geographically distant from salivary glands, it is unclear if the biomarkers that are detectable in the saliva are a by product of the tumor or a specific component of the pathophysiological process.

One mechanism in which these biomarkers may migrate/traffic into the saliva is through extracellular vesicles (EVs) transportation. Traditionally, EVs are organized and distributed into three subgroups based on size: exosomes, microvesicles, and apoptotic bodies (Kalra et al., 2012). Exosomes range from 30 to 100 nm in size and 1.13–1.19 g/mL in density (Théry, Amigorena, Raposo, & Clayton, 2006) and were first isolated from saliva in 2008 (Ogawa, Kanai-Azuma, Akimoto, Kawakami, & Yanoshita, 2008). Traditionally, exosomes are isolated through density gradient or sucrose cushion by ultracentrifugation at 100,000 × g (Théry et al., 2006). Additional methods, however, such as polymer-assisted precipitation (Niu, Pang, Liu, Cheng, & WSB, 2017), immunoaffinity-based capture beads (Sharma et al., 2018), immunoaffinity-based microfluidic chips (He & Zeng, 2016), and acoustic fluidic chips (Wu et al., 2017) have surfaced with promising capabilities.

In order to reduce the aforementioned variabilities of saliva during collection, it may be ideal to focus on studying the isolated EVs secreted by cancer cells present in the saliva. The term *saliva-exosomics* is used to describe the integration of various -omics approaches (e.g. genomics, transcriptomics, or proteomics) to examine the presence and function of salivary exosomes and exosome-related biomarkers that emerge in the setting of oral and systemic diseases (Nonaka & Wong, 2017).

Sharma, Gillespie, Palanisamy, and Gimzewski (2011) showed direct release of exosomes from oral cancer cells in saliva. In pursuit of the mechanistic tumor salivary exosomal axis, an orthotopic pancreatic cancer mouse model was developed by injecting a pancreatic cancer cell line (Panc02) into the pancreas of a syngeneic mouse (Lau et al., 2013). It was observed that exosome biogenesis could be inhibited by stable transfection of the dominant-negative form of GTPase Rab11 (DN-Rab11). When exosome biogenesis is suppressed by knocking down Rab11 in the Panc02 cells, the salivary transcriptome signature of mice injected with those genetically modified cells was ablated, compared to control tumor-bearing mice. This suggests a mechanistic link between pancreatic exosome biogenesis and the transportation/trafficking/migration of the exosomes to salivary glands and presentation of the disease exRNA profile in saliva.

Yang et al. developed a human lung cancer cell (H460) model expressing hCD63-GFP (an exosome marker) by injecting H460 cells into the chest cavity of immunocompromised mice (Yang, Wei, Schafer, & Wong, 2014). Human GAPDH mRNA was identified in hCD63 + GFP+ exosome-like microvesicles in saliva of the tumor-bearing mice. This finding suggests that human tumor cell-specific mRNA encased in exosome-like microvesicle can be transported from the organ of pathology (lung) to the salivary gland and into saliva.

More recently, Katsiougiannis et al. used a rodent model to show that saliva possess immunoregulatory properties (Katsiougiannis, Chia, Kim, Singh, & Wong, 2017). Saliva from Panc02-injected mice was collected and orally administered into non-tumor bearing control mice. NK cell activation markers, CD69 and NKG2D, were shown to significantly decrease when tumor saliva was gavaged into non-tumor bearing mice. Contrastingly, when saliva collected from mice injected with exosome biogenesis-suppressed Panc02 cells (DN-Rab11 transfected) was orally administered to non-tumor bearing mice, the NK activation markers were not affected. These findings depict an immunological relevance to exosome migration from the primary tumor site.

These physiological studies begin to shine light on the relevance of salivary exosomes and their capability to alter and regulate remote sites in the body (Fig. 8.1). Saliva-exosomics presents a new landscape and new horizon of saliva biology that is just being explored.

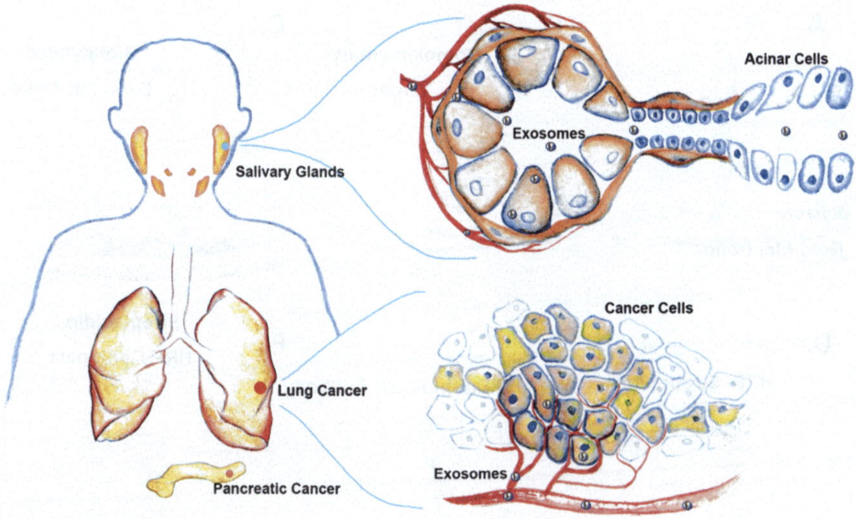


**Fig. 8.1** Potential biological mechanism for exosomes trafficking between salivary gland and distal tumor sites. Exosomes (not drawn to scale) are released by distant cancer cells (e.g., in the lung or pancreas) through multivesicular body (MVB) and enter the circulation. Exosomes may be up taken into salivary gland cells through endocytosis or membrane fusion and later released into the saliva. Exosomes have been shown to contain cancer-derived miRNAs, mRNAs, genomic DNAs, and proteins which can be isolated and assayed in saliva

## 8.5 EFIRM and Saliva Liquid Biopsy

An impactful engagement of saliva liquid biopsy is in the intensely research landscape of liquid biopsy to detect actionable mutation in biofluids, where lung cancer is the major organ site being explored where three actionable mutations (L858R, Exon 19 deletion, and T790M) in the epidermal growth factor receptor (*EGFR*) gene can be drugged and impact on the progress free survival of NSCLC patients (Wei, Yang, & Wong, 2013).

Our group has developed a novel liquid biopsy technology termed electric field-induced release and measurement method (EFIRM). This non-PCR-based electrochemical platform utilizes a capture probe that is complementary to a ctDNA target is designed and then immobilized on the surface of a gold electrode by encapsulating it in a conducting polymer matrix. After the immobilization of the capture probe on the surface of the electrode, the saliva specimen is placed on the surface of the electrode and a cyclic square wave (CSW) is applied. This CSW is designed to specifically lyse the exosomal structure that encapsulates the ctDNA sequence and aids in the DNA hybridization process. EFIRM can disrupt exosomes to release mRNAs and proteins in a similar manner as triton lysis (Wei et al., 2013). Following the incubation of the target sequence to the capture probe, a detector probe that is also complementary to the ctDNA is hybridized. This detector probe is biotinylated at its terminal end, which is then complexed to a streptavidin-horseradish peroxidase

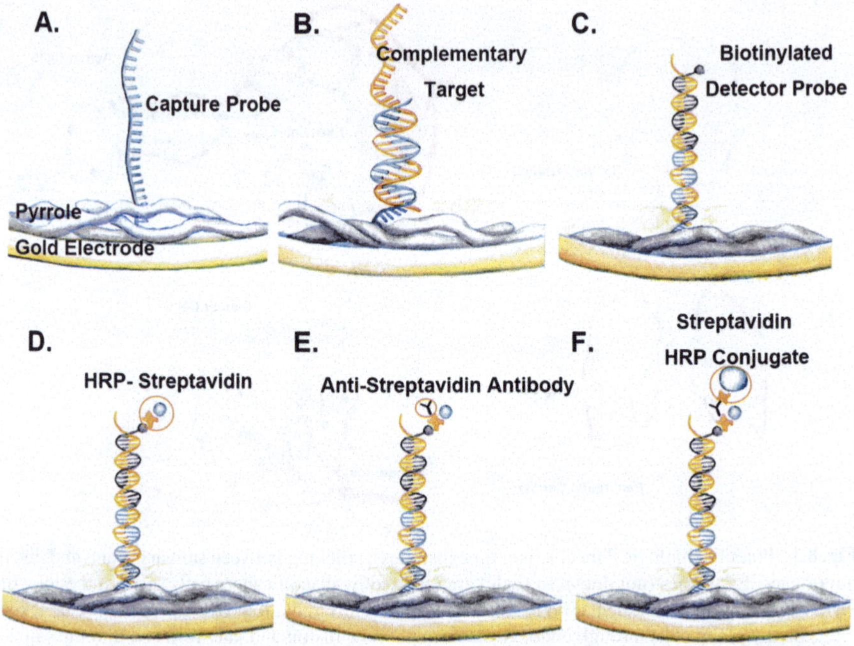


**Fig. 8.2** Schematic steps of the EFIRM Assay. (**a**) An electric field is applied polymerizing pyrrole in order to embed a capture probe specific for a cancer biomarker onto a gold electrode. (**b**) Complementary biomarker target is added and hybridizes with capture probe. (**c**) Complementary biotinylated detector probe hybridizes with target. (**d**) HRP (Horseradish peroxidase)-conjugated streptavidin binds to biotin on detector probe. (**e**) and (**f**) A subsequent layer of anti-streptavidin antibody and streptavidin-HRP amplify the signal. 3,3′,5,5′-Tetramethylbenzidine (TMB) substrate is added to generate a current through a reaction with HRP. The current is read by the gold electrode which represents the relative biomarker abundance

(HRP). The final output signal is determined by measuring a current generated by a tetramethylbenzadine-HRP reaction which proportionally reflects the amount of detector probe and target present in the saliva sample (Fig. 8.2).

Current clinical practice to detect signature *EGFR* ctDNA for NSCLC is ddPCR and NGS with performance ranges from 60–80% concordance with biopsy genotyping (Cohen et al., 2018; Newman et al., 2014; Phallen et al., 2017). The EFIRM technology detected signature oncogenic *EGFR* mutations in plasma and saliva of NSCLC patients, in two blinded clinical studies, with near-perfect concordance with biopsy genotyping (96–100%) (Pu et al., 2016; Wei et al., 2014).

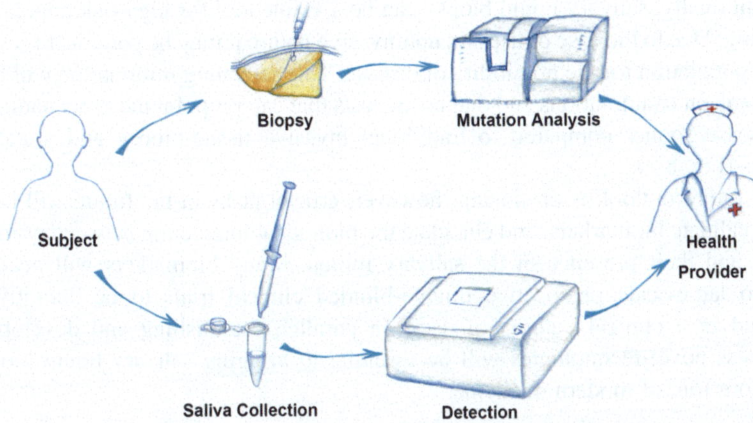


**Fig. 8.3** Potential of liquid biopsy in clinical practice. Sampling of biofluids for liquid biopsy can provide monitoring of patient disease progression or efficacy of pharmaceutical interventions. Point-of-care monitoring of liquid biopsy provides clinical information to the health provider with lower turn around time compared to traditional invasive tissue-biopsy and mutation analysis

## 8.6 Consideration and Future Outlook

Scientific and translational advances have established saliva as a sample type for omics-based biomarker development and liquid biopsy applications. Salivaomics and saliva-exosomics have mechanistically connected salivary glands with networking of personalized omics constituents from systemic origins. At this time, while many discovered salivary biomarkers remain to be definitively validated, it is foreseeable that as reliability and robustness of salivary biomarkers continue to be strengthened and their biological relevance uncovered, saliva liquid biopsy will become a major factor in many clinical applications.

Liquid biopsy will be a valid drug selection tool, treatment monitoring, and detection of acquire resistance mutations in cancer patients. In HNSCC, analyzing both saliva and plasma may be optimal for effective screening and monitoring of cancer (Nonaka & Wong, 2018). In NSCLC, detecting (L858R and Exon19 deletion) ctDNA targets with EFIRM could potentially be used to guide therapeutic decisions for specific tyrosine kinases inhibitors (Wei et al., 2014). The two-mutation assay has been shown to be feasible and therefore expanding the number of targets is needed. For example, T790M, another *EGFR* mutation, is responsible for 50% of acquired resistance to first-generation tyrosine kinase inhib itors if detected, its presence would indicate the use of a third-generation osimertinib (Inal, Yilmaz, Piperdi, Perez-Soler, & Cheng, 2015). Recently, patients been presenting ostimertinib resistant-associated C797S mutation which at this time has no therapeutic options (Wang, Tsui, Liu, Song, & Liu, 2016). Accordingly, if salivary liquid biopsy detects the levels of different mutations (e.g. L858R to T790M to C797S), it can monitor the progression of disease.

Additionally, salivary liquid biopsy can be a viable tool for high-risk population screening. Due to the ease of use and noninvasive nature it may be possible to screen a large population for the possibility of disease. This screening information will lead to subsequent examinations and diagnostic tests that can pinpoint the exact nature of the disease sooner compared to traditional invasive tissue-biopsy and mutation analysis (Fig. 8.3).

The future outlook is promising, however, crucial tasks in the future will be to verify, validate biomarkers, and elucidate the biological interaction between primary tumors and their presence in the salivary milieu. Many biomarkers will need to undergo large-scale prospective double-blinded clinical trials to be definitively validated in a clinical context of uses. In parallel, establishing and developing robustness novel technologies will be essential to bringing salivary liquid biopsy to the forefront of modern medicine.

**Acknowledgments** This work was supported by Grants (UH2/UH3 CA206126; U01 CA233370; UO1 DE017790) from the National Institutes of Health (NIH) to David T. W. Wong and the Canadian Institute of Health (CIHR) Doctoral Foreign Student Award and Tobacco Related Disease Research Program (TRDRP) Predoctoral Fellowship to Jordan Cheng.

**Disclosures** David T.W. Wong is a consultant to GlaxoSmithKlein, Wrigley and Colgate-Palmolive.

## References

Ai, J., Smith, B., & Wong, D. T. (2010). Saliva ontology: An ontology-based framework for a salivaomics knowledge base. *BMC Bioinformatics, 11*, 302. https://doi.org/10.1186/1471-2105-11-302

Bahn, J. H., Zhang, Q., Li, F., Chan, T. M., Lin, X., Kim, Y., . . . Xiao, X. (2015). The landscape of microRNA, Piwi-interacting RNA, and circular RNA in human saliva. *Clinical Chemistry, 61*, 221–230. https://doi.org/10.1373/clinchem.2014.230433

Bandhakavi, S., Stone, M. D., Onsongo, G., Van Riper, S. K., & Griffin, T. J. (2009). A dynamic range compression and three-dimensional peptide fractionation analysis platform expands proteome coverage and the diagnostic potential of whole saliva. *Journal of Proteome Research, 8*, 5590–5600. https://doi.org/10.1021/pr900675w

Barra, G. B., Santa Rita, T. H., Vasques J de, A., Chianca, C. F., Nery, L. F., & Santana Soares Costa, S. (2015). EDTA-mediated inhibition of DNases protects circulating cell-free DNA from ex vivo degradation in blood samples. *Clinical Biochemistry, 48*, 976–981. https://doi.org/10.1016/j.clinbiochem.2015.02.014

Bettegowda, C., Sausen, M., Leary, R. J., Kinde, I., Wang, Y., Agrawal, N., . . . Diaz, L. A., Jr. (2014). Detection of circulating tumor DNA in early- and late-stage human malignancies. *Science Translational Medicine, 6*, 224ra24. https://doi.org/10.1126/scitranslmed.3007094

Bonne, N. J., & Wong, D. T. (2012). Salivary biomarker development using genomic, proteomic and metabolomic approaches. *Genome Medicine, 4*, 82. https://doi.org/10.1186/gm383

Brooks, M. N., Wang, J., Li, Y., Zhang, R., Elashoff, D., & Wong, D. T. (2008). Salivary protein factors are elevated in breast cancer patients. *Molecular Medicine Reports, 1*, 375–378.

Burnham, P., Kim, M. S., Agbor-Enoh, S., Luikart, H., Valantine, H. A., Khush, K. K., & De Vlaminck, I. (2016). Single-stranded DNA library preparation uncovers the origin and diversity of ultrashort cell-free DNA in plasma. *Scientific Reports, 6*, 27859. https://doi.org/10.1038/srep27859

Chatterjee, D. (2016). Liquid biopsies: Handle with care. *Journal of Pathology and Laboratory Medicine, 1*, 3–6.

Chen, X., Liang, H., Zhang, J., Zen, K., & Zhang, C. Y. (2012). Secreted microRNAs: A new form of intercellular communication. *Trends in Cell Biology, 22*, 125–132. https://doi.org/10.1016/j.tcb.2011.12.001

Chen, D. X., Schwartz, P. E., & Li, F. Q. (1990). Saliva and serum CA 125 assays for detecting malignant ovarian tumors. *Obstetrics and Gynecology, 75*, 701–704.

Cohen, J. D., Li, L., Wang, Y., Thoburn, C., Afsari, B., Danilova, L., . . . Papadopoulos, N. (2018). Detection and localization of surgically resectable cancers with a multi-analyte blood test. *Science, 359*(6378), 926–930. https://doi.org/10.1126/science.aar3247

Denny, P., Hagen, F. K., Hardt, M., Liao, L., Yan, W., Arellanno, M., . . . Fisher, S. J. (2008). The proteomes of human parotid and submandibular/sublingual gland salivas collected as the ductal secretions. *Journal of Proteome Research, 7*, 1994–2006. https://doi.org/10.1021/pr700764j

Diaz, L. A., & Bardelli, A. (2014). Liquid biopsies: Genotyping circulating tumor DNA. *Journal of Clinical Oncology, 32*, 579–586. https://doi.org/10.1200/JCO.2012.45.2011

Dowling, P., Wormald, R., Meleady, P., Henry, M., Curran, A., & Clynes, M. (2008). Analysis of the saliva proteome from patients with head and neck squamous cell carcinoma reveals differences in abundance levels of proteins associated with tumour progression and metastasis. *Journal of Proteomics, 71*, 168–175. https://doi.org/10.1016/j.jprot.2008.04.004

Du, J., & Zhang, L. (2017). Analysis of salivary microRNA expression profiles and identification of novel biomarkers in esophageal cancer. *Oncology Letters, 14*, 1387–1394. https://doi.org/10.3892/ol.2017.6328

Duffy, M. J., Shering, S., Sherry, F., McDermott, E., & O'Higgins, N. (2000). CA 15-3: A prognostic marker in breast cancer. *The International Journal of Biological Markers, 15*, 330–333.

Elashoff, D., Zhou, H., Reiss, J., Wang, J., Xiao, H., Henson, B., . . . Wong, D. T. (2012). Prevalidation of salivary biomarkers for oral cancer detection. *Cancer Epidemiology, Biomarkers & Prevention, 21*, 664–672. https://doi.org/10.1158/1055-9965.EPI-11-1093

Füzéry, A. K., Levin, J., Chan, M. M., & Chan, D. W. (2013). Translation of proteomic biomarkers into FDA approved cancer diagnostics: Issues and challenges. *Clinical Proteomics, 10*, 13. https://doi.org/10.1186/1559-0275-10-13

Gallo, A., Tandon, M., Alevizos, I., & Illei, G. G. (2012). The majority of microRNAs detectable in serum and saliva is concentrated in exosomes. *PLoS One, 7*, e30679. https://doi.org/10.1371/journal.pone.0030679

García-Olmo, D. C., Domínguez, C., García-Arranz, M., Anker, P., Stroun, M., García-Verdugo, J. M., & García-Olmo, D. (2010). Cell-free nucleic acids circulating in the plasma of colorectal cancer patients induce the oncogenic transformation of susceptible cultured cells. *Cancer Research, 70*, 560–567. https://doi.org/10.1158/0008-5472.CAN-09-3513

Grassl, N., Kulak, N. A., Pichler, G., Geyer, P. E., Jung, J., Schubert, S., . . . Mann, M. (2016). Ultra-deep and quantitative saliva proteome reveals dynamics of the oral microbiome. *Genome Medicine, 8*, 44. https://doi.org/10.1186/s13073-016-0293-0

Gröschl, M. (2008). Current status of salivary hormone analysis. *Clinical Chemistry, 54*, 1759–1769. https://doi.org/10.1373/clinchem.2008.108910

Ha, M., & Kim, V. N. (2014). Regulation of microRNA biogenesis. *Nature Reviews. Molecular Cell Biology, 15*, 509–524. https://doi.org/10.1038/nrm3838

Han, Y.-N., Li, Y., Xia, S.-Q., Zhang, Y. Y., Zheng, J. H., & Li, W. (2017). PIWI proteins and PIWI-interacting RNA: Emerging roles in cancer. *Cellular Physiology and Biochemistry, 44*, 1–20. https://doi.org/10.1159/000484541

Han, X., Wang, J., & Sun, Y. (2017). Circulating tumor DNA as biomarkers for cancer detection. *Genomics, Proteomics & Bioinformatics, 15*, 59–72. https://doi.org/10.1016/j.gpb.2016.12.004

Han, N., Lu, H., Zhang, Z., Ruan, M., Yang, W., & Zhang, C. (2018). Comprehensive and in-depth analysis of microRNA and mRNA expression profile in salivary adenoid cystic carcinoma. *Gene, 678*, 349–360. https://doi.org/10.1016/j.gene.2018.08.023

He, M., & Zeng, Y. (2016). Microfluidic exosome analysis toward liquid biopsy for cancer. *Journal of Laboratory Automation, 21*, 599–608. https://doi.org/10.1177/2211068216651035

Helmerhorst, E. J., & Oppenheim, F. G. (2007). Saliva: A dynamic proteome. *Journal of Dental Research, 86*, 680–693. https://doi.org/10.1177/154405910708600802

Henson, B. S., & Wong, D. T. (2010). Collection, storage, and processing of saliva samples for downstream molecular applications. *Methods in Molecular Biology (Clifton NJ), 666*, 21–30. https://doi.org/10.1007/978-1-60761-820-1_2

Hsu, C. M., Lin, P. M., Wang, Y. M., Chen, Z. J., Lin, S. F., & Yang, M. Y. (2012). Circulating miRNA is a novel marker for head and neck squamous cell carcinoma. *Tumor Biology, 33*(6), 1933–1942.

Hu, S., Arellano, M., Boontheung, P., Wang, J., Zhou, H., Jiang, J., . . . Wong, D. T. (2008). Salivary proteomics for oral cancer biomarker discovery. *Cancer Research, 14*, 6246–6252. https://doi.org/10.1158/1078-0432.CCR-07-5037

Hui, A. B. Y., Lenarduzzi, M., Krushel, T., Waldron, L., Pintilie, M., Shi, W., . . . Liu, F. F. (2010). Comprehensive MicroRNA profiling for head and neck squamous cell carcinomas. *Cancer Research, 16*, 1129–1139. https://doi.org/10.1158/1078-0432.CCR-09-2166

Ilié, M., & Hofman, P. (2016). Pros: Can tissue biopsy be replaced by liquid biopsy? *Translational Lung Cancer Research, 5*, 420–423. https://doi.org/10.21037/tlcr.2016.08.06

Inal, C., Yilmaz, E., Piperdi, B., Perez-Soler, R., & Cheng, H. (2015). Emerging treatment for advanced lung cancer with EGFR mutation. *Expert Opinion on Emerging Drugs, 20*, 597–612. https://doi.org/10.1517/14728214.2015.1058778

Jahr, S., Hentze, H., Englisch, S., Hardt, D., Fackelmayer, F. O., Hesch, R. D., & Knippers, R. (2001). DNA fragments in the blood plasma of cancer patients: Quantitations and evidence for their origin from apoptotic and necrotic cells. *Cancer Research, 61*, 1659–1665.

Kalra, H., Simpson, R. J., Ji, H., Aikawa, E., Altevogt, P., Askenase, P., . . . Mathivanan, S. (2012). Vesiclepedia: A compendium for extracellular vesicles with continuous community annotation. *PLoS Biology, 10*, e1001450. https://doi.org/10.1371/journal.pbio.1001450

Katsiougiannis, S., Chia, D., Kim, Y., Singh, R. P., & Wong, D. T. (2017). Saliva exosomes from pancreatic tumor–bearing mice modulate NK cell phenotype and antitumor cytotoxicity. *The FASEB Journal, 31*, 998–1010. https://doi.org/10.1096/fj.201600984R

Kawahara, R., Bollinger, J. G., Rivera, C., Ribeiro, A. C., Brandão, T. B., Paes Leme, A. F., & MJ, M. C. (2016). A targeted proteomic strategy for the measurement of oral cancer candidate biomarkers in human saliva. *Proteomics, 16*, 159–173. https://doi.org/10.1002/pmic.201500224

Lau, C., Kim, Y., Chia, D., Spielmann, N., Eibl, G., Elashoff, D., . . . Wong, D. T. (2013). Role of pancreatic cancer-derived exosomes in salivary biomarker development. *The Journal of Biological Chemistry, 288*, 26888–26897. https://doi.org/10.1074/jbc.M113.452458

Lee, Y.-H., Kim, J. H., Zhou, H., Kim, B. W., & Wong, D. T. (2012). Salivary transcriptomic biomarkers for detection of ovarian cancer: For serous papillary adenocarcinoma. *Journal of Molecular Medicine, 90*, 427–434. https://doi.org/10.1007/s00109-011-0829-0

Li, Y., St. John, M. A. R., Zhou, X., Kim, Y., Sinha, U., Jordan, R. C., . . . Wong, D. T. (2004). Salivary transcriptome diagnostics for oral cancer detection. *Cancer Research, 10*, 8442–8450. https://doi.org/10.1158/1078-0432.CCR-04-1167

Li, Y., Zhou, X., St. John, M. A. R., & Wong, D. T. W. (2004). RNA profiling of cell-free saliva using microarray technology. *Journal of Dental Research, 83*, 199–203. https://doi.org/10.1177/154405910408300303

Lin, S., & Gregory, R. I. (2015). MicroRNA biogenesis pathways in cancer. *Nature Reviews. Cancer, 15*, 321–333. https://doi.org/10.1038/nrc3932

Lin, S. C., Liu, C. J., Lin, J. A., Chiang, W. F., Hung, P. S., & Chang, K. W. (2010). miR-24 up-regulation in oral carcinoma: Positive association from clinical and in vitro analysis. *Oral Oncology, 46*(3), 204–208.

Liu, C. J., Kao, S. Y., Tu, H. F., Tsai, M. M., Chang, K. W., & Lin, S. C. (2010). Increase of microRNA miR-31 level in plasma could be a potential marker of oral cancer. *Oral Diseases, 16* (4), 360–364.

Liu, X., Yu, H., Qiao, Y., Yang, J., Shu, J., Zhang, J., ... Li, Z. (2018). Salivary glycopatterns as potential biomarkers for screening of early-stage breast cancer. *eBioMedicine, 28*, 70–79. https://doi.org/10.1016/j.ebiom.2018.01.026

Loo, J. A., Yan, W., Ramachandran, P., & Wong, D. T. (2010). Comparative human salivary and plasma proteomes. *Journal of Dental Research, 89*, 1016–1023. https://doi.org/10.1177/0022034510380414

Looi, M.-L., Zakaria, H., Osman, J., & Jamal, R. (2012). Quantity and quality assessment of DNA extracted from saliva and blood. *Clinical Laboratory, 58*, 307–312.

Matse, J. H., Yoshizawa, J., Wang, X., Elashoff, D., Bolscher, J. G., Veerman, E. C., ... Wong, D. T. (2013). Discovery and prevalidation of salivary extracellular microRNA biomarkers panel for the noninvasive detection of benign and malignant parotid gland tumors. *Clinical Cancer Research, 19*, 3032–3038. https://doi.org/10.1158/1078-0432.CCR-12-3505

National Cancer Institute. (2011). NCI dictionary of cancer terms. *Journal of National Cancer Institute*. Accessed November 25, 2018, from https://www.cancer.gov/publications/dictionaries/cancer-terms

Navarro, M. A., Mesía, R., Díez-Gibert, O., Rueda, A., Ojeda, B., & Alonso, M. C. (1997). Epidermal growth factor in plasma and saliva of patients with active breast cancer and breast cancer patients in follow-up compared with healthy women. *Breast Cancer Research and Treatment, 42*, 83–86. https://doi.org/10.1023/A:1005755928831

Newman, A. M., Bratman, S. V., To, J., Wynne, J. F., Eclov, N. C., Modlin, L. A., ... Diehn, M. (2014). An ultrasensitive method for quantitating circulating tumor DNA with broad patient coverage. *Nature Medicine, 20*, 548–554. https://doi.org/10.1038/nm.3519

Niu, Z., Pang, R. T. K., Liu, W., Cheng, R., & WSB, Y. (2017). Polymer-based precipitation preserves biological activities of extracellular vesicles from an endometrial cell line. *PLoS One, 12*, e0186534. https://doi.org/10.1371/journal.pone.0186534

Nonaka, T., & Wong, D. T. W. (2017). Saliva-exosomics in cancer: Molecular characterization of cancer-derived exosomes in saliva. *The Enzymes, 42*, 125–151. https://doi.org/10.1016/bs.enz.2017.08.002

Nonaka, T., & Wong, D. T. W. (2018). Liquid biopsy in head and neck cancer: Promises and challenges. *Journal of Dental Research, 97*, 701–708. https://doi.org/10.1177/0022034518762071

Ogawa, Y., Kanai-Azuma, M., Akimoto, Y., Kawakami, H., & Yanoshita, R. (2008). Exosome-like vesicles with dipeptidyl peptidase IV in human saliva. *Biological & Pharmaceutical Bulletin, 31*, 1059–1062.

Ohshiro, K., Rosenthal, D. I., Koomen, J. M., Streckfus, C. F., Chambers, M., Kobayashi, R., & El-Naggar, A. K. (2007). Pre-analytic saliva processing affect proteomic results and biomarker screening of head and neck squamous carcinoma. *International Journal of Oncology, 30*, 743–749.

Park, N. J., Li, Y., Yu, T., Brinkman, B. M., & Wong, D. T. (2006). Characterization of RNA in saliva. *Clinical Chemistry, 52*, 988–994. https://doi.org/10.1373/clinchem.2005.063206

Park, N. J., Zhou, H., Elashoff, D., Henson, B. S., Kastratovic, D. A., Abemayor, E., & Wong, D. T. (2009). Salivary microRNA: Discovery, characterization, and clinical utility for oral cancer detection. *Cancer Research, 15*, 5473–5477. https://doi.org/10.1158/1078-0432.CCR-09-0736

Park, N. J., Zhou, X., Yu, T., Brinkman, B. M. N., Zimmerman, B., Palaniswamy, V., & Wong, D. T. (2007). Characterization of salivary RNA by cDNA library analysis. *Archives of Oral Biology, 52*, 30–35. https://doi.org/10.1016/j.archoralbio.2006.08.014

Pepe, M. S., Etzioni, R., Feng, Z., Potter, J. D., Thompson, M. L., Thornquist, M., … Yasui, Y. (2001). Phases of biomarker development for early detection of cancer. *Journal of the National Cancer Institute, 93*, 1054–1061.

Phallen, J., Sausen, M., Adleff, V., Leal, A., Hruban, C., White, J., … Velculescu, V. E. (2017). Direct detection of early-stage cancers using circulating tumor DNA. *Science Translational Medicine, 9*, eaan2415. https://doi.org/10.1126/scitranslmed.aan2415

Pu, D., Liang, H., Wei, F., Akin, D., Feng, Z., Yan, Q., … Zhou, Q. (2016). Evaluation of a novel saliva-based epidermal growth factor receptor mutation detection for lung cancer: A pilot study. *Thoracic Cancer, 7*, 428–436. https://doi.org/10.1111/1759-7714.12350

Rai, B., Kaur, J., Jacobs, R., & Anand, S. C. (2011). Adenosine deaminase in saliva as a diagnostic marker of squamous cell carcinoma of tongue. *Clinical Oral Investigations, 15*, 347–349. https://doi.org/10.1007/s00784-010-0404-z

Schulz, B. L., Cooper-White, J., & Punyadeera, C. K. (2013). Saliva proteome research: Current status and future outlook. *Critical Reviews in Biotechnology, 33*, 246–259. https://doi.org/10.3109/07388551.2012.687361

Sharma, S., Gillespie, B. M., Palanisamy, V., & Gimzewski, J. K. (2011). Quantitative nanostructural and single-molecule force spectroscopy biomolecular analysis of human-saliva-derived exosomes. *Langmuir, 27*(23), 14394–14400.

Sharma, P., Ludwig, S., Muller, L., Hong, C. S., Kirkwood, J. M., Ferrone, S., & Whiteside, T. L. (2018). Immunoaffinity-based isolation of melanoma cell-derived exosomes from plasma of patients with melanoma. *Journal of Extracellular Vesicles, 7*, 1435138. https://doi.org/10.1080/20013078.2018.1435138

Siegel, R. L., Miller, K. D., & Jemal, A. (2018). Cancer statistics, 2018. *CA: A Cancer Journal for Clinicians, 68*, 7–30. https://doi.org/10.3322/caac.21442

Spielmann, N., & Wong, D. T. (2011). Saliva: Diagnostics and therapeutic perspectives. *Oral Diseases, 17*, 345–354. https://doi.org/10.1111/j.1601-0825.2010.01773.x

Streckfus, C., Bigler, L., Dellinger, T., Dai, X., Kingman, A., & Thigpen, J. T. (2000). The presence of soluble c-erbB-2 in saliva and serum among women with breast carcinoma: A preliminary study. *Cancer Research, 6*, 2363–2370.

Streckfus, C., Bigler, L., Tucci, M., & Thigpen, J. T. (2000). A preliminary study of CA15-3, c-erbB-2, epidermal growth factor receptor, cathepsin-D, and p53 in saliva among women with breast carcinoma. *Cancer Investigation, 18*(2), 101–109.

Stroun, M., Anker, P., Maurice, P., & Gahan, P. B. (1977). Circulating nucleic acids in higher organisms. *International Review of Cytology, 51*, 1–48.

Stroun, M., Anker, P., Maurice, P., Lyautey, J., Lederrey, C., & Beljanski, M. (1989). Neoplastic characteristics of the DNA found in the plasma of cancer patients. *Oncology, 46*, 318–322. https://doi.org/10.1159/000226740

Summerer, I., Unger, K., Braselmann, H., Schuettrumpf, L., Maihoefer, C., Baumeister, P., … Zitzelsberger, H. (2015). Circulating microRNAs as prognostic therapy biomarkers in head and neck cancer patients. *British Journal of Cancer, 113*, 76–82. https://doi.org/10.1038/bjc.2015.111

Sun, Y., Huo, C., Qiao, Z., Shang, Z., Uzzaman, A., Liu, S., … Xiao, H. (2018). Comparative proteomic analysis of exosomes and microvesicles in human saliva for lung cancer. *Journal of Proteome Research, 17*, 1101–1107. https://doi.org/10.1021/acs.jproteome.7b00770

Tajmul, M., Parween, F., Singh, L., Mathur, S. R., Sharma, J. B., Kumar, S., … Yadav, S. (2018). Identification and validation of salivary proteomic signatures for non-invasive detection of ovarian cancer. *International Journal of Biological Macromolecules, 108*, 503–514. https://doi.org/10.1016/j.ijbiomac.2017.12.014

Théry, C., Amigorena, S., Raposo, G., & Clayton, A. (2006). Isolation and characterization of exosomes from cell culture supernatants and biological fluids. *Current Protocols in Cell Biology, Chapter 3*, Unit 3.22. https://doi.org/10.1002/0471143030.cb0322s30

Thierry, A. R., El Messaoudi, S., Gahan, P. B., Anker, P., & Stroun, M. (2016). Origins, structures, and functions of circulating DNA in oncology. *Cancer Metastasis Reviews, 35*, 347–376. https://doi.org/10.1007/s10555-016-9629-x

Wang, Y., Springer, S., Mulvey, C. L., Silliman, N., Schaefer, J., Sausen, M., & Agrawal, N. (2015). Detection of somatic mutations and HPV in the saliva and plasma of patients with head and neck squamous cell carcinomas. *Science Translational Medicine, 7*, 293ra104. https://doi.org/10.1126/scitranslmed.aaa8507

Wang, S., Tsui, S. T., Liu, C., Song, Y., & Liu, D. (2016). EGFR C797S mutation mediates resistance to third-generation inhibitors in T790M-positive non-small cell lung cancer. *Journal of Hematology and Oncology, 9*, 59. https://doi.org/10.1186/s13045-016-0290-1

Wei, F., Lin, C.-C., Joon, A., Feng, Z., Troche, G., Lira, M. E., … Wong, D. T. (2014). Noninvasive saliva-based EGFR gene mutation detection in patients with lung cancer. *American Journal of Respiratory and Critical Care Medicine, 190*, 1117–1126. https://doi.org/10.1164/rccm.201406-1003OC

Wei, F., Yang, J., & Wong, D. T. W. (2013). Detection of exosomal biomarker by electric field-induced release and measurement (EFIRM). *Biosensors & Bioelectronics, 44*, 115–121. https://doi.org/10.1016/j.bios.2012.12.046

Wood, N., & Streckfus, C. F. (2015). The expression of lung resistance protein in saliva: A novel prognostic indicator protein for carcinoma of the breast. *Cancer Investigation, 33*, 510–515. https://doi.org/10.3109/07357907.2015.1081920

Wu, M., Ouyang, Y., Wang, Z., Zhang, R., Huang, P. H., Chen, C., … Huang, T. J. (2017). Isolation of exosomes from whole blood by integrating acoustics and microfluidics. *Proceedings of the National Academy of Sciences of the United States of America, 114*, 10584–10589. https://doi.org/10.1073/pnas.1709210114

Wu, Z.-Z., Wang, J.-G., & Zhang, X.-L. (2009). Diagnostic model of saliva protein finger print analysis of patients with gastric cancer. *World Journal of Gastroenterology, 15*, 865–870. https://doi.org/10.3748/wjg.15.865

Xiao, H., & Wong, D. T. W. (2012). Method development for proteome stabilization in human saliva. *Analytica Chimica Acta, 722*, 63–69. https://doi.org/10.1016/j.aca.2012.02.017

Xiao, H., Zhang, Y., Kim, Y., Kim, S., Kim, J. J., Kim, K. M., … Wong, D. T. (2016). Differential proteomic analysis of human saliva using tandem mass tags quantification for gastric cancer detection. *Scientific Reports, 6*, 22165. https://doi.org/10.1038/srep22165

Xiao, H., Zhang, L., Zhou, H., Lee, J. M., Garon, E. B., & Wong, D. T. (2012). Proteomic analysis of human saliva from lung cancer patients using two-dimensional difference gel electrophoresis and mass spectrometry. *Molecular & Cellular Proteomics, 11*, M111.012112. https://doi.org/10.1074/mcp.M111.012112

Yan, W., Apweiler, R., Balgley, B. M., Boontheung, P., Bundy, J. L., Cargile, B. J., … Wong, D. T. (2009). Systematic comparison of the human saliva and plasma proteomes. *Proteomics. Clinical Applications, 3*, 116–134. https://doi.org/10.1002/prca.200800140

Yang, J., Wei, F., Schafer, C., & Wong, D. T. W. (2014). Detection of tumor cell-specific mRNA and protein in exosome-like microvesicles from blood and saliva. *PLoS One, 9*, e110641. https://doi.org/10.1371/journal.pone.0110641

Zhang, L., Farrell, J. J., Zhou, H., Elashoff, D., Akin, D., Park, N. H., … Wong, D. T. (2010). Salivary transcriptomic biomarkers for detection of resectable pancreatic cancer. *Gastroenterology, 138*, 949–957.e1–7. https://doi.org/10.1053/j.gastro.2009.11.010

Zhang, J., Li, S., Li, L., Li, M., Guo, C., Yao, J., & Mi, S. (2015). Exosome and Exosomal MicroRNA: Trafficking, sorting, and function. *Genomics Proteomics Bioinformatics, 13*, 17–24. https://doi.org/10.1016/j.gpb.2015.02.001

Zhang, L., Xiao, H., Karlan, S., Zhou, H., Gross, J., Elashoff, D., … Wong, D. T. (2010). Discovery and preclinical validation of salivary transcriptomic and proteomic biomarkers for the non-invasive detection of breast cancer. *PLoS One, 5*, e15573. https://doi.org/10.1371/journal.pone.0015573

# Chapter 9
# Salivary Bioscience, Immunity, and Inflammation

**Jenna L. Riis, Michelle L. Byrne, Lisa M. Hernández, and Theodore F. Robles**

**Abstract** The study of immune and inflammatory markers in saliva has gained increased attention in recent years with the advancements in assay technology and a heightened focus on cross-systems biology and psychoneuroimmunology. Salivary immune markers are important for the study of both oral and systemic health. Salivary inflammation, in particular, has been widely examined across many fields as both an area of interest and a source of confounding variance. In this chapter, we discuss the opportunities and challenges of studying immune markers in saliva and review the current state of knowledge regarding the study of salivary immune biomeasures, including salivary cytokines, C-reactive protein, and immunoglobulins. Analysis and interpretation issues particularly important for studying immune-related analytes, such as the impact of oral and systemic health, the interpretation of the serum–saliva correlation, and multisystem measurement and analysis techniques, are discussed. Finally, we discuss future directions for the study of salivary immune markers and applications of this research to clinical care and health monitoring and surveillance programs.

**Keywords** Inflammation · Salivary immunoglobulins · Cytokines · C-reactive protein · MMP

---

J. L. Riis (✉)
Department of Psychological Science, School of Social Ecology, University of California, Irvine, CA, USA

Institute for Interdisciplinary Salivary Bioscience Research, University of California, Irvine, CA, USA
e-mail: jriis@uci.edu

M. L. Byrne
Department of Psychology, University of Oregon, Eugene, OR, USA

L. M. Hernández
Leidos, Inc., San Diego, CA, USA

T. F. Robles
Department of Psychology, University of California, Los Angeles, CA, USA



D. A. Granger, M. K. Taylor (eds.), *Salivary Bioscience*,
https://doi.org/10.1007/978-3-030-35784-9_9

## 9.1 Introduction

The ability to measure immune markers in saliva provides unique opportunities in the study of inflammation, health, development, and disease risk. Using saliva as a biospecimen in health-related research and clinical programs may help broaden participation in research, increase our ability to monitor and track disease risk, progression, and treatment, and improve the ecological validity of assessments of immune function. The promise of salivary immune markers to advance our understanding of the biopsychosocial factors affecting health and development has garnered the attention of researchers from various fields, including public health, psychology, and medicine. Despite exciting advances in our understanding of the correlates of salivary immune markers, there remain critical validity, reliability, and conceptual questions that need to be addressed regarding the interpretation and implications of our findings.

Salivary immune markers have potential value for measuring oral, mucosal, and systemic immune function, though limitations in our ability to isolate these sources of variability contributes to uncertainty in interpreting salivary immune marker findings. The immune system in the oral cavity includes the functioning of resident immune cells in the mouth (e.g., in the salivary glands), migrating immune cells from systemic circulation, and the mucosal immune system (Bergmeier, 2018). With bidirectional connections to both the gut mucosal and the systemic immune systems, oral immune processes may provide exciting opportunities to use salivary immune markers as indices of immune function (Bergmeier, 2018). The value of salivary immune markers for informing health, however, has not been fully examined in the literature. Researchers in the fields of oral biology, dentistry, periodontology, and oral cancer examine immune markers in saliva to study risk of oral health problems such as dental caries, periodontitis, gingivitis, and oral cancer (Belstrøm, Damgaard, Könönen, Gürsoy, & Gürsoy, 2017; Rhodus, Ho, Miller, Myers, & Ondrey, 2005; St. John et al., 2004; Teles, Likhari, Socransky, & Haffajee, 2009; Zhang et al., 2016) (see Chaps. 8 and 18 for discussions of salivary biomeasures and oral cancers and periodontal medicine). Researchers in fields such as psychoneuroendocrinology and public health typically use salivary immune markers to gain insight into how specific exposures, such as psychosocial stress and environmental toxins, are related to systemic health and well-being. While converging evidence from across these fields point to the value of salivary immune markers in the study and understanding of oral, physical, and mental health, there has been little cross-disciplinary work linking traditional oral biology studies with studies that use salivary immune markers to make inferences about systemic health.

In this chapter, we will discuss the current state of knowledge regarding salivary immune markers in the study of overall health and development. To provide context concerning the study of salivary immune markers in biobehavioral and health research, we discuss the history of the field of psychoneuroimmunology (PNI) and its adoption of salivary bioscience methods to study the biopsychosocial processes underlying health. We also provide recommendations for the investigation and interpretation of salivary immune markers in biobehavioral and health studies and discuss exciting research and clinical opportunities afforded by advancing the study of salivary immune markers.

## 9.2 History of Psychoneuroimmunology and the Integration with Salivary Bioscience

While the origins of PNI date back to 1964 (Solomon & Moos, 1964), the field truly blossomed in the 1970s and 1980s (Ader, 2000) through research on brain–immune connections and their implications for adaptive immunity—the slow-acting but highly specific arm of the immune system that hinges on T- and B-cell activation. Throughout the 1980s and 1990s, psychological stress was widely viewed as immunosuppressive (Segerstrom & Miller, 2004). Toward the end of the 1990s and early 2000s, research on immune-to-brain communication (Maier & Watkins, 1998), age-related increases in innate immunity (Papanicolaou, Wilder, Manolagas, & Chrousos, 1998), and depression-related alterations in immune function (Miller & Raison, 2016) steered the field toward studying innate immune inflammation (Kiecolt-Glaser, McGuire, Robles, & Glaser, 2002). Since that time, the body's rapid response to infection and injury has dominated research in PNI—from the impact of stress on inflammation (Marsland, Walsh, Lockwood, & John-Henderson, 2017; Rohleder, 2014), to immune system contributors to neural development and function (Bilbo & Schwarz, 2012; DiSabato, Quan, & Godbout, 2016), and the clinical implications of all of the above. Throughout these research areas, plasma and serum markers of innate immune inflammation, including pro-inflammatory cytokines (interleukin(IL)-1β, IL-6, and tumor necrosis factor(TNF)-α) and the acute phase C-reactive protein (CRP), dominated inflammation assessments from the measurement of stimulated cytokine levels in vitro to large-scale epidemiologic research.

The salivary compartment may be favorable for studying links between psychosocial factors and inflammation for several reasons, many of which were articulated in a systematic review by Slavish and colleagues on salivary inflammation and stress (2015b). First, saliva allows us to collect biospecimens in ecologically valid contexts with minimal disruption to normal, everyday activities. We can also collect multiple saliva samples across time and with a high sampling frequency while minimizing participant burden. In addition, many inflammatory markers, such as IL-1β and IL-8, are present at high concentrations in saliva and are highly detectable (Byrne et al., 2013).

Interestingly, the history of using salivary immune markers in human psychosocial research has unfolded in a manner similar to the broader field of PNI. In the mid-1990s and early 2000s much of the emphasis was on demonstrating changes in immune markers during stress exposures in the laboratory (Segerstrom & Miller, 2004). However, as the clinical relevance of short-term immune changes in the laboratory began to receive more scrutiny (Kiecolt-Glaser, Cacioppo, Malarkey, & Glaser, 1992), the emphasis shifted to using salivary markers that are more ecologically valid. The earliest reports of using salivary assays of cytokines in the context of psychological factors emerged in the late 1990s across independent groups (Dugue (1996) included IL-2; Nishanian, Aziz, Chung, Detels, and Fahey (1998) studied soluble TNFα). Studies in early 2000s focused on changes in salivary immune markers as a function of stress exposures, including social (Dickerson, Kemeny, Aziz, Kim, & Fahey, 2004) and physical stressors [exercise—(Ilardo et al., 2001; Minetto et al., 2007); sauna—(Dugue, 1996)]. As the decade advanced, research interests expanded to include associations between salivary immune markers and naturalistic stress exposures [police officers' shifts—(Zefferino et al., 2006); exams—(Lester, Brown, Aycock, Grubbs, & Johnson, 2010)]; neural activity (Master et al., 2009; Slavich, Way, Eisenberger, & Taylor, 2010); emotional and interpersonal experiences (Chiang, Eisenberger, Seeman, & Taylor, 2012; Moons, Eisenberger, & Taylor, 2010); and development (Riis, Granger, Dipietro, Bandeen-Roche, & Johnson, 2015).

Modern PNI research has increasingly focused on how signals from the brain are transmitted to the immune system via neuroendocrine pathways, as well as the impact of immune system signals on neural functioning. For example, previous studies have shown peripheral inflammation is correlated with task-related activation of neural regions associated social-affective processes (Brydon, Harrison, Walker, Steptoe, & Critchley, 2008; Eisenberger, Inagaki, Rameson, Mashal, & Irwin, 2009; Muscatell et al., 2015; Prossin et al., 2011). Some research suggests that similar associations can be found with salivary inflammatory markers (O'Connor, Irwin, & Wellisch, 2009; Slavich et al., 2010). This may be partly due to a process whereby the brain receives information from the local trigeminal nerve during acute oral inflammation (Navarro, Iyomasa, Leite-Panissi, Almeida, & Branco, 2006). Integrating salivary immune processes with neural activity involves assessing multisystem functioning, such as examining whether changes in autonomic and hypothalamic–pituitary–adrenal (HPA) axis measures mediate links between psychosocial factors and immune changes; honing in on intracellular mechanisms linking neuroendocrine signals and changes in immune cells; and examining how immune signals influence critical cellular processes such as cell metabolism and repair. Developments in multiplex technology are now allowing PNI researchers to examine such cross-system relations with more analytes available for assay from single biospecimens. The clinical relevance of these neuroendocrine–immune studies for the health of individuals and groups is of ever-increasing interest.

## 9.3 Current State of Knowledge

This section synthesizes the current research examining salivary immune markers in humans including the study of cytokines, CRP, matrix metalloproteinases, and markers of humoral immune processes. These biomeasures are either directly produced by or act synergistically with the immune system in the oral compartment. While not an exhaustive list of salivary immune-related analytes, the analytes discussed below represent the most commonly examined immune markers in saliva.

## *9.3.1 Functions of Salivary Immune Markers*

**Salivary Cytokines** Cytokines are key signaling molecules of the immune system. While many types of cells may secrete cytokines, the primary sources of cytokine secretion are lymphoid cells that are involved in the initiation, amplification, or attenuation of immune activity. Cytokines are categorized into different families such as interleukins, interferons, growth factors, and tumor necrosis factors, and they differ in their sources of secretion and biologic activities [for review, see Granger, Granger, and Granger (2006)]. Although they are often grouped into pro-inflammatory (e.g., IL-1, TNF-α) and anti-inflammatory (e.g., IL-4, IL-10) categories, many cytokines can have both pro- and anti-inflammatory effects (e.g., IL-6) (Cavaillon, 2001). Cytokine receptors have been identified on central nervous and endocrine system cells [e.g., Besedovsky et al. (1983) and Miyake (2012)], meaning that cytokine concentrations may effect multiple biological systems, the brain, and human behavior (Maier & Watkins, 1998).

Cytokines in saliva have varying origins; they may be expressed by the salivary glands or resident immune cells in the mouth (e.g., in the gingiva), secreted from lymphoid cells that have migrated into the oral mucosa, or come from serum constituents that pass into the oral fluid from the general circulation (Brennan & Fox, 2010; Gröschl, 2009; Moutsopoulos & Konkel, 2018). Salivary cytokines secreted from cells in the oral compartment are thought to be involved in the coordination and mobilization of local oral immune processes (Brennan & Fox, 2010; Moutsopoulos & Konkel, 2018). Whereas cytokines from serum constituents that are found in saliva may, to some extent, reflect systemic immune processes.

**Salivary C-Reactive Protein** CRP is an acute phase protein primarily synthesized by the liver in response to inflammatory cytokines (Sproston & Ashworth, 2018). CRP activates the complement system, which assists the immune system in killing and clearing pathogens from the body (Sproston & Ashworth, 2018). As such, it is part of the body's systemic inflammatory response and increases quickly after tissue damage or an infection (Sproston & Ashworth, 2018). Both pro- and anti-inflammatory effects of CRP have been described (Sproston & Ashworth, 2018).

CRP in saliva is assumed to be an overflow from the blood compartment, perhaps entering saliva through the inflammatory exudate of gingival tissues (Giannobile et al., 2009), i.e., gingival crevicular fluid (Megson et al., 2010). CRP and other acute phase proteins can also pass from blood to saliva via diffusion through the porous capillaries around the salivary glands, or through a process called ultrafiltration, which is filtration through the spaces between salivary gland cells (Pfaffe, Cooper-white, Beyerlein, Kostner, & Punyadeera, 2011). Recent reports also find evidence of CRP being produced locally by the gingiva (Lu & Jin, 2010; Maekawa et al., 2011). Presumably, CRP has a similar role in saliva as it does in blood—to trigger an immune response. However, no studies to date have examined this explicitly; rather most research has focused on salivary CRP as a biomarker of systemic inflammation.

**Matrix Metalloproteinases** Matrix metalloproteinases (MMPs) are a family of enzymes that activate leukocytes and help regulate the immune and inflammatory response in the oral cavity (Hannas, Pereira, Granjeiro, & Tjäderhane, 2007; Smigiel & Parks, 2017). MMPs are involved in tissue degradation and restructuring, as well as cell proliferation, migration, and apoptosis (Hannas et al., 2007). The study of salivary MMPs has primarily focused on their ability to index oral health problems. Salivary MMP-8 may be a central biomarker for periodontal disease (Hannas et al., 2007; Zhang, Li, Yan, & Huang, 2018) and has been associated with oral HPV infection (Haukioja, Tervahartiala, Sorsa, & Syrjänen, 2017). In addition, MMP-9 may be a biomarker of malignant disorders (Venugopal & Maheswari, 2016). MMP inhibitors (e.g., TIMP-1) have also been examined in conjunction with salivary MMPs as a measure of MMP regulation [e.g., with HPV; (Haukioja et al., 2017)].

**Humoral Immune Markers** Immunoglobulins, also known as antibodies, are secreted by B cells and are key aspects of humoral immunity. Immunoglobulins bind to and neutralize specific pathogens, including viruses, toxins, bacteria, and fungi. In salivary research, virus-specific immunoglobulins (e.g., for the Epstein–Barr virus and cytomegalovirus) are often examined as indices of viral exposure or viral load. Chapter 13 reviews salivary antibodies as indicies of pathogen exposure and infection. In this chapter, we focus on total concentrations of select salivary immunoglobulins.

Total salivary concentrations of immunoglobulins A and G are commonly examined in salivary bioscience research and provide important information about humoral immunity in the oral cavity. Secretory IgA (SIgA) and IgG are the two most common immunoglobulins in saliva (Brandtzaeg, 2013). IgG enters saliva primarily from circulation through crevicular fluid (Brandtzaeg, 2013), and some IgG is produced within the oral cavity by the salivary glands and gums (Brandtzaeg, 2013). IgG protects the body from infection by a variety of viral, bacterial, and fungal pathogens. For these reasons, whole saliva total IgG may be a valuable indicator of oral mucosal inflammatory disease, such as periodontitis (Taubman & Smith, 1993).

SIgA, the secreted form of the IgA antibody, is integral to the immune function of mucous membranes and plays major role in gut immunity and mucosal homeostasis (Mantis, Rol, & Corthésy, 2011). Salivary SIgA is produced by plasma cells in the salivary glands and is secreted via exocytosis (Brandtzaeg, 2013). As part of the adaptive immune system, salivary SIgA is more dynamic than other salivary proteins and its secretion from the parotid gland is reportedly more stable than its oral concentration (Brandtzaeg, 2013). SIgA in the mouth plays an important role in protecting epithelial cells and teeth from bacteria, toxins, and viruses (Teeuw, Bosch, Veerman, & Nieuw Amerongen, 2004).

### 9.3.2 Do Salivary Immune Markers Reflect Systemic Immune Processes?

The strength of the associations between immune and inflammatory markers measured in blood and whole saliva differ depending on the marker. Below we summarize the current understanding of the extent to which key salivary immune markers represent systemic immune functioning, and we offer suggestions regarding the interpretation of salivary immune marker concentrations.

**Salivary Cytokines** There is limited evidence supporting significant associations between concentrations of cytokines measured in saliva and blood. Importantly, these cross-specimen relations also vary considerably by cytokine. For example, Williamson and colleagues (2012) examined 27 cytokines in saliva and plasma samples from healthy adults and found that only three, interferon (IFN)-ɣ, IL-6, and macrophage inflammatory protein-1β, showed significant associations across biospecimen (at medium effect sizes—correlation coefficients = 0.31 – 0.34). A significant serum–saliva correlation for IFN-ɣ (as well as IL-2 and IL-12p70) was also found in a study of adolescents (Byrne et al., 2013). However, these relations were only significant when non-detectable cytokine concentrations (>80% of serum concentrations) were replaced with zero, rather than excluding them (Byrne et al., 2013). A significant and moderate saliva to blood correlation was also reported for IL-6 in a sample of young men (Nam, Kim, Chang, & Kho, 2019), and Fernandez-Botran and colleagues (2011) also reported a modest plasma–saliva correlation ($r = 0.29$, $p < 0.05$) for IL-6 in a sample of post-menopausal women. However, this correlation was only significant for biospecimens collected on one of two study visits; the correlation between plasma and salivary IL-6 collected during the second visit was only 0.10 ($p = 0.41$) (Fernandez-Botran et al., 2011). Other studies have also failed to find an association between salivary and plasma levels of IL-6 (e.g., Cullen et al. 2015) including another study of post-menopausal women which found that GM-CSF and IL-5 demonstrated significant serum–saliva correlations, but levels of IFN-ɣ, IL-10, IL-1β, IL-2, IL-4, IL-6, IL-8, and TNF-α did not (Browne et al., 2013). Furthermore, Riis and colleagues (2014) measured GM-CSF, IFN-ɣ, IL-1β, IL-2, IL-6, IL-8, IL-10, IL-12p70, and TNF-α in serum and saliva samples from adolescent girls and found that only IL-1β showed a significant serum–saliva correlation. However, this association was weak and not consistent across time (Riis et al., 2014). It is important to note that the latter study statistically corrected for multiple comparisons, which is not yet standard in salivary bioscience studies but may be appropriate in studies examining multiple immune markers simultaneously. Given the current evidence, salivary cytokine concentrations should generally be interpreted as measures of oral immune processes, rather than systemic immune function.

**C-Reactive Protein** Of all the immune markers discussed in this chapter, salivary CRP has the strongest evidence supporting a significant and meaningful correlation between systemic and salivary concentrations. Several studies have shown significant correlations between blood and salivary concentrations of CRP (Browne et al., 2013; Byrne et al., 2013; La Fratta et al., 2018; Ouellet-Morin, Danese, Williams, & Arseneault, 2011; Out, Hall, Granger, Page, & Woods, 2012; Pay & Shaw, 2019; Punyadeera, Dimeski, Kostner, Beyerlein, & Cooper-White, 2011). These studies find medium to large effect sizes for correlations between serum or plasma and salivary CRP (correlation coefficients = 0.38 – 0.92). Significant serum–saliva correlations for CRP are not, however, a universal finding [e.g., Dillon et al. (2010) and Pay and Shaw (2019)]. Although additional research is needed in healthy and clinical populations, the extant literature suggests that CRP may be a salivary immune marker that has significant associations with systemic inflammatory processes and measures of physical and psychological health.

**Salivary MMPs and Immunoglobulins** Salivary MMPs and immunoglobulins are typically used to index oral mucosal immune processes (Lahdentausta et al., 2018; Rathnayake et al., 2013; Salminen et al., 2014) and are not generally examined in relation to serum levels. SIgA is secreted from the mucous membranes, so associations with systemic IgA are not generally examined. While most salivary IgG leaks into the oral compartment from the blood, salivary IgG concentrations are thought to correlate with oral health and periodontal problems (Brandtzaeg, 2013; Taubman & Smith, 1993). Therefore, the primary utility of these indices is likely as measures of local oral and/or mucosal, rather than systemic health.

**Recommendations** The current literature does not support the use of most salivary immune markers as indices of systemic health. Most of these studies, however, examined relations within healthy samples and some used assay kits not validated for use with saliva. It is also important to note that the strength and nature of associations between immune markers in saliva and blood may depend on the oral and/or physical health of the population studied. Acute illness or inflammation of either the oral compartment (e.g., periodontitis) or systemically (e.g., infection) may affect the extent to which salivary and serum immune markers are correlated. For example, in a sample of healthy adolescents with no signs of oral disease, the serum–saliva correlation for CRP varied by level of serum CRP (Byrne et al., 2013). Among adolescents with higher levels of systemic inflammation (higher serum CRP), there were similarly high levels of salivary CRP, and CRP was significantly associated across biospecimen (serum–saliva correlation: $r = 0.62$, $p < 0.01$) (Byrne et al., 2013). In contrast, among adolescents with lower levels of systemic inflammation (lower serum CRP), salivary CRP did not correlate with serum CRP (serum–saliva correlation: $r = 0.11$, $p = 0.79$) (Byrne et al., 2013). Similar findings have been found in other studies of salivary and serum CRP (Pay & Shaw, 2019).

In addition to oral and physical health, age and developmental stage may also influence the strength and nature of serum–saliva correlations for salivary immune markers. For example, very early in development, before the emergence of teeth, and during old age, when there is typically an increase in medication use and/or disease pathologies, may represent periods of life when salivary immune markers are more or less related to systemic immune function. Variation in serum–saliva correlations by age and health/disease status, however, has not yet been fully examined in the literature.

It is also important to consider the coordinated and synergistic nature of immune processes in the oral cavity when interpreting serum–saliva associations. Salivary immune markers tend to be highly intercorrelated and more variable than immune markers in serum (Riis et al., 2014). The dynamic local immune environment of the oral cavity therefore likely influences salivary concentrations of immunosensitive markers more so than serum concentrations. For example, adiponectin, an immunosensitive biomeasure of metabolic function, exhibits a stronger serum–saliva correlation after accounting for variance related to oral inflammation (salivary cytokines and MMP-8) (Riis et al., 2017). Similar processes likely affect the serum–saliva associations of immune markers such as CRP and cytokines. Recent findings that CRP is produced within the oral compartment (Lu & Jin, 2010; Maekawa et al., 2011) highlight the importance of controlling for local oral immune processes even when significant serum–saliva associations are found. Future studies that are able to parse the variance in salivary immune markers due to oral and systemic health will provide important information about the extent to which immune measures in saliva may reflect systemic immune processes. Furthermore, rigorous examinations of salivary immune markers that consider oral and systemic disease states and include diverse samples are needed to fully understand the utility of salivary immune markers as indices or oral and systemic health.

### 9.3.3 Are Salivary Immune Markers Sensitive to Stress?

The opportunity to examine neuroendocrine–immune relations with salivary biomeasures has been an exciting advancement for biobehavioral and health researchers, especially those in the field of PNI. Encouraging findings from salivary immune marker studies illustrate the bidirectional relationship between stress, stress-related pathologies, and immunity. For example, salivary measures of immunity (i.e., salivary SIgA, SIgA1, SIgA2, and secretory component) and early-life inflammatory events (indexed by salivary SIgA) have been correlated with various pathologies including anxiety and post-traumatic stress disorder (Ulmer-Yaniv et al., 2018). Others have used salivary immune markers to show a link between immunocompetence and psychological stress [e.g., Engeland et al. (2016)].

Despite a relatively deep and growing literature examining neuroendocrine–immune relations in saliva and the stress sensitivity of salivary immune markers, the interpretation of these findings is complicated by the study of a wide range of populations and stressor paradigms in a variety of laboratory and real-world settings. It is also important to note that the sensitivity of an individual's immune and inflammatory responses to stress are also influenced by several factors, including age and developmental stage, history of chronic and acute stress exposure, the nature of their neuroendocrine response, and the nature and intensity of the stressor (acute vs. chronic). Methodological oversights, such as inadequate adjustment for oral health and salivary flow rate, also make cross-study comparisons difficult. With these caveats in mind, we present the current findings of stress-related change in salivary cytokines, CRP, MMPs, and immunoglobulins below.

**Cytokines** Although studies are limited, acute laboratory stress has been somewhat consistently associated with increased levels of inflammatory cytokines in saliva including IL-1β, TNF-α, and IL-6 [for review, see Slavish, Graham-Engeland, Smyth, and Engeland (2015a)]. Furthermore, there is evidence from one study that salivary inflammatory reactivity to acute social-evaluative stress is positively associated with neural activity during a social rejection task (Slavich et al., 2010).

In assessing the strength and nature of salivary cytokine stress responses, it is important to consider how individual and environmental factors may influence the immune response. For example, there is some evidence that age and sex may moderate the salivary cytokine response to acute stress. One study of adults found greater increases in salivary cytokine (IL-6, IL-8, IL-10, and IL-4) concentrations after a pain stressor task among older, compared to younger, adults (mean age of older adults = 68.3; mean age of younger adults = 21.4 years) (Sorenson et al., 2017). The use of a pain stressor task in this study is important to note, as physical stressor tasks may not elicit the same type of neuroendocrine stress response as social-evaluative, emotional, and/or socio-cognitive stressor tasks employed by other studies. However, consistent with a pattern of age-related increases in salivary cytokine stress reactivity, two studies of preschool-aged children found stable and/or decreasing pro-inflammatory salivary cytokine trajectories across a series of emotional and cognitive lab stressor tasks (Riis et al., 2015; Tyrka, Parade, Valentine, Eslinger, & Seifer, 2015). While a study of elementary school-aged children (mean age = 8.75) found mixed stress-related trajectories for salivary IL-6 with 55% decreasing and 45% increasing in IL-6 concentrations across the study tasks (El-Sheikh, Buckhalt, Granger, Erath, & Acebo, 2007). Two of these studies of cytokine stress reactivity among children also found stress-related pro-inflammatory salivary cytokine trajectories varied by sex (El-Sheikh et al., 2007; Riis et al., 2015).

Beyond sample demographics, participant characteristics and their subjective approach to the lab stressor tasks may further moderate stress-related salivary cytokine concentrations. For example, individual affect, cognitive control, and attention to the task, as well as history of trauma and level of perceived discrimination, have all been shown to moderate pro-inflammatory cytokine stress responses in the lab (Maydych, Claus, Watz, & Kleinsorge, 2018; Newton et al., 2017; Shields, Kuchenbecker, Pressman, Sumida, & Slavich, 2016; Szabo, Fernandez-Botran, & Newton, 2019)

Outside the laboratory, several studies find associations between real-world stressors and levels of pro-inflammatory cytokines. Early-life adversity, fear of deportation, and chronic family stress have been associated with elevated baseline concentrations of inflammatory cytokines in saliva (Martínez, Ruelas, & Granger, 2018; Tyrka et al., 2015). Perceived discrimination stress has also been associated with baseline salivary IL-6 concentrations among homosexual adults (Doyle & Molix, 2016). However, echoing the importance of accounting for individual differences, these relations varied by sex and individual expression of identity (Doyle & Molix, 2016). Associations between real-life acute stress and salivary markers of inflammation are similarly complex. La Fratta and colleagues found that salivary IL-1β and IL-8 were higher in anticipation of an academic exam and decreased after, while IL-6 increased from pre- to post-exam (2018). Importantly, these patterns of stress-related change in salivary cytokines were mirrored in plasma cytokine levels (La Fratta et al., 2018). Similar findings were reported from a study of undergraduate students performing a real-life public speaking task; salivary IL-1β decreased from pre- to post-task (Auer, Calvi, Jordan, Schrader, & Byrd-Craven, 2018). Despite an overall declining IL-1β trajectory, participants with higher levels of social anxiety had greater increases in IL-1β concentrations across the task (Auer et al., 2018). These real-world examinations of acute stress and inflammation in the oral compartment may seem contradictory to lab-based study findings of increases in pro-inflammatory cytokines in response to stressor tasks. However, pre-stressor levels in real-world studies likely reflect anticipatory stress levels, rather than true baseline levels (as measured in the lab).

**CRP** There are only a few studies examining acute stress-related changes in salivary CRP, and these studies point to a complex stress response model for salivary CRP. In a series of studies of African American adults, Lucas and colleagues found that changes in salivary CRP across the Trier Social Stress Test (TSST) were related to self-reported subjective stress, stress appraisal, affect, and level of perceived discrimination during the task (Laurent, Lucas, Pierce, Goetz, & Granger, 2016; Lucas et al., 2016, 2017). Furthermore, relations between stress-related levels of salivary CRP and cortisol also varied by individual stress appraisals during the task (Laurent et al., 2016). Given the complexities demonstrated in the Lucas studies, it is not surprising that other studies have failed to find significant changes in salivary CRP across the TSST [e.g., Campisi, Bravo, Cole, and Gobeil (2012)]. Outside of the laboratory, a study of acute socio-cognitive exam stress also found no significant stress-related changes in salivary nor serum CPR among young adult men (La Fratta et al., 2018).

Studies of chronic stress and adversity, however, report relatively consistent associations between life stressors and salivary CRP. For example, poor parental monitoring and fewer positive parenting behaviors have been associated with higher levels of salivary CRP in children and adolescents (Byrne, Badcock, et al., 2017; Byrne, Horne, et al., 2017). Similarly, early-life adversity has been linked to higher levels of salivary CRP among infants (David, Measelle, Ostlund, & Ablow, 2017; Measelle, David, & Ablow, 2017). However, similar relations were not found when examined among young children [e.g., Tyrka et al. (2015)]. Underscoring the complexity of the salivary CRP stress relations and their implications for health and well-being, Maldonado and colleagues found that salivary CRP moderated the relation between acculturative stress and anxiety in a sample of adult Latino Americans (2018).

**Immunoglobulins** SIgA secretion is under neuroendocrine control and salivary concentrations are sensitive to acute and chronic stress (Brandtzaeg, 2013; Teeuw et al., 2004). There is evidence that SIgA increases in response to acute stress and is suppressed during chronic stress (Birkett, Johnson, & Gelety, 2017; Brandtzaeg, 2013; Laurent, Stroud, Brush, D'Angelo, & Granger, 2015; Ohira, 2005; Phillips et al., 2006). Chronic psychological stressors have been associated with lower concentrations of the IgA1 subtype in particular (Engeland et al., 2016). However, neuroendocrine stress responses, affect, externalizing problems, and emotional support experienced during the acute stressor may moderate SIgA changes in response to acute stress (Laurent et al., 2015; Ohira, 2005).

Results from studies of real-life stressors are mixed. A large study of adults found that SIgA secretion rate was inversely associated with life stress load (Phillips et al., 2006). However, results from a study of mother–child dyads found higher SIgA concentrations among war-exposed women and adolescents compared to controls, even a decade post trauma (Ulmer-Yaniv, Djalovski, Priel, Zagoory-Sharon, & Feldman, 2018; Yirmiya, Djalovski, Motsan, Zagoory-Sharon, & Feldman, 2018). In the same sample, SIgA concentrations were positively associated with symptoms of depression among mothers, and internalizing, externalizing, and anxiety symptoms among their children (Ulmer-Yaniv, Djalovski, Priel, et al., 2018; Yirmiya et al., 2018). Byrne and colleagues found complementary results with higher child salivary SIgA concentrations associated with poor parental monitoring (2017).

**Exercise Stress** Both immune (e.g., SIgA and IgG) and inflammatory (e.g., CRP) markers are implicated in the healthy adaptation to exercise stress and training. Inflammation is a necessary component of muscle regeneration, and it has been shown that use of non-steroidal anti-inflammatory drugs can attenuate muscle protein synthesis (Bondesen, Mills, Kegley, & Pavlath, 2004). Findings in salivary immune markers are mixed. Some studies of salivary SIgA in response to acute exercise find no change in analyte concentrations and secretion rates, while others find increases or decreases across the exercise and recovery period (Campbell & Turner, 2018; Keaney, Kilding, Merien, & Dulson, 2018). The effects of longer physical training periods among athletes have found reductions in SIgA during high training periods (Keaney et al., 2018). Although, these findings are not universal, and the implications for health are unclear (Campbell & Turner, 2018).

Reports on salivary CRP responses to exercise stress are limited. A pilot study of men who completed a maximal effort exercise test found no changes in salivary CRP across the task (Hernandez, Fuller, Stone, Carpenter, & Taylor, 2016). In ultra-endurance athletes, there was an increase in salivary CRP after a long-distance run, but not after an open water swim (Tauler, Martinez, Moreno, Martínez, & Aguilo, 2014). Exercise-related increases in salivary cytokines, including IL-1β, IL-6, TNF-α, and IL-8, have been found among adults (Hayashida, 2016; Slavish et al., 2015b). However, these effects are not universal. Exercise-related changes in IL-6 are especially varied and may depend on the type of exercise and the sample collection protocol (e.g., swab vs. passive drool) (Cullen et al., 2015; Hayashida, 2016; Slavish et al., 2015b).

**Recommendations** Several studies have shown that salivary immune markers differ with respect to the sensitivity and directionality of changes in response to stress and these responses vary by the type of stress exposure (Slavish et al., 2015b). Therefore, while many of the salivary immune markers of interest exhibit synergistic physiologic effects, hypotheses about salivary immune marker stress reactivity should consider each analyte individually as well as part of a coordinated immune response in the oral cavity.

Relations between the immune markers and biomeasures of other physiologic systems, such as the HPA and autonomic nervous system, are also particularly important to consider in the context of the stress response. The neuroendocrine stress response may be associated with the magnitude and/or timing of the immune response. There is very little research examining the timing of immune and inflammatory stress responses in the oral compartment and how these stress response profiles vary by individual and stressor characteristics. For example, salivary cortisol research has shown that basal and stress-related hormone profiles may be confounded by personal history of stress exposure, so it is important to control for these confounds or focus on within-person changes when evaluating these hormone levels (Lucas et al., 2016; Slavish et al., 2015a). Similar confounds may affect salivary immune marker concentrations and dynamics. Future studies should directly address these sources of variability.

### *9.3.4 Do Salivary Immune Markers Exhibit Developmental Trajectories?*

The immune system begins developing in utero and continues throughout early childhood (Veru, Laplante, Luheshi, & King, 2014). While adult levels of immunocompetence are typically achieved around age five (Veru et al., 2014), the immune system continues to adapt throughout life to support tissue repair and growth, as well as protect the body against toxic exposures, pathogens, and disease. The flexibility of the immune system in the oral cavity may be particularly important for protecting and maintaining health in the context of a dynamic and variable antigen environment. There are also developmental changes specific to the oral cavity, such as the growth and loss of teeth and the progression of oral diseases, which may influence oral immune function.

Age-related changes in resting concentrations of many salivary immune markers, including cytokines and MMPs, have not been rigorously examined. Many immunoglobulin concentrations in saliva vary across age group. Salivary immunoglobulin levels may be particularly variable during infancy when salivary gland development, breastfeeding, and immunization exposures likely influence their measured concentrations (Brandtzaeg, 2013). SIgA is detectable in early infancy and may increase with age; although this is not a universal finding [e.g. Evans et al. (2000) and Ewing et al. (2010)] and age trajectories may vary by SIgA subtype (Jafarzadeh, Sadeghi, Karam, & Vazirinejad, 2010; Weemaes et al., 2003).

Salivary CRP concentrations may also increase with age. Looking at salivary CRP concentrations across studies, CRP in saliva tends to be higher in adults compared to children and adolescents. Average or median salivary CRP concentrations in studies of adults typically range from approximately 100–8500 pg/ml (Browne et al., 2013; Laurent et al., 2016; Mohamed, Campbell, Cooper-White, Dimeski, & Punyadeera, 2012; Ouellet-Morin et al., 2011; Out et al., 2012), while levels in adolescents and children are generally lower with mean or median levels in these studies ranging from approximately 7–2030 pg/ml (Byrne, Badcock, et al., 2017; Byrne, Horne, et al., 2017; Byrne et al., 2013; Cullen et al., 2017; Goodson et al., 2014; Naidoo, Konkol, Biccard, Dubose, & McKune, 2012; Shields, Slavich, Perlman, Klein, & Kotov, 2019). Importantly, these summary statistics are drawn from studies of samples across a wide range of ages and body compositions with varying saliva sample collection times, and some measurements were adjusted for flow rate while others were not. Other factors, such as health status and smoking behaviors, may also be confounding the observed age differences in salivary CRP. Additional research is needed to understand the underlying causes and correlates of developmental differences in salivary CRP.

**Recommendations** Our current understanding of age-related changes in salivary immune markers is largely based on cross-sectional studies and cohort effects. Rigorous, longitudinal examinations of developmental changes in salivary immune markers are needed to further our understanding of age-related changes in the oral immune environment and to help us interpret age-related differences in salivary immune markers. These studies should consider oral and systemic health factors, such as oral health and disease, obesity, diet, and chronic stress exposure, which may influence immune marker concentrations and contribute to age-related differences in analyte concentrations.

## 9.4 Methodological Issues

In some ways, the widespread incorporation of salivary immune markers in biobehavioral and health research has pushed the study of salivary immune analytes out beyond our methodological understanding of these biomeasures. In this section, we provide our recommendations for current and future research to advance our understanding of salivary immune markers in studies of health and well-being.

### *9.4.1 Saliva Collection and Processing Methods*

In the absence of rigorous research examining the impact of collection technique on each salivary immune marker, the collection of whole unstimulated saliva via passive drool is the recommended "universal" method for saliva collection. Collecting whole saliva allows for the examination of the widest range of salivary analytes and avoids inconsistencies in concentrations of some immune markers that may depend on the idiosyncrasies in oral fluids produced by different salivary glands (Ruhl et al., 2004). SIgA may be particularly sensitive to sample collection and processing procedures (Brandtzaeg, 2007). Investigators should review the existing literature on their specific analyte and consult with their assay manufacturer and laboratory staff when determining sample collection and processing protocols.

With approximately 30,000 neutrophils entering the gingival crevice from the blood every minute and cytokine expression across various oral glands and tissues (Gröschl, 2009; Moutsopoulos & Konkel, 2018), differences in immune marker concentrations across biospecimen type (e.g., whole saliva vs. crevicular fluid) is expected. For example, IgG concentrations will be higher in crevicular fluid than in whole saliva as IgG enters the oral compartment primarily via crevicular fluid and may be diluted in whole saliva (Brandtzaeg, 2013). Biospecimen type may also influence the strength of the association between oral immune markers and those measured in serum (Nishanian et al., 1998).

Salivary flow rate is another important factor to consider when examining salivary immune marker concentrations. Analytes that are produced locally, such as SIgA, and those that are brought into saliva from circulation via the crevicular fluid, such as salivary CRP and total IgG, have known associations with flow rate (Brandtzaeg, 2013; Pay & Shaw, 2019). The impact of flow rate on other salivary immune markers, such as salivary cytokines, is less clear. Given the lack of rigorous investigation of flow rate and immune markers in saliva, investigators should record saliva collection duration and sample weight in studies that will examine salivary immune markers. This will allow researchers to assess the impact of flow rate on analyte concentrations. Chapter 4 discusses flow rate calculations and statistical adjustments for flow rate.

### *9.4.2 Statistical Considerations and Cautions*

Salivary immune markers typically display a strong positive skew with a long tail of very high concentrations. Some markers, like IL-10, are typically present at very low concentrations and may even be undetectable with current assay technology (Browne et al., 2013; Byrne et al., 2013; Riis et al., 2014; Shields et al., 2019). Others, like IL-8 and IL-1β, are generally present at very high concentrations in saliva (Browne et al., 2013; Byrne et al., 2013; Riis et al., 2014, 2015, 2017; Shields et al., 2019). The interpretation and statistical handling of such varying levels of salivary immune markers is complicated by a lack of established norms and cutoffs. We refer investigators to Chap. 4 for a review of how to address kurtotic and skewed distributions. The pre-analytic processing of salivary immune marker data is especially important given their unique distribution and ranges which may preclude the use of raw analyte data in parametric statistical models. We recommend that researchers report their data processing steps and the results from both raw and cleaned (e.g., extrapolated, replaced, or transformed) data in their manuscripts.

The immunoassay and multiplex assay technology used to measure immune markers in saliva is very advanced. Commercially available tests consistently report detection limits in the very low pg/mL range, and modern multiplexing assays have log-scale dynamic ranges. These advances in technology allow for the simultaneous testing of a large number of analytes. While this provides a wealth of information, highly correlated determinations of salivary immune markers from each individual may require advanced statistical modeling to parse the variance and individual effects of each immune marker alone and as part of a coordinated immune response. Investigators should also be particularly aware of issues related to collinearity among salivary immune markers. Chapter 4 reviews statistical approaches and methods used in salivary biomeasure studies. We also present code for an interactive case study with mock salivary immune marker data here (https://github.com/michellebyrne1/immune-multiplex). This web link allows the reader to explore issues related to salivary immune marker multiplexing, collinearity, and data cleaning, and how differences in data processing steps may affect study results. It is important to note that the data cleaning procedures included in our web link serve as simplified examples of steps investigators have traditionally taken to help visualize and process their salivary immune marker data. Some of these approaches are now outdated, and we recommend researchers read Chap. 4 for a detailed review of and guidelines for salivary biomeasure data cleaning and analysis.

While there are generally high correlations among salivary inflammatory markers, analytes should be examined both together, as an overall index of inflammation, and as separate analytes with potentially unique mechanisms of action and associations with other analytes and health conditions. The production of salivary immune markers and their individual roles in both local and systemic immune processes may be unique. For example, IL-8 is released by neutrophils, the most common immune cell entering the mouth, and is involved in neutrophil migration and inflammation in the oral compartment (Hasturk, Kantarci, & Van Dyke, 2012; Moutsopoulos & Konkel, 2018). With concentrations of IL-8 in saliva generally higher than many other cytokines [e.g., Browne et al. (2013), Byrne et al. (2013), La Fratta et al. (2018), Riis et al. (2014, 2015, 2017), and Shields et al. (2019)], salivary IL-8 may play an important and specific role in oral health and disease [e.g., Belstrøm et al. (2017), Finoti et al. (2017), Sahibzada et al. (2017), and St. John et al. (2004)]. Despite robust correlations between IL-8 and other inflammatory cytokines, IL-8 may have unique physiologic effects worthy of independent investigation.

Finally, rapid technological advancements in our ability to study salivary immune markers should be paired with rigorous and disciplined scientific study designs and analytic testing. Unless research studies are explicitly exploratory, specific hypotheses for each immune marker and research question should be stated a priori. Researchers can preregister these hypotheses on a platform such as the Open Science Framework. Even when hypotheses are exploratory, researchers should correct for multiple comparisons by using a Bonferroni correction, or similar method for conducting multiple statistical tests, in order to reduce Type I error [e.g., Riis et al. (2014)].

### 9.4.3 Confounding Factors and Covariates

**Smoking Behavior and Tobacco Exposure** Smoking is an important confounder in the measurement of salivary immune markers. Smoking may have acute and chronic effects on the oral immune environment. Levels of salivary cytokines, CRP, immunoglobulins, and MMPs have been associated with smoking and environmental tobacco smoke exposure (Azar & Richard, 2011; Evans et al., 2000; Ewing et al., 2010; Haukioja et al., 2017; Lira-Junior, Åkerman, Gustafsson, Klinge, & Boström, 2017; Riis et al., 2014, 2015). Oral health and disease are also strongly associated with smoking behavior and environmental tobacco smoke exposure (Hanioka et al., 2019; Hasmun et al., 2017; Javed, Ahmed, & Romanos, 2014). In the absence of self-reported smoking data, salivary cotinine can be assessed to test for exposure to nicotine. However, cotinine levels will only reflect recent exposure (approximately exposure in the past 16 h) and will not provide information about the duration of smoking behavior which may be important for understanding the impact of smoking on the oral immune system (Jarvis, Russell, Benowitz, & Feyerabend, 1988).

**Blood Leakage into Saliva** Given their role in maintaining oral health and their potential transport into saliva via crevicular fluid, many of the salivary immune markers discussed in this chapter may be influenced by blood leakage into saliva. Associations between immune makers and blood leakage may be due to increased passage of serum immune markers into saliva and/or increased local production of salivary immune markers in response to injury in the oral compartment. As such, blood leakage may affect both salivary immune maker levels and the observed serum–saliva correlation.

Saliva samples should be visually examined for contamination by blood and flagged if there is evidence of contamination [see Kivlighan et al. (2004) for details]. Blood leakage in the oral compartment can also be assessed with salivary transferrin and used as a statistical control.

**Oral Health** In dentistry, periodontology, oral cancer, and public oral health research, oral immune markers, such as MMPs, immunoglobulins, and cytokines, have been associated with oral inflammation, disease, and health (Rangbulla, Nirola, Gupta, & Batra, 2017; Rathnayake et al., 2013; Salminen et al., 2014; Taba, Kinney, Kim, & Giannobile, 2005; Taubman & Smith, 1993). Nearly half of all adults in the USA have periodontitis and rates of periodontitis vary by racial/ethnic group (Eke et al., 2018). Biobehavioral and health researchers, however, often fail to appropriately assess and adjust for oral health in their studies of salivary immune markers. Unexamined associations between oral health and salivary immune marker concentrations may have considerable impacts on study findings that aim to use salivary immune markers as indices of systemic health. In conceptualizing the role of oral health in these studies, it is important to consider the many ways in which oral immune function may be related to systemic health. For example, oral immune function may: (1) initiate a systemic immune response and/or influence systemic health; (2) be initiated and/or influenced by systemic immune function; and (3) reflect a number of health and immune-related factors that are related to both systemic and mucosal health. Building our understanding of the complex, bidirectional relations among oral, mucosal, and systemic health and immune function requires a thoughtful, interdisciplinary approach to study planning and hypothesis generation, as well as careful measurement and adjustment for potential confounding factors.

The study of cardiovascular disease provides an interesting case study in our developing understanding of oral–systemic immune relations. Atherosclerosis is the primary contributor to cardiovascular disease (CVD). The early phase of atherosclerosis includes the induction of an inflammatory response which eventually leads to arterial plaque development (Berliner et al., 1995). Periodontal disease (PD) is a bacteria-induced infection which also elicits an inflammatory response (Bascones-Martínez et al., 2009). An initial study in 1989 reported that oral health status was associated with myocardial infarctions (MI), commonly known as heart attacks (Mattila et al., 1989). While a causal relation between PD and atherosclerotic vascular disease has not been established, the evidence to date supports a modest association between these two diseases (Lockhart et al., 2012; Scannapieco, Bush, & Paju, 2003). These relations may hint at important cross-system connections between oral disease and other physical and mental health disorders as dysregulated inflammation is considered a pathophysiologic mechanism for a number of health conditions, including diabetes, depression, and Alzheimer's Disease (Dai, Golembiewska, Lindholm, & Stenvinkel, 2017; Manthiram, Zhou, Aksentijevich, & Kastner, 2017; Saltiel & Olefsky, 2017).

Covariation among oral health problems and systemic health and disease risk is an interesting and important area of research to which salivary bioscience investigators can make substantial contributions via the study of salivary immune markers. While aiming to disentangle the effects of oral, systemic, and mucosal immune function on health, this line of research will also need to examine covariation in oral and systemic health from a broader, social perspective. In the USA, there are vast overlapping disparities in health and access to care for both oral and systemic health; in many cases, individuals at higher risk for oral health problems are also at higher risk for physical and mental health problems (U.S. Department of Health and Human Services, 2000). This covariation is especially important in studies that use salivary immune markers to study the impact of environmental and psychosocial exposures on health. Failing to adjust for oral health in these studies may considerably affect the interpretation and implications of the findings.

The most common approach to assessing oral health in biobehavioral and health studies is via self-report questionnaires. There are several self-report tools and approaches available to assess participants' oral health [e.g., Fisher-Owens et al. (2007), Liu et al. (2016), Nirmal, Ramachandiran, Anand, Sathiamurthy, and Sekaran (2005), and World Health Organization (2013)]. Associations between responses on these questionnaires and levels of salivary analytes related to oral inflammation and disease (cytokines, MMPs, immunoglobulins) have not been fully examined, and, in some cases, no relations have been found [e.g., Riis et al. (2017)]. Concentrations of salivary immune markers, such as MMPs, cytokines, and immunoglobulins may also be used to statistically control for variance associated with oral health (Riis et al., 2017); however, these approaches have also not yet been validated and may not be appropriate when the focus of investigation is inflammation. With marked disparities in oral health and access to oral health care, conceptual overlap among health behaviors for oral and physical health, and known associations between oral, mucosal, physical, and mental health, investigators should carefully consider the role of oral health and inflammation in their research questions and statistical models.

**Sex** Sexual differentiation of the immune system begins in utero and differences in inflammatory and immune responses by sex are seen across the lifespan (Bouman, Jan Heineman, & Faas, 2005; Klein & Flanagan, 2016; Taneja, 2018). Sex hormones, such as estradiol, may also influence immune marker levels in saliva (Teeuw et al., 2004), and secretion rates of salivary SIgA may vary by sex (Rutherfurd-Markwick, Starck, Dulson, & Ali, 2017). Sex-related differences in salivary immune markers can be seen in children as young as 5 years old (El-Sheikh et al., 2007; Riis et al., 2015, 2016). Sex may influence the levels of immune markers in saliva, as well as their stress sensitivity and associations with other biomeasures (Riis et al., 2015, 2016; Teeuw et al., 2004). Sex may also interact with other important covariates, such as age and developmental stage. For example, the developmental trajectories and the effects of pubertal status and timing on salivary acute phase proteins and cytokines are not yet known. Investigations using salivary immune markers should examine the effects of sex, and sample size calculations may need to anticipate sex stratified analyses.

**Other Health-Related Variables** Other factors including mood, psychosocial stressors, sleep, diet, medication use, body mass index, menstrual phase, pregnancy, current physical health, and physical activity [for review, see O'Connor et al. (2009)] should be considered as possible covariates in studies examining salivary immune markers. Factors related to systemic health and functioning have the potential to confound salivary immune measures as well as their relations with health conditions of interest. For example, a meta-analysis examining the methodological rigor in studies of serum CRP and its relations with depression found that only a small number of studies controlled for age, sex, obesity, medical conditions, substance/medication use, and psychosocial factors (Horn et al., 2018). Importantly, the effect size of the association between serum CRP and depression was small in the studies that controlled for these confounders (Horn et al., 2018). Similar examinations are needed for the appropriate interpretation of salivary immune marker study findings.

**Time Since Waking and Diurnal Patterns** Circadian and diurnal rhythms are essential for health and wellness. Salivary immune markers may provide insight into these patterns in ways that would not be possible with traditional blood-based assessments of immune function. Unfortunately, to date, few studies have examined the variability in salivary immune markers across the day.

Diurnal patterns have been reported for SIgA (Hucklebridge, Clow, & Evans, 1998; Li & Gleeson, 2004; Pritchard, Stanton, Lord, Petocz, & Pepping, 2017), salivary IL-6 (Izawa, Miki, Liu, & Ogawa, 2013; Reinhardt, Fernandes, Markus, & Fischer, 2019), and salivary CRP (Hernandez & Taylor, 2017; Out et al., 2012; Rudnicka, Rumley, Lowe, & Strachan, 2007). These markers share a common general pattern with higher levels in the morning and lower levels in the afternoon. Specifically, the diurnal cycle for SIgA begins with a peak upon awakening, and then an immediate decline to stable concentrations across the rest of the day (Hucklebridge et al., 1998; Pritchard et al., 2017). Salivary CRP levels are also elevated upon awakening and decline thereafter (Hernandez & Taylor, 2017; Izawa et al., 2013). Interestingly, the diurnal pattern for CRP may be specific to the oral compartment as studies of serum CRP have failed to find a significant pattern in diurnal serum CRP concentrations and it has been reported to be stable over a 24-h period (Meier-Ewert et al., 2001; Rudnicka et al., 2007). Compared to SIgA and CRP, the decline of salivary IL-6 after awakening is muted. There may also be an increase in concentrations proximal to bedtime (Izawa et al., 2013; Sjögren, Leanderson, Kristenson, & Ernerudh, 2006). This pattern is different than those observed in serum and plasma IL-6 which have been found to exhibit a "morning trough" without a morning or evening peak (Nilsonne, Lekander, Åkerstedt, Axelsson, & Ingre, 2016).

There is limited research examining the diurnal pattern of other immune markers in saliva. However, given the overlapping functions of many of the immune analytes of interest, investigators should consider time of waking and time of day when examining immune markers in saliva. It is also important to note that the physiologic mechanisms underlying the diurnal patterns in salivary immune markers may reflect processes specific to the oral environment, and this may be especially true for analytes with diurnal patterns that are unique to salivary concentrations and not observed in blood measurements. On the other hand, diurnal patterns in salivary immune markers may also reflect coordinated action across multiple physiologic systems within the body. For example, marked diurnal changes in HPA axis activity, reflected in salivary cortisol concentrations, may influence concentrations of immune markers in saliva across the day, and, given cortisol's potent anti-inflammatory properties, these effects may vary by analyte.

Beyond a methodologic confound, the physiological significance of a biomeasure's diurnal pattern has potential implications for health and disease outcomes. For example, sleep quality, duration, and disorders have been associated with variation in some salivary immune markers [e.g., El-Sheikh et al. (2007), Nizam,

Basoglu, Tasbakan, Nalbantsoy, and Buduneli (2014), and Reinhardt, Fernandes, Markus, and Fischer (2016)]. Given sleep's crucial role in repair and restorative processes, studies of the diurnal patterns of salivary immune markers may provide important information about health and disease risk. Closely monitoring changes in immune marker values across the day and understanding the relationships between these patterns and biobehavioral correlates (e.g., activity, diet, and blood pressure) may provide insight into optimizing health and reducing disease risk.

### 9.4.4 Salivary Immune Marker Stability and Reliability

Establishing measurement reliability is a key component of validating new biomeasures for the study of health and development. Measurement reliability for salivary immune markers at the level of the assay is generally strong and has improved considerably with the development of new assay technology. Coefficients of variation for these assays are typically low and within the accepted range for salivary bioscience studies (<5% for intra-assay CV; <15% for inter-assay CV). In contrast, the within-person short- and long-term stability of most salivary immune markers and acute phase proteins is largely unknown.

Given their diurnal patterns, the stability of some salivary immune marker concentrations likely varies by sampling time. For example, in a study of healthy young adults, concentrations of salivary IL-6 across two consecutive days showed weak to strong correlations depending on the sample timing ($r$'s $= 0.21 - 0.90$) (Izawa et al., 2013). The strongest associations in salivary IL-6 concentrations across the two days were found at night and early in the morning (Izawa et al., 2013). In the same study, salivary CRP concentrations were correlated across two consecutive days of sampling ($r$'s $= 0.23 - 0.96$), and, as seen with IL-6, the strength of the associations varied by sampling time (Izawa et al., 2013). The weakest across-day associations for salivary CRP were found in the afternoon and early evening (Izawa et al., 2013). For saliva samples collected on the same day, one study of adolescent girls found salivary IL-1β, IL-6, IL-8, TNF-α, IL-18, and CRP displayed moderate/strong correlations ($r$'s $= 0.38 - 0.81$; $p$'s $< 0.05$) across a 120 min waiting period (Shields et al., 2019).

Few studies have examined the long term stability of salivary immune markers. One study examining salivary cytokines in healthy adolescent girls across a 3-year study period found that across-year correlations varied by cytokine and were generally weak to moderate in strength ($r$'s $= 0.02 - 0.46$) (Riis et al., 2014). These findings were largely confirmed by another study of salivary IL-1β, IL-6, IL-8, TNF-α, IL-18, and CRP in adolescent girls which found similarly weak correlations across an 18-month period ($r$'s $= 0.04 - 0.31$) (Shields et al., 2019). In this study, salivary CRP stood out as one of the more stable immune markers with a correlation of 0.31 ($p < 0.001$) across the 18-month study period (Shields et al., 2019). More robust long-term stability in salivary CRP was also found in a study of adult women with salivary CRP showing moderate stability across 3 years of assessments ($r$'s = 0.46 – 0.61, $p$ <0.01) (Out et al., 2012).

The oral compartment is exposed to the outside world and therefore vulnerable to a wide range of threats from which immune processes in the blood are protected. The oral cavity is also home to extremely dense and diverse communities of microbes which influence oral, and potentially systemic, health (see Chap. 7 for a discussion of the oral microbiome). Oral and physical health may therefore rely on a local, oral immune response that is inherently more variable than that observed in blood. Baseline concentrations and changes in immune markers in saliva may represent immune processes responding to extrinsic factors, such as exposure to environmental pollutants and antigens (e.g., tobacco smoke and pollen), intrinsic factors related to oral health (e.g., dental caries and bacterial load), as well as intrinsic factors related to systemic health (e.g., infection or fever). Researchers should take these factors into consideration when interpreting between and within-person differences in salivary immune marker concentrations.

## 9.5 Future Directions and Opportunities

The integration of salivary immune markers into biobehavioral studies of health has provided insight into the complex mechanisms underlying health and development. These studies have allowed us to measure relations between salivary immune markers and the family environment (Byrne, Badcock, et al., 2017; Byrne, Horne, et al., 2017), social stress (Lucas et al., 2016, 2017; Slavich et al., 2010; Slavish et al., 2015a), sleep (El-Sheikh et al., 2007; Zheng et al., 2014), and depression and other mental health symptomatology (Cicchetti, Handley, & Rogosch, 2015; Delany et al., 2016; Keller, El-Sheikh, Vaughn, & Granger, 2010; Riis et al., 2016), as well as neuroendocrine–immune variation in response to environmental and psychosocial stressors (O'Connor, Irwin, & Wellisch, 2009; Riis et al., 2015; Slavish et al., 2015b). Despite this progress, we believe that furthering the study of salivary immune markers requires rigorous methodological investigations as well as an interdisciplinary, cross-systems approach to the study of health and development. Through this deliberate advancement of salivary immune marker research, we hope to gain a deeper, more nuanced, and meaningful understanding of the processes influencing health.

**Multisystem Assessments** The oral immune environment is complex with multiple analytes being secreted, expressed, and transported into the oral fluid from different areas of the mouth and body. Therefore, while many immune markers may work synergistically in the oral compartment and salivary concentrations often strongly covary within individuals, single marker assessments of immune function may not be useful for understanding oral or systemic immune processes as a whole and should not be interpreted as such. Instead, models that measure multiple analytes can highlight the synergistic effects that occur within the immune system, such as ratios between pro- and anti-inflammatory cytokines, as well as associations between inflammatory markers and other markers of immune activation such as SIgA. A critical aspect of examining oral immune processes may also lie in the oral microbial environment. While a relatively new area of study, advancing understanding of how the oral immune system interacts with these microbial communities, and how these interactions effect oral and physical health is a growing and important area of research (see Chap. 7 for a review of the oral microbiome). Finally, multisystem studies that use salivary biomeasures to examine how the immune system interacts with other biological processes, such as gonadal and neuroendocrine function, will be especially important for advancing our understanding of overall health and development.

It is also vital to measure immune system functioning along other analytes that are immune sensitive, such as cortisol. Several salivary bioscience studies have reported neuroendocrine–immune associations in behavioral science (Laurent et al., 2016; Riis et al., 2015, 2016). Neuroendocrine–immune relations are also implicated in the pathophysiology of many diseases. For example, chronic inflammation may trigger a cycle of glucocorticoid resistance or negative feedback associated with depression (Pariante & Miller, 2001). One study using serum biomeasures found that a lower cortisol to CRP ratio, suggesting an overproduction of inflammatory markers in relation to glucocorticoid release, was associated with depression, specifically for women (Suarez, Sundy, & Erkanli, 2015). Salivary bioscience is poised to contribute to this growing field of multisystem health and pathophysiology. Saliva is generally easier and cheaper to collect than other biospecimens and modern multiplex assay technologies, adapted for use with saliva, allow for the measurement of several analytes from a single aliquot.

**Clinical Populations and Applications** Of all the immune markers discussed in this chapter, based on the current state of knowledge, salivary SIgA and CRP have the most promise for clinical applications for systemic health. SIgA has the potential to help identify individualized alternative therapies and targeted interventions for mood disorders (Kreutz, Bongard, Rohrmann, Hodapp, & Grebe, 2004) and anxiety (Ma, Serbin, & Stack, 2018). With respect to precision medicine, SIgA is now a proposed marker of patient status in a variety of diseases such as oral cancer (Zhang et al., 2017) and multiple sclerosis (Kaplan et al., 2018). Such discoveries may provide noninvasive, complementary tools for providers to help improve patient outcomes.

In the last few decades, SIgA has also been used as a marker of over-training and impending upper respiratory infections (URI) in athletic populations who experience high levels of physical stress. However, the recent discovery of SIgA's diurnal pattern may impact the findings of historical studies in which SIgA may have been sampled at sub-optimal or inconsistent times (Pritchard et al., 2017). Still, URI are the most common, noninjury-related reason that athletes seek medical attention, and reductions in SIgA values may correlate with URI onset (Gleeson & Pyne, 2016). SIgA levels are also highly associated with training load (Engels, Kendall, Fahlman, Gothe, & Bourbeau, 2018) which could help coaches and athletic trainers mitigate decrements in sports performance due to over-training or nonfunctional overreaching. These studies point to great promise for using salivary SIgA to easily monitor a variety of clinical patients (Kaplan et al., 2018; Zhang et al., 2017), athletes, and at-risk populations like children (Ma et al., 2018) and older adults (Jiang, Yin, Li, Chen, & Gu, 2018).

Salivary CRP is the inflammatory maker with the most consistent serum–saliva correlation, making it a candidate for potential systemic health applications. Studies have shown that salivary levels of CRP are associated with measures of physical (Goodson et al., 2014; Naidoo et al., 2012) and psychological (Cicchetti et al., 2015) health. Salivary CRP has also been related to pediatric health conditions with higher levels associated with pediatric obesity, poor cardiorespiratory health (Naidoo et al., 2012), and allergic asthma (Krasteva et al., 2010). Higher levels of salivary CRP have also been linked with active and passive smoking (Azar & Richard, 2011), subacute thyroiditis (Rao et al., 2010), and reduced cognitive function in childhood (Cullen et al., 2017).

The American Heart Association and the US Centers for Disease Control and Prevention have established thresholds for high sensitivity CRP in blood and cardiovascular disease risk (Grundy et al., 2000). Unfortunately, clinically relevant reference ranges for salivary CRP have not yet been identified. Given generally low concentrations of CRP in saliva, more sensitive detection techniques may be warranted. However, a preliminary study by Out and colleagues found that salivary CRP concentrations reliably differentiated participants with high vs. low plasma CRP levels suggesting that salivary CRP could be a potential indicator of CVD risk (2012). With additional research and technological advances in detection and collection methods, salivary CRP has potential clinical applications for heart disease and a range of other inflammation-related conditions (Pay & Shaw, 2019).

**Large-Scale Monitoring, Assessment, and Treatment** As potential windows into immune functioning, salivary immune markers hold great potential for use in large-scale health programs and interventions. Point-of-care testing for salivary immunoglobulins, cytokines, CRP, and MMPs have been developed and are being tested to determine their utility in oral and physical health screening and treatment programs (Herr et al., 2007; Khan, Khurshid, & Yahya Ibrahim Asiri, 2017; Rathnayake, Gieselmann, Heikkinen, Tervahartiala, & Sorsa, 2017). The large-scale application of these easy-to-use and rapid tests of immune function would create new opportunities to assess, treat, and track community health using objective, biologic measures in the field. For example, salivary immunoglobulins may be useful indices of infection exposure and response to vaccination (Heaney, Phillips, Carroll, & Drayson, 2018; Lim, Garssen, & Sandalova, 2016; Pisanic et al., 2017, 2018). Chapter 30 provides additional discussion of the potential for salivary biomeasures to influence community and public health.

## 9.6 Concluding Comments

There is vast untapped potential for the study of salivary immune markers to advance our understanding of health and development. Realizing this potential will require both basic methodological studies, as well as studies that inform the reshaping of our conceptual understanding of salivary immune markers. Future investigators should also work toward establishing a set of reporting standards for the collection and processing of saliva samples, assay results (e.g., percent CVs, percent undetectable), data cleaning techniques, and study paradigms (e.g., TSST vs. physical stressor, timing of the sample collections). Adding this level of transparency into salivary immune marker studies will help us understand the differences observed across studies and move the field toward a set of standard protocols.

Methodologic studies that examine the validity, reliability, stability, correlates, and confounders of salivary immune markers in healthy and clinical samples are needed to provide the foundation for future salivary immune marker research. Basic methodologic studies examining the influence of typical salivary bioscience confounders, such as individual characteristics, and sample timing, collection, and processing methods, are needed to improve our ability to attribute changes in salivary immune markers to changes in functioning rather than unaccounted for biases and/or confounding in the data. Additionally, we recommend experimental research that includes the induction of inflammatory states through therapeutic drug administration, vaccines, and oral inflammatory activation to further our understanding of the within-person interactions among oral, mucosal, and systemic inflammation, and how changes in one compartment influences function across the body. Finally, to develop our understanding of between-person differences in salivary immune markers, we recommend expanding research into varying populations across sexes, developmental stages, races, ethnicities, and health and disease states.

Beyond methodologic studies, we believe interdisciplinary research that bridges across fields such as PNI, oral health and periodontology, public health, human performance, and medicine, and recognizes the importance of understanding complex, multisystem biologic processes will be key to furthering salivary bioscience and health research. Rather than viewing the associations between salivary immune markers and oral, mucosal, and systemic health as a confound to be measured and parceled out in study designs or statistical models, we believe this complexity may make salivary immune markers uniquely valuable biomarkers of health and well-being.

## References

Ader, R. (2000). On the development of psychoneuroimmunology Robert. *European Journal of Pharmacology, 405*, 167–176.

Auer, B. J., Calvi, J. L., Jordan, N. M., Schrader, D., & Byrd-Craven, J. (2018). Communication and social interaction anxiety enhance interleukin-1 beta and cortisol reactivity during high-stakes public speaking. *Psychoneuroendocrinology, 94*, 83–90. https://doi.org/10.1016/j.psyneuen.2018.05.011

Azar, R., & Richard, A. (2011). Elevated salivary C-reactive protein levels are associated with active and passive smoking in healthy youth: A pilot study. *Journal of Inflammation, 8*, 37. https://doi.org/10.1186/1476-9255-8-37

Bascones-Martínez, A., Muñoz-Corcuera, M., Noronha, S., Mota, P., Bascones-Ilundain, C., & Campo-Trapero, J. (2009). Host defence mechanisms against bacterial aggression in periodontal disease: Basic mechanisms. *Medicina Oral, Patología Oral y Cirugía Bucal, 14*, 10–15. https://doi.org/10.4317/medoral.14.e680

Belstrøm, D., Damgaard, C., Könönen, E., Gürsoy, M., & Gürsoy, U. K. (2017). Salivary cytokine levels in early gingival inflammation. *Journal of Oral Microbiology, 9*, 1364101. https://doi.org/10.1080/20002297.2017.1364101

Bergmeier, L. A. (2018). Immunology of the oral mucosa. In L. Bergmeier (Ed.), *Oral Mucosa in Health and Disesae. A Concise Handbook* (pp. 53–67). Cham: Springer. https://doi.org/10.1007/978-3-319-56065-6_4

Berliner, J., Navab, M., Fogelman, A., Frank, J., Demer, L., Edwards, P., . . . Lusis, A. (1995). Atherosclerosis: Basic mechanisms. Oxidation, inflammation, and genetics. *Circulation, 91*, 2488–2496.

Besedovsky, H., del Rey, A., Sorkin, E., Da Prada, M., Burri, R., & Honegger, C. (1983). The immune response evokes changes in brain noradrenergic neurons. *Science, 221*, 564–566.

Bilbo, S., & Schwarz, J. (2012). The immune system and developmental programming of brain and behavior. *Frontiers in Neurendocrinology, 33*, 267–286. https://doi.org/10.1016/j.yfrne.2012.08.006

Birkett, M., Johnson, L., & Gelety, C. (2017). Investigation of sex differences in sIgA response to the trier social stress test. *Stress and Health, 33*, 158–163. https://doi.org/10.1002/smi.2680

Bondesen, B. A., Mills, S. T., Kegley, K. M., & Pavlath, G. K. (2004). The COX-2 pathway is essential during early stages of skeletal muscle regeneration. *American Journal of Physiology. Cell Physiology, 287*, C475–C483. https://doi.org/10.1152/ajpcell.00088.2004

Bouman, A., Jan Heineman, M., & Faas, M. M. (2005). Sex hormones and the immune response in humans. *Human Reproduction Update, 11*, 411–423. https://doi.org/10.1093/humupd/dmi008

Brandtzaeg, P. (2007). Do salivary antibodies reliably reflect both mucosal and systemic immunity? *Annals of the New York Academy of Sciences, 1098*, 288–311. https://doi.org/10.1196/annals.1384.012

Brandtzaeg, P. (2013). Secretory immunity with special reference to the oral cavity. *Journal of Oral Microbiology, 5*, 1–24. https://doi.org/10.3402/jom.v5i0.20401

Brennan, M., & Fox, P. (2010). Cytokine mRNA expression in the labial salivary glands of healthy volunteers. *Oral Diseases, 6*, 222–226. https://doi.org/10.1111/j.1601-0825.2000.tb00117.x

Browne, R. W., Kantarci, A., LaMonte, M. J., Andrews, C. A., Hovey, K. M., Falkner, K. L., . . . Wactawski-Wende, J. (2013). Performance of multiplex cytokine assays in serum and saliva among community-dwelling postmenopausal women. *PLoS One, 8*, e59498. https://doi.org/10.1371/journal.pone.0059498

Brydon, L., Harrison, N. A., Walker, C., Steptoe, A., & Critchley, H. D. (2008). Peripheral inflammation is associated with altered substantia nigra activity and psychomotor slowing in humans. *Biological Psychiatry, 63*, 1022–1029. https://doi.org/10.1016/j.biopsych.2007.12.007

Byrne, M., Badcock, P., Simmons, J., Whittle, S., Pettitt, A., Olsson, C., . . . Allen, N. (2017). Self-reported parenting style is associated with children's inflammation and immune activation. *Journal of Family Psychology, 31*, 374.

Byrne, M., Horne, S., O'Brien-Simpson, N., Walsh, K., Reynolds, E., Schwartz, O., . . . Allen, N. (2017). Associations between observed parenting behavior and adolescent inflammation two and a half years later in a community sample. *Health Psychology, 36*, 641.

Byrne, M. L., O'Brien-Simpson, N. M., Reynolds, E. C., Walsh, K. A., Laughton, K., Waloszek, J. M., . . . Allen, N. B. (2013). Acute phase protein and cytokine levels in serum and saliva: A comparison of detectable levels and correlations in a depressed and healthy adolescent sample. *Brain, Behavior, and Immunity, 34*, 164–175. https://doi.org/10.1016/j.bbi.2013.08.010

Campbell, J. P., & Turner, J. E. (2018). Debunking the myth of exercise-induced immune suppression: Redefining the impact of exercise on immunological health across the lifespan. *Frontiers in Immunology, 9*, 1–21. https://doi.org/10.3389/fimmu.2018.00648

Campisi, J., Bravo, Y., Cole, J., & Gobeil, K. (2012). Acute psychosocial stress differentially influences salivayr endocrine and immune measures in undergraduate students. *Physiology and Behavior, 107*, 317–321. https://doi.org/10.1016/j.physbeh.2012.09.003

Cavaillon, J. (2001). Pro- versus anti-inflammatory cytokines: myth or reality. *Cellular and Molecular Biology, 47*, 695–702.

Chiang, J. J., Eisenberger, N. I., Seeman, T. E., & Taylor, S. E. (2012). Negative and competitive social interactions are related to heightened proinflammatory cytokine activity. *Proceedings of the National Academy of Sciences of the United States of America, 109*, 1878–1882. https://doi.org/10.1073/pnas.1120972109

Cicchetti, D., Handley, E. D., & Rogosch, F. A. (2015). Child maltreatment, inflammation, and internalizing symptoms: Investigating the roles of C-reactive protein, gene variation, and neuroendocrine regulation. *Development and Psychopathology, 27*, 553–566. https://doi.org/10.1017/S0954579415000152

Cullen, A. E., Tappin, B. M., Zunszain, P. A., Dickson, H., Roberts, R. E., Nikkheslat, N., … Laurens, K. R. (2017). The relationship between salivary C-reactive protein and cognitive function in children aged 11–14 years: Does psychopathology have a moderating effect? *Brain, Behavior, and Immunity, 66*, 221–229. https://doi.org/10.1016/j.bbi.2017.07.002

Cullen, T., Thomas, A. W. A., Web, R., Hughes, M. G., Webb, R., & Hughes, M. G. (2015). The relationship between interleukin-6 in saliva, venous and capillary plasma, at rest and in response to exercise. *Cytokine, 71*(2), 397–400. https://doi.org/10.1016/j.cyto.2014.10.011

Dai, L., Golembiewska, E., Lindholm, B., & Stenvinkel, P. (2017). End-stage renal disease, inflammation and cardiovascular outcomes. *Contributions to Nephrology, 191*, 32–43. https://doi.org/10.1159/000479254

David, J., Measelle, J., Ostlund, B., & Ablow, J. (2017). Association between early life adversity and inflammation during infancy. *Developmental Psychobiology, 59*, 696–702.

Delany, F. M., Byrne, M. L., Whittle, S., Simmons, J. G., Olsson, C., Mundy, L. K., … Allen, N. B. (2016). Depression, immune function, and early adrenarche in children. *Psychoneuroendocrinology, 63*, 228–234. https://doi.org/10.1016/j.psyneuen.2015.10.003

Dickerson, S. S., Kemeny, M. E., Aziz, N., Kim, K. H., & Fahey, J. L. (2004). Immunological effects of induced shame and guilt. *Psychosomatic Medicine, 66*, 124–131. https://doi.org/10.1097/01.PSY.0000097338.75454.29

Dillon, M., Opris, D., Kopanczyk, R., Lickliter, J., Cornwell, H., Bridges, E., … Bridges, K. (2010). Detection of homocysteine and C-reactive protein in the saliva of healthy adults: Comparison with blood levels. *Biomarker Insights, 5*, 57–61.

DiSabato, D., Quan, N., & Godbout, J. (2016). Neuroinflammation: The devil is in the details. *Journal of Neurochemistry, 139*, 136–153. https://doi.org/10.1111/jnc.13607

Doyle, D., & Molix, L. (2016). Minority stress and inflammatory mediators: Covering moderates associations between perceived discrimination and salivary interleukin-6 in gay men. *Journal of Behavioral Medicine, 39*, 782–792. https://doi.org/10.1097/CCM.0b013e31823da96d. Hydrogen

Dugue, B. (1996). Cytokines in saliva. Basal concentrations and the effect of high ambient heat (sauna). *Stress Medicine, 12*, 193–197. https://doi.org/10.1002/(SICI)1099-1700(199607)12:3<193::AID-SMI701>3.0.CO;2-Z

Eisenberger, N. I., Inagaki, T. K., Rameson, L. T., Mashal, N. M., & Irwin, M. R. (2009). An fMRI study of cytokine-induced depressed mood and social pain: The role of sex differences. *NeuroImage, 47*, 881–890. https://doi.org/10.1016/j.neuroimage.2009.04.040

Eke, P. I., Thornton-Evans, G. O., Wei, L., Borgnakke, W. S., Dye, B. A., & Genco, R. J. (2018). Periodontitis in US adults: National Health and Nutrition Examination Survey 2009–2014.

*Journal of the American Dental Association, 149*, 576–588.e6. https://doi.org/10.1016/j.adaj.2018.04.023

El-Sheikh, M., Buckhalt, J. A., Granger, D. A., Erath, S. A., & Acebo, C. (2007). The association between children's sleep disruption and salivary interleukin-6. *Journal of Sleep Research, 16*, 188–197. https://doi.org/10.1111/j.1365-2869.2007.00593.x

Engeland, C. G., Hugo, F. N., Hilgert, J. B., Nascimento, G. G., Junges, R., Lim, H. J., . . . Bosch, J. A. (2016). Psychological distress and salivary secretory immunity. *Brain, Behavior, and Immunity, 52*, 11–17. https://doi.org/10.1016/j.bbi.2015.08.017

Engels, H., Kendall, B., Fahlman, M., Gothe, N., & Bourbeau, K. (2018). Salivary immunoglobulin A in healthy adolescent females: Effects of maximal exercise, physical activity, body composition and diet. *The Journal of Sports Medicine and Physical Fitness, 58*, 1096–1101.

Evans, P., Der, G., Ford, G., Hucklebridge, F., Hunt, K., & Lambert, S. (2000). Social class, sex, and age differences in mucosal immunity in a large community sample. *Brain, Behavior, and Immunity, 14*, 41–48. https://doi.org/10.1006/brbi.1999.0571

Ewing, P., Otczyk, D., Occhipinti, S., Kyd, J., Gleeson, M., & Cripps, A. (2010). Developmental profiles of mucosal immunity in pre-school children. *Clinical and Developmental Immunology, 2010*, 196785. https://doi.org/10.1155/2010/196785

Fernandez-Botran, R., Miller, J. J., Burns, V. E., & Newton, T. L. (2011). Correlations among inflammatory markers in plasma, saliva and oral mucosal transudate in post-menopausal women with past intimate partner violence. *Brain, Behavior, and Immunity, 25*, 314–321. https://doi.org/10.1016/j.bbi.2010.09.023

Finoti, L. S., Nepomuceno, R., Pigossi, S. C., Corbi, S. C., Secolin, R., & Scarel-Caminaga, R. M. (2017). Association between interleukin-8 levels and chronic periodontal disease. *Medicine (Baltimore), 96*, e6932. https://doi.org/10.1097/MD.0000000000006932

Fisher-Owens, S. A., Gansky, S. A., Platt, L. J., Weintraub, J. A., Soobader, M.-J., Bramlett, M. D., & Newacheck, P. W. (2007). Influences on children's oral health: A conceptual model. *Pediatrics, 120*, e510–e520. https://doi.org/10.1542/peds.2006-3084

Giannobile, W., Beikler, T., Kinney, J., Ramseier, C., Morelli, T., & Wong, D. (2009). Saliva as a diagnostic tool for periodontal disease: Current state and future directions. *Periodontology 2000, 2000*(50), 52–64. https://doi.org/10.4103/2468-6360.186487

Gleeson, M., & Pyne, D. B. (2016). Respiratory inflammation and infections in high-performance athletes. *Immunology and Cell Biology, 94*, 124–131. https://doi.org/10.1038/icb.2015.100

Goodson, J. M., Kantarci, A., Hartman, M. L., Denis, G. V., Stephens, D., Hasturk, H., . . . Welty, F. (2014). Metabolic disease risk in children by salivary biomarker analysis. *PLoS One, 9*, e98799. https://doi.org/10.1371/journal.pone.0098799

Granger, D. A., Granger, G. A., & Granger, S. W. (2006). Immunology and developmental psychopathology. In D. Cicchetti & D. Cohen (Eds.), *Developmental neuroscience* (pp. 677–709). New York, NY: Wiley.

Gröschl, M. (2009). The physiological role of hormones in saliva. *Bioessays, 31*, 843–852. https://doi.org/10.1002/bies.200900013

Grundy, S., Bazzarre, T., Cleeman, J., D'Agostino, R., Hill, M., Houston-Miller, N., . . . Wood, D. (2000). Prevention conference V: Beyond secondary prevention: Identifying the high-risk patient for primary prevention: Medical office assessment: Writing Group I. *Circulation, 101*, E3–E11.

Hanioka, T., Morita, M., Yamamoto, T., Inagaki, K., Wang, P.-L., Ito, H., . . . Ogawa, H. (2019). Smoking and periodontal microorganisms. *Japanese Dental Science Review, 55*, 88–94. https://doi.org/10.1016/j.jdsr.2019.03.002

Hannas, A. R., Pereira, J. C., Granjeiro, J. M., & Tjäderhane, L. (2007). The role of matrix metalloproteinases in the oral environment. *Acta Odontologica Scandinavica, 65*, 1–13. https://doi.org/10.1080/00016350600963640

Hasmun, N., Drummand, B., Milne, T., Cullinan, M., Meldrum, A., & Coates, D. (2017). Effects of environmental tobacco smoke on the oral health of preschool children. *European Archives of Paediatric Dentistry, 18*, 393–398.

Hasturk, H., Kantarci, A., & Van Dyke, T. E. (2012). Oral inflammatory diseases and systemic inflammation: Role of the macrophage. *Frontiers in Immunology, 3*, 1–17. https://doi.org/10.3389/fimmu.2012.00118

Haukioja, A., Tervahartiala, T., Sorsa, T., & Syrjänen, S. (2017). Persistent oral human papillomavirus (HPV) infection is associated with low salivary levels of matrix metalloproteinase 8 (MMP-8). *Journal of Clinical Virology, 97*, 4–9. https://doi.org/10.1016/j.jcv.2017.10.011

Hayashida, H. (2016). Effects of the menstrual cycle and acute aerobic exercise on cytokine levels. *Journal of Sports Medicine and Doping Studies, 06*, 6–10. https://doi.org/10.4172/2161-0673.1000173

Heaney, J. L. J., Phillips, A. C., Carroll, D., & Drayson, M. T. (2018). The utility of saliva for the assessment of anti-pneumococcal antibodies: investigation of saliva as a marker of antibody status in serum. *Biomarkers, 23*, 115–122. https://doi.org/10.1080/1354750X.2016.1265009

Hernandez, L., Fuller, S., Stone, M., Carpenter, J., & Taylor, M. (2016). Catabolic/anabolic hormone and inflammatory marker responses to a graded exercise test in military men. *Medicine and Science in Sports and Exercise, 48*, 273–274.

Hernandez, L., & Taylor, M. (2017). Diurnal pattern of salivary C-reactive protein and associations with biobehavioral correlates in military men. In *Medicine and science in sports and exercise*. Philadelphia, PA: Lippincott Williams & Wilkins.

Herr, A. E., Hatch, A. V., Throckmorton, D. J., Tran, H. M., Brennan, J. S., Giannobile, W. V., & Singh, A. K. (2007). Microfluidic immunoassays as rapid saliva-based clinical diagnostics. *Proceedings of the National Academy of Sciences, 104*, 5268–5273. https://doi.org/10.1073/pnas.0607254104

Horn, S., Long, M., Nelson, B., Allen, N., Fisher, P., & Byrne, M. (2018). Replication and reproducibility issues in the relationship between C-reactive protein and depression: A systematic review and focused meta-analysis. *Brain, Behavior, and Immunity, 73*, 85–114.

Hucklebridge, F., Clow, A., & Evans, P. (1998). The relationship between salivary secretory immunoglobulin A and cortisol: Neuroendocrine response to awakening and the diurnal cycle. *International Journal of Psychophysiology, 31*, 69–76. https://doi.org/10.1016/S0167-8760(98)00042-7

Ilardo, C., Toniolo, A., Aimone-gastin, I., Abdelmouttaleb, I., Gueant, J., & Desor, D. (2001). Effects of psycho-physical stress (competitive rafting) on saliva interleukin-1 beta. *Stress and Health, 17*, 9–15.

Izawa, S., Miki, K., Liu, X., & Ogawa, N. (2013). The diurnal patterns of salivary interleukin-6 and C-reactive protein in healthy young adults. *Brain, Behavior, and Immunity, 27*, 38–41. https://doi.org/10.1016/j.bbi.2012.07.001

Jafarzadeh, A., Sadeghi, M., Karam, G. A., & Vazirinejad, R. (2010). Salivary IgA and IgE levels in healthy subjects: Relation to age and gender. *Brazilian Oral Research, 24*, 21–27. https://doi.org/10.1590/S1806-83242010000100004

Jarvis, M., Russell, M., Benowitz, N., & Feyerabend, C. (1988). Elimination of cotinine from body fluids: Implications for noninvasive measurement of tobacco smoke exposure. *American Journal of Public Health, 78*, 696–698. https://doi.org/10.2105/AJPH.78.6.696

Javed, F., Ahmed, H., & Romanos, G. (2014). Association between environmental tobacco smoke and periodontal disease: A systematic review. *Environmental Research, 133*, 117–122.

Jiang, M., Yin, Z., Li, S., Chen, X., & Gu, J. (2018). Effect of positive psychological intervention on well-being, 2-week illness prevalence, and salivary immunoglobulin A. *Western Journal of Nursing Research, 40*, 889–906. https://doi.org/10.1177/0193945917690733

Kaplan, B., Golderman, S., Ganelin-Cohen, E., Miniovitch, A., Korf, E., Ben-Zvi, I., … Flechter, S. (2018). Immunoglobulin free light chains in saliva: A potential marker for disease activity in multiple sclerosis. *Clinical and Experimental Immunology, 192*, 7–17. https://doi.org/10.1111/cei.13086

Keaney, L. C., Kilding, A. E., Merien, F., & Dulson, D. K. (2018). The impact of sport related stressors on immunity and illness risk in team-sport athletes. *Journal of Science and Medicine in Sport, 21*, 1192–1199. https://doi.org/10.1016/j.jsams.2018.05.014

Keller, P. S., El-Sheikh, M., Vaughn, B., & Granger, D. A. (2010). Relations between mucosal immunity and children's mental health: The role of child sex. *Physiology and Behavior, 101*, 705–712. https://doi.org/10.1016/j.physbeh.2010.08.012

Khan, R., Khurshid, Z., & Yahya Ibrahim Asiri, F. (2017). Advancing point-of-care (PoC) testing using human saliva as liquid biopsy. *Diagnostics (Basel), 7*, E39. https://doi.org/10.3390/diagnostics7030039

Kiecolt-Glaser, J., Cacioppo, J., Malarkey, W., & Glaser, R. (1992). Acute psychological stressors and short-term immune changes: What, why, for whom, and to what extent? *Psychosomatic Medicine, 54*, 680–685.

Kiecolt-Glaser, J., McGuire, L., Robles, T., & Glaser, R. (2002). Emotions, morbidity, and mortality: New perspectives from psychoneuroimmunology. *Annual Review of Psychology, 53*, 83–107.

Kivlighan, K., Granger, D. A., Schwartz, E., Nelson, V., Curran, M., & Shirtcliff, E. (2004). Quantifying blood leakage into the oral mucosa and its effects on the measurement of cortisol, dehydroepiandrosterone, and testosterone in saliva Katie. *Hormones and Behavior, 46*, 39–46. https://doi.org/10.1016/j.yhbeh.2005.10.005

Klein, S. L., & Flanagan, K. L. (2016). Sex differences in immune responses. *Nature Reviews. Immunology, 16*, 626–638. https://doi.org/10.1038/nri.2016.90

Krasteva, A., Perenovska, P., Ivanova, A., Altankova, I., Bocheva, T., & Kisselova, A. (2010). Alteration in salivary components of children with allergic asthma. *Biotechnology and Biotechnological Equipment, 24*, 1866–1869. https://doi.org/10.2478/V10133-010-0050-2

Kreutz, G., Bongard, S., Rohrmann, S., Hodapp, V., & Grebe, D. (2004). Effects of choir singing or listening on secretory immunoglobulin A, cortisol, and emotional state. *Journal of Behavioral Medicine, 27*, 623–635.

La Fratta, I., Tatangelo, R., Campagna, G., Rizzuto, A., Franceschelli, S., Ferrone, A., ... Pesce, M. (2018). The plasmatic and salivary levels of IL-1β, IL-18 and IL-6 are associated to emotional difference during stress in young male. *Scientific Reports, 8*, 1–10. https://doi.org/10.1038/s41598-018-21474-y

Lahdentausta, L., Paju, S., Mäntylä, P., Buhlin, K., Tervahartiala, T., Pietiäinen, M., ... Pussinen, P. J. (2018). Saliva and serum biomarkers in periodontitis and coronary artery disease. *Journal of Clinical Periodontology, 45*(9), 1045–1055. https://doi.org/10.1111/jcpe.12976

Laurent, H. K., Stroud, L. R., Brush, B., D'Angelo, C., & Granger, D. A. (2015). Secretory IgA reactivity to social threat in youth: Relations with HPA, ANS, and behavior. *Psychoneuroendocrinology, 59*, 81–90. https://doi.org/10.1016/j.psyneuen.2015.04.021

Laurent, H. K., Lucas, T., Pierce, J., Goetz, S., & Granger, D. A. (2016). Coordination of cortisol response to social evaluative threat with autonomic and inflammatory responses is moderated by stress appraisals and affect. *Biological Psychology, 118*, 17–24. https://doi.org/10.1016/j.biopsycho.2016.04.066.Coordination

Lester, S., Brown, J., Aycock, J., Grubbs, S., & Johnson, R. (2010). Use of saliva for assessment of stress and its effect on the immune system prior to gross anatomy practical examinations. *Anatomical Sciences Education, 3*, 160–167. https://doi.org/10.1002/ase.161

Li, T., & Gleeson, M. (2004). The effect of single and repeated bouts of prolonged cycling and circadian variation on saliva flow rate, immunoglobulin A and alpha-amylase responses. *Journal of Sports Sciences, 22*, 1015–1024.

Lim, P. W., Garssen, J., & Sandalova, E. (2016). Potential use of salivary markers for longitudinal monitoring of inflammatory immune responses to vaccination. *Mediators of Inflammation, 2016*, 6958293. https://doi.org/10.1155/2016/6958293

Lira-Junior, R., Åkerman, S., Gustafsson, A., Klinge, B., & Boström, E. A. (2017). Colony stimulating factor-1 in saliva in relation to age, smoking, and oral and systemic diseases. *Scientific Reports, 7*, 1–8. https://doi.org/10.1038/s41598-017-07698-4

Liu, H., Hays, R. D., Marcus, M., Coulter, I., Maida, C., Ramos-gomez, F., ... Crall, J. (2016). Patient-Reported oral health outcome measurement for children and adolescents. *BMC Oral Health, 16*. https://doi.org/10.1186/s12903-016-0293-x

Lockhart, P. P. B., Bolger, A. A. F., Papapanou, P. P. N., Osinbowale, O., Trevisan, M., Levison, M. M. E., … Baddour, L. M. L. (2012). Periodontal disease and atherosclerotic vascular disease: Does the evidence support an independent association? A scientific statement from the American heart association. *Circulation, 125*, 2520–2544. https://doi.org/10.1161/CIR.0b013e31825719f3

Lu, Q., & Jin, L. (2010). Human gingiva is another site of C-reactive protein formation. *Journal of Clinical Periodontology, 37*, 789–796. https://doi.org/10.1111/j.1600-051X.2010.01600.x

Lucas, T., Lumley, M., Flack, J., Wegner, R., Pierce, J., & Goetz, S. (2016). A preliminary experimental examination of worldview verification, perceived racism, and stress reactivity in African Americans. *Health Psychology, 35*, 366–375. https://doi.org/10.1037/hea0000284

Lucas, T., Wegner, R., Pierce, J., Lumley, M. A., Laurent, H. H. K. H. H. K., & Granger, D. D. A. D. A. (2017). Perceived discrimination, racial identity, and multisystem stress response to social evaluative threat among African American men and women. *Psychosomatic Medicine, 79*, 293–305. https://doi.org/10.1097/PSY.0000000000000406

Ma, D., Serbin, L. A., & Stack, D. M. (2018). Children's anxiety symptoms and salivary immunoglobulin A: A mutual regulatory system? *Developmental Psychobiology, 60*, 202–215. https://doi.org/10.1002/dev.21590

Maekawa, T., Tabeta, K., Kajita-Okui, K., Nakajima, T., Yamazaki, K., Tomoki, M., … Kazuhisa, Y. (2011). Increased expression of C-reactive protein gene in inflamed gingival tissues could be derived from endothelial cells stimulated with interleukin-6. *Archives of Oral Biology, 56*, 1312–1318. https://doi.org/10.1016/j.archoralbio.2011.04.010

Maier, S., & Watkins, L. (1998). Cytokines for psychologists: Implications of bidirectional immune-to- brain communication for understanding behavior, mood, and cognition. *Psychological Review, 105*, 83–107.

Maldonado, A., Preciado, A., Buchanan, M., Pulvers, K., Romero, D., & D'Anna-Hernandez, K. (2018). Acculturative stress, mental health symptoms, and the role of salivary inflammatory markers among a Latino sample. *Cultural Diversity and Ethnic Minority Psychology, 24*, 277–283.

Manthiram, K., Zhou, Q., Aksentijevich, I., & Kastner, D. L. (2017). The monogenic autoinflammatory diseases define new pathways in human innate immunity and inflammation. *Nature Immunology, 18*, 832–842. https://doi.org/10.1038/ni.3777

Mantis, N. J., Rol, N., & Corthésy, B. (2011). Secretory IgA's complex roles in immunity and mucosal homeostasis in the gut. *Mucosal Immunology, 4*, 603–611. https://doi.org/10.1038/mi.2011.41

Marsland, A., Walsh, C., Lockwood, K., & John-Henderson, N. (2017). The effects of acute psychological stress on circulating and stimulated inflammatory markers: A systematic review and meta-analysis. *Brain, Behavior, and Immunity, 64*, 208–219. https://doi.org/10.1016/j.bbi.2017.01.011

Martínez, A. D., Ruelas, L., & Granger, D. A. (2018). Household fear of deportation in relation to chronic stressors and salivary proinflammatory cytokines in Mexican-origin families post-SB 1070. *SSM-Population Health, 5*, 188–200.

Master, S., Amodio, D., Stanon, A., Yee, C., Hilmert, C., & Taylor, S. (2009). Neurobiogical correlates of coping through emotional approach. *Brain, Behavior, and Immunity, 23*, 27–35.

Mattila, K. J., Nieminen, M. S., Valtonen, V. V., Rasi, V. P., Kesäniemi, Y. A., Syrjälä, S. L., … Jokinen, M. J. (1989). Association between dental health and acute myocardial infarction. *BMJ, 298*, 779–781.

Maydych, V., Claus, M., Watz, C., & Kleinsorge, T. (2018). Attention to emotional information is associated with cytokine responses to psychological stress. *Frontiers in Neuroscience, 12*, 687. https://doi.org/10.3389/fnins.2018.00687

Measelle, J. R., David, J., & Ablow, J. C. (2017). Increased levels of inflammation among infants with disorganized histories of attachment. *Behavioural Brain Research, 325*, 260–267. https://doi.org/10.1016/j.bbr.2016.12.001

Megson, E., Fitzsimmons, T., Charmapatni, K., Bartold, P., Dharmapatni, K., & Bartold, P. (2010). C-reactive protein in gingival crevicular fluid may be indicative of systemic inflammation. *Journal of Clinical Periodontology, 37*, 797–804.

Meier-Ewert, H. K., Ridker, P. M., Rifai, N., Price, N., Dinges, D. F., & Mullington, J. M. (2001). Absence of diurnal variation of C-reactive protein concentrations in healthy human subjects. *Clinical Chemistry, 47*, 426–430.

Miller, A., & Raison, C. (2016). The role of inflammation in depression: From evolutionary imperative to modern treatment target. *Nature Reviews. Immunology, 16*, 22–34. https://doi.org/10.1038/nri.2015.5

Minetto, M. A., Gazzoni, M., Lanfranco, F., Baldi, M., Saba, L., Pedrola, R., . . . Rainoldi, A. (2007). Influence of the sample collection method on salivary interleukin-6 levels in resting and post-exercise conditions. *European Journal of Applied Physiology, 101*, 249–256. https://doi.org/10.1007/s00421-007-0484-x

Miyake, S. (2012). Mind over cytokines: Crosstalk and regulation between the neuroendocrine and immune systems. *Clinical Experimental Neuroimmunology, 3*, 1–15. https://doi.org/10.1111/j.1759-1961.2011.00023.x

Mohamed, R., Campbell, J.-L., Cooper-White, J., Dimeski, G., & Punyadeera, C. (2012). The impact of saliva collection and processing methods on CRP, IgE, and Myoglobin immunoassays. *Clinical and Translational Medicine, 1*, 19. https://doi.org/10.1186/2001-1326-1-19

Moons, W. G., Eisenberger, N. I., & Taylor, S. E. (2010). Anger and fear responses to stress have different biological profiles. *Brain, Behavior, and Immunity, 24*, 215–219. https://doi.org/10.1016/j.bbi.2009.08.009

Moutsopoulos, N. M., & Konkel, J. E. (2018). Tissue-specific immunity at the oral mucosal barrier. *Trends in Immunology, 39*, 276–287. https://doi.org/10.1016/j.it.2017.08.005

Muscatell, K. A., Dedovic, K., Slavich, G. M., Jarcho, M. R., Breen, E. C., Bower, J. E., . . . Eisenberger, N. I. (2015). Greater amygdala activity and dorsomedial prefrontal—Amygdala coupling are associated with enhanced inflammatory responses to stress. *Brain, Behavior, and Immunity, 43*, 46–53. https://doi.org/10.1016/j.bbi.2014.06.201

Naidoo, T., Konkol, K., Biccard, B., Dubose, K., & McKune, A. (2012). Elevated salivary C-reactive protein predicted by low cardio-respiratory fitness and being overweight in African children: Cardiovascular topic. *Cardiovascular Journal of Africa, 23*, 501–506. https://doi.org/10.5830/CVJA-2012-058

Nam, Y., Kim, Y. Y., Chang, J. Y., & Kho, H. S. (2019). Salivary biomarkers of inflammation and oxidative stress in healthy adults. *Archives of Oral Biology, 97*, 215–222. https://doi.org/10.1016/j.archoralbio.2018.10.026

Navarro, V. P., Iyomasa, M. M., Leite-Panissi, C. R., Almeida, M. C., & Branco, L. G. (2006). New role of the trigeminal nerve as a neuronal pathway signaling brain in acute periodontitis: Participation of local prostaglandins. *Pflugers Archiv: European Journal of Physiology, 453*, 73–82.

Newton, T., Fernandez-Botran, R., Lyle, K., Szabo, Y., Miller, J., & Warnecke, A. (2017). Salivary cytokine response in the aftermath of stress: An emotion regulation perspective. *Emotion, 17*, 1007–1020.

Nilsonne, G., Lekander, M., Åkerstedt, T., Axelsson, J., & Ingre, M. (2016). Diurnal variation of circulating interleukin-6 in humans: A meta-analysis. *PLoS One, 11*, 1–17. https://doi.org/10.1371/journal.pone.0165799

Nirmal, B., Ramachandiran, H., Anand, V., Sathiamurthy, A., & Sekaran, P. (2005). Tools for evaluating oral health and quality of life. *Journal of Pharmacy and Bioallied Sciences, 7*, S414–S419.

Nishanian, P., Aziz, N., Chung, J., Detels, R., & Fahey, J. L. (1998). Oral fluids as an alternative to serum for measurement of markers of immune activation. *Clinical and Diagnostic Laboratory Immunology, 5*, 507–512.

Nizam, N., Basoglu, O. K., Tasbakan, M. S., Nalbantsoy, A., & Buduneli, N. (2014). Salivary cytokines and the association between obstructive sleep apnea syndrome and periodontal disease. *Journal of Periodontology, 85*, e251–e258. https://doi.org/10.1902/jop.2014.130579

O'Connor, M., Bower, J., Cho, H., Creswell, J., Dimitrov, S., Hamby, M., . . . Irwin, M. (2009). To assess, to control, to exclude: Effects of biobehavioral factors on circulating inflammatory markers. *Brain, Behavior, and Immunity, 23*, 887–897.

O'Connor, M. F., Irwin, M. R., & Wellisch, D. K. (2009). When grief heats up: Pro-inflammatory cytokines predict regional brain activation. *NeuroImage, 47*, 891–896. https://doi.org/10.1016/j.neuroimage.2009.05.049

Ohira, H. (2005). Social support and salivary secretory immunoglobulin A response in women to stress of making a public speech. *Perceptual and Motor Skills, 98*, 1241. https://doi.org/10.2466/pms.98.4.1241-1250

Ouellet-Morin, I., Danese, A., Williams, B., & Arseneault, L. (2011). Validation of a high-sensitivity assay for C-reactive protein in human saliva. *Brain, Behavior, and Immunity, 25*, 640–646. https://doi.org/10.1016/j.bbi.2010.12.020

Out, D., Hall, R. J., Granger, D. A., Page, G. G., & Woods, S. J. (2012). Assessing salivary C-reactive protein: Longitudinal associations with systemic inflammation and cardiovascular disease risk in women exposed to intimate partner violence. *Brain, Behavior, and Immunity, 26*, 543–551. https://doi.org/10.1021/nl061786n.Core-Shell

Papanicolaou, D., Wilder, R., Manolagas, S., & Chrousos, G. (1998). The pathophysiologic roles of interleukin-6 in human disease. *Annals of Internal Medicine, 128*, 127–137.

Pariante, C. M. C., & Miller, A. A. H. (2001). Glucocorticoid receptors in major depression: Relevance to pathophysiology and treatment. *Biological Psychiatry, 49*, 391–404. https://doi.org/10.1016/S0006-3223(00)01088-X

Pay, J. B., & Shaw, A. M. (2019). Towards salivary C-reactive protein as a viable biomarker of systemic inflammation. *Clinical Biochemistry, 68*, 1–8. https://doi.org/10.1016/j.clinbiochem.2019.04.006

Pfaffe, T., Cooper-white, J., Beyerlein, P., Kostner, K., & Punyadeera, C. (2011). Diagnostic potential of saliva: Current state and future applications. *Clinical Chemistry, 57*, 675–687. https://doi.org/10.1373/clinchem.2010.153767

Phillips, A. C., Carroll, D., Evans, P., Bosch, J. A., Clow, A., Hucklebridge, F., & Der, G. (2006). Stressful life events are associated with low secretion rates of immunoglobulin a in saliva in the middle aged and elderly. *Brain, Behavior, and Immunity, 20*, 191–197. https://doi.org/10.1016/j.bbi.2005.06.006

Pisanic, N., Ballard, S.-B., Colquechagua, F. D., François, R., Exum, N., Yori, P. P., . . . Kosek, M. N. (2018). Minimally invasive saliva testing to monitor norovirus infection in community settings. *The Journal of Infectious Diseases, 219*(8), 1234–1242. https://doi.org/10.1093/infdis/jiy638

Pisanic, N., Rahman, A., Saha, S. K., Labrique, A. B., Nelson, K. E., Granger, D. A., . . . Heaney, C. D. (2017). Development of an oral fluid immunoassay to assess past and recent hepatitis E virus (HEV) infection. *Journal of Immunological Methods, 448*, 1–8. https://doi.org/10.1016/j.jim.2017.04.012

Pritchard, B. T., Stanton, W., Lord, R., Petocz, P., & Pepping, G. J. (2017). Factors affecting measurement of salivary cortisol and secretory immunoglobulin a in field studies of athletes. *Frontiers in Endocrinology (Lausanne), 8*, 1–7. https://doi.org/10.3389/fendo.2017.00168

Prossin, A. R., Koch, A. E., Campbell, P. L., McInnis, M. G., Zalcman, S. S., & Zubieta, J. K. (2011). Association of plasma interleukin-18 levels with emotion regulation and mu-opioid neurotransmitter function in major depression and healthy volunteers. *Biological Psychiatry, 69*, 808–812. https://doi.org/10.1016/j.biopsych.2010.10.014

Punyadeera, C., Dimeski, G., Kostner, K., Beyerlein, P., & Cooper-White, J. (2011). One-step homogeneous C-reactive protein assay for saliva. *Journal of Immunological Methods, 373*, 19–25. https://doi.org/10.1016/j.jim.2011.07.013

Rangbulla, V., Nirola, A., Gupta, M., & Batra, P. (2017). Salivary IgA, interleukin-1b and MMP-8 as salivary biomarkers in chronic periodontitis patients. *The Chinese Journal of Dental Research, 20*, 43–51.

Rao, N. L., Shetty, S., Upadhyaya, K., Prasad, R. M., Lobo, E. C., Kedilaya, H. P., & Prasad, G. (2010). Salivary C-reactive protein in Hashimoto's thyroiditis and subacute thyroiditis. *International Journal of Inflammation, 2010*, 1–5. https://doi.org/10.4061/2010/514659

Rathnayake, N., Åkerman, S., Klinge, B., Lundegren, N., Jansson, H., Tryselius, Y., . . . Gustafsson, A. (2013). Salivary biomarkers of oral health—A cross-sectional study. *Journal of Clinical Periodontology, 40*, 140–147. https://doi.org/10.1111/jcpe.12038

Rathnayake, N., Gieselmann, D.-R., Heikkinen, A., Tervahartiala, T., & Sorsa, T. (2017). Salivary diagnostics—Point-of-care diagnostics of MMP-8 in dentistry and medicine. *Diagnostics, 7*, 7. https://doi.org/10.3390/diagnostics7010007

Reinhardt, É. L., Fernandes, P. A. C. M., Markus, R. P., & Fischer, F. M. (2016). Short sleep duration increases salivary IL-6 production. *Chronobiology International, 33*, 780–782. https://doi.org/10.3109/07420528.2016.1167710

Reinhardt, E., Fernandes, P., Markus, R., & Fischer, F. (2019). Night work effects on salivary cytokines TNF, IL-1β and IL-6. *Chronobiology International, 36*, 11–26.

Rhodus, N., Ho, V., Miller, C., Myers, S., & Ondrey, F. (2005). NF-κB dependent cytokine levels in saliva of patients with oral preneoplastic lesions and oral squamous cell carcinoma. *Cancer Detection and Prevention, 29*, 42–45.

Riis, J. L., Bryce, C. I., Ha, T., Hand, T., Stebbins, J. L., Matin, M., . . . Granger, D. A. (2017). Adiponectin: Serum-saliva associations and relations with oral and systemic markers of inflammation. *Peptides, 91*, 58–64. https://doi.org/10.1016/j.peptides.2017.03.006

Riis, J. L., Granger, D. A., Dipietro, J. A., Bandeen-Roche, K., & Johnson, S. B. (2015). Salivary cytokines as a minimally-invasive measure of immune functioning in young children: Correlates of individual differences and sensitivity to laboratory stress. *Developmental Psychobiology, 57*, 153–167. https://doi.org/10.1002/cncr.27633.Percutaneous

Riis, J. L., Granger, D. A., Minkovitz, C. S., Bandeen-Roche, K., DiPietro, J. A., & Johnson, S. B. (2016). Maternal distress and child neuroendocrine and immune regulation. *Social Science and Medicine (1982), 151*, 206–214. https://doi.org/10.1016/j.socscimed.2015.12.043

Riis, J., Out, D., Dorn, L., Beal, S., Denson, L., Pabst, S., . . . Granger, D.A. (2014). Salivary cytokines in healthy adolescent girls: Intercorrelations, stability, and associations with serum cytokines, age, and pubertal stage. *Developmental Psychobiology, 56*, 797–811. https://doi.org/10.1002/dev.21149

Rohleder, N. (2014). Stimulation of systemic low-grade inflammation by psychosocial stress. *Psychosomatic Medicine, 76*, 181–189. https://doi.org/10.1097/PSY.0000000000000049

Rudnicka, A., Rumley, A., Lowe, G., & Strachan, D. (2007). Clinical perspective. *Circulation, 115*, 996–1003.

Ruhl, S., Homberaer, S., Betz, R., Sukkar, T., Schmalz, G., Seymour, R. A., . . . Thomason, J. M. (2004). Salivary proteins and cytokines in drug-induced gingival overgrowth. *Journal of Dental Research, 83*, 322–326. https://doi.org/10.1177/154405910408300410

Rutherfurd-Markwick, K., Starck, C., Dulson, D., & Ali, A. (2017). Salivary diagnostic markers in males and females during rest and exercise. *Jounal of the International Society of Sports Nutrition, 14*, 27. https://doi.org/10.1186/s12970-017-0185-8

Sahibzada, H. A., Khurshid, Z., Khan, R. S., Naseem, M., Siddique, K. M., Mali, M., & Zafar, M. S. (2017). Salivary IL-8, IL-6 and TNF-α as potential diagnostic biomarkers for oral cancer. *Diagnostics, 7*, 21. https://doi.org/10.3390/diagnostics7020021

Salminen, A., Gursoy, U. K., Paju, S., Hyvärinen, K., Mäntylä, P., Könönen, E., . . . Pussinen, P. J. (2014). Salivary biomarkers of bacterial burden, inflammatory response, and tissue destruction in periodontitis. *Journal of Clinical Periodontology, 41*, 442–450. https://doi.org/10.1111/jcpe.12234

Saltiel, A. R., & Olefsky, J. M. (2017). Inflammarory linking obesity and metabolic disease and metabolic disease. *The Journal of Clinical Investigation, 127*, 1–4. https://doi.org/10.1172/JCI92035

Scannapieco, F., Bush, R., & Paju, S. (2003). Associations between periodontal disease and risk for atherosclerosis, cardiovascular disease, and stroke. A systematic review. *Annals of Periodontology, 8*, 38–53.

Segerstrom, S., & Miller, G. (2004). Psychological stress and the human immune system: A meta-analytic study of 30 years of inquiry. *Psychological Bulletin, 130*, 601–630. https://doi.org/10.1038/mp.2011.182

Shields, G. S., Kuchenbecker, S. Y., Pressman, S., Sumida, K. D., & Slavich, G. M. (2016). Better cognitive control of emotional information is associated with reduced pro-inflammatory cytokine reactivity to emotional stress. *Stress, 19*, 63–68. https://doi.org/10.3109/10253890.2015.1121983

Shields, G. S., Slavich, G. M., Perlman, G., Klein, D. N., & Kotov, R. (2019). The short-term reliability and long-term stability of salivary immune markers. *Brain, Behavior, and Immunity, 81*, 650–654. https://doi.org/10.1016/j.bbi.2019.06.007

Sjögren, E., Leanderson, P., Kristenson, M., & Ernerudh, J. (2006). Interleukin-6 levels in relation to psychosocial factors: Studies on serum, saliva, and in vitro production by blood mononuclear cells. *Brain, Behavior, and Immunity, 20*, 270–278. https://doi.org/10.1016/j.bbi.2005.08.001

Slavich, G. M., Way, B. M., Eisenberger, N. I., & Taylor, S. E. (2010). Neural sensitivity to social rejection is associated with inflammatory responses to social stress. *Proceedings of the National Academy of Sciences, 107*, 14817–14822. https://doi.org/10.1073/pnas.1009164107

Slavish, D., Graham-Engeland, J., Smyth, J., & Engeland, C. (2015a). Salivary markers of inflammation in response to acute stress. *Brain, Behavior, and Immunity, 44*, 253–269.

Slavish, D. C., Graham-Engeland, J. E., Smyth, J. M., & Engeland, C. G. (2015b). Salivary markers of inflammation in response to acute stress. *Brain, Behavior, and Immunity, 44*, 253–269. https://doi.org/10.1016/j.bbi.2014.08.008

Smigiel, K., & Parks, W. (2017). Chapter five—Matrix metalloproteinases and leukocyte activation. *Science, 147*, 167–195.

Solomon, G., & Moos, R. (1964). Emotions, immunity, and disease. *Archives of General Psychiatry, 11*, 657–674. https://doi.org/10.1016/j.humimm.2009.07.015

Sorenson, H., Cruz-Almeida, Y., Tighe, P., Wallet, S. M., Riley, J. L., Aguirre, M., … Riley, J. L. (2017). Age differences in salivary markers of inflammation in response to experimental pain: Does venipuncture matter? *Journal of Pain Research, 10*, 2365–2372. https://doi.org/10.2147/jpr.s138460

Sproston, N. R., & Ashworth, J. J. (2018). Role of C-reactive protein at sites of inflammation and infection. *Frontiers in Immunology, 9*, 1–11. https://doi.org/10.3389/fimmu.2018.00754

St. John, M., Li, Y., Zhou, X., Denny, P., Ho, C.-M., Montemagno, C., … Wong, D. (2004). Interleukin 6 and interleukin 8 as potential biomarkers for oral cavity and oropharyngeal squamous cell carcinoma. *Archives of Otolaryngology—Head and Neck Surgery, 130*, 929–935. https://doi.org/10.1001/archotol.130.8.929

Suarez, E. C., Sundy, J. S., & Erkanli, A. (2015). Depressogenic vulnerability and gender-specific patterns of neuro-immune dysregulation: What the ratio of cortisol to C-reactive protein can tell us about loss of normal regulatory control. *Brain, Behavior, and Immunity, 44*, 137–147. https://doi.org/10.1016/j.bbi.2014.09.008

Szabo, Y. Z., Fernandez-Botran, R., & Newton, T. L. (2019). Cumulative trauma, emotion reactivity and salivary cytokine levels following acute stress in healthy women. *Anxiety, Stress and Coping, 32*, 82–94. https://doi.org/10.1080/10615806.2018.1524377

Taba, M., Kinney, J., Kim, A. S., & Giannobile, W. V. (2005). Diagnostic biomarkers for oral and periodontal diseases. *Dental Clinics of North America, 49*, 551–571. https://doi.org/10.1016/j.cden.2005.03.009

Taneja, V. (2018). Sex hormones determine immune response. *Frontiers in Immunology, 9*, 1–5. https://doi.org/10.3389/fimmu.2018.01931

Taubman, M., & Smith, D. (1993). Significance of salivary antibody in dental diseases. *Annals of the New York Academy of Sciences, 694*, 202–215. https://doi.org/10.1196/annals.1340.018

Tauler, P., Martinez, S., Moreno, C., Martínez, P., & Aguilo, A. (2014). Changes in salivary hormones, immunoglobulin A, and C-reactive protein in response to ultra-endurance exercises. *Applied Physiology, Nutrition, and Metabolism, 39*, 560–565. https://doi.org/10.1139/apnm-2013-0466

Teeuw, W., Bosch, J. A., Veerman, E. C. I., & Nieuw Amerongen, A. V. (2004). Neuroendocrine regulation of salivary IgA synthesis and secretion: Implication for oral health. *Biological Chemistry, 385*, 1137–1146. https://doi.org/10.1515/BC.2004.147

Teles, R., Likhari, V., Socransky, S., & Haffajee, A. (2009). Salivary cytokine levels in subjects with chronic periodontitis and in periodontally healthy individuals: a cross-sectional study. *Journal of Periodontal Research, 44*, 411–417.

Tyrka, A. R., Parade, S. H., Valentine, T., Eslinger, N. M., & Seifer, R. (2015). Adversity in preschool-aged children: Effects on salivary interleukin-1β. *Development and Psychopathology, 27*, 567–576. https://doi.org/10.1017/S0954579415000164

U.S. Department of Health and Human Services. (2000). *Oral health in America: A report of the surgeon general* (Vol. 44, pp. Rockville, MD, 16–Rockville, 28). National Institute of Dental and Craniofacial Research.. https://doi.org/10.1089/pop.2013.0038

Ulmer-Yaniv, A., Djalovski, A., Priel, A., Zagoory-Sharon, O., & Feldman, R. (2018). Maternal depression alters stress and immune biomarkers in mother and child. *Depression and Anxiety, 35*, 1145–1157. https://doi.org/10.1002/da.22818

Ulmer-Yaniv, A., Djalovski, A., Yirmiya, K., Halevi, G., Zagoory-Sharon, O., & Feldman, R. (2018). Maternal immune and affiliative biomarkers and sensitive parenting mediate the effects of chronic early trauma on child anxiety. *Psychological Medicine, 48*, 1020–1033. https://doi.org/10.1017/S0033291717002550

Venugopal, A., & Maheswari, T. (2016). Expression of matrix metalloproteinase-9 in oral potentially malignant disorders: A systematic review. *Journal of Oral and Maxillofacial Pathology, 20*, 474–479.

Veru, F., Laplante, D. P., Luheshi, G., & King, S. (2014). Prenatal maternal stress exposure and immune function in the offspring. *Stress, 17*, 133–148. https://doi.org/10.3109/10253890.2013.876404

Weemaes, C., Klasen, I., Göertz, J., Beldhuis-Valkis, M., Olafsson, O., & Haraldsson, A. (2003). Development of immunoglobulin a in infancy and childhood. *Scandinavian Journal of Immunology, 58*, 642–648. https://doi.org/10.1111/j.1365-3083.2003.01344.x

Williamson, S., Munro, C., Pickler, R., Grap, M. J., & Elswick, R. K. (2012). Comparison of biomarkers in blood and saliva in healthy adults. *Nursing Research and Practice, 2012*, 1–4. https://doi.org/10.1155/2012/246178

World Health Organization. (2013). *Oral health surveys: Basic methods* (5th ed.). Geneva, World Health Organization.

Yirmiya, K., Djalovski, A., Motsan, S., Zagoory-Sharon, O., & Feldman, R. (2018). Stress and immune biomarkers interact with parenting behavior to shape anxiety symptoms in trauma-exposed youth. *Psychoneuroendocrinology, 98*, 153–160. https://doi.org/10.1016/j.psyneuen.2018.08.016

Zefferino, R., Facciorusso, A., Lasalvia, M., Narciso, M., Nuzzaco, A., Lucchini, R., & L'Abbate, N. (2006). Salivary markers of work stress in an emergency team of urban police (1 degree step). *Giornale Italiano Medicina del Lavoro ed Ergonomia, 28*, 472–477.

Zhang, C., Cheng, X., Li, J., Zhang, P., Yi, P., Xu, X., & Zhou, X. (2016). Saliva in the diagnosis of diseases. *International Journal of Oral Science, 8*, 133–137. https://doi.org/10.1038/ijos.2016.38

Zhang, L., Li, X., Yan, H., & Huang, L. (2018). Salivary matrix metalloproteinase (MMP)-8 as a biomarker for periodontitis. *Medicine (Baltimore), 97*, 1–6. https://doi.org/10.1097/MD.0000000000009642

Zhang, S., Zhang, X., Yin, K. E., Li, T., Bao, Y., & Chen, Z. (2017). Variation and significance of secretory immunoglobulin A, interleukin 6 and dendritic cells in oral cancer. *Oncology Letters, 13*, 2297–2303. https://doi.org/10.3892/ol.2017.5703

Zheng, H., Li, R., Zhang, J., Zhou, S., Ma, Q., Zhou, Y., … Lin, J. (2014). Salivary biomarkers indicate obstructive sleep apnea patients with cardiovascular diseases. *Scientific Reports, 4*, 1–7. https://doi.org/10.1038/srep07046

# Chapter 10
# Salivary Bioscience and Human Development

**Leah C. Hibel, Kristine Marceau, and Andrea C. Buhler-Wassmann**

**Abstract** Human Development is an interdisciplinary field with the goal of understanding the complex biological, psychological, social, and contextual interactions that underlie individual development across the life span. Each phase of development (e.g., infancy, middle childhood, and adolescence) is uniquely characterized by differential experiences, developmental capabilities, and physiological functioning. Yet even within a particular phase of life, individual differences in behavioral expression exist, reflecting dynamic responses to experiences occurring in the past years, months, weeks, or even minutes. Further, human physiology is multifaceted, with multiple biological agents acting and interacting to influence behavior. Minimally invasive salivary assessments enable human development researchers to carefully disentangle these complex features of behavior by collecting an array of biological markers, across ecologically meaningful events, at appropriately timed intervals, starting on the first day of life. This chapter will review salivary bioscience's contributions in furthering the field of human development. The chapter will focus on birth through adolescence, with special attention to the unique developmental experiences that characterize each broad phase of life.

**Keywords** Salivary biomarkers · Development · Infants · Children · Adolescents

L. C. Hibel (✉) · A. C. Buhler-Wassmann
Department of Human Ecology, University of California, Davis, Davis, CA, USA
e-mail: lchibel@ucdavis.edu

K. Marceau
Department of Human Development and Family Studies, Purdue University, West Lafayette, IN, USA



D. A. Granger, M. K. Taylor (eds.), *Salivary Bioscience*,
https://doi.org/10.1007/978-3-030-35784-9_10

## 10.1 History of Salivary Bioscience in Human Development

The science of human development is by definition interdisciplinary, focusing on understanding the dynamic interplay among biological, cognitive, behavioral, and social–emotional domains. It aims to take a holistic approach and inherently acknowledges that humans exist and develop within social contexts (e.g., relationships and culture), physical contexts (e.g., home, school, and workplace), and temporal contexts (e.g., historic and age). Developmental theorists have created conceptual frameworks that illustrate individuals as agents within their environments, acting in and being acted on by the human ecology (Bronfenbrenner, 1977; Sroufe, 1979). For example, mothers shape their children's biology through parent–child interactions; however, maternal behavior is largely determined by cultural norms and early caregiving experiences (e.g., Sameroff, 2009). Over time, as individuals develop into adolescents, their biological systems and developmental goals change (Booth, Carver, & Granger, 2000). For adolescents, peers and romantic relationships play an increasingly important role on behavioral and biological development, yet broader social expectations continue to influence experiences (Crockett & Crouter, 2014). The field of human development acknowledges these proximal biobehavioral transactions, as well as the more distal influences of the individuals' social and cultural contexts, that shape developmental trajectories and outcomes.

Biological systems are responsive to context and assist in the behavioral response to contextual demands. Scientists study biological markers (e.g., hormones) as a conduit for understanding how external factors influence these physiological systems, and how these physiological systems, in turn, influence internal functioning and development (D'Onofrio & Lahey, 2010; Hostinar & Gunnar, 2013). Importantly, advances in salivary bioscience have allowed for the noninvasive and inexpensive measurement of many physiological processes through saliva (e.g., Granger et al., 2012), making it possible to test human development frameworks and illuminate the transactions between environmental influences and biology (Booth et al., 2000).

Links among behavior, environment, and hormones were first demonstrated in nonhuman animal and primate research (e.g., Bernstein, Rose, & Gordon, 1974), and in the 1980s, there was a rise in studies investigating the physiological correlates of human behavioral development (e.g., Gunnar, 1986, 1989). Many relied on salivary bioscience as a practical approach to examine individual differences in biological functioning (Laudat et al., 1988; Vining, McGinley, Maksvytis, & Ho, 1983). Thus, examining saliva has enabled scientists to test core human development theories. For example, attachment theory has assumed that primary caregivers fulfill infants' biological needs by acting as external regulators, and in analyzing saliva, scientists have found evidence that a secure attachment relationship reduces physiological reactivity (Bernard & Dozier, 2010; Gunnar, Brodersen, Nachmias, Buss, & Rigatus, 1996). Likewise, salivary bioscience gives insight into how conditions during pregnancy program infant physiology (Rash et al., 2016; see Chap. 25), the

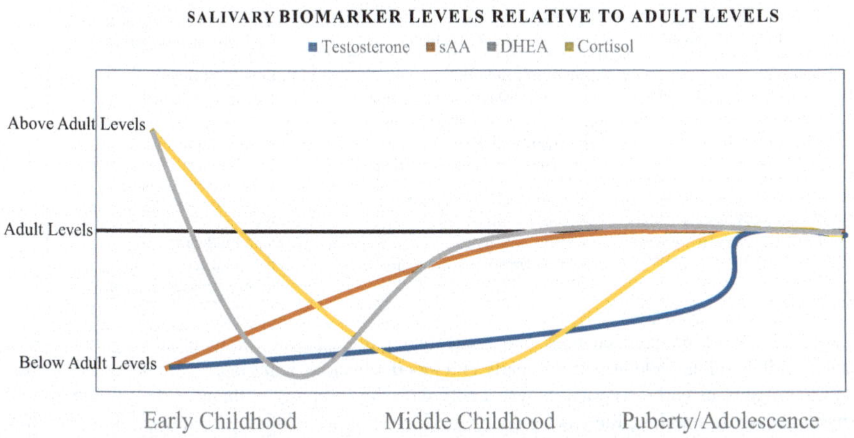


**Fig. 10.1** Approximate trajectories of testosterone (blue), salivary alpha-amylase (sAA; orange), dehydroepiandrosterone (DHEA; gray), and cortisol (yellow) from infancy through puberty and adolescence. Trajectories are graphed in relation to adult levels. Adult levels are represented by the horizontal line across the middle of the graph. Both cortisol and DHEA begin with relatively high levels compared to adults, followed by a decrease to middle childhood and an increase into puberty. Salivary AA and testosterone begin low, with sAA gradually increasing through middle childhood, and testosterone sharply increasing at puberty

mechanisms through which early experiences are linked to later outcomes (Sullivan, Hawes, Winchester, & Miller, 2008), and how youths' peer experiences shape biobehavioral development (Kornienko & Granger, 2018).

In this chapter, we will discuss the ways in which salivary bioscience has allowed scientists to better understand human development and the connections among biology, behavior, and context. Collecting and studying saliva grants access to a wide array of analytes, and we will focus on those most commonly incorporated into studies of developmental processes and behavior, namely cortisol, salivary alpha-amylase (sAA), testosterone, and dehydroepiandrosterone (DHEA; note Chap. 9 focuses on immunological markers and thus will not be detailed here). These markers provide a window into the functioning of autonomic nervous system (via sAA), the hypothalamic–pituitary–adrenal (HPA) axis (via cortisol and DHEA), and the hypothalamic–pituitary–gonadal (HPG) axis (via testosterone and DHEA). Though the field of human development covers the full life span, other chapters in this volume will discuss the prenatal period (see Chap. 25) and adulthood and aging (see Chap. 27); therefore, this chapter will examine these processes during infancy and early childhood (0–6 years), middle childhood (6 years–puberty), and puberty and adolescence (puberty–18 years). Developmental changes and contextual influences on these salivary biomarkers will be explored during each phase of childhood and adolescence. In Fig. 10.1 we provide the approximate developmental trajectories of these main salivary biomarkers from infancy through puberty. Figure 10.2 illustrates the primary developmental changes and main contextual influences at each phase.

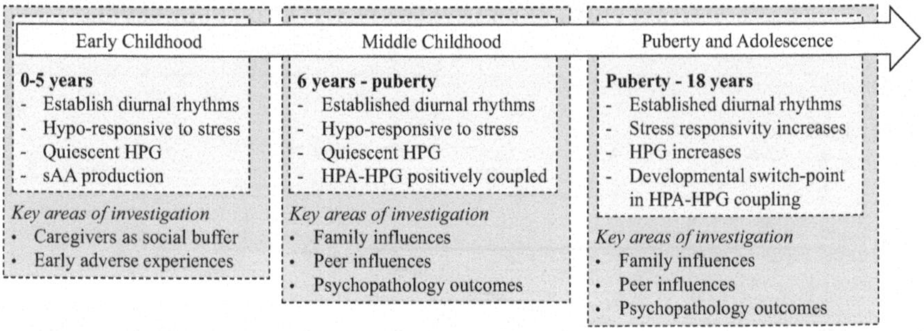


**Fig. 10.2** Developmental changes and contextual influences on common salivary biomarkers across childhood and adolescence. In early childhood, physiological systems are being established and the quality of early experiences, especially with the caregiver, influences developing diurnal rhythms. During middle childhood, physiological systems are relatively hyporesponsive, yet they undergo changes in preparation for puberty and begin to be implicated in internalizing and externalizing problems. In puberty and adolescence, increases in hormone levels and responsivity and changes in co-activation of physiological systems equip individuals for developmentally appropriate goals. Families and peers affect the continued organization of hormonal activity, which often precedes or is associated with psychopathology

## 10.2 Current Status of Knowledge in Human Development

### *10.2.1 Infancy and Early Childhood (0–5 years)*

***Developmental Changes*** Beginning during mid-gestation, environmental stimuli evoke HPA axis responses and influence future adrenocortical functioning (Gitau, Fisk, & Glover, 2004; Levine, 2005; see Chap. 25). The use of salivary biomarkers in newborns, however, has also illuminated the immaturity of certain physiological systems at birth. For instance, although children, adolescents, and adults exhibit a reliable cortisol diurnal rhythm with high cortisol in the morning and the lowest levels in the evening (see Chap. 5), this pattern is not developed until 3–6 months of age (Antonini, Jorge, & Moreira, 2000; de Weerth, Zijl, & Buitelaar, 2003). Further, average child cortisol levels are significantly higher than in adults, slowly decreasing into middle childhood (Finegood, Rarick, Blair, and Family Life Project Investigators, 2017; Laurent, Harold, Leve, Shelton, & Van Goozen, 2016; Tollenaar, Jansen, Beijers, Riksen-Walraven, & de Weerth, 2010). Salivary AA production is also underdeveloped in infancy. Specifically, only small amounts of sAA are produced at birth, though production undergoes a sharp increase across the second year of life, reaching maximum levels by 5–6 years of age (O'Donnell & Miller, 1980). Salivary AA also exhibits diurnal fluctuations, but in an opposite pattern as cortisol, with lower morning levels that increase over the course of the day (Miller et al., 2015). However, relatively less research has been conducted on the development of the sAA diurnal rhythm, compared to cortisol. Thus, while these systems are active, neither sAA nor cortisol secretion during early childhood reflects adult levels or patterns (see Figs. 10.1 and 10.2).

Across the 30–45 min post waking, cortisol exhibits a sharp increase (named the cortisol awakening response; CAR) while sAA exhibits a sharp decrease (see Chap. 5; Clow, Thorn, Evans, & Hucklebridge, 2004; Nater, Rohleder, Schlotz, Ehlert, & Kirschbaum, 2007). The CAR appears to develop in the first 6 months as sleep consolidates, although to the best of our knowledge, comparable studies have not been conducted with sAA (Bright, Frick, Out, & Granger, 2014; Stalder et al., 2013; Tegethoff, Knierzinger, Meyer, & Meinlschmidt, 2013). Across the second year of life, children exhibit either a cortisol or sAA awakening response only about 50% of mornings (Bright et al., 2014). However, implementing rigorous compliance inclusion criteria (e.g., less than 15 min difference between self-report and actigraphy-verified wake time; less than 15 min difference between self-report collection time and electronic time stamp of collection time) in a study of the CAR in 12–87 month old children suggests this percentage could be as high as 80% (Baumler, Kirschbaum, Kliegel, Alexander, & Stalder, 2013). Further, the degree of cortisol change in the awakening response exhibits stability across early childhood (Baumler et al., 2013). Lastly, individual differences in children's awakening responses have been found to be sensitive to family experiences (Hibel, Trumbell, & Mercado, 2014) and associated with health outcomes (Miller et al., 2015).

Environmentally challenging or difficult situations prompt the activation of various neurophysiological systems, with increases in cortisol and sAA, to aid in children's immediate behavioral response, as well as long-term adaptation (e.g., Lupien, McEwen, Gunnar, & Heim, 2009). However, during early childhood, attachment figures provide a strong buffer to physiological responses to stress, rendering stress paradigms less successful at eliciting increases in cortisol (Gunnar, 2017; Gunnar, Brodersen, Krueger, & Rigatuso, 1996; Gunnar, Talge, & Herrera, 2009; Lewis & Ramsay, 1995). Despite this period of hypo-responsivity, individual differences in reactivity (i.e., stress-induced increases) and recovery (i.e., return to baseline) to physical and psychological stressors are thought to index sensitivity to context (Del Giudice, Ellis, & Shirtcliff, 2011), coping ability (e.g., Obradović, 2012), and inform research on the development of self-regulation (e.g., Blair & Raver, 2012).

***Family Influences*** Caregiver interactions profoundly influence the child's developing biobehavioral systems. When confronted with a stressor, infants and young toddlers rely on their immediate caregivers to help them cope with the stressor (e.g., Hofer, 2006). When distressed, a sensitive and responsive caregiver perceives the child's cues, and quickly and accurately responds to the child's needs. These supportive behaviors not only regulate behavioral distress, but physiological arousal as well. For example, sensitive caregiving and higher maternal caregiving quality are associated with overall lower levels of adrenocortical and sAA output (Clowtis, Kang, Padhye, Rozmus, & Barratt, 2016; Hatfield, Hestenes, Kintner-Duffy, & O'Brien, 2013; Kaplan, Evans, & Monk, 2008; Pendry & Adam, 2007; Taylor et al., 2013) and reactivity (Hibel, Granger, Blair, Cox, and Family Life Project Key Investigators, 2011; Laurent et al., 2016; Mörelius, Nelson, & Gustafsson,

2007; Spanglar, Schieche, Ilg, Maier, & Ackermann, 1994), and greater cortisol recovery (Albers, Riksen-Walraven, Sweep, & Weerth, 2008). In sum, it appears that young children with secure attachment relationships and more sensitive caregivers show healthier response profiles in the face of stress.

***Influence of Adversity*** Numerous early adverse experiences have been found to strongly shape the developing child's biological functioning. For example, child maltreatment is a profound stressor, characterized by extreme physical punishment or neglect of the child's emotional and/or physical needs. Over time these toxic interactions are thought to shift the levels and responsivity of the child's HPA axis toward reduced morning cortisol, and overall flatter diurnal slopes via altered feedback regulation (Bruce, Fisher, Pears, & Levine, 2009; Cicchetti, Rogosch, Gunnar, & Toth, 2010; Cicchetti, Rogosch, Toth, & Sturge-Apple, 2011; Dozier et al., 2006; Kuras et al., 2017). Similarly, poverty is a multifaceted stressor associated with harsh parenting, chaotic households, substandard housing, and other risky exposures (Evans & Kim, 2007; Yoshikawa, Aber, & Beardslee, 2012). Thus, as expected, early exposure to poverty has also been related to flattened cortisol diurnal rhythms (Zalewski, Lengua, Kiff, & Fisher, 2012).

These early adverse experiences can heighten vigilance and perception of threat (for review see Pakulak, Stevens, & Neville, 2018), and studies have found that exposure to marital violence (Hibel et al., 2011), maternal depression (Azar, Paquette, Zoccolillo, Baltzer, & Tremblay, 2007; Dougherty, Klein, Rose, & Laptook, 2011), and cumulative risk (as indexed by maternal demographic characteristics and family income; Holochwost et al., 2017) during infancy and toddlerhood can heighten cortisol reactivity (see Hunter, Minnis, & Wilson, 2011, for review). Similarly, young children exposed to war have higher levels of sAA than unexposed children (Feldman, Vengrober, Eidelman-Rothman, & Zagoory-Sharon, 2013). Importantly, physiology at this point is amenable to change, and when quality of care is increased, by either removal from a maltreating home into supportive foster care (Bernard, Butzin-Dozier, Rittenhouse, & Dozier, 2010) or through parenting interventions aimed at increasing nurturance and decreasing problematic parenting behaviors (Bernard, Hostinar, & Dozier, 2015), cortisol levels are lowered and rhythms are normalized.

Individual differences in physiological reactivity and recovery from stress are thought to mediate the relationship between early adversity and mental and physical health (e.g., Loman & Gunnar, 2010; Zalewski et al., 2012). For example, early exposure to poverty has been related to lower sAA levels at 12 months which in turn predicted higher levels of internalizing behaviors at 36 months (Hill-Soderlund et al., 2015). Similarly, exposure to intimate partner violence in utero most strongly predicted internalizing behaviors when infants exhibited asymmetrical HPA–SNS (i.e., high-cortisol and low-sAA) reactivity, compared to symmetrical HPA–SNS (i.e., low-cortisol and low-sAA or high-cortisol and high-sAA) reactivity (Martinez-Torteya, Bogat, Lonstein, Granger, & Levendosky, 2017). Salivary studies provide evidence of the early dysregulation of stress physiology within the context of adversity, as a mechanism increasing the potential for pathology.

In summary, early childhood presents a time of physiological development, and diurnal rhythm consolidation. This developmental process is aided by caregivers' regulatory control over child physiology through the quality of their caregiving behavior. Likewise, adverse childhood experiences, such as child maltreatment, shape individuals' physiological reactivity and recovery, providing the biological underpinnings of certain pathological conditions (see Figs. 10.1 and 10.2).

### *10.2.2 Middle Childhood (6 years–Puberty)*

***Developmental Changes*** In a potential continuation of the stress hyporesponsive period, middle childhood is generally considered a period of relative quiescence in the overall scope of salivary HPA and HPG activity (Kamin & Kertes, 2017; Shirtcliff et al., 2012). Toward the end of middle childhood, salivary cortisol and sAA reactivity increases, likely with the onset of adrenarche, an early stage of puberty (Auchus & Rainey, 2004; Leppert, Kushner, Smith, Lemay, & Dougherty, 2016; Strahler, Skoluda, Kappert, & Nater, 2017; Törnhage, 2002). However, during this developmental life stage, the relative quiescence of salivary hormone levels does not equate to a lack of individual differences. For example, at the age of 7 years, cortisol has been shown to be responsive to fear and frustration (Lopez-Duran, Hajal, Olson, Felt, & Vazquez, 2009), and this stress responsivity has been shown to increase from early childhood (e.g., ages 3–5 years) to middle childhood (e.g., ages 6–10 years) with substantial individual differences (Leppert et al., 2016). Further, increasing levels of DHEA drive adrenarche, ushering in the first pubertal changes (Havelock, Auchus, & Rainey, 2004). However, longitudinal changes in *salivary* DHEA with adrenarche have yet to be documented.

***Family and Peer Influences*** During middle childhood, the family environment is still a strong regulator of HPA activity (Gunnar & Hostinar, 2015). Most often, research has examined the effect of maternal depression on cortisol levels and reactivity. For example, paternal depression was associated with increased cortisol reactivity, whereas maternal depression strengthened the association of children's positive emotionality with lower cortisol reactivity (Mackrell et al., 2014). The importance of the family context is not limited to cortisol; poorer family environments have been linked to earlier adrenarche as assessed by salivary DHEA concentrations (Ellis & Essex, 2007). Middle childhood marks an important shift in social development, whereby peers become more influential than earlier in childhood. Although more pronounced in adolescence, there is evidence that peers begin to play an important role in stress regulation in middle childhood. For example, fourth graders who experienced peer rejection had higher levels of cortisol at school and a blunted diurnal rhythm, but having better friendships or more friends attenuated this association (Peters, Riksen-Walraven, Cillessen, & de Weerth, 2011). Importantly, an earlier study found less evidence of cortisol or sAA responsivity to a peer rejection task in 7–12 year olds (Stroud et al., 2009). Together, these studies suggest that peer influences on biological stress responding are more complex than simply experiencing rejection.

***Influence of Adversity*** A substantial body of literature has examined the influence of early life stress on salivary cortisol during middle childhood. In particular, a history of child maltreatment, and early institutional care have been shown to predict dysregulated cortisol functioning (Doom, Cicchetti, & Rogosch, 2014; Flannery et al., 2017; Puetz et al., 2016). Importantly, the nature of the dysregulation varies widely, with some studies showing increased responsivity or indexes of diurnal cortisol and others decreased. It is likely that further contextual influences, like the availability of social buffers like supportive parents play a key role. For example, the effects of early life stress on middle childhood cortisol reactivity are also exacerbated by more recent traumatic events (Jaffee et al., 2015). Further, adrenarche may play a contextual role: Early life stress was associated with lower cortisol awakening response among 9–13 year olds prior to puberty, but with higher cortisol awakening response among youth who had begun puberty (King et al., 2017).

***Psychopathology: Internalizing*** One of the major foci of salivary bioscience research has been to identify biomarkers of risk for psychopathology. Most of the literature examining how salivary biomarkers are associated with psychopathology during childhood have focused on cortisol. However, findings within this literature are mixed (Guerry & Hastings, 2011). For example, lower cortisol levels (Badanes, Watamura, & Hankin, 2011) and blunted CAR (McGinnis, Lopez-Duran, Martinez-Torteya, Abelson, & Muzik, 2016) have been associated with internalizing problems. Conversely, higher cortisol levels (in the context of low autonomic arousal; El-Sheikh, Erath, Buckhalt, Granger, & Mize, 2008) and higher cortisol reactivity (Martinez-Torteya, Bogat, Levendosky, & Eye, 2016) have also been associated with internalizing problems. Increasingly, studies have begun to examine moderators of these associations (El-Sheikh et al., 2008; Martinez-Torteya et al., 2016), as well as longitudinal associations (Ruttle et al., 2011) in order to attempt to clarify whether internalizing problems during childhood predict altered cortisol later in development or vice versa. The limited evidence available suggests that internalizing problems in middle childhood precede lower diurnal cortisol in adolescence (Ruttle et al., 2011).

***Psychopathology: Externalizing*** The literature on externalizing problems is similarly focused on cortisol, also with mixed findings. Meta-analytic evidence suggests that in middle childhood, lower basal cortisol is linked to externalizing problems, the opposite direction of effects found among preschoolers (Alink et al., 2008). This association of low-activity dysregulation with externalizing problems has also been found for the cortisol awakening response (Freitag et al., 2009). Similar to internalizing problems, contextual factors including parenting (Wagner et al., 2017) have been shown to moderate cortisol-externalizing associations. Further, emerging longitudinal work suggests that externalizing problems may precede a dampening of the HPA axis, at least as indexed by diurnal cortisol, across middle childhood (Salis, Bernard, Black, Dougherty, & Klein, 2016).

In sum, studies of middle childhood reveal salivary hormones to be relatively quiescent in response to stressors, though studies using salivary bioscience have found individual differences in the reactivity of stress hormones. Family context remains important, but there is a substantial increase in peer influence. Further, individual differences in physiological levels, rhythms, and reactivity have been linked to behavior and psychopathology in middle childhood (see Figs. 10.1 and 10.2).

### 10.2.3 Puberty and Adolescence (Puberty–18 years)

***Developmental Changes*** There are clear increases in HPG hormones during puberty, as a hallmark of pubertal development is the increase in testosterone (for males especially, but also females; Matchock, Dorn, & Susman, 2007), and estradiol for females (Grumbach, 2002; Styne & Grumbach, 2002). There is also substantial evidence that the HPA axis "reactivates" (e.g., HPA activity increases again) during adolescence after the relative quiescence during childhood (Gunnar, Wewerka, Frenn, Long, & Griggs, 2009; Romeo, 2013). For example, salivary levels of cortisol and DHEA increase across puberty (Netherton, Goodyer, Tamplin, & Herbert, 2004; Shirtcliff, Zahn-Waxler, Klimes-Dougan, & Slattery, 2007).

There is also evidence that pubertal hormones may also exhibit increases in stress reactivity across puberty (Eatough, Shirtcliff, Hanson, & Pollak, 2009; Marceau, Ruttle, Shirtcliff, Hastings, et al., 2015). For example, age and pubertal status were both positively related to salivary testosterone increases in response to exercise in adolescent boys (Di Luigi et al., 2006). Cortisol (Ji, Negriff, Kim, & Susman, 2016), DHEA (Marceau et al., 2014; Shirtcliff et al., 2007), and sAA (Strahler et al., 2017) have all been shown to increase in response to stressors during adolescence.

A growing body of research suggests that HPA and HPG hormones change together, or are "coupled" in response to stressors (Han, Miller, Cole, Zahn-Waxler, & Hastings, 2015; Marceau et al., 2014; Turan, Tackett, Lechtreck, & Browning, 2015), across the day (Harden et al., 2016; Marceau, Ruttle, Shirtcliff, Hastings, et al., 2015), and across development (Ruttle, Shirtcliff, Armstrong, Klein, & Essex, 2015). Although coupling is positive in middle childhood and adolescence, there is thought to be a developmental switch-point in late adolescence (Ruttle et al., 2015; Shirtcliff et al., 2015) where the relation of HPA and HPG hormones transitions to a mutual inhibition (see Fig. 10.3). This initial positive coupling likely serves a developmental purpose by allowing adolescents to exhibit more autonomy in dealing with stressors while also undergoing reproductive system changes during early and middle adolescence.

***Family and Peer Influences*** The role of the family context changes substantially during adolescence, as youth develop autonomy and the parent–child relationship is repeatedly renegotiated (Steinberg & Silk, 2002), and parents play less of a role for

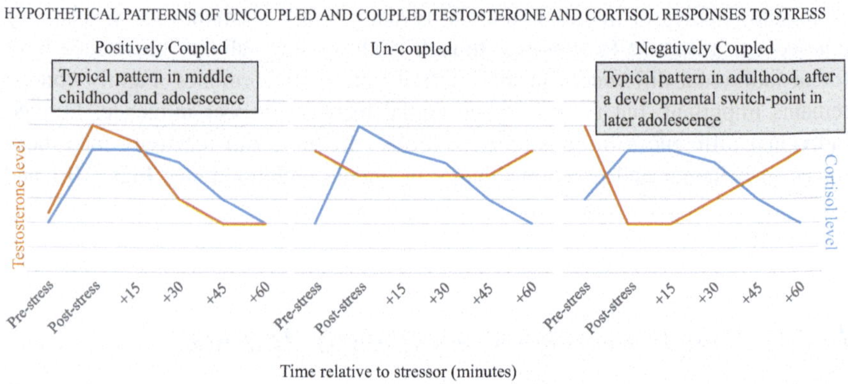


**Fig. 10.3** Hypothetical patterns of the coupling and uncoupling of cortisol and testosterone stress reactivity in middle childhood and adolescence. Middle childhood is marked by positive cortisol/testosterone coupling in response to stress; adolescence (and adulthood) is marked by a negative coupling of cortisol and testosterone

regulating children's hormonal responses during adolescence as compared to earlier in childhood (Hostinar, Johnson, & Gunnar, 2015). During adolescence, higher testosterone levels have been associated with lower quality of mother– son and father–son relationships (Booth, Johnson, Granger, Crouter, & McHale, 2003). Thus, the role of the family for the development of hormone regulation, for example as a social buffer, may change with pubertal maturational stage (Ellis, Boyce, Belsky, Bakermans-Kranenburg, & IJzendoorn, 2011; Zhang et al., 2016), although this body of work is small and mixed with regard to the nature of HPA–HPG–family interplay.

A hallmark of adolescence is the shift in youths' social spheres, whereby adolescents begin to spend more time with peers of their choosing, and peers become increasingly important to hormonal responses to stress (Stroud et al., 2009). For example, post-pubertal girls demonstrated higher sAA responses to peer rejection tasks, whereas pre-pubertal girls were more responsive to social performance (Stroud, Papandonatos, D'Angelo, Brush, & Lloyd-Richardson, 2017). Peer influences have been shown to be both a source of increased stress and a social buffer, depending on the aspect of peer relationships under study (Kornienko & Granger, 2018). Specifically, cortisol levels have been shown to increase with delinquent peer involvement among youth with disruptive behavior disorders, but decrease among healthy controls (Dorn et al., 2009). There is also some research to suggest that motivation for social dominance during adolescence is linked to basal levels of pubertal hormones in girls (e.g., testosterone; Cardoos et al., 2017) and boys (Schaal, Tremblay, Soussignan, & Susman, 1996). Together, advances in research using salivary bioscience have shown that the changes in pubertal hormones occurring during adolescence may play an important role in the shift from parents to peers as the key socializers during this developmental period.

***Influence of Adversity*** Early life adversity, including trauma and abuse but also less severe forms of adversity, has been shown to play a critical role in the development of cortisol function during adolescence, as in earlier developmental periods (Lupien et al., 2009). Across studies, the direction of associations are somewhat unclear, with higher levels of early life stress being associated with both higher reactivity (Halligan, Herbert, Goodyer, & Murray, 2004; Rao, Hammen, Ortiz, Chen, & Poland, 2008) and lower levels of cortisol (Stroud, Chen, Doane, & Granger, 2016). Examining HPA and HPG function together in order to elucidate the role of early life stress may help to clarify these associations. For example, there is evidence from a study of incarcerated boys that experiencing early life adversity may lead to more strongly positively coupled cortisol and testosterone across the day (Dismukes, Johnson, Vitacco, Iturri, & Shirtcliff, 2015). Further, pubertal maturation itself may moderate associations of early life stress with cortisol function, with early adversity blunting, and later adversity heightening (King et al., 2017). However, other studies did not find differences based on pubertal status (McLaughlin et al., 2015).

***Psychopathology: Internalizing*** Although the cross-sectional literature is mixed (with higher and lower levels and reactivity of cortisol being associated with internalizing phenotypes), longitudinal work on children and adolescents suggest that higher levels of cortisol and cortisol reactivity often precede the development of internalizing problems (Burke, Davis, Otte, & Mohr, 2005; Marceau, Ruttle, Shirtcliff, Essex, & Susman, 2015). Recent meta-analyses suggest that flatter diurnal slopes (Adam et al., 2017) are associated with internalizing phenotypes during adolescence, although associations with cortisol awakening responses are mixed with both positive and negative associations found (Chida & Steptoe, 2009).

Basal levels of DHEA and testosterone have also been associated with both higher and lower levels of internalizing symptoms during adolescence, which also points to the importance of contextual moderators (Marceau, Ruttle, Shirtcliff, Essex, et al., 2015). Further, one study found that salivary DHEA responsivity to a social stressor was associated with higher internalizing problems (Shirtcliff et al., 2007). Because of evidence that DHEA may protect against some of the deleterious effects of cortisol, a substantial body of research has examined high cortisol-to-low DHEA ratios in association with internalizing problems during adolescence (Goodyer, Park, Netherton, & Herbert, 2001). Generally, cross-sectional work suggests that high cortisol-to-DHEA ratios are associated with internalizing problems (Kamin & Kertes, 2017).

***Psychology: Externalizing*** Theoretically, low arousal in certain brain regions lead youth to engage in stimulating behaviors (e.g., sensation seeking; Zuckerman, 1979), which often fall into the category of externalizing behaviors (e.g., delinquency, aggression, rule breaking, substance use). An increase in risk-taking and sensation seeking has been linked to levels of testosterone and estradiol during adolescence (Harden et al., 2018; Peper, Braams, Blankenstein, Bos, & Crone, 2018). Indeed, testosterone levels and responsivity are often associated with externalizing problems, particularly aggression, during adolescence (Yildirim & Derksen,

2012), although there are ample null findings and most studies are cross-sectional (Book, Starzyk, & Quinsey, 2001; Duke, Balzer, & Steinbeck, 2014). Although a number of studies suggest that higher levels of DHEA are associated with externalizing phenotypes during adolescence (Marceau, Ruttle, Shirtcliff, Essex, et al., 2015), meta-analytic evidence finds no association between cortisol and externalizing behaviors in adolescence (Alink et al., 2008).

HPA–HPG interplay has also been investigated with respect to externalizing problems. There is some evidence that high levels of testosterone and low levels of cortisol together (either assessed via ratios or statistical interactions) predict externalizing psychopathology during adolescence, at least for boys (Dabbs, Jurkovic, & Frady, 1991; Popma et al., 2007; Yu & Shi, 2009). In a recent longitudinal study, statistical interactions (as opposed to ratios) have suggested that low cortisol reactivity was associated with externalizing phenotypes in the context of low but not high diurnal testosterone in boys, and to a lesser extent, girls (Susman, Peckins, Bowes, & Dorn, 2017).

In summary, puberty is marked by increases in cortisol, estradiol, testosterone, DHEA, and sAA activity and reactivity. During this time, parents or peers can act as either stressors or buffers on physiology, based on the quality of the relationships. Further, salivary studies continue to highlight the importance of physiology in understanding adolescent behaviors and psychopathology (see Figs. 10.1 and 10.2).

## 10.3 Methodological Issues, Challenges, and Considerations

Saliva can provide a large array of biological data and the ease of collection has resulted in its rapid incorporation into developmental studies. Historically, studying basic physiological processes in nonclinical children was difficult due to parents' and children's unwillingness to participate in research with invasive biological collections such as blood draws. Salivary collections circumvent this problem, but that does not mean that salivary measures are without challenges or methodological considerations (see Chaps. 3 and 4). For example, as described in the previous sections, physiology changes across the life span, as do the collection methods and study designs. Therefore, the collection and interpretation of salivary biomarkers must be couched within an understanding of human development. Additionally, humans live complex and varied lives, and these diverse experiences play a fundamental role in physiological levels and trajectories across an event, a day, a year, or a life span. Therefore, understanding salivary biomarkers requires a nuanced understanding of the social and cultural environment. Though other methodological considerations exist, these will be our focus.

***Developmental Considerations*** Developmental abilities and changes introduce a number of methodological challenges. A benefit of salivary collections is that they are not confined to laboratory settings, and present the opportunity to collect in real-world settings (e.g., school, home, and playground). To do this, participants are trained on salivary collection techniques, such as how (and how much) to collect, when to collect, how to store the sample, abstaining from eating and drinking, etc. However, young children are not able to collect their own saliva and must rely on others to ensure compliance. This magnifies the potential for noncompliance in that not only must the participant (i.e., the child) be compliant, the caregiver must also be compliant. Importantly, if compliance is related to developmental or contextual variables of interest (e.g., emotion regulation, poverty, and maltreatment), these systematic biases may influence the ability to detect differences in physiological functioning (Valentino, Alba, Hibel, Fondren, & McDonnell, 2017).

Study methods may also need to vary to account for developmental differences. For example, many studies attempt to elicit a stress response to determine individual differences in physiological (e.g., cortisol, sAA, and DHEA) reactivity to stress. This is problematic because not all stressors induce the same physiological response, and responses depend on an individual's perception of the stressor. Dickerson and Kemeny (2004) identified effective stress paradigms as characterized by unpredictability, uncontrollability, and social-evaluative threat. Yet before 3 months of age physical stimuli (e.g., physical examinations, and blood draws), not psychological, are most appropriate (e.g., Mörelius, He, & Shorey, 2016). Around 3 months, infants begin to show stress responses to interactions with insensitive caregivers. Between 6 and 9 months, physiological reactivity can be triggered by short parental separations; however, by 1 year, only longer separations from the parent, like spending the day at child care, increase salivary stress physiology (Gunnar, Talge, & Herrera, 2009). Self-referent emotions (e.g., embarrassment) emerge by preschool; therefore, by preschool age, tasks with parents or peers provoking negative self-evaluation induce increases in stress hormones (Gunnar, Talge, & Herrera, 2009). Relatively fewer studies of early childhood have incorporated sAA into stress paradigms with appropriate attention to its faster timeline of change; sAA does appear to be responsive to physical (Davis & Granger, 2009), emotional (Laurent, Ablow, & Measelle, 2012; Spinrad et al., 2009; Taylor et al., 2013), and cognitive (Miller et al., 2015) stressors in young children. Though not many studies have been successful in eliciting physiological reactivity in middle childhood, after puberty, children are physiologically reactive to public speaking tasks like the Trier Social Stress Test for children (TSST-C; Buske-Kirschbaum et al., 1997; Gunnar, Talge, & Herrera, 2009).

Further, one's developmental abilities (e.g., cognitive abilities, emotion regulation, and brain development) are likely to influence the ways in which experiences impact their physiological functioning. For example, earlier in development, early life adversity is associated with higher cortisol production, whereas later in development, the early life and prolonged adversity is associated with lower cortisol production (Tarullo & Gunnar, 2006). This developmental change may be explained by a transition into a state of hypo-responsivity to protect from the potentially deleterious effects of prolonged cortisol exposure on the brain (Juster, McEwen, & Lupien, 2010). This notion is also in part supported by evidence that the timing of early adversity matters for the association with adolescent cortisol function. For example, prenatal adversity and adversity from ages 6–11 years (but not 0–5 years) predicted increased cortisol reactivity and higher cortisol levels at age 16 years, whereas more proximal adversity during age 12–13 and 14–15 years predicted low cortisol levels at age 16 years (Bosch et al., 2012).

***Contextual Considerations*** Current theoretical approaches in the field of human development acknowledge that individual differences in developmental processes arise out of complex transactions within and between multiple levels of influence, from cellular to societal (Booth et al., 2000; Bronfenbrenner, 1977; see Fig. 10.2). Physiological systems do not work in isolation; numerous co-activating systems provide a physiological context by which biological substrates interact and work together to increase or decrease the probability that a behavior will occur, depending on the social context (Sapolsky, 1997; Shirtcliff et al., 2015). For example, incorporating the role of oxytocin and early life stress is expected to further clarify and advance our understanding of how cortisol and testosterone jointly influence aggression during adolescence (Fragkaki, Cima, & Granic, 2018). Likewise, exposure to a stressor results in the coordinated activation of multiple physiological systems including components of the autonomic, neuroendocrine, metabolic, and immune system (Lupien et al., 2009). Importantly, examination of inter-subsystem relations may provide unique information on the health of the individual, above and beyond the examination of each system in isolation (Bauer, Quas, & Boyce, 2002; Granger et al., 2012). Thus, although to date most human development studies have tended to incorporate only one biological marker, human development theories and biological realities necessitate attention to the larger physiological context.

Moment-to-moment changes in daily experiences (i.e., intra-individual difference) such as eating a big meal, napping, waking up late, or rushing to get to a scheduled event all have the potential to change physiological analytes of interest in saliva (e.g., Gibson et al., 1999; Robles et al., 2012; Stalder et al., 2016) and add unwanted (and perhaps uncontrolled) variance into saliva studies. Importantly, experiences differentially impact different analytes, with some analytes being more acutely susceptible than others. For example, cortisol appears to be highly sensitive to fluctuations in context with large diurnal and event reactivity-related variability within person (Tomarken, Han, & Corbett, 2015), whereas sAA diurnal profiles remain far more stable in the face of daily changes (Out, Granger, Sephton, & Segerstrom, 2013). To complicate this issue further, not all individuals are uniformly sensitive to external events (Hibel, Senguttuvan, & Bauer, 2013; Out et al., 2013). Thus, the real-world collection of salivary biomarkers necessitates detailed logs of daily routines, events, and stressors to control for, or examine, everyday contexts on individuals' physiology.

Broader individual differences in social experiences also have the potential to amplify or suppress biobehavioral associations (Chen, Raine, & Granger, 2018). Specifically, during early childhood, high-risk contexts and negative parent–child relationships have been found to influence biobehavioral associations, such that physiology more strongly predicts, or is predicted by, poor health and development (Boyce et al., 2006; Obradović, 2012; Wolf, Nicholls, & Chen, 2008). Likewise, during adolescence parent–child relationships, peer influences, and neighborhood quality each have been shown to moderate testosterone- and/or cortisol- externalizing and internalizing associations (Booth et al., 2003; Dorn et al., 2009; Steeger, Cook, & Connell, 2017; Yu et al., 2016). In the future, human development models are expected to become even more complex in an attempt to better capture the theorized interplay of endogenous (multiple physiological systems) and exogenous (e.g., social and cultural) contexts for hormone–behavior associations, across multiple time scales.

## 10.4 Future Directions and Opportunities

Before addressing the future directions and opportunities, it is important to acknowledge the promises realized by salivary biomarker work in the field of human development. The biobehavioral research of the 1980s was marked by attempts to clarify the scientific concept of stress (Engel, 1985), determine the most accurate biological marker of stress (Kagan Reznick, & Snidman, 1987; Levine, Winer, Coe, Bayart, & Hayashi, 1987) and provide theoretical explanations for dissociations between behavior and biology (e.g., Frankenhaeuser, 1980; Gunnar, Isensee, & Fust, 1987; Henry, 1980; Levine, 1985). The incorporation of salivary biomarkers into studies of human development in the time since has allowed researchers to move beyond establishing foundational definitions and processes to addressing complex, longitudinal, developmental, and multisystem questions. For example, current research often examines how early life experiences shaped the trajectory of physiological development (e.g., Del Giudice et al., 2011; Gunnar, 2016); how relationships buffer stress responses (e.g., Doom et al., 2015; Gunnar, 2017), or how experiences are translated into mental and physical health (e.g., Koss & Gunnar, 2018; Miller, Chen, & Parker, 2011; Shonkoff et al., 2012). While there is certainly still work to be done in these areas, significant advances have been made in our understanding of biopsychosocial transactions. Importantly, salivary biomarkers have played a leading role in these discoveries.

Salivary biomarkers will continue to provide crucial physiological data that informs human developmental science. This dynamic biofluid expands human development research and models at multiple levels. At the methodological level, saliva has allowed researchers to more accurately capture life as it is lived, with collections occurring in people's natural environments and on time scales meaningful to the behavioral, contextual, or physiological process. At the individual level, new markers continue to be uncovered allowing for novel, and potentially more accurate, understandings of biobehavioral associations across the life span. For example, tests for salivary uric acid and oxytocin have recently been developed or improved (Martin, Kagerbauer, Gempt, Podtschaske, Hapfelmeier, & Schneider, 2018). Further, at the population level, utilizing saliva has allowed researchers to expand the populations examined. Requiring participants to travel to large universities, or medical centers to receive blood draws necessarily narrowed the pool of eligibility. Saliva opens up the possibility of assessing individuals and groups who are disenfranchised, under-studied, or in remote locations, and whose experiences are vastly different than white, middle class, educated participants who tend to dominate the scientific literature.

***Emerging Methods*** For the first two decades of salivary bioscience in human development, methods largely mirrored laboratory designs, substituting saliva for more invasively collected tissues and fluids (e.g., blood). These studies provided crucial information on individual differences in physiological responses to highly controlled paradigms. Occasionally laboratory like methods were conducted in the home, but for the most part studies failed to capitalize on the flexibility of salivary collections. In the past decade the types of studies utilizing salivary biomarkers expanded to include examinations of naturally occurring day-to-day fluctuations, physiological changes to developmentally meaningful transitions [development of sleep cycles (Gribbin, Watamura, Cairns, Harsh, & LeBourgeois, 2012), puberty (Stroud et al., 2017)] and responses to ecologically valid experiences [beginning school (Yang, Lamb, Kappler, & Ahnert, 2017), maltreatment (Valentino et al., 2015), problems with peers (Bai, Robles, Reynolds, & Repetti, 2017)]. The next generation of salivary biomarker studies will more fully capture the lived human experience by examining life in vivo.

Further, salivary collections can be conducted on much larger groups of people than studies confined to small laboratory settings. For example, recent studies have collected samples from multiple members of a family (e.g., Hibel & Mercado, 2017), large peer networks (e.g., Kornienko & Granger, 2018), or even entire classrooms of children (Spray, Floyd, Littleton, Trnka, & Mattison, 2018). Humans are highly social animals and emotions and behaviors have been found to be highly "contagious" across groups (Butler, 2011; Hatfield, Cacioppo, & Rapson, 1993). Saliva facilitated the discovery that humans not only coordinate behaviors, but also experience linkage or attunement in physiological reactivity (e.g., Laurent et al., 2012; Ruttle, Serbin, Stack, Schwartzman, & Shirtcliff, 2011) and rhythms (Hibel, Mercado, & Valentino, 2019; Papp, Pendry, & Adam, 2009). Future studies will continue to uncover the situations that encourage (or discourage) group level attunement, and the health and behavioral implications of these associations.

***Emerging Markers*** Though most studies of human development have focused on the biological markers discussed in this chapter, the National Institute for Cranofacial and Dental Research funded a systematic examination and characterization of the salivary proteome. This effort resulted in the identification of more than 1000 analytes available for assay in saliva (Hu, Loo, & Wong, 2007). Clearly there are too many candidates to discuss, but we will identify a few emerging biomarkers with the potential to provide a new understanding of human development. Recent studies show promise for the incorporation of immune markers and uric acid, into studies of human behavior.

The immune system is responsible for coordinating a network of responses to insults such as viruses, bacteria, and parasites, and therefore plays a prominent role in physical health (see Chap. 9). The immune system is also intimately tied with the nervous and endocrine systems (i.e., neuroendocrine-immune network, NEI), and as such is calibrated by early life adversity and experiences of stress (Johnson, Riley, Granger, & Riis, 2013). Incorporation of salivary biomarkers of immunity open up new opportunities to gain a better understanding of NEI functioning in children and how contextual experiences modulate stress-related damage within the NEI network. However, it is important to note that immune factors in saliva (e.g., immunoglobulins, cytokines) are actively produced and excreted from the salivary glands, and therefore do not reflect peripheral levels found in blood, or even fluids from other mucus membranes (e.g., tears). Further, as detailed in Chap. 9, there is limited evidence suggesting that salivary immune markers serve as systemic indices of health.

Uric acid is produced during the breakdown of purine nucleotides, and heightened levels have been associated with a host of chronic illnesses such as hypertension, metabolic syndrome, obesity, and Type-2 diabetes (Biscaglia, Ceconi, Malagu, Pavasini, & Ferrari, 2015; Kushiyama, Tanaka, Hara, & Kawazu, 2014; Soukup et al., 2012). Recently it was shown that salivary uric acid levels are highly stable, and robustly correlated with serum levels (Riis et al., 2018). As obesity and type-2 diabetes rates continue to climb worldwide, especially in children, salivary uric acid might prove to be an important marker of health in future studies. These emerging biomarkers, along with advances in genome/epigenome analysis (e.g., single-nucleotide polymorphisms, genome-wide association studies, telomeres), and other biological markers of health and development will play important roles in the next generation of developmental research (see Chap. 26 for a discussion on salivary omics in pediatric health).

***Emerging Populations*** Salivary bioscience opens the possibility of assessing biomarkers of stress and adaptation in at-risk or remote environments (Masten, 2014), including trailer parks after Hurricane Katrina (e.g., Vigil, Geary, Granger, & Flinn, 2010), homeless shelters (e.g., Cutuli, Wiik, Herbers, Gunnar, & Masten, 2010), foster homes (e.g., Fisher, Van Ryzin, & Gunnar, 2011), low-income and rural family homes in the USA (e.g., Blair et al., 2013), and rural areas in Nepal (e.g., Worthman & Panter-Brick, 2008). It is important to study people from diverse backgrounds and take an intersectional approach in understanding how one's identities within the social environment affect physiology and health (Doane, Sladek, & Adam, 2018; Parra & Hastings, 2018). Individuals who identify as a sexual, ethnic, racial, and/or gender minority live at a higher risk of experiencing stressful life events (e.g., discrimination) that impact physiological functioning. Studying the saliva of marginalized people opens the door to understanding the mechanisms through which social stressors influence health and perpetuate inequalities. For Latinx adolescents, perceived racial or ethnic discrimination is associated with elevations in diurnal cortisol (Zeiders, Doane, & Roosa, 2012). Furthermore, aspects of one's identity may serve as protective factors toward healthful adaptation to challenging contexts. For example, Latina pregnant women who endorsed more acculturation (i.e., greater adaptation of the dominant culture) had more blunted diurnal cortisol slopes, as measured through saliva, than Latina pregnant women who were less acculturated, and these cortisol patterns affected their babies' birth weights (D'Anna et al., 2012). Thus, having a stronger tie to one's ethnic group is associated with more positive physiological and developmental outcomes. Human development research using salivary bioscience has endorsed that people's diverse experiences, influenced by their multifaceted identities and settings, impact their physiology in different ways. By assessing these experiences and populations using saliva, scientists are acknowledging the vast array of human experiences and uncovering the developmental processes through which individuals interact with their social context.

To better understand individuals' full lived experiences and different physiological profiles, scientists are integrating salivary biomarkers into studies with greater attention to diverse contexts, in participants' authentic ecological settings. Groups who have been historically disenfranchised are underrepresented in research, and some disenfranchisement even stems from abuse of power within the research community (e.g., Brandt, 1978). Institutional Review Board oversight and researchers/research partners with a shared cultural and/or linguistic background with participants reduce the likelihood of future abuses, and potentially increase minority group participation in research. Likewise, the use of salivary biomarkers also has the potential to increase the reach of developmental science. For example, using salivary collections removes the reliance on hospital or university clinicians/phlebotomists and allows trusted community members to be employed as part of the research team to train participants on saliva collections. Similarly, when participants are able to collect in the comfort of their own settings (e.g., school and home), barriers such as requiring participant transportation (which can be overly costly, time consuming, or logistically not possible without a car), or requiring participants to navigate predominantly white, privileged institutions are reduced. Lastly, the minimally invasive nature of saliva is inherently more welcoming and acceptable to a larger percentage of the population than other painful or restrictive procedures or devices. By opening up the possibilities of where physiological research can take place and who can participate, salivary bioscience is creating opportunities for a more accurate representation of the human experience.

## References

Adam, E. K., Quinn, M. E., Tavernier, R., McQuillan, M. T., Dahlke, K. A., & Gilbert, K. E. (2017). Diurnal cortisol slopes and mental and physical health outcomes: A systematic review and meta-analysis. *Psychoneuroendocrinology, 83*, 25–41. https://doi.org/10.1016/j.psyneuen.2017.05.018

Albers, E. M., Riksen-Walraven, J., Sweep, F. C., & Weerth, C. D. (2008). Maternal behavior predicts infant cortisol recovery from a mild everyday stressor. *Journal of Child Psychology and Psychiatry, 49*(1), 97–103. https://doi.org/10.1111/j.1469-7610.2007.01818.x

Alink, L. R., van Ijzendoorn, M. H., Bakermans-Kranenburg, M. J., Mesman, J., Juffer, F., & Koot, H. M. (2008). Cortisol and externalizing behavior in children and adolescents: Mixed meta-analytic evidence for the inverse relation of basal cortisol and cortisol reactivity with externalizing behavior. *Developmental Psychobiology, 50*(5), 427–450. https://doi.org/10.1002/dev.20300

Antonini, S. R., Jorge, S. M., & Moreira, A. C. (2000). The emergence of salivary cortisol circadian rhythm and its relationship to sleep activity in preterm infants. *Clinical Endocrinology, 52*(4), 423–426. https://doi.org/10.1111/j.1365-2265.2000.tb00561.x

Auchus, R. J., & Rainey, W. E. (2004). Adrenarche–physiology, biochemistry and human disease. *Clinical Endocrinology, 60*(3), 288–296. https://doi.org/10.1046/j.1365-2265.2003.01858.x

Azar, R., Paquette, D., Zoccolillo, M., Baltzer, F., & Tremblay, R. E. (2007). The association of major depression, conduct disorder, and maternal overcontrol with a failure to show a cortisol buffered response in 4-month-old infants of teenage mothers. *Biological Psychiatry, 62*(6), 573–579. https://doi.org/10.1016/j.biopsych.2006.11.009

Badanes, L. S., Watamura, S. E., & Hankin, B. L. (2011). Hypocortisolism as a potential marker of allostatic load in children: Associations with family risk and internalizing disorders. *Development and Psychopathology, 23*(3), 881–896. https://doi.org/10.1017/S095457941100037X

Bai, S., Robles, T. F., Reynolds, B. M., & Repetti, R. L. (2017). Children's diurnal cortisol responses to negative events at school and home. *Psychoneuroendocrinology, 83*, 150–158. https://doi.org/10.1016/j.psyneuen.2017.05.027

Bauer, A. M., Quas, J. A., & Boyce, W. T. (2002). Associations between physiological reactivity and children's behavior: Advantages of a multisystem approach. *Journal of Developmental and Behavioral Pediatrics, 23*(2), 102–113.

Baumler, D., Kirschbaum, C., Kliegel, M., Alexander, N., & Stalder, T. (2013). The cortisol awakening response in toddlers and young children. *Psychoneuroendocrinology, 38*, 2485–2492. https://doi.org/10.1016/j.psyneuen.2013.05.008

Bernard, K., Butzin-Dozier, Z., Rittenhouse, J., & Dozier, M. (2010). Cortisol production patterns in young children living with birth parents vs children placed in foster care following involvement of Child Protective Services. *Archives of Pediatrics and Adolescent Medicine, 164*(5), 438–443. https://doi.org/10.1001/archpediatrics.2010.54

Bernard, K., & Dozier, M. (2010). Examining infants' cortisol responses to laboratory tasks among children varying in attachment disorganization: Stress reactivity or return to baseline? *Developmental Psychology, 46*(6), 1771–1778. https://doi.org/10.1037/a0020660

Bernard, K., Hostinar, C. E., & Dozier, M. (2015). Intervention effects on diurnal cortisol rhythms of Child Protective Services–referred infants in early childhood: Preschool follow-up results of a randomized clinical trial. *JAMA Pediatrics, 169*(2), 112–119. https://doi.org/10.1001/jamapediatrics.2014.2369

Bernstein, I. S., Rose, R. M., & Gordon, T. P. (1974). Behavioral and environmental events influencing primate testosterone levels. *Journal of Human Evolution, 3*(6), 517–525. https://doi.org/10.1016/0047-2484(74)90012-8

Biscaglia, S., Ceconi, C., Malagu, M., Pavasini, R., & Ferrari, R. (2015). Uric acid and coronary artery disease: An elusive link deserving further attention. *International Journal of Cardiology, 213*, 28–32. https://doi.org/10.1016/j.ijcard.2015.08.086

Blair, C., Berry, D., Mills-Koonce, R., Granger, D. A., & The FLP Investigators. (2013). Cumulative effects of early poverty on cortisol in young children: Moderation by autonomic nervous system activity. *Psychoneuroendocrinology, 38*(11), 2666–2675. https://doi.org/10.1016/j.psyneuen.2013.06.025

Blair, C., & Raver, C. C. (2012). Child development in the context of adversity: Experiential canalization of brain and behavior. *American Psychologist, 67*, 309–318. https://doi.org/10.1037/a0027493

Book, A. S., Starzyk, K. B., & Quinsey, V. L. (2001). The relationship between testosterone and aggression: A meta analysis. *Aggression and Violent Behavior, 6*(6), 579–599. https://doi.org/10.1016/S1359-1789(00)00032-X

Booth, A., Carver, K., & Granger, D. A. (2000). Biosocial perspectives on the family. *Journal of Marriage and Family, 62*(4), 1018–1034. https://doi.org/10.1111/j.1741-3737.2000.01018.x

Booth, A., Johnson, D. R., Granger, D. A., Crouter, A. C., & McHale, S. (2003). Testosterone and child and adolescent adjustment: The moderating role of parent-child relationships. *Developmental Psychology, 39*(1), 85–98. https://doi.org/10.1037/0012-1649.39.1.85

Bosch, N. M., Riese, H., Reijneveld, S. A., Bakker, M. P., Verhulst, F. C., Ormel, J., & Oldehinkel, A. J. (2012). Timing matters: Long term effects of adversities from prenatal period up to adolescence on adolescents' cortisol stress response. The TRAILS study. *Psychoneuroendocrinology, 37*(9), 1439–1447. https://doi.org/10.1016/j.psyneuen.2012.01.013

Boyce, W. T., Essex, M. J., Alkon, A., Goldsmith, H. H., Kraemer, H. C., & Kupfer, D. J. (2006). Early father involvement moderates biobehavioral susceptibility to mental health problems in middle childhood. *Journal of the American Academy of Child and Adolescent Psychiatry, 45* (12), 1510–1520. https://doi.org/10.1097/01.chi.0000237706.50884.8b

Brandt, A. M. (1978). Racism and research: The case of the Tuskegee Syphilis Study. *Hastings Center Report, 8*(6), 21–29. https://doi.org/10.2307/3561468

Bright, M. A., Frick, J. E., Out, D., & Granger, D. A. (2014). Individual differences in the cortisol and salivary α-amylase awakening responses in early childhood: Relations to age, sex, and sleep. *Developmental Psychobiology, 56*(6), 1300–1315. https://doi.org/10.1002/dev.21209

Bronfenbrenner, U. (1977). Toward an experimental ecology of human development. *American Psychologist, 32*(7), 513–531. https://doi.org/10.1037/0003-066X.32.7.513

Bruce, J., Fisher, P. A., Pears, K. C., & Levine, S. (2009). Morning cortisol levels in preschool-aged foster children: Differential effects of maltreatment type. *Developmental Psychobiology, 51*(1), 14–23. https://doi.org/10.1002/dev.20333

Burke, H. M., Davis, M. C., Otte, C., & Mohr, D. C. (2005). Depression and cortisol responses to psychological stress: A meta-analysis. *Psychoneuroendocrinology, 30*(9), 846–856. https://doi.org/10.1016/j.psyneuen.2005.02.010

Buske-Kirschbaum, A., Jobst, S., Wustmans, A., Kirschbaum, C., Rauh, W., & Hellhammer, D. (1997). Attenuated free cortisol response to psychosocial stress in children with atopic dermatitis. *Psychosomatic Medicine, 59*(4), 419–426. https://doi.org/10.1097/00006842-199707000-00012

Butler, E. A. (2011). Temporal Interpersonal Emotion Systems the "TIES" that form relationships. *Personality and Social Psychology Review, 15*(4), 367–393. https://doi.org/10.1177/1088868311411164

Cardoos, S. L., Suleiman, A. B., Johnson, M., van den Bos, W., Hinshaw, S. P., & Dahl, R. E. (2017). Social status strategy in early adolescent girls: Testosterone and value-based decision making. *Psychoneuroendocrinology, 81*, 14–21. https://doi.org/10.1016/j.psyneuen.2017.03.013

Chen, F. R., Raine, A., & Granger, D. A. (2018). Testosterone and proactive-reactive aggression in youth: The moderating role of harsh discipline. *Journal of Abnormal Child Psychology, 46*, 1–14. https://doi.org/10.1007/s10802-018-0399-5

Chida, Y., & Steptoe, A. (2009). Cortisol awakening response and psychosocial factors: A systematic review and meta-analysis. *Biological Psychology, 80*(3), 265–278. https://doi.org/10.1016/j.biopsycho.2008.10.004

Cicchetti, D., Rogosch, F. A., Gunnar, M. R., & Toth, S. L. (2010). The differential impacts of early physical and sexual abuse and internalizing problems on daytime cortisol rhythm in school-aged children. *Child Development, 81*(1), 252–269. https://doi.org/10.1111/j.1467-8624.2009.01393.x

Cicchetti, D., Rogosch, F. A., Toth, S. L., & Sturge-Apple, M. L. (2011). Normalizing the development of cortisol regulation in maltreated infants through preventive interventions. *Development and Psychopathology, 23*(3), 789–800. https://doi.org/10.1017/S0954579411000307

Clow, A., Thorn, L., Evans, P., & Hucklebridge, F. (2004). The awakening cortisol response: Methodological issues and significance. *Stress, 7*(1), 29–37. https://doi.org/10.1080/10253890410001667205

Clowtis, L. M., Kang, D. H., Padhye, N. S., Rozmus, C., & Barratt, M. S. (2016). Biobehavioral factors in child health outcomes: The roles of maternal stress, maternal–child engagement,

salivary cortisol, and salivary testosterone. *Nursing Research, 65*(5), 340–351. https://doi.org/10.1097/NNR.0000000000000172
Crockett, L. J., & Crouter, A. C. (2014). *Pathways through adolescence: Individual development in relation to social contexts*. New York: Psychology Press.
Cutuli, J. J., Wiik, K. L., Herbers, J. E., Gunnar, M. R., & Masten, A. S. (2010). Cortisol function among early school-aged homeless children. *Psychoneuroendocrinology, 35*, 833–845. https://doi.org/10.1016/j.psyneuen.2009.11.008
D'Anna, K. L., Hoffman, M. C., Zerbe, G. O., Coussons-Read, M., Ross, R. G., & Laudenslager, M. L. (2012). Acculturation, maternal cortisol and birth outcomes in women of Mexican descent. *Psychosomatic Medicine, 74*(3), 296–304. https://doi.org/10.1097/PSY.0b013e318244fbde
D'Onofrio, B. M., & Lahey, B. B. (2010). Biosocial influences on the family: A decade review. *Journal of Marriage and Family, 72*(3), 762–782. https://doi.org/10.1111/j.1741-3737.2010.00729.x
Dabbs, J. M., Jurkovic, G. J., & Frady, R. L. (1991). Salivary testosterone and cortisol among late adolescent male offenders. *Journal of Abnormal Child Psychology, 19*(4), 469–478. https://doi.org/10.1007/BF00919089
Davis, E. P., & Granger, D. A. (2009). Developmental differences in infant salivary alpha-amylase and cortisol responses to stress. *Psychoneuroendocrinology, 34*(6), 795–804. https://doi.org/10.1016/j.psyneuen.2009.02.001
de Weerth, C., Zijl, R. H., & Buitelaar, J. K. (2003). Development of cortisol circadian rhythm in infancy. *Early Human Development, 73*(1–2), 39–52. https://doi.org/10.1016/S0378-3782(03)00074-4
Del Giudice, M., Ellis, B. J., & Shirtcliff, E. A. (2011). The adaptive calibration model of stress responsivity. *Neuroscience and Biobehavioral Reviews, 35*(7), 1562–1592. https://doi.org/10.1016/j.neubiorev.2010.11.007
Di Luigi, L., Baldari, C., Gallotta, M. C., Perroni, F., Romanelli, F., Lenzi, A., & Guidetti, L. (2006). Salivary steroids at rest and after a training load in young male athletes: Relationship with chronological age and pubertal development. *International Journal of Sports Medicine, 27*(09), 709–717. https://doi.org/10.1055/s-2005-872931
Dickerson, S. S., & Kemeny, M. E. (2004). Acute stressors and cortisol responses: A theoretical integration and synthesis of laboratory research. *Psychological Bulletin, 130*(3), 355–391. https://doi.org/10.1037/0033-2909.130.3.355
Dismukes, A. R., Johnson, M. M., Vitacco, M. J., Iturri, F., & Shirtcliff, E. A. (2015). Coupling of the HPA and HPG axes in the context of early life adversity in incarcerated male adolescents. *Developmental Psychobiology, 57*(6), 705–718. https://doi.org/10.1002/dev.21231
Doane, L. D., Sladek, M. R., & Adam, E. (2018). An introduction to cultural neurobiology: Evidence from physiological stress systems. In J. M. Causadias, E. H. Telzer, & N. A. Gonzalez (Eds.), *The handbook of culture and biology* (pp. 227–254). Hoboken, NJ: Wiley.
Doom, J. R., Cicchetti, D., & Rogosch, F. A. (2014). Longitudinal patterns of cortisol regulation differ in maltreated and nonmaltreated children. *Journal of the American Academy of Child and Adolescent Psychiatry, 53*(11), 1206–1215. https://doi.org/10.1016/j.jaac.2014.08.006
Doom, J. R., Hostinar, C. E., Van Zomeren-Dohm, A. A., & Gunnar, M. R. (2015). The roles of puberty and age in explaining the diminished effectiveness of parental buffering of HPA reactivity and recovery in adolescence. *Psychoneuroendocrinology, 59*, 102–111. https://doi.org/10.1016/j.psyneuen.2015.04.024
Dorn, L. D., Kolko, D. J., Susman, E. J., Huang, B., Stein, H., Music, E., & Bukstein, O. G. (2009). Salivary gonadal and adrenal hormone differences in boys and girls with and without disruptive behavior disorders: Contextual variants. *Biological Psychology, 81*(1), 31–39. https://doi.org/10.1016/j.biopsycho.2009.01.004
Dougherty, L. R., Klein, D. N., Rose, S., & Laptook, R. S. (2011). Hypothalamic-pituitary-adrenal axis reactivity in the preschool-age offspring of depressed parents: Moderation by early parenting. *Psychological Science, 22*(5), 650–658. https://doi.org/10.1177/0956797611404084

Dozier, M., Manni, M., Gordon, M. K., Peloso, E., Gunnar, M. R., Stovall-McClough, K. C., ... Levine, S. (2006). Foster children's diurnal production of cortisol: An exploratory study. *Child Maltreatment, 11*(2), 189–197. https://doi.org/10.1177/1077559505285779

Duke, S. A., Balzer, B. W., & Steinbeck, K. S. (2014). Testosterone and its effects on human male adolescent mood and behavior: A systematic review. *Journal of Adolescent Health, 55*(3), 315–322. https://doi.org/10.1016/j.jadohealth.2014.05.007

Eatough, E. M., Shirtcliff, E. A., Hanson, J. L., & Pollak, S. D. (2009). Hormonal reactivity to MRI scanning in adolescents. *Psychoneuroendocrinology, 34*(8), 1242–1246. https://doi.org/10.1016/j.psyneuen.2009.03.006

Ellis, B. J., Boyce, W. T., Belsky, J., Bakermans-Kranenburg, M. J., & Van IJzendoorn, M. H. (2011). Differential susceptibility to the environment: An evolutionary–neurodevelopmental theory. *Development and Psychopathology, 23*(1), 7–28. https://doi.org/10.1017/S0954579410000611

Ellis, B. J., & Essex, M. J. (2007). Family environments, adrenarche, and sexual maturation: A longitudinal test of a life history model. *Child Development, 78*(6), 1799–1817. https://doi.org/10.1111/j.1467-8624.2007.01092.x

El-Sheikh, M., Erath, S. A., Buckhalt, J. A., Granger, D. A., & Mize, J. (2008). Cortisol and children's adjustment: The moderating role of sympathetic nervous system activity. *Journal of Abnormal Child Psychology, 36*(4), 601–611. https://doi.org/10.1007/s10802-007-9204-6

Engel, B. T. (1985). Stress is a noun! No, a verb! No, an adjective! In T. M. Field, P. M. McCabe, & N. Schneiderman (Eds.), *Stress and coping* (pp. 3–12). Hillsdale, NJ: Erlbaum.

Evans, G. W., & Kim, P. (2007). Childhood poverty and health: Cumulative risk exposure and stress dysregulation. *Psychological Science, 18*(11), 953–957. https://doi.org/10.1111/j.1467-9280.2007.02008.x

Feldman, R., Vengrober, A., Eidelman-Rothman, M., & Zagoory-Sharon, O. (2013). Stress reactivity in war-exposed young children with and without posttraumatic stress disorder: Relations to maternal stress hormones, parenting, and child emotionality and regulation. *Development and Psychopathology, 25*(4pt1), 943–955. https://doi.org/10.1017/S0954579413000291

Finegood, E. D., Rarick, J. R., Blair, C., & Family Life Project Investigators. (2017). Exploring longitudinal associations between neighborhood disadvantage and cortisol levels in early childhood. *Development and Psychopathology, 29*(5), 1649–1662. https://doi.org/10.1017/S0954579417001304

Fisher, P. A., Van Ryzin, M. J., & Gunnar, M. R. (2011). Mitigating HPA axis dysregulation associated with placement changes in foster care. *Psychoneuroendocrinology, 36*, 531–539. https://doi.org/10.1016/j.psyneuen.2010.08.007

Flannery, J. E., Gabard-Durnam, L. J., Shapiro, M., Goff, B., Caldera, C., Louie, J., ... Tottenham, N. (2017). Diurnal cortisol after early institutional care—Age matters. *Developmental Cognitive Neuroscience, 25*, 160–166. https://doi.org/10.1016/j.dcn.2017.03.006

Fragkaki, I., Cima, M., & Granic, I. (2018). The role of trauma in the hormonal interplay of cortisol, testosterone, and oxytocin in adolescent aggression. *Psychoneuroendocrinology, 88*, 24–37. https://doi.org/10.1016/j.psyneuen.2017.11.005

Frankenhaeuser, M. (1980). Psychobiological aspects of life stress. In *Coping and health* (pp. 203–223). Boston, MA: Springer.

Freitag, C. M., Hänig, S., Palmason, H., Meyer, J., Wüst, S., & Seitz, C. (2009). Cortisol awakening response in healthy children and children with ADHD: Impact of comorbid disorders and psychosocial risk factors. *Psychoneuroendocrinology, 34*(7), 1019–1028. https://doi.org/10.1016/j.psyneuen.2009.01.018

Gibson, E. L., Checkley, S., Papadopoulos, A., Poon, L., Daley, S., & Wardle, J. (1999). Increased salivary cortisol reliably induced by a protein-rich midday meal. *Psychosomatic Medicine, 61* (2), 214–224. https://doi.org/10.1097/00006842-199903000-00014

Gitau, R., Fisk, N. M., & Glover, V. (2004). Human fetal and maternal corticotrophin releasing hormone responses to acute stress. *Archives of Disease in Childhood. Fetal and Neonatal Edition, 89*(1), F29–F32. https://doi.org/10.1136/fn.89.1.F29

Goodyer, I. M., Park, R. J., Netherton, C. M., & Herbert, J. (2001). Possible role of cortisol and dehydroepiandrosterone in human development and psychopathology. *The British Journal of Psychiatry, 179*(3), 243–249. https://doi.org/10.1192/bjp.179.3.243

Granger, D. A., Fortunato, C. K., Beltzer, E. K., Virag, M., Bright, M. A., & Out, D. (2012). Focus on methodology: Salivary bioscience and research on adolescence: An integrated perspective. *Journal of Adolescence, 35*(4), 1081–1095. https://doi.org/10.1016/j.adolescence.2012.01.005

Gribbin, C. E., Watamura, S. E., Cairns, A., Harsh, J. R., & LeBourgeois, M. K. (2012). The cortisol awakening response (CAR) in 2-to 4-year-old children: Effects of acute nighttime sleep restriction, wake time, and daytime napping. *Developmental Psychobiology, 54*(4), 412–422. https://doi.org/10.1002/dev.20599

Grumbach, M. M. (2002). The neuroendocrinology of human puberty revisited. *Hormone Research in Pædiatrics, 57*(Suppl 2), 2–14. https://doi.org/10.1159/000058094

Guerry, J. D., & Hastings, P. D. (2011). In search of HPA axis dysregulation in child and adolescent depression. *Clinical Child and Family Psychology Review, 14*(2), 135–160. https://doi.org/10.1007/s10567-011-0084-5

Gunnar, M. R. (1986). Human developmental psychoneuroendocrinology: A review on research on neuroendocrine responses to challenge and threat in infancy and childhood. In M. Lamb, A. Brown, & B. Rogoff (Eds.), *Advances in developmental psychology* (Vol. 4, pp. 51–103). Hillsdale, NJ: Erlbaum.

Gunnar, M. R. (1989). Studies of the human infant's adrenocortical response to potentially stressful events. *New Directions for Child and Adolescent Development, 1989*(45), 3–18. https://doi.org/10.1002/cd.23219894503

Gunnar, M. R. (2016). Early life stress: What is the human chapter of the mammalian story? *Child Development Perspectives, 10*(3), 178–183. https://doi.org/10.1111/cdep.12182

Gunnar, M. R. (2017). Social buffering of stress in development: A career perspective. *Perspectives on Psychological Science, 12*(3), 355–373. https://doi.org/10.1177/1745691616680612

Gunnar, M., Isensee, J., & Fust, L. (1987). Adrenocortical activity and the brazelton neonatal assessment scale: Moderating effects of the newborn's biomedical status. *Child Development, 58*(6), 1448–1458. https://doi.org/10.2307/1130684

Gunnar, M. R., Brodersen, L., Krueger, K., & Rigatuso, J. (1996). Dampening of adrenocortical responses during infancy: Normative changes and individual differences. *Child Development, 67*(3), 877–889. https://doi.org/10.1111/j.1467-8624.1996.tb01770.x

Gunnar, M. R., Brodersen, L., Nachmias, M., Buss, K., & Rigatuso, J. (1996). Stress reactivity and attachment security. *Developmental Psychobiology, 29*(3), 191–204. https://doi.org/10.1002/(SICI)1098-2302(199604)29:3<191::AID-DEV1>3.0.CO;2-M

Gunnar, M. R., & Hostinar, C. E. (2015). The social buffering of the hypothalamic–pituitary–adrenocortical axis in humans: Developmental and experiential determinants. *Social Neuroscience, 10*(5), 479–488. https://doi.org/10.1080/17470919.2015.1070747

Gunnar, M. R., Talge, N. M., & Herrera, A. (2009). Stressor paradigms in developmental studies: What does and does not work to produce mean increases in salivary cortisol. *Psychoneuroendocrinology, 34*(7), 953–967. https://doi.org/10.1016/j.psyneuen.2009.02.010

Gunnar, M. R., Wewerka, S., Frenn, K., Long, J. D., & Griggs, C. (2009). Developmental changes in hypothalamus–pituitary–adrenal activity over the transition to adolescence: Normative changes and associations with puberty. *Development and Psychopathology, 21*(1), 69–85. https://doi.org/10.1017/S0954579409000054

Halligan, S. L., Herbert, J., Goodyer, I. M., & Murray, L. (2004). Exposure to postnatal depression predicts elevated cortisol in adolescent offspring. *Biological Psychiatry, 55*(4), 376–381. https://doi.org/10.1016/j.biopsych.2003.09.013

Han, G., Miller, J. G., Cole, P. M., Zahn-Waxler, C., & Hastings, P. D. (2015). Adolescents' internalizing and externalizing problems predict their affect-specific HPA and HPG axes reactivity. *Developmental Psychobiology, 57*(6), 769–785. https://doi.org/10.1002/dev.21268

Harden, K. P., Mann, F. D., Grotzinger, A. D., Patterson, M. W., Steinberg, L., Tackett, J. L., & Tucker-Drob, E. M. (2018). Developmental differences in reward sensitivity and sensation seeking in adolescence: Testing sex-specific associations With gonadal hormones and pubertal development. *Journal of Personality and Social Psychology, 115*(1), 161–178. https://doi.org/10.1037/pspp0000172

Harden, K. P., Wrzus, C., Luong, G., Grotzinger, A., Bajbouj, M., Rauers, A., … Riediger, M. (2016). Diurnal coupling between testosterone and cortisol from adolescence to older adulthood. *Psychoneuroendocrinology, 73*, 79–90. https://doi.org/10.1016/j.psyneuen.2016.07.216

Hatfield, E., Cacioppo, J. T., & Rapson, R. L. (1993). Emotional contagion. *Current Directions in Psychological Science, 2*(3), 96–100. https://doi.org/10.1111/1467-8721.ep10770953

Hatfield, B. E., Hestenes, L. L., Kintner-Duffy, V., & O'Brien, M. (2013). Classroom emotional support predicts differences in preschool children's cortisol and alpha-amylase levels. *Early Childhood Research Quarterly, 28*(2), 347–356. https://doi.org/10.1016/j.ecresq.2012.08.001

Havelock, J. C., Auchus, R. J., & Rainey, W. E. (2004). The rise in adrenal androgen biosynthesis: Adrenarche. *Seminars in Reproductive Medicine, 22*, 337–347.

Henry, J. P. (1980). Present concept of stress theory. In *Catecholamines and stress: Recent advances* (pp. 557–571). New York: Elsevier.

Hibel, L. C., Granger, D. A., Blair, C., Cox, M. J., & Family Life Project Key Investigators. (2011). Maternal sensitivity buffers the adrenocortical implications of intimate partner violence exposure during early childhood. *Development and Psychopathology, 23*(2), 689–701. https://doi.org/10.1017/S0954579411000010

Hibel, L. C., & Mercado, E. (2017). Marital conflict predicts mother-to-infant adrenocortical transmission. *Child Development*, 1–16. https://doi.org/10.1111/cdev.13010

Hibel, L. C., Senguttuvan, U., & Bauer, N. S. (2013). Do state factors moderate the relationship between depressive symptoms and morning cortisol? *Hormones and Behavior, 63*(3), 484–490. https://doi.org/10.1016/j.yhbeh.2012.12.012

Hibel, L. C., Trumbell, J. M., & Mercado, E. (2014). Work/non-workday differences in mother, child, and mother–child morning cortisol in a sample of working mothers and their children. *Early Human Development, 90*(1), 1–7. https://doi.org/10.1016/j.earlhumdev.2013.11.007

Hibel, L. C., Mercado, E., & Valentino, K. (2019). Child maltreatment and mother–child transmission of stress physiology. *Child Maltreatment, 24*(4), 340–352. https://doi.org/10.1177/1077559519826295

Hill-Soderlund, A. L., Holochwost, S. J., Willoughby, M. T., Granger, D. A., Gariépy, J. L., Mills-Koonce, W. R., & Cox, M. J. (2015). The developmental course of salivary alpha-amylase and cortisol from 12 to 36 months: Relations with early poverty and later behavior problems. *Psychoneuroendocrinology, 52*, 311–323. https://doi.org/10.1016/j.psyneuen.2014.08.011

Hofer, M. A. (2006). Psychobiological roots of early attachment. *Current Directions in Psychological Science, 15*(2), 84–88. https://doi.org/10.1111/j.0963-7214.2006.00412.x

Holochwost, S. J., Gariépy, J. L., Mills-Koonce, W. R., Propper, C. B., Kolacz, J., & Granger, D. A. (2017). Individual differences in the activity of the hypothalamic pituitary adrenal axis: Relations to age and cumulative risk in early childhood. *Psychoneuroendocrinology, 81*, 36–45. https://doi.org/10.1016/j.psyneuen.2017.03.023

Hostinar, C. E., & Gunnar, M. R. (2013). The developmental psychobiology of stress and emotion in childhood. In I. B. Weiner, D. K. Freedheim, & R. M. Lerner (Eds.), *Handbook of psychology* (2nd ed., pp. 121–141). Hoboken, NJ: Wiley.

Hostinar, C. E., Johnson, A. E., & Gunnar, M. R. (2015). Parent support is less effective in buffering cortisol stress reactivity for adolescents compared to children. *Developmental Science, 18*(2), 281–297. https://doi.org/10.1111/desc.12195

Hu, S., Loo, J. A., & Wong, D. T. (2007). Human saliva proteome analysis. *Annals of the New York Academy of Sciences, 1098*(1), 323–329. https://doi.org/10.1196/annals.1384.015

Hunter, A. L., Minnis, H., & Wilson, P. (2011). Altered stress responses in children exposed to early adversity: A systematic review of salivary cortisol studies. *Stress, 14*(6), 614–626. https://doi.org/10.3109/10253890.2011.577848

Jaffee, S. R., McFarquhar, T., Stevens, S., Ouellet-Morin, I., Melhuish, E., & Belsky, J. (2015). Interactive effects of early and recent exposure to stressful contexts on cortisol reactivity in middle childhood. *Journal of Child Psychology and Psychiatry, 56*(2), 138–146. https://doi.org/10.1111/jcpp.12287

Ji, J., Negriff, S., Kim, H., & Susman, E. J. (2016). A study of cortisol reactivity and recovery among young adolescents: Heterogeneity and longitudinal stability and change. *Developmental Psychobiology, 58*(3), 283–302. https://doi.org/10.1002/dev.21369

Johnson, S. B., Riley, A. W., Granger, D. A., & Riis, J. (2013). The science of early life toxic stress for pediatric practice and advocacy. *Pediatrics, 131*(2), 319–327. https://doi.org/10.1542/peds.2012–0469

Juster, R. P., McEwen, B. S., & Lupien, S. J. (2010). Allostatic load biomarkers of chronic stress and impact on health and cognition. *Neuroscience and Biobehavioral Reviews, 35*(1), 2–16. https://doi.org/10.1016/j.neubiorev.2009.10.002

Kagan, J., Reznick, J., & Snidman, N. (1987). The physiology and psychology of behavioral inhibition in children. *Child Development, 58*(6), 1459–1473. https://doi.org/10.2307/1130685

Kamin, H. S., & Kertes, D. A. (2017). Cortisol and DHEA in development and psychopathology. *Hormones and Behavior, 89*, 69–85. https://doi.org/10.1016/j.yhbeh.2016.11.018

Kaplan, L. A., Evans, L., & Monk, C. (2008). Effects of mothers' prenatal psychiatric status and postnatal caregiving on infant biobehavioral regulation: Can prenatal programming be modified? *Early Human Development, 84*(4), 249–256. https://doi.org/10.1016/j.earlhumdev.2007.06.004

King, L. S., Colich, N. L., LeMoult, J., Humphreys, K. L., Ordaz, S. J., Price, A. N., & Gotlib, I. H. (2017). The impact of the severity of early life stress on diurnal cortisol: The role of puberty. *Psychoneuroendocrinology, 77*, 68–74. https://doi.org/10.1016/j.psyneuen.2016.11.024

Kornienko, O., & Granger, D. A. (2018). Peer networks, psychobiology of stress response, and adolescent development. In *The Oxford handbook of evolution, biology, and society*. New York, NY: Oxford University Press, pp. 327–353. https://doi.org/10.1093/oxfordhb/9780190299323.013.25

Koss, K. J., & Gunnar, M. R. (2018). Annual research review: Early adversity, the hypothalamic–pituitary–adrenocortical axis, and child psychopathology. *Journal of Child Psychology and Psychiatry, 59*(4), 327–346. https://doi.org/10.1111/jcpp.12784

Kuras, Y. I., Assaf, N., Thoma, M. V., Gianferante, D., Hanlin, L., Chen, X., … Rohleder, N. (2017). Blunted diurnal cortisol activity in healthy adults with childhood adversity. *Frontiers in Human Neuroscience, 11*, 1–8. https://doi.org/10.3389/fnhum.2017.00574

Kushiyama, A., Tanaka, K., Hara, S., & Kawazu, S. (2014). Linking uric acid metabolism to diabetic complications. *World Journal of Diabetes, 5*(6), 787–795. https://doi.org/10.4239/wjd.v5.i6.787

Laudat, M. H., Cerdas, S., Fournier, C., Guiban, D., Guilhaume, B., & Luton, J. P. (1988). Salivary cortisol measurement: A practical approach to assess pituitary-adrenal function. *Journal of Clinical Endocrinology and Metabolism, 66*(2), 343–348. https://doi.org/10.1210/jcem-66-2-343

Laurent, H. K., Ablow, J. C., & Measelle, J. (2012). Taking stress response out of the box: Stability, discontinuity, and temperament effects on HPA and SNS across social stressors in mother–infant dyads. *Developmental Psychology, 48*(1), 35–45. https://doi.org/10.1037/a0025518

Laurent, H. K., Harold, G. T., Leve, L., Shelton, K. H., & Van Goozen, S. H. (2016). Understanding the unfolding of stress regulation in infants. *Development and Psychopathology, 28*(4pt2), 1431–1440. https://doi.org/10.1017/S0954579416000171

Leppert, K. A., Kushner, M., Smith, V. C., Lemay, E. P., & Dougherty, L. R. (2016). Children's cortisol responses to a social evaluative laboratory stressor from early to middle childhood. *Developmental Psychobiology, 58*(8), 1019–1033. https://doi.org/10.1002/dev.21435

Levine, S. (1985). Animal stress. In *A definition of stress?* (pp. 51–69). New York, NY: Springer.

Levine, S. (2005). Developmental determinants of sensitivity and resistance to stress. *Psychoneuroendocrinology, 30*(10), 939–946. https://doi.org/10.1016/j.psyneuen.2005.03.013

Levine, S., Wiener, S., Coe, C., Bayart, F. E. S., & Hayashi, K. (1987). Primate vocalization: A psychobiological approach. *Child Development, 58*(6), 1408–1419. https://doi.org/10.2307/1130681

Lewis, M., & Ramsay, D. S. (1995). Developmental change in infants' responses to stress. *Child Development, 66*(3), 657–670. https://doi.org/10.1111/j.1467-8624.1995.tb00896.x

Loman, M. M., & Gunnar, M. R. (2010). Early experience and the development of stress reactivity and regulation in children. *Neuroscience and Biobehavioral Reviews, 34*(6), 867–876. https://doi.org/10.1016/j.neubiorev.2009.05.007

Lopez-Duran, N. L., Hajal, N. J., Olson, S. L., Felt, B. T., & Vazquez, D. M. (2009). Individual differences in cortisol responses to fear and frustration during middle childhood. *Journal of Experimental Child Psychology, 103*(3), 285–295. https://doi.org/10.1016/j.jecp.2009.03.008

Lupien, S. J., McEwen, B. S., Gunnar, M. R., & Heim, C. (2009). Effects of stress throughout the lifespan on the brain, behaviour and cognition. *Nature Reviews Neuroscience, 10*(6), 434–445. https://doi.org/10.1038/nrn2639

Mackrell, S. V., Sheikh, H. I., Kotelnikova, Y., Kryski, K. R., Jordan, P. L., Singh, S. M., & Hayden, E. P. (2014). Child temperament and parental depression predict cortisol reactivity to stress in middle childhood. *Journal of Abnormal Psychology, 123*(1), 106–116. https://doi.org/10.1037/a0035612

Marceau, K., Ruttle, P. L., Shirtcliff, E. A., Essex, M. J., & Susman, E. J. (2015). Developmental and contextual considerations for adrenal and gonadal hormone functioning during adolescence: Implications for adolescent mental health. *Developmental Psychobiology, 57*(6), 742–768. https://doi.org/10.1002/dev.21214

Marceau, K., Ruttle, P. L., Shirtcliff, E. A., Hastings, P. D., Klimes-Dougan, B., & Zahn-Waxler, C. (2015). Within-person coupling of changes in cortisol, testosterone, and DHEA across the day in adolescents. *Developmental Psychobiology, 57*(6), 654–669. https://doi.org/10.1002/dev.21173

Marceau, K., Shirtcliff, E. A., Hastings, P. D., Klimes-Dougan, B., Zahn-Waxler, C., Dorn, L. D., & Susman, E. J. (2014). Within-adolescent coupled changes in cortisol with DHEA and testosterone in response to three stressors during adolescence. *Psychoneuroendocrinology, 41*, 33–45. https://doi.org/10.1016/j.psyneuen.2013.12.002

Martin, J., Kagerbauer, S. M., Gempt, J., Podtschaske, A., Hapfelmeier, A., & Schneider, G. (2018). Oxytocin levels in saliva correlate better than plasma levels with concentrations in the cerebrospinal fluid of patients in neurocritical care. *Journal of Neuroendocrinology, 30*(5), e12596. https://doi.org/10.1111/jne.12596

Martinez-Torteya, C., Bogat, G. A., Levendosky, A. A., & Von Eye, A. (2016). The influence of prenatal intimate partner violence exposure on hypothalamic–pituitary–adrenal axis reactivity and childhood internalizing and externalizing symptoms. *Development and Psychopathology, 28*(1), 55–72. https://doi.org/10.1017/S0954579415000280

Martinez-Torteya, C., Bogat, G. A., Lonstein, J. S., Granger, D. A., & Levendosky, A. A. (2017). Exposure to intimate partner violence in utero and infant internalizing behaviors: Moderation by salivary cortisol-alpha amylase asymmetry. *Early Human Development, 113*, 40–48. https://doi.org/10.1016/j.earlhumdev.2017.07.014

Masten, A. S. (2014). Global perspectives on resilience in children and youth. *Child Development, 85*(1), 6–20. https://doi.org/10.1111/cdev.12205

Matchock, R. L., Dorn, L. D., & Susman, E. J. (2007). Diurnal and seasonal cortisol, testosterone, and DHEA rhythms in boys and girls during puberty. *Chronobiology International, 24*(5), 969–990. https://doi.org/10.1080/07420520701649471

McGinnis, E. W., Lopez-Duran, N., Martinez-Torteya, C., Abelson, J. L., & Muzik, M. (2016). Cortisol awakening response and internalizing symptoms across childhood: Exploring the role of age and externalizing symptoms. *International Journal of Behavioral Development, 40*(4), 289–295. https://doi.org/10.1177/0165025415590185

McLaughlin, K. A., Sheridan, M. A., Tibu, F., Fox, N. A., Zeanah, C. H., & Nelson, C. A. (2015). Causal effects of the early caregiving environment on development of stress response systems in children. *Proceedings of the National Academy of Sciences, 112*(18), 5637–5642. https://doi.org/10.1073/pnas.1423363112

Miller, G. E., Chen, E., & Parker, K. J. (2011). Psychological stress in childhood and susceptibility to the chronic diseases of aging: Moving toward a model of behavioral and biological mechanisms. *Psychological Bulletin, 137*(6), 959–997. https://doi.org/10.1037/a0024768

Miller, A. L., Sturza, J., Rosenblum, K., Vazquez, D. M., Kaciroti, N., & Lumeng, J. C. (2015). Salivary alpha amylase diurnal pattern and stress response are associated with body mass index in low-income preschool-aged children. *Psychoneuroendocrinology, 53*, 40–48. https://doi.org/10.1016/j.psyneuen.2014.12.011

Mörelius, E., He, H. G., & Shorey, S. (2016). Salivary cortisol reactivity in preterm infants in neonatal intensive care: An integrative review. *International Journal of Environmental Research and Public Health, 13*(3), 337. https://doi.org/10.3390/ijerph13030337

Mörelius, E., Nelson, N., & Gustafsson, P. A. (2007). Salivary cortisol response in mother–infant dyads at high psychosocial risk. *Child: Care, Health and Development, 33*(2), 128–136. https://doi.org/10.1111/j.1365-2214.2006.00637.x

Nater, U. M., Rohleder, N., Schlotz, W., Ehlert, U., & Kirschbaum, C. (2007). Determinants of the diurnal course of salivary alpha-amylase. *Psychoneuroendocrinology, 32*(4), 392–401. https://doi.org/10.1016/j.psyneuen.2007.02.007

Netherton, C., Goodyer, I., Tamplin, A., & Herbert, J. (2004). Salivary cortisol and dehydroepiandrosterone in relation to puberty and gender. *Psychoneuroendocrinology, 29*(2), 125–140. https://doi.org/10.1016/S0306-4530(02)00150-6

O'Donnell, M. D., & Miller, N. J. (1980). Plasma pancreatic and salivary-type amylase and immunoreactive trypsin concentrations: Variations with age and reference ranges for children. *Clinica Chimica Acta, 104*(3), 265–273. https://doi.org/10.1016/0009-8981(80)90384-8

Obradović, J. (2012). How can the study of physiological reactivity contribute to our understanding of adversity and resilience processes in development? *Development and Psychopathology, 24*(2), 371–387. https://doi.org/10.1017/S0954579412000053

Out, D., Granger, D. A., Sephton, S. E., & Segerstrom, S. C. (2013). Disentangling sources of individual differences in diurnal salivary α-amylase: Reliability, stability and sensitivity to context. *Psychoneuroendocrinology, 38*(3), 367–375. https://doi.org/10.1016/j.psyneuen.2012.06.013

Pakulak, E., Stevens, C., & Neville, H. (2018). Neuro-, cardio-, and immunoplasticity: Effects of early adversity. *Annual Review of Psychology, 69*, 131–156. https://doi.org/10.1146/annurev-psych-010416-044115

Papp, L. M., Pendry, P., & Adam, E. K. (2009). Mother-adolescent physiological synchrony in naturalistic settings: Within-family cortisol associations and moderators. *Journal of Family Psychology, 23*, 882–894. https://doi.org/10.1037/a0017147

Parra, L. A., & Hastings, P. D. (2018). Integrating the neurobiology of minority stress with an intersectionality framework for LGBTQ-Latinx populations. *New Directions for Child and Adolescent Development, 161*, 91–108. https://doi.org/10.1002/cad.20244

Pendry, P., & Adam, E. K. (2007). Associations between parents' marital functioning, maternal parenting quality, maternal emotion and child cortisol levels. *International Journal of Behavioral Development, 31*(3), 218–231. https://doi.org/10.1177/0165025407074634

Peper, J. S., Braams, B. R., Blankenstein, N. E., Bos, M. G., & Crone, E. A. (2018). Development of multifaceted risk taking and the relations to sex steroid hormones: A longitudinal study. *Child Development, 89*(5), 1887–1907. https://doi.org/10.1111/cdev.13063

Peters, E., Riksen-Walraven, J. M., Cillessen, A. H., & de Weerth, C. (2011). Peer rejection and HPA activity in middle childhood: Friendship makes a difference. *Child Development, 82*(6), 1906–1920. https://doi.org/10.1111/j.1467-8624.2011.01647.x

Popma, A., Vermeiren, R., Geluk, C. A., Rinne, T., van den Brink, W., Knol, D. L., . . . Doreleijers, T. A. (2007). Cortisol moderates the relationship between testosterone and aggression in delinquent male adolescents. *Biological Psychiatry, 61*(3), 405–411. https://doi.org/10.1016/j.biopsych.2006.06.006

Puetz, V. B., Viding, E., Palmer, A., Kelly, P. A., Lickley, R., Koutoufa, I., . . . McCrory, E. J. (2016). Altered neural response to rejection-related words in children exposed to maltreatment. *Journal of Child Psychology and Psychiatry, 57*(10), 1165–1173. https://doi.org/10.1111/jcpp.12595

Rao, U., Hammen, C., Ortiz, L. R., Chen, L. A., & Poland, R. E. (2008). Effects of early and recent adverse experiences on adrenal response to psychosocial stress in depressed adolescents. *Biological Psychiatry, 64*(6), 521–526. https://doi.org/10.1016/j.biopsych.2008.05.012

Rash, J. A., Thomas, J. C., Campbell, T. S., Letourneau, N., Granger, D. A., Giesbrecht, G. F., & APrON Study Team. (2016). Developmental origins of infant stress reactivity profiles: A multisystem approach. *Developmental Psychobiology, 58*(5), 578–599. https://doi.org/10.1002/dev.21403

Riis, J. L., Bryce, C. I., Matin, M. J., Stebbins, J. L., Kornienko, O., van Huisstede, L., & Granger, D. A. (2018). The validity, stability, and utility of measuring uric acid in saliva. *Biomarkers in Medicine, 12*(6). https://doi.org/10.2217/bmm-2017-0336

Robles, T. F., Sharma, R., Park, K. S., Harrell, L., Yamaguchi, M., & Shetty, V. (2012). Utility of a salivary biosensor for objective assessment of surgery-related stress. *Journal of Oral and Maxillofacial Surgery, 70*(10), 2256–2263. https://doi.org/10.1016/j.joms.2012.03.005

Romeo, R. D. (2013). The teenage brain: The stress response and the adolescent brain. *Current Directions in Psychological Science, 22*(2), 140–145. https://doi.org/10.1177/0963721413475445

Ruttle, P. L., Serbin, L. A., Stack, D. M., Schwartzman, A. E., & Shirtcliff, E. A. (2011). Adrenocortical attunement in mother–child dyads: Importance of situational and behavioral characteristics. *Biological Psychology, 88*, 104–111. https://doi.org/10.1016/j.biopsycho.2011.06.014

Ruttle, P. L., Shirtcliff, E. A., Armstrong, J. M., Klein, M. H., & Essex, M. J. (2015). Neuroendocrine coupling across adolescence and the longitudinal influence of early life stress. *Developmental Psychobiology, 57*(6), 688–704. https://doi.org/10.1002/dev.21138

Ruttle, P. L., Shirtcliff, E. A., Serbin, L. A., Fisher, D. B. D., Stack, D. M., & Schwartzman, A. E. (2011). Disentangling psychobiological mechanisms underlying internalizing and externalizing behaviors in youth: Longitudinal and concurrent associations with cortisol. *Hormones and Behavior, 59*(1), 123–132. https://doi.org/10.1016/j.yhbeh.2010.10.015

Salis, K. L., Bernard, K., Black, S. R., Dougherty, L. R., & Klein, D. (2016). Examining the concurrent and longitudinal relationship between diurnal cortisol rhythms and conduct problems during childhood. *Psychoneuroendocrinology, 71*, 147–154. https://doi.org/10.1016/j.psyneuen.2016.05.021

Sameroff, A. (2009). *The transactional model.* Washington, DC: American Psychological Association.

Sapolsky, R. (1997). *The trouble with testosterone: And other essays on the biology of the human predicament.* New York: Scribner.

Schaal, B., Tremblay, R. E., Soussignan, R., & Susman, E. J. (1996). Male testosterone linked to high social dominance but low physical aggression in early adolescence. *Journal of the American Academy of Child and Adolescent Psychiatry, 35*(10), 1322–1330. https://doi.org/10.1097/00004583-199610000-00019

Shirtcliff, E. A., Allison, A. L., Armstrong, J. M., Slattery, M. J., Kalin, N. H., & Essex, M. J. (2012). Longitudinal stability and developmental properties of salivary cortisol levels and circadian rhythms from childhood to adolescence. *Developmental Psychobiology, 54*(5), 493–502. https://doi.org/10.1002/dev.20607

Shirtcliff, E. A., Dismukes, A. R., Marceau, K., Ruttle, P. L., Simmons, J. G., & Han, G. (2015). A dual-axis approach to understanding neuroendocrine development. *Developmental Psychobiology, 57*(6), 643–653. https://doi.org/10.1002/dev.21337

Shirtcliff, E., Zahn-Waxler, C., Klimes-Dougan, B., & Slattery, M. (2007). Salivary dehydroepiandrosterone responsiveness to social challenge in adolescents with internalizing problems.

*Journal of Child Psychology and Psychiatry, 48*(6), 580–591. https://doi.org/10.1111/j.1469-7610.2006.01723.x

Shonkoff, J. P., Garner, A. S., Siegel, B. S., Dobbins, M. I., Earls, M. F., McGuinn, L., . . . Committee on Early Childhood, Adoption, and Dependent Care. (2012[3]). The lifelong effects of early childhood adversity and toxic stress. *Pediatrics, 129*(1), e232–e246. https://doi.org/10.1542/peds.2011-2663

Soukup, M., Biesiada, I., Henderson, A., Idowu, B., Rodeback, D., Ridpath, L., . . . Bridges, K. G. (2012). Salivary uric acid as a noninvasive biomarker of metabolic syndrome. *Diabetology and Metabolic Syndrome, 4*(1), 14. https://doi.org/10.1186/1758-5996-4-14

Spanglar, G., Schieche, M., Ilg, U., Maier, U., & Ackermann, C. (1994). Maternal sensitivity as an external organizer for biobehavioral regulation in infancy. *Developmental Psychobiology: The Journal of the International Society for Developmental Psychobiology, 27*(7), 425–437. https://doi.org/10.1002/dev.420270702

Spinrad, T. L., Eisenberg, N., Granger, D. A., Eggum, N. D., Sallquist, J., Haugen, R. G., . . . Hofer, C. (2009). Individual differences in preschoolers' salivary cortisol and alpha-amylase reactivity: Relations to temperament and maladjustment. *Hormones and Behavior, 56*(1), 133–139. https://doi.org/10.1016/j.yhbeh.2009.03.020

Spray, J., Floyd, B., Littleton, J., Trnka, S., & Mattison, S. (2018). Social group dynamics predict stress variability among children in a New Zealand classroom. *Homo, 69*(1–2), 50–61. https://doi.org/10.1016/j.jchb.2018.03.005

Sroufe, L. A. (1979). The coherence of individual development: Early care, attachment, and subsequent developmental issues. *American Psychologist, 34*(10), 834–841. https://doi.org/10.1037/0003-066X.34.10.834

Stalder, T., Bäumler, D., Miller, R., Alexander, N., Kliegel, M., & Kirschbaum, C. (2013). The cortisol awakening response in infants: Ontogeny and associations with development-related variables. *Psychoneuroendocrinology, 38*(4), 552–559. https://doi.org/10.1016/j.psyneuen.2012.07.015

Stalder, T., Kirschbaum, C., Kudielka, B. M., Adam, E. K., Pruessner, J. C., Wüst, S., . . . Miller, R. (2016). Assessment of the cortisol awakening response: Expert consensus guidelines. *Psychoneuroendocrinology, 63*, 414–432. https://doi.org/10.1016/j.psyneuen.2015.10.010

Steeger, C. M., Cook, E. C., & Connell, C. M. (2017). The interactive effects of stressful family life events and cortisol reactivity on adolescent externalizing and internalizing behaviors. *Child Psychiatry and Human Development, 48*(2), 225–234. https://doi.org/10.1007/s10578-016-0635-6

Steinberg, L., & Silk, X. (2002). Parenting adolescents. In M. H. Bornstein (Ed.), *Handbook of Parenting: Vol. 1: Children and parenting* (pp. 103–133). Mahwah, NJ: Erlbaum.

Strahler, J., Skoluda, N., Kappert, M. B., & Nater, U. M. (2017). Simultaneous measurement of salivary cortisol and alpha-amylase: Application and recommendations. *Neuroscience and Biobehavioral Reviews, 83*, 657–677. https://doi.org/10.1016/j.neubiorev.2017.08.015

Stroud, C. B., Chen, F. R., Doane, L. D., & Granger, D. A. (2016). Individual differences in early adolescents' latent trait cortisol (LTC): Relation to early adversity. *Developmental Psychobiology, 58*(6), 700–713. https://doi.org/10.1002/dev.21410

Stroud, L. R., Foster, E., Papandonatos, G. D., Handwerger, K., Granger, D. A., Kivlighan, K. T., & Niaura, R. (2009). Stress response and the adolescent transition: Performance versus peer rejection stressors. *Development and Psychopathology, 21*(1), 47–68. https://doi.org/10.1017/S0954579409000042

Stroud, L. R., Papandonatos, G. D., D'Angelo, C. M., Brush, B., & Lloyd-Richardson, E. E. (2017). Sex differences in biological response to peer rejection and performance challenge across development: A pilot study. *Physiology and Behavior, 169*, 224–233. https://doi.org/10.1016/j.physbeh.2016.12.005

Styne, D. M., & Grumbach, M. M. (2002). Puberty in boys and girls. *Hormones, Brain and Behavior, 4*, 661–716. https://doi.org/10.1016/B978-012532104-4/50078-0

Sullivan, M. C., Hawes, K., Winchester, S. B., & Miller, R. J. (2008). Developmental origins theory from prematurity to adult disease. *Journal of Obstetric, Gynecologic and Neonatal Nursing, 37* (2), 158–164. https://doi.org/10.1111/j.1552-6909.2008.00216.x

Susman, E. J., Peckins, M. K., Bowes, J. L., & Dorn, L. D. (2017). Longitudinal synergies between cortisol reactivity and diurnal testosterone and antisocial behavior in young adolescents. *Development and Psychopathology, 29*(4), 1353–1369. https://doi.org/10.1017/S0954579416001334

Tarullo, A. R., & Gunnar, M. R. (2006). Child maltreatment and the developing HPA axis. *Hormones and Behavior, 50*(4), 632–639. https://doi.org/10.1016/j.yhbeh.2006.06.010

Taylor, Z. E., Spinrad, T. L., VanSchyndel, S. K., Eisenberg, N., Huynh, J., Sulik, M. J., & Granger, D. A. (2013). Sociodemographic risk, parenting, and effortful control: Relations to salivary alpha-amylase and cortisol in early childhood. *Developmental Psychobiology, 55*(8), 869–880. https://doi.org/10.1002/dev.21079

Tegethoff, M., Knierzinger, N., Meyer, A. H., & Meinlschmidt, G. (2013). Cortisol awakening response in infants during the first six postnatal months and its relation to birth outcome. *Psychoneuroendocrinology, 38*(5), 629–637. https://doi.org/10.1016/j.psyneuen.2012.08.002

Tollenaar, M. S., Jansen, J., Beijers, R., Riksen-Walraven, J. M., & de Weerth, C. (2010). Cortisol in the first year of life: Normative values and intra-individual variability. *Early Human Development, 86*(1), 13–16. https://doi.org/10.1016/j.earlhumdev.2009.12.003

Tomarken, A. J., Han, G. T., & Corbett, B. A. (2015). Temporal patterns, heterogeneity, and stability of diurnal cortisol rhythms in children with autism spectrum disorder. *Psychoneuroendocrinology, 62*, 217–226. https://doi.org/10.1016/j.psyneuen.2015.08.016

Törnhage, C. J. (2002). Reference values for morning salivary cortisol concentrations in healthy school-aged children. *Journal of Pediatric Endocrinology and Metabolism, 15*(2), 197–204. https://doi.org/10.1515/JPEM.2002.15.2.197

Turan, B., Tackett, J. L., Lechtreck, M. T., & Browning, W. R. (2015). Coordination of the cortisol and testosterone responses: A dual axis approach to understanding the response to social status threats. *Psychoneuroendocrinology, 62*, 59–68. https://doi.org/10.1016/j.psyneuen.2015.07.166

Valentino, K., De Alba, A., Hibel, L. C., Fondren, K., & McDonnell, C. G. (2017). Adherence to diurnal cortisol sampling among mother–child dyads from maltreating and nonmaltreating families. *Child Maltreatment, 22*(4), 286–294. https://doi.org/10.1177/1077559517725208

Valentino, K., Hibel, L. C., Cummings, E. M., Nuttall, A. K., Comas, M., & McDonnell, C. G. (2015). Maternal elaborative reminiscing mediates the effect of child maltreatment on behavioral and physiological functioning. *Development and Psychopathology, 27*(4pt2), 1515–1526. https://doi.org/10.1017/S0954579415000917

Vigil, J. M., Geary, D. C., Granger, D. A., & Flinn, M. V. (2010). Sex differences in salivary cortisol, alpha-amylase, and psychological functioning following Hurricane Katrina. *Child Development, 81*, 1227–1239. https://doi.org/10.1111/j.1467-8624.2010.01464.x

Vining, R. F., McGinley, R. A., Maksvytis, J. J., & Ho, K. Y. (1983). Salivary cortisol: A better measure of adrenal cortical function than serum cortisol. *Annals of Clinical Biochemistry, 20*(6), 329–335. https://doi.org/10.1177/000456328302000601

Wagner, N. J., Mills-Koonce, W. R., Willoughby, M. T., Cox, M. J., Family Life Project Key Investigators, Vernon-Feagans, L., … Garrett-Peters, P. (2017). Parenting and cortisol in infancy interactively predict conduct problems and callous–unemotional behaviors in childhood. *Child Development, 90*(1), 279–297. https://doi.org/10.1111/cdev.12900

Wolf, J. M., Nicholls, E., & Chen, E. (2008). Chronic stress, salivary cortisol, and α-amylase in children with asthma and healthy children. *Biological Psychology, 78*(1), 20–28. https://doi.org/10.1016/j.biopsycho.2007.12.004

Worthman, C. M., & Panter-Brick, C. (2008). Homeless stress children in Nepal: Use of allostatic load to assess the burden of childhood adversity. *Development and Psychopathology, 20*, 233–255. https://doi.org/10.1017/S0954579408000114

Yang, P. J., Lamb, M. E., Kappler, G., & Ahnert, L. (2017). Children's diurnal cortisol activity during the first year of school. *Applied Developmental Science, 21*(1), 30–41. https://doi.org/10.1080/10888691.2016.1140578

Yildirim, B. O., & Derksen, J. J. (2012). A review on the relationship between testosterone and life-course persistent antisocial behavior. *Psychiatry Research, 200*(2–3), 984–1010. https://doi.org/10.1016/j.psychres.2012.07.044

Yoshikawa, H., Aber, J. L., & Beardslee, W. R. (2012). The effects of poverty on the mental, emotional, and behavioral health of children and youth: Implications for prevention. *American Psychologist, 67*(4), 272–284. https://doi.org/10.1037/a0028015

Yu, R., Nieuwenhuis, J., Meeus, W., Hooimeijer, P., Koot, H. M., & Branje, S. (2016). Biological sensitivity to context: Cortisol awakening response moderates the effects of neighbourhood density on the development of adolescent externalizing problem behaviours. *Biological Psychology, 120*, 96–107. https://doi.org/10.1016/j.biopsycho.2016.08.004

Yu, Y. Z., & Shi, J. X. (2009). Relationship between levels of testosterone and cortisol in saliva and aggressive behaviors of adolescents. *Biomedical and Environmental Sciences, 22*(1), 44–49. https://doi.org/10.1016/S0895-3988(09)60021-0

Zalewski, M., Lengua, L. J., Kiff, C. J., & Fisher, P. A. (2012). Understanding the relation of low income to HPA-axis functioning in preschool children: Cumulative family risk and parenting as pathways to disruptions in cortisol. *Child Psychiatry and Human Development, 43*(6), 924–942. https://doi.org/10.1007/s10578-012-0304-3

Zeiders, K. H., Doane, L. D., & Roosa, M. W. (2012). Perceived discrimination and diurnal cortisol: Examining relations among Mexican American adolescents. *Hormones and Behavior, 61*(4), 541–548. https://doi.org/10.1016/j.yhbeh.2012.01.018

Zhang, J., Lam, S.-P., Kong, A. P., Ma, R. C., Li, S. X., Chan, J. W., ... Wing, Y.-K. (2016). Family conflict and lower morning cortisol in adolescents and adults: Modulation of puberty. *Scientific Reports, 6*, 22531. https://doi.org/10.1038/srep22531

Zuckerman, M. (1979). Attribution of success and failure revisited, or: The motivational bias is alive and well in attribution theory. *Journal of Personality, 47*(2), 245–287. https://doi.org/10.1111/j.1467-6494.1979.tb00202.x

# Part III
# Biomedical Research and Related Applications

**Elizabeth A. Thomas, Section Editor**

# Chapter 11
# Biomedical Research and Related Applications: Current Assay Methods and Quality Requirements in Oral Fluid Diagnostics Applications

**Steve W. Granger and Supriya Gaitonde**

**Abstract** Salivary diagnostics has been a field of great promise for several decades, yet few salivary assays have been FDA approved and adopted by the medical community. Why has't the momentum driving salivary diagnostics translated into more groundbreaking in vitro diagnostics (IVD) products? This chapter titled *Salivary Biomedical Research and Related Applications* will cover a brief history; the status of knowledge based on discoveries; the methodological issues, challenges, and considerations for use of salivary bioscience in biomedical research and clinical diagnostics; and future directions and opportunities in the field. Saliva represents a challenging diagnostic bodily fluid. In general, published findings using assays that associate salivary marker levels with serum using large sample numbers to calculate meaningful correlations have difficulty translating to the next level where the rigorous predictive requirements for an assay to achieve regulatory approval must have high specificity and sensitivity to minimize the risk of false-negatives or false-positives. In other words, approval requires an assay to correctly diagnose an individual in contrast to achieving a positive correlation in a large group. Likewise, oral fluid must achieve consistent results, a difficult task in saliva where analyte levels are low and composition and viscosity are variable when compared to serum and plasma. In addition, even the best assays require rigorous validation and costly clinical trials for FDA diagnostic approval. Success stories lie with salivary cortisol for Cushing's disease, Point of Care Tests (POCTs) for HIV, Streptococcus infections, and Drugs of Abuse. Several future assays show promise and the field is evolving toward new methods that may bring salivary diagnostics directly to consumers.

S. W. Granger (✉) · S. Gaitonde
Salimetrics Technology and Development Center, Carlsbad, CA, USA
e-mail: sgranger@salimetrics.com



D. A. Granger, M. K. Taylor (eds.), *Salivary Bioscience*,
https://doi.org/10.1007/978-3-030-35784-9_11

## 11.1 Brief History

Since the mid-1980s, oral fluid-based diagnostics have been described as a paradigm changing solution for patient monitoring and of great interest to the medical community. Modern saliva testing began using immunoassays mainly in the radioimmunoassay (RIA) and traditional enzyme-linked immunosorbent assay (ELISA) formats due to their high sensitivity, specificity, and throughput. Assays that were originally designed for the measurement of analytes in blood were used for proof-of-principle testing in saliva when an advantage to salivary testing for the molecule of interest appealed to the investigator. The attraction and advantages of saliva collection when compared to drawing blood were mainly due to the ease of noninvasive self-collection, and the ability to provide multiple samples through the course of the day. Another advantage was that the collection expertise of a trained phlebotomist was not necessary and instead the individual could be easily trained to accurately collect with limited instruction. In addition, since saliva testing began measuring mainly hormones, the benefit of selectively assessing the functionally relevant, unbound fraction found in saliva was a selling point, making saliva arguably a more relevant sample type compared to a blood sample which measures both free and bound hormones (Read, Greene, & Katzenellenbogen, 1989).

In the beginning, analytes that provided high serum to saliva correlations, such as observed with cortisol, became attractive candidates to diagnose systemic conditions. From this the term "mirror or window of the body" emerged as a novel concept for saliva testing and several direct to consumer methodologies were the focus of immediate attention and government-funded research (Wong, 2006). Salivary cortisol was soon adopted as one of the first FDA cleared diagnostic salivary assays for Cushing's disease and was FDA 510(k) cleared due to its high correlation with blood levels and substantial equivalence to serum immunoassays (Raff, 2009; Raff & Findling, 1989; Raff, Homar, & Skoner, 2003). As the field evolved, since collection occurs in a less regulated environment, such as at home or as part of a field study, the responsibility of proper sample handling and maintenance of the cold chain became a critical part of instruction and training.

From a technology perspective, various immunoassay platforms remain the methodology of choice for diagnostics when central laboratories are used for testing or for rapid on-site tests. The health care industry in general has made considerable advances in the testing of serum or plasma for a wide array of markers. In developed countries, most hospitals have incorporated automated magnetic bead-based immunoassays, or similar technologies, as a standard of diagnostic care. Roche, Abbott, Siemens, Beckman Coulter, and Alere lead the market for immunoassay-based in vitro diagnostics for clinical use (IVDs). Transfer of these specific technologies into the salivary diagnostic segment has generally not occurred and to date the majority of salivary assays remain in the traditional 96-well immunoassay format and are mostly used in research. Analytical methods using mass spectrometry have also been adopted in some diagnostic labs, but costs, highly trained users, and rigorous method development have resulted in a high barrier for entry. On the other hand, rapid paper-based immunoassays that are low cost and can be used directly by the consumer, have been developed for drugs of abuse and infectious disease when answers are qualitative and require less sensitivity. These and other direct-to-consumer methodologies are an attractive angle for salivary diagnostics and current smart phone-enabled personalized mobile health care tools are becoming a very popular goal for developers in this market segment.

The first oral fluid FDA approved direct-to-consumer diagnostic test for the detection of HIV was in 2002. The OraQuick oral fluid rapid HIV test detects antibodies to HIV as an indication of prior infection. In this case, oral fluid is collected by swiping the gums and test results are obtained after 20 min with a serum follow-up required for positive results. The HIV test falls into the category of infectious disease diagnostics and several have since been developed but not FDA approved for the detection of antibodies to Hepatitis A, B, and C (Cameron & Carman, 2005; Parry, Perry, Panday, & Mortimer, 1989). A rapid oral fluid-based anti-Hepatitis C virus test has been approved in Europe but not yet approved in the USA (Lee et al., 2011).

Additionally, assays for drugs of abuse have entered the market. In 2009, Pink et al. reported that saliva testing had become so widespread that it had begun to replace urine testing as the standard for detecting illicit drugs and prescription medications (Pink et al., 2009). One such device is the Oratect® III Oral Fluid Drug Screen Device from Alere Toxicology/Branan Medical Corporation which simultaneously tests for six drugs, Marijuana (THC), Cocaine, Opiates, Amphetamine, Methamphetamine including MDMA (Ecstasy), and Phencyclidine (PCP). In either case for HIV or testing for illicit drugs, these assays provide a preliminary analytical result with the recommendation to have alternate chemical methods and sample type used for confirmation.

Salivary assays will likely be significantly utilized in the detection and diagnosis of periodontal disease where the tests are measuring analytes that originate from the oral cavity, as well as other important infectious and noninfectious diseases (Rossomando, Kousvelari, Janicki, & Tabak, 2001; Streckfus & Bigler, 2002; Tabak, 2001). Additional advancements in the field of salivary diagnostics are outlined in Table 11.1 (Luppa, Sokoll, & Chan, 2001; Vashist & Luong, 2016; Vashist, Luppa, Yeo, Ozcan, & Luong, 2015).

Rapid tests such as lateral flow assays are in development to evaluate oral fluids for disease diagnosis and monitoring. These tests have the advantage of frequent monitoring using an easily accessible biospecimen and could noninvasively improve health care outcomes for patients; however, none of these technologies currently have FDA clearance for clinical use. Oral fluid or saliva-based biomarkers have been investigated for diagnosing or monitoring disease conditions such as diabetes, cardiovascular disease, cystic fibrosis, Sjogren's syndrome, and various forms of cancer.

**Table 11.1** Examples of commercially marketed oral fluid tests

| Manufacturer/Test | Intended use |
|---|---|
| 23andMe®/Health + Ancestry | Identifies genetic health risks, physical traits, carrier status, and ancestry |
| OralDNA® Labs/MyPerioPath®, Celsus One®, OraRisk® HPV | For determining early warning of oral pathogens, gene markers related to inflammatory response, and increased risk for HPV-related oral cancers, respectively |
| Vigilant Biosciences®/OncAlert Oral Cancer RAPID Test | Measures CD44 and total protein levels, associated with oral cancer |
| OraSure Technologies, Inc./OraQuick® HIV Self-Test/ OraQuick® Ebola Test | Detects antibodies to both HIV-1 and HIV-2 in oral fluid, detects Ebola Zaire virus in cadaveric oral fluid |
| Alere/ Oratect® Saliva Drug Test | Detects methamphetamine, marijuana, cocaine, amphetamine, opiate, and phencyclidine |
| PeriRx, LLC/SaliMark™ OSCC Test | Early detection of oral squamous cell carcinoma |

## 11.2 The Methodological Issues, Challenges, and Considerations for Use in Biomedical Research and Clinical Diagnostics

The challenges of precise testing in Oral Fluids. Saliva is uniquely heterogeneous when compared to other bodily fluids. One of the most problematic methodological issues facing oral fluids relates to the irregularity of the salivary matrix. Saliva components vary significantly between individuals and over time within the same individual. One of the most prominent variables is the viscosity of the sample. Mucins, a type of glycoprotein found in saliva, gives saliva its visco-elastic properties. Depending on the concentration of specific mucins, especially MUC5B and MUC7, which have the highest gel-forming capacity, there can be a wide range in salivary viscosity (Prodan et al., 2015) leading to interference in multiple immunoassay formats. The heterogeneity of salivary components are also influenced by diet and unexplained individual physiology which translates to unique properties.

In addition, saliva has many unique constituents that are not found in other commonly sampled bodily fluids mainly due to the challenging physiologic functional requirements of saliva. Coating and lubricating food for swallowing is one major salivary function. The swallowing process is aided by chewing but also involves enzymatic digestion. Saliva must coat foods with very different properties that range from dry and bread like to fat rich. Thus, salivary components are inherently suited to adhere to food items and this stickiness is often troublesome for assays (Celebioglu, Lee, & Chronakis, 2019). Another salivary function is antimicrobial (Frenkel & Ribbeck, 2015; Marsh, Do, Beighton, & Devine, 2016). To appreciate the microbial challenges saliva must overcome, imagine a bag that is maintained at body temperature, always coated in liquid and opened and closed thousands of times a day. When you add the fact that non-sterile food items pass through it many times a day and that the bag is closed for most of the night, it is a wonder that our mouths are not overrun with fungal and bacterial growth. Salivary mucins and innate immune factors play a critical role in maintaining the health of the mouth and minimizing bacterial and fungal growth.

With these difficulties in mind, preprocessing of saliva is a common way to minimize assay interference for saliva diagnostics. Often the goal of salivary preprocessing techniques help produce a more uniformly behaving solution with reduced, more uniform viscosity. In direct to consumer tests, diluents that facilitate flow and include a sieving filtration step are often incorporated. For samples that are accumulated, stored at a central site and tested in bulk in a laboratory setting, freeze thaw and centrifugation are common practice to remove mucins, food, and cellular debris, and other inhibitory factors. Other preprocessing steps include sample concentration for example, by solid phase extraction and/or lyophilization when necessitated by very low analyte levels.

Saliva production can be a challenge for some. In individuals with xerostomia or dry mouth, a condition defined by reduced or no saliva production, obtaining a saliva sample can be difficult. There are several reasons, temporary or permanent, that can lead to xerostomia, including dry mouth as a contraindication for specific medication use which is often temporary, to more permanent situations such as radiation for head and neck cancers, which can affect salivary gland function. Sjogren's syndrome, an autoimmune condition in which the body mounts an immune response to tear and salivary glands, both of which markedly reduce saliva production and make it close to impossible to provide a sufficient volume of saliva for testing. Other factors that can affect saliva production include aging, smoking, excessive alcohol use, recreational drug use especially methamphetamine, and nerve damage.

Another common challenge and consideration facing saliva is the requirement for high sensitivity assays to detect low-abundance analytes. Some very high-profile analytes measured in blood are present at 40–100-fold lower levels in saliva requiring sensitivity in the low picogram per milliliter range. This is especially true for steroid hormones which are present in blood at higher concentrations but typically bound to sex-hormone binding globulin or albumin, with only a small fraction that is bio-active or "free." Since only the free or unbound fraction enters saliva, it is considered the more relevant sample type; however, these low concentrations of free hormone concentrations require assay technologies to be of higher sensitivity (Estrada & Orlander, 2011; Vining, McGinley, Maksvytis, & Ho, 1983).

Sensitivity is not a limiting factor when measuring analytes that are locally produced in the mouth, salivary glands, or oral cavity, such as alpha-amylase, secretory IgA. While some of these analytes are locally produced, the measurement of some other analytes, such as IL-6 and CRP can be confusing since the source of these cytokines may either be due to local production by circulating neutrophils due to oral inflammation, or due to systemic inflammation and originate in the blood. These considerations are discussed in a separate chapter but it is important to consider the source of the analyte being measured, and its origin.

Advances in assay types and the emergence of novel detection technology are easing the path forward to increased sensitivity in saliva-based testing. Methods used for oral fluid testing can be classified into two general categories depending on if the samples are tested at the point of care and rapid results are needed or at a centralized testing facility.

The enzyme immunoassay (EIA) and enzyme-linked immunosorbent assay (ELISA) are platforms well suited to measure these low analyte levels and have been the historic methods of choice used at centralized testing sites; however, the quality of immunoassays varies depending on how rigorously the assays have been optimized for performance in saliva. These assays can range in signal type from radioactivity, colorimetry, chemiluminescence, and electrochemiluminescence, with each signal type offering additional advantages in terms of sensitivity, increased dynamic range, and specificity. Currently, salivary cortisol measurements to screen for Cushing's disease are performed in reference laboratories using EIA-based testing and this testing is insurance reimbursable.

Lateral flow immunoassays conventionally have had limited capabilities in saliva mainly due to the nanogram per milliliter minimum sensitivity limits imposed by shortened exposure times on the test strip between antibodies and their targets. For a qualitative application such as in infectious diseases, or as a screening tool, lateral flow assays (LFAs) using saliva are very attractive. Use of innovative detection technologies such as fluorescent labels or nanotechnology, and the implementation of devices, such as desktop or on-cassette disposable readers make these methodologies very attractive for increased sensitivity and the application as a direct-to-consumer rapid test. Instrument-fortified LFAs have demonstrated quantification and enhanced sensitivity and specificity; however, no commercially available product is available that uses saliva or oral fluid as a sample type at this time. This will change as a number of companies are currently working on developing lateral flow immunoassays for the quantitative measurement of analytes in saliva, and with increased sophistication in signal generation, it is inevitable that rapid tests using saliva as a sample type will be commercially available either directly to the consumer or as a point of care device used in a physician's or dentist's office.

Commercial examples of quantitative lateral flow assay output, such as Lepu Medical's Leccurate and comparable technologies from Quidel Triage system, Abbott i-STAT, Abaxis and Gyrolab centrifugal CD systems were recently reviewed in the context of immunoassays for cardiovascular diseases but these technologies are lab-confined and restricted to blood-based assays (Morin et al., 2018).

Other more quantitative analytical methods, such as mass spectrometry, can be prohibitively expensive when considering per test costs and they require high-level operators and expensive high-sensitivity equipment. These methods are often used to benchmark other less expensive techniques, but none the less require rigorous validation and implementation of internal controls to be considered gold standard methodologies. Although accuracy is seldom questioned, precision and hence precise quantification can be dependent on the quality of the laboratory, the rigor in the method, and skill of the operator. Mass spectrometry-based quantitation of cortisol is used in reference laboratories when screening for Cushing's disease using saliva as the sample type.

Other methodologies exist that use colorimetric, and/or enzymatic reactions for rapid assessments of salivary components such as for uric acid and nitric oxide. The measurement is used to screen or monitor levels of these analytes in saliva and are designed to be semi-quantitative, similar to a pH strip as a direct-to-consumer over-the-counter offering. Wearable biosensors have also gained popularity and to this end, mouth wearable sensors have been described in the literature for monitoring uric acid levels with data showing promise in the ability to track levels in subjects with hyperuricemia; however, no commercially available product is currently available.

## 11.3 Methods and Levels of Assay Validation

Salivary assays have several levels of validation or qualification regarding regulatory requirements that vary depending on the intended use of the assay. These include research use only (RUO), laboratory-developed tests (LDTs), 510k-cleared and FDA-approved diagnostics. At an absolute minimum, assay developers must demonstrate with certainty that the assay measures the molecule of choice specifically, accurately, and precisely with sensitivity suitable for the range of analyte in the solution and that the assay can reproduce these results with the intended precision. Several resources exist for guidance about requirements and technological method development for each level of assay classification. These include books, review articles, and regulatory guidance documents (Andreasson et al., 2015; Tiwari & Tiwari, 2010; Bioanalytical method validation: guidance for industry US FDA, CDER, CVM, dept. health and human services May 2018; Immunoassay handbook 2013 edition). In some cases, an assay that has been validated for use in another sample type such as serum or plasma is cross-validated or validated at a fit–for-purpose level, which requires a partial validation that does not include some parameters such as the stability of reagents provided in the kit. As an example of a *practical fit for purpose validation*, Table 11.2 lists commonly used tests from the practical guide to immunoassay method validation.

A more comprehensive document that describes recommendations that are compliant with the FDA is the *FDA guidelines for Bioanalytical Method Validation Guidance for Industry* (May 2018) and defines critical bioanalytical parameters. Major topics that are described in detail include: (1) Reference Standards and critical reagents; (2) Calibration Curve; (3) Quality Control Samples; (4) Selectivity and Specificity; (5) Sensitivity; (6) Accuracy, Precision, and Recovery; (7) Stability; (8) Dilution Effects; and (9) Partial and cross validations. In the document, each of these sections are outlined and In-Study Analysis recommendations with acceptance criteria for each included. Also included is a helpful glossary to clarify validation terminology. Lot-to-lot variability and reagent qualification and assay component stability are critical for kit manufacturers, but often underappreciated by assays built and used for research studies.

In general, RUO salivary immunoassays are often originally built for academic or commercial use to measure analytes in serum or plasma and later adapted for saliva use. These RUO assays require fit-for-purpose validation to show that matrix effects

**Table 11.2** Short description of the validation parameters for which SOPs are presented (Andreasson et al., 2015)

| | Parameter | Definition |
|---|---|---|
| 1 | Robustness | The ability of a method to remain unaffected by small variations in method parameters |
| 2 | Precision | The closeness of agreement between independent test results obtained under stipulated conditions |
| 3 | Trueness | The closeness of agreement between the average value obtained from a large series of test results and an accepted reference value |
| 4 | Uncertainty | A parameter associated with the result of a measurement, that characterizes the dispersion of the values could reasonably be attributed to the measurand |
| 5 | Limits of quantification | Highest and lowest concentrations of analyte that have been demonstrated to be measurable with acceptable levels of precision and accuracy |
| 6 | Dilutional linearity | Dilutional linearity is performed to demonstrate that a sample with a spiked concentration above the ULOQ can be diluted to a concentration within the working range and still give a reliable result |
| 7 | Parallelism | Relative accuracy from recovery tests on the biological matrix or diluted matrix against the calibrators in a substitute matrix |
| 8 | Recovery | The recovery of an analyte in an assay is the detector response obtained from an amount of the analyte added to and extracted from the biological matrix, compared to the detector response obtained for the true concentration of the analyte in the solvent |
| 9 | Selectivity | The ability of the bioanalytical method to measure and differentiate the analytes in the presence of components that may be expected to be present |
| 10 | Sample stability | The chemical stability of an analyte in a given matrix under specific conditions for given time intervals |

of saliva are minimized by testing sample spike and recovery with acceptance often set at 80–120% recovery at the minimum required dilution (MRD). When analytes are at low levels, the flexibility to dilute samples to overcome matrix effects is limited, so often matrix issues are solved by the composition of complex assay running buffers or diluents. Sensitivity is optimized at the same time to assure the sample range fits in the linear portion of the calibration curve. Dilution linearity tests specificity along with cross reactivity testing directly against closely related molecules. Precision is critical when compared to the absolute difference in test samples. This defines the signal-to-noise ratio of the assay. The highest precision is optimal, but if anticipated differences in sample values are several fold over the precision tolerance, then confident conclusions can be determined. For infectious disease, high signal is frequently observed when individuals are exposed to a pathogen relative to non-exposure, while some hormones are operating in regions of the calibration curve where very low variation in assay precision are tolerated. In sum, an optimal assay will show required sensitivity for the sample range, high precision ($CV < 5–15\%$), high specificity, and low cross reactivity.

Suggested references and guidance documents for validation of *RUO assays*:

- Andreasson U, et al. A Practical Guide to Immunoassay Method Validation. Front Neurol. 2015 Aug 19; 6:179.
- US FDA, CDER, CVM. Bioanalytical method validation: guidance for industry (2018). http://www.fda.gov/downloads/drugs/guidances/ucm070107.

Suggested references and guidance documents for validation of *FDA cleared assays*:

- US FDA, CDER, CVM. Bioanalytical method validation: guidance for industry (2018). https://www.fda.gov/regulatory-information/search-fda-guidance-documents/bioanalytical-method-validation-guidance-industry.

(Allinson, 2018).

*FDA-regulated assays require another level of validation.* 510k-cleared assays are approved for diagnostic use, but require submission to the FDA and require more rigorous control of calibrators used and precision of measurements. Diagnostic kits are evaluated on performance as RUO assays, but require precision and accuracy measures that evaluate the possibilities of false-negatives and false-positives. The predictive value of an assay and the risk of an error are assay qualities that are important for the FDA.

## 11.4 Medical Device Classification

As clarification, there are two pathways to gain entry into the market from the FDA perspective—the section 510(k) process, where a device is "cleared" for distribution, or the Premarket Approval (PMA) process, where a device is "approved" by the FDA. In either case, approved or cleared devices are equally legal authorizations to market and export a medical device. The FDA classifies medical devices into three regulatory categories: Class I (low risk), for example, elastic bandages; Class II (higher risk), for example, pregnancy test kits; and Class III (highest risk), for example, implantable devices such as pacemakers. As device classification increases from Class I through to Class III, the regulatory controls also increase, where Class III devices are subject to the highest level of regulatory control. Class I products must register with the FDA and are often marketed as FDA listed or registered. Class II products must show equivalence to a product that the FDA has already cleared and involves a premarket notification (PMN) or 510 (k) submission which notifies the FDA of intent to market a device. Class III products are subject to a much more rigorous review by the FDA to gain FDA approved status which includes a premarket approval (PMA) submission. Class III devices are those that support or sustain human life, are of substantial importance in preventing impairment of human health, or which present a potential, unreasonable risk of illness or injury. In the event that a device is not found to be substantially equivalent to a Class I or Class II predicate device through the 510(k) process, it is required to go through a PMA submission, which is a very involved process. PMA submissions require scientific evidence that the benefits to health from the intended use of a device outweigh the possible risks and that the device will significantly help a large portion of the target population. More recently, the FDA has included two new categories: "De Novo" and "Humanitarian Device Exemption" where the "de novo" pathway was added to address novel devices with low to moderate risk that do not have a predicate device, and "humanitarian device exemptions" are intended for devices for orphan diseases affecting fewer than 4000 patients per year in the USA.

PMA approval is based on scientific evidence providing a reasonable assurance that the device is safe and effective for its intended use or uses. For IVDs, there is a unique link between safety and effectiveness since the safety of the device is not generally related to contact between the device and patient. For IVD products, the safety of the device relates to the impact of the device's performance, and in particular on the impact of false-negative and false-positive results, on patient health.

## 11.5 Requirements for FDA-Cleared or FDA-Approved Diagnostics

*For salivary diagnostics that are FDA approved or cleared*, the ability of the measurement method to predict an outcome with certainty for an individual becomes an important requirement to consider. In general, an assay must precisely and accurately measure an analyte and that measurement will clearly indicate something to a physician that will diagnose a condition that leads to a corrective action by the physician. This is very clear for infectious diseases. In this case, the term "Sensitivity" is also called the true positive rate or probability of detection and is reporting the actual number of positives that are correctly determined. The term "specificity", or true-negative rate, is a measure of the actual negatives that are correctly determined. The sensitivity and specificity are two components that are used to define the validity of a diagnostic test for use in the clinic. So, to have a true diagnostic, the risk of having a false-positive or false-negative result when an individual sample is tested must be evaluated and the consequences of each tie into whether the diagnostic is effective. High sensitivity and high specificity are critical to diagnose a single patient, while in research, large populations can identify trends or correlations and may tolerate some incorrect diagnoses. An example for an infectious disease test to illustrate these terms is as follows. If an assay is 100% sensitive, then all the individuals who are infected are correctly identified as infected, while 100% specific would mean no uninfected individuals would be incorrectly identified as infected. In all cases, large clinical studies are important for regulatory submission and approval of a diagnostic assay in the affected target population of interest.

$$\text{sensitivity} = \frac{\text{number of true positives}}{\text{number of true positives} + \text{number of false negatives}}$$

$$= \frac{\text{number of true positives}}{\text{total number of sick individuals in population}}$$

$$= \text{probability of a positive test given that the patiert has the disease}$$

$$\text{specificity} = \frac{\text{number of true negatives}}{\text{number of true negatives} + \text{number of false positives}}$$

$$= \frac{\text{number of true negatives}}{\text{total number of well individuals in population}}$$

$$= \text{probability of a negative test given that the patient is well}$$

These properties can be captured by analyzing data using a Receiver Operating Characteristics (ROC) curve (Zou, Liu, Bandos, & Ohno-Machado, 2011) to show the relationship between sensitivity (true-positive) and one minus the specificity (true-negative) in determining the inherent predictive value of a diagnostic test. The ROC curve can be used to define a clinical cutoff to make a clinical discrimination. The cutoff point is determined to minimize misclassification and the area under the curve evaluates the validity of a diagnostic test.

Source:
Sprawls, P. Image Characteristics and Quality. In: Sprawls Educational Foundation. Open Resources for Learning and Teaching: The Physical Principles of Medical Imaging [online textbook]. Available at: http://www.sprawls.org/ppmi2/IMGCHAR/. Accessed August 11, 2015.

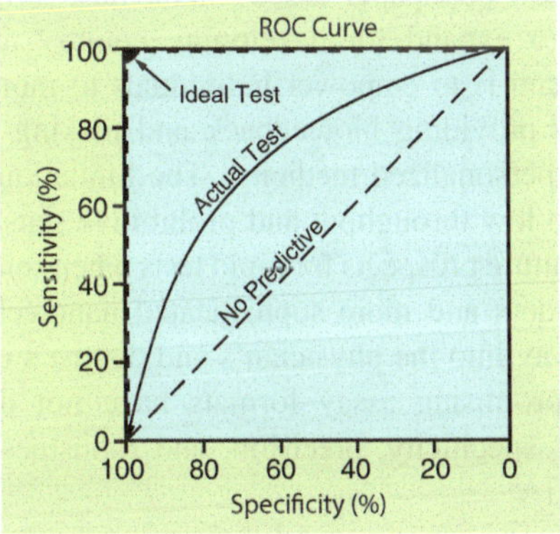


ROC curves are also important for finding an optimal cutoff point to create the lowest likelihood of misdiagnosing diseased or non-diseased subjects, evaluating the ability of a test to correctly pick diseased and non-diseased subjects; comparing the ability of different tests to assess the same disease; and comparing two or more observers measuring the same test.

Clinical trials to generate data to meet these requirements for regulatory submission are usually sponsored by larger companies and partnership is often part of the path forward for smaller companies or university laboratories that developed the original test. This stage of development is often where a great idea may disappear and programs can be canceled in the context of overarching priorities of larger companies' business plan. Another method to help support the large financial burden of clinical trials is the SBIR (Small Business Innovation Research) Grant. In either case, the clinical unmet needs and market valuation of the opportunity must justify the high cost of bringing a diagnostic through the phases of clinical trials and regulatory submission for approval.

## 11.6 Future Directions and Opportunities

The future of salivary diagnostics will benefit from new technologies that may enable multiple classes of direct-to-consumer applications. There has been considerable interest in developing cell phone readers and disposable low-cost tests to provide helpful biofeedback for wellness, infectious disease for public health, diagnostics in the dental and physician's office and for law enforcement. Saliva has yet to be accepted into the automated immunoassay space of hospital laboratories. In the future we predict that innovative and breakthrough formats will continue to emerge for immunoassays, bioanalytical and biosensing platforms, and detection methods. Cost is a driver for the methods of choice for direct to consumer or rapid point of care products, which is often predicated by the setting under which the values are obtained. Low-cost-paper-based lateral flow strips with mobile phone readers are expected to expand for developing country and direct-to-consumer products. A general trend is to empower individuals to monitor and manage their health and lifestyle by providing biofeedback and making informed decisions to promote wellness and personalized medicine. The limitations of these lateral flow immunoassays relate to low throughput and prohibitive sensitivity when compared to the EIA platforms. Future prospects for rapid tests when sensitivity is necessary lie in more expensive readers and more sophisticated nanotechnology or biosensors which may find their way into the physician's and dentist's offices and hospitals.

Most of the new promising assay formats have not demonstrated the high throughput, sensitivity, specificity, precision, and robustness that are satisfied by ELISA or automated EIA.

## References

Allinson, J. L. (2018). Clinical biomarker validation. *Bioanalysis, 10*(12), 957–968.

Andreasson, U., Perret-Liaudet, A., Van Waalwijk van Doorn, L. J., Blennow, K., Chiasserini, D., Engelborghs, S., ... Teunissen, C. E. (2015). A practical guide to immunoassay method validation. *Frontiers in Neurology, 6*, 179.

Cameron, S. O., & Carman, W. F. (2005). The use of the OraSure collection device for hepatitis virus testing in health care settings. *Journal of Clinical Virology, 34*(Suppl 1), S22–S28.

Celebioglu, H. Y., Lee, S., & Chronakis, I. S. (2019). Interactions of salivary mucins and saliva with food proteins: A review. *Critical Reviews in Food Science and Nutrition*, 1–20.

Estrada, Y. M. R. M., & Orlander, P. R. (2011). Salivary cortisol can replace free serum cortisol measurements in patients with septic shock. *Chest, 140*(5), 1216–1222.

Frenkel, E. S., & Ribbeck, K. (2015). Salivary mucins in host defense and disease prevention. *Journal of Oral Microbiology, 7*, 29759.

Lee, S. R., Kardos, K. W., Schiff, E., Berne, C. A., Mounzer, K., Banks, A. T., ... Roehler, M. (2011). Evaluation of a new, rapid test for detecting HCV infection, suitable for use with blood or oral fluid. *Journal of Virological Methods, 172*(1–2), 27–31.

Luppa, P. B., Sokoll, L. J., & Chan, D. W. (2001). Immunosensors—Principles and applications to clinical chemistry. *Clinica Chimica Acta; International Journal of Clinical Chemistry, 314*(1–2), 1–26.

Marsh, P. D., Do, T., Beighton, D., & Devine, D. A. (2016). Influence of saliva on the oral microbiota. *Periodontology 2000, 70*(1), 80–92.

Morin, T. J., McKenna, W. L., Shropshire, T. D., Wride, D. A., Deschamps, J. D., Liu, X., ... Dunbar, W. B. (2018). A handheld platform for target protein detection and quantification using disposable nanopore strips. *Scientific Reports, 8*(1), 14834.

Parry, J. V., Perry, K. R., Panday, S., & Mortimer, P. P. (1989). Diagnosis of hepatitis A and B by testing saliva. *Journal of Medical Virology, 28*(4), 255–260.

Pink, R., Simek, J., Vondrakova, J., Faber, E., Michl, P., Pazdera, J., & Indrak, K. (2009). Saliva as a diagnostic medium. *Biomedical Papers of the Medical Faculty of the University Palacky, Olomouc, Czech Republic, 153*(2), 103–110.

Prodan, A., Brand, H. S., Ligtenberg, A. J., Imangaliyev, S., Tsivtsivadze, E., van der Weijden, F., ... Veerman, E. C. (2015). Interindividual variation, correlations, and sex-related differences in the salivary biochemistry of young healthy adults. *European Journal of Oral Sciences, 123*(3), 149–157.

Raff, H. (2009). Utility of salivary cortisol measurements in Cushing's syndrome and adrenal insufficiency. *The Journal of Clinical Endocrinology and Metabolism, 94*(10), 3647–3655.

Raff, H., & Findling, J. W. (1989). A new immunoradiometric assay for corticotropin evaluated in normal subjects and patients with Cushing's syndrome. *Clinical Chemistry, 35*(4), 596–600.

Raff, H., Homar, P. J., & Skoner, D. P. (2003). New enzyme immunoassay for salivary cortisol. *Clinical Chemistry, 49*(1), 203–204.

Read, L. D., Greene, G. L., & Katzenellenbogen, B. S. (1989). Regulation of estrogen receptor messenger ribonucleic acid and protein levels in human breast cancer cell lines by sex steroid hormones, their antagonists, and growth factors. *Molecular Endocrinology, 3*(2), 295–304.

Rossomando, E. F., Kousvelari, E., Janicki, B. W., & Tabak, L. A. (2001). Improvement of oral health in the postgenomic era: Opportunities for government/industry partnerships. *The Compendium of Continuing Education in Dentistry, 22*(7), 570–572. 4.

Streckfus, C. F., & Bigler, L. R. (2002). Saliva as a diagnostic fluid. *Oral Diseases, 8*(2), 69–76.

Tabak, L. A. (2001). A revolution in biomedical assessment: The development of salivary diagnostics. *Journal of Dental Education, 65*(12), 1335–1339.

Tiwari, G., & Tiwari, R. (2010). Bioanalytical method validation: An updated review. *Pharmaceutical Methods, 1*(1), 25–38.

Vashist, S. K., & Luong, J. H. (2016). Trends in in vitro diagnostics and mobile healthcare. *Biotechnology Advances, 34*(3), 137–138.

Vashist, S. K., Luppa, P. B., Yeo, L. Y., Ozcan, A., & Luong, J. H. T. (2015). Emerging technologies for next-generation point-of-care testing. *Trends in Biotechnology, 33*(11), 692–705.

Vining, R. F., McGinley, R. A., Maksvytis, J. J., & Ho, K. Y. (1983). Salivary cortisol: A better measure of adrenal cortical function than serum cortisol. *Annals of Clinical Biochemistry, 20* (Pt 6), 329–335.

Wong, D. T. (2006). Towards a simple, saliva-based test for the detection of oral cancer 'oral fluid (saliva), which is the mirror of the body, is a perfect medium to be explored for health and disease surveillance'. *Expert Review of Molecular Diagnostics, 6*(3), 267–272.

Zou, K. H., Liu, B., Bandos, A. I., & Ohno-Machado, H. E. (2011). *Statistical evaluation of diagnostic performance: Topics in ROC analysis.*

# Chapter 12
# Salivary Biomarkers and Neurodegenerative Conditions

**Elizabeth A. Thomas**

**Abstract** A major drawback for biomarker research on neurodegenerative diseases is that the pathologically affected tissue, i.e., the brain, cannot easily be accessed for sampling or biopsy. Hence, researchers have turned to peripheral tissues as biospecimens for such studies. The most commonly used biofluid is blood (serum or plasma), although urine and sweat have also been studied. This chapter will focus on the use of saliva as a biofluid for biomarker research studies on neurodegenerative diseases. Saliva is known to contain an abundance of hormones, proteins, and nucleic acid components that reflect physiological function, including several neurodegenerative disease-related proteins such as tau, amyloid beta, alpha-synuclein, and the huntingtin protein. Levels of these proteins in saliva have been proposed to represent useful biomarkers for their indicated diseases. This chapter will review studies demonstrating the presence of central nervous system proteins in saliva and the potential for salivary proteins to serve as biomarkers for neurological and neurodegenerative disorders, with a focus on three of the most common neurodegenerative diseases: Alzheimer's disease, Parkinson's disease, and Huntington's disease.

**Keywords** Central nervous system · Biomarker · Neurodegenerative disease

## 12.1 Background

Neurodegenerative diseases represent a broad family of diseases afflicting both the central and peripheral nervous systems. A primary feature of all neurodegenerative diseases is a progressive loss of certain classes of neurons that affect either motor function or memory and cognition. The degeneration of these neurons can be due to several molecular mechanisms that promote cell death including excitotoxicity,

---

E. A. Thomas (✉)
Department of Neuroscience, The Scripps Research Institute, La Jolla, CA, USA
e-mail: bthomas@scripps.edu



D. A. Granger, M. K. Taylor (eds.), *Salivary Bioscience*,
https://doi.org/10.1007/978-3-030-35784-9_12

mitochondrial dysfunction, and intracellular inclusions or extracellular aggregation of toxic molecules (Dong, Wang, & Qin, 2009; Ross & Poirier, 2004; Trushina & McMurray, 2007). For genetic neurodegenerative diseases, such as triplet repeat disorders, diagnoses can be confirmed by genetic testing. However, for most other neurodegenerative diseases, diagnosis can be challenging and is usually based on symptom presentation. In either case, however, early detection is crucial for improved prognosis and optimal therapeutic outcomes.

There is a great need to identify biomarkers for these disorders for several reasons, including determining disease risk and onset, assessing the severity of symptoms, predicting disease outcomes and to track therapeutics. However, because sampling of central nervous system tissue is not possible, researchers have increasingly focused on other sources for biomarker studies. Cerebral spinal fluid (CSF) is thought to represent the biofluid most like the brain environment, however, CSF collection is a highly invasive technique that requires a lumbar puncture, which can be painful and lead to side effects and complications (Evans, 1998) (Table 12.1).

**Table 12.1** Different types of biofluids used for biomarker research in neurodegenerative conditions

| Biomarker type | Advantages | Disadvantages |
|---|---|---|
| CSF-based biomarkers | Reflects disease and pathological changes in the brain | Requires invasive lumbar puncture |
| | Does not require special laboratory equipment for analysis | Not suitable for all patients (i.e., elderly and those with advanced disease) |
| | | Variability of analytical procedures across laboratories |
| | | Normal biomarker range difficult to establish |
| Blood-based biomarkers | Easily available for analysis | Requires trained personnel to obtain sample |
| | Minimally invasive | Still invasive |
| | Cost-effective assays involved | Large starting volumes typically required |
| | Multiple assessments possible | Normal biomarker range difficult to establish |
| | Broad applications for diagnostics and monitoring therapeutic outcomes | Variability of analytical procedures across diagnostic laboratories |
| Saliva-based biomarkers | Noninvasive | Disease analytes present at low levels |
| | Does not require specially trained personnel | Possible matrix effects |
| | Can collect samples in any setting; minimal sample processing | Variability of analytical procedures across diagnostic laboratories |
| | Broad applications for diagnostics and monitoring therapeutic outcomes | |

The advantages and disadvantages of each biofluid are summarized
*CSF* cerebrospinal fluid

With regards to peripheral sources, investigations in blood have dominated the field for decades; however, blood sampling is also an invasive technique that has several drawbacks (Table 12.1), notably the need for large starting volumes and the need for trained personnel to collect the blood samples. These drawbacks necessitate the pursuit for more advanced and less invasive testing options and have opened the doors for other biofluids, such as saliva, to serve as a source for biomarker analyses.

Like blood, saliva contains a multitude of constituents, including hormones, proteins, and nucleic acids that reflect biological functions (Yan et al., 2009). Unlike blood, which is typically similar in composition throughout the body, saliva is a composite of oral fluids secreted from many different sources with most originating from the major salivary glands, which include the parotid, submandibular, and sublingual glands (Baum, 1993; Granger et al., 2012) (Fig. 12.1). The minor salivary glands, which comprise ~600–1000 glands distributed throughout the oral cavity, are also important components of saliva secretion (Baum, 1993; Granger et al., 2012) (Fig. 12.1).

In general, human salivary glands produce ~750 ml of serous and mucinous saliva daily. This output, which is considered "whole saliva," consists of water, salts, and an abundance of molecules from the blood, as well as salivary proteins in the oral cavity (Humphrey & Williamson, 2001). Saliva secretion is influenced by many factors, which include the diurnal cycle, autonomic nervous system activity, exercise, and chewing. Medications, other treatments, and various medical conditions can also affect saliva secretion (Granger et al., 2012).

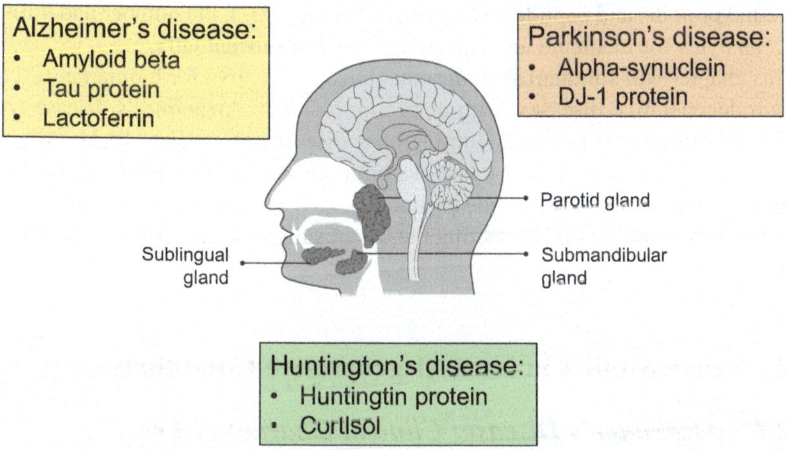


**Fig. 12.1** Summary of the most promising salivary biomarkers for Alzheimer's disease, Parkinson's disease, and Huntington's disease. Biomarkers for each disease are listed. Saliva secretion occurs in the mouth by the actions of the major salivary glands, the parotid gland, submandibular gland, and sublingual gland, which are shown in the cartoon. The minor salivary glands (not shown) also contribute to saliva content and are most concentrated along the buccal, labial, and lingual mucosa, as well as in the soft and hard palates of the mouth

The use of saliva as a biofluid has several advantages over blood sampling (Kaczor-Urbanowicz et al., 2017; Pfaffe, Cooper-White, Beyerlein, Kostner, & Punyadeera, 2011). Most importantly, whole saliva is easy to collect in a noninvasive way. Further, compared to blood sampling, whole saliva collection requires no specially trained personnel, and is easier to process because it does not coagulate like blood. Further, providing a saliva sample reduces discomfort and anxiety for the patient and simplifies serial sample collection over long periods of time, such as in a clinical trial or testing at home (Table 12.1). Saliva collection can be considered safer than blood collection with regards to the risk for hepatitis and HIV (Campo et al., 2006; Wormwood et al., 2015), because needles are not used.

As a diagnostic fluid, saliva has been assessed in a growing number of studies for several pathological conditions, such as celiac disease (Lenander-Lumikari, Ihalin, & Lahteenoja, 2000), rheumatoid arthritis (Helenius et al., 2005), HIV (Holmstrom, Syrjanen, Laine, Valle, & Suni, 1990; Matsuda, Oka, Honda, Takebe, & Takemori, 1993), diabetes mellitus (Belazi, Galli-Tsinopoulou, Drakoulakos, Fleva, & Papanayiotou, 1998; Lopez et al., 2003), breast cancer (Streckfus & Bigler, 2005), Sjögren's syndrome (Ryu, Atkinson, Hoehn, Illei, & Hart, 2006), as well as for therapeutic drug monitoring (Drobitch & Svensson, 1992) (see Chap. 17). However, the use of saliva for biomarker studies on CNS disorders is a relatively new field, despite a report dating back to 1980 suggesting that salivary levels may reflect changes in CSF (Scherber, Richter, & Schaps, 1980). Later studies demonstrating the expression of CNS disease-related proteins in salivary epithelial cells and salivary glands helped promote the use of saliva for biomarker research for these disorders (Oh & Turner, 2006; Sousa, do Amaral, Guimaraes, & Saraiva, 2005). With the characterization of the salivary proteome, which was found to contain more than 2000 proteins and peptides (Hu, Loo, & Wong, 2007), the number and utility of saliva analytes for medicine are expected to expand substantially.

This chapter will summarize studies that have used saliva for biomarker research in neurodegenerative diseases, with a specific focus on Alzheimer's disease (AD), Parkinson's disease (PD), and Huntington's disease (HD) (Fig. 12.1). Although saliva bioscience research as it relates to brain conditions is a field in its infancy, the goal of this chapter is to provide a summary, to date, of the advances made in this area of research in hopes of promoting the use of saliva for nervous system research.

## 12.2 Saliva Studies in Neurodegenerative Conditions

### *12.2.1 Alzheimer's Disease: Clinical Characteristics and Pathology*

Alzheimer's disease (AD) is the most common neurodegenerative disorder that currently afflicts approximately 2% of the general population, with the risk substantially increasing in individuals 70 years or older (Scheltens et al., 2016). Because life expectancies are increasing in human populations, it is estimated that the number of individuals afflicted with AD will double in the next 10–15 years, becoming one of the leading causes of disability and death among the elderly (Alzheimer's Association, 2010; Ferri et al., 2005), not to mention the enormous economical cost to society. AD is characterized by the gradual loss of memory, inability to learn new things, deficits in speech and language, and in recognition of people and objects. Additionally, depression, delusions, and other psychiatric conditions can be present. AD is progressive in nature and eventually leads to a complete incapacity and death of the patient within 8–10 years of diagnosis (Magalingam, Radhakrishnan, Ping, & Haleagrahara, 2018). Major degeneration of neurons in AD occurs particularly in the basal forebrain, hippocampus, and other cortical areas of the brain, and dysfunction in these areas accounts for a majority of the symptoms associated with this disease.

AD can be classified into familial and sporadic forms. Familial AD comprises about 5% of AD cases and is due to mutations in the Amyloid beta precursor protein (*APP*) gene, Presenilin-1 (*PSEN1*), and Presenilin-2 genes (*PSEN2*) genes (Lanoiselee et al., 2017). The gene products of *PSEN1/PSEN2*, the presenilins, are components of the gamma-secretase multicomplex, which is responsible for the cleavage of the APP protein and the formation of amyloid beta peptides. Sporadic forms of AD show no family history, but some cases may be associated with the expression of apolipoprotein E4, a protein involved in the transport of lipids, cholesterol, and other hydrophobic molecules into the brain (Raber, Huang, & Ashford, 2004). In addition, brain changes associated with depressive episodes that compromise the ability of the brain to cope with stress may constitute risk factors for the development of AD (Aznar & Knudsen, 2011). Despite enormous progress in AD research over the past few decades, the precise cause of AD is not completely understood, and currently, there is no effective treatment for the disease.

### 12.2.2 AD Disease-Related Proteins: Amyloid Beta and Tau

Two main pathological hallmarks of AD are extracellular plaques, consisting of insoluble aggregates of the amyloid beta protein, and intracellular neurofibrillary tangles, composed of hyperphosphorylated tau protein (Magalingam et al., 2018; Scheltens et al., 2016). Although the exact nature of how these plaques and tangles are formed is not entirely clear, one well-established and detrimental consequence is neuronal loss and synaptic dysfunction in the AD brain. The normal function of the APP protein is not clear but it is believed to play important roles in memory formation, lipid homeostasis, and the regulation of neuronal activity and neurite outgrowth (Mitsuyama et al., 2009). Amyloid beta pathology arises from the abnormal cleavage of the APP protein, resulting in amyloid beta monomers and oligomers that aggregate ultimately forming amyloid beta fibrils and plaques (Lanoiselee et al., 2017). Dysregulated amyloid beta homeostasis is thought to represent an early event in AD pathology, as studies have shown that plaques can accumulate for up to 10 years before the onset of AD symptoms (Masters et al., 1985). Another argument in favor of a role for amyloid beta in AD pathology is that both familial and sporadic forms of AD are associated with increased levels of this peptide (Dorszewska, Prendecki, Oczkowska, Dezor, & Kozubski, 2016).

The second core pathology is the formation of neurofibrillary tangles, which result from the hyperphosphorylation of tau, a microtubule-associated protein that functions in the cytoskeletal network and maintains proper neuronal structure and intracellular transport (Richter-Landsberg, 2016; Scholz & Mandelkow, 2014). In AD, phosphorylation of tau at multiple sites results in its removal from the microtubule causing disruption of microtubule structures, which leads to dysregulation of a number of neuronal processes ranging from protein trafficking to cellular morphology (Ulrich et al., 2018; Weingarten, Lockwood, Hwo, & Kirschner, 1975). The aggregation and toxic deposition of tau, and the ensuing formation of the characteristic neurofibrillary tangles, leads to compromised cellular function and ultimately, neuronal death (Braak & Braak, 1991; Maeda et al., 2007).

To date, CSF measures of amyloid beta, total tau, and phosphorylated tau proteins represent the most promising biomarkers for AD. Specifically, decreased levels of beta amyloid and increased levels of tau and phosphorylated tau in the CSF are the most reproducible biomarkers for AD diagnosis (Jack et al., 2018; Scheltens et al., 2016) (Table 12.2). Previous studies have successfully measured these in blood samples from AD patients, showing results generally consistent with the CSF findings (Jack et al., 2018; Scheltens et al., 2016) (Table 12.2). In addition to measurements of these disease-related proteins, biomarkers identifying complex

**Table 12.2** Comparison of neurodegenerative biomarker measures in CSF, blood, and saliva

| Disease | Biomarker | Effect in CSF | Effect in plasma/serum | Effect in saliva |
|---|---|---|---|---|
| Alzheimer's disease | | | | |
| | Tau-total | ↑ | ↑ | ↔ |
| | Tau-phospho | ↑ | ↑ | ↑ |
| | Aβ42 | ↓ | ↔ | ↑ or ↔ |
| | Aβ40 | ↔ | ↔ | ↔ |
| Huntingon's disease | | | | |
| | Htt-total | Not tested | ↓ or ↔ | ↑ |
| | Htt-mutant | ↑ | ↑ | Not tested |
| Parkinson's disease | | | | |
| | Alpha-syn-total | ↓ | ↔ | ↓ or ↔ |
| | Alpha-syn-oligo | ↑ | ↑ | ↑ |
| | DJ-1 | ↓ | ↔ | ↑ or ↔ |

Up and down arrows depict the overall change in levels of the biomarker in patients with the disease compared to normal controls, based on previous studies. ↔, not significantly different between patients and controls. Data summarized from Devic et al. (2011), Lin et al. (2012), Weiss et al. (2012), Massai et al. (2013), Al-Nimer et al. (2014), Masters et al. (2015), Southwell et al. (2015), Wild et al. (2015), Vivacqua et al. (2016), Kang et al. (2016), Hijioka et al. (2017), Pchelina et al. (2017), Tatebe et al. (2017), Goldman et al. (2018), An et al. (2018), Corey-Bloom et al. (2018) and from https://www.alzforum.org/alzbiomarker

*Htt* huntingtin, *Alpha-syn* alpha-synuclein, *Alpha-syn-oligo* alpha-synuclein oligomeric, *Aβ* amyloid beta

pathways contributing to pathology in AD would be highly useful, especially for early AD patients. Because of the aforementioned advantages of saliva as a biospecimen, several studies have measured salivary levels of amyloid beta and tau in AD patients. These will be discussed below. Further, lactoferrin is another important inflammation-related protein that has been measured in saliva, which is also discussed below. Such noninvasive biomarkers could also have therapeutic applications, given that several AD therapeutic approaches have targeted the amyloid beta peptide and tau pathways (Barten & Albright, 2008).

### *12.2.3 Salivary Biomarkers in AD*

#### 12.2.3.1 Salivary Amyloid Beta Studies

The amyloid beta protein is typically a 40 amino acid long peptide, but lengths can range from 38 to 43 amino acids, with different forms being measured in different studies. The first study to measure amyloid beta in saliva used an enzyme-linked immunosorbent assay (ELISA) to compare amyloid beta 40 (Aβ40) and amyloid beta 42 (Aβ42) levels between AD patients and two groups of controls: normal healthy subjects and Parkinson's disease patients (Bermejo-Pareja, Antequera, Vargas, Molina, & Carro, 2010). Results showed a small, but statistically significant, increase in salivary Aβ42 levels in mild and moderate AD patients compared to both of the control groups, but no differences in saliva concentrations of Aβ42 in patients with severe AD compared to the control groups (Bermejo-Pareja et al., 2010). In contrast, no changes in salivary levels of Aβ40 were detected among the three cohorts of subjects. Further, this study showed that the association between saliva Aβ42 levels and AD was dependent on gender and associated with Total Functional Capacity clinical scores, but independent of known risk factors, such as age or the *APOE4* genotype (Bermejo-Pareja et al., 2010). In another study using an antibody-based magnetic nanoparticle immunoassay, in addition to an ELISA method, increases in both Aβ40 and Aβ42 were detected in AD patients (Kim, Choi, Song, & Song, 2014). However, there was no difference in Aβ42 concentrations with disease progression, ranging from mild cognitive impairment to severe AD symptomatology (Kim et al., 2014), an effect also observed by Bermejo Pareja and colleagues (Bermejo-Pareja et al., 2010). The fact that PD patients showed no significant difference in Aβ42 levels compared to healthy controls suggests specificity for salivary Aβ42 to serve as a biomarker for AD patients.

Another recent study, also using ELISA to measure Aβ42, found similar results whereby, salivary Aβ42 levels were significantly higher in a cohort of 15 AD patients compared to normal controls (Sabbagh et al., 2018). Interestingly, other studies using a mass spectrometry approach did not detect amyloid beta in saliva of human patients at all (Shi et al., 2011). Other studies aimed at optimizing salivary amyloid beta measurements by adding thioflavin S to prevent its aggregation, and sodium azide to inhibit bacterial growth in saliva (Lee, Guo, Kennedy, McGeer, &

McGeer, 2017). Both of these technical steps acted to prevent sample degradation and improve sample quality and subsequent Aβ42 detection by ELISA (Lee et al., 2017). Aβ42 detection using this method found that the concentration of Aβ42 was about twice as high in AD patients (~40 pg/ml) compared to healthy controls (~20 pg/ml) (Lee et al., 2017). To test the efficiency of their method, these authors compared their in-house ELISA to the commercially available ELISA (Invitrogen) used in the early study and found that it only detected 25% of Aβ42 in the sample when compared to their method (Lee et al., 2017). These findings suggest that adding thioflavin S and sodium azide can dramatically improve salivary Aβ42 measures. This study did not report any differences in Aβ42 concentrations according to disease stage (mild, moderate, or severe) within AD patients, nor did they measure Aβ40 with their method (Lee et al., 2017). Overall, these studies had suggested that saliva testing could be a promising method for detecting AD during its critical early stages and could be useful in tracking therapeutic strategies aimed at reducing amyloid beta levels (Madav, Wairkar, & Prabhakar, 2019).

#### 12.2.3.2 Measures of Tau in Saliva

The first study to measure tau in saliva utilized a mass spectrometry approach, which showed unequivocally that tau is present in human saliva (Shi et al., 2011). In that study, a Luminex method and a modified protocol to achieve minimal matrix effects (see below Sect. 12.3.3) were used to quantify salivary total tau (t-tau) and phosphorylated tau (p-tau) levels in 21 AD patients and 38 healthy controls. Results showed that t-tau levels were slightly decreased in AD patients compared to controls, but the difference diminished after normalizing to the total protein levels in the sample (Shi et al., 2011). There was a clear trend for increased p-tau levels in AD patients, whether or not the values were normalized to saliva total protein (Shi et al., 2011). Significant increases were also found in the p-tau/t-tau ratios in patients with AD, though the differences in t-tau and p-tau levels between patients and controls did not reach statistical significance. These findings suggest that p-tau or p-tau/t-tau ratios might be useful as a possible biomarker for AD.

Another study also quantified salivary t-tau levels using a different method, ultrasensitive Single molecule array technology, in three groups of subjects: 53 AD patients, 68 mild cognitive impairment, and 160 healthy elderly controls (Ashton et al., 2018). However, in that study, no significant differences in salivary t-tau levels were found in AD patients compared to mild cognitive impairment or healthy elderly controls (Ashton et al., 2018). Further, there was no association between salivary t-tau concentrations and clinical assessments or structural magnetic resonance imaging data (Ashton et al., 2018). Hence, more studies are needed to determine whether salivary tau could be a relevant biomarker for this disease.

#### 12.2.3.3 Salivary Lactoferrin in AD

A growing body of literature supports the notion that immune system dysfunction plays a major role in the pathophysiology of AD (Akiyama et al., 2000; Kinney et al., 2018). Although several immune-related biomarkers have been measured in saliva, recent studies have focused on lactoferrin, an antimicrobial peptide which functions in the modulation of immune reactions and inflammatory pathways (Carro et al., 2017). In this study, both mass spectrometry and ELISA methodologies were utilized to measure lactoferrin in AD patients compared to healthy controls, and also subjects with PD, with results showing significantly reduced levels of lactoferrin in AD patients versus both control groups (Carro et al., 2017). Presumably, lower lactoferrin levels would allow for uncontrolled inflammation and immune signaling. Further, significant correlations between salivary lactoferrin and *APOE4* allele status, Mini Mental State Examination (MMSE) scores, CSF amyloid beta-42, and CSF tau were also demonstrated (Carro et al., 2017), providing convincing evidence for lactoferrin as a relevant salivary biomarker for early detection and diagnosis in AD. Importantly, lactoferrin may represent a useful salivary biomarker for other neurodegenerative diseases, many of which have been related to dysregulated immune and inflammatory mechanisms.

### *12.2.4 Huntington's Disease (HD): Clinical Characteristics and Pathology*

HD is an inherited, progressive autosomal-dominant, neurodegenerative disorder that affects approximately one person per 10,000 people worldwide (Huntington's Disease Collaborative Research Group, 1993). It is classified as a movement disorder with the most characteristic symptom of HD being chorea, which is uncontrolled and involuntary movement of the limbs and face. In addition to motor dysfunction, which defines the clinical onset of the disease, other symptoms are commonly present. These include cognitive and psychiatric deficits, which are often detectable before the appearance of motor abnormalities. HD is typically diagnosed in one's late 30s or early 40s, with death following approximately 15 years after diagnosis, from complications such as aspiration pneumonia or cardiac failure (Heemskerk & Roos, 2012). A juvenile form of HD occurs in approximately 5% of cases (Nance & Myers, 2001; van Dijk, van der Velde, Roos, & Bruyn, 1986). Juvenile forms progress more rapidly and show more stiffness of movement rather than chorea. Seizures are also common with this form of the disease (Nance & Myers, 2001). Symptoms arise from prominent neuronal cell loss in the striatum of the brain, which is a region controlling movement and motor function. However, cell death and atrophy in cortical regions also occur.

The HD gene mutation is a CAG repeat expansion in exon 1 of the Huntington (*HTT*) gene (Huntington's Disease Collaborative Research Group, 1993). The length of the CAG repeat expansion determines whether an individual will inherit the disease. Repeat mutations of 40 or greater lead are pathogenic, while repeat lengths <35 are generally considered to be non-pathological. In between mutations, those between 36 and 39 repeats, show unpredictable penetrance (Kremer et al., 1994; Myers et al., 1993). Although HD is a single gene disease, with genetic testing readily available, there is still substantial variability in the onset and severity of disease symptoms, even in patients with the same CAG repeat mutation (Andresen et al., 2007; Andrew et al., 1993; Wexler et al., 2004). For example, studies have shown that HD patients with CAG repeat lengths between 40 and 44 can have an age of onset that differs by up to 20 years (Andrew et al., 1993). The nature and severity of disease symptoms, as well as the course of illness, can also vary among patients, highlighting the need for biomarkers to predict and monitor these features. There are only two FDA-approved treatments available for HD (Richard & Frank, 2019; Yero & Rey, 2008). However, the most promising therapeutic approaches involve *HTT* gene silencing and knockdown technologies, which are currently in different pre-clinical and clinical stages of development.

### 12.2.5 The Huntingtin (Htt) Protein

The *HTT* gene encodes the huntingtin protein (Htt), a large protein consisting of 3144 amino acids that is ubiquitously expressed throughout the brain and periphery (Huntington's Disease Collaborative Research Group, 1993; Marques Sousa & Humbert, 2013). For decades, the exact function(s) of the Htt protein was elusive, but now a growing body of literature shows that Htt is an essential protein (knockout mice are embryonic-lethal) and is involved in a variety of cellular functions including gene transcription, vesicle transport, and energy metabolism (Marques Sousa & Humbert, 2013; Nasir et al., 1995). The primary pathological hallmark of HD is the formation of insoluble Htt aggregates, which are found in both the nucleus and cytoplasm, not only in neurons but other CNS and non-CNS cell types (Sathasivam et al., 1999). The expression of the polyglutamine tract in the mutated Htt protein leads to protein misfolding, and the formation of toxic soluble protein oligomers and insoluble aggregates/inclusion bodies that contributes to the disruption of many intracellular pathways (Arrasate & Finkbeiner, 2012). Although HD is primarily considered to be a CNS disease, patients with HD also exhibit a wide range of peripheral changes, including skeletal muscle dysfunction and peripheral immune system abnormalities (van der Burg, Bjorkqvist, & Brundin, 2009).

Given that levels of mutant Htt correlate with the severity of HD symptoms, and that Htt is a primary target for HD therapeutics, an obvious biomarker for HD is the disease protein itself, Htt. Accordingly, many researchers have turned to measuring Htt in CSF (Wild et al., 2015) and blood (Massai et al., 2013; Weiss et al., 2012) to assess its ability to serve as a biomarker. However, both CSF collection and blood drawing are invasive procedures, with several additional drawbacks (see Table 12.1). Further, CSF levels of Htt are known to be very low (Wild et al.,

2015) and blood measurements of Htt are confounded by the cell-type heterogeneity of blood and the requirement for large starting volumes (Table 12.1). Hence, salivary measurements of the Htt protein would be an ideal alternative. Below, advances in salivary measures of Htt, as well as other studies that have investigated salivary cortisol in HD, will be discussed.

## 12.2.6 Salivary Biomarkers in HD

### 12.2.6.1 Salivary Htt as a Biomarker for HD

Recent studies by Corey-Bloom and colleagues tested whether Htt protein was present in saliva from human subjects and whether salivary levels of Htt might represent a peripheral biomarker for HD. After showing that Htt protein was present in saliva of normal individuals using Western blot methods, saliva samples were collected from 146 HD patients and matched controls to determine if levels differed according to diagnosis (Corey-Bloom et al., 2018). Using ELISA methodology, which detected both normal and mutant forms of the Htt protein (total Htt; tHtt), these authors showed that levels of tHtt were significantly higher in saliva from HD patients versus matched, normal control subjects (Corey-Bloom et al., 2018). Further, there was a nonsignificant trend toward increased levels of tHtt in pre-symptomatic HD patients compared to normal controls, suggesting that Htt levels might accumulate prior to the onset of symptoms (Corey-Bloom et al., 2018). Similar findings were reproduced in a validation cohort of HD patients and normal controls (Corey-Bloom et al., 2018). Importantly, covariable analyses showed no gender effects on salivary tHtt, but significant positive correlations to age in both HD patients and normal controls. Levels of salivary tHtt were not found to be correlated with CAG mutation length, nor age of onset of disease symptoms (Corey-Bloom et al., 2018).

That study also assessed whether tHtt levels were correlated with the severity of disease symptoms at different stages of HD patients using various clinical scores, which included the Mental State Examination (MMSE), the Unified Huntington's Disease Rating Scale (UHDRS), and Total Functional Capacity (TFC). Results showed that salivary tHtt was significantly positively correlated to the UHDRS score and significantly negatively correlated to the TFC score, with no associations with the other measures (Corey-Bloom et al., 2018). These results indicated that salivary tHtt concentrations could have clinical relevance. One drawback of this study was the potential effect of medications, given that ~50% of the HD patients were taking antidepressants, such as Paxil, Fluoxetine, Wellbutrin, and Zoloft. How medications affect saliva production is a potential overall concern, which is addressed below. Nonetheless, a major implication of this work related to HD therapeutics, where salivary levels of Htt might serve as a biomarker to monitor the effects of Htt-lowering therapies, several of which are currently in clinical trials.

#### 12.2.6.2 Salivary Cortisol in HD

Dysfunction of the Hypothalamic–pituitary–adrenal (HPA) axis, which traditionally is seen as the body's stress system, has been linked to learning and memory deficits in a variety of neuropsychiatric and neurodegenerative conditions, including HD (Aziz et al., 2009; Du & Pang, 2015; Hubers et al., 2015). Further, evidence supports a link between HPA dysfunction and depressive symptoms in HD, as well as AD and PD, (Aziz et al., 2009; Du & Pang, 2015; Hubers et al., 2015), however the degree to which HPA dysfunction contributes to these symptoms is unknown. Depression and cognitive deficits are among the most common of the non-motor symptoms that occur in HD, with an estimated 20–50% of HD gene carriers showing depressive symptoms and/or cognitive dysfunction (Gargiulo et al., 2009). These symptoms typically occur prior to the onset of overt motor symptoms. Several studies have measured salivary cortisol in HD patients to investigate the association between the HPA axis and presence of HD symptoms in pre-symptomatic patients and those with early-stage HD (Aziz et al., 2009; Du & Pang, 2015; Hubers et al., 2015; Shirbin et al., 2013; van Duijn et al., 2010). In one study, cortisol concentrations were measured in saliva samples from HD subjects diurnally across a single day in combination with a verbal memory performance task. It was found that the severity of motor and memory retrieval symptoms was associated with higher levels of evening cortisol (Shirbin, Chua, Churchyard, Hannan, et al., 2013). Further, this study showed a trend toward higher levels of salivary cortisol in pre-symptomatic patients suggesting that HPA dysfunction and hypercortisolism could begin much earlier that the diagnostic onset of disease but memory deficits associated with HPA axis abnormalities may only manifest once motor signs are present (Shirbin, Chua, Churchyard, Hannan, et al., 2013).

In contrast, another study showed that pre-symptomatic HD patients, who were not depressed, had significantly lower morning cortisol levels relative to early-stage HD patients and controls (Shirbin et al., 2013). Further, it was shown that the cortisol awakening response was elevated in HD patients who did show depressive-like symptoms (Shirbin, Chua, Churchyard, Lowndes, et al., 2013). Similar studies have been carried out in large animal models of HD, such as the Libechov minipig; however, no significant differences in cortisol response were detected in the HD minipigs versus normal minipigs (Schuldenzucker et al., 2018). Such studies in animal models may improve translational reliability and provide insight into stress reactivity and behavioral correlations and neurodegenerative diseases.

### *12.2.7 Parkinson's Disease: Clinical Characteristics and Pathology*

Parkinson's disease (PD) affects approximately 1–2% of the population over the age of 60 years (Thomas & Beal, 2007). Roughly 60,000 new cases of PD present in the

USA every year and an estimated 10 million people have been diagnosed with the disease worldwide, making PD the second most common neurodegenerative disease (Thomas & Beal, 2007). PD is characterized by muscle rigidity, slowness of movement, resting tremor and other features such as impaired posture and balance and loss of autonomic movements and speech. These symptoms are progressive in nature and caused by loss of dopaminergic neurons in a small region of the midbrain known as the substantia nigra (Dauer & Przedborski, 2003). PD appears to have no strong impact on life expectancy, with most people living up to 20 years after their diagnosis.

Although most PD is sporadic in nature, there are also genetic forms. Several genes have been identified in which mutations have been shown to cause an early-onset form of the disease (Dawson & Dawson, 2003; Gasser, 2007; Kitada et al., 1998; Pramstaller et al., 2005). These findings have greatly accelerated research progress for this disease. Mutations in the alpha-synuclein gene (*SNCA*) cause autosomal-dominant PD via a toxic gain of function of the encoded mutant protein (Campelo & Silva, 2017). In contrast, other early-onset forms of PD can be caused by mutations in the genes encoding the Parkinson protein 2, E3 ubiquitin protein ligase (Parkin; *PARK2*), Parkinsonism associated deglycase (PARK7; a.k.a. DJ-1) or PTEN induced putative kinase 1 (*PINK1*) (Deng, Wang, & Jankovic, 2018; Saito, 2017; Valente et al., 2004). These mutations likely result in pathological effects due to loss-of-function mechanisms. The cause and pathogenesis of the selective loss of dopamine neurons in the substantia nigra of the brain in PD also remain unclear, but accumulating evidence implicates roles for oxidative stress and mitochondrial dysfunction (Al Shahrani, Heales, Hargreaves, & Orford, 2017). To date, no treatment has been identified that represses or slows the death of dopaminergic neurons in PD, however, there are FDA-approved options to treat symptoms, with Levodopa (also called L-dopa) being the most commonly prescribed medicine for PD.

### *12.2.8 PD Disease-Related Proteins: Alpha-Synuclein and DJ-1*

The main pathological hallmark of adult-onset PD is the Lewy body, an insoluble inclusion body localized in the cytoplasm of neurons in the brain. There are also aggregates found in neurites, referred to as Lewy neurites. A major constituent of Lewy bodies is the aggregated form of the alpha-synuclein protein. It is therefore not surprising that mutations in the gene encoding this protein, as well as other genes in the proteasomal pathway, are observed in PD.

Alpha-synuclein is a 140 amino acid protein and member of the synuclein family. It is a neuron-specific protein that is abundantly found throughout the brain and localized in the presynaptic nerve terminals. Several functions for α-synuclein have been proposed including synaptic vesicle release and vesicle trafficking, fatty acid binding, as well as roles in neuronal survival (Dev, Hofele, Barbieri, Buchman, &

van der Putten, 2003). These roles can depend on whether the protein exists in its soluble cytosolic form, or its membrane-bound form. Alpha-synuclein pathology in PD is caused by overexpression of the *SNCA* gene resulting in increased levels of the protein in the brain. In addition to gene overexpression, other mutations in the *SNCA* gene sequence have been identified in familial PD that affects aggregation states of the alpha-synuclein protein (Narhi et al., 1999; Polymeropoulos et al., 1997). There are two major forms of alpha-synuclein aggregation that occur in PD, oligomeric aggregation, which is linked to multiple organelle dysfunction and deficits in the axonal transport system (Hsu et al., 2000), and fibrillar insoluble aggregation that leads to the formation of Lewy Bodies (Baba et al., 1998; Spillantini, Crowther, Jakes, Hasegawa, & Goedert, 1998). Both forms of alpha-synuclein have been measured in CSF and plasma from PD patients in biomarker discovery research studies (Goldman et al., 2018; Tokuda et al., 2010; Vivacqua et al., 2016).

One of the causative genes of familial Parkinson's disease, *PARK7*, encodes a 189 amino acid protein, called Parkinson protein 7, more commonly known as DJ-1. DJ-1 is thought to act as a neuroprotective antioxidant, a transcriptional regulator, and as a molecular chaperone in protein degradation (Hijioka, Inden, Yanagisawa, & Kitamura, 2017; Saito, 2017). DJ-1 is typically located in the cytoplasm of dopaminergic neurons that are not only destined to die in this disease but also can be found in the nucleus mitochondria. DJ-1 levels have been shown to be altered in the CSF and plasma from patients with PD (Lin et al., 2012).

PD lacks robust diagnostic and prognostic biomarkers; however studies on CSF and blood have been widely employed for biomarker identification studies. The diagnosis of PD currently depends on the presence of specific clinical features. However, these features appear only years after the degeneration of dopaminergic neurons. Hence, there is a growing need for biomarkers that might predict early detection of the disease as well as provide a means to monitor disease progression. Alpha-synuclein has generated enormous interest as a biomarker, not only because of its role in disease pathology but also because many therapeutic efforts have focused on reducing the aggregated form of alpha-synuclein (Goldman et al., 2018; Tokuda et al., 2010; Vivacqua et al., 2016). The identification of α-Syn and DJ-1 in human saliva, two proteins that are critically involved in both familial and sporadic PD, suggests that saliva could be a potentially important biofluid for PD biomarker studies. These will be discussed below.

## *12.2.9 Salivary Biomarkers in PD*

### 12.2.9.1 Measures of Alpha-Synuclein in Saliva

In efforts to establish alpha-synuclein as a relevant salivary biomarker of PD diagnosis, studies from 2011 were the first to show the existence of alpha-synuclein in human saliva (Devic et al., 2011). They found that alpha-synuclein concentrations were significantly decreased in the saliva of PD patients as compared to healthy controls (Devic et al., 2011). These findings were replicated in other studies using ELISA methodologies (Al-Nimer, Mshatat, & Abdulla, 2014; Vivacqua et al., 2016), one of which examined different forms of alpha-synuclein in saliva (Vivacqua et al., 2016). In that study, the authors detected a significant increase in the oligomeric form of alpha-synuclein, as well as in the oligomeric alpha-synuclein/total alpha-synuclein ratios in the saliva of PD patients compared to healthy controls (Vivacqua et al., 2016). However, levels of total alpha-synuclein were significantly decreased in the saliva of PD patients when compared to healthy controls (Vivacqua et al., 2016). In that study, it was suggested that the differences in levels of the total and oligomeric forms of the protein were due to the oligomerization of monomeric alpha-synuclein in saliva, which can lead to the lowering of the total alpha-synuclein concentrations. Importantly, within the PD population, correlations were detected between salivary alpha-synuclein levels and the severity of motor symptoms, as assessed by the Unified Parkinson's Disease Rating Scale (UPDRS), as well as other measures of disease progression, stages of illness, and cognitive impairment scores (Vivacqua et al., 2016). Overall, these findings suggest that salivary alpha-synuclein might provide a relevant means for predicting disease progression, whereby lower concentrations would reflect early stages of disease and higher concentrations indicative of a more progressive stage of illness (Vivacqua et al., 2016). A similar study found increases in oligomeric synuclein in PD patients compared to normal controls in saliva (Kang et al., 2016), as well as plasma (Pchelina et al., 2017), but no difference in total alpha-synuclein between the two groups (Kang et al., 2016). In contrast to these highly interesting findings, another study did not find differences in levels of alpha-synuclein in PD patients and controls in either saliva or plasma, nor did alpha-synuclein significantly correlate among biofluids, including CSF (Goldman et al., 2018).

#### 12.2.9.2 DJ-1 in Saliva of PD Patients

Several studies have measured DJ-1 in saliva to assess its potential to serve as a peripheral biomarker for PD, albeit with mixed results. In one pilot study of $n = 16$ PD patients and $n = 22$ matched controls, an increase in both total protein concentration and DJ-1 concentrations were found in PD patients, but there was no difference in salivary DJ-1 after correctly for total protein concentrations in the samples (Masters, Noyce, Warner, Giovannoni, & Proctor, 2015). However, the adjusted DJ-1 levels did correlate with disease symptoms as measured by the UPDRS scores (Masters et al., 2015). In contrast, in a larger study of $n = 74$ PD patients, salivary levels of DJ-1 were higher in later stages of PD compared to early-stage patients and normal controls (Kang et al., 2014). Although in that study, salivary DJ-1 levels were not found to be correlated with UPDRS scores (Kang et al., 2014). Another study used an established Luminex assay with in-house modifications in order to achieve improved accuracy with minimal matrix effect in saliva (see below discussion on matrix effects) (Devic et al., 2011). Differences in DJ-1 levels between patients and controls did not reach statistical significance, regardless of whether or not protein concentrations were normalized, but a trend toward increased levels of DJ-1 was observed in PD patients compared to controls (Devic et al., 2011). Again, no correlation was observed between total DJ-1 levels and UPDRS motor scores (Devic et al., 2011). In assessing DJ-1 levels specifically in buccal cells of saliva, no change was found in DJ-1 levels in PD patients compared to normal controls (Stewart et al., 2014), consistent with studies in plasma (An, Pu, Xiao, & Zhang, 2018).

## 12.3 Challenges and Considerations

### *12.3.1 Origins of Salivary Analytes*

One important consideration in salivary bioscience research relating to neurological disorders is the origin of the analyte/protein being measured. While several neurodegenerative disease-related proteins are undoubtedly present in saliva, the origin of these proteins is unclear and could derive from several sources, or a combination of sources. Constituents from the blood can enter into the saliva via transcellular transport, passive intracellular diffusion, or active transport, giving credence to the notion that some salivary proteins could reflect those circulating levels in the body. Additional studies directly comparing saliva and blood levels will be necessary to test this notion.

Additionally, salivary analytes could come from the salivary glands themselves. Salivary glands are known to express several disease proteins, including the Htt protein, the amyloid beta precursor protein and tau (Conrad, Vianna, Freeman, & Davies, 2002; Marques Sousa & Humbert, 2013; Oh & Turner, 2006), which would be secreted into the whole saliva. It is also possible that the nerves innervating salivary glands could release proteins into the saliva. Parasympathetic innervation to the submandibular glands is achieved by the superior salivatory nucleus via the VIIth cranial nerve, a branch of the facial nerve, while sympathetic innervation of the salivary glands takes place via preganglionic nerves located in the intermediolateral nucleus of spinal cord (Silvers & Som, 1998). It will be critical for future investigations to explore the precise sources and contributions of these salivary biomarkers, not only for understanding the fundamental mechanisms involved in the transportation of these proteins in disease states, but also for controlling the effects of potential covariables if salivary biomarkers are to be used in the clinic.

### *12.3.2 Medication Considerations*

Most patients with neurodegenerative diseases take several different types of medications and these could affect measurements of biomarkers in saliva. In particular, the rate of saliva secretion can be influenced by medications and other treatments.

The presence of other drugs in the body can affect accurate measures of a drug of interest. For example, antidepressant drugs, antihistamines, antipsychotics, sedatives, methyldopa, and diuretics are all known to lead to a low saliva volume or hyposalivation in many, but not all, patients. The presence of these drugs may not directly reduce saliva production, but may lead to a loss of saliva volume secondary to dehydration. This is the case, for example, of diuretics. In addition, cholinergic agents, such as those used in PD patients, can cause dry mouth or xerostomia. Moreover, patient intrinsic factors such as salivary pH and salivary protein composition vary among individuals and can be affected by the disease state itself (Kiang & Ensom, 2016).

### 12.3.3 Matrix Effects

Although ELISA assays designed for blood samples are often used for other biological fluids, it is important to consider the saliva matrix, which can have a confounding effect on immunoassay results. Interactions between the protein of interest and other constituents in the saliva (i.e., the saliva matrix) can result in erroneous readings, typically by affecting the binding of antibody to the protein of interest, or by altering the signal-to-noise ratio. Most matrix effects can be attributed to hydrophobic substances in the sample, such as the phospholipids and carbohydrates that make up mucins, which are abundantly present in saliva and account for its viscous nature. In addition, matrix effects can be due to the pH of the sample, protein–protein interactions, and/or differential salt and ion concentrations in the saliva. One way to assess the extent of the matrix effect is to spike-in different amounts of standards into pooled saliva samples. This has been carried out in several of the aforementioned studies with varying results. For DJ-1, the authors carefully addressed the saliva matrix effect and used a protocol that accomplished 100% recovery of DJ-1 in saliva (Devic et al., 2011). In other studies, it was found that the average recovery rate for tau proteins was between 50 and 70% (Shi et al., 2011), while that of the Htt protein was higher, at 91% (Corey-Bloom et al., 2018). Unfortunately, most studies did not mention their recovery of the measured disease protein in saliva, nor whether matrix effects were addressed.

## 12.4 Conclusions

There is an urgent need for the development of noninvasive and easily accessible biomarkers for neurodegenerative diseases, which are severely debilitating, not only for patients but also their family members and caregivers as well. Saliva is a biological fluid that shows significant potential for the development of noninvasive testing of disease biomarkers as summarized in this chapter. This is a particularly innovative area of research for neurodegenerative diseases, although there are still several challenges that need to be addressed. Because relevant biomarkers might be present at very low concentrations in saliva, there is a need for more specific and sensitive analytical methods to identify and quantify these disease proteins in saliva. Finally, many of the studies on candidate biomarkers in saliva summarized in this chapter will require extensive validation in larger cohorts, but this does not diminish the exciting potential for saliva to be translated into clinical diagnostic, prognostic, and screening efforts for neurological disorders.

## References

Akiyama, H., Barger, S., Barnum, S., Bradt, B., Bauer, J., Cole, G. M., … Wyss-Coray, T. (2000). Inflammation and Alzheimer's disease. *Neurobiology of Aging, 21*, 383–421.

Al Shahrani, M., Heales, S., Hargreaves, I., & Orford, M. (2017). Oxidative stress: Mechanistic insights into inherited mitochondrial disorders and Parkinson's disease. *Journal of Clinical Medicine, 6*(11), 100.

Al-Nimer, M. S., Mshatat, S. F., & Abdulla, H. I. (2014). Saliva alpha-synuclein and a high extinction coefficient protein: A novel approach in assessment biomarkers of Parkinson's disease. *North American Journal of Medical Sciences, 6*, 633–637.

Alzheimer's Association. (2010). 2010 Alzheimer's disease facts and figures. *Alzheimers Dement, 6*, 158–194.

An, C., Pu, X., Xiao, W., & Zhang, H. (2018). Expression of the DJ-1 protein in the serum of Chinese patients with Parkinson's disease. *Neuroscience Letters, 665*, 236–239.

Andresen, J. M., Gayan, J., Djousse, L., Roberts, S., Brocklebank, D., Cherny, S. S., … Wexler, N. S. (2007). The relationship between CAG repeat length and age of onset differs for Huntington's disease patients with juvenile onset or adult onset. *Annals of Human Genetics, 71*, 295–301.

Andrew, S. E., Goldberg, Y. P., Kremer, B., Telenius, H., Theilmann, J., Adam, S., … Hayden, M. R. (1993). The relationship between trinucleotide (CAG) repeat length and clinical features of Huntington's disease. *Nature Genetics, 4*, 398–403.

Arrasate, M., & Finkbeiner, S. (2012). Protein aggregates in Huntington's disease. *Experimental Neurology, 238*, 1–11.

Ashton, N. J., Ide, M., Scholl, M., Blennow, K., Lovestone, S., Hye, A., & Zetterberg, H. (2018). No association of salivary total tau concentration with Alzheimer's disease. *Neurobiology of Aging, 70*, 125–127.

Aziz, N. A., Pijl, H., Frolich, M., van der Graaf, A. W., Roelfsema, F., & Roos, R. A. (2009). Increased hypothalamic-pituitary-adrenal axis activity in Huntington's disease. *The Journal of Clinical Endocrinology and Metabolism, 94*, 1223–1228.

Aznar, S., & Knudsen, G. M. (2011). Depression and Alzheimer's disease: Is stress the initiating factor in a common neuropathological cascade? *Journal of Alzheimer's Disease, 23*, 177–193.

Baba, M., Nakajo, S., Tu, P. H., Tomita, T., Nakaya, K., Lee, V. M., … Iwatsubo, T. (1998). Aggregation of alpha-synuclein in Lewy bodies of sporadic Parkinson's disease and dementia with Lewy bodies. *The American Journal of Pathology, 152*, 879–884.

Barten, D. M., & Albright, C. F. (2008). Therapeutic strategies for Alzheimer's disease. *Molecular Neurobiology, 37*, 171–186.

Baum, B. J. (1993). Principles of saliva secretion. *Annals of the New York Academy of Sciences, 694*, 17–23.

Belazi, M. A., Galli-Tsinopoulou, A., Drakoulakos, D., Fleva, A., & Papanayiotou, P. H. (1998). Salivary alterations in insulin-dependent diabetes mellitus. *International Journal of Paediatric Dentistry, 8*, 29–33.

Bermejo-Pareja, F., Antequera, D., Vargas, T., Molina, J. A., & Carro, E. (2010). Saliva levels of Abeta1-42 as potential biomarker of Alzheimer's disease: A pilot study. *BMC Neurology, 10*, 108.

Braak, H., & Braak, E. (1991). Neuropathological stageing of Alzheimer-related changes. *Acta Neuropathologica, 82*, 239–259.

Campelo, C., & Silva, R. H. (2017). Genetic variants in SNCA and the risk of sporadic Parkinson's disease and clinical outcomes: A review. *Parkinson's Disease, 2017*, 4318416.

Campo, J., Perea, M. A., del Romero, J., Cano, J., Hernando, V., & Bascones, A. (2006). Oral transmission of HIV, reality or fiction? An update. *Oral Diseases, 12*, 219–228.

Carro, E., Bartolome, F., Bermejo-Pareja, F., Villarejo-Galende, A., Molina, J. A., Ortiz, P., ... Orive, G. (2017). Early diagnosis of mild cognitive impairment and Alzheimer's disease based on salivary lactoferrin. *Alzheimers Dement (Amst), 8*, 131–138.

Conrad, C., Vianna, C., Freeman, M., & Davies, P. (2002). A polymorphic gene nested within an intron of the tau gene: Implications for Alzheimer's disease. *Proceedings of the National Academy of Sciences of the United States of America, 99*, 7751–7756.

Corey-Bloom, J., Haque, A. S., Park, S., Nathan, A. S., Baker, R. W., & Thomas, E. A. (2018). Salivary levels of total huntingtin are elevated in Huntington's disease patients. *Scientific Reports, 8*, 7371.

Dauer, W., & Przedborski, S. (2003). Parkinson's disease: Mechanisms and models. *Neuron, 39*, 889–909.

Dawson, T. M., & Dawson, V. L. (2003). Rare genetic mutations shed light on the pathogenesis of Parkinson disease. *The Journal of Clinical Investigation, 111*, 145–151.

Deng, H., Wang, P., & Jankovic, J. (2018). The genetics of Parkinson disease. *Ageing Research Reviews, 42*, 72–85.

Dev, K. K., Hofele, K., Barbieri, S., Buchman, V. L., & van der Putten, H. (2003). Part II: Alpha-synuclein and its molecular pathophysiological role in neurodegenerative disease. *Neuropharmacology, 45*, 14–44.

Devic, I., Hwang, H., Edgar, J. S., Izutsu, K., Presland, R., Pan, C., ... Zhang, J. (2011). Salivary alpha-synuclein and DJ-1: Potential biomarkers for Parkinson's disease. *Brain, 134*, e178.

Dong, X. X., Wang, Y., & Qin, Z. H. (2009). Molecular mechanisms of excitotoxicity and their relevance to pathogenesis of neurodegenerative diseases. *Acta Pharmacologica Sinica, 30*, 379–387.

Dorszewska, J., Prendecki, M., Oczkowska, A., Dezor, M., & Kozubski, W. (2016). Molecular basis of familial and sporadic Alzheimer's disease. *Current Alzheimer Research, 13*, 952–963.

Drobitch, R. K., & Svensson, C. K. (1992). Therapeutic drug monitoring in saliva. An update. *Clinical Pharmacokinetics, 23*, 365–379.

Du, X., & Pang, T. Y. (2015). Is dysregulation of the HPA-axis a core pathophysiology mediating co-morbid depression in neurodegenerative diseases? *Frontiers in Psychiatry, 6*, 32.

Evans, R. W. (1998). Complications of lumbar puncture. *Neurologic Clinics, 16*, 83–105.

Ferri, C. P., Prince, M., Brayne, C., Brodaty, H., Fratiglioni, L., Ganguli, M., ... Alzheimer's Disease International. (2005). Global prevalence of dementia: A Delphi consensus study. *Lancet, 366*, 2112–2117.

Gargiulo, M., Lejeune, S., Tanguy, M. L., Lahlou-Laforet, K., Faudet, A., Cohen, D., ... Durr, A. (2009). Long-term outcome of presymptomatic testing in Huntington disease. *European Journal of Human Genetics, 17*, 165–171.

Gasser, T. (2007). Update on the genetics of Parkinson's disease. *Movement Disorders, 22*(Suppl 17), S343–S350.

Goldman, J. G., Andrews, H., Amara, A., Naito, A., Alcalay, R. N., Shaw, L. M., ... Kang, U. J. (2018). Cerebrospinal fluid, plasma, and saliva in the BioFIND study: Relationships among biomarkers and Parkinson's disease Features. *Movement Disorders, 33*, 282–288.

Granger, D. A., Fortunato, C. K., Beltzer, E. K., Virag, M., Bright, M. A., & Out, D. (2012). Focus on methodology: Salivary bioscience and research on adolescence: An integrated perspective. *Journal of Adolescence, 35*, 1081–1095.

Heemskerk, A. W., & Roos, R. A. (2012). Aspiration pneumonia and death in Huntington's disease. *PLoS Current, 4*, RRN1293.
Helenius, L. M., Meurman, J. H., Helenius, I., Kari, K., Hietanen, J., Suuronen, R., ... Lindqvist, C. (2005). Oral and salivary parameters in patients with rheumatic diseases. *Acta Odontologica Scandinavica, 63*, 284–293.
Hijioka, M., Inden, M., Yanagisawa, D., & Kitamura, Y. (2017). DJ-1/PARK7: A new therapeutic target for neurodegenerative disorders. *Biological & Pharmaceutical Bulletin, 40*, 548–552.
Holmstrom, P., Syrjanen, S., Laine, P., Valle, S. L., & Suni, J. (1990). HIV antibodies in whole saliva detected by ELISA and western blot assays. *Journal of Medical Virology, 30*, 245–248.
Hsu, L. J., Sagara, Y., Arroyo, A., Rockenstein, E., Sisk, A., Mallory, M., ... Masliah, E. (2000). Alpha-synuclein promotes mitochondrial deficit and oxidative stress. *American Journal of Pathology, 157*, 401–410.
Hu, S., Loo, J. A., & Wong, D. T. (2007). Human saliva proteome analysis and disease biomarker discovery. *Expert Review of Proteomics, 4*, 531–538.
Hubers, A. A., van der Mast, R. C., Pereira, A. M., Roos, R. A., Veen, L. J., Cobbaert, C. M., ... Giltay, E. J. (2015). Hypothalamic-pituitary-adrenal axis functioning in Huntington's disease and its association with depressive symptoms and suicidality. *Journal of Neuroendocrinology, 27*, 234–244.
Humphrey, S. P., & Williamson, R. T. (2001). A review of saliva: Normal composition, flow, and function. *The Journal of Prosthetic Dentistry, 85*, 162–169.
Huntington's Disease Collaborative Research Group. (1993). A novel gene containing a trinucleotide repeat that is expanded and unstable on Huntington's disease chromosomes. The Huntington's Disease Collaborative Research Group. *Cell, 72*, 971–983.
Jack, C. R., Jr., Bennett, D. A., Blennow, K., Carrillo, M. C., Dunn, B., Haeberlein, S. B., ... Sperling, R. (2018). NIA-AA research framework: Toward a biological definition of Alzheimer's disease. *Alzheimer's & Dementia, 14*, 535–562.
Kaczor-Urbanowicz, K. E., Martin Carreras-Presas, C., Aro, K., Tu, M., Garcia-Godoy, F., & Wong, D. T. (2017). Saliva diagnostics—Current views and directions. *Experimental Biology and Medicine (Maywood, N.J.), 242*, 459–472.
Kang, W., Chen, W., Yang, Q., Zhang, L., Zhang, L., Wang, X., ... Liu, J. (2016). Salivary total alpha-synuclein, oligomeric alpha-synuclein and SNCA variants in Parkinson's disease patients. *Scientific Reports, 6*, 28143.
Kang, W. Y., Yang, Q., Jiang, X. F., Chen, W., Zhang, L. Y., Wang, X. Y., ... Chen, S. D. (2014). Salivary DJ-1 could be an indicator of Parkinson's disease progression. *Frontiers in Aging Neuroscience, 6*, 102.
Kiang, T. K., & Ensom, M. H. (2016). A qualitative review on the pharmacokinetics of antibiotics in saliva: Implications on clinical pharmacokinetic monitoring in humans. *Clinical Pharmacokinetics, 55*, 313–358.
Kim, C. B., Choi, Y. Y., Song, W. K., & Song, K. B. (2014). Antibody-based magnetic nanoparticle immunoassay for quantification of Alzheimer's disease pathogenic factor. *Journal of Biomedical Optics, 19*, 051205.
Kinney, J. W., Bemiller, S. M., Murtishaw, A. S., Leisgang, A. M., Salazar, A. M., & Lamb, B. T. (2018). Inflammation as a central mechanism in Alzheimer's disease. *Alzheimers Dement (N Y), 4*, 575–590.
Kitada, T., Asakawa, S., Hattori, N., Matsumine, H., Yamamura, Y., Minoshima, S., ... Shimizu, N. (1998). Mutations in the parkin gene cause autosomal recessive juvenile parkinsonism. *Nature, 392*, 605–608.
Kremer, B., Goldberg, P., Andrew, S. E., Theilmann, J., Telenius, H., Zeisler, J., ... Hayden, M. R. (1994). A worldwide study of the Huntington's disease mutation. The sensitivity and specificity of measuring CAG repeats. *The New England Journal of Medicine, 330*, 1401–1406.
Lanoiselee, H. M., Nicolas, G., Wallon, D., Rovelet-Lecrux, A., Lacour, M., Rousseau, S., ... Collaborators of the CNR-MAJ Project. (2017). APP, PSEN1, and PSEN2 mutations in early-

onset Alzheimer disease: A genetic screening study of familial and sporadic cases. *PLoS Medicine, 14*, e1002270.

Lee, M., Guo, J. P., Kennedy, K., McGeer, E. G., & McGeer, P. L. (2017). A method for diagnosing Alzheimer's disease based on salivary amyloid-beta protein 42 levels. *Journal of Alzheimer's Disease, 55*, 1175–1182.

Lenander-Lumikari, M., Ihalin, R., & Lahteenoja, H. (2000). Changes in whole saliva in patients with coeliac disease. *Archives of Oral Biology, 45*, 347–354.

Lin, X., Cook, T. J., Zabetian, C. P., Leverenz, J. B., Peskind, E. R., Hu, S. C., . . . Shi, M. (2012). DJ-1 isoforms in whole blood as potential biomarkers of Parkinson disease. *Scientific Reports, 2*, 954.

Lopez, M. E., Colloca, M. E., Paez, R. G., Schallmach, J. N., Koss, M. A., & Chervonagura, A. (2003). Salivary characteristics of diabetic children. *Brazilian Dental Journal, 14*, 26–31.

Madav, Y., Wairkar, S., & Prabhakar, B. (2019). Recent therapeutic strategies targeting beta amyloid and tauopathies in Alzheimer's disease. *Brain Research Bulletin, 146*, 171–184.

Maeda, S., Sahara, N., Saito, Y., Murayama, M., Yoshiike, Y., Kim, H., . . . Takashima, A. (2007). Granular tau oligomers as intermediates of tau filaments. *Biochemistry, 46*, 3856–3861.

Magalingam, K. B., Radhakrishnan, A., Ping, N. S., & Haleagrahara, N. (2018). Current concepts of neurodegenerative mechanisms in Alzheimer's disease. *BioMed Research International, 2018*, 3740461.

Marques Sousa, C., & Humbert, S. (2013). Huntingtin: Here, there, everywhere! *Journal of Huntington's Disease, 2*, 395–403.

Massai, L., Petricca, L., Magnoni, L., Rovetini, L., Haider, S., Andre, R., . . . Bernocco, S. (2013). Development of an ELISA assay for the quantification of soluble huntingtin in human blood cells. *BMC Biochemistry, 14*, 34.

Masters, J. M., Noyce, A. J., Warner, T. T., Giovannoni, G., & Proctor, G. B. (2015). Elevated salivary protein in Parkinson's disease and salivary DJ-1 as a potential marker of disease severity. *Parkinsonism & Related Disorders, 21*, 1251–1255.

Masters, C. L., Simms, G., Weinman, N. A., Multhaup, G., McDonald, B. L., & Beyreuther, K. (1985). Amyloid plaque core protein in Alzheimer disease and Down syndrome. *Proceedings of the National Academy of Sciences of the United States of America, 82*, 4245–4249.

Matsuda, S., Oka, S., Honda, M., Takebe, Y., & Takemori, T. (1993). Characteristics of IgA antibodies against HIV-1 in sera and saliva from HIV-seropositive individuals in different clinical stages. *Scandinavian Journal of Immunology, 38*, 428–434.

Mitsuyama, F., Futatsugi, Y., Okuya, M., Karagiozov, K., Peev, N., Kato, Y., . . . Koide, T. (2009). Amyloid beta: A putative intra-spinal microtubule-depolymerizer to induce synapse-loss or dentritic spine shortening in Alzheimer's disease. *Italian Journal of Anatomy and Embryology, 114*, 109–120.

Myers, R. H., MacDonald, M. E., Koroshetz, W. J., Duyao, M. P., Ambrose, C. M., Taylor, S. A., . . . Gusella, J. F. (1993). De novo expansion of a (CAG)n repeat in sporadic Huntington's disease. *Nature Genetics, 5*, 168–173.

Nance, M. A., & Myers, R. H. (2001). Juvenile onset Huntington's disease—Clinical and research perspectives. *Mental Retardation and Developmental Disabilities Research Reviews, 7*, 153–157.

Narhi, L., Wood, S. J., Steavenson, S., Jiang, Y., Wu, G. M., Anafi, D., . . . Citron, M. (1999). Both familial Parkinson's disease mutations accelerate alpha-synuclein aggregation. *The Journal of Biological Chemistry, 274*, 9843–9846.

Nasir, J., Floresco, S. B., O'Kusky, J. R., Diewert, V. M., Richman, J. M., Zeisler, J., . . . Hayden, M. R. (1995). Targeted disruption of the Huntington's disease gene results in embryonic lethality and behavioral and morphological changes in heterozygotes. *Cell, 81*, 811–823.

Oh, Y. S., & Turner, R. J. (2006). Effect of gamma-secretase inhibitors on muscarinic receptor-mediated calcium signaling in human salivary epithelial cells. *American Journal of Physiology Cell Physiology, 291*, C76–C82.

Pchelina, S., Emelyanov, A., Baydakova, G., Andoskin, P., Senkevich, K., Nikolaev, M., … Zakharova, E. (2017). Oligomeric alpha-synuclein and glucocerebrosidase activity levels in GBA-associated Parkinson's disease. *Neuroscience Letters, 636*, 70–76.

Pfaffe, T., Cooper-White, J., Beyerlein, P., Kostner, K., & Punyadeera, C. (2011). Diagnostic potential of saliva: Current state and future applications. *Clinical Chemistry, 57*, 675–687.

Polymeropoulos, M. H., Lavedan, C., Leroy, E., Ide, S. E., Dehejia, A., Dutra, A., … Nussbaum, R. L. (1997). Mutation in the alpha-synuclein gene identified in families with Parkinson's disease. *Science, 276*, 2045–2047.

Pramstaller, P. P., Schlossmacher, M. G., Jacques, T. S., Scaravilli, F., Eskelson, C., Pepivani, I., … Klein, C. (2005). Lewy body Parkinson's disease in a large pedigree with 77 Parkin mutation carriers. *Annals of Neurology, 58*, 411–422.

Raber, J., Huang, Y., & Ashford, J. W. (2004). ApoE genotype accounts for the vast majority of AD risk and AD pathology. *Neurobiology of Aging, 25*, 641–650.

Richard, A., & Frank, S. (2019). Deutetrabenazine in the treatment of Huntington's disease. *Neurodegenerative Disease Management, 9*(1), 31–37.

Richter-Landsberg, C. (2016). Protein aggregate formation in oligodendrocytes: Tau and the cytoskeleton at the intersection of neuroprotection and neurodegeneration. *Biological Chemistry, 397*, 185–194.

Ross, C. A., & Poirier, M. A. (2004). Protein aggregation and neurodegenerative disease. *Nature Medicine, 10*(Suppl), S10–S17.

Ryu, O. H., Atkinson, J. C., Hoehn, G. T., Illei, G. G., & Hart, T. C. (2006). Identification of parotid salivary biomarkers in Sjogren's syndrome by surface-enhanced laser desorption/ionization time-of-flight mass spectrometry and two-dimensional difference gel electrophoresis. *Rheumatology (Oxford, England), 45*, 1077–1086.

Sabbagh, M. N., Shi, J., Lee, M., Arnold, L., Al-Hasan, Y., Heim, J., & McGeer, P. (2018). Salivary beta amyloid protein levels are detectable and differentiate patients with Alzheimer's disease dementia from normal controls: Preliminary findings. *BMC Neurology, 18*, 155.

Saito, Y. (2017). DJ-1 as a biomarker of Parkinson's disease. *Advances in Experimental Medicine and Biology, 1037*, 149–171.

Sathasivam, K., Hobbs, C., Turmaine, M., Mangiarini, L., Mahal, A., Bertaux, F., … Bates, G. P. (1999). Formation of polyglutamine inclusions in non-CNS tissue. *Human Molecular Genetics, 8*, 813–822.

Scheltens, P., Blennow, K., Breteler, M. M., de Strooper, B., Frisoni, G. B., Salloway, S., & Van der Flier, W. M. (2016). Alzheimer's disease. *Lancet, 388*, 505–517.

Scherber, A., Richter, K., & Schaps, P. (1980). Distribution of antiepileptic drugs between plasma, plasma water, cerebrospinal fluid, saliva and brain. *Monographs in Neural Sciences, 5*, 208–212.

Scholz, T., & Mandelkow, E. (2014). Transport and diffusion of Tau protein in neurons. *Cellular and Molecular Life Sciences, 71*, 3139–3150.

Schuldenzucker, V., Schubert, R., Muratori, L. M., Freisfeld, F., Rieke, L., Matheis, T., … Reilmann, R. (2018). Behavioral assessment of stress compensation in minipigs transgenic for the Huntington gene using cortisol levels: a proof-of-concept study. *Journal of Huntington's Disease, 7*, 151–161.

Shi, M., Sui, Y. T., Peskind, E. R., Li, G., Hwang, H., Devic, I., … Zhang, J. (2011). Salivary tau species are potential biomarkers of Alzheimer's disease. *Journal of Alzheimer's Disease, 27*, 299–305.

Shirbin, C. A., Chua, P., Churchyard, A., Hannan, A. J., Lowndes, G., & Stout, J. C. (2013). The relationship between cortisol and verbal memory in the early stages of Huntington's disease. *Journal of Neurology, 260*, 891–902.

Shirbin, C. A., Chua, P., Churchyard, A., Lowndes, G., Hannan, A. J., Pang, T. Y., … Stout, J. C. (2013). Cortisol and depression in pre-diagnosed and early stage Huntington's disease. *Psychoneuroendocrinology, 38*, 2439–2447.

Silvers, A. R., & Som, P. M. (1998). Salivary glands. *Radiologic Clinics of North America, 36*, 941–66, vi.

Sousa, M. M., do Amaral, J. B., Guimaraes, A., & Saraiva, M. J. (2005). Up-regulation of the extracellular matrix remodeling genes, biglycan, neutrophil gelatinase-associated lipocalin, and matrix metalloproteinase-9 in familial amyloid polyneuropathy. *FASEB Journal, 19*, 124–126.

Southwell, A. L., Smith, S. E., Davis, T. R., Caron, N. S., Villanueva, E. B., Xie, Y., ... Hayden, M. R. (2015). Ultrasensitive measurement of huntingtin protein in cerebrospinal fluid demonstrates increase with Huntington disease stage and decrease following brain huntingtin suppression. *Scientific Reports, 5*, 12166.

Spillantini, M. G., Crowther, R. A., Jakes, R., Hasegawa, M., & Goedert, M. (1998). Alpha-synuclein in filamentous inclusions of Lewy bodies from Parkinson's disease and dementia with Lewy bodies. *Proceedings of the National Academy of Sciences of the United States of America, 95*, 6469–6473.

Stewart, T., Sui, Y. T., Gonzalez-Cuyar, L. F., Wong, D. T., Akin, D. M., Tumas, V., ... Zhang, J. (2014). Cheek cell-derived alpha-synuclein and DJ-1 do not differentiate Parkinson's disease from control. *Neurobiology of Aging, 35*, 418–420.

Streckfus, C., & Bigler, L. (2005). The use of soluble, salivary c-erbB-2 for the detection and post-operative follow-up of breast cancer in women: The results of a five-year translational research study. *Advances in Dental Research, 18*, 17–24.

Tatebe, H., Kasai, T., Ohmichi, T., Kishi, Y., Kakeya, T., Waragai, M., ... Tokuda, T. (2017). Quantification of plasma phosphorylated tau to use as a biomarker for brain Alzheimer pathology: Pilot case-control studies including patients with Alzheimer's disease and down syndrome. *Molecular Neurodegeneration, 12*, 63.

Thomas, B., & Beal, M. F. (2007). Parkinson's disease. *Human Molecular Genetics, 16*(Spec No. 2), R183–R194.

Tokuda, T., Qureshi, M. M., Ardah, M. T., Varghese, S., Shehab, S. A., Kasai, T., ... El-Agnaf, O. M. (2010). Detection of elevated levels of alpha-synuclein oligomers in CSF from patients with Parkinson disease. *Neurology, 75*, 1766–1772.

Trushina, E., & McMurray, C. T. (2007). Oxidative stress and mitochondrial dysfunction in neurodegenerative diseases. *Neuroscience, 145*, 1233–1248.

Ulrich, G., Salvade, A., Boersema, P., Cali, T., Foglieni, C., Sola, M., ... Paganetti, P. (2018). Phosphorylation of nuclear Tau is modulated by distinct cellular pathways. *Scientific Reports, 8*, 17702.

Valente, E. M., Abou-Sleiman, P. M., Caputo, V., Muqit, M. M., Harvey, K., Gispert, S., ... Wood, N. W. (2004). Hereditary early-onset Parkinson's disease caused by mutations in PINK1. *Science, 304*, 1158–1160.

van der Burg, J. M., Bjorkqvist, M., & Brundin, P. (2009). Beyond the brain: Widespread pathology in Huntington's disease. *Lancet Neurology, 8*, 765–774.

van Dijk, J. G., van der Velde, E. A., Roos, R. A., & Bruyn, G. W. (1986). Juvenile Huntington disease. *Human Genetics, 73*, 235–239.

van Duijn, E., Selis, M. A., Giltay, E. J., Zitman, F. G., Roos, R. A., van Pelt, H., & van der Mast, R. C. (2010). Hypothalamic-pituitary-adrenal axis functioning in Huntington's disease mutation carriers compared with mutation-negative first-degree controls. *Brain Research Bulletin, 83*, 232–237.

Vivacqua, G., Latorre, A., Suppa, A., Nardi, M., Pietracupa, S., Mancinelli, R., ... Berardelli, A. (2016). Abnormal salivary total and oligomeric alpha-synuclein in Parkinson's disease. *PLoS One, 11*, e0151156.

Weingarten, M. D., Lockwood, A. H., Hwo, S. Y., & Kirschner, M. W. (1975). A protein factor essential for microtubule assembly. *Proceedings of the National Academy of Sciences of the United States of America, 72*, 1858–1862.

Weiss, A., Trager, U., Wild, E. J., Grueninger, S., Farmer, R., Landles, C., ... Tabrizi, S. J. (2012). Mutant huntingtin fragmentation in immune cells tracks Huntington's disease progression. *The Journal of Clinical Investigation, 122*, 3731–3736.

Wexler, N. S., Lorimer, J., Porter, J., Gomez, F., Moskowitz, C., Shackell, E., … U.S.-Venezuela Collaborative Research Project. (2004). Venezuelan kindreds reveal that genetic and environmental factors modulate Huntington's disease age of onset. *Proceedings of the National Academy of Sciences of the United States of America, 101*, 3498–3503.

Wild, E. J., Boggio, R., Langbehn, D., Robertson, N., Haider, S., Miller, J. R., … Weiss, A. (2015). Quantification of mutant huntingtin protein in cerebrospinal fluid from Huntington's disease patients. *The Journal of Clinical Investigation, 125*, 1979–1986.

Wormwood, K. L., Aslebagh, R., Channaveerappa, D., Dupree, E. J., Borland, M. M., Ryan, J. P., … Woods, A. G. (2015). Salivary proteomics and biomarkers in neurology and psychiatry. *Proteomics Clinical Applications, 9*, 899–906.

Yan, W., Apweiler, R., Balgley, B. M., Boontheung, P., Bundy, J. L., Cargile, B. J., … Wong, D. T. (2009). Systematic comparison of the human saliva and plasma proteomes. *Proteomics Clin Appl, 3*, 116–134.

Yero, T., & Rey, J. A. (2008). Tetrabenazine (xenazine), an FDA-approved treatment option for Huntington's disease-related chorea. *P T, 33*, 690–694.

# Chapter 13
# The Utility of Antibodies in Saliva to Measure Pathogen Exposure and Infection

**Pranay R. Randad, Kyla Hayford, Richard Baldwin, Lindsay Avolio, Nora Pisanic, William J. Moss, Douglas A. Granger, and Christopher D. Heaney**

**Abstract** Saliva contains pathogen-specific antibodies that can provide quantitative information on the type and temporality of infection. As such, saliva as a biospecimen can be used to noninvasively assess seroconversion to infectious

---

In the interest of full disclosure, Douglas A. Granger is the founder and chief scientific and strategy advisor of Salimetrics LLC and Salivabio LLC (Carlsbad, CA) and the nature of these relationships is managed by the policies of the committees on conflict of interest at Johns Hopkins University School of Medicine and the University of California at Irvine.

---

P. R. Randad (✉) · L. Avolio · N. Pisanic
Department of Environmental Health and Engineering, Johns Hopkins Bloomberg School of Public Health, Johns Hopkins University, Baltimore, MD, USA
e-mail: prandad1@jhmi.edu

K. Hayford · W. J. Moss
Department of International Health, International Vaccine Access Center, Johns Hopkins Bloomberg School of Public Health, Johns Hopkins University, Baltimore, MD, USA

R. Baldwin
NanoComposix, San Diego, CA, USA
e-mail: Richard.Baldwin@nanocomposix.com

D. A. Granger
Institute for Interdisciplinary Salivary Bioscience Research, University of California at Irvine, Irvine, CA, USA

Johns Hopkins University, Baltimore, MD, USA
e-mail: dagrange@uci.edu

C. D. Heaney (✉)
Department of Environmental Health and Engineering, Johns Hopkins Bloomberg School of Public Health, Johns Hopkins University, Baltimore, MD, USA

Department of International Health, Bloomberg School of Public Health, Johns Hopkins University, Baltimore, MD, USA

Department of Epidemiology, Bloomberg School of Public Health, Johns Hopkins University, Baltimore, MD, USA
e-mail: cheaney1@jhu.edu



D. A. Granger, M. K. Taylor (eds.), *Salivary Bioscience*,
https://doi.org/10.1007/978-3-030-35784-9_13

pathogens in the clinical setting as well as for epidemiological surveillance, improving understanding of the epidemiology and natural history of infectious diseases. While saliva is not ideal for analysis of all infectious diseases, it remains a valuable means for data collection across populations varying in age, geographic location, and health status. Developments in the field could allow for rapid testing of exposure to infectious agents through point-of-care assays in areas where blood-based assays or laboratory analysis would be impractical. Here we review the advantages and disadvantages of using salivary assays to assess infectious disease exposure and infection in different settings and explore future directions.

## 13.1 Introduction

Saliva contains pathogen-specific antibodies that can provide quantitative information on the type and temporality of pathogen exposure and infection. As such, saliva can serve as a noninvasive biospecimen to monitor recent or historical exposure to pathogens in both clinical and epidemiological settings. In clinical settings, saliva-based diagnostic testing can inform clinical decision-making around infectious disease treatment, management, and vaccination. In epidemiological surveys, saliva-based infectious disease detection methods can be used to assess seroprevalence and seroconversion among high-risk populations; gain epidemiological inference on risk factors associated with pathogen exposure, infection, and transmission; and facilitate self-collection and testing in national- and community-level public health infectious disease surveillance programs. Saliva-based methods for infectious disease detection have a unique and critical role in the context of vaccine preventable diseases (VPDs), where they can be used to assess disease burden and improve case detection, estimate population immunity, identify immunity gaps, and assess the impact of vaccination campaigns and routine immunization.

Most of the studies discussed in this chapter utilize "oral fluids" specifically. We thus use the term oral fluids throughout the chapter unless stated otherwise. In this chapter, we discuss the utility of antibodies in oral fluids to measure pathogen exposure and infection in two major infectious disease contexts: (1) clinical diagnostic testing and (2) epidemiological research and population-based surveillance, with a special emphasis on VPDs. We first discuss the utility of oral fluid-based methods in clinical infectious disease diagnostic testing. In this section, we review oral fluid-based HIV testing as an example of a US Food and Drug Administration (FDA) approved and CLIA-waived oral fluid-based test for diagnosis of infection. Second, we discuss the utility of oral fluid-based methods in epidemiological research and the inference that can be gained from using such tools in cross-sectional and longitudinal study designs. We further discuss and provide examples of national- and community-level public health programs for population-based infectious disease surveillance where oral fluid-based self-collection and testing has been implemented. We end this section with a discussion of the utility of oral fluid-based methods in the context of VPDs. Following these two sections, we address the limitations of oral fluid-based methods compared to blood-based methods, such as diagnostic accuracy. We conclude this chapter with a section on future directions for the field of oral fluid-based infectious disease surveillance and diagnostic testing, highlighting point-of-care (POC) development for VPD-related applications, the identification of asymptomatic subpopulations in population-based studies, and a more nuanced argument on the potential utility of saliva-based secretory IgA (SIgA) biomarkers of infectious disease. Through this chapter, the reader will gain an understanding of the different potential uses of oral fluid-based methods for infectious disease testing in clinical- and population-based settings, the current status of pathogen-specific oral fluid-based methods and assays (Table 13.1), and advantages and disadvantages of oral fluid-based infectious disease testing.

## 13.2 Utility of Oral Fluid in Clinical Infectious Disease Diagnostic Testing

Serological diagnostic testing has proven to be an indispensable tool in the clinical setting for the diagnosis and management of infectious diseases. Serological diagnostic testing is ideal for the diagnosis of infectious diseases where active infection can be identified by pathogen-specific antibody levels and early clinical identification could initiate treatment interventions and infection control practices. Human immunodeficiency virus (HIV) is one of the few pathogens that has an oral fluid-based clinical diagnostic test that is approved by the US FDA and is Clinical Laboratory Improvement Amendments (CLIA)-waived for diagnosis of infection. In this section, we discuss oral fluid-based diagnostic testing for HIV in the context of its clinical evaluation and FDA approval for diagnostic and point-of-care (POC) purposes.

### *13.2.1 Oral Fluid-Based Human Immunodeficiency Virus Diagnostic Testing*

Identification of infection with HIV is critical to inform clinical decision-making, including initiation of antiretroviral therapy or pre-exposure prophylaxis (PrEP), and to prevent transmission. The Center for Disease Control and Prevention (CDC) recommends that everyone between the ages of 13 and 64 get tested for HIV at least once, and at least once a year for higher risk populations ("HIV/AIDS testing," 2019). A number of studies have shown that IgG antibodies against HIV type 1 (HIV-1) can be detected in the oral fluid of seropositive subjects (Hodinka, Nagashunmugam, & Malamud, 1998). These reports garnered considerable interest and prompted the further development of oral mucosal transudate (OMT)-based methods for serological detection of HIV. Advantages of an OMT-based method to detect HIV-specific antibodies include greater safety when collecting and handling HIV-positive specimens due to low concentrations of HIV antigens in saliva (Major et al., 1991), increased patient compliance (Major et al., 1991; Spielberg et al.,

**Table 13.1** Studies regarding salivary pathogen-specific antibody biomarkers of pathogen exposure and/or infection

| Pathogen | | Antibody biomarker | References |
|---|---|---|---|
| Virus | Dengue virus (DENV) | IgM, IgA, and IgG antibody | Andries et al. (2015), Balmaseda et al. (2008), Spoorthi (2014), Vázquez et al. (2007) |
| | Ebola | IgM and IgG antibody | Lambe et al. (2016) |
| | Human immuno-deficiency virus (HIV) | IgG antibody | Delaney et al. (2006), Delaney et al. (2011), Figueroa et al. (2018), Gallo (1997), Hodinka et al. (1998), Wesolowski et al. (2006), Reid et al. (2008) |
| | Hepatitis A virus (HAV) | IgM and IgG antibody | Amado, Villar, de Paula, de Almeida, and Gaspar (2006), Laufer et al. (1995), Morris-Cunnington et al. (2004), Ochnio, Scheifele, Ho, and Mitchell (1997), Parry, Perry, Panday, and Mortimer (1989), Quoilin et al. (2007), Tourinho et al. (2015) |
| | Hepatitis B virus (HBV) | IgM, IgG anti-body, and viral antigen | Fisker, Georgsen, Stolborg, Khalil, and Christensen (2002), Hutse et al. (2005), Nokes et al. (2001), O'Connell et al. (2000), Piacentini, Thieme, Beller, and Davidson (1993), Quoilin et al. (2007), Simani et al. (2008), Thornton et al. (2000), Weild et al. (2000) |
| | Hepatitis C virus (HCV) | IgM and IgG antibody | De Cock et al. (2004), Judd et al. (2003), Thornton et al. (2000), Weild et al. (2000) |
| | Hepatatis E virus (HEV) | IgG and IgA antibody | Pisanic et al. (2017) |
| | Measles | IgM and IgG antibody | Goyal et al. (2009), Hayford, Al-Emran, et al. (2013), Hübschen et al. (2018), Jani et al. (2008), Kremer and Muller (2005), Nigatu et al. (2008), Ohuma et al. (2009), Sheikh et al. (2011), Shonhai et al. (2015), Thieme et al. (1994), Vainio et al. (2006) |
| | Mumps | IgM and IgG antibody | Hübschen et al. (2018), Jin et al. (2004), Reid et al. (2008), Thieme et al. (1994), Warrener et al. (2011) |
| | Norovirus | IgA and IgG antibodies | Griffin et al. (2011, 2015), Lindesmith et al. (2003), Lindesmith et al. (2005), Moe et al. (2004), Pisanic et al. (2018), Wade et al. (2019) |
| | Pertussis | IgG antibody | Kara et al. (2017), Wang et al. (2014) |
| | Rotavirus | IgM, IgA, and IgG antibody | Aiyar et al. (1990), Friedman, Entin, Zedaka, and Dagan (1996), Grimwood et al. (1988), Stals, Walther, and Bruggeman (1984) |
| | Rubella | IgM and IgG antibody | Ben Salah et al. (2003), Hübschen et al. (2018), Nokes et al. (1998), Nokes et al. (2001), Thieme et al. (1994) |

(continued)

**Table 13.1** (continued)

| Pathogen | | Antibody biomarker | References |
|---|---|---|---|
| | Zika virus | IgM, IgA, and IgG antibody | Zhao et al. (2019) |
| Bacteria | *Escherichia coli* O157 | IgM and IgA antibodies | Chart and Jenkins (1998), Chart et al. (2003), Ludwig et al. (2002) |
| | *Salmonella Typhii* | IgA antibody | Herath (2003), Zaka-ur-Rab et al. (2012) |
| | *Campylobacter* spp. | IgG and IgA antibody | Cawthraw et al. (2002) |
| | *Shigella* spp. | IgA antibody | Schultsz et al. (1992) |
| | *Vibrio cholerae* | IgA antibody | Jertborn, Svennerholm, and Holmgren (1986) |
| Parasite | *Plasmodium falciparum* | IgG antibody | Chidi et al. (2011), Estévez et al. (2011) |
| | *Giardia* spp. | IgG and IgA (including SIgA) antibody | El-Gebaly et al. (2012), Rodríguez et al. (2004) |
| | *Cryptosporidium* spp. | IgG and IgA antibody | Cozon, Biron, Jeannin, Cannella, and Revillard (1994), Egorov et al. (2010); Griffin et al. (2011); Moss et al. (2019) |

2000), and the potential to improve access to testing and results (Martin, Williams, Ferguson, & Read, 2018; Nangendo et al., 2017).

Testing for HIV using OMT was enabled by a number of products that have been approved by the US FDA. The OraSure oral-specimen collection device has been licensed by the FDA to use in conjunction with a laboratory-based enzyme immunoassay (EIA) antibody test, the Oral Fluid Vironostika® HIV-1 Microelisa system (approved in 1994) ("Approval of The First U.S. HIV Test System Using Oral Fluid Samples," , 1994). The OraSure® collection apparatus was designed specifically to enhance the collection of OMT rich in IgG antibodies derived from OMT at the gingival crevice (Brandtzaeg, 2013). Two separate large-scale clinical trials determined that OMT-based testing for HIV-1 antibodies using this system was a highly accurate alternative to serum testing. The system performed with a sensitivity and specificity of 99.2% (Granade et al., 1998) to 99.9% (Gallo, 1997). The oral fluid Vironostika HIV-1 Microelisa System has since been replaced and is currently marketed as the Avioq HIV-1 Microelisa System (Food and Drug Administration, 2009). It is important to note that the FDA requires a confirmatory test to rule out false positives (Food and Drug Administration, 2009). The OraSure HIV-1 Western blot kit (Epitope Inc.) has been licensed by the FDA as a more specific method to perform a requisite confirmation of oral fluid specimens positive for HIV-1 specific antibodies (Gallo, 1997), which was approved in 1996. FDA licensure of the OraSure® OMT collection device with an associated EIA method now provides an accurate alternative to routine, blood-based, serological HIV-1 testing.

Rapid, POC tools are critical in meeting national goals to improve coverage of HIV testing and access to HIV status results. The OraQuick® *ADVANCE* Rapid

HIV-1/2 Antibody Test, for the qualitative detection of antibodies to both HIV-1 and HIV-2 in human oral fluids, was approved by the FDA and CLIA-waived in 2004 (Food and Drug Administration, 2004). A direct comparison of six FDA approved rapid HIV antibody tests concluded that this OMT-based rapid test performed similarly to other blood-based rapid tests (Delaney et al., 2011). This same system was repackaged as the OraQuick® In-Home HIV Test, and was approved as an over-the-counter, POC, test in 2012 (Food and Drug Administration, 2012b). Most recently, the FDA approved the Dual Path Platform® (DPP) HIV-1/2 POC test in 2012, which can utilize oral fluid as input specimens (Food and Drug Administration, 2012a). While both blood- and oral fluid-based rapid HIV antibody tests conducted via self-testing provide accurate HIV results compared to when conducted by trained personnel, the sensitivity and specificity have been shown to be higher for self-testing using blood-based rapid tests compared with oral fluid-based rapid tests (Figueroa et al., 2018). Such POC tools can be used in HIV screening programs to identify patients that require intervention, improve access to HIV testing results, and ensure that negative patients are truly negative prior to initiating PrEP.

The CDC continues to strongly advocate for the most sensitive rapid HIV test (Bernard et al., 2014). An evaluation of four separate CDC studies showed that the OraQuick® ADVANCE Rapid HIV-1/2 Antibody test performed with a sensitivity and specificity >99% for HIV antibody in oral fluid specimens in diverse clinical and nonclinical settings (Delaney et al., 2006). Post-marketing surveillance of the rapid OraQuick® test in 368 testing sites affiliated with 17 state and city health departments concluded that oral fluid-based testing performed with a median specificity of 99.89% (99.44–100%) and positive predictive value of 90.00% (range: 50.00–100%) (Wesolowski et al., 2006).

There are several important limitations of OMT-based HIV testing. Because antibody levels are lower in oral fluids, rapid HIV tests using oral fluids are less sensitive when HIV-specific antibody levels are low, i.e., during the early stages of infection (CDC, 2008; Mortimer & Parry, 1991; Pant Pai et al., 2007). These limitations also apply to population-based studies, where OMT-based HIV diagnostic testing was shown to be less sensitive for the early detection of HIV antibodies compared to blood-based methods (Luo et al., 2013). Despite these limitations, a large body of evidence supports the utility of OMT-based testing in widespread screening for HIV infection, including within national-, state-, and community-level public health programs.

## 13.3 Utility of Oral Fluid in Population-Based Infectious Disease Epidemiology and Surveillance

Saliva can serve as a tool in population-based epidemiological surveillance to improve understanding of the epidemiology and natural history of exposure and infection. Pathogen-specific antibodies are unique biomarkers in their ability to identify both historical and recent exposure to a diversity of pathogens, including bacteria, viruses, and parasites. Built into epidemiological research, serological methods provide an objective and sensitive method of elucidating population-level seroprevalence, identifying risk factors for exposure and transmission, and monitoring and evaluating preventative intervention strategies.

Blood-based serological methods have several limitations for surveillance and epidemiological studies that can often be mitigated by the use of oral fluid samples. First, the collection of venous blood is invasive and requires phlebotomists. Due to the ease of oral fluid collection, oral fluid-based serology facilitates self-collection and mail-in methods which can be advantageous in both field-based epidemiological studies (Wade et al., 2018) and in national-level population-based infectious disease screening programs (Morris-Cunnington, Edmunds, Miller, & Brown, 2004; Quoilin et al., 2007). Furthermore, self-testing can be reliable and accurate compared to testing performed by trained personnel (Figueroa et al., 2018). Because phlebotomists, transport for researchers and participants, and blood collection consumables are not needed, oral fluid-based testing can also have cost advantages (Morris-Cunnington et al., 2004). Second, blood samples must be centrifuged and frozen relatively quickly, which can be difficult in low-resource settings without electricity and laboratory infrastructure. Provided that oral fluid samples are buffered or stored properly, antibody integrity in oral fluid samples can be maintained for several days, prior to processing. Third, blood is a biohazardous material increasing risk to personnel and adding complexity to shipping. In the case of HIV-positive biospecimens, oral fluids are generally thought to be safer to handle than blood due to lower virus levels (Major et al., 1991). Fourth, social and cultural taboos, religion, and hesitancy toward blood collection can bias participation. Oral fluid-based methods may improve coverage and reach populations not reached by blood collection due to hesitancy or lack of personnel or facilities (Manikkavasagan, Bukasa, Brown, Cohen, & Ramsay, 2010; Wang et al., 2014). By increasing coverage, oral fluid-based methods can reduce case detection biases and improve understanding of the epidemiology of transmission and risk (Manikkavasagan et al., 2010; Wang et al., 2014). Acceptability of oral fluid-based testing is generally high (Krause, Subklew-Sehume, Kenyon, & Colebunders, 2013), often due to a preference for less painful methods that do not require blood collection (Nangendo et al., 2017). Oral fluids also uniquely allow for the study of the secretory or mucosal immune responses. This is especially advantageous for enteric diseases, where salivary IgA has been shown to serve as a noninvasive proxy for intestinal immune induction (Aase et al., 2016). Limitations of oral fluid-based methods compared to blood-based methods are discussed in a later section.

In this section we focus on three aspects of oral fluid-based methods in population-based epidemiological research: (1) development of oral fluid-based methods to evaluate infectious diseases in population-based epidemiological research, including critical parameters to address during each stage of the development process; (2) application of oral fluid-based methods in population-based cross-sectional and longitudinal epidemiological research; and (3) use of oral fluids for epidemiological studies and surveillance of vaccine preventable diseases (VPD).

### 13.3.1 Development of Oral Fluid-Based Multiplex EIA Tools Intended for Use in Epidemiological Research and Surveillance

While it may not be necessary to seek FDA approval for oral fluid-based methods for epidemiological research, a rigorous assessment of critical assay parameters must be addressed prior to implementing and drawing inference from oral fluid-based infectious disease diagnostic tools in population-based epidemiological studies. The development of oral fluid-based immunoassays can be simplified into two stages, optimization and validation. The US Environmental Protection Agency's (EPA) development of a multiplex EIA to detect salivary IgA and IgG antibodies against potential waterborne pathogens (Augustine et al., 2015; Augustine et al., 2016; Griffin, Chen, Fout, Wafe, & Egorov, 2011), and its subsequent application in population-based studies (Augustine et al., 2017; Griffin et al., 2015; Wade et al., 2018), provides an illustrative example of the rigorous upstream optimization and validation required of an oral fluid-based multiplex EIA intended for use in epidemiological research and surveillance.

During the optimization phase, critical parameters to address include optimizing coupling concentrations, identifying and mitigating protein–reagent interferences, and evaluating cross-reactivity. For the US EPA multiplex EIA for waterborne pathogens, reagent conditions, including pH, protein concentrations, and antibody concentrations, were first optimized for the coupling and confirmation of each unique pathogen-specific antigen to unique Luminex™ xMAP microsphere bead-sets (Augustine et al., 2015; Augustine et al., 2016; Griffin et al., 2011). A Design of Experiments (DOE) approach, a statistical method that can systematically identify factors affecting the output of a process, enabled the group to determine which antigens could be utilized in the multiplex with little to no cross-reactivity, and which ones to exclude from the multiplex EIA pending further studies (Augustine et al., 2015).

During the validation phase, critical parameters to address include establishing the sensitivity and specificity of the oral fluid-based method compared to a gold-standard diagnostic method and defining cutoff values to determine seropositivity and seroconversion. The optimized waterborne pathogen multiplex EIA was validated using matched serum and saliva samples from a longitudinal cohort (Griffin et al., 2011) and plasma samples from a cross-sectional cohort (Augustine et al., 2015) of diagnostically characterized individuals. Defining and validating cutoffs for oral fluid assays to assess seropositivity can be complex, depending on the distribution of immunological response data (discussed briefly in the next section). The US EPA employed finite mixed modeling, a statistical method that can be used to predict exposure subgroup distributions based on antibody levels, to establish cutoff values to differentiate positive and negative individuals in a population for each pathogen included in the optimized multiplex EIA (Augustine et al., 2015; Augustine et al., 2016).

Irrespective of FDA approval, this rigorous process of development, including optimization and validation, provided the proof-of-concept for implementing this oral fluid-based multiplex EIA into population-based studies to investigate swimming-related exposure to waterborne pathogens (discussed in the following sections). This example provides a framework for the development, optimization, and validation of other oral fluid-based multiplex EIAs intended for use in population-based epidemiological research and surveillance.

### *13.3.2 Application of Oral Fluid-Based Methods in Cross-Sectional Surveys to Estimate Seroprevalence*

Most infections result in mild or asymptomatic disease presentation, meaning that surveillance based on clinical manifestation of symptomatic disease represents only the "tip of the iceberg" in terms of the true burden of exposure. Because antibody levels typically reflect historical exposure over a period of months to years, cross-sectional surveys of seroprevalence are rich with information about prior pathogen exposure (Arnold, Scobie, Priest, & Lammie, 2018). Furthermore, multiplex assays create an opportunity to efficiently expand resources beyond single-disease testing (Arnold et al., 2018). Integrating multiplexed serosurveillance using oral fluid samples into field-based cross-sectional surveys could enable an unprecedented evaluation of multi-pathogen seroprevalence and improve estimates on burden of disease and exposure.

The utility of rigorously optimized and validated oral fluid-based multiplex immunoassays in field-based cross-sectional surveys of high-risk populations can be illustrated through the US EPA's use of an oral fluid-based multiplex EIA for waterborne pathogens in an epidemiological study of visitors of Boqueron Beach, Puerto Rico (Augustine et al., 2017). The identification of waterborne pathogen seroprevalence is a necessary first step in linking acute gastrointestinal illness (AGI) to contaminated water, identifying waterborne pathogens associated with AGI, and assessing and managing contaminated water associated with AGI risks in humans (Exum et al., 2016). While blood-based methods can provide these measures, in this study oral fluid-based methods allowed for field-based sample collection in a low-resource setting. Oral fluid samples were collected at the beach site for all participants and probed for pathogen-specific IgG antibodies against *Campylobacter jejuni* (*C. jejuni*), *Helicobacter pylori* (*H. pylori*), *Toxoplasma gondii* (*T. gondii*), hepatitis A virus (HAV), and noroviruses (NoV) GI.1 and GII.4 at a single time point. The study revealed that more than two-thirds of beachgoers were previously infected by at least one of the included waterborne pathogens (Augustine et al., 2017). Over 60% had evidence of prior exposure to NoV, while over 20% had evidence of previous exposure to *H. pylori* and HAV (Augustine et al., 2017). Antibodies against *T. gondii* and *C. jejuni* were less common (Augustine et al., 2017). Interestingly, many of the beachgoers had evidence of detectable antibodies to two or more pathogen-specific antigens, demonstrating an advantage of simultaneously multiplexing for multiple pathogens (Augustine et al., 2017).

Immunoassay data from cross-sectional studies can be difficult to interpret due to the wide range of antibody reactivity, including among recently infected, historically infected, previously vaccinated, or unexposed individuals. This complicates the critical step of establishing a cutoff value for discriminating positive and negative individuals in a population-based cross-sectional survey (Augustine et al., 2015). In some cases, cross-sectional studies may allow for cutoff values to be defined for classification of seropositivity. A cutoff can easily be established in the case of a bimodal distribution of serological response data. Data that does not display a bimodal distribution may be more difficult to interpret. Longitudinal studies allow for cutoff values to be defined for classification of seroconversion (i.e., incident infection). Cutoff values for saliva-based serological surveys should be decided upon based on the research question or goal of the cross-sectional survey. For example, less stringent criteria may be applied for population-based screening of saliva samples requiring confirmatory clinical diagnostic testing. In the case of the US EPA study, where the immunological status of subjects was not known or the data were not normally distributed, more stringent criteria may be applied to reduce false positives (Augustine et al., 2017). A conservative cutoff, however, has the limitation of increasing false negatives. One strategy is to employ multiple cutoff values to examine the effects of conservative and liberal cutoff values on the interpretation of results from cross-sectional seroprevalence studies.

### *13.3.3 Application of Oral Fluid-Based Methods in Longitudinal Studies to Estimate Seroconversion and Incidence of Infection*

Integrated multi-pathogen serosurveillance in longitudinal study designs can facilitate the detection of seroconversion to inform estimates of incident infection in population-based settings. In a time series of two or more saliva samples, a change from an antibody-negative to an antibody-positive sample, or a fourfold increase in pathogen-specific antibody titer between samples, can be employed to measure incident, acute cases of infection or exposure in a defined population over a defined period of time (Exum et al., 2016). Built into longitudinal studies, oral fluid-based methods can inform the incidence of infections, identify risk factors for transmission, and serve as tools in monitoring and evaluating population-based prevention strategies.

The US EPA's application of an optimized multiplex EIA for the detection of salivary IgA and IgG against two NoV genotypes (GI.1 and GII.4) in an NoV challenge trial, and subsequently in a longitudinal population-based study provides a example of the value of oral fluid-based serological methods in longitudinal epidemiological studies of infectious disease (Griffin et al., 2015; Wade et al.,

2018). Because NoV-specific antibodies typically increase after NoV infection (Graham et al., 1994), seroconversion can be used as a serological biomarker of recent infection. To first determine whether NoV-specific IgA and IgG seroconversion could be detected in oral fluids, the optimized NoV EIA was applied to pre- and post-challenge oral fluid samples collected from participants of an NoV (GI.1) challenge study (Griffin et al., 2015). Using a fourfold increase in salivary antibody response as a threshold for seroconversion, NoV-specific IgG in saliva correctly identified all infected and noninfected individuals in the challenge study (Griffin et al., 2015). NoV-specific IgG tests had higher sensitivity than IgA, which displayed weaker fold increases associated with NoV challenge. While participants in a challenge study may not be representative of the general population, these results suggested that IgG antibody responses in oral fluids can be used to detect seroconversion to NoV infection in longitudinal population-based studies.

Subsequently, the optimized multiplex EIA was employed in a field-based longitudinal cohort of asymptomatic beachgoers in Puerto Rico to evaluate the incidence of NoV infection, the risk of NoV transmission, and swimming-related risk factors of NoV infection associated with contaminated recreational waterbodies (Wade et al., 2018). Measuring NoV GI.1- and GII.4-specific IgG levels in a time series of three oral fluid samples the US EPA study determined that 2.6% of asymptomatic participants seroconverted to the two NoV genotypes within 3 weeks of the initial beach visit (Wade et al., 2018). Risk factor analysis revealed that seroconversion rates were approximately five times higher among beachgoers who immersed their heads in beach water compared to non-swimmers and swimmers who did not immerse their heads in the water (Wade et al., 2018). Interestingly, nearly all seroconversion events were unaccompanied by AGI symptoms, supporting an important role for oral fluid-based serological methods in identifying asymptomatic source populations, where clinical symptom presentation would not appropriately capture the incidence of infection (Wade et al., 2018). While a similar inference could have potentially been drawn from blood-based methods, longitudinal blood collection may not have been feasible in this field-based study. By providing beachgoers oral fluid collection kits following the initial beach visit, oral fluid-based self-collection and mail-in testing enabled the collection of longitudinal samples in this field-based study, allowing for the evaluation of incident NoV infection and waterborne transmission of NoV associated with a recreational waterbody.

A separate multiplex EIA developed by investigators at Johns Hopkins Bloomberg School of Public Health was aimed at measuring IgG responses against virus like particles (VLP) of five common NoV genotypes (GI.1, GII.2, GII.4, GII.6, and GII.17) in oral fluid collected from children in an endemic setting (Pisanic et al., 2018). Using a longitudinal cohort of children, under the age of 5 years living in Peru, the study revealed that NoV infections elicit a genotype-specific IgG response that can be measured in oral fluids and correctly determined recent, PCR confirmed, NoV infections (Pisanic et al., 2018). Compared to PCR-diagnosed NoV infection, the oral fluid-based method performed with a sensitivity of 71% and specificity of 96% across the evaluated genotypes (Pisanic et al., 2018). An important observation in this study was that the median NoV-specific IgG signal in oral fluid was similar among children with or without an NoV positive stool sample, suggesting that most children in this population had been exposed to NoV in the past (Pisanic et al., 2018). Thus, in highly endemic settings it may not be feasible to apply anti-NoV IgG signal cutoff values to discriminate seropositive from seronegative individuals based on a single time point. Although it may be possible to develop a salivary NoV assay to measure the acute phase IgA response at a single time point, without the need for repeat sampling, additional studies are needed.

### 13.3.4 Oral Fluid-Based Serological Methods Facilitate Self-Collection in Population-Based Postal Surveys for Infectious Disease

Understanding the proportion of a population that is immune or has experienced prior infection with certain pathogens has important epidemiologic applications, including identifying susceptible subpopulations, and monitoring and evaluating preventative intervention strategies (e.g., vaccination programs) (Morris-Cunnington et al., 2004). Blood-based serological surveillance is an accurate method for monitoring population immunity but presents limitations for widespread use in population-based settings as previously described. Oral fluid-based serosurveillance provides tremendous value in population-based surveillance of infectious disease at the national, state, and community level by enabling self-collection (Egorov et al., 2010; Morris-Cunnington et al., 2004; Quoilin et al., 2007), reducing costs (Morris-Cunnington et al., 2004), and potentially increasing response rates compared to blood-based methods.

Pathogen-specific oral fluid-based self-collection and serological testing have been successfully employed in population-based postal surveys of infectious disease screening at the national level, such as HAV in England and Wales (Morris-Cunnington et al., 2004) and HAV, hepatitis B virus (HBV) and hepatitis C virus (HCV) in Belgium (Quoilin et al., 2007), and at the community level, such as *Cryptosporidium hominis* (*C. hominis*) in the USA (Egorov et al., 2010). The HAV study in England and Wales demonstrated three major findings regarding the implementation of oral fluid-based national-level HAV surveillance: (1) oral fluid-based self-collection and serology was logistically feasible; (2) HAV-specific IgG provided accurate estimates of population-level exposure compared to previous population-based studies using blood-based methods; and (3) coupled with extensive demographic and social data, oral fluid-based methods were able to uncover unexpected epidemiological associations with age, ethnicity, and socioeconomic status (Morris-Cunnington et al., 2004). The authors also noted a cost advantage for oral fluid-based methods compared to blood-based methods. Because specially trained personnel were not employed to collect serum sample, no costs were incurred for salaries for trained personnel or for the transport of researchers or participants. In terms of equipment required for specimen collection, the authors estimated $1.14 per self-collected oral fluid sample compared to $4.00 per blood sample (Morris-Cunnington et al., 2004). Given the success of this program, the Immunisation, Hepatitis and Blood Safety Department of Public Health United Kingdom created and implemented a nationwide hepatitis A outbreak investigation program ("Hepatitis A: Oral Fluid Testing for Household Contacts," 2015). Population-based postal surveillance for HAV, HBV, and HCV in Belgium has also shown value in estimating national-level disease burden and is addressed in the next section where VPD disease surveillance to estimate disease burden is discussed. At the community-level, a study in the USA demonstrated the utility of oral fluid-based self-collection in a longitudinal postal survey of community-acquired *C. hominis* cases (Egorov et al., 2010), but additional studies are needed before incorporating oral fluid-based testing for diarrheal disease into population-based screening programs. Taken together, oral fluid-based self-collection methods have the potential to be incorporated into national- and community- level public health programs as an accurate, inexpensive, and logistically feasible alternative to blood-based serological methods.

### *13.3.5 Uses of Oral Fluid for VPD Epidemiologic Studies and Surveillance*

The acceptability and ease of oral fluid self-collection has facilitated a unique role in the surveillance of vaccine preventable diseases (VPDs). This section reviews uses of oral fluid testing for VPDs, which are defined as all infectious diseases with a licensed vaccine.

Although there are no FDA approved or CLIA-waived oral fluid-based diagnostic tests for VPDs, many studies have reported the development, adaptation, or optimization of commercial or noncommercial assays for use with oral fluid. Diagnostic accuracy studies of oral fluid-based antibody detection have been published on the following VPDs with variable sensitivity and specificity: cholera, dengue, Ebola, hepatitis A, hepatitis B, hepatitis E, Haemophilus influenzae type B (Hib) disease, human papillomavirus (HPV), influenza, measles, meningococcal disease, mumps, pertussis, pneumococcal disease, polio, rubella, tetanus, typhoid, and varicella (Hayford et al., under review; Holroyd et al., under review; Lambe et al., 2016; Pisanic et al., 2017). In addition, POC lateral flow devices for measles and mumps antibody detection using oral fluid are in development (Warrener et al., 2011; Warrener, Slibinskas, Brown, Sasnauskas, & Samuel, 2010).

Despite generally lower diagnostic accuracy for VPDs, oral fluid-based testing is increasingly recognized as an alternative or complementary tool to assess disease burden, exposure to and protection against many VPDs at the population level in order to inform disease control strategies and guide immunization programs. VPD serological surveillance is used primarily in three ways: (1) disease surveillance and enhanced case detection, (2) assessing population immunity and identifying immunity gaps, and (3) assessing the impact of vaccination campaigns and routine immunization.

#### 13.3.5.1 VPD Disease Surveillance to Estimate Disease Burden and Improve Case Detection

For VPDs, oral fluid testing has been used primarily to: (1) estimate disease burden at the population level or (2) confirm infection among suspected cases. Specific IgM detection is most commonly used for acute disease surveillance but detection of specific IgG, pathogen antigens, or multiple markers may be required for case detection for some VPDs such as hepatitis B or pertussis.

Estimating the burden of VPDs is a core function of public health systems and important for informing disease control strategies and preventing outbreaks. Measuring the prevalence of VPDs can be complicated due to nonspecific clinical case definitions, "tip of the iceberg" case detection, and challenges of conducting high-quality and unbiased representative surveys with biomarker collection (MacNeil, Lee, & Dietz, 2014). Large-scale oral fluid surveys have been used to estimate past exposure and burden of hepatitis A in England and Wales (described above) (Morris-Cunnington et al., 2004) and hepatitis A, B, and C in Flanders, Belgium (Quoilin et al., 2007). Results from the Belgium cross-sectional oral fluid IgG survey in 2003 were compared to a representative serosurvey 10 years earlier. Substantial declines in IgG seroprevalence suggested the burden of hepatitis A and hepatitis C was substantially lower than estimated in 1993. The prevalence of hepatitis B surface antigen (HBsAg), an indicator of active infection, remained stable and below 1% over the 10 year period, indicating that incident infections had not increased. Results affirmed the impact of routine hepatitis B vaccination, which was introduced in 1999 for infants and 12-year-old children, on disease burden. Like other studies discussed above, the study in Belgium showed that population-based surveys with self-collection of oral fluid are acceptable and feasible although achieving high response rates remains a challenge. Other studies have estimated the burden of hepatitis B and C and infections using oral fluid in settings with high disease transmission risk, such as prisons (Weild et al., 2000).

Oral fluid-based testing has also been used to improve measles, mumps, rubella, and pertussis case detection during outbreak investigations and characterize outbreak epidemiology in order to better guide disease control and immunization strategies (Campbell et al., 2014; Manikkavasagan et al., 2010; Ramsay et al., 2003; Reid et al., 2008; Thieme, Piacentini, Davidson, & Steingart, 1994). The public health agency in England and Wales, Public Health England (PHE, formerly HPA) has tested self-collected oral fluid samples from suspected measles, mumps, and rubella cases since 1994 and for pertussis since 2013 as an additional surveillance tool to identify laboratory-confirmed cases (Campbell et al., 2014; Manikkavasagan et al., 2010; Ramsay et al., 2003; Thieme et al., 1994). When a suspected case is identified by a health practitioner, PHE ships an oral fluid collection kit to the primary care physician or patient. Oral fluid samples are self- or physician-collected with an Oracol swab and returned in the mail at ambient temperature to PHE for testing. The addition of oral fluid-based testing began, in part, to improve case detection as the country moved closer toward elimination of these diseases. The clinical case definition for measles, rubella, and pertussis had poor sensitivity and specificity, which was worsened as disease burden continued to decline. In addition, PHE suspected laboratory confirmation with serum did not identify all cases due to the invasiveness and challenges of blood collection. For all four diseases, oral fluid testing both substantially improved case detection and revealed new information about the epidemiology of outbreaks. For rubella, the incidence of rubella cases increased from 0.5 to 0.77 per 1,000,000 population when oral fluid testing was added. The increase was attributed to improved case ascertainment, particularly among children and adult men, and helped the UK assess progress toward rubella elimination goals. Oral fluid-based surveillance also revealed that the clinical case definition overestimated rubella incidence, especially in children (Manikkavasagan et al., 2010). For measles surveillance, IgM-based oral fluid testing improved the diagnostic accuracy of clinical case detection and significantly increased case detection rates. 56% of confirmed measles cases were identified by an oral fluid sample alone, suggesting blood-based surveillance failed to detect all cases and therefore underestimated the size of outbreaks. For pertussis, case detection with oral fluid was rolled out in England and Wales in 2013 and expanded in 2018 following the same model as measles and rubella outbreak surveillance. Pertussis antitoxin IgG surveillance in oral fluid also improved case detection, particularly among nonhospitalized and milder cases. The age distribution of cases detected in oral fluid revealed a high burden in young infants and that transmission occurred primarily within the household from the mother to infant (Kara et al., 2017; Wang et al., 2014). These findings informed the decision by PHE to vaccinate pregnant women during the 2012 pertussis outbreak.

Another advantage of serological sampling during outbreaks is that DNA or RNA can be detected to improve case detection and genotyped from positive samples to characterize the pathogen strain. Like serum or dried blood spots, genotyping is routinely conducted on oral fluid swabs to identify measles, mumps, and rubella (Abernathy et al., 2009; Chibo, Riddell, Catton, & Birch, 2005; Jin, Brown, Litton, & White, 2004; Jin, Vyse, & Brown, 2002; Ramsay et al., 2003; Reid et al., 2008). Adding nucleic acid testing to serology can improve case detection, particularly if samples are collected quickly after onset of infection prior to a detectable increase of specific IgM in oral fluid. A study of suspected rubella cases in Peru found that IgM serology and viral RNA detection from oral fluid samples detected more cases than blood-based serology alone (Abernathy et al., 2009). After the mass measles vaccination campaign in the UK in 1994, IgM surveillance and viral genotyping with oral fluid and serum samples revealed that the campaign had successfully interrupted endemic measles transmission and new outbreaks were caused primarily by imported cases (Ramsay, Brugha, & Brown, 1997; Ramsay et al., 2003).

#### 13.3.5.2 Estimating Population Immunity and Identifying Immunity Gaps to VPDs

Estimating the proportion of a population exposed to or immune to VPDs enables public health agencies to target disease control and immunization strategies. Because measles, rubella, varicella and, to a lesser degree, mumps viruses are antigenically stable and confer lifelong immunity from natural infection or vaccination, they are good pathogens to assess previous exposure and population immunity across all age groups. For other VPDs like tetanus or hepatitis B where antibody levels wane below detectable levels over time, antibody surveillance can be used among limited age groups or shortly after exposure or vaccination. Many oral fluid-based IgG antibody serosurveys have been conducted to estimate population immunity at national and subnational levels throughout Europe, North America, Africa, and South Asia for tetanus (Tapia et al., 2006), rubella (Ben Salah et al., 2003; Nokes et al., 2001; Thieme et al., 1994), measles (Goyal, Shaikh, Kinikar, & Wairakgar, 2009; Hayford et al., 2013; Kremer & Muller, 2005; Nigatu et al., 2008; Ohuma et al., 2009; Sheikh et al., 2011; Thieme et al., 1994; Vainio et al., 2006), and hepatitis B (Nokes et al., 2001; O'Connell et al., 2000; Simani et al., 2008). A nationally representative postal survey in Ireland revealed exposure to hepatitis B virus was very low based on 0.51% prevalence of antibodies to hepatitis B core antigen in oral fluid (O'Connell et al., 2000). In rural Ethiopia, measles and rubella IgG testing in oral fluid had high sensitivity and specificity and demonstrated that antibody surveillance with oral fluid is feasible and accurate when protocols for collection and testing are rigorous (Nokes et al., 2001). For other VPDs like mumps and HPV, oral fluid testing was not adequately sensitive to accurately estimate population immunity (Buchinsky et al., 2006; Passmore et al., 2007; Rowhani-Rahbar et al., 2009; Thieme et al., 1994; Vainio et al., 2006).

Although the majority of VPD seroprevalence studies are community-based surveys, nonrepresentative study designs have been used to assess population immunity and identify immunity gaps in high-risk settings such as prisons and refugee settlements (Gill, Aston, Vyse, White, & Greenwood, 2002). Oral fluid testing among refugees and asylum seekers in Luxembourg revealed low levels of immunity to measles, mumps, and rubella and guided subsequent vaccination policy (Hübschen, Charpentier, Weicherding, & Muller, 2018). Identifying acceptable and feasible tools like oral fluid sampling to assess population immunity to outbreak prone VPDs is a priority in refugee camps due to high density, elevated disease risk, and the lack of medical records and vaccination history for many displaced people.

#### 13.3.5.3 Assessing Impact of Vaccination Campaigns and Routine Immunization

Vaccination campaigns are conducted to significantly increase population immunity and interrupt disease transmission but they can be expensive and logistically challenging to roll out. Oral fluid testing enables governments and funders to assess the impact of the campaigns without drawing blood. For example, pre- and post-campaign cross-sectional oral fluid serosurveys were conducted in Ethiopia and Kenya to assess the impact of measles vaccination campaign on population immunity (Nigatu et al., 2008; Ohuma et al., 2009). In Ethiopia, overall seroprevalence increased by 39% but results revealed immunity gaps among children above the target age of the campaign. Age-specific seroprevalence results in the pre-campaign period could have been used to justify increasing the campaign age range to maximize impact. If it is not feasible to collect samples before the campaign, oral fluid studies from the post-campaign period only can provide information on population immunity and remaining immunity gaps. A cross-sectional rubella serosurvey with oral fluid found that 95% of children were seropositive 4 months after a national vaccination campaign (de Azevedo Neto et al., 1995). In contrast, a cross-sectional oral fluid serosurvey of HIV and measles IgG antibodies in Lusaka, Zambia, found that measles population immunity was insufficient to interrupt measles transmission 3 years after a measles vaccination campaign (Lowther et al., 2009). Lastly, to evaluate the impact of a 20-year policy to selectively vaccinate injection drug users (IDUs) in England, an oral fluid survey found vaccination was associated with lower rates of hepatitis B and a continuation of the policy was recommended (Judd et al., 2003). Longitudinal studies with oral fluid collection have demonstrated high seroconversion rates after measles vaccination in Mozambique, Norway, and the USA (Jani et al., 2008; Thieme et al., 1994; Vainio et al., 2006). However, most VPD immunogenicity studies are still conducted with blood.

Although many studies have attempted to validate vaccination history with oral fluid serology, this body of research is often conflicting and can be difficult to interpret, in large part due to expected discrepancies in vaccination status and serostatus caused by primary and secondary vaccine failure, circulation of natural infection, as well as lower diagnostic accuracy of oral fluid assays (Hayford et al., 2013; Jenkins et al., 2009; MacNeil et al., 2014).

## 13.4 Limitations of Oral Fluid-Based Serological Methods Compared to Blood-Based Methods

Oral fluid-based serological methods have several limitations compared to blood-based methods. First, the diagnostic accuracy of oral fluid-based serology is typically lower than that of blood-based methods but depends on the antigen, assay type, and quality of the collection and testing procedures. Because antibody levels in oral fluid are typically lower than levels observed in sera or plasma, the sensitivity of oral fluid-based testing may be impaired. For example, in studies comparing rubella or measles IgG in serum and oral fluid, many participants with low antibody levels in serum were systematically misclassified as negative by oral fluid (Hayford, Al-Emran, et al., 2013; Helfand et al., 1999; Nokes et al., 2001; Vainio et al.,

2006). In addition, variability in antibody levels in oral fluid related to, for example, time of collection, sample contamination, or hydration and health of participant can increase non-systematic error in oral fluid results. The implications of systematic and non-systematic errors are most significant when a quantitative result is needed or when levels are near the cutoff resulting in qualitative misclassifications. Thus, populations with bimodal antibody distributions of clear positives and negatives are best suited for use with oral fluid testing. Although some reductions in sensitivity can be mitigated by adjusting cutoff values, the tradeoff with specificity must be weighed against the scientific goal. Second, there is currently a lack of standardization in oral fluid collection, storage, and processing methods used across studies and settings. One review of oral fluid-based serological testing for VPDs reported that a majority of studies did not provide details on the collection, transport, storage, or volume of oral fluid samples (Holroyd et al., under review), complicating the ability to compare between studies. Each of these technical considerations is critical for maintaining sample integrity and ensuring high-quality results. Although oral fluid antibodies are not rapidly degraded by enzymes, cold-chain during transport and buffering can reduce degradation during long-term storage. Third, oral fluid antibody levels are highly variable. Along with differences in collection and processing, factors contributing to the variability in IgA levels in oral fluids include age, geography, time of day, and secretion flow rate (Brandtzaeg, 2013). Various stressors and diseases have also been reported to influence oral fluid antibody levels in different ways (Brandtzaeg, 2013). Fourth, similar to blood collection, oral fluid collection is difficult in young children, often resulting in low volumes of diluted saliva. Oral fluid collection methods for young children need to be optimized for size, taste, and texture, while maintaining the ability to recover adequate volumes of oral fluid and the intended analytes. Safety is an especially important issue in young children, due to the risk of choking.

## 13.5 Future Directions

As we look forward in the field, oral fluid-based methods have the power to not only support individual patient diagnostic testing but also broader applications in population-based epidemiological research. Future directions for oral fluid-based serology include expanding the pathogen panel to comprise emerging infectious diseases of critical priority, developing rapid testing formats such as POC tests, and applying both established and new oral fluid-based EIAs in different clinical and population-based settings. In this section we outline future directions in oral fluid-based serology of infectious diseases that could be of high impact including identifying pathogens for which oral fluid-based detection methods are urgently needed, and new settings where we see the fruitful application of oral fluid-based methods.

### *13.5.1 Oral Fluid-Based Point-of-Care Diagnostic Tools for Infectious Diseases*

It is now well established that oral fluid-based infectious disease diagnostic POC tools add tremendous value in clinical settings, epidemiological research, and national health programs. This has been clearly demonstrated in the case of HIV, described previously. Oral fluid-based POC tools for measles diagnosis are currently being evaluated (Warrener et al., 2010), and in the context of VPDs will provide unprecedented ability to epidemiologically characterize population immunity, identify susceptible populations, and inform vaccination strategies. As more oral fluid-based infectious disease diagnostic tools are advanced into POC development, several critical challenges must be addressed to successfully support this technological transition while maintaining diagnostic accuracy. Three major challenges are: (1) oral fluid matrix effects, (2) interindividual variability in oral fluid composition, and (3) lateral flow platform optimization. Each of these challenges will require methodological advances to develop, optimize, and apply POC infectious disease diagnostic tools in clinical and population-based epidemiological settings.

Oral fluid is a complex matrix, composed of a variety of macromolecules, and exogenous substances such as food particles, many of which can interfere with antibody reactivity in oral fluid-based immunoassays. In laboratory settings, oral fluid samples can be centrifuged to clarify the sample, and diluted to minimize inhibitors of antibody reactivity, prior to application in an ELISA or EIA format. Such laboratory techniques may not be afforded in the case of oral fluid-based POC infectious disease diagnostic tools, intended for rapid identification of infectious diseases in field-based settings. Field-based methods to process oral fluid are urgently needed to support oral fluid-based POC infectious disease diagnostic development and application. One solution may include the addition of portable centrifuge devices to the workflow but this would introduce additional costs and time in the POC protocol. Another solution may involve the use of nanoparticles that can target and concentrate the intended analytes in the oral fluid sample, prior to processing.

While methodological advances are necessary to optimize diagnostic accuracy, oral fluid-based POC infectious disease diagnostic tools are urgently needed in several VPD-related applications. In this section, we identify two examples where oral fluid-based POC tools are needed to inform vaccination decision-making and VPD outbreak investigation.

#### 13.5.1.1 The Role of Oral Fluid-Based Companion Diagnostic Tools to Inform Vaccination Decision-Making

An important takeaway from Sanofi-Pasteur's dengue virus (DENV) vaccine clinical trials was that seronegative vaccine recipients had a higher risk of severe dengue and hospitalization than vaccine recipients who had been previously exposed to DENV

(Aguiar, Stollenwerk, & Halstead, 2016). Secondary infections with a heterologous DENV serotype can lead to hemorrhagic fever or dengue shock syndrome (severe dengue) (Durbin, 2016; Halstead, 2015). As a result, understanding DENV serostatus of any individual prior to receiving a DENV vaccination is critical in endemic regions and for travelers returning from endemic regions. Oral fluid-based diagnostic tests that can reveal an individual's DENV serostatus prior to vaccination could inform vaccination decision-making and avert vaccine-mediated adverse events. For the case of DENV vaccination, rapid POC tools will be needed to support real-time, individual-level, decision-making in both field-based and clinical vaccination settings.

Multiple longitudinal studies, conducted in different patient populations and settings, provide strong support for the utility of oral fluids in discriminating primary and secondary DENV infection. While three separate longitudinal studies demonstrated that oral fluid-based methods for the detection of DENV-specific IgM and IgG were less sensitive than blood-based methods, all of the studies determined that DENV-specific IgG in oral fluids could potentially serve as a biomarker to discriminate primary and secondary infection (Andries et al., 2015; Balmaseda et al., 2008; Vázquez et al., 2007). Considering the fatal outcomes of heterologous DENV infection and the recent licensure of Dengvaxia®, rigorous, FDA approved, oral fluid-based serological methods to determine DENV serostatus prior to vaccine administration are urgently needed. Moreover, oral fluid-based DENV diagnostic testing will need to be developed into rapid POC tools to be effective in field-based applications.

Similarly, oral fluid-based epidemiological tools will be critical in guiding vaccination strategy in endemic regions by informing population-level seroprevalence of DENV exposure and infection. Because of the special considerations around secondary DENV infection, the World Health Organization (WHO) has made the recommendation that the vaccine only be used in endemic areas defined by high seroprevalence (World Health Organization (WHO), 2016). Seroprevalence thresholds are considered the best approach to identify target populations for vaccination (World Health Organization (WHO), 2017). While Dengvaxia® has not been implemented in any countrywide programs to date, its recent licensure demands noninvasive methods for population-based screening to target populations for vaccination.

#### 13.5.1.2 The Role of Oral Fluid-Based POC Tools for VPD Outbreak Investigation

As countries move toward elimination of polio, measles, rubella, and hepatitis B, public health systems need rapid and nimble strategies to confirm suspected cases and identify susceptible populations, both of which can be done with oral fluid-based testing (Ramsay, Brugha, Brown, Cohen, & Miller, 1998). Serological confirmation is needed because clinical case detection becomes less sensitive and specific as incidence declines but the challenges of blood collection and testing can undermine the intended goals of a comprehensive and nimble case detection system.

The WHO Global Measles and Rubella Laboratory network representing thousands of laboratories in 191 countries has played a crucial role in confirming suspected measles and rubella cases and characterizing viral genotypes and epidemiologic patterns of outbreaks. However, laboratory confirmation with blood samples consistently underestimates the number and size of measles and rubella outbreaks, especially in hard-to-reach, remote, and under-resourced settings (World Health Organization (WHO), 2019). An oral fluid-based POC test for measles and rubella IgM would overcome many of these challenges and potentially improve case detection rate and turnaround time for case confirmation. A POC lateral flow device for measles IgM antibodies had 90% sensitivity and 96% specificity in preliminary studies and 75% and 96% in a field study in Zimbabwe, respectively. Further refinements and field testing of the POC test are needed (Shonhai et al., 2015; Warrener et al., 2011).

Improved assays for VPD-specific IgG detection in oral fluid are also needed to enhance epidemiological surveillance. If diagnostic accuracy can be improved and the impact of variability in specimen quality can be minimized, an oral fluid test on a lab-based or POC device would be an appealing and accessible tool for estimating population immunity to VPDs. Improvements in diagnostic accuracy observed with bead-based or microfluidic assays for other infectious diseases need to be explored for measles, rubella, pertussis, and other VPDs.

### *13.5.2 Improving Estimates of Asymptomatic Exposure and Infection in Endemic Regions*

Understanding asymptomatic exposure and infection is critical for numerous infectious diseases. For example, malaria parasitemia is commonly missed or underreported in passive symptom-based case detection surveillance. It is well established that undetected, clinically immune, individuals serve as reservoirs of ongoing malaria transmission in hypoendemic regions (Bousema, Okell, Felger, & Drakeley, 2014). Because malaria-specific antibodies can persist for months to years after infection, they may have utility as proxy measures of malaria transmission in low-transmission settings (Drakeley et al., 2005). Previous studies have shown that age-specific anti-malaria antibody seroconversion rates are an effective tool to assess endemicity and public health burden of malaria (Corran, Coleman, Riley, & Drakeley, 2007; Kobayashi et al., 2019; Williams et al., 2009). Blood samples are typically used for serological estimation but present challenges in communities with blood-related taboos. Thus, the strategic eradication of malaria in low-transmission regions demands the development and widespread application of noninvasive, low-resource friendly, screening tools for asymptomatic malaria reservoirs.

Two studies have provided a proof-of-concept for the utility of oral fluid in detecting *Plasmodium falciparum* (*P. falciparum*)-specific antibodies in epidemiological studies in Gambia, Tanzania, and Zambia. (Chidi et al., 2011; Estévez et al., 2011). The Gambia and Tanzania study demonstrated that antimalarial IgG antibodies against *P. falciparium* antigens merozoite surface protein-1 (MSP-$1_{19}$) and apical membrane antigen (AMA-1) could be detected in oral fluid and correlated strongly with levels in plasma (Estévez et al., 2011). A separate cross-sectional study in Tanzania observed a statistically significant correlation between IgG antibody levels to whole, asexual stage *P. falciparium* antigens in oral fluids and dried blood spots (DBS) (Chidi et al., 2011). In this study, the oral fluid-based assay performed with sensitivity and specificity of 100%, correctly identifying all DBS confirmed positive and negative samples (Chidi et al., 2011).

Two limitations were apparent from these studies. First, IgG antibody titers to *P. falciparum* antigens were significantly lower in oral fluid compared to blood samples. Second, both studies noted a difficulty in assessing whether an adequate sample was collected and a need for standardization in quantity and quality during collection. One potential strategy to evaluate improper collection of oral fluid samples is to test for total IgG concentration. Nevertheless, the results of this study open a pathway for more extensive studies to assess the utility of oral fluid in the detection of antimalarial antibodies, which could be particularly useful for population-based surveys in malaria hypoendemic regions. Further research could also help determine if antimalarial antibodies in oral fluid could be extended to study exposure to other malaria species, such as *Plasmodium vivax*.

### *13.5.3 Distinguishing Between Secretory IgA and IgA in Population-Based Studies of Enteric Disease*

A hallmark of the mucosal immune system is epithelial export of locally produced IgA polymers (mainly dimers) that are transported to the lumen as secretory IgA (SIgA) antibodies (Brandtzaeg, 2009). Pathogen-specific SIgA provides critical immune functions against enteric pathogens at mucosal sites (Brandtzaeg, 2009). Salivary IgA has demonstrated potential as a noninvasive proxy for intestinal immune induction by enteric pathogens (Brandtzaeg, 2007). While pathogen-specific salivary IgA biomarkers have been developed and validated in patient populations for multiple enteric pathogens, including *Campylobacter* spp., *Giardia* spp., *Escherichia coli* O157 (*E. coli* O157), *Salmonella* serotype *Typhii*, *Shigella* spp., *Vibrio cholerae*, Norovirus, and Rotavirus, only a few studies have implemented saliva-based IgA detection into population-based epidemiological studies of enteric disease.

Cross-sectional studies of children with microscopically confirmed *Giardia duodenalis* (*G. duodenalis*) infections have shown that *G. duodenalis*-specific IgA is elevated in the saliva of parasitized children compared to their non-parasitized counterparts (El-Gebaly, Halawa, Moussa, Rabia, & Abu-Zekry, 2012; Rodríguez et al., 2004). While the sensitivity of the *G. duodenalis*-specific IgA ELISAs were not ideal compared to stool-based microscopic diagnosis, longitudinal studies, incorporating sequential saliva sampling, have shown improved sensitivity in the detection of infection with other enteric pathogens. In a longitudinal study of children with culture confirmed *E. coli* O157 associated hemolytic uremic syndrome (infectious HUS), a saliva-based *E. coli* O157 LPS-specific ELISA performed with a sensitivity of 92% for IgA and 100% for IgM compared to blood-based serology when both initial and follow-up saliva specimens were included in the analysis (Ludwig et al., 2002)—far greater than that observed in cross-sectional studies of children with infectious HUS (Chart & Jenkins, 1998; Chart, Perry, Willshaw, & Cheasty, 2003). Longitudinal studies of children with clinically confirmed typhoid have evaluated *Salmonella* serotype *Typhii* LPS-specific salivary IgA for early diagnosis typhoid infection and determined that diagnostic accuracy is greatest during the second and third weeks of illness (Herath, 2003; Zaka-ur-Rab, Abqari, Shahab, Islam, & Shukla, 2012), reaching a sensitivity of 100% during the second and third week of illness and dropping to 0% by the fifth week (Zaka-ur-Rab et al., 2012) Although limited by a small sample size, longitudinal studies of Shigellosis and Campylobacteriosis have also demonstrated that *Shigella*- and *C. jejuni*-specific IgA are elevated in saliva of clinically confirmed patients compared to healthy controls (Cawthraw, Feldman, Sayers, & Newell, 2002; Schultsz, Qadri, Hossain, Ahmed, & Ciznar, 1992), providing a proof-of-concept for the further development and application of these pathogen-specific saliva-based tools in epidemiological studies. Due to the similar clinical presentation of enteric infections, multiplexing for enteric pathogens would provide additional value in identifying etiologic agents and risk factors associated with enteric disease in population-based studies.

Although studies show promise in the utility of SIgA in epidemiological studies of enteric disease, saliva-based SIgA detection has two important limitations. First, it is challenging to distinguish between polymeric SIgA containing a secretory component (SC) produced locally in the mucosa vs. monomeric IgA devoid of an SC circulating in whole blood. Only one of the studies of enteric disease mentioned above specified a SIgA signal in their ELISA methodology, by using a detect antibody against the IgA SC (Rodríguez et al., 2004). The remaining studies did not discriminate between mucosally produced SIgA and IgA that is devoid of SC, thus conflating the SIgA signal with IgA. The ability to distinguish between SIgA and systemic IgA signal would improve saliva-based infectious disease detection methods, particularly for enteric pathogens whose pathogenicity involves the mucosal immune system.

A second challenge is that the transient and acute nature of the IgA (including SIgA) immune response limits its diagnostic accuracy. In the *C. jejuni* study mentioned above, for example, IgA and IgG levels in saliva declined over time, but IgG levels remained significantly higher in convalescent phases (Cawthraw et al., 2002). This was also the case in a NoV challenge study where NoV-specific salivary IgA peaked at 14 days post-challenge, whereas NoV-specific salivary IgG continued to rise during 21 days post-challenge (Moe, Sair, Lindesmith, Estes, & Jaykus, 2004). Combining NoV-specific IgA and IgG assay results improved the detection of seroconversion to 100% of infected participants, compared to only 83%

when either IgA or IgG was used alone (Moe et al., 2004). While IgA may be advantageous for detecting the acute phase of enteric disease, the larger window of IgG kinetics may allow for increased sensitivity in the detection of historical exposure to enteric pathogens. Salivary IgG has also been used in a *Cryptosporidium parvum* (*C. parvum*) foodborne-outbreak investigation, where *C. parvum*-specific salivary IgG was able to confirm *C. parvum* exposure in all cryptosporidiosis cases (Moss et al., 2019). Additional longitudinal studies of salivary IgA (including SIgA) and IgG responses among patients with clinically confirmed enteric disease, and analytical techniques that combine different salivary antibody isotype results, are needed.

## 13.6 Conclusions

Throughout this chapter, we have discussed the utility of pathogen-specific antibodies in oral fluid to evaluate pathogen exposure and infection. Oral fluid has utility in monitoring recent and/or historical infection in both clinical and epidemiological settings. In clinical settings, oral fluid-based diagnostic testing can inform clinical decision making around infectious disease treatment, management, and vaccination. In epidemiological surveys, oral fluid-based serological methods can be used to assess seroprevalence and seroconversion among high-risk populations and gain epidemiological inference on risk factors associated with pathogen exposure, infection, and transmission. Due to the ease of oral fluid collection, saliva-based methods facilitate self-collection and testing in national- and community-level public health infectious disease surveillance programs. Furthermore, oral fluid-based methods for infectious disease detection have a unique role in the context of VPDs, where they can be used to assess disease burden and improve case detection, measure population immunity and identify immunity gaps, and assess the impact of vaccination campaigns and routine immunization. While oral fluid has shown promise as a non-invasive proxy for blood for the detection of pathogen exposure and infectious disease, oral fluid-based methods pose several limitations compared to blood-based serological methods, including reduced sensitivity. Additional research is needed to both improve the diagnostic accuracy of oral fluid-based serological methods and develop oral fluid-based POC tools for infectious disease screening and diagnosis.

## References

Aase, A., Sommerfelt, H., Petersen, L. B., Bolstad, M., Cox, R. J., Langeland, N., . . . Brandtzaeg, P. (2016). Salivary IgA from the sublingual compartment as a novel noninvasive proxy for intestinal immune induction. *Mucosal Immunology, 9*(4), 884–893. https://doi.org/10.1038/mi.2015.107

Abernathy, E., Cabezas, C., Sun, H., Zheng, Q., Chen, M. H., Castillo-Solorzano, C., . . . Icenogle, J. (2009). Confirmation of rubella within 4 days of rash onset: Comparison of rubella virus RNA detection in oral fluid with immunoglobulin M detection in serum or oral fluid. *Journal of Clinical Microbiology, 47*(1), 182–188. https://doi.org/10.1128/JCM.01231-08

Aguiar, M., Stollenwerk, N., & Halstead, S. B. (2016). The impact of the newly licensed dengue vaccine in endemic countries. *PLoS Neglected Tropical Diseases, 10*(12), 1–23. https://doi.org/10.1371/journal.pntd.0005179

Aiyar, J., Bhan, M. K., Bhandari, N., Kumar, R., Raj, P., & Sazawal, S. (1990). Rotavirus-specific antibody response in saliva of infants with rotavirus diarrhea. *Journal of Infectious Diseases, 162*(6), 1383–1384. https://doi.org/10.1093/infdis/162.6.1383

Amado, L. A., Villar, L. M., de Paula, V. S., de Almeida, A. J., & Gaspar, A. M. C. (2006). Detection of hepatitis A, B, and C virus-specific antibodies using oral fluid for epidemiological studies. *Memórias Do Instituto Oswaldo Cruz, 101*(2), 149–155. https://doi.org/10.1590/S0074-02762006000200006

Andries, A. C., Duong, V., Ly, S., Cappelle, J., Kim, K. S., Lorn Try, P., … Buchy, P. (2015). Value of routine dengue diagnostic tests in urine and saliva specimens. *PLoS Neglected Tropical Diseases, 9*(9). https://doi.org/10.1371/journal.pntd.0004100

Approval of The First U.S. HIV Test System Using Oral Fluid Samples. (1994). Retrieved July 26, 2019, from https://aidsinfo.nih.gov/news/267/approval-of-the-first-u-s%2D%2Dhiv-test-system-using-oral-fluid-samples

Arnold, B. F., Scobie, H. M., Priest, J. W., & Lammie, P. J. (2018). Integrated serologic surveillance of population immunity and disease transmission. *Emerging Infectious Diseases, 24*(7), 1188–1194. https://doi.org/10.3201/eid2407.171928

Augustine, S. A. J., Eason, T. N., Simmons, K. J., Curioso, C. L., Griffin, S. M., Ramudit, M. K. D., & Plunkett, T. R. (2016). Developing a salivary antibody multiplex immunoassay to measure human exposure to environmental pathogens. *Journal of Visualized Experiments*, (115), 1–7. https://doi.org/10.3791/54415

Augustine, S. A. J., Simmons, K. J., Eason, T. N., Curioso, C. L., Griffin, S. M., Wade, T. J., … Wymer, L. J. (2017). Immunoprevalence to six waterborne pathogens in beachgoers at boquerón beach, puerto rico: Application of a microsphere-based salivary antibody multiplex immunoassay. *Frontiers in Public Health, 5*(May). https://doi.org/10.3389/fpubh.2017.00084

Augustine, S. A. J., Simmons, K. J., Eason, T. N., Griffin, S. M., Curioso, C. L., Wymer, L. J., … Dufour, A. (2015). Statistical approaches to developing a multiplex immunoassay for determining human exposure to environmental pathogens. *Journal of Immunological Methods, 425*, 1–9. https://doi.org/10.1016/j.jim.2015.06.002

Balmaseda, A., Saborio, S., Tellez, Y., Mercado, J. C., Pérez, L., Hammond, S. N., … Harris, E. (2008). Evaluation of immunological markers in serum, filter-paper blood spots, and saliva for dengue diagnosis and epidemiological studies. *Journal of Clinical Virology, 43*(3), 287–291. https://doi.org/10.1016/j.jcv.2008.07.016

Ben Salah, A., Zaâtour, A., Pomery, L., Cohen, B. J., Brown, D. W. G., & Andrews, N. (2003). Validation of a modified commercial assay for the detection of rubella-specific IgG in oral fluid for use in population studies. *Journal of Virological Methods, 114*(2), 151–158.

Bernard, M. B., Michele Owen, S., Laura, G. W., Berry, B., Barbara, G. W., Kelly, E. W., & Michael, A. P. (2014). *Laboratory testing for the diagnosis of HIV infection: Updated recommendations*. https://doi.org/10.15620/cdc.23447

Bousema, T., Okell, L., Felger, I., & Drakeley, C. (2014). Asymptomatic malaria infections: Detectability, transmissibility and public health relevance. *Nature Reviews Microbiology, 12*(12), 833–840. https://doi.org/10.1038/nrmicro3364

Brandtzaeg, P. (2007). Do salivary antibodies reliably reflect both mucosal and systemic immunity? *Annals of the New York Academy of Sciences, 1098*, 288–311. https://doi.org/10.1196/annals.1384.012

Brandtzaeg, P. (2009). Mucosal immunity: Induction, dissemination, and effector functions. *Scandinavian Journal of Immunology, 70*(6), 505–515. https://doi.org/10.1111/j.1365-3083.2009.02319.x

Brandtzaeg, P. (2013). Secretory immunity with special reference to the oral cavity. *Journal of Oral Microbiology, 5*(2013), 1–24. https://doi.org/10.3402/jom.v5i0.20401

Buchinsky, F. J., Carter, J. J., Wipf, G. C., Hughes, J. P., Koutsky, L. A., & Galloway, D. A. (2006). Comparison of oral fluid and serum ELISAs in the determination of IgG response to natural human papillomavirus infection in university women. *Journal of Clinical Virology, 35*(4), 450–453. https://doi.org/10.1016/j.jcv.2005.09.014

Campbell, H., Amirthalingam, G., Fry, N. K., Litt, D., Harrison, T. G., Wagner, K., . . . Miller, E. (2014). Oral fluid testing for pertussis, England and Wales, June 2007–August 2009. *Emerging Infectious Diseases, 20*(6), 968–975.

Cawthraw, S. A., Feldman, R. A., Sayers, A. R., & Newell, D. G. (2002). Long-term antibody responses following human infection with Campylobacter jejuni. *Clinical and Experimental Immunology, 130*(1), 101–106. https://doi.org/10.1046/j.1365-2249.2002.01966.x

CDC. (2008). False-positive oral fluid rapid HIV tests—New York City, 2005–2008. *MMWR, 57* (24), 660–665.

Chart, H., & Jenkins, C. (1998). Salivary antibodies to lipopolysaccharide antigens of *Escherichia coli* 0157. *The Lancet, 352*(9125), 371.

Chart, H., Perry, N. T., Willshaw, G. A., & Cheasty, T. (2003). Analysis of saliva for antibodies to the LPS of *Escherichia coli* O157 in patients with serum antibodies to *E. coli* O157 LPS. *Journal of Medical Microbiology, 52*(7), 569–572. https://doi.org/10.1099/jmm.0.05126-0

Chibo, D., Riddell, M. A., Catton, M. G., & Birch, C. J. (2005). Applicability of oral fluid collected onto filter paper for detection and genetic characterization of measles virus strains. *Journal of Clinical Microbiology, 43*(7), 3145–3149. https://doi.org/10.1128/JCM.43.7.3145-3149.2005

Chidi, A. P., Chishimba, S., Kobayashi, T., Hamapumbu, H., Mharakurwa, S., Thuma, P. E., & Moss, W. J. (2011). Validation of oral fluid samples to monitor serological changes to Plasmodium falciparum: An observational study in southern Zambia. *Malaria Journal, 10*(1), 162. https://doi.org/10.1186/1475-2875-10-162

Corran, P., Coleman, P., Riley, E., & Drakeley, C. (2007). Serology: A robust indicator of malaria transmission intensity? *Trends in Parasitology, 23*(12), 575–582. https://doi.org/10.1016/j.pt.2007.08.023

Cozon, G., Biron, F., Jeannin, M., Cannella, D., & Revillard, J.-P. (1994). Secretory IgA antibodies to Cryptosporidium parvum in AIDS patients with chronic cryptosporidiosis. *Journal of Infectious Diseases, 169*(3), 696–699. https://doi.org/10.1093/infdis/169.3.696

de Azevedo Neto, R., Richards, A., Nokes, D., Silveira, A., Cohen, B., Passos, S., . . . Massad, E. (1995). Salivary antibody detection in epidemiological surveys: A pilot study after a mass vaccination campaign against rubella in São Paulo, Brazil. *Transactions of the Royal Society of Tropical Medicine and Hygiene, 89*(1), 115–118.

De Cock, L., Hutse, V., Verhaegen, E., Quoilin, S., Vandenberghe, H., & Vranckx, R. (2004). Detection of HCV antibodies in oral fluid. *Journal of Virological Methods, 122*(2), 179–183. https://doi.org/10.1016/j.jviromet.2004.09.001

Delaney, K., Branson, B., Uniyal, A., Kerndt, P., Keenan, P., Jafa, K., . . . Bulterys, M. (2006). Performance of an oral fluid rapid HIV-1/2 test: Experience from four CDC studies. *AIDS, 20* (12), 1655–1660.

Delaney, K. P., Branson, B. M., Uniyal, A., Phillips, S., Candal, D., Owen, S. M., & Kerndt, P. R. (2011). Evaluation of the performance characteristics of 6 rapid HIV antibody tests. *Clinical Infectious Diseases, 52*(2), 257–263. https://doi.org/10.1093/cid/ciq068

Drakeley, C. J., Corran, P. H., Coleman, P. G., Tongren, J. E., McDonald, S. L. R., Carneiro, I., . . . Riley, E. M. (2005). Estimating medium- and long-term trends in malaria transmission by using serological markers of malaria exposure. *Proceedings of the National Academy of Sciences, 102* (14), 5108–5113. https://doi.org/10.1073/pnas.0408725102

Durbin, A. P. (2016). Vaccine development for Zika virus—timelines and strategies. *Seminars in Reproductive Medicine, 34*(5), 299–304.

Egorov, A. I., Montuori Trimble, L. M., Ascolillo, L., Ward, H. D., Levy, D. A., Morris, R. D., . . . Griffiths, J. K. (2010). Recent diarrhea is associated with elevated salivary IgG responses to cryptosporidium in residents of an eastern massachusetts community. *Infection, 38*(2), 117–123. https://doi.org/10.1007/s15010-009-9323-4

El-Gebaly, N. S. M., Halawa, E. F., Moussa, H. M. E., Rabia, I., & Abu-Zekry, M. (2012). Saliva and sera IgA and IgG in Egyptian Giardia-infected children. *Parasitology Research, 111*(2), 571–575. https://doi.org/10.1007/s00436-012-2869-y

Estévez, P. T., Satoguina, J., Nwakanma, D. C., West, S., Conway, D. J., & Drakeley, C. J. (2011). Human saliva as a source of anti-malarial antibodies to examine population exposure to Plasmodium falciparum. *Malaria Journal, 10*, 1–8. https://doi.org/10.1186/1475-2875-10-104

Exum, N. G., Pisanic, N., Granger, D. A., Schwab, K. J., Detrick, B., Kosek, M., . . . Heaney, C. D. (2016). Use of pathogen-specific antibody biomarkers to estimate waterborne infections in population-based settings. *Current Environmental Health Reports, 3*(3), 322–334. https://doi.org/10.1007/s40572-016-0096-x

Figueroa, C., Johnson, C., Ford, N., Sands, A., Dalal, S., Meurant, R., . . . Baggaley, R. (2018). Reliability of HIV rapid diagnostic tests for self-testing compared with testing by health-care workers: a systematic review and meta-analysis. *The Lancet HIV, 5*(6), e277–e290. https://doi.org/10.1016/S2352-3018(18)30044-4

Fisker, N., Georgsen, J., Stolborg, T., Khalil, M. R., & Christensen, P. B. (2002). Low hepatitis B prevalence among pre-school children in Denmark: saliva anti-HBc screening in day care centres. *Journal of Medical Virology, 68*(4), 500–504. https://doi.org/10.1002/jmv.10242

Food and Drug Administration. (2004). *OraQuick® ADVANCE Rapid HIV-1/2 Antibody Test Summary of Safety and Effectiveness Data.*

Food and Drug Administration. (2009). *Avioq HIV-1 Microelisa System.*

Food and Drug Administration. (2012a). *Chembio DPP® HIV 1/2 Assay Summary of Safety and Effectiveness*. Retrieved from http://weekly.cnbnews.com/news/article.html?no=124000

Food and Drug Administration. (2012b). *OraQuick® In-Home HIV Test Summary of Safety and Effectiveness.*

Friedman, M. G., Entin, N., Zedaka, R., & Dagan, R. (1996). Subclasses of IgA antibodies in serum and saliva samples of newborns and infants immunized against rotavirus. *Clinical and Experimental Immunology, 103*(2), 206–211. https://doi.org/10.1046/j.1365-2249.1996.d01-620.x

Gallo, D. (1997). Evaluation of a system using oral mucosal transudate for HIV-1 antibody screening and confirmatory testing. OraSure HIV Clinical Trials Group. *JAMA, 277*(3), 254–258. https://doi.org/10.1001/jama.277.3.254

Gill, J., Aston, R., Vyse, A. J., White, J. M., & Greenwood, A. (2002). Susceptibility of young offenders to measles and rubella: an antibody prevalence study using oral fluid samples. *Communicable Disease and Public Health, 5*(4), 314–317.

Goyal, A., Shaikh, N., Kinikar, A., & Wairakgar, N. (2009). Oral fluid, a substitute for serum to monitor measles IgG antibody? *Indian Journal of Medical Microbiology, 27*(4), 351–353.

Graham, D. Y., Jiang, X., Tanaka, T., Opekun, A. R., Madore, H. P., & Estes, M. K. (1994). Norwalk virus infection of volunteers: New insights based on improved assays. *Journal of Infectious Diseases, 170*(1), 34–43. https://doi.org/10.1093/infdis/170.1.34

Granade, T. C., Phillips, S. K., Parekh, B., Lee-Thomas, S., Gomez, P., Oleander, H., . . . Charles, W. (1998). Detection of antibodies to human immunodeficiency virus type 1 in oral fluids: A large-scale evaluation of immunoassay performance. *Clinical and Diagnostic Laboratory Immunology, 5*(2), 171–175.

Griffin, S. M., Chen, I. M., Fout, G. S., Wade, T. J., & Egorov, A. I. (2011). Development of a multiplex microsphere immunoassay for the quantitation of salivary antibody responses to selected waterborne pathogens. *Journal of Immunological Methods, 364*(1–2), 83–93.

Griffin, S. M., Converse, R. R., Leon, J. S., Wade, T. J., Jiang, X., Moe, C. L., & Egorov, A. I. (2015). Application of salivary antibody immunoassays for the detection of incident infections with Norwalk virus in a group of volunteers. *Journal of Immunological Methods, 424*, 53–63. https://doi.org/10.1016/j.jim.2015.05.001

Grimwood, K., Lund, J. C., Coulson, B. S., Hudson, I. L., Bishop, R. F., & Barnes, G. L. (1988). Comparison of serum and mucosal antibody responses following severe acute rotavirus gastroenteritis in young children. *Journal of Clinical Microbiology, 26*(4), 732–738.

Halstead, S. B. (2015). Pathogenesis of dengue: Dawn of a new era. *F1000Research, 4*(0), 1353. https://doi.org/10.12688/f1000research.7024.1

Hayford, K. T., Al-Emran, H. M., Moss, W. J., Shomik, M. S., Bishai, D., & Levine, O. S. (2013). Validation of an anti-measles virus-specific IgG assay with oral fluid samples for immunization surveillance in Bangladesh. *Journal of Virological Methods, 193*(2), 512–518. https://doi.org/10.1016/j.jviromet.2013.07.014

Hayford, K. T., Holroyd, T. A., Schiaffino, F., Chang, R. H., Wanyiri, J. W., Gross, M., . . . Moss, W. J. (under review). Systematic review and meta-analysis of diagnostic accuracy of measles and rubella IgG and IgM antibodies in oral fluid. *Vaccine.*

Hayford, K. T., Levine, O. S., Shomik, M. S., Al-Emran, H. M., Moss, W. J., & Bishai, D. (2013). Measles vaccination coverage estimates from surveys, clinic records, and immune markers in oral fluid and blood: A population-based cross-sectional study. *BMC Public Health, 13*(1). https://doi.org/10.1186/1471-2458-13-1211

Helfand, R. F., Kebede, S., Mercader, S., Gary, H. E., Beyene, H., & Bellini, W. J. (1999). The effect of timing of sample collection on the detection of measles-specific IgM in serum and oral fluid samples after primary measles vaccination. *Epidemiology and Infection, 123*(3), 451–455. https://doi.org/10.1017/S0950268899002988

Hepatitis A: Oral Fluid Testing for Household Contacts. (2015). Retrieved from https://www.gov.uk/government/publications/hepatitis-a-oral-fluid-testing-for-household-contacts

Herath, H. M. T. U. (2003). Early diagnosis of typhoid fever by the detection of salivary IgA. *Journal of Clinical Pathology, 56*(9), 694–698. https://doi.org/10.1136/jcp.56.9.694

HIV/AIDS testing. (2019). Retrieved July 26, 2019, from https://www.cdc.gov/hiv/basics/testing.html

Hodinka, R. L., Nagashunmugam, T., & Malamud, D. (1998). Detection of human immunodeficiency virus antibodies in oral fluids. *Clinical and Diagnostic Laboratory Immunology, 5*(4), 419–426.

Holroyd, T. A., Schiaffino, F., Chang, R. H., Wanyiri, J. W., Saldanha, I., Gross, M., . . . Hayford, K. T. (under review). Overview of diagnostic accuracy of saliva and feasibility of saliva collection for use in estimating prevalence and immunity to vaccine-preventable diseases. *Journal of Virological Methods.*

Hübschen, J. M., Charpentier, E., Weicherding, P., & Muller, C. P. (2018). IgG antibody prevalence suggests high immunization needs in newcomers to Luxembourg, 2012. *Vaccine, 36*(6), 899–905. https://doi.org/10.1016/j.vaccine.2017.12.041

Hutse, V., Verhaegen, E., De Cock, L., Quoilin, S., Vandenberghe, H., Horsmans, Y., . . . Oyen, H. Van (2005). Oral fluid as a medium for the detection of hepatitis B surface antigen. *Journal of Medical Virology, 77*(1), 53–56. https://doi.org/10.1002/jmv.20413

Jani, J. V., Holm-Hansen, C., Mussá, T., Zango, A., Manhiça, I., Bjune, G., & Jani, I. V. (2008). Assessment of measles immunity among infants in Maputo City, Mozambique. *BMC Public Health, 8*, 1–8. https://doi.org/10.1186/1471-2458-8-386

Jenkins, M. M., Reed-Gross, E., Rasmussen, S. A., Barfield, W. D., Prue, C. E., Gallagher, M. L., & Honein, M. A. (2009). Maternal attitudes toward DNA collection for gene-environment studies: A qualitative research study. *American Journal of Medical Genetics Part A, 149A*(11), 2378–2386. https://doi.org/10.1002/ajmg.a.33043

Jertborn, M., Svennerholm, A. M., & Holmgren, J. (1986). Saliva, breast milk, and serum antibody responses as indirect measures of intestinal immunity after oral cholera vaccination or natural disease. *Journal of Clinical Microbiology, 24*(2), 203–209.

Jin, L., Brown, D. W. G., Litton, P. A., & White, J. M. (2004). Genetic diversity of mumps virus in oral fluid specimens: Application to mumps epidemiological study. *The Journal of Infectious Diseases, 189*(6), 1001–1008. https://doi.org/10.1086/382134

Jin, L., Vyse, A., & Brown, D. W. G. (2002). The role of RT-PCR assay of oral fluid for diagnosis and surveillance of measles, mumps and rubella. *Bulletin of the World Health Organization, 80*(1), 76–77.

Judd, A., Parry, J., Hickman, M., McDonald, T., Jordan, L., Lewis, K., … Nelson, M. (2003). Evaluation of a modified commercial assay in detecting antibody to hepatitis C virus in oral fluids and dried blood spots. *Journal of Medical Virology, 71*(1), 49–55. https://doi.org/10.1002/jmv.10463

Kara, E. O., Campbell, H., Ribeiro, S., Fry, N., Litt, D., Eletu, S., & Amirthalingam, G. (2017). Survey of household contacts of infants with laboratory-confirmed pertussis infection during a national pertussis outbreak in England and Wales. *The Pediatric Infectious Disease Journal, 36*(2), 140–145.

Kobayashi, T., Jain, A., Liang, L., Obiero, J. M., Hamapumbu, H., Stevenson, J. C., … Moss, W. J. (2019). Distinct antibody signatures associated with different malaria transmission intensities in Zambia and Zimbabwe. *MSphere, 4*(2). https://doi.org/10.1128/mSphereDirect.00061-19

Krause, J., Subklew-Sehume, F., Kenyon, C., & Colebunders, R. (2013). Acceptability of HIV self-testing: A systematic literature review. *BMC Public Health, 13*(1), 735. https://doi.org/10.1186/1471-2458-13-735

Kremer, J. R., & Muller, C. P. (2005). Evaluation of commercial assay detecting specific immunoglobulin g in oral fluid for determining measles immunity in vaccinees. *Clinical and Vaccine Immunology, 12*(5), 668–670. https://doi.org/10.1128/cdli.12.5.668-670.2005

Lambe, T., Rampling, T., Samuel, D., Bowyer, G., Ewer, K. J., Venkatraman, N., … Gilbert, S. C. (2016). Detection of vaccine-induced antibodies to Ebola virus in oral fluid. *Open Forum Infectious Diseases, 3*(1), 2–4. https://doi.org/10.1093/ofid/ofw031

Laufer, D. S., Hurni, W., Watson, B., Miller, W., Ryan, J., Nalin, D., & Brown, L. (1995). Saliva and serum as diagnostic media for antibody to hepatitis A virus in adults and in individuals who have received an inactivated hepatitis A vaccine. *Clinical Infectious Diseases, 20*(4), 868–871. https://doi.org/10.1093/clinids/20.4.868

Lindesmith, L., Moe, C., LePendu, J., Frelinger, J. A., Treanor, J., & Baric, R. S. (2005). Cellular and humoral immunity following snow mountain virus challenge. *Journal of Virology, 79*(5), 2900–2909. https://doi.org/10.1128/JVI.79.5.2900-2909.2005

Lindesmith, L., Moe, C., Marionneau, S., Ruvoen, N., Jiang, X., Lindblad, L., … Baric, R. (2003). Human susceptibility and resistance to Norwalk virus infection. *Nature Medicine, 9*(5), 548–553. https://doi.org/10.1038/nm860

Lowther, S. A., Curriero, F. C., Kalish, B. T., Shields, T. M., Monze, M., & Moss, W. J. (2009). Population immunity to measles virus and the effect of HIV-1 infection after a mass measles vaccination campaign in Lusaka, Zambia: a cross-sectional survey. *The Lancet, 373*(9668), 1025–1032. https://doi.org/10.1016/S0140-6736(09)60142-2

Ludwig, K., Grabhorn, E., Bitzan, M., Bobrowski, C., Kemper, M. J., Sobottka, I., … Müller-Wiefel, D. E. (2002). Saliva IgM and IgA are a sensitive indicator of the humoral immune response to *Escherichia coli* O157 lipopolysaccharide in children with enteropathic hemolytic uremic syndrome. *Pediatric Research, 52*(2), 307–313. https://doi.org/10.1203/00006450-200208000-00026

Luo, W., Masciotra, S., Delaney, K. P., Charurat, M., Croxton, T., Constantine, N., … Owen, S. M. (2013). Comparison of HIV oral fluid and plasma antibody results during early infection in a longitudinal Nigerian cohort. *Journal of Clinical Virology, 58*, e113–e118. https://doi.org/10.1016/j.jcv.2013.08.017

MacNeil, A., Lee, C. w., & Dietz, V. (2014). Issues and considerations in the use of serologic biomarkers for classifying vaccination history in household surveys. *Vaccine, 32*(39), 4893–4900. https://doi.org/10.1016/j.vaccine.2014.07.005

Major, C. J., Read, S. E., Coates, R. A., Francis, A., McLaughlin, B. J., Millson, M., … O'Shaughnessy, M. V. (1991). Comparison of saliva and blood for human immunodeficiency virus prevalence testing. *Journal of Infectious Diseases, 163*(4), 699–702. https://doi.org/10.1093/infdis/163.4.699

Manikkavasagan, G., Bukasa, A., Brown, K. E., Cohen, B. J., & Ramsay, M. E. (2010). Oral fluid testing during 10 years of rubella elimination, england and wales. *Emerging Infectious Diseases, 16*(10), 1532–1538. https://doi.org/10.3201/eid1610.100560

Martin, I. B., Williams, V., Ferguson, D., & Read, S. (2018). Performance of and preference for oral rapid HIV testing in The Bahamas. *Journal of Infection and Public Health, 11*(1), 126–129. https://doi.org/10.1016/j.jiph.2017.06.005

Moe, C. L., Sair, A., Lindesmith, L., Estes, M. K., & Jaykus, L.-A. (2004). Diagnosis of norwalk virus infection by indirect enzyme immunoassay detection of salivary antibodies to recombinant norwalk virus antigen. *Clinical and Vaccine Immunology, 11*(6), 1028–1034. https://doi.org/10.1128/CDLI.11.6.1028-1034.2004

Morris-Cunnington, M. C., Edmunds, W. J., Miller, E., & Brown, D. W. G. (2004). A population-based seroprevalence study of hepatitis A virus using oral fluid in England and Wales. *American Journal of Epidemiology, 159*(8), 786–794. https://doi.org/10.1093/aje/kwh107

Mortimer, P. P., & Parry, J. V. (1991). Non-invasive virological diagnosis: Are saliva and urine specimens adequate substitutes for blood? *Reviews in Medical Virology, 1*(2), 73–78. https://doi.org/10.1002/rmv.1980010204

Moss, D. M., Montgomery, J. M., Newland, S. V., Priest, J. W., Moss, D. M., Montgomery, J. M., ... Lammie, P. J. (2019). Detection of cryptosporidium antibodies in sera and oral fluids using multiplex bead assay. *Journal of Parasitology, 90*(2), 397–404.

Nangendo, J., Obuku, E. A., Kawooya, I., Mukisa, J., Nalutaaya, A., Musewa, A., ... Kalyango, J. N. (2017). Diagnostic accuracy and acceptability of rapid HIV oral testing among adults attending an urban public health facility in Kampala, Uganda. *PLoS One, 12*(8), e0182050. https://doi.org/10.1371/journal.pone.0182050

Nigatu, W., Samuel, D., Cohen, B., Cumberland, P., Lemma, E., Brown, D. W. G., & Nokes, J. (2008). Evaluation of a measles vaccine campaign in Ethiopia using oral-fluid antibody surveys. *Vaccine, 26*(37), 4769–4774. https://doi.org/10.1016/j.vaccine.2008.07.005

Nokes, D. J., Enquselassie, F., Nigatu, W., Vyse, A. J., Cohen, B. J., Brown, D. W., & Cutts, F. T. (2001). Has oral fluid the potential to replace serum for the evaluation of population immunity levels? A study of measles, rubella and hepatitis B in rural Ethiopia. *Bulletin of the World Health Organization, 79*(7), 588–595.

Nokes, D. J., Nigatu, W., Abebe, A., Messele, T., Dejene, A., Enquselassie, F., ... Cutts, F. T. (1998). A comparison of oral fluid and serum for the detection of rubella-specific antibodies in a community study in Addis Ababa, Ethiopia. *Tropical Medicine and International Health, 3*(4), 258–267.

O'Connell, T., Thornton, L., O'Flanagan, D., Staines, A., Connell, J., Dooley, S., & McCormack, G. (2000). Prevalence of hepatitis B anti-core antibody in the Republic of Ireland. *Epidemiology and Infection, 125*(3), 701–704. https://doi.org/10.1017/S095026880000491X

Ochnio, J. J., Scheifele, D. W., Ho, M., & Mitchell, L. A. (1997). New, ultrasensitive enzyme immunoassay for detecting vaccine- and disease-induced hepatitis A virus-specific immunoglobulin G in saliva. *Journal of Clinical Microbiology, 35*(1), 98–101.

Ohuma, E. O., Okiro, E. A., Bett, A., Abwao, J., Were, S., Samuel, D., ... Nokes, D. J. (2009). Evaluation of a measles vaccine campaign by oral-fluid surveys in a rural Kenyan district: Interpretation of antibody prevalence data using mixture models. *Epidemiology and Infection, 137*(2), 227–233. https://doi.org/10.1017/S0950268808000848

Pant Pai, N., Joshi, R., Dogra, S., Taksande, B., Kalantri, S. P., Pai, M., ... Reingold, A. L. (2007). Evaluation of diagnostic accuracy, feasibility and client preference for rapid oral fluid-based diagnosis of HIV infection in rural India. *PLoS One*, *2*(4), e367. https://doi.org/10.1371/journal.pone.0000367

Parry, J., Perry, K., Panday, S., & Mortimer, P. (1989). Diagnosis of hepatitis A and B by testing saliva. *Journal of Medical Virology, 28*(4), 255–260.

Passmore, J.-A. S., Marais, D. J., Sampson, C., Allan, B., Parker, N., Milner, M., ... Williamson, A.-L. (2007). Cervicovaginal, oral, and serum IgG and IgA responses to human papillomavirus type 16 in women with cervical intraepithelial neoplasia. *Journal of Medical Microbiology, 79*, 1375–1380. https://doi.org/10.1002/jmv

Piacentini, S. C., Thieme, T. R., Beller, M., & Davidson, S. L. (1993). Diagnosis of Hepatitis A, B, and C using oral samples. *Annals of the New York Academy of Sciences, 694*, 334–336. https://doi.org/10.1111/j.1749-6632.1993.tb18382.x

Pisanic, N., Ballard, S.-B., Colquechagua, F. D., François, R., Exum, N., Yori, P. P., … Kosek, M. N. (2018). Minimally invasive saliva testing to monitor norovirus infection in community settings. *The Journal of Infectious Diseases, 219*(8), 1234–1242. https://doi.org/10.1093/infdis/jiy638

Pisanic, N., Rahman, A., Saha, S. K., Labrique, A. B., Nelson, K. E., Granger, D. A., … Heaney, C. D. (2017). Development of an oral fluid immunoassay to assess past and recent hepatitis E virus (HEV) infection. *Journal of Immunological Methods, 448*, 1–8. https://doi.org/10.1016/j.jim.2017.04.012

Quoilin, S., Hutse, V., Vandenberghe, H., Claeys, F., Verhaegen, E., De Cock, L., … Van Oyen, H. (2007). A population-based prevalence study of hepatitis A, B and C virus using oral fluid in Flanders, Belgium. *European Journal of Epidemiology, 22*(3), 195–202. https://doi.org/10.1007/s10654-007-9105-6

Ramsay, M., Brugha, R., & Brown, D. (1997). Surveillance of measles in England and Wales: Implications of a national saliva testing programme. *Bulletin of the World Health Organization, 75*(6), 515–521.

Ramsay, M. E., Brugha, R., Brown, D. W. G., Cohen, B. J., & Miller, E. (1998). Salivary diagnosis of rubella: A study of notified cases in the United Kingdom, 1991–4. *Epidemiology and Infection, 120*(3), 315–319. https://doi.org/10.1017/S0950268898008838

Ramsay, M. E., Jin, L., White, J., Litton, P., Cohen, B., & Brown, D. (2003). The elimination of indigenous measles transmission in England and Wales. *The Journal of Infectious Diseases, 187* (s1), S198–S207. https://doi.org/10.1086/368024

Reid, F., Hassan, J., Irwin, F., Waters, A., Hall, W., & Connell, J. (2008). Epidemiologic and diagnostic evaluation of a recent mumps outbreak using oral fluid samples. *Journal of Clinical Virology, 41*(2), 134–137. https://doi.org/10.1016/j.jcv.2007.10.009

Rodríguez, O. L., Hagel, I., González, Y., Roque, M. E., Vásquez, N., López, E., & Di Prisco, M. C. (2004). Secretory IgA antibody responses in Venezuelan children infected with Giardia duodenalis. *Journal of Tropical Pediatrics, 50*(2), 68–72. https://doi.org/10.1093/tropej/50.2.68

Rowhani-Rahbar, A., Carter, J. J., Hawes, S. E., Hughes, J. P., Weiss, N. S., Galloway, D. A., & Koutsky, L. A. (2009). Antibody responses in oral fluid after administration of prophylactic human papillomavirus vaccines. *The Journal of Infectious Diseases, 200*(9), 1452–1455. https://doi.org/10.1086/606026

Schultsz, C., Qadri, F., Hossain, S. A., Ahmed, F., & Ciznar, I. (1992). Shigella -specific IgA in saliva of children with bacillary dysentery. *FEMS Microbiology Letters, 89*(2), 65–72. https://doi.org/10.1111/j.1574-6968.1992.tb04972.x

Sheikh, S., Ali, A., Zaidi, A. K. M., Agha, A., Khowaja, A., Allana, S., … Azam, I. (2011). Measles susceptibility in children in Karachi, Pakistan. *Vaccine, 29*(18), 3419–3423. https://doi.org/10.1016/j.vaccine.2011.02.087

Shonhai, A., Warrener, L., Mangwanya, D., Slibinskas, R., Brown, K., Brown, D., … Samuel, D. (2015). Investigation of a measles outbreak in Zimbabwe, 2010: Potential of a point of care test to replace laboratory confirmation of suspected cases. *Epidemiology and Infection, 143*(16), 3442–3450. https://doi.org/10.1017/S0950268815000540

Simani, O., Fracois, G., Burnett, R., Meheus, A., Basetse, H., & Mphahlele, M. (2008). Oral fluid detection of hepatitis B vaccine-induced antibodies can improve vaccination programmes. *South African Medical Journal, 98*(8), 608–609.

Spielberg, F., Critchlow, C., Vittinghoff, E., Coletti, A. S., Sheppard, H., Mayer, K. H., … Gross, M. (2000). Home collection for frequent HIV testing: Acceptability of oral fluids, dried blood spots and telephone results. *AIDS, 14*(12), 1819–1828. https://doi.org/10.1097/00002030-200008180-00018

Spoorthi, B. R. (2014). Diagnostic Efficacy of saliva for dengue – A reality in near future? A piloting initiative. *Journal of Clinical and Diagnostic Research, 8*(3), 229–232. https://doi.org/10.7860/JCDR/2014/7521.4169

Stals, F., Walther, F. J., & Bruggeman, C. A. (1984). Faecal and pharyngeal shedding of rotavirus and rotavirus IgA in children with diarrhoea. *Journal of Medical Virology, 14*(4), 333–339.

Tapia, M. D., Pasetti, M. F., Cuberos, L., Sow, S. O., Doumbia, M. N., Bagayogo, M., . . . Levine, M. M. (2006). Measurement of tetanus antitoxin in oral fluid. *The Pediatric Infectious Disease Journal, 25*(9), 819–825. https://doi.org/10.1097/01.inf.0000232629.72160.bb

Thieme, T., Piacentini, S., Davidson, S., & Steingart, K. (1994). Determination of measles, mumps, and rubella immunization status using oral fluid samples. *JAMA: The Journal of the American Medical Association, 272*(3), 219–221. https://doi.org/10.1001/jama.1994.03520030061029

Thornton, L., Barry, J., Long, J., Allwright, S., Bradley, F., & Parry, J. (2000). Comparison between self-reported hepatitis B, hepatitis C, and HIV antibody status and oral fluid assay results in Irish prisoners. *Communicable Disease and Public Health2, 3*(4), 253–255.

Tourinho, R., de Almeida, A., Villar, L., Murat, P., Capelin, G., Castro, A., & de Paula, V. (2015). Cross-sectional study of hepatitis A virus infection in the pantanal population before vaccine implementation in Brazil: usage of non-invasive specimen collection. *International Journal of Environmental Research and Public Health, 12*(7), 7357–7369. https://doi.org/10.3390/ijerph120707357

Vainio, K., Samdal, H. H., Ånestad, G., Skutlaberg, D. H., Bransdal, K. T., Mundal, R., & Aaberge, I. (2006). Seroprevalence of measles among Norwegian military conscripts in 2004. *European Journal of Clinical Microbiology and Infectious Diseases, 26*(3), 217–220. https://doi.org/10.1007/s10096-007-0258-8

Vázquez, S., Cabezas, S., Pérez, A. B., Pupo, M., Ruiz, D., Calzada, N., . . . Guzmán, M. G. (2007). Kinetics of antibodies in sera, saliva, and urine samples from adult patients with primary or secondary dengue 3 virus infections. *International Journal of Infectious Diseases, 11*(3), 256–262. https://doi.org/10.1016/j.ijid.2006.05.005

Wade, T. J., Sams, E. A., Egorov, A. I., Eason, T. N., Augustine, S. A. J., Griffin, S. M., . . . Simmons, K. J. (2018). Asymptomatic norovirus infection associated with swimming at a tropical beach: A prospective cohort study. *PLoS One, 13*(3), 1–14. https://doi.org/10.1371/journal.pone.0195056

Wade, T. J., Griffin, S. M., Egorov, A. I., et al. (2019). Application of a multiplex salivary immunoassay to detect sporadic incident norovirus infections. *Scientific Reports, 9*(1), 19576. https://doi.org/10.1038/s41598-019-56040-7

Wang, K., Fry, N. K., Campbell, H., Amirthalingam, G., Harrison, T. G., Mant, D., & Harnden, A. (2014). Whooping cough in school age children presenting with persistent cough in UK primary care after introduction of the preschool pertussis booster vaccination: prospective cohort study. *BMJ, 348*(jun24 18), g3668–g3668. https://doi.org/10.1136/bmj.g3668

Warrener, L., Slibinskas, R., Brown, D., Sasnauskas, K., & Samuel, D. (2010). Development and evaluation of a rapid immunochromatographic test for mumps-specific IgM in oral fluid specimens and use as a matrix for preserving viral nucleic acid for RT-PCR. *Journal of Medical Virology, 82*, 485–493. https://doi.org/10.1002/jmv

Warrener, L., Slibinskas, R., Chua, K. B., Nigatu, W., Brown, K. E., Sasnauskas, K., . . . Brown, D. (2011). A point-of-care test for measles diagnosis: Detection of measles-specific igM antibodies and viral nucleic acid. *Bulletin of the World Health Organization, 89*(9), 675–682. https://doi.org/10.2471/BLT.11.088427

Weild, A., Gill, O., Bennett, D., Livingstone, S., Parry, J., & Curran, L. (2000). Prevalence of HIV, hepatitis B, and hepatitis C antibodies in prisoners in England and Wales: A national survey. *Communicable Disease and Public Health, 3*(2), 121–126.

Wesolowski, L. G., MacKellar, D. A., Facente, S. N., Dowling, T., Ethridge, S. F., Zhu, J. H., & Sullivan, P. S. (2006). Post-marketing surveillance of OraQuick whole blood and oral fluid rapid HIV testing. *Aids, 20*(12), 1661–1666. https://doi.org/10.1097/01.aids.0000238413.13442.ed

Williams, G. S., Mweya, C., Stewart, L., Mtove, G., Reyburn, H., Cook, J., … Drakeley, C. J. (2009). Immunophoretic rapid diagnostic tests as a source of immunoglobulins for estimating malaria sero-prevalence and transmission intensity. *Malaria Journal*, *8*(1), 1–11. https://doi.org/10.1186/1475-2875-8-168

World Health Organization (WHO). (2016). *Dengue vaccine: WHO position paper – July 2016. Weekly epidemiological record, 91*. https://doi.org/10.1371/jour

World Health Organization (WHO). (2017). *Informing vaccination programs: A guide to the design and conduct of dengue serosurveys*. Retrieved from http://www.who.int/about/licensing

World Health Organization (WHO). (2019). Chapter 2: The global measles and rubella laboratory network. In *Manual for the Laboratory-based Surveillance of Measles, Rubella, and Congenital Rubella Syndrome* (pp. 3–5)

Zaka-ur-Rab, Z., Abqari, S., Shahab, T., Islam, N., & Shukla, I. (2012). Evaluation of salivary anti-Salmonella typhi lipopolysaccharide IgA ELISA for serodiagnosis of typhoid fever in children. *Archives of Disease in Childhood, 97*(3), 236–238. https://doi.org/10.1136/adc.2011.300622

Zhao, L.-Z., Hong, W.-X., Wang, J., Yu, L., Hu, F.-Y., Qiu, S., … Zhang, F.-C. (2019). Kinetics of antigen-specific IgM/IgG/IgA antibody responses during Zika virus natural infection in two patients. *Journal of Medical Virology, 91*(5), 872–876. https://doi.org/10.1002/jmv.25366

# Chapter 14
# Salivary Bioscience and Pain

**Laura A. Payne and Michelle A. Fortier**

**Abstract** Salivary biomarkers and analytes represent an important advancement in the field of pain by suggesting the possibility of identifying an objective measurement of individual differences in the experience of acute and chronic pain. In adults and children, the most widely studied salivary pain biomarkers are salivary alpha-amylase (sAA) and cortisol. Existing data suggest that, whereas in the acute pain context there are elevations in the production of salivary biomarkers, in chronic pain there is often a blunting effect. These findings support the overlap between acute pain and stress responses, as well as a chronic stress model where the experience of persistent pain is associated with sustained changes in stress biomarkers. However, individual factors and mechanisms contributing to this relationship remain unclear. Additionally, more research is needed to determine the specific and interactive factors and effects of individual salivary measures.

**Keywords** Cortisol · Salivary alpha-amylase · Acute pain · Chronic pain

## 14.1 History of Salivary Bioscience and Pain

As noted by the International Association for the Study of Pain (IASP), pain is a complex phenomenon involving "an unpleasant sensory and emotional experience associated with actual or potential tissue damage, or described in terms of such damage" (IASP Task Force on Taxonomy, 2004). Because the experience of pain varies across individuals and is influenced by not just actual or potential tissue damage, but emotional, cognitive, familial, cultural, and environmental factors,

L. A. Payne (✉)
McLean Hospital/Harvard Medical School, Belmont, MA, USA
e-mail: LPayne@mclean.harvard.edu

M. A. Fortier
University of California at Irvine Center on Stress and Health, Sue and Bill Gross School of Nursing at University of California at Irvine, Irvine, CA, USA



D. A. Granger, M. K. Taylor (eds.), *Salivary Bioscience*,
https://doi.org/10.1007/978-3-030-35784-9_14

assessment of a broad range of both subjective and objective markers of pain can inform effective interventions. Although self-report has historically been considered the gold standard for pain assessment and has largely been used to guide pain treatment decisions, critics of this approach argue that self-report oversimplifies the pain phenomenon and in fact, may contradict alternative measures of pain (Twycross, Voepel-Lewis, Vincent, Franck, & Von Baeyer, 2015). Accordingly, it is important to examine other individual or contextual factors associated with pain.

Salivary biomarkers prove one meaningful avenue to further understand the pain experience. Pain is conceptualized as a stressor that activates the autonomic nervous system (ANS) and subsequently the sympathetic nervous system (SNS), which triggers the changes that occur in physiological functioning during the fight-or-flight response. Several hormones are produced during this response and have been assessed as a part of the pain response, including cortisol and alpha-amylase. Cortisol production following a stressor is delayed and thus assessment must occur 15–20 min following the experience of pain; whereas alpha-amylase increases more quickly and increases are seen more proximal to the stressor. Pain biomarkers are effectively biological "signatures" that may represent specific phenotypic characteristics associated with either acute or chronic pain, specific chronic pain conditions, or someone at risk for transitioning from acute or recurrent to chronic pain. Identification of specific biomarkers is thought to be an important window into mechanisms of pain, which could allow for the development of novel therapies to address these mechanisms (Reckziegel et al., 2019). Additionally, salivary biomarkers, in particular salivary secretory immunoglobulin-A (s-IgA) and soluble tumor necrosis factor—alpha (tnf-alpha), demonstrate good reliability across time (Sobas et al., 2016). Salivary bioscience is critical to identifying biomarkers, as many different components of functioning, from gonadal hormones to salivary proteins, can be measured noninvasively.

## 14.2 Salivary Bioscience and Pain in Adult Populations

### *14.2.1 Clinical Pain Syndromes*

#### 14.2.1.1 Fibromyalgia

Fibromyalgia (FM) is a complex pain syndrome defined by chronic, widespread pain and tenderness in multiple body locations (Wolfe et al., 1990). Like other pain conditions, it is associated with deficits in central pain processing, although the exact mechanisms by which these deficits occur are not well understood (Eller-Smith, Nicol, & Christianson, 2018). A long-standing hypothesis is that alterations in the functioning of the hypothalamic–pituitary–adrenal axis (HPA axis) may be a significant contributor to deficiencies in pain processing; thus, salivary bioscience investigations have primarily focused on examining cortisol levels in this population. Because cortisol reactivity can be affected by so many possible factors (presence of chronic pain, pain levels at the time of sampling), the data are hard to interpret. One early study examined cortisol in individuals with FM and found higher daily average cortisol levels than healthy controls, despite no differences in stress (Catley, Kaell, Kirschbaum, & Stone, 2000). Yet, hypocortisolism has also been demonstrated in women with FM (Riva, Mork, Westgaard, Rø, & Lundberg, 2010). For example, women with FM were shown to have significantly lower levels of cortisol compared to women with shoulder and neck pain and healthy women with no pain (Riva, Mork, Westgaard, & Lundberg, 2012). These studies either did not obtain current pain ratings or did not compare cortisol levels to pain ratings specifically, which makes it difficult to tease apart possible mechanisms contributing to differences in cortisol reactivity in FM.

A subsequent study did find a positive relationship between salivary cortisol levels and pain ratings at waking and one-hour post-waking, with cortisol levels explaining 38% and 14% of the variation in pain ratings, respectively (McLean et al., 2005). These data suggest that the relationship between cortisol and FM pain is dynamic and fluctuating throughout the day. A more recent study confirmed these findings in a sample of women with FM who provided both pain and stress ratings, as well as saliva samples six times per day for 14 consecutive days (Fischer et al., 2016). Results showed that stress predicted pain ratings, but not vice versa; additionally, cortisol levels were significantly positively related to pain levels only. However, additional research has found only slightly elevated cortisol levels in women with FM compared to healthy controls, despite no relationship of cortisol to laboratory-induced pain via pressure and heat pain thresholds (Wingenfeld, Nutzinger, Kauth, Hellhammer, & Lautenbacher, 2010). Yet, the opposite pattern was demonstrated in a separate study that found no differences in cortisol upon awakening or throughout the day, but significantly elevated cortisol in response to pressure pain threshold measurement (Geiss, Rohleder, & Anton, 2012). Social stressors, on the other hand, may result in attenuated cortisol responses in women with FM (Wingenfeld et al., 2008). These data highlight the potential utility of cortisol as a pain-specific biomarker in fibromyalgia, but also suggest the need for continued investigation of other factors that may alter the cortisol-pain relationship.

One possible pathway that the HPA axis becomes dysregulated in FM is through childhood maltreatment and neglect. A number of studies have demonstrated a relationship between childhood trauma and chronic pain (Davis, Luecken, & Zautra, 2005; Jones, Power, & Macfarlane, 2009), but the mechanisms by which childhood experiences exert their influences on physical health have not been elucidated. Disruption of HPA axis pathways during critical developmental periods may be one explanation; as a result, examining cortisol levels in individuals with childhood trauma and chronic pain may provide some answers. In support of this notion, cortisol has been shown to mediate the relationship between childhood neglect and current pain symptoms in women with FM (Yeung, Davis, & Ciaramitaro, 2016). Other investigations have shown significantly lower levels of cortisol in women with FM and a history of child abuse compared to women with FM and no history of abuse (McLean et al., 2005).

Other potential biomarkers, such as salivary alpha-amylase (sAA) have not been evaluated extensively. In women with FM, levels of sAA showed no relationship to self-reported daily pain (Fischer et al., 2016). There is some evidence that glucocorticoid sensitivity of inflammatory cytokine production is reduced in FM, and a pro-inflammatory cytokine, interleukin-6, is elevated following a pressure pain task and was significantly correlated with ratings of pressure pain and fatigue (Geiss et al., 2012).

Another potential utility of salivary bioscience in this population is to identify potential biomarkers that would facilitate diagnosis and disease prognosis. Differences in expressions of salivary proteins between those with FM and those without have been shown (Bazzichi et al., 2009). The authors attempted to further refine the identification of salivary proteins specific to FM by comparing them to individuals with rheumatoid arthritis, migraine, or healthy controls (Ciregia et al., 2019). Although expressions of specific salivary proteins were able to differentiate those with FM from the other groups in this specific study, additional research for salivary biomarkers is warranted.

#### 14.2.1.2 Dental, Mouth, and Jaw Pain

Like in FM, salivary cortisol has been a widely used biomarker of pain in individuals with various forms of dental/mouth pain. Temporomandibular disorders (TMD) are chronic conditions of the mouth and jaw, with a common symptom of orofacial pain. Cortisol may be a useful biomarker for evaluating patients with TMD to develop new treatment approaches (Wadhwa & Kapila, 2008). Cross-sectional studies have found elevated levels of morning cortisol in TMD patients compared to controls (Chinthakanan et al., 2018; Vrbanović et al., 2018), although other studies have found no group differences and no relationship of self-reported pain to cortisol levels (Jo et al., 2016; Nilsson & Dahlstrm, 2010; Quartana et al., 2010). One study compared salivary cortisol levels in patients with chronic orofacial pain, acute orofacial pain, and healthy controls. This unique study design allows for comparison of duration of pain (chronic versus acute), which may help better determine whether alterations in salivary analyte levels are related to the ongoing experience of pain. Interestingly, no differences in cortisol levels were present between any of the groups, even though the chronic orofacial pain group reported significantly higher levels of pain and stress (Jasim, Louca, Christidis, & Ernberg, 2014). However, pain catastrophizing (which describes the tendency to describe pain in an exaggerated way) may help explain these findings. Pain catastrophizing has been linked to elevated salivary cortisol, *regardless of group* (i.e., TMD versus healthy control) in response to laboratory pain tasks (Quartana et al., 2010).

Patients with recurrent aphthous stomatitis (a relatively common condition where painful ulcers appear in the oral cavity) have shown hypocortisolism compared to controls, regardless of whether they were experiencing active lesions or if the lesions had healed (Rezaei, Aminian, & Raygani, 2017), and cortisol was also unrelated to pain severity. However, acute dental pain due to pulpal or periapical inflammation has been associated with increased cortisol levels compared to controls (Haug & Marthinussen, 2018).

Additional salivary analytes such as dehydroepiandrosterone (DHEA; a steroid secreted by the adrenal glands) have not shown any group differences (Jo et al., 2016), although the research in these areas is still in its infancy. In a sample of individuals with tooth pain, sAA was related to the severity of self-reported pain (Ahmadi-Motamayel et al., 2013), and inflammatory cytokines have also been shown to be elevated in those with acute orofacial pain compared to controls (Haug & Marthinussen, 2018). Salivary total antioxidant capacity [TAC; a biomarker of the antioxidant potential of bodily fluids (Peluso & Raguzzini, 2016)] studies have not produced consistent findings. Two studies found lower levels of salivary TAC in patients with TMD (de Almeida & Amenábar, 2016; Rodríguez de Sotillo, Velly, Hadley, & Fricton, 2011), but two recent studies found significantly higher levels in patients with TMD compared to controls (de Almeida & Amenábar, 2016; Vrbanović et al., 2018). One possible explanation for this difference is that the participants in the study found that elevated TAC had documented chronic pain for at least 6 months; duration of pain was not reported in the other two studies. With regard to other salivary analytes, one recent study found salivary opiorphin (an endogenous opioid peptide) levels were significantly positively correlated to pretreatment tooth pain in participants who received root canals (Ozdogan et al., 2019), although a separate study found no significant relationship of s-IgA (an antibody located in the oral mucosa of the mouth) to self-reported pain during an archwire insertion (José da Silva Campos, César Souza Alves, Rezende Barbosa Raposo, Paula Ferreira, & Willer Farinazzo Vitral, 2010). These data highlight the importance of assessing dynamic changes occurring in salivary measures as a function of continued pain experience, which naturally introduces another potential variable and adds to the challenges in drawing significant conclusions.

### 14.2.1.3 Heterogeneous Chronic Pain

Cortisol has also been widely studied in heterogenous populations of chronic pain, with general trends toward evidence of hypocortisolism in patients with chronic pain. This population includes individuals with widespread musculoskeletal pain and FM. Cortisol has been found to follow a similar trajectory as reported pain intensity following a 4-week intensive pain management program (Evans, Douglas, Bruce, & Drummond, 2008). It also appears as though cortisol levels are negatively associated with degree of chronic pain, such that individuals with generalized chronic pain were three times as likely to have cortisol levels in the lowest third of a group of individuals that also included those "at risk" for chronic pain (as evidenced by high levels of somatization), and healthy controls. Similarly, the individuals in the "at risk" group were 1.8 times more likely than controls to have cortisol levels in the lowest third (Mcbeth et al., 2005). In a follow-up study by the same authors, those in the "at risk" group with new-onset chronic widespread pain showed significantly lower morning salivary cortisol and higher evening salivary cortisol (indicative of a "blunting" cortisol response), suggesting that HPA axis dysfunction may be a predictor of who will develop chronic pain in those with psychological risk factors (McBeth et al., 2007). This evidence of at least morning hypocortisolism in individuals with chronic pain is a finding that has been supported by several other studies (Generaal et al., 2014; Turner-Cobb, Osborn, da Silva, Keogh, & Jessop, 2010).

Only one study to date has examined other salivary measures associated with chronic pain (in this case neuropathic pain) (Kallman, Ghafouri, & Bäckryd, 2018). The authors compared salivary concentrations of beta-endorphins and substance P in individuals with chronic neuropathic pain to healthy controls and found no significant differences in either salivary peptide.

#### 14.2.1.4 Pelvic Pain

No significant differences have been found in salivary cortisol concentrations between women with menstrual pain and healthy women during menstruation, although this study did not assess the correlation of cortisol with pain ratings and also did not specify what time of day the samples were obtained (Park & Watanuki, 2005). Men with chronic prostatitis/chronic pelvic pain (CPP) appear to demonstrate elevated cortisol awakening responses compared to healthy controls (Anderson, Orenberg, Chan, Morey, & Flores, 2008), and diurnal salivary cortisol levels have been shown to be normal to low in women with CPP (Heim, Hanker, & Hellhammer, 1998) and women with endometriosis (Petrelluzzi, Garcia, Petta, Grassi-Kassisse, & Spadari-Bratfisch, 2008), despite women with endometriosis having higher levels of perceived stress (Petrelluzzi et al., 2008). This same pattern was found in another study of women with endometriosis compared to healthy women, although both groups in this study had similar levels of perceived stress (Quiñones, Urrutia, Torres-Reverón, Vincent, & Flores, 2015). Additionally, incapacitating pain was the strongest predictor of hypocortisolism.

In response to a corticotropin releasing factor (CRF; a neuropeptide that regulates activity in the HPA axis) stimulation test or dexamethasone suppression test (DST; a test that measures the responsiveness of the adrenal glands), women with CPP have decreased cortisol levels compared to healthy controls (Heim et al., 1998; Heim, Ehlert, Hanker, & Hellhammer, 1999). Yet, women with CPP showed identical cortisol response patterns as healthy women and women with fibromyalgia in response to a social stress task or an adrenocorticotropic hormone (ACTH; a hormone that also stimulates the adrenal glands to produce cortisol) 1–24 stimulation task (Wingenfeld et al., 2008). These data highlight the complexities of assessing cortisol reactivity.

Measuring prostaglandin levels is a novel application of salivary bioscience and is particularly relevant for women with dysmenorrhea, as elevated prostaglandin levels are believed to be a causal factor for menstrual pain. Only one study to date has examined salivary prostaglandin levels in women with menstrual migraines, each of whom received either a placebo pill or a pill containing sumatriptan succinate and naproxen sodium (Durham et al., 2010). Results indicated elevated levels of prostaglandins following the migraine attack in women who received the pill placebo; women who received the active pill did not show these elevations. It is conceivable that salivary prostaglandin levels may be an indicator of treatment response.

One other area of physiological assessment in CPP is measuring levels of secretory immunoglobulin-A (s-IgA). S-IgA—an antibody that helps prevent adherence of potentially harmful microorganisms to healthy cells (Corthésy, 2010)—is a marker of immune function and may play a role in the experience of pain (Willemsen et al., 1998). A single study examined this measure in women with severe menstrual pain and found higher concentrations of s-IgA in women with menstrual pain compared to healthy controls during menses only; there were no group differences in non-menstrual phases of the cycle (Park & Watanuki, 2005).

#### 14.2.1.5 Neck and Back Pain

In a study comparing the cortisol awakening responses of patients with acute low back pain to those with chronic low back pain, the authors found no group differences in cortisol profiles (Sudhaus et al., 2009). However, within the chronic low back pain group, those with elevations on psychometric measures such as avoidance of social activities, depression, and fatigue showed blunted or weakened cortisol responses compared to those who scored low on these measures. This relationship was not evident in the acute low back pain group. Lower cortisol has also been associated with increased health complaints, pain, and fatigue in patients with low back pain (Sveinsdottir, Eriksen, Ursin, Hansen, & Harris, 2016). Interestingly, one study examined gender-related salivary cortisol levels in both patients with chronic neck, back, or shoulder pain and healthy controls. Cortisol levels were significantly *higher* in the pain group, but for men only (Schell, Theorell, Hasson, Arnetz, & Saraste, 2008).

Salivary cortisol can also be a useful indicator of treatment response. A recent study manipulated treatment expectations for physical therapy (positive, neutral, and negative) in patients with neck pain (Malfliet, Lluch Girbés, Pecos-Martin, Gallego-Izquierdo, & Valera-Calero, 2018). Increased cortisol was observed in patients in the neutral and negative expectations groups, although not in the positive expectation group. Additionally, salivary cortisol was not related to pain intensity report. Another study found that significant increases in both morning and evening cortisol levels in patients with low back pain who were treated using Mindfulness-Based Stress Reduction (Ardito et al., 2017; Kabat-Zinn, 2006). Similarly, salivary cortisol may be a useful tool for identifying individuals who will transition from acute pain to chronic pain (Nees, Löffler, Usai, & Flor, 2019).

Laboratory studies of pain sensitivity have also revealed differences in salivary cortisol between patients with low back pain and healthy controls. Another study found significantly lower salivary cortisol (measured as "area under the curve," AUC) in chronic low back pain patients compared to patients with depression and healthy controls, following experimental pain testing using heat stimuli (Muhtz et al., 2013). However, the opposite pattern has also been demonstrated—individuals with chronic low back pain show similar cortisol reactivity to healthy individuals following heat pain testing (Vachon-Presseau et al., 2013).

Like in other pain conditions, sAA has also been shown to be a possible biomarker for pain intensity in this population, with sAA levels positively correlating with reports of pain intensity using a visual analog scale (VAS) (Shirasaki et al., 2007). Similarly, substance P has provided some interesting results in its relationship to pain. There are data to indicate substance P levels are lower in patients with chronic pain compared to healthy controls (Parris, Kambam, Naukam, & Rama Sastry, 2006; Parris, Rama Sastry, Kambam, Naukam, & Johnson, 2006). However, no recent studies have continued this line of investigation with regard to this specific biomarker.

#### 14.2.1.6 Headache/Migraine

Headache and migraine studies are unique in that they are recurrent pain conditions (typically not chronic), so that each individual experiences intervals of pain and no pain. Interestingly, only one study has examined salivary cortisol in chronic headache/migraine. Chronic migraine has been associated with increased diurnal cortisol levels compared to controls (Patacchioli et al., 2006). Two other studies have evaluated sAA as a potential biomarker for sympathetic arousal in this population. One study found that women with frequent, episodic tension-type headache had significantly higher sAA levels compared to healthy women, and sAA levels were significantly positively correlated with pain ratings in the headache group (Vahedi et al., 2018). A separate study went beyond evaluating headache/control group differences and compared sAA levels in four groups: individuals currently experiencing a migraine attack, those in post-attack, those who were in a pain-free interval between migraines, and healthy controls (Bugdayci, Yildiz, Altunrende, Yildiz, & Alkoy, 2010). Compared to the control group, sAA levels were significantly *lower* during a migraine attack and significantly *higher* in post-attack. The pain-free interval and control groups did not show any differences in sAA levels. Pain ratings were not correlated to sAA levels in any group.

A small group of additional salivary analytes have also been explored in this population. Salivary magnesium levels, which have been thought to contribute to the onset of headache, are lower in patients with migraine or tension-type headache compared to controls (Gallai et al., 1992). Nerve growth factor, Substance P, and calcitonin gene-related peptide (CGRP) levels have been shown to be higher in individuals with chronic migraine compared to healthy controls, with Substance P and CGRP highly positively correlated with reported pain intensity (Jang, Park, Kho, Chung, & Chung, 2011). Salivary measures of testosterone and dehydroepiandrosterone-sulfate have not shown group differences when comparing chronic migraine and healthy controls (Patacchioli et al., 2006). However, oxytocin and interleukin—1 beta did show elevations in migraineurs compared to controls, with changes in those measures appearing to reflect degree of clinical improvement following vagal nerve stimulation (Boström et al., 2019).

#### 14.2.1.7 Cancer Pain

To our knowledge, only one study has evaluated cancer-related pain and salivary biomarkers in adults. Thirty-eight patients with cancer were asked to rate their overall cancer pain intensity and provide a saliva sample for analysis. Levels of sAA were significantly positively correlated with pain intensity in this sample (Arai et al., 2009). Although it is difficult to draw conclusions about this population from this single study, the results support other findings of the positive relationship between sAA and indices of pain intensity.

#### 14.2.1.8 Other Pain Syndromes

Other pain syndromes have not been studied widely, but some general patterns have revealed the possibility of using salivary biomarkers for diagnostic assessment. Alpha-amylase levels have been shown to successfully discriminate those with myocardial infarction from those with acute pain (and no myocardial infarction) (Shen et al., 2012) and women with and without carpal tunnel syndrome (Fernández-de-las-Peñas et al., 2014). Cortisol and sAA was also related to the severity of carpal tunnel syndrome. In patients with interstitial cystitis and chronic myogenous facial pain, there have been no differences found in cortisol levels compared to controls (Galli et al., 2008; Lutgendorf et al., 2002). However, cortisol levels do appear to be related to disease severity in a number of pain conditions, with hypocortisolism associated with greater symptom severity (Lutgendorf et al., 2002; Nierop et al., 2006), although in osteoarthritis (Carlesso, Sturgeon, & Zautra, 2016) and rheumatoid arthritis (Kim, Jeon, Koh, Park, & Suh, 2016), the opposite has been true.

Additional studies have found elevations in free radicals (peroxidase, superoxide dismutase) and uric acid and total antioxidant status in patients with Complex Regional Pain Syndrome—Type 1 (Eisenberg et al., 2008) and salivary exosomes in inflammatory bowel disease (Zheng et al., 2017).

### *14.2.2 Nonclinical Populations*

Salivary bioscience has also been evaluated in nonclinical populations to determine normative levels of various analytes either in experimental or naturalistic settings.

#### 14.2.2.1 Experimental Pain

Thermal Pain

The cold pressor test (CPT) involves participants putting their entire hand and forearm in a large container of cold water, typically around 12 °C. The CPT has been shown to be associated with increased cortisol responses and decreased tnf-alpha (a measure of inflammation) reactivity (Goodin, Smith, Quinn, King, & McGuire, 2012). However, positive social support can possibly attenuate cortisol reactivity to the CPT (Roberts, Klatzkin, & Mechlin, 2015). In response to a CPT, women were found to report higher levels of pain display more overt behavioral indicators of pain (i.e., jumping and cursing) compared to men; however, these differences were not accounted for by salivary testosterone levels (Archey, Goldey, Crockett, & Boyette-Davis, 2018). The CPT was also used to evaluate cortisol reactivity in healthy adults with different styles of coping with pain. People who reported being fearful of and avoiding pain were more likely to show an increase in cortisol responses at baseline and following the CPT, compared to people who reported using humor or distraction or persisting despite pain who tended to show lower cortisol baseline levels and decreased cortisol responses following the cold pressor (Sudhaus et al., 2015). Similarly, cortisol responses following a DST have been linked with pain sensitivity, with cortisol suppression associated with higher pain tolerance ratings during the CPT and impaired pain inhibition (Godfrey et al., 2017). The CAR is a significant predictor of postoperative pain (Lautenbacher et al., 2009). Studies have also investigated how the CAR may be related to cortisol and tnf-alpha reactivity following cold and hot water pain tasks. Increased CAR was associated with higher pain intensity and pain unpleasantness; however, CAR was not related to cortisol or tnf-alpha reactivity to the pain tasks (Goodin, Quinn, et al., 2012). However, the authors found the opposite results in a previous study where suppression of CAR was associated with increased pain intensity and unpleasantness during the CPT (Fabian et al., 2009). In support of these findings, a study by the same authors found that baseline cortisol and tnf-alpha levels (obtained prior to administering the CPT) were not related to the change in CPT pain responses following either a brief hypnosis session or no intervention, *even though hypnosis was associated with lower pain intensity and unpleasantness* (Goodin, Quinn, et al., 2012). Poor sleep has also been shown to be predictive of greater cortisol reactivity to the CPT and mediate the relationship between poor sleep quality and pain severity (Goodin, Smith, et al., 2012).

In a sample of healthy adults, sAA has been used to measure responses to heat pain stimuli. Even though sAA is a marker of sympathetic activity in response to stress, it can also be a potentially useful biomarker for pain reactivity in healthy populations. Levels of sAA have been shown to be related to self-reported pain intensity and unpleasantness of heat stimuli, but not pain tolerance (Wittwer, Krummenacher, La Marca, Ehlert, & Folkers, 2016). Sex differences in responses to thermal pain can also shed light on individual patterns of reactivity.

Other Models of Experimental Pain

Other studies have used different methods for delivering experimental pain. Although the exact relationship between cortisol and pain reactivity to electrical stimulation was not reported, one study found that pain ratings of electrical pain were higher when participants were under stress, and cortisol levels were also higher during stress (Choi, Chung, & Lee, 2012). Salivary melatonin also appears to be reactive to acute electrical pain stimuli; although the pattern of melatonin levels during recovery may show time-dependent both decreases and increases (Nelson, Farr, Ebadi, & Nelson, 2001).

#### 14.2.2.2 Naturalistic Pain Settings

Evaluating physiological reactivity and nonexperimental stressors provides a unique method of assessing "real life" reactivity. One study evaluated women in labor by assessing salivary levels of chromogranin A (CgA), which is a marker of psychological stress, separate from physical stress. Results indicated that CgA levels were significantly higher when the sample was taken when the participant showed 10 cm of cervical dilation, as compared to 4–6 cm cervical dilation and immediately after birth (Iizuka, Masaoka, & Ohashi, 2018). In a similar population, there were no differences in salivary substance P between women in active labor, pregnant women not in labor, nonpregnant women without pain, nonpregnant women with acute pain due to postoperative hysterectomy (Dalby et al., 1997). However, sAA levels have not been shown to be related to pain during dental extraction (Lee & Bassiur, 2017). Salivary melatonin levels have been measured in relation to gastrointestinal symptoms in patients seeking psychiatric care (Söderquist et al., 2019). Melatonin levels after lunch were significantly related to report of gastrointestinal symptoms, particularly gastrointestinal pain and bloating. Other studies have shown elevated pro- and anti-inflammatory cytokine levels following an experimental cold pressor test as compared to an experimental warm water test, but no elevations were found following venipuncture (Cruz-Almeida et al., 2017). Additionally, blunted CARs have been shown during menses in women with premenstrual syndrome, as compared to other phases of the menstrual cycle. Menses was also associated with the highest pain intensity, although there were no significant correlations between cortisol measures and pain intensity (Ozgocer, Ucar, & Yildiz, 2017).

### *14.2.3 Methodological Considerations*

With regard to adult pain, there have been mixed findings across both chronic pain diagnoses and healthy samples. One relatively consistent finding is that widespread pain conditions, such as FM, do appear to be associated with a blunting of salivary cortisol responses. Attempting to identify specific salivary biomarkers related to adult pain presents a number of challenges. Varied methodologies, including the time of day the sample was obtained and the analytical method (e.g., using AUC or raw levels) have made it difficult to draw conclusions across studies, as well as across pain diagnoses (Sobas et al., 2016). Variations in sample collection methods (e.g., unstimulated whole saliva versus stimulated sublingual saliva, etc.) have also likely contributed to inconsistent findings across biomarkers and in different populations (Jasim, Carlsson, Hedenberg-Magnusson, Ghafouri, & Ernberg, 2018). Additionally, the number of potential configurations for comparing groups, such as chronic pain versus healthy, chronic pain versus healthy in response to acute pain, comparing specific pain conditions, etc., presents many challenges. Likely, inaccurate or nonspecific phenotyping of various pain conditions has contributed to the challenges of identifying specific salivary pain biomarkers.

## 14.3 Salivary Bioscience and Pain in Pediatric Populations

### *14.3.1 Neonatal Pain*

A number of studies have examined salivary biomarkers in the context of procedural pain and have largely focused on salivary cortisol as a biomarker for pain as a result of early exposure to painful procedures in neonates. For example, the earliest study we identified documented increased salivary cortisol in a cross-sectional study of 2-, 4-, and 6-month-old infants in response to immunization and showed correlation in behavioral responses indicative of pain and cortisol levels (Lewis & Thomas, 1990). A significant concentration of the literature has also been on cortisol in preterm infants, with early work focused on the feasibility of salivary sampling for cortisol assay in infants in response to heel stick—a common painful procedure for neonates. This work demonstrated elevations in cortisol following the heel stick, although unlike the previous study, did not find correlations between cortisol levels and behavioral response to pain (Herrington, Olomu, & Geller, 2004). Additional work focused on older infants showed similar results, with 6-month old infants demonstrating increases in cortisol following routine immunizations, but no associations between cortisol and behavioral reactivity (Ramsay & Lewis, 2003). The inconsistency in findings related to cortisol and behavioral response in infants may reflect differences in observational measures of pain used in these studies. These discrepancies highlight the importance of relying on multimodal assessment of pain, particularly in infants.

Further evaluation of how exposure to distressing and painful procedures early in life may impact HPA axis functioning in preterm infants has resulted in some mixed findings, but examination of cortisol response in this population supports potential alterations in the stress response system. For example, early work showed that extremely low gestational age (ELGA) infants had altered cortisol response, including higher basal and more sustained cortisol levels, compared to very low gestational age (VLGA) infants, suggesting that increased exposure to painful procedures in

ELGA compared to VLGA infants may lead to alterations in the stress response system (Grunau, Weinberg, & Whitfield, 2004). Additional studies focused on preterm infants documented a blunted cortisol response compared to full-term or near-term infants, and this response was associated with the number of painful procedures the infant experienced (Provenzi et al., 2016) and sex, with male preterm infants showing more blunted response compared to female preterm infants (Grunau et al., 2010). Consistent with this later work, school-age boys who were born preterm who had higher neonatal procedural distress showed lower cortisol levels compared to full-term boys (Brummelte et al., 2015). Thus, early exposure to pain may lead to alterations in the HPA axis, as demonstrated by assessment of cortisol in preterm infants.

In addition to cortisol, salivary alpha-amylase, a salivary biomarker of the sympathetic nervous system component of the stress response, has also been examined in the context of painful procedures in infants (Davis & Granger, 2009). This work suggests that evidence of the sympathetic nervous system via sAA develops somewhere between 2 and 6 months of age and that infant sAA levels correlate with maternal levels (Davis & Granger, 2009). Although a number of investigators have examined sAA in neonates as a marker of stress, to date we were unable to locate any studies that included sAA as a biomarker of pain in infants. However, as will be presented later in the chapter, sAA in response to pain has been examined in school-aged children.

### *14.3.2 Children with Clinical Pain Syndromes*

Chronic pain conditions are prevalent in children and although estimates vary widely, it is suggested that nearly 40% of children may experience chronic pain (Huguet & Miró, 2008). Recent research documents that depending upon the type of pain encountered, prevalence rates may even be over 80% for some pain conditions, such as headache (King et al., 2011). Chronic pain in children is associated with a wide array of negative sequelae, including difficulties with school attendance and academic performance, impaired sleep and social and role functioning (Haraldstad, Sørum, Eide, Natvig, & Helseth, 2011), and comorbid emotional disorders such as depression and anxiety (Kashikar-Zuck, Goldschneider, Powers, Vaught, & Hershey, 2001; Knook et al., 2011).

#### 14.3.2.1 Functional Abdominal Pain

Functional abdominal pain (FAP) is common in children, with epidemiological research suggesting worldwide prevalence rates of 13.5% (Korterink, Diederen, Benninga, & Tabbers, 2015) and perhaps even as high as 53% (King et al., 2011). Although the body of literature on salivary bioscience and chronic pain in children is small, several studies have focused on FAP and such studies have examined cortisol and sAA in children with FAP compared to healthy controls or have examined these biomarkers in children with FAP in response to a stressor. In examinations of children with FAP compared to healthy children, morning and total cortisol levels were found to be higher, although the authors noted that the cortisol results may have reflected comorbid depression in the FAP sample (Tornhage & Alfven, 2006; Törnhage & Alfvén, 2015). A study by Dorn and colleagues comparing cortisol response to a stressful procedure in children with FAP and those with anxiety disorders found no differences in baseline or maximum cortisol levels between groups (Dorn et al., 2003). Finally, children with FAP were found to have a blunted cortisol response to a stressful laboratory task compared to healthy children (Gulewitsch et al., 2017). Thus, based upon this small body of literature, it is difficult to draw conclusions of how FAP may impact response to pain as assessed by cortisol.

#### 14.3.2.2 Juvenile Primary Fibromyalgia Syndrome

The only additional pediatric chronic pain article we located that included salivary biomarkers focused on children with juvenile primary fibromyalgia syndrome (JPFMS), which is an amplified pain syndrome associated with musculoskeletal pain in conjunction with fatigue and disruptions in sleep and emotional functioning (Anthony & Schanberg, 2001). In a comparison of salivary cortisol levels in children with JPFMS to a group of children with arthritis and a group of healthy controls in response to venipuncture, no significant differences were found between groups (Conte, Walco, & Kimura, 2003). However, the authors noted the procedure may not have induced a significant stress response in children and that the baseline saliva sampling may have been too temporally proximate to the procedure. Additionally, circadian rhythm and baseline cortisol levels were not taken into account.

### *14.3.3 Pain Associated with Medical Disease*

#### 14.3.3.1 Cancer

Salivary biomarkers of pain have been examined in children with cancer, a pediatric population that is subjected to repeated painful medical procedures and experiences chronic and recurrent pain associated with cancer and its treatment (Miller, Jacob, & Hockenberry, 2011). Two studies have examined sensitivity to pain and physiological pain responses using salivary cortisol, connecting psychosocial and physiological aspects of pain in children diagnosed with cancer (Chen, Craske, Katz, Schwartz, & Zeltzer, 2000; Firoozi & Rostami, 2012). Both studies showed that children who had lower levels of pain tolerance demonstrated increases in cortisol following a medical procedure, including a lumbar puncture (Chen et al., 2000) and administration of chemotherapy (Firoozi & Rostami, 2012). In examining a range of self-report, observational, and physiological indices of pain during a lumbar puncture, including salivary cortisol, low associations between assessment modalities were found (Walco, Conte, Labay, Engel, & Zeltzer, 2005). For example, cortisol increased significantly over the course of the painful procedure, even when observational and/or self-report assessments did not show elevations in pain. Thus, it is important to note that these various modalities of pain assessment should not be used in isolation or to the exclusion of other modalities.

Jenkins and colleagues examined sAA in response to an experimental pain task in children undergoing treatment for cancer (Jenkins et al., 2018). Experimental pain tasks allow for safe induction of pain that allows for methodological standardization in order to better isolate aspects of pain response, including tolerance and sensitivity. In this study, children ages 8 and up were randomized to one of three emotional regulation conditions (reassurance, reappraisal, or distraction) and participated in the CPT, which involves submerging the nondominant hand in 7 °C water until the child can no longer withstand the pain. Results demonstrated that certain emotion regulation strategies seemed to reduce the stress response to pain as reflected by lower levels of sAA during the CPT. Specifically, children in the distraction condition demonstrated greater sAA reactivity (levels that continued to rise following the CPT) compared to children in the other two conditions (Jenkins et al., 2018).

### *14.3.4 Pain Associated with Neurodevelopmental Disorders*

Assessment of pain in children with neurodevelopmental disorders can be challenging as evidence suggests such children are at high risk for experiencing pain and many children with neurodevelopmental disorders are unable to verbally report pain (Oberlander, O'Donnell, & Montgomery, 1999). Accordingly, strategies that focus on nonverbal modes of pain assessment, such as observational and physiological measures are important. The Paediatric Pain Profile (PPP) is one such measure that was developed for use in children with severe neurological disability (Hunt et al., 2007). In efforts to validate the PPP, salivary cortisol was analyzed in relation to observational pain ratings; however the authors did not find significant correlations between cortisol levels and pain severity (Hunt et al., 2007). This may be due to methodological issues, such as the timing of the sampling, which was not necessarily conducted in the context of painful events, and may also be a result of the finding that salivary cortisol concentrations in this population were lower when compared to data on healthy controls.

#### 14.3.4.1 Epilepsy

Evidence suggests that children with epilepsy may be at higher risk for the experience of pain syndromes given the high rates of depression found in people with epilepsy, the high comorbidity of chronic pain and depression, and higher rates of pain in two conditions related to epilepsy—traumatic brain injury (TBI) and migraine headaches (Finocchi, Villani, & Casucci, 2010; Kanner et al., 2012). We identified one investigation of pain response via sAA in response to a venous blood draw in children with epilepsy compared with healthy controls (Ferrara et al., 2013). Results showed that children with epilepsy experienced greater sAA reactivity compared to healthy controls, despite no differences in self-reported pain scores in the epilepsy group over the course of the procedure. Conversely, the self-reported pain severity of the healthy control group increased significantly over the course of the procedure. The authors could not rule out possible interactions between anti-seizure medications and alpha-amylase; nonetheless, this study represents an important potential avenue for assessment of pain in neurologically impaired children.

#### 14.3.4.2 Attention-Deficit/Hyperactivity Disorder

Typical cortisol levels follow a circadian rhythm, with baseline levels at their lowest in the morning followed by a peak that gradually decreases throughout the afternoon and evening. Children with attention-deficit/hyperactivity disorder (ADHD) have been found to have atypical cortisol patterns, including decreased awakening cortisol levels and blunted cortisol levels in response to stressors (Kariyawasam, Zaw, & Handley, 2002; Randazzo, Dockray, & Susman, 2008). Thus, McCarthy and colleagues examined salivary cortisol levels in response to an intravenous (IV) catheter insertion in children with and without ADHD (McCarthy et al., 2011). In this study, children with ADHD demonstrated higher (although not significant) baseline cortisol levels at home prior to the painful procedure but significantly lower cortisol levels before and after the IV insertion compared to children without ADHD. This study provides some evidence that physiological response to stressful and painful procedures may be altered in children with neurodevelopmental disorders such as ADHD.

#### 14.3.4.3 Cerebral Palsy

Pain in individuals with cerebral palsy (CP), the most common cause of severe neurological impairment in childhood, is frequently present as a result of spasticity, contractures, and bony deformities (Surveillance of Cerebral Palsy in Europe, 2000). In children with CP, pain is frequently as issue due to early developmental interventions that involve physical manipulation (Zhao, Chen, Du, Li, & Li, 2015). Two studies identified have focused on salivary cortisol response to pain in children with CP. To document the feasibility of saliva collection and cortisol as a possible marker to discriminate pain versus no pain in children with CP, Symons and colleagues examined cortisol response to a medical procedure in children with CP with and without pain (Symons et al., 2015). Findings showed that cortisol levels were higher in children with pain compared to those without. Zhao and colleagues examined cortisol response to neurodevelopmental intervention programs and found significant increases in cortisol in response to the majority of interventions conducted with children ages 1 to 4 years (Zhao et al., 2015). Thus, although this body of literature is scant, there is some evidence that cortisol may be a useful salivary biomarker of pain in children with CP.

### 14.3.5 Pain in Children Without Medical or Developmental Disorders

A few studies located have focused on salivary biomarkers of pain in healthy children or children without any specified disease or condition. One of these represented a descriptive study to provide normative data on cortisol response to IV placement in children and included a large sample ($N = 384$) of children ages 4 to 10 years (McCarthy et al., 2009). Baseline samples were collected from children at home prior to the procedure and then again on the day of the procedure prior to and following their IV placement. Results showed that baseline cortisol levels reflected the expected pattern, peaking in early morning and then gradually decreasing throughout the day. On the day of the painful procedure, cortisol levels were higher than baseline and increased significantly at IV insertion, suggesting that cortisol may be a valid biomarker of pain in healthy children (McCarthy et al., 2009).

Allen and colleagues examined cortisol response to experimentally induced pain in healthy children and specifically focused on sex differences, given sex differences found in adults (Allen, Lu, Tsao, Worthman, & Zeltzer, 2009). Participants included 235 healthy children ages 8–18 years who completed three laboratory pain tasks, including pressure, heat, and CPT. Salivary cortisol samples were taken at baseline, after completion of all tasks, and 20 min post-completion. Cortisol levels did not increase over the course of the study, in fact, the highest cortisol level in both boys and girls was the baseline sample, suggesting that as children became familiar with study procedures, stress levels decreased. Boys' baseline cortisol levels were higher, although not statistically, compared to girls'; thus, it remains unclear whether sex differences in cortisol response to pain emerges in childhood (Allen et al., 2009).

Finally, Payne and colleagues examined sAA and cortisol in response to experimentally induced pain in healthy children with and without social anxiety (Payne, Hibel, Granger, Tsao, & Zeltzer, 2014). Their sample included 231 children ages 8 to 18 who completed three laboratory pain tasks, including cold, pressure, and heat. Saliva samples were obtained at baseline, after completion of all tasks, and 20 min post-completion. Overall, sAA did not change across time and cortisol decreased from time one to time two and then remained stable (Payne et al., 2014). Children with social anxiety had higher levels of sAA, but not cortisol at each assessment point. Thus, although cortisol reactivity has been demonstrated in a study of healthy children undergoing a painful medical procedure, neither cortisol or sAA increased in experimentally induced pain, suggesting that laboratory pain models may not be ideal for salivary biomarkers of pain in healthy children.

### 14.3.6 Methodological Considerations

Saliva sampling is often a preferred approach to assessment of biomarkers, particularly in children, because it is simple, noninvasive, and painless. In the context of multimodal pain assessment, salivary biomarkers are useful to provide a more complete picture of the pain experience in children and are particularly helpful in nonverbal populations, such as infants and cognitively impaired children. However, it is important to keep in mind that although biomarkers such as cortisol and alpha-amylase may increase in the context of pain, they reflect the stress response more broadly and thus should not be used as a sole indicator of pain. There is also no standardization in interpreting changes in these biomarkers in a clinically meaningful manner. That is, although self-report scores of 0–10 are often accepted as the gold standard for assessment of pain severity and treatment decisions, with numeric ratings reflecting categories of pain severity (e.g., $\geq 3$ = mild pain, 4–6 = moderate pain, etc.) (Boonstra et al., 2016) there are no such standardized tools that include clinically meaningful interpretations of changes in physiological parameters of pain.

In terms of sampling, although saliva is objectively noninvasive, as noted by Granger and colleagues, there are some populations in which saliva sampling may pose particular challenges (Granger et al., 2007). For example, collecting saliva from preterm and newborn infants can be difficult due to the inability to collect a sufficient volume of saliva. This is particularly concerning due to evidence that insufficient sample volume, particularly when absorbent collection devices are used, can influence the accuracy of assay results (Harmon, Hibel, Rumyantseva, & Granger, 2007). In addition, it is not well established when neonates establish diurnal cortisol patterns and data regarding normative development of basal cortisol levels in infants is lacking (Tryphonopoulos, Letourneau, & Azar, 2014). In older infants, additional challenges can hamper collection, including sleep-wake cycles, residual food and liquids in the mouth, and the simple unwillingness of an infant with stranger anxiety to allow a researcher to collect saliva (Granger et al., 2007). Nonetheless, there are alternative sample collection strategies (e.g., microsponge, paper) that are proving promising in aiding in sample collection in infants (Granger et al., 2007; Tryphonopoulos et al., 2014; Voegtline & Granger, 2014). Jessop and Turner-Cobb (2008) published a review of factors that can impact the accuracy of cortisol assessment in children and noted that age, sex, body mass index, conditions of sampling, units of measurement, assay conditions, and compliance are all important factors that can lead to variations in the precision and accuracy of salivary cortisol assessment (Jessop & Turner-Cobb, 2008). These authors suggest that rather than focusing on the establishment of normative basal cortisol levels, it is important to understand how these factors can impact the diurnal cortisol pattern so we can begin to understand atypical cortisol responses and their sequelae (Jessop & Turner-Cobb, 2008).

## 14.4 Future Directions

More accurate phenotyping of participants is the next critical step in identifying candidate salivary biomarkers for pain. The mere presence of a pain diagnosis as the primary inclusion criterion results in tremendous variability within the group, which is likely obscuring individual differences that could identify subtypes. Assessing the duration of pain, the contribution of additional body pain sites (that may not necessarily be chronic pain), and the influence of gonadal hormones and the menstrual cycle, are just some of the important variables to consider when attempting to identify different pain signatures associated with salivary biomarkers. Related to this, perhaps moving away from comparing pain diagnoses and instead focusing on those with central sensitization will result in a more accurate and generalizable phenotype (Woolf, 2011). Newly identified salivary biomarkers of pain will hopefully shed additional light on how pain can affect various bodily functions. Using markers of uric acid levels to assess painful conditions that are not typically thought of as chronic pain, such as gout, may provide additional answers about the role of pain in chronic medical conditions. Salivary oxytocin, which is beginning to be used intranasally as an anesthetic, (Paloyelis et al., 2016) and salivary opiorphin also show promise as pain biomarkers (Rock, Kataoka, & Lai, 2013). Future work in the area of salivary bioscience and pain should also focus on standardizing methodological approaches. For example, many studies have used the method of analyzing the AUC for salivary cortisol across time (e.g., Muhtz et al., 2013), but this is not entirely consistent. Using a standardized analytical approach will facilitate the generalizability of results.

## References


Ahmadi-Motamayel, F., Shahriari, S., Goodarzi, M. T., Moghimbeigi, A., Jazaeri, M., & Babaei, P. (2013). The relationship between the level of salivary alpha amylase activity and pain severity in patients with symptomatic irreversible pulpitis. *Restorative Dentistry and Endodontics, 38*(3), 141–145. https://doi.org/10.5395/rde.2013.38.3.141

Allen, L. B., Lu, Q., Tsao, J. C. I., Worthman, C. M., & Zeltzer, L. K. (2009). Sex differences in the association between cortisol concentrations and laboratory pain responses in healthy children. *Gender Medicine, 6*(2), 193–207. https://doi.org/10.1016/j.genm.2009.03.001

Anderson, R. U., Orenberg, E. K., Chan, C. A., Morey, A., & Flores, V. (2008). Psychometric profiles and hypothalamic-pituitary-adrenal axis function in men with chronic prostatitis/chronic pelvic pain syndrome. *Journal of Urology, 179*(3), 956–960. https://doi.org/10.1016/j.juro.2007.10.084

Anthony, K. K., & Schanberg, L. E. (2001). Juvenile primary fibromyalgia syndrome. *Current Rheumatology Reports, 3*(2), 165–171.

Arai, Y. C., Matsubara, T., Shimo, K., Osuga, T., Ushida, T., Suzuki, C., … Arakawa, M. (2009). Small correlation between salivary alpha-amylase activity and pain intensity in patients with cancer pain. *Acta Anaesthesiol Scand, 53*(3), 408. https://doi.org/10.1111/j.1399-6576.2008.01833.x

Archey, M., Goldey, K., Crockett, E., & Boyette-Davis, J. (2018). An investigation of the effects of testosterone and behavioral expressions of pain on sex/gender differences in pain perception. *Psychological Reports, 0*(0), 1–15. https://doi.org/10.1177/0033294118781320

Ardito, R. B., Pirro, P. S., Re, T. S., Bonapace, I., Menardo, V., Bruno, E., & Gianotti, L. (2017). Mindfulness-based stress reduction program on chronic low-Back pain: A study investigating the impact on endocrine, physical, and psychologic functioning. *The Journal of Alternative and Complementary Medicine, 23*(8), 615–623. https://doi.org/10.1089/acm.2016.0423

Bazzichi, L., Ciregia, F., Giusti, L., Baldini, C., Giannaccini, G., Giacomelli, C., … Lucacchini, A. (2009). Detection of potential markers of primary fibromyalgia syndrome in human saliva. *Proteomics – Clinical Applications, 3*(11), 1296–1304. https://doi.org/10.1002/prca.200900076

Boonstra, A. M., Stewart, R. E., Köke, A. J. A., Oosterwijk, R. F. A., Swaan, J. L., Schreurs, K. M. G., & Schiphorst Preuper, H. R. (2016). Cut-off points for mild, moderate, and severe pain on the numeric rating scale for pain in patients with chronic musculoskeletal pain: Variability and influence of sex and Catastrophizing. *Frontiers in Psychology, 7*, 1466. https://doi.org/10.3389/fpsyg.2016.01466

Boström, A., Scheele, D., Stoffel-Wagner, B., Hönig, F., Chaudhry, S. R., Muhammad, S., … Kinfe, T. M. (2019). Saliva molecular inflammatory profiling in female migraine patients responsive to adjunctive cervical non-invasive vagus nerve stimulation: The MOXY study. *Journal of Translational Medicine, 17*, 53. https://doi.org/10.1186/s12967-019-1801-y

Brummelte, S., Chau, C. M. Y., Cepeda, I. L., Degenhardt, A., Weinberg, J., Synnes, A. R., & Grunau, R. E. (2015). Cortisol levels in former preterm children at school age are predicted by neonatal procedural pain-related stress. *Psychoneuroendocrinology, 51*, 151–163. https://doi.org/10.1016/j.psyneuen.2014.09.018

Bugdayci, G., Yildiz, S., Altunrende, B., Yildiz, N., & Alkoy, S. (2010). Salivary alpha amylase activity in migraine patients. *Autonomic Neuroscience, 155*(1–2), 121–124. https://doi.org/10.1016/j.autneu.2010.01.011

Carlesso, L. C., Sturgeon, J. A., & Zautra, A. J. (2016). Exploring the relationship between disease-related pain and cortisol levels in women with osteoarthritis. *Osteoarthritis and Cartilage, 24*(12), 2048–2054. https://doi.org/10.1016/j.joca.2016.06.018

Catley, D., Kaell, A. T., Kirschbaum, C., & Stone, A. A. (2000). A naturalistic evaluation of cortisol secretion in persons with fibromyalgia and rheumatoid arthritis. *Arthritis and Rheumatism, 13*(1), 51–61.

Chen, E., Craske, M. G., Katz, E. R., Schwartz, E., & Zeltzer, L. K. (2000). Pain-sensitive temperament: Does it predict procedural distress and response to psychological treatment among children with cancer? *Journal of Pediatric Psychology, 25*(4), 269–278.

Chinthakanan, S., Laosuwan, K., Boonyawong, P., Kumfu, S., Chattipakorn, N., & Chattipakorn, S. C. (2018). Reduced heart rate variability and increased saliva cortisol in patients with TMD. *Archives of Oral Biology, 90*, 125–129. https://doi.org/10.1016/j.archoralbio.2018.03.011

Choi, J. C., Chung, M. I., & Lee, Y. D. (2012). Modulation of pain sensation by stress-related testosterone and cortisol. *Anaesthesia, 67*(10), 1146–1151. https://doi.org/10.1111/j.1365-2044.2012.07267.x

Ciregia, F., Giacomelli, C., Giusti, L., Boldrini, C., Piga, I., Pepe, P., … Bazzichi, L. (2019). Putative salivary biomarkers useful to differentiate patients with fibromyalgia. *Journal of Proteomics, 190*, 44–54. https://doi.org/10.1016/j.jprot.2018.04.012

Conte, P. M., Walco, G. A., & Kimura, Y. (2003). Temperament and stress response in children with juvenile primary fibromyalgia syndrome. *Arthritis and Rheumatism, 48*(10), 2923–2930.

Corthésy, B. (2010). Role of secretory immunoglobulin A and secretory component in the protection of mucosal surfaces. *Future Microbiology, 5*(5), 817–829. https://doi.org/10.2217/fmb.10.39

Cruz-Almeida, Y., Aguirre, M., Sorenson, H., Tighe, P., Wallet, S., & Riley, J. (2017). Age differences in salivary markers of inflammation in response to experimental pain: Does venipuncture matter? *Journal of Pain Research, 10*, 2365–2372. https://doi.org/10.2147/jpr.s138460

Dalby, P. L., Ramanathan, S., Roy, L., Aber, A., Rudy, T. E., & Amenta, J. S. (1997). Plasma and saliva substance P levels: The effects of acute pain in pregnant and non-pregnant women. *Pain, 69*, 263–267. https://doi.org/10.1016/s0304-3959(96)03258-7

Davis, E. P., & Granger, D. A. (2009). Developmental differences in infant salivary alpha-amylase and cortisol responses to stress. *Psychoneuroendocrinology, 34*(6), 795–804. https://doi.org/10.1016/j.psyneuen.2009.02.001

Davis, D. A., Luecken, L. J., & Zautra, A. J. (2005). Are reports of childhood abuse related to the experience of chronic pain in adulthood? A meta-analytic review of the literature. *Clinical Journal of Pain, 21*(5), 398–405. https://doi.org/10.1097/01.ajp.0000149795.08746.31

de Almeida, C., & Amenábar, J. M. (2016). Changes in the salivary oxidative status in individuals with temporomandibular disorders and pain. *Journal of Oral Biology and Craniofacial Research, 6*(Suppl 1), S1–S4. https://doi.org/10.1016/j.jobcr.2016.10.006

Dorn, L. D., Campo, J. C., Thato, S., Dahl, R. E., Lewin, D., Chandra, R., & Di Lorenzo, C. (2003). Psychological comorbidity and stress reactivity in children and adolescents with recurrent abdominal pain and anxiety disorders. *Journal of the American Academy of Child and Adolescent Psychiatry, 42*(1), 66–75. https://doi.org/10.1097/01.CHI.0000024897.60748.43

Durham, P. L., Vause, C. V., Derosier, F., McDonald, S., Cady, R., & Martin, V. (2010). Changes in salivary prostaglandin levels during menstrual migraine with associated dysmenorrhea. *Headache, 50*(5), 844–851. https://doi.org/10.1111/j.1526-4610.2010.01657.x

Eisenberg, E., Shtahl, S., Geller, R., Reznick, A. Z., Sharf, O., Ravbinovich, M., . . . Nagler, R. M. (2008). Serum and salivary oxidative analysis in complex regional pain syndrome. *Pain, 138*(1), 226–232. https://doi.org/10.1016/j.pain.2008.04.019

Eller-Smith, O. C., Nicol, A. L., & Christianson, J. A. (2018). Potential mechanisms underlying centralized pain and emerging therapeutic interventions. *Frontiers in Cellular Neuroscience, 12*, 35. https://doi.org/10.3389/fncel.2018.00035

Evans, K. D., Douglas, B., Bruce, N., & Drummond, P. D. (2008). An exploratory study of changes in salivary cortisol, depression, and pain intensity after treatment for chronic pain. *Pain Medicine, 9*(6), 752–758. https://doi.org/10.1111/j.1526-4637.2006.00285.x

Fabian, L. A., McGuire, L., Page, G. G., Goodin, B. R., Edwards, R. R., & Haythornthwaite, J. (2009). The association of the cortisol awakening response with experimental pain ratings. *Psychoneuroendocrinology, 34*(8), 1247–1251. https://doi.org/10.1016/j.psyneuen.2009.03.008

Fernández-de-las-Peñas, C., Díaz-Rodríguez, L., Salom-Moreno, J., Galiano-Castillo, N., Valverde-Herreros, L., Martínez-Martín, J., & Pareja, J. A. (2014). Activation in the hypothalamic-pituitary-adrenocortical axis and sympathetic nervous system in women with carpal tunnel syndrome. *Pain Medicine (United States), 15*(8), 1373–1378. https://doi.org/10.1111/pme.12453

Ferrara, P., Bottaro, G., Angeletti, S., Gatto, A., Vitelli, O., Battaglia, D., . . . Dicuonzo, G. (2013). Salivary alpha-amylase: A new non-invasive biomarker for assessment of pain perception in epileptic children. *Acta Neurologica Belgica, 113*(3), 279–283. https://doi.org/10.1007/s13760-013-0180-z

Finocchi, C., Villani, V., & Casucci, G. (2010). Therapeutic strategies in migraine patients with mood and anxiety disorders: Clinical evidence. *Neurological Sciences, 31*(Suppl 1), S95–S98. https://doi.org/10.1007/s10072-010-0297-2

Firoozi, M., & Rostami, R. (2012). Sensitivity to pain in children with acute lymphoblastic leukemia (ALL). *Iranian Journal of Cancer Prevention, 5*(2), 74–80.

Fischer, S., Doerr, J. M., Strahler, J., Mewes, R., Thieme, K., & Nater, U. M. (2016). Stress exacerbates pain in the everyday lives of women with fibromyalgia syndrome – The role of cortisol and alpha-amylase. *Psychoneuroendocrinology, 63*, 68–77. https://doi.org/10.1016/j.psyneuen.2015.09.018

Gallai, V., Sarchielli, P., Coata, G., Firenze, C., Morucci, P., & Abbritti, G. (1992). Serum and salivary magnesium levels in migraine. Results in a group of juvenile patients. *Headache: The*

*Journal of Head and Face Pain, 32*(3), 132–135. https://doi.org/10.1111/j.1526-4610.1992.hed3203132.x

Galli, U., Gaab, J., Ettlin, D., Ruggia, F., Ehlert, U., & Palla, S. (2008). Enhanced negative feedback sensitivity of the hypothalamus-pituitary-adrenal axis in chronic myogenous facial pain. *European Journal of Pain, 13*(6), 600–605. https://doi.org/10.1016/j.ejpain.2008.07.010

Geiss, A., Rohleder, N., & Anton, F. (2012). Evidence for an association between an enhanced reactivity of interleukin-6 levels and reduced glucocorticoid sensitivity in patients with fibromyalgia. *Psychoneuroendocrinology, 37*(5), 671–684. https://doi.org/10.1016/j.psyneuen.2011.07.021

Generaal, E., Vogelzangs, N., Macfarlane, G. J., Geenen, R., Smit, J. H., Penninx, B. W., & Dekker, J. (2014). Reduced hypothalamic-pituitary-adrenal axis activity in chronic multi-site musculoskeletal pain: Partly masked by depressive and anxiety disorders. *BMC Musculoskeletal Disorders, 15*, 277. https://doi.org/10.1186/1471-2474-15-227

Godfrey, K. M., Herbert, M., Strachan, E., Mostoufi, S., Crofford, L. J., Buchwald, D., . . . Afari, N. (2017). Dexamethasone-suppressed salivary cortisol and pain sensitivity in female twins. *Clinical Journal of Pain, 33*(3), 246–253. https://doi.org/10.1097/AJP.0000000000000398

Goodin, B. R., Quinn, N. B., King, C. D., Page, G. G., Haythornthwaite, J. A., Edwards, R. R., . . . McGuire, L. (2012). Enhanced cortisol increase upon awakening is associated with greater pain ratings but not salivary cortisol or soluble tumor necrosis factor-α receptor II responses to acute pain. *Clinical Journal of Pain, 28*(4), 291–299. https://doi.org/10.1097/AJP.0b013e31822cf542

Goodin, B. R., Quinn, N. B., Kronfli, T., King, C. D., Page, G. G., Haythornthwaite, J. A., . . . McGuire, L. (2012). Experimental pain ratings and reactivity of cortisol and soluble tumor necrosis factor-α receptor II following a trial of hypnosis: Results of a randomized controlled pilot study. *Pain Medicine, 13*, 29–44. https://doi.org/10.1111/j.1526-4637.2011.01293.x

Goodin, B. R., Smith, M. T., Quinn, N. B., King, C. D., & McGuire, L. (2012). Poor sleep quality and exaggerated salivary cortisol reactivity to the cold pressor task predict greater acute pain severity in a non-clinical sample. *Biological Psychology, 91*(1), 36–41. https://doi.org/10.1016/j.biopsycho.2012.02.020

Granger, D. A., Kivlighan, K. T., Fortunato, C., Harmon, A. G., Hibel, L. C., Schwartz, E. B., & Whembolua, G. L. (2007). Integration of salivary biomarkers into developmental and behaviorally-oriented research: Problems and solutions for collecting specimens. *Physiology and Behavior, 92*(4), 583–590. https://doi.org/10.1016/j.physbeh.2007.05.004

Grunau, R. E., Tu, M. T., Whitfield, M. F., Oberlander, T. F., Weinberg, J., Yu, W., . . . Scheifele, D. (2010). Cortisol, behavior, and heart rate reactivity to immunization pain at 4 months corrected age in infants born very preterm. *Clinical Journal of Pain, 26*(8), 698–704. https://doi.org/10.1097/AJP.0b013e3181e5bb00

Grunau, R. E., Weinberg, J., & Whitfield, M. F. (2004). Neonatal procedural pain and preterm infant cortisol response to novelty at 8 months. *Pediatrics, 114*(1), e77–e84. https://doi.org/10.1542/peds.114.1.e77

Gulewitsch, M. D., Weimer, K., Enck, P., Schwille-Kiuntke, J., Hautzinger, M., & Schlarb, A. A. (2017). Stress reactivity in childhood functional abdominal pain or irritable bowel syndrome. *European Journal of Pain (United Kingdom), 21*, 166–177. https://doi.org/10.1002/ejp.914

Haraldstad, K., Sørum, R., Eide, H., Natvig, G. K., & Helseth, S. (2011). Pain in children and adolescents: Prevalence, impact on daily life, and parents' perception, a school survey. *Scandinavian Journal of Caring Sciences, 25*(1), 27–36. https://doi.org/10.1111/j.1471-6712.2010.00785.x

Harmon, A. G., Hibel, L. C., Rumyantseva, O., & Granger, D. A. (2007). Measuring salivary cortisol in studies of child development: Watch out – What goes in may not come out of saliva collection devices. *Developmental Psychobiology, 49*, 495–500. https://doi.org/10.1002/dev.20231

Haug, S. R., & Marthinussen, M. C. (2018). Acute dental pain and salivary biomarkers for stress and inflammation in patients with pulpal or periapical inflammation. *Journal of Oral & Facial Pain and Headache, 33*, 1–8. https://doi.org/10.11607/ofph.2007

Heim, C., Ehlert, U., Hanker, J. P., & Hellhammer, D. H. (1999). Psychological and endocrine correlates of chronic pelvic pain associated with adhesions. *Journal of Psychosomatic Obstetrics and Gynaecology, 20*(1), 11–20. https://doi.org/10.3109/01674829909075572

Heim, C., Hanker, J. P., & Hellhammer, D. H. (1998). Abuse-related posttraumatic stress disorder and alterations of the hypothalamic-pituitary-adrenal axis in women with chronic pelvic pain. *Psychosomatic Medicine, 60*(3), 309–318.

Herrington, C. J., Olomu, I. N., & Geller, S. M. (2004). Salivary cortisol as indicators of pain in preterm infants: A pilot study. *Clinical Nursing Research, 13*(1), 53–68. https://doi.org/10.1177/1054773803259665

Huguet, A., & Miró, J. (2008). The severity of chronic pediatric pain: An epidemiological study. *The Journal of Pain, 9*(3), 226–236. https://doi.org/10.1016/j.jpain.2007.10.015

Hunt, A., Wisbeach, A., Seers, K., Goldman, A., Crichton, N., Perry, L., & Mastroyannopoulou, K. (2007). Development of the paediatric pain profile: Role of video analysis and saliva cortisol in validating a tool to assess pain in children with severe neurological disability. *Journal of Pain and Symptom Management, 33*(3), 276–289. https://doi.org/10.1016/j.jpainsymman.2006.08.011

IASP Task Force on Taxonomy. (2004). *Classification of chronic pain* (2nd ed.). Seattle: IASP Press.

Iizuka, Y., Masaoka, N., & Ohashi, K. (2018). Pain in labor assessed from two discomfort aspects: Physical pain intensity and psychological stress state. *Journal of Obstetrics and Gynaecology Research, 44*(7), 1243–1251. https://doi.org/10.1111/jog.13660

Jang, M. U., Park, J. W., Kho, H. S., Chung, S. C., & Chung, J. W. (2011). Plasma and saliva levels of nerve growth factor and neuropeptides in chronic migraine patients. *Oral Diseases, 17*, 187–193. https://doi.org/10.1111/j.1601-0825.2010.01717.x

Jasim, H., Carlsson, A., Hedenberg-Magnusson, B., Ghafouri, B., & Ernberg, M. (2018). Saliva as a medium to detect and measure biomarkers related to pain. *Scientific Reports, 8*(1), 3220. https://doi.org/10.1038/s41598-018-21131-4

Jasim, H., Louca, S., Christidis, N., & Ernberg, M. (2014). Salivary cortisol and psychological factors in women with chronic and acute oro-facial pain. *Journal of Oral Rehabilitation., 41*, 122. https://doi.org/10.1111/joor.12118

Jenkins, B. N., Granger, D. A., Roemer, R. J., Martinez, A., Torres, T. K., & Fortier, M. A. (2018). Emotion regulation and positive affect in the context of salivary alpha-amylase response to pain in children with cancer. *Pediatric Blood and Cancer, 65*(6), e26973. https://doi.org/10.1002/pbc.26973

Jessop, D. S., & Turner-Cobb, J. M. (2008). Measurement and meaning of salivary cortisol: A focus on health and disease in children. *Stress, 11*(1), 1–14. https://doi.org/10.1080/10253890701365527

Jo, K. B., Lee, Y. J., Lee, I. G., Lee, S. C., Park, J. Y., & Ahn, R. S. (2016). Association of pain intensity, pain-related disability, and depression with hypothalamus-pituitary-adrenal axis function in female patients with chronic temporomandibular disorders. *Psychoneuroendocrinology, 69*, 106–115. https://doi.org/10.1016/j.psyneuen.2016.03.017

Jones, G. T., Power, C., & Macfarlane, G. J. (2009). Adverse events in childhood and chronic widespread pain in adult life: Results from the 1958 British birth cohort study. *Pain, 143*(1), 92–96. https://doi.org/10.1016/j.pain.2009.02.003

José da Silva Campos, M., César Souza Alves, C., Rezende Barbosa Raposo, N., Paula Ferreira, A., & Willer Farinazzo Vitral, R. (2010). Influence of salivary secretory immunoglobulin A level on the pain experienced by orthodontic patients. *Medical Science Monitor, 16*(9), CR405–CR409.

Kabat-Zinn, J. (2006). Review of full catastrophe living: Using the wisdom of your body and mind to face stress, pain, and illness. *Contemporary Psychology: A Journal of Reviews, 37*, 609. https://doi.org/10.1037/032287

Kallman, T. F., Ghafouri, B., & Bäckryd, E. (2018). Salivary beta-endorphin and substance P are not biomarkers of neuropathic chronic pain propensity. *Heliyon, 4*, e00718. https://doi.org/10.1016/j.heliyon.2018

Kanner, A. M., Schachter, S. C., Barry, J. J., Hersdorffer, D. C., Mula, M., Trimble, M., . . . Gilliam, F. (2012). Depression and epilepsy, pain and psychogenic non-epileptic seizures: Clinical and therapeutic perspectives. *Epilepsy and Behavior, 24*(2), 169–181. https://doi.org/10.1016/j.yebeh.2012.01.008

Kariyawasam, S. H., Zaw, F., & Handley, S. L. (2002). Reduced salivary cortisol in children with comorbid attention deficit hyperactivity disorder and oppositional defiant disorder. *Neuro Endocrinology Letters, 23*(1), 45–48.

Kashikar-Zuck, S., Goldschneider, K. R., Powers, S. W., Vaught, M. H., & Hershey, A. D. (2001). Depression and functional disability in chronic pediatric pain. *The Clinical Journal of Pain, 17* (4), 341–349. https://doi.org/10.1097/00002508-200112000-00009

Kim, H. A., Jeon, J. Y., Koh, B. R., Park, S. B., & Suh, C. H. (2016). Salivary cortisol levels, but not salivary α-amylase levels, are elevated in patients with rheumatoid arthritis irrespective of depression. *International Journal of Rheumatic Diseases., 19*, 172. https://doi.org/10.1111/1756-185X.12224

King, S., Chambers, C. T., Huguet, A., MacNevin, R. C., McGrath, P. J., Parker, L., & MacDonald, A. J. (2011). The epidemiology of chronic pain in children and adolescents revisited: A systematic review. *Pain, 152*(12), 2729–2738. https://doi.org/10.1016/j.pain.2011.07.016

Knook, L. M. E., Konijnenberg, A. Y., van der Hoeven, J., Kimpen, J. L. L., Buitelaar, J. K., van Engeland, H., & de Graeff-Meeder, E. R. (2011). Psychiatric disorders in children and adolescents presenting with unexplained chronic pain: What is the prevalence and clinical relevancy? *European Child and Adolescent Psychiatry, 20*(1), 39–48. https://doi.org/10.1007/s00787-010-0146-0

Korterink, J. J., Diederen, K., Benninga, M. A., & Tabbers, M. M. (2015). Epidemiology of pediatric functional abdominal pain disorders: A meta-analysis. *PLoS One, 10*(5), e0126982. https://doi.org/10.1371/journal.pone.0126982

Lautenbacher, S., Huber, C., Kunz, M., Parthum, A., Weber, P. G., Griessinger, N., & Sittl, R. (2009). Hypervigilance as predictor of postoperative acute pain: Its predictive potency compared with experimental pain sensitivity, cortisol reactivity, and affective state. *The Clinical Journal of Pain, 25*(2), 92–100. https://doi.org/10.1097/AJP.0b013e3181850dce

Lee, K. C., & Bassiur, J. P. (2017). Salivary alpha amylase, dental anxiety, and extraction pain: A pilot study. *Anesthesia Progress, 64*, 22. https://doi.org/10.2344/anpr-63-03-02

Lewis, M., & Thomas, D. (1990). Cortisol release in infants in response to inoculation. *Child Development, 61*(1), 50–59.

Lutgendorf, S. K., Kreder, K. J., Rothrock, N. E., Hoffman, A., Kirschbaum, C., Sternberg, E. M., . . . Ratliff, T. L. (2002). Diurnal cortisol variations and symptoms in patients with interstitial cystitis. *Journal of Urology, 167*(3), 1338–1343. https://doi.org/10.1016/S0022-5347(05)65295-0

Malfliet, A., Lluch Girbés, E., Pecos-Martin, D., Gallego-Izquierdo, T., & Valera-Calero, A. (2018). The influence of treatment expectations on clinical outcomes and cortisol levels in patients with chronic neck pain: An experimental study. *Pain Practice, 19*(4), 370–381. https://doi.org/10.1111/papr.12749

Mcbeth, J., Chiu, Y. H., Silman, A. J., Ray, D., Morriss, R., Dickens, C., . . . Macfarlane, G. J. (2005). Hypothalamic-pituitary-adrenal stress axis function and the relationship with chronic widespread pain and its antecedents. *Arthritis Research and Therapy, 7*(5), 992–1000. https://doi.org/10.1186/ar1772

McBeth, J., Silman, A. J., Gupta, A., Chiu, Y. H., Ray, D., Morriss, R., . . . Macfarlane, G. J. (2007). Moderation of psychosocial risk factors through dysfunction of the hypothalamic-pituitary-adrenal stress axis in the onset of chronic widespread musculoskeletal pain: Findings of a population-based prospective cohort study. *Arthritis and Rheumatism, 56*(1), 360–371. https://doi.org/10.1002/art.22336

McCarthy, A. M., Hanrahan, K., Kleiber, C., Zimmerman, M. B., Lutgendorf, S., & Tsalikian, E. (2009). Normative salivary cortisol values and responsivity in children. *Applied Nursing Research, 22*, 54–62. https://doi.org/10.1016/j.apnr.2007.04.009

McCarthy, A. M., Hanrahan, K., Scott, L. M., Zemblidge, N., Kleiber, C., & Zimmerman, M. B. (2011). Salivary cortisol responsivity to an intravenous catheter insertion in children with attention-deficit/hyperactivity disorder. *Journal of Pediatric Psychology, 36*(8), 902–910. https://doi.org/10.1093/jpepsy/jsr012

McLean, S. A., Williams, D. A., Harris, R. E., Kop, W. J., Groner, K. H., Ambrose, K., . . . Clauw, D. J. (2005). Momentary relationship between cortisol secretion and symptoms in patients with fibromyalgia. *Arthritis and Rheumatism, 52*(11), 3660–3669. https://doi.org/10.1002/art.21372

Miller, E., Jacob, E., & Hockenberry, M. J. (2011). Nausea, pain, fatigue, and multiple symptoms in hospitalized children with cancer. *Oncology Nursing Forum, 38*(5), E382–E393. https://doi.org/10.1188/11.ONF.E382-E393

Muhtz, C., Rodriguez-Raecke, R., Hinkelmann, K., Moeller-Bertram, T., Kiefer, F., Wiedemann, K., . . . Otte, C. (2013). Report cortisol response to experimental pain in patients with chronic low back pain and patients with major depression. *Pain Medicine, 14*, 498–503. https://doi.org/10.1111/j.1526-4637.2012.01514.x

Nees, F., Löffler, M., Usai, K., & Flor, H. (2019). Hypothalamic-pituitary-adrenal axis feedback sensitivity in different states of back pain. *Psychoneuroendocrinology, 101*, 60–66. https://doi.org/10.1016/j.psyneuen.2018.10.026

Nelson, F. A., Farr, L. A., Ebadi, M., & Nelson, F. A. (2001). Salivary melatonin response to acute pain stimuli. *Journal of Pineal Research, 30*(1), 206–212. https://doi.org/10.1034/j.1600-079X.2001.300403.x

Nierop, A., Bratsikas, A., Klinkenberg, A., Nater, U. M., Zimmermann, R., & Ehlert, U. (2006). Prolonged salivary cortisol recovery in second-trimester pregnant women and attenuated salivary alpha-amylase responses to psychosocial stress in human pregnancy. *Journal of Clinical Endocrinology and Metabolism, 91*(4), 1329–1335. https://doi.org/10.1210/jc.2005-1816

Nilsson, A. M., & Dahlstrm, L. (2010). Perceived symptoms of psychological distress and salivary cortisol levels in young women with muscular or disk-related temporomandibular disorders. *Acta Odontologica Scandinavica, 68*(5), 284–288. https://doi.org/10.3109/00016357.2010.494620

Oberlander, T. F., O'Donnell, M. E., & Montgomery, C. J. (1999). Pain in children with significant neurological impairment. *Journal of Developmental and Behavioral Pediatrics, 20*(4), 235–243.

Ozdogan, M. S., Gungormus, M., Yusufoglu, S. I., Ertem, S. Y., Sonmez, C., & Orhan, M. (2019). Salivary opiorphin in dental pain: A potential biomarker for dental disease. *Archives of Oral Biology, 99*, 15–21. https://doi.org/10.1016/j.archoralbio.2018.12.006

Ozgocer, T., Ucar, C., & Yildiz, S. (2017). Cortisol awakening response is blunted and pain perception is increased during menses in cyclic women. *Psychoneuroendocrinology, 77*, 158–164. https://doi.org/10.1016/j.psyneuen.2016.12.011

Paloyelis, Y., Krahé, C., Maltezos, S., Williams, S. C., Howard, M. A., & Fotopoulou, A. (2016). The analgesic effect of oxytocin in humans: A double-blind, placebo-controlled cross-over study using laser-evoked potentials. *Journal of Neuroendocrinology, 28*(4). https://doi.org/10.1111/jne.12347

Park, M.-K., & Watanuki, S. (2005). Specific physiological responses in women with severe primary dysmenorrhea during the menstrual cycle. *Journal of Physiological Anthropology and Applied Human Science, 24*(6), 601–609. https://doi.org/10.2114/jpa.24.601

Parris, W. C. V., Kambam, J. R., Naukam, R. J., & Rama Sastry, B. V. (2006). Immunoreactive substance P is decreased in saliva of patients with chronic Back pain syndromes. *Anesthesia and Analgesia, 70*(1), 63–67. https://doi.org/10.1213/00000539-199001000-00010

Parris, W. C., Rama Sastry, B. V., Kambam, J. R., Naukam, R. J., & Johnson, B. W. (2006). Immunoreactive substance P in human saliva. *Annals of the New York Academy of Sciences, 694* (1 Saliva as a D), 308–310. https://doi.org/10.1111/j.1749-6632.1993.tb18373.x

Patacchioli, F. R., Monnazzi, P., Simeoni, S., De Filippis, S., Salvatori, E., Coloprisco, G., & Martelletti, P. (2006). Salivary cortisol, dehydroepiandrosterone-sulphate (DHEA-S) and testosterone in women with chronic migraine. *Journal of Headache and Pain, 7*(2), 90–94. https://doi.org/10.1007/s10194-006-0274-6

Payne, L., Hibel, L., Granger, D. A., Tsao, J., & Zeltzer, L. (2014). Relationship of salivary alpha amylase and cortisol to social anxiety in healthy children undergoing laboratory pain tasks. *Journal of Child and Adolescent Behaviour, 2*, 129. https://doi.org/10.4172/jcalb.1000129

Peluso, I., & Raguzzini, A. (2016). Salivary and urinary total antioxidant capacity as biomarkers of oxidative stress in humans. *Pathology Research International, 2016*, 1. https://doi.org/10.1155/2016/5480267

Petrelluzzi, K. F. S., Garcia, M. C., Petta, C. A., Grassi-Kassisse, D. M., & Spadari-Bratfisch, R. C. (2008). Salivary cortisol concentrations, stress and quality of life in women with endometriosis and chronic pelvic pain. *Stress, 11*(5), 390–397. https://doi.org/10.1080/10253890701840610

Provenzi, L., Giusti, L., Fumagalli, M., Tasca, H., Ciceri, F., Menozzi, G., ... Montirosso, R. (2016). Pain-related stress in the neonatal intensive care unit and salivary cortisol reactivity to socio-emotional stress in 3-month-old very preterm infants. *Psychoneuroendocrinology, 72*, 161–165. https://doi.org/10.1016/j.psyneuen.2016.07.010

Quartana, P. J., Buenaver, L. F., Edwards, R. R., Klick, B., Haythornthwaite, J. A., & Smith, M. T. (2010). Pain catastrophizing and salivary cortisol responses to laboratory pain testing in temporomandibular disorder and healthy participants. *Journal of Pain, 11*(2), 186–194. https://doi.org/10.1016/j.jpain.2009.07.008

Quiñones, M., Urrutia, R., Torres-Reverón, A., Vincent, K., & Flores, I. (2015). Anxiety, coping skills and hypothalamus-pituitary-adrenal (HPA) axis in patients with endometriosis. *Journal of Reproductive Biology and Health, 3*, 2. https://doi.org/10.7243/2054-0841-3-2

Ramsay, D., & Lewis, M. (2003). Reactivity and regulation in cortisol and behavioral responses to stress. *Child Development, 74*(2), 456–464. https://doi.org/10.1111/1467-8624.7402009

Randazzo, W. T., Dockray, S., & Susman, E. J. (2008). The stress response in adolescents with inattentive type ADHD symptoms. *Child Psychiatry and Human Development, 39*(1), 27–38. https://doi.org/10.1007/s10578-007-0068-3

Reckziegel, D., Vachon-Presseau, E., Petre, B., Schnitzer, T. J., Baliki, M. N., & Apkarian, A. V. (2019). Deconstructing biomarkers for chronic pain: Context- and hypothesis-dependent biomarker types in relation to chronic pain. *Pain, 160*(Suppl 1), S37–S48. https://doi.org/10.1097/j.pain.0000000000001529

Rezaei, F., Aminian, M., & Raygani, A. (2017). Evaluation of salivary cortisol changes and psychological profiles in patients with recurrent aphthous stomatitis. *Contemporary Clinical Dentistry, 8*(2), 259. https://doi.org/10.4103/ccd.ccd_165_17

Riva, R., Mork, P. J., Westgaard, R. H., & Lundberg, U. (2012). Comparison of the cortisol awakening response in women with shoulder and neck pain and women with fibromyalgia. *Psychoneuroendocrinology, 37*(2), 299–306. https://doi.org/10.1016/j.psyneuen.2011.06.014

Riva, R., Mork, P. J., Westgaard, R. H., Rø, M., & Lundberg, U. (2010). Fibromyalgia syndrome is associated with hypocortisolism. *International Journal of Behavioral Medicine, 17*, 223–233. https://doi.org/10.1007/s12529-010-9097-6

Roberts, M. H., Klatzkin, R. R., & Mechlin, B. (2015). Social support attenuates physiological stress responses and experimental pain sensitivity to cold Pressor pain. *Annals of Behavioral Medicine, 49*(1), 557–569. https://doi.org/10.1007/s12160-015-9686-3

Rock, K. L., Kataoka, H., & Lai, J.-J. (2013). Uric acid as a danger signal in gout and its comorbidities. *Nature Reviews. Rheumatology, 9*(1), 13–23. https://doi.org/10.1038/nrrheum.2012.143

Rodríguez de Sotillo, D., Velly, A. M., Hadley, M., & Fricton, J. R. (2011). Evidence of oxidative stress in temporomandibular disorders: A pilot study. *Journal of Oral Rehabilitation, 38*(10), 722–728. https://doi.org/10.1111/j.1365-2842.2011.02216.x

Schell, E., Theorell, T., Hasson, D., Arnetz, B., & Saraste, H. (2008). Stress biomarkers' associations to pain in the neck, shoulder and back in healthy media workers: 12-month prospective follow-up. *European Spine Journal, 17*(3), 393–405. https://doi.org/10.1007/s00586-007-0554-0

Shen, Y. S., Chen, W. L., Chang, H. Y., Kuo, H. Y., Chang, Y. C., & Chu, H. (2012). Diagnostic performance of initial salivary alpha-amylase activity for acute myocardial infarction in patients with acute chest pain. *Journal of Emergency Medicine, 43*(4), 553–560. https://doi.org/10.1016/j.jemermed.2011.06.040

Shirasaki, S., Fujii, H., Takahashi, M., Sato, T., Ebina, M., Noto, Y., & Hirota, K. (2007). Correlation between salivary alpha-amylase activity and pain scale in patients with chronic pain. *Regional Anesthesia and Pain Medicine, 32*(2), 120–123. https://doi.org/10.1016/j.rapm.2006.11.008

Sobas, E. M., Reinoso, R., Cuadrado-Asensio, R., Fernández, I., Maldonado, M. J., & Pastor, J. C. (2016). Reliability of potential pain biomarkers in the saliva of healthy subjects: Inter-individual differences and intersession variability. *PLoS One, 11*(12), e0166976. https://doi.org/10.1371/journal.pone.0166976

Söderquist, F., Sundberg, I., Ramklint, M., Widerström, R., Hellström, P. M., & Cunningham, J. L. (2019). The relationship between daytime salivary melatonin and gastrointestinal symptoms in young adults seeking psychiatric care. *Psychosomatic Medicine, 81*(1), 51–56. https://doi.org/10.1097/PSY.0000000000000644

Sudhaus, S., Fricke, B., Stachon, A., Schneider, S., Klein, H., von Düring, M., & Hasenbring, M. (2009). Salivary cortisol and psychological mechanisms in patients with acute versus chronic low back pain. *Psychoneuroendocrinology, 34*(4), 513–522. https://doi.org/10.1016/j.psyneuen.2008.10.011

Sudhaus, S., Held, S., Schoofs, D., Bültmann, J., Dück, I., Wolf, O. T., & Hasenbring, M. I. (2015). Associations between fear-avoidance and endurance responses to pain and salivary cortisol in the context of experimental pain induction. *Psychoneuroendocrinology, 52*(1), 195–199. https://doi.org/10.1016/j.psyneuen.2014.11.011

Surveillance of Cerebral Palsy in Europe. (2000). Surveillance of cerebral palsy in Europe: A collaboration of cerebral palsy surveys and registers. Surveillance of Cerebral Palsy in Europe (SCPE). *Developmental Medicine and Child Neurology, 42*(12), 816–824.

Sveinsdottir, V., Eriksen, H. R., Ursin, H., Hansen, Å. M., & Harris, A. (2016). Cortisol, health, and coping in patients with nonspecific low back pain. *Applied Psychophysiology Biofeedback, 41*(1), 9–16. https://doi.org/10.1007/s10484-015-9300-2

Symons, F. J., Elghazi, I., Reilly, B. G., Barney, C. C., Hanson, L., Panoskaltsis-Mortari, A., … Wilcox, G. L. (2015). Can biomarkers differentiate pain and no pain subgroups of nonverbal children with cerebral palsy? A preliminary investigation based on noninvasive saliva sampling. *Pain Medicine (United States), 16*, 249–256. https://doi.org/10.1111/pme.12545

Tornhage, C. J., & Alfven, G. (2006). Diurnal salivary cortisol concentration in school-aged children: Increased morning cortisol concentration and Total cortisol concentration negatively correlated to body mass index in children with recurrent abdominal pain of psychosomatic origin. *Journal of Pediatric Endocrinology and Metabolism, 19*(6), 843–854.

Törnhage, C. J., & Alfvén, G. (2015). Children with recurrent psychosomatic abdominal pain display increased morning salivary cortisol and high serum cortisol concentrations. *Acta Paediatrica, International Journal of Paediatrics, 104*, e577–e580. https://doi.org/10.1111/apa.13206

Tryphonopoulos, P. D., Letourneau, N., & Azar, R. (2014). Approaches to salivary cortisol collection and analysis in infants. *Biological Research for Nursing, 16*(4), 398–408. https://doi.org/10.1177/1099800413507128

Turner-Cobb, J. M., Osborn, M., da Silva, L., Keogh, E., & Jessop, D. S. (2010). Sex differences in hypothalamic–pituitary–adrenal axis function in patients with chronic pain syndrome. *Stress, 13*(4), 293–301. https://doi.org/10.3109/10253890903524785

Twycross, A., Voepel-Lewis, T., Vincent, C., Franck, L. S., & Von Baeyer, C. L. (2015). A debate on the proposition that self-report is the gold standard in assessment of pediatric pain intensity. *Clinical Journal of Pain, 31*(8), 707–712. https://doi.org/10.1097/AJP.0000000000000165

Vachon-Presseau, E., Martel, M.-O., Roy, M., Caron, E., Albouy, G., Marin, M.-F., … Rainville, P. (2013). Acute stress contributes to individual differences in pain and pain-related brain activity in healthy and chronic pain patients. *Journal of Neuroscience, 33*(16), 6826–6833. https://doi.org/10.1523/jneurosci.4584-12.2013

Vahedi, M., Mazdeh, M., Hajilooi, M., Farhadian, M., Barakian, Y., & Sadr, P. (2018). The relationship between salivary alpha amylase activity and score of mcgill pain questionnaire in patients with tension type headache. *Basic and Clinical Neuroscience, 9*(1), 59–64. https://doi.org/10.29252/nirp.bcn.9.1.59

Voegtline, K. M., & Granger, D. A. (2014). Dispatches from the interface of salivary bioscience and neonatal research. *Frontiers in Endocrinology, 5*(25), 1–8. https://doi.org/10.3389/fendo.2014.00025

Vrbanović, E., Alajbeg, I. Z., Vuletić, L., Lapić, I., Rogić, D., Andabak Rogulj, A., … Alajbeg, I. (2018). Salivary oxidant/antioxidant status in chronic temporomandibular disorders is dependent on source and intensity of pain – A pilot study. *Frontiers in Physiology, 9*. https://doi.org/10.3389/fphys.2018.01405

Wadhwa, S., & Kapila, S. (2008). TMJ disorders: Future innovations in diagnostics and therapeutics. *Journal of Dental Education, 72*(8), 930–947.

Walco, G. A., Conte, P. M., Labay, L. E., Engel, R., & Zeltzer, L. K. (2005). Procedural distress in children with cancer: Self-report, behavioral observations, and physiological parameters. *The Clinical Journal of Pain, 21*(6), 484–490.

Willemsen, G., Ring, C., Carroll, D., Evans, P., Clow, A., & Hucklebridge, F. (1998). Secretory immunoglobulin A and cardiovascular reactions to mental arithmetic and cold pressor. *Psychophysiology, 35*(3), 252–259.

Wingenfeld, K., Heim, C., Schmidt, I., Wagner, D., Meinlschmidt, G., & Hellhammer, D. H. (2008). HPA axis reactivity and lymphocyte glucocorticoid sensitivity in fibromyalgia syndrome and chronic pelvic pain. *Psychosomatic Medicine, 70*(1), 65–72. https://doi.org/10.1097/PSY.0b013e31815ff3ce

Wingenfeld, K., Nutzinger, D., Kauth, J., Hellhammer, D. H., & Lautenbacher, S. (2010). Salivary cortisol release and hypothalamic pituitary adrenal axis feedback sensitivity in fibromyalgia is associated with depression but not with pain. *Journal of Pain, 11*(11), 1195–1202. https://doi.org/10.1016/j.jpain.2010.02.011

Wittwer, A., Krummenacher, P., La Marca, R., Ehlert, U., & Folkers, G. (2016). Salivary alpha-amylase correlates with subjective heat pain perception. *Pain Medicine (United States)., 17*(6), 1131–1136.

Wolfe, F., Smythe, H. A., Yunus, M. B., Bennett, R. M., Bombardier, C., Goldenberg, D. L., … Clark, P. (1990). The American College of Rheumatology 1990 criteria for the classification of fibromyalgia. Report of the Multicenter Criteria Committee. *Arthritis and Rheumatism, 33*(2), 160–172.

Woolf, C. J. (2011). Central sensitization: Implications for the diagnosis and treatment of pain. *Pain, 152*(Supplement), S2–S15. https://doi.org/10.1016/j.pain.2010.09.030

Yeung, E. W., Davis, M. C., & Ciaramitaro, M. C. (2016). Cortisol profile mediates the relation between childhood neglect and pain and emotional symptoms among patients with fibromyalgia. *Annals of Behavioral Medicine, 50*(1), 87–97. https://doi.org/10.1007/s12160-015-9734-z

Zhao, X., Chen, M., Du, S., Li, H., & Li, X. (2015). Evaluation of stress and pain in young children with cerebral palsy during early developmental intervention programs: A descriptive study. *American Journal of Physical Medicine and Rehabilitation, 94*(3), 169–175. https://doi.org/10.1097/PHM.0000000000000252

Zheng, X., Chen, F., Zhang, Q., Liu, Y., You, P., Sun, S., … Chen, N. (2017). Salivary exosomal PSMA7: A promising biomarker of inflammatory bowel disease. *Protein and Cell, 8*(9), 686–695. https://doi.org/10.1007/s13238-017-0413-7

# Chapter 15
# Salivary Bioscience and Environmental Exposure Assessment

**Parinya Panuwet, Priya E. D'Souza, Emily R. Phillips, P. Barry Ryan, and Dana Boyd Barr**

**Abstract** Human exposure to toxic, environmental chemicals can be assessed by measuring levels of parent chemicals or their metabolites in biological matrices such as urine, plasma, and whole blood. This technique is known as biomonitoring. If these measurements are accurate, the internal dose of the target chemicals in humans can be estimated. Biomonitoring has become an important tool for epidemiological studies linking exposures to toxic, environmental chemicals to health outcomes. Selection of the appropriate biological matrices for biomonitoring is crucial and requires a comprehensive understanding of the physiochemical properties of the target chemicals, their toxicokinetic properties, as well as the biological properties of the matrix. Different matrices may provide different information on the magnitude and frequency of dose, toxicity, and biological endpoint. Saliva has been used as a matrix for biomonitoring of toxic, environmental chemicals, but with far less frequency than urine or plasma. The success of using saliva as a matrix for exposure assessment has been proven for cotinine, a principal metabolite of nicotine. However, a number of limitations and challenges remain for other chemicals such as plasticizers and heavy metals. In this chapter, the current knowledge on salivary biomonitoring is provided, including the associated challenges researchers face. More importantly, practical recommendations are provided with the intention of supporting future research using saliva as a matrix to assess exposure to toxic, environmental chemicals.

**Keywords** Saliva · Biomonitoring · Environmental exposure · Toxic chemicals · Exposure assessment

P. Panuwet (✉) · P. E. D'Souza · E. R. Phillips · P. B. Ryan · D. B. Barr
Laboratory of Exposure Assessment and Development for Environmental Research (LEADER), Department of Environmental Health, Rollins School of Public Health, Emory University, Atlanta, GA, USA
e-mail: ppanuwe@emory.edu



D. A. Granger, M. K. Taylor (eds.), *Salivary Bioscience*,
https://doi.org/10.1007/978-3-030-35784-9_15

## 15.1 History of Salivary Bioscience in Environmental Exposure

At present, in addition to naturally occurring inorganic and organic chemicals, humans are exposed to a wide variety of anthropogenic chemicals through their environment (e.g., air, water, soil, etc.). This is a result of industrial growth, urbanization, industrial and nonindustrial product consumption and substitution, and other human practices. Product substitution refers to the replacement of a key chemical ingredient that was proven toxic to humans with a newly developed substance for which toxicological evidence in humans is not yet available. For instance, bisphenol A has been phased out and replaced with bisphenol F or bisphenol S in consumable plastic materials (Lehmler, Liu, Gadogbe, & Bao, 2018).

Examples of chemicals to which humans are commonly exposed include heavy metals/metalloids, dioxins, pesticides, plasticizers (phthalates, bisphenols), cosmetic and pharmaceutical preservatives (parabens), flame retardants (polybrominated diphenyl ethers or PBDEs, organophosphate flame-retardants), polycyclic aromatic hydrocarbons, gasoline emissions and combustion products, per- and poly-fluoroalkyl substances, tobacco smoke, environmental phenols, volatile organic compounds, and perchlorate. Some of these chemicals enter the food chain and are ingested by humans, while others become suspended in the air or attached to particulate matter or dust particles and are inhaled into human lungs. Other environmental chemicals dissolve in water and are ingested or absorbed through human skin. Some chemicals remain in the soil and enter the human body via dermal absorption or ingestion. Other chemicals (i.e., key ingredients in cosmetic products) are specifically designed to be applied to human skin and therefore are readily absorbed by the human body.

Some environmental chemicals, such as pesticides, heavy metals/metalloids, and tobacco smoke have well-defined toxicological endpoints that are associated with health outcomes in humans (Damalas & Koutroubas, 2016; IARC, 2012; Kakkar & Jaffery, 2005; Wallace, 2012). However, the toxicological endpoints for many chemicals are ambiguous, particularly when exposures to them are chronic and low level. Epidemiological and risk assessment studies continue to play an important role in quantifying the associated health risks or in investigating the relationships between exposures to these environmental chemicals and health outcomes. It is undeniable that a key factor in these studies is the ability to determine accurate and precise exposure levels in the study population. While there are other tools (i.e., questionnaires and ecological and environmental assessments) that can be used to estimate exposure levels, *biomonitoring* is a key analytic process in defining exposure-disease risk/outcomes, largely due to the biological significance of internal dose measurements (Needham et al., 2005). In short, biomonitoring refers to the measurement of parent compounds, their metabolites, or reaction products, in biological matrices such as urine, blood, and serum as well as other collectible biological matrices (Angerer, Ewers, & Wilhelm, 2007). Biomonitoring and biological measurement of exposure can be used interchangeably.

Upon entering the human body, a chemical undergoes four complex steps in the pharmacokinetic process: absorption, distribution, metabolism, and excretion (ADME). In order to assess human exposure to a given chemical, biological measurements of the chemical can be made after the absorption step or during each of the subsequent steps of ADME. Measurement of the concentration of a chemical involves the selection of appropriate biological matrices. The concentrations in each matrix depend on the amount of the intake and uptake of the absorbed chemical, the pharmacokinetics of the chemical, and the exposure scenario, including the timing of the exposure (Sexton, Callahan, & Bryan, 1995). Biomonitoring data integrate all routes of exposure and are independent of the exposure pathway. Ideally, measurements of the biologically effective dose, the dose at the target site that causes an adverse health effect, are preferred. However, in most circumstances, the target organ is unknown or unavailable for collection. As such, the chemical is measured in a proximal matrix, typically blood or urine, to estimate the internal dose (Barr, Wang, & Needham, 2005; Pirkle, Needham, & Sexton, 1995).

Selection of the most appropriate matrix for biomonitoring depends upon the different classes of environmental chemicals to be monitored. For example, some persistent organic pollutants, such as dioxins and PBDEs, are nonpolar and lipophilic and thus they tend to partition in lipid stores in the body (CDC, 2017). Therefore, biological matrices that are rich in lipid content (e.g., serum, adipose tissue, brain, and breast milk) are likely to be appropriate matrices for biological measurements of these chemicals (Barr et al., 2005). For biological measurements of environmental chemicals, blood or urine is often the matrix of choice. Blood has inherent, toxicological advantages for biomonitoring because the majority of chemicals must be absorbed by the bloodstream and circulated to the tissues to have an effect. Additionally, measurements are automatically normalized because blood volume is constant per given body weight, allowing for the estimation of total toxicant body burden. Urine is the other most commonly used biological matrix. Several environmental chemicals are nonpersistent (i.e., they have short environmental and biological half-lives) and are excreted in the urine in their original form, as metabolites, or both, depending on their pharmacokinetic characteristics. Thus, urine is considered an excellent matrix for measuring those compounds (Needham et al., 2005).

Saliva has been used as a matrix for the biological measurement of exposure to environmental chemicals, mainly those found in tobacco smoke, for more than 38 years (Greenberg, Haley, Etzel, & Loda, 1984; Jarvis et al., 1985; Peyton III, Wilson, & Benowitz, 1981; Wall, Johnson, Jacob, & Benowitz, 1988). Historically, saliva was used, together with serum and/or urine, to quantify nicotine and its major metabolite, cotinine, using gas chromatographic methods (Feyerabend & Russell, 1990; Peyton III et al., 1981). A major limitation of these methods was low selectivity of the analytical instrumentation available at the time, making the elimination of chemical interferences impossible. This also led to difficulties in quantifying low levels of nicotine and cotinine resulting from second- or third-hand smoke exposure (also collectively called environmental tobacco smoke (ETS) exposure). In the mid- to late-1990s, the advance of efficient and effective interfaces between liquid chromatography and mass spectrometry enabled better separation and detection of cotinine and nicotine in biological samples, allowing for the detection of cotinine in serum and saliva resulting from ETS exposure (Bernert et al., 1997).

Since then, saliva has been widely used as the matrix of choice for biological measurement of cotinine to assess exposure to ETS in adults. Because the collection procedure is minimally invasive and relatively easy in comparison to other biological matrices (e.g., blood), saliva has also been explored as a suitable matrix for the analysis of other environmental chemicals (either in the form of metabolites or the original parent compounds) for exposure assessment purposes. However, concentrations of the environmental chemicals in saliva are usually very low and may not be in equilibrium with concentrations in plasma. This is a major factor that limits the application of saliva as a matrix for exposure assessment of environmental chemicals.

## 15.2 Current Status of Knowledge in Environmental Exposure

Saliva is considered a nontraditional matrix for biomonitoring of most environmental chemicals. As reviewed in Hold, de Boer, and Zuidema (1999), saliva is produced by a number of specialized glands and then discharged into the oral cavity of mammalian vertebrates. The major salivary glands (i.e., parotid, submandibular, and sublingual) are responsible for producing most of the saliva. A small portion of saliva also comes from the small labial, buccal, and palatal glands that line the mouth. In adults, the total volume of saliva produced each day is 1000–1500 mL (Humphrey & Williamson, 2001).

There are three primary mechanisms by which a given biomolecule or small chemical circulating in plasma can enter or be transported into saliva. These mechanisms include passive diffusion, active transport against a concentration gradient, and ultrafiltration through pores in the membrane. Some biomolecules, such as steroid hormones, can pass through the membrane of the glandular epithelial cells via a passive diffusion process because their molecules contain fatty acids, which permits their transcellular diffusion through the lipophilic layer of the epithelial membrane. An active transport mechanism is responsible for the presence of electrolytes, proteins such as immunoglobulin A, as well as some drugs in saliva. For environmental chemicals, the primary transport mechanism is thought to be ultrafiltration. Due to the hydrophilic properties of these chemicals, they are unlikely to pass through to the saliva compartment via passive transcellular diffusion or active transcellular transport. More detailed information on the transport of biomolecules from blood to saliva can be found in the article by Pfaffe, Cooper-White, Beyerlein, Kostner, and Punyadeera (2011). It is possible that the free form of environmental chemicals that pass into the saliva compartment will later bind to proteins that are present in saliva (Fig. 15.1).

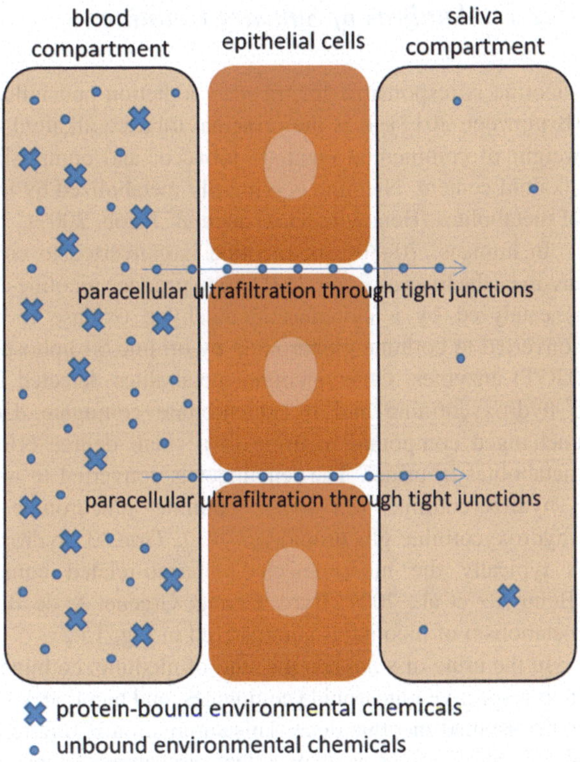


**Fig. 15.1** Transport of environmental chemicals from the blood compartment to saliva. Note: Information was extracted from several published articles (Cooper, 2000; Hold et al., 1999; Pfaffe et al., 2011)

In comparison to plasma, saliva contains more water (it is 97–99.5% water) and less protein, with a range of 0.15–0.64 g protein/100 mL saliva (which is less than 1% of the protein found in plasma). The pH of saliva is between 5.3 and 7.8, depending on the level of stimulation. Due to the low protein content, protein-bound molecules are unlikely to be transported into saliva. Therefore, the unbound fraction of any compound is likely to partition into saliva (Hold et al., 1999; Humphrey & Williamson, 2001).

Due to its unique characteristics and the ease by which it is collected, saliva has been explored for its use as a matrix for biomonitoring. So far, saliva is a proven reliable matrix for biomonitoring of exposure to nicotine found in tobacco smoke (Avila-Tang et al., 2013; Demkowska, Polkowska, & Namiesnik, 2011). Other applications of saliva in biomonitoring are for the analysis of some heavy metals, phthalate metabolites, and perchlorate, although these applications still have some limitations.

### *15.2.1 Analysis of Salivary Cotinine*

Nicotine is responsible for tobacco addiction and influences tobacco use patterns (Bruijnzeel, 2012). It is the principal tobacco alkaloid, making up about 1.5% by weight of commercial cigarette tobacco, and comprising about 95% of the total alkaloid content. Nicotine is primarily metabolized by liver enzymes into a number of metabolites (Benowitz, Hukkanen, & Jacob, 2009).

In humans, 70–80% of nicotine is converted to cotinine. This transformation involves the liver enzyme CYP2A6 to produce nicotine-$\Delta 1'$ (5′)-iminium ion, which is catalyzed by a cytoplasmic aldehyde oxidase to cotinine. Cotinine is then converted to cotinine glucuronide by uridine 5′-diphospho-glucuronosyltransferase (UGT) enzymes. Other nicotine metabolites detected in smoker urine are *trans*-3′-hydroxycotinine and its glucuronide conjugate. Cotinine is excreted as an unchanged compound in urine to a small degree (10–15% of the nicotine and metabolites in urine). The remainder is converted to metabolites, primarily *trans*-3′-hydroxycotinine (33–40%), cotinine glucuronide (12–17%), and *trans*-3-′-hydroxycotinine glucuronide (7–9%). *Trans*-3′-hydroxycotinine, in its free form, is typically the most detected tobacco-related compound in smokers' urine (Benowitz et al., 2009; Byrd, Chang, Greene, & de Bethizy, 1991). The primary metabolism of nicotine is summarized in Fig. 15.2.

In the urine of smokers, the sum of nicotine, cotinine, *trans*-3′-hydroxycotinine, their respective glucuronide conjugates, and nicotine *N*-oxide accounts for >85% of the consumed nicotine dose. This summation is referred to as total nicotine equivalents, which serves as an excellent biomarker of tobacco exposure (Murphy et al., 2017; Wang, Liang, Mendes, & Sarkar, 2011). A lack of UGT-2B10 enzyme activity (as a result of a high frequency of nonfunctional alleles) responsible for the conversion of cotinine into cotinine glucuronide largely influences the level of cotinine in serum and saliva in an individual. African Americans are more likely to have no active UGT-2B10 enzyme, leading to higher cotinine concentrations in their plasma and lower cotinine glucuronide levels in their urine. This could mediate the toxicological mechanisms and biological outcomes of nicotine exposure. In addition, when comparing cotinine levels in serum or saliva, across different races, genetic polymorphisms should be characterized and taken into account before interpreting the data (Fowler et al., 2015; Murphy et al., 2017).

According to a recent research study using a highly sensitive and selective quadrupole-time of flight tandem mass spectrometer, nicotine, cotinine, and *trans*-3′-hydroxycotinine were found in saliva samples (Carrizo, Nerin, Domeno, Alfaro, & Nerín, 2016). In this study, only the free forms of nicotine, cotinine, and *trans*-3′-hydroxycotinine were found in the salivary fluid. The conjugated forms were not detected in the saliva samples. This may suggest that the conjugated forms of the primary nicotine metabolites are unable to cross into salivary fluid, despite the fact that cotinine glucuronide and *trans*-3′-hydroxycotinine glucuronide were detected in human plasma samples from individuals exposed to tobacco smoke (de Leon et al., 2002). On the other hand, the conjugate forms may partition in saliva before

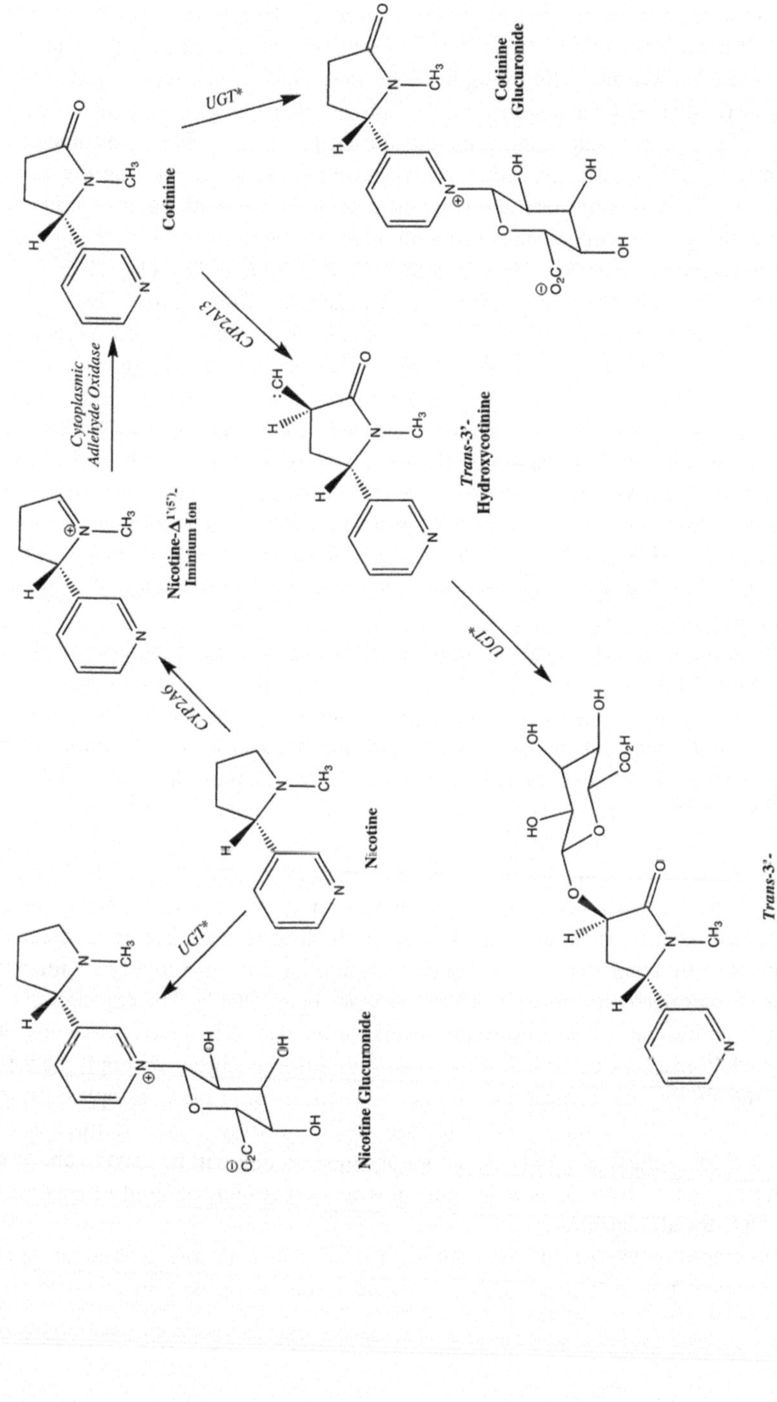


**Fig. 15.2** Primary metabolism of nicotine to cotinine, *trans*-3′-hydroxycotinine, and their conjugated metabolites

glucuronidase enzyme breaks the glucuronide bond and liberates the free forms (Chauncey, Lionetti, Winer, &Lisanti, 1954).

Concentrations of cotinine vary across different biological matrices. A comparative study conducted by Wall et al. (1988) found that concentrations of cotinine found in serum, saliva, and urine of active smokers ($\leq$ 10 cigarettes per day) were 78 ng/mL, 66.9 ng/mL, and 673.4 ng/mL, respectively (Wall et al., 1988). Another study, where paired urine–saliva samples from pregnant women who reported not smoking were analyzed for cotinine levels, showed that the geometric mean concentration of cotinine in saliva and urine are similar ($n = 52$, during third trimester, 3.06 ng/mL vs. 3.07 ng/mL, respectively). However, among the smoking pregnant women ($n = 17$, third trimester), the geometric mean concentrations of cotinine in saliva and urine were drastically different (128.45 ng/mL vs. 341.72 ng/mL) (Stragierowicz, Mikołajewska, Zawadzka-Stolarz, Polańska, & Ligocka, 2013).

A study that analyzed paired serum-saliva samples for cotinine levels also indicated that, on average, the concentration of salivary cotinine is approximately 27% higher than the concentration in serum. The difference in concentrations between the two matrices was much more pronounced in ETS-exposed individuals than active smokers. When the blood compartment becomes saturated with cotinine, a portion of the cotinine is transported into the saliva compartment, making the ratio of cotinine in blood versus saliva close to 1:1. However, saliva fluid content is largely dependent on pH and the amount of protein that can bind to cotinine (Bernert Jr, McGuffey, Morrison, & Pirkle, 2000). Overall, these results suggest that salivary cotinine concentrations are closer to those found in serum but much less than those found in urine, particularly for smokers.

A correlation of cotinine concentrations in serum (or plasma), urine, and saliva is of great interest because it suggests that chemical equilibrium can be established across these biological matrices. This equilibrium is important to determine if salivary concentrations can be used as an indicator of exposures to nicotine or to predict exposure magnitudes. According to a study by Bernert Jr et al. (2000), a simple, linear relationship between serum and saliva cotinine concentrations can be seen in an individual exposed to nicotine. This study found that this relationship can be used to reasonably estimate the serum cotinine concentration in an individual given his or her salivary cotinine result (Bernert Jr et al., 2000). Another study concluded that urinary cotinine concentrations (which were not creatinine-adjusted to account for differences in urinary concentration/dilution) were highly correlated with salivary cotinine concentrations (Stragierowicz et al., 2013). The correlation of cotinine concentrations across serum (or plasma), urine, and saliva means that any of these biological matrices can be used to assess nicotine exposures, although the low nicotine metabolite concentrations found in serum or saliva samples from ETS-exposed individuals may limit detection ability, leading to low frequency of detection and low statistical power. A laboratory method that will be used to analyze these samples should have a suitable limit of detection (LOD) or limit of quantification (LOQ) for all matrices.

Success in the analysis of salivary cotinine has led to its widespread application in research studies. For instance, salivary cotinine levels were used to evaluate the association between Spanish smoking legislation and second-hand smoke exposure in a nonsmoking adult cohort in Barcelona, Spain. This study analyzed 397 saliva samples from nonsmokers before and after the implementation of national smoking bans. The results showed a decrease in the frequency of detection of salivary cotinine (from 92.4% to 64.2%) among the study subjects following the implementation of the national smoking bans (LOQ = 0.05 ng/mL). The geometric mean of salivary cotinine concentration significantly decreased from 0.98 ng/mL to 0.12 ng/mL after the implementation of the smoke-free legislation. This decrease was independent of sociodemographic variables (Lidón-Moyano et al., 2017). Another study used salivary cotinine levels to evaluate the impact of ETS exposure during pregnancy on child neurodevelopment within the first 2 years of life. Saliva samples from 461 non-smoking pregnant women were analyzed for cotinine levels. The results indicated that ETS exposures in the first and second trimesters of pregnancy were associated with a decrease in child language function at the age of 1 year and 2 years. A negative association was found for cotinine level in all trimesters of pregnancy and child motor abilities at 2 years of age. This study, using saliva samples, could confirm that ETS exposure during pregnancy could have a negative impact on child psychomotor development within the first 2 years of life (Polanska et al., 2017).

### 15.2.2 *Analysis of Phthalate Metabolites in Saliva*

Phthalates, often called plasticizers, are a group of chemicals used to increase the flexibility and durability of plastic or to hold color, shine, or fragrance in various consumer products. They are used in vinyl flooring, adhesives, detergents, lubricating oils, automotive plastics, plastic clothes (raincoats), and personal care products (soaps, shampoos, hair sprays, and nail polishes). Phthalates are key ingredients in plastic packaging film and sheets, household products, children's toys, and medical materials. Once phthalates enter the human body, they are quickly metabolized and excreted in urine (CDC, 2017). Examples of phthalates are di-2-ethylhexyl phthalate (DEHP), dibutyl phthalate (DBP), and diethyl phthalate (DEP). Figure 15.3 summarizes the primary metabolism of phthalates.

Phthalate metabolites have been reportedly detected in saliva samples (Silva et al., 2005). In Silva et al. (2005), salivary concentrations of phthalate metabolites were measured in 39 anonymous adult volunteers using high-performance liquid chromatography–tandem mass spectrometry. Seven out of the 14 measured phthalate metabolites were detected. These include phthalic acid, monomethyl phthalate (MMP), monoethyl phthalate (MEP), mono-n-butyl phthalate (MnBP), monoisobutyl phthalate (MiBP), monobenzyl phthalate (MBzP), and mono-2-ethylhexyl phthalate (MEHP). The detection frequencies ranged from 8% (for MMP) to 85% (for MnBP). The concentrations of these metabolites varied. The highest concentration, 353.6 ng/mL, was observed for MBzP. The frequency of detection and the salivary levels of each phthalate monoester in this study population

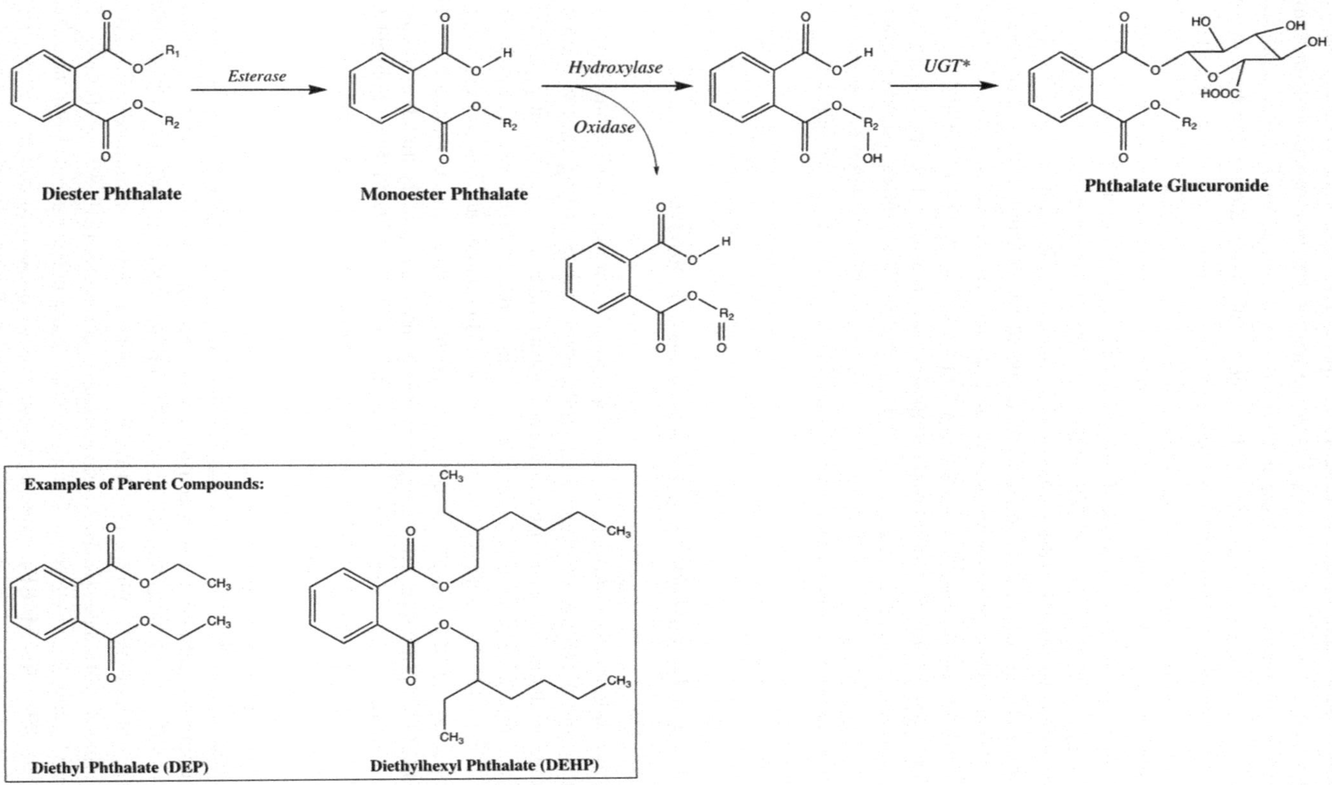


**Fig. 15.3** Primary metabolism of phthalates. Note: *UGT∗ = UGT isoenzymes*

were lower than those found in urine for the same monoester in the general US population (Silva et al., 2005).

In the study of Silva et al. (2005), the metabolites detected in saliva samples were non-oxidative metabolites normally formed by hydrolysis enzymes (esterases) in serum, breast milk, and saliva (Calafat, Slakman, Silva, Herbert, & Needham, 2004; Kato et al., 2003; Silva et al., 2005). The levels of these metabolites, for example, MEHP, in these matrices can be a result of DEHP contamination. To prevent this issue, an appropriate post-collection treatment to denature these enzymes is needed (Calafat et al., 2004; Kato et al., 2003).

It is known that the oxidative metabolites are better biomarkers of phthalate exposure, as they are the metabolic products of specific liver enzymes (Choi et al., 2012). Thus, the concentrations of oxidative metabolites are not subject to contamination of the parent compounds (Silva et al., 2006). However, the concentrations of oxidative metabolites, mostly present as glucuronide-bound metabolites, are much lower than the non-oxidative metabolites. In addition, the concentrations of glucuronide-bound, oxidative metabolites in serum are much lower than those found in urine (Kato et al., 2004).

Another study has attempted to investigate the relationships among phthalate metabolites across biological matrices such as breast milk, urine, saliva, and serum collected from 33 US lactating women. Using the gold-standard liquid chromatographic–tandem mass spectrometric method, this study only detected two phthalate metabolites that are oxidative metabolites: mono-(3-carboxypropyl) phthalate (MCPP) (2.2 μg/L) and mono-(2-ethyl-5-carboxypentyl) phthalate (MECPP) (2.3 μg/L). This detection was from a single saliva sample from one woman on one visit. All other phthalate metabolites were not detected (Hines, Calafat, Silva, Mendola, & Fenton, 2009).

The results from both studies (Silva et al., 2005; Hines et al., 2009) are not in agreement in terms of the detection of oxidative metabolites of phthalates. In addition, the detection frequencies for these metabolites vary greatly. No detection of some oxidative metabolites, such as mono-(2-ethyl-5-hydroxyhexyl) phthalate (MEHHP), mono-(2-ethyl-5-oxohexyl) phthalate (MEOHP), may indicate no transport of these metabolites from blood plasma to saliva fluids (lack of chemical equilibrium), or insufficient sensitivity of the employed method. If chemical equilibrium can only be established for some metabolites between blood plasma and saliva and the likelihood of pre-analytic and analytic contamination is high, saliva will likely be an unsuitable matrix for exposure assessment of phthalates. Besides, for a comprehensive assessment of phthalate exposures, a panel of metabolites must be analyzed altogether (CDC, 2017).

### *15.2.3 Analysis of Toxic Heavy Metals in Saliva*

There is enough scientific evidence to indicate that toxic heavy metals can also be detected in saliva, similarly to electrolytes. A recent study showed that arsenic (As), chromium (Cr), and lead (Pb) were detected in the saliva of healthy individuals (Dame et al., 2015).

In the study by Bhowmick et al., (2013), 50 saliva samples were collected from male residents of Nadia district in West Bengal, India, where the groundwater is reported to be heavily contaminated with manganese (Mn) and As. In this study, Mn, As, Cr, and Pb were detected in saliva samples with mean concentrations of 5.4 μg/L (range 0.69–22 μg/L), 6.3 μg/L (range 0.70–29 μg/L), 0.78 μg/L (range < LOD-5.9 μg/L), and 0.94 μg/L (range < LOD-4.2 μg/L), respectively. Because the water source was contaminated predominantly by Mn and As, a significant correlation between Mn and As was observed in these saliva samples (Bhowmick et al., 2013).

Wang et al. (2017) reported the concentrations of total As and its species in urine and saliva samples of 70 residents exposed to As from drinking water in Shanxi, China. The result showed that the total As concentration in saliva was relatively lower than in urine samples (mean concentrations, 12.31 μg/L in saliva and 124.93 μg/g in urine). Salivary total As concentrations were positively correlated with total urinary As and drinking water As concentrations. This study also found that in saliva, most As species were not methylated. The major species in saliva was $As^{III}$ and $As^{V}$ (76.18% of total), followed by dimethylarsinate (13.08%) and methylarsonic acid (9.13%). This study suggests that the methylated As species have a limited transfer capability from blood plasma into saliva fluid (Wang et al., 2017).

In an occupational study by Gil et al. (2011), where cadmium (Cd), Cr, Mn, nickel (Ni), and Pb levels in whole blood, urine, and saliva (axillary hair included) from 178 individuals were analyzed, no correlation was found for Mn concentrations across all biological matrices analyzed. Salivary Ni concentrations did not correlate with blood or urinary concentrations. Salivary Cd concentrations did not correlate with blood Cd or urinary Cd concentrations, but significantly correlated with concentrations in hair. Salivary Cr concentrations were only correlated with urinary concentrations, but not with those found in other matrices. Salivary Pb concentrations were inversely, significantly correlated with blood Pb concentrations. No correlation was found between salivary Pb and urinary Pb concentrations. Overall, the results of this study indicate a lack of uniform correlations between salivary metal concentrations and those found in other biological matrices (Gil et al., 2011).

In a study by Nriagu, Burt, Linder, Ismail, and Sohn (2006), a total of 904 pairs of blood and saliva samples were collected and analyzed for Pb levels. The population was composed of low-income adults who lived in Michigan, USA. The average blood Pb and salivary Pb levels were 2.7 μg/dL (or 27 μg/L) and 2.4 μg/L, respectively. The log-transformed Pb concentrations derived from both matrices were statistically correlated, but the regression coefficient was very weak ($r = 0.135$) (Nriagu et al., 2006).

A recent study from Staff et al. (2014) demonstrated the limited use of saliva as an alternative to whole blood for the biomonitoring of inorganic Pb. In their study, 105 pairs of blood and saliva samples were obtained from occupationally exposed workers. For the paired samples, the median whole blood Pb level was 6.00 μg/dL (or 60 μg/L) and the median salivary Pb level was 17.1 μg/L. Although a significant correlation between blood and salivary Pb levels was observed, it was weak (Pearson's correlation coefficient = 0.457; 95% C.I. 0.291–0.596). The authors concluded that saliva could only be effective as a surrogate for whole blood in highly exposed populations (Staff et al., 2014). This conclusion is in agreement with the article published previously by Koh et al. (2003) that examined the relationship between salivary Pb and whole blood Pb levels in 82 exposed adults. In this article, the authors concluded that the use of salivary Pb for biomonitoring in individuals with whole blood Pb levels ranging from 10 to 50 μg/dL is not recommended.

Moreover, in a study by Barbosa et al. (2006), salivary Pb levels were compared to blood Pb and plasma Pb levels. A total of 88 subjects participated in this study and provided all samples analyzed. Significant, albeit weak, correlations were found between salivary Pb concentrations, blood Pb concentrations, and plasma Pb concentrations, respectively. However, the salivary Pb to plasma Pb ratio was highly variable (from 0.20 to 18.0), suggesting a high degree of variation in Pb concentrations in both matrices, thus suggesting a lack of chemical equilibrium. Hence, it was concluded that salivary Pb may not be used as a biomarker to indicate Pb exposure nor can it be used to predict plasma Pb levels in a low to moderately exposed population (Barbosa et al., 2006).

The study by de Souza Guerra et al. (2015) investigated whole blood, serum, and saliva Pb levels from the same populations (children and adolescents) at two time points (12 months apart) to assess the effectiveness of the mitigation measures undertaken by the Brazilian authorities to reduce Pb exposures. They found no significant correlations between the salivary Pb and whole blood Pb levels as well as the salivary Pb and serum Pb levels in both time points. Note that the numbers of pairs varied from $n = 50$ to $n = 57$ (de Souza Guerra et al., 2015).

The above information, including the results from Costa de Almeida et al. (2010) and Wilhelm et al. (2002), which were not mentioned in detail here, shows that heavy metal concentrations in saliva and other biological matrices are not always correlated. This suggests that, in some cases, equilibrium may not be established across human compartments for some metals. This may complicate the use of saliva as a matrix to assess exposures to those metals.

### 15.2.4 Analysis of Perchlorate in Saliva

Perchlorate ion is a negatively charged group of atoms consisting of a central chlorine atom bonded to four oxygen atoms. The molecular formula of perchlorate is $ClO_4^-$. Perchlorates can form naturally in the atmosphere, leading to trace levels of perchlorate in precipitation. Perchlorates are also manufactured in large amounts for its use in rocket fuels, explosives, temporary adhesives, electrolysis baths, batteries, airbags, drying agents, etching agents, cleaning agents and bleach, and oxygen generating systems. Perchlorate has gained public health concern due to its ability to inhibit partially the thyroid's uptake of iodine. It is anticipated that people exposed to excessive amounts of perchlorate for a long period of time may develop a diminished capacity to produce thyroid hormones (ATSDR, 2008). A developing fetus or infant, whose normal neurodevelopment depends on adequate iodine intake for the production of thyroid hormones, may be affected upon exposure to perchlorate (Leung, Pearce, & Braverman, 2010).

An exploratory study analyzed a subset of 13 paired saliva–serum samples for perchlorate using a liquid chromatographic–tandem mass spectrometric method. The concentrations of perchlorate in serum correlated well with the concentrations of perchlorate in saliva. Although the sample size was low, the mean saliva–serum concentration ratio of perchlorate was 14:1, suggesting that perchlorate concentrations in saliva are generally higher than those in serum. Although salivary perchlorate concentrations are generally lower than those in urine; this research may lead to the use of salivary perchlorate in epidemiological investigations aiming to assess the associations between environmental exposure to perchlorate and health outcomes (Oldi & Kannan, 2009).

## 15.3 Methodology Issues, Challenges, and Considerations

Several environmental chemicals, to which humans are commonly exposed, can be measured in saliva in order to provide a snapshot of the internal dose at the time of collection. However, for other chemicals, salivary concentrations are too low to be reliably quantified by current analytical instruments, leading to the limited interpretation of the results. This factor prevents the widespread application of saliva in human biomonitoring of environmental chemicals.

Nonetheless, for the successful utilization of saliva in human biomonitoring of environmental chemicals, several analytical issues must be considered.

### *15.3.1 Sample Collection*

Saliva is easier to collect than blood. Like urine, it can be collected multiple times in a given time period, offering an excellent opportunity for longitudinal assessment of exposure levels.

Saliva can be collected via a self-spitting technique or a variety of commercial devices. The spitting technique provides oral fluid that is relatively viscous and non-uniform. The commercial devices provide advantages such as decreased collection time and ease of use. These devices may be used together with stimulators such as citric acid and chewing gum to increase the volume of saliva (Drummer, 2008). However, the use of such collection devices may not be suitable for use with children because of potential choking hazards.

When the flow rate of saliva increases, the chemical composition changes. The pH of saliva also changes which affects the ability of acidic, e.g., citric acid, and basic chemicals to partition into the oral fluid. Thus, the use of stimulators can control the salivary pH and concentrations of some chemicals in the oral fluid. Devices that contain absorbent pads must be centrifuged to release the fluid from the pad. The collection process is an important element that can affect the concentrations of target chemicals (Drummer, 2008).

Because the buccal cavity can be contaminated by components derived from previous ingestions, it is recommended that saliva is collected a few hours after drinking or eating (Bessonneau, Pawliszyn, & Rappaport, 2017). In addition, the salivary flow rate and concentrations vary according to the circadian clock (Dawes, 1972). As such, collection time is an important factor in determining the concentrations of target chemicals or metabolites in saliva (Dallmann, Viola, Tarokh, Cajochen, & Brown, 2012). It is important that saliva samples are collected at the same time for repeat measurements (Bessonneau et al., 2017).

### 15.3.2 Extraction Method and Analysis Technique

Usually, exposure to a given environmental chemical results in a low, circulating concentration in plasma. Only a portion of this chemical (whether it is the parent form or metabolite) is transferred into saliva, while the rest is excreted via urine or feces. Another portion may partition in adipose tissues or bind to macromolecules, depending on its lipophilicity (Barr et al., 2005). Even in the case of exposure to nicotine, the concentrations of cotinine found in saliva can be quite low, especially for those of ETS-exposed individuals. Thus, salivary cotinine cannot be used to estimate more precisely the level of exposure to tobacco smoke in nonsmokers (Stragierowicz et al., 2013).

Environmental chemical concentrations in saliva are relatively low, therefore it is crucial that the appropriate extraction technique and analytical instruments are selected to enable the quantification of these chemicals or their metabolites. The majority of chemicals that partition in saliva are highly polar and may not bind to proteins (Hold et al., 1999). To extract these chemicals, solid-phase extraction or liquid–liquid extraction techniques are normally used, with or without prior protein precipitation. Extraction recovery may vary across different chemicals, but the extractants are clean, as they contain fewer chemicals than in urine or blood plasma (Bessonneau et al., 2017).

The analysis of target environmental chemicals in saliva requires the use of highly sensitive and selective techniques such as mass spectrometry (Barr et al., 2005). Mass spectrometry offers unprecedented capabilities over other techniques such as colorimetric analyses or spectrophotometry; it is able to separate and analyze chemicals based on their masses (i.e., mass-to-charge ratio). When mass spectrometry is operated in tandem, by combining at least two mass spectrometric analyzers into one instrument, it is even more sensitive and is able to analyze chemicals at very low concentrations. However, mass spectrometry requires that the target chemicals be initially separated with liquid or gas chromatography to achieve optimal sensitivity and selectivity during the analysis.

Mass spectrometric methods are subject to certain limitations as well. Matrix effects, especially ion suppression, can lead to low sensitivity of mass analyzers when used with liquid chromatography to analyze polar chemicals. Ion suppression is caused by co-eluting chemicals that reduce the ionization efficiency of target analytes, therefore allowing a lower volume of the target analytes to enter the mass analyzer (Panuwet et al., 2016). As such, the samples that contain fewer chemical components will likely be free from ion suppression. This is true for saliva samples, as they were found to lack interference from matrix effects during the analysis of environmental pollutants (Russo, Barbato, Mita, & Grumetto, 2019). This may facilitate wider use of saliva as a matrix for chemical analyses in the future.

### *15.3.3 Lack of Reference Materials*

It is important to determine the accuracy of any given analytical method used for human biomonitoring, in order to allow for direct comparisons of the results with other studies. This can be achieved via the analysis of standard reference materials of known concentrations such as Standard Reference Materials produced by the National Institutes of Standards and Technology (NIST). Currently, there is no standard reference material available for the determination of environmental chemicals in saliva.

## 15.4 Future Directions and Opportunities

With the continued advancement of analytical capabilities, biomonitoring will play a critical role in understanding chemical exposures that have personal and public health significance (Dennis et al., 2017). With the emergence of a new concept, the human exposome, the exposure assessment paradigm has shifted. Saliva may play a significant role in this development.

The concept of the human exposome was proposed in 2005 by molecular epidemiologist Dr. Christopher P. Wild, to draw attention to the need to improve environmental exposure assessment to complement the progress made on the human genome characterization effort. In his view, there was a desperate need to develop methods with the same precision for characterizing an individual's environmental exposures, as there was for an individual's genome. Initially, the concept was defined as, "*...life-course environmental exposures (including lifestyle factors), from the prenatal period, onwards*" (Wild, 2005). The original definition was strictly centered on exposure assessment. However, a new definition proposed in 2014 defines the exposome as, "*The cumulative measure of environmental influences and associated biological responses throughout the lifespan, including exposures* *from the environment, diet, behavior, and endogenous processes.*" This definition allows the inclusion of three new concepts: cumulative biological responses, human behaviors, and endogenous processes. All of these components have a profound role in understanding the etiology of disease development (Miller & Jones, 2014). The human exposome concept emphasizes the fact that the sources of toxic chemicals are both exogenous (such as air, water, diet, drugs, and radiation) and endogenous (from inflammation, lipid peroxidation, oxidative stress, existing diseases, infections, and gut flora) (Rappaport, 2011).

Thus, in the era of the human exposome, exposure assessment of environmental chemicals should not only aim to quantify the concentrations of chemicals but also to understand the resulting biological responses. Some of the biological responses may result in a series of changes in endogenous metabolites. By looking at these changes, subtle alterations in biological pathways that underlie various physiological conditions and aberrant processes, including diseases, can be revealed (Johnson, Ivanisevic, & Siuzdak, 2016). The changes in metabolites can be characterized using metabolomics. Metabolomics was developed to enable the comprehensive measurement of all metabolites (both endogenous and exogenous) and low-molecular weight molecules in a biological specimen (Scalbert et al., 2014). Endogenous metabolites are small molecules synthesized by enzymes or our microflora while exogenous metabolites have their sources from the environment, food, drugs, or other consumable products (Wishart et al., 2013). In a comprehensive metabolomics study, advanced analytical platforms are required to uncover a large set of metabolites. These usually include liquid chromatography–mass spectrometry, nuclear magnetic resonance spectroscopy, gas chromatography–mass spectrometry, and inductively coupled plasma mass spectrometry. The sum of metabolites in a given biological matrix or from a particular origin is called the metabolome (Scalbert et al., 2014). Note that the extraction procedure and analytical method of choice largely influence the types of metabolites detected in each measurement (Johnson et al., 2016).

The saliva metabolome has been characterized to provide a solid foundation for future studies that seek to understand the etiology of the disease (Dame et al., 2015; Takeda et al., 2009). The recent characterization of the saliva metabolome was reported by Dame et al. (2015), in which a large set of metabolites was determined using multiple analytical platforms. In their report, there were at least 853 detectable endogenous and exogenous metabolites (corresponding to 1237 probable chemical species) in human saliva. Approximately 300 of them have been quantified (Dame et al., 2015). Although the saliva metabolome has considerable chemical diversity, saliva contains fewer chemicals than urine, which has been confirmed to contain more than 3000 chemicals and serum, which has more than 4000 chemicals (Bouatra et al., 2013; Psychogios et al., 2011; Wishart et al., 2013). In addition, the saliva metabolome may differ according to gender, sample collection strategy (stimulated versus unstimulated), as well as smoking status (Takeda et al., 2009).

Nevertheless, because saliva is likely in equilibrium with whole blood, and both whole blood and saliva temporally represent diverse groups of small molecules in the body, there is great potential for the use of saliva as a biological matrix in studies of the human exposome. In a recent study, the saliva metabolome was evaluated for its use in exposome-wide association studies, or E-WAS, to link human metabolic pathways to the development of diseases. This study found that the saliva exposome represents at least 14 metabolic pathways, including amino acid metabolism, the citrate cycle, gluconeogenesis, glutathione metabolism, and butanoate metabolism. These metabolic pathways are connected to human metabolic diseases, central nervous system diseases, and neoplasms (Bessonneau et al., 2017). Therefore, saliva may be useful to discover exposure-risk factors for several chronic diseases. This study also emphasized that saliva specimens offer a more practical way to investigate longitudinal individual exposomes due to the simplicity and noninvasiveness of specimen collection and the ability to collect multiple samples from many individuals at any given time.

With the continued improvement of analytical technologies and instrumentation, the ability to detect and quantify low-level metabolites will be achieved. More metabolites or environmental chemicals in saliva will be uncovered in the future. Standardization of biomonitoring methods will permit a direct comparison between studies, allowing exposure levels to be compared across different populations. It is anticipated that saliva will continue to offer a unique advantage in human biomonitoring studies aiming at linking exposures to environmental chemicals to the development of chronic diseases in humans.

## References

Angerer, J., Ewers, U., & Wilhelm, M. (2007). Human biomonitoring: State of the art. *International Journal of Hygiene and Environmental Health, 210*(3–4), 201–228.

ATSDR. (2008). *Toxicological profile for perchlorates*. Retrieved Dec 12, 2018, from https://www.atsdr.cdc.gov/phs/phs.asp?id=892&tid=181#top

Avila-Tang, E., Al-Delaimy, W. K., Ashley, D. L., Benowitz, N., Bernert, J. T., Kim, S., . . . Hecht, S. S. (2013). Assessing secondhand smoke using biological markers. *Tobacco Control, 22*, 164–171.

Barbosa, F., Corrêa Rodrigues, M. H., Buzalaf, M. R., Krug, F. J., Gerlach, R. F., & Tanus-Santos, J. E. (2006). Evaluation of the use of salivary lead levels as a surrogate of blood lead or plasma lead levels in lead exposed subjects. *Archives of Toxicology, 80*(10), 633–637.

Barr, D. B., Wang, R. Y., & Needham, L. L. (2005). Biologic monitoring of exposure to environmental chemicals throughout the life stages: Requirements and issues for consideration for the National Children's Study. *Environmental Health Perspectives, 113*(8), 1083–1091.

Benowitz, N. L., Hukkanen, J., & Jacob, P. (2009). Nicotine chemistry, metabolism, kinetics and biomarkers. *Handbook of Experimental Pharmacology, 192*, 29–60.

Bernert, J. T., Jr., McGuffey, J. E., Morrison, M. A., & Pirkle, J. L. (2000). Comparison of serum and salivary cotinine measurements by a sensitive high-performance liquid chromatography-tandem mass spectrometry method as an indicator of exposure to tobacco smoke among smokers and nonsmokers. *Journal of Analytical Toxicology, 24*(5), 333–339.

Bernert, J. T., Turner, W. E., Pirkle, J. L., Sosnoff, C. S., Akins, J. R., Waldrep, M. K., . . . Sampson, E. J. (1997). Development and validation of sensitive method for determination of serum cotinine in smokers and nonsmokers by liquid chromatography/atmospheric pressure ionization tandem mass spectrometry. *Clinical Chemistry, 43*(12), 2281–2291.

Bessonneau, V., Pawliszyn, J., & Rappaport, S. M. (2017). The saliva exposome for monitoring of individuals' health trajectories. *Environmental Health Perspectives, 125*, 1–10.

Bhowmick, S., Halder, D., Kundu, A. K., Saha, D., Iglesias, M., Nriagu, J., ... Chatterjee, D. (2013). Is saliva a potential biomarker of arsenic exposure? A case-control study in West Bengal, India. *Environmental Science and Technology, 47*, 3326–3332.

Bouatra, S., Aziat, F., Mandal, R., Guo, A. C., Wilson, M. R., Knox, C., ... Wishart, D. S. (2013). The human urine metabolome. *PLoS One, 8*(9), e73076.

Bruijnzeel, A. W. (2012). Tobacco addiction and the dysregulation of brain stress systems. *Neuroscience and Biobehavioral Reviews, 36*, 1418–1441.

Byrd, G. D., Chang, K. M., Greene, J. M., & de Bethizy, J. D. (1991). Evidence for urinary excretion of glucuronide conjugates of nicotine, cotinine, and trans-3′-hydroxycotinine in smokers. *Drug Metabolism and Disposition, 20*(2), 192–197.

Calafat, A. M., Slakman, A. R., Silva, M. J., Herbert, A. R., & Needham, L. L. (2004). Automated solid phase extraction and quantitative analysis of human milk for 13 phthalate metabolites. *Journal of Chromatography B, 805*(1), 49–56.

Carrizo, D., Nerin, I., Domeno, C., Alfaro, P., & Nerín, C. (2016). Direct screening of tobacco indicators in urine and saliva by atmospheric pressure solid analysis probe coupled to quadrupole-time of flight mass spectrometry (ASAP-MS-Q-TOF-). *Journal of Pharmaceutical and Biomedical Analysis, 124*, 149–156.

Chauncey, H. H., Lionetti, F., Winer, R. A., & Lisanti, V. F. (1954). Enzymes of human saliva: I. the determination, distribution, and origin of whole saliva enzymes. *Journal of Dental Research, 33*(3), 321–334.

Choi, K., Joo, H., Campbell, J. L., Clewell, R. A., Andersen, M. E., & Clewell, H. J., 3rd. (2012). In vitro metabolism of di(2-ethylhexyl) phthalate (DEHP) by various tissues and cytochrome P450s of human and rat. *Toxicology In Vitro, 26*(2), 315–322.

Cooper, G. M. (2000). *Transport of small molecules. The cell: A molecular approach* (2nd ed.). Sunderland, MA: Sinauer Associates. www.ncbi.nlm.nih.gov/books/NBK9847/

Costa de Almeida, G. R., de Freitas Tavares, C. F., de Souza, A. M., Sampaio de Sousa, T., Rodrigues Funayama, C. A., Barbosa, F., Jr., ... Gerlach, R. F. (2010). Whole blood, serum, and saliva lead concentrations in 6- to 8-year-old children. *The Science of the Total Environment, 408*(1), 1551–1556.

Dallmann, R., Viola, A. U., Tarokh, L., Cajochen, C., & Brown, S. A. (2012). The human circadian metabolome. *Proceedings of the National Academy of Sciences, 109*, 2625–2629.

Damalas, C., & Koutroubas, S. (2016). Farmers' exposure to pesticides: Toxicity types and ways of prevention. *Toxics, 4*(1), 1.

Dame, Z. T., Aziat, F., Mandal, R., Krishnamurthy, R., Bouatra, S., Borzouie, S., ... Wishart, D. S. (2015). The human saliva metabolome. *Metabolomics, 11*, 1864–1883.

Dawes, C. (1972). Circadian rhythms in human salivary flow rate and composition. *The Journal of Physiology, 220*, 529–545.

de Leon, J., Diaz, F. J., Rogers, T., Browne, D., Dinsmore, L., Ghosheh, O. H., ... Crooks, P. A. (2002). Total cotinine in plasma: A stable biomarker for exposure to tobacco smoke. *Journal of Clinical Psychopharmacology, 22*(5), 496–501.

de Souza Guerra, C., Magna, G. A. M., Machado, S. L., De Capitani, E. M., Ramos, J., Peixoto, I. T., ... de Fátima Carvalho, M. (2015). Lead concentrations in whole blood, serum, saliva and house dust in samples collected at two time points (12 months apart) in Santo Amaro, BA, Brazil. *Environmental Research, 142*(1), 337–344.

Demkowska, I., Polkowska, Z., & Namiesnik, J. (2011). Non-invasive biological fluid matrices analysed to assess exposure to environmental tobacco smoke. *Journal of Exposure Science & Environmental Epidemiology, 21*(6), 656–661.

Dennis, K. K., Marder, E., Balshaw, D. M., Cui, Y., Lynes, M. A., Patti, G. J., ... Barr, D. B. (2017). Biomonitoring in the era of the exposome. *Environmental Health Perspectives, 125*(4), 502–510.

Drummer, O. H. (2008). Introduction and review of collection techniques and applications of drug testing of oral fluid. *Therapeutic Drug Monitoring, 30*(2), 203–206.

Feyerabend, C., & Russell, M. A. (1990). A rapid gas-liquid chromatographic method for the determination of cotinine and nicotine in biological fluids. *The Journal of Pharmacy and Pharmacology, 42*(6), 450–452.

Fowler, S., Kletzl, H., Finel, M., Manevski, N., Schmid, P., Tuerck, D., . . . Iglesias, V. A. (2015). A UGT2B10 splicing polymorphism common in African populations may greatly increase drug exposure. *The Journal of Pharmacology and Experimental Therapeutics, 352*, 358–367.

Gil, F., Hernández, A. F., Márquez, C., Femia, P., Olmedo, P., López-Guarnido, O., & Pla, A. (2011). Biomonitorization of cadmium, chromium, manganese, nickel and lead in whole blood, urine, axillary hair and saliva in an occupationally exposed population. *The Science of the Total Environment, 409*, 1172–1180.

Greenberg, R. A., Haley, N. J., Etzel, R. A., & Loda, F. A. (1984). Measuring the exposure of infants to tobacco smoke. Nicotine and cotinine in urine and saliva. *The New England Journal of Medicine, 310*(17), 1075–1078.

Hines, E. P., Calafat, A. M., Silva, M. J., Mendola, P., & Fenton, S. E. (2009). Concentrations of phthalate metabolites in milk, urine, saliva, and serum of lactating North Carolina women. *Environmental Health Perspectives, 117*(1), 86–92.

Hold, K., de Boer, D., & Zuidema, J. (1999). Saliva as an analytical tool in toxicology. *International Journal of Drug Testing, 1*, 1–36.

Humphrey, S. P., & Williamson, R. T. (2001). A review of saliva: Normal composition, flow, and function. *The Journal of Prosthetic Dentistry, 85*, 162–169.

International Agency for Research on Cancer. (2012). *Personal habits and indoor combustions*. WHO (100E):541.

Jarvis, M. J., Russell, M. A., Feyerabend, C., Eiser, J. R., Morgan, M., Gammage, P., & Gray, E. M. (1985). Passive exposure to tobacco smoke: Saliva cotinine concentrations in a representative population sample of non-smoking school children. *BMJ, 291*(6500), 927–929.

Johnson, C. H., Ivanisevic, J., & Siuzdak, G. (2016). Metabolomics: Beyond biomarkers and towards mechanisms. *Nature Reviews. Molecular Cell Biology, 17*, 451–459.

Kakkar, P., & Jaffery, F. N. (2005). Biological markers for metal toxicity. *Environmental Toxicology and Pharmacology, 19*(2), 335–349.

Kato, K., Silva, M. J., Brock, J. W., Reidy, J. A., Malek, N. A., Hodge, C. C., . . . Barr, D. B. (2003). Quantitative detection of nine phthalate metabolites in human serum using reversed-phase high-performance liquid chromatography-electrospray ionization-tandem mass spectrometry. *Journal of Analytical Toxicology, 27*(5), 284–289.

Kato, K., Silva, M. J., Reidy, J. A., Hurtz, D., 3rd, Malek, N. A., Needham, L. L., . . . Calafat, A. M. (2004). Mono(2-ethyl-5-hydroxyhexyl) phthalate and mono-(2-ethyl-5-oxohexyl) phthalate as biomarkers for human exposure assessment to di-(2-ethylhexyl) phthalate. *Environmental Health Perspectives, 112*(3), 327–330.

Koh, D., Ng, V., Chua, L. H., Yang, Y., Ong, H., & Chia, S. (2003). Can salivary lead be used for biological monitoring of lead exposed individuals? *Occupational and Environmental Medicine, 60*, 696–698.

Lehmler, H.-J., Liu, B., Gadogbe, M., & Bao, W. (2018). Exposure to bisphenol A, bisphenol F, and bisphenol S in U.S. adults and children: The National Health and Nutrition Examination Survey 2013–2014. *ACS Omega, 3*, 6523–6532.

Leung, A. M., Pearce, E. N., & Braverman, L. E. (2010). Perchlorate, iodine and the thyroid. *Best Practice & Research. Clinical Endocrinology & Metabolism, 24*(1), 133–141.

Lidón-Moyano, C., Fu, M., Perez-Ortuño, R., Ballbè, M., Sampedro-Vida, M., Martín-Sánchez, J. C., . . . Martínez-Sánchez, J. M. (2017). Assessment of salivary cotinine concentration among general non-smokers population: Before and after Spanish smoking legislations. *Cancer Epidemiology, 51*, 87–91.

Miller, G. W., & Jones, D. P. (2014). The nature of nurture: Refining the definition of the exposome. *Toxicological Sciences, 137*(1), 1–2.

Murphy, S. E., Sipe, C. J., & Choi, K., Raddatz, L. M., Koopmeiners, J. S., Donny, E. C. , & Hatsukami, D. K. (2017). Low cotinine glucuronidation results in higher serum and saliva cotinine in African American compared to white smokers. *Cancer Epidemiology, Biomarkers & Prevention, 26*, 1093–1099.

Needham, L. L., Özkaynak, H., Whyatt, R. M., Barr, D. B., Wang, R. Y., Naeher, L., . . . Zartarian, V. (2005). Exposure assessment in the National Children's Study: Introduction. *Environmental Health Perspectives, 113*(8), 1076–1082.

Nriagu, J., Burt, B., Linder, A., Ismail, A., & Sohn, W. (2006). Lead levels in blood and saliva in a low-income population of Detroit, Michigan. *International Journal of Hygiene and Environmental Health, 209*(1), 109–121.

Oldi, J. F., & Kannan, K. (2009). Perchlorate in human blood serum and plasma: Relationship to concentrations in saliva. *Chemosphere, 77*(1), 43–47.

Panuwet, P., Hunter, R. E., D'Souza, P. E., Chen, X., Radford, S. A., Cohen, J. R., . . . Barr, D. B. (2016). Biological matrix effects in quantitative tandem mass spectrometry-based analytical methods: Advancing biomonitoring. *Critical Reviews in Analytical Chemistry, 46*, 93–105.

Peyton, J., III, Wilson, M., & Benowitz, N. L. (1981). Improved gas chromatographic method for the determination of nicotine and cotinine in biologic fluids. *Journal of Chromatography. B, Biomedical Sciences and Applications, 222*(1), 61–70.

Pfaffe, T., Cooper-White, J., Beyerlein, P., Kostner, K., & Punyadeera, C. (2011). Diagnostic potential of saliva: Current state and future applications. *Clinical Chemistry, 57*, 675–687.

Pirkle, J. L., Needham, L. L., & Sexton, K. (1995). Improving exposure assessment by monitoring human tissues for toxic chemicals. *Journal of Exposure Analysis and Environmental Epidemiology, 5*(3), 405–424.

Polanska, K., Krol, A., Merecz-Kot, D., Ligocka, D., Mikolajewska, K., Mirabella, F., . . . Hanke, W. (2017). Environmental tobacco smoke exposure during pregnancy and child neurodevelopment. *International Journal of Environmental Research and Public Health, 14* (7), 796.

Psychogios, N., Hau, D. D., Peng, J., Guo, A. C., Mandal, R., Bouatra, S., . . . Wishart, D. S. (2011). The human serum metabolome. *PLoS One, 6*(2), e16957.

Rappaport, S. M. (2011). Implications of the exposome for exposure science. *Journal of Exposure Science & Environmental Epidemiology, 21*(1), 5–9.

Russo, G., Barbato, F., Mita, D. G., & Grumetto, L. (2019). Simultaneous determination of fifteen multiclass organic pollutants in human saliva and serum by liquid chromatography-tandem ultraviolet/fluorescence detection: A validated method. *Biomedical Chromatography, 33*, e4427.

Scalbert, A., Brennan, L., Manach, C., Andres-Lacueva, C., Dragsted, L. O., Draper, J., . . . Wishart, D. S. (2014). The food metabolome: A window over dietary exposure. *The American Journal of Clinical Nutrition, 99*, 1286–1308.

Sexton, K., Callahan, M. A., & Bryan, E. F. (1995). Estimating exposure and dose to characterize health risks: The role of human issue monitoring in exposure assessment. *Environmental Health Perspectives, 103*, 17.

Silva, M. J., Reidy, J. A., Preau, J. L., Samandar, E., Needham, L. L., & Calafat, A. M. (2006). Measurement of eight urinary metabolites of di(2-ethylhexyl) phthalate as biomarkers for human exposure assessment. *Biomarkers, 11*(1), 1–13.

Silva, M. J., Reidy, J. A., Samandar, E., Herbert, A. R., Needham, L. L., & Calafat, A. M. (2005). Detection of phthalate metabolites in human saliva. *Archives of Toxicology, 79*, 647–652.

Staff, J. F., Harding, A.-H., Morton, J., Jones, K., Guice, E. A., & McCormick, T. (2014). Investigation of saliva as an alternative matrix to blood for the biological monitoring of inorganic lead. *Toxicology Letters, 231*(1), 270–276.

Stragierowicz, J., Mikołajewska, K., Zawadzka-Stolarz, M., Polańska, K., & Ligocka, D. (2013). Estimation of cutoff values of cotinine in urine and saliva for pregnant women in Poland. *BioMed Research International, 1*, 1–11.

Takeda, I., Stretch, C., Barnaby, P., Bhatnager, K., Rankin, K., Fu, H., … Slupsky, C. (2009). Understanding the human salivary metabolome. *NMR in Biomedicine, 22*, 577–584.

U.S. Centers for Disease Control and Prevention. (2017). *Environmental chemicals*. Retrieved Dec 12, 2018 from https://www.cdc.gov/biomonitoring/environmental_chemicals.html

Wall, M. A., Johnson, J., Jacob, P., & Benowitz, N. L. (1988). Cotinine in the serum, saliva, and urine of nonsmokers, passive smokers, and active smokers. *American Journal of Public Health, 78*(6), 699–701.

Wallace, A. D. (2012). Toxic endpoints in the study of human exposure to environmental chemicals. *Progress in Molecular Biology and Translational Science*, 89–115.

Wang, J., Liang, Q., Mendes, P., & Sarkar, M. (2011). Is 24h nicotine equivalents a surrogate for smoke exposure based on its relationship with other biomarkers of exposure? *Biomarkers, 16* (2), 144–154.

Wang, D., Shimoda, Y., Wang, S., Wang, Z., Liu, J., Liu, X., … An, Y. (2017). Total arsenic and speciation analysis of saliva and urine samples from individuals living in a chronic arsenicosis area in China. *Environmental Health and Preventive Medicine, 22*(45), 1–9.

Wild, C. P. (2005). Complementing the genome with an "Exposome": The outstanding challenge of environmental exposure measurement in molecular epidemiology. *Cancer Epidemiology, Biomarkers & Prevention, 14*(8), 1847–1850.

Wilhelm, M., Pesch, A., Rostek, U., Begerow, J., Schmitz, N., Idel, H., & Ranft, U. (2002). Concentrations of lead in blood, hair and saliva of German children living in three different areas of traffic density. *The Science of the Total Environment, 297*, 109–118.

Wishart, D. S., Jewison, T., Guo, A. C., Wilson, M., Knox, C., Liu, Y., … Scalbert, A. (2013). HMDB 3.0—The human metabolome database in 2013. *Nucleic Acids Research, 41*, D801–D807.

# Chapter 16
# Saliva and Drugs of Abuse

**Mahvash Navazesh and Azadeh Ahmadieh**

**Abstract** The drug overdose epidemic resulted in approximately 630,000 deaths from 1999 to 2016 according to the Center for Disease Control and Prevention (CDC). In 2015, 52,404 drug overdose deaths occurred in the USA in which 63.1% involved an opioid; whereas in 2016, there were 63,632 drug overdose deaths reported, opioids accounted for 66.4% of them. On average, 115 Americans die every day from an opioid overdose. National Institute of Drug Abuse (NIDA) confirmed this data by reporting that drug-related deaths have more than doubled since 2000. There are more deaths, oral and systemic complications and disabilities from substance use than from any other preventable health condition. Today, one in four deaths is attributable to alcohol, tobacco, and illicit or prescription drug use. NIDA 5 pertains to the five drugs of abuse, which should be tested in relation to a drug-free workplace, by the National Institute on Drug Abuse. The NIDA 5 drugs that need to be tested are *Cannabinoids (hashish, marijuana, THC or tetrahydrocannabinol), phencyclidine (PCP), opiates (opium, codeine, heroin, and morphine), amphetamines (methamphetamine), and cocaine (benzoylecognine, cocaethylene).*

Saliva or oral fluid has become increasingly area of focus for, detection, recognition, and diagnosis of potential drug abuse in the workplace, clinical toxicology, and driving under the influence of drugs (DUID). This is due to multiple factors such as simplicity and rapidity of collection and storage, cost-effectiveness, noninvasiveness, and low biohazard considerations, no need for specialized collection by medical personnel, and the possibility of second sample collection for laboratory confirmation. As only unbound or free drugs are excreted into the oral fluid, there is evidence that oral fluid drug concentrations correlate with free drug plasma levels; thus, oral fluid can reflect a recent drug use. Limitations associated

M. Navazesh (✉)
Division of Periodontology, Diagnostic Sciences & Dental Hygiene, Herman Ostrow School of Dentistry of University of Southern California, Los Angeles, CA, USA
e-mail: navazesh@usc.edu

A. Ahmadieh
Division of Biomedical Sciences, Herman Ostrow School of Dentistry of University of Southern California, Los Angeles, CA, USA
e-mail: ahmadieh@usc.edu



D. A. Granger, M. K. Taylor (eds.), *Salivary Bioscience*,
https://doi.org/10.1007/978-3-030-35784-9_16

with oral fluid testing may include the difficulty in collecting a proper volume, possible oral cavity contamination after drug administration, and dry mouth following the drug use. Due to several advantages of oral fluid testing, quite recently this matrix has been introduced in place of blood or urine testing.

Oral fluid application in detection of drugs of abuse is reviewed in this chapter regarding five main categories of illicit drugs (*cannabinoids, opioids, cocaine, amphetamine, and benzodiazepines*). Relevant scientific and evidence-based publications from 2000 to 2018 were identified, reviewed, and will be presented in the chapter. The efficacy of the main saliva collection kits (*Intercept, StatSure, Quantisal, and Oral-Eze, Draeger cassette)* and different saliva assessment methods (*Liquid chromatography–tandem mass spectrometry, SPE 2D-GCMS assay, Direct ELISA, Surface-enhanced Raman spectroscopy, Smartphone, (UCNPs)-Based paper device, Headspace solid-phase microextraction and gas chromatography–mass spectrometry, Radioimmunoassay, Triage kit, Aptasensor, Point of collection testing devices (POCT))* pertaining to drugs of abuse will be discussed. Validity of saliva/oral fluid in comparison to urine and blood-based testing will also be mentioned.

## 16.1 Introduction

The drug overdose epidemic resulted in approximately 630,000 deaths from 1999 to 2016 according to the Center for Disease Control and Prevention (CDC). In 2015, there were 52,404 drug overdose deaths in the USA in which 63.1% of them involved an opioid, whereas, in 2016, there were 63,632 drug overdose deaths reported in which opioids accounted for 66.4% of them. On average, 115 Americans die every day from an opioid overdose. The National Institute of Drug Abuse (NIDA) confirmed this data by reporting that drug-related deaths have more than doubled since 2000 (More than 72,000 Americans died from drug overdoses in 2017, including illicit drugs and prescription opioids). There are more deaths, oral and systemic complications, and disabilities from substance use than from any other preventable health condition. Today, one in four deaths is attributable to alcohol, tobacco, and illicit or prescription drug use (CDC, 2017; NIDA, 2018).

The Substance Abuse and Mental Health Services Administration (SAMHSA) conducts the annual National Survey on Drug Use and Health (NSDUH), a major source of information on substance use, abuse, and dependence among Americans 12 years and older. The most recent published data, from 2013, show that illicit drug use in the USA has been increasing. In 2013, an estimated 24.6 million Americans aged 12 or older—9.4% of the population—had used an illicit drug in the past month. This number is up from 8.3% in 2002. The increase mostly reflects a recent rise in the use of *marijuana*, the most commonly used illicit drug. In 2013, there were 19.8 million current marijuana users—about 7.5% of people aged 12 or older—up from 14.5 million (5.8%) in 2007. In 2013, 6.5 million Americans aged 12 or older (or 2.5%) had used prescription drugs nonmedically in the past month.

Prescription drugs include pain relievers, tranquilizers, stimulants, and sedatives. Methamphetamine use was higher in 2013, with 595,000 current users, compared with 353,000 users in 2010. *Most people use drugs for the first time when they are teenagers. More than half of new illicit drug users begin with marijuana* (SAMHSA, 2015).

The National Institute on Drug Abuse (NIDA) is the prime federal agency that supports scientific studies and research on drug use and abuse and the inevitable consequences. NIDA 5 pertains to the five drugs of abuse, which should be tested in relation to a drug-free workplace, by the National Institute on Drug Abuse. The NIDA 5 drugs that need to be tested are *Cannabinoids (hashish, marijuana, THC or tetrahydrocannabinol), phencyclidine (PCP) opiates (opium, codeine, heroin, and morphine), amphetamines (methamphetamine), and cocaine (benzoylecognine, cocaethylene).* ADHD medications and other prescription drugs may be detected in a NIDA 5 panel drug test, but other semi-synthetic and synthetic drugs will not (NIDA, 2018).

***Driving Under the Influence of Drugs (DUID)*** A control case study based on data from nine European countries estimated a moderately increased relative risk (RR 2-10) of serious injury when driving under the influence of medicinal or illicit opioids (Busardo et al., 2018). In another study, the odds for involvement in fatal traffic accidents for different substances or combinations of substances were in increasing order: single drug < multiple drugs < alcohol only < alcohol + drugs (Gjerde, Normann, Christophersen, Samuelsen, & Mørland, 2011). The drugs found most frequently were zopiclone, benzodiazepines, codeine, THC or tetrahydrocannabinol, and amphetamines. Psychoactive medicinal drugs were more prevalent among females than males, among drivers stopped on working days rather than weekends, and among those who reported annual driving distances less than 16,000 km (10,000 miles) (Gjerde et al., 2008). The World Health Organization (WHO) report on road safety showed that amphetamine is responsible for 51% of drug-related traffic deaths, followed by cannabis (22%), cocaine (14%), and opioids (13%) (WHO, 2016).

***Saliva as a Matrix to Detect Drugs of Abuse*** Since the 1970s, saliva has been studied as an alternative matrix to disclose current consumption of psychotropic drugs (Bosker & Huestis, 2009), and recently it has been used at roadsides to identify drivers under the influence of drugs. The main advantages of oral fluid are the simplicity and rapidity of collection and storage, cost-effectiveness, noninvasiveness, low biohazard considerations, no need for specialized collection by medical personnel, and the possibility of second sample collection for laboratory confirmation. As only unbound or free drugs are excreted into the oral fluid, there is evidence that oral fluid drug concentrations correlate with free drug plasma levels; thus, oral fluid can reflect recent drug use and provide a better correlation with pharmacodynamic effects, such as impaired performance. Limitations associated with oral fluid testing may include the difficulty in collecting a proper volume, possible oral cavity contamination after drug administration, and dry mouth following cannabis or stimulant drug use. Due to several advantages of oral fluid testing, quite recently this matrix has been introduced in place of blood or urine testing (Busardo et al., 2018).

Oral fluid application in the detection of drugs of abuse is reviewed in this chapter in regard to five main categories of illicit drugs (cannabinoids, opioids, cocaine, amphetamine, and benzodiazepines). Relevant publications from January 2000 to April 2018 were identified using the following key words in PubMed: "Saliva," Oral fluid," and "Drugs of abuse." 133 articles found initially were reviewed; 66 of them were included (papers were excluded if the saliva was assessed in cigarette or alcohol consumers). Main saliva collection kits and different saliva assessment methods pertaining to drugs of abuse will be discussed briefly in this chapter.

## 16.2 Cannabinoids (Hashish, Marijuana, and THC)

Cannabis is a natural drug, consisting of the dried flowering, fruiting tops, and leaves of the *Cannabis sativa* plant. Hashish is the dried resinous secretion of the plant, and cannabis oil is a solvent extract of cannabis (Busardo et al., 2018). Cannabis remains the most commonly used illicit drug worldwide (Swortwood et al., 2017). The main psychoactive compound in cannabis is $\Delta^9$-tetrahydrocannabinol (THC), which is highly lipophilic and can be distributed widely in the body. THC was detected in 12.6% of US weekend nighttime drivers' blood or oral fluid samples (Berning, Compton, & Wochinger, 2015). Other cannabinoid metabolites include 11-hydroxy-THC (11-OH-THC), 11-nor-9-carboxy-THC (THC-COOH), tetrahydrocannabivarin (THCV), cannabidiol (CBD), and cannabigerol (CBG). Acute and chronic use of cannabis has been shown to impair psychomotor functions, memory and attention, and an increased risk of a traffic accident, often in a dose dependent manner. Epidemiological studies have suggested that THC blood concentrations of 2–5 ng/ml are generally associated with an increased risk of an accident. Proposed cutoffs for THC blood concentration range from 1 to 5 ng/ml, but several countries have adopted a zero-tolerance policy that does not allow the presence of any amount of THC in blood while driving (Busardo et al., 2018). An experimental study on performance impairment as a function of THC in serum and oral fluid by Ramaekers and colleagues showed a strong and linear correlation between THC in serum and oral fluid. Serum THC concentration between 2 and 5 ng/ml was concluded to represent the lower and upper range of the THC limit for impairment (Ramaekers et al., 2006). Gjerde and Verstraete investigated the possibility of predicting the blood THC concentrations above the chosen limits by analyzing the oral fluid. Equivalent cutoff thresholds could be estimated in a regression model with concentration percentiles in oral fluid as dependent variables and the corresponding concentration percentiles in blood as independent variables. The authors highlighted that the regression formula could be influenced by factors such as the number of samples investigated, the oral fluid sampling method, and the time between cannabis smoking and sampling (Gjerde & Verstraete, 2011). In the study by Vindense and colleagues on multiple drugs of abuse, the use of cannabis was confirmed by detecting THC in oral fluid and THC-COOH in urine. In 34 of 46 cases the use of cannabis was confirmed both in oral fluid and urine. The use of cannabis was confirmed by a positive finding in only urine in 11 cases and in only oral fluid in one case. All the drug groups detected in blood were also found in oral fluid (Vindenes et al., 2012). Another study was performed to acquire urine, serum, and oral fluid samples in cases of suspected driving under the influence of drugs of abuse. Oral fluid was collected using a novel sampling/testing device (Dräger DrugTest System). The aim of the study was to evaluate oral fluid and urine as a predictor of blood samples positive for drugs of abuse including, cannabinoids, amphetamine and its derivatives, opiates, and cocaine. Accuracy in correlating drug detection in oral fluid and serum was > 90% for all substances and also >90% in urine and serum except for THC (71.0%). Of the cases with oral fluid positive for any drug 97.1% of corresponding serum samples were also positive for at least one drug; of drug-positive urine samples this were only 82.4%. Of the cases with drugs detected in serum, 19.1% appeared not impaired, which were the same with drug-positive oral fluid while more persons with drug-positive urine samples appeared uninfluenced (32.7%). *The data demonstrate that oral fluid is superior to urine in correlating with serum analytical data and impairment symptoms of drivers under the influence of drugs of abuse* (Toennes, Kauert, Steinmeyer, & Moeller, 2005). The relationship of drug concentrations between oral fluid and whole blood was assessed by studying the linear correlation of concentrations and calculating the oral fluid to blood concentration ratios for different substances by Langel and colleagues. Study showed that for most substances the correlation between the oral fluid and whole blood concentrations was observed expect for THC and lorazepam; for THC, the median oral fluid to blood ratio reported was 14 (Langel et al., 2014).

Significantly greater cannabinoid metabolites were observed after smoked and vaporized cannabis compared to oral cannabis in frequent smokers only, by (Swortwood et al., 2017). Stability for cannabinoid products was determined after storage at 4 °C by Scheidweiler and colleagues using the Quantisal collection method. They concluded that Quantisal™-collected oral fluid specimens should be stored at 4 °C for no more than 2 months to assure accurate THC, THC-COOH, THCV, CBD, and CBG quantitative results (Scheidweiler, Andersson, Swortwood, Sempio, & Huestis, 2017).

Hartman assessed subjective effects of cannabis with and without alcohol, relative to blood and oral fluid. Subjective effects persisted through 3.3–4.3 h, with *alcohol potentiating the duration of the cannabis effects*. Effect-versus-THC concentration and effect-versus-alcohol concentration hysteresis were counterclockwise and clockwise, respectively. Oral fluid to blood and oral fluid to plasma THC significantly correlated, but variability was high. Vaporized cannabis subjective effects were similar to those previously reported after smoking, with duration extended by concurrent alcohol. Cannabis intake was identified by oral fluid testing, but oral fluid concentration variability limited interpretation. They concluded that the blood THC concentrations were more consistent and accurate across subjects (Hartman et al., 2016). Odell and colleagues performed a study on the time period for detection of THC. Subjects volunteered to provide once-daily blood, urine, and oral fluid samples for seven consecutive days following admission involving cessation and abstinence from all cannabis use. Blood and oral fluid specimens were analyzed for THC and urine specimens for the metabolite THC-COOH. In some subjects THC was detectable in blood for at least 7 days, and oral fluid specimens were positive for THC up to 78 h after admission to the unit. Urinary THC-COOH concentrations exceeded 1000 ng/mL for some subjects 129 h after the last use (Odell, Frei, Gerostamoulos, Chu, & Lubman, 2015). As the number of cannabis cigarettes smoked increased over the study days, oral fluid THC, CBN, and THC-COOH also increased with a significant effect of time since last smoking in a study by Lee and colleagues; concentrations on Day 4 were significantly lower than those on days 32 and 46 but not day 18. Cannabinoid disposition in oral fluid was highly influenced by Δtime and the composition of smoked cannabis. Furthermore, cannabinoid oral fluid concentrations increased over ad libitum smoking days, in parallel with increased cannabis self-administration, possibly reflecting the development of increased cannabis tolerance (Lee and colleagues, 2015). Lee and colleagues, in another study showed that lack of measurable THC, CBD, and CBN in oral fluid following oral THC, and high oral fluid CBD/THC ratios after Sativex distinguish oral and sublingual drug delivery routes from cannabis smoking. Sativex is a whole plant cannabis extract, contains nearly equivalent THC and CBD concentrations delivered via spray onto the oral mucosa to improve bioavailability. Sativex is an approved medication in Canada for multiple sclerosis (MS) neuropathic and opioid-resistant cancer pain, and in the UK, Spain, New Zealand, Germany, and Denmark to treat MS-related spasticity. Low THC-COOH/THC ratios suggest recent Sativex and smoked cannabis exposure. These data indicate that oral fluid cannabinoid monitoring can document compliance with Sativex pharmacotherapy, and identify relapse to smoked cannabis during oral THC medication but not Sativex treatment, unless samples were collected shortly after smoking (Lee et al., 2013). THC was quantifiable in 95.2%, cannabidiol in 69.3%, cannabinol in 62.3%, and THC-COOH in 94.7% of specimens in the study by Milman and colleagues, *THC-COOH in oral fluid suggests no passive contamination, and CBD and CBN suggest recent cannabis smoking* (Milman, Schwope, Gorelick, & Huestis, 2012). In the study by Lee and colleagues, (2012), maximum THC, CBD, and CBN concentrations *occurred within 0.5 h.* All samples were THC positive at 6 h and all samples were positive to 6 h for THC-COOH (Lee, Schwope, et al., 2012). In the study performed by Huestis and Cone, plasma specimens by (GC–MS) and paired oral fluid specimens were analyzed by radioimmunoassay (RIA) and GC–MS. These data represent simultaneous collections of oral fluid and plasma specimens in subjects following controlled cannabis dosing. The first oral fluid specimen, collected from one subject at 0.2 h following initiation smoking, contained a THC concentration of 5800 ng/mL (GC–MS). By 0.33 h, the THC concentration in oral fluid had fallen to 81 ng/mL. THC ratio of oral fluid to plasma THC concentrations was 1.18. Within 12 h, both oral fluid and plasma THC concentrations generally declined below 1 ng/mL (Huestis & Cone, 2004). Another study was done on samples collected with the Quantisal method. THC-COOH was the most prevalent metabolite. In contrast, 11-OH-THC was not identified in any sample; cannabidiol and THC were quantified in few samples. Measurement of THC-COOH in oral fluid not only identifies cannabis exposure, but also minimizes the possibility of passive inhalation. *Authors suggested that THC-COOH might be a better finding to detect cannabis use* (Milman et al., 2010). Choi and colleagues reported that the concentrations of THC and THC-COOH in oral fluid showed a large variation, and the results from oral fluid and urine samples from cannabis abusers did not show any correlation. Thus, detailed information about the time interval between drug use and sample collection is needed to interpret the oral fluid results properly. In addition, further investigation about the detection time window of THC and THC-COOH in oral fluid is required to substitute oral fluid for urine in drug testing (Choi et al., 2009).

## 16.3 Opiates (Opium, Codeine, Heroin, and Morphine)

Opioids are primarily used as licit drugs in the treatment of moderate to severe pain. At the same time, natural, synthetic, and new synthetic opioids (e.g., fentanyl and derivatives) are a class of psychotropic substances that are widely misused. In North America (the USA and Canada), there is an epidemic of the abuse of prescribed opioids, e.g., fentanyl and oxycodone or hydrocodone (Busardo et al., 2018). In the study by Fierro I, the prevalence of opioids in Spanish drivers was 1.8%. Polydrug detection was common (56.2%): of these, in two out of three cases, two opioids were detected, and cocaine was also detected in 86% of the cases. Morphine was always detected with 6-acetylmorphine (6-AM) (Fierro, Colás, González-Luque, & Álvarez, 2017). Another study by Herrera-Gomez and colleagues, on 65, 244 Spanish drivers, showed that opioids were detected in 8.6% of positive cases, while 7.2% were positives for 6-AM, 6.5% positive for morphine, 5.4% for codeine, and 4.1% for methadone. The majority of the confirmed tests for morphine, codeine, and methadone were also positive for 6-AM (heroin use). Confirmed positive tests for morphine, codeine, and methadone had also consumed heroin and illicit drugs, such as cocaine and/or THC, and at a relevant amount (Herrera-Gómez, García-Mingo, Colás, González-Luque, & Álvarez, 2018).

Toennes and colleagues found more than 90% accuracy in correlating opioids' detection in oral fluid and serum (Toennes et al., 2005). In the study by Gray and colleagues oral fluid specimens from methadone-maintained patients exceeded the DRUID guideline, and the 20 ng/mL cutoff appears to be sensitive enough to detect daily methadone exposure; however, additional indicators of behavioral and/or motor impairment would be necessary to provide evidence (Gray, Dams, Choo, Jones, & Huestis, 2011). Vindense and colleagues suggested that oral fluid can be used in driving-under-the-influence cases instead of urine. They found a satisfactory result in correlation between the findings in urine and oral fluid for multiple drugs including opiates. Cocaine and the heroin marker (6-MAM) were more frequently detected in the oral fluid than in urine according to this study (Vindenes et al., 2012). In another work, by Langel and collaborators, the median oral fluid to blood ratios for codeine (4.8), methadone (1.8), morphine (6.4), and tramadol (1.1) were found to be close to theoretical values based on the physicochemical properties of the drugs, and a statistically significant correlation between oral fluid and blood concentrations was observed (Langel et al., 2014).

## 16.4 Cocaine (Benzoylecognine, Cocaethylene)

Cocaine, the psychoactive drug extracted from the leaves of *Erythroxylum Coca* shrub, is the most popular drug of abuse in Europe after cannabis. It belongs to the group of "stimulant drugs," enhancing mood, feelings of well-being, energy, and alertness (Busardo et al., 2018). In South America cocaine is still one of the drugs of most concern, mainly because its consumption continues to increase as opposed to most either regions worldwide—where the consumption remains stable or decreases (Fiorentin et al., 2017). In a meta-analysis carried out by Elvik and colleagues in 2013, the best estimate of the relative risk of accident involvement with cocaine was 2.96 (Elvik, 2013). Proposed cutoff values for cocaine in the blood range from 10 to 80 ng/mL, depending on different international studies as reported by Busardo and collaborators (Busardò, Pichini, & Pacifici, 2017). The correlation between oral fluid and blood cocaine concentrations is still not well established. Cocaine is a weak base and is subject to oral fluid ion trapping; however, the use of cocaine leads to reduced salivary volume (dry mouth), and abuse of smoked crack cocaine, insufflation of cocaine hydrochloride, and oral cocaine abuse can lead to a contamination of the oral cavity resulting in high initial oral fluid levels compared to concentrations that occur after intravenous cocaine administration. In controlled administration studies, cocaine was identified in oral fluid after smoking, intravenous, intranasal, and oral administration. After subcutaneous administration, cocaine was identified in oral fluid, 0.08–0.32 h after dosing, with a half-life of 1.1–3.8 h; the correlation between cocaine oral fluid and blood concentrations was significant, although no correlation with clinical symptoms was described (Scheidweiler et al., 2010). A cocaine oral fluid to blood ratio of 22 was reported in a sample of people driving under suspicion of drugs and a similar oral fluid to blood ratio value was also obtained by Langel and collaborators (2014). These data indicate that oral fluid is a good alternative biological matrix to test cocaine use roadside, or in drivers involved in car crashes, providing immediate evidence of driving under the influence of this psychostimulant drug (Busardo et al., 2018). However, in the study by Fiorentin and colleagues in cocaine and/or crack cocaine users benzoylecgonine (BZE) was the most prevalent analyte, followed by cocaine (COC), anhydroecgonine (AEC), cocaethylene (CE), and anhydroecgonine methyl ester (AEME). In addition, the concentration of benzoylecgonine in urine was higher compared to oral fluid and plasma samples (Fiorentin et al., 2017). In another study by the same author, the relationship between drug concentrations in the oral fluid to plasma and oral fluid to urine ratios was evaluated. Significant correlations were found for cocaine and BE. Many factors contribute to the variability of drug correlation ratios in studies involving random samples, including uncertainty about the time of last administration and dosage. They found significant R2 values for COC and BZE in the oral fluid to plasma and oral fluid to urine ratios, but not for AEC. Despite the good correlations found in some cases, especially for BZE, the large variation in drug concentrations seen in this work suggests that oral fluid concentrations should not be used to estimate concentrations of COC, BZE, or AEC in plasma and/or urine (Fiorentin et al., 2018). 11 healthy, cocaine-using adults received 25 mg intravenous cocaine in the study by Ellefsen and colleagues; physiological and subjective effects (visual analog scales), and plasma were collected 1 h prior, and up to 21 h post-dose peak subjective effects ("Rush," "Good drug effect," and "Bad drug effect") occurred prior to peak oral fluid cocaine concentration, whereas observed peak plasma concentrations and subjective measures occurred simultaneously, most likely due to significantly earlier plasma Tmax compared to oral fluid Tmax (time to maximum concentration). Tlast (time of last observed concentration) was generally longer in the oral fluid than plasma (Ellefsen, Concheiro, Pirard, Gorelick, & Huestis, 2016a). In the study by Scheidweiler and colleagues, cocaine ratios increased after cocaine administration; this may be helpful for interpreting the time of last use. Comparison of oral fluid collection through citric acid candy-stimulated expectoration, citric acid-treated Salivette, and neutral cotton Salivette devices did *not reveal significant* differences between devices for areas under the curve for cocaine, BE, or ecgonine methyl ester (EME) (Scheidweiler et al., 2010).

## 16.5 Amphetamines

Amphetamine and amphetamine-type substances such as methamphetamine and 3, 4-methylendioxymethamphetamine (MDMA, ecstasy) are most commonly abused due to their stimulant effect on the central nervous system that results in sociability enhancement, increased mood, auditory and/or visual perceptions, and energy boosting as reported by recreational users during controlled administrations or recreational settings. Acute or chronic use of amphetamines has been implicated in increased dangerous driving, with an estimated relative risk of 8.67 for crashes resulting in property damage, 6.19 for injury accidents, and 5.17 for fatal accidents (Elvik, 2013). In contrast with the above-reported data, some experimental studies showed that amphetamines like MDMA, methamphetamine and dexamphetamine, could improve certain aspects of cognitive/driving performances. Skills such as tracking, impulse control, and reaction time can be generally improved, whereas cognitive functions such as working memory and movement perception can be impaired (Silber, Croft, Papafotiou, & Stough, 2006). Blood cutoff values for amphetamines have been recently proposed, ranging from 20 to 600 ng/mL for amphetamine, from 20 to 200 ng/mL for methamphetamine, and from 20 to 300 ng/mL for MDMA (Busardò et al., 2017). In the study by Langel and colleagues, the relationship of drug concentrations between oral fluid and whole blood was evaluated by studying the linear correlation of concentrations and calculating the oral fluid to blood concentration ratios for different substances. A significant correlation was observed for most drugs including amphetamine (Langel et al., 2014). In the study performed by Schepers and colleagues methamphetamine and amphetamine pharmacokinetics in oral fluid and plasma after controlled oral methamphetamine administration were evaluated. After the first oral dose, initial plasma METH detection was within 0.25–2 h. In oral fluid, METH was detected as early as 0.08–2 h. Neutral cotton swab collection yielded significantly higher METH and AMP concentrations than citric acid candy-stimulated expectoration. The plasma and oral fluid 24 h METH detection rates were 54 and 60%, respectively. After four administrations, METH was measurable for 36–72 h (Schepers et al., 2003). Another study by Engblom and colleagues investigated amphetamine concentrations in both oral fluid and whole blood samples taken from individuals suspected of driving under the influence of drugs. The total number of cases positive for amphetamine in the oral fluid was 100 out of 114. Therefore, the authors concluded that oral fluid could be used as a testing matrix for amphetamine when driving under the influence is suspected (Engblom, Gunnar, Rantanen, & Lillsunde, 2007). Gjerde and Verstraete suggested that Monte Carlo simulations might give better accuracies than simple calculations if good data on the oral fluid to blood ratios are available (The study was done on amphetamine and THC). They found that if using only 20 randomly selected oral fluid to blood ratios, a Monte Carlo simulation gave the best accuracy but not the best precision. None of the methods gave acceptable accuracy if the prevalence of high blood drug concentrations was less than 15% (Gjerde & Verstraete, 2010). In addition, equivalent cutoff thresholds could be estimated using a regression model with concentration percentiles in the oral fluid as the response variable and the corresponding concentration percentiles in blood as the predictor variable; the same authors (Gjerde and Verstraete) validated a method for determining equivalent drug cutoff concentrations for THC and amphetamine in blood and oral fluid, which ensures that the drug prevalence in samples of blood and oral fluid taken simultaneously is equal.

## 16.6 Benzodiazepines

Nordal and colleagues did the assessment on Diazepam, Clonazepam, and Alprazolam in oral fluid and urine: The maximum detection times for diazepam and N-desmethyldiazepam in oral fluid were 7 and 9 days. For clonazepam and 7-aminoclonazepam, the maximum detection times in oral fluid were 5 and 6 days, respectively. The maximum detection time for alprazolam in the oral fluid was 2.5 days (Nordal, Øiestad, Enger, Christophersen, & Vindenes, 2015). According to the study done by Gjerde and colleagues in 2014, the concentration of benzodiazepines in the oral fluid can neither be used to accurately estimate the concentrations in blood nor to correctly identify patients with blood drug concentrations below or above recommended therapeutic levels (Gjerde, Langel, Favretto, & Verstraete, 2014).

## 16.7 The Impact of Saliva (Oral Fluid) Collection Methods in Drugs of Abuse Studies

***Intercept*** Drug concentration ratios between oral fluid collected with the Intercept device (see Table 16.1 for kit elements and instructions) and whole blood was determined by Gjerde and colleagues. The distributions of drug concentration ratios were wide for most drugs and did not allow reliable estimations of drug concentrations in blood using concentrations in oral fluid. The median oral fluid to blood drug concentration ratios for the most prevalent drugs was 0.036 diazepam, 0.027 nordiazepam, 7.1 amphetamine, 2.9 methamphetamine, 5.4 codeine, 1.9 morphine, and 4.7 THC (Gjerde et al., 2010).

***StatSure Versus Intercept*** Zopiclone in the oral fluid was generally higher when using the Intercept compared to the Statsure device; the median oral fluid to whole blood concentration ratios were 3.8 and 1.9, respectively. The correlation between zopiclone concentrations in oral fluid collected with the two devices was fairly poor. The results indicate that the type of sampling device may significantly affect the analytical result for zopiclone in sampled oral fluid (Gjerde et al., 2010).

***Quantisal*** The study by Lee showed that all Quantisal oral fluid cannabinoid concentrations were stable for 1 week at 4 °C. After 4 weeks at 4 °C, as well as 4 and 24 weeks at −20 °C, THC was stable in 90%, 80%, and 80% and THC-COOH in 89%, 40%, and 50% of Quantisal samples, respectively. Cannabinoids in the expectorated oral fluid were less stable than in Quantisal samples when refrigerated or frozen. After 4 weeks at 4 and − 20 °C, CBD and CBN were stable in 33–100% of Quantisal and expectorated samples; by 24 weeks at −20 °C, CBD and CBN were stable in $\leq$44% (Lee et al., 2012).

***Intercept Versus Quantisal Kits*** In a comparison between Intercept and Quantisal oral fluid kits, better recoveries and fewer matrix effects were observed for some compounds using Quantisal. The method is sensitive and robust for its purposes and has been used successfully since February 2015 for the analysis of Intercept oral fluid samples from 2600 cases in a 12-month period (Valen, Leere Øiestad, Strand, Skari, & Berg, 2017).

***Oral-Eze*** Data by Newmeyer and colleagues illustrate the effectiveness of the Oral-Eze® device for oral fluid collection, the impact of self-administered smoked cannabis history on oral fluid cannabinoid results, and the ability to improve interpretation and tailor oral fluid cannabinoid cutoffs to fulfill the detection window needs of a given program (Newmeyer, Desrosiers, & Lee, 2014).

***Oral-Eze Versus Quantisal*** In the study by Newmeyer and colleagues 16 participants received 7 Vicks VapoInhaler doses according to manufacturer's recommendations. Specimens were collected before and up to 32 h after the first dose with the Oral-Eze and Quantisal devices. No d-methamphetamine or d-amphetamine was detected in any sample. All participants had measurable oral

**Table 16.1** Saliva collection kits summary instructions

| Collection Kit | Manufacture | Kit elements and instructions |
|---|---|---|
| Intercept[a] | OraSure Technologies, Bethlehem, PA | The donor places the collection pad (an absorbent foam pad with diluent) in his or her cheek and gum for at least 2 min. Once the absorbent collection pad is saturated, it is placed in a vial, the handle of the collection device is snapped off at the rim of the vial, the vial is sealed, and the donor initials the seal. The entire process takes just 5 min. |
| Oral Eze[b] | Quest Diagnostics | The Oral-Eze collection consists of an absorbent pad and plastic shield mounted on a plastic handle for the collection of oral fluid and a collection tube for the stabilization and transportation of the specimen. The collection is complete when the sample adequacy window turns blue. This typically occurs within 5 min. Once the Oral-Eze collector is placed in the mouth, the donor is asked to refrain from swallowing and talking, instead directing saliva toward the device. |
| StatSure[a] | StatSur Diagnostic Systems, Inc. Framingham, MA | The StatSure device consists of an absorptive cellulose pad with a volume adequacy indicator and a plastic tube containing a buffer solution. The window in the stem of the collection pad turns blue when oral fluid is collected. The supplier states that the volume of the buffer solution is 1 mL. For sample collection, the level of buffer solution must be checked against a line on the side of the tube before collection begins. Oral fluid is allowed to gather in the mouth and the collection pad is placed under the tongue. When the indicator window turns completely blue, the pad is removed from the mouth and placed into the collection/transport tube. In the laboratory, the collection pad is disconnected from the stem and dropped to the bottom of the tube, and a filter is inserted into the tube to recover the oral fluid buffer solution for testing. |
| Dräger DrugTest System[c] | Dräger Company | The DrugTest 5000 test cassette is equipped with a polymeric noncompressible pad for oral fluid collection. Samples are collected by swiping the test cassette on the tongue and sides of the cheeks. |
| Salivette[a] | Salivette Sarstedt AG & Co., Numbrecht, Germany | The Salivette device consists of a cotton wool swab in a plastic tube with an insert and a cap. For sample collection, the swab is placed into the mouth and chewed for approximately 45 s. It is removed, placed into the insert in the collection tube, and shipped to the laboratory. In the laboratory it is centrifuged and the insert with the swab is removed to recover clear oral fluid. |
| Quantisal[a] | Immunalysis, Pomona, CA | The Quantisal device consists of an absorptive cellulose pad with a polypropylene stem and a |

(continued)

**Table 16.1** (continued)

| Collection Kit | Manufacture | Kit elements and instructions |
|---|---|---|
| | | plastic tube containing a buffer solution. The collection pad has a volume adequacy indicator based on a blue dye that is dissolved when the aqueous medium (oral fluid) migrates along with the cellulose pad by capillary action. When 1 mL + 10% of the fluid is collected, the dye becomes visible in the window of the stem. The supplier states that the volume of the buffer solution is 3 mL, and it contains a non-azide preservative. For sample collection, the collection pad is placed under the tongue. When the indicator window turns blue, the pad is removed from the mouth and placed into the collection/transport tube. |
| Cozart[a] | Cozart Bioscience, Oxfordshire, UK | The Cozart device consists of an oral fluid collection pad with a volume adequacy indicator and a collection/transport tube with a preservative buffer. The indicator area in the stem of the collection pad turns blue once 1 mL of OF has been collected. The volume of the buffer solution is stated to be 2 mL, and the solution contains salts, preservatives, and detergents. For sample collection, the collector is actively swabbed around the gums, the tongue, and the inside of the cheek, and then held inside the mouth until the sample presence indicator turns blue. The pad is placed into the collection tube bud end first. |
| Greiner[a] | Greiner Bio-One GmbH, Kremsmünster, Austria | The Greiner device consists of a rinsing solution tube (6 mL), an oral fluid extraction solution tube (4 mL), a collection beaker with a cap, and two vacuum transfer tubes that contain stabilizers and preservatives. The extraction solution contains a yellow food dye, tartrazine, which serves as an internal standard enabling spectrophotometric quantification of collected oral fluid. It also contains a citrate buffer at a certain pH to increase salivation. The Quantification Kit contains five calibrators, two controls, oral fluid extraction solution, and artificial oral fluid for spectrophotometric measurements. |
| OraCol test kit[a] | Malvern Medical Developments, Worcester, UK | The OraCol device consists of a centrifuge tube with a cap and an absorbent foam swab designed to collect up to 1 mL of oral fluid. For sample collection, the swab is firmly rubbed along the gum at the base of the teeth until it is wet. This should take about 1 min. The swab is then removed, placed into the collection tube, and shipped to the laboratory. In the laboratory, the tube is centrifuged to extract the sample from the swab. |

(continued)

**Table 16.1** (continued)

| Collection Kit | Manufacture | Kit elements and instructions |
|---|---|---|
| Oratube[a] | Varian OraTubeTM | The OraTube device consists of a collection pad and a plastic collection tube with an expresser. For sample collection, the foam collector is kept in the mouth until it is thoroughly soaked (up to 3 rain). The collector is then placed into the expresser in the tube and pushed to the bottom of the expresser so that the sample flows to the bottom of the tube. Finally, the collector and the expresser are thrown away and the sample is shipped to the laboratory. |
| Salicule[a] | Acro Biotech Salicule | The Salicule device is a collection vial with an expectoration straw and a two-piece cap. For sample collection, the cap of the collection vial is removed to pull out the expectoration straw. Oral fluid is expectorated via the straw into the vial until enough fluid is collected (there is a scale on the side of the vial). In the laboratory, the top cap is removed, and the sample can be poured into a test tube for analysis. |

[a]Langel and colleagues (2008)
[b]Newmeyer and colleagues (2015)
[c]Ellefsen and colleagues (2016c)

fluid l-methamphetamine with median maximum concentrations 14.8 and 16.1 μg/L in Quantisal™ and Oral-Eze® devices, respectively, after a median of five doses. There were no positive DrugTest® 5000 results. No significant differences in the performance between the two oral fluid collection devices were observed (Newmeyer et al., 2015).

***Oral-Eze® Versus StatSure*** To ensure oral fluid cannabinoid concentration accuracy, data by Anizan suggest analysis within 4 weeks at 4 °C storage for Oral-Eze collection and within 4 weeks at 4 °C or 24 weeks at −20 °C for StatSure collection (Anizan et al., 2015). Another study by Desrosiers NA and colleagues showed that THC and CBD concentrations were not significantly different between devices (StatSure and Oral-Eze) in consecutively collected samples. However, StatSure and Oral-Eze THC-COOH and CBN concentrations differed significantly by device, with THC-COOH and CBN concentrations being generally higher in Oral-Eze samples. Given the low THC-COOH detection rate in occasional smokers, the authors compared THC-COOH concentrations between devices for frequent smokers only; concentrations remained significantly different (Desrosiers et al., 2014). In the study by Ellefsen and colleagues, ten cocaine-using adults provided oral fluid, collected with Oral-Eze® and StatSure Saliva Sampler™ devices, an hour prior to and up to 69 h after 25 mg intravenous (IV) cocaine administration. Cocaine and benzoylecgonine (BE) were quantified by a validated 2D-GC–MS method. There were no significant pharmacokinetic differences between Oral-Eze and StatSure oral fluid collection devices, except cocaine half-life was significantly shorter in StatSure oral fluid specimens. This difference could be attributed to differences in stabilizing buffers present in oral fluid collection devices, which may affect cocaine stability in oral fluid specimens or decreased recovery from collection pads. Both Oral-Eze and StatSure oral fluid collection devices were effective in monitoring cocaine and metabolite concentrations with similar detection windows (Ellefsen, Concheiro, Pirard, Gorelick, & Huestis, 2016b).

***Draeger Cassette, Oral-Eze, and StatSure*** Cocaine test-strip median Tlast for screening only results were 6.5 h with Oral-Eze® and 4 h for StatSure oral fluid confirmation for cocaine and/or BE at 1, 8, and 10 μg/L; sensitivity, specificity, and efficiency ranged from 85.5 to 100% and 83.3 to 100% for cocaine only confirmation at 8 and 10 μg/L. For the BE test-strip, median Tlast was 12.5 h for screening only and confirmation for cocaine and/or BE at all three cutoffs; sensitivity, specificity, and efficiency ranged from 85.5 to 97.5% and 78.4 to 97.4% with cocaine and/or BE confirmation at 8 and 10 μg/L cutoffs, respectively. The Draeger cocaine test-strip with cocaine only confirmation offers a useful option for monitoring the acute intoxication phase of DUID (Driving Under the Influence of Drugs) (Ellefsen, Concheiro, Pirard, Gorelick, & Huestis, 2016c).

Nine different oral fluid (OF) collection devices (Greiner Bio-One, Orasure, Interceptl, Quantisal™, StatSure Saliva Sampler™, Cozart Sarstedt Salivette, Malvern Medical OraCol, Acro Biotech Salicule, and VarianOraTube™) were studied to evaluate their suitability for collecting samples for drug analysis. The volume of collected OF was quantified for samples collected both in vitro and from volunteers. Drug recovery was studied by collecting OF fortified at 1000 ng/mL with amphetamine, 3,4-methylenedioxymethamphetamine, cocaine, A9-tetrahydrocannabinol, morphine, codeine, diazepam, and alprazolam with the devices in vitro and analyzing the samples with gas chromatography–mass spectrometry. The stability of drugs in the samples was studied by analyzing the samples after 0, 14, and 28 days storage. The study shows that there are substantial differences between the OF collection devices on the market. Some are well suited for collecting samples for toxicological analysis, but some give quite poor results.

## 16.8 Available Saliva (Oral Fluid) Assessment Techniques to Detect Drugs of Abuse

***Liquid Chromatography–Tandem Mass Spectrometry*** LC MS combines the physical separation abilities of liquid chromatography with the mass analysis abilities of mass spectrometry (MS), this technique is especially important to detect low levels of drugs in saliva (Chap. 17). Dams in 2003, proved that this method can be used for the evaluation of oral fluid as an alternative matrix to urine for monitoring illicit drug use and for determining oral fluid methadone concentrations in pregnant opiate and/or cocaine addicts (Dams et al., 2003). Identification and quantitation of amphetamines, cocaine, opiates, phencyclidine, and other illicit drugs in the oral fluid were assessed by this method in the study by Fritch and colleagues. This validated method is suitable for simultaneous measurement of 21 licit and illicit drugs and metabolites in oral fluid (Fritch et al., 2009). Desrosiers and colleagues developed and validated a sensitivity and specificity of LC–MS/MS method for cannabinoid and its metabolite; they showed that this method of evaluating oral fluid to detect cannabis will be useful in the workplace, criminal justice cases, drug treatment, and driving under the influence of cannabis (Desrosiers, Scheidweiler, & Huestis, 2015). The study by Barnes showed that GC–MS for the evaluation of THC-COOH in oral fluid reduces analysis time by 9 min per sample compared with the 2-dimensional gas chromatography–mass spectrometry method. This method was applied to the analysis of oral fluid specimens collected from individuals participating in controlled cannabis studies and will be effective for distinguishing passive environmental contamination from active cannabis smoking (Barnes, Scheidweiler, & Huestis, 2014). The first LC–MS/MS method for methamphetamine and amphetamine assessment performed by Newmeyer, using the Oral-Eze® and Quantisal™ saliva collection devices, showed that this method is sensitive, selective, economic, and improves methamphetamine result interpretation (Newmeyer et al., 2014).

***SPE 2D-GCMS Assay*** This method achieved efficient quantification of five cannabinoids in oral fluid, including pg/mL concentrations of THC-COOH by combining differential elution, 2D-GCMS with electron ionization and negative chemical ionization. This method will be applied to the quantification of cannabinoids in oral fluid specimens from individuals participating in controlled cannabis and Sativex (50% THC and 50% CBD) administration studies, and during cannabis withdrawal. This new analytical method simultaneously identifies and quantifies THC, CBD, CBN, 11-OH-THC, and THC-COOH in a single extraction of oral fluid collected with the Quantisal™ device. This validated method provides specific and accurate results over an analyte concentration range that is consistent with expected oral fluid concentrations following oral THC administration. Enhanced analytical sensitivity with improved S/N (signal to noise) and detection limits for cannabinoids was achieved with 2D-GCMS with cold trapping. This method also is useful for quantification of cannabinoids after cannabis smoking, when parent analytes may be present in higher concentrations than after ingestion; however, appropriate specimen dilutions may be required (Milman et al., 2010).

***Direct ELISA*** Cannabinoid metabolites Immunoanalysis Sweat/oral fluid THC Direct ELISA and confirmed by 2-dimensional GC–MS (collected with Quantisal); results showed 89.5% diagnostic specificity and 90.6% diagnostic efficiency for this method (Schwope, Milman, & Huestis, 2010). Another study by Barnes on sensitivity, specificity, and efficiency of the opiate ELISA showed the lowest cutoff of sensitivity, specificity, and efficiency to be 91.5%, 88.6%, and 89.3%. Application of the cutoff used in the UK yielded sensitivity, specificity, and efficiency results of 79.7%, 99.0%, and 95.4% and similar results of 76.7%, 99.1%, and 95.1% when applying the SAMHSA criteria. These data indicate that the Opiate ELISA efficiently detects oral codeine use (Barnes et al., 2003).

***Surface-Enhanced Raman Spectroscopy (SERS)*** A method was successfully developed by Inscore and colleagues that allowed consistent detection of five different drugs in saliva at 1 ppm or less within 10 min. In a focused study, ROC curves were used to establish that cocaine could be measured at 50 ppb (ng/mL) more than 90% of the time. This concentration is 5000 times more sensitive than previously reported SERS measurements of cocaine in saliva (Inscore, Shende, & Sengupta, 2011).

***ELISA Versus GC-MS*** Due to the small sample size, no significant differences in the measured performance of onsite- and laboratory-based tests were seen in the study by Veitenheimer and Wagner. Study results indicate that oral fluid testing is a viable option both at the roadside and in a laboratory setting (Veitenheimer & Wagner, 2017).

***Smartphone*** An innovative, low-cost, portable approach for the rapid interpretation of lateral flow saliva test results for drug-of-abuse detection is based on the use of a smartphone device (Carrio, Sampedro, Sanchez-Lopez, Pimienta, & Campoy, 2015). The test reader is able to obtain the same result if not better, than a trained operator, therefore reducing the subjectivity of the analysis by standardizing the test interpretation according to the study by Meagher and Kousvelari. The system is automatic, allowing one to systematize the collection and analysis of the data in real time, removing the risks of managing manual results, and allowing for centralized processing. This method can be used in places where there is no qualified staff and rapid detection is needed (Meagher & Kousvelari, 2018).

***(UCNPs)-Based Paper Device*** He and colleagues published the first portable upconversion nanoparticles (UCNPs)-based paper device to detect cocaine in oral fluid. This device can give quantitative results in a short time with high sensitivity using only a smartphone as the apparatus; this device is applicable in human saliva and blood samples to monitor the changes in the cocaine content (He, Li, Ge, & Liu, 2016).

***Headspace Solid-Phase Micro Extraction and Gas Chromatography-Mass Spectrometry (HS-SPME–GC–MS)*** Procedure for the simultaneous detection of methylene-dioxyamphetamine (MDA), methylene-dioxymethamphetamine (MDMA), methylene-dioxyethamphetamine (MDE), and N-methyl-1-(1,3-benzodioxol-5-yl)-2-butanamine (MBDB) in hair has been developed. This method is suitable for the separation of primary and secondary amines, is reproducible, is not time consuming, and requires small quantities of samples. It provides sufficient sensitivity and specificity. This method is suitable for routine clinical, epidemiological, and forensic purposes and can be used for the preliminary screening of many other substances (amphetamine, methamphetamine, ketamine, ephedrine, nicotine, phencyclidine, and methadone) in hair and other biological matrices such as saliva, urine, and blood (Gentili, Torresi, Marsili, Chiarotti, & Macchia, 2002).

***Triage Kit Versus GC–MS*** Results were positive for methadone in 9 saliva and 14 urine specimens, for opiates in 2 and 10, respectively, and for barbiturates in

2 specimens. Saliva specimens were negative for cannabis, THC, benzodiazepines, and tricyclic antidepressants, although the GC–MS analysis revealed low concentrations of these drugs in the saliva (Lo Muzio et al., 2005).

***Radioimmunoassay (RIA) Versus GC–MS*** RIA analyses of oral fluid specimens for six subjects demonstrated the same pattern of GC–MS: initial high levels of contamination immediately after smoking, followed by rapid clearing, and a slower decline over 12 h (Huestis & Cone, 2004).

***Aptasensor*** Used to assess human plasma, serum, saliva, and urine. The cocaine solutions in biological fluids were prepared by mixing 7.8 μM cocaine (in T-buffer) and 50% biological fluid (in T-buffer) with a ratio of 1:1. The final concentration of cocaine was 3.8 μM in 25% biological fluids (Du et al., 2010).

***Point of Collection Testing Devices (POCT)*** Scherer and colleagues published a systematic review on the reliability of the point of collection devices for drugs of abuse in oral fluid. They concluded that the DrugWipe test (Securetec, Germany) was the most evaluated device among the studies, followed by the Drug Test 5000 (Drager Safety AG & CO), the Rapiscan (Cozart Bioscience Ltd., UK), and the Rapid Stat (Mavand Solutions, Germany). The meta-analysis on five main classes of drugs (amphetamine, cocaine, benzodiazepine, cannabinoid, and opioids) using the listed devices showed sensitivities ranging from 78.6 to 85.2% and specificities from 81.3 to 94.9% (Scherer et al., 2017).

## 16.9 Future Directions

This chapter reviews the literature in regard to the use of saliva as a substitute matrix for blood and urine to assess drugs of abuse. As mentioned in most of the included studies, saliva collection is simple, fast, and does not require privacy or specific expertise; these advantages can be very helpful to use saliva, especially in roadside settings. Although many advantages have been reviewed, there are many areas that need to be addressed in future studies. Most of the available publications have used a small number of samples, so more large-scale studies are needed to validate these findings. It is imperative to know the exact saliva collection method and the time window between the sample collection and sample study (which is missing in most studies). It is also significant to know how long after taking a drug, the saliva was collected. Another significant factor that may affect study results, and is missing in the available publications, is saliva contamination (possibly with blood) due to factors such as an individual's poor oral hygiene. Available studies have not focused on the exact source of collected saliva—major versus minor salivary glands. Furthermore, the type of collected saliva is not clear (stimulated versus non-stimulated, spitting, or drooling). In the absence of all this information, it is difficult to compare saliva with other matrixes such as blood or urine. This chapter has focused on the five main categories of drugs including, cannabinoids, opioids, cocaine, amphetamine, and benzodiazepines. PCP, as one of the major drugs according to the NIDA recommended panel drug tests, should be validated in future studies. Traditional and evolving technologies were referred to in this chapter. Given the rapid development of new technologies and scientific discoveries, saliva will continue to be a fountain of opportunities for the future development of cost-effective, efficient, and noninvasive screening and/or diagnostic tests for drugs of abuse.

## References

Anizan, S., Bergamaschi, M. M., Barnes, A. J., Milman, G., Desrosiers, N., Lee, D., … Huestis, M. A. (2015). Impact of oral fluid collection device on cannabinoid stability following smoked cannabis. *Drug Testing and Analysis, 7*(2), 114–120.

Barnes, A. J., Kim, I., Schepers, R., Moolchan, E. T., Wilson, L., Cooper, G., … Huestis, M. A. (2003). Sensitivity, specificity, and efficiency in detecting opiates in oral fluid with the Cozart Opiate Microplate EIA and GC-MS following controlled codeine administration. *Journal of Analytical Toxicology, 27*(7), 402–407.

Barnes, A. J., Scheidweiler, K. B., & Huestis, M. A. (2014). Quantification of 11-nor-9-carboxy-δ9 tetrahydrocannabinol in human oral fluid by gas chromatography-tandem mass spectrometry. *Therapeutic Drug Monitoring, 36*(2), 225–233.

Berning, A., Compton, R., & Wochinger, K. (2015). *Results of the 2013–2014 national roadside survey of alcohol and drug use by drivers*. Washington, DC: National Highway Traffic Safety Administration. Retrieved from https://www.nhtsa.gov/sites/nhtsa.dot.gov/files/812118-roadside_survey_2014.pdf

Bosker, W. M., & Huestis, M. A. (2009). Oral fluid testing for drugs of abuse. *Clinical Chemistry, 55*(11), 1910–1931.

Busardò, F. P., Pichini, S., & Pacifici, R. (2017). Driving under the influence of drugs: Looking for reasonable blood cutoffs and realistic analytical values. *Clinical Chemistry, 63*(3), 781–783.

Busardo, F. P., Pichini, S., Pellegrini, M., Montana, A., Fabrizio Lo Faro, A., Zaami, S., & Graziano, S. (2018). Correlation between blood and oral fluid psychoactive drug concentrations and cognitive impairment in driving under the influence of drugs. *Current Neuropharmacology, 16*(1), 84–96.

Carrio, A., Sampedro, C., Sanchez-Lopez, J. L., Pimienta, M., & Campoy, P. (2015). Automated low-cost smartphone-based lateral flow saliva test reader for drugs-of-abuse detection. *Sensors (Basel), 15*(11), 29569–29593.

Center for Disease Control and Prevention. (2017). *Opioid overdose: Understanding the epidemic*. Retrieved from https://www.cdc.gov/drugoverdose/epidemic/index.html

Choi, H., Baeck, S., Kim, E., Choi, H., Baeck, S., Kim, E., … Chung, H. (2009). Analysis of cannabis in oral fluid specimens by GC-MS with automatic SPE. *Science & Justice, 49*(4), 242–246.

Dams, R., Murphy, C. M., Choo, R. E., Lambert, W. E., De Leenheer, A. P., & Huestis, M. A. (2003). LC-atmospheric pressure chemical ionization-MS/ MS analysis of multiple illicit drugs, methadone, and their metabolites in oral fluid following protein precipitation. *Analytical Chemistry, 75*(4), 798–804.

Desrosiers, N. A., Milman, G., Mendu, D. R., Lee, D., Barnes, A. J., Gorelick, D. A., & Huestis, M. A. (2014). Cannabinoids in oral fluid by on-site immunoassay and by GC-MS using two different oral fluid collection devices. *Analytical and Bioanalytical Chemistry, 406*(17), 4117–4128.

Desrosiers, N. A., Scheidweiler, K. B., & Huestis, M. A. (2015). Quantification of six cannabinoids and metabolites in oral fluid by liquid chromatography-tandem mass spectrometry. *Drug Testing and Analysis, 7*(8), 684–694.

Du, Y., Chen, C., Yin, J., Li, B., Zhou, M., Dong, S., & Wang, E. (2010). Solid-state probe based electrochemical aptasensor for cocaine: A potentially convenient, sensitive, repeatable, and integrated sensing platform for drugs. *Analytical Chemistry, 82*(4), 1556–1563.

Ellefsen, K. N., Concheiro, M., Pirard, S., Gorelick, D. A., & Huestis, M. A. (2016a). Pharmacodynamic effects and relationships to plasma and oral fluid pharmacokinetics after intravenous cocaine administration. *Drug and Alcohol Dependence, 163*, 116–125.

Ellefsen, K. N., Concheiro, M., Pirard, S., Gorelick, D. A., & Huestis, M. A. (2016b). Oral fluid cocaine and benzoylecgonine concentrations following controlled intravenous cocaine administration. *Forensic Science International, 260*, 95–101.

Ellefsen, K. N., Concheiro, M., Pirard, S., Gorelick, D. A., & Huestis, M. A. (2016c). Cocaine and benzoylecgonine oral fluid on-site screening and confirmation. *Drug Testing and Analysis, 8* (3–4), 296–303.

Elvik, R. (2013). Risk of road accident associated with the use of drugs: A systematic review and meta-analysis of evidence from epidemiological studies. *Accident; Analysis and Prevention, 60*, 254–267.

Engblom, C., Gunnar, T., Rantanen, A., & Lillsunde, P. (2007). Driving under the influence of drugs—amphetamine concentrations in oral fluid and whole blood samples. *Journal of Analytical Toxicology, 31*(5), 276–280.

Fierro, I., Colás, M., González-Luque, J. C., & Álvarez, F. J. (2017). Roadside opioid testing of drivers using oral fluid: The case of a country with a zero tolerance law, Spain. *Substance Abuse Treatment, Prevention, and Policy, 12*(1), 22.

Fiorentin, T. R., D'Avila, F. B., Comiran, E., Zamboni, A., Scherer, J. N., Pechansky, F., ... Limberger, R. P. (2017). Simultaneous determination of cocaine/crack and its metabolites in oral fluid, urine and plasma by liquid chromatography-mass spectrometry and its application in drug users. *Journal of Pharmacological and Toxicological Methods, 86*, 60–66.

Fiorentin, T. R., Scherer, J. N., Marcelo, M. C. A., Sousa, T. R. V., Pechansky, F., Ferrão, M. F., & Limberger, R. P. (2018). Comparison of cocaine/crack biomarkers concentrations in oral fluid, urine and plasma simultaneously collected from drug users. *Journal of Analytical Toxicology, 42*(2), 69–76.

Fritch, D., Blum, K., Nonnemacher, S., Haggerty, B. J., Sullivan, M. P., & Cone, E. J. (2009). Identification and quantitation of amphetamines, cocaine, opiates, and phencyclidine in oral fluid by liquid chromatography-tandem mass spectrometry. *Journal of Analytical Toxicology, 33*(9), 569–577.

Gentili, S., Torresi, A., Marsili, R., Chiarotti, M., & Macchia, T. (2002). Simultaneous detection of amphetamine-like drugs with headspace solid-phase microextraction and gas chromatography-mass spectrometry. *Journal of Chromatography. B, Analytical Technologies in the Biomedical and Life Sciences, 780*(1), 183–192.

Gjerde, H., Langel, K., Favretto, D., & Verstraete, A. G. (2014). Estimation of equivalent cutoff thresholds in blood and oral fluid for drug prevalence studies. *Journal of Analytical Toxicology, 38*(2), 92–98.

Gjerde, H., Normann, P. T., Christophersen, A. S., Samuelsen, S. O., & Mørland, J. (2011). Alcohol, psychoactive drugs and fatal road traffic accidents in Norway: A case-control study. *Accident; Analysis and Prevention, 43*(3), 1197–1203.

Gjerde, H., Normann, P. T., Pettersen, B. S., Assum, T., Aldrin, M., Johansen, U., ... Mørland, J. (2008). Prevalence of alcohol and drugs among Norwegian motor vehicle drivers: A roadside survey. *Accident; Analysis and Prevention, 40*(5), 1765–1772.

Gjerde, H., Øiestad, E. L., Øiestad, Å. M., Langødegård, M., Gustavsen, I., Hjelmeland, K., ... Christophersen, A. S. (2010). Comparison of zopiclone concentrations in oral fluid sampled with Intercept® oral specimen collection device and Statsure Saliva Sampler™ and concentrations in blood. *Journal of Analytical Toxicology, 34*(9), 590–593.

Gjerde, H., & Verstraete, A. (2010). Can the prevalence of high blood drug concentrations in a population be estimated by analysing oral fluid? A study of tetrahydrocannabinol and amphetamine. *Forensic Science International, 195*(1–3), 153–159.

Gjerde, H., & Verstraete, A. G. (2011). Estimating equivalent cutoff thresholds for drugs in blood and oral fluid using prevalence regression: A study of tetrahydrocannabinol and amphetamine. *Forensic Science International, 212*(1–3), e26–e30.

Gray, T. R., Dams, R., Choo, R. E., Jones, H. E., & Huestis, M. A. (2011). Methadone disposition in oral fluid during pharmacotherapy for opioid-dependence. *Forensic Science International, 206* (1–3), 98–102.

Hartman, R. L., Brown, T. L., Milavetz, G., Spurgin, A., Gorelick, D. A., Gaffney, G., & Huestis, M. A. (2016). Controlled vaporized cannabis, with and without alcohol: Subjective effects and oral fluid-blood cannabinoid relationships. *Drug Testing and Analysis, 8*(7), 690–701.

He, M., Li, Z., Ge, Y., & Liu, Z. (2016). Portable upconversion nanoparticles-based paper device for field testing of drug abuse. *Analytical Chemistry, 88*(3), 1530–1534.

Herrera-Gómez, F., García-Mingo, M., Colás, M., González-Luque, J. C., & Álvarez, F. J. (2018). Opioids in oral fluid of Spanish drivers. *Drug and Alcohol Dependence, 187*, 35–39.

Huestis, M. A., & Cone, E. J. (2004). Relationship of Delta 9-tetrahydrocannabinol concentrations in oral fluid and plasma after controlled administration of smoked cannabis. *Journal of Analytical Toxicology, 28*(6), 394–399.

Inscore, F., Shende, C., Sengupta, A., Huang, H., & Farquharson, S. (2011). Detection of drugs of abuse in saliva by surface-enhanced Raman spectroscopy (SERS). *Applied Spectroscopy, 65*(9), 1004–1008.

Langel, K., Engblom, C., Pehrsson, A., Gunnar, T., Ariniemi, K., & Lillsunde, P. (2008). Drug testing in oral fluid-evaluation of sample collection devices. *Journal of Analytical Toxicology, 32*(6), 393–401.

Langel, K., Gjerde, H., Favretto, D., Lillsunde, P., Øiestad, E. L., Ferrara, S. D., & Verstraete, A. G. (2014). Comparison of drug concentrations between whole blood and oral fluid. *Drug Testing and Analysis, 6*(5), 461–471.

Lee, D., Karschner, E. L., Milman, G., Barnes, A. J., Goodwin, R. S., & Huestis, M. A. (2013). Can oral fluid cannabinoid testing monitor medication compliance and/or cannabis smoking during oral THC and oromucosal Sativex administration? *Drug and Alcohol Dependence, 130*(1–3), 68–76.

Lee, D., Milman, G., Schwope, D. M., Barnes, A. J., Gorelick, D. A., & Huestis, M. A. (2012). Cannabinoid stability in authentic oral fluid after controlled cannabis smoking. *Clinical Chemistry, 58*(7), 1101–1109.

Lee, D., Schwope, D. M., Milman, G., Barnes, A. J., Gorelick, D. A., & Huestis, M. A. (2012). Cannabinoid disposition in oral fluid after controlled smoked cannabis. *Clinical Chemistry, 58* (4), 748–756.

Lee, D., Vandrey, R., Mendu, D. R., Murray, J. A., Barnes, A. J., & Huestis, M. A. (2015). Oral fluid cannabinoids in chronic frequent cannabis smokers during ad libitum cannabis smoking. *Drug Testing and Analysis, 7*(6), 494–501.

Lo Muzio, L., Falaschini, S., Rappelli, G., Bambini, F., Baldoni, A., Procaccini, M., & Cingolani, M. (2005). Saliva as a diagnostic matrix for drug abuse. *International Journal of Immunopathology and Pharmacology, 18*(3), 567–573.

Meagher, R. J., & Kousvelari, E. (2018). Mobile oral heath technologies based on saliva. *Oral Diseases, 24*(1–2), 194–197.

Milman, G., Barnes, A. J., Schwope, D. M., Schwilke, E. W., Darwin, W. D., Goodwin, R. S., . . . Huestis, M. A. (2010). Disposition of cannabinoids in oral fluid after controlled around-the-clock oral THC administration. *Clinical Chemistry, 56*(8), 1261–1269.

Milman, G., Schwope, D. M., Gorelick, D. A., & Huestis, M. A. (2012). Cannabinoids and metabolites in expectorated oral fluid following controlled smoked cannabis. *Clinica Chimica Acta, 413*(7–8), 765–770.

National Institute on Drug Abuse. (2018). *Overdose death rates*. Retrieved from https://www.drugabuse.gov/related-topics/trends-statistics/overdose-death-rates

Newmeyer, M. N., Concheiro, M., da Costa, J. L., Flegel, R., Gorelick, D. A., & Huestis, M. A. (2015). Oral fluid with three modes of collection and plasma methamphetamine and amphetamine enantiomer concentrations after controlled intranasal l-methamphetamine administration. *Drug Testing and Analysis, 7*(10), 877–883.

Newmeyer, M. N., Desrosiers, N. A., & Lee, D. (2014). Cannabinoid disposition in oral fluid after controlled cannabis smoking in frequent and occasional smokers. *Drug Testing and Analysis, 6* (10), 1002–1010.

Nordal, K., Øiestad, E. L., Enger, A., Christophersen, A. S., & Vindenes, V. (2015). Detection times of diazepam, clonazepam, and alprazolam in oral fluid collected from patients admitted to detoxification, after high and repeated drug intake. *Therapeutic Drug Monitoring, 37*(4), 451–460.

Odell, M. S., Frei, M. Y., Gerostamoulos, D., Chu, M., & Lubman, D. I. (2015). Residual cannabis levels in blood, urine and oral fluid following heavy cannabis use. *Forensic Science International, 249*, 173–180.

Ramaekers, J. G., Moeller, M. R., van Ruitenbeek, P., Theunissen, E. L., Schneider, E., & Kauert, G. (2006). Cognition and motor control as a function of Delta9-THC concentration in serum and oral fluid: Limits of impairment. *Drug and Alcohol Dependence, 85*(2), 114–122.

Scheidweiler, K. B., Andersson, M., Swortwood, M. J., Sempio, C., & Huestis, M. A. (2017). Long-term stability of cannabinoids in oral fluid after controlled cannabis administration. *Drug Testing and Analysis, 9*(1), 143–147.

Scheidweiler, K. B., Spargo, E. A., Kelly, T. L., Cone, E. J., Barnes, A. J., & Huestis, M. A. (2010). Pharmacokinetics of cocaine and metabolites in human oral fluid and correlation with plasma concentrations after controlled administration. *Therapeutic Drug Monitoring, 32*(5), 628–637.

Schepers, R. J., Oyler, J. M., Joseph, R. E., Jr., Cone, E. J., Moolchan, E. T., & Huestis, M. A. (2003). Methamphetamine and amphetamine pharmacokinetics in oral fluid and plasma after controlled oral methamphetamine administration to human volunteers. *Clinical Chemistry, 49* (1), 121–132.

Scherer, J. N., Fiorentin, T. R., Borille, B. T., Pasa, G., TRV, S., von Diemen, L., … Pechansky, F. (2017). Reliability of point-of-collection testing devices for drugs of abuse in oral fluid: A systematic review and meta-analysis. *Journal of Pharmaceutical and Biomedical Analysis, 143*, 77–85.

Schwope, D. M., Milman, G., & Huestis, M. A. (2010). Validation of an enzyme immunoassay for detection and semi quantification of cannabinoids in oral fluid. *Clinical Chemistry, 56*(6), 1007–1014.

Silber, B. Y., Croft, R. J., Papafotiou, K., & Stough, C. (2006). The acute effects of d-amphetamine and methamphetamine on attention and psychomotor performance. *Psychopharmacology, 187* (2), 154–169.

Substance Abuse and Mental Health Service Administration. (2015). *Age- and gender-based populations*. Retrieved from https://www.samhsa.gov/specific-populations/age-gender-based

Swortwood, M. J., Newmeyer, M. N., Andersson, M., Abulseoud, O. A., Scheidweiler, K. B., & Huestis, M. A. (2017). Cannabinoid disposition in oral fluid after controlled smoked, vaporized, and oral cannabis administration. *Drug Testing and Analysis, 9*(6), 905–915.

Toennes, S. W., Kauert, G. F., Steinmeyer, S., & Moeller, M. R. (2005). Driving under the influence of drugs—evaluation of analytical data of drugs in oral fluid, serum and urine, and correlation with impairment symptoms. *Forensic Science International, 152*(2–3), 149–155.

Valen, A., Leere Øiestad, Å. M., Strand, D. H., Skari, R., & Berg, T. (2017). Determination of 21 drugs in oral fluid using fully automated supported liquid extraction and UHPLC-MS/MS. *Drug Testing and Analysis, 9*(5), 808–823.

Veitenheimer, A. M., & Wagner, J. R. (2017). Evaluation of oral fluid as a specimen for DUID. *Journal of Analytical Toxicology, 41*(6), 517–522.

Vindenes, V., Lund, H. M., Andresen, W., Gjerde, H., Ikdahl, S. E., Christophersen, A. S., & Øiestad, E. L. (2012). Detection of drugs of abuse in simultaneously collected oral fluid, urine and blood from Norwegian drug drivers. *Forensic Science International, 219*(1–3), 165–171.
World Health Organization. (2016). *Drug use and road safety: a policy brief.* Retrieved from http://www.who.int/substance_abuse/drug_use_road_safety/en

# Chapter 17
# Therapeutic Drug Monitoring in Saliva

**Elizabeth A. Thomas**

**Abstract** Blood monitoring of therapeutic drugs is frequently used in clinical settings for the purpose of optimizing drug therapy. Psychotropic and psychiatric medications are used to treat a variety of mental health conditions, including depression, anxiety, bipolar disorder, and schizophrenia. Saliva is gaining increasing attention as a bioanalytical sample matrix for drug monitoring, including monitoring of psychotropic medications. Much of the advantages of saliva sampling come from the easy and noninvasive nature of saliva collection, which has clear advantages over blood draws, especially in pediatric populations or when frequent monitoring is required. Great improvements in the development of accurate and reliable methods for sample collection and highly sensitive and specific analytical methods for saliva testing of drugs have been observed in recent years. This chapter will review studies supporting the applicability of saliva for monitoring of psychotropic drugs, including antidepressants, mood stabilizers, and antipsychotic drugs. Further, this chapter will cover the main technological advances over the last decade, which have made oral fluid analysis of drugs possible.

**Keywords** Treatment · Therapy · Psychotropic · Lithium · Anti-psychotic · Monitoring

## 17.1 Background

### *17.1.1 Therapeutic Drug Monitoring*

Psychotropic medications, including antidepressants, mood stabilizers, and antipsychotics, are used to treat a variety of psychiatric disorders. Therapeutic drug monitoring of these medications is necessary for the optimal management of a patient's

E. A. Thomas (✉)
Department of Neuroscience, The Scripps Research Institute, La Jolla, CA, USA
e-mail: bthomas@scripps.edu



D. A. Granger, M. K. Taylor (eds.), *Salivary Bioscience*,
https://doi.org/10.1007/978-3-030-35784-9_17

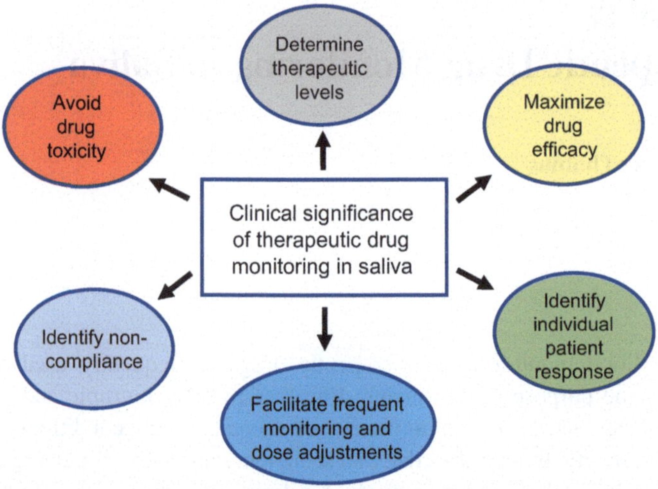


**Fig. 17.1** A summary of the clinical significance of therapeutic drug monitoring in saliva. The ovals depict six main necessities for the clinical monitoring of drugs in biological fluids, which can be conducted in saliva, among other fluids

drug regimen for many reasons. These include determining the effective dose for an individual, avoiding medical complications and drug toxicity, and identifying non-responsiveness and noncompliance (Fig. 17.1). Historically, the routine use of therapeutic drug monitoring developed out of the recognition that certain drugs have a narrow therapeutic range or window. That is, in concentrations above the upper limit of the range, the drug can be toxic, while concentrations below the lower limit of the range, the drug can be ineffective (Fig. 17.2). Further, it is now realized that not all patients have the same response to a medication, even at the same dose. Poor adherence is also a common issue in psychiatric patients. For example, rates of noncompliance in patients with bipolar disorder are typically 20–40% in euthymic patients (Murru et al., 2013), and up to 40–64% among acute manic patients (Montoya et al., 2007). Poorly adherent patients suffer from greater chronicity, worse functioning, an increased number of visits to emergency units and increased and longer acute admissions, with subsequent higher health expenses.

Typically, the drug monitoring process requires the drawing of blood, which remains the gold standard matrix. Therapeutic ranges for various drugs have been based on blood levels, despite the fact that the majority of therapeutic targets lie outside of this matrix (i.e., the brain or other peripheral organs). However, obtaining blood from human patients is an invasive procedure with inherent risks, including pain at the needle site and an increased risk of infection. It is also associated with increased healthcare costs as the process of acquiring blood requires trained personnel and laboratory equipment that are commonly only available at a hospital, laboratory, or institutional setting. Additionally, some patient populations are not well suited for blood sampling, because they may have a limited blood supply (e.g., pediatric subjects) or difficult veins (e.g., elderly subjects). Additionally, in patients

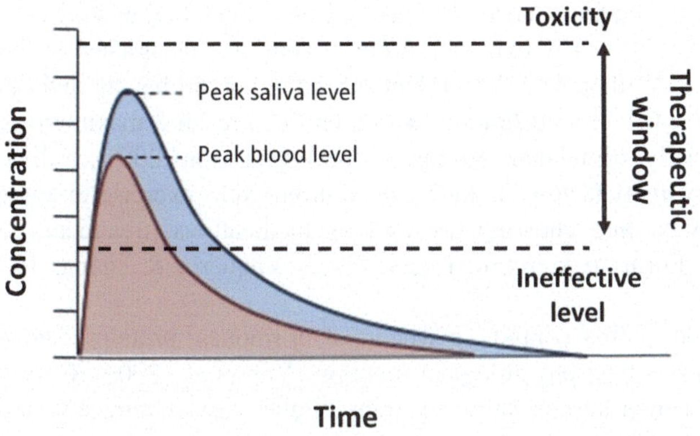


**Fig. 17.2** Hypothetical depiction of the therapeutic efficacy of psychotropic drugs. Drugs, such as lithium medications, show higher peak concentrations in saliva compared to blood, although this depends on the timing of the last dose. Many drugs for mood disorders have narrow therapeutic windows, (i.e., the difference between ineffective and toxic doses), exemplifying the need for drug monitoring

with psychiatric disorders, anxiety and discomfort in relation to blood sampling is an issue. These factors have provided motivation for finding an alternative matrix for conducting therapeutic drug monitoring and clinical pharmacokinetic research.

Saliva is considered as an alternative matrix for drug monitoring and is increasingly being used for testing of chemotherapeutic agents (Belostotsky, Adaway, Keevil, Cohen, & Webb, 2011; Greenaway, Ratnaraj, Sander, & Patsalos, 2011) and illicit substance testing (Huestis & Cone, 2007). The main advantages of saliva come from the ease of sampling and noninvasive nature of saliva collection, which does not induce patient discomfort or require trained personnel (Liu & Duan, 2012). Several psychotropic drugs, including mood stabilizers, antidepressants, and antipsychotic drugs, have been measured in saliva and these findings will be reviewed herein. Notable improvements in the development of accurate, reliable, and highly sensitive methods for saliva testing of drugs have been observed in recent years and these efforts will hopefully expand the use of saliva for monitoring of other therapeutic drugs. Saliva monitoring of drugs of abuse is covered in Chap. 16.

### *17.1.2 Saliva as a Clinical Matrix*

Saliva in the oral cavity originates mainly from three major salivary glands: the parotid, submandibular, and sublingual glands (Humphrey & Williamson, 2001). Minor salivary glands, including the buccal, labial, palatal, palatoglossal, and lingual glands, also contribute to salivary output, as well as gingival crevicular fluid, and fluids stemming from epithelial cells, erythrocytes, and leukocytes. In general, these salivary glands produce ~750 ml of serous and mucinous saliva daily by combining water, salts, and an abundance of molecules from the blood and salivary proteins in the oral cavity to give rise to the multi-constituent fluid designated as "whole saliva" (Humphrey & Williamson, 2001). Under resting conditions, an individual secretes approximately 0.1–0.3 mL/min of saliva, and can reach a maximum of 7 mL/min when artificially stimulated (Humphrey & Williamson, 2001). Saliva secretion i influenced by many factors, including the diurnal cycle, exercise, autonomic nervous system activity, and chewing, as well as medications, treatments, and various medical conditions (Chicharro, Lucia, Perez, Vaquero, & Urena, 1998; Granger et al., 2012).

Like blood, saliva contains a wealth of hormones, proteins, and nucleic acid molecules that reflect physiological function (Yan et al., 2009). Constituents from the blood can enter into the saliva via transcellular, passive intracellular diffusion, or active transport. Hence, many proteins in saliva reflect those levels circulating in the body. In addition, many factors can alter the distribution of blood versus salivary molecules. The physiology of drug distribution into saliva can depend on the drug's solubility, molecular weight, and degree of protein binding (Kiang & Ensom, 2016). Moreover, patient intrinsic factors such as salivary flow rate, salivary pH, and salivary protein composition, which can be altered by concomitant drugs or pathophysiological states, can also affect drug distribution into the saliva (Kiang & Ensom, 2016). Hence, saliva drug monitoring may not be ideal for all drugs and needs to be evaluated on a case-by-case basis.

## 17.2 Studies Measuring Psychotropic Drugs in Saliva

### *17.2.1 Mood Stabilizers*

Mood stabilizers are psychiatric medications used to treat mood disorders, such as bipolar disorder (type I or II), mania and hypomania, borderline personality disorder and schizoaffective disorder. These types of drugs have the added benefit of preventing recurrence and worsening of symptoms and are also used for the maintenance of mood and as a prophylaxis for abnormal mood swings. The prevalence of bipolar disorder has been estimated to be 1–2% worldwide, although rates as high as 4.4% have been cited when considering bipolar threshold (defined as having a lifetime history of 2 subthreshold hypomanic episodes) (Kessler et al., 1994). Lifetime prevalence rates of major depressive disorder have been estimated to be as high as 16% (Kessler et al., 1994). Given the relatively high, and possibly increasing, prevalence of these disorders, managing and optimizing treatment options for these types of drugs is essential. Several mood-stabilizing drugs have been studied in saliva, including lithium, valproic acid, carbamazepine, and lamotrigine and these will be discussed below. Lithium is the first-choice mood-stabilizing drug, but blood levels must be constantly monitored to ensure the lithium level is high enough to be effective but not high enough to be toxic. Hence, a more detailed summary of salivary lithium studies is provided.

#### 17.2.1.1 Lithium

Lithium medications, typically in the form of lithium carbonate (trade names: Lithium Carbonate ER, Eskalith, Lithobid, Lithotabs), are the prototypical mood stabilizers and commonly used for the treatment of bipolar mania and major depressive disorder (Thomas, Stansifer, & Findling, 2011). These drugs have been shown to decrease the risk of suicide in patients with mood disorders by nearly 80% (Benard, Vaiva, Masson, & Geoffroy, 2016; Zalsman et al., 2016). The use of lithium in psychiatry actually dates back to the mid-nineteenth century; however, John Cade is credited with introducing lithium into psychiatry for bipolar mania in 1949 (Cade, 1949). Although physicians in the USA were prescribing lithium medications in the late 1960s, it was not approved by the Food and Drug Administration in the USA until 1970, due to concerns about its safety (Shorter, 2009). Lithium therapy is one of the most effective and affordable long-term treatments for bipolar disorder but less than 25% of patients with bipolar disorders are prescribed with lithium drugs, due in large part to its very narrow therapeutic window (blood levels of 0.5–1.5 mM, between ineffectiveness or very severe side effects) (see Fig. 17.2). Lithium's narrow therapeutic index, intolerable side effects, and toxicity are likely the reasons it is less utilized in modern practice despite being considered a first-line agent, when compared to other agents, such as valproate or second-generation antipsychotics.

Patients who are prescribed lithium require frequent dose adjustments and regular laboratory monitoring, which involves repeated drawing of blood samples from the patient. This process can be cumbersome, expensive, and uncomfortable. Moreover, blood draws are difficult to time in a clinic setting and may not capture the full range of drug exposure over the course of the day, which can lead to errors in the estimation of drug levels in the patient. This monitoring of lithium levels in the blood is necessary to avoid toxic effects, which can include nausea, diarrhea, tremors, dizziness, confusion, and seizures. Both acute and chronic toxicity can occur due to dehydration, interactions with other medications, compromised kidney function, and several other factors/conditions that can affect how lithium is handled in the body. Although patients with serum lithium levels greater than 1.5 mM have a greater risk of developing toxicity, patients can have signs of toxicity below this range as well. Like other pharmacologic treatments, signs and symptoms of toxicity vary based on individual patient sensitivity and tolerability (Danielyan & Kowatch, 2005). Managing the inherent risk and maximizing the benefit of lithium therapy are essential for optimal treatment of patients with bipolar disorder.

#### 17.2.1.2 Lithium Monitoring in Saliva

With regards to lithium monitoring, there is substantial precedence for the idea that saliva can be used in place of blood; in fact, a large body of work dating back >30 years has suggested that salivary lithium might replace blood for patient monitoring of lithium levels (Ben-Aryeh et al., 1984; Bowden, Houston, Shulman, & Clothier, 1982; Khare, Sankaranarayanan, Goel, Khandelwal, & Srinivasa Murthy, 1983; McKeage & Maling, 1989; Nataraj & Bhat, 1981; Obach et al., 1988; Rosman, Sczupak, & Pakes, 1980; Shetty, Desai, Patil, & Nayak, 2012). Most of these studies have utilized atomic absorption spectrometry (AAS) or a similar method. Early studies dating back to 1983 conducted on 20 normal individuals who ingested single and multiple doses of lithium medications found significant correlations between saliva and serum levels of lithium (Doshi, Rout, & Kulkarni, 1983). Studies carried out one year later utilized actual patients ($n = 22$) diagnosed with bipolar disorder who were on lithium therapy. This study also reported a significant correlation between saliva and serum lithium concentrations (Ben-Aryeh et al., 1984). Similar findings were reported in several additional studies over the next 20 years that demonstrated significant correlations between serum lithium and salivary lithium in adult patients being treated with lithium carbonate (Ben-Aryeh et al., 1984; Khare et al., 1983; Murru et al., 2017; Rosman et al., 1980; Shetty et al., 2012), as well as in children (Vitiello et al., 1988). However, other work, notably older studies carried out >20 years ago, found only weaker associations between blood and saliva levels of lithium (Bowden et al., 1982; McKeage & Maling, 1989; Nataraj & Bhat, 1981; Obach et al., 1988; Rosman et al., 1980), which suggested that saliva would not serve as a suitable surrogate for blood.

The reported poor correlations between serum and salivary levels of lithium are likely related to techniques of obtaining and processing saliva samples and a lack of consideration of other factors that could affect saliva pharmacokinetics; these are discussed below (see "Challenges and Considerations") (Table 17.1). However, one principal issue worth mentioning here has to do with saliva processing. Saliva consists of water and mucopolysaccharide parts, whereby lithium is stored mainly in the water part. Centrifugation of saliva samples, to remove the mucins or dialysis of saliva, was found to improve the quality of lithium measurements and to increase the observed serum to saliva correlations (El-Mallakh, Linder, Valdes, & Looney, 2004; Murru et al., 2017).

Aside from examining serum–saliva correlations in patients, other important aspects of these studies included identifying the ratio of the lithium concentrations found in blood to that detectable in saliva. The cumulative data from the literature reveal higher concentrations of lithium in saliva compared to serum (Khare et al., 1983; Moody, 1999; Serdarevic, Kozjek, & Malesic, 2006; Shetty et al., 2012) (Fig. 17.2). This can be explained by the fact that lithium ions are eliminated slower from saliva than from serum as well as by the active transport of lithium into saliva (Serdarevic et al., 2006). In one of the first studies to examine saliva–serum ratios came from Khare et al. (1983), who utilized a group of 60 patients with bipolar disorder on lithium therapy and found that the saliva–serum ratios varied from 1.77 to 6.68 with a majority of the samples being between 3 and 3.99 (Khare et al., 1983). Similar variability in the saliva–serum ratios among individual adult patients was also observed in other studies (Khare et al., 1983; Moody, 1999; Shetty et al., 2012), highlighting the need to standardize salivary measures of lithium, in order for it to be reliable clinical reference.

**Table 17.1** Factors that can account for the variations in saliva lithium measurements reported in previous studies

| Issue | Addressed in previous publications | | |
|---|---|---|---|
| | Yes | No | Not reported |
| Separation of cellular components/mucins | Nataraj and Bhat (1981), Bowden et al. (1982), Ben-Aryeh et al. (1984), Obach et al. (1988), Vitiello et al. (1988), El-Mallakh et al. (2004), Shetty et al. (2012), Murru et al. (2017) | McKeage and Maling (1989), Spencer et al. (1990), Moody (1999), Serdarevic et al. (2006), Shetty et al. (2012) | Khare et al. (1983), Sankaranarayanan et al. (1985) |
| Use of stimulated saliva | Nataraj and Bhat (1981), Bowden et al. (1982), Khare et al. (1983), Obach et al. (1988), Spencer et al. (1990) | Ben-Aryeh et al. (1984), Moody (1999), El-Mallakh et al. (2004), Serdarevic et al. (2006), Shetty et al. (2012) | Sankaranarayanan et al. (1985), McKeage and Maling (1989), Murru et al. (2017) |
| Consideration of other medications | Bowden et al. (1982), McKeage and Maling (1989), Khare et al. (1983), Moody (1999), Shetty et al. (2012), Murru et al. (2017) | Nataraj and Bhat (1981), Ben-Aryeh et al. (1984), Obach et al. (1988), Spencer et al. (1990) | Sankaranarayanan et al. (1985), El-Mallakh et al. (2004), Serdarevic et al. (2006) |
| Timing of collection (Varied between 2 and 24 h after last dose) | Nataraj and Bhat (1981), Bowden et al. (1982), Khare et al. (1983), Ben-Aryeh et al. (1984), Obach et al. (1988), McKeage and Maling (1989), Spencer et al. (1990), El-Mallakh et al. (2004), Serdarevic et al. (2006), Shetty et al. (2012) | | Sankaranarayanan et al. (1985), Spencer et al. (1990), Moody (1999), Murru et al. (2017) |
| Information about xerostomy, dehydration, or saliva dilulteness | Khare et al. (1983), Ben-Aryeh et al. (1984), Obach et al. (1988), Shetty et al. (2012), Murru et al. (2017) | Rosman et al. (1980), Nataraj and Bhat (1981), Bowden et al. (1982), Khare et al. (1983), Ben-Aryeh et al. (1984), Obach et al. (1988), McKeage and Maling (1989), Spencer et al. (1990), Moody (1999), El-Mallakh et al. (2004), Serdarevic et al. (2006), Shetty et al. (2012) | |

The citations in the different columns represent whether or not a particular issue was considered in that study

Regarding gender differences, levels of lithium ion concentrations were not found to be different between men and women in most studies. However, in one study, a higher correlation between lithium levels in saliva and serum was found in women (Shetty et al., 2012), suggesting the possibility of gender differences. Finally, consideration to the different galenic forms of lithium taken by the patient also should be noted, as it may influence the results and clinical interpretation, as suggested previously (Obach et al., 1988).

#### 17.2.1.3 Valproic Acid

Valproic acid (sodium valproate, Convulex, Depakote, and Epilim) is an 8-carbon 2-chain fatty acid that is metabolized by the liver and processed at a variable rate based on the patient's liver function and age. Valproic acid is a broad-spectrum drug, and like lithium, it has a narrow therapeutic window. By the mid-1990s, valproic acid had replaced lithium as the mood stabilizer of choice, most likely due to its slightly larger therapeutic window and less stringent blood monitoring requirements. Valproic acid is also often used in combination with an atypical antipsychotic, which can provide synergistic mood-stabilizing and antidepressant activity, which is useful in controlling the extreme changes in mood characteristics of bipolar disorder, as well as antipsychotic activity in patients with schizoaffective disorder.

It has been suggested that the efficacy of valproic acid in bipolar disorder results from enhanced γ-aminobutyric acid (GABA) neurotransmission and the inhibition of enzymes involved in GABA metabolism. Valproic acid also inhibits histone deacetylase enzymes at therapeutic serum levels. The therapeutic range for total valproic acid is 50–125 μg/mL in blood. The toxic level in blood is greater than 150 μg/mL. Within that therapeutic range, most patients experience therapeutic drug effects without excessive side effects. Even within the recommended therapeutic range, however, results vary between individual patients and some may experience adverse effects at the low end of the spectrum. The range of effectiveness and adverse reactions seen with this drug require monitoring of the drug levels to ensure optimal dosing for each patient. Commonly reported side effects of valproic acid include abdominal pain, asthenia, drowsiness, nausea, tremor, vomiting, alopecia, diarrhea, dizziness, flu-like symptoms, thrombocytopenia, and anorexia.

#### 17.2.1.4 Monitoring Valproic Acid in Saliva

Correlations between serum and saliva levels of valproic acid were initially determined in patients with epilepsy (Dwivedi et al., 2015). In that study, a total of 59 paired serum and saliva samples were assayed from patients subjected to valproic acid for at least 3 months and levels were quantified using high-performance liquid chromatography (HPLC). It was found that salivary valproic acid concentrations were significantly correlated with serum free valproic acid concentration (Dwivedi et al., 2015). Levels of the drug were lower in saliva compared to serum, with the mean ratio of saliva to serum levels of valproic acid being 0.68 (Dwivedi et al., 2015). These findings suggested that valproic acid measures in saliva could also be useful for patients with bipolar disorder who are also taking the same medication. A later study using a similar method to measure valproic acid in saliva supported these findings (Tonic-Ribarska et al., 2012). However, another study using a different method, an fluorescence polarization immunoassay found a poor correlation between salivary and serum free concentrations, with a mean ratio of salivary to serum free concentration being lower than previously reported, with a value of 0.48 (al Za'abi, Deleu, & Batchelor, 2003). Similarly, another study using an immunoassay for valproic acid also did not show a good correlation between saliva and serum levels (Murru et al., 2017). Overall these findings suggest that monitoring and evaluating valproic acid levels in saliva from patients with bipolar disorder or other psychiatric conditions are feasibly provided that the correct methodology is used.

#### 17.2.1.5 Carbamazepine and Lamotrigine

Carbamazepine (Tegretol, Equetro) and lamotrigine (Lamictal) were first developed to treat epilepsy but were later found to help bipolar disorder patients (Post et al., 1998). Currently, they are primarily used to prevent and/or control seizures in adults and children, although the US Food and Drug Administration has also approved these drugs for the treatment of depression, mania, anxiety, bipolar, and borderline personality disorder (Post et al., 1998). A standard initiation and titration practice for lamotrigine involves a daily dose of 25 mg and increasing this to 50 mg per day after 1–2 weeks, then doubling the dose every 1–2 weeks until a dose of 200 mg per day is reached (Ng, Hallam, Lucas, & Berk, 2007). For carbamazepine, the dosage range is 400–1200 mg/day. Blood levels of carbamazepine are typically monitored weekly for the first 4 weeks to ensure that adequate levels are achieved to produce a therapeutic response. Hence, salivary monitoring of these medications would facilitate determinations of optimal dosing.

#### 17.2.1.6 Salivary Measures of Carbamazepine and Lamotrigine

Both carbamazepine and lamotrigine have been measured in saliva, but most studies have focused on patients with epilepsy, not bipolar disorder. Several early studies comparing carbamazepine in blood and saliva from patients with epilepsy found significant correlations between the two biofluids although levels were much lower in saliva compared to serum (MacKichan, Duffner, & Cohen, 1981; Rosenthal et al., 1995; Schramm, Annesley, Siegel, Sackellares, & Smith, 1991). Further, it was shown that stimulated saliva had no effect on salivary levels and that both salivary and serum carbamazepine levels were not affected by storing the samples for 7 days at room temperature (Rosenthal et al., 1995).

In a more recent study, serum and salivary levels of carbamazepine were measured in 116 persons with epilepsy using high-performance liquid chromatography

(HPLC) (Dwivedi, Singh, Kaleekal, Gupta, & Tripathi, 2016). The levels of carbamazepine showed a significant correlation ($p < 0.05$) between serum and saliva for those patients taking monotherapy or bi-therapy; however, there was no significant association in the case of poly-therapy (Dwivedi et al., 2016). This is an important finding, given that current psychiatric drug regimens often include poly-therapy.

Similarly, Serum and salivary lamotrigine concentrations were determined by high-performance liquid chromatography, also in patients with epilepsy, and good correlations between salivary and serum concentrations were detected (Ryan et al., 2003; Tsiropoulos, Kristensen, & Klitgaard, 2000). However, there was wide interpatient variability in the saliva:serum ratios (Ryan et al., 2003; Tsiropoulos et al., 2000). A study on normal healthy populations given single oral doses of 200 mg lamotrigine also found that plasma and stimulated saliva levels were significantly correlated with one another (Incecayir, Agabeyoglu, & Gucuyener, 2007).

### *17.2.2 Antidepressant Medications*

Therapeutic drug monitoring is also often used in antidepressant pharmacotherapy to improve dose titrations, efficacy, and avoiding side effects (Laux, Baumann, & Hiemke, 2007). Antidepressant medications are given to relieve the symptoms of depression, such as low mood, anxiety, and worthlessness. Antidepressants are classified into several different types depending on their structure and the way that they work. Of the five major classes of antidepressants, selective serotonin reuptake inhibitors (SSRIs) and serotonin and norepinephrine reuptake inhibitors (SNRIs) are the most commonly prescribed, particularly in first-line treatment.

#### 17.2.2.1 Antidepressant Drug Monitoring in Saliva

A few studies have explored the use of saliva for antidepressant drug monitoring. In one study, a single dose of desipramine, a tricyclic antidepressant, was given to healthy controls and saliva samples were analyzed at different time points. At any given time at least 24 h after the dose, there were significant relationships between saliva and plasma levels across individuals. However, comparing levels across subjects, there was wide variability in predicting plasma levels from saliva (Pi et al., 1991). A similar study on healthy controls was done with desmethylimipramine whereby considerable interindividual variation in the saliva/plasma desmethylimipramine ratio was observed, although this ratio within individual subjects was found to be stable (Cooper, Bark, & Simpson, 1981).

Another study compared blood and saliva levels of venlafaxine, an SSNRI, and citalopram, an SSRI, from psychiatric patients (Ebert et al., 2018). Results from that study showed significant correlations between concentrations of venlafaxine, and its metabolite O-desmethylvenlafaxine, as well as citalopram in serum and in saliva

(Ebert et al., 2018). These correlations were similar to another study measuring venlafaxine in saliva by HPLC (Dziurkowska & Wesolowski, 2013).

### *17.2.3 Antipsychotic Medications*

Overall, the presently available antipsychotic medications reduce symptomatology and prevent relapse in a sizeable percentage of patients. However, 20–30% of patients do not respond to antipsychotic therapy and most patients require several rounds of "trial and error," before finding a suitable medication to relieve their psychiatric symptoms. The two main classes of drugs used are the "typical" and "atypical" antipsychotics. Typical antipsychotic drugs, which include haloperidol, chlorpromazine, and fluphenazine, are primarily dopamine $D_2$ receptor antagonists. The "atypical" antipsychotics, such as clozapine, risperidone, quetiapine, and olanzapine, have a range of affinities for several different neurotransmitter receptors in addition to those for dopamine (Bymaster et al., 1996; Jann, 1991; Kerwin & Taylor, 1996). Atypical and typical antipsychotics are generally thought to be equivalent for the treatment of the positive symptoms of psychiatric disorders, however studies have shown that clozapine exhibits superior clinical efficacy over typical antipsychotics (Iqbal et al., 2003). The newer atypical antipsychotic medications are also usually preferred over older typical antipsychotic medications due to their more favorable side effect profiles. Compared to the typical antipsychotics, atypical drugs are associated with a lower incidence of extrapyramidal side effects and tardive dyskinesia. Unfortunately, despite this benefit, many atypical drugs (especially clozapine) are associated with metabolic side effects such as weight gain, hyperglycemia, and hypertriglyceridemia (Baptista, 1999; Baptista, Kin, Beaulieu, & de Baptista, 2002; Gupta, Droney, Al-Samarrai, Keller, & Frank, 1999). These issues can greatly reduce patient compliance, which is already very poor in psychiatric patients, with rates being as high as 40–64% (Montoya et al., 2007). Frequent monitoring of patients for drug levels in saliva would dramatically improve doctors' abilities to evaluate patient compliance, which is essential in order to know if a particular antipsychotic drug is working or not.

#### 17.2.3.1 Atypical Antipsychotics Measures in Saliva

One comprehensive study has examined saliva, plasma, and whole blood from patients prescribed any one of the following atypical antipsychotic drugs: amisulpride, aripiprazole, clozapine, quetiapine, risperidone, or sulpiride (Fisher, van Schalkwyk et al., 2013). These drugs were measured in Matched saliva, plasma, and whole-blood samples using liquid chromatography–tandem mass spectrometry (LC–MS/MS) from 90 patients taking these medications (Fisher, van Schalkwyk et al., 2013). Significant associations between the plasma and whole-blood concentrations ($R > 0.95$) of these drugs were observed, but poorer correlations were detected between saliva and serum, with r-values ranging from 0.3 to 0.7, which were thought to be due to saliva pH variation among the different patients (Fisher, van Schalkwyk et al., 2013). This study also compared the levels of drugs in the two biofluids. Amisulpride levels in plasma and saliva were found to be similar, whereas salivary aripiprazole concentrations were only approximately 8% of those found in the plasma. This was thought to reflect the strong plasma protein binding properties of aripiprazole. For the other drugs, plasma concentrations were 2–4 times higher than salivary concentrations. These findings concluded that measures of aripiprazole and its metabolite would probably not be feasible in saliva. However, the relationship between plasma and saliva levels for amisulpride, clozapine, quetiapine, and risperidone could be possible Fisher, van Schalkwyk et al., 2013).

Another study measured olanzapine and its major metabolite desmethylclozapine in saliva samples from a group of 34 inpatients or outpatients with a diagnosis of schizophrenia (Dumortier et al., 1998) for the purpose of assessing compliance. In this case, the exact correlations between saliva and blood levels would not be as important. Using HPLC it was shown that both clozapine and desmethylclozapine could be robustly detected in saliva (Dumortier et al., 1998), suggesting the use of salivary monitoring for checking compliance to treatment, in particular, among outpatients.

#### 17.2.3.2 Salivary Measures of Typical Antipsychotics: Haloperidol

The clinical utility of monitoring haloperidol in salivary samples has been assessed based on plasma-to-saliva correlations in a few previous studies. Early studies dating back to 1981 were the first to report significant correlations between saliva and serum levels of haloperidol measured by radioimmunoassay in human subjects (Yamazumi & Miura, 1981), findings that were supported by later studies examining both unstimulated whole saliva and gustatory stimulated parotid gland saliva (Dysken et al., 1992). Later studies using a more sensitive HPLC method were carried out on matched plasma and saliva samples from 105 patients with schizophrenia (Jain et al., 2011). These results also found statistically significant correlations of haloperidol levels in saliva and plasma, with strong linear relationships [$r = 0.93$, $p < 0.0001$] (Jain et al., 2011). In another study using rat models, a significant correlation between plasma and submandibular saliva elicited by parasympathetic, sympathetic, and pilocarpine stimulation was demonstrated using ELISA-based commercial kits called "Markit-M Haloperidol kit" (Takai, Eto, Uchihashi, Yamaguchi, & Nishikawa, 2006). Although it is generally considered that the salivary flow rate is a significant factor influencing salivary drug concentrations, that study found that the haloperidol concentrations in saliva were not affected by the salivary flow rate (Takai et al., 2006).

Additionally, the most striking finding of the Takai and colleagues' study was the demonstration of significant correlations of haloperidol levels between saliva and the brain (Takai et al., 2006), the target organ of antipsychotic drugs. And that the correlation of haloperidol between saliva and brain was higher than that between plasma and brain (Takai et al., 2006). While it is generally agreed that there is some correlation between plasma levels of the drug levels and clinical response, measuring drug levels in the brain is not possible. These findings suggest that monitoring salivary concentrations of haloperidol could be a clinically beneficial substitute for those levels present in the brain.

## 17.3 Methodologies, Challenges, and Considerations

### *17.3.1 Methodologies and Techniques Used for Drug Monitoring in Saliva*

#### 17.3.1.1 Identification of Drugs Levels

Great improvements in the development of highly sensitive and specific analytical methods for saliva testing of drugs have been observed in the last decade. The general methods for analyzing psychotropic drugs in different biological fluids are typically based on a combination of efficient separation method with a sensitive detection technique. A summary of the different methods used to analyze psychotropic drugs in saliva is listed in Table 17.2.

In the past, numerous separation techniques, including high-performance liquid chromatography (HPLC) (Xiong, Ruan, Cai, & Tang, 2009), gas chromatography (GC) (Huang et al., 2007; Sasaki et al., 2013), and capillary electrophoresis (Dziomba, Kowalski, & Baczek, 2012; Li, Zhao, & Ju, 2006) have been employed for the analysis of psychotropic drugs. Among those methods, HPLC is probably the most popular separation technique in analytical chemistry, with advantages including convenience, high separation ability, and simple operation. It is mainly used to separate components and to identify and to quantify each component in a mixture. Earlier, this method was known as high-pressure liquid chromatography, as it depended on a pump to introduce a pressurized liquid solvent containing the sample mixture through a column packed with a solid adsorbent material. Each constituent in the sample could be separated according to their different flow rates.

However, without mass spectrometry-based analytical methods, such as liquid chromatography coupled to mass spectrometry (LC–MS), tandem mass spectrometry (LC–MS/MS), or gas chromatography–mass spectrometry (GC–MS), the desired sensitivity of many drug measures would not be sufficient. This is due to the potentially low amounts of drugs often present in saliva and the small volumes of saliva. LC–MS combines the physical separation abilities of liquid chromatography with the mass analysis abilities of mass spectrometry (MS). LC–MS is a dominant analytical technique that has very high accuracy, sensitivity, and specificity compared to HPLC. Hence, this technique is especially important to detect low levels of drugs in saliva.

**Table 17.2** List of psychotropic drugs that have been measured in saliva and the methods used

| Drug class and names | Method for salivary measurements | References |
|---|---|---|
| *Mood stabilizers:* | | |
| Lithium | AA, ISE, Flame photometer, dry-slide analysis | Doshi et al. (1983), Sankanarayanan et al. (1985), Vitiello et al. (1988), Obach et al. (1988), McKeage and Maling (1989), Moody et al. (1999), Spencer et al. (1990), Serdarevic et al. (2006), Shetty et al. (2012), Murru et al. (2017) |
| Valproic acid | HPLC; Immunoassay | Dwivedi et al. (2015), Murru et al. (2017) |
| Carbamazepine | HPLC | MacKichan et al. (1981), Schramm et al. (1991), Rosenthal et al. (1995), Dwivedi et al. (2016) |
| Lamotrigine | LC, HPLC | Tsiropoulos et al. (2000), Ryan et al. (2003), Incecayir et al. (2007) |
| *Antidepressants:* | | |
| Venlafaxine | HPLC, LC–MS/MS | Dziurkowska and Wesolowski (2013), Ebert et al. (2018) |
| Quetiapine | LC–MS/MS | Ebert et al. (2018) |
| *Typical antipsychotics:* | | |
| Haloperidol | RIA, LC, ELISA, GC–MS | Yamazumi and Miura (1981), Jain et al. (1988), Takai et al. (2006), Pujadas et al. (2007) |
| Chlorpromazine | GC–MS | Pujadas et al. (2007) |
| Fluphenazine | GC–MS | Pujadas et al. (2007) |
| Sulpiride | LC–MS/MS | Fisher, Partridge, Handley, Couchman, et al. (2013), Fisher, Partridge, Handley, and Flanagan (2013), Fisher, van Schalkwyk, et al. (2013) |
| *Atypical antipsychotics:* | | |
| Amisulpride | LC–MS/MS | Fisher, Partridge, Handley, Couchman, et al. (2013), Fisher, Partridge, Handley, and Flanagan (2013), Fisher, van Schalkwyk, et al. (2013) |
| Aripiprazole | LC–MS/MS, HPLC–DAD | Fisher, Partridge, Handley, Couchman, et al. (2013), Fisher, Partridge, Handley, and Flanagan (2013), Fisher, van Schalkwyk, et al. (2013), Petruczynik et al. (2016) |
| Clozapine | LC–MS/MS, HPLC–DAD | Fisher, Partridge, Handley, Couchman, et al. (2013), Fisher, Partridge, Handley, and Flanagan (2013), Fisher, van Schalkwyk, et al. (2013), Petruczynik et al. (2016) |
| Quetiapine | LC–MS/MS, HPLC–DAD | Fisher, Partridge, Handley, Couchman, et al. (2013), Fisher, Partridge, Handley, and Flanagan (2013), Fisher, van Schalkwyk, et al. (2013), Petruczynik et al. (2016) |
| Risperidone | LC–MS/MS, HPLC–DAD, MEPS–LC–UV | Fisher, Partridge, Handley, Couchman, et al. (2013), Fisher, Partridge, Handley, and Flanagan (2013), Fisher, van Schalkwyk, et al. (2013), Petruczynik et al. (2016), Mandrioli et al. (2011) |
| Sulpiride | LC–MS/MS | Fisher, Partridge, Handley, Couchman, et al. (2013), Fisher, Partridge, Handley, and Flanagan (2013), Fisher, van Schalkwyk, et al. (2013) |

(continued)

**Table 17.2** (continued)

| Drug class and names | Method for salivary measurements | References |
|---|---|---|
| Citalopram | HPLC, HPLC–DAD | Petruczynik et al. (2016), Ebert et al. (2018) |
| Perazine | HPLC–DAD | Petruczynik et al. (2016) |
| Levomepromazine | HPLC–DAD | Petruczynik et al. (2016) |

*AA* atomic absorption, *ISE* ion selective electrode, *RIA* radioimmunoassay, *ELISA* enzyme-linked immunosorbent assay, *LC–MS/MS* liquid chromatography–tandem mass spectrometry, *HPLC–DAD* high-performance liquid chromatography with diode array detection, *GC–MS* gas chromatography–mass spectrometry, *LC* liquid chromatography, *MEPS/LC–UV* microextraction packed sorbent and liquid chromatography with UV detection

#### 17.3.1.2 Methods for Lithium Measurements

Lithium is a chemical element and a group I alkali metal; hence, detection of lithium levels in biological fluids can be achieved by a variety of chemical analyses that do not require sophisticated separations methods, such as HPLC. Many different methods have been utilized for serum lithium determination. In the late 1980s, ion selective electrodes (ISEs) were developed for lithium and used for salivary measures (Spencer et al., 1990), and even a colorimetric method has been developed and deemed suitable to detect lithium (Chapoteau, Czech, Zazulak, & Kumar, 1992). A "dry-slide" technique, involving measuring an increase in absorbance of a dye at 600 nm reflecting the concentration of lithium in the sample, has also been used for salivary measures (Serdarevic et al., 2006). But by far, the most common method used in past studies comparing levels of lithium in saliva and blood is atomic absorption spectrometry (AAS) (Table 17.2). Newer techniques, including inductively coupled plasma optical emission spectrometry (ICP-OES) and inductively coupled plasma mass spectrometry (ICP-MS) have been developed which have several advantages over AAS. These include: (1) Increased sensitivity and dynamic range for most elements; (2) The ability to analyze multiple elements at the same time; (3) No requirement for flammable gasses; (4) Easier to teach and maintain. So it is expected that studies in the future will utilize these more advanced methods.

#### 17.3.1.3 Simultaneously Measurements of Multiple Drugs and Their Metabolites

Because a significant percentage of patients who are treated with psychotropic drugs are treated with more than one drug, it is advantageous to use an assay that is suitable for the determination of multiple antipsychotic drugs in a single run. Because the concentrations of many drugs in biological samples are often very low, the analytical methods used for identification have to be highly sensitive and selective for accurate and precise quantification. Several of the above methods, or variations therein, have been utilized for this purpose (Fisher, Partridge, Handley, Couchman, et al., 2013; Fisher, Partridge, Handley, & Flanagan, 2013; Fisher, van Schalkwyk, et al., 2013;

Petruczynik et al., 2016; Pujadas et al., 2007). For example, LC–MS/MS has been used for the simultaneous analysis of amisulpride, aripiprazole and dehydroaripiprazole, clozapine and norclozapine, olanzapine, quetiapine, risperidone and 9-hydroxyrisperidone, and sulpiride in small (200 μL) volumes of plasma as well as saliva (Fisher, Partridge, Handley, Couchman, et al., 2013).

A simple and rapid of HPLC with a diode array detector (HPLC–DAD) method has been developed and validated for the simultaneous determination of seven psychotropic drugs (risperidone, citalopram, clozapine, quetiapine, levomepromazine, perazine, and aripiprazole) and their metabolites in saliva (Petruczynik et al., 2016). Including the detection of metabolites can require measuring concentrations of drugs over a substantially wider range. The HPLC–DAD method was compared to an HPLC–MS method to verify the presence of drugs in serum and saliva samples from the same patients (Petruczynik et al., 2016). Measurements of drug metabolites are especially important in some cases. For example, risperidone, which is one of the most frequently prescribed atypical antipsychotic drugs, has a main active metabolite, 9-hydroxyrisperidone, and both of these contribute significantly to the observed therapeutic effects. Further, because this metabolite possesses pharmacological properties similar to those of risperidone, any individual differences in how the liver metabolizes this drug can lead to toxicity. Hence, therapeutic drug monitoring can help in assessing the most effective dose, considering both the parent drug and its metabolite, not relying solely on the oral dose given.

Another study described a different analytical method to simultaneously measure risperidone and its metabolite, in plasma, urine, and saliva using HPLC coupled to an original sample pretreatment procedure based on micro-extraction by packed sorbent (Mandrioli, Mercolini, Lateana, Boncompagni, & Raggi, 2011). This study was successful in utilizing the same method for different matrices, namely plasma, urine, and saliva, with high accuracy and no interference (Mandrioli et al., 2011).

### *17.3.2 Challenges and Considerations*

The studies described above demonstrate the strong potential for saliva to be used for therapeutic drug monitoring. However, there are several challenges and considerations associated with the use of saliva for these purposes. These include: (1) Lack of standardization in saliva collection procedures; (2) Inconsistencies in saliva processing; (3) Lack of consideration of other medications; (4) Considerations of other variables, such as dehydration and sodium intake, etc. Overall, it is expected that consideration and standardization of these factors will improve serum-saliva associations and subsequent use of saliva for lithium monitoring in the clinic.

#### 17.3.2.1 Saliva Collection

Saliva as a biological material requires specific sample collecting methods. Two primary sample collection techniques are the "passive drool" method and the use of a "swab/salivette," which involves placing an absorbent material within the cheek. Whole saliva collected by passive drool is the gold standard when collecting saliva for biological testing, as it avoids localized secretions from specific salivary glands, providing a more consistent sample (Granger et al., 2012). In contrast, absorbent devices may collect localized saliva rather than whole saliva, which could affect results for some analytes (Mohamed, Campbell, Cooper-White, Dimeski, & Punyadeera, 2012).

Another key factor is whether saliva is stimulated or unstimulated for the collection procedure. Unstimulated saliva involves passively drooling into the collection vial. In contrast, saliva can be stimulated as a way to increase the flow rate and subsequent volume collected. Mechanically stimulated saliva involves the participants chewing onto a piece of tubing for 1 minute and then expectorating the first component while keeping the tubing in the mouth. Saliva can also be stimulated saliva using citric acid. Many reports of salivary drug monitoring have used both, which would certainly affect the quality and diluteness of the sample (see Table 17.1). Variation in time and frequency of sampling compared to the last dose of medication is also a variable that is not always consistently documented. This is especially important given that physicians typically consider trough levels of the drugs, defined as the levels right before the next dose is due, however, patients do not always have their levels monitored exactly at that time. Thus, standardization of collection methods and timing are needed to improve the quality of the data generated from saliva samples.

#### 17.3.2.2 Inconsistencies in Saliva Processing

Saliva consists of both aqueous watery component and a lipid-rich component which contains cellular components and mucins. Separation of these components can be easily done by centrifugation. Many past studies using whole saliva did not separate out these two components, which can negatively impact the accuracy of analysis (Table 17.1). One example of the major difference this step can make comes from a study showing that separating out the mucins from the soluble saliva fraction dramatically improved serum-saliva correlations for lithium measurements (El-Mallakh et al., 2004). However, it becomes clear how discrepancies can arise when only roughly half of the prior studies measuring lithium performed this separation step (Table 17.2).

#### 17.3.2.3 Drug Stability in Saliva

Researchers have also used highly sensitive chromatography methodologies to assess the stability of antipsychotic drugs in saliva at different temperatures. The stability of amisulpride, aripiprazole, and dehydroaripiprazole, clozapine and norclozapine, olanzapine, quetiapine, risperidone, and 9-hydroxyrisperidone, and sulpiride was assessed in human saliva using LC–MS/MS. Analyte instability was defined as a deviation of 15% or greater from the expected concentration. With the exception of olanzapine, all analytes were found to be stable for at least 2 days at ambient temperature, 1 week at 2–8 °C, and 2 months at −20 °C (Fisher, Partridge, Handley, Couchman, et al., 2013). The stability of olanzapine was found to be improved in both whole blood and saliva with the prior addition of ascorbic acid (Fisher, Partridge, Handley, & Flanagan, 2013)

#### 17.3.2.4 Consideration of Other Drugs

The rate of saliva secretion can be influenced by medications, treatments, and medical conditions that affect salivary gland function and oral health. The presence of other drugs in the body can affect accurate measures of a drug of interest. In particular, the use of polydrug therapy is common with psychiatric patients can confound measurements of single drugs.

Further, drugs like antidepressants, antihistamines, antipsychotics, sedatives, methyldopa, and diuretics are known to contribute to a low saliva volume or hyposalivation. Sometimes these drugs do not directly affect saliva production but in the case of diuretics, for example, the loss of fluid leads to a secondary reduction in saliva production. In contrast, cholinergic agents and clozapine can enhance the flow rate. Excessive salivation, especially during sleep is a common side effect, which appears in 23% of patients treated by clozapine and could explain the variations observed in plasma/saliva ratios (Dumortier et al., 1998). Moreover, patient intrinsic factors such as salivary flow rate, salivary pH, and salivary protein composition, which can be altered by concomitant drugs or pathophysiological states, can also affect drug distribution into the saliva (Kiang & Ensom, 2016). Hence, saliva drug monitoring may not be ideal for all drugs and needs to be evaluated on a case-by-case basis.

#### 17.3.2.5 Other Factors

Serum and/or salivary levels of medications can be influenced by dehydration, liver, and kidney function as well as sodium/potassium intake. Dehydration can occur for several reasons. Most of the time it is due to fluid and electrolyte loss from vomiting and copious diarrhea as is seen with gastroenteritis. Excessive sweating with inadequate fluid replenishment is another cause as is blood loss. Sometimes dehydration arises with the use of certain drugs that promote fluid loss (diuretics). A dry mouth (xerostomia) is the most obvious symptom of hyposalivation. It is important to note that dry mouth is not always due to hyposalivation. Mouth breathing, for example, can lead to dryness of the mouth due to the airflow although the daily saliva production is within the normal range. Finally, blood leakage into saliva may confound the analysis of blood versus saliva drug concentrations.

## 17.4 Conclusions

Therapeutic drug monitoring is carried out to optimize the management of patients receiving drug therapy to alleviate symptoms associated with a variety of neurological disorders. This is especially important in psychiatry with the use of psychotropic drugs for three main reasons: (1) Medications often need to be tailored to an individual's response in order to receive the maximum benefit; (2) Some psychiatric medications have narrow therapeutic ranges, which can lead to unexpected toxicity; (3) Patient noncompliance is high with psychiatric medications. The value of saliva as a key biological fluid for drug monitoring is emerging due to its many advantages over blood or serum, in particular, its noninvasive nature and ease of collection. However, for psychiatry, in particular, the use of saliva for drug monitoring could represent a major advance in the individualized treatment of patients. Frequent monitoring of patients for drug levels in saliva would dramatically improve doctors' abilities to evaluate patient compliance, optimize therapeutic dosing, and avoid complications and drug toxicity. The studies summarized above demonstrate the enormous potential of saliva to serve as a lead matrix for therapeutic drug monitoring.

## References


al Za'abi, M., Deleu, D., & Batchelor, C. (2003). Salivary free concentrations of anti-epileptic drugs: An evaluation in a routine clinical setting. *Acta Neurologica Belgica, 103*(1), 19–23.

Baptista, T. (1999). Body weight gain induced by antipsychotic drugs: Mechanisms and management. *Acta Psychiatrica Scandinavica, 100*(1), 3–16.

Baptista, T., Kin, N. M., Beaulieu, S., & de Baptista, E. A. (2002). Obesity and related metabolic abnormalities during antipsychotic drug administration: Mechanisms, management and research perspectives. *Pharmacopsychiatry, 35*(6), 205–219.

Belostotsky, V., Adaway, J., Keevil, B. G., Cohen, D. R., & Webb, N. J. (2011). Measurement of saliva tacrolimus levels in pediatric renal transplant recipients. *Pediatric Nephrology, 26*(1), 133–138.

Benard, V., Vaiva, G., Masson, M., & Geoffroy, P. A. (2016). Lithium and suicide prevention in bipolar disorder. *Encephale, 42*(3), 234–241.

Ben-Aryeh, H., Laor, R., Szargel, R., Gutman, D., Naon, H., Pascal, M., & Hefetz, A. (1984). Saliva for monitoring of patients with primary affective disorders. *Israel Journal of Medical Sciences, 20*(3), 197–201.

Bowden, C. L., Houston, J. P., Shulman, R. S., & Clothier, J. M. (1982). Clinical utility of salivary lithium concentration. *International Pharmacopsychiatry, 17*(2), 104–113.

Bymaster, F. P., Calligaro, D. O., Falcone, J. F., Marsh, R. D., Moore, N. A., Tye, N. C., . . . Wong, T. (1996). Radioreceptor binding profile of the atypical antipsychotic olanzapine. *Neuropsychopharmacology, 14*, 87–96.

Cade, J. F. (1949). Lithium salts in the treatment of psychotic excitement. *The Medical Journal of Australia, 2*(10), 349–352.

Chapoteau, E., Czech, B. P., Zazulak, W., & Kumar, A. (1992). First practical colorimetric assay of lithium in serum. *Clinical Chemistry, 38*(9), 1654–1657.

Chicharro, J. L., Lucia, A., Perez, M., Vaquero, A. F., & Urena, R. (1998). Saliva composition and exercise. *Sports Medicine, 26*(1), 17–27.

Cooper, T. B., Bark, N., & Simpson, G. M. (1981). Prediction of steady state plasma and saliva levels of desmethylimipramine using a single dose, single time point procedure. *Psychopharmacology, 74*(2), 115–121.

Danielyan, A., & Kowatch, R. A. (2005). Management options for bipolar disorder in children and adolescents. *Paediatric Drugs, 7*(5), 277–294.

Doshi, B. S., Rout, S. J., & Kulkarni, R. D. (1983). Absorption & urinary & salivary excretion of lithium after oral administration. *The Indian Journal of Medical Research, 78*, 708–712.

Dumortier, G., Lochu, A., Zerrouk, A., Van Nieuwenhuyse, V., Colen de Melo, P., Roche Rabreau, D., & Degrassat, K. (1998). Whole saliva and plasma levels of clozapine and desmethylclozapine. *Journal of Clinical Pharmacy and Therapeutics, 23*(1), 35–40.

Dwivedi, R., Gupta, Y. K., Singh, M., Joshi, R., Tiwari, P., Kaleekal, T., & Tripathi, M. (2015). Correlation of saliva and serum free valproic acid concentrations in persons with epilepsy. *Seizure, 25*, 187–190.

Dwivedi, R., Singh, M., Kaleekal, T., Gupta, Y. K., & Tripathi, M. (2016). Concentration of antiepileptic drugs in persons with epilepsy: A comparative study in serum and saliva. *The International Journal of Neuroscience, 126*(11), 972–978.

Dysken, M. W., Johnson, S. B., Holden, L., Kuskowski, M., Ofstehage, J., Skare, S., & Vatassery, G. (1992). Haloperidol and reduced haloperidol in saliva and blood. *J Clin Psychopharmacol, 12*(3), 186-190.

Dziomba, S., Kowalski, P., & Baczek, T. (2012). Micelle to solvent stacking of tricyclic psychiatric drugs in capillary electrophoresis. *Journal of Pharmaceutical and Biomedical Analysis, 62*, 149–154.

Dziurkowska, E., & Wesolowski, M. (2013). Simultaneous quantitation of venlafaxine and its main metabolite, O-desmethylvenlafaxine, in human saliva by HPLC. *Journal of Separation Science, 36*(11), 1726–1733.

Ebert, K., Maurice, E., Lukacin, R., Fleischhaker, C., Schulz, E., Ebert, D., & Clement, H. W. (2018). Serum and saliva concentrations of venlafaxine, O-Desmethylvenlafaxine, Quetiapine, and citalopram in psychiatric patients. *Therapeutic Drug Monitoring, 40*(3), 351–355.

El-Mallakh, R. S., Linder, M., Valdes, R., & Looney, S. (2004). Dialysis of saliva improves accuracy of saliva lithium determinations. *Bipolar Disorders, 6*(1), 87–89.

Fisher, D. S., Partridge, S. J., Handley, S. A., Couchman, L., Morgan, P. E., & Flanagan, R. J. (2013). LC-MS/MS of some atypical antipsychotics in human plasma, serum, oral fluid and haemolysed whole blood. *Forensic Science International, 229*(1–3), 145–150.

Fisher, D. S., Partridge, S. J., Handley, S. A., & Flanagan, R. J. (2013). Stability of some atypical antipsychotics in human plasma, haemolysed whole blood, oral fluid, human serum and calf serum. *Forensic Science International, 229*(1–3), 151–156.

Fisher, D. S., van Schalkwyk, G. I., Seedat, S., Curran, S. R., & Flanagan, R. J. (2013). Plasma, oral fluid, and whole-blood distribution of antipsychotics and metabolites in clinical samples. *Therapeutic Drug Monitoring, 35*(3), 345–351.

Granger, D. A., Fortunato, C. K., Beltzer, E. K., Virag, M., Bright, M. A., & Out, D. (2012). Focus on methodology: Salivary bioscience and research on adolescence: An integrated perspective. *Journal of Adolescence, 35*(4), 1081–1095.

Greenaway, C., Ratnaraj, N., Sander, J. W., & Patsalos, P. N. (2011). Saliva and serum lacosamide concentrations in patients with epilepsy. *Epilepsia, 52*(2), 258–263.

Gupta, S., Droney, T., Al-Samarrai, S., Keller, P., & Frank, B. (1999). Olanzapine: Weight gain and therapeutic efficacy. *Journal of Clinical Psychopharmacology, 19*(3), 273–275.

Huang, H. C., Liu, C. H., Lan, T. H., Hu, T. M., Chiu, H. J., Wu, Y. C., & Tseng, Y. L. (2007). Detection and quantification of aripiprazole and its metabolite, dehydroaripiprazole, by gas chromatography-mass spectrometry in blood samples of psychiatric patients. *Journal of Chromatography. B, Analytical Technologies in the Biomedical and Life Sciences, 856*(1–2), 57–61.

Huestis, M. A., & Cone, E. J. (2007). Methamphetamine disposition in oral fluid, plasma, and urine. *Annals of the New York Academy of Sciences, 1098*, 104–121.

Humphrey, S. P., & Williamson, R. T. (2001). A review of saliva: Normal composition, flow, and function. *The Journal of Prosthetic Dentistry, 85*(2), 162–169.

Incecayir, T., Agabeyoglu, I., & Gucuyener, K. (2007). Comparison of plasma and saliva concentrations of lamotrigine in healthy volunteers. *Arzneimittel-Forschung, 57*(8), 517–521.

Iqbal, M. M., Rahman, A., Husain, Z., Mahmud, S. Z., Ryan, W. G., & Feldman, J. M. (2003). Clozapine: A clinical review of adverse effects and management. *Annals of Clinical Psychiatry, 15*(1), 33–48.

Jain, A. K., Kelwala, S., & Gershon, S. (1988). Antipsychotic drugs in schizophrenia: current issues. *International Clinical Psychopharmacology, 3*, 1–30.

Jain, T., Bhandari, A., Ram, V., Sharma, S., Parakh, M., & Parakh, M. C. (2011). Correlation of haloperidol levels between saliva and plasma of acutely ill schizophrenic patients. *Clinical Biochemistry, 44*(8–9), 675–680.

Jann, M. W. (1991). Clozapine. *Pharmacotherapy, 11*, 179–195.

Kerwin, R., & Taylor, D. (1996). Antipsychotics – a review of the current status and clinical potential. *CNS Drugs, 6*, 71–82.

Kessler, R. C., McGonagle, K. A., Zhao, S., Nelson, C. B., Hughes, M., Eshleman, S., . . . Kendler, K. S. (1994). Lifetime and 12-month prevalence of DSM-III-R psychiatric disorders in the United States. Results from the National Comorbidity Survey. *Archives of General Psychiatry, 51*(1), 8–19.

Khare, C. B., Sankaranarayanan, A., Goel, A., Khandelwal, S. K., & Srinivasa Murthy, R. (1983). Saliva lithium levels for monitoring lithium prophylaxis of manic depressive psychosis. *International Journal of Clinical Pharmacology, Therapy, and Toxicology, 21*(9), 451–453.

Kiang, T. K., & Ensom, M. H. (2016). A qualitative review on the pharmacokinetics of antibiotics in saliva: Implications on clinical pharmacokinetic monitoring in humans. *Clinical Pharmacokinetics, 55*(3), 313–358.

Laux, G., Baumann, P., & Hiemke, C. (2007). Therapeutic drug monitoring of antidepressants – clinical aspects. *Journal of Neural Transmission (Suppl), 72*, 261–267.

Li, J., Zhao, F., & Ju, H. (2006). Simultaneous determination of psychotropic drugs in human urine by capillary electrophoresis with electrochemiluminescence detection. *Analytica Chimica Acta, 575*(1), 57–61.

Liu, J., & Duan, Y. (2012). Saliva: a potential media for disease diagnostics and monitoring. *Oral Oncology, 48*(7), 569–577.

MacKichan, J. J., Duffner, P. K., & Cohen, M. E. (1981). Salivary concentrations and plasma protein binding of carbamazepine and carbamazepine 10,11-epoxide in epileptic patients. *British Journal of Clinical Pharmacology, 12*(1), 31–37.

Mandrioli, R., Mercolini, L., Lateana, D., Boncompagni, G., & Raggi, M. A. (2011). Analysis of risperidone and 9-hydroxyrisperidone in human plasma, urine and saliva by MEPS-LC-UV. *Journal of Chromatography. B, Analytical Technologies in the Biomedical and Life Sciences, 879*(2), 167–173.

McKeage, M. J., & Maling, T. J. (1989). Saliva lithium: A poor predictor of plasma and erythrocyte levels. *The New Zealand Medical Journal, 102*(878), 559–560.

Mohamed, R., Campbell, J. L., Cooper-White, J., Dimeski, G., & Punyadeera, C. (2012). The impact of saliva collection and processing methods on CRP, IgE, and myoglobin immunoassays. *Clinical and Translational Medicine, 1*(1), 19.

Montoya, A., Perez Sanchez Toledo, J., Gilaberte, I., Gonzalez-Pinto, A., Haro, J. M., Vieta, E., & Tohen, M. (2007). Patterns of drug treatment for manic episode in the clinical practice. Outcomes of the Spanish sample in the EMBLEM study. *Actas Españolas de Psiquiatría, 35*(5), 315–322.

Moody, J. P. (1999). Biologic variation of serum and salivary lithium. *Therapeutic Drug Monitoring, 21*(1), 97–101.

Murru, A., Pacchiarotti, I., Amann, B. L., Nivoli, A. M., Vieta, E., & Colom, F. (2013). Treatment adherence in bipolar I and schizoaffective disorder, bipolar type. *Journal of Affective Disorders, 151*(3), 1003–1008.

Murru, A., Torra, M., Callari, A., Pacchiarotti, I., Romero, S., Gonzalez de la Presa, B., . . . Colom, F. (2017). A study on the bioequivalence of lithium and valproate salivary and blood levels in the treatment of bipolar disorder. *European Neuropsychopharmacology, 27*(8), 744–750.

Nataraj, G. S., & Bhat, V. K. (1981). Reliability of saliva lithium level-a prospective study. *Indian Journal of Psychiatry, 23*(2), 115–119.

Ng, F., Hallam, K., Lucas, N., & Berk, M. (2007). The role of lamotrigine in the management of bipolar disorder. *Neuropsychiatric Disease and Treatment, 3*(4), 463–474.

Obach, R., Borja, J., Prunonosa, J., Valles, J. M., Torrent, J., Izquierdo, I., & Jane, F. (1988). Lack of correlation between lithium pharmacokinetic parameters obtained from plasma and saliva. *Therapeutic Drug Monitoring, 10*(3), 265–268.

Petruczynik, A., Wroblewski, K., Szultka-Mlynska, M., Buszewski, B., Karakula-Juchnowicz, H., Gajewski, J., . . . Waksmundzka-Hajnos, M. (2016). Determination of some psychotropic drugs in serum and saliva samples by HPLC-DAD and HPLC MS. *Journal of Pharmaceutical and Biomedical Analysis, 127*, 68–80.

Pi, E. H., Tran-Johnson, T. K., Gray, G. E., Walker, N. R., Suckow, R. F., & Cooper, T. B. (1991). Saliva and plasma desipramine levels in Asian and Caucasian volunteers. *Psychopharmacology Bulletin, 27*(3), 281–284.

Post, R. M., Denicoff, K. D., Frye, M. A., Dunn, R. T., Leverich, G. S., Osuch, E., & Speer, A. (1998). A history of the use of anticonvulsants as mood stabilizers in the last two decades of the 20th century. *Neuropsychobiology, 38*(3), 152–166.

Pujadas, M., Pichini, S., Civit, E., Santamarina, E., Perez, K., & de la Torre, R. (2007). A simple and reliable procedure for the determination of psychoactive drugs in oral fluid by gas chromatography-mass spectrometry. *Journal of Pharmaceutical and Biomedical Analysis, 44* (2), 594–601.

Rosenthal, E., Hoffer, E., Ben-Aryeh, H., Badarni, S., Benderly, A., & Hemli, Y. (1995). Use of saliva in home monitoring of carbamazepine levels. *Epilepsia, 36*(1), 72–74.

Rosman, A. W., Sczupak, C. A., & Pakes, G. E. (1980). Correlation between saliva and serum lithium levels in manic-depressive patients. *American Journal of Hospital Pharmacy, 37*(4), 514–518.

Ryan, M., Grim, S. A., Miles, M. V., Tang, P. H., Fakhoury, T. A., Strawsburg, R. H., . . . Baumann, R. J. (2003). Correlation of lamotrigine concentrations between serum and saliva. *Pharmacotherapy, 23*(12), 1550–1557.

Sankaranarayanan, A., Goel, A., & Pant, V. L. (1985). Variation in the relationship between serum and saliva lithium levels. *International Journal of Clinical Pharmacology, Therapy, and Toxicology, 23*, 365–366.

Sasaki, C., Shinozuka, T., Murakami, C., Irie, W., Maeda, K., Watanabe, T., . . . Kurihara, K. (2013). Simultaneous determination of 5 psychotropic drugs of various types in an autopsy case of acute multiple drug poisoning. *Forensic Science International, 227*(1–3), 90–94.

Schramm, W., Annesley, T. M., Siegel, G. J., Sackellares, J. C., & Smith, R. H. (1991). Measurement of phenytoin and carbamazepine in an ultrafiltrate of saliva. *Therapeutic Drug Monitoring, 13*(5), 452–460.

Serdarevic, N., Kozjek, F., & Malesic, I. (2006). Saliva and serum lithium monitoring in hospitalized patients and possibility to replace serum to saliva. *Bosnian Journal of Basic Medical Sciences, 6*(4), 32–35.

Shetty, S. J., Desai, P. B., Patil, N. M., & Nayak, R. B. (2012). Relationship between serum lithium, salivary lithium, and urinary lithium in patients on lithium therapy. *Biological Trace Element Research, 147*(1–3), 59–62.

Shorter, E. (2009). The history of lithium therapy. *Bipolar Disorders, 11*(Suppl 2), 4–9.

Spencer, E. K., Campbell, M., Adams, P., Perry, R., Choroco, M. C., Padron-Gayol, M., & Small, A. M. (1990). Saliva and serum lithium monitoring in hospitalized children. *Psychopharmacology Bulletin, 26*(2), 239–243.

Takai, N., Eto, K., Uchihashi, K., Yamaguchi, M., & Nishikawa, Y. (2006). Correlation of haloperidol levels between submandibular saliva and brain in the rat. *Archives of Oral Biology, 51*(7), 567–572.

Thomas, T., Stansifer, L., & Findling, R. L. (2011). Psychopharmacology of pediatric bipolar disorders in children and adolescents. *Pediatric Clinics of North America, 58*(1), 173–187. xii.

Tonic-Ribarska, J., Haxhiu, A., Sterjev, Z., Kiteva, G., Suturkova, L., & Trajkovic-Jolevska, S. (2012). Development and validation of a bioanalytical LC-UV method with solid-phase extraction for determination of valproic acid in saliva. *Acta Pharmaceutica, 62*(2), 211–220.

Tsiropoulos, I., Kristensen, O., & Klitgaard, N. A. (2000). Saliva and serum concentration of lamotrigine in patients with epilepsy. *Therapeutic Drug Monitoring, 22*(5), 517–521.

Vitiello, B., Behar, D., Malone, R., Delaney, M. A., Ryan, P. J., & Simpson, G. M. (1988). Pharmacokinetics of lithium carbonate in children. *Journal of Clinical Psychopharmacology, 8*(5), 355-359.

Xiong, C., Ruan, J., Cai, Y., & Tang, Y. (2009). Extraction and determination of some psychotropic drugs in urine samples using dispersive liquid-liquid microextraction followed by high-performance liquid chromatography. *Journal of Pharmaceutical and Biomedical Analysis, 49*(2), 572–578.

Yamazumi, S., & Miura, S. (1981). Haloperidol concentrations in saliva and serum: Determination by the radioimmunoassay method. *International Pharmacopsychiatry, 16*(3), 174–183.

Yan, W., Apweiler, R., Balgley, B. M., Boontheung, P., Bundy, J. L., Cargile, B. J., . . . Wong, D. T. (2009). Systematic comparison of the human saliva and plasma proteomes. *Proteomics Clinical Applications, 3*(1), 116–134.

Zalsman, G., Hawton, K., Wasserman, D., van Heeringen, K., Arensman, E., Sarchiapone, M., . . . Zohar, J. (2016). Suicide prevention strategies revisited: 10-year systematic review. *Lancet Psychiatry, 3*(7), 646–659.

# Chapter 18
# Salivary Bioscience and Periodontal Medicine

**Chen Xuan Wei, Yanbao Yu, Gajender Aleti, Manny Torralba, Anna Edlund, Karen E. Nelson, and Marcelo Freire**

**Abstract** Mammalian saliva carries a precise developmental and biological program that defines the function of multiple organs including oral and periodontal tissues (tissues surrounding the tooth). Saliva is at the interface of dynamic interactions between microbial ecosystems and oral immune cells. Dysregulation of saliva physiology has an impact on oral and systemic health. Elevated inflammatory cytokines in saliva, for example, activates dysbiosis (disease) of periodontal tissues, feeding forward tissue inflammation, leading to oral tissue loss. Complex host–microbiome interactions modulate pathological transitions including gingivitis, periodontal diseases, and oral cancer. However, mechanisms involving these interactions are poorly understood at the molecular level. Hence, the overall understanding of salivary biosciences, in the context of periodontal health and disease, is crucial to oral medicine, microbiology, translational sciences, and human health. High-throughput screening techniques, presented here, allow for the identification and quantification of salivary molecular information and its functional kinetics. With current available "multi-omic" tools, we highlight the importance of saliva in

C. X. Wei
Faculty of Dentistry, Department of Prosthodontics, The University of Hong Kong, Sai Ying Pun, Hong Kong
e-mail: sanhin@connect.hku.hk

Y. Yu
Infectious Disease, J. Craig Venter Institute, La Jolla, CA, USA
e-mail: yayu@jcvi.org

G. Aleti
Mammalian Genomics, J. Craig Venter Institute, La Jolla, CA, USA

M. Torralba · A. Edlund · K. E. Nelson
Genomic Medicine, J. Craig Venter Institute, La Jolla, CA, USA
e-mail: mtorralb@jcvi.org; aedlund@jcvi.org; kenelson@jcvi.org

M. Freire (✉)
Infectious Disease, J. Craig Venter Institute, La Jolla, CA, USA

Genomic Medicine, J. Craig Venter Institute, La Jolla, CA, USA
e-mail: mfreire@jcvi.org



D. A. Granger, M. K. Taylor (eds.), *Salivary Bioscience*,
https://doi.org/10.1007/978-3-030-35784-9_18

precision dental medicine and in influencing dynamic interactions of the periodontium. This chapter reviews fundamental roles of saliva in periodontal health with particular focus given to oral microbiome, tissue inflammation, resolution signals, local-systemic circuitry, and periodontal dysbiosis.

**Keywords** Saliva · Inflammation · Oral microbiome · Periodontal diseases · Omics

## 18.1 History of Salivary Biosciences in Periodontal Medicine

Periodontal tissues are supporting tissues surrounding the tooth comprising of periodontal ligament, connective tissue, and epithelium. While dental caries historically developed as a result of forming into society and emerging civilizations, paleontological studies revealed that "gum diseases" and "loose teeth" are as old as humanity. Periodontal diseases occur in humans, domesticated animals, and wild animals and are one of the most common diseases afflicting mammalian dentition (Löe, 1993). There are records of Egyptian and Chinese civilizations describing signs and symptoms of periodontal diseases as an inflammatory condition. In "*Le chirugien dentist*," Pierre Fauchard provided the first pathological description of the periodontal lesion and new treatment instructions that conceptually are used to this day by periodontists and dental medicine (Fauchard, 1728). Later, John Hunter, a physician–dentist and scientist, wrote *The Natural History of Human Teeth* (Hunter, 1778), describing the association of tooth loss and calculus accumulation.

Recent evidence defined periodontal diseases as a group of complex and destructive lesions affecting oral supporting tissues (Armitage, 2000, 2004), which develop through chronic inflammation induced by pathogenic microbes (Haffajee & Socransky, 1994; Socransky et al., 1984). The landmark study by Harold Loe et al. (1965) demonstrated that bacteria possess a plethora of virulence factors, which induce inflammatory mediator production by host cells at the gingival level. While pathogenic biofilms are traditionally associated with the reversible stage of inflammation (in gingivitis), the disease progression to irreversible tissue loss (in periodontitis) requires a susceptible host (Offenbacher, 1996). For example, patients presenting poor oral hygiene have demonstrated variable response and sometimes a healthy phenotype, implying the role of the host immunity and saliva in control of pathophysiology. The localized nature of gingival sulcus/periodontal pocket infection, which receives and delivers molecules to saliva, can lead to systemic inflammation beyond the periodontium (Van Dyke & van Winkelhoff, 2013). A larger number of studies have investigated the potential associations between periodontal diseases and systemic conditions (Linden et al., 2013) including diabetes (Chapple et al., 2013), cardiovascular conditions (Tonetti & Dyke, 2013), and adverse pregnancy outcomes (Sanz et al., 2013). Periodontitis is reported to be highly prevalent worldwide, affecting children and adults (Papapanou, 1996). Globally, half the adults have a form of periodontal disease (Eke et al., 2016; Patel, 2012), with different distributions according to severity (slight, moderate, and severe). The prevalence of periodontitis is more in elderly (Holtfreter, Kocher, Hoffmann, Desvarieux, & Micheelis, 2010; Mattila, Niskanen, Vehkalahti, Nordblad, & Knuuttila, 2010) and in indigenous populations (Brothwell & Ghiabi, 2009).

Periodontal health and disease etiopathogenesis and severity are partly modulated by saliva biology. For instance, some of the most abundant saliva proteins, proline-rich proteins (PRPs), amylase, lysozyme, lactoferrin, and peroxidases have been identified possessing antimicrobial functions (Gorr, 2009; van't Hof, Veerman, Nieuw Amerongen, & Ligtenberg, 2014) in control of periodontal severity. Also, macromolecule proteins, such as mucins, function in lubrication, food cleaning, facilitation of oral microbial adherence and/or aggregation. Monitoring periodontal diseases and response to treatment through saliva has provided novel understanding of how specific markers react in different stages of periodontal diseases. Aspartate and alanine aminotransferases have been measured locally as new biomarkers for periodontal pocket progression (Totan et al., 2006). Immunologic proteins, such as immunoglobulins and complement components facilitate host immune response against microbes and subsequent inflammation (Humphrey & Williamson, 2001). Periodontal diseases elicit high salivary concentration of immunoglobulins (IgA, IgG and IgM) specific to periodontal pathogens when compared to healthy subjects (Seemann et al., 2004). While periodontal disease is influenced by saliva molecular content, periodontal treatment also influences saliva protein and host response (Nomura et al., 2012). Thus, the study of saliva metabolism, in the context of periodontal tissues, holds important promises for the future understanding of periodontal health and disease initiation, progression, and prevention/treatment modalities. New biomarker evidence in saliva can predict the risk of local tissue inflammation progression and certain distant tissues, organs, and diseases (Podzimek, Vondrackova, Duskova, Janatova, & Broukal, 2016). From the initial insights of "oral focal infection" to this date, saliva bioscience has continuously advanced, becoming an important part of clinical dentistry and laboratory armamentarium in predicting, measuring, and treating periodontal and systemic diseases. Regarding the aforementioned salivary diagnostic markers, immunoglobulins, enzymatic constituents from gingival crevicular fluid, bacterial components or metabolites and to some extent volatile compounds serve as potential diagnostic markers for early detection of oral and also systemic health.

## 18.2 The Role of Saliva in Periodontal Health and Diseases

### *18.2.1 Saliva and Periodontal Homeostasis*

Periodontal tissues undergo constant remodeling and are located in a highly dynamic microenvironment with broad internal (i.e., hormone, blood pressure) and external (i.e., alimentary, chemical, mechanical, microbial) stimuli (Luan et al., 2017;

Newman et al., 2018). In this sense, understanding the role of saliva and its components in periodontal homeostasis is essential for maintaining tissue integrity. Saliva modulates periodontal tissues by controlling host–microbiome interactions, regulating tissue hydration, maintaining chemical balance, and controlling periodontal homeostasis (Dodds, Johnson, & Yeh, 2005). Thus, specific molecular compositions, such as lipids, urea, and innate immune cells, proteins, amino acids, enzymes and microbial-derived metabolites have been characterized to regulate oral homeostasis and periodontal tissue integrity.

Saliva lipids are chemically and functionally diverse group of compounds including cholesterol, triglycerides, high-density lipoprotein cholesterol, low-density lipoprotein cholesterol, and very low-density lipoprotein cholesterol (Singh et al., 2014). Lipid oxidation and modification in saliva could reflect on enhanced superoxide anion production and leaking from gingival crevicular fluid (GCF, a serum-like fluid within the gingival pockets) during the interaction with periodontal pathogens, and neutrophils within periodontal tissues or pockets. The composition and quantity of saliva lipids vary in different pathological states, and lipid peroxidation is a major consequence of oxidative stress. This could be evaluated via malondialdehyde levels that are positively correlated with elevated levels in chronic periodontitis (Guentsch et al., 2008; Nguyen et al., 2016; Trivedi, Lal, Mahdi, Singh, & Pandey, 2015) and reduction upon treatment. In vivo detection of lipid peroxidation could be achieved through measuring prostaglandin-like compounds, namely 8-epi-PGF2α, that is produced by the action of reactive oxygen species (ROS) on arachidonic acid (Singh et al., 2014; Su et al., 2009). Although reactive oxygen species such as hydrogen peroxide, superoxide and hydroxyl ions are produced by the host to defend against bacterial pathogens, overproduction of ROS may lead to oxidation of DNA, lipids, and proteins contributing to tissue destruction. Molecules associated with the resolution of inflammation lipid mediators are also present in saliva, counteracting the role of pro-inflammatory signals. A new genus of eicosanoids are involved in resolution activation: *resolvins, lipoxins, maresins*, and *protectins* (Freire & Van Dyke, 2013). The ratio of leukotriene-b4 and resolvin D1, for example, has been predictive of vascular health and disease (Thul, Labat, Temmar, Benetos, & Bäck, 2017). Increasing exogenous amounts of oral eicosanoid lipoxins has demonstrated to promote periodontal regeneration in animal models (Van Dyke et al., 2015).

In addition to lipid mediators that control initiation of inflammation and resolution in periodontal pathology, urea generated mainly from protein catabolism serves as one of the buffer factors in saliva, that could release ammonia and carbon dioxide to neutralize acids produced by bacteria, and maintain the pH in saliva and dental plaque when high saliva flow is present (Humphrey & Williamson, 2001). Higher salivary urea levels could potentially protect the teeth from demineralization but on the other hand induce calculus formation, which suggests patients with higher salivary urea level, such as in chronic and terminal renal failure, receive special periodontal care (Bots et al., 2006). The by-products of urea metabolism, i.e., ammonia, could be cytotoxic to gingival tissues, increase the sulcular epithelium permeability, and may play an important role in gingivitis onset (de Almeida, Gregio, Machado, De Lima, & Azevedo, 2008).

Salivary leukocytes have previously been shown elevated in gingivitis (Aps et al., 2002) and periodontitis (Bender et al., 2006). As the most abundant leukocytes in saliva, it is estimated that 30,000 polymorphonuclear neutrophils (PMNs) cells (neutrophils) transit through periodontal tissue every minute (Darveau, 2010; Tonetti, Imboden, & Lang, 1998; Vidović et al., 2012). In the presence of inflammation, the increased number of PMNs could be examined in the periodontal tissue, GCF, and saliva (Akalın, Baltacıoğlu, Alver, & Karabulut, 2007; Ozmeric, 2004). The major biological functions of saliva PMNs include chemotaxis and generation of reactive oxygen species, and phagocytosis of bacterial pathogens (Hajishengallis, 2014). PMNs could form a barrier between the junctional epithelium and the subgingival biofilm, preventing the apical migration of the bacteria, which is essential in periodontal homeostasis (Delima & Van Dyke, 2003). Evidence implicated PMNs as important mediators of the host immune response against proliferating pathogenic microorganisms during inflammatory processes, and subsequently generated antimicrobial factors, such as ROS, in inflamed periodontal tissues (Canakci, Cicek, & Canakci, 2005). When periodontal homeostasis is interrupted, the related alteration of PMNs activities including quantity and functions could be examined as an important indicator of periodontal disease status (Landzberg, Doering, Aboodi, Tenenbaum, & Glogauer, 2015; Ozmeric, 2004). Hence, since the 1970s, salivary PMNs are often adapted in evaluating periodontal disease status and treatment outcomes (Bhadbhade, Acharya, & Thakur, 2012; Raeste & Aura, 1978).

Saliva provides an optimal buffering system for microbial acidic metabolites (Scannapieco, 1994), maintaining microbial community homeostasis in a neutral pH and a constant supply of nutrients to the environment. However, microorganisms are sensitive to minor changes in the local environment, including pH, nutrients, oxygen, moisture, and host immune responses. Changes in saliva flow rate (i.e., in dry mouth patient) or pH (i.e., during soft drink consumption) may dysregulate its buffering capacity, and over time may result in shifts in the microbial composition, and some pathogenic bacteria may become dominant in their local community (Cho & Blaser, 2012; Zarco, Vess, & Ginsburg, 2012). Additionally, while many salivary components are known to affect bacterial growth, e.g., pH modulating molecules such as bicarbonates, phosphates, and urea, salivary proteins contribute to processes associated with microbial metabolism, aggregation, and attachment of biofilms (Kolenbrander, 2011; Scannapieco, 1994). Saliva also provides necessary moisture and slightly acidic to neutral pH (6–7) for microbial growth and biofilm development (Humphrey & Williamson, 2001; Scannapieco, 1994). Hence, decreased individual salivary flow rates may promote bacterial growth, decreased fermentable products clearance around the dental plaque, and increased risks of periodontal bacterial infections (Dodds et al., 2005; Navazesh, Mulligan, Kipnis, Denny, & Denny, 1992).

### 18.2.2 Saliva and Periodontal Dysbiosis

Bacteria that colonize subgingival sites are associated with periodontal disease. Although their specific roles and homeostasis mechanisms have been a matter of debate, human microbiome studies in preclinical models suggest that periodontitis is a dysbiotic disease (i.e., imbalance in the microbial numbers) rather than infection caused by a single bacterium. As the major point of entry to the human body bacteria, archaea, protists and viruses the oral cavity hosts approximately 150 taxa per mouth and 750 taxa have been identified. Thus, saliva could provide valid information of the overall taxonomic and functional capacities of the oral microbiome (Quinque, Kittler, Kayser, Stoneking, & Nasidze, 2006). Poor oral hygiene leads to oral diseases, including mineralized tissues (caries) and to supporting tissues (periodontal tissues). Extensive studies have linked the pathology of gingivitis and periodontitis (Darveau, 2010) to certain subgingival microbial taxa, such as *Aggregatibacter actinomycetemcomitans, Bacteroides forsythus, Campylobacter rectus, Fusobacterium nucleatum, Porphyromonas gingivalis, Prevotella intermedia*, and *Treponema denticola* where *P. gingivalis* is considered as the most virulent pathogen causing chronic periodontitis (Paster et al., 2001; Pihlstrom, Michalowicz, & Johnson, 2005; Socransky, Haffajee, Cugini, Smith, & Kent, 1998; Ximénez-Fyvie, Haffajee, & Socransky, 2000). These pathobionts can induce specific host immune defenses, which may further damage the oral epithelia and periodontal connective tissues, while simultaneously inactivating repair systems from the host response (Battino, Bullon, Wilson, & Newman, 1999). The etiology of periodontal disease is not only attributed to bacterial virulence factors but also to host genetic factors and systemic conditions (Albandar, 2002). For instance, subjects with type 2 diabetes are highly susceptible to gingivitis and periodontitis (Mealey, 2006; Taylor & Borgnakke, 2008). In addition, gastric precancerous lesions are also positively correlated with the colonization of periodontal pathogens (Salazar et al., 2012). Also, herpesvirus can promote the onset of periodontal infections by elevated levels of periodontal pathogens with concomitant immunosuppression signals (Contreras & Slots, 2000; Kamma, Contreras, & Slots, 2001).

Periodontal diseases are positively correlated with increased oxidative modification of salivary DNA, lipids, and proteins in both stimulated and unstimulated saliva (Diab-Ladki, Pellat, & Chahine, 2003; Sculley & Langley-Evans, 2003; Su et al., 2009). Increased ROS levels, produced by neutrophils, modulate tissue permeability allowing salivary pathogens to invade the mucosal barrier (Canakci et al., 2005; Chapple & Matthews, 2007). 8-Hydroxy-2′-deoxyguanosine (8-OHdG) is a principal stable marker of oxidative DNA damage in saliva (Helbock, Beckman, & Ames, 1999; Su et al., 2009), which is shown to be elevated in periodontal diseases. Beyond genetic factors, low antioxidant capacity is influenced by environmental factors such as smoking cigarettes, chronic excessive exercise, and poor nutritional status (Canakci et al., 2005; Sculley & Langley-Evans, 2003). In fact, high salivary ROS could induce osteoclastic bone resorption, and lead to excessive lipid peroxides, inflammatory mediator, and oxidized proteins in periodontal tissues, which could further feed forward inflammation via host cells (Chapple, Brock, Eftimiadi, & Matthews, 2002; Chapple & Matthews, 2007; Dahiya et al., 2013).

## 18.3 Methodological Issues, Challenges, and Considerations

### *18.3.1 Evaluating Salivary Host–Microbiome Interactions*

The oral cavity represents synergistic polymicrobial networks spanning different oral biogeographic niches. The interconnected network of microbes that live in the oral cavity (oral microbiota) constitutes major portion of the total bacterial load and it is evident that the microbial signatures are host specific and are stable over time, providing unique opportunities for advancing prognosis, diagnosis, and treatment. Indeed, deeper understanding of the oral microbiome and their interaction with the host will enable revolution in personalized health care and to promote targeted clinical approaches. Commensal oral microbes are closely interlinked with health since they prevent colonization by external microbes, some of which would be potentially pathogenic. The characterization of complex host–microbe interactions is essential to gain a deeper understanding of both disease development, and mechanisms that are important in maintaining health. Interactions between the molecular components of saliva and the oral microbiome are important determinants of healthy host tissues. While biomechanical and biochemical cues control initial adhesion of microbial biofilms on hard and soft tissue surfaces, multiple biological molecules facilitate and/or protect microbial colonization (van't Hof et al., 2014). Salivary proteins (i.e., saliva pellicle) and minerals are responsible for the health-associated remineralization, buffering capacity, and hydration of the soft tissue (Lindh, Aroonsang, Sotres, & Arnebrant, 2014). Salivary mucins and proline-rich proteins help initiate bacterial colonization and biofilm formation (Douglas, 1994; Tabak, 1995). Salivary mucins are also able to aggregate and clear the bacteria from oral surfaces (Tabak, 1995). Colonizing pathogens can influence the periodontium by disrupting the epithelia, which promotes inflammation, and further stimulates a host immune response. The latter is initially aimed to respond to invading pathogens; however, if pathogen clearance is unresolved, the response can cause toxic damages inducing periodontal pathogenesis. Thus, unraveling complex host–microbe interactions responsible for regulating health versus periodontal diseases require integration of novel tools in genomics, proteomics, and bioinformatics.

Bacteria are introduced to saliva via multiple pathways, e.g., our diet, shedding from microbial biofilms from various intraoral tissues including dental surfaces, tongue, and buccal mucosa (Li et al., 2013; Zaura et al., 2009). Several groups of bacteria belonging to the *Streptococcus*, *Gemella*, *Granulicatella*, *Neisseria*, *Prevotella*, and *Veillonella* genera are shared between intraoral sites. However, some taxa are site specific, showing unique niche adaptations (Segata et al., 2012).

Studies have also reported the differentiation of composition of salivary microbiota between periodontal health and disease (Belstrøm et al., 2014; Paju et al., 2009). By using metagenomics and metatranscriptomics approaches, Belstrøm et al. (2017) characterized the salivary microbiota in subjects with periodontitis and compared to healthy controls (Fig. 18.1) and confirmed that traditional periodontal pathogens, such as *Porphyromonas gingivalis*, *Tannerella forsythia*, *Parvimonas micra*, and *Filifactor alocis* also were transcriptionally active in saliva. These findings suggest that anaerobic periodontal pathogens, whose natural habitat is in the gum pocket, also have the capacity to interact with other intraoral sites and actively spread to other tissues. Furthermore, high metabolic activity of *Filifactor alocis* and *Neisseria* sp. was associated with active periodontal host response. The most predominant and transcriptionally active plaque community members in the same study belonged to the *Streptococcus* genus, which suggests a potential role in periodontitis. Our current knowledge of periodontal pathogens is derived from only a few bacterial taxa, which likely represent only a small fraction of the true complexity. This gap of knowledge was further explored in a recent study by employing parallel whole community shotgun sequencing and machine learning on DNA samples obtained from subjects with severe periodontitis (Torres, Thompson, McLean, Kelley, & Edlund, 2018). By mapping 272 metagenomes from various human body sites against the *Candidatus Bacteroides periocalifornicus* genome, Torres and collaborators showed ubiquitous distribution of *Candidatus Bacteroides periocalifornicus* bacterium in dental plaque, as well as its strong association with the pathogenic red complex bacteria

In previous studies, the rich diversity of bacteria associated with various intraoral sites were shown to differ significantly between healthy and disease states, suggesting an overall important role of host–microbiome interactions (Ahn et al., 2012; Curtis et al., 2011; Peterson et al., 2011). The mechanism by which microbiota differ between subjects with oral squamous cell carcinoma (OSCC) and periodontal disease also showed unique taxonomic signatures associated with disease (Gholizadeh et al., 2016; Saxena et al., 2017; Schmidt et al., 2014). Thus far, such complex associations have not yet been fully elucidated beyond the characterization of the microbial community. Future studies, including influencing factors such as diet and host genetics will be required to elucidate these associations further. *For advanced genetics and epigenics relationships please see details in Chap. 6.*

### *18.3.2 Genomics Tools to Investigate the Salivary Microbiome*

Although the oral cavity is constantly subjected to external and internal stimuli, the oral microbiome remains stable over a long period of time. Indeed, changes in the microbiome profile may provide insights into mechanisms of disease initiation and progression. The advent of next-generation sequencing (NGS) technologies has revolutionized the field of microbiology, and enables the detection of myriads of uncultivable microorganisms and their genetic potentials. One of the most common

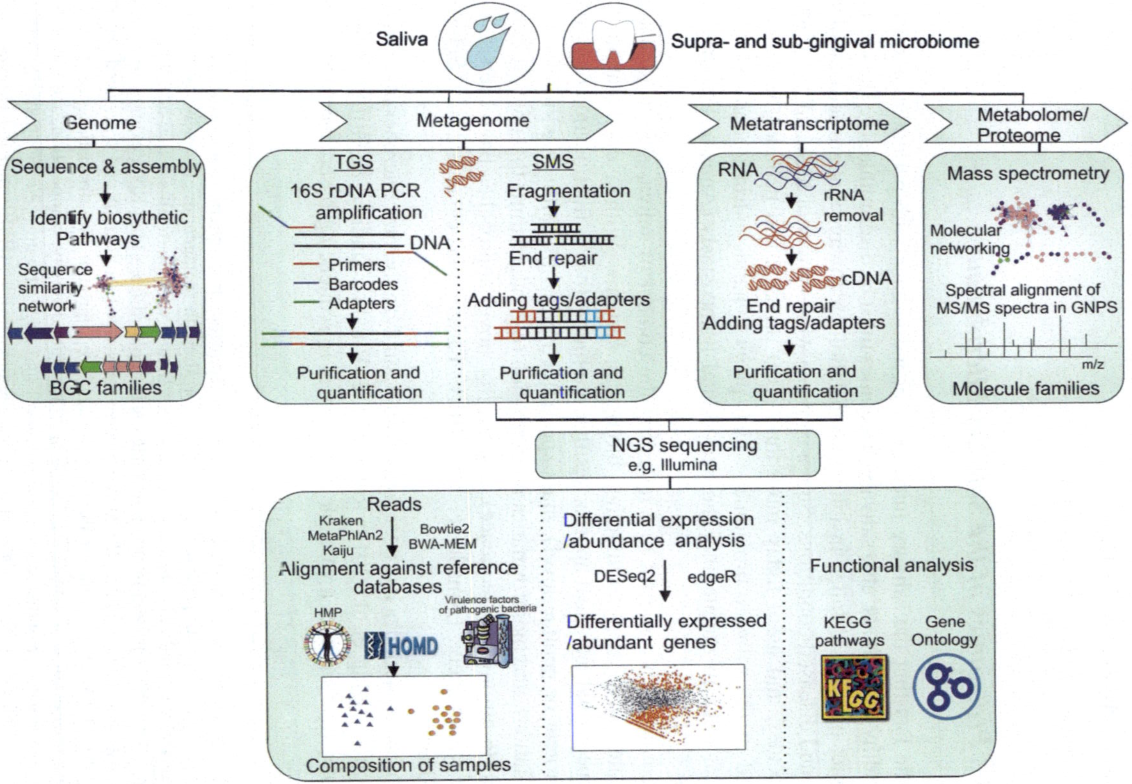


**Fig. 18.1** Omics approaches and bioinformatics tools for microbial community analysis. For target gene sequencing (TGS), i.e., sequencing of 16S rRNA gene fragments from DNA extracts of saliva and/or dental plaque. Gene fragments can be amplified with conserved 16S rRNA gene primers barcoded with short

culture-independent techniques for profiling microbial communities is based on sequencing the bacterial and archaeal taxonomic 16S rRNA marker gene and whole community shotgun sequencing of DNA (metagenomics) and mRNA (metatranscriptomics). The latter methodologies have the capacity to provide valuable snapshots of the genetic content and gene activities in any given environment (Zhou et al., 2015) (Fig. 18.1).

### *18.3.3 Proteomics of the Salivary Microbiome*

Given the fact that the salivary microbiota of individuals with periodontitis differ from healthy individuals, also salivary proteomes must be discriminatory. Therefore, proteomics provides detailed insights into metabolic pathways that are active under healthy and diseased conditions. Saliva carries a wide variety of host and microbial biomolecules, and as of today more than 3000 proteins and peptides have been identified in the salivary proteome, which plays dynamic intrinsic and extrinsic biological roles in maintaining the periodontal homeostasis (Amado, Ferreira, & Vitorino, 2013).

Proteomics is a powerful tool revealing the content of saliva, able to globally investigate saliva along with other biological systems. With a combination of multidimensional technical approaches that are innovative in fractionation (e.g., on either protein or peptide level), mass spectrometric identification (e.g., ionization, fragmentation), and bioinformatics interpretation (e.g., algorithms that convert mass spectrum to amino acid sequence), a variety of novel approaches have been summarized extensively (Amado et al., 2013; Castagnola et al., 2012; Schulz et al., 2013).

MS-based proteomics has been employed to analyze the oral microbes and saliva microbiome (Jagtap et al., 2012; Rudney, Xie, Rhodus, Ondrey, & Griffin, 2010). Metaproteomics provides a valuable complement to sequencing-based approaches, as

---

**Fig. 18.1** (continued) oligonucleotides, as well as sequencing adapters to facilitate multiple samples for sequencing (Caporaso et al., 2012). For shotgun metagenomic sequencing (SMS), community DNA is randomly sheared, end-repaired, ligated with platform-specific adapters. For sequencing of mRNA (metatranscriptomics), rRNA is depleted from total RNA extracts to enrich mRNA transcripts. The remaining RNAs are reverse transcribed into cDNA and sequenced (Zhou et al., 2015). DNA/RNA sequence libraries are analyzed through bioinformatics pipelines to address specific questions, for instance microbial community diversity, composition, and function. Typically, sequences are subjected to quality control and aligned with reference genes/genomes obtained from databases such as the Human Microbiome Project (HMP) and the Human Oral Microbiome Database (HOMD). Differential gene expression/abundance analyses, gene ontology and pathway enrichment analyses are performed using, for example, the DESeq normalization algorithm (Anders & Huber, 2010) and the Kyoto Encyclopedia of Genes and Genomes (KEGG) pathways tools. For comparative metabolomics analysis, mass spectrometry (MS)-based network infrastructures are used (e.g., the Global Natural Products Social, GNPS) (Wang et al., 2016)

it is possible to simultaneously characterize microbe as well as host-specific differences (Belstrøm et al., 2016). For example, from the teeth biofilm 7771 bacterial and 853 human proteins were identified by using a metaproteomic approach. Of them, seven bacterial and five human proteins showed 96% sensitivity to be able to discriminate healthy versus caries-affected dental plaques (Belda-Ferre et al., 2015). They have also observed that inter-person variability was significantly high, specially when evaluating before and after eating and prior to oral hygiene (Starr et al., 2018). Over 5500 human proteins along with over 2000 microbial proteins derived from 50 bacterial genera were identified and quantified (Grassl et al., 2016).

### *18.3.4 Proteomics of the Salivary Host Response*

Human saliva proteome is characterized by a high complex of proteins derived from a variety of sources (Castagnola et al., 2017), including secreted proteins by oral glands, shed epithelium cells by the oral cavity, blood proteins from wounds in oral tissue, dietary proteins from food, and microbial species residing in the oral cavity, all of which form the so-called metaproteome. Meanwhile, saliva contains DNAs, RNAs, lipids, and metabolites delivered by exosomes from distal organs (Majem, Rigau, Reventós, & Wong, 2015). In addition, saliva proteome was found to have broad dynamic range. A few dominant proteins (e.g., amylases, albumin, and IgGs) would interfere with the detection of low-abundance proteins on mass spectrometer (Amado et al., 2013). Saliva exhibits high proteolytic activity, and rapid protein degradation may occur at room temperature (Amado et al., 2013; Esser et al., 2008; Thomadaki et al., 2011), bringing great challenges to proteomic sample preparation. In addition, salivary proteins are susceptible to undergoing various posttranslational modifications (PTMs) before secretion (Sondej et al., 2009), such as glycosylation, phosphorylation, protease cleavage, sulfation, pyro-Glu conversion, and unusual N-acetylhexosamine modification which might be cancer associated (Vitorino et al., 2010). Considering the complexities mentioned above, analysis of saliva proteome is extremely challenging. Early attempts toward global analysis of human saliva by using 2D-gel electrophoresis (2-DE) and MS-based approaches only obtained less than 200 proteins (Guo et al., 2006). As biomarker proteins of many diseases tend to have notoriously low expression levels in saliva, high proteome coverage is critical for recognizing its full diagnostic potential. In this regard, multidimensional fractionation was later adapted to reduce the complexity and increase the coverage of saliva proteome. This led to the identification of over 3000 proteins (Amado et al., 2013). Secretory IgA, the dominant immunoglobulin in saliva, functions as a first line of defense in periodontal defense processes, and is claimed to reduce bacterial adherence to gingival surfaces (Gorr, 2009; Kaufman & Lamster, 2000; Seemann et al., 2004). IgM was found to work in a similar fashion as IgA, while IgG is mainly entering saliva via GCF and has been found to inhibit colonization of bacterial pathogens colonization on periodontal tissues (Brandtzaeg et al., 1997). Previous studies show that mucins work with secretory IgA and bind bacterial pathogens with greater affinity than either molecule working alone (Marcotte & Lavoie, 1998). Mucin-7 is a well-documented bactericidal and antifungal protein that binds to a wide range of oral microorganisms (Lis, Liu, Barker, Rogers, & Bobek, 2010). Cystatin C and cystatin S were proved to be able to inhibit the growth of *P. gingivalis* growth (Blankenvoorde et al., 1998). Salivary enzymes, such as lactoferrin, lysozyme, peroxidase, and amylase protect oral surfaces against microbial invasion (Scannapieco, 1994). For instance, lysozyme is the first discovered antibacterial protein in saliva that breaks down bacterial cell walls and promotes clearance (Fábián, Hermann, Beck, Fejérdy, & Fábián, 2012). Novel methods in proteomics allow for identification and quantification of novel proteins.

Using simple in-solution digestion, 2-h Liquid Chromatography (LC) gradient, and Q Exactive HF instrument, an Orbitrap-based mass spectrometer, current studies (Belstrøm et al., 2016) were able to identify 1946 proteins from saliva in the context of oral diseases, including periodontal and dental caries diseases. Similarly, new evidence has identified 3835 proteins by single LC-MS run and 5563 proteins by fractionation, revealing the unexpected complexity of saliva proteome (Grassl et al., 2016). New technological trends for saliva proteomics, including shotgun proteomic profiling, can improve in-depth proteome coverage. In addition, minor technical improvements such as ultra-long LC column separation ($\geq$40 cm), double digestion (e.g., LysC and Trypsin), and tip-based sample processing with minimal sample loss have also shown significant contributions to the enhanced proteomics analyses (Glatter et al., 2012; Kulak, Pichler, Paron, Nagaraj, & Mann, 2014; Thakur et al., 2011). Thus, proteomics is a technology-driven science, yet new advancements to improve range and specificity are needed (Lee, 2001).

#### 18.3.4.1 Comparison of Saliva and Plasma Proteome

While saliva is considered as an ultrafiltrate of blood plasma, only 27% of the proteomes are overlapping (Loo et al., 2010), thus arguing the uniqueness of saliva and the possibility of using it as a diagnostic alternative to blood tests. With the availability of the most comprehensive saliva proteome (Grassl et al., 2016) and the recent advances in plasma proteomics (Keshishian et al., 2015), we have revisited the two proteomes and provided VENN diagram comparisons where the gingival crevicular fluid (GCF) proteome also is included (see Fig. 18.2).

The GCF proteome was included in this analysis because it is a major component of human saliva, and most of GCF contents are originated from blood. We extracted the GCF proteome from one of the most comprehensive studies (Rody et al., 2014) and to ensure high confidence, we only included the proteins that were identified by at least two unique peptides from the three studies. The data indicated that 2195 (~43%) proteins are shared by saliva and plasma proteome, much higher than previously thought (Loo et al., 2010). However, such significant overlap in protein content does not correlate well to the quantitative protein levels. The estimated protein abundances in the two proteomes showed little overall correlation ($R^2 < 0.2$) (Grassl et al., 2016), which raises the concern that saliva may not be used directly to determine plasma protein levels. On the other hand, saliva proteins tend to have slightly more even distribution than plasma or urine proteins. The

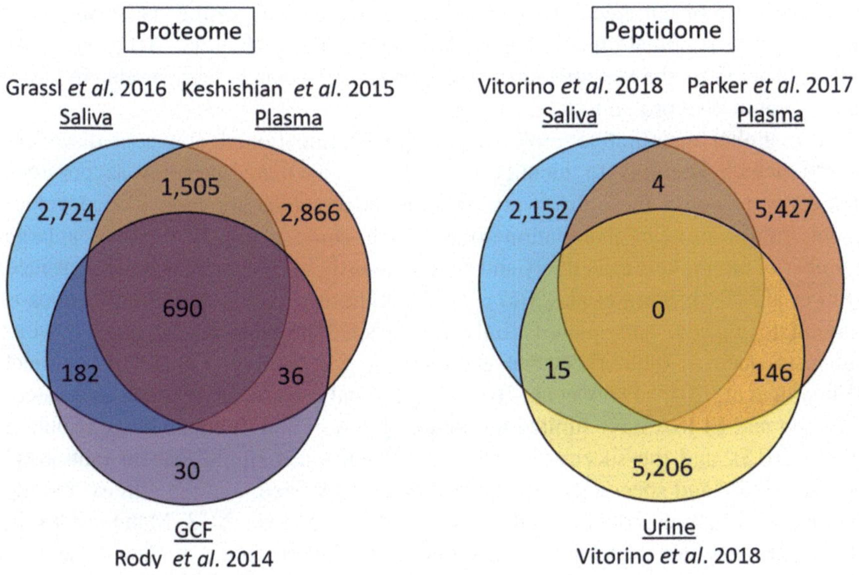


**Fig. 18.2** Venn diagram comparisons of proteomes and peptidomes of human body fluids. (Left) Comparison of the proteomes of human saliva, plasma, and gingival crevicular fluid (GCF). The data sets derived from three comprehensive studies that utilized high-resolution mass spectrometry (Orbitrap or Triple TOF) and stringent criteria (≥2 unique peptides per protein) for protein identification were extracted and reanalyzed. Protein IDs were mapped to the latest release (02/2018) of UniProt Knowledge Base. Some of the IDs were *Obsolete*, *Deleted*, or *Unmapped* after ID Mapping. Therefore, the total number of proteins used in the analysis is slightly different from the original publication. (Right) Comparison of peptidomes of saliva, plasma, and urine

dynamic range of the saliva proteome spans six orders of magnitude (Grassl et al., 2016), much less dramatic than plasma proteome ($>10^{12}$) (Anderson & Anderson, 2002). Therefore, saliva may hold the advantage over blood for the identification of less abundant proteins by mass spectrometry (Wang et al., 2017).

Quantitative proteomics holds the potential to perform unbiased characterization of diverse protein signatures in periodontal and other oral diseases, thus pinpointing to candidate biomarkers with diagnostic and prognostic values. Large-scale proteome quantitation includes two typical strategies: label based and label free. Label-based quantitation includes metabolic (e.g., stable isotope labeling using amino acids in cell culture-SILAC, $N^{15}$) and chemical labeling approaches (e.g., isotope code affinity tag—ICAT, isobaric tags for relative and absolute quantification—iTRAQ, etc.) Label-free quantitation includes spectral counting and peak area-based methods (such as MaxLFQ) (Cox et al., 2014). These methods and their application in clinical proteomics have been summarized extensively in recent articles (Ahrné et al., 2013; Megger et al., 2013; Schubert, Röst, Collins, Rosenberger, & Aebersold, 2017).

One of the advantages of label-free method is that the number of samples that can be compared is unlimited, whereas labeling quantitation has limited multiplexing capability. Only a certain number of samples can be analyzed with iTRAQ or TMT labeling strategy (Rauniyar & Yates, 2014). In clinical proteomics, a large cohort with the scale of hundreds of samples is desirable to achieve the statistical power necessary for candidate biomarker identification (Cominetti et al., 2016; Yu et al., 2014). Therefore, the versatile label-free proteomics seems to be favorable to large-scale quantitative proteomics.

The global quantitative proteomics approach mentioned above is discovery based, which attempts to identify and quantify as many proteins as possible. However, to verify these discovery findings, in particular for biomarker development, robust protein quantitation tools which could precisely measure a large number of biomarkers effectively and simultaneously in large sample sets are needed (Shi et al., 2016; Song et al., 2017). The hypothesis-driven targeted proteomics is aiming to analyze and quantify only pre-selected proteins or the ones of your interests, such as biomarkers that are found in the discovery phase of a project (Ebhardt et al., 2015; Percy et al., 2017). Traditional immunoassays such as Western Blotting and ELISA have limited multiplexing power and dynamic range (Ebhardt et al., 2015), and the success is solely dependent on a single reagent (antibody) whose quality and specificity may be poorly characterized, and the affinity and the epitope are often unknown or undisclosed (Aebersold et al., 2013; Mann, 2008). In this regard, targeted proteomics offers a promising alternative as it allows quantitation of multiple candidate markers across a wide range of samples. The approaches often refer to single, multiple, or parallel reaction monitoring (SRM/MRM/PRM). There is another quantitation strategy termed SWATH (sequential windowed acquisition of all theoretical product ion mass spectra) which combines data-independent acquisition (DIA) with targeted MS/MS data extraction approach (Gillet et al., 2012).

### *18.3.5 Saliva Peptidomics*

In recent years, tremendous efforts have been made to systematically survey the salivary peptidome in different pathological conditions in order to map disease-associated molecular pathways. More than 2000 peptides have been identified so far which constitute the salivary peptidome, of which approximately 600 are derived from salivary glands. A large number of endogenous peptides (usually <15 kDa) remain undetectable using conventional proteomic approaches. These peptides are predicted to account for approximately 40~50% of the total secreted proteins (Amado, Lobo, Domingues, Duarte, & Vitorino, 2010). However, two recent saliva proteomics studies show that peptides in saliva were found to contribute to only around 10% of the total proteome (Belstrøm et al., 2016; Grassl et al., 2016).

The peptides might be originated directly from salivary gland secretion, or from proteolysis of proteins of different sources. For instance, the microorganisms inhabiting the human mouth are known to produce a variety of enzymes that proteolytically cleave the host proteins and generate a complex of peptides (Kuboniwa et al., 2012). Previous studies estimate that the salivary peptidome consists of approximately 2000 endogenous peptides (Vitorino, 2018; Vitorino et al., 2009), which suggest that peptides are an essential component of whole saliva and may play significant roles in anti-bacterial, anti-viral, anti-pain, and wound healing (Amado et al., 2010; Jönsson, 2017; Trindade et al., 2015). This makes sense as the oral cavity is one of the main gateways for microbes to invade the human (or host) body. Consequently, saliva antimicrobial properties are important to maintain oral and bodily homeostasis (Trindade et al., 2015).

Analyzing the salivary peptidome is more technically challenging as compared to conventional shotgun proteomics analysis because endogenous peptides vary dramatically in length and physicochemical properties. In contrast to classical shotgun proteomics where every tryptic peptide is ended with a lysine (K) or arginine (R), about 29.1% of endogenous salivary peptides do not carry a positive charge (Fig. 18.3, left). Therefore, these peptides are neutral in principle and will unlikely be detected by MS. Meanwhile, around 22.1% peptides carry three or more charges (Fig. 18.3, right), which will not be easily fragmented (or MS/MS sequenced) by conventional collision methods (CID or HCD). Alternative fragmentation methods such as electron-transfer dissociation (ETD) could be explored to improve peptide identification (Mommen et al., 2014). Naturally occurring peptides will be extremely time consuming and have very high false discovery rate (FDR), especially when considering the unknown type and number of PTMs that may occur on the peptides (Menschaert et al., 2010; Schrader et al., 2014).

Although human body fluids share a significant number of proteins (as shown in Fig. 18.1, left), their peptidomes are highly unique (Fig. 18.1, right). Around 43% of the total saliva and plasma proteins overlap, whereas only four peptides are common to the two peptidomes (Parker et al., 2017). One of the possible reasons is that distinct peptides may be cleaved from different regions of the same protein (Vitorino, 2018). The data also highlight the uniqueness of the endogenous peptides in specific body fluids. Additional investigations are required to figure the enzymes that produce the peptides. Their regulations in health and disease are to be determined. For instance, a study by Cabras et al. (2010) identified the decreased level of

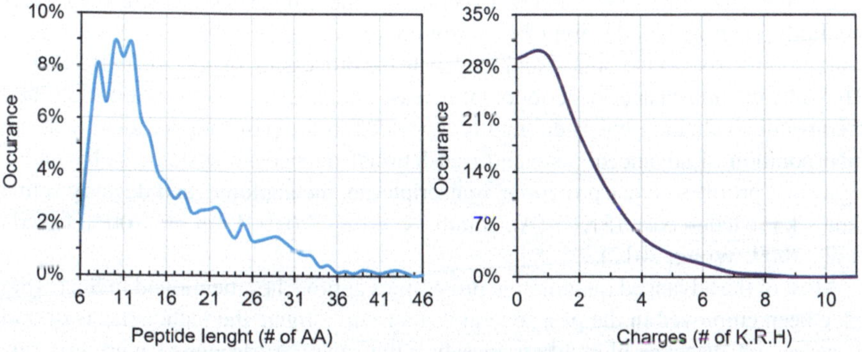


**Fig. 18.3** The length (left) and charge (right) distribution of salivary peptides. The data were retrieved from the article by Vitorino (Proteomics, 2018), and reanalyzed for the plots. In total, 2209 peptides were used for the analysis. K, R, and H stand for lysine, arginine, and histidine residues, respectively

statherin, a proline-rich peptide, and histatins in the saliva of type 1 diabetic (T1D) children as compared to control subjects. The underlying mechanism of their association with the incidence of dental and T1D diseases are still unknown. Nevertheless, it is apparent that the salivary peptidome data complements the proteomics knowledge. With the solid improvement of MS-based technologies, it is now time to combine our current knowledge to elucidate the diagnostics potential of the salivary peptides and proteins.

Increase of sensitivity and specificity of potential salivary biomarkers can provide clinical information regarding periodontal diseases. Lack of large clinical studies for downstream validation may have hampered the translation of those candidate marker proteins to clinical applications. Efficient sample preparation method that can help minimize sample loss, maximize sample recovery, and help detect low concentrations of markers with clinical relevance would be highly desirable. As in the case of exosomes released by a variety of cells, including tumor cells, for the purpose of intercellular communication, they tend to have extremely low abundance while harboring source-specific protein contents (Cocucci & Meldolesi, 2015). Therefore, to fully explore the potential of exosomal biomarkers for disease diagnosis, highly effective approaches that can deliver in-depth proteome coverage would play important roles (Keller, Ridinger, Rupp, Janssen, & Altevogt, 2011). In addition, to obtain statistically significant biomarker candidates, either targeted or untargeted large-scale proteome quantitation (for example, of hundreds of samples), with high effectiveness and sensitivity are still under development (Cominetti et al., 2016; Rosenberger et al., 2017).

## 18.4 Future Directions and Opportunities

### *18.4.1 Salivary Biomarkers*

The major advantages of saliva in experiential and clinical assessment over other biofluids, such as blood, might be the minimally invasive collection, easy access, with minimal risk to the donor, and low costs for processing (Miočević et al., 2017). The salivary bioscience in periodontal disease diagnosis, clinical monitoring, and therapeutic evaluation has been greatly advanced in the past two decades, with the incorporation of advanced basic and translational sciences to systematically investigate and profile salivary proteome, transcriptome, metabolome, and develop saliva omics knowledge base (SKB) (Ai, Smith, & Wong, 2010; Li et al., 2004; Takeda et al., 2009; Wong, 2012).

Most of the advanced quantitative proteomics approaches mentioned in Sect. 18.3 have been employed in the past several years to investigate the pathogenesis of oral diseases and discover biomarkers in saliva that may be associated with systemic diseases. For instance, Xiao et al. (2016) investigated gastric cancer by globally comparing the salivary proteome with matched control subjects. The isobaric labeling (tandem mass tag, TMT) quantitation was applied. Among the over 500 proteins that were quantified, 48 of them showed variation with significance ($p < 0.05$). Three proteins (e.g., cystatin B, triosephosphate isomerase, and deleted in malignant brain tumors 1 protein) were successfully verified by ELISA and were suggested to be potential noninvasive diagnostic biomarkers of gastric cancer with 85% sensitivity, 80% specificity, and 0.93 accuracies. Mertens and co-workers directly used MRM approach to assess 35 plasma biomarkers of periodontitis in saliva (Mertens et al., 2018). Three proteins (e.g., hemopexin, plasminogen, and α-fibrinogen) have shown significant association with periodontitis, and for the first time, Apolipoprotein H seemed to be able to differentiate chronic versus aggressive periodontitis. Therefore, the study suggested the clinical value of saliva as a noninvasive diagnostic sample of periodontitis disease. Similarly, Chen et al. evaluated the possibility of 56 salivary proteins that are known to be associated with human cancers as potential oral squamous cell carcinoma (OSCC) biomarkers (Mertens et al., 2018). The authors found that 25 of the 54 proteins were significantly different between the two groups, and thus are potential markers of oral cancer. In another study, Wu and co-workers also investigated the OSCC saliva proteome with a global label-free approach and found 33 proteins that were differentially expressed when compared with control subjects (Wu, Chu, Hsu, Chang, & Liu, 2015). However, regarding OSCC biomarker studies, only a single protein (named vitronectin) was overlapped as a common marker protein, implicating the highly complex nature of biomarker discovery studies. Future studies require to include host–pathogen interactions to identify novel relationships between the dynamic microbiome and immune systems.

#### 18.4.1.1 Salivary Biomarkers in Periodontium

Salivary biomarkers are salivary components, proteins, enzymes, and other biochemical elements, that could serve as indicators for a normal or pathogenic process, categories of disease or disease process of interest, or response to therapeutic and/or pharmacologic interventions, and are gaining popularity providing insights to periodontal etiology, pathogenesis, prognosis, and treatment outcomes (Naylor, 2003; Streckfus, 2015). As a diagnostic specimen in the dental clinic, saliva has many advantages in terms of collection, storage, shipping, and voluminous sampling, and all of these processes can be carried out economically (Giannobile, 2012). Additionally, there are compelling reasons for adapting saliva as a diagnostic tool over GCF, such as easier access, less technique sensitive in sample collection, less time consuming, and not site specific.

While many periodontal diseases often go undiagnosed until primary or secondary symptoms are recognized by the patients, the high sensitivity of oral microorganisms should forecast subtle changes in the health status and potentially serve as an indicator for early detection of diseases even before primary disease symptoms are noticed. Paired with other host saliva biomarkers (increased likelihood), the oral microbiome presents a novel noninvasive diagnostic tool for monitoring changes in human physiology and early disease detection (Streckfus, 2015). Investigations toward periodontal disease detected pathogenic bacteria paired with biomarkers in saliva showed that the bacterial composition shift along with salivary biomarkers during periodontitis disease progression (Ramseier et al., 2009; Streckfus, 2015; Zhang, Henson, Camargo, & Wong, 2009). A recent example investigating possible connections between microbial composition and disease progression revealed positive associations between periodontal pathogens (*Porphyromonas gingivalis, Prevotella intermedia, Tannerella forsythia, Fusobacterium nucleatum, Treponema denticola*, and *Campylobacter rectus*) quantity and progressed periodontitis in 82% of patients, which indicated that microbial biomarkers offer potential for the identification of periodontal disease states (Kinney et al., 2011).

Advances in high-throughput omic analysis also provided novel insights into salivary biomarkers' discovery in periodontal disease onset, progression, and therapeutic outcomes, for instance, beta-2-microglobulin found in both saliva and GCF, causing the agglutination of the bacterial pathogen, and was considered one of the indicators for periodontal inflammation activity (Ericson, 1984; Loos & Tjoa, 2005). Other proteins such as actin, profilin, hemoglobin, plastins, alpha-amylase, matrix metalloproteinases, histones, annexins, and antimicrobial peptides including histatins, S-100A8, cathelicidin-related peptides, human neutrophil peptides, statherin, and cystatins are also commonly found overrepresented in periodontal disease (Table 18.1) in proteomic analysis and thus may serve as biomarkers (Sakellari, 2017). Combining the information gained from salivary biomarkers would also provide opportunities to comprehensively profile and reveal disease status, and would be more synergistic for early disease detection and monitoring (Spielmann & Wong, 2011). *For further details on biomarkers, please see Chap. 12.*

### *18.4.2 Precision Periodontal Medicine*

In the last decade, salivary diagnostics for periodontal diseases has achieved great progress as a result of the advancement and incorporations of omics sciences, that could provide great amount of evidence and information regarding both oral microbial and human nuclear acid in dynamic physiological and pathological conditions, providing insight into the patient's genetic predisposition to periodontal diseases even before diseases onset, or if the treatment has suppressed certain etiological agent(s) (Wong, 2012). In one example, Gomes et al. (2006) adapted a salivary transcriptomic approach to detect inflammatory mRNA in type II diabetic patients' saliva to evaluate periodontal disease status. Currently, some laboratory saliva tests are available to examine the presence of certain periodontal pathogens and genetic predisposition, such as MyPerioPath® and MyPerioID PST® from OralDNA Labs (Brentwood, Tenn) that are designed to provide dental professions further information regarding their patients' periodontal status. MyPerioPath® is a salivary test that evaluates periodontal status of a patient and identifies the type and concentration of specific periodontal pathogenic bacteria in a periodontitis setting, whereas MyPerioID PST® is a salivary test that identifies individual genetic susceptibility of developing periodontal disease.

**Table 18.1** Salivary biomarkers identified in periodontal diseases

| Pathological condition | Type of sample | Methodology | Major results | Refs. |
|---|---|---|---|---|
| Aggressive periodontitis | Unstimulated WS | 2DE-ESI-MS/MS | ↑ Albumin; amylase; Ig α(alpha)2 chain C region; Ig γ(gamma)2 chain C region; vitamin D-binding protein; zinc α(alpha)2 glycoprotein ↓ Lactotransferrin; elongation factor 2; 14-3-3 sigma; SPLUNC-2; carbonic anhydrase VI | Dowling et al. (2008) |
| Chronic periodontitis | Unstimulated WS | 2DE-MALDI-MS/MS; LC-ESI-MS/MS 2DE-MALDI-MS/MS; 2DE-ESI-MS/MS; Enzyme-linked immunosorbent assay | ↑ Albumin; hemoglobin; amylase; Ig α(alpha)2 chain C region; Ig γ(gamma)2 chain C region; MMP-8; IL-1β(beta); S100A8; S100A9; S100A6; vitamin D-binding protein; zinc α(alpha)2 glycoprotein ↓ Cystatin, lactotransferrin, elongation factor 2, 14-3-3 sigma, SPLUNC-2; carbonic anhydrase VI | Dowling et al. (2008), Haigh et al. (2010), Miller, King, Langub, Kryscio, and Thomas (2006) |
| Gingivitis | Unstimulated WS | 2DE-MALDI-MS/MS; LC-ESI-MS/MS | ↑ Albumin; amylase | Gonçalves et al. (2011) |

Hence, with further improvements of salivary diagnostics precision with the advance of novel omics and bioinformatics techniques, dental professions would be offered great opportunities to provide individualized periodontal therapy (Urdea et al., 2011). Validated salivary biomarkers for periodontal disease diagnostics would become a valuable tool for the low-cost, noninvasive, and easy-to-use screening method for customization of treatment decisions, practices, and health products being tailored to individuals in periodontal applications.

## 18.5 Final Remarks

Here, we have discussed the importance of saliva in periodontal health and disease, and the importance of saliva as a source for biomarkers in precision medicine/ dentistry. Likewise, we presented novel concepts in salivary biology and the overall role of dynamic changes controlling periodontium ecology and host responses. Recent developments in point-of-care tests (POC, standardized saliva collection with functional lateral flow immunochromatographic test strips and results delivered in real time) allow for rapid precision salivary diagnostics, treatment outcome monitoring, with faster results and improvemed costs (Urdea et al., 2011). The design and development of saliva-based POC devices with minimum sample requirements are also expected to be adapted for population-wide screening programs in the future (Pfaffe, Cooper-White, Beyerlein, Kostner, & Punyadeera, 2011). One example would be (Oncescu et al., 2013) designed smartphone-based device to monitor salivary pH change over different diets. Also, "Lab-on-a-chip" revolutionized saliva-based biosensor technology, which incorporated nanotechnology, clinical chemistry, bioinformatics, microfluidics, optics, image analysis, and pattern recognition to generate an integrated measurement approach in a small device footprint, that allows for personal and private diagnosis outside laboratories, and applying it to saliva will enhance health care delivery, reduce health disparities, and improve access to care. Oral-based assays have been designed for psychological purposes, including cortisol levels as well as cardiac diagnostics, which is considered a pioneer next-generation lab-on-a-chip system with ultrasensitive multi-analyte detection capabilities in a miniaturized format (Christodoulides et al., 2005), and the related applications also include tests for periodontitis.

Finally, the rapid development of salivary molecular libraries reveals the fundamental physiological and pathological role of saliva in multiple states of health and disease. In the future, microbiome and immune connectivities embedded in saliva will shed light on the understanding of pathological pathways. The importance of salivary proteomics and peptidomics is already leading to new discoveries of reliable biomarkers with potential applicability to precision diagnostics. Integration of omic studies is now pursuing saliva as the most noninvasive and informative biofluid for medical and dental investigations. Novel dynamic, robust, rapid, portable, and user-friendly saliva diagnostic devices with the capacity to simultaneously screen multiple biomarkers will offer accurate information. The promising future of saliva bioscience in both understanding periodontal disease etiology and delivering treatments brings hope to the development of novel diagnostics and therapeutics.

## References

Aebersold, R., Burlingame, A. L., & Bradshaw, R. A. (2013). Western blots versus selected reaction monitoring assays: Time to turn the tables? *Molecular & Cellular Proteomics, 12*, 2381–2382.

Ahn, J., Chen, C. Y., & Hayes, R. B. (2012). Oral microbiome and oral and gastrointestinal cancer risk. *Cancer Causes & Control, 23*, 399–404.

Ahrné, E., Molzahn, L., Glatter, T., & Schmidt, A. (2013). Critical assessment of proteome-wide label-free absolute abundance estimation strategies. *Proteomics, 13*, 2567–2578.

Ai, J., Smith, B., & Wong, D. T. (2010). Saliva ontology: An ontology-based framework for a Salivaomics Knowledge Base. *BMC Bioinformatics, 11*(1), 302.

Akalın, F. A., Baltacıoğlu, E., Alver, A., & Karabulut, E. (2007). Lipid peroxidation levels and total oxidant status in serum, saliva and gingival crevicular fluid in patients with chronic periodontitis. *Journal of Clinical Periodontology, 34*(7), 558–565.

Albandar, J. M. (2002). Global risk factors and risk indicators for periodontal diseases. *Periodontology 2000, 29*(1), 177–206.

Amado, F. M., Ferreira, R. P., & Vitorino, R. (2013). One decade of salivary proteomics: Current approaches and outstanding challenges. *Clinical Biochemistry, 46*(6), 506–517.

Amado, F., Lobo, M. J. C., Domingues, P., Duarte, J. A., & Vitorino, R. (2010). Salivary peptidomics. *Expert Review of Proteomics, 7*, 709–721.

Anders, S., & Huber, W. (2010). Differential expression analysis for sequence count data. *Genome Biology, 11*, R106.

Anderson, N. L., & Anderson, N. G. (2002). The human plasma proteome: History, character, and diagnostic prospects. *Molecular & Cellular Proteomics, 1*, 845–867.

Aps, J. K. M., Van den Maagdenberg, K., Delanghe, J. R., & Martens, L. C. (2002). Flow cytometry as a new method to quantify the cellular content of human saliva and its relation to gingivitis. *Clinica Chimica Acta, 321*, 35–41.

Armitage, G. C. (2000). Development of a classification system for periodontal diseases and conditions. *Northwest Dentistry, 79*, 31–35.

Armitage, G. C. (2004). Periodontal diagnoses and classification of periodontal diseases. *Periodontology 2000, 34*, 9–21.

Battino, M., Bullon, P., Wilson, M., & Newman, H. (1999). Oxidative injury and inflammatory periodontal diseases: The challenge of anti-oxidants to free radicals and reactive oxygen species. *Critical Reviews in Oral Biology & Medicine, 10*(4), 458–476.

Belda-Ferre, P., Williamson, J., Simón-Soro, Á., Artacho, A., Jensen, O. N., & Mira, A. (2015). The human oral metaproteome reveals potential biomarkers for caries disease. *Proteomics, 15*, 3497–3507.

Belstrøm, D., Constancias, F., Liu, Y., Yang, L., Drautz-Moses, D. I., Schuster, S. C., . . . Givskov, M. (2017). Metagenomic and metatranscriptomic analysis of saliva reveals disease-associated microbiota in patients with periodontitis and dental caries. *NPJ Biofilms and Microbiomes, 3*, 23.

Belstrøm, D., Fiehn, N.-E., Nielsen, C. H., et al. (2014). Differences in bacterial saliva profile between periodontitis patients and a control cohort. *Journal of Clinical Periodontology, 41*, 104–112.

Belstrøm, D., Jersie-Christensen, R. R., Lyon, D., Damgaard, C., Jensen, L. J., Holmstrup, P., & Olsen, J. V. (2016). Metaproteomics of saliva identifies human protein markers specific for individuals with periodontitis and dental caries compared to orally healthy controls. *PeerJ, 4*, e2433.

Bender, J. S., Thang, H., & Glogauer, M. (2006). Novel rinse assay for the quantification of oral neutrophils and the monitoring of chronic periodontal disease. *Journal of Periodontal Research, 41*, 214–220.

Bhadbhade, S. J., Acharya, A. B., & Thakur, S. (2012). Correlation between probing pocket depth and neutrophil counts in dental plaque, saliva, and gingival crevicular fluid. *Quintessence International, 43*(2), 111–117.

Blankenvoorde, M. F. J., Van't Hof, W., Walgreen-Weterings, E., Van Steenbergen, T. J. M., Brand, H. S., Veerman, E. C. I., & Nieuw Amerongen, A. V. (1998). Cystatin and cystatin-derived peptides have antibacterial activity against the pathogen *Porphyromonas gingivalis*. *Biological Chemistry, 379*(11), 1371–1376.

Bots, C. P., Poorterman, J. H. G., Brand, H. S., Kalsbeek, H., van Amerongen, B. M., Veerman, E. C., & Nieuw Amerongen, A. V. (2006). The oral health status of dentate patients with chronic renal failure undergoing dialysis therapy. *Oral Diseases, 12*, 176–180.

Brandtzaeg, P., Berstad, A. E., Farstad, I. N., Haraldsen, G., Helgeland, L., Jahnsen, F. L., ... Rugtveit, J. (1997). Mucosal immunity—A major adaptive defence mechanism. *Behring Institute Mitteilungen, 98*, 1–23.

Brothwell, D., & Ghiabi, E. (2009). Periodontal health status of the Sandy Bay First Nation in Manitoba, Canada. *International Journal of Circumpolar Health, 68*, 23–33.

Cabras, T., Pisano, E., Mastinu, A., Denotti, G., Pusceddu, P. P., Inzitari, R., ... Messana, I. (2010). Alterations of the salivary secretory peptidome profile in children affected by type 1 diabetes. *Molecular & Cellular Proteomics, 9*, 2099–2108.

Canakci, C., Cicek, Y., & Canakci, V. (2005). Reactive oxygen species and human inflammatory periodontal diseases. *Biochemistry (Moscow), 70*(6), 619–628.

Caporaso, J. G., Lauber, C. L., Walters, W. A., Berg-Lyons, D., Huntley, J., Fierer, N., ... Knight, R. (2012). Ultra-high-throughput microbial community analysis on the Illumina HiSeq and MiSeq platforms. *The ISME Journal, 6*, 1621–1624.

Castagnola, M., Cabras, T., Iavarone, F., Fanali, C., Nemolato, S., Peluso, G., ... Messana, I. (2012). The human salivary proteome: A critical overview of the results obtained by different proteomic platforms. *Expert Review of Proteomics, 9*, 33–46.

Castagnola, M., Scarano, E., Passali, G. C., Messana, I., Cabras, T., Iavarone, F., ... Paludetti, G. (2017). Salivary biomarkers and proteomics: Future diagnostic and clinical utilities. *Acta Otorhinolaryngologica Italica: Organo Ufficiale Della Societa Italiana Di Otorinolaringologia E Chirurgia Cervico-Facciale, 37*(2), 94–101.

Chapple, I., Brock, G., Eftimiadi, C., & Matthews, J. (2002). Glutathione in gingival crevicular fluid and its relation to local antioxidant capacity in periodontal health and disease. *Molecular Pathology, 55*(6), 367.

Chapple, I. L. C., Genco, R., & Working Group 2 of the Joint EFP/AAP Workshop. (2013). Diabetes and periodontal diseases: Consensus report of the Joint EFP/AAP Workshop on Periodontitis and Systemic Diseases. *Journal of Periodontology, 84*, S106–S112.

Chapple, I. L., & Matthews, J. B. (2007). The role of reactive oxygen and antioxidant species in periodontal tissue destruction. *Periodontology 2000, 43*(1), 160–232.

Cho, I., & Blaser, M. J. (2012). The human microbiome: At the interface of health and disease. *Nature Reviews Genetics, 13*(4), 260–270.

Christodoulides, N., Floriano, P. N., Acosta, S. A., Ballard, K. L. M., Weigum, S. E., Mohanty, S., ... McDevitt, J. T. (2005). Toward the development of a lab-on-a-chip dual-function leukocyte and C-reactive protein analysis method for the assessment of inflammation and cardiac risk. *Clinical Chemistry, 51*(12), 2391–2395.

Cocucci, E., & Meldolesi, J. (2015). Ectosomes and exosomes: Shedding the confusion between extracellular vesicles. *Trends in Cell Biology, 25*, 364–372.

Cominetti, O., Núñez Galindo, A., Corthésy, J., Oller Moreno, S., Irincheeva, I., Valsesia, A., ... Dayon, L. (2016). Proteomic biomarker discovery in 1000 human plasma samples with mass spectrometry. *Journal of Proteome Research, 15*, 389–399.

Contreras, A., & Slots, J. (2000). Herpesviruses in human periodontal disease. *Journal of Periodontal Research, 35*(1), 3–16.

Cox, J., Hein, M. Y., Luber, C. A., Paron, I, Nagaraj, N., & Mann, M. (2014). Accurate proteome-wide label-free quantification by delayed normalization and maximal peptide ratio extraction, termed MaxLFQ. *Molecular & Cellular Proteomics, 13*, 2513–2526.

Curtis, M. A., Zenobia, C., & Darveau, R. P. (2011). The relationship of the oral microbiota to periodontal health and disease. *Cell Host & Microbe, 10*, 302–306.

Dahiya, P., Kamal, R., Gupta, R., Bhardwaj, R., Chaudhary, K., & Kaur, S. (2013). Reactive oxygen species in periodontitis. *Journal of Indian Society of Periodontology, 17*(4), 411–416.

Darveau, R. P. (2010). Periodontitis: A polymicrobial disruption of host homeostasis. *Nature Reviews Microbiology, 8*(7), 481–490.

de Almeida, P. D. V., Gregio, A., Machado, M., De Lima, A., & Azevedo, L. R. (2008). Saliva composition and functions: A comprehensive review. *The Journal of Contemporary Dental Practice, 9*(3), 72–80.

Delima, A. J., & Van Dyke, T. E. (2003). Origin and function of the cellular components in gingival crevice fluid. *Periodontology 2000, 31*(1), 55–76.

Diab-Ladki, R., Pellat, B., & Chahine, R. (2003). Decrease in the total antioxidant activity of saliva in patients with periodontal diseases. *Clinical Oral Investigations, 7*(2), 103–107.

Dodds, M. W., Johnson, D. A., & Yeh, C. K. (2005). Health benefits of saliva: A review. *Journal of Dentistry, 33*(3), 223–233.

Douglas, C. (1994). Bacterial-protein interactions in the oral cavity. *Advances in Dental Research, 8* (2), 254–262.

Dowling, P., Wormald, R., Meleady, P., Henry, M., Curran, A., & Clynes, M. (2008). Analysis of the saliva proteome from patients with head and neck squamous cell carcinoma reveals differences in abundance levels of proteins associated with tumour progression and metastasis. *Journal of Proteomics, 71*(2), 168–175.

Ebhardt, H. A., Root, A., Sander, C., & Aebersold, R. (2015). Applications of targeted proteomics in systems biology and translational medicine. *Proteomics, 15*, 3193–3208.

Eke, P. I., Wei, L., Borgnakke, W. S., Thornton-Evans, G., Zhang, X., Lu, H., … Genco, R. J. (2016). Periodontitis prevalence in adults $\geq$ 65 years of age, in the USA. *Periodontology 2000, 72*(1), 76–95.

Ericson, D. (1984). Agglutination of Streptococcus mutans by low-molecular-weight salivary components: Effect of beta 2-microglobulin. *Infection and Immunity, 46*(2), 526–530.

Esser, D., Alvarez-Llamas, G., de Vries, M. P., Weening, D., Vonk, R. J., & Roelofsen, H. (2008). Sample stability and protein composition of saliva: Implications for its use as a diagnostic fluid. *Biomarker Insights, 3*(February), 25–27.

Fábián, T. K., Hermann, P., Beck, A., Fejérdy, P., & Fábián, G. (2012). Salivary defense proteins: Their network and role in innate and acquired oral immunity. *International Journal of Molecular Sciences, 13*(4), 4295–4320.

Fauchard, P. (1728). *Le chirurgien dentiste, ou traité des dents… par Pierre Fauchard.*

Freire, M. O., & Van Dyke, T. E. (2013). Natural resolution of inflammation. *Periodontology 2000, 63*, 149–164.

Giannobile, W. V. (2012). Salivary diagnostics for periodontal diseases. *The Journal of the American Dental Association, 143*, 6S–11S.

Gholizadeh, P., Eslami, H., Yousefi, M., Asgharzadeh, M., Aghazadeh, M., & Kafil, H. S. (2016). Role of oral microbiome on oral cancers, a review. *Biomedicine & Pharmacotherapy, 84*, 552–558.

Gillet, L. C., Navarro, P., Tate, S., Röst, H., Selevsek, N., Reiter, L., … Aebersold, R. (2012). Targeted data extraction of the MS/MS spectra generated by data-independent acquisition: A new concept for consistent and accurate proteome analysis. *Molecular & Cellular Proteomics, 11*, O111.016717.

Glatter, T., Ludwig, C., Ahrné, E., Aebersold, R., Heck, A. J. R., & Schmidt, A. (2012). Large-scale quantitative assessment of different in-solution protein digestion protocols reveals superior cleavage efficiency of tandem Lys-C/trypsin proteolysis over trypsin digestion. *Journal of Proteome Research, 11*, 5145–5156.

Gomes, M. A., Rodrigues, F. H., Afonso-Cardoso, S. R., Buso, A. M., Silva, A. G., Favoreto, S., & Souza, M. A. (2006). Levels of immunoglobulin A1 and messenger RNA for interferon γ and tumor necrosis factor α in total saliva from patients with diabetes mellitus type 2 with chronic periodontal disease. *Journal of Periodontal Research, 41*(3), 177–183.

Gonçalves, L. d. R., Soares, M. R., Nogueira, F. C., Garcia, C. H., Camisasca, D. R., Domont, G., … Alves, G. (2011). Analysis of the salivary proteome in gingivitis patients. *Journal of Periodontal Research, 46*(5), 599–606.

Gorr, S. U. (2009). Antimicrobial peptides of the oral cavity. *Periodontology 2000, 51*(1), 152–180.

Grassl, N., Kulak, N. A., Pichler, G., Geyer, P. E., Jung, J., Schubert, S., ... Mann, M. (2016). Ultra-deep and quantitative saliva proteome reveals dynamics of the oral microbiome. *Genome Medicine, 8*, 44.

Guentsch, A., Preshaw, P. M., Bremer-Streck, S., Klinger, G., Glockmann, E., & Sigusch, B. W. (2008). Lipid peroxidation and antioxidant activity in saliva of periodontitis patients: Effect of smoking and periodontal treatment. *Clinical Oral Investigations, 12*, 345–352.

Guo, T., Rudnick, P. A., Wang, W., Lee, C. S., Devoe, D. L., & Balgley, B. M. (2006). Characterization of the human salivary proteome by capillary isoelectric focusing/nanoreversed-phase liquid chromatography coupled with ESI-tandem MS. *Journal of Proteome Research, 5*, 1469–1478.

Haffajee, A. D., & Socransky, S. S. (1994). Microbial etiological agents of destructive periodontal diseases. *Periodontology 2000, 5*, 78–111.

Haigh, B. J., Stewart, K. W., Whelan, J. R., Barnett, M. P., Smolenski, G. A., & Wheeler, T. T. (2010). Alterations in the salivary proteome associated with periodontitis. *Journal of Clinical Periodontology, 37*(3), 241–247.

Hajishengallis, G. (2014). Immunomicrobial pathogenesis of periodontitis: Keystones, pathobionts, and host response. *Trends Immunology, 35*(1), 3–11. https://doi.org/10.1016/j.it.2013.09.001

Helbock, H. J., Beckman, K. B., & Ames, B. N. (1999). 8-Hydroxydeoxyguanosine and 8-hydroxyguanine as biomarkers of oxidative DNA damage. *Methods in Enzymology, 300*, 156–166.

Holtfreter, B., Kocher, T., Hoffmann, T., Desvarieux, M., & Micheelis, W. (2010). Prevalence of periodontal disease and treatment demands based on a German dental survey (DMS IV). *Journal of Clinical Periodontology, 37*, 211–219.

Humphrey, S. P., & Williamson, R. T. (2001). A review of saliva: Normal composition, flow, and function. *The Journal of Prosthetic Dentistry, 85*(2), 162–169.

Hunter, J. (1778). *The natural history of the human teeth: Explaining their structure, use, formation, growth, and diseases*. Illustrated with Copper-plates. By John Hunter. J. Johnson.

Jagtap, P., McGowan, T., Bandhakavi, S., Tu, Z. J., Seymour, S., Griffin, T.J., & Rudney, J. D. (2012). Deep metaproteomic analysis of human salivary supernatant. *Proteomics, 12*, 992–1001.

Jönsson, D. (2017). Antimicrobial peptides: Roles in periodontal health and disease. In *Pathogenesis of periodontal diseases* (pp. 97–110).

Kamma, J. J., Contreras, A., & Slots, J. (2001). Herpes viruses and periodontopathic bacteria in early onset periodontitis. *Journal of Clinical Periodontology, 28*(9), 879–885.

Kaufman, E., & Lamster, I. B. (2000). Analysis of saliva for periodontal diagnosis. *Journal of Clinical Periodontology, 27*(7), 453–465.

Keller, S., Ridinger, J., Rupp, A.-K., Janssen, J. W., & Altevogt, P. (2011). Body fluid derived exosomes as a novel template for clinical diagnostics. *Journal of Translational Medicine, 9*, 86.

Keshishian, H., Burgess, M. W., Gillette, M. A., Mertins, P., Clauser, K. R., Mani, D. R., ... Carr, S. A. (2015). Multiplexed, quantitative workflow for sensitive biomarker discovery in plasma yields novel candidates for early myocardial injury. *Molecular & Cellular Proteomics, 14*, 2375–2393.

Kinney, J. S., Morelli, T., Braun, T., Ramseier, C. A., Herr, A. E., Sugai, J. V., ... Giannobile, W. V. (2011). Saliva/pathogen biomarker signatures and periodontal disease progression. *Journal of Dental Research, 90*(6), 752–758.

Kolenbrander, P. E. (2011). Multispecies communities: Interspecies interactions influence growth on saliva as sole nutritional source. *International Journal of Oral Science, 3*(2), 49–54.

Kuboniwa, M., Tribble, G. D., Hendrickson, E. L., Amano, A., Lamont, R. J., & Hackett, M. (2012). Insights into the virulence of oral biofilms: Discoveries from proteomics. *Expert Review of Proteomics, 9*, 311–323.

Kulak, N. A., Pichler, G., Paron, I., Nagaraj, N., & Mann, M. (2014). Minimal, encapsulated proteomic-sample processing applied to copy-number estimation in eukaryotic cells. *Nature Methods, 11*, 319–324.

Landzberg, M., Doering, H., Aboodi, G., Tenenbaum, H., & Glogauer, M. (2015). Quantifying oral inflammatory load: Oral neutrophil counts in periodontal health and disease. *Journal of Periodontal Research, 50*(3), 330–336.

Lee, K. H. (2001). Proteomics: A technology-driven and technology-limited discovery science. *Trends in Biotechnology, 19*, 217–222.

Li, K., Bihan, M., & Methé, B. A. (2013). Analyses of the stability and core taxonomic memberships of the human microbiome. *PLoS One, 8*, e63139.

Li, Y., St John, M. A., Zhou, X., Kim, Y., Sinha, U., Jordan, R. C., . . . Wong, D. T. (2004). Salivary transcriptome diagnostics for oral cancer detection. *Clinical Cancer Research, 10*(24), 8442–8450.

Linden, G. J., Herzberg, M. C., & Working Group 4 of Joint EFP/AAP Workshop. (2013). Periodontitis and systemic diseases: A record of discussions of Working Group 4 of the Joint EFP/AAP Workshop on Periodontitis and Systemic Diseases. *Journal of Clinical Periodontology, 40*(Suppl 14), S20–S23.

Lindh, L., Aroonsang, W., Sotres, J., & Arnebrant, T. (2014). Salivary pellicles. *Monographs in Oral Science, 24*, 30–39.

Lis, M., Liu, T. T., Barker, K. S., Rogers, P. D., & Bobek, L. A. (2010). Antimicrobial peptide MUC7 12-mer activates the calcium/calcineurin pathway in *Candida albicans*. *FEMS Yeast Research, 10*(5), 579–586.

Loos, B. G., & Tjoa, S. (2005). Host-derived diagnostic markers for periodontitis: Do they exist in gingival crevice fluid? *Periodontology 2000, 39*(1), 53–72.

Löe, H. (1993). Periodontal diseases: A brief historical perspective. *Periodontology 2000, 2*, 7–12.

Loe, H., Theilade, E., & Jensen, S. B. (1965). Experimental gingivitis in man. *Journal of Periodontology, 36*, 177–187.

Loo, J. A., Yan, W., Ramachandran, P., & Wong, D. T. (2010). Comparative human salivary and plasma proteomes. *Journal of Dental Research, 89*, 1016–1023.

Luan, X., Zhou, X., Trombetta-eSilva, J., Gaharwar, A. K., Atsawasuwan, P., & Diekwisch, T. G. H. (2017). MicroRNAs and periodontal homeostasis. *Journal of Dental Research, 96*, 491–500.

Majem, B., Rigau, M., Reventós, J., & Wong, D. T. (2015). Non-coding RNAs in saliva: Emerging biomarkers for molecular diagnostics. *International Journal of Molecular Sciences, 16*(4), 8676–8698.

Mann, M. (2008). Can proteomics retire the western blot? *Journal of Proteome Research, 7*, 3065.

Marcotte, H., & Lavoie, M. C. (1998). Oral microbial ecology and the role of salivary immunoglobulin A. *Microbiology and Molecular Biology Reviews, 62*(1), 71–109.

Mattila, P. T., Niskanen, M. C., Vehkalahti, M. M., Nordblad, A., & Knuuttila, M. L. (2010). Prevalence and simultaneous occurrence of periodontitis and dental caries. *Journal of Clinical Periodontology, 37*, 962–967.

Mealey, B. L. (2006). Periodontal disease and diabetes: A two-way street. *The Journal of the American Dental Association, 137*, S26–S31.

Megger, D. A., Bracht, T., Meyer, H. E., & Sitek, B. (2013). Label-free quantification in clinical proteomics. *Biochimica et Biophysica Acta, 1834*, 1581–1590.

Menschaert, G., Vandekerckhove, T. T. M., Baggerman, G., Schoofs, L., Luyten, W., & Van Criekinge, W. (2010). Peptidomics coming of age: A review of contributions from a bioinformatics angle. *Journal of Proteome Research, 9*, 2051–2061.

Mertens, B., Orti, V., Vialaret, J., Gibert, P., Relaño-Ginés, A., Lehmann, S., . . . Hirtz, C. (2018). Assessing a multiplex-targeted proteomics approach for the clinical diagnosis of periodontitis using saliva samples. *Bioanalysis, 10*, 35–45.

Miller, C. S., King, C. P., Langub, M. C., Kryscio, R. J., & Thomas, M. V. (2006). Salivary biomarkers of existing periodontal disease: A cross-sectional study. *The Journal of the American Dental Association, 137*(3), 322–329.

Miočević, O., Cole, C. R., Laughlin, M. J., Buck, R. L., Slowey, P. D., & Shirtcliff, E. A. (2017). Quantitative lateral flow assays for salivary biomarker assessment: A review. *Frontiers in Public Health, 5*, 133.

Mommen, G. P. M., Frese, C. K., Meiring, H. D., van Gaans-van den Brink, J., de Jong, A. P., van Els, C. A., & Heck, A. J. (2014). Expanding the detectable HLA peptide repertoire using electron-transfer/higher-energy collision dissociation (EThcD). *Proceedings of the National Academy of Sciences of the United States of America, 111*, 4507–4512.

Navazesh, M., Mulligan, R., Kipnis, V., Denny, P., & Denny, P. (1992). Comparison of whole saliva flow rates and mucin concentrations in healthy Caucasian young and aged adults. *Journal of Dental Research, 71*(6), 1275–1278.

Naylor, S. (2003). Biomarkers: Current perspectives and future prospects. *Expert Review of Molecular Diagnostics, 3*(5), 525–529. Taylor & Francis.

Newman, M. G., Takei, H., Carranza, F. A., & Klokkevold, P. R. (2018). *Newman and Carranza's clinical periodontology*. Amsterdam: Elsevier - Health Sciences Division.

Nguyen, T. T., Ngo, L. Q., Promsudthi, A., & Surarit, R. (2016). Salivary lipid peroxidation in patients with generalized chronic periodontitis and acute coronary syndrome. *Journal of Periodontology, 87*, 134–141.

Nomura, Y., Shimada, Y., Hanada, N., Numabe, Y., Kamoi, K., Sato, T., ... Yoshie, H. (2012). Salivary biomarkers for predicting the progression of chronic periodontitis. *Archives of Oral Biology, 57*, 413–420.

Offenbacher, S. (1996). Periodontal diseases: Pathogenesis. *Annals of Periodontology, 1*, 821–878.

Oncescu, V., O'Dell, D., & Erickson, D. (2013). Smartphone based health accessory for colorimetric detection of biomarkers in sweat and saliva. *Lab on a Chip, 13*, 3232–3238.

Ozmeric, N. (2004). Advances in periodontal disease markers. *Clinica Chimica Acta, 343*(1), 1–16.

Paster, B. J., Boches, S. K., Galvin, J. L., Ericson, R. E., Lau, C. N., Levanos, V. A., ... Dewhirst, F. E. (2001). Bacterial diversity in human subgingival plaque. *Journal of Bacteriology, 183*(12), 3770–3783.

Paju, S., Pussinen, P. J., Suominen-Taipale, L., Hyvönen, M., Knuuttila, M., & Könönen, E. (2009). Detection of multiple pathogenic species in saliva is associated with periodontal infection in adults. *Journal of Clinical Microbiology, 47*, 235–238.

Papapanou, P. N. (1996). Periodontal diseases: Epidemiology. *Annals of Periodontology, 1*, 1–36.

Parker, B. L., Burchfield, J. G., Clayton, D., Geddes, T. A., Payne, R. J., Kiens, B., ... James, D. E. (2017). Multiplexed temporal quantification of the exercise-regulated plasma peptidome. *Molecular & Cellular Proteomics, 16*, 2055–2068.

Patel, R. (2012). *The state of oral health in Europe*. Report commissioned by the Platform for Better Oral Health in Europe 2012.

Percy, A. J., Hardie, D. B., Jardim, A., Yang, J., Elliott, M. H., Zhang, S., ... Borchers, C. H. (2017). Multiplexed panel of precisely quantified salivary proteins for biomarker assessment. *Proteomics, 17*. https://doi.org/10.1002/pmic.201600230

Peterson, S. N., Snesrud, E., Schork, N. J., & Bretz, W. A. (2011). Dental caries pathogenicity: A genomic and metagenomic perspective. *International Dental Journal, 61*(Suppl 1), 11–22.

Pfaffe, T., Cooper-White, J., Beyerlein, P., Kostner, K., & Punyadeera, C. (2011). Diagnostic potential of saliva: Current state and future applications. *Clinical Chemistry, 57*(5), 675–687.

Pihlstrom, B. L., Michalowicz, B. S., & Johnson, N. W. (2005). Periodontal diseases. *The Lancet, 366*(9499), 1809–1820.

Podzimek, S., Vondrackova, L., Duskova, J., Janatova, T., & Broukal, Z. (2016). Salivary markers for periodontal and general diseases. *Disease Markers, 2016*, 9179632.

Quinque, D., Kittler, R., Kayser, M., Stoneking, M., & Nasidze, I. (2006). Evaluation of saliva as a source of human DNA for population and association studies. *Analytical Biochemistry, 353*(2), 272–277.

Raeste, A., & Aura, A. (1978). Rate of migration of oral leukocytes in patients with periodontitis. *European Journal of Oral Sciences, 86*(1), 43–51.

Ramseier, C. A., Kinney, J. S., Herr, A. E., Braun, T., Sugai, J. V., Shelburne, C. A., ... Giannobile, W. V. (2009). Identification of pathogen and host-response markers correlated with periodontal disease. *Journal of Periodontology, 80*(3), 436–446.

Rauniyar, N., & Yates, J. R., 3rd. (2014). Isobaric labeling-based relative quantification in shotgun proteomics. *Journal of Proteome Research, 13*, 5293–5309.
Rody, W. J., Jr., Holliday, L. S., McHugh, K. P., Wallet, S. M., Spicer, V., & Krokhin, O. (2014). Mass spectrometry analysis of gingival crevicular fluid in the presence of external root resorption. *American Journal of Orthodontics and Dentofacial Orthopedics, 145*, 787–798.
Rosenberger, G., Liu, Y., Röst, H. L., Ludwig, C., Buil, A., Bensimon, A., . . . Aebersold, R. (2017). Inference and quantification of peptidoforms in large sample cohorts by SWATH-MS. *Nature Biotechnology, 35*, 781–788.
Rudney, J. D., Xie, H., Rhodus, N. L., Ondrey, F. G., & Griffin, T. J. (2010). A metaproteomic analysis of the human salivary microbiota by three-dimensional peptide fractionation and tandem mass spectrometry. *Molecular Oral Microbiology, 25*, 38–49.
Sakellari, D. (2017). Proteomics of periodontal pocket. *Current Oral Health Reports, 4*(4), 271–277.
Salazar, C. R., Francois, F., Li, Y., Corby, P., Hays, R., Leung, C., . . . Chen, Y. (2012). Association between oral health and gastric precancerous lesions. *Carcinogenesis, 33*(2), 399–403.
Sanz, M., Kornman, K., & Working Group 3 of the Joint EFP/AAP Workshop. (2013). Periodontitis and adverse pregnancy outcomes: Consensus report of the Joint EFP/AAP Workshop on Periodontitis and Systemic Diseases. *Journal of Periodontology, 84*, S164–S169.
Saxena, S., Sankhla, B., Sundaragiri, K. S., & Bhargava, A. (2017). A review of salivary biomarker: A tool for early oral cancer diagnosis. *Advanced Biomedical Research, 6*, 90.
Scannapieco, F. A. (1994). Saliva-bacterium interactions in oral microbial ecology. *Critical Reviews in Oral Biology & Medicine, 5*(3), 203–248.
Schmidt, B. L., Kuczynski, J., Bhattacharya, A., Huey, B., Corby, P. M., Queiroz, E. L., . . . Albertson, D. G. (2014). Changes in abundance of oral microbiota associated with oral cancer. *PLoS One, 9*, e98741.
Schrader, M., Schulz-Knappe, P., & Fricker, L. D. (2014). Historical perspective of peptidomics. *EuPA Open Proteomics, 3*, 171–182.
Schubert, O. T., Röst, H. L., Collins, B. C., Rosenberger, G., & Aebersold, R. (2017). Quantitative proteomics: Challenges and opportunities in basic and applied research. *Nature Protocols, 12*, 1289–1294.
Schulz, B. L., Cooper-White, J., & Punyadeera, C. K. (2013). Saliva proteome research: Current status and future outlook. *Critical Reviews in Biotechnology, 33*, 246–259.
Sculley, D. V., & Langley-Evans, S. C. (2003). Periodontal disease is associated with lower antioxidant capacity in whole saliva and evidence of increased protein oxidation. *Clinical Science, 105*(2), 167–172.
Seemann, R., Hägewald, S., Sztankay, V., Drews, J., Bizhang, M., & Kage, A. (2004). Levels of parotid and submandibular/sublingual salivary immunoglobulin A in response to experimental gingivitis in humans. *Clinical Oral Investigations, 8*(4), 233–237.
Segata, N., Haake, S. K., Mannon, P., Lemon, K. P., Waldron, L., Gevers, D., . . . Izard, J. (2012). Composition of the adult digestive tract bacterial microbiome based on seven mouth surfaces, tonsils, throat and stool samples. *Genome Biology, 13*, R42.
Shi, T., Song, F., Nie, S., Rodland, K. D., Liu, T., Qian, W. J., & Smith, R. D. (2016). Advances in targeted proteomics and applications to biomedical research. *Proteomics, 16*, 2160–2182.
Singh, S., Ramesh, V., Oza, N., Balamurali, P. D., Prashad, K. V., & Balakrishnan, P. (2014). Evaluation of serum and salivary lipid profile: A correlative study. *Journal of Oral and Maxillofacial Pathology: JOMFP, 18*(1), 4.
Socransky, S., Haffajee, A., Cugini, M., Smith, C., & Kent, R. (1998). Microbial complexes in subgingival plaque. *Journal of Clinical Periodontology, 25*(2), 134–144.
Socransky, S. S., Haffajee, A. D., Goodson, J. M., & Lindhe, J. (1984). New concepts of destructive periodontal disease. *Journal of Clinical Periodontology, 11*, 21–32.
Sondej, M., Denny, P. A., Xie, Y., Ramachandran, P., Si, Y., Takashima, J., . . . Denny, P. C. (2009). Glycoprofiling of the human salivary proteome. *Clinical Proteomics, 5*, 52–68.

Song, E., Gao, Y., Wu, C., Shi, T., Nie, S., Fillmore, T. L., … Liu, T. (2017). Targeted proteomic assays for quantitation of proteins identified by proteogenomic analysis of ovarian cancer. *Scientific Data, 4*, 170091.

Spielmann, N., & Wong, D. T. (2011). Saliva: Diagnostics and therapeutic perspectives. *Oral Diseases, 17*(4), 345–354.

Starr, A. E., Deeke, S. A., Li, L., Zhang, X., Daoud, R., Ryan, J., … Figeys, D. (2018). Proteomic and metaproteomic approaches to understand host-microbe interactions. *Analytical Chemistry, 90*, 86–109.

Streckfus, C. F. (2015). *Advances in salivary diagnostics*. Berlin: Springer.

Su, H., Gornitsky, M., Velly, A. M., Yu, H., Benarroch, M., & Schipper, H. M. (2009). Salivary DNA, lipid, and protein oxidation in nonsmokers with periodontal disease. *Free Radical Biology and Medicine, 46*(7), 914–921.

Tabak, L. A. (1995). In defense of the oral cavity: Structure, biosynthesis, and function of salivary mucins. *Annual Review of Physiology, 57*(1), 547–564.

Takeda, I., Stretch, C., Barnaby, P., Bhatnager, K., Rankin, K., Fu, H., … Slupsky, C. (2009). Understanding the human salivary metabolome. *NMR in Biomedicine, 22*(6), 577–584.

Taylor, G. W., & Borgnakke, W. (2008). Periodontal disease: Associations with diabetes, glycemic control and complications. *Oral Diseases, 14*(3), 191–203.

Thakur, S. S., Geiger, T., Chatterjee, B., Bandilla, P., Fröhlich, F., Cox, J., & Mann, M. (2011). Deep and highly sensitive proteome coverage by LC-MS/MS without prefractionation. *Molecular & Cellular Proteomics, 10*, M110.003699.

Thomadaki, K., Helmerhorst, E., Tian, N., Sun, X., Siqueira, W., Walt, D., & Oppenheim, F. (2011). Whole-saliva proteolysis and its impact on salivary diagnostics. *Journal of Dental Research, 90*(11), 1325–1330.

Thul, S., Labat, C., Temmar, M., Benetos, A., & Bäck, M. (2017). Low salivary resolvin D1 to leukotriene B4 ratio predicts carotid intima media thickness: A novel biomarker of non-resolving vascular inflammation. *European Journal of Preventive Cardiology, 24*, 903–906.

Tonetti, M. S., & Dyke, T. E. (2013). Periodontitis and atherosclerotic cardiovascular disease: Consensus report of the Joint EFP/AAP Workshop on Periodontitis and Systemic Diseases. *Journal of Clinical Periodontology, 40*(s14).

Tonetti, M. S., Imboden, M. A., & Lang, N. P. (1998). Neutrophil migration into the gingival sulcus is associated with transepithelial gradients of interleukin-8 and ICAM-1. *Journal of Periodontology, 69*(10), 1139–1147.

Torres, P. J., Thompson, J., McLean, J. S., Kelley, S. T., & Edlund, A. (2018). Discovery of a novel periodontal disease-associated bacterium. *Microbial Ecology*. https://doi.org/10.1007/s00248-018-1200-6

Totan, A., Greabu, M., Totan, C., & Spinu, T. (2006). Salivary aspartate aminotransferase, alanine aminotransferase and alkaline phosphatase: Possible markers in periodontal diseases? *Clinical Chemistry and Laboratory Medicine, 44*, 612–615.

Trindade, F., Amado, F., Pinto da Costa, J., Ferreira, R., Maia, C., Henriques, I. … Vitorino, R. (2015). Salivary peptidomic as a tool to disclose new potential antimicrobial peptides. *Journal of Proteomics, 115*, 49–57.

Trivedi, S., Lal, N., Mahdi, A. A., Singh, B., & Pandey, S. (2015). Association of salivary lipid peroxidation levels, antioxidant enzymes, and chronic periodontitis. *International Journal of Periodontics and Restorative Dentistry, 35*, e14–e19.

Urdea, M., Neuwald, P., Greenberg, B., Glick, M., Galloway, J., Williams, D., & Wong, D. (2011). Saliva, diagnostics, and dentistry. *Advances in Dental Research, 23*(4), 353–359.

Van Dyke, T. E., Hasturk, H., Kantarci, A., Freire, M. O., Nguyen, D., Dalli, J., & Serhan, C. N. (2015). Proresolving nanomedicines activate bone regeneration in periodontitis. *Journal of Dental Research, 94*, 148–156.

Van Dyke, T. E., & van Winkelhoff, A. J. (2013). Infection and inflammatory mechanisms. *Journal of Clinical Periodontology, 40*(Suppl 14), S1–S7.

van't Hof, W., Veerman, E. C. I., Nieuw Amerongen, A. V., & Ligtenberg, A. J. M. (2014). Antimicrobial defense systems in saliva. In A. J. M. Ligtenberg & E. C. I. Veerman (Eds.), *Saliva: Secretion and functions* (Vol. 24, pp. 40–51). Basel: Karger.

Vidović, A., Vidović Juras, D., Vučićević Boras, V., Lukač, J., Grubišić-Ilić, M., Rak, D., & Sabioncello, A. (2012). Determination of leucocyte subsets in human saliva by flow cytometry. *Archives of Oral Biology, 57*, 577–583.

Vitorino, R. (2018). Digging deep into peptidomics applied to body fluids. *Proteomics*, 18. https://doi.org/10.1002/pmic.201700401

Vitorino, R., Alves, R., Barros, A., Caseiro, A., Ferreira, R., Lobo, M. C., . . . Amado, F. L. (2010). Finding new posttranslational modifications in salivary proline-rich proteins. *Proteomics, 10*, 3732–3742.

Vitorino, R., Barros, A., Caseiro, A., Domingues P, Duarte J, & Amado, F. (2009). Towards defining the whole salivary peptidome. *Proteomics - Clinical Applications, 3*, 528–540.

Wang, M., Carver, J. J., Phelan, V. V., Sanchez, L. M., Garg, N., Peng, Y., . . . Bandeira, N. (2016). Sharing and community curation of mass spectrometry data with Global Natural Products Social Molecular Networking. *Nature Biotechnology, 34*, 828–837.

Wang, X., Kaczor-Urbanowicz, K. E., & Wong, D. T. W. (2017). Salivary biomarkers in cancer detection. *Medical Oncology, 34*, 7.

Wong, D. T. (2012). Salivaomics. *The Journal of the American Dental Association, 143*, 19S–24S.

Wu, C.-C., Chu, H.-W., Hsu, C.-W., Chang, K. P., & Liu, H. P. (2015). Saliva proteome profiling reveals potential salivary biomarkers for detection of oral cavity squamous cell carcinoma. *Proteomics, 15*, 3394–3404.

Xiao, H., Zhang, Y., Kim, Y., Kim, S., Kim, J. J., Kim, K. M., . . . Wong, D. T. (2016). Differential proteomic analysis of human saliva using tandem mass tags quantification for gastric cancer detection. *Scientific Reports, 6*, 22165.

Ximénez-Fyvie, L. A., Haffajee, A. D., & Socransky, S. S. (2000). Microbial composition of supra- and subgingival plaque in subjects with adult periodontitis. *Journal of Clinical Periodontology, 27*(10), 722–732.

Yu, Y., Suh, M.-J., Sikorski, P., Kwon, K., Nelson, K. E., & Pieper, R. (2014). Urine sample preparation in 96-well filter plates for quantitative clinical proteomics. *Analytical Chemistry, 86*, 5470–5477.

Zarco, M., Vess, T., & Ginsburg, G. (2012). The oral microbiome in health and disease and the potential impact on personalized dental medicine. *Oral Diseases, 18*(2), 109–120.

Zaura, E., Keijser, B. J. F., Huse, S. M., & Crielaard, W. (2009). Defining the healthy "core microbiome" of oral microbial communities. *BMC Microbiology, 9*, 259.

Zhang, L., Henson, B. S., Camargo, P. M., & Wong, D. T. (2009). The clinical value of salivary biomarkers for periodontal disease. *Periodontology 2000, 51*(1), 25–37.

Zhou, J., He, Z., Yang, Y., Deng, Y., Tringe, S. G., & Alvarez-Cohen, L. (2015). High-throughput metagenomic technologies for complex microbial community analysis: Open and closed formats. *MBio, 6*. https://doi.org/10.1128/mBio.02288-14

# Chapter 19
# Salivary Bioscience and Cancer

**Mahvash Navazesh and Sibel Dincer**

**Abstract** Saliva is a fountain of opportunities. It plays a significant role in health and disease due to its versatile composition and distinct function. With the advancements of technology and research methodologies, saliva is gaining larger recognition for its vital role in risk assessment and disease prevention.

Recently salivary molecular, microbial and immunologic biomarkers have been evaluated and tested as potential screening and diagnostic tools for sundry of infectious, neoplastic, inflammatory, and hereditary conditions. Due to its noninvasive and cost-effective nature, many scholars have begun to focus on the utilization of saliva as a potential diagnostic tool for systemic diseases.

Saliva is considered to be an ultra-filtrate of the blood and its composition is impacted by different diseases. Salivary proteome, transcriptome, metabolome, and microbiome have been introduced in relevant literature in recent years.

According to World Health Organization (WHO) cancer is the second leading cause of death globally. Lung, prostate, colorectal, stomach, and liver cancer are the most common types of cancer in men, while breast, colorectal, lung, cervix, and thyroid cancer are the most common among women. One of the major reasons for high mortality rate associated with cancer is failure to make early diagnosis. As bimolecular changes precede clinical signs of cancer, identification of cancer-specific biomarkers in bodily fluids such as serum and saliva is crucial for early diagnosis, closer monitoring of the treatment process, and better cancer surveillance. Salivary biomarkers have received more visibility in recent years for risk assessment of and susceptibility to different types of cancer. Changes in certain salivary metabolite levels, presence of aberrant noncoding RNAs (lncRMA, miRNA, mRNA), circulating tumor DNAs (ctDNA), aberrant DNA methylation, and changes

M. Navazesh (✉)
Division of Periodontology, Diagnostic Sciences & Dental Hygiene, Herman Ostrow School of Dentistry of University of Southern California, Los Angeles, CA, USA
e-mail: navazesh@usc.edu

S. Dincer
Division of Biomedical Sciences, Herman Ostrow School of Dentistry of University of Southern California, Los Angeles, CA, USA
e-mail: dincer@usc.edu

 
D. A. Granger, M. K. Taylor (eds.), *Salivary Bioscience*,
https://doi.org/10.1007/978-3-030-35784-9_19

in peptide expressions have been studied for better understanding of their relation to cancer pathophysiology and for their utilization as salivary biomarkers for early diagnosis, monitoring of the treatment, and the prognosis of many different types of cancer.

This chapter will focus on a systematic review of the available scientific literature and evidence-based information, published between 2000 and 2018 on salivary biomarkers and cancer. Outcomes of 104 clinical investigations, demonstrating the abundance of available information on potential salivary biomarkers for head and neck, breast, pancreatic and lung cancers, and evolving information on less-studied salivary biomarkers of ovarian, prostate, brain, hematological, as well as hepatocellular cancers will be presented. The specificity and sensitivity, as well as diagnostic clinical validity of studied biomarkers will be discussed. Available and evolving technologies are also briefly discussed, and the future needs are assessed.

## 19.1 Introduction

Known for its diverse and unique composition, saliva has been considered as a "mirror of the body" (Mandel, 1993). Constituents of saliva are being studied for their potential use for diagnostic and screening purposes, because saliva can be obtained easily, repeatedly, noninvasively, and cost effectively (Slavkin, 1998).

Saliva was mainly utilized for monitoring the concentration and presence of drugs of misuse, steroid hormones, antibodies, and antigens before the 1980s. Interest toward better understanding and assessing the value of saliva as a potential diagnostic aid increased rapidly in the late 1980s and early 1990s in a wide array of disciplines like dentistry, biochemistry, clinical chemistry, molecular biology, and immunodiagnostics (Malamud & Tabak, 1993).

In 1993 the National Institute of Dental Research supported the first broad-based meeting, organized by the New York Academy of Sciences, where researchers from different disciplines evaluated the past progress in salivary research and set future directions and recommendations (Mandel, 1993).

With the rapid development of technology and research methodology, the number of basic, translational, and clinical science research on the characterization of salivary content, salivary biomarkers, salivary physiology, and potential utilization of saliva for screening and diagnostics skyrocketed in the early 2000s. A brand new ontology of "saliva omics" (Yan et al., 2008) was developed. This term includes genomics and epigenomics (studying genes and their methylation), transcriptomics (studying RNAs), metabolomics (studying metabolites), proteomics (studying proteins), and microbiomics (studying pathogens). Characterizations of salivary transcriptome, RNA, and proteome were accomplished through multicenter research cooperation (Denny et al., 2008; Helmerhorst & Oppenheim, 2008; Hu et al., 2008; Li, St John, & Zhou, 2004, Li, Zhou, St John, & Wong, 2004; Spielman & Wong, 2011).

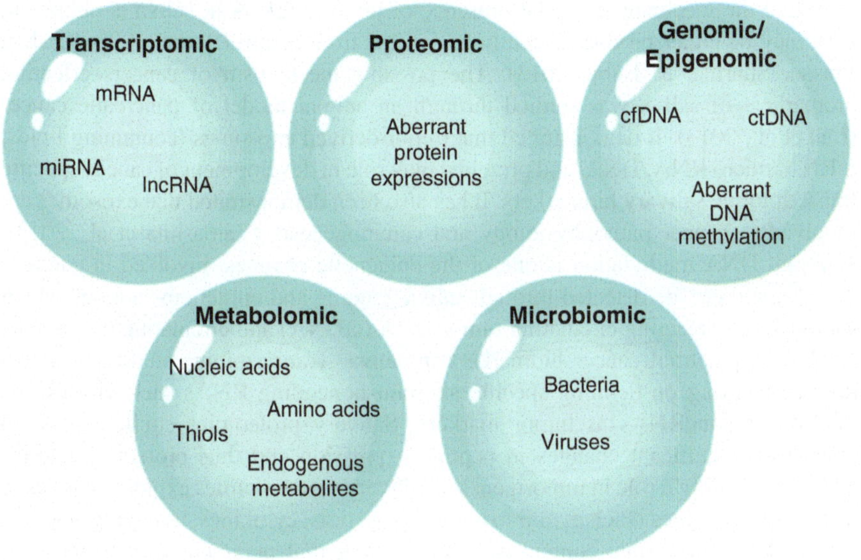


**Fig. 19.1** Salivary biomarkers

Furthermore, collaborative research among teams of engineers and scientists with expertise in nanotechnology microfluidic and molecular biology and biochemistry focused on the development of portable point of care diagnostic platforms for rapid detection and analysis of oral biomarkers and potential advancements in salivary diagnostics and screening for systemic diseases such as tuberculosis, malaria, HIV, chronic obstructive pulmonary disease, acute myocardial infarction, cystic fibrosis, and oral cancer (Malamud, 2011a, 2011b).

On the other hand, profound lessons were learned on further understanding and treating human disease, for example cancer, through the impact of the Human Genome Project that was completed in 2003. The evolving scientific information significantly influenced the way cancer is researched and understood (Wheeler, 2013), opening new horizons in cancer early diagnostics and monitoring. Greater interest in utilization of salivary biomarkers in cancer diagnostics and screening emerged simultaneously, due to saliva's potential as a liquid biopsy, targeting early diagnostics and prognostic screening for improvement of cancer survival.

Cancer is a major cause of morbidity and mortality, based on the latest World Cancer Report (Stewart & Wild, 2014), presenting GLOBOCAN study data reports, revealing an increase of cancer. The incidence of cancer has increased from 12.7 million in 2008 to 14.1 million in 2012 and is projected to reach to 25 million cases over the next two decades.

Since bio molecular and genetic alterations precede clinical symptoms of cancer, many cancer cases are diagnosed in late stages of the disease. Discovering, validating, and testing bio molecular markers in body fluids such as saliva (Fig. 19.1) will significantly contribute to early detection, recognition, and management of cancer.

While the mechanism for the presence of tumor markers in saliva is unknown, they may be either produced locally or derived from serum (Baum, 1993; Krishna Prasad, Sharma, & Babu, 2013). The possible mechanism of cancers effecting biomarkers of saliva was studied through an animal model of pancreatic cancer (Lau et al., 2013). It demonstrated that tumor-derived exosomes (containing lipids, mRNA, micro RNA, DNA, and proteins) play role in development of cancer-specific transcriptomic salivary biomarkers. It has also been demonstrated that exosomes are involved in cancer pathophysiology and carcinogenesis (Yamashita et al., 2013). Aberrant DNA methylation is one of the epigenetic changes, involved in carcinogenesis, and can be detected through saliva genome and epigenome studies which also work on detecting circulating tumor DNA (ctDNA) and circulating tumor cells (CTCs) as potential cancer biomarkers in saliva. Transcriptome studies of saliva involve working on defining specific aberrant noncoding RNAs such as miRNA, mRNA, and lncRNAs as tumor markers. Salivary proteomics studies work on identifying significant changes in peptide expression and thus protein structures, which are playing a role in tumorigenesis. Salivary metabolomics explore changes in levels of metabolites (such as amino acids, lipids, tiols, cytokines, immunoglobulins, etc.) and the metabolite content of saliva as potential parameters to diagnose or monitor cancer. Sugimoto, Wong, Hirayama, Soga, and Tomita (2010) conducted detailed salivary metabolite analysis in patients with oral, pancreatic or breast cancer as compared with healthy controls. Through this study, 57 salivary metabolites were found to have predictive value for specific diseases. Salivary microbiomics studies are designed to identify bacteria, viruses, and microorganisms involved in the pathogenesis of systemic diseases including cancers.

In this chapter, we present the current state of knowledge accumulated through research on potential utilization of salivary biomarkers in cancer diagnostic sciences. For this purpose, we conducted a PubMed review of clinical research published on salivary biomarkers of cancers between 2000 and 2018 in English language. We present available information, related to each group of cancers, based on the cancer categories presented by the latest WHO classification of cancers, organized by organs. Our review demonstrates the abundance of available information on potential salivary biomarkers for head and neck, breast, pancreatic and lung cancers, and evolving information on less-studied cancers such as ovarian, prostate, brain, hematological, as well as hepatocellular cancers (Fig. 19.2).

## 19.2 Brain Cancer

Suma et al. (2010) studied ferric reducing antioxidant power (FRAP) and protein thiol as indicators of oxidative damage and thus potentially useful in monitoring brain cancer prognosis through unstimulated whole saliva analysis. The mean FRAP values were found to be significantly lower while the mean protein thiol values were significantly higher in all patients with brain tumors. No significant difference in salivary FRAP or protein thiol were observed.

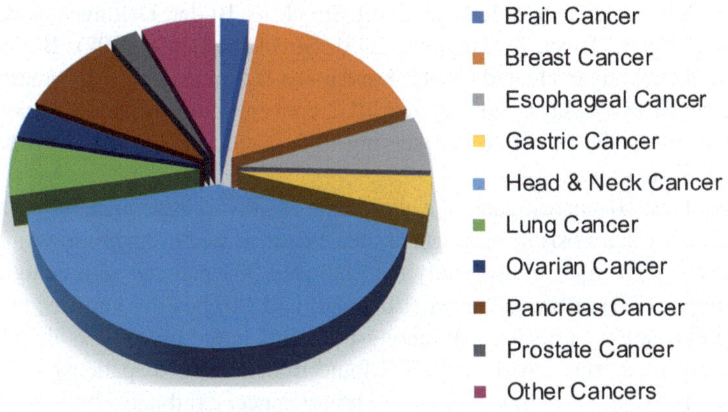


**Fig. 19.2** Proportion of salivary biomarker studies

Fischer, Attenhofer, Gultekin, Ross, and Heinimann (2017) investigated the prevalence of *TRPS1* gene alterations in subependymomas by investigating saliva and tumor tissue samples of two unrelated subependymoma patients with autosomal dominant TRPS and six patients with sporadic solitary subependymomas; the study results indicated a prominent role of *TRPS1 gene alterations* in subependymoma development.

Daubenbuchel et al. (2016) performed oxytocin analysis in the saliva of childhood-onset craniopharyngioma survivors and healthy controls. This study reveals that patients with hypothalamic surgical lesions grade 1 present with a significant lower fasting level of oxytocin.

## 19.3 Breast Cancer

Salivary markers evaluated in breast cancer patients include proteins (Bigler et al., 2002; Brooks et al., 2008; Laidi et al., 2014; Streckfus et al., 2009; Wood & Streckfus, 2015; We, Gong et al. 2015), métabolites (Jinno, Murata, Sunamura, & Sugimoto, 2015; Ozturk, Emekli-Alturfan, & Kasikci, 2011; Sugimoto et al., 2010; Zhong, Cheng, Lu, Duan, & Wang, 2016), and carbohydrates (Liu et al., 2018). CA 15-3 and c-erb-B-2 are the most frequently assessed salivary biomarkers. Protein markers are the most frequently assessed type of markers, followed by metabolites and carbohydrates. CA 15-3 is a breast cancer marker that has been found to be correlated with tumor size and could predict bone and liver metastases (Frenette et al., 1994).

CA 15-3 is also a biomarker, approved by FDA for monitoring the metastasis of breast cancer (Füzéry, Levin, Chan, & Chan, 2013). Saliva concentrations of protein CA 15-3 were significantly higher in breast cancer patients than controls (Agha-

Hosseini, Mirzaii-Dizgah, & Rahimi, 2009; Streckfus, Bigler, Dellinger, et al., 2000; Streckfus, Bigler, Tucci, & Thigpen, 2000; Streckfus et al., 2008). Bigler et al. (2002) in their study evaluated CA 15-3 and c-erb-B-2 in patients with breast cancer performing ELISA essays. Salivary c-erbB-2 concentrations were significantly different indicating the possibility of serving as a useful aid in evaluating patient's responses to treatment.

Liu et al. in 2018 investigated the alterations of salivary glycopatterns as possible biomarkers for diagnosis of early-stage breast cancer. Lectin microarrays and blotting analysis were used in this study that included 337 patients with benign breast cyst or tumor (BB) or breast cancer (I/II stage) and 110 healthy humans, providing critical information to distinguish amongst patients based on precise alterations in salivary glycopatterns, which may contribute to screening of patients with early-stage BC. The diagnostic model of stage I breast cancer exhibited a high accuracy of 0.902 in a double-blind cohort.

We, Gong et al. in 2015 explored the diagnostic value of salivary protein markers in detecting benign and malignant breast tumors by applying surface-enhanced Raman spectroscopy (SERS).The diagnostic accuracy rates of 92.78% for healthy, 95.87% for benign, and 88.66% for malignant breast tumors were obtained.

Studies exploring free amino acids and metabolomics were also performed with the intention of finding early breast cancer diagnosis markers. Zhong et al. in 2016 analyzed entire salivary metabolome for potential breast cancer diagnosis and staging biomarkers. LysoPC (18:1), LysoPC (22:6), and MG (0:0/14:0/0:0) were found to display the AUC values of 0.920, 0.920, and 0.929.

Cheng, Wang, Huang, Duan, and Wang in 2015 reported significant differences in 15 salivary free amino acids in 27 individuals with breast cancer as compared to 28 healthy adults.

## 19.4 Esophageal Cancer

Multiple transcriptomic studies are designed and conducted in order to specify the roles of different miRNA in esophageal cancer pathogenesis. Ye, Ye, Zhang, Rao, and Xie (2014) studied miRNA21 expression and found miRNA21 to be significantly upregulated in esophageal cancer groups compared to control groups. Li, Yu, Shi, Li, and Fu (2015) reported that higher levels of miRNA are correlated with poor prognosis in patients with esophageal carcinoma. Wu et al. (2013) found miR144 levels in saliva samples of patients with esophageal carcinoma to be significantly higher compared to the control group. Xie et al. (2013) found upregulated expressions of six miRNA (miR144, miR-10b, miR-451, miR486-5p, miR-634, and miR-21). After validation with RT-qPCR MiR-144, miR-10b, and miR-451 were found to be significantly upregulated in whole saliva whereas miR-10b, miR-144, miR-21, and miR-451 were significantly upregulated in saliva supernatant. Du and Zhang (2017) studied 18 differentially expressed salivary miRNAs of EC patients. Findings of this investigation revealed that miR-144, miR-451, miR-98, miR-10b, and miR-363 may be involved in esophageal cancer through regulating target genes *NOTCH*, brinogen, *AKT1*, *MMPs*, *PPARA*, *KLF4*, MYC, *E2F1*, *E2F2*, and *CDKN1A*.

## 19.5 Gastric Cancer

Proteomic, transcriptomic, genomic, and microbiomic salivary markers were studied in patients with gastric cancer. Wu, Wang, and Zhang in 2009 identified four discriminatory salivary proteins (1472.78, 2936.49, 6556.81, and 7081.17 Da) using mass spectrometry with 95.65% sensitivity and 100% specificity.

Xiao et al. (2016) identified and quantified more than 500 proteins; among those 48 showed significant expression in gastric cancer patients compared to controls. The presence of a combination of three of those specific proteins led to 85% sensitivity and 80% specificity with accuracy of 0.93.

Yao et al. (2017) conducted a genomics study to observe genetic variations of Epstein Barr Virus strains from patients diagnosed with carcinoma and healthy donors and identified genetic variants and genes associated with gastric carcinoma.

In their metabolomics study, Zilberman and Sonkusale (2015) identified clinically significant levels of $CO_2$ and $NH_3$ at ppm levels. Both $CO_2$ and $NH_3$, being Pylori metabolites, provide room for consideration on the potential contribution of *H. pylori* in gastric cancer pathophysiology and points toward the need for further microbiomic or genomic studies to be performed.

## 19.6 Head and Neck Cancer

Many studies have focused on squamous cell carcinoma biomarkers in bodily fluids such as plasma, serum, oral rinses, and saliva. Salivary tumor markers of head and neck cancers include genomic and epigenomic markers, such as mutated tumor suppressor genes and oncogenes, viral-induced DNA alterations resulting from high-risk HPV and EBV involvement in tumorigenesis, tumor-specific DNA (ctDNA), and DNA methylation profiles (Ahn et al., 2014; Carvalho et al., 2011; Guerrero-Preston et al., 2011; Lim et al., 2016; Ovchinnikov et al., 2012; Rettori et al., 2013; Righini et al., 2007; Sun et al., 2012; Wang et al., 2015). Transcriptomic markers of head and neck cancer include coding RNAs (mRNA) (Bu, Bu, Liu, Chen, & Chen, 2015; Chai et al., 2016; Elashoff et al., 2012; Li, St John, et al., 2004; Michailidou et al., 2016), noncoding RNAs such as miRNAs (Duz et al., 2016; Liu et al., 2010; Momen-Heravi, Trachtenberg, Kuo, & Cheng, 2014; Park et al., 2009 Salazar et al., 2014; Wiklund et al., 2011), and long-noncoding RNAs (Tang, Wu, Zhang, & Su, 2013). Salivary protein biomarkers of head and neck cancer consist of nucleic acids and proteinaceous analytes (Allegra et al., 2014; Aggarwal et al., 2015; Camisasca et al., 2017; Chen et al., 2017; Jarai et al., 2012; Kawahara et al.,

2016; Krapfenbauer, Drucker, & Thurnher, 2014; Li, Yang, Jin, Cai, & Sun, 2016; Pereira et al., 2016; Vidotto, Henrique, Raposo, Maniglia, & Tajara, 2010; Wu, Chu et al., 2015; Yu et al., 2016). Microbiomic markers include inhabitants of the oral microflora, possibly predisposing the site to disease, considered as diagnostic or prognostic indicators (Guerrero-Preston et al. 2017; Mager et al., 2005). Salivary biomarkers of head and neck cancers have been studied for their potential value in diagnosis of head and neck cancers and for their association with the prognosis, disease monitoring, and monitoring of the treatment efficacy.

Li, St John, et al. (2004) identified seven cancer-related mRNA biomarkers that exhibited at least 3.5-fold elevations in saliva of oral squamous cell carcinoma patients, the combination of four of those salivary mRNAs (OAZ, SAT, IL8, and IL1B) had 91% specificity and 91% sensitivity for oral cancer identification. Elashoff et al. in 2012 evaluated 395 subjects and found that expression of same seven mRNAs (IL8, IL1B, DUSP1, HA3, OAZ1, S100P, and SAT) was increased in OSCC patients compared to controls with elevations of IL-8 and SAT showing the best sensitivity and specificity. Michailidou et al. (2016) evaluated the presence of extracellular IL-8, IL-1B, OAZ, and SAT miRNAs in 54 patients with primary OSCC, oral leukoplakia and dysplasia, and healthy control. The combination of those four mRNAs had predictive ability of 80% for patients with OSCC. The combination of SAT and IL8 demonstrated a predictive ability of 75.5%, confirming the value of those markers as a prediction model for OSCC. Park et al. in 2009 found statistically significant downregulation of miR-125a and miR-200a in patients with oral cancer. On the other hand, Wiklund et al. (2011) described a panel of aberrant miR-375 and miR-200a expression and miR-200c-141 methylation being significantly present in saliva and oral rinse of oral cancer patients compared to healthy individuals. More than 700 salivary miRNA were tested by Momen-Haravi et al. in 2014 using NanoString nCounter technology. Eleven miRNA were significantly downregulated and two miRNAs were found to be significantly upregulated in OCC patients as compared to the control group. Salazar et al. in 2014 tested five miRNAs in saliva of HNSCC patients and healthy controls and reported that miR-9, miR-191, and miR-134 can serve as salivary biomarkers for HNSCC with great discriminatory capacity.

Tumor suppressor gene-promoter DNA methylation panel of DAPK1, P16MSP, and RASF1A was able to detect tumor presence with 81% accuracy in saliva of HNSCC patients and sensitivity of 94% and specificity of 87% compared to healthy nonsmoker controls (Ovchinnikov et al., 2012). Guerrero-Preston et al. (2011) conducted a Phase I Biomarker Development Trial identifying differentially methylated genes HOXA9 and NID2 as highly discriminative for OSCC. Lim and colleagues (Lim et al., 2016) studied DNA methylation levels of a five suppressor genes panel and reported high methylation levels in HPV(–) HNSCC patients and low methylation levels in HPV(+) HSCC patients. Wang et al. in 2015 tested saliva samples for somatic DNA mutations or HPV genes (HPV16 and HPV18). Tumor DNA was found in saliva samples of 100% of patients with oral cancer and 47–70% of patients with cancers of other sites in head and neck. Furthermore, tumor DNA was found in three patients before their clinical recurrence diagnoses after curative treatment, but not found in patients without recurrence.

Krapfenbauer et al. (2014) identified 25 protein expression levels to be altered in saliva and validated GAL-7 as a potential screening marker with 90% specificity and 80% sensitivity. Another salivary proteome study was able to validate the RETN protein as salivary OSCC marker showing RETN levels to be significantly high in OSCC patients compared to controls. Elevated levels of RETN were also found to be correlated with late stage primary tumors, advanced overall stage, and lymph node metastasis (Wu, Chu et al. 2015). Liu et al. (2010) analyzed the levels of miR-31 in saliva of patients with OSCC, oral premalignant lesions, and healthy controls and found those levels to be significantly increased in all stage OSCC patients compared to controls and patients with premalignant oral lesions. MiR-31 levels were observed to be remarkably reduced upon surgical removal of OSCC lesions, suggesting this marker to be tumor associated. Duz et al. (2016) found expression levels of miR-139-5p to be significantly reduced in OSCC patients compared to the controls, and restored upon surgical removal of the primary tumor. Carvalho et al. (2011) was able to show significant association of DNA methylation and poor local control and overall poor survival, analyzing a pretreatment salivary DNA methylation panel (CCNA1, DAPK, DCC, MGMT, MINT31). Promoter hypermethylation of TIMP3 in oral rinse has been found to be significantly associated with recurrence-free survival in HNSCC patients than in those without hypermethylation (Sun et al., 2012).

## 19.7 Lung Cancer

Genomics, transcriptomic, and proteomic studies have focused on upregulated genes, aberrant mRNA, and characteristic proteins in the saliva of lung cancer patients. An electrochemical sensor was developed for EGFR mutation detection (Wei et al., 2014). Xiao et al. (2012) detected three upregulated proteins (HP, AZGP1, and human Cal protein) to be significantly higher in lung cancer patients than in healthy controls with 88.5% sensitivity and 92.3% specificity with an AUC of 0.90. Li, Yang, and Lin (2012) found nine bimolecular peaks presenting significant differences between cancer patients and controls. Those amino acids and nucleic acids detected with 78% specificity and 83% sensitivity represented potential for salivary diagnosis. Zhang et al. (2012) performed transcriptomics analysis for pre-validation of seven upregulated genes.

Sun, Liu, and Qiao (2017) conducted a proteomics study aiming to systematically compare the protein profiling in saliva and serum exosomes, and 319 and 994 exosomal proteins were identified from saliva and serum, respectively. Results revealed that around 80% of saliva exosomal proteins were shared with serum exosomes. In particular, 11 potential candidates were coincidently discovered in both body fluids of lung cancer patients. Four of those proteins: Alpha-1-acid glycoprotein 1(A1AG1), Saliva acidic proline-rich phosphoprotein 1/2(PRPC), Aquaporin-5(AQP5), and Mucin-5B (MUC5B) were found to be highly associated with lung cancer. Sun et al. (2018) also conducted salivary extracellular vesicle protein profiling in normal subjects and lung cancer patients. 25 and 40 proteins, originally from distal organ cells, were found in the salivary exosomes and micro vesicles of lung cancer patients, respectively. In particular, 5 out of 25 and 9 out of 40 were lung-related proteins. BPIFA1, CRNN, MUC5B, and IQGAP were confirmed either in salivary microvesicles or in exosomes of lung cancer patients.

## 19.8 Ovarian Cancer

Chen, Schwartz, and Li (1990) compared potential diagnostic value of salivary and serum CA-125 in patients with pelvic tumors. A serum CA-125 value greater than 65 U/mL and a saliva CA-125 value greater than 3000 U/mL were defined as the positivity criteria. The sensitivities of the saliva and serum CA-125 assays in 16 patients with epithelial ovarian cancer were 81.3% and 93.8%, respectively. The false-positive rates of serum CA 125 in patients with endometriomas and pelvic tuberculosis were 72.7% and 80%, respectively, and the false-positive rates for saliva were only 13.6% and 10%. It was, therefore, concluded that the saliva CA-125 assay had a better diagnostic value than serum.

Lee et al. (2013) performed a salivary transcriptome analysis and clinically validated seven downregulated mRNA markers through RT-qPCR. The combination of five validated biomarkers (AGPAT1, B2M, BASP2, IER3, and IL1B) significantly discriminated ovarian cancer patients from the healthy controls, yielding a receiver operating characteristic plot area under the curve value of 0.909 with 85.7% sensitivity and 91.4% specificity. This study demonstrated that salivary RNA can be used as ovarian cancer biomarkers with high sensitivity and specificity.

Zermeño-Nava et al. (2018) tested the overexpression of sialic acid (SA) in saliva of 52 patients with a diagnosis of adnexal tumor. Results showed a sensitivity/specificity of 80%/100% with a cutoff to distinguish between benign/cancer cases of SA 15.5 mg/dL, as established from an ROC analysis. This result presents the value of salivary SA testing as an adjunct in diagnosing ovarian cancer by distinguishing between adnexal tumor and ovarian cancer.

## 19.9 Pancreatic Cancer

Transcriptome (Gao, Chen, Wang, Liu, & Chen, 2014; Humeau et al., 2015; Lau et al., 2013; Machida et al., 2016; Xie et al., 2015; Zhang et al., 2010), metabolome (Sugimoto et al., 2010), and microbiome (Farrell, Zhang, & Zhou, 2012; Torres et al., 2015) studies were conducted in order to identify pancreatic cancer-specific upregulated and/or downregulated miRNA, mRNA, oncogenes, pancreatic cancer-related metabolites and variations in pancreatic cancer patients oral microbiome, and identify their value as salivary biomarkers. Sugimoto et al. (2010) performed a metabolomics study through which a panel of five cancer-specific metabolites was identified with ROC curve AUC = 0.944 for pancreatic cancer. Farrell et al. (2012) compared salivary microbiomes of patients with pancreatic cancer to the control healthy group and found *Neiseria elongata* and *Streptococcus mitis* species that showed significant difference yielding 96.45% sensitivity and 82.1% specificity with an AU = 0.90. Torres et al. (2015), on the other hand, found a significantly higher ratio of *Leptotrichia* sp. to *Porphyromonas* sp. in pancreatic cancer patients, compared to healthy individuals. A salivary transcriptomic study found four mRNAs (KRAS, MBD3L2, ACRV1, and DPM10) to be significantly associated with pancreatic cancer with 90% sensitivity and 95% specificity, suggesting that those could be utilized in a screening test for pancreatic cancer (Zhang et al., 2010). A transcriptomic study, conducted by Xie et al. in 2015, evaluating potentially pancreatic cancer-specific miRNAs, found miR-367 to be downregulated and miR-940 upregulated between cancer patients and control group with 72.5% sensitivity and 70% specificity. Multiple transcriptomic studies (Humeau et al., 2015; Machida et al., 2016; Wang et al., 2009, 2014), evaluating pancreatic cancer-specific miRNA discovered various combinations of significantly upregulated miRNAs in patients with pancreatic cancer compared to healthy controls with sensitivity ratios varying from 64 to 85.7% and specificity ratio from 70 to 100%. Despite the fact that those miRNA combinations were significantly upregulated in cancer patients, some of those miRNAs (miR-21, miR-155, miR-210) were frequently found to be a part of those combinations. Humeau et al. (2015) found miR-23a and miR23b to be expressed in patients with pancreatic precursor lesions, whereas miR-216 was upregulated in patients with pancreatic cancers, which is valuable information for the possible applications of those markers in cancer screening.

## 19.10 Prostate Cancer

Prostate-specific antigen (PSA) is the most studied saliva and serum tumor marker for prostate adenocarcinoma (PA) in clinical studies. Shiiki et al. (2011) reported an association between saliva and serum PSA. Turan et al. (2000) used microparticle immunoassay technology and did not find an association between the serum and saliva PSA levels. Further studies are needed for better identification. Isaacsson Velho et al. (2018) retrospectively evaluated 150 prostate patients with recurrent and metastatic prostate cancer for germline DNA repair mutations utilizing next-generation sequencing from saliva. The presence of intraductal/ductal histology and lymphovascular invasion was found to be associated with pathogenic germline DNA-repair gene mutations. There were no associations between germline mutations and age, tumor stage, Gleason sum, or family history; mutation-positive patients had lower median PSA levels at diagnosis (5.5 vs. 8.6 ng/mL, $P = 0.01$) and unique pathologic features. The results of this study demonstrated that identification of clinical features associated with pathogenic germ line mutations might help to select patients for germ line testing.

## 19.11 Other Cancers

### *19.11.1 Hepatocellular Carcinoma*

Xie et al. (2018) studied the diagnostic value of salivary lncRNA in HCC. Salivary lnc-PCDH9-13:1 was significantly elevated in patients with HCC compared to healthy controls and groups of several benign liver diseases and showed a better diagnostic value than that of serum alpha-fetoprotein; it was also found to predict the HCC recurrence, and thus it was defined as a novel hypersensitive salivary HCC marker for early diagnosis of HCC.

Recently Qin et al. (2017) compared unstimulated whole saliva of 50 healthy volunteers (HV), to those 40 HBV-infected patients (HB), 50 cirrhosis patients (HC), and 60 hepatocellular carcinoma patients (HCC). A total of 40, 47, 29, and 33 N-glycan peaks were identified from HV, HB, HC, and HCC groups, respectively. The proportion of fucosylated N-glycans was increased in the HCC group (84.8%) more than any other groups (73.1% $\pm$ 0.01); however, the proportion of sialylated N-glycans was decreased in the HCC group (12.1%) less than any of the other groups (17.23% $\pm$ 0.003). The results of this study provided important insight for the ability to differentiate viral hepatitis and cirrhosis based on alteration of salivary N-linked glycans.

### *19.11.2 Leukemias*

Chen et al. (2014) analyzed leukemia-specific fusion gene transcripts and reported that leukemia-specific BCR-ABL, PML-RARα, and AML-ETO fusion genes were detected in all whole saliva samples of leukemia patients. This study developed a composition for stable storage of saliva RNA at room temperature and confirmed that saliva can be used as a sampling source for the detection of leukemic fusion transcripts.

Pre-, peri-, and post-intervention (chemotherapy or marrow transplantation) genomic, proteomics and metabolomics studies were also performed (Gershan et al., 2015, Guerra et al., 2012; Streckfus, Romaguera, & Guajardo-Streckfus, 2013). These studies found positive correlations in changes of concentration range of salivary neuropeptides, urea, glucose, insulin, and in salivary protein profile, reinforcing the value of saliva in monitoring the therapy and disease progression in leukemia patients.

## 19.12 Future Opportunities

The number of cross-sectional studies on salivary diagnostics is growing rapidly. Many cancer biomarkers have been discovered, attempts are made to enhance the specificity and sensitivity of available biomarker-based tests and to better understand their roles in onset, diagnosis, progression, and management of cancers. Our review revealed an abundance of available publications based on salivary biomarkers in cancers. We have reported here outcomes of 104 clinical investigations. The number of investigations per cancer varied from 2 to 36.

Heterogeneity of tumors in reference to tumor stage, cite, biology, and etiology, along with standardization of saliva sample collection, processing and storage, as well as molecular capturing, extraction, and quantification techniques all need to be addressed when detecting new and validating current salivary biomarkers to enhance sensitivity and specificity of study results.

Targeted approaches, incorporating the knowledge and understanding of unique pathophysiology of each type and stage of oncogenesis, with the use of advanced biomolecular analysis techniques and developments in point-of-care biomarker detection tools, will lead to even better accuracy in identifying combined cancer biomarker panels. Well-structured prospective large cohort studies of different populations should be conducted for further validation and to test reproducibility, accuracy, and applicability of promising salivary biomarker panel findings and thus utilization of saliva as a true diagnostic fluid.

Altogether, our review affirms the proposition of saliva as a novel liquid biopsy application. We emphasize the promising opportunity for future utilization of saliva as a diagnostic and screening tool for early diagnosis and personalized monitoring of treatment and prognosis of cancers, with the ultimate intention of urgent improvement in cancer surveillance.

## References

Aggarwal, S., Sharma, S. C., & Das, S. N. (2015). Galectin-1 and galectin-3: Plausible tumour markers for oral squamous cell carcinoma and suitable targets for screening high-risk population. *Clinica Chimica Acta, 442*, 13–21.

Agha-Hosseini, F., Mirzaii-Dizgah, I., & Rahimi, A. (2009). Correlation of serum and salivary CA15-3 levels in patients with breast cancer. *Medicina Oral, Patología Oral y Cirugía Bucal, 14*(10), e521–e524.

Ahn, S. M., Chan, J. Y., Zhang, Z., Wang, H., Khan, Z., Bishop, J. A., … Califano, J. A. (2014). Saliva and plasma quantitative polymerase chain reaction-based detection and surveillance of human papillomavirus-related head and neck cancer. *JAMA Otolaryngology–Head & Neck Surgery, 140*(9), 846–854.

Allegra, E., Trapasso, S., La Boria, A., Aragona, T., Pisani, D., Belfiore, A., & Garozzo, A. (2014). Prognostic role of salivary CD44sol levels in the follow-up of laryngeal carcinomas. *Journal of Oral Pathology & Medicine, 43*(4), 276–281.

Baum, B. J. (1993). Principles of saliva secretion. *Annals of the New York Academy of Sciences, 694*, 17–23.

Bigler, L. R., Streckfus, C. F., Copeland, L., Burns, R., Dai, X., Kuhn, M., ... Bigler, S. A. (2002). The potential use of saliva to detect recurrence of disease in high-risk breast cancer patients. *Journal of Oral Pathology & Medicine, 31*(7), 421–431.

Brooks, M. N., Wang, J., Li, Y., Zhang, R., Elashoff, D., & Wong, D. T. (2008). Salivary protein factors are elevated in breast cancer patients. *Molecular Medicine Reports, 1*(3), 375–378.

Bu, J., Bu, X., Liu, B., Chen, F., & Chen, P. (2015). Increased expression of tissue/salivary transgelin mRNA predicts poor prognosis in patients with oral squamous cell carcinoma (OSCC). *Medical Science Monitor, 21*, 2275–2281.

Camisasca, D. R., Da Ros Goncalves, L., Soares, M. R., Sandim, V., Nogueira, F. C., Garcia, C. H., ... Lourenço, S. Q. (2017). A proteomic approach to compare saliva from individuals with and without oral leukoplakia. *Journal of Proteomics, 151*, 43–52.

Carvalho, A. L., Henrique, R., Jeronimo, C., Nayak, C. S., Reddy, A. N., Hoque, M. O., ... Califano, J. A. (2011). Detection of promoter hypermethylation in salivary rinses as a biomarker for head and neck squamous cell carcinoma surveillance. *Clinical Cancer Research, 17*(14), 4782–4789.

Chai, R. C., Lim, Y., Frazer, I. H., Wan, Y., Perry, C., Jones, L., ... Punyadeera, C. (2016). A pilot study to compare the detection of HPV-16 biomarkers in salivary oral rinses with tumour p16 (INK4a) expression in head and neck squamous cell carcinoma patients. *BMC Cancer, 16*, 178.

Chen, Y. T., Chen, H. W., Wu, C. F., Chu, L. J., Chiang, W. F., Wu, C. C., ... Ou Yang, W. T. (2017). Development of a multiplexed liquid chromatography multiple-reaction-monitoring mass spectrometry (LC-MRM/MS) method for evaluation of salivary proteins as oral cancer biomarkers. *Molecular & Cellular Proteomics, 16*(5), 799–811.

Chen, D. X., Schwartz, P. E., & Li, F. Q. (1990). Saliva and serum CA 125 assays for detecting malignant ovarian tumours. *Obstetrics & Gynecology, 75*, 701–704.

Chen, D., Song, N., Ni, R., Zhao, J., Hu, J., Lu, Q., & Li, Q. (2014). Saliva as a sampling source for the detection of leukemic fusion transcripts. *Journal of Translational Medicine, 12*, 321.

Cheng, F., Wang, Z., Huang, Y., Duan, Y., & Wang, X. (2015). Investigation of salivary free amino acid profile for early diagnosis of breast cancer with ultra performance liquid chromatography-mass spectrometry. *Clinica Chimica Acta, 447*, 23–31.

Daubenbuchel, A. M., Hoffmann, A., Eveslage, M., Özyurt, J., Lohle, K., Reichel, J., ... Müller, H. L. (2016). Oxytosin in survivors of childhood onset craniopharyngioma. *Endocrine, 54*(2), 524–531.

Denny, P., Hagen, F. K., Hardt, M., Liao, L., Yan, W., Arellanno, M., ... Fisher, S. J. (2008). The proteomes of human parotid and submandibular/sublingual gland salivas collected as the ductal secretions. *Journal of Proteome Research, 7*(5), 1994–2006.

Du, J., & Zhang, L. (2017). Analysis of salivary microRNA expression profiles and identification of novel biomarkers in oesophageal cancer. *Oncology Letters, 14*(2), 1378–1394.

Duz, M. B., Karatas, O. F., Guzel, E., Turgut, N. F., Yilmaz, M., Creighton, C. J., & Ozen, M. (2016). Identification of miR-139-5p as a saliva biomarker for tongue squamous cell carcinoma: A pilot study. *Cellular Oncology (Dordrecht, Netherlands), 39*(2), 187–193.

Elashoff, D., Zhou, H., Reiss, J., Wang, J., Xiao, H., Henson, B., ... Wong, D. T. (2012). Prevalidation of salivary biomarkers for oral cancer detection. *Cancer Epidemiology, Biomarkers & Prevention, 21*(4), 664–672.

Farrell, J. J., Zhang, L., & Zhou, H. (2012). Variations of oral microbiota are associated with pancreatic diseases including pancreatic cancer. *Gut, 61*(4), 582–588.

Fischer, S. B., Attenhofer, M., Gultekin, S. H., Ross, D. A., & Heinimann, K. (2017). TRPS1 gene alterations in human subependymoma. *Journal of Neuro-Oncology, 134*(1), 133–138.

Frenette, P. S., Thirlwell, M. P., Trudeau, M., Thomson, D. M., Joseph, L., & Shuster, J. S. (1994). The diagnostic value of CA 27-29, CA 15-3, mucin-like carcinoma antigen, carcinoembryonic antigen and CA 19-9 in breast and gastrointestinal malignancies. *Tumour Biology, 15*(5), 247–254.

Füzéry, A. K., Levin, J., Chan, M. M., & Chan, D. W. (2013). Translation of proteomic biomarkers into FDA approved cancer diagnostics: Issues and challenges. *Clinical Proteomics, 10*(1), 13.

Gao, S., Chen, L. Y., Wang, P., Liu, L. M., & Chen, Z. (2014). MicroRNA expression in salivary supernatant of patients with pancreatic cancer and its relationship with ZHENG. *BioMed Research International, 2014*, 756347.

Gershan, L. A., Durham, P. L., Skidmore, J., Shimizu, J., Cady, R. J., Sheng, X., & Maloney, C. G. (2015). The role of salivary neuropeptides in pediatrics: Potential biomarkers for integrated therapies. *European Journal of Integrative Medicine, 7*, 372–377.

Guerra, R. N., Oliveira-Junior, J. J., Mouchrek-Filho, J. C. E., Liberio, S. A., Lima, M. V., Paim, D. B., . . . Pereira, A. L. (2012). Salivary evaluation of pediatric patients with cancer, before and after antineoplastic treatment. *Journal of Oral Pathology & Medicine, 41*(7), 527–532.

Guerrero-Preston, R., Soudry, E., Acero, J., Orera, M., Moreno-López, L., Macía-Colón, G., . . . Sidransky, D. (2011). NID2 and HOXA9 promoter hypermethylation as biomarkers for prevention and early detection in oral cavity squamous cell carcinoma tissues and saliva. *Cancer Prevention Research (Philadelphia, Pa.), 4*(7), 1061–1072.

Guerrero-Preston, R., White, J. R., Godoy-Vitorino, F., Rodríguez-Hilario, A., Navarro, K., González, H., . . . Sidransky, D. (2017). High-resolution microbiome profiling uncovers fusobacterium nucleatum, Lactobacillus gasseri/johnsonii, and Lactobacillus vaginalis associated to oral and oropharyngeal cancer in saliva form HPV positive and HPV negative patients treated with surgery and chemo-radiation. *Oncotarget, 8*(67), 110931–110948.

Helmerhorst, E. J., & Oppenheim, F. G. (2008). Saliva: A dynamic proteome. *Journal of Dental Research, 86*(8), 680–693.

Hu, Z., Zimmermann, B. G., Zhou, H., Wang, J., Henson, B. S., Yu, W., . . . Wong, D. T. (2008). Exon-level expression profiling: A comprehensive transcriptome analysis of oral fluids. *Clinical Chemistry, 54*(5), 824–832.

Humeau, M., Vignolle-Vidoni, A., Sicard, F., Martins, F., Bournet, B., Buscail, L., . . . Cordelier, P. (2015). Salivary microRNA in pancreatic cancer patients. *PLoS One, 10*(6), e0130996.

Isaacsson Velho, P., Silberstein, J. L., Markowski, M. C., Luo, J., Lotan, T. L., Isaacs, W. B., & Antonarakis, E. S. (2018). Intraductal/ductal histology and lymphovascular invasion are associated with germline DNA-repair gene mutations in prostate cancer. *Prostate, 78*(5), 401–407.

Jarai, T., Maasz, G., Burian, A., Bona, A., Jambor, E., Gerlinger, I., & Mark, L. (2012). Mass spectrometry-based salivary proteomics for the discovery of head and neck squamous cell carcinoma. *Pathology Oncology Research, 18*(3), 623–628.

Jinno, H., Murata, T., Sunamura, M., & Sugimoto, M. (2015). Investigation of potential salivary biomarkers for the diagnosis of breast cancer. *ASCO Annual Meeting Proceedings, 33*, 145.

Kawahara, R., Bollinger, J. G., Rivera, C., Ribeiro, A. C., Brandão, T. B., Paes Leme, A. F., & MacCoss, M. J. (2016). A targeted proteomic strategy for the measurement of oral cancer candidate biomarkers in human saliva. *Proteomics, 16*(1), 159–173.

Krapfenbauer, K., Drucker, E., & Thurnher, D. (2014). Identification of tumour-related proteins as potential screening markers by proteome analysis-protein profiles of human saliva as a predictive and prognostic tool. *The EPMA Journal, 5*(1), 20.

Krishna Prasad, R. B., Sharma, A., & Babu, H. M. (2013). An insight into salivary markers in oral cancer. *Dental Research Journal (Isfahan), 10*(3), 287–295.

Laidi, F., Bouziane, A., Lakhdar, A., Khabouze, S., Amrani, M., Rhrab, B., & Zaoui, F. (2014). Significant correlation between salivary and serum Ca 15–3 in healthy women and breast cancer patients. *Asian Pacific Journal of Cancer Prevention, 15*(11), 4659–4662.

Lau, C., Kim, Y., Chia, D., Spielmann, N., Eibl, G., Elashoff, D., . . . Wong, D. T. (2013). Role of pancreatic cancer-derived exosomes in salivary biomarker development. *The Journal of Biological Chemistry, 288*(37), 26888–26897.

Lee, J. Y., Kim, H. S., Suh, D. H., Kim, M. K., Chung, H. H., & Song, Y. S. (2013). Ovarian cancer biomarker discovery based on genomic approaches. *Journal of Cancer Prevention, 18*(4), 298–312.

Li, Y., St John, M. A., & Zhou, X. (2004). Salivary transcriptome diagnostics for oral cancer detection. *Clinical Cancer Research, 10*(24), 8442–8450.

Li, S. X., Yang, Y. Q., Jin, L. J., Cai, Z. G., & Sun, Z. (2016). Detection of survivin, carcinoembryonic antigen and ErbB2 level in oral squamous cell carcinoma patients. *Cancer Biomarkers, 17*(4), 377–382.

Li, X., Yang, T., & Lin, J. (2012). Spectral analysis of human saliva for detection of lung cancer using surface-enhanced Raman spectroscopy. *Journal of Biomedical Optics, 17*, 037003.

Li, B. X., Yu, Q., Shi, Z. L., Li, P., & Fu, S. (2015). Circulating microRNAs in esophageal squamous cell carcinoma: Association with locoregional staging and survival. *International Journal of Clinical and Experimental Medicine, 8*(5), 7241–7250.

Li, Y., Zhou, X., St John, M. A., & Wong, D. T. (2004). RNA profiling of cell-free saliva using microarray technology. *Journal of Dental Research, 83*(3), 199–203.

Lim, Y., Wan, Y., Vagenas, D., Ovchinnikov, D. A., Perry, C. F., Davis, M. J., & Punyadeera, C. (2016). Salivary DNA methylation panel to diagnose HPV-positive and HPV-negative head and neck cancers. *BMC Cancer, 16*(1), 749.

Liu, C. J., Kao, S. Y., Tu, H. F., Tsai, M. M., Chang, K. W., & Lin, S. C. (2010). Increase of microRNA miR-31 level in plasma could be a potential marker of oral cancer. *Oral Diseases, 16* (4), 360–364.

Liu, X., Yu, H., Qiao, Y., Yang, J., Shu, J., Zhang, J., . . . Li, Z. (2018). Salivary glycopatterns as potential biomarkers for screening of early-stage breast cancer. *eBioMedicine, 28*, 70–79.

Machida, T., Tomofuji, T., Maruyama, T., Yoneda, T., Ekuni, D., Azuma, T., . . . Morita, M. (2016). MiR-1246 and miR-4644 in salivary exosome as potential biomarkers for pancreatobiliary tract cancer. *Oncology Reports, 36*(4), 2375–2381.

Mager, D. L., Haffajee, A. D., Devlin, P. M., Norris, C. M., Posner, M. R., & Goodson, J. M. (2005). The salivary microbiota as a diagnostic indicator of oral cancer: A descriptive, non-randomized study of cancer free and oral squamous cell carcinoma subjects. *Journal of Translational Medicine, 3*, 27.

Malamud, D. (2011a). Saliva as a diagnostic fluid. *Dental Clinics of North America, 55*(1), 159–178.

Malamud, D. (2011b). Point-of-care oral-based diagnostics. *Oral Diseases, 17*(8), 745–752.

Malamud, D., & Tabak, L. (Eds.). (1993). *Saliva as a diagnostic fluid* (Annals of the New York Academy of Sciences) (Vol. 694). New York: New York Academy of Sciences.

Mandel, I. D. (1993). Salivary diagnosis: Promises, promises. *Annals of the New York Academy of Sciences, 694*, 1–10.

Michailidou, E., Tzimagiorgis, G., Chatzopoulou, F., Vahtsevanos, K., Antoniadis, K., Kouidou, S., . . . Antoniades, D. (2016). Salivary mRNA markers having the potential to detect oral squamous cell carcinoma segregated from oral leukoplakia with dysplasia. *Cancer Epidemiology, 43*, 112–118.

Momen-Heravi, F., Trachtenberg, A. J., Kuo, W. P., & Cheng, Y. S. (2014). Genomewide study of salivary microRNAs for detection of oral cancer. *Journal of Dental Research, 93*(7 Suppl), 86S–93S.

Ovchinnikov, D. A., Cooper, M. A., Pandit, P., Coman, W. B., Cooper-White, J. J., Keith, P., . . . Punyadeera, C. (2012). Tumor-suppressor gene promoter hypermethylation in saliva of head and neck cancer patients. *Translational Oncology, 5*(5), 321–326.

Ozturk, L. K., Emekli-Alturfan, E., & Kasikci, E. (2011). Salivary total sialic acid levels increase in breast cancer patients: A preliminary study. *Medicinal Chemistry, 7*(5), 443–444.

Park, N. J., Zhou, H., Elashoff, D., Henson, B. S., Kastratovic, D. A., Abemayor, E., & Wong, D. T. (2009). Salivary microRNA: Discovery, characterization, and clinical utility for oral cancer detection. *Clinical Cancer Research, 15*(17), 5473–5477.

Pereira, L. H., Reis, I. M., Reategui, E. P., Gordon, C., Saint-Victor, S., Duncan, R., . . . Franzmann, E. J. (2016). Risk stratification system for oral cancer screening. *Cancer Prevention Research (Philadelphia, PA), 9*(6), 445–455.

Qin, Y., Zhong, Y., Ma, T., Zhang, J., Yang, G., Guan, F., . . . Li, B. (2017). A pilot study of salivary N-glycome in HBV-induced chronic hepatitis, cirrhosis, and hepatocellular carcinoma. *Glycoconjugate Journal, 34*(4), 523–535

Rettori, M. M., De Carvalho, A. C., Bomfim Longo, A. L., de Oliveira, C. Z., Kowalski, L. P., Carvalho, A. L., & Vettore, A. L. (2013). Prognostic significance of TIMP3 hypermethylation in post-treatment salivary rinse from head and neck squamous cell carcinoma patients. *Carcinogenesis, 34*(1), 20–27.

Righini, C. A., De Fraipont, F., Timsit, J. F., Faure, C., Brambilla, E., Reyt, E., & Favrot, M. C. (2007). Tumor-specific methylation in saliva: A promising biomarker for early detection of head and neck cancer recurrence. *Clinical Cancer Research, 13*(4), 1179–1185.

Salazar, C., Nagadia, R., Pandit, P., Cooper-White, J., Banerjee, N., Dimitrova, N., … Punyadeera, C. (2014). A novel saliva-based microRNA biomarker panel to detect head and neck cancers. *Cellular Oncology (Dordrecht), 37*(5), 331–338.

Shiiki, N., Tokuyama, S., Sato, C., Kondo, Y., Saruta, J., Mori, Y., … Tsukinoki, K. (2011). Association between saliva PSA and serum PSA in conditions with prostate adenocarcinoma. *Biomarkers, 16*(6), 498–503.

Slavkin, H. C. (1998). Toward molecularly based diagnostics for the oral cavity. *Journal of the American Dental Association (1939), 129*(8), 1138–1143.

Spielman, N., & Wong, D. T. (2011). Saliva: Diagnostics and therapeutic perspectives. *Oral Diseases, 17*(4), 345–354.

Stewart, B. W., Wild, C. P. (Eds.) (2014). *World Cancer Report 2014*. Lyon, France: International Agency for Research on Cancer.

Streckfus, C., Bigler, L., Dellinger, T., Dai, X., Kingman, A., & Thigpen, J. T. (2000). The presence of soluble c-erbB-2 in saliva and serum among women with breast carcinoma: A preliminary study. *Clinical Cancer Research, 6*(6), 2363–2370.

Streckfus, C., Bigler, L., Tucci, M., & Thigpen, J. T. (2000). A preliminary study of CA15-3, c-erbB-2, epidermal growth factor receptor, cathepsin-D, and p53 in saliva among women with breast carcinoma. *Cancer Investigation, 18*(2), 101–109.

Streckfus, C. F., Mayorga-Wark, O., Arreola, D., Edwards, C., Bigler, L., & Dubinsky, W. P. (2008). Breast cancer related proteins are present in saliva and are modulated secondary to ductal carcinoma in situ of the breast. *Cancer Investigation, 26*(2), 159–167.

Streckfus, C. F., Romaguera, J., & Guajardo-Streckfus, C. (2013). The use of salivary protein secretions as an in vivo model to study Mantel cell lymphoma progression and treatment. *Cancer Investigation, 31*(7), 494–499.

Streckfus, C. F., Storthz, K. A., Bigler, L., & Dubinsky, W. P. (2009). A comparison of the proteomic expression in pooled saliva specimens from individuals diagnosed with ductal carcinoma of the breast with and without lymph node involvement. *Journal of Oncology, 2009*, 737619.

Sugimoto, M., Wong, D. T., Hirayama, A., Soga, T., & Tomita, M. (2010). Capillary electrophoresis mass spectrometry-based saliva metabolomics identified oral, breast and pancreatic cancer-specific profiles. *Metabolomics, 6*(1), 78–95.

Suma, H. R., Prabhu, K., Shenoy, R. P., Annaswamy, R., Rao, S., & Rao, A. (2010). Estimation of salivary protein thiols and total antioxidant power of saliva in brain tumour patients. *Journal of Cancer Research and Therapeutics, 6*(3), 278–281.

Sun, Y., Huo, C., Quao, Z., Shang, Z., Uzzaman, A., Liu, S., … Xiao, H. (2018). Comparative proteomic analysis of exosomes and microvesicles in human saliva for lung cancer. *Journal of Proteome Research, 17*(3), 1101–1107.

Sun, Y., Liu, S., Qiao, Z., Shang, Z., Xia, Z., Niu, X., … Xiao, H. (2017). Systematic comparison of exosomal proteomes from human saliva and serum for the detection of lung cancer. *Analityca Chimica Acta, 982*, 84–95.

Sun, W., Zaboli, D., Wang, H., Liu, Y., Arnaoutakis, D., Khan, T., … Califano, J. A. (2012). Detection of TIMP3 promoter hypermethylation in salivary rinse as an independent predictor of local recurrence-free survival in head and neck cancer. *Clinical Cancer Research, 18*(4), 1082–1091.

Tang, H., Wu, Z., Zhang, J., & Su, B. (2013). Salivary lncRNA as a potential marker for oral squamous cell carcinoma diagnosis. *Molecular Medicine Reports, 7*(3), 761–766.

Torres, P. J., Fletcher, E. M., Gibbons, S. M., Bouvet, M., Doran, K. S., & Kelley, S. T. (2015). Characterization of the salivary microbiome in patients with pancreatic cancer. *PeerJ, 3*, e1373.

Turan, T., Demir, S., Aybek, H., Atahan, O., Tuncay, O. L., & Aybek, Z. (2000). Free and total prostate-specific antigen levels in saliva and the comparison with serum levels in men. *European Urology, 38*, 550–554.

Vidotto, A., Henrique, T., Raposo, L. S., Maniglia, J. V., & Tajara, E. H. (2010). Salivary and serum proteomics in head and neck carcinomas: Before and after surgery and radiotherapy. *Cancer Biomarkers, 8*(2), 95–107.

Wang, J., Chen, J., Chang, P., LeBlanc, A., Li, D., Abbruzzesse, J. L., ... Sen, S. (2009). MicroRNAs in plasma of pancreatic ductal adenocarcinoma patients as novel blood-based biomarkers of disease. *Cancer Prevention Research, 2*, 807–813.

Wang, J., Raimondo, M., Guha, S., Chen, J., Diao, L., Dong, X., ... Sen, S. (2014). Circulating microRNAs in pancreatic juice as candidate biomarkers of pancreatic cancer. *Journal of Cancer, 5*, 696–705.

Wang, Y., Springer, S., Mulvey, C. L., Silliman, N., Schaefer, J., Sausen, M., ... Agrawal, N. (2015). Detection of somatic mutations and HPV in the saliva and plasma of patients with head and neck squamous cell carcinomas. *Science Translational Medicine, 7*(293), 293ra104.

Wei, F., Lin, C. C., Joon, A., Feng, Z., Troche, G., Lira, M. E., ... Wong, D. T. (2014). Noninvasive saliva-based EGFR gene mutation detection in patients with lung cancer. *American Journal of Respiratory and Critical Care Medicine, 190*, 1117–1126.

Wheeler, D. A. (2013). From human genome to cancer genome: The first decade. *Genome Research, 23*(7), 1054–1062.

Wiklund, E. D., Gao, S., Hulf, T., Sibbritt, T., Nair, S., Costea, D. E., ... Kjems, J. (2011). MicroRNA alterations and associated aberrant DNA methylation patterns across multiple sample types in oral squamous cell carcinoma. *PLoS One, 6*(11), e27840.

Wood, N., & Streckfus, C. F. (2015). The expression of lung resistance protein in saliva: A novel prognostic indicator protein for carcinoma of the breast. *Cancer Investigation, 33*(10), 510–515.

Wu, C. C., Chu, H. W., Hsu, C. W., Chang, K. P., & Liu, H. P. (2015). Saliva proteome profiling reveals potential salivary biomarkers for detection of oral cavity squamous cell carcinoma. *Proteomics, 15*(19), 3394–3404.

Wu, W., Gong, H., Liu, M., Chen, G., Chen, R. (2015). Noninvasive breast tumors detection based on saliva protein surface enhanced Raman spectroscopy and regularized multinomial regression. In *2015 8th International conference on biomedical engineering and informatics (BMEI)* (pp. 214–218). IEEE.

Wu, W., Hou, W., Wu, Z., Wang, Y., Yi, Y., & Lin, W. (2013). miRNA-144 in the saliva is a genetic marker for early diagnosis of esophageal cancer. *Nan Fang Yi Ke Da Xue Bao, 33*(12), 1783–1786.

Wu, Z. Z., Wang, J. G., & Zhang, X. L. (2009). Diagnostic model of saliva protein finger print analysis of patients with gastric cancer. *World Journal of Gastroenterology, 15*(7), 865–870.

Xiao, H., Zhang, Y., Kim, Y., Kim, S., Kim, J. J., Kim, K. M., ... Wong, D. T. (2016). Differential proteomic analysis of human saliva using tandem mass tags quantification for gastric cancer detection. *Scientific Reports, 6*, 22165.

Xiao, H., Zhang, L., Zhou, H., Lee, J. M., Garon, E. B., & Wong, D. T. (2012). Proteomic analysis of human saliva from lung cancer patients using two-dimensional difference gel electrophoresis and mass spectrometry. *Molecular & Cellular Proteomics, 11*(2), M111.012112.

Xie, Z., Chen, G., Zhang, X., Li, D., Huang, J., Yang, C., ... Li, Z. (2013). Salivary microRNAs as promising biomarkers for detection of esophageal cancer. *PLoS One*. 2013, *8*(4), e57502.

Xie, Z., Yin, X., Gong, B., Nie, W., Wu, B., Zhang, X., ... Li, Z. (2015). Salivary microRNAs show potential as a noninvasive biomarker for detecting resectable pancreatic cancer. *Cancer Prevention Research (Philadelphia, PA), 8*(2), 165–173.

Xie, Z., Zhou, F., Yang, Y., Li, L., Lei, Y., Lin, X., ... Wu, B. (2018). Lnc-PCDH9-13:1 is a hypersensitive and specific biomarker for early hepatocellular carcinoma. *eBioMedicine, 33*, 57–67.

Yamashita, T., Kamada, H., Kanasaki, S., Maeda, Y., Nagano, K., Abe, Y., ... Tsunoda, S. (2013). Epidermal growth factor receptor localized to exosome membranes as a possible biomarker for lung cancer diagnosis. *Pharmazie, 68*(12), 969–973.

Yan, W., Yu, W., Than, S., Hu, Z., Zhou, H., & Wong, D. T. (2008, April 5). *Salivaomics knowledge base (SKB) (abstract 1179)*. Presented at the 37th annual meeting and exhibition of the American Association for Dental Research, Dallas, TX.

Yao, Y., Xu, M., Liang, L., Zhang, H., Xu, R., Feng, Q., ... Zeng, Y. X. (2017). Genome-wide analysis of Epstein-Barr virus identifies variants and genes associated with gastric carcinoma and population structure. *Tumour Biology, 39*(10), 1010428317714195.

Ye, M., Ye, P., Zhang, W., Rao, J., & Xie, Z. (2014). Diagnostic values of salivary versus and plasma microRNA-21 for early esophageal cancer. *Nan Fang Yi Ke Da Xue Bao, 34*(6), 885–889.

Yu, J. S., Chen, Y. T., Chiang, W. F., Hsiao, Y. C., Chu, L. J., See, L. C., ... Hartwell, L. H. (2016). Saliva protein biomarkers to detect oral squamous cell carcinoma in a high-risk population in Taiwan. *Proceedings of the National Academy of Sciences of the United States of America, 113* (41), 11549–11554.

Zermeño-Nava, J. J., Martínez-Martínez, M. U., Rámirez-de-Ávila, A. L., Hernández-Arteaga, A. C., García-Valdivieso, M. G., Hernández-Cedillo, A., ... Navarro-Contreras, H. R. (2018). Determination of sialic acid in saliva by means of surface-enhanced Raman spectroscopy as a marker in adnexal mass patients: Ovarian cancer vs benign cases. *Journal of Ovarian Research, 11*(1), 61.

Zhang, L., Farrell, J. J., Zhou, H., Elashoff, D., Akin, D., Park, N. H., ... Wong, D. T. (2010). Salivary transcriptomic biomarkers for detection of resectable pancreatic cancer. *Gastroenterology, 138*(3), 949–957.

Zhang, L., Xiao, H., Zhou, H., Santiago, S., Lee, J. M., Garon, E. B., ... Wong, D. T. W. (2012). Development of transcriptomic biomarker signature in human saliva to detect lung cancer. *Cellular and Molecular Life Sciences, 69*(19), 3341–3350.

Zhong, L., Cheng, F., Lu, X., Duan, Y., & Wang, X. (2016). Untargeted saliva metabonomics study of breast cancer based on ultra performance liquid chromatography coupled to mass spectrometry with HILIC and RPLC separations. *Talanta, 158*, 351–360.

Zilberman, Y., & Sonkusale, S. R. (2015). Microfluidic optoelectronic sensor for salivary diagnostics of stomach cancer. *Biosensors & Bioelectronics, 67*, 465–471.

# Part IV
# Psychosocial Research and Applications

**Kate Ryan Kuhlman, Section Editor**

# Chapter 20
# Salivary Bioscience in Clinical Psychology and Psychiatry

**Kate Ryan Kuhlman, Nestor L. Lopez-Duran, and Zahra Mousavi**

**Abstract** The development of biomarkers of psychiatric disease has been among the highest priorities in mental health research over the past century. The ideal biomarkers in this context would serve one of several purposes, such as (1) distinguish individuals with and without psychiatric disease, (2) provide evidence that treatments mitigate not only the behavioral symptoms of disorders but also the biological underpinnings, and (3) elucidate future risk for development of a disorder. To date, salivary markers have been used in each of these contexts. In this chapter, we summarize the history of the use of salivary biomarkers in research aimed at identifying, characterizing, and treating psychiatric disorders, such as mood and anxiety disorders, psychotic disorders, and neurodevelopmental disorders. Finally, this chapter reviews the strengths and limitations of the knowledge gained to date and introduces how salivary bioscience can contribute to the next generation of research on the course, causes, and treatment for psychiatric disorders.

**Keywords** Clinical psychology · Psychiatry · Treatment · Diagnosis · Psychopathology

K. R. Kuhlman (✉)
Department of Psychological Science, School of Social Ecology, University of California Irvine, Irvine, CA, USA

Institute for Interdisciplinary Salivary Bioscience Research, University of California, Irvine, CA, USA
e-mail: krkuhl@uci.edu

N. L. Lopez-Duran
Department of Psychology, University of Michigan, Ann Arbor, MI, USA

Z. Mousavi
Department of Psychological Science, School of Social Ecology, University of California Irvine, Irvine, CA, USA



D. A. Granger, M. K. Taylor (eds.), *Salivary Bioscience*,
https://doi.org/10.1007/978-3-030-35784-9_20

## 20.1 History of Salivary Bioscience in Psychiatry and Clinical Psychology

Psychiatric disorders account for 7.4% of all disability-adjusted life years (DALYs) globally (Whiteford et al., 2013). Thus, efforts to reliably identify individuals at risk for and who have already developed psychiatric disorders across the population are among the highest priorities to public health. While it had been a focus of some psychiatric research for decades, the field of psychiatry prioritized efforts to identify reliable and objective biological markers, or biomarkers, that distinguished individuals with and without psychopathology in the 1980s (Singh & Rose, 2009). Biomarkers of psychiatric disease can be used to identify individuals at risk for psychopathology, to characterize the processes that lead to the development of psychiatric illnesses, to determine the severity of psychiatric illness, and to determine the effectiveness of treatments for psychiatric illness (Davis et al., 2015). Biomarkers of psychiatric disease have taken many forms such as the structure and function of neural substrates, autonomic nervous system activity, DNA methylation, and measuring concentrations of hormones, proteins, and peptides in blood, urine, feces, cerebrospinal fluid, and saliva. Through these methods, a wealth of information has emerged about the pathogenesis of psychiatric disorders and markers of risk. Yet, the gold standard for diagnosing psychiatric disorders remains subjective self-report or clinicians' observations of symptoms, and no biomarkers have been adopted to inform diagnoses of psychiatric disorders.

As is the case with many aspects of science, innovation is closely tied to the development of technology, and dissemination is closely tied to the cost of that technology to health care providers. Thus, the emergence of neuroimaging technology led to a burst of new information on the neural underpinnings of psychiatric disease in research, but its clinical utility has been limited as there remain significant validity, reliability, feasibility, and cost-related barriers to using neuroimaging in health care settings for psychiatric patients (Kapur, Phillips, & Insel, 2012). In contrast, salivary biomarkers are cost-effective and involve minimal supervision for collection in ambulatory settings. Thus, salivary biomarkers of psychiatric disease are among the most versatile and cost-effective tools available to us today. In this chapter, we review the current state of knowledge on the ability of salivary biomarkers to distinguish individuals with and without several of the most common and impairing psychiatric diseases. In addition, we review the state of the literature using salivary biomarkers to determine the effectiveness of both pharmacological and behavioral interventions for psychiatric disorders.

## 20.2 Use of Salivary Biomarkers in Psychiatric Diagnoses

### *20.2.1 Major Depressive Disorder (MDD) and Bipolar Disorder (BD)*

Perhaps the largest literature using salivary biomarkers in psychiatry exists for depressive disorders. Because up to 80% of depressive episodes occur in the wake of major life stressors, the search for salivary biomarkers has focused largely on stress-related hormones. Stress activates several interdependent physiological processes, including the hypothalamic–pituitary–adrenal axis (HPA axis). In the context of stress, the HPA axis activates a hormonal cascade that redistributes energy throughout the body to enable survival in the short term. The final product of HPA axis activation are cortisol, dehydroepiandrosterone (DHEA), and epinephrine/norepinephrine in humans. Both cortisol and DHEA can be reliably measured in saliva. These biomarkers are complex and can reflect the intensity of stress exposure, the health and sensitivity of physiological stress response systems, and be impacted by cognitive, contextual, and biological moderators.

Due to the role of stress in the pathogenesis of depression, a predominant theory in clinical psychology and psychiatry is that stressful experiences upregulate the HPA axis, resulting in elevated and even chronic exposure to cortisol (de Kloet, Joëls, & Holsboer, 2005; Pariante & Lightman, 2008). High levels of endogenous and exogenous cortisol have profound effects on cognition and behavior (Juster, McEwen, & Lupien, 2010; Lupien, McEwen, Gunnar, & Heim, 2009), and it is through this pathway that stress may lead to depressive disorders. However, it is important to note that the findings of the relationship between cortisol and depression remain mixed (e.g., Knight, Avery, Janssen, & Powell, 2010; Mangold, Marino, & Javors, 2011; Schuler et al., 2017; Sjögren, Leanderson, & Kristenson, 2006). Cortisol and DHEA are thought to have complementary effects on one another, such that cortisol can have neurotoxic effects on the brain at high and sustained concentrations, while simultaneous increases in DHEA are thought to be neuroprotective against the putative consequences of cortisol (Maninger, Wolkowitz, Reus, Epel, & Mellon, 2009). Thus, it is also posited that depletion of DHEA over time may lead to depression by making the central nervous system more vulnerable to the effects of cortisol (Jin et al., 2016).

Cortisol concentrations are highest in the morning and lowest in the evening among healthy individuals. This decline across the day is referred to as the diurnal cortisol slope. Cortisol values increase by about 50% within an hour of waking (Wüst et al., 2000), a phenomenon called the cortisol awakening responses (CAR) (Clow, Hucklebridge, Stalder, Evans, & Thorn, 2010; Stalder et al., 2016). Deviations from these patterns are often thought to indicate underlying disease processes. Indeed, several studies have shown that increases in daily experiences of stress correspond to increases in daily levels of salivary cortisol (Adam & Gunnar, 2001; Adam, Hawkley, Kudielka, & Cacioppo, 2006; Kuhlman, Repetti, Reynolds, & Robles, 2016, 2018). Further, there are several diurnal cortisol markers that have been prospectively linked to depression. Individuals with a family history of depression exhibit elevated salivary cortisol concentrations, particularly CAR (Mannie, Harmer, & Cowen, 2007). In adolescents, elevated cortisol at 8 am predicted MDD one year later (Goodyer, Herbert, Tamplin, & Altham, 2000) and increases in depressive symptoms 3 years later (Halligan, Herbert, Goodyer, & Murray, 2007). CAR prospectively predicted the onset of depressive episodes during late adolescence one year (Adam et al., 2010) and 2.5 years later (Vrshek-Schallhorn et al., 2013). In adult women, cortisol at 8 am predicted MDD 13 months later (Harris et al., 2000), and greater CAR immediately following breast cancer diagnosis was associated with increases in depressive symptoms during treatment and recovery

(Kuhlman et al., 2017). Finally, elevated CAR also predicts the recurrence of MDD (Hardeveld et al., 2014). This is particularly notable given that CAR is a unique index of HPA axis functioning. Several studies have shown that CAR can be upregulated, in other words the production of cortisol between waking and 30 minutes post-waking is increased, from one day to the next based on daily demands and stressors (Fries, Dettenborn, & Kirschbaum, 2009; Pruessner, Hellhammer, Pruessner, & Lupien, 2003; Schlotz, Hellhammer, Schulz, & Stone, 2004). Further, preclinical experiments have shown that CAR is likely an index of adrenal sensitivity to ACTH (Clow et al., 2010). That being said, studies vary in their approach to calculating CAR (e.g., slope of increase from waking, magnitude of change over time, or total cortisol production via AUC following waking) which precludes our having a nuanced understanding of the underlying neurobiology indexed by CAR (See Stalder et al., 2016).

Salivary indicators of HPA axis functioning can also distinguish between individuals with and without MDD. Depressed individuals tend to have higher salivary cortisol concentrations than control participants (Cubała, Landowski, Dziadziuszko, Chrzanowska, & Wielgomas, 2016; Wei et al., 2016), particularly in the morning (Bhagwagar, Hafizi, & Cowen, 2005). Having elevated cortisol concentrations throughout the day has been interpreted in many ways, such as chronic synthesis of glucocorticoids in the context of ongoing stress, impaired negative feedback of the HPA axis that fails to regulate or decrease glucocorticoid secretion, or dysregulated circadian regulation of the neuroendocrine system within the central nervous system (De Kloet, 1991; de Kloet et al., 2005). Depressed individuals also tend to have prolonged elevations in salivary cortisol during recovery from laboratory stress (Burke, Davis, Otte, & Mohr, 2005; Lopez-Duran, Kovacs, & George, 2009), lending support to the potential role of psychosocial stress to diurnal overproduction of glucocorticoids. Similarly, individuals with depression also exhibit attenuated DHEA concentrations 50 minutes after acute laboratory exposure (Jiang et al., 2017), as well as throughout the day (Michael, Jenaway, Paykel, & Herbert, 2000).

While much has been learned in the past 2 decades by carefully interrogating the role of the HPA axis and its by-products in the pathogenesis of depression, there are some promising, non-HPA salivary biomarkers that warrant further investigation. For example, salivary testosterone appears to be lower in females with an anxiety or depressive disorder than in non-affected females (Giltay et al., 2012). Given the profound sex differences in the onset of depression that correspond with pubertal development (Kessler et al., 2005), pursuit of the role of gonadal hormones in the pathogenesis of depression is important (Angold & Costello, 2006). Along these lines, another adrenal hormone, aldosterone, appears to be correlated with the length and severity of a depressive episode in women (Segeda, Izakova, Hlavacova, Bednarova, & Jezova, 2017).

Further, sleep and circadian dysregulation is emerging as a critical component to the onset and maintenance of affective disorders. Indeed, depressed young women exhibit alterations to salivary melatonin which may be a useful indicator of dysregulated sleep–wake cycles (Frey et al., 2012). Finally, there is a single nucleotide polymorphism, spindle and kinetochore-associated protein 2 (SKA2), which is linked to suicidal ideation. Recently, a blood and saliva-based stress-biosignature was derived that interacts with SKA2 to predict suicidal ideation across multiple, large clinical samples (Clive et al., 2016). This line of research is critical to our ability as health care providers to noninvasively identify individuals at the greatest risk of harm to themselves and others.

Far less salivary biomarker research has been conducted within the context of Bipolar Disorder; however, a few studies suggest some promising future directions. In contrast with the studies in MDD, it appears that individuals with Bipolar Disorder exhibit a blunted CAR compared with control participants (Huang et al., 2017). Given what we know about the underlying neurobiology of CAR as a biomarker (Clow et al., 2010), this may suggest a decrease in neuroendocrine flexibility or adaptability in the context of life events. Perhaps the most compelling application of salivary bioscience within Bipolar Disorder research has been the validation of lithium assays. Briefly, lithium is a long-standing and effective pharmacological treatment for Bipolar Disorder (Miura et al., 2014; Severus et al., 2014). However, one challenge to lithium treatment is its narrow therapeutic window. At low concentrations in the body, lithium does not affect mood and behavior, but at high concentrations, it can be toxic. Thus, participants receiving lithium treatment for Bipolar Disorder must regularly provide blood samples to monitor the levels of lithium in their system. The invasive nature of these repeated blood samples can be a barrier to prescribing and taking this medication, especially in vulnerable populations such as children and adolescents. Lithium concentrations can also be measured noninvasively in saliva, which may provide a less invasive alternative to blood sampling (Ben-Aryeh, Naon, Szargel, Gutman, & Hefetz, 1980). However, the correspondence between blood-based and salivary measures of lithium appears to be stronger for some individuals than others (Shetty, Desai, Patil, & Nayak, 2012). This may be the result of individual differences in metabolism of the drug, timing of the salivary sample collection, and assay technology. The field of salivary bioscience is working to establish validated practice guidelines for effective monitoring of lithium in saliva to increase access of this effective medicine to larger populations (Gałuszko-Węgielnik et al., 2017; Khare, Sankaranarayanan, Goel, Khandelwal, & Srinivasa Murthy, 1983). Salivary biomarkers have similarly been employed to monitor therapeutic and toxic levels of antipsychotic medications, such as Haloperidol (Dysken et al., 1992). Chapter 17 of this book details the progress and potential of using salivary biomarkers in therapeutic drug monitoring.

### 20.2.2 *Anxiety Disorders*

Anxiety disorders can manifest in several different ways. Generalized Anxiety Disorder (GAD) is the persistence of apprehensive expectation and difficulty controlling worry; Panic Disorder (PD) is characterized by reoccurring, unexpected panic attacks; Obsessive-Compulsive Disorder (OCD) describes a phenomenon where a person is distressed and impaired by uncontrollable, reoccurring thoughts (*obsessions*) and behaviors (*compulsions*) that he or she feels the urge to repeat over and over; Social Anxiety Disorder (SAD) is intense anxiety or fear of being judged, negatively evaluated, or rejected in a social or performance situation; and Post-traumatic Stress Disorder (PTSD) is a disorder that develops in some people who have experienced a shocking, scary, or dangerous event who then go on to suffer from symptoms that make them feel stressed or frightened even when they are not in danger (American Psychiatric Association, 2013).

Similar to the research conducted in mood disorders, salivary bioscience in the context of anxiety disorders have also largely focused on peptides associated with stress physiology, such as salivary α-amylase, a biomarker of sympathetic nervous system activation and responds to physical and psychological stress (Granger, Kivlighan, el-Sheikh, Gordis, & Stroud, 2007; Nater & Rohleder, 2009), and cortisol. However, there also appears to be a burgeoning literature looking at genetic vulnerability to anxiety disorders measured in saliva.

#### 20.2.2.1 Generalized Anxiety Disorder

As is seen with depression, GAD appears to be associated with an exaggerated physiological stress response system, and can be observed in populations across the lifespan, although indexed by different salivary analytes (e.g., α-amylase, cortisol, and inflammation). Pediatric GAD patients with comorbid depression have elevated α-amylase throughout the day, and greater symptom severity was correlated with elevated salivary cortisol and α-amylase reactivity to acute lab stress (Funke et al., 2017). Elevated salivary cortisol has been observed in older adults (age 60+) with GAD as well, and cortisol concentrations are positively correlated with symptom severity (Mantella et al., 2008). Notably, elevated stress hormones (α-amylase) were also associated with salivary inflammatory cytokines, salivary IL-1 and IL-6, in asthmatic adults with GAD (Yang, Liu, Xu, Shi, & Du, 2017). Cytokines are soluble proteins secreted from immune cells that allow immune cells to communicate with one another; a subset of these proteins are considered inflammatory. These cytokines are thought to indicate upregulation of the inflammatory arm of the immune system which, if prolonged, may contribute to both physical and mental health problems. These elevated markers of stress physiology are more evident when looking at aggregated measures of HPA axis functioning (e.g., diurnal slope) rather than at specific times of the day. Specifically, children with GAD showed no differences in cortisol at bedtime compared with children without GAD, despite differences in sleep latency and sleep architecture (Alfano, Reynolds, Scott, Dahl, & Mellman, 2013). These preliminary observations indicate that the differences in HPA axis functioning between individuals with and without GAD are subtle, but may reflect general disruption to the diurnal regulation of both glucocorticoids and sleep; which are both modulated by similar regions of the central nervous system.

#### 20.2.2.2 Social Anxiety Disorder

The association between salivary biomarkers of stress physiology in patients with SAD is less consistent. In perhaps the most comprehensive study of SAD biomarkers to date, individuals with SAD exhibited greater salivary cortisol responses to acute laboratory stress, greater amygdala responses to social phobia relevant words, and decreased methylation of the oxytocin receptor gene (measured in whole blood) (Ziegler et al., 2015). Oxytocin is a neuropeptide and hormone produced by the pituitary gland. It is known to facilitate childbirth and offspring caretaking (e.g., nursing) in mammals, but is also thought to be generally associated with prosocial behaviors, attachment, and bonding beyond the parent–offspring relationship (MacDonald & MacDonald, 2010). However, it is important to note that investigators do not agree that oxytocin is a useful biomarker for psychiatric disorders (Rutigliano et al., 2016).

With respect to physiological stress response systems, one study reported that patients with SAD show a blunted cortisol response to acute lab stress compared with healthy controls (Furlan, DeMartinis, Schweizer, Rickels, & Lucki, 2001). However, several other studies have observed no differences in salivary measures of α-amylase (García-Rubio, Espín, Hidalgo, Salvador, & Gómez-Amor, 2017; Krämer et al., 2012) or cortisol following acute laboratory stress (Klumbies, Braeuer, Hoyer, & Kirschbaum, 2014; Krämer et al., 2012; Martel et al., 1999) among those with and without SAD.

These inconsistencies raise the possibility that alterations to the functioning of the HPA axis are better indicators of risk for SAD rather than indicators of the presence of the disorder. Further supporting this possibility, children exhibit a within-person increase in salivary cortisol in the morning and afternoon when school begins, but children of mothers with SAD exhibit persisting increases in salivary cortisol at bedtime as well (Russ et al., 2012). This suggests that children of mothers with SAD, who are presumably at greater risk of developing SAD, may experience a global alteration to the circadian rhythm in HPA axis activity. The other possible explanation is that those with SAD, following chronic activation of HPA axis prior to the onset of the disorder, eventually exhibit chronic attenuation of the HPA axis as evidenced by blunted salivary cortisol.

#### 20.2.2.3 Obsessive-Compulsive Disorder

Very little salivary bioscience research has been conducted in populations with OCD, although the few studies that have been published suggest increased cortisol concentrations in the morning that correlate with symptom severity and comorbidity. Pediatric patients with OCD, compared with adult patients with OCD and healthy controls, were observed to have greater methylation of high expressing loci of the serotonin transporter gene (5-HTTLPR + rs25531) measured in saliva, which correlated positively with greater salivary cortisol levels at waking (Grünblatt et al.,

2018). Methylation describes the phenomenon where environmental influences can inhibit the expression of a particular gene, thus reducing its influence on an organism's phenotype. Methylation of the serotonin transporter (5-HTTLPR) is associated with reduced inhibitory influence of serotonin on the HPA axis and other systems. This is consistent with another study of children and adolescents with OCD, which documented elevated salivary cortisol concentrations in the morning, as well as blunted cortisol responses to acute laboratory stress (Gustafsson, Gustafsson, Ivarsson, & Nelson, 2008). Also, patients with OCD and high trait anxiety in the same sample exhibited larger CARs and less cortisol suppression following a dose of dexamethasone (Labad et al., 2018), suggesting impaired regulatory feedback of the HPA axis. Dexamethasone is a synthetic steroid that binds to glucocorticoid receptors to inhibit the release of ACTH from the pituitary, subsequently decreasing the release of cortisol as a result, in healthy individuals. Finally, patients with comorbid OCD and MDD exhibit a flatter diurnal cortisol slope compared to healthy controls or patients with either OCD or MDD alone (Labad et al., 2018). Further research is needed to support the association between symptom severity and comorbidity of OCD with alterations in salivary markers of neuroendocrine functioning and other physiological systems.

#### 20.2.2.4 Panic Disorder

Panic Disorder does not appear to be consistently related to differences in the functioning of physiological systems measured in saliva, at least not in trait-measures of functioning. One study observed that patients with Panic Disorder exhibit an attenuated salivary cortisol response to acute lab stress compared to healthy controls (Petrowski, Wintermann, Schaarschmidt, Bornstein, & Kirschbaum, 2013). Other studies have observed no difference in salivary cortisol (Wintermann, Kirschbaum, & Petrowski, 2016) or salivary α-amylase (Petrowski, Wintermann, Kirschbaum, & Strahler, 2016; Tanaka et al., 2012) responses to acute lab stress between patients with and without Panic Disorder; nor have studies observed differences in salivary cortisol responses to electrical stimulation (Tanaka et al., 2012) or cortisol responses to a 200 mg dose of oral L-5-hydroxytryptophan (Schruers, van Diest, Nicolson, & Griez, 2002). L-5-hydroxytryptophan is a natural amino acid and contributes to the biosynthesis of serotonin, thus providing some insight into the complex interplay between serotonin and stress physiology in Panic Disorder patients. Some of these mixed observations in response to lab procedures may occur because patients with Panic Disorder may not exhibit the same sensitivity to the novel laboratory context that other populations do (Stones, Groome, Perry, Hucklebridge, & Evans, 1999). However, there is preliminary evidence that patients with Panic Disorder exhibit elevated 3-methoxy-4-hydroxyphenylglycol (MHPG), a salivary norepinephrine metabolite marker (Yamada et al., 2000), suggesting that altered stress reactivity may occur further "upstream" in individuals with Panic Disorder, such as within the central rather than the peripheral nervous system.

#### 20.2.2.5 Post-traumatic Stress Disorder

Exposure to trauma is ubiquitous, more than 50% of children will be exposed to at least one potentially traumatic event by the age of 18 (Breslau, 2009; Kuhlman, Robles, Bower, & Carroll, 2018). Yet, only 10% of exposed individuals will develop PTSD. Thus, the focus of biomarker research in populations with PTSD has been on identifying markers of risk and resilience. For example, there is preliminary evidence that adolescent females whose mothers had PTSD have larger CARs and greater cortisol concentrations throughout the day than adolescent females with mothers with no history of PTSD (Liu et al., 2016). Additionally, both exposure to childhood maltreatment and having the Met allele of the catechol-O-methyltransferase (COMT) genetic polymorphism are linked to risk for PTSD. Having the COMT Met allele, measured in saliva, and a history of childhood maltreatment was associated with elevated PTSD symptoms and decreased activation of the amygdala during an inhibition task (van Rooij et al., 2016). Similarly, mothers exposed to intimate partner violence (IPV) during adulthood may be more at risk for parenting stress and PTSD symptoms with less methylation of the promoter region of NR3C1, the glucocorticoid receptor gene (Schechter et al., 2015). Further, adolescents girls with and without exposure to recent sexual abuse did not exhibit differences in diurnal regulation of salivary α-amylase, but among the sexually abused girls, having PTSD symptoms was associated with elevated morning concentrations (Keeshin, Strawn, Out, Granger, & Putnam, 2015). Within survivors of medical surgical intensive care units, homozygosity for the corticotrophin-releasing hormone binding protein, CRHBP rs10055255, T allele was associated with significantly fewer post-ICU PTSD symptoms, while carrying the CRH receptor type 1, CRHR1 rs1876831, C allele was associated with more post-ICU depressive symptoms (Davydow et al., 2014).

Suppressed production of glucocorticoids appears to be characteristic of PTSD. One common theoretical model of this phenomenon is that chronic and repeated activation of stress physiology contributes to allostatic load and, ultimately exhaustion, of these systems (Juster et al., 2010; McEwen, 2000). For example, combat veterans with PTSD have lower CAR and lower cortisol throughout the day compared to combat veterans without PTSD (Wahbeh & Oken, 2013). Importantly, SKA2 DNA methylation, measured in both saliva and blood, is hypothesized to suppress the production of glucocorticoids following stress and interacts with trauma exposure to predict PTSD as well as suicide attempts (Kaminsky et al., 2015). This suggests that there are salivary markers that can be used to predict risk for PTSD following exposure to trauma as well as severity of distress during the course of the disorder. However, it is important to note that both hyper- and hyporeactivity of the HPA axis have been observed in the context of PTSD.

Research into risk for and biological underpinnings of PTSD have also branched out into other salivary biomarkers. For example, male police officers with PTSD have lower basal salivary oxytocin levels compared to male officers without PTSD, whereas there were no differences in salivary oxytocin between affected and non-affected female officers (Frijling et al., 2015). There is also some evidence that PTSD is associated with immune alterations. Veterans with PTSD exhibit immune alterations, specifically an imbalance of cytokines toward more pro-inflammatory and Th1 salivary markers compared with veterans without PTSD (Wang, Mandel, Levingston, & Young, 2016). Further, women with current PTSD had elevated salivary IL-6 upon lab arrival compared to women with remitted or no history of PTSD (Newton, Fernandez-Botran, Miller, & Burns, 2014).

### *20.2.3 Substance Use Disorders*

Salivary bioscience within populations of patients suffering from substance use disorders have been limited to date and have largely focused on either the impact of the used substance on physiological processes or biomarkers of substance use that have the potential to obviate self-report measures. For example, alcohol-dependent patients exhibit altered salivary α-amylase, clusterin, haptoglobin, heavy and light chains of immunoglobulins, and transferrin when compared with healthy control participants (Kratz et al., 2014). Alterations to these biomarkers may be used in monitoring the damage to multiple physiological systems that are attributed to substance use and dependence. Chronic cannabis users exhibit attenuated psychological and salivary cortisol responses to acute stress (Cuttler et al., 2017). Additionally, smokers with greater nicotine use exhibit attenuated cortisol responses to stress, high concentrations of salivary cotinine (the predominant metabolite of nicotine), and lower concentrations of salivary carbon monoxide. The altered cortisol responses and salivary cotinine are related to greater salivary α-amylase responses to acute stress (Morris, Mielock, & Rao, 2016). In fact, salivary cotinine appears to be a good salivary index for nicotine exposure (Chen et al., 2017; Fu et al., 2012) that could be used widely to understand the impact of nicotine exposure on public health and human development. Additionally, oral fluids are a valid method of assessing the presence of behavior-altering narcotics, such as opioids, in ambulatory and field settings (Herrera-Gómez, García-Mingo, Colás, González-Luque, & Álvarez, 2018). Though these findings are promising, they indicate that use of salivary bioscience in the context of risk for and consequences of substance use disorders is currently in the very early stages with great potential for impact and innovation.

### *20.2.4 Eating Disorders*

There are two predominant eating disorders in psychiatry: anorexia nervosa and bulimia nervosa (American Psychiatric Association, 2013). The use of salivary bioscience in this field of psychiatry and clinical psychology research has been limited. One thrust of this research area has focused on differentiating trait-like biomarkers of risk for eating disorders from state-like biomarkers of current disease status and severity. For example, women with current anorexia nervosa have exaggerated CARs when compared to both healthy control women and weight-restored women with a history of anorexia nervosa (Monteleone et al., 2016). Consistent with this, women with anorexia nervosa also exhibit elevated cortisol concentrations throughout the day prior to inpatient treatment (Shibuya et al., 2011). Further, cortisol concentrations in anorexia nervosa patients also appear to be less stable from day to day compared with healthy controls (Wild et al., 2016). Despite these consistent observations of elevated cortisol in the morning, women with eating disorders appear to exhibit blunted salivary cortisol and α-amylase responses to acute lab stress (Het et al., 2015).

While all of these studies have focused on markers of stress physiology, interest in other physiological markers is also growing. For example, gonadal hormones including estradiol and testosterone play organizing roles in the development of eating disorders, although only some of this work has been conducted using salivary as opposed to blood-based biomarkers (Klump et al., 2006). Additionally, uric acid has recently been validated as a reliable and meaningful salivary biomarker (Riis et al., 2018). Uric acid is a normal component of urine that is the product of purine nucleotide metabolism and reflects functioning across multiple physiological systems. Uric acid may be bidirectionally involved in anorexia nervosa; alterations in uric acid not only play a role in the pathogenesis of the disease but also may be perpetuated by the symptoms and course of the illness (Simeunovic Ostojic & Maas, 2018). That being said, it is important to acknowledge the limitations of salivary bioscience research in the context of eating disorders given the impact of eating disorders on oral health. These include inflammation of the gums and oral cavity, and introduction of additional bacteria into the salivary fluid. Chapters 2, 3, and 4 in this book provide information on the role of oral health in salivary biomarkers and how to effectively address them.

### *20.2.5 Autism Spectrum Disorder*

Within Autism Spectrum Disorders (ASD) research, there has been a focus on identifying genetic markers of risk in both blood and saliva. In saliva, 14 microRNAs are differentially expressed in individuals with ASD compared to typically developing children, and these microRNAs were also correlated with functional impairment (Hicks, Ignacio, Gentile, & Middleton, 2016). Among pediatric populations, and populations with ASD specifically, there has also been interest in validating more salivary biomarkers to be used instead of invasive measures like blood and urine (Galiana-Simal, Muñoz-Martinez, Calero-Bueno, Vela-Romero, & Beato-Fernandez, 2018), including examining neuroactive amino acids, such as glutamate, tryptophan, and GABA that may be involved in symptom development and maintenance (Zheng, Wang, Li, Rauw, & Baker, 2017). Indeed, proteomic analysis of salivary samples comparing individuals with and without ASD found elevations in several salivary proteins among individuals with ASD (Ngounou Wetie et al., 2015).

Children with ASD appear to exhibit elevated diurnal cortisol throughout the day with a flat pattern of diurnal regulation (Tomarken, Han, & Corbett, 2015; Tordjman et al., 2014). And further, elevated salivary cortisol in children under 12 appears to be positively correlated with stereotypic behaviors, a phenomenon that is no longer apparent in adolescents (Bitsika, Sharpley, Agnew, & Andronicos, 2015). Some developmental research has also shown that the synchrony of parent–child stress hormones (e.g., cortisol) is weaker in dyads including a child with ASD (Saxbe et al., 2017). Further, youth with ASD exhibit blunted cortisol responses to acute lab stress but only when they also had a comorbid anxiety disorder (Hollocks, Howlin, Papadopoulos, Khondoker, & Simonoff, 2014). On the other hand, children with ASD appear to exhibit elevated cortisol during social interactions solicited from another child, and concentrations of cortisol were negatively correlated with social communication (Corbett et al., 2014). Importantly, elevated hormones in ASD extend well beyond cortisol and have been observed for several adrenal hormones (Majewska et al., 2014). One important issue is that individuals with ASD may have poorer oral health (Bhandary & Hari, 2017; Blomqvist, Bejerot, & Dahllöf, 2015; Rai, Hegde, & Jose, 2012), which may confound the results associated with the diagnosis and should be carefully examined and accounted for.

### *20.2.6 Externalizing Behavior Disorders*

Attention-deficit/hyperactivity disorder (ADHD) is a neurodevelopmental disorder that is typically diagnosed in childhood and includes pervasive externalizing behavior problems in the domains of inattention (difficulty deploying and sustaining attention), hyperactivity (excessive movement that is not fitting to the setting), and impulsivity (difficulty with inhibitory control and executive planning) (American Psychiatric Association, 2013). ADHD is a heterogeneous disorder such that some patients can predominantly suffer from symptoms of inattention, others may suffer from symptoms of hyperactivity, and others from a combination of both symptom clusters. Correlates of ADHD with stress physiology appear to be contingent upon the predominant symptoms the individual exhibits. For example, hyperactivity symptoms have been related to attenuated salivary cortisol stress reactivity while inattentive symptoms have been associated with robust stress hormone reactivity (Maldonado, Trianes, Cortés, Moreno, & Escobar, 2009; van West, Claes, & Deboutte, 2009). These findings may be specific to the type of laboratory stressor used. One study found that children with ADHD exhibit larger salivary cortisol responses to an academic stressor (Palma, Fernandes, Muszkat, & Calil, 2012) while another found that children with ADHD have more attenuated responses to a catheter insertion (McCarthy et al., 2011). Individuals with higher ADHD symptoms may exhibit an exaggerated CAR overall (Vogel et al., 2017), however, both inattentive and combined (mixed hyperactive and inattentive symptoms) types of ADHD have been linked to lower cortisol values across the day as measured by several diurnal indices (Angeli et al., 2018; Isaksson, Nilsson, Nyberg, Hogmark, & Lindblad, 2012; Pinto et al., 2016). Further, another study of 6–12 year old children observed that cortisol values were no different in children with ADHD compared to controls, but that salivary DHEA was lower, as well as the DHEA/cortisol ratio (Wang et al., 2011). These patterns may reverse in adulthood where ADHD has been associated with both more robust salivary cortisol reactivity to acute stress (Raz & Leykin, 2015) and normalized responses to stress (Corominas-Roso et al., 2015). Salivary markers of oxidative stress have also been implicated in the pathogenesis of ADHD, such that children with ADHD exhibited elevated thiols and pseudocholinesterase, and attenuated magnesium compared to healthy controls (Archana et al., 2012).

While patterns of stress hormone reactivity have dominated this area of research, several promising findings have been published that indicate other important salivary biomarkers of ADHD. For example, salivary melatonin has been used to show alterations to circadian rhythms in children with ADHD (Nováková et al., 2011), and elevated salivary melatonin in adults with ADHD have been linked to expression of circadian clock genes, BMAL1 and PER2 (Baird, Coogan, Siddiqui, Donev, & Thome, 2012). Higher DNA methylation of the dopamine receptor, DRD4, inflammatory processes, VIPR2, and monamine cholinergic neurotransmission, MYT1L, all measured in saliva was associated with either more severe ADHD or diagnostic status (Dadds, Schollar-Root, Lenroot, Moul, & Hawes, 2016; Wilmot et al., 2016). Finally, similar to monitoring of lithium toxicity in patients with Bipolar Disorder, therapeutic drug monitoring of methylphenidate can also be conducted in saliva instead of blood which presents an opportunity to significantly improve the dissemination of this pharmacological treatment to the populations in need (Preiskorn et al., 2018). Again, Chap. 17 of this book provides more details on the potential for salivary biomarkers to contribute to therapeutic drug monitoring.

It is also important to note whether ADHD is linked to differences in oral health that can influence the integrity of salivary biomarkers. To our knowledge, only one study has directly addressed this issue and found that youth with ADHD (both medicated and unmedicated) have slower salivary flow rates and more plaque than youth without ADHD (Hidas et al., 2011).

Male youth with externalizing behavior problems, heterogeneous for oppositional defiant disorder (ODD), ADHD, and Conduct Disorder problems, did not exhibit any differences in diurnal regulation of cortisol, but male youth with conduct problems exhibited lower salivary α-amylase (Angyal et al., 2016). ODD is an externalizing behavior disorder in children and adolescents characterized by defiance of authority (American Psychiatric Association, 2013). Similar patterns were observed when comparing salivary cortisol concentrations and their reactivity to acute lab stress between individuals with ADHD alone and comorbid ADHD and Conduct Disorder (Northover, Thapar, Langley, Fairchild, & van Goozen, 2016). One factor that may be particularly important to consider in youth with externalizing behavior problems are callous-unemotional traits, which are an endophenotype for Conduct Disorder. The antisocial behaviors (e.g., rule breaking, violence, and destruction of others' property) are thought to be related to general hypo-arousal of physiological systems (Scarpa & Raine, 1997). Indeed, youth (ages 8–14) with ADHD and high callous-unemotional traits exhibit attenuated salivary cortisol responses to stress but this pattern was not seen in youth with ADHD in the absence of these traits (Stadler et al., 2011). Similarly, children with comorbid ADHD and ODD/CD exhibit a blunted CAR compared to youth with ADHD alone or control children (Freitag et al., 2009), and salivary cortisol overall appears to be lower in children with comorbid ADHD and ODD compared with ADHD alone (Kariyawasam, Zaw, & Handley, 2002).

A number of studies have looked at salivary biomarkers that offer insight into callous-unemotional traits. First, salivary measures of DNA have been used to show that functional single nucleotide polymorphisms from the serotonin receptor, HTR1b, are associated with callous-unemotional traits (Moul, Dobson-Stone, Brennan, Hawes, & Dadds, 2013, 2015). There also appears to be a consensus that male youth with high callous-unemotional traits exhibit lower circulating cortisol than comparison youth (Loney, Butler, Lima, Counts, & Eckel, 2006; von Polier et al., 2013) and lower DHEA (Kimonis, Goulter, Hawes, Wilbur, & Groer, 2017), although the findings have been somewhat mixed (See Moul, Hawes, & Dadds, 2018 for review). Typically, salivary markers of cortisol, DHEA, and testosterone are correlated within individuals across time. However, higher callous-unemotional traits in adolescents appear to be related to greater asynchrony of testosterone with cortisol and DHEA (Johnson et al., 2014), suggesting that the best biomarkers of these traits are indicators of coordination across the HPA and HPG axes, rather than any single analyte on its own.

## *20.2.7 Schizophrenia*

Salivary bioscience research on schizophrenia has broadly focused much of its effort on identifying genetic markers of risk. The salivary bioscience research in this area is no exception. Salivary DNA has been used to identify single nucleotide polymorphisms, rs7004633 (MMP16), to understand genetic risk for schizophrenia (Morton et al., 2017), as well as hypermethylation of the serotonin transporter gene (Abdolmaleky et al., 2014), and hypomethylation of the MB-COMT gene promoter (Nohesara et al., 2011), and the serotonin receptor (HTR2A) (Ghadirivasfi et al., 2011).

Salivary biomarkers of stress hormone regulation, such as CAR, appear to be attenuated in individuals with schizophrenia and other psychiatric disorders (Berger et al., 2016), although one recent study observed an exaggerated CAR in patients with first-episode psychosis who were exposed to child abuse (Ciufolini et al., 2018). Otherwise, very few differences have been observed between individuals with a psychotic disorder and healthy controls in the diurnal regulation or tonic concentrations of cortisol (Chaumette et al., 2016; Vaessen et al., 2018), although some studies have reported salivary cortisol concentrations to be lower (Hempel et al., 2010; Tobolska et al., 2016). Some studies have also explored whether individuals with schizophrenia demonstrate different physiological responses to acute stress, however so far, findings have been mixed and largely inconclusive (Brenner et al., 2009; Jansen et al., 1998; Lange et al., 2017; Zorn et al., 2017).

Salivary measures of the kynurenine pathway may also be important to examine in the context of schizophrenia. Kynurenic acid is a neuroactive product of tryptophan metabolism and may be associated with neurophysiological and neuropathological processes. Elevated kynurenic acid has been related to the cognitive symptoms observed in patients with schizophrenia (e.g., confusion and psychotic symptoms) by contributing to excessive inhibition of glutamatergic and acetylcholine neurotransmission (Schwarcz & Hunter, 2007). Individuals with schizophrenia also exhibit larger increases in salivary kynurenic acid compared to healthy controls (Chiappelli et al., 2014, 2018). This may be consistent with other neuroendocrine-immune interaction differences observed in individuals with schizophrenia. For example, cortisol inhibits the release of pro-inflammatory cytokines from immune cells. Thus, rising concentrations of cortisol should track with declines in inflammatory cytokines over time. This is observed in the saliva of healthy control participants after a psychological stressor, but not in individuals with schizophrenia suggesting that their immune cells may be glucocorticoid resistant (Chiappelli et al., 2016). Indeed, proteomics of whole saliva in patients with serious mental illness, specifically Bipolar Disorder and schizophrenia, have been used to show that serious mental illness is associated with widespread immune regulation differences (Iavarone et al., 2014). Alternatively, these neuroendocrine-immune differences could indicate greater sympathetic nervous system activation in individuals with schizophrenia, which directly influences the production of inflammatory cytokines. For example, individuals with schizophrenia have shown greater salivary α-amylase concentrations compared to healthy controls (Inagaki et al., 2010; Monteleone et al., 2015).

There is also limited but promising evidence that individuals with schizophrenia may have compromised endogenous sleep-onset physiology. In healthy controls, higher salivary melatonin was associated with shorter sleep-onset latency while such a relationship was not present in patients with schizophrenia (Afonso, Figueira, & Paiva, 2011).

Very little research has been conducted to date on the oral health of patients with schizophrenia, although the studies that do exist suggest a need for this research to better understand how to ensure the integrity of saliva samples in this population and protect its comparability with other populations. For example, the use of different antipsychotic medications has implications for oral health (Eltas, Kartalcı, Eltas, Dündar, & Uslu, 2013).

## 20.3 Use of Salivary Biomarkers in Behavioral and Pharmacological Treatments for Psychiatric Disorders

One of the most exciting new frontiers in the use of salivary biomarkers has been in the evaluation of pharmacological and behavioral interventions for patients with psychiatric disorders. Rather than evaluating intervention effectiveness on symptoms alone, use of salivary biomarkers can offer insight into physiological mechanisms through which these interventions work as well as the potentially pleiotropic impact interventions have on behavior, cognition, and physiological processes.

For example, decreases in salivary cortisol concentrations following antidepressant treatment for generalized anxiety disorder were correlated with symptom improvement (Lenze et al., 2012). Treatment with antidepressants may also regulate HPA axis functioning in individuals with depression (Hinkelmann et al., 2012; Ruhé et al., 2015). Also, one study has shown that supplementation with curcumin decreased salivary cortisol in a randomized, double-blind, placebo-controlled study among depressed males (Yu, Pei, Zhang, Wen, & Yang, 2015), although this observation is inconsistent across studies (Lopresti et al., 2015).

Fewer studies using behavioral interventions have incorporated salivary biomarkers into their outcome measures, although the findings are promising so far. For example, salivary cortisol increased progressively across sessions of prolonged exposure in a veteran population with PTSD, and the pattern of cortisol change across sessions predicted treatment response (Rauch, King, Liberzon, & Sripada, 2017). Mindfulness-based interventions for stress reduction in patients with depressive symptoms and chronic physical illness are associated with decreases in salivary cytokines (Stefanaki et al., 2015; Walsh, Eisenlohr-Moul, & Baer, 2016) and salivary cortisol (Christopher et al., 2018).

Finally, there have been some findings from nontraditional therapies. Light therapy increases salivary melatonin in the evening and decreases salivary cortisol in patients with depression (Lieverse et al., 2011). Salivary markers of circadian genes can be used to show improvements in circadian rhythm over the course of treatment for mood disorders (Moon et al., 2016). And finally, having a service animal was associated with lower concentrations of salivary cortisol, such as CAR and AUCi, in veterans with PTSD (Rodriguez, Bryce, Granger, & O'Haire, 2018). Salivary biomarkers have great clinical potential; they offer noninvasive assessment of individual differences in functioning across multiple physiological systems and could be more widely used in evaluation of the effectiveness of treatments and designing new therapies in the service of precision medicine. Chapter 29 of this book goes into further detail on the potential for salivary biomarkers to contribute to advances in precision medicine.

## 20.4 Methodological Considerations: Oral Health

Oral health is an important consideration in salivary bioscience research on populations with or at risk for psychiatric disorders. For example, oral health can be impacted by different factors such as antipsychotic medications, and researchers should carefully consider the role of psychiatric medications on these biomarkers in their study designs and analyses. Lifestyle factors can also be an important consideration in psychiatric populations, such as limited access to regular medical and dental care, difficulties engaging in activities of daily living, and symptoms of the disorders themselves. In particular, concerns about the role of oral health in studies appear to be more important in populations with eating disorders, Autism Spectrum Disorders, ADHD and Schizophrenia. Chapters 2, 3, and 4 provide detailed explanations and strategies for study design, analysis, and consideration of the potential role of oral health when collecting salivary biomarkers.

## 20.5 Conclusions and Future Directions

Salivary biomarkers have now been incorporated into research across psychiatric disorders and treatment modalities. This research has predominantly focused on the regulation of HPA axis stress hormones, such as cortisol and DHEA. However, very promising results have emerged from the measurement of gonadal hormones, DNA, melatonin, and other proteins that allow us to less invasively, and more comprehensively understand the biological underpinnings of psychiatric disease. We see several areas of opportunity for salivary bioscience to drive clinically meaningful progress in the field of clinical psychology and psychiatry. First, this research has focused predominantly on case-control designs characterizing individuals with a psychiatric disorder when compared with healthy control participants. As a result, our understanding of the specificity of any salivary biomarkers to any one psychiatric disorder remains unknown. If one of the purposes of identifying biomarkers of psychiatric disease is to more accurately identify individuals in need of treatment, characterizing the specificity of these salivary biomarkers to disorders is an important next step. Along these lines, this chapter focused on what were previously known as Axis I disorders (American Psychological Association, 2000), and does not address research looking at the salivary biomarkers observed in the context of personality disorders or symptoms. Individuals with borderline personality disorder (BPD), for example, exhibit different cortisol and α-amylase reactivity in response to social evaluative stress (Scott, Levy, & Granger, 2013). As psychiatric research becomes less categorically focused, and more dedicated to dimensional models of symptoms and domains of impairment, the application of knowledge to persistent disorders such as personality disorders will be more and more apparent and relevant. A second area of critical need at the intersection of psychiatric research and salivary bioscience is more widespread examination of the sensitivity of disease-specific biomarkers to effective treatment. Indeed, biomarkers are intended to reflect the underlying neurobiology of psychiatric disease, in which case effective treatments should mitigate not only the behavioral manifestation of the disorder but also the underlying neurobiology. Research to this end has been extremely limited and represents an important new frontier for research in clinical psychology and psychiatry. Historically, this may be the case because biomarkers were largely assessed through the collection of blood and therefore more invasive and difficult to collect and store. The feasibility of biomarker collection in clinical and ambulatory settings through saliva essentially eliminates that barrier. Finally, it is important to keep in mind that our approach to summarizing the research linking salivary biomarkers to the diagnosis and treatment of psychiatric disorders was largely descriptive. There are numerous theories driving progress in the psychobiological processes underlying psychiatric diseases, yet given the scope of this chapter, we have only acknowledged a select few. As more studies are conducted in this area and more biomarkers validated and used across laboratories, synthesizing the strength of support for these theories will be of the utmost importance in informing patient care and policy.

## References

Abdolmaleky, H. M., Nohesara, S., Ghadirivasfi, M., Lambert, A. W., Ahmadkhaniha, H., Ozturk, S., . . . Thiagalingam, S. (2014). DNA hypermethylation of serotonin transporter gene promoter in drug naïve patients with schizophrenia. *Schizophrenia Research, 152*(2–3), 373–380. https://doi.org/10.1016/j.schres.2013.12.007

Adam, E. K., Doane, L. D., Zinbarg, R. E., Mineka, S., Craske, M. G., & Griffith, J. W. (2010). Prospective prediction of major depressive disorder from cortisol awakening responses in adolescence. *Psychoneuroendocrinology, 35*(6), 921–931. https://doi.org/10.1016/j.psyneuen.2009.12.007

Adam, E. K., & Gunnar, M. R. (2001). Relationship functioning and home and work demands predict individual differences in diurnal cortisol patterns in women. *Psychoneuroendocrinology, 26*(2), 189–208. https://doi.org/10.1016/S0306-4530(00)00045-7

Adam, E. K., Hawkley, L. C., Kudielka, B. M., & Cacioppo, J. T. (2006). Day-to-day dynamics of experience—cortisol associations in a population-based sample of older adults. *Proceedings of the National Academy of Sciences, 103*(45), 17058–17063. https://doi.org/10.1073/pnas.0605053103

Afonso, P., Figueira, M. L., & Paiva, T. (2011). Sleep-promoting action of the endogenous melatonin in schizophrenia compared to healthy controls. *International Journal of Psychiatry in Clinical Practice, 15*(4), 311–315. https://doi.org/10.3109/13651501.2011.605954

Alfano, C. A., Reynolds, K., Scott, N., Dahl, R. E., & Mellman, T. A. (2013). Polysomnographic sleep patterns of non-depressed, non-medicated children with generalized anxiety disorder. *Journal of Affective Disorders, 147*(1–3), 379–384. https://doi.org/10.1016/j.jad.2012.08.015

American Psychiatric Association. (2013). *Diagnostic and statistical manual of mental disorders*. https://doi.org/10.1176/appi.books.9780890425596.744053

American Psychological Association. (2000). *Diagnostic and statistical manual of mental disorders, fourth edition, text revision (DSM-IV-TR).* Washington, DC.

Angeli, E., Korpa, T., Johnson, E. O., Apostolakou, F., Papassotiriou, I., Chrousos, G. P., & Pervanidou, P. (2018). Salivary cortisol and alpha-amylase diurnal profiles and stress reactivity in children with Attention Deficit Hyperactivity Disorder. *Psychoneuroendocrinology, 90*, 174–181. https://doi.org/10.1016/j.psyneuen.2018.02.026

Angold, A., & Costello, E. J. (2006). Puberty and depression. *Child and Adolescent Psychiatric Clinics of North America, 15*(4), 919–937, ix. https://doi.org/10.1016/j.chc.2006.05.013

Angyal, N., Halasz, J., Meszaros, G., Kovacs, J. K., Kruk, E., & Nemoda, Z. (2016). Potential salivary biomarkers and their genetic effects in a pilot study of adolescent boys with externalizing problems. *Neuropsychopharmacologia Hungarica: A Magyar Pszichofarmakologiai Egyesulet Lapja = Official Journal of the Hungarian Association of Psychopharmacology, 18* (4), 173–179.

Archana, E., Pai, P., Prabhu, B. K., Shenoy, R. P., Prabhu, K., & Rao, A. (2012). Altered biochemical parameters in saliva of pediatric attention deficit hyperactivity disorder. *Neurochemical Research, 37*(2), 330–334. https://doi.org/10.1007/s11064-011-0616-x

Baird, A. L., Coogan, A. N., Siddiqui, A., Donev, R. M., & Thome, J. (2012). Adult attention-deficit hyperactivity disorder is associated with alterations in circadian rhythms at the behavioural, endocrine and molecular levels. *Molecular Psychiatry, 17*(10), 988–995. https://doi.org/10.1038/mp.2011.149

Ben-Aryeh, H., Naon, H., Szargel, R., Gutman, D., & Hefetz, A. (1980). Salivary lithium concentration—a tool for monitoring psychiatric patients. *Oral Surgery, Oral Medicine, and Oral Pathology, 50*(2), 127–129.

Berger, M., Kraeuter, A. K., Romanik, D., Malouf, P., Amminger, G. P., & Sarnyai, Z. (2016). Cortisol awakening response in patients with psychosis: Systematic review and meta-analysis. *Neuroscience and Biobehavioral Reviews, 68*, 157–166. https://doi.org/10.1016/j.neubiorev.2016.05.027

Bhagwagar, Z., Hafizi, S., & Cowen, P. J. (2005). Increased salivary cortisol after waking in depression. *Psychopharmacology, 182*(1), 54–57. https://doi.org/10.1007/s00213-005-0062-z

Bhandary, S., & Hari, N. (2017). Salivary biomarker levels and oral health status of children with autistic spectrum disorders: A comparative study. *European Archives of Paediatric Dentistry: Official Journal of the European Academy of Paediatric Dentistry, 18*(2), 91–96. https://doi.org/10.1007/s40368-017-0275-y

Bitsika, V., Sharpley, C. F., Agnew, L. L., & Andronicos, N. M. (2015). Age-related differences in the association between stereotypic behaviour and salivary cortisol in young males with an Autism Spectrum Disorder. *Physiology & Behavior, 152*(Pt A), 238–243. https://doi.org/10.1016/j.physbeh.2015.10.010

Blomqvist, M., Bejerot, S., & Dahllöf, G. (2015). A cross-sectional study on oral health and dental care in intellectually able adults with autism spectrum disorder. *BMC Oral Health, 15*(81). https://doi.org/10.1186/s12903-015-0065-z

Brenner, K., Liu, A., Laplante, D. P., Lupien, S., Pruessner, J. C., Ciampi, A., ... King, S. (2009). Cortisol response to a psychosocial stressor in schizophrenia: Blunted, delayed, or normal? *Psychoneuroendocrinology, 34*(6), 859–868. https://doi.org/10.1016/j.psyneuen.2009.01.002

Breslau, N. (2009). The epidemiology of trauma, PTSD, and other posttrauma disorders. *Trauma, Violence, & Abuse, 10*(3), 198–210. https://doi.org/10.1177/1524838009334448

Burke, H. M., Davis, M. C., Otte, C., & Mohr, D. C. (2005). Depression and cortisol responses to psychological stress: A meta-analysis. *Psychoneuroendocrinology, 30*(9), 846–856. https://doi.org/10.1016/j.psyneuen.2005.02.010

Chaumette, B., Kebir, O., Mam-Lam Fook, C., Morvan, Y., Bourgin, J., Godsil, B. P., ... Krebs, M.-O. (2016). Salivary cortisol in early psychosis: New findings and meta-analysis. *Psychoneuroendocrinology, 63*, 262–270. https://doi.org/10.1016/j.psyneuen.2015.10.007

Chen, A., Krebs, N. M., Zhu, J., Sun, D., Stennett, A., & Muscat, J. E. (2017). Sex/gender differences in cotinine levels among daily smokers in the Pennsylvania Adult Smoking Study. *Journal of Women's Health (2002), 26*(11), 1222–1230. https://doi.org/10.1089/jwh.2016.6317

Chiappelli, J., Pocivavsek, A., Nugent, K. L., Notarangelo, F. M., Kochunov, P., Rowland, L. M., ... Hong, E, L. (2014). Stress-induced increase in kynurenic acid as a potential biomarker for patients with schizophrenia and distress intolerance. *JAMA Psychiatry, 71*(7), 761–768. https://doi.org/10.1001/jamapsychiatry.2014.243

Chiappelli, J., Rowland, L. M., Notarangelo, F. M., Wijtenburg, S. A., Thomas, M. A. R., Pocivavsek, A., … Hong, L. E. (2018). Salivary kynurenic acid response to psychological stress: Inverse relationship to cortical glutamate in schizophrenia. *Neuropsychopharmacology: Official Publication of the American College of Neuropsychopharmacology*. https://doi.org/10.1038/s41386-018-0072-2

Chiappelli, J., Shi, Q., Kodi, P., Savransky, A., Kochunov, P., Rowland, L. M., … Hong, L. E. (2016). Disrupted glucocorticoid—Immune interactions during stress response in schizophrenia. *Psychoneuroendocrinology, 63*, 86–93. https://doi.org/10.1016/j.psyneuen.2015.09.010

Christopher, M. S., Hunsinger, M., Goerling, L. R. J., Bowen, S., Rogers, B. S., Gross, C. R., … Pruessner, J. C. (2018). Mindfulness-based resilience training to reduce health risk, stress reactivity, and aggression among law enforcement officers: A feasibility and preliminary efficacy trial. *Psychiatry Research, 264*, 104–115. https://doi.org/10.1016/j.psychres.2018.03.059

Ciufolini, S., Gayer-Anderson, C., Fisher, H. L., Marques, T. R., Taylor, H., Di Forti, M., … Mondelli, V. (2018). Cortisol awakening response is decreased in patients with first-episode psychosis and increased in healthy controls with a history of severe childhood abuse. *Schizophrenia Research*. https://doi.org/10.1016/j.schres.2018.05.002

Clive, M. L., Boks, M. P., Vinkers, C. H., Osborne, L. M., Payne, J. L., Ressler, K. J., … Kaminsky, Z. (2016). Discovery and replication of a peripheral tissue DNA methylation biosignature to augment a suicide prediction model. *Clinical Epigenetics, 8*, 113. https://doi.org/10.1186/s13148-016-0279-1

Clow, A., Hucklebridge, F., Stalder, T., Evans, P., & Thorn, L. (2010). The cortisol awakening response: More than a measure of HPA axis function. *Neuroscience & Biobehavioral Reviews, 35*(1), 97–103. https://doi.org/10.1016/j.neubiorev.2009.12.011

Corbett, B. A., Swain, D. M., Newsom, C., Wang, L., Song, Y., & Edgerton, D. (2014). Biobehavioral profiles of arousal and social motivation in autism spectrum disorders. *Journal of Child Psychology and Psychiatry, and Allied Disciplines, 55*(8), 924–934. https://doi.org/10.1111/jcpp.12184

Corominas-Roso, M., Palomar, G., Ferrer, R., Real, A., Nogueira, M., Corrales, M., … Ramos-Quiroga, J. A. (2015). Cortisol response to stress in adults with attention deficit hyperactivity disorder. *The International Journal of Neuropsychopharmacology, 18*(9). https://doi.org/10.1093/ijnp/pyv027

Cubała, W. J., Landowski, J., Dziadziuszko, M., Chrzanowska, A., & Wielgomas, B. (2016). Magnesium, C-reactive protein, and cortisol in drug-naïve patients with short illness-duration, first episode major depressive disorder: Possible immunomodulatory role for magnesium. *Magnesium Research, 29*(4), 169–174. https://doi.org/10.1684/mrh.2016.0413

Cuttler, C., Spradlin, A., Nusbaum, A. T., Whitney, P., Hinson, J. M., & McLaughlin, R. J. (2017). Blunted stress reactivity in chronic cannabis users. *Psychopharmacology, 234*(15), 2299–2309. https://doi.org/10.1007/s00213-017-4648-z

Dadds, M. R., Schollar-Root, O., Lenroot, R., Moul, C., & Hawes, D. J. (2016). Epigenetic regulation of the DRD4 gene and dimensions of attention-deficit/hyperactivity disorder in children. *European Child & Adolescent Psychiatry, 25*(10), 1081–1089. https://doi.org/10.1007/s00787-016-0828-3

Davis, J., Maes, M., Andreazza, A., McGrath, J. J., Tye, S. J., & Berk, M. (2015). Towards a classification of biomarkers of neuropsychiatric disease: From encompass to compass. *Molecular Psychiatry, 20*(2), 152–153. https://doi.org/10.1038/mp.2014.139

Davydow, D. S., Kohen, R., Hough, C. L., Tracy, J. H., Zatzick, D., & Katon, W. J. (2014). A pilot investigation of the association of genetic polymorphisms regulating corticotrophin-releasing hormone with posttraumatic stress and depressive symptoms in medical-surgical intensive care unit survivors. *Journal of Critical Care, 29*(1), 101–106. https://doi.org/10.1016/j.jcrc.2013.08.016

De Kloet, E. R. (1991). Brain corticosteroid receptor balance and homeostatic control. *Frontiers in Neuroendocrinology, 12*(2), 95–164.

de Kloet, E. R., Joëls, M., & Holsboer, F. (2005). Stress and the brain: From adaptation to disease. *Nature Reviews Neuroscience, 6*(6), 463–475. https://doi.org/10.1038/nrn1683

Dysken, M. W., Johnson, S. B., Holden, L., Kuskowski, M., Ofstehage, J., Skare, S., & Vatassery, G. (1992). Haloperidol and reduced haloperidol in saliva and blood. *Journal of Clinical Psychopharmacology, 12*(3), 186–190.

Eltas, A., Kartalcı, S., Eltas, S. D., Dündar, S., & Uslu, M. O. (2013). An assessment of periodontal health in patients with schizophrenia and taking antipsychotic medication. *International Journal of Dental Hygiene, 11*(2), 78–83. https://doi.org/10.1111/j.1601-5037.2012.00558.x

Freitag, C. M., Hänig, S., Palmason, H., Meyer, J., Wüst, S., & Seitz, C. (2009). Cortisol awakening response in healthy children and children with ADHD: Impact of comorbid disorders and psychosocial risk factors. *Psychoneuroendocrinology, 34*(7), 1019–1028. https://doi.org/10.1016/j.psyneuen.2009.01.018

Frey, S., Birchler-Pedross, A., Hofstetter, M., Brunner, P., Götz, T., Münch, M., … Cajochen, C. (2012). Young women with major depression live on higher homeostatic sleep pressure than healthy controls. *Chronobiology International, 29*(3), 278–294. https://doi.org/10.3109/07420528.2012.656163

Fries, E., Dettenborn, L., & Kirschbaum, C. (2009). The cortisol awakening response (CAR): Facts and future directions. *International Journal of Psychophysiology, 72*(1), 67–73. https://doi.org/10.1016/j.ijpsycho.2008.03.014

Frijling, J. L., van Zuiden, M., Nawijn, L., Koch, S. B. J., Neumann, I. D., Veltman, D. J., & Olff, M. (2015). Salivary oxytocin and vasopressin levels in police officers with and without post-traumatic stress disorder. *Journal of Neuroendocrinology, 27*(10), 743–751. https://doi.org/10.1111/jne.12300

Fu, M., Martínez-Sánchez, J. M., Agudo, A., Pascual, J. A., Ariza, C., Moncada, A., … DCOT Study Investigators. (2012). Nicotine dependence and salivary cotinine concentration in daily smokers. *European Journal of Cancer Prevention: The Official Journal of the European Cancer Prevention Organisation (ECP), 21*(1), 96–102. https://doi.org/10.1097/CEJ.0b013e32834a7e59

Funke, R., Eichler, A., Distler, J., Golub, Y., Kratz, O., & Moll, G. H. (2017). Stress system dysregulation in pediatric generalized anxiety disorder associated with comorbid depression. *Stress and Health: Journal of the International Society for the Investigation of Stress, 33*(5), 518–529. https://doi.org/10.1002/smi.2736

Furlan, P. M., DeMartinis, N., Schweizer, E., Rickels, K., & Lucki, I. (2001). Abnormal salivary cortisol levels in social phobic patients in response to acute psychological but not physical stress. *Biological Psychiatry, 50*(4), 254–259.

Galiana-Simal, A., Muñoz-Martinez, V., Calero-Bueno, P., Vela-Romero, M., & Beato-Fernandez, L. (2018). Towards a future molecular diagnosis of autism: Recent advances in biomarkers research from saliva samples. *International Journal of Developmental Neuroscience: The Official Journal of the International Society for Developmental Neuroscience, 67*, 1–5. https://doi.org/10.1016/j.ijdevneu.2018.03.004

Gałuszko-Węgielnik, M., Jakuszkowiak-Wojten, K., Wiglusz, M. S., Szarmach, J., Włodarczyk, A., Herstowska, M., & Cubała, W. J. (2017). Lithium therapeutic dose monitoring in human saliva. *Psychiatria Danubina, 29*(Suppl 3), 361–364.

García-Rubio, M. J., Espín, L., Hidalgo, V., Salvador, A., & Gómez-Amor, J. (2017). Autonomic markers associated with generalized social phobia symptoms: Heart rate variability and salivary alpha-amylase. *Stress (Amsterdam, Netherlands), 20*(1), 44–51. https://doi.org/10.1080/10253890.2016.1265939

Ghadirivasfi, M., Nohesara, S., Ahmadkhaniha, H.-R., Eskandari, M.-R., Mostafavi, S., Thiagalingam, S., & Abdolmaleky, H. M. (2011). Hypomethylation of the serotonin receptor type-2A Gene (HTR2A) at T102C polymorphic site in DNA derived from the saliva of patients with schizophrenia and bipolar disorder. *American Journal of Medical Genetics. Part B,*

*Neuropsychiatric Genetics: The Official Publication of the International Society of Psychiatric Genetics, 156B*(5), 536–545. https://doi.org/10.1002/ajmg.b.31192

Giltay, E. J., Enter, D., Zitman, F. G., Penninx, B. W. J. H., van Pelt, J., Spinhoven, P., & Roelofs, K. (2012). Salivary testosterone: Associations with depression, anxiety disorders, and antidepressant use in a large cohort study. *Journal of Psychosomatic Research, 72*(3), 205–213. https://doi.org/10.1016/j.jpsychores.2011.11.014

Goodyer, I. M., Herbert, J., Tamplin, A., & Altham, P. M. E. (2000). Recent life events, cortisol, dehydroepiandrosterone and the onset of major depression in high-risk adolescents. *British Journal of Psychiatry, 177*, 499–504. https://doi.org/10.1192/bjp.177.6.499

Granger, D. A., Kivlighan, K. T., el-Sheikh, M., Gordis, E. B., & Stroud, L. R. (2007). Salivary alpha-amylase in biobehavioral research: Recent developments and applications. *Annals of the New York Academy of Sciences, 1098*, 122.

Grünblatt, E., Marinova, Z., Roth, A., Gardini, E., Ball, J., Geissler, J., ... Walitza, S. (2018). Combining genetic and epigenetic parameters of the serotonin transporter gene in obsessive-compulsive disorder. *Journal of Psychiatric Research, 96*, 209–217. https://doi.org/10.1016/j.jpsychires.2017.10.010

Gustafsson, P. E., Gustafsson, P. A., Ivarsson, T., & Nelson, N. (2008). Diurnal cortisol levels and cortisol response in youths with obsessive-compulsive disorder. *Neuropsychobiology, 57*(1–2), 14–21. https://doi.org/10.1159/000123117

Halligan, S. L., Herbert, J., Goodyer, I., & Murray, L. (2007). Disturbances in morning cortisol secretion in association with maternal postnatal depression predict subsequent depressive symptomatology in adolescents. *Biological Psychiatry, 62*(1), 40–46. https://doi.org/10.1016/j.biopsych.2006.09.011

Hardeveld, F., Spijker, J., Vreeburg, S. A., Graaf, R. D., Hendriks, S. M., Licht, C. M. M., ... F, A. T. (2014). Increased cortisol awakening response was associated with time to recurrence of major depressive disorder. *Psychoneuroendocrinology, 50*, 62–71. https://doi.org/10.1016/j.psyneuen.2014.07.027

Harris, T. O., Borsanyi, S., Messari, S., Stanford, K., Brown, G. W., Cleary, S. E., ... Herbert, J. (2000). Morning cortisol as a risk factor for subsequent major depressive disorder in adult women. *The British Journal of Psychiatry, 177*(6), 505–510. https://doi.org/10.1192/bjp.177.6.505

Hempel, R. J., Tulen, J. H. M., van Beveren, N. J. M., Röder, C. H., de Jong, F. H., & Hengeveld, M. W. (2010). Diurnal cortisol patterns of young male patients with schizophrenia. *Psychiatry and Clinical Neurosciences, 64*(5), 548–554. https://doi.org/10.1111/j.1440-1819.2010.02121.x

Herrera-Gómez, F., García-Mingo, M., Colás, M., González-Luque, J. C., & Álvarez, F. J. (2018). Opioids in oral fluid of Spanish drivers. *Drug and Alcohol Dependence, 187*, 35–39. https://doi.org/10.1016/j.drugalcdep.2018.02.016

Het, S., Vocks, S., Wolf, J. M., Hammelstein, P., Herpertz, S., & Wolf, O. T. (2015). Blunted neuroendocrine stress reactivity in young women with eating disorders. *Journal of Psychosomatic Research, 78*(3), 260–267. https://doi.org/10.1016/j.jpsychores.2014.11.001

Hicks, S. D., Ignacio, C., Gentile, K., & Middleton, F. A. (2016). Salivary miRNA profiles identify children with autism spectrum disorder, correlate with adaptive behavior, and implicate ASD candidate genes involved in neurodevelopment. *BMC Pediatrics, 16*, 52. https://doi.org/10.1186/s12887-016-0586-x

Hidas, A., Noy, A. F., Birman, N., Shapira, J., Matot, I., Steinberg, D., & Moskovitz, M. (2011). Oral health status, salivary flow rate and salivary quality in children, adolescents and young adults with ADHD. *Archives of Oral Biology, 56*(10), 1137–1141. https://doi.org/10.1016/j.archoralbio.2011.03.018

Hinkelmann, K., Moritz, S., Botzenhardt, J., Muhtz, C., Wiedemann, K., Kellner, M., & Otte, C. (2012). Changes in cortisol secretion during antidepressive treatment and cognitive improvement in patients with major depression: A longitudinal study. *Psychoneuroendocrinology, 37*(5), 685–692. https://doi.org/10.1016/j.psyneuen.2011.08.012

Hollocks, M. J., Howlin, P., Papadopoulos, A. S., Khondoker, M., & Simonoff, E. (2014). Differences in HPA-axis and heart rate responsiveness to psychosocial stress in children with autism spectrum disorders with and without co-morbid anxiety. *Psychoneuroendocrinology, 46*, 32–45. https://doi.org/10.1016/j.psyneuen.2014.04.004

Huang, M.-C., Chuang, S.-C., Tseng, M.-C. M., Chien, Y.-L., Liao, S.-C., Chen, H.-C., & Kuo, P.-H. (2017). Cortisol awakening response in patients with bipolar disorder during acute episodes and partial remission: A pilot study. *Psychiatry Research, 258*, 594–597. https://doi.org/10.1016/j.psychres.2017.08.084

Iavarone, F., Melis, M., Platania, G., Cabras, T., Manconi, B., Petruzzelli, R., … Castagnola, M. (2014). Characterization of salivary proteins of schizophrenic and bipolar disorder patients by top-down proteomics. *Journal of Proteomics, 103*, 15–22. https://doi.org/10.1016/j.jprot.2014.03.020

Inagaki, T., Miyaoka, T., Okazaki, S., Yasuda, H., Kawamukai, T., Utani, E., … Tsuji, S. (2010). High salivary alpha-amylase levels in patients with schizophrenia: A pilot study. *Progress in Neuro-Psychopharmacology & Biological Psychiatry, 34*(4), 688–691. https://doi.org/10.1016/j.pnpbp.2010.03.024

Isaksson, J., Nilsson, K. W., Nyberg, F., Hogmark, A., & Lindblad, F. (2012). Cortisol levels in children with attention-deficit/hyperactivity disorder. *Journal of Psychiatric Research, 46*(11), 1398–1405. https://doi.org/10.1016/j.jpsychires.2012.08.021

Jansen, L. M., Gispen-de Wied, C. C., Gademan, P. J., De Jonge, R. C., van der Linden, J. A., & Kahn, R. S. (1998). Blunted cortisol response to a psychosocial stressor in schizophrenia. *Schizophrenia Research, 33*(1–2), 87–94.

Jiang, X., Zhong, W., An, H., Fu, M., Chen, Y., Zhang, Z., & Xiao, Z. (2017). Attenuated DHEA and DHEA-S response to acute psychosocial stress in individuals with depressive disorders. *Journal of Affective Disorders, 215*, 118–124. https://doi.org/10.1016/j.jad.2017.03.013

Jin, R. O., Mason, S., Mellon, S. H., Epel, E. S., Reus, V. I., Mahan, L., … Wolkowitz, O. M. (2016). Cortisol/DHEA ratio and hippocampal volume: A pilot study in major depression and healthy controls. *Psychoneuroendocrinology, 72*, 139–146. https://doi.org/10.1016/j.psyneuen.2016.06.017

Johnson, M. M., Dismukes, A. R., Vitacco, M. J., Breiman, C., Fleury, D., & Shirtcliff, E. A. (2014). Psychopathy's influence on the coupling between hypothalamic-pituitary-adrenal and -gonadal axes among incarcerated adolescents. *Developmental Psychobiology, 56*(3), 448–458. https://doi.org/10.1002/dev.21111

Juster, R.-P., McEwen, B. S., & Lupien, S. J. (2010). Allostatic load biomarkers of chronic stress and impact on health and cognition. *Neuroscience & Biobehavioral Reviews, 35*(1), 2–16. https://doi.org/10.1016/j.neubiorev.2009.10.002

Kaminsky, Z., Wilcox, H. C., Eaton, W. W., Van Eck, K., Kilaru, V., Jovanovic, T., … Smith, A. K. (2015). Epigenetic and genetic variation at SKA2 predict suicidal behavior and post-traumatic stress disorder. *Translational Psychiatry, 5*, e627. https://doi.org/10.1038/tp.2015.105

Kapur, S., Phillips, A. G., & Insel, T. R. (2012). Why has it taken so long for biological psychiatry to develop clinical tests and what to do about it? *Molecular Psychiatry, 17*(12), 1174.

Kariyawasam, S. H., Zaw, F., & Handley, S. L. (2002). Reduced salivary cortisol in children with comorbid Attention deficit hyperactivity disorder and oppositional defiant disorder. *Neuro Endocrinology Letters, 23*(1), 45–48.

Keeshin, B. R., Strawn, J. R., Out, D., Granger, D. A., & Putnam, F. W. (2015). Elevated salivary alpha amylase in adolescent sexual abuse survivors with posttraumatic stress disorder symptoms. *Journal of Child and Adolescent Psychopharmacology, 25*(4), 344–350. https://doi.org/10.1089/cap.2014.0034

Kessler, R. C., Berglund, P., Demler, O., Jin, R., Merikangas, K. R., & Walters, E. E. (2005). Lifetime prevalence and age-of-onset distributions of DSM-IV disorders in the National Comorbidity Survey Replication. *Archives of General Psychiatry, 62*(6), 593.

Khare, C. B., Sankaranarayanan, A., Goel, A., Khandelwal, S. K., & Srinivasa Murthy, R. (1983). Saliva lithium levels for monitoring lithium prophylaxis of manic depressive psychosis. *International Journal of Clinical Pharmacology, Therapy, and Toxicology, 21*(9), 451–453.

Kimonis, E. R., Goulter, N., Hawes, D. J., Wilbur, R. R., & Groer, M. W. (2017). Neuroendocrine factors distinguish juvenile psychopathy variants. *Developmental Psychobiology, 59*(2), 161–173. https://doi.org/10.1002/dev.21473

Klumbies, E., Braeuer, D., Hoyer, J., & Kirschbaum, C. (2014). The reaction to social stress in social phobia: Discordance between physiological and subjective parameters. *PloS One, 9*(8), e105670. https://doi.org/10.1371/journal.pone.0105670

Klump, K. L., Gobrogge, K. L., Perkins, P. S., Thorne, D., Sisk, C. L., & Breedlove, S. M. (2006). Preliminary evidence that gonadal hormones organize and activate disordered eating. *Psychological Medicine, 36*(4), 539.

Knight, J. M., Avery, E. F., Janssen, I., & Powell, L. H. (2010). Cortisol and depressive symptoms in a population-based cohort of midlife women. *Psychosomatic Medicine, 72*(9), 855.

Krämer, M., Seefeldt, W. L., Heinrichs, N., Tuschen-Caffier, B., Schmitz, J., Wolf, O. T., & Blechert, J. (2012). Subjective, autonomic, and endocrine reactivity during social stress in children with social phobia. *Journal of Abnormal Child Psychology, 40*(1), 95–104. https://doi.org/10.1007/s10802-011-9548-9

Kratz, E. M., Waszkiewicz, N., Kałuża, A., Szajda, S. D., Zalewska-Szajda, B., Szulc, A., ... Ferens-Sieczkowska, M. (2014). Glycosylation changes in the salivary glycoproteins of alcohol-dependent patients: A pilot study. *Alcohol and Alcoholism, 49*(1), 23–30. https://doi.org/10.1093/alcalc/agt152

Kuhlman, K. R., Irwin, M. R., Ganz, P. A., Crespi, C. M., Petersen, L., Asher, A., & Bower, J. E. (2017). Cortisol awakening response as a prospective risk factor for depressive symptoms in women after treatment for breast cancer. *Psychosomatic Medicine, 79*(7), 763–769. https://doi.org/10.1097/PSY.0000000000000499

Kuhlman, K. R., Repetti, R. L., Reynolds, B. M., & Robles, T. F. (2016). Change in parent-child conflict and the HPA-axis: Where should we be looking and for how long? *Psychoneuroendocrinology, 68*, 74–81. https://doi.org/10.1016/j.psyneuen.2016.02.029

Kuhlman, K. R., Repetti, R. L., Reynolds, B. M., & Robles, T. F. (2018). Interparental conflict and child HPA-axis responses to acute stress: Insights using intensive repeated measures. *Journal of Family Psychology*. https://doi.org/10.1037/fam0000437

Kuhlman, K. R., Robles, T. F., Bower, J. E., & Carroll, J. E. (2018). Screening for childhood adversity: The what and when of identifying individuals at risk for lifespan health disparities. *Journal of Behavioral Medicine., 41*(4), 516–527.

Labad, J., Soria, V., Salvat-Pujol, N., Segalàs, C., Real, E., Urretavizcaya, M., ... Menchón, J. M. (2018). Hypothalamic-pituitary-adrenal axis activity in the comorbidity between obsessive-compulsive disorder and major depression. *Psychoneuroendocrinology, 93*, 20–28. https://doi.org/10.1016/j.psyneuen.2018.04.008

Lange, C., Deutschenbaur, L., Borgwardt, S., Lang, U. E., Walter, M., & Huber, C. G. (2017). Experimentally induced psychosocial stress in schizophrenia spectrum disorders: A systematic review. *Schizophrenia Research, 182*, 4–12. https://doi.org/10.1016/j.schres.2016.10.008

Lenze, E. J., Dixon, D., Mantella, R. C., Dore, P. M., Andreescu, C., Reynolds, C. F., ... Butters, M. A. (2012). Treatment-related alteration of cortisol predicts change in neuropsychological function during acute treatment of late-life anxiety disorder. *International Journal of Geriatric Psychiatry, 27*(5), 454–462. https://doi.org/10.1002/gps.2732

Lieverse, R., Van Someren, E. J. W., Nielen, M. M. A., Uitdehaag, B. M. J., Smit, J. H., & Hoogendijk, W. J. G. (2011). Bright light treatment in elderly patients with nonseasonal major depressive disorder: A randomized placebo-controlled trial. *Archives of General Psychiatry, 68*(1), 61–70. https://doi.org/10.1001/archgenpsychiatry.2010.183

Liu, K., Ruggero, C. J., Goldstein, B., Klein, D. N., Perlman, G., Broderick, J., & Kotov, R. (2016). Elevated cortisol in healthy female adolescent offspring of mothers with posttraumatic stress disorder. *Journal of Anxiety Disorders, 40*, 37–43. https://doi.org/10.1016/j.janxdis.2016.04.003

Loney, B. R., Butler, M. A., Lima, E. N., Counts, C. A., & Eckel, L. A. (2006). The relation between salivary cortisol, callous-unemotional traits, and conduct problems in an adolescent non-referred sample. *Journal of Child Psychology and Psychiatry, and Allied Disciplines, 47* (1), 30–36. https://doi.org/10.1111/j.1469-7610.2005.01444.x

Lopez-Duran, N. L., Kovacs, M., & George, C. J. (2009). Hypothalamic-pituitary-adrenal axis dysregulation in depressed children and adolescents: A meta-analysis. *Psychoneuroendocrinology, 34*(9), 1272–1283. https://doi.org/10.1016/j.psyneuen.2009.03.016

Lopresti, A. L., Maes, M., Meddens, M. J. M., Maker, G. L., Arnoldussen, E., & Drummond, P. D. (2015). Curcumin and major depression: A randomised, double-blind, placebo-controlled trial investigating the potential of peripheral biomarkers to predict treatment response and antidepressant mechanisms of change. *European Neuropsychopharmacology: The Journal of the European College of Neuropsychopharmacology, 25*(1), 38–50. https://doi.org/10.1016/j.euroneuro.2014.11.015

Lupien, S. J., McEwen, B. S., Gunnar, M. R., & Heim, C. (2009). Effects of stress throughout the lifespan on the brain, behaviour and cognition. *Nature Reviews Neuroscience, 10*(6), 434–445. https://doi.org/10.1038/nrn2639

MacDonald, K., & MacDonald, T. M. (2010). The peptide that binds: A systematic review of oxytocin and its prosocial effects in humans. *Harvard Review of Psychiatry, 18*(1), 1–21. https://doi.org/10.3109/10673220903523615

Majewska, M. D., Hill, M., Urbanowicz, E., Rok-Bujko, P., Bieńkowski, P., Namysłowska, I., & Mierzejewski, P. (2014). Marked elevation of adrenal steroids, especially androgens, in saliva of prepubertal autistic children. *European Child & Adolescent Psychiatry, 23*(6), 485–498. https://doi.org/10.1007/s00787-013-0472-0

Maldonado, E. F., Trianes, M. V., Cortés, A., Moreno, E., & Escobar, M. (2009). Salivary cortisol response to a psychosocial stressor on children diagnosed with attention-deficit/hyperactivity disorder: Differences between diagnostic subtypes. *The Spanish Journal of Psychology, 12*(2), 707–714.

Mangold, D., Marino, E., & Javors, M. (2011). The cortisol awakening response predicts subclinical depressive symptomatology in Mexican American adults. *Journal of Psychiatric Research, 45*(7), 902–909.

Maninger, N., Wolkowitz, O. M., Reus, V. I., Epel, E. S., & Mellon, S. H. (2009). Neurobiological and neuropsychiatric effects of dehydroepiandrosterone (DHEA) and DHEA sulfate (DHEAS). *Frontiers in Neuroendocrinology, 30*(1), 65–91. https://doi.org/10.1016/j.yfrne.2008.11.002

Mannie, Z. N., Harmer, C. J., & Cowen, P. J. (2007). Increased waking salivary cortisol levels in young people at familial risk of depression. *American Journal of Psychiatry, 164*(4), 617–621. https://doi.org/10.1176/ajp.2007.164.4.617

Mantella, R. C., Butters, M. A., Amico, J. A., Mazumdar, S., Rollman, B. L., Begley, A. E., … Lenze, E. J. (2008). Salivary cortisol is associated with diagnosis and severity of late-life generalized anxiety disorder. *Psychoneuroendocrinology, 33*(6), 773–781.

Martel, F. L., Hayward, C., Lyons, D. M., Sanborn, K., Varady, S., & Schatzberg, A. F. (1999). Salivary cortisol levels in socially phobic adolescent girls. *Depression and Anxiety, 10*(1), 25–27.

McCarthy, A. M., Hanrahan, K., Scott, L. M., Zemblidge, N., Kleiber, C., & Zimmerman, M. B. (2011). Salivary cortisol responsivity to an intravenous catheter insertion in children with attention-deficit/hyperactivity disorder. *Journal of Pediatric Psychology, 36*(8), 902–910. https://doi.org/10.1093/jpepsy/jsr012

McEwen, B. S. (2000). Allostasis and allostatic load: Implications for neuropsychopharmacology. *Neuropsychopharmacology: Official Publication of the American College of Neuropsychopharmacology, 22*(2), 108.

Michael, A., Jenaway, A., Paykel, E. S., & Herbert, J. (2000). Altered salivary dehydroepiandrosterone levels in major depression in adults. *Biological Psychiatry, 48*(10), 989–995.

Miura, T., Noma, H., Furukawa, T. A., Mitsuyasu, H., Tanaka, S., Stockton, S., … Kanba, S. (2014). Comparative efficacy and tolerability of pharmacological treatments in the maintenance treatment of bipolar disorder: A systematic review and network meta-analysis. *The Lancet. Psychiatry, 1*(5), 351–359. https://doi.org/10.1016/S2215-0366(14)70314-1

Monteleone, A. M., Monteleone, P., Di Filippo, C., Pellegrino, F., Grillo, F., & Maj, M. (2015). Investigation of the endogenous stress response system in patients with chronic schizophrenia. *Neuropsychobiology, 72*(1), 1–7. https://doi.org/10.1159/000437437

Monteleone, A. M., Monteleone, P., Serino, I., Amodio, R., Monaco, F., & Maj, M. (2016). Underweight subjects with anorexia nervosa have an enhanced salivary cortisol response not seen in weight restored subjects with anorexia nervosa. *Psychoneuroendocrinology, 70*, 118–121. https://doi.org/10.1016/j.psyneuen.2016.05.004

Moon, J.-H., Cho, C.-H., Son, G. H., Geum, D., Chung, S., Kim, H., … Lee, H.-J. (2016). Advanced circadian phase in mania and delayed circadian phase in mixed mania and depression returned to normal after treatment of bipolar disorder. *EBioMedicine, 11*, 285–295. https://doi.org/10.1016/j.ebiom.2016.08.019

Morris, M. C., Mielock, A. S., & Rao, U. (2016). Salivary stress biomarkers of recent nicotine use and dependence. *The American Journal of Drug and Alcohol Abuse, 42*(6), 640–648. https://doi.org/10.1080/00952990.2016.1202263

Morton, S. E., O'Hare, K. J. M., Maha, J. L. K., Nicolson, M. P., Machado, L., Topless, R., … Linscott, R. J. (2017). Testing the validity of taxonic schizotypy using genetic and environmental risk variables. *Schizophrenia Bulletin, 43*(3), 633–643. https://doi.org/10.1093/schbul/sbw108

Moul, C., Dobson-Stone, C., Brennan, J., Hawes, D., & Dadds, M. (2013). An exploration of the serotonin system in antisocial boys with high levels of callous-unemotional traits. *PLoS One, 8*(2), e56619. https://doi.org/10.1371/journal.pone.0056619

Moul, C., Dobson-Stone, C., Brennan, J., Hawes, D. J., & Dadds, M. R. (2015). Serotonin 1B receptor gene (HTR1B) methylation as a risk factor for callous-unemotional traits in antisocial boys. *PLoS One, 10*(5), e0126903. https://doi.org/10.1371/journal.pone.0126903

Moul, C., Hawes, D. J., & Dadds, M. R. (2018). Mapping the developmental pathways of child conduct problems through the neurobiology of empathy. *Neuroscience and Biobehavioral Reviews, 91*, 34–50. https://doi.org/10.1016/j.neubiorev.2017.03.016

Nater, U. M., & Rohleder, N. (2009). Salivary alpha-amylase as a non-invasive biomarker for the sympathetic nervous system: Current state of research. *Psychoneuroendocrinology, 34*(4), 486–496. https://doi.org/10.1016/j.psyneuen.2009.01.014

Newton, T. L., Fernandez-Botran, R., Miller, J. J., & Burns, V. E. (2014). Interleukin-6 and soluble interleukin-6 receptor levels in posttraumatic stress disorder: Associations with lifetime diagnostic status and psychological context. *Biological Psychology, 99*, 150–159. https://doi.org/10.1016/j.biopsycho.2014.03.009

Ngounou Wetie, A. G., Wormwood, K. L., Russell, S., Ryan, J. P., Darie, C. C., & Woods, A. G. (2015). A pilot proteomic analysis of salivary biomarkers in autism spectrum disorder. *Autism Research: Official Journal of the International Society for Autism Research, 8*(3), 338–350. https://doi.org/10.1002/aur.1450

Nohesara, S., Ghadirivasfi, M., Mostafavi, S., Eskandari, M.-R., Ahmadkhaniha, H., Thiagalingam, S., & Abdolmaleky, H. M. (2011). DNA hypomethylation of MB-COMT promoter in the DNA derived from saliva in schizophrenia and bipolar disorder. *Journal of Psychiatric Research, 45*(11), 1432–1438. https://doi.org/10.1016/j.jpsychires.2011.06.013

Northover, C., Thapar, A., Langley, K., Fairchild, G., & van Goozen, S. H. M. (2016). Cortisol levels at baseline and under stress in adolescent males with attention-deficit hyperactivity disorder, with or without comorbid conduct disorder. *Psychiatry Research, 242*, 130–136. https://doi.org/10.1016/j.psychres.2016.05.052

Nováková, M., Paclt, I., Ptáček, R., Kuželová, H., Hájek, I., & Sumová, A. (2011). Salivary melatonin rhythm as a marker of the circadian system in healthy children and those with attention-deficit/hyperactivity disorder. *Chronobiology International, 28*(7), 630–637. https://doi.org/10.3109/07420528.2011.596983

Palma, S. M. M., Fernandes, D. R. M., Muszkat, M., & Calil, H. M. (2012). The response to stress in Brazilian children and adolescents with attention deficit hyperactivity disorder. *Psychiatry Research, 198*(3), 477–481. https://doi.org/10.1016/j.psychres.2011.10.016

Pariante, C. M., & Lightman, S. L. (2008). The HPA axis in major depression: Classical theories and new developments. *Trends in Neurosciences, 31*(9), 464–468.

Petrowski, K., Wintermann, G.-B., Kirschbaum, C., & Strahler, J. (2016). Salivary alpha-amylase response following repeated psychosocial stress in patients with panic disorder. *Journal of Anxiety Disorders, 37*, 54–63. https://doi.org/10.1016/j.janxdis.2015.10.001

Petrowski, K., Wintermann, G.-B., Schaarschmidt, M., Bornstein, S. R., & Kirschbaum, C. (2013). Blunted salivary and plasma cortisol response in patients with panic disorder under psychosocial stress. *International Journal of Psychophysiology: Official Journal of the International Organization of Psychophysiology, 88*(1), 35–39. https://doi.org/10.1016/j.ijpsycho.2013.01.002

Pinto, R., Rijsdijk, F., Ouellet-Morin, I., Asherson, P., McLoughlin, G., & Kuntsi, J. (2016). The aetiological association between the dynamics of cortisol productivity and ADHD. *Journal of Neural Transmission (Vienna, Austria: 1996), 123*(8), 991–1000. https://doi.org/10.1007/s00702-016-1534-5

Preiskorn, J., Studer, S., Rauh, R., Lukačin, R., Geffert, C., Fleischhaker, C., ... Biscaldi, M. (2018). Inter- and intraindividual variation of methylphenidate concentrations in serum and saliva of patients with attention-deficit/hyperactivity disorder. *Therapeutic Drug Monitoring*. https://doi.org/10.1097/FTD.0000000000000520

Pruessner, M., Hellhammer, D. H., Pruessner, J. C., & Lupien, S. J. (2003). Self-reported depressive symptoms and stress levels in healthy young men: Associations with the cortisol response to awakening. *Psychosomatic Medicine, 65*(1), 92–99. https://doi.org/10.1097/01.PSY.0000040950.22044.10

Rai, K., Hegde, A. M., & Jose, N. (2012). Salivary antioxidants and oral health in children with autism. *Archives of Oral Biology, 57*(8), 1116–1120. https://doi.org/10.1016/j.archoralbio.2012.03.006

Rauch, S. A. M., King, A. P., Liberzon, I., & Sripada, R. K. (2017). Changes in salivary cortisol during psychotherapy for posttraumatic stress disorder: A pilot study in 30 veterans. *The Journal of Clinical Psychiatry, 78*(5), 599–603. https://doi.org/10.4088/JCP.15m10596

Raz, S., & Leykin, D. (2015). Psychological and cortisol reactivity to experimentally induced stress in adults with ADHD. *Psychoneuroendocrinology, 60*, 7–17. https://doi.org/10.1016/j.psyneuen.2015.05.008

Riis, J. L., Bryce, C. I., Matin, M. J., Stebbins, J. L., Kornienko, O., Huisstede, L. V., & Granger, D. A. (2018). The validity, stability, and utility of measuring uric acid in saliva. *Biomarkers in Medicine* 12(6):583-596

Rodriguez, K. E., Bryce, C. I., Granger, D. A., & O'Haire, M. E. (2018). The effect of a service dog on salivary cortisol awakening response in a military population with posttraumatic stress disorder (PTSD). *Psychoneuroendocrinology*. https://doi.org/10.1016/j.psyneuen.2018.04.026

Ruhé, H. G., Khoenkhoen, S. J., Ottenhof, K. W., Koeter, M. W., Mocking, R. J. T., & Schene, A. H. (2015). Longitudinal effects of the SSRI paroxetine on salivary cortisol in major depressive disorder. *Psychoneuroendocrinology, 52*, 261–271. https://doi.org/10.1016/j.psyneuen.2014.10.024

Russ, S. J., Herbert, J., Cooper, P., Gunnar, M. R., Goodyer, I., Croudace, T., & Murray, L. (2012). Cortisol levels in response to starting school in children at increased risk for social phobia. *Psychoneuroendocrinology, 37*(4), 462–474. https://doi.org/10.1016/j.psyneuen.2011.07.014

Rutigliano, G., Rocchetti, M., Paloyelis, Y., Gilleen, J., Sardella, A., Cappucciati, M., ... Fusar-Poli, P. (2016). Peripheral oxytocin and vasopressin: Biomarkers of psychiatric disorders? A

comprehensive systematic review and preliminary meta-analysis. *Psychiatry Research, 241*, 207–220. https://doi.org/10.1016/j.psychres.2016.04.117
Saxbe, D. E., Golan, O., Ostfeld-Etzion, S., Hirschler-Guttenberg, Y., Zagoory-Sharon, O., & Feldman, R. (2017). HPA axis linkage in parent-child dyads: Effects of parent sex, autism spectrum diagnosis, and dyadic relationship behavior. *Developmental Psychobiology, 59*(6), 776–786. https://doi.org/10.1002/dev.21537
Scarpa, A., & Raine, A. (1997). Psychophysiology of anger and violent behavior. *The Psychiatric Clinics of North America, 20*(2), 375–394.
Schechter, D. S., Moser, D. A., Paoloni-Giacobino, A., Stenz, L., Gex-Fabry, M., Aue, T., … Rusconi Serpa, S. (2015). Methylation of NR3C1 is related to maternal PTSD, parenting stress and maternal medial prefrontal cortical activity in response to child separation among mothers with histories of violence exposure. *Frontiers in Psychology, 6*, 690. https://doi.org/10.3389/fpsyg.2015.00690
Schlotz, W., Hellhammer, J., Schulz, P., & Stone, A. A. (2004). Perceived work overload and chronic worrying predict weekend-weekday differences in the cortisol awakening response. *Psychosomatic Medicine, 66*(2), 207–214.
Schruers, K., van Diest, R., Nicolson, N., & Griez, E. (2002). L-5-hydroxytryptophan induced increase in salivary cortisol in panic disorder patients and healthy volunteers. *Psychopharmacology, 161*(4), 365–369. https://doi.org/10.1007/s00213-002-1072-8
Schuler, K. L., Ruggero, C. J., Goldstein, B. L., Perlman, G., Klein, D. N., & Kotov, R. (2017). Diurnal cortisol interacts with stressful events to prospectively predict depressive symptoms in adolescent girls. *The Journal of Adolescent Health: Official Publication of the Society for Adolescent Medicine, 61*(6), 767–772. https://doi.org/10.1016/j.jadohealth.2017.06.005
Schwarcz, R., & Hunter, C. A. (2007). Toxoplasma gondii and Schizophrenia: Linkage through astrocyte-derived kynurenic acid? *Schizophrenia Bulletin, 33*(3), 652.
Scott, L. N., Levy, K. N., & Granger, D. A. (2013). Biobehavioral reactivity to social evaluative stress in women with borderline personality disorder. *Personality Disorders, 4*(2), 91.
Segeda, V., Izakova, L., Hlavacova, N., Bednarova, A., & Jezova, D. (2017). Aldosterone concentrations in saliva reflect the duration and severity of depressive episode in a sex dependent manner. *Journal of Psychiatric Research, 91*, 164–168. https://doi.org/10.1016/j.jpsychires.2017.04.011
Severus, E., Taylor, M. J., Sauer, C., Pfennig, A., Ritter, P., Bauer, M., & Geddes, J. R. (2014). Lithium for prevention of mood episodes in bipolar disorders: Systematic review and meta-analysis. *International Journal of Bipolar Disorders, 2*(15). https://doi.org/10.1186/s40345-014-0015-8
Shetty, S. J., Desai, P. B., Patil, N. M., & Nayak, R. B. (2012). Relationship between serum lithium, salivary lithium, and urinary lithium in patients on lithium therapy. *Biological Trace Element Research, 147*(1), 59–62. https://doi.org/10.1007/s12011-011-9295-3
Shibuya, I., Nagamitsu, S., Okamura, H., Komatsu, H., Ozono, S., Yamashita, Y., & Matsuishi, T. (2011). Changes in salivary cortisol levels as a prognostic predictor in children with anorexia nervosa. *International Journal of Psychophysiology: Official Journal of the International Organization of Psychophysiology, 82*(2), 196–201. https://doi.org/10.1016/j.ijpsycho.2011.08.008
Simeunovic Ostojic, M., & Maas, J. (2018). Anorexia nervosa and uric acid beyond gout: An idea worth researching. *The International Journal of Eating Disorders, 51*(2), 97–101. https://doi.org/10.1002/eat.22817
Singh, I., & Rose, N. (2009). Biomarkers in psychiatry [Comments and opinion]. *Nature*. https://doi.org/10.1038/460202a
Sjögren, E., Leanderson, P., & Kristenson, M. (2006). Diurnal saliva cortisol levels and relations to psychosocial factors in a population sample of middle-aged Swedish men and women. *International Journal of Behavioral Medicine, 13*(3), 193–200. https://doi.org/10.1207/s15327558ijbm1303_2

Stadler, C., Kroeger, A., Weyers, P., Grasmann, D., Horschinek, M., Freitag, C., & Clement, H.-W. (2011). Cortisol reactivity in boys with attention-deficit/hyperactivity disorder and disruptive behavior problems: The impact of callous unemotional traits. *Psychiatry Research, 187*(1–2), 204–209. https://doi.org/10.1016/j.psychres.2010.05.004

Stalder, T., Kirschbaum, C., Kudielka, B. M., Adam, E. K., Pruessner, J. C., Wüst, S., ... Hellhammer, D. H. (2016). Assessment of the cortisol awakening response: Expert consensus guidelines. *Psychoneuroendocrinology, 63*, 414–432.

Stefanaki, C., Bacopoulou, F., Livadas, S., Kandaraki, A., Karachalios, A., Chrousos, G. P., & Diamanti-Kandarakis, E. (2015). Impact of a mindfulness stress management program on stress, anxiety, depression and quality of life in women with polycystic ovary syndrome: A randomized controlled trial. *Stress (Amsterdam, Netherlands), 18*(1), 57–66. https://doi.org/10.3109/10253890.2014.974030

Stones, A., Groome, D., Perry, D., Hucklebridge, F., & Evans, P. (1999). The effect of stress on salivary cortisol in panic disorder patients. *Journal of Affective Disorders, 52*(1–3), 197–201.

Tanaka, Y., Ishitobi, Y., Maruyama, Y., Kawano, A., Ando, T., Imanaga, J., ... Akiyoshi, J. (2012). Salivary α-amylase and cortisol responsiveness following electrical stimulation stress in panic disorder patients. *Neuroscience Research, 73*(1), 80–84. https://doi.org/10.1016/j.neures.2012.01.006

Tobolska, D., Wilczyński, K. M., Lorek, M., Mazgaj, E., Krysta, K., Gawlik, A., & Krzystanek, M. (2016). Evaluation of the cortisol concentrations in patients with schizophrenia. *Psychiatria Danubina, 28*(Suppl-1), 162–164.

Tomarken, A. J., Han, G. T., & Corbett, B. A. (2015). Temporal patterns, heterogeneity, and stability of diurnal cortisol rhythms in children with autism spectrum disorder. *Psychoneuroendocrinology, 62*, 217–226. https://doi.org/10.1016/j.psyneuen.2015.08.016

Tordjman, S., Anderson, G. M., Kermarrec, S., Bonnot, O., Geoffray, M.-M., Brailly-Tabard, S., ... Touitou, Y. (2014). Altered circadian patterns of salivary cortisol in low-functioning children and adolescents with autism. *Psychoneuroendocrinology, 50*, 227–245. https://doi.org/10.1016/j.psyneuen.2014.08.010

Vaessen, T., Kasanova, Z., Hernaus, D., Lataster, J., Collip, D., van Nierop, M., & Myin-Germeys, I. (2018). Overall cortisol, diurnal slope, and stress reactivity in psychosis: An experience sampling approach. *Psychoneuroendocrinology, 96*, 61–68. https://doi.org/10.1016/j.psyneuen.2018.06.007

van Rooij, S. J. H., Stevens, J. S., Ely, T. D., Fani, N., Smith, A. K., Kerley, K. A., ... Jovanovic, T. (2016). Childhood trauma and COMT genotype interact to increase hippocampal activation in resilient individuals. *Frontiers in Psychiatry, 7*, 156. https://doi.org/10.3389/fpsyt.2016.00156

van West, D., Claes, S., & Deboutte, D. (2009). Differences in hypothalamic-pituitary-adrenal axis functioning among children with ADHD predominantly inattentive and combined types. *European Child & Adolescent Psychiatry, 18*(9), 543–553. https://doi.org/10.1007/s00787-009-0011-1

Vogel, S. W. N., Bijlenga, D., Verduijn, J., Bron, T. I., Beekman, A. T. F., Kooij, J. J. S., & Penninx, B. W. J. H. (2017). Attention-deficit/hyperactivity disorder symptoms and stress-related biomarkers. *Psychoneuroendocrinology, 79*, 31–39. https://doi.org/10.1016/j.psyneuen.2017.02.009

von Polier, G. G., Herpertz-Dahlmann, B., Konrad, K., Wiesler, K., Rieke, J., Heinzel-Gutenbrunner, M., ... Vloet, T. D. (2013). Reduced cortisol in boys with early-onset conduct disorder and callous-unemotional traits. *BioMed Research International, 2013*, 349530. https://doi.org/10.1155/2013/349530

Vrshek-Schallhorn, S., Doane, L. D., Mineka, S., Zinbarg, R. E., Craske, M. G., & Adam, E. K. (2013). The cortisol awakening response predicts major depression: Predictive stability over a 4-year follow-up and effect of depression history. *Psychological Medicine, 43*(03), 483–493. https://doi.org/10.1017/S0033291712001213

Wahbeh, H., & Oken, B. S. (2013). Salivary cortisol lower in posttraumatic stress disorder. *Journal of Traumatic Stress, 26*(2), 241–248. https://doi.org/10.1002/jts.21798

Walsh, E., Eisenlohr-Moul, T., & Baer, R. (2016). Brief mindfulness training reduces salivary IL-6 and TNF-α in young women with depressive symptomatology. *Journal of Consulting and Clinical Psychology, 84*(10), 887–897. https://doi.org/10.1037/ccp0000122

Wang, L.-J., Huang, Y.-S., Hsiao, C.-C., Chiang, Y.-L., Wu, C.-C., Shang, Z.-Y., & Chen, C.-K. (2011). Salivary dehydroepiandrosterone, but not cortisol, is associated with attention deficit hyperactivity disorder. *The World Journal of Biological Psychiatry, 12*(2), 99–109. https://doi.org/10.3109/15622975.2010.512090

Wang, Z., Mandel, H., Levingston, C. A., & Young, M. R. I. (2016). An exploratory approach demonstrating immune skewing and a loss of coordination among cytokines in plasma and saliva of Veterans with combat-related PTSD. *Human Immunology, 77*(8), 652–657. https://doi.org/10.1016/j.humimm.2016.05.018

Wei, K., Xue, H.-L., Guan, Y.-H., Zuo, C.-T., Ge, J.-J., Zhang, H.-Y., … Du, Y.-J. (2016). Analysis of glucose metabolism of (18)F-FDG in major depression patients using PET imaging: Correlation of salivary cortisol and α-amylase. *Neuroscience Letters, 629*, 52–57. https://doi.org/10.1016/j.neulet.2016.06.039

Whiteford, H. A., Degenhardt, L., Rehm, J., Baxter, A. J., Ferrari, A. J., Erskine, H. E., … Vos, T. (2013). Global burden of disease attributable to mental and substance use disorders: Findings from the Global Burden of Disease Study 2010. *The Lancet, 382*(9904), 1575–1586. https://doi.org/10.1016/S0140-6736(13)61611-6

Wild, B., Stadnitski, T., Wesche, D., Stroe-Kunold, E., Schultz, J.-H., Rudofsky, G., … Friederich, H.-C. (2016). Temporal relationships between awakening cortisol and psychosocial variables in inpatients with anorexia nervosa—A time series approach. *International Journal of Psychophysiology: Official Journal of the International Organization of Psychophysiology, 102*, 25–32. https://doi.org/10.1016/j.ijpsycho.2016.03.002

Wilmot, B., Fry, R., Smeester, L., Musser, E. D., Mill, J., & Nigg, J. T. (2016). Methylomic analysis of salivary DNA in childhood ADHD identifies altered DNA methylation in VIPR2. *Journal of Child Psychology and Psychiatry, and Allied Disciplines, 57*(2), 152–160. https://doi.org/10.1111/jcpp.12457

Wintermann, G.-B., Kirschbaum, C., & Petrowski, K. (2016). Predisposition or side effect of the duration: The reactivity of the HPA-axis under psychosocial stress in panic disorder. *International Journal of Psychophysiology: Official Journal of the International Organization of Psychophysiology, 107*, 9–15. https://doi.org/10.1016/j.ijpsycho.2016.06.008

Wüst, S., Wolf, J., Hellhammer, D. H., Federenko, I., Schommer, N., & Kirschbaum, C. (2000). The cortisol awakening response-normal values and confounds. *Noise & Health, 2*(7), 79.

Yamada, S., Yamauchi, K., Yajima, J., Hisadomi, S., Maeda, H., Toyomasu, K., & Tanaka, M. (2000). Saliva level of free 3-methoxy-4-hydroxyphenylglycol (MHPG) as a biological index of anxiety disorders. *Psychiatry Research, 93*(3), 217–223.

Yang, C.-J., Liu, D., Xu, Z.-S., Shi, S.-X., & Du, Y.-J. (2017). The pro-inflammatory cytokines, salivary cortisol and alpha-amylase are associated with generalized anxiety disorder (GAD) in patients with asthma. *Neuroscience Letters, 656*, 15–21. https://doi.org/10.1016/j.neulet.2017.07.021

Yu, J.-J., Pei, L.-B., Zhang, Y., Wen, Z.-Y., & Yang, J.-L. (2015). Chronic supplementation of curcumin enhances the efficacy of antidepressants in major depressive disorder: A randomized, double-blind, placebo-controlled pilot study. *Journal of Clinical Psychopharmacology, 35*(4), 406–410. https://doi.org/10.1097/JCP.0000000000000352

Zheng, H.-F., Wang, W.-Q., Li, X.-M., Rauw, G., & Baker, G. B. (2017). Body fluid levels of neuroactive amino acids in autism spectrum disorders: A review of the literature. *Amino Acids, 49*(1), 57–65. https://doi.org/10.1007/s00726-016-2332-y

Ziegler, C., Dannlowski, U., Bräuer, D., Stevens, S., Laeger, I., Wittmann, H., … Domschke, K. (2015). Oxytocin receptor gene methylation: Converging multilevel evidence for a role in

social anxiety. *Neuropsychopharmacology: Official Publication of the American College of Neuropsychopharmacology, 40*(6), 1528–1538. https://doi.org/10.1038/npp.2015.2
Zorn, J. V., Schür, R. R., Boks, M. P., Kahn, R. S., Joëls, M., & Vinkers, C. H. (2017). Cortisol stress reactivity across psychiatric disorders: A systematic review and meta-analysis. *Psychoneuroendocrinology, 77*, 25–36. https://doi.org/10.1016/j.psyneuen.2016.11.036

# Chapter 21
# Salivary Bioscience Research in Health Psychology and Behavioral Medicine

**Michael A. Hoyt and Caroline F. Zimmermann**

**Abstract** Since their inception, the fields of health psychology and behavioral medicine have largely adopted a biopsychosocial approach to understanding health and the manner in which psychological and social factors "get under the skin." Facilitated by advances in salivary bioscience, great strides over the past several decades have been made in understanding the biological processes by which such factors influence health and disease. Health psychology and behavioral medicine research have integrated advanced clinical and laboratory assessments of relevant immune system and neuroendocrine markers in saliva to identify mechanisms, stress processes, and evaluate the impact of clinical intervention on physiological systems. This chapter highlights contributions of salivary bioscience to health psychology and behavioral medicine with an emphasis on research related to understanding adjustment to chronic illness and the influence of psychological and social factors on disease processes. Research utilizing salivary markers of hypothalamic–pituitary–adrenal axis activity, sympathetic nervous system activation, as well as other neuroendocrine and immune processes has greatly contributed to our understanding of psychological adaptation to illness and the composition of clusters of adjustment-related symptoms such as fatigue, pain, and depression. In addition, documented changes in salivary levels of health-relevant biomarkers in response to behavioral interventions have contributed to a new definition of intervention efficacy. This broad synthesis of the literature emphasizes a more integrated biopsychosocial framework for understanding health and well-being.

M. A. Hoyt (✉)
Population Health and Disease Prevention, University of California, Irvine, CA, USA

Institute for Interdisciplinary Salivary Bioscience Research, University of California, Irvine, CA, USA
e-mail: mahoyt@uci.edu

C. F. Zimmermann
Department of Psychology, Hunter College, City University of New York, New York, NY, USA

The Graduate Center, City University of New York, New York, NY, USA



D. A. Granger, M. K. Taylor (eds.), *Salivary Bioscience*,
https://doi.org/10.1007/978-3-030-35784-9_21

**Keywords** Salivary bioscience · Health psychology · Behavioral medicine · Chronic illness · Stress · Cortisol

## 21.1 Salivary Bioscience Research in Health Psychology and Behavioral Medicine

The quest to understand the interactions between psychological and physiological factors in human health can be traced across ancient and modern civilizations. Catalyzed by advances in salivary bioscience, the past several decades have brought significant progress in psychosomatic medicine, psychoneuroimmunology, neuroscience, and in our understanding of the health impact of psychological and social stress. The fields of health psychology and behavioral medicine have made vital contributions to this knowledge base and have increasingly become more interdisciplinary as a more complete biopsychosocial model actualizes. This chapter highlights contributions of salivary bioscience to health psychology and behavioral medicine with an emphasis on research related to understanding adjustment to chronic illness and the influence of psychological and social factors on disease processes. As a nexus between health psychology and behavioral medicine, the focus will be on pertinent studies of conditions that comprise significant causes of morbidity and mortality such as cardiovascular and pulmonary diseases, cancer, HIV/AIDS, and rheumatic diseases. The emphasis is on building from empirically supported conceptualizations of psychological adaptation to these conditions, to present an expanded biobehavioral model that emerges, in part, from contributions from salivary bioscience. This includes understanding the management of illness across the disease trajectory as well as influences on its progression. Although this chapter concentrates on chronic illness, it should be noted that salivary bioscience also exerts influence across a variety of areas in health psychology and behavioral medicine including studies of general health promotion and prevention (e.g., Tomiyama et al., 2014), experimental health psychology (e.g., Dickerson & Kemeny, 2004), social processes and relationships (e.g., Kornienko, Schaefer, Pressman, & Granger, 2018; Robles, Slatcher, Trombello, & McGinn, 2014; see also Chap. 22), health disparities (e.g., Le-Scherban et al., 2018), and basic stress processes (see Nater, Skoluda, & Starhler, 2013). We call upon mostly studies of adults with chronic illness; however, relevant literatures exist with a focus on the health of children and adolescents (e.g., Jessop & Turner-Cobb, 2008; see also Chap. 26).

Noncommunicable disease accounts for 70% of worldwide mortality (World Health Organization [WHO], 2017). At the same time, compared with past decades, individuals are living long periods of time following a diagnosis (Centers for Disease Control and Prevention [CDC], 2017). There are thousands of studies to show that the experience of chronic illness can cause significant disruption across life domains. Thus, health psychologists and others have devoted intense empirical and clinical attention to identifying psychosocial and biobehavioral contributors to, and consequences of, chronic illness and to developing approaches to reduce physical and psychological morbidity (see Hoyt & Stanton, 2018). Hoyt and Stanton (2018) articulated a model of psychological adjustment to chronic illness in which interrelated contexts of risk and resilience (e.g., disease/treatment factors, social processes, personality, socioeconomic factors, and gender) exert influence on various domains of psychological adjustment by way of cognitive, affective, and behavioral responses that include coping strategies.

Although this and other (see Helgeson & Zajdel, 2017) models of disease adjustment possess significant utility, more recent research utilizing salivary markers of hypothalamic–pituitary–adrenal axis activity, sympathetic nervous system activation, as well as other neuroendocrine and immune processes has greatly added to our understanding of the biobehavioral pathways by which psychological processes influence adaptation to illness; relate to the composition of clusters of adjustment-related symptoms such as fatigue (e.g., Bower, Ganz, & Aziz, 2005), pain (e.g., Fischer et al., 2016), and depression (e.g., Weinrib et al., 2010); and possess potential to influence progression of disease (e.g., Spiegel, 2012). For instance, emotional responses to acute and sustained stressors contribute to the modulation of neuroendocrine activity that may provide a link between stress responses, suppressed immunity, and disease-related processes (Fagundes, Murdock, Chirinos, & Green, 2017).

## 21.2 Measuring Stress to Understand Adjustment

Since the advent of the field, health psychologists and behavioral medicine professionals have been searching to understand how stressful experiences impact human health (Friedman & Adler, 2007). The growth of biological knowledge by the 1970s and a new demand for rigorous research methods that incorporated both behavioral as well as biological science gave way to a new research zeitgeist which included understanding links between external stressors and organic illness. However, new methods of inquiry that integrated and measured biological processes were needed for behavioral scientists.

Salivary bioscience has provided an accessible window into physiological stress processes. This has included access to a broad range of salivary biomarkers, with health psychologists and behavioral medicine professionals being largely interested in understanding how psychosocial experiences relate to physiological stress. Stress is the dynamic response that takes place in both the central nervous system and the periphery of the body that one experiences when demands are perceived to exceed one's personal resources. The involvement of two distinct physiological systems has been identified. The sympathetic adrenomedullary system prepares the body's *flight or fight* response through activation of the sympathetic nervous system. This involves the secretion of epinephrine and norepinephrine by the adrenal glands leading to an increase in heart rate and more rapid metabolization of glucose. The hypothalamic–pituitary–adrenal (HPA) axis is also involved in the stress response. Activation of this system involves increased secretion of corticotrophin-releasing hormone and stimulation of the pituitary gland. The pituitary gland secretes adrenocorticotrophic hormone inducing an increased release of glucocorticoids by the adrenal cortex. In humans, the glucocorticoid cortisol prepares the body for responding to stressful demands and has been implicated in delaying the body's innate immune response. These systems work in tandem to direct energy from nonessential processes to those necessary to overcome stressors, such as increases in cardiac and respiratory activity, energy consumption, and mental activity.

Although a broader array of salivary analytes is gaining prominence in the health psychology and behavioral medicine literature, salivary cortisol (as a marker of HPA activity) has been a dominative focus. Consistent with findings in individuals who experience chronic stressors (e.g., informal caregivers; Leggett, Liu, Klein, & Zarit, 2016), relationships of both blunted (e.g., flatter diurnal cortisol slope) and elevated cortisol (e.g., higher daily cortisol output or greater area under the curve) responses have been documented in chronic illness groups. Salivary cortisol also has documented associations to health outcomes known to greatly impact quality of life including functional disability (Weinrib et al., 2010), pain exacerbations (Yeung, Davis, & Ciaramitaro, 2016), fatigue (Bower et al., 2005), and sexual dysfunction (Hoyt, Gaffey, Wang, & Lawsin, 2018). The HPA system is often considered to be a mediating system of acute and chronic psychosocial states and disease processes. It is also shown to be particularly sensitive to laboratory stressors (Dickerson & Kemeny, 2004).

Increasingly researchers have been utilizing measures of salivary alpha-amylase (sAA) as a reliable marker of autonomic nervous system (ANS) activity (Granger, Kivlighan, El-Sheikh, Gordis, & Stroud, 2008; Nater & Rohleder, 2009). sAA increases with ANS activation via a release by acinar cells which are innervated by both sympathetic and parasympathetic activity. During periods of psychological stress, sAA level is predominantly influenced by SNS activity in the cervical sympathetic pathway, and sAA levels rise in response to stress (Bosch, de Geus, Veerman, Hoogstraten, & Nieuw Amerongen, 2003; Nater, Rohleder, Schlotz, Ehlert, & Kirschbaum, 2007). Studies that have included measures of both salivary cortisol and sAA have been able to distinguish associations between HPA and somatic nervous system involvement. SAA shows distinct patterns from salivary cortisol in response to stress and might better capture anxious versus depressive reactivity (Yoon & Weierich, 2016). For instance, in fibromyalgia patients Fischer and colleagues (2016) found that cortisol, but not alpha-amylase, impacted momentary pain in a 14-day ambulatory monitoring study.

## 21.3 Salivary Bioscience and Adjustment to Chronic Illness

### *21.3.1 Depressive Symptoms After Chronic Illness*

The onset or worsening of depressive symptoms following the diagnosis or treatment of chronic illness is an important indicator of adjustment to chronic disease. In fact, although heterogeneity exists, large prospective studies demonstrate that individuals with chronic illness are at elevated risk for depressive symptoms (Polsky et al., 2005). Cancer survivors have the highest risk of depressive symptoms within 2 years after diagnosis (hazard ratio [HR] = 3.55), followed by chronic lung disease (HR = 2.21) and heart disease (HR = 1.45), versus those with no incident disease. Researchers have been successful in identifying psychological and social predictors of symptom onset; however, research to understand the neurobiological risk factors and mechanisms is emerging. Relationships of dysregulation in salivary cortisol and depressive symptoms have been documented. Studies outside of chronic illness groups have pointed to dysregulation in HPA axis as underpinning elevations in depressive symptoms (Bhagwagar & Cowen, 2008; Stetler & Miller, 2011). It appears that HPA axis over-activation, at least initially, dampens prefrontal inhibition of a sustained stress response, which yields adrenal overstimulation in a feedback loop that terminates in anhedonic states (e.g., Gold, 2015), though notable variation exists across groups and individuals.

Depressive symptoms have been associated with dysregulation in multiple indices of salivary cortisol including morning levels, evening values, and the pattern of cortisol levels across the diurnal cycle in samples of individuals with cancer (Jehn et al., 2006; Lutgendorf et al., 2008; Sephton et al., 2009), cardiovascular disease (Nikkheslat et al., 2015), and HIV (Barroso, Burrage, Carlson, & Carlson, 2006). In some cases, salivary cortisol is more strongly related to depressive symptoms than are treatment factors (see Bower, 2008). However, more longitudinal research is necessary to establish a mechanistic role of cortisol dysregulation. For instance, Kuhlman et al. (2017) found that a higher cortisol awakening response predicted greater increases in depressive symptoms across 6 months in women with early-stage breast cancer.

It is notable that the HPA axis is implicated in regulating inflammation. In depressed patients, cortisol's role in the negative feedback loop leading to anti-inflammatory effects is impaired (Pariante & Miller, 2001) and depressed individuals can experience chronically elevated nocturnal cortisol (Deuschle et al., 1997). Thus, excessive inflammation resulting from impaired glucocorticoid receptor sensitivity is observed with greater depressive symptomatology (e.g., Carvalho et al., 2015). In a study of coronary heart disease patients, Nikkheslat et al. (2015) document that in the presence of low salivary cortisol output, depression is accompanied by elevated levels of inflammation and glucocorticoid resistance. Such findings provide insight into the role of inflammation and salivary cortisol in the etiology of depression in chronic illness patients, and evidence of possible inflammatory pathways to the progression of the disease.

### *21.3.2 Coping with Chronic Illness*

Evidence exists that diurnal cortisol rhythm may be influenced by individual coping strategies in chronic illness groups (e.g., Diaz, Aldridge-Gerry, & Spiegel, 2014).

However, relatively few studies have sought to identify the possibility that specific strategies for coping with illness-related stressors can shape salivary biomarkers (e.g., diurnal cortisol). In this context, coping constitutes behavioral and cognitive efforts to manage the illness-related demands (Lazarus & Folkman, 1984). Coping responses, often categorized as either approach-oriented (e.g., active planning) or avoidance-oriented (e.g., disengagement), modulate the negative consequences of illness-related stressors and impact adjustment and quality of life (Baldwin, Kellerman, & Christensen, 2010). In prostate cancer patients, cancer-related avoidance-oriented coping, and not approach-oriented coping, is associated with more blunted cortisol slope over time (Hoyt et al., 2014).

Identification of the mechanistic salivary biomarkers that influence adjustment and overall health outcomes holds promise for the identification of modifiable coping behaviors useful for the development of tailored interventions. A strong example comes from work on coping processes relevant to goal pursuit. As disease-related demands are perceived as threatening to central goals, the greater the perception of stress, and the more coping processes are engaged. Self-regulation theories highlight the importance of perceived goal blockage in shaping coping and adjustment (e.g., Leventhal, Halm, Horowitz, Leventhal, & Ozakinci, 2005). To the extent that an individual expects those goals are obtainable despite illness, or to the extent that one perceives the ability to identify and engage in alternative goal pursuit, then initiation of approach-oriented coping strategies is likely (Hoyt, Gamarel, Saigal, & Stanton, 2016). However, if a person expects unremitting goal blockage and does not engage in new goals, disengagement might ensue. Likewise, continued pursuit of unobtainable goals will likely exert a negative effect on adjustment (Wrosch, Scheier, Miller, Schulz, & Carver, 2003). The ability to disengage from unattainable goals is associated with better self-reported health and more normative patterns of diurnal cortisol secretion (Wrosch, Miller, Scheier, & de Pontet, 2007). In breast cancer patients, reengagement in meaningful goals following goal blockage appears to buffer against elevated daily cortisol secretion associated with negative affect (Castonguay, Wrosch, & Sabiston, 2017).

Research utilizing salivary biomarkers can also uncover patterns of resilience. In a study of 111 cancer survivors, Costanzo, Stawski, Ryff, Coe, and Almeida (2012) observed that relative to a matched non-cancer comparison group, cancer survivors showed less pronounced changes in cortisol output in response to interpersonal conflict. Similarly, higher mean daily cortisol is associated with poor disease-related social support (Turner-Cobb, Sephton, Koopmanm, Blake-Mortimer, & Spiegel, 2000). Observations that salivary markers are sensitive to variation in coping behaviors, resilience factors, and coping resources provide a strong rationale for targeted intervention development.

### *21.3.3 Behavioral Medicine Interventions and Salivary Bioscience*

The many associations between salivary biomarkers and health indicators suggest that modulation of salivary biomarkers via behavioral interventions may play an important role in influencing health and long-term adjustment to disease. Targeted interventions that aim to modulate HPA axis function or autonomic nervous system regulation hold significant promise, though more randomized controlled trials (RCTs) are necessary. Cortisol is the most widely studied salivary marker in behavioral medicine intervention research. However, reviews of this literature are inconclusive in determining whether such interventions render meaningful changes on salivary diurnal cortisol (Ryan, Booth, Spathis, Mollart, & Clow, 2016). This is in large due to significant heterogeneity across studies in regard to saliva collection procedures and reported cortisol parameters (i.e., cortisol awakening response, area under the curve, diurnal slope, etc.).

Broad variation exists in the design and results of behavioral medicine interventions with identified salivary biomarkers outcomes. Several studies have focused on the use of mindfulness-based stress reduction (MBSR). In an RCT with 271 distressed breast cancer survivors, Carlson et al. (2013) observed more flattened cortisol slopes over time in their control group, but the maintenance of slope in those receiving MBSR plus yoga (as well as those receiving supportive-expressive therapy). However, MBSR was not associated with changes in salivary melatonin in breast and prostate cancer patients (Carlson, Speca, Patel, & Goodey, 2004). However, null findings of cortisol parameters have been documented after mindfulness-based intervention trials including fibromyalgia patients (Cash et al., 2016) and cancer patients with sleep disturbance (Lipschitz, Kuhn, Kinney, Donaldson, & Nakamura, 2013). Although Lipschitz et al. (2013) did not observe effects on cortisol, they found lower sAA activity in the morning after a sleep hygiene education group compared to those in a sleep-focused mind–body intervention.

Group-based cognitive behavioral stress management (CBSM) interventions have also demonstrated a salutary impact on HPA regulation. In a study of HIV+ gay men, baseline cortisol levels decreased across 10-week CBSM participation (Cruess, Antoni, Kumar, & Schneiderman, 2000). Moreover, cortisol reductions had associations decreases in distress, perceived stress, and more frequent at-home relaxation practice. Salivary biomarkers have also been used in pharmacologic trials (e.g., Chaborski, Bitterlich, Alteheld, Parsi, & Metzner, 2015) and integrative medicine trials (e.g., Tornhage et al., 2013).

As discussed by Ryan et al. (2016), little guidance exists on the use of salivary measures in intervention research. To avoid bias and error in RCT design, researchers must consider their hypothesized mechanism of change when identifying *primary* intervention outcomes a priori. For instance, the diurnal profile of various salivary markers renders a host of measurement parameters (e.g., cortisol awakening response versus diurnal cortisol slope) that signal different aspects of the biological mechanism of interest. They also warn that researchers should consider long- and short-term stability and reliability of the marker of interest when considering intervention interval and follow-up periods.

## 21.4 Salivary Biomarkers, Disease Processes, and Progression

In addition to the contribution of salivary bioscience to identifying the biological pathways to domains of disease adjustment, the past several decades have yielded groundbreaking strides in our understanding of the multiple biological pathways by which psychosocial and behavioral factors can also affect the progression and symptoms of chronic illness. Long-term epidemiological studies demonstrate that higher cortisol levels (in men) and higher evening cortisol (in women) are associated with increased mortality risk and chronic disease onset (Schoorlemmer, Peeters, van Schoor, & Lips, 2009).

To the degree that salivary bioscience allows researchers to measure aspects of physiological stress patterns, it facilitates the discovery of how stress precipitates, exacerbates, or maintains physical symptoms. As discussed, cortisol dysregulation is linked to pain (Crofford et al., 2004) and mood disturbance (Deuschle et al., 1997), with some evidence that cortisol dysregulation may precede symptoms. However, despite that psychological models have emphasized its role in physical health, evidence linking cortisol to specific disease outcomes has been mixed. As evidence builds for HPA dysregulation underlying physical symptoms such as fatigue, results across studies are not always consistent. More research examining the contextual moderators of cortisol's impact on physical symptoms is warranted. Wrosch, Miller, Lupien, and Pruessner (2008) showed that higher cortisol levels were associated with increases in physical symptoms in older adults, but only among those who reported relatively high negative affect and poor sleep.

Strides are being made to not only understand influences on physical symptoms but also on disease progression. This association has been evidenced across chronic disease and illness populations including cancer, cardiovascular disease, asthma, HIV, and pain conditions. For instance, asthmatics with poorly controlled asthma (a signifier of progressed disease) presented with significantly lower levels of morning salivary cortisol than those with greater symptom control (Shin et al., 2014). Similarly, in cardiovascular disease patients, intima media thickness (a measure of atherosclerosis or arterial plaque buildup) is associated with a greater cortisol awakening response in women (Eller, Netterstrøm, & Allerup, 2005). In the case of HIV, elevated cortisol levels (measured in saliva and serum) have been linked to markers of disease progression, including increased viral load (see Ironson et al., 2015; Schneiderman, Ironson, & Siegel, 2005). Flatter diurnal cortisol slopes evidenced in patients with metastatic breast cancer (Abercrombie et al., 2004) have been linked to early mortality (Cohen et al., 2012; Schrepf et al., 2015; Sephton et al., 2013) leaving researchers to discover in what ways cortisol might be influencing tumor dynamics. Early work in animal models suggested that experimentally induced impairments in cortisol feedback enhance tumor growth (Sapolsky & Donnelly, 1985). Similar patterns are evident in human studies. Blunted diurnal slopes of salivary cortisol observed prior to surgery have been associated with markers of inflammation in the tumor microenvironment in ovarian cancer patients (Lutgendorf et al., 2008; Weinrib et al., 2010) and appear to play a role in the upregulation of tumor inflammation (Schrepf et al., 2015).

Research has found that sleep is vital to slow disease progression and conversely, that without adequate sleep an individual with a chronic health condition may be at risk for worsened disease progression. Moreover, the assessment of salivary measures has helped to identify the mechanisms of the impact of sleep on disease progression. Although Cash et al. (2015) did not find support for an association of psychological distress and tumor progression among breast cancer patients ($n = 43$), they did find that elevated cortisol awakening responses and circadian activity rhythm disruption (i.e., poor sleep quality) were related to tumor progression and immunosuppression markers (i.e., VEGF, MMP-9, and TGF-b). It may be that increased nighttime cortisol impairs adequate biological restoration during sleep.

Although melatonin is known to modulate the sleep–wake cycle (Zhdanova, Lynch, & Wurtman, 1997), few behavioral medicine studies have examined salivary melatonin, particularly in the context of chronic illness. More research examining salivary melatonin in chronic illness groups might help to better understand changes in sleep quality, as well as the impact of sleep disruption on disease outcomes. For instance, work in patients with chronic renal failure suggests that disease-related disruptions in nocturnal production of melatonin might underscore sleep problems (Pinto, da Silva, & Pinato, 2016). Research in this area is limited; however, similar patterns have been documented in pregnant women with gestational hypertension or diabetes (Shimada, Seki, Samejima, Hayase, & Shirai, 2016).

Identifying mechanisms of how stress-related pathways influence the disease process is a new frontier for research in health psychology and behavioral medicine, but more longitudinal studies in this domain are warranted. Most studies to date highlight significant differences between groups at different stages of disease severity. However, a stronger contribution will be made with within-subject studies that track changes in salivary markers overtime. Further, more research that address salivary biomarker's predictive capabilities of disease progression will help elucidate causal chains.

## 21.5 Conclusions and Future Directions

This chapter called upon research on psychological adaptation to chronic illness to highlight the contributions and potential of salivary bioscience to develop new and expanded conceptual models that underscore research and practice in health psychology and behavioral medicine. What emerges is an expanded model of adjustment to chronic illness that includes biological mechanisms and physiological pathways connecting stress and psychosocial determinants of adaptation, as well as a framework for inclusion of relationships with disease processes and progression. Future research should continue to test and refine this emergent biobehavioral model.

The progression of knowledge about biological stress processes is intertwined with the advent and growth of behavioral medicine and health psychology. Understanding how external factors impact health via stress-mediated physiological pathways has marked progress in these fields for decades. It follows that research to date has centered on the assessment of salivary cortisol and alpha-amylase. This work has made vital contributions to the field, and now more studies are emerging examining other salivary biomarkers. These new frontiers hold promise for strengthening the biobehavioral model. For instance, salivary oxytocin has been linked with enhanced feelings of empathy and social connectedness, anxiety reduction, better pain tolerance, and increased attention to positive stimuli in depressed patients (Cochran, Fallon, Hill, & Frazier, 2013; Domes, Normann, & Heinrichs, 2016; Rash, Aguirre-Camacho, & Campbell, 2014) and has been shown to be sensitive to behavioral intervention (Fancourt et al., 2016). Despite these observations, studies of those with chronic illness are only beginning to emerge. In a prospective study, craniopharyngioma patients showed a blunted salivary oxytocin-release compared to controls and this was associated with higher anxiety (Gebert et al., 2018). Continual advances in the measurement of neurotransmitters, cytokines, and analytes that signal such processes as sleep (e.g., melatonin) and metabolic activity (e.g., uric acid) will yield more sophisticated biobehavioral research and more targeted behavioral interventions.

Notably, research designed to understand the mechanistic potential of pro-inflammatory cytokine activation in unique relationships of psychosocial factors and disease processes has been fruitful (see Irwin & Slavich, 2017). The preponderance of this research has relied on the measurement of circulating levels of immune markers in the blood. Acquisition of blood in health psychology and behavioral medicine research impairs ecological assessment and presents logistical challenges and undue participant burden in studies that rely on observations across time or at multiple moments across the day. Salivary measures of inflammation offer the promise of measuring inflammation *in the moment*. However, salivary cytokine studies should be interpreted with significant caution. Across the limited number of empirical studies, salivary markers of inflammation (i.e., IL-1b, TNF-a, and IL-6) have been reliably measured; however, associations with measures of circulating blood levels have been inconsistent (Slavish, Graham-Engeland, Smyth, & Engeland, 2015). Salivary measures of inflammatory markers appear to better reflect activity in the oral mucosa immune environment rather than systemic inflammation (Riis, Granger, DiPietro, Bandeen-Roche, & Johnson, 2015).

Several challenges remain that challenge the progression of salivary bioscience in health psychology and behavioral medicine. Foremost is a need for training across disciplines. Few health psychologists and behavioral medicine professionals receive training in laboratory procedures or research design related to the assessment of salivary biomarkers. In addition to pairing with professionals across disciplines, researchers are encouraged to seek training experiences and guidance when designing sample collection and analysis protocols (see Adam & Kumari, 2009; Nicolson, 2008). It is imperative that investigators understand the dynamics of measured markers, including any diurnal rhythm, sampling strategy considerations, storage and handling recommendations, and potential within- and between-individual factors confounding measurement. Finally, researchers will need to ensure that all laboratory results are inspected for quality assurance.

Behavioral science researchers should apply principles of high-quality research design used in studies of psychosocial variables to studies including salivary biomarkers. This begins with having clearly defined study objectives and hypotheses. Health psychology and behavioral medicine research are predicated on theory-driven hypotheses. The integration of salivary biomarkers should fit within sound conceptual underpinnings. Studies that relegate biomarkers to exploratory measures run a greater risk of Type I error. Relatedly, studies should be adequately powered for planned analyses. The collection of saliva from a mere sample subset might render inadequate statistical power or contain systematic bias.

Research in salivary bioscience continues to expand our collective knowledge of biobehavioral processes and health. Perhaps more than ever before we are unlocking the secrets of mind–body relationships. However, this can only be achieved when researchers adhere to fundamental principles of the careful study design including vigilant quality control and utilization of informed laboratory procedures, careful data analysis, caution in the interpretation of research findings, and sufficiently powered replication of research results.

## References

Abercrombie, H. C., Giese-Davis, J., Sephton, S., Epel, E. S., Turner-Cobb, J. M., & Spiegel, D. (2004). Flattened cortisol rhythms in metastatic breast cancer patients. *Psychoneuroendocrinology, 29*, 1082–1092.

Adam, E. K., & Kumari, M. (2009). Assessing salivary cortisol in large scale, epidemiological research. *Psychoneuroendocrinology, 34*, 1423–1436.

Baldwin, A. S., Kellerman, Q. D., & Christensen, A. J. (2010). Coping with chronic illness. In J. M. Suls, K. W. Davidson, & R. M. Kaplan (Eds.), *Handbook of health psychology and behavioral medicine* (pp. 494–507). New York: Guilford Press.

Barroso, J., Burrage, J., Carlson, J., & Carlson, B. W. (2006). Salivary cortisol values in HIV-positive people. *Journal of the Association of Nurses in AIDS Care, 17*, 29–36.

Bhagwagar, Z., & Cowen, P. J. (2008). 'It's not over when it's over': Persistent neurobiological abnormalities in recovered depressed patients. *Psychological Medicine, 38*, 307–313.

Bosch, J. A., de Geus, E. J., Veerman, E. C., Hoogstraten, J., & Nieuw Amerongen, A. V. (2003). Innate secretory immunity in response to laboratory stressors that evoke distinct patterns of cardiac autonomic activity. *Psychosomatic Medicine, 65*, 245–258.

Bower, J. E. (2008). Behavioral symptoms in patients with breast cancer and survivors. *Journal of Clinical Oncology, 26*, 768–777.

Bower, J. E., Ganz, P. A., & Aziz, N. (2005). Altered cortisol response to psychologic stress in breast cancer survivors with persistent fatigue. *Psychosomatic Medicine, 67*, 277–280.

Carlson, L. E., Doll, R., Stephen, J., Faris, P., Tamagawa, R., Drysdale, E., & Speca, M. (2013). Randomized controlled trial of mindfulness-based cancer recovery versus supportive expressive group therapy for distressed survivors of breast cancer. *Journal of Clinical Oncology, 31*, 3119–3126.

Carlson, L. E., Speca, M., Patel, K. D., & Goodey, E. (2004). Mindfulness-based stress reduction in relation to quality of life, mood, symptoms of stress and levels of cortisol, dehydroepiandrosterone sulfate (DHEAS) and melatonin in breast and prostate cancer outpatients. *Psychoneuroendocrinology, 29*, 448–474.

Carvalho, L. A., Urbanova, L., Hamer, M., Hackett, R. A., Lazzarino, A. I., & Steptoe, A. (2015). Blunted glucocorticoid and mineralocorticoid sensitivity to stress in people with diabetes. *Psychoneuroendocrinology, 51*, 09–218.

Cash, E., Salmon, P., Weissbecker, I., Rebholz, W. N., Bayley-Veloso, R., Zimmaro, L. A., ... Sephton, S. E. (2016). Mindfulness meditation alleviates fibromyalgia symptoms in women: Results of a randomized clinical trial. *Annals of Behavioral Medicine, 49*, 319–330.

Cash, E., Sephton, S. E., Chagpar, A. B., Spiegel, D., Rebholz, W. N., Zimmaro, L. A., ... Dhabar, F. S. (2015). Circadian disruption and biomarkers of tumor progression in breast cancer patients awaiting surgery. *Brain, Behavior, and Immunity, 48*, 102–114.

Castonguay, A. L., Wrosch, C., & Sabiston, C. M. (2017). The roles of negative affect and goal adjustment capacities in breast cancer survivors: Associations with physical activity and diurnal cortisol secretion. *Health Psychology, 36*, 320–331.

Centers for Disease Control and Prevention, U.S. Department of Health and Human Services. (2017). Deaths: Final data for 2015. *National Vital Statistics Report, 66*, 1–73.

Chaborski, K., Bitterlich, N., Alteheld, B., Parsi, E., & Metzner, C. (2015). Placebo-controlled dietary intervention of stress-induced neurovegetative disorders with a specific amino acid composition: A pilot-study. *Nutrition Journal, 14*, 43.

Cochran, D. M., Fallon, D., Hill, M., & Frazier, J. A. (2013). The role of oxytocin in psychiatric disorders: A review of biological and therapeutic research findings. *Harvard Review of Psychiatry, 21*, 219–247.

Cohen, L., Cole, S. W., Sood, A. K., Prinsloo, S., Kirchbaum, C., Arevalo, J. M. G., ... Pisters, L. (2012). Depressive symptoms and cortisol rhythmicity predict survival in patients with renal cell carcinoma: Role of inflammatory signaling. *PLoS One, 7*, s42324.

Costanzo, E. S., Stawski, R. S., Ryff, C. D., Coe, C. L., & Almeida, D. M. (2012). Cancer survivors' responses to daily stressors: Implications for quality of life. *Health Psychology, 31*, 360–370.

Crofford, L. J., Young, E. A., Engleberg, N. C., Korszun, A., Brucksch, C. B., McClure, L. A., ... Demitrack, M. A. (2004). Basal circadian and pulsatile ACTH and cortisol secretion in patients with fibromyalgia and/or chronic fatigue syndrome. *Brain, Behavior and Immunity, 18*, 314–325.

Cruess, D. G., Antoni, M. H., Kumar, M., & Schneiderman, N. (2000). Reductions in salivary cortisol are associated with mood improvement during relaxation training among HIV-seropositive men. *Journal of Behavioral Medicine, 23*, 107–122.

Deuschle, M., Schweiger, U., Weber, B., Gotthardt, U., Körner, A., Schmider, J., ... Heuser, J. (1997). Diurnal activity and pulsatility of the hypothalamus-pituitary-adrenal system in male depressed patients and healthy controls. *Journal of Clinical Endocrinology and Metabolism, 82*, 234–238.

Diaz, M., Aldridge-Gerry, A., & Spiegel, D. (2014). Posttraumatic growth and diurnal cortisol slope among women with metastatic breast cancer. *Psychoneuroendocrinology, 44*, 83–87.

Dickerson, S. S., & Kemeny, M. E. (2004). Acute stressors and cortisol responses: A theoretical integration of laboratory research. *Psychological Bulletin, 130*, 355–391.

Domes, G., Normann, C., & Heinrichs, M. (2016). The effect of oxytocin on attention to angry and happy faces in chronic depression. *BMC Psychiatry, 16*, 92.

Eller, N. H., Netterstrøm, B., & Allerup, P. (2005). Progression in intima media thickness -the significance of hormonal biomarkers of chronic stress. *Psychoneuroendocrinology, 30*, 715–723.

Fagundes, C. P., Murdock, K. W., Chirinos, D. A., & Green, P. A. (2017). Biobehavioral pathways to cancer incidence, progression, and quality of life. *Current Directions in Psychological Science, 26*, 548–553.

Fancourt, D., Williamon, A., Cavalho, L. A., Steptoe, A., Dow, R., & Lewis, I. (2016). Singing modulates mood, stress, cortisol and neuropeptide activity in cancer patients and carers. *Ecancermedicalscience, 10*, 631.

Fischer, S., Doerr, J. M., Strahler, J., Mewes, R., Thieme, K., & Nater, U. M. (2016). Stress exacerbates pain in the everyday lives of women with fibromyalgia syndrome: The role of cortisol and alpha-amylase. *Psychoneuroendocrinology, 63*, 68–77.

Friedman, H. S., & Adler, N. E. (2007). The history and background of health psychology. In H. S. Friedman & R. C. Silver (Eds.), *Foundations of health psychology* (pp. 3–18). New York, NY: Oxford University Press.

Gebert, D., Auer, M. K., Steig, M. R., Freitag, M. T., Lahne, M., Fuss, J., … Kopczak, A. (2018). De-masking oxytocin-deficiency in craniopharyngioma and assessing its link with affective function. *Psychoneuroendocrinology, 88*, 61–69.

Gold, P. W. (2015). The organization of the stress system and its dysregulation in depressive illness. *Molecular Psychiatry, 20*, 32–47.

Granger, D. A., Kivlighan, K. T., El-Sheikh, M., Gordis, E. B., & Stroud, L. R. (2008). Assessment of salivary α-amylase in biobehavioral research. In L. Luecken & L. C. Gallo (Eds.), *Handbook of physiological research methods in health psychology* (pp. 95–114). Thousand Oaks, CA: Sage.

Helgeson, V. S., & Zajdel, M. (2017). Adjusting to chronic health conditions. *Annual Review of Psychology, 68*, 545–571.

Hoyt, M. A., Gamarel, K. E., Saigal, C. S., & Stanton, A. L. (2016). Goal navigation, approach-oriented coping, and cancer adjustment in young men with testicular cancer. *Annals of Behavioral Medicine, 50*, 572–581.

Hoyt, M. A., Gaffey, A. E., Wang, A. W., & Lawsin, C. J. (2018). Sexual well-being and diurnal cortisol after prostate cancer treatment. *Journal of Health Psychology*. https://doi.org/10.1177/1359105318772655

Hoyt, M. A., Marin-Chollom, A. M., Bower, J. E., Thomas, K. S., Irwin, M. R., & Stanton, A. L. (2014). Approach and avoidance coping: Diurnal cortisol rhythm in prostate cancer survivors. *Psychoneuroendocrinology, 49*, 182–186.

Hoyt, M. A., & Stanton, A. L. (2018). Adjustment to chronic illness. In T. A. Revenson & R. A. R. Gurung (Eds.), *Handbook of health psychology* (3rd ed.). New York, NY: Taylor & Francis.

Ironson, G., O'Cleirigh, C. O., Kumar, M., Kaplan, L., Balbin, E., Kelsch, C. B., … Shneiderman, N. (2015). Psychosocial and neurohormonal predictors of HIV disease progression (CD4 cells and viral load): A 4 year prospective study. *AIDS & Behavior, 19*, 1388–1397.

Irwin, M. R., & Slavich, G. M. (2017). In J. T. Cacioppo, L. G. Tassinary, & G. G. Berntson (Eds.), *Handbook of psychophysiology* (pp. 377–397). New York: Cambridge University Press.

Jehn, C. F., Kuehnhardt, D., Bartholomae, A., Pfeiffer, S., Krebs, M., Regierer, A. C., … Flath, B. C. (2006). Biomarkers of depression in cancer patients. *Cancer, 107*, 2723–2729.

Jessop, D. S., & Turner-Cobb, J. M. (2008). Measurement and meaning of salivary cortisol: A focus on health and disease in children. *Stress, 11*, 1–14.

Kornienko, O., Schaefer, D. R., Pressman, S. D., & Granger, D. A. (2018). Associations between secretory immunoglobulin a and social network structure. *International Journal of Behavioral Medicine, 25*, 669–681.

Kuhlman, K. R., Irwin, M. R., Ganz, P. A., Crespi, C. M., Petersen, M. S., Asher, A., & Bower, J. E. (2017). Cortisol awakening response as a prospective risk factor for depressive symptoms in women after treatment for breast cancer. *Psychosomatic Medicine, 79*, 763–769.

Lazarus, R. S., & Folkman, S. (1984). *Stress, appraisal, and coping*. New York, NY: Springer.

Leggett, A. N., Liu, Y., Klein, L. C., & Zarit, S. H. (2016). Sleep duration and the cortisol awakening response in dementia caregivers utilizing adult day services. *Health Psychology, 35*, 465–473.

Le-Scherban, F., Brenner, A., Hicken, M. T., Needham, B. L., Seeman, T., Sloan, R. P., … Diez Roux, A. V. (2018). Child and adult socioeconomic status and the cortisol response to acute stress: Evidence from the multi-ethnic study of atherosclerosis. *Psychosomatic Medicine, 80*, 184–192.

Leventhal, H., Halm, E., Horowitz, C., Leventhal, E. A., & Ozakinci, G. (2005). Living with chronic illness: A contextualized, self-regulation approach. In S. Sutton, A. Baum, & M. Johnston (Eds.), *The Sage handbook of health psychology* (pp. 197–240). Thousand Oaks, CA: Sage.

Lipschitz, L., Kuhn, R., Kinney, Y., Donaldson, W., & Nakamura, Y. (2013). Reduction in salivary α-amylase levels following a mind–body intervention in cancer survivors—an exploratory study. *Psychoneuroendocrinology, 38*, 1521–1531.

Lutgendorf, S. K., Weinrib, A. Z., Penedo, F., Russell, D., DeGeest, K., Costanzo, E. S., … Lubaroff, D. M. (2008). Interleukin-6, cortisol, and depressive symptoms in ovarian cancer patients. *Journal of Clinical Oncology, 26*, 4820–4827.

Nater, U. M., & Rohleder, N. (2009). Salivary alpha-amylase as a noninvasive biomarker for the sympathetic nervous system: Current state of research. *Psychoneuroendocrinology, 34*, 486–496.

Nater, U. M., Rohleder, N., Schlotz, W., Ehlert, U., & Kirschbaum, C. (2007). Determinants of the diurnal course of salivary alpha-amylase. *Psychoneuroendocrinology, 32*, 392–401.

Nater, U. M., Skoluda, N., & Starhler, J. (2013). Biomarkers of stress in behavioral medicine. *Current Opinion in Psychiatry, 260*, 440–445.

Nicolson, N. A. (2008). Measurement of cortisol. In L. Luecken & L. C. Gallo (Eds.), *Handbook of physiological research methods in health psychology* (pp. 37–74). Thousand Oaks, CA: Sage.

Nikkheslat, N., Zunszain, P. A., Horowitz, M. A., Barbosa, I. G., Parker, J. A., Myint, A. M., … Pariante, C. M. (2015). Insufficient glucocorticoid signaling and elevated inflammation in coronary heart disease patients with comorbid depression. *Brain, Behavior, and Immunity, 48*, 8–18.

Pariante, C. M., & Miller, A. H. (2001). Glucocorticoid receptors in major depression: Relevance to pathophysiology and treatment. *Biological Psychiatry, 49*, 391–404.

Pinto, A. R., da Silva, N. C., & Pinato, L. (2016). Analyses of melatonin, cytokines, and sleep in chronic renal failure. *Sleep & Breathing, 20*, 339–334.

Polsky, D., Doshi, J. A., Marcus, S., Oslin, D., Rothbard, A., Thomas, N., & Thompson, C. L. (2005). Long-term risk for depressive symptoms after a medical diagnosis. *Archives of Internal Medicine, 165*, 1260–1266.

Rash, J. A., Aguirre-Camacho, A., & Campbell, T. S. (2014). Oxytocin and pain: A systematic review and synthesis of findings. *The Clinical Journal of Pain, 30*, 453–462.

Riis, J. L., Granger, D. A., DiPietro, J. A., Bandeen-Roche, K., & Johnson, S. B. (2015). Salivary cytokines as a minimally-invasive measure of immune functioning in young children: Correlates of individual differences and sensitivity to laboratory stress. *Developmental Psychobiology, 57*, 153–167.

Robles, T. F., Slatcher, R. B., Trombello, J. M., & McGinn, M. M. (2014). Marital quality and health: A meta-analytic review. *Psychological Bulletin, 140*, 140–187.

Ryan, R., Booth, S., Spathis, A., Mollart, S., & Clow, A. (2016). Use of salivary diurnal cortisol as an outcome measure in randomised controlled trials: A systematic review. *Annals of Behavioral Meidicne, 50*, 210–236.

Sapolsky, R. M., & Donnelly, T. M. (1985). Vulnerability to stress-induced tumor growth increases with age in rats: Role of glucocorticoids. *Endocrinology, 117*, 662–666.

Schneiderman, N., Ironson, G., & Siegel, S. (2005). Stress and health: Psychological, behavioral and biological determinants. *Annual Review of Clinical Psychology, 1*, 607–628.

Schoorlemmer, R. M., Peeters, G. M., van Schoor, N. M., & Lips, P. (2009). Relationships between cortisol level, mortality and chronic diseases in older persons. *Clinical Endocrinology, 71*, 779–786.

Schrepf, A., Thaker, P. H., Goodheart, M. J., Bender, D., Slavich, G. M., Dahmoush, L., … Lutgendorf, S. K. (2015). Diurnal cortisol and survival in epithelial ovarian cancer. *Psychoneuroendocrinology, 53*, 256–267.

Sephton, S. E., Dhabhar, F. S., Keuroghlian, A. S., Giese-Davis, J., McEwen, B. S., Ionan, A. C., & Spiegel, D. (2009). Depression, cortisol, and suppressed cell-mediated immunity in metastatic breast cancer. *Brain, Behavior, and Immunity, 23*, 1148–1155.

Sephton, S. E., Lush, E., Dedert, E. A., Floyd, A. R., Rebholz, W. N., Dhabhar, F. S., … Salmon, P. (2013). Diurnal cortisol rhythm as a predictor of lung cancer survival. *Brain, Behavior, and Immunity, 30*, S163–S170.

Shimada, M., Seki, H., Samejima, M., Hayase, M., & Shirai, F. (2016). Salivary melatonin levels and sleep-wake rhythms in pregnant women with hypertensive and glucose metabolic disorders: A prospective analysis. *Bioscience Trends, 10*, 34–41.

Shin, Y. S., Liu, J. N., Kim, J., Nam, Y., Choi, G. S., & Park, H. (2014). The impact of asthma control on salivary cortisol level in adult asthmatics. *Allergy, Asthma and Immunology Research, 6*, 463–466.

Slavish, D. C., Graham-Engeland, J. E., Smyth, J. M., & Engeland, C. G. (2015). Salivary markers of inflammation in response to acute stress. *Brain, Behavior, and Immunity, 44*, 253–269.

Spiegel, D. (2012). Mind matters in cancer. *Psycho-Oncology, 21*, 588–593.

Stetler, C., & Miller, G. E. (2011). Depression and hypothalamic-pituitary-adrenal activation: A quantitative summary of four decades of research. *Psychosomatic Medicine, 73*, 114–126.

Tomiyama, A. J., Epel, E. S., McClatchey, T. M., Poelke, G., Kemeny, M. E., McCoy, S. K., & Daubenmier, J. (2014). Associations of weight stigma with cortisol and oxidative stress independent of adiposity. *Health Psychology, 33*, 862–867.

Tornhage, C. J., Skogar, Ö., Borg, A., Larrson, B., Robertsson, L., Ansersson, L., … Lökk, J. (2013). Short- and long-term effects of tactile massage on salivary cortisol concentrations in Parkinson's disease: A randomised controlled pilot study. *BMC Complementary and Alternative Medicine, 13*, 357.

Turner-Cobb, J., Sephton, S., Koopmanm, C., Blake-Mortimer, J., & Spiegel, D. (2000). Social support and salivary cortisol in women with metastatic breast cancer. *Psychosomatic Medicine, 62*, 337–345.

Weinrib, A. Z., Sephton, S. E., DeGeest, K., Penedo, F., Bender, D., Zimmerman, B., … Lutgendorf, S. K. (2010). Dirunal cortisol dysregulation, functional disability, and depression in women with ovarian cancer. *Cancer, 116*, 4410–4419.

World Health Organization. (2017). *Noncommunicable diseases*. Accessed June, 2017 from http://www.who.int/mediacentre/factsheets/fs355/en/

Wrosch, C., Miller, G. E., Lupien, S., & Pruessner, J. (2008). Diurnal cortisol secretion and 2-year changes in older adults' physical symptoms: The moderating roles of negative affect and sleep. *Health Psychology, 27*, 685–693.

Wrosch, C., Miller, G. E., Scheier, M. F., & de Pontet, S. B. (2007). Giving up on unattainable goals: Benefits for health? *Personality and Social Psychology Bulletin, 33*, 251–265.

Wrosch, C., Scheier, M. F., Miller, G. E., Schulz, R., & Carver, C. S. (2003). Adaptive self-regulation of unattainable goals. Goal disengagement, goal reengagement, and subjective well-being. *Personality and Social Psychology Bulletin, 29*, 1494–1508. https://doi.org/10.1177/0146167203256921

Yeung, E. W., Davis, M. C., & Ciaramitaro, M. C. (2016). Cortisol profile mediates the relation between childhood neglect and pain and emotional symptoms among patients with fibromyalgia. *Annals of Behavioral Medicine, 50*, 87–97.

Yoon, S. A., & Weierich, M. R. (2016). Salivary biomarkers of neural hypervigilance in trauma-exposed women. *Psychoneuroendocrinology, 63*, 17–25.

Zhdanova, I. V., Lynch, H. J., & Wurtman, R. J. (1997). Melatonin: A sleep-promoting hormone. *Sleep, 20*, 899–907.

# Chapter 22
# The Neurobiology of Human Social Behavior: A Review of How Testosterone and Cortisol Underpin Competition and Affiliation Dynamics

**Joey T. Cheng and Olga Kornienko**

**Abstract** The brain, behavior, and neuroendocrine system have coevolved to support human group living. Recent developments in behavioral endocrinology over last several decades increasingly point to the powerful role of social experiences in influencing and being influenced by hormones. Here, we review the accumulated empirical developments that link two hormones—testosterone and cortisol—to social competition and affiliation. We suggest that testosterone and cortisol both influence and reflect the dynamics of human social behavior in domains of competition and affiliation, albeit in very different ways. The evidence supports the notion that testosterone may function as a competition hormone that calibrates psychological systems to current social standing and adaptively guide status-seeking efforts. As for cortisol, much evidence reveals that cortisol modulates affiliative behaviors in ways that appear to be adaptive; cortisol is elevated during times of social threat, social isolation, and loneliness, possibly to mobilize responses geared toward seeking coping and support, but is dampened when individuals gain social control and affiliative support. Still, more work is needed to unpack the complex interplay between neurobiology and human sociality. We end with a number of methodological recommendations on how using salivary bioscience methods may ultimately lead to a richer understanding of the complex reciprocal ties between biology and human social behavior.

**Keywords** Testosterone · Cortisol · Competition · Affiliation · Social status · Stress

J. T. Cheng (✉)
Department of Psychology, York University, Toronto, ON, Canada
e-mail: chengjt@yorku.ca

O. Kornienko
Department of Psychology, George Mason University, Fairfax, VA, USA
e-mail: okornien@gmu.edu



D. A. Granger, M. K. Taylor (eds.), *Salivary Bioscience*,
https://doi.org/10.1007/978-3-030-35784-9_22

## 22.1 History of Salivary Bioscience in Social Competition and Affiliation Dynamics

The study of hormones and social behavior in primates using salivary methods, which began a little over two decades ago, has since become an indispensable tool kit for understanding the interplay between neuroendocrinology and the social faculties of diverse species. Most notably, this work has made substantial contributions to our understanding of the social correlates of stress and the role of androgens in competition. In the earliest studies, salivary cortisol was sampled to study the effects of social interactions on stress in infant rhesus monkeys (Boyce, Champoux, Suomi, & Gunnar, 1995), and then later in squirrel monkeys (Fuchs, Kirschbaum, Benisch, & Bieser, 1997), tree shrews (Ohl, Kirschbaum, & Fuchs, 1999), adult rhesus monkeys (Lutz, Tiefenbacher, Jorgensen, Meyer, & Novak, 2000), and humans (Davis & Emory, 1995). In contrast, the application of salivary testosterone to studying primate social behavior, most notably aggression and competitive behavior, is comparably more recent, despite early validation work undertaken with rhesus monkeys (Arslan, Akhtar, & Nieschlag, 1984), and is more widely adopted in studies of humans than other primates (Dabbs, 1993), for which measures from urinary and fecal samples are widely used (Anestis, 2006; Beehner, Bergman, Cheney, Seyfarth, & Whitten, 2006; Behringer, Deschner, Deimel, Stevens, & Hohmann, 2014; Kutsukake et al., 2009).

The popularity of salivary cortisol and testosterone is driven by the many advantages they present over other methods. For example, approaches that rely on blood plasma sampling often require capture, restraint, and perhaps even sedation, thus inducing substantial stress. Moreover, the collection of urine, feces, or hair is often limited by availability, making difficult repeated sampling within short spans of time, such as in investigations of hormone reactivity or diurnal rhythm. On the contrary, salivary cortisol and testosterone are relatively easy to collect and store, and can be assessed repeatedly insofar as subjects can be trained to suck or chew on an absorbent material. Compared to their alternatives, salivary methods are less invasive and stress-inducing (though in nonhuman primates restraint is still sometimes required), and provide temporally sensitive measurements within a short intersampling interval (Behringer & Deschner, 2017; Kutsukake et al., 2009; Novak, Hamel, Kelly, Dettmer, & Meyer, 2013).

## 22.2 Current Status of Knowledge in Testosterone and Human Social Competition

We begin this review by surveying the extant evidence on the steroid hormone testosterone as a proximate mediator of human competitive social behavior. We propose that this evidence is best viewed in light of theorizing that emphasizes how testosterone (T), a principal androgen regulated by the hypothalamic–pituitary–

gonadal (HPG) axis, functions as a "competition hormone," readying individuals for and facilitating the attainment of social status. In constructing this empirical review, we rely on large assemblies of empirical studies, and, when available, meta-analyses to identify central areas of insight stemming from studies on T. Using this approach, we identified five key insights into T and competitive social behavior in humans:

- Does T cause aggression? A straightforward, one-to-one association between T and aggression is unlikely.
- Nevertheless, there is mounting evidence that T underpins a range of competitive motivation, behaviors, strategies, and propensities in humans.
- Conversely, T also responds to social contexts and experiences.
- T rises during the anticipation of competition, in order to prepare the organism for challenge.
- T responds to competitive outcomes, rising following victory and falling following defeat.

### 22.2.1 *Is Testosterone a Causal Agent of Aggression?*

Converging lines of animal research appear to suggest that aggressive behavior across diverse nonhuman species is, in part, facilitated by T. The earliest evidence for the inductive effect of T on aggression comes, for instance, from studies of rodents and red deer stags that reveal how individuals whose circulating T is suppressed or removed by castration show an absence of agonistic behavior; however, after T is supplemented and restored, fighting resumes (Beeman, 1947; Lincoln, Guinness, & Short, 1972). Conversely, aggression also influences T levels. This regularity is captured by the now well-supported challenge hypothesis (Wingfield, 2017; Wingfield, Hegner, Dufty, & Ball, 1990; Wingfield et al., 2000), which proposes that T fluctuates in concert with challenges—rising during the mating season when aggression is most intense in sexual competition for mates, but falling when physical contests are infrequent (e.g., during non-mating season or periods marked by paternal care or social stability). Consistent with this, early evidence in male birds point to seasonal variation in aggression coinciding robustly with seasonal variation in T, with a high peak of T during the height of intrasexual aggressive competition for females (Beletsky, Orians, & Wingfield, 1992; Vleck & Brown, 1999). Though the challenge hypothesis was initially conceived to explain patterns of androgenic activity birds, the last decades have seen substantial evidence confirming the key predictions that stem from this theorizing in diverse species (Archer, 2006; Hirschenhauser & Oliveira, 2006). However, it must be noted that substantial variation exists both within- and across-species and these effects appear to depend substantially on social context and environmental influences, many of which are still largely unknown (Wingfield, Ball, Dufty, Hegner, & Ramenofsky, 1987).

Does this link between T and aggression in nonhuman animals generalize to humans? The prevailing consensus is yes, but with a caveat: The link is weak at best

(Archer, 2006). Over the years, a series of meta-analyses of the numerous studies have been performed; all converge on the conclusion that the link between T and aggression in humans is weak but positive (in the range of $r = 0.08$ to 0.14), but is also inconsistent and highly variable (Archer, Graham-Kevan, & Davies, 2005; Book, Starzyk, & Quinsey, 2001). What explains the observed empirical inconsistency that plagues this literature? A key issue may be that aggression in humans is complex and non-unidimensional (Brain & Haug, 1992; Carré & Olmstead, 2015; Wrangham, 2018). While some forms of aggression are direct and physical (the variety studied in other mammals), others are indirect and nonphysical (Archer, 2004; Archer & Coyne, 2005). The clearest evidence on physical aggression comes from a large-scale study showing that male prisoners with higher salivary T are more likely to have a history of violent crime (such as homicide, assault, robbery, and rape), but less likely to have a record involving nonviolent crimes (such as theft and drugs; Dabbs, Carr, Frady, & Riad, 1995, Dabbs, Frady, Carr, & Besch, 1987). Similar patterns are found in female prison inmates (Dabbs & Hargrove, 1997). Nevertheless, evidence on physical aggression (and its relations to T) remains limited still. After all, human physical aggression is relatively rare (compared to other species) and conflict is often resolved without escalation (Cant, English, Reeve, & Field, 2006), both of which restrict opportunities for research.

In part out of necessity then, the bulk of other available evidence addressing T and aggression in human relies on indirect aggression or aggressive motivation, rather than threat postures and actual fights (as in studies of nonhuman animals). But even so, the evidence based on these measures still appears mixed. Stronger positive associations are sometimes obtained in studies using peer reports (in contrast to self-reports) of aggressive intent or hostility (Archer, 1991; Assari, Caldwell, & Zimmerman, 2014; Persky, Smith, & Basu, 1971), and salivary T compared to other sampling methods (Archer, Birring, & Wu, 1998). Meanwhile, laboratory-based studies that attempt to simulate and capture actual aggressive behavior (such as using willingness to inflict harm by blasting an opponent with aversive sound) have similarly produced mixed results, reporting positive, null, or sometimes even *negative* associations with salivary T (Buades-Rotger et al., 2016; Carré, McCormick, & Hariri, 2011). Perhaps most problematically, a causal effect of T cannot be confirmed given emerging null evidence of T induction in these laboratory situations (Boksem et al., 2013; Eisenegger, Naef, Snozzi, Heinrichs, & Fehr, 2010; Zethraeus et al., 2009). Overall, given these conceptual and methodological differences, it is unsurprising that studies often yield different estimates of the link between T and aggressiveness. Another issue is that existing studies of T are bedeviled by small samples that hinder the search for firm conclusions (Geniole, Bird, Ruddick, & Carré, 2017). In summary, though the evidence is plentiful, the results are mixed and indicate, at best, a weak positive link between T and aggression in humans (Carré & Olmstead, 2015).[1]

---

[1]Note that although steroid hormones have long been a focus in the study of aggression, it is important to keep in mind that aggressive behavior is ultimately determined by complex interactions

### 22.2.2 Testosterone Underpins Human Competitive Behavior

In light of this controversial, and at best minor, influence of T on human aggression, where does this leave our understanding of the role of T in social behavior? In recent years, an emerging new perspective is that T functions as a "competition hormone" that readies and calibrates an organism's psychology for the pursuit and maintenance of status and by doing so maximizes fitness across competitive contexts (Booth, Granger, Mazur, & Kivlighan, 2006; Eisenegger, Haushofer, & Fehr, 2011; Knight & Mehta, 2014; Mazur & Booth, 1998). In contrast to the early (and perhaps overly simplistic) claim of a one-to-one link between T and human aggression, this revised notion emphasizes the role of T in activating a suite of competitive motivation, behaviors, and strategies that altogether coordinate a complex, integrated behavioral repertoire that facilitates the pursuit and maintenance of status. Unlike the case for aggression, this newly emerging integrated and nuanced view is finding broad and robust support in diverse empirical research programs. In the remainder of this section, we provide a brief review of the relevant evidence, which can be parsed into three interconnected literatures: (a) the effects of T on status-enhancing motivation and behaviors; (b) an anticipatory T increase before an impending competition; and (c) the modulation of T by the outcome of competitions (success and defeat).

#### 22.2.2.1 Testosterone Propels Interlocking Motivation, Behaviors, and Strategies that Enhance Social Status

Diverse lines of research are converging on the notion that T facilitates status attainment in humans by propelling a repertoire of competitive psychology and behavior. First, evincing the key role of T in status-seeking efforts, endogenous T (or, baseline levels of T)—which in many existing studies are assessed using a single saliva sample—are positively correlated with a range of cognitive states and behaviors that increase the success and competitiveness of an individual in competitive situations and conflict. This collection of cognitive states includes those that directly and indirectly increase one's odds of prevailing in conflict—such as implicit power motivation (Schultheiss, Wirth, & Stanton, 2004; Schultheiss et al., 2005; Stanton & Schultheiss, 2009), risk-taking in economic domains (Apicella, Carré, & Dreber, 2015; Apicella et al., 2008; Coates, Gurnell, & Sarnyai, 2010; Sapienza, Zingales, & Maestripieri, 2009), overconfidence (Johnson et al., 2006; Ronay, Tybur, van Huijstee, & Morssinkhof, 2017), intuitive (rather than deliberate) and "hawkish" decision-making (Mehta, Lawless DesJardins, van Vugt, & Josephs, 2017; Nave,

---

between genes (e.g., MAOA genotype), biological signals (e.g., dopamine receptors, steroid hormones including T and estrogen), neural circuits (e.g., amygdala, frontal cortex suppression), and gene-environmental interactions. Hormones do not function in isolation (Batrinos, 2012; Nelson & Trainor, 2007).

Nadler, Zava, & Camerer, 2017), persistence (Andrew & Rogers, 1972; Archer, 1977), willingness to enter competitive interactions (Carré & McCormick, 2008; Coates, Gurnell, & Rustichini, 2009; Mehta & Josephs, 2006), and reduced sensitivity to threat (Hermans et al., 2007; Hermans, Putman, Baas, Koppeschaar, & Honk, 2006; van Honk & Schutter, 2007; van Honk et al., 1999). Behaviorally, higher T is associated with conspicuous consumption to increase perceived status (Nave et al., 2018; Wu, Eisenegger, Sivanathan, Crockett, & Clark, 2017), calling the bluffs of opponents (van Honk et al., 2016), and making more threats, confrontations, and rule infractions in prison (Dabbs et al., 1995; Dabbs & Hargrove, 1997).

Of these results, the effects of T in increasing concern for status and reputation are particularly well illustrated by three sets of laboratory findings. First, Josephs, Sellers, Newman, and Mehta (2006) assigned individuals to high- or low-status treatments to compare how the physiological, emotional, and cognitive states elicited might differ across people with different levels of T. Their results reveal that placing high T individuals in a low-status position created much distress, including inducing negative affect and physiological arousal, heightening their mental preoccupation with status cues, and suppressing their cognitive functioning and performance. Second, building on this evidence, Mehta, Jones, and Josephs (2008) further demonstrated that the consequences of low status vary across men with different levels of T. They found that cortisol rises sharply among high T men following a loss of status (defeat in a competition) but drops in high T men who gain status by winning. Highlighting the role of T in status concerns, in low T men, no changes in cortisol were observed after victory or defeat. Third, further illustrating the link between T and status-seeking, more recent behavioral evidence shows that high T individuals show more characteristic dominant ethological displays, including selfish and forceful gestures and verbal statements, and disproportionate claims to shared resources (Mehta et al., 2017; Slatcher, Mehta, & Josephs, 2011). In sum, there is strong and abundant empirical support for the notion that androgen levels predict a well-coordinated repertoire of motivations and behaviors that regulate and increase one's status and influence.

#### 22.2.2.2 The Modulation of Testosterone by Context: Competition and the Outcome of Conflicts (Winning and Losing)

Above, we have seen how T is a potent proximate mechanism that contributes to regulating competitive behavior. However, T not only propels behavior but also responds to them, meaning that the social environment, in turn, also affects T levels (Mazur & Booth, 1998; van Anders & Watson, 2006). Since its inception over half a century ago, a long-standing interest in the field of behavioral endocrinology entails establishing precisely how endocrine systems interact with social stimuli to jointly regulate the expression of behavior (Beach, 1948; Ford & Beach, 1951). While earlier work has tended to focus on the response of the hypothalamic-pituitary-adrenal (HPA) axis to stressors (Kudielka & Kirschbaum, 2005; Tsigos & Chrousos, 2002), only more recently has research turned to exploring how the HPG responds to social environments. Before proceeding, let us consider why, theoretically, T might be expected to vary in response to social situations.

Despite the advantages of high T in enhancing success in competitions (and hence success in mating, territorial defense, resource acquisition, and so forth), prolonged T elevation presents substantial costs, including high energetic demands, depressed immune function, increased risk of parasitic infestation and mortality, and suppressed investment in the care of offspring (Folstad & Karter, 1992; Lynn, 2016; Oliveira, 2004; Wingfield, Lynn, & Soma, 2001). As a result, males in many species face a trade-off between competitive inclinations (due to its facilitation of mating effort) and parental effort. Possibly owing to this trade-off, rather than remain persistently elevated, in many species T levels respond flexibly to context (Harding, 1981). The challenge hypothesis predicts long-term (seasonal) and short-term patterns of T modulation that correlate with mating and parenting efforts (Wingfield et al., 1990). This means that T should rise in response to challenges when increased competitiveness is particularly advantageous, but dampen during periods when care of offspring is paramount. In this section, we review two major lines of empirical evidence supporting these context effects on T, and (given our particular interest here on competitive behavior) with a focus on work showing that T responds to situations involving (a) social challenge, such as in male–male competition; and (b) social victory and defeat.[2]

## Testosterone Responds to Competition and Social Challenges

One of the most compelling lines of evidence for context effects is that, across diverse species, androgens are modulated by competition. If, as discussed above, T responsiveness is favored by intrasexual selection to turn on or off androgen-dependent behaviors to facilitate mating efforts, a similar T response should also

[2]One of the first published reports documenting an effect of T modulation by social experience appeared in somewhat unusual circumstances. In 1970, an anonymous author published a report in the journal *Nature* (Anonymous, 1970). This communication, entitled "Effects of sexual activity on beard growth in men," reports the author's study of his own personal experience. Having lived in isolation on a remote island for two years, the author noticed what appeared to be a correlation between his beard growth and the timing of visit to the mainland where his fiancée resided. He noticed that when in isolation, his beard grew slowly, but just before his visit to the mainland, it would grow quickly. Inspired by this initial observation, he carried out a detailed study, meticulously quantifying his beard growth by collecting and weighing his beard shavings daily. Confirming his hunch, the data reveal a progressive increase in beard growth in the days that led up to his visit. In fact, he noticed that the longer the period of abstinence, the more obvious the anticipatory response. He concluded that the mere expectation of sexual activity served as a stimulus for his accelerated beard growth. He surmised that beard growth acts as a proxy for androgenic activity, and that the anticipation of sexual activity triggers androgen production, in turn stimulating beard growth. As we summarize in this chapter, subsequent research, some of which might well have been inspired by this anonymous author's experimentation, reveals qualitatively similar patterns that point to the role of sexual activity and its anticipation in inducing T production in men.

emerge in a range of situations involving competition with other males, given that male–male conflicts represent a primary means through which males compete, both directly and indirectly, for access to females. Evolutionary logic proposes that T responses to competition should operate in two principle ways (Archer, 2006; Mazur & Booth, 1998): (a) T should rise in anticipation of competition to prepare the organism for impending contest; and (b) following the competition, T should be elevated in winners but suppressed in losers to adaptively modulate future competitive motivation. We discuss these two ways in which T responds to competition next.

*Testosterone Rises in Anticipation of Competition*

An anticipatory androgen responsiveness to competition allows an individual to better adjust and regulate its subsequent behavioral output to the current context. That is, given the physiological effects of androgens on competitive ability, ranging from muscular development to rapid dampening of anxiety responses (Aikey, Nyby, Anmuth, & James, 2002; Celec, Ostatníková, & Hodosy, 2015), an anticipatory androgen response to contest may facilitate subsequent competitive behavior expressed in territorial defense, mate-guarding, and status contests (Carré & Olmstead, 2015).

Indeed, numerous studies performed on a range of species confirm an anticipatory T effect. These studies show that challenges from conspecific males in competitive encounters lead to a spike in T production in males, including primates (Cavigelli & Pereira, 2000; Harding, 1981; Loren Buck & Barnes, 2003; Muller & Wrangham, 2004; Rose, Holaday, & Bernstein, 1971). Interestingly, this effect may even operate on bystanders uninvolved in the fight but in whom aggressive motivation is merely primed (Clotfelter & Paolino, 2003; Oliveira, Lopes, Carneiro, & Canário, 2001). In the case of humans, similar evidence of an anticipatory T effect is observed in studies that track endocrine activity in the moments that precede the competition, many of which in the form of competitive sports and contrived laboratory competitions that attempt to simulate real-life human dominance contests. In their now classic tennis study, Booth, Shelley, Mazur, Tharp, and Kittok (1989) found that players displayed an anticipatory rise of T, as evidenced in their higher salivary T measured 15 min before the match compared to a baseline T assessed the day prior. Similar patterns of T rise before the competitive encounter have been observed in judo, wrestling, hockey, chess, and video game tournaments (Booth, Mazur, & Dabbs, 1993; Mazur, Booth, & Dabbs, 1992; Mazur, Susman, & Edelbrock, 1997; Salvador, Suay, González-Bono, & Serrano, 2003; Salvador, Suay, Martinez-Sanchis, Simon, & Brain, 1999), though note that the use of small samples remains as a key limitation of many of these studies.

*Testosterone Responds to Contest Outcomes: Rising After Success and Falling After Defeat*

Beyond an anticipatory T response, another hormonal adaptation that may enhance success in competitive interactions is a flexible T response that takes into account the current competitive standing of the self vis-à-vis the rival, foremost in the form of a T rise during victory and T decline during defeat. This highly specific pattern of T response may act as an adaptive mechanism to adjust future behavior, such that future competitive behavior is facilitated in winners to prepare for future challenges, but dampened in losers whose withdrawal from future challenges minimizes costs from further injury and loss in status (Carré et al., 2011; Carré & Olmstead, 2015; Mazur & Booth, 1998; Mehta & Josephs, 2006; Zilioli & Bird, 2017). These outcome-dependent T changes may thus be functionally similar to other status-dependent changes in morphology (e.g., alterations in facial coloration and other sexual adornments, testicular size) and behavior (e.g., sociality) that occur when adult males rise or fall in social contests. That is, these changes may be considered part of a broader suite of behavioral and physiological adaptations for calibrating ongoing and future behavior in intrasexual competition (Setchell & Dixson, 2001; Wingfield et al., 1990).

Evidence indicates that a wide range of social mammals calibrate their T levels to wins and losses (Mazur & Booth, 1998). Winners of status contests generally show a rapid increase in circulating T relative to pre-competition or losers in physically demanding competitions, such as wrestling, rowing, and tennis (Booth et al., 1989; Elias, 1981; Longman, Surbey, Stock, & Wells, 2018; Mazur & Lamb, 1980), as well as nonphysical competitions with sanctioned competitors, such as chess, domino, and video game matches (Flinn, Ponzi, & Muehlenbein, 2012; Mazur et al., 1992; Zilioli & Watson, 2012). This victory-induced T effect is particularly pronounced when the stakes of competition (and thus competitive motivation) are especially high, such as when status concerns are hyper-salient and the domain of competition has high self-importance (Edwards, Wetzel, & Wyner, 2006; Schultheiss et al., 2005; Vongas & Al Hajj, 2017), the competition venue is in own territory rather than away (i.e., "home advantage;" Carré, 2009; Fuxjager, Mast, Becker, & Marler, 2009; Neave & Wolfson, 2003), or the defeated rival is from an antagonistic out-group rather than in-group (Flinn et al., 2012; Oxford, Ponzi, & Geary, 2010).

A key issue in these earlier studies on victory-induced T effects, however, is that causality cannot be firmly established. A rising T profile may be a cause (rather than a consequence) of winning, thus leaving the possibility of reciprocal causality on the table. Recent laboratory experiments, however, have dramatically clarified the causal basis. Using rigged competitions to manipulate wins versus losses, a number of studies confirm a causal effect of winning on elevated T (Gladue, Boechler, & McCaul, 1989; Josephs, Newman, Brown, & Beer, 2003; Josephs et al., 2006; Longman et al., 2018; Newman, Sellers, & Josephs, 2005; Schultheiss et al., 2005; Zilioli & Watson, 2012, 2014; but see Wu et al., 2017). These results are obtained even when the outcome is knowingly chance-based and independent of ability (e.g., coin-tosses; McCaul, Gladue, & Joppa, 1992), and victory is experiential rather than personal, such as for fans of a winning soccer team and supporters of a winning political candidate (Bernhardt, Dabbs, Fielden, & Lutter, 1998; Stanton, Beehner, Saini, Kuhn, & LaBar, 2009). This causal effect of winning on T can be sizable; for example, a 40% T increase is induced by simply watching one's previous victory on video (Carré & Putnam, 2010). Taken together, these lines of evidence from a range of competitive contexts indicate that winners display a greater rise in T than losers, supporting the notion that these context-dependent T responses provide organisms with feedback on prior social experiences to effectively calibrate future competitive efforts.

In summary, converging lines of evidence indicate that T predicts as well as responds to competitive interactions; it is both a cause and a consequence of social challenge.

### *Testosterone Facilitates the Competition for Social Status Based on Both Dominance (Agonistic Contests Based on Fear) and Prestige (Non-agonistic Contests Based on Respect)*

Humans can gain social rank in different ways. In most other primates and social mammals, rank structure within groups is principally organized around dominance—rank differences established on the basis of competitive interactions involving agonism, force, aggression, intimidation, and violence (Bernstein, 1981; Hinde, 1974). In humans, however, the social organization of a collection of individuals is not simply an extension of these dominance hierarchies. For instance, we often seek out and defer—out of personal choice—to people who are particularly successful, skilled, and knowledgeable in locally valued domains (Boyd & Richerson, 1985; Boyd, Richerson, & Henrich, 2011; Henrich & Gil-White, 2001). These rank differences do not appear to be products of any agonistic coercion, fear, or imposition (as they are in dominance), but rather results from non-agonistic persuasion that is freely conferred. Emerging theoretical and empirical work delineates the distinction between these two forms of rank that operate in human societies. This work proposes that, unlike in other social mammals whose social organization is principally based on dominance (coercive capacity that derives from strength, threat, and intimidation), humans possess a separate pathway to social rank termed *prestige* (persuasive capacity that derives from valued skills, abilities, and knowledge; Cheng & Tracy, 2014; Cheng, Tracy, Foulsham, Kingstone, & Henrich, 2013; Henrich, 2016; Henrich & Gil-White, 2001; Maner, 2017). Empirically confirming this distinction, a substantial body of laboratory and field evidence indicates that prestige and dominance (a) can be distinguished by their ethological displays (e.g., postural and vocal cues and signals; Cheng, Tracy, & Henrich, 2010; Cheng, Tracy, Ho, & Henrich, 2016), motivational profiles (Case & Maner, 2014; Maner & Mead, 2010; Mead & Maner, 2012), and affective responses (Cheng et al., 2010); and (b) coexist to influence group decision-making and attention patterns in laboratory small groups (Cheng et al., 2013), naturalistic groups and teams in the field (Cheng et al., 2010; Redhead, Cheng, Driver, Foulsham, & O'Gorman, 2018), and even within the communities of people living in small-scale societies typified by highly egalitarian social norms (Garfield & Hagen, in press); and (c) lead to higher fitness outcomes, but via different mechanisms (Snyder, Kirkpatrick, & Barrett, 2008; von Rueden, Gurven, & Kaplan, 2011).

Recognizing this duality of social rank in human life opens up to a new vista of questions, including foremost: Does T regulate the pursuit and expression of both forms of social rank? The bulk of prior studies exploring the connection between T and social rank treat status as a unidimensional construct, operationalizing it as a confusing mix of both prestige and dominance. As discussed above, in these studies T effects are often explored in physically taxing sports competitions (e.g., wrestling, tennis) that emphasize both physical prowess (e.g., size and strength, agility, and endurance)—which may induce submission via perceived force and coercion (dominance)—and other game-relevant skills and abilities (e.g., planning, strategizing, technique) that may attract deference based on earned respect for perceived success and achievement (prestige). Even in studies of skill-based contests that lack physical confrontation (e.g., chess, tetris), this issue persists because direct competition with a sanctioned rival or enemy is likely to evoke a dominance psychology based on domination and subordination fueled by animosity and hostility, in addition to prestige stemming from greater intellectual skill.

Recent empirical work, however, is rectifying this conceptual ambiguity, and the latest findings from these efforts, when considered in conjunction with other existing evidence, are beginning to create a new picture that points to a role of T in both prestige and dominance rank competitions. Given that the focus on strictly prestige and T represents a stark departure from existing work on nonhumans (with its focus on agonistic dominance), here we focus on sketching the evidence that points to a link between T and prestige. The first line of evidence demonstrates that T levels predict the expression of behaviors that facilitate the pursuit of prestige (such as generosity; rather than dominance and aggression). That is, in sharp contrast to the early (but likely inaccurate) view that T causes hostile and aggressive behavior in humans, emerging evidence suggests that T may fuel prosocial preferences and behaviors, especially in contexts in which a prestige-based avenue to rank appears more viable or profitable (in terms of fitness gains). For example, in female communities—where highly dominant women may evoke particularly strong anti-dominance sentiments from subordinates (Benenson, 2013; Cashdan, 1995), thus making dominance a precarious long-term strategy in this context (Redhead et al., 2018)—a rise in T actually predicts greater affiliative interactions with other women (Casto & Edwards, 2016). In fact, directly linking T to prestige-based status, male and female athletes with higher T are not only seen as more skilled by teammates but also enjoy greater social popularity and connectedness—two proxies of high prestige (Edwards et al., 2006). In another study, salivary T is linked to reduced aggression and reactivity to angry faces in laboratory studies (Buades-Rotger et al., 2016). These findings dovetail with results from T administration studies, which reveal that an acute dose of T increases fairness in a bargaining game (Eisenegger et al., 2010), generosity toward those who are prosocial with them (reciprocity) (Boksem et al.,

2013), and willingness to sanction norm-violators at a personal cost (Dreher et al., 2016).

The second line of evidence, which complements the research program (described above) on how T responds to contest outcomes, addresses whether and how T is modulated by experiences of gaining prestige (that is, winning and losing contests based solely on prestige, absent of dominance). In this vein, a recent study by our team examined T responses to changes in prestige ranking within a highly cooperative community. To unambiguously distinguish prestige from dominance, we focused on how T changes among individuals who attract substantial respect and admiration from members of their community, as a result of their earned merit and achievement, devoid of dominance, fear, or antagonism. Results show that men who achieve top ranks of the prestige hierarchy in the initial weeks of the group's formation show a rise in testosterone over the subsequent 2 months, whereas men with low prestige show a decline or little change in testosterone (Cheng, Kornienko, & Granger, 2018). These results converge with prior work demonstrating how winning competitions modulates T, but supply novel evidence that winning prestige, devoid of any dominance or antagonism, is sufficient to raise T levels. This hints at a possible role of T in facilitating the emergence and maintenance of prestige hierarchies.

In sum, recently emerging evidence suggests that T may be a candidate physiological mechanism that orchestrates both prestige- and dominance-seeking efforts (Eisenegger et al., 2011; Gray, McHale, & Carré, 2017). When the environment conspires against dominance and offers incentives for a prestige-based route to rank (such as in many contemporary workplaces), T may propel emotions (e.g., pride), motivations (e.g., affiliation versus aggression), and behaviors (e.g., generosity) that help sustain or increase an individual's prestige and influence (Cheng et al., 2010; Henrich, Chudek, & Boyd, 2015), all the time while aggression and a general inclination toward coercive tactics remain suppressed. Although this work is still in its infancy and a much larger database is needed, the empirical patterns available are consistent with the notion that T may, in some contexts, facilitate competition in prestige-based rank contests.

## 22.3 Current Status of Knowledge on Cortisol and Human Social Affiliation

In the next section, we shift our focus to cortisol (C) and affiliative relationships. Volumes of research have focused on the neuroendocrine underpinnings of the stress response, and the role of HPA axis and its end product cortisol in readying and facilitating the adaptation to chronic, unpredictable, and long-lasting stressors (e.g., Del Giudice, Ellis, & Shirtcliff, 2011; Gunnar & Quevedo, 2007; McEwen & Gianaros, 2010). Emerging conceptual and empirical efforts have been directed at describing the role of C in modulating social behavior. Here, we review evidence on associations between social stress and salivary C in humans and propose that this evidence is best viewed in light of theorizing that emphasizes how C relates to social affiliation behavior that is directed at adaptation to stress, which among human primates (as well as nonhuman primates and other species, (Raulo & Dantzer, 2018), although the latter is beyond the scope of this review) unfolds as intricately connected with social interactions and embedded within social context. In constructing this empirical review, we again rely on extant reviews of empirical studies, and, when available, on meta-analyses to identify central themes from studies using salivary C. Using this approach, we identified five key insights into associations between C and social behavior. The major take-home points from this section of the review are that:

- Elevated C is associated with inhibition and withdrawal from social relationships and anxiety, loneliness, social isolation, social rejection, and social status threat.
- But elevated C can also increase social affiliation as a means of stress reduction, and serve as a social buffering mechanism within the context of parental relationships.
- Distinguishing between stressors that are endogenous (i.e., internal) and exogenous (i.e., external) to social relationship may help delineate patterns of cortisol–social behavior links.
- C response is dampened in the presence of conspecifics who provide social buffering effects when the stressor is exogenous.
- The social nature and intensity of stressors moderate C response and its association with social behavior.

### 22.3.1 *Cortisol Underpins Social Behavior as a Part of the Stress Response*

C, a glucocorticoid, is the primary end product of activity of the HPA axis in humans (Chrousos & Gold, 1992; see Chap. 5). Cortisol levels trend higher when individuals (a) appraise a situation to be challenging, uncertain, and intense, and (b) experience rumination and social status threat (Denson, Spanovich, & Miller, 2009; Dickerson, 2008). Short-term elevation of cortisol is considered to be adaptive in novel or dynamic social environments (Sapolsky et al., 2000), whereas prolonged activation has the potential to translate into cumulative wear and tear on many biological systems with downstream consequences for health (McEwen & Gianaros, 2010).

According to the integrated specificity model of stress (Kemeny, 2003; Lazarus & Folkman, 1984; Weiner, 1992), specific stressors and their cognitive appraisals initiate a psychobiological stress response, including the mobilization of emotional, motivational, and physiological systems. Perceived control over stressors instantiates *defeat* and *defense* strategies as a part of the integrated stress response profile. Threats to one's well-being that are appraised as uncontrollable and outside of the scope of one's coping resources may initiate *defeat or disengagement responses.*

Defeat responses drive HPA axis activity and associated distress, withdrawal, and depressed affect (Chrousos & Gold, 1992). Research has shown that individual differences in C levels are linked with social withdrawal, inhibition, and anxiety (e.g., Kagan, Reznick, & Snidman, 1987; Shoal, Giancola, & Kirillova, 2003; Smider et al., 2002).

Stressors appraised as controllable and within one's coping resources lead to *defense or engagement responses* including fight or flight responses, which are mediated by the activity of autonomic nervous system cascades involving respective upregulation of sympathetic and downregulation of parasympathetic branches (Koolhaas & Bohus, 1989). Stressful conditions, in which competition and dominance dynamics are apparent, especially under conditions of instability of a social hierarchy, have been linked to increased T levels when an individual engages in active coping with a challenge; by contrast, T levels tend to plummet when an individual is passively or reactively coping with a challenge (Archer, 2006; Mehta & Josephs, 2010; Salvador & Costa, 2009; Schoofs & Wolf, 2011). Another line of research suggests that activation of HPA axis is associated with the mobilization of coping resources and increased sensitivity to social cues and feedback (Del Giudice et al., 2011), which may be beneficial for an individual who occupies a position of high social status, in part because being at the top of the group hierarchy may require the capacity to detect and respond to threats to one's social standing.

Yet another pattern through which cortisol might be associated with social behavior involves the "*tend-and-befriend*" model as an alternative coping response to stress. Affiliation with others under stress is a protective mechanism to restore safety and avert threats (Taylor, 2011). Given the prevalence of gender-based division of labor in human history, in which men were primarily responsible for hunting and group protection and women for gathering and child-rearing, women's stress response may have evolved to not only protect oneself but one's offspring during times of stress. Taylor et al. (2000) and Taylor (2006) proposed that, particularly for women, fight or flight may not be the most adaptive response compared to tend or befriend. Here, "tending" involves nurturant activities designed to promote safety and reduce the distress of the self and offspring, whereas "befriending" is the existence and use of social networks that may aid in these processes (Taylor et al., 2000). The "tend-and-befriend" model posits that individuals, especially women, may form tight and/or extensive social ties and seek out friends in times of stress, who provide them with social support and help buffer against the deleterious effects of stress. Indeed, tending and befriending, as in turning to others for support and help, have been shown to be an effective stress coping strategy (e.g., Tamres, Janicki, & Helgeson, 2002). But it is noteworthy that both men and women turn to others for help under stress, and other lines of research guided by evolutionary logic suggest that men are also likely to create more extensive networks and befriend more non-kin peers given their greater reliance on coalitional building during hunting and group protection (Benenson, 2014).

#### 22.3.1.1 Social Context Moderates Stress Response and Adaptation

Adapting to the trials and tribulations of everyday life may sometimes occur in social isolation, but, among inherently social humans, it is also common for these processes to unfold in the context of networked social relationships (i.e., family, friends). Living in social groups affords individuals many advantages and theorists suggest that the nature and complexity of components of the human central nervous system have evolved to meet the demands of living in large groups (e.g., processing ambiguous social information; Chang et al., 2013; Seyfarth & Cheney, 2013; Silk, 2007). Not surprisingly, individual differences in the activity and regulation of the HPA axis are associated with subjective experiences created throughout social interactions with group members (e.g., social evaluation and status threat, novelty, unpredictability; Denson et al., 2009; Dickerson & Kemeny, 2004). The biobehavioral associations between hormones and social behavior are shaped not only by the nature of a stressor (status-related, requiring an extended effort) and its cognitive appraisal (Denson et al., 2009), but also by its social context (e.g., quality of family relationships; Booth, Johnson, Granger, Crouter, & McHale, 2003).

When considering the adaption to stressors within a context of social relationships, it is useful to distinguish between whether the stressor to which an individual is responding occurs within the context of a social relationship (i.e., stressor is endogenous to a relationship), or if the stressor occurs outside of the context of a social relationship (i.e., stressor is exogenous to a relationship). Social support or buffering effects on HPA activity are typically documented with stressors that are exogenous to one's relationship (i.e., public speaking or some other external event). For example, one study showed that the presence of a parent or a friend was associated with faster HPA axis recovery after a public speaking task, suggesting a social buffering effect (Hostinar, Johnson, & Gunnar, 2015). However, a closer look at the accumulating evidence on the moderating role of social context hints at how potent stressors may emerge *as a function of one's social relationships* (i.e., endogenous stressors), with interpersonal stressors of poor relationship quality, high degree of conflict and negativity, and social rejection exerting an influence on neuroendocrine processes and social behavior. We next briefly consider these two types of stressors.

##### Stressors that Are Endogenous to Social Relationships

**Social Rejection and Social Status Threat** As previously discussed, higher C levels are linked to fear of social rejection and losing social status, acceptance, and esteem. These patterns have been robustly established among children (Gunnar, Sebanc, Tout, Donzella, & van Dulmen, 2003), adolescents (Adam, 2006; Blackhart, Eckel, & Tice, 2007), and adults (Dickerson, 2008; Dickerson & Kemeny, 2004; Dickerson & Zoccola, 2013).

**Loneliness** Being isolated from the social group is a major stressor, and neuroendocrine and physiological systems link loneliness with heightened activity of the HPA axis (Adam, Hawkley, Kudielka, & Cacioppo, 2006; Cacioppo, Capitanio, & Cacioppo, 2014). Prolonged experiences of loneliness have detrimental health outcomes, including diminished cardiovascular health, immunity, sleep quality, and mental health (Pressman et al., 2005; Valtorta, Kanaan, Gilbody, Ronzi, & Hanratty, 2016). In contrast, transient feelings of loneliness serve an adaptive signaling function and promote the restoration of the basic belongingness need by seeking and renewing social relationships (Cacioppo et al., 2014). It is thought that feelings of loneliness activate the self-preservation system and its integral component—heightened vigilance to social threat (Cacioppo et al., 2014). A critical physiological marker of social threat is the activity of HPA axis and its end product C (Dickerson, 2008). Indeed, research has linked loneliness to increased HPA activity (Adam et al., 2006; Cacioppo et al., 2002; Doane & Adam, 2010; Glaser, Kiecolt-Glaser, Speicher, & Holliday, 1985; Steptoe, Owen, Kunz-Ebrecht, & Brydon, 2004). Whereas cortisol is a stress hormone, it can serve as an adaptive function of energy mobilization (Del Giudice et al., 2011), which could support the development of social connection for a lonely individual.

**Poor Relationship Quality, Negativity, and Conflict** Social relationships may sometimes be the source of conflict, strain, and discord, which qualifies them to be salient psychosocial stressors with negative consequences for health (e.g., Newsom, Mahan, Rook, & Krause, 2008). Indeed, empirical evidence suggests that negativity and aversive interactions within close relationships are significant stressors (Rook, 1984, 2001) and have detrimental effects on health (Kiecolt-Glaser & Newton, 2001) and stress physiology (Timmons, Margolin, & Saxbe, 2015). A recent review of 18 studies reveals cortisol coregulation within romantic or marital dyads, and moreover joint increases in cortisol levels predict poorer relationship quality and higher levels of conflict (Timmons et al., 2015).

**Attunement on Cortisol in Social Relationships** Group living and social ties that connect individual members within a group are viewed as crucial adaptations contributing to the fitness of human and nonhuman primates (e.g., Silk, 2007). Researchers have proposed that this adaptive nature of groups emerges through processes of *bio-behavioral synchrony* (Feldman, 2015) or *attunement* (Granger et al., 2012), defined as the temporal, ongoing coordination of biological, and behavioral processes among members of a social group. Empirical evidence documents physiological attunement for a variety of dyadic social relationships, including parent–child bonds (Booth et al., 2003; Sethre-Hofstad, Stansbury, & Rice, 2002), married or dating couples (Timmons et al., 2015), and friendship dyads (Rankin, Swearingen-Stanborough, Granger, & Byrd-Craven, 2018). Moving beyond the dyad as the unit of analysis, physiological attunement has been reported even among small family groups (Booth et al., 2000). A recent meta-analysis of 29 laboratory and 16 ecological studies of romantic and family dyads suggests robust adrenocortical attunement among dyad members (Ha et al., 2018).

### Stressors that Are Exogenous to Social Relationships

When adapting to exogenous stressors—external to one's social relationship—individuals may draw on social connections, which provide social buffering effects on adrenocortical activity. In such circumstances, as a prominent model—the social buffering hypothesis—posits, social support derived from one's social relationships serves protective effects by attenuating the physiological stress response (Carter, 1998; Feldman, 2015; Gunnar & Hostinar, 2015; Hostinar, Sullivan, & Gunnar, 2014). This hypothesis has received some support within the context of parent–child relationships (e.g., Gunnar & Quevedo, 2007). However, a recent meta-analysis of 38 studies failed to find an overall non-zero effect for the association between parental warmth and offspring HPA axis basal levels, reactivity, or recovery (Hackman, O'Brien, & Zalewski, 2018). Upon examination of moderators, it appeared that parental warmth was protective against reactivity but not recovery in laboratory tasks in which social status threat was acute. Hackman et al. concluded that this might have implications for children's appraisal of stressors.

Interestingly, emerging evidence suggests that parental support does not effectively promote recovery of the HPA axis after a public speaking task during adolescence, whereas friend support does (Hostinar et al., 2015). Furthermore, the presence of a friend appears to have dampening effects on the reactivity of the HPA axis to negative social experiences and exclusion in naturalistic ecologies (Adams, Santo, & Bukowski, 2011; Peters, Riksen-Walraven, Cillessen, & de Weerth, 2011) and the laboratory (Calhoun et al., 2014).

In summary, the extensive body of theoretical and empirical work points to the vital role that C may play in the modulation of social behavior. The emerging patterns of this association are not linear and straightforward. On the one hand, elevated C is associated with reduced social affiliation—social inhibition, withdrawal from social relationships, social isolation, social rejection, and social status threat. On the other hand, elevated C can also increase social affiliation as a means of stress reduction, where social support and buffering mechanisms operate in the context of a social relationship (e.g., parent–child relationship) and increased sociality is associated with downregulation of HPA axis activity. Another theme that is apparent across the vast corpus of evidence on cortisol-social affiliation link is that features of the social context moderate C response and its association with social behavior.

## 22.4 Methodological Opportunities, Challenges, and Considerations in the Study of Salivary Bioscience and Human Social Behavior

As reviewed above, in the last several decades, a large scientific enterprise has emerged to lay the foundation for understanding endocrinology and its relation with social relationships in humans and nonhuman animals. Despite the progress made, however, many key puzzles in human social behavioral endocrinology remain to be solved and a number of important discoveries are, in all likelihood, yet to be made. Here we propose three methodological recommendations to facilitate a greater understanding of the complex interplay between neurobiology and social behavior: (1) conduct field studies; (2) sample diverse, non-WEIRD populations; and (3) extend focus to social networks and communities, beyond individuals and dyads.

### *22.4.1 Go Wild! The Importance of Studying Neurobiology and Human Social Behavior in the Field and Outside of the Laboratory*

As our review above highlights, scientific inquiry into the neurobiological underpinnings of human social dynamics has proliferated from an exceptionally strong integration of extensive field and laboratory evidence. Insofar that the combined use of both laboratory and field studies will continue to be paramount for developing comprehensive models of neurobiology and social dynamics, one important direction for future work is to further expand the existing corpus of field studies. In the branch of work on cortisol (Saxbe, 2008), for instance, giant strides have been made over the last two decades, owing in part to an increasing shift in focus from laboratory-based stress induction methodologies (reviewed in Dickerson & Kemeny, 2004) to field research on everyday stress and coping. The tremendous promise and value of studying neurobiology "in the wild" is strikingly demonstrated by a recent meta-analysis of T effects. After surveying the last 35 years of research on the effects of competition outcome on T responses, the authors concluded that "the strength of the 'winner-loser' effect depended on the location of the competition, whereby the effect was much stronger in studies conducted outside the lab... compared to studies conducted in the lab" (Geniole et al., 2017, p. 47). "Much stronger" is perhaps an understatement; the average effect is *six times* stronger in the field as it is in the laboratory. Although the basis for why these T effects are larger in the field is not well understood, a primary factor, that is suspect is the greater potency of the social experience in the field (e.g., greater psychological investment in winning in real life compared to a laboratory game; Geniole et al., 2017; Oliveira, 2004). Put simply, there are simply few laboratory analogs capable of simulating the powerful experience of human social interactions as in real life. There are, of course, an array of other important reasons for undertaking field research—from establishing ecological validity to exploring questions that cannot be answered in the laboratory (e.g., How does social rank alter T longitudinally? How does reproductive behavior change hormones? What effect does basal HPA axis activity have on primate health? Cheng et al., 2018; Gray et al., 2017; Sapolsky, 2005). Nevertheless, on a practical front the unparalleled potency of human social relationships and experiences in real life, and their powerful effects on endocrinology, provide unrivaled research opportunities.

### 22.4.2 *Sample Diverse Populations, Especially non-WEIRD Individuals*

Across the behavioral sciences there is growing interest in identifying aspects of human nature that are universal as well as those that are culturally variable, using comparative research with diverse societies, beyond the traditional WEIRD (Western Educated Industrialized Rich Democratic) samples on which much of existing empirical foundations is built (Apicella & Barrett, 2016; Henrich, Heine, & Norenzayan, 2010). Behavioral endocrinology researchers, too, are responding to this call; an emerging generation of comparative research programs is beginning to shed new light on the ways in which the interplay between hormones and behavior are similar or variable across cultures. Among these efforts, for example, is Trumble et al.' (2012) recent study of the Tsimane, forager-horticulturalists of the Bolivian Amazon, which reveals evidence of the same pattern of a competition-induced T increase in men typically found in WEIRD samples.

These results contribute a crucial data point to the existing empirical database that is heavily skewed toward sampling men in industrialized societies, who, in fact, have an unusually high level of T across all ages (Bribiescas, 1996), possibly owing to the low energetic and pathogenic stress typified by industrialized settings. Thus, evidence of a qualitatively similar competition-induced increase in men in nonindustrialized societies, where investment in exaggerated T-related faculties may be too costly and is thus reflected in a lower basal T level in males, is crucial and offers some suggestive preliminary evidence that perhaps the social modulation of T effects that are relatively well established in industrialized contexts may, in fact, generalize beyond the WEIRD contexts studied and apply species-wide. Other similar efforts include work that investigates, for instance, how T responds to experiences of challenge, such as hunting excursions and actual kills in Tsimane and !Kung San men (Jaeggi, Trumble, Kaplan, & Gurven, 2015; Trumble, Smith, O'Connor, Kaplan, & Gurven, 2014; Worthman & Konner, 1987), and paternal caregiving among East African foragers and pastoralists (Muller, Marlowe, Bugumba, & Ellison, 2009). As these efforts illustrate, comparative research on behavioral endocrinology with diverse societies, while challenging, will likely occupy an increasingly central role for generating broad insights into the complex nature of hormone-behavior interactions in our species.

### *22.4.3 Contributions of Social Network Analysis to Study the Social System in Its Entirety, Beyond Individuals and Dyads*

Research to date has advanced our understanding of hormone regulation of social behavior considered at the level of an individual, dyads, and social groups. A key limitation of this research when it comes to examining social groups and communities is that it has predominantly relied on aggregate composites of group processes and ignoring the role that social network structure and dynamics play for hormone-social behavior associations (for an exception see research with nonhuman primate dominance hierarchies; Sapolsky, 2005). Social networks represent the structures and dynamics of group living and social connections among individuals in a group are central to understanding social behavior and context because they govern the ways in which relational provisions such as information, resources, and support are distributed (for reviews, see Borgatti, Mehra, Brass, & Labianca, 2009; Kadushin, 2012). As such social networks emerge as a result of an individual's social behavior in a group and influence psychological and neuroendocrine processes (for reviews, see Berkman, Glass, Brissette, & Seeman, 2000; Crosier, Webster, & Dillon, 2012).

Social network theory and analytical tools enable pursuing several innovative research questions by examining social systems in their entirety (e.g., communities and networks), beyond individuals and dyads. The first set of questions focuses on the understanding of hormone associations with network position, which is derived from nominations collected from individuals (i.e., egos) and their group members (i.e., alters) within a defined social group. Complete network data, referring to multi-informant assessment of network ties (O'Mailey & Marsden, 2008; Wasserman & Faust, 1994), allow the consideration of directed and mutual relationships within a social system. An individual's outgoing ties depict social network activity or gregariousness, whereas incoming ties describe social popularity and status (Wasserman & Faust, 1994). Social network density, describing the degree of interconnectedness among one's friends, has implications for the flow of social support and stress-buffering processes (Walker, Wasserman, & Wellman, 1993) and, therefore, may have direct and moderating effects on associations between social stressors and cortisol. An individual's network centrality is measured by outgoing direct and indirect ties to all other members of a community and has implications for access to resources and information (Borgatti et al., 2009) and likely serves a stress-buffering effect.

Emerging studies of nonhuman and human primates suggest that hormones are associated with social network position and structure. Specifically, high-ranking free-ranging female macaques were shown to have lower glucocorticoid levels when their association networks were smaller and more focused, as indexed by a lower number of outgoing connections (Brent, Semple, Dubuc, Heistermann, & MacLarnon, 2011). Among female nursing students (Kornienko, Clemans, Out, & Granger, 2013; Kornienko, Clemans, Out, & Granger, 2014) and male rugby players (Ponzi, Zilioli, Mehta, Maslov, & Watson, 2016), salivary C levels were also inversely associated with gregariousness levels; these findings are in line with prior research documenting the impact of social isolation on HPA axis activity.

Because social ties within a group are not formed and dissolved randomly, the second set of questions that could be addressed with the use of SNA approaches focuses on examining how hormones are associated with how social network structures are being created. Social network selection refers to understanding changes in networks based on factors that include (a) individual characteristics (e.g., hormone levels) that affect the tendency to form ties and (b) network structural processes (e.g., popularity, transitivity), reflecting how connections between individuals depend on the nature of their ties with other members of a group. Networks research has shown that to obtain unbiased estimates of how hormone concentrations contribute to network selection; thus, we need to statistically control for alternative network structural processes using social network modeling approaches (Snijders, 2011).

A recent investigation explored the role of cortisol and testosterone levels in predicting friendship tie maintenance and creation over time in a social network of members of a collegiate marching band (Kornienko, Schaefer, Weren, Hill, & Granger, 2016). The findings revealed that over time, individuals with lower cortisol levels were more likely to maintain friendships, and those with higher cortisol levels were more likely to create new ties. In contrast, individuals with higher testosterone levels were more likely to maintain existing friendships and less likely to create new relationships. The reader is referred to Figs. 22.1 and 22.2 for illustrations of contemporaneous associations between social network structure and salivary levels of cortisol and testosterone in a large mixed-gender organization of a marching band. This work points to exciting new directions to discover how hormones are associated with social behavior through which social relationships among group members are initiated, maintained, and lost in order to advance our understanding the structure and function of human social ecology for individual behavior and adaptation (for a review, see Pinter-Wollman et al., 2014).

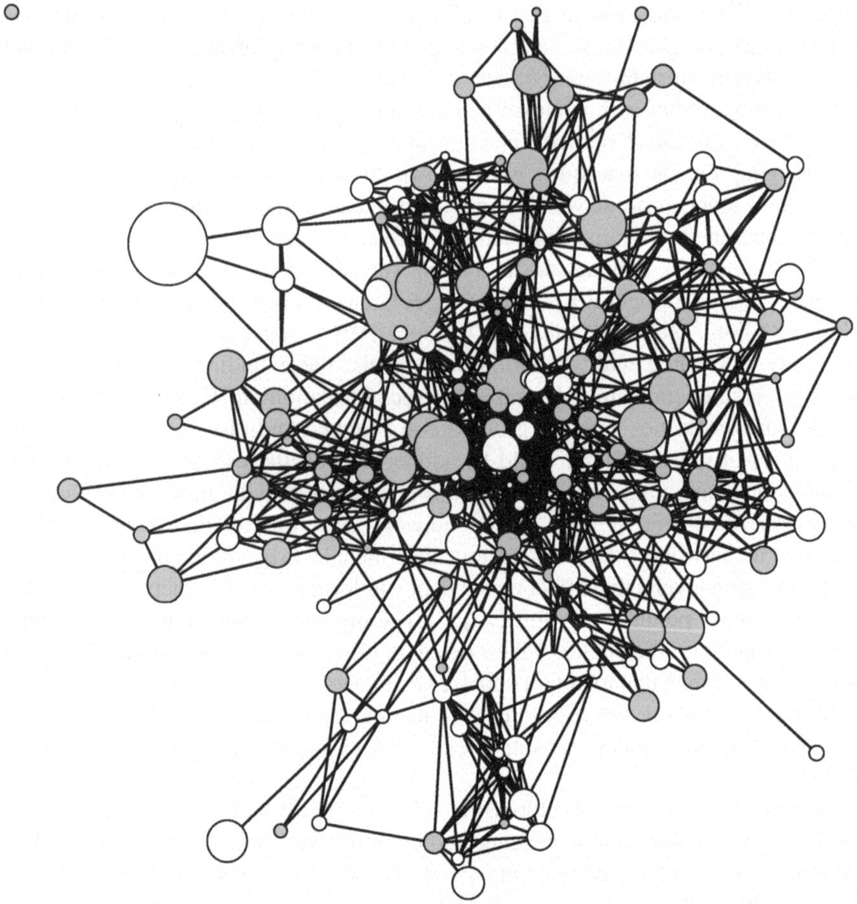

**Fig. 22.1** Visual representation of friendship network and salivary cortisol levels. Links between nodes represent directed friendship ties, arrows omitted to improve visual layout. Node size corresponds to individual's salivary C level, which was multiplied by a constant of 7.5 (larger nodes = higher levels of C). Node color denotes gender (white = male; gray = female)

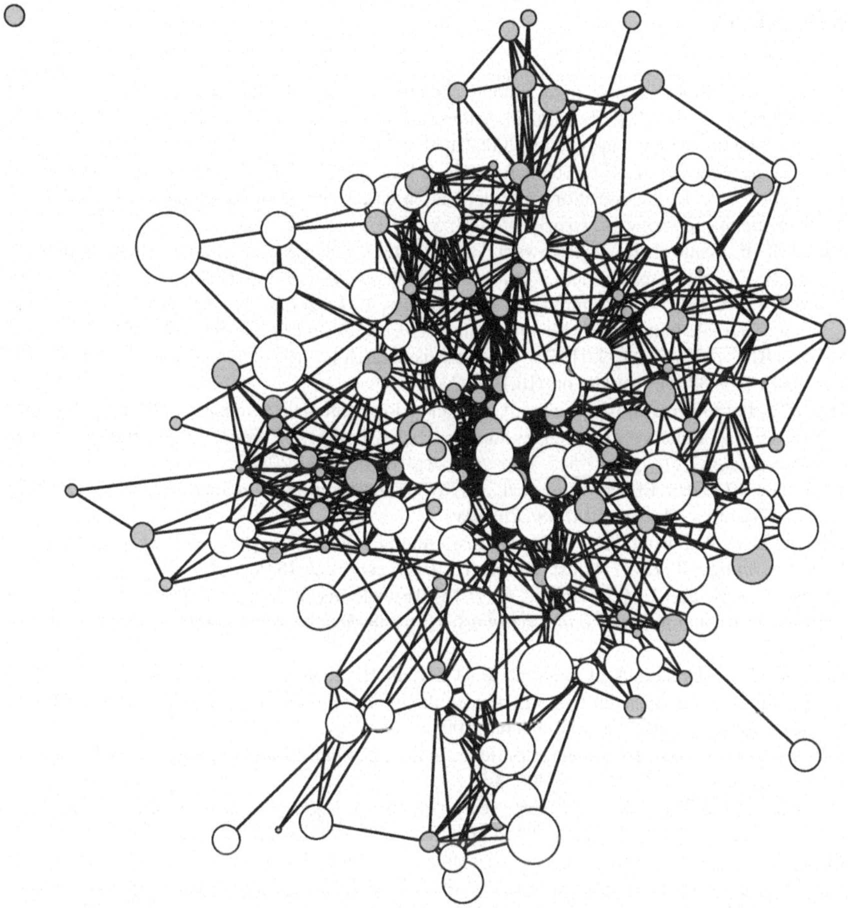

**Fig. 22.2** Visual representation of friendship network and salivary testosterone levels. Links between nodes represent directed friendship ties, arrows omitted to improve visual layout. Node size corresponds to individual's salivary T level, which was divided by a constant of 50 (larger nodes = higher levels of T). Node color denotes gender (white = male; gray = female)

## References

Adam, E. K. (2006). Transactions among adolescent trait and state emotion and diurnal and momentary cortisol activity in naturalistic settings. *Psychoneuroendocrinology, 31*, 664–679. https://doi.org/10.1016/j.psyneuen.2006.01.010

Adam, E. K., Hawkley, L. C., Kudielka, B. M., & Cacioppo, J. T. (2006). Day-to-day dynamics of experience–cortisol associations in a population-based sample of older adults. *Proceedings of the National Academy of Sciences of the U S A, 103*, 17058–17063.

Adams, R. E., Santo, J. B., & Bukowski, W. M. (2011). The presence of a best friend buffers the effects of negative experiences. *Developmental Psychology, 47*(6), 1786.

Aikey, J. L., Nyby, J. G., Anmuth, D. M., & James, P. J. (2002). Testosterone rapidly reduces anxiety in male house mice (Mus musculus). *Hormones and Behavior, 42*(4), 448–460.

Andrew, R. J., & Rogers, L. J. (1972). Testosterone, search behaviour and persistence. *Nature, 237* (5354), 343–345. https://doi.org/10.1038/237343a0

Anestis, S. F. (2006). Testosterone in juvenile and adolescent male chimpanzees (Pan troglodytes): Effects of dominance rank, aggression, and behavioral style. *American Journal of Physical Anthropology, 130*(4), 536–545. https://doi.org/10.1002/ajpa.20387

Anonymous. (1970). Effects of sexual activity on beard growth in man. *Nature, 226*(5248), 869–870. https://doi.org/10.1038/226869a0

Apicella, C. L., & Barrett, H. C. (2016). Cross-cultural evolutionary psychology. *Current Opinion in Psychology, 7*, 92–97. https://doi.org/10.1016/j.copsyc.2015.08.015

Apicella, C. L., Carré, J. M., & Dreber, A. (2015). Testosterone and economic risk taking: A review. *Adaptive Human Behavior and Physiology, 1*(3), 358–385. https://doi.org/10.1007/s40750-014-0020-2

Apicella, C. L., Dreber, A., Campbell, B., Gray, P. B., Hoffman, M., & Little, A. C. (2008). Testosterone and financial risk preferences. *Evolution and Human Behavior, 29*(6), 384–390. https://doi.org/10.1016/j.evolhumbehav.2008.07.001

Archer, J. (1977). Testosterone and persistence in mice. *Animal Behaviour, 25*(2), 479–488. https://doi.org/10.1016/0003-3472(77)90023-9

Archer, J. (1991). The influence of testosterone on human aggression. *British Journal of Psychology, 82*(1), 1–28. https://doi.org/10.1111/j.2044-8295.1991.tb02379.x

Archer, J. (2004). Sex differences in aggression in real-world settings: A meta-analytic review. *Review of General Psychology, 8*(4), 291–322. https://doi.org/10.1037/1089-2680.8.4.291

Archer, J. (2006). Testosterone and human aggression: An evaluation of the challenge hypothesis. *Neuroscience and Biobehavioral Reviews, 30*(3), 319–345. https://doi.org/10.1016/j.neubiorev.2004.12.007

Archer, J., Birring, S. S., & Wu, F. C. W. (1998). The association between testosterone and aggression among young men: Empirical findings and a meta-analysis. *Aggressive Behavior, 24*(6), 411–420. https://doi.org/10.1002/(SICI)1098-2337(1998)24:6<411::AID-AB2>3.0.CO;2-9

Archer, J., & Coyne, S. M. (2005). An integrated review of indirect, relational, and social aggression. *Personality and Social Psychology Review, 9*(3), 212–230. https://doi.org/10.1207/s15327957pspr0903_2

Archer, J., Graham-Kevan, N., & Davies, M. (2005). Testosterone and aggression: A reanalysis of Book, Starzyk, and Quinsey's (2001) study. *Aggression and Violent Behavior, 10*(2), 241–261. https://doi.org/10.1016/j.avb.2004.01.001

Arslan, M., Akhtar, F. B., & Nieschlag, E. (1984). Salivary testosterone levels in normal and testosterone treated monkeys. *Journal of Steroid Biochemistry, 21*, 461–464. https://doi.org/10.1016/0022-4731(84)90312-1

Assari, S., Caldwell, C. H., & Zimmerman, M. A. (2014). Sex differences in the association between testosterone and violent behaviors. *Trauma Monthly, 19*(3), e18040. https://doi.org/10.5812/traumamon.18040

Batrinos, M. L. (2012). Testosterone and aggressive behavior in man. *International Journal of Endocrinology and Metabolism, 10*(3), 563–568. https://doi.org/10.5812/ijem.3661

Beach, F. A. (1948). *Hormones and behavior*. New York, NY: Paul Hoeber.

Beehner, J. C., Bergman, T. J., Cheney, D. L., Seyfarth, R. M., & Whitten, P. L. (2006). Testosterone predicts future dominance rank and mating activity among male chacma baboons. *Behavioral Ecology and Sociobiology, 59*(4), 469–479. https://doi.org/10.1007/s00265-005-0071-2

Beeman, E. A. (1947). The relation of the interval between castration and first encounter to the aggressive behavior of mice. *The Anatomical Record, 99*(4), 570.

Behringer, V., & Deschner, T. (2017). Non-invasive monitoring of physiological markers in primates. *Hormones and Behavior, 91*, 3–18. https://doi.org/10.1016/j.yhbeh.2017.02.001

Behringer, V., Deschner, T., Deimel, C., Stevens, J. M. G., & Hohmann, G. (2014). Age-related changes in urinary testosterone levels suggest differences in puberty onset and divergent life history strategies in bonobos and chimpanzees. *Hormones and Behavior, 66*(3), 525–533. https://doi.org/10.1016/j.yhbeh.2014.07.011

Beletsky, L. D., Orians, G. H., & Wingfield, J. C. (1992). Year-to-year patterns of circulating levels of testosterone and corticosterone in relation to breeding density, experience, and reproductive success of the polygynous red-winged blackbird. *Hormones and Behavior, 26*(3), 420–432. https://doi.org/10.1016/0018-506X(92)90011-J

Benenson, J. F. (2013). The development of human female competition: Allies and adversaries. *Philosophical Transactions of the Royal Society B, 368*(1631), 20130079. https://doi.org/10.1098/rstb.2013.0079

Benenson, J. F. (2014). *Warriors and worriers: The survival of the sexes*. Oxford: Oxford University Press.

Berkman, L. F., Glass, T., Brissette, I., & Seeman, T. E. (2000). From social integration to health: Durkheim in the new millennium. *Social Science and Medicine, 51*, 843–857.

Bernhardt, P. C., Dabbs, J. M., Jr., Fielden, J. A., & Lutter, C. D. (1998). Testosterone changes during vicarious experiences of winning and losing among fans at sporting events. *Physiology and Behavior, 65*(1), 59–62. https://doi.org/10.1016/S0031-9384(98)00147-4

Bernstein, I. S. (1981). Dominance: The baby and the bathwater. *Behavioral and Brain Sciences, 4* (3), 419–429.

Blackhart, G. C., Eckel, L. A., & Tice, D. M. (2007). Salivary cortisol in response to acute social rejection and acceptance by peers. *Biological Psychology, 75*(3), 267–276.

Boksem, M. A. S., Mehta, P. H., den Bergh, B. V., van Son, V., Trautmann, S. T., Roelofs, K., ... Sanfey, A. G. (2013). Testosterone inhibits trust, but promotes reciprocity. *Psychological Science, 24*(11), 2306–2314. https://doi.org/10.1177/0956797613495063

Booth, A., Carver, K., & Granger, D. A. (2000). Biosocial perspectives on the family. *Journal of Marriage and Family, 62*(4), 1018–1034. https://doi.org/10.1111/j.1741-3737.2000.01018.x

Book, A. S., Starzyk, K. B., & Quinsey, V. L. (2001). The relationship between testosterone and aggression: A meta-analysis. *Aggression and Violent Behavior, 6*(6), 579–599. https://doi.org/10.1016/S1359-1789(00)00032-X

Booth, A., Granger, D. A., Mazur, A., & Kivlighan, K. T. (2006). Testosterone and social behavior. *Social Forces, 85*(1), 167–191. https://doi.org/10.1353/sof.2006.0116

Booth, A., Johnson, D. R., Granger, D. A., Crouter, A. C., & McHale, S. (2003). Testosterone and child and adolescent adjustment: The moderating role of parent-child relationships. *Developmental Psychology, 39*(1), 85.

Booth, A., Mazur, A. C., & Dabbs, J. M. (1993). Endogenous testosterone and competition: The effect of "fasting". *Steroids, 58*(8), 348–350. https://doi.org/10.1016/0039-128X(93)90036-M

Booth, A., Shelley, G., Mazur, A., Tharp, G., & Kittok, R. (1989). Testosterone, and winning and losing in human competition. *Hormones and Behavior, 23*(4), 556–571. https://doi.org/10.1016/0018-506X(89)90042-1

Borgatti, S. P., Mehra, A., Brass, D. J., & Labianca, G. (2009). Network analysis in the social sciences. *Science, 323*, 892–895. https://doi.org/10.1126/science.1165821

Boyce, W. T., Champoux, M., Suomi, S. J., & Gunnar, M. R. (1995). Salivary cortisol in nursery-reared rhesus monkeys: Reactivity to peer interactions and altered circadian activity. *Developmental Psychobiology, 28*(5), 257–267. https://doi.org/10.1002/dev.420280502

Boyd, R., & Richerson, P. J. (1985). *Culture and the evolutionary process*. Chicago, IL: University of Chicago Press.

Boyd, R., Richerson, P. J., & Henrich, J. (2011). The cultural niche: Why social learning is essential for human adaptation. *Proceedings of the National Academy of Sciences, 108*, 10918–10925. https://doi.org/10.1073/pnas.1100290108

Brain, P. F., & Haug, M. (1992). Hormonal and neurochemical correlates of various forms of animal "aggression". *Psychoneuroendocrinology, 17*(6), 537–551. https://doi.org/10.1016/0306-4530(92)90014-X

Brent, L. J. N., Semple, S., Dubuc, C., Heistermann, M., & MacLarnon, A. (2011). Social capital and physiological stress levels in free-ranging adult female rhesus macaques. *Physiology and Behavior, 102*, 76–83. https://doi.org/10.1016/j.physbeh.2010.09.022

Bribiescas, R. G. (1996). Testosterone levels among Aché hunter-gatherer men. *Human Nature, 7* (2), 163–188. https://doi.org/10.1007/BF02692109

Buades-Rotger, M., Engelke, C., Beyer, F., Keevil, B. G., Brabant, G., & Krämer, U. M. (2016). Endogenous testosterone is associated with lower amygdala reactivity to angry faces and reduced aggressive behavior in healthy young women. *Scientific Reports, 6*, 38538. https://doi.org/10.1038/srep38538

Cacioppo, S., Capitanio, J. P., & Cacioppo, J. T. (2014). Toward a neurology of loneliness. *Psychological Bulletin, 140*, 1464–1504. https://doi.org/10.1037/a0037618

Cacioppo, J. T., Hawkley, L. C., Berntson, G. G., Ernst, J. M., Gibbs, A. C., Stickgold, R., & Hobson, J. A. (2002). Do lonely days invade the nights? Potential social modulation of sleep efficiency. *Psychological Science, 13*(4), 384–387.

Calhoun, C. D., Helms, S. W., Heilbron, N., Rudolph, K. D., Hastings, P. D., & Prinstein, M. J. (2014). Relational victimization, friendship, and adolescents' hypothalamic–pituitary–adrenal axis responses to an in vivo social stressor. *Development and Psychopathology, 26*, 605–618. https://doi.org/10.1017/S0954579414000261

Cant, M. A., English, S., Reeve, H. K., & Field, J. (2006). Escalated conflict in a social hierarchy. *Proceedings of the Royal Society of London B: Biological Sciences, 273*(1604), 2977–2984. https://doi.org/10.1098/rspb.2006.3669

Carré, J. M. (2009). No place like home: Testosterone responses to victory depend on game location. *American Journal of Human Biology, 21*(3), 392–394. https://doi.org/10.1002/ajhb.20867

Carré, J. M., & McCormick, C. M. (2008). Aggressive behavior and change in salivary testosterone concentrations predict willingness to engage in a competitive task. *Hormones and Behavior, 54* (3), 403–409. https://doi.org/10.1016/j.yhbeh.2008.04.008

Carré, J. M., McCormick, C. M., & Hariri, A. R. (2011). The social neuroendocrinology of human aggression. *Psychoneuroendocrinology, 36*(7), 935–944. https://doi.org/10.1016/j.psyneuen.2011.02.001

Carré, J. M., & Olmstead, N. A. (2015). Social neuroendocrinology of human aggression: Examining the role of competition-induced testosterone dynamics. *Neuroscience, 286*, 171–186. https://doi.org/10.1016/j.neuroscience.2014.11.029

Carré, J. M., & Putnam, S. K. (2010). Watching a previous victory produces an increase in testosterone among elite hockey players. *Psychoneuroendocrinology, 35*(3), 475–479. https://doi.org/10.1016/j.psyneuen.2009.09.011

Carter, C. S. (1998). Neuroendocrine perspective on social attachment and love. *Psychoneuroendocrinology, 23*, 779–818.

Case, C. R., & Maner, J. K. (2014). Divide and conquer: When and why leaders undermine the cohesive fabric of their group. *Journal of Personality and Social Psychology, 107*(6), 1033–1050. https://doi.org/10.1037/a0038201

Cashdan, E. (1995). Hormones, sex, and status in women. *Hormones and Behavior, 29*(3), 354–366. https://doi.org/10.1006/hbeh.1995.1025

Casto, K. V., & Edwards, D. A. (2016). Testosterone and reconciliation among women: After-competition testosterone predicts prosocial attitudes towards opponents. *Adaptive Human Behavior and Physiology, 2*(3), 220–233. https://doi.org/10.1007/s40750-015-0037-1

Cavigelli, S. A., & Pereira, M. E. (2000). Mating season aggression and fecal testosterone levels in male ring-tailed lemurs (Lemur catta). *Hormones and Behavior, 37*(3), 246–255. https://doi.org/10.1006/hbeh.2000.1585

Celec, P., Ostatníková, D., & Hodosy, J. (2015). On the effects of testosterone on brain behavioral functions. *Frontiers in Neuroscience, 9*, 12. https://doi.org/10.3389/fnins.2015.00012

Chang, S. W. C., Brent, L. J. N., Adams, G. K., Klein, J. T., Pearson, J. M., Watson, K. K., & Platt, M. L. (2013). Neuroethology of primate social behavior. *PNAS, 110*(S2), 10387–10394.

Cheng, J. T., Kornienko, O., & Granger, D. A. (2018). Prestige in a large-scale social group predicts longitudinal changes in testosterone. *Journal of Personality and Social Psychology, 114*(6), 924–944. https://doi.org/10.1037/pspi0000126

Cheng, J. T., & Tracy, J. L. (2014). Toward a unified science of hierarchy: Dominance and prestige are two fundamental pathways to human social rank. In J. T. Cheng, J. L. Tracy, & C. Anderson (Eds.), *The psychology of social status* (pp. 3–27). New York, NY: Springer. https://doi.org/10.1007/978-1-4939-0867-7_1

Cheng, J. T., Tracy, J. L., Foulsham, T., Kingstone, A., & Henrich, J. (2013). Two ways to the top: Evidence that dominance and prestige are distinct yet viable avenues to social rank and influence. *Journal of Personality and Social Psychology, 104*(1), 103–125. https://doi.org/10.1037/a0030398

Cheng, J. T., Tracy, J. L., & Henrich, J. (2010). Pride, personality, and the evolutionary foundations of human social status. *Evolution and Human Behavior, 31*(5), 334–347. https://doi.org/10.1016/j.evolhumbehav.2010.02.004

Cheng, J. T., Tracy, J. L., Ho, S., & Henrich, J. (2016). Listen, follow me: Dynamic vocal signals of dominance predict emergent social rank in humans. *Journal of Experimental Psychology: General, 145*(5), 536–547. https://doi.org/10.1037/xge0000166

Chrousos, G. P., & Gold, P. W. (1992). The concepts of stress and stress system disorders: Overview of physical and behavioral homeostasis. *Journal of American Medical Association, 267*, 1244–1252. https://doi.org/10.1001/jama.1992.03480090092034

Clotfelter, E. D., & Paolino, A. D. (2003). Bystanders to contests between conspecifics are primed for increased aggression in male fighting fish. *Animal Behaviour, 66*(2), 343–347. https://doi.org/10.1006/anbe.2003.2227

Coates, J. M., Gurnell, M., & Rustichini, A. (2009). Second-to-fourth digit ratio predicts success among high-frequency financial traders. *Proceedings of the National Academy of Sciences, 106*(2), 623–628. https://doi.org/10.1073/pnas.0810907106

Coates, J. M., Gurnell, M., & Sarnyai, Z. (2010). From molecule to market: Steroid hormones and financial risk-taking. *Philosophical Transactions of the Royal Society of London B: Biological Sciences, 365*(1538), 331–343. https://doi.org/10.1098/rstb.2009.0193

Crosier, B. S., Webster, G. D., & Dillon, H. M. (2012). Wired to connect: Evolutionary psychology and social networks. *Review of General Psychology, 16*(2), 230–239. https://doi.org/10.1037/a0027919

Dabbs, J. M., Jr. (1993). Salivary testosterone measurements in behavioral studies. *Annals of the New York Academy of Sciences, 694*(1), 177–183. https://doi.org/10.1111/j.1749-6632.1993.tb18351.x

Dabbs, J. M., Jr., Carr, T. S., Frady, R. L., & Riad, J. K. (1995). Testosterone, crime, and misbehavior among 692 male prison inmates. *Personality and Individual Differences, 18*(5), 627–633. https://doi.org/10.1016/0191-8869(94)00177-T

Dabbs, J. M., Jr., Frady, R. L., Carr, T. S., & Besch, N. F. (1987). Saliva testosterone and criminal violence in young adult prison inmates. *Psychosomatic Medicine, 49*(2), 174–182. https://doi.org/10.1097/00006842-198703000-00007

Dabbs, J. M., Jr., & Hargrove, M. F. (1997). Age, testosterone, and behavior among female prison inmates. *Psychosomatic Medicine, 59*(5), 477.

Davis, M., & Emory, E. (1995). Sex differences in neonatal stress reactivity. *Child Development, 66* (1), 14–27. https://doi.org/10.1111/j.1467-8624.1995.tb00852.x

Del Giudice, M., Ellis, B. J., & Shirtcliff, E. A. (2011). The adaptive calibration model of stress responsivity. *Neuroscience and Biobehavioral Reviews, 35*, 1562–1592. https://doi.org/10.1016/j.neubiorev.2010.11.007

Denson, T. F., Spanovich, M., & Miller, N. (2009). Cognitive appraisals and emotions predict cortisol and immune responses: A meta-analysis of acute laboratory social stressors and emotion inductions. *Psychological Bulletin, 135*, 823–853. https://doi.org/10.1037/a0016909

Dickerson, S. S. (2008). Emotional and physiological responses to social-evaluative threat. *Social and Personality Psychology Compass, 2*(3), 1362–1378. https://doi.org/10.1111/j.1751-9004.2008.00095.x

Dickerson, S. S., & Kemeny, M. E. (2004). Acute stressors and cortisol responses: A theoretical integration and synthesis of laboratory research. *Psychological Bulletin, 130*(3), 355–391. https://doi.org/10.1037/0033-2909.130.3.355

Dickerson, S. S., & Zoccola, P. M. (2013). Cortisol responses to social exclusion. In C. N. DeWall (Ed.), *The Oxford handbook of social exclusion* (pp. 143–151). Oxford, UK: Oxford University Press.

Doane, L. D., & Adam, E. K. (2010). Loneliness and cortisol: Momentary, day-to-day, and trait associations. *Psychoneuroendocrinology, 35*(3), 430–441.

Dreher, J.-C., Dunne, S., Pazderska, A., Frodl, T., Nolan, J. J., & O'Doherty, J. P. (2016). Testosterone causes both prosocial and antisocial status-enhancing behaviors in human males. *Proceedings of the National Academy of Sciences, 113*(41), 11633–11638. https://doi.org/10.1073/pnas.1608085113

Edwards, D. A., Wetzel, K., & Wyner, D. R. (2006). Intercollegiate soccer: Saliva cortisol and testosterone are elevated during competition, and testosterone is related to status and social connectedness with teammates. *Physiology and Behavior, 87*(1), 135–143. https://doi.org/10.1016/j.physbeh.2005.09.007

Eisenegger, C., Haushofer, J., & Fehr, E. (2011). The role of testosterone in social interaction. *Trends in Cognitive Sciences, 15*(6), 263–271. https://doi.org/10.1016/j.tics.2011.04.008

Eisenegger, C., Naef, M., Snozzi, R., Heinrichs, M., & Fehr, E. (2010). Prejudice and truth about the effect of testosterone on human bargaining behaviour. *Nature, 463*(7279), 356–359. https://doi.org/10.1038/nature08711

Elias, M. (1981). Serum cortisol, testosterone, and testosterone-binding globulin responses to competitive fighting in human males. *Aggressive Behavior, 7*(3), 215–224.

Feldman, R. (2015). The neurobiology of mammalian parenting and the biosocial context of human caregiving. *Hormones and Behavior, 77*, 3–17. https://doi.org/10.1016/j.yhbeh.2015.10.001

Flinn, M. V., Ponzi, D., & Muehlenbein, M. P. (2012). Hormonal mechanisms for regulation of aggression in human coalitions. *Human Nature, 23*(1), 68–88. https://doi.org/10.1007/s12110-012-9135-y

Folstad, I., & Karter, A. J. (1992). Parasites, bright males, and the immunocompetence handicap. *American Naturalist, 139*, 603–622.

Ford, C. S., & Beach, F. A. (1951). *Patterns of sexual behavior*. Oxford: Harper and Paul B. Hoeber.

Fuchs, E., Kirschbaum, C., Benisch, D., & Bieser, A. (1997). Salivary cortisol: A non-invasive measure of hypothalamo-pituitary-adrenocortical activity in the squirrel monkey, Saimiri sciureus. *Laboratory Animals, 31*(4), 306–311. https://doi.org/10.1258/002367797780596077

Fuxjager, M. J., Mast, G., Becker, E. A., & Marler, C. A. (2009). The 'home advantage' is necessary for a full winner effect and changes in post-encounter testosterone. *Hormones and Behavior, 56*(2), 214–219. https://doi.org/10.1016/j.yhbeh.2009.04.009

Garfield, Z. H., & Hagen, E. H. (in press). Investigating evolutionary models of leadership among recently settled Ethiopian hunter-gatherers. *The Leadership Quarterly*. https://doi.org/10.1016/j.leaqua.2019.03.005

Geniole, S. N., Bird, B. M., Ruddick, E. L., & Carré, J. M. (2017). Effects of competition outcome on testosterone concentrations in humans: An updated meta-analysis. *Hormones and Behavior, 92*, 37–50. https://doi.org/10.1016/j.yhbeh.2016.10.002

Gladue, B. A., Boechler, M., & McCaul, K. D. (1989). Hormonal response to competition in human males. *Aggressive Behavior, 15*(6), 409–422.

Glaser, R., Kiecolt-Glaser, J. K., Speicher, C. E., & Holliday, J. E. (1985). Stress, loneliness, and changes in herpesvirus latency. *Journal of Behavioral Medicine, 8*(3), 249–260.

Granger, D. A., Fortunato, C. K., Beltzer, E. B., Virag, M., Bright, M., & Out, D. (2012). Salivary bioscience and research on adolescence: An integrated perspective. *Journal of Adolescence, 32*, 1081–1095. https://doi.org/10.1016/j.adolescence.2012.01.005

Gray, P. B., McHale, T. S., & Carré, J. M. (2017). A review of human male field studies of hormones and behavioral reproductive effort. *Hormones and Behavior, 91*, 52–67. https://doi.org/10.1016/j.yhbeh.2016.07.004

Gunnar, M. R., & Hostinar, C. E. (2015). The social buffering of the hypothalamic–pituitary–adrenocortical axis in humans: Developmental and experiential determinants. *Social Neuroscience, 10*, 479–488. https://doi.org/10.1080/17470919.2015.1070747

Gunnar, M. R., & Quevedo, K. (2007). The neurobiology of stress and development. *Annual Review of Psychology, 58*, 145–173. https://doi.org/10.1146/annurev.psych.58.110405.085605

Gunnar, M. R., Sebanc, A. M., Tout, K., Donzella, B., & van Dulmen, M. M. H. (2003). Peer rejection, temperament, and cortisol activity in preschoolers. *Developmental Psychobiology, 43*, 346–368. https://doi.org/10.1002/dev.10144

Ha, T., Rogers, A. A., Hibel, L. C., Saxbe, D. E., Laurent, H. K., & Granger, D. A. (2018). *Interpersonal attunement in the activity of the HPA axis: A review and meta-analysis*. Poster presented at the ISPNE Conference, Newport Beach, CA.

Hackman, D. A., O'Brien, J. R., & Zalewski, M. (2018). Enduring association between parenting and cortisol: A Meta-analysis. *Child Development, 89*(5), 1485–1503.

Harding, C. F. (1981). Social modulation of circulating hormone levels in the male. *Integrative and Comparative Biology, 21*(1), 223–231. https://doi.org/10.1093/icb/21.1.223

Henrich, J. (2016). *The secret of our success: How culture is driving human evolution, domesticating our species, and making us smarter*. Princeton, NJ: Princeton University Press.

Henrich, J., Chudek, M., & Boyd, R. (2015). The Big Man Mechanism: How prestige fosters cooperation and creates prosocial leaders. *Philosophical Transactions of The Royal Society B Biological Sciences, 370*(1683), 20150013. https://doi.org/10.1098/rstb.2015.0013

Henrich, J., & Gil-White, F. J. (2001). The evolution of prestige: Freely conferred deference as a mechanism for enhancing the benefits of cultural transmission. *Evolution and Human Behavior, 22*(3), 165–196. https://doi.org/10.1016/S1090-5138(00)00071-4

Henrich, J., Heine, S. J., & Norenzayan, A. (2010). The weirdest people in the world? *Behavioral and Brain Sciences, 33*(2–3), 61–83. https://doi.org/10.1017/S0140525X0999152X

Hermans, E. J., Putman, P., Baas, J. M., Gecks, N. M., Kenemans, J. L., & van Honk, J. (2007). Exogenous testosterone attenuates the integrated central stress response in healthy young women. *Psychoneuroendocrinology, 32*(8–10), 1052–1061. https://doi.org/10.1016/j.psyneuen.2007.08.006

Hermans, E. J., Putman, P., Baas, J. M., Koppeschaar, H. P., & Honk, J. v. (2006). A single administration of testosterone reduces fear-potentiated startle in humans. *Biological Psychiatry, 59*(9), 872–874. https://doi.org/10.1016/j.biopsych.2005.11.015

Hinde, R. A. (1974). *Biological bases of human social behaviour*. New York, NY: McGraw-Hill.

Hirschenhauser, K., & Oliveira, R. F. (2006). Social modulation of androgens in male vertebrates: Meta-analyses of the challenge hypothesis. *Animal Behaviour, 71*(2), 265–277. https://doi.org/10.1016/j.anbehav.2005.04.014

Hostinar, C. E., Johnson, A. E., & Gunnar, M. R. (2015). Parent support is less effective in buffering cortisol stress reactivity for adolescents compared to children. *Developmental Science, 18*, 281–297. https://doi.org/10.1111/desc.12195

Hostinar, C. E., Sullivan, R. M., & Gunnar, M. R. (2014). Psychobiological mechanisms underlying the social buffering of the hypothalamic–pituitary–adrenocortical axis: A review of animal models and human studies across development. *Psychological Bulletin, 140*, 256–282. https://doi.org/10.1037/a0032671

Jaeggi, A. V., Trumble, B. C., Kaplan, H. S., & Gurven, M. (2015). Salivary oxytocin increases concurrently with testosterone and time away from home among returning Tsimane' hunters. *Biology Letters, 11*(3), 20150058. https://doi.org/10.1098/rsbl.2015.0058

Johnson, D. D. P., McDermott, R., Barrett, E. S., Cowden, J., Wrangham, R., McIntyre, M. H., & Rosen, S. P. (2006). Overconfidence in wargames: Experimental evidence on expectations, aggression, gender and testosterone. *Proceedings of the Royal Society B: Biological Sciences, 273*(1600), 2513–2520.

Josephs, R. A., Newman, M. L., Brown, R. P., & Beer, J. M. (2003). Status, testosterone, and human intellectual performance: Stereotype threat as status concern. *Psychological Science, 14* (2), 158–163. https://doi.org/10.1111/1467-9280.t01-1-01435

Josephs, R. A., Sellers, J. G., Newman, M. L., & Mehta, P. H. (2006). The mismatch effect: When testosterone and status are at odds. *Journal of Personality and Social Psychology, 90*(6), 999–1013. https://doi.org/10.1037/0022-3514.90.6.999

Kadushin, C. (2012). *Understanding Social Networks*. New York, NY: Oxford University Press.

Kagan, J., Reznick, J. S., & Snidman, N. (1987). The physiology and psychology of behavioral inhibition in children. *Child Development, 58*(6), 1459–1473.

Kemeny, M. E. (2003). The psychobiology of stress. *Current Directions in Psychological Science, 12*(4), 124–129.

Kiecolt-Glaser, J. K., & Newton, T. L. (2001). Marriage and health: His and hers. *Psychological Bulletin, 127*, 472–503. https://doi.org/10.1037/0033-2909.127.4.472

Knight, E. L., & Mehta, P. H. (2014). Hormones and hierarchies. In J. T. Cheng, J. L. Tracy, & C. Anderson (Eds.), *The psychology of social status* (pp. 269–301). New York, NY: Springer. https://doi.org/10.1007/978-1-4939-0867-7_13

Koolhaas, J., & Bohus, B. (1989). Social control in relation to neuroendocrine and immunological responses. In A. Steptoe & A. Appels (Eds.), *Stress, personal control and health* (pp. 295–305). Chichester: Wiley.

Kornienko, O., Clemans, K., Out, D., & Granger, D. A. (2013). Friendship network position and salivary cortisol levels. *Social Neuroscience, 8*, 385–396. https://doi.org/10.1080/17470919.2013.795500

Kornienko, O., Clemans, K. H., Out, D., & Granger, D. A. (2014). Hormones, behavior, and social network analysis: Exploring associations between cortisol, testosterone, and network structure. *Hormones and Behavior, 66*, 534–544. https://doi.org/10.1016/j.yhbeh.2014.07.009

Kornienko, O., Schaefer, D. R., Weren, S., Hill, G. W., & Granger, D. A. (2016). Cortisol and testosterone associations with social network dynamics. *Hormones and Behavior, 80*, 92–102. https://doi.org/10.1016/j.yhbeh.2016.01.013

Kudielka, B. M., & Kirschbaum, C. (2005). Sex differences in HPA axis responses to stress: A review. *Biological Psychology, 69*(1), 113–132. https://doi.org/10.1016/j.biopsycho.2004.11.009

Kutsukake, N., Ikeda, K., Honma, S., Teramoto, M., Mori, Y., Hayasaka, I., … Hasegawa, T. (2009). Validation of salivary cortisol and testosterone assays in chimpanzees by liquid chromatography-tandem mass spectrometry. *American Journal of Primatology, 71*(8), 696–706. https://doi.org/10.1002/ajp.20708

Lazarus, R. S., & Folkman, S. (1984). Coping and adaptation. In W. D. Gentry (Ed.), *The handbook of behavioral medicine* (pp. 282–325). New York: Guilford.

Lincoln, G. A., Guinness, F., & Short, R. V. (1972). The way in which testosterone controls the social and sexual behavior of the red deer stag (Cervus elaphus). *Hormones and Behavior, 3*(4), 375–396. https://doi.org/10.1016/0018-506X(72)90027-X

Longman, D. P., Surbey, M. K., Stock, J. T., & Wells, J. C. K. (2018). Tandem androgenic and psychological shifts in male reproductive effort following a manipulated "win" or "loss" in a sporting competition. *Human Nature, 29*, 1–28. https://doi.org/10.1007/s12110-018-9323-5

Loren Buck, C., & Barnes, B. M. (2003). Androgen in free-living arctic ground squirrels: Seasonal changes and influence of staged male-male aggressive encounters. *Hormones and Behavior, 43* (2), 318–326. https://doi.org/10.1016/S0018-506X(02)00050-8

Lutz, C. K., Tiefenbacher, S., Jorgensen, M. J., Meyer, J. S., & Novak, M. A. (2000). Techniques for collecting saliva from awake, unrestrained, adult monkeys for cortisol assay. *American Journal of Primatology, 52*(2), 93–99. https://doi.org/10.1002/1098-2345(200010)52:2<93::AID-AJP3>3.0.CO;2-B

Lynn, S. E. (2016). Endocrine and neuroendocrine regulation of fathering behavior in birds. *Hormones and Behavior, 77*, 237–248. https://doi.org/10.1016/j.yhbeh.2015.04.005

Maner, J. K. (2017). Dominance and prestige: A tale of two hierarchies. *Current Directions in Psychological Science, 26*(6), 526–531. https://doi.org/10.1177/0963721417714323

Maner, J. K., & Mead, N. L. (2010). The essential tension between leadership and power: When leaders sacrifice group goals for the sake of self-interest. *Journal of Personality and Social Psychology, 99*(3), 482–497. https://doi.org/10.1037/a0018559

Mazur, A., & Booth, A. (1998). Testosterone and dominance in men. *Behavioral and Brain Sciences, 21*(03), 353–363.

Mazur, A., Booth, A., & Dabbs, J. M., Jr. (1992). Testosterone and chess competition. *Social Psychology Quarterly, 55*(1), 70–77. https://doi.org/10.2307/2786687

Mazur, A., & Lamb, T. A. (1980). Testosterone, status, and mood in human males. *Hormones and Behavior, 14*(3), 236–246. https://doi.org/10.1016/0018-506X(80)90032-X

Mazur, A., Susman, E. J., & Edelbrock, S. (1997). Sex difference in testosterone response to a video game contest. *Evolution and Human Behavior, 18*(5), 317–326. https://doi.org/10.1016/S1090-5138(97)00013-5

McCaul, K. D., Gladue, B. A., & Joppa, M. (1992). Winning, losing, mood, and testosterone. *Hormones and Behavior, 26*(4), 486–504. https://doi.org/10.1016/0018-506X(92)90016-O

McEwen, B. S., & Gianaros, P. J. (2010). Central role of the brain in stress and adaptation: Links to socioeconomic status, health, and disease. *Annals of the New York Academy of Sciences, 1186*, 190–222. https://doi.org/10.1111/j.1749-6632.2009.05331.x

Mead, N. L., & Maner, J. K. (2012). On keeping your enemies close: Powerful leaders seek proximity to ingroup power threats. *Journal of Personality and Social Psychology, 102*(3), 576–591. https://doi.org/10.1037/a0025755

Mehta, P. H., & Josephs, R. A. (2006). Testosterone change after losing predicts the decision to compete again. *Hormones and Behavior, 50*(5), 684–692. https://doi.org/10.1016/j.yhbeh.2006.07.001

Mehta, P. H., Jones, A. C., & Josephs, R. A. (2008). The social endocrinology of dominance: Basal testosterone predicts cortisol changes and behavior following victory and defeat. *Journal of Personality and Social Psychology, 94*(6), 1078–1093. https://doi.org/10.1037/0022-3514.94.6.1078

Mehta, P. H., & Josephs, R. A. (2010). Testosterone and cortisol jointly regulate dominance: Evidence for a dual-hormone hypothesis. *Hormones and Behavior, 58*(5), 898–906.

Mehta, P. H., Lawless DesJardins, N. M., van Vugt, M., & Josephs, R. A. (2017). Hormonal underpinnings of status conflict: Testosterone and cortisol are related to decisions and satisfaction in the hawk-dove game. *Hormones and Behavior, 92*, 141–154. https://doi.org/10.1016/j.yhbeh.2017.03.009

Muller, M. N., Marlowe, F. W., Bugumba, R., & Ellison, P. T. (2009). Testosterone and paternal care in East African foragers and pastoralists. *Proceedings of the Royal Society of London B: Biological Sciences, 276*(1655), 347–354. https://doi.org/10.1098/rspb.2008.1028

Muller, M. N., & Wrangham, R. W. (2004). Dominance, aggression and testosterone in wild chimpanzees: A test of the 'challenge hypothesis'. *Animal Behaviour, 67*(1), 113–123. https://doi.org/10.1016/j.anbehav.2003.03.013

Nave, G., Nadler, A., Dubois, D., Zava, D., Camerer, C., & Plassmann, H. (2018). Single-dose testosterone administration increases men's preference for status goods. *Nature Communications, 9*(1), 2433. https://doi.org/10.1038/s41467-018-04923-0

Nave, G., Nadler, A., Zava, D., & Camerer, C. (2017). Single-dose testosterone administration impairs cognitive reflection in men. *Psychological Science, 28*(10), 1398–1407. https://doi.org/10.1177/0956797617709592

Neave, N., & Wolfson, S. (2003). Testosterone, territoriality, and the 'home advantage'. *Physiology and Behavior, 78*(2), 269–275. https://doi.org/10.1016/S0031-9384(02)00969-1

Nelson, R. J., & Trainor, B. C. (2007). Neural mechanisms of aggression. *Nature Reviews Neuroscience, 8*(7), 536–546. https://doi.org/10.1038/nrn2174

Newman, M. L., Sellers, J. G., & Josephs, R. A. (2005). Testosterone, cognition, and social status. *Hormones and Behavior, 47*(2), 205–211. https://doi.org/10.1016/j.yhbeh.2004.09.008

Newsom, J. T., Mahan, T. L., Rook, K. S., & Krause, N. (2008). Stable negative social exchanges and health. *Health Psychology, 27*(1), 78.

Novak, M. A., Hamel, A. F., Kelly, B. J., Dettmer, A. M., & Meyer, J. S. (2013). Stress, the HPA axis, and nonhuman primate well-being: A review. *Applied Animal Behaviour Science, 143*(2), 135–149. https://doi.org/10.1016/j.applanim.2012.10.012

O'Mailey, A. J., & Marsden, P. V. (2008). The analysis of social networks. *Health Services and Outcomes Research Methodology, 8*, 222–269. https://doi.org/10.1007/s10742-008-0041-z.

Ohl, F., Kirschbaum, C., & Fuchs, E. (1999). Evaluation of hypothalamo-pituitary-adrenal activity in the tree shrew (Tupaia belangeri) via salivary cortisol measurement. *Laboratory Animals, 33*(3), 269–274. https://doi.org/10.1258/002367799780578237

Oliveira, R. F. (2004). Social modulation of androgens in vertebrates: Mechanisms and function. *Advances in the Study of Behavior, 34*, 165–239.

Oliveira, R. F., Lopes, M., Carneiro, L. A., & Canário, A. V. M. (2001). Watching fights raises fish hormone levels. *Nature, 409*(6819), 475–475. https://doi.org/10.1038/35054128

Oxford, J., Ponzi, D., & Geary, D. C. (2010). Hormonal responses differ when playing violent video games against an ingroup and outgroup. *Evolution and Human Behavior, 31*(3), 201–209. https://doi.org/10.1016/j.evolhumbehav.2009.07.002

Persky, H., Smith, K. D., & Basu, G. K. (1971). Relation of psychologic measures of aggression and hostility to testosterone production in man. *Psychosomatic Medicine, 33*(3), 265–277. https://doi.org/10.1097/00006842-197105000-00007

Peters, E., Riksen-Walraven, J. M., Cillessen, A. H., & de Weerth, C. (2011). Peer rejection and HPA activity in middle childhood: Friendship makes a difference. *Child Development, 82*, 1906–1920.

Pinter-Wollman, N., Hobson, E. A., Smith, J. E., Edelman, A. J., Shizuka, D., de Silva, S., … McDonald, D. (2014). The dynamics of animal social networks: Analytical, conceptual, and theoretical advances. *Behavioral Ecology, 25*, 242–255. https://doi.org/10.1093/beheco/art047

Ponzi, D., Zilioli, S., Mehta, P. H., Maslov, A., & Watson, N. V. (2016). Social network centrality and hormones: The interaction of testosterone and cortisol. *Psychoneuroendocrinology, 68*, 6–13. https://doi.org/10.1016/j.psyneuen.2016.02.014

Pressman, S. D., Cohen, S., Miller, G. E., Barkin, A., Rabin, B. S., & Treanor, J. (2005). Loneliness, social network size, and immune response to influenza vaccination in college freshmen. *Health Psychology, 24*, 297–306. https://doi.org/10.1037/0278-6133.24.3.297

Rankin, A., Swearingen-Stanborough, C., Granger, D. A., & Byrd-Craven, J. (2018). The role of co-rumination and adrenocortical attunement in young women's close friendships. *Psychoneuroendocrinology, 98*, 61–66.

Raulo, A., & Dantzer, B. (2018). Associations between glucocorticoids and sociality across a continuum of vertebrate social behavior. *Ecology and Evolution, 8*(15), 7697–7716.

Redhead, D. J., Cheng, J. T., Driver, C., Foulsham, T., & O'Gorman, R. (2018). On the dynamics of social hierarchy: A longitudinal investigation of the rise and fall of prestige, dominance, and social rank in naturalistic task groups. *Evolution and Human Behavior, 40*, 222–234.

Ronay, R., Tybur, J. M., van Huijstee, D., & Morssinkhof, M. (2017). Embodied power, testosterone, and overconfidence as a causal pathway to risk-taking. *Comprehensive Results in Social Psychology, 2*(1), 28–43. https://doi.org/10.1080/23743603.2016.1248081

Rook, K. S. (1984). The negative side of social interaction: Impact on psychological well-being. *Journal of Personality and Social Psychology, 46*(5), 1097.

Rook, K. S. (2001). Emotional health and positive versus negative social exchanges: A daily diary analysis. *Applied Developmental Science, 5*(2), 86–97.

Rose, R. M., Holaday, J. W., & Bernstein, I. S. (1971). Plasma testosterone, dominance rank and aggressive behaviour in male rhesus monkeys. *Nature, 231*(5302), 366–368. https://doi.org/10.1038/231366a0

Salvador, A., & Costa, R. (2009). Coping with competition: Neuroendocrine responses and cognitive variables. *Neuroscience and Biobehavioral Reviews, 33*(2), 160–170.

Salvador, A., Suay, F., González-Bono, E., & Serrano, M. A. (2003). Anticipatory cortisol, testosterone and psychological responses to judo competition in young men. *Psychoneuroendocrinology, 28*(3), 364–375. https://doi.org/10.1016/S0306-4530(02)00028-8

Salvador, A., Suay, F., Martinez-Sanchis, S., Simon, V. M., & Brain, P. F. (1999). Correlating testosterone and fighting in male participants in judo contests. *Physiology and Behavior, 68*(1–2), 205–209. https://doi.org/10.1016/S0031-9384(99)00168-7

Sapienza, P., Zingales, L., & Maestripieri, D. (2009). Gender differences in financial risk aversion and career choices are affected by testosterone. *Proceedings of the National Academy of Sciences, 106*(36), 15268–15273. https://doi.org/10.1073/pnas.0907352106

Sapolsky, R. M. (2005). The influence of social hierarchy on primate health. *Science, 308*(5722), 648–652. https://doi.org/10.1126/science.1106477

Sapolsky, R. M., Romero, L. M., & Munck, A. U. (2000). How do glucocorticoids influence stress responses? Integrating permissive, suppressive, stimulatory, and preparative actions. *Endocrine Reviews, 21*(1), 55–89. https://doi.org/10.1210/edrv.21.1.0389

Saxbe, D. E. (2008). A field (researcher's) guide to cortisol: Tracking HPA axis functioning in everyday life. *Health Psychology Review, 2*(2), 163–190. https://doi.org/10.1080/17437190802530812

Schoofs, D., & Wolf, O. T. (2011). Are salivary gonadal steroid concentrations influenced by acute psychosocial stress? A study using the Trier Social Stress Test (TSST). *International Journal of Psychophysiology, 80*(1), 36–43. https://doi.org/10.1016/j.ijpsycho.2011.01.008

Schultheiss, O. C., Wirth, M. M., & Stanton, S. J. (2004). Effects of affiliation and power motivation arousal on salivary progesterone and testosterone. *Hormones and Behavior, 46*(5), 592–599. https://doi.org/10.1016/j.yhbeh.2004.07.005

Schultheiss, O. C., Wirth, M. M., Torges, C. M., Pang, J. S., Villacorta, M. A., & Welsh, K. M. (2005). Effects of implicit power motivation on men's and women's implicit learning and testosterone changes after social victory or defeat. *Journal of Personality and Social Psychology, 88*(1), 174–188. https://doi.org/10.1037/0022-3514.88.1.174

Setchell, J. M., & Dixson, A. F. (2001). Changes in the secondary sexual adornments of male Mandrills (Mandrillus sphinx) are associated with gain and loss of alpha status. *Hormones and Behavior, 39*(3), 177–184. https://doi.org/10.1006/hbeh.2000.1628

Sethre-Hofstad, L., Stansbury, K., & Rice, M. A. (2002). Attunement of maternal and child adrenocortical response to child challenge. *Psychoneuroendocrinology, 27*(6), 731–747.

Seyfarth, R. M., & Cheney, D. L. (2013). Affiliation, empathy, and the origins of theory of mind. *Proceedings of the National Academy of Sciences, 110*(Supplement 2), 10349–10356.

Shoal, G. D., Giancola, P. R., & Kirillova, G. P. (2003). Salivary cortisol, personality, and aggressive behavior in adolescent boys: A 5-year longitudinal study. *Journal of the American Academy of Child and Adolescent Psychiatry, 42*(9), 1101–1107.

Silk, J. B. (2007). Social components of fitness in primate groups. *Science, 317*, 1347–1351.

Slatcher, R. B., Mehta, P. H., & Josephs, R. A. (2011). Testosterone and self-reported dominance interact to influence human mating behavior. *Social Psychological and Personality Science, 2*(5), 531–539. https://doi.org/10.1177/1948550611400099

Smider, N. A., Essex, M. J., Kalin, N. H., Buss, K. A., Klein, M. H., Davidson, R. J., & Goldsmith, H. H. (2002). Salivary cortisol as a predictor of socioemotional adjustment during kindergarten: A prospective study. *Child Development, 73*(1), 75–92.

Snijders, T. A. B. (2011). Statistical models for social networks. *Annual Review of Sociology, 37*, 131–153. https://doi.org/10.1146/annurev.soc.012809.102709

Snyder, J. K., Kirkpatrick, L. A., & Barrett, H. C. (2008). The dominance dilemma: Do women really prefer dominant mates? *Personal Relationships, 15*(4), 425–444.

Stanton, S. J., Beehner, J. C., Saini, E. K., Kuhn, C. M., & LaBar, K. S. (2009). Dominance, politics, and physiology: Voters' testosterone changes on the night of the 2008 United States presidential election. *PLoS One, 4*(10), e7543. https://doi.org/10.1371/journal.pone.0007543

Stanton, S. J., & Schultheiss, O. C. (2009). The hormonal correlates of implicit power motivation. *Journal of Research in Personality, 43*(5), 942–949. https://doi.org/10.1016/j.jrp.2009.04.001

Steptoe, A., Owen, N., Kunz-Ebrecht, S. R., & Brydon, L. (2004). Loneliness and neuroendocrine, cardiovascular, and inflammatory stress responses in middle-aged men and women. *Psychoneuroendocrinology, 29*, 593–611. https://doi.org/10.1016/S0306-4530(03)00086-6

Tamres, L. K., Janicki, D., & Helgeson, V. S. (2002). Sex differences in coping behavior: A meta-analytic review and an examination of relative coping. *Personality and Social Psychology Review, 6*(1), 2–30.

Taylor, S. E. (2006). Tend and befriend biobehavioral bases of affiliation under stress. *Current Directions in Psychological Science, 15*, 273–277.

Taylor, S. E. (2011). Tend and befriend theory. In P. A. M. Van Lange, A. W. Kruglanski, & E. T. Higgins (Eds.), *Handbook of theories of social psychology: Collection* (Vol. 1 & 2, p. 32). Thousand Oaks, CA: Sage.

Taylor, S. E., Klein, L. C., Lewis, B. P., Gruenewald, T. L., Gurung, R. A., & Updegraff, J. A. (2000). Biobehavioral responses to stress in females: Tend-and-befriend, not fight-or-flight. *Psychological Review, 107*(3), 411.

Timmons, A. C., Margolin, G., & Saxbe, D. E. (2015). Physiological linkage in couples and its implications for individual and interpersonal functioning: A literature review. *Journal of Family Psychology, 29*(5), 720–731.

Trumble, B. C., Cummings, D., von Rueden, C., O'Connor, K. A., Smith, E. A., Gurven, M., & Kaplan, H. (2012). Physical competition increases testosterone among Amazonian forager-horticulturalists: A test of the 'challenge hypothesis'. *Proceedings of the Royal Society of London B: Biological Sciences, 279*(1739), 2907–2912. https://doi.org/10.1098/rspb.2012.0455

Trumble, B. C., Smith, E. A., O'Connor, K. A., Kaplan, H. S., & Gurven, M. D. (2014). Successful hunting increases testosterone and cortisol in a subsistence population. *Proceedings of the Royal Society of London B: Biological Sciences, 281*(1776), 20132876. https://doi.org/10.1098/rspb.2013.2876

Tsigos, C., & Chrousos, G. P. (2002). Hypothalamic–pituitary–adrenal axis, neuroendocrine factors and stress. *Journal of Psychosomatic Research, 53*(4), 865–871. https://doi.org/10.1016/S0022-3999(02)00429-4

Valtorta, N. K., Kanaan, M., Gilbody, S., Ronzi, S., & Hanratty, B. (2016). Loneliness and social isolation as risk factors for coronary heart disease and stroke: Systematic review and meta-analysis of longitudinal observational studies. *Heart, 102*(13), 1009–1016.

van Anders, S. M., & Watson, N. V. (2006). Social neuroendocrinology: Effects of social contexts and behaviors on sex steroids in humans. *Human Nature, 17*(2), 212–237. https://doi.org/10.1007/s12110-006-1018-7

van Honk, J., & Schutter, D. J. L. G. (2007). Testosterone reduces conscious detection of signals serving social correction: Implications for antisocial behavior. *Psychological Science, 18*(8), 663–667. https://doi.org/10.1111/j.1467-9280.2007.01955.x

van Honk, J., Tuiten, A., Verbaten, R., van den Hout, M., Koppeschaar, H., Thijssen, J., & de Haan, E. (1999). Correlations among salivary testosterone, mood, and selective attention to threat in humans. *Hormones and Behavior, 36*(1), 17–24. https://doi.org/10.1006/hbeh.1999.1521

van Honk, J., Will, G.-J., Terburg, D., Raub, W., Eisenegger, C., & Buskens, V. (2016). Effects of testosterone administration on strategic gambling in poker play. *Scientific Reports, 6*, 18096. https://doi.org/10.1038/srep18096

Vleck, C. M., & Brown, J. L. (1999). Testosterone and social and reproductive behaviour in Aphelocoma jays. *Animal Behaviour, 58*(5), 943–951. https://doi.org/10.1006/anbe.1999.1226

von Rueden, C., Gurven, M., & Kaplan, H. (2011). Why do men seek status? Fitness payoffs to dominance and prestige. *Proceedings of the Royal Society B: Biological Sciences, 278*(1715), 2223.

Vongas, J. G., & Al Hajj, R. (2017). The effects of competition and implicit power motive on men's testosterone, emotion recognition, and aggression. *Hormones and Behavior, 92*, 57–71. https://doi.org/10.1016/j.yhbeh.2017.04.005

Walker, M. E., Wasserman, S., & Wellman, B. (1993). Statistical models for social support networks. *Sociological Methods Research, 22*, 71–98. https://doi.org/10.1177/0049124193022001004

Wasserman, S., & Faust, K. (1994). *Social network analysis: Methods and applications*. Cambridge: Cambridge University Press.

Weiner, H. (1992). *Perturbing the organism: The biology of stressful experience*. Chicago, IL: University of Chicago Press.

Wingfield, J. C. (2017). The challenge hypothesis: Where it began and relevance to humans. *Hormones and Behavior, 92*, 9–12. https://doi.org/10.1016/j.yhbeh.2016.11.008

Wingfield, J. C., Ball, G. F., Dufty, A. M., Hegner, R. E., & Ramenofsky, M. (1987). Testosterone and aggression in birds. *American Scientist, 75*(6), 602–608.

Wingfield, J. C., Hegner, R. E., Dufty, A. M., & Ball, G. F. (1990). The "Challenge Hypothesis": Theoretical implications for patterns of testosterone secretion, mating systems, and breeding strategies. *The American Naturalist, 136*(6), 829–846.

Wingfield, J. C., Jacobs, J. D., Tramontin, A. D., Perfito, N., Meddle, S., Maney, D. L., & Soma, K. (2000). Toward an ecological basis of hormone–behavior interactions in reproduction of birds. In K. Wallen & J. E. Schneider (Eds.), *Reproduction in context. Social and environmental influences on reproduction* (pp. 85–128). Cambridge, MA: MIT Press.

Wingfield, J. C., Lynn, S. E., & Soma, K. K. (2001). Avoiding the 'costs' of testosterone: Ecological bases of hormone-behavior interactions. *Brain, Behavior and Evolution, 57*(5), 239–251. https://doi.org/10.1159/000047243

Worthman, C. M., & Konner, M. J. (1987). Testosterone levels change with subsistence hunting effort inU+202F.!Kung San Men. *Psychoneuroendocrinology, 12*(6), 449–458. https://doi.org/10.1016/0306-4530(87)90079-5

Wrangham, R. W. (2018). Two types of aggression in human evolution. *Proceedings of the National Academy of Sciences, 115*(2), 245–253. https://doi.org/10.1073/pnas.1713611115

Wu, Y., Eisenegger, C., Sivanathan, N., Crockett, M. J., & Clark, L. (2017). The role of social status and testosterone in human conspicuous consumption. *Scientific Reports, 7*(1), 11803. https://doi.org/10.1038/s41598-017-12260-3

Zethraeus, N., Kocoska-Maras, L., Ellingsen, T., von Schoultz, B., Hirschberg, A. L., & Johannesson, M. (2009). A randomized trial of the effect of estrogen and testosterone on economic behavior. *Proceedings of the National Academy of Sciences, 106*(16), 6535–6538. https://doi.org/10.1073/pnas.0812757106

Zilioli, S., & Bird, B. M. (2017). Functional significance of men's testosterone reactivity to social stimuli. *Frontiers in Neuroendocrinology, 47*, 1–18. https://doi.org/10.1016/j.yfrne.2017.06.002

Zilioli, S., & Watson, N. V. (2012). The hidden dimensions of the competition effect: Basal cortisol and basal testosterone jointly predict changes in salivary testosterone after social victory in men. *Psychoneuroendocrinology, 37*(11), 1855–1865. https://doi.org/10.1016/j.psyneuen.2012.03.022

Zilioli, S., & Watson, N. V. (2014). Testosterone across successive competitions: Evidence for a 'winner effect' in humans? *Psychoneuroendocrinology, 47*, 1–9. https://doi.org/10.1016/j.psyneuen.2014.05.001

# Chapter 23
# Salivary Studies of the Social Neuroscience of Human–Animal Interaction

**Patricia Pendry and Jaymie L. Vandagriff**

**Abstract** This chapter focuses on salivary bioscience applications in Human–Animal Interaction (HAI) research, an expanding field of interdisciplinary inquiry examining the mutual, dynamic relationships between people and animals and the ways in which these interactions may affect health and well-being in both species. Using a social neuroscience perspective, this chapter provides a snapshot of research on the physiological impacts of animal companionship, pet ownership, and the intentional inclusion of animals as part of a therapeutic or ameliorative process or milieu, Animal Assisted Interventions (AAIs), in both humans and animals. While considering a wide range of contexts, populations, species, interactions, and therapeutic outcomes, this chapter will explore empirical evidence aiming to capture the bidirectional links between HAI and the activity of biological systems through examining a variety of salivary analytes. Given the focus of existing empirical work in the field, we will emphasize activity of the Hypothalamic–Pituitary–Adrenal (HPA) axis and its main marker, cortisol, although other analytes including salivary alpha amylase (sAA), secretory immunoglobulin A (SIgA), salivary Oxytocin (sOT), and Nerve Growth Factor (sNGF) will be discussed. After an overview of relevant theoretical perspectives and the introduction of a model describing links between HAI and HPA axis activity, we provide an empirical review organized by intervention approach, population, and species, including dyadic approaches. Methodological challenges, implications for practice, and future research are discussed.

P. Pendry (✉) · J. L. Vandagriff
Department of Human Development, Washington State University, Pullman, WA, USA
e-mail: ppendry@wsu.edu; jaymie.vandagriff@wsu.edu



D. A. Granger, M. K. Taylor (eds.), *Salivary Bioscience*,
https://doi.org/10.1007/978-3-030-35784-9_23

## 23.1 History of Salivary Bioscience in Human–Animal Interaction

### *23.1.1 Human–Animal Interaction Research: The Emergence of the Field*

Throughout history, humans and animals have enjoyed a unique and enduring bond. Humans have depended on wild and agricultural animals as a source of food, safety and materials, a means of transportation, and an instrument for cultural and religious practices. In addition, keeping animals as pets for enjoyment and companionship has been practiced throughout human history in both Western and non-Western civilizations. A recent survey (GfK, 2016) indicates a high prevalence of pet ownership worldwide, polled at 57% in 22 countries, with the US ranking fifth overall (70%), and Argentina topping the list with 82% of its populace owning pets. This is of interest from a public health standpoint as early work on HAI has provided intriguing evidence that pets may promote better physical and psychological health in their owners, support socio-emotional development of children, enhance learning in the classroom, facilitate more active, healthy lifestyles, help improve or maintain functionality in elderly adults, and provide emotional support in a wide range of challenging contexts.

### *23.1.2 Increasing Popularity of Animal Assisted Interventions (AAIs)*

Furthermore, there has been a significant increase in the practice of including animals in interventions designed to prevent or treat physical and psychological symptoms and disorders. For example, Animal Assisted Therapy (AAT) intentionally incorporates animals as part of a goal-directed, personalized treatment plan aiming to provide measurable improvements in human physical, social, emotional, or cognitive functioning (International Association for Human Animal Interaction Organizations [IAHAIO], 2014). Although dogs are the species most commonly incorporated in AATs, the use of horses in Equine-Assisted Psychotherapy (EFP)[1] and Hippotherapy[2] is quite common as well, and AATs with cats, Guinea pigs, goats, sheep, and camels, while less common, are not unheard of. Similarly, Animal Assisted Activities (AAAs) are becoming increasingly prevalent. While lacking the

---

[1]EFP is defined as an interactive process in which a licensed mental health professional working as or with an appropriately credentialed equine specialists to address psychotherapy goals set forth by the therapists and client.

[2]Hippotherapy is conducted by physical, occupational, and speech therapists who use the motion of the horse to facilitate clients' improvements in motor skills, neurological functioning, and ability to process sensory input.

formal therapeutic aim and context of AATs, AAAs incorporate animals in interventions to improve individuals' quality of life and well-being by reducing stress or distress, providing social support, increasing motivation, creating a sense of community, and facilitating social interactions, as well as facilitating behavioral change or the acquisition of new skills (IAHAIO, 2014). One of the most common examples is the use of Animal Visitation Programs (AVPs) featuring cats and dogs, which have become increasingly popular in various settings such as college campuses, classrooms, hospitals, rehabilitation and care facilities, professional and community venues. Another example is the use of equines to assist individuals with cognitive, physical, and emotional impairments through Therapeutic Riding (TR) or Equine Facilitated Learning (EFL), an experiential learning approach designed to foster various competencies in typical and atypically developing youth and adults. Today, Animal Assisted Interventions (AAIs) are practiced with people at all stages of life in every conceivable medical, therapeutic, rehabilitation, and educational setting. Interestingly, despite the increase in use and popularity of AAIs and the prevalence of pet ownership, scientific evidence demonstrating their causal impact, and the mechanisms that may underlie these effects is just beginning to emerge.

### *23.1.3 Expansion of HAI Research*

Focused on developing an understanding of the extent to which—and ways through which—HAI may affect biobehavioral indices of health and well-being across species, the field of HAI has enjoyed a period of unprecedented expansion. The impetus for this expansion was undoubtedly informed by the formation of a public–private partnership in 2008 between the WALTHAM® Centre for Pet Nutrition and the Eunice Kennedy Shriver National Institute of Child Health and Human Development, which has supported the field of HAI of through a variety of activities and resources. In response to a call in the literature to increase the number of scientific studies, including the integration of valid and repeatable biobehavioral measures of short- and long-term effects (e.g., Esposito, Mccune, Griffin, & Maholmes, 2011), the consortium sponsored workshops and meetings connecting researchers and practitioners from a wide range of disciplines and backgrounds with the goal of fostering rigorous, innovative investigations within and across species. Supported by several million dollars in HAI research funds, these efforts have resulted in the formation of unprecedented collaborations resulting in a seminal publication on the social neuroscience of HAI (Freund, McCune, Esposito, Gee, & McCardle, 2016) and an expansion of empirical work using salivary bioscience approaches.

This chapter reviews Human–Animal Interaction (HAI) research that aims to capture the physiological impacts of animal companionship, pet ownership, and AAIs as measured through salivary analytes in humans and animals. While we considered a range of contexts, approaches and analytes, the emphasis of research in the field leads us to focus on links between HAI and activity of the Hypothalamic–Pituitary–Adrenal (HPA) axis and its main marker, cortisol, although links between

HAI and SIgA, Nerve Growth Factor (NGF), Testosterone (T), Alpha-Amylase (sAA), and Oxytocin (sOT) are discussed as well. After discussing relevant theoretical perspectives, we propose an integrative model that can be used to guide future inquiries linking HAI with biobehavioral indices. Following this, we provide a selective empirical review. This starts with research examining the physiological effects of HAI as a stress reducer in humans of various age groups. Next, we discuss the potential of HAI as a stress buffer against acute or subsequent stressors in humans. Then, we examine effects of HAI from an animal welfare perspective before reviewing studies that have taken a dyadic approach by simultaneously examining biomarkers in the context of mutual dynamic interactions. Last, we discuss the methodological strengths and limitations of the current state of the field and provide suggestions for practice and future research.

## 23.2 Theoretical Rationale for the Effects of HAI on Stress

### *23.2.1 Absence of a Single Overarching Framework*

In Fine's (2015) 4th edition of the *Handbook of Animal Assisted Therapy,* a recurrent theme is the "absence of a single overarching framework for understanding HAI" (p. 179), lending a great deal of opportunity for theoretical grounding and clarity in HAI research. It is not uncommon for clinicians and practitioners working in AAT to take their own individual theoretical approach (Fine, 2015), as is true for professionals developing and implementing AAAs (VanFleet, Fine, O'Callaghan, Mackintosh, & Gimeno, 2015). That said, the field of HAI is critically aware of the need to guide, frame, and support research and practice and as such has begun to apply and test unifying theories from other fields to develop HAI specific theories, paradigms and perspectives to help explain potential mechanisms underlying effects of HAI on humans and animals and vice versa.

For the purposes of this chapter, we will discuss theoretical models we deem most relevant to salivary bioscience in HAI research and describe how together they inform bidirectional pathways between HAI and HPA axis activity. This includes theoretical underpinnings aligning with more general models of stress involving social support theories (Cohen & McKay, 1984; Cohen & Wills, 1985; Guéguen & Ciccotti, 2008; Wood, Giles-Corti, & Bulsara, 2005), models of appraisal, stress and coping (Lazarus & Folkman, 1984, 1987; Schneider & Harley, 2006; Wells & Perrine, 2001), and the oxytocin hypothesis (Beetz, Unväs-Moberg, Julius, & Kotrschal, 2012). Although a detailed and nuanced explanation of the HPA axis and its implications for human development more generally is beyond the scope of this chapter (see Chap. 5 for a detailed review), we will briefly summarize the characteristics of the HPA and its functioning as it relates to stress and coping in the context of HAI, and provide examples of research, in HAI and general human development, that lend support to these proposed pathways.

### 23.2.2 HPA Axis Activity and HAI

It is well known that the HPA axis is composed of limbic system and endocrine structures in the body starting with the amygdala, hypothalamus, and pituitary gland in the limbic region of the brain to the adrenal glands at the top of the kidneys. It is a component of the stress system that is not only acutely activated in times of heightened stress, but is also utilized in daily endocrine functioning to maintain homeostasis (Nicolaides, Charmandari, Chrousos, & Kino, 2014). As part of daily functioning, cortisol output follows a circadian pattern, or diurnal rhythm, that is consistent from day to day. In a given saliva sample, approximately 62–72% of variation in cortisol levels is explained by diurnal patterns, with the remaining portion thought to reflect momentary influences (Adam, 2005; Adam & Gunnar, 2001). A typical, healthy diurnal rhythm for cortisol is observed as a moderate output slightly before wake up that rises to a peak within 30 min post-wake up, followed by a rapid decline in the next few hours that slows to a steady decrease until nighttime, when levels reach a nadir in the first few hours of sleep (Kirschbaum & Hellhammer, 1989; Pruessner et al., 1997). When the HPA axis is activated by appraisal of stress—a phenomenon commonly known as the HPA cascade (Gunnar & Quevedo, 2007)—the hypothalamus stimulates cells that release corticotrophin-releasing hormone (CRH) and arginine vasopressin (AVP) to flow through blood vessels to the pituitary gland, which thereby triggers release of adrenocorticotropic hormone (ACTH), signaling the adrenal glands to release glucocorticoids such as cortisol in blood, urine, and saliva (Stratakis & Chrousos, 1995).

It is thought that HPA activation is dependent not only on the nature and frequency of daily stressors encountered but also on appraisal processes (Herman & Cullinan, 1997). These include both the individual's interpretation of stressors, as well as the assessment of the availability—and adequacy—of coping resources including social support, along with the individual's ability to regulate arousal in times of stress (Cohen & Wills, 1985; Lazarus & Folkman, 1987). Thus, relevant for salivary bioscience research on HAI is the role that animals may play in modulating the way humans appraise the animal as a coping resource by providing social support. This may occur both in the initial appraisal of a potentially stressful event, thereby preventing overwhelm before it begins, as well as through one's "use" of the animal as a catalyst lending ameliorative effects through downregulation of physiological arousal and cortisol production via the feedback loop that constitutes the HPA axis.

### 23.2.3 Social Support, Appraisal, and Attachment

The notion that interaction with animals constitutes an instrumental source of social support conducive to relieving stress in humans is widely supported by HAI researchers and clinicians. Many individuals perceive that their pets consistently provide emotional support, perhaps more readily than fellow humans (McNicholas & Collis, 2006). In addition, since a sense of belonging is thought to lead to increased perceptions of social support (Cohen & McKay, 1984), companionship with one's pet, interactions with animals generally, and the social facilitation of human interaction that may co-occur in such contexts, are likely to contribute to humans feeling supported (McNicholas & Collis, 2006). Similarly, HAI researchers, pet owners, and clinicians alike have described sustained relationships with companion animals and working pairs in terms of attachment and security, core concepts of attachment theory (Ainsworth, 1963; Bowlby, 1969) known to inform internal working models around social support, as well as aspects of HPA axis activity (Hane & Fox, 2016). In addition, physical contact—a common feature of HAI—is regarded as an *expression* of social support (Ditzen et al., 2007), which, when combined with emotional support has been found to be most effective to reduce autonomic and endocrinological stress responses (Demakis & McAdams, 1994; Ditzen et al., 2007). As such, it is reasonable to suggest that HAI may influence individuals' appraisal of stressors as less threatening, which is an important modulator of HPA axis activity.

### *23.2.4 Expectancy*

The perception of animals' capacity in providing social support and effective stress relief may in itself play an important role in modulating HPA axis activity in the context of AAIs through *expectancy*. Since appraisal of stressors is a key component in the process of coping with stress (Cohen & Wills, 1985; Lazarus & Folkman, 1984, 1987), the belief that animals provide an effective coping source may encourage individuals in stressful situations to appraise their stress as less threatening or aid management of emotions in response to stress. Furthermore, the presence of animals is found to affect humans' appraisal of other humans, such as faculty and professors (Wells & Perrine, 2001) and psychotherapists (Schneider & Harley, 2006). In fact, research suggests that participants are willing to engage in AAIs and believe the experience will be helpful to them (Adamle, Riley, & Carlson, 2009; Rabbitt, Kazdin, & Hong, 2014). Moreover, evidence in mental health treatment literature suggests that expectancy plays a critical role in the adoption of therapeutic interventions (Constantino, Arnkoff, Glass, Ametrano, & Smith, 2011; Greenberg, Constantino, & Bruce, 2006) and adherence (Calvo et al., 2016).

This notion is exemplified in college-based Animal Assisted Activities (AAAs), which are commonly attended by college students expecting to relieve high levels of perceived stress (Crossman & Kazdin, 2015; Vandagriff, 2017). Causal efficacy trials employing self-report measures indeed find that participating in brief, 5–20-min college-based AVPs results in higher ratings of momentary positive emotions and a reduction in stress-related negative emotions, including irritability and anxiety (Pendry, Carr, Roeter, & Vandagriff, 2018), lower perceived stress (Barker, Barker, McCain, & Schubert, 2016; Binfet, 2017; Binfet, Passmore, Cebry, Struik, & McKay, 2017; Crump & Derting, 2015), and improvements in mood (Crossman,

Kazdin, & Knudson, 2015; Grajfoner, Harte, Potter, & McGuigan, 2017), results which could be informed by expectancy. While the role of expectancy as a mechanism for program effects has not been formally examined in AAIs, it is possible that expectations play a role, consciously or unconsciously, in tandem with effects conferred by the physical interactions such as stroking and petting and the physiological effects these are thought to impart.

### 23.2.5 HAI, Oxytocin, and HPA Axis Activity

There is evidence to suggest that HAI evokes several interrelated physiological responses likely to inform HPA activation. The oxytocin hypothesis posed by Beetz et al. (2012) serves as a useful, complimentary perspective for examining implications of HAI on HPA axis activity. Oxytocin (OT) is a neuropeptide produced in the hypothalamus that is released into the circulatory system and brain in response to sensory stimulation (for an overview see Beetz, Kotrschal, Hediger, Turner, & Unväs-Moberg, 2011). While OT is involved in many aspects of physiological functioning, its proposed role in promoting human–animal bonding (Beetz & Bales, 2016), and decreasing stress through suppression of cortisol (Cardoso, Ellenbogen, Orlando, Bacon, & Joober, 2013; Heinrichs et al., 2003), blood pressure (Petersson, 2002), anxiety and depression (Cochran, Fallon, Hill, & Frazier, 2013), thereby linking HAI to stress-related outcomes, are of most interest to HAI researchers. In fact, several HAI investigators (Odendaal, 2000; Odendaal & Meintjes, 2003; Handlin et al., 2011) have noted significant increases in OT levels in human plasma after as little as 3 min of stroking a dog, with effects depending on the relationship quality within the human–dog pair.

Recent studies suggest *canine* OT can be reliably assayed in saliva. In addition to studies validating the assessment of canine sOT (MacLean et al., 2018), research has found significant increases in canine salivary and plasma OT in response to brief HAI proportional to levels of affiliative behavior in the interaction (MacLean et al., 2017a). Moreover, higher levels of endogenous OT can be observed among AAA dogs compared to pet dogs (MacLean et al., 2017b). In contrast, studies with human populations have been less successful in assaying sOT reliably (McCullough, Churchland, & Mendez, 2013; Quintana et al., 2018). Concentrations in human plasma are often quite low, and compounds that transmit to saliva have lower concentrations often below possible levels of detection. Though some researchers have detected either endogenous sOT (Carter et al., 2007; Daughters Manstead, Hubble, Rees, Thapar, & van Goozen, 2015; White-Traut, Watanabe, Pournajafi-Nazarloo, Schwertz, Bell, & Carter, 2009) or exogenous sOT administered intranasally or intravenously (Franke, Li, Menden, Lee, & Lai, 2019; Javor, Reidl, Kindermann, Brandstätter, Ransmayr, & Gabriel, 2014; Martin et al., 2018; Quintana et al., 2018; van IJzendoorn, Bhandari, van der Veen, Grewen, & Bakermans-Kranenburg, 2012; Weisman, Zagoory-Sharon, & Feldman, 2012) in humans, associations with plasma OT levels often fail to reach statistical significance. Likewise, considerable debate remains about sOT's practical significance (McCullough et al., 2013).

Since increases in plasma OT are associated with decreases in plasma cortisol levels in humans and animals (Legros, Chiodera, Geenen, & von Frenckell, 1987; Neumann, Krömer, Toschi, & Ebner, 2000; Petersson, Lundeberg, & Unväs-Moberg, 1999), the notion that engagement in AAIs may exert a stress-relieving effect through suppressing HPA activity via affiliative experiences including stroking, petting, and grooming is nevertheless reasonable. Given that cortisol levels in plasma and saliva are known to be highly correlated and thought to reflect levels in circulation, we suggest that empirical links between HPA axis activity and HAI reflect a valuable representation of stress-relieving effects of HAI in their own right. Given the recognition that OT is an important variable in HAI as well as many other fields, it appears that laboratories are rapidly advancing assaying methodologies to accommodate the growing interest in measuring sOT in various settings. However, until a reliable salivary assay is developed, results based on sOT in human populations need to be interpreted with caution.

### *23.2.6 The HAI-HPA Transactional Model*

There is a strong rationale for incorporating the aforementioned constructs in the newly proposed HAI-HPA Transactional model presented here. Based on the observed links between periods of heightened stress, elevations in basal cortisol, dysregulation of diurnal patterns, and the emergence of psychopathology, heightened diurnal HPA axis activity is hypothesized to play a role in the development of stress-related disorders (Grant et al., 2014; Lupien, McEwen, Gunnar, & Heim, 2009). As such, aiming to affect HPA axis activity and facilitate adaptive regulation of diurnal patterns through HAI is a sound goal. We suggest that this goal may be achieved in the interim by providing short-term exposure to HAI activities that affect cortisol levels momentarily, which, when repeated regularly, may over time facilitate a gradual change in individuals' diurnal and basal HPA axis activity through the biological stress system's tendency toward homeostasis (Goldstein & Kopin, 2007). We point out that downregulation of basal HPA axis activity should not always be a preferred goal, as this should depend on the target population and outcome. For example, it is well known that individual differences in activation and regulation exist based on developmental disorder and physical and mental health conditions, potentially leading to unexpected reactivity—or lack thereof—to intervention attempts. For example, Isaksson, Nilsson, Nyberg, Hogmark, and Lindblad (2012) reported that children aged 10–17 with Attention Deficit Hyperactivity Disorder (ADHD), when compared to controls, display lower levels of salivary cortisol in the morning and evening regardless of subtype or comorbidity, gender, or ADHD medications, suggesting under-arousal, which may counteract perceived benefits of downregulation. As such, the design of HAI activities and AAIs should be informed by carefully examining links between specific HAI behaviors promoted and their

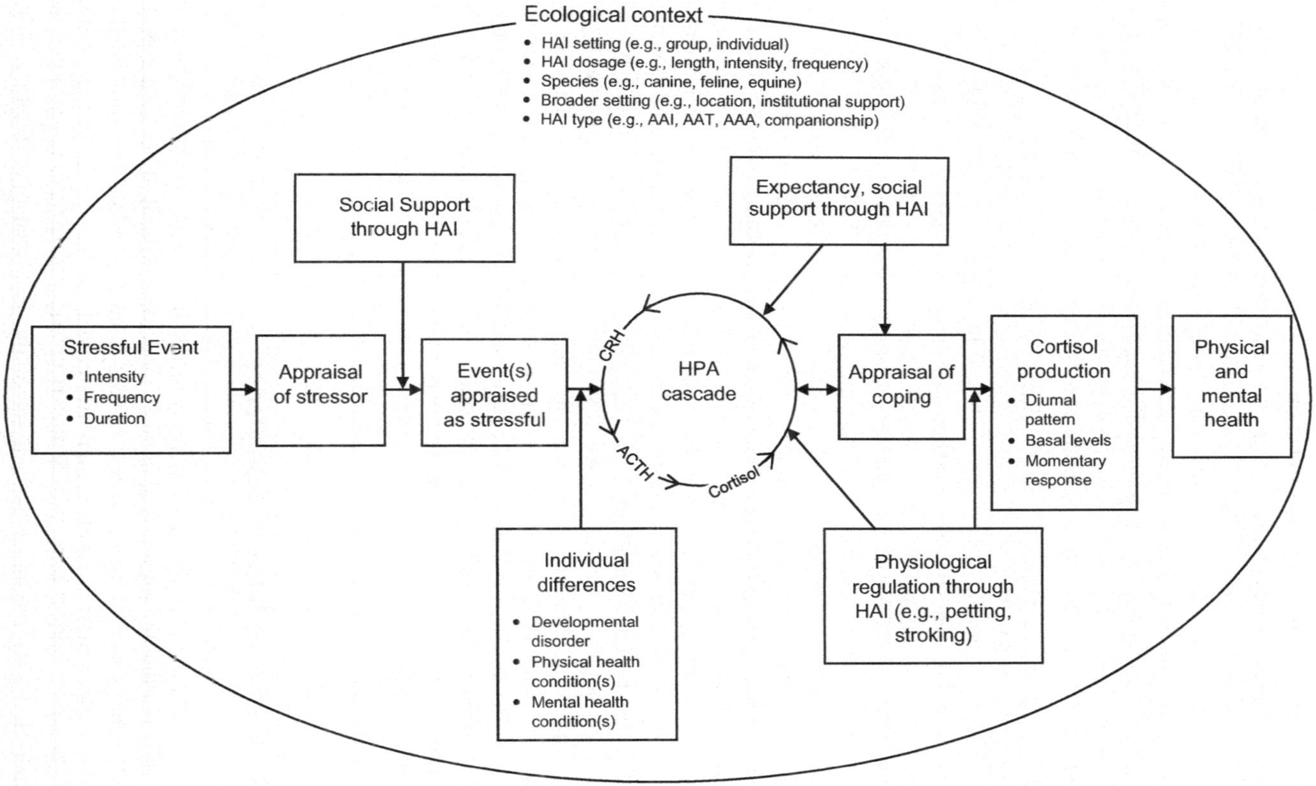


**Fig. 23.1** Proposed HAI–HPA transactional model. *HAI* Human–animal interaction, *HPA* Hypothalamic–pituitary–adrenal, *CRH* corticotropin releasing hormone, *ACTH* adrenocorticotropin hormone

proposed effects on HPA axis activity, based on the known features of HPA axis regulation—and dysregulation thereof—in the specific populations targeted and the stress-related outcomes one aims to target.

Drawing from existing stress-coping and social support frameworks in the human stress literature (Cohen & Wills, 1985; Lazarus & Folkman, 1987), we propose that HAI affects the activity of the HPA axis through the following mechanisms (see Fig. 23.1). It is likely that HAI informs activation and cessation of the HPA cascade through perceiving a stressor, followed by cognitive appraisal of that stressor as informed by perceptions of social support provided by the animal. Although activation of the HPA axis may occur, appraisal and reappraisal may aid in the evaluation of stressors as less overwhelming and coping resources as adequate as informed by expectancy, through the belief that the presence of the animal enhances coping skills, attenuating overwhelm. Regulation of arousal in response to the resulting cascade activation is informed by co-occurring physiological responses through physical contact with the animals or while engaging in rhythmic, sustained petting and stroking through OT production and subsequent downregulation of cortisol, which may attenuate the cascade by interrupting the production of CRH and ACTH. At the same time, being in the presence of animals exerts a direct physiological impact which suppresses HPA activation while simultaneously conferring the aforementioned psychological processes of appraisal and expectancy. Clearly, theoretical notions postulating links between HAI and HPA axis activity will not be complete without considering the role of specific dyadic interaction behavior in humans and animals, or the relevant developmental and treatment considerations and moderating variables in special populations often targeted in AAIs. Individual differences in perceptions of stress, appraisal of social support, expectancy, or ability to regulate arousal are likely to influence the extent to which HAI informs HPA activity, as the ecology in which HAI occurs, such as the species and behavior of the animal; the nature of the interaction; length, intensity, and frequency of exposure; and the broader social context in which the interaction occurs (e.g., group based, individual). Given that included constructs and pathways are measurable, future investigations should be conducted to examine the utility of the proposed model linking HAI and HPA.

## 23.3 Current Status of Knowledge

There is a substantial body of non-salivary research on the stress-buffering effects of HAI showing that the presence of animals moderates arousal of selected aspects of the ANS component of the biological stress system (e.g., blood pressure, heart rate, and skin temperature) in adults (Allen, Blascovich, & Mendes, 2002) and children (Havener et al., 2001). Building on this, HAI researchers have begun to examine whether HAI may modulate the activity of the Hypothalamic–Pituitary–Adrenal (HPA) axis by measuring changes in cortisol in response to a wide range of HAIs by considering both group and one-on-one interactions lasting various time periods (one 3-min exposure to 11 weeks of 90-min weekly exposures), examining interactions in various settings from laboratory studies to real-life interventions, as well as considering various species (e.g., dogs, cats, and equines), and a variety of human populations (e.g., child, college students, adults, and elderly), including special populations (e.g., Veterans with Post Traumatic Stress Disorder (PTSD), and children with Autism Spectrum Disorder (ASD). While cortisol is the most frequently researched analyte and thus emphasized in this review, we also include studies examining other salivary indices of stress system functioning including salivary alpha amylase (sAA), secretory immunoglobulin A (SIgA), and Nerve Growth Factor (sNGF). Given that the use of salivary methodologies is in its infancy in the field of HAI, the aim of this section is to present a variety of empirical work, including studies with null or difficult to interpret results due to small sample sizes and limitations in design and methods. We begin our overview by describing studies that have examined the effects of various types of HAI on momentary reactivity parameters of stress system functioning.

### *23.3.1 College-Based Animal Assisted Activities (AAAs)*

One type of AAI that has received ample attention from HAI researchers is the use of AAAs on university campuses. In the US alone, close to a thousand universities offer on campus AVPs, which are brief AAAs promoted as stress reduction programs (Crossman & Kazdin, 2015). Most AAAs focus on providing students with an opportunity to engage in small group interactions that consist of 5-to 30-min petting sessions (Vandagriff, 2017) conducted in collaboration with therapy animal organizations (e.g., Pet Partners) or animal shelters, who bring dogs and/or cats to campus in days leading up to final exams. While there is promising evidence to suggest that students taking part in college-based AAAs experience short-term increases in positive emotion and reductions in negative emotion including irritability and anxiety (Pendry, Carr, Roeter, & Vandagriff, 2018), lower perceived stress (Barker et al., 2016; Binfet, 2017; Crump & Derting, 2015), greater coping (Barker et al., 2017), less homesickness, and higher sense of belonging in school (Binfet et al., 2017), studies examining physiological indices of stress provide inconsistent results. A selection of current studies is summarized below.

In a randomized controlled trial (RCT), Pendry & Vandagriff (2019) randomly assigned undergraduate college students ($N = 249$) to one of four 10-min treatment conditions: the *hands-on* condition ($n = 73$) spent 10 min petting cats and dogs in a small group setting; the *observation* condition ($n = 62$) spent 10 min watching others' pet animals while waiting in line, unable to touch the animals; the *slideshow* condition ($n = 57$) viewed images of the same animals for 10 min; and the *waitlist* condition ($n = 57$) spent 10 min without verbal, physical, or electronic activities in a waiting room. These comparison conditions were explicitly designed to test the effects of isolated features common to college-based AVPs, i.e. physically interacting with animals, waiting in line in view of the animals, and socializing with peers. Participants collected three cortisol samples through passive drool: a basal sample taken at home immediately upon waking, and two more samples timed exactly 25 min after the beginning and end of the 10-min intervention under the direction of researchers. Controlling for basal wake-up cortisol levels, wake-to-pretest cortisol slope, and total time awake at the start of the interaction, cortisol levels were significantly lower in the hands-on interaction condition than in the slideshow (+27 %), waitlist (+28 %), and observation condition (+29 %). These findings suggest that a 10-min, hands-on interaction with dogs and cats provided significant stress relief.

In an RCT, Charnetski, Riggers, and Brennan (2004) assessed changes in SIgA, an antibody used to assess an individual's susceptibility to contracting infection, in response to 18 min of petting a live dog ($n = 19$) compared to petting a replica stuffed dog ($n = 17$) or sitting on a couch ($n = 19$). Saliva samples were taken immediately before and after the intervention. SIgA levels significantly increased in the live dog group, but not other groups, which was interpreted as a potential stress-relieving effect.

While promising, causal effects of short-term exposure to animals during college-based AVPs on salivary indices are not found consistently. In an RCT of first-year female undergraduate students, Crump and Derting (2015) assessed the effects of a group-based AVP on salivary cortisol immediately before a 30-min AVP session ($n = 34$) or 30 min of independent activity ($n = 28$), and once more 20 min after the intervention. Statistically significant effects were not observed on salivary cortisol. Stress relief effects of a one-on-one, 15-min interaction with a therapy dog or an attention-control condition were also examined on sNGF and sAA in a randomized crossover study of students ($N = 78$) by Barker et al. (2016). Salivary NGF levels were undetectable in this sample (see Matin et al., 2016) and no between-group differences in sAA emerged. In sum, while evidence of stress-relieving effects on IgA and cortisol have been demonstrated in RCTs featuring one-on-one HAI and group-based AVPs, studies using noncausal paradigms and/or featuring other salivary analytes have failed to show robust effects.

## 23.3.2 AAAs in Adult Populations

Studies examining potential stress-relieving effects of AAAs in adult populations also show inconsistent results. In a sample of health-care professionals ($N = 20$), Barker, Knisely, McCain, and Best (2005) used a within-subjects design to compare effects of 20 min of quiet rest to 5 and 20 min of unstructured interaction with a therapy dog-handler team on various biosalivary outcomes. Saliva samples were collected at baseline, 5, 15, 30, 45, and 60 min post-condition. Although cortisol and IgA decreased across time for all three conditions, no consistent, statistically significant group differences were found. Other studies have explored whether individual differences in physiological responding or lack thereof may be explained by dog ownership. Also using a within-person design, Krause-Parello, Tychowski,

Gonzalez, and Boyd (2012) compared changes in cortisol and IgA levels of pet owning ($n = 16$) and non-pet owning adults ($n = 17$) collected before and after a 20-min canine-related movie and a 20-min canine interaction 1 week later. While cortisol significantly decreased among non-pet owners in response to canine interaction by 14.1% and increased insigificantly in response to viewing the video in both groups, decreases in cortisol of pet owners in response to canine interaction did not reach significance. No changes in IgA in response to canine interaction were observed in either condition or group.

### *23.3.3 Diurnal HPA Parameters and HAI*

While most of the work reviewed thus far examined short-term responses to relatively brief exposures, several studies have examined the effects of sustained HAI exposure on diurnal indices of HPA activity. Rodriguez, Bryce, Granger, and O'Haire (2018) examined the role of service dogs on the salivary cortisol awakening response (CAR) among military veterans with PTSD. Participants ($N = 73$) were selected from a national sample of veterans approved to receive a PTSD service dog, including 45 members with a service dog, and 28 members actively awaiting their service dog. Salivary cortisol was sampled immediately upon waking and again 30 min later over two consecutive days to measure CAR. Findings suggest the presence of a service dog was associated with higher CAR and higher area under the curve (AUC) increase in cortisol, which was interpreted as improved health and well-being among this population.

In contrast, *reductions* in the CAR in response to sustained HAIs have been observed in children. Viau et al. (2010) examined basal cortisol levels and CAR among children with ASD ($N = 42$, $M_{age} = 7.1 \pm 3.1$ years) by sampling once weekly, 3 times per day over an 8-week period, including 2 weeks before introduction of a service dog, 4 weeks in which the dog remained in the home, and 2 weeks after the dog was temporarily removed from the home. Results demonstrated the presence of the service dog did not relate to children's average basal cortisol levels. However, the presence of the service dog was related to the children's CAR as demonstrated by a statistically significant decrease while the dog was present (10%), compared to before the dog was introduced (58%) and following dog removal (48%).

### *23.3.4 Effects of Human–Equine Interaction on Diurnal HPA Parameters*

While the studies reviewed thus far examined exposure to HAI featuring dogs, and one featuring dogs and cats (Pendry & Vandagriff, 2019), there is causal evidence to suggest that engaging with equines in the context of EFL affects diurnal functioning of HPA axis activity in humans. In an RCT ($N = 113$), Pendry, Smith, and Roeter (2014) examined the effects of participation in an EFL program on adolescents' diurnal cortisol. Participants were randomly assigned to an 11-week EFL program, attending once-weekly 90-min sessions ($n = 53$; $M_{age} = 11.3$ years), or to a waitlist control group ($n = 60$; $M_{age} = 11.4$ years). Passive drool saliva samples were collected at home immediately upon wake up, at 4 PM in the afternoon, and immediately before going bed over two consecutive days, resulting in six saliva samples collected before random assignment to condition (pretest) and once more after the 11-week program's completion (posttest). Results showed that at posttest, compared to the control condition, youth who engaged in the EFL program had significantly lower afternoon cortisol levels ($d = 0.48$) and significantly lower diurnal cortisol concentrations (Area Under the Curve -AUC) during waking hours ($d = 0.46$).

While the aforementioned findings (Pendry et al., 2014) highlight the efficacy of the 11-week EFL program, investigators also embedded a study to explore the role of participants' ($N = 59$; $M_{age} = 11.63$) affective and physiological regulatory abilities commonly associated with risk status in shaping their moment-to-moment experiences and behavioral responses to specific EAL activities. About halfway through the program, on the day they experienced their first horseback ride, participants provided three additional salivary samples at the end of their regular school day, and 25 and 35 min after their first mount. While indices of basal levels and potential dysregulation of cortisol diurnal patterns were controlled, regression analyses showed that greater *cortisol reactivity* in response to riding was linked to higher negative and lower positive (self-reported) emotion as well as increased negative behavior observed by three independent raters (Pendry, Carr, & Vandagriff, 2018). Findings suggest that adolescents' physiological and affective arousal *during* HAI influenced their perception and behavior during the intervention, which may influence the quality of HAI during the intervention. Evidence suggesting that physiological and emotional states may affect behavioral responses during equine-assisted activities was also observed in an equine AAT context among children and adolescents with ASD (Pan, Granger, Guérin, Shoffner, & Gabriels, 2018). Few researchers have found significant changes in diurnal or momentary parameters of equine salivary cortisol in response to equine-facilitated activities. For instance, McKinney, Mueller & Frank (2015) found salivary cortisol parameters in equines do not differ by activity type (e.g., rest, activity, therapeutic riding), which was interpreted as not inducing stress in the equine. Further, salivary cortisol activity in horses does not appear to be affected by the presence of clients presenting with PTSD compared to healthy controls (Merkies, McKechnie, & Zakraisek, 2018).

### 23.3.5 Buffering Acute Stress Through HAI

Studies in naturalistic settings examining the impact of HAI on acute stress have produced mixed results. Examples of this work include studies investigating stressors such as hospitalization, which have not generated evidence to suggest short-term exposure to HAI influences salivary cortisol responses in children (Branson, Boss, Padhye, Trötscher, & Ward, 2017; Calcaterra et al., 2015). On the other hand, research examining the effect of experimentally induced stressors such as the Trier Social Stress Test (TSST; Foley & Kirschbaum, 2010) has shown various effects depending on the population examined. Kertes and colleagues (2017) conducted an RCT with typically developing 7–12-year-old children ($N = 101$) examining cortisol reactivity to the TSST-C in the presence of a pet dog, parent, or no support figure present. Saliva was collected via passive drool 30 min after arriving the lab, and once more at +15, 25, 35, and 45 min after the start of the TSST-C. Results showed that while perceived stress from baseline to stress response (+15 sample) was significantly lower for children with a pet present, cortisol levels did not differ among groups across sampling time points. However, children in the pet condition who engaged in more child-directed petting, and whose dogs engaged less in spontaneous proximity-seeking behavior, exhibited a lower cortisol response, indicating physiological changes in children's stress response occurring independently of their perceived stress. Similarly, Beetz et al. (2011) randomized fourth grade boys with insecure-disorganized attachment ($N = 31$) to receive social support during the TSST-C from a live dog ($n = 11$), toy dog ($n = 9$), or friendly human ($n = 11$). Samples were taken at five time points prior to, during, and after the TSST-C, with the final sample taken 30 min post-TSST. Results showed that children's cortisol levels were significantly lower in the real dog condition than in the toy dog or human conditions with longer durations of stroking significantly linked to a less pronounced stress response. Although further inquiry is needed before implications for practice can be made, findings of these studies may inform our thinking about the types of child behaviors we may want to encourage (petting) and the behavior we may want to limit in therapy animals (proximity seeking) when the opportunity to engage with a live animal during acute stress exposures is made available to buffer against stress.

Although no causal work has examined effects in college students taking final exams, recent research in a similarly stressful context is promising. In an RCT, Polheber and Matchock (2014) assigned undergraduate students to 40 min of relaxing with a human friend ($n = 16$), interacting with a novel dog ($n = 16$), or sitting quietly alone ($n = 18$) before engaging in the TSST, during which the dog or friend remained present. Samples were taken immediately before, 7 and 30 min post-TSST. Participants in the dog group had significantly lower cortisol than those in the friend and waiting groups at all three time points, suggesting a stress-buffering effect.

Similarly, focusing on young adults, ($N = 120$; $M_{age} = 21.8$ years) Jacobson (2014) conducted an RCT collecting saliva immediately before and after the TSST, as well as 10, 20 and 35 and 60 min after its completion. Participants who performed the TSST immediately after an unstructured, 3-min HAI interaction with a therapy dog showed an attenuated cortisol response to stress compared to subjects who completed the TSST before the HAI and those receiving no HAI, with group differences peaking 10–20 min after the TSST. Overall, these results suggest a brief 3–7-min HAI interaction with a therapy dog immediately preceding exposure to an acute stressor may have a significant stress-buffering effect.

### 23.3.6 *Effects of AAI Participation on Canine Salivary Cortisol*

Given the bidirectional dynamic processes inherent in HAI, it makes sense that we closely examine the effects of HAI on animal health and well-being, particularly in their role as partners providing AAIs. The majority of this work was conducted with certified therapy animals whose cortisol levels were measured in response to engaging in AAIs compared to nonworking days. To investigate the impact of regular AAI visits on trained therapy dogs ($N = 26$) in a pediatric oncology setting, McCullough and colleagues (2018) assessed dogs' salivary cortisol levels before, during, and after regular hospital visits and compared them to salivary samples at dog awakening, midday, 20 min after an AAI "trigger" (e.g., taking out therapy vest), and bedtime on non-visit days. Based on 411 valid samples, results showed no significant differences between any of the five baseline samples and the post-AAI session samples. Cortisol levels remained consistent over time with their involvement in the 3-year study.

Further, to explore the effect of college-based AAA participation on therapy dog-handler teams, Ng et al. (2014) examined salivary cortisol and behavior responses to a standardized 60-min session across three settings: a college-based AAA, a novel room, and home. Saliva samples were taken every 30 min before, during, and after each session. Results showed that cortisol levels were significantly higher in the novel setting compared to the AAA and home settings at the 30-min sampling time point and continued to be higher than in the AAA setting at the 60-min time point. Similar findings were corroborated by Pirrone et al (2017) in that canine cortisol levels did not differ between AAA and home settings. However, it is possible that habituation to novelty may play a role. In a pilot study of experienced therapy dogs ($N = 5$) utilized in AAI sessions, Glenk et al. (2014) compared pre- and post-session cortisol levels over 5 weekly group therapy sessions 55–60 min in length. In the first three sessions, cortisol levels marginally decreased from pre-to post-session, and in sessions 4 and 5, post-session cortisol levels were significantly lower than pre-session levels. While pre- and post-session cortisol levels did not significantly differ from levels taken at the same time on a nonworking day at home, dog stress behaviors observed during the fifth session, (e.g., lip licking, body shake) were positively associated with declines in cortisol, suggesting a clear link between animal behavior and cortisol. On the other hand, recent studies have shown that the frequency of therapy dog visits in AAA contexts is negatively associated with cortisol concentrations in dogs (Clark, Smidt, & Bauer, 2019), suggesting further evaluation of potential stress in therapy dogs is warranted.

While novelty and context appear important for canine cortisol activation and possibly behavior, the type of approach utilized during AAI appears relevant as well. In a study examining the effects of HAI on certified and in-training therapy dogs ($N = 21$) during therapy sessions at an adult in-patient mental facility, Glenk, Stetina, Kepplinger, and Baran (2013) found that therapy dogs trained to serve off-leash demonstrated significantly lower levels of cortisol during sessions compared to dogs trained to serve on-leash.

Other investigators too have found that the types of activities in which dogs are asked to engage may impact their cortisol levels. In a study exploring canine cortisol response to various activities, Colussi, Stefanon, Adorini, and Sandri (2018) found that working dogs trained for AAAs experienced a decrease in cortisol during a 1.5-h AAA activity. Compared to dogs engaged in other working activities (i.e., pointing hunting, blood tracking, and agility training), cortisol levels in AAA dogs were elevated immediately prior to the start of their activity; however, only AAA dogs experienced a significant decrease in cortisol and return to evening at-home levels following their intervention, whereas other-trained dogs in their respective activities experienced non-changes or increases in cortisol from pre- to post-session. This contrasts with findings from Melco et al (2018) who did not find significant differences in dog salivary cortisol levels across five different AAA activities (e.g., social skills training, reading, etc.) conducted with children diagnosed with ADHD.

While cortisol has been the most frequent analyte of interest, recent developments in assaying technology have emerged to allow examination of canine sOT as mentioned above. In a study by MacLean et al. (2017a), assistance dogs in training ($N = 38$) were evaluated for salivary and plasma sOT and vasopressin levels following HAI. Dogs who spent 10-min interacting with a friendly experimenter demonstrated significantly higher salivary and plasma sOT levels compared to dogs who spent 10 min relaxing. No significant differences in vasopressin levels were noted between the experimental and control conditions.

### *23.3.7 Physiological Synchrony between Humans and Animals*

Research examining associations between salivary analytes in both species simultaneously is underdeveloped, yet emerging. For example, in a pilot study of 132 owner-dog dyads, Schöberl, Wedl, Beetz, and Kotrschal (2017) assessed the intraindividual cortisol variability of dogs and their owners during several laboratory-based challenges while investigating owner and dog social characteristics. Results showed that dogs whose owners scored high in Neuroticism, reported insecure-ambivalent attachment to their dogs, showed human-directed separation anxiety, or expressed a strong desire for independence exhibited a low coefficient of variance of cortisol, which was interpreted as a decreased ability to adapt to challenging circumstances. The authors concluded that dogs and owners both influence dyadic cortisol variability, suggesting human partners are more influential than the dog in this context.

Although some researchers have collected data to allow for examination of physiological synchrony of salivary cortisol within handler-dog dyads in an AAA context (Pirrone et al., 2017) and of plasma OT and cortisol among dog-owner pairs in a laboratory setting (Petersson et al., 2017), analyses of interspecies synchrony were not conducted. Instead, synchrony work can be found on dog-handler pairs engaged in agility competitions. To investigate the reciprocal hormonal influence among agility dog-handler teams, Jones and Josephs (2006) assessed changes in salivary Testosterone (T) in male handlers and salivary cortisol in their dogs ($n = 83$ teams). Samples were taken 1.5 h prior to competition and 20 min after results were announced to all teams. Only among losing teams, male handlers' pre-competition T was positively associated with changes in dogs' cortisol from pre-to-post competition, predicting 50.9% of the variance in dogs' cortisol change. Greater decreases in handler T also predicted greater increases in dog cortisol to a lesser extent. Similarly, only among losing teams were post-competition behaviors correlated with handler T or dog cortisol, such that pre-competition handler T positively predicted time spent in affiliative behaviors (e.g., petting, playing) and negatively predicted time spent in punitive behaviors (e.g., physically pushing or yelling at the dog). Finally, more time spent in affiliative behaviors was associated with a lesser increase in dog cortisol, while the opposite was true for punitive behaviors.

Only one pilot study of synchrony was conducted in a therapeutic riding context. In a sample of four children aged 8–10 years with PTSD, Yorke et al. (2013) employed a multiple baseline, single case design assessing effects of a 12-day intervention, consisting of once-weekly 50-min sessions, 6 of which involved riding. To establish a pre-intervention baseline, saliva samples were collected by parents three times per day, 1 day per week, for 3 weeks, upon awakening, midday, after school, and before bed. Horse ($N = 4$) baseline samples were collected simultaneously. During the 12-day intervention, saliva samples were collected every 20–30 min. At the end of the intervention, diurnal samples were once again collected according to baseline procedures. The results showed a mild correlation between horse and child cortisol levels across the sessions, which authors acknowledged as difficult to interpret. Overall, understanding how species respond to each other neurobiologically may provide some insight into how the effects of AAIs can be optimized while safeguarding animal well-being.

## 23.4 Methodological Issues, Challenges, and Considerations for Use of Salivary Bioscience in HAI

As biosalivary approaches in the field have expanded, they have garnered great enthusiasm from HAI researchers and funders in the hopes that incorporating these methods will boost the development of a coherent body of knowledge and theory on HAI. However, it is clear from reviewing salivary HAI research to date that significant variation exists with regards to the use and documentation of sound, theoretical, and empirically informed research design, use of recommended sampling paradigms, storage and assaying methods, calculation and interpretation of relevant salivary parameters, and use of sensitive, suitable analytic strategies. While salivary bioscience has the potential to profoundly impact the field, it is important that investigators understand the potential applications and limitations of these methods and adhere to sound, rigorous methodological and analytic procedures.

The most common shortcoming of HAI work incorporating salivary biomarkers to date is the lack of sound research design, thoughtful sampling schemes, and comprehensive analytic strategies. While the so-called gold standard of clinical research, the double-blind placebo-controlled trial, is not feasible in this line of work, more consistent efforts should be made to employ causal designs with adequate sample sizes that include comparisons to meaningful control conditions, particularly when aiming to make causal inferences on the effects of AAIs.

Although the choice of design is important, of even greater importance from a salivary bioscience perspective is the choice of sampling scheme. HAI researchers interested in examining effects of specific, discrete HAI events on momentary HPA activity should aim to incorporate collection of basal levels, and ideally, diurnal patterns. This is important so that potential aspects of HPA axis dysregulation, which may be present given the populations targeted in AAIs, are considered. Considering these parameters and other potentially confounding variables (e.g., time of day; medication and/or substance use; indices of diurnal dysregulation) is likely to reveal rather than obscure existing treatment effects.

Overall, whether examining effects of momentary reactivity, a stress-buffering effect, or treatment effects of longer term HAIs on diurnal indices, the exact timing of sample collection should be informed by empirical and theoretical consideration of the specific analyte under examination. This is important as the timeframe in which an analyte peaks in saliva varies among analytes. Moreover, some analytes are not expected to change within a short timeframe, are only reactive in response to extreme stress, or do not have known links to relevant developmental outcomes, and as such are difficult to interpret within the context of AAIs. In addition, special consideration should be given to the research question, the tolerance of human and animal participants to collect multiple samples without fundamentally changing the experience of the interaction under study, as well as logistical and practical issues of salivary sampling paradigms.

Last, but not least, while sampling of oral fluids is perceived as quick, uncomplicated, and minimally invasive, HAI investigators too often ignore key facts about oral biology and the nature of saliva as a biospecimen to prevent compromising the value of the information obtained. This has distinct significance when measuring salivary outcomes in animals, as the structural characteristics and functions of saliva vary across species (Hofmann, Streich, Fickel, Hummel, & Clauss, 2008; Tucker, 2010), and validation of sampling paradigms varies across biomarker and species, Chap. 28, for a comprehensive review). Too many HAI studies still use stimulants to stimulate saliva production, a methodology that, when not used both minimally and consistently, has been demonstrated to manipulate analyte performance (Schwartz, Granger, Susman, Gunnar, & Laird, 1998), which poses barriers to validity in human and animal studies (Dreschel & Granger, 2009). Furthermore, many studies utilize cotton-based swabs in human and animal populations where other, less invasive methods could be used. Passive drool sampling is identified (Granger et al., 2007) as the preferred method in humans minimizing threats to measurement validity; it comes as no surprise that studies employing this method seem to have the most robust results.

Most importantly, given that we collect data across a multitude of fields and settings, it is important to more explicitly *describe* procedural paradigms employed and parameter calculation methods used. Information should be collected about behavior, activities, and events to potentially exclude samples or statistically control for their potential influence. Additionally, criteria and decisions made about inclusion and exclusion of samples should be clearly described. Similarly, storing procedures on the day of sampling and thereafter, along with shipping and assay procedures, are rarely described in adequate detail. We must continue to foster work employing sound research design and highlight the credentials of such work by explicitly conveying study details in succeeding manuscripts.

## 23.5 Closing Comments

From the perspective of more established disciplines, the field of HAI represents an eclectic mix of researchers and clinicians who apply an equally eclectic range of theories and methods toward the establishment of a coherent body of knowledge and theory describing the purported benefits of animal companionship and AAIs. Rather than perceiving this characteristic as a shortcoming, we should first and foremost consider the exciting aspects of engaging in novel, interdisciplinary research and continue to draw like-minded investigators to join in focused, rigorous scientific research incorporating salivary approaches. Moving forward, we suggest that the field of HAI will most benefit from incorporating multifaceted approaches through examining a variety of relevant analytes associated with positive and negative indicators of behavior, health, well-being, and development in human and animal populations.

## References


Adam, E. (2005). Momentary emotions and physiological stress levels in the everyday lives of working parents. In B. Schneider & J. Waite (Eds.), *Being together, working apart: Dual-career families and the work-life balance* (pp. 105–133). Cambridge: Cambridge University Press.

Adam, E. K., & Gunnar, M. R. (2001). Relationship functioning and home and work demands predict individual differences in diurnal cortisol patterns in women. *Psychoneuroendocrinology, 26*(2), 189–208. https://doi.org/10.1016/S0306-4530(00)00045-7

Adamle, K., Riley, T., & Carlson, T. (2009). Evaluating college student interest in pet therapy. *Journal of American College Health, 57*(5), 545–548. https://doi.org/10.3200/JACH.57.5.545-548

Ainsworth, M. D. S. (1963). The development of infant-mother interaction among the Ganda. In B. M. Foss (Ed.), *Determinants of infant behavior* (pp. 67–104). New York, NY: Wiley.

Allen, K., Blascovich, J., & Mendes, W. B. (2002). Cardiovascular reactivity and the presence of pets, friends, and spouses: The truth about cats and dogs. *Psychosomatic Medicine, 64*, 727–739.

American College Health Association. (2018). *American College Health Association-National College Health Assessment II: Reference group executive summary fall 2017.* Hanover, MD: American College Health Association.

Barker, S. B., Barker, R. T., McCain, N. L., & Schubert, C. M. (2016). A randomized cross-over exploratory study of the effect of visiting therapy dogs on college student stress before final exams. *Anthrozoös, 29*(1), 35–46. https://doi.org/10.1080/08927936.2015.1069988

Barker, S. B., Barker, R. T., McCain, N. L., & Schubert, C. M. (2017). The effect of a canine-assisted activity on college student perceptions of family supports and current stressors. *Anthrozoös, 30*(4), 595–606. https://doi.org/10.1080/08927936.2017.1370227

Barker, S. B., Knisely, J. S., McCain, N. L., & Best, A. M. (2005). Measuring stress and immune responses in healthcare professionals following interaction with a therapy dog: A pilot study. *Psychological Reports, 96*, 713–729.

Beetz, A., & Bales, K. (2016). Affiliation in human-animal interaction. In L. S. Freund, S. McCune, L. Esposito, N. R. Gee, & P. McCardle (Eds.), *The social neuroscience of human-animal interaction* (pp. 107–125). Washington, DC: American Psychological Association. https://doi.org/10.1037/14856-000

Beetz, A., Kotrschal, K., Hediger, K., Turner, D., & Unväs-Moberg, K. (2011). The effect of a real dog, toy dog and friendly person on insecurely attached children during a stressful task: An exploratory study. *Anthrozoös, 24*, 349–368.

Beetz, A., Unväs-Moberg, K., Julius, H., & Kotrschal, K. (2012). Psychosocial and psychophysiological effects of human-animal interactions: The possible role of oxytocin. *Frontiers in Psychology, 3*(Jul), 1–15. https://doi.org/10.3389/fpsyg.2012.00234

Binfet, J. T. (2017). The effects of group-administered canine therapy on university students' wellbeing: A randomized controlled trial. *Anthrozoös, 30*(3), 397–414. https://doi.org/10.1080/08927936.2017.1335097

Binfet, J. T., Passmore, H. A., Cebry, A., Struik, K., & McKay, C. (2017). Reducing university students' stress through a drop-in canine-therapy program. *Journal of Mental Health, 27*(3), 193–196. https://doi.org/10.1080/09638237.2017.1417551

Bowlby, J. (1969). Disruption of affectional bonds and its effects on behavior. *Canada's Mental Health Supplement, 69*, 1–17.

Branson, S. M., Boss, L., Padhye, N. S., Trötscher, T., & Ward, A. (2017). Effects of animal-assisted activities on biobehavioral stress responses in hospitalized children: A randomized controlled study. *Journal of Pediatric Nursing, 36*, 84–91. https://doi.org/10.1016/j.pedn.2017.05.006

Calcaterra, V., Veggiotti, P., Palestrini, C., De Giorgis, V., Raschetti, R., Tumminelli, M., … Pelizzo, G. (2015). Post-operative benefits of animal-assisted therapy in pediatric surgery: A randomized study. PLoS One, 10(6), e0125813. https://doi.org/10.1371/journal.pone.0125813

Calvo, P., Fortuny, J. R., Guzmán, S., Macías, C., Bowen, J., García, M. L., … Fatjó, J. (2016). Animal assisted therapy (AAT) program as a useful adjunct to conventional psychosocial rehabilitation for patients with schizophrenia: Results of a small-scale randomized controlled trial. *Frontiers in Psychology, 7*(May), 1–13. https://doi.org/10.3389/fpsyg.2016.00631

Cardoso, C., Ellenbogen, M. A., Orlando, M. A., Bacon, S. L., & Joober, R. (2013). Intranasal oxytocin attenuates the cortisol response to physical stress: A dose-response study. *Psychoneuroendocrinology, 38*(3), 399–407. https://doi.org/10.1016/j.psyneuen.2012.07.013

Carter, C. S., Pournajavi-Nazarloo, H., Kramer, K. M., Zeigler, T. E., White-Traut, R., Bello, D., & Schwertz, D. (2007). Oxytocin: Behavioral associations and potential as a salivary biomarker. *Annals of the New York Academy of Sciences, 1098*(1), 312–322. https://doi.org/10.1196/annals.1384.006

Charnetski, C. J., Riggers, S., & Brennan, F. X. (2004). Effect of petting a dog on immune system function. *Psychological Reports, 95*, 1087–1091.

Clark, S. D., Smidt, J. M., & Bauer, B. A. (2019). Welfare considerations: Salivary cortisol concentrations on the frequency of therapy dogs visits in an outpatient hospital setting: A pilot study. *Journal of Veterinary Behavior, 30*, 88–91. https://doi.org/10.1016/j.jveb.2018.12.002

Cochran, D. M., Fallon, D., Hill, M., & Frazier, J. A. (2013). The role of oxytocin in psychiatric disorders: A review of biological and therapeutic research findings. *Harvard Review of Psychiatry, 21*(5), 219–247. https://doi.org/10.1097/HRP.0b013e3182a75b7d

Cohen, S., & McKay, G. (1984). Social support, stress, and the buffering hypothesis: A theoretical analysis. In A. Baum, S. E. Taylor, & J. E. Singer (Eds.), *Handbook of psychology and health* (4th ed.). Hillsdale, NJ: Psychology Press. Retrieved from http://www.psy.cmu.edu/~scohen/buffer84.pdf

Cohen, S., & Wills, T. A. (1985). Stress, social support, and the buffering hypothesis. *Psychological Bulletin, 98*(2), 310–357. https://doi.org/10.1037/0033-2909.98.2.310

Colussi, A., Stefanon, B., Adorini, C., & Sandri, M. (2018). Variations of salivary cortisol in dogs exposed to different cognitive and physical activities. *Italian Journal of Animal Science, 0*(0), 1–8. https://doi.org/10.1080/1828051X.2018.1453756

Constantino, M. J., Arnkoff, D. B., Glass, C. R., Ametrano, R. M., & Smith, J. A. Z. (2011). Expectations. *Journal of Clinical Psychology, 67*(2), 184–192. https://doi.org/10.1002/jclp.20754

Crossman, M. K., & Kazdin, A. E. (2015). Animal visitation programs in colleges and universities: An efficient model for reducing student stress. In A. H. Fine (Ed.), *Handbook on animal-assisted therapy* (4th ed., pp. 333–337). San Diego, CA: Elsevier Academic Press. https://doi.org/10.1016/B978-0-12-801292-5.00024-9

Crossman, M. K., Kazdin, A. E., & Knudson, K. (2015). Brief unstructured interaction with a dog reduces distress. *Anthrozoös, 28*(4), 649–659. https://doi.org/10.1080/08927936.2015.1070008

Crump, C., & Derting, T. L. (2015). Effects of pet therapy on the psychological and physiological stress levels of first-year female undergraduates. *North American Journal of Psychology, 17*(3), 575.

Daughters, K., Manstead, A. S., Hubble, K., Rees, A., Thapar, A., & van Goozen, S. H. M. (2015). Salivary oxytocin concentrations in males following intranasal administration of oxytocin: A double-blind, cross-over study. *PLoS One, 10*(12). https://doi.org/10.1371/journal.pone.0145104

Demakis, G. J., & McAdams, D. P. (1994). Personality, social support and well-being among first year college students. *College Student Journal, 28*, 235–243.

Ditzen, B., Neumann, I. D., Bodenmann, G., von Dawans, B., Turner, R. A., Ehlert, U., & Heinrichs, M. (2007). Effects of different kinds of couple interaction on cortisol and heart rate responses to stress in women. *Psychoneuroendocrinology, 32*(5), 565–574. https://doi.org/10.1016/j.psyneuen.2007.03.011

Dreschel, N. A., & Granger, D. A. (2009). Methods of collection for salivary cortisol measurement in dogs. *Hormones and Behavior, 55*(1), 163–168. https://doi.org/10.1016/j.yhbeh.2008.09.010

Esposito, L., Mccune, S., Griffin, J. A., & Maholmes, V. (2011). Directions in human-animal interaction research: Child development, health, and therapeutic interventions. *Child Development Perspectives, 5*(3), 205–211. https://doi.org/10.1111/j.1750-8606.2011.00175.x

Franke, A. A., Li, X., Menden, A., Lee, M. R., & Lai, J. F. (2019). Oxytocin analysis from human serum, urine, and saliva by orbitrap liquid chromatography-mass spectrometry. *Drug Testing and Analysis, 11*(1), 199–128. https://doi.org/10.1002/dta.2475

Fine, A. H. (Ed.). (2015). *Handbook on animal assisted therapy: Theoretical foundations and guidelines for practice* (4th ed.). London: Elsevier Academic Press.

Foley, P., & Kirschbaum, C. (2010). Human hypothalamus-pituitary-adrenal axis responses to acute psychosocial stress in laboratory settings. *Neuroscience and Biobehavioral Reviews, 35*(1), 91–96. https://doi.org/10.1016/j.neubiorev.2010.01.010

Freund, L. S., McCune, S., Esposito, L., Gee, N. R., & McCardle, P. (Eds.). (2016). *The social neuroscience of human-animal interaction*. Washington, DC: American Psychological Association.

GfK. (2016, May). *Pet ownership: Global GfK survey.* Retrieved from https://www.gfk.com/global-studies/global-studies-pet-ownership/

Glenk, L. M., Kothgassner, O. D., Stetina, B. U., Palme, R., Kepplinger, B., & Baran, H. (2014). Salivary cortisol and behavior in therapy dogs during animal-assisted interventions: A pilot study. *Journal of Veterinary Behavior: Clinical Applications and Research, 9*(3), 98–106. https://doi.org/10.1016/j.jveb.2014.02.005

Glenk, L. M., Stetina, B. U., Kepplinger, B., & Baran, H. (2013). Therapy dogs' salivary cortisol levels vary during animal-assisted interventions. *Animal Welfare, 22*(3), 369–378.

Goldstein, D. S., & Kopin, I. J. (2007). Evolution of concepts of stress. *Stress, 10*(2), 109–120. https://doi.org/10.1080/10253890701288935

Grajfoner, D., Harte, E., Potter, L. M., & McGuigan, N. (2017). The effect of dog-assisted intervention on student well-being, mood, and anxiety. *International Journal of Environmental Research and Public Health, 14*(5), 483. https://doi.org/10.3390/ijerph14050483

Granger, D. A., Kivlighan, K. T., Fortunato, C., Harmon, A. G., Hibel, L. C., Schwartz, E. B., & Whembolua, G. L. (2007). Integration of salivary biomarkers into developmental and behaviorally-oriented research: Problems and solutions for collecting specimens. *Physiology and Behavior, 92*(4), 583–590. https://doi.org/10.1016/j.physbeh.2007.05.004

Grant, K. E., McMahon, S. D., Carter, J. S., Carleton, R. A., Adam, E. K., & Chen, E. (2014). The influence of stressors on the development of psychopathology. In M. Lewis & K. D. Rudolph (Eds.), *Handbook of developmental psychopathology* (pp. 205–223). New York, NY: Springer Science and Business Media New York. https://doi.org/10.1007/978-1-4614-9608-3

Greenberg, R. P., Constantino, M. J., & Bruce, N. (2006). Are patient expectations still relevant for psychotherapy process and outcome? *Clinical Psychology Review, 26*(6), 657–678. https://doi.org/10.1016/j.cpr.2005.03.002

Guéguen, N., & Ciccotti, S. (2008). Domestic dogs as facilitators in social interaction: An evaluation of helping and courtship behaviors. *Anthrozoös, 21*(4), 339–349. https://doi.org/10.2752/175303708X371564

Gunnar, M., & Quevedo, K. (2007). The neurobiology of stress and development. *Annual Review of Psychology, 58*(1), 145–173. https://doi.org/10.1146/annurev.psych.58.110405.085605

Handlin, L., Hydbring-Sandberg, E., Nilsson, A., Ejdebäck, M., Jansson, A., & Uvnäs-Moberg, K. (2011). Short-term interaction between dogs and their owners: Effects on oxytocin, cortisol, insulin and heart rate-an exploratory study. *Anthrozoös, 24*(3), 301–315. https://doi.org/10.2752/175303711X13045914865385

Hane, A. A., & Fox, N. A. (2016). Studying the biology of human attachment. In J. Cassidy & P. R. Shaver (Eds.), *Handbook of attachment: Theory, research, and clinical applications* (3rd ed., pp. 223–241). New York, NY: Guilford Press.

Havener, L., Gentes, L., Thaler, B., Megel, M. E., Baun, M. M., Driscoll, F. A., ... Agrawal, S. (2001). The effects of a companion animal on distress in children undergoing dental procedures. *Issues in Comprehensive Pediatric Nursing, 24*(2), 137–152.

Heinrichs, M., Baumgartner, T., Kirschbaum, C., & Ehlert, U. (2003). Social support and oxytocin interact to suppress cortisol and subjective responses to psychosocial stress. *Biological Psychiatry, 54*(12), 1389–1398. https://doi.org/10.1016/S0006-3223(03)00465-7

Herman, J. P., & Cullinan, W. E. (1997). Neurocircuitry of stress: Central control of the hypothalamo- pituitary-adrenocortical axis. *Trends in Neurosciences, 20*(2), 78–84. https://doi.org/10.1016/S0166-2236(96)10069-2

Hofmann, R. R., Streich, W. J., Fickel, J., Hummel, J., & Clauss, M. (2008). Convergent evolution in feeding types: Salivary gland mass differences in wild ruminant species. *Journal of Morphology, 269*(2), 240–257. https://doi.org/10.1002/jmor.10580

International Association of Human-Animal Interaction Organizations. (2014). *IAHAIO WHITE PAPER 2014: The IAHAIO definitions for animal assisted intervention and guidelines for wellness of animals involved.* Seattle, WA: IAHAIO. Retrieved from http://iahaio.org/best-practice/white-paper-on-animal-assisted-interventions/

Isaksson, J., Nilsson, K. W., Nyberg, F., Hogmark, Å., & Lindblad, F. (2012). Cortisol levels in children with attention-deficit/hyperactivity disorder. *Journal of Psychiatric Research, 46*(11), 1398–1405. https://doi.org/10.1016/j.jpsychires.2012.08.021

Jacobson, K. C. (2014). *The effects of HAI on cortisol reactivity in response to psychosocial stress* (pp. 66–67). Vienna: International Society for Anthrozoölogy. Retrieved from www.isaz.net/conferences/Abstractbook2014.pdf

Javor, A., Reidl, R., Kindermann, H., Brandstätter, W., Ransmayr, G., & Gabriel, M. (2014). Correlation of plasma and salivary oxytocin in healthy young men – experimental evidence. *Neuroendocrinology Letters, 35*(6), 470–473.

Jones, A. C., & Josephs, R. A. (2006). Interspecies hormonal interactions between man and the domestic dog (*Canis familiaris*). *Hormones and Behavior, 50*(3), 393–400. https://doi.org/10.1016/j.yhbeh.2006.04.007

Kertes, D. A., Liu, J., Hall, N. J., Hadad, N. A., Wynne, C. D. L., & Bhatt, S. S. (2017). Effect of pet dogs on children's perceived stress and cortisol stress response. *Social Development, 26*(2), 382–401. https://doi.org/10.1111/sode.12203

Kirschbaum, C., & Hellhammer, D. H. (1989). Salivary cortisol in psychobiological research: An overview. *Neuropsychobiology, 22*, 150–169.

Krause-Parello, C. A., Tychowski, J., Gonzalez, A., & Boyd, Z. (2012). Human-canine interaction: Exploring stress indicator response patterns of salivary cortisol and Immunoglobin A. *Research and Theory for Nursing Practice, 26*(1), 25–40.

Lazarus, R. S., & Folkman, S. (1984). *Stress, appraisal and coping*. New York, NY: Springer.

Lazarus, R., & Folkman, S. (1987). Transactional theory and research on emotion and coping. *European Journal of Personality, 1*, 141–169. https://doi.org/10.1002/per.2410010304

Legros, J., Chiodera, P., Geenen, V., & von Frenckell, R. (1987). Confirmation of the inhibitory influence of exogenous oxytocin on cortisol and ACTH in man: Evidence of reproducibility. *Acta Endocrinologica, 114*(3), 345–349.

Lupien, S. J., McEwen, B. S., Gunnar, M. R., & Heim, C. (2009). Effects of stress throughout the lifespan on the brain, behaviour and cognition. *Nature Reviews Neuroscience, 10*(6), 434–445. https://doi.org/10.1038/nrn2639

MacLean, E. L., Gesquiere, L. R., Gee, N. R., Levy, K., Martin, W. L., & Carter, S. (2017a). Effects of affiliative human–animal interaction on dog salivary and plasma oxytocin and vasopressin. *Frontiers in Psychology, 8*, 1606–1615. https://doi.org/10.3389/fpsyg.2017.01606

MacLean, E. L., Gesquiere, L. R., Gruen, M. E., Sherman, B. L., Martin, W. L., & Carter, C. S. (2017b). Endogenous oxytocin, vasopressin, and aggression in domestic dogs. *Frontiers in Psychology, 8*, 1613. https://doi.org/10.3389/fpsyg.2017.01613

MacLean, E. L., Gesquiere, L. R., Gee, N., Levy, K., Martin, W. L., & Carter, C. S. (2018). Validation of salivary oxytocin and vasopressin as biomarkers in domestic dogs. *Journal of Neuroscience Methods, 293*, 67–76. https://doi.org/10.1016/j.jneumeth.2017.08.033

Matin, M. J., Li. D., Peterson, J., Taylor, M. K., Laurent, H. K., Lucas, T., . . . Granger, S. W. (2016). Measuring nerve growth factor in saliva by immunoassay: A cautionary note. *Psychoneuroendocrinology, 63*, 235–237.

Martin, J., Kagerbauer, S. M., Gempt, J., Podtscaske, A., Hapfelmeier, A., & Schneider, G. (2018). Oxytocin levels in saliva correlate better than plasma levels with concentrations in the cerebrospinal fluid of patients in neurocritical care. *Journal of Neuroendocrinology, 30*(5), e12596. https://doi.org/10.1111/jne.12596

McCullough, M. E., Churchland, P. S., & Mendez, A. J. (2013). Problems with measuring peripheral oxytocin: can the data on oxytocin and human behavior be trusted? *Neuroscience and Behavioral Reviews, 37*(8), 1485–1492. https://doi.org/10.1016/j.neubiorev.2013.04.018

McCullough, A., Jenkins, M. A., Ruerhdanz, A., Gilmer, M. J., Olson, J., Pawar, A., . . . O'Haire, M. E. (2018). Physiological and behavioral effects of animal-assisted interventions on therapy dogs in pediatric oncology settings. Applied Animal Behaviour Science, 200, 86–95. doi:https://doi.org/10.1016/j.applanim.2017.11.014

McKinney, C., Mueller, M. K., & Frank, N. (2015). Effects of therapeutic riding on measures of stress in horses. *Journal of Equine Veterinary Science, 35*(11–12), 922–928. https://doi.org/10.1016/j.jevs.2015.08.015

McNicholas, J., & Collis, G. (2006). Animals as social supports: Insights for understanding animal-assisted therapy. In A. H. Fine (Ed.), *Handbook on animal assisted therapy: Theoretical foundations and guidelines for practice* (pp. 49–71). New York: Academic Press.

Melco, A. L., Goldman, L., Fine, A. H., & Peralta, J. M. (2018). Investigation of physiological and behavioral responses in dogs participating in animal-assisted therapy with children diagnosed with attention-deficit hyperactivity disorder. *Journal of Applied Animal Welfare Science.* https://doi.org/10.1080/10888705.2018.1536979

Melson, G. F., & Fine, A. H. (2015). Animals in the lives of children. In A. H. Fine (Ed.), *Handbook on animal assisted therapy: Theoretical foundations and guidelines for practice* (4th ed., pp. 179–194). San Diego, CA: Elsevier Academic Press.

Merkies, K., McKechnie, M. J., & Zakrajsek, E. (2018). Behavioral and physiological responses of therapy horses to mentally traumatized humans. *Applied Animal Behaviour Science, 205*, 61–67. https://doi.org/10.1016/j.applanim.2018.05.019

Neumann, I. D., Krömer, S. A., Toschi, N., & Ebner, K. (2000). Brain oxytocin inhibits the (re) activity of the hypothalamo-pituitary-adrenal axis in male rats: Involvement of hypothalamic and limbic brain regions. *Regulatory Peptides, 96*(1–2), 31–38. https://doi.org/10.1016/S0167-0115(00)00197-X

Ng, Z. Y., Pierce, B. J., Otto, C. M., Buechner-Maxwell, V. A., Siracusa, C., & Werre, S. R. (2014). The effect of dog-human interaction on cortisol and behavior in registered animal-assisted activity dogs. *Applied Animal Behaviour Science, 159*, 69–81. https://doi.org/10.1016/j.applanim.2014.07.009

Nicolaides, N. C., Charmandari, E., Chrousos, G. P., & Kino, T. (2014). Circadian endocrine rhythms: The hypothalamic–pituitary–adrenal axis and its actions. *Annals of the New York Academy of Sciences, 1318*, 71–80. https://doi.org/10.1111/nyas.12464

Odendaal, J. S. J. (2000). Animal-assisted therapy: Magic or medicine? *Journal of Psychosomatic Research, 49*(4), 275–280. https://doi.org/10.1016/S0022-3999(00)00183-5

Odendaal, J. S. J., & Meintjes, R. A. (2003). Neurophysiological correlates of affiliative behaviour between humans and dogs. *The Veterinary Journal, 165*(3), 296–301. https://doi.org/10.1016/S1090-0233(02)00237-X

Pan, Z., Granger, D. A., Guérin, N. A., Shoffner, A., & Gabriels, R. L. (2018). Replication pilot trial of therapeutic horseback riding and cortisol collection with children on the autism spectrum. *Frontiers in Veterinary Science, 5*. https://doi.org/10.3389/fvets.2018.00312

Pendry, P., & Vandagriff, J. L. (2019). Animal visitation program (AVP) reduces cortisol levels of university students: A randomized controlled trial. *AERA Open, 5*(2), 233285841985259. https://doi.org/10.1177/2332858419852592

Pendry, P., Carr, A. M., Roeter, S. M., & Vandagriff, J. L. (2018). Experimental trial demonstrates effects of animal-assisted stress prevention program on college students' positive and negative emotion. *Human-Animal Interaction Bulletin, 6*(1), 81–97.

Pendry, P., Carr, A. M., & Vandagriff, J. L. (2018). Adolescents' affective and physiological regulation shape negative behavior during challenging equine assisted learning activities. *Frontiers in Veterinary Science, 5*, 300. https://doi.org/10.3389/fvets.2018.00300

Pendry, P., Smith, A. N., & Roeter, S. M. (2014). Randomized trial examines effects of equine facilitated learning on adolescents' basal cortisol levels. *Human Animal Interaction Bulletin, 2* (1), 80–95.

Petersson, M. (2002). Cardiovascular effects of oxytocin. *Progress in Brain Research, 139*, 281–288. https://doi.org/10.1016/S0079-6123(02)39024-1

Petersson, M., Lundeberg, T., & Unväs-Moberg, K. (1999). Short-term increase and long-term decrease of blood pressure in response to oxytocin-potentiating effect of female steroid hormones. *Journal of Cardiovascular Pharmacology, 33*(1), 102–108.

Petersson, M., Unväs-Moberg, K., Nilsson, A., Gustafson, L., Hydbring-Sandberg, E., & Handlin, L. (2017). Oxytocin and cortisol levels in dog owners and their dogs are associated with behavioral patterns: An exploratory study. *Frontiers in Psychology, 8*, 1796. https://doi.org/10.3389/fpsyg.2017.01796

Pirrone, F., Ripamonti, A., Garoni, E. C., Stradiotti, S., & Albertini, M. (2017). Measuring social synchrony and stress in the handler-dog dyad during animal assisted activities: A pilot study. *Journal of Veterinary Behavior, 21*, 45–52. https://doi.org/10.1016/j.jveb.2017.07.004

Polheber, J. P., & Matchock, R. L. (2014). The presence of a dog attenuates cortisol and heart rate in the trier social stress test compared to human friends. *Journal of Behavioral Medicine, 37*(5), 860–867. https://doi.org/10.1007/s10865-013-9546-1

Pruessner, J. C., Wolf, O. T., Hellhammer, D. H., Buske-Kirschbaum, A., Von Auer, K., Jobst, S., . . . Kirschbaum, C. (1997). Free cortisol levels after awakening: A reliable biological marker for the assessment of adrenocortical activity. *Life Sciences, 61*(26), 2539–2549. https://doi.org/10.1016/S0024-3205(97)01008-4

Quintana, D. S., Westlye, L. T., Smerud, K. T., Mahmoud, R. A., Andreassen, O. A., & Djupesland, P. G. (2018). Saliva oxytocin measures do not reflect peripheral plasma concentrations after intranasal oxytocin administration in men. *Hormones and Behavior, 102*, 85–92. https://doi.org/10.1016/j.yhbeh.2018.05.004

Rabbitt, S. M., Kazdin, A. E., & Hong, J. (2014). Acceptability of animal-assisted therapy: Attitudes toward AAT, psychotherapy, and medication for the treatment of child disruptive behavioral problems. *Anthrozoös, 27*(3), 335–350. https://doi.org/10.2752/175303714X13903827487881

Rodriguez, K. E., Bryce, C. I., Granger, D. A., & O'Haire, M. E. (2018). The effect of a service dog on salivary cortisol awakening response in a military population with posttraumatic stress disorder (PTSD). *Psychoneuroendocrinology, 98*, 202. https://doi.org/10.1016/j.psyneuen.2018.04.026

Schneider, M. S., & Harley, L. P. (2006). How dogs influence the evaluation of psychotherapists. *Anthrozoös, 19*(2), 128–142. https://doi.org/10.2752/089279306785593784

Schöberl, I., Wedl, M., Beetz, A., & Kotrschal, K. (2017). Psychobiological factors affecting cortisol variability in human-dog dyads. *PLoS One, 12*(2), 11–13. https://doi.org/10.1371/journal.pone.0170707

Schwartz, E. B., Granger, D. A., Susman, E. J., Gunnar, M. R., & Laird, B. (1998). Assessing salivary cortisol in studies of child development. *Child Development, 69*(6), 1503–1513. https://doi.org/10.1111/j.1467-8624.1998.tb06173.x

Stratakis, C. A., & Chrousos, G. P. (1995). Neuroendocrinology and pathophysiology of the stress system. *Annals of the New York Academy of Sciences, 771*, 1–18.

Tucker, A. S. (2010). Salivary gland adaptations: Modification of the glands for novel uses. In *Salivary glands* (pp. 21–31). Basel: KARGER. https://doi.org/10.1159/000313705

Vandagriff, J. L. (2017, June). *Considering the implementation of animal visitation programs (AVPs) on college campuses: A review of causal studies*. Poster presented at the annual meeting of the International Society for Anthrozoölogy, Davis, CA.

VanFleet, R., Fine, A. H., O'Callaghan, D., Mackintosh, T., & Gimeno, J. (2015). Application of animal-assisted interventions in professional settings: An overview of alternatives. In A. H. Fine (Ed.), *Handbook on animal assisted therapy: Theoretical foundations and guidelines for practice* (4th ed., pp. 157–177). London: Elsevier Academic Press. https://doi.org/10.1016/B978-0-12-801292-5.00012-2

van IJzendoorn, M. H., Bhandari, R., van der Veen, R., Grewen, K. M., & Bakermans-Kranenburg, M. J. (2012). Elevated salivary levels of oxytocin persist more than 7 h after intranasal administration. *Frontiers in Neuroscience, 6*. https://doi.org/10.3389/fnins.2012.00174

Viau, R., Arsenault-Lapierre, G., Fecteau, S., Champagne, N., Walker, C. D., & Lupien, S. (2010). Effect of service dogs on salivary cortisol secretion in autistic children. *Psychoneuroendocrinology, 35*(8), 1187–1193. https://doi.org/10.1016/j.psyneuen.2010.02.004

Weisman, O., Zagoory-Sharon, O., & Feldman, R. (2012). Intranasal oxytocin administration is reflected in human saliva. *Psychoneuroendocrinology, 37*, 1582–1586. https://doi.org/10.1016/j.psyneuen.2012.02.014

Wells, M., & Perrine, R. (2001). Pets go to college: The influence of pets on students' perceptions of faculty and their offices. *Anthrozoös, 14*(3), 161–168. https://doi.org/10.2752/089279301786999472

White-Traut, R., Watanabe, K., Pournajafi-Nazarloo, H., Schwertz, D., Bell, A., & Carter, C. S. (2009). Detection of salivary oxytocin levels in lactating women. *Developmental Psychobiology, 51*(4), 367–373. https://doi.org/10.1002/dev.20376

Wood, L., Giles-Corti, B., & Bulsara, M. (2005). The pet connection: Pets as a conduit for social capital? *Social Science and Medicine, 61*(6), 1159–1173. https://doi.org/10.1016/j.socscimed.2005.01.017

Yorke, J., Nugent, W., Strand, E., Bolen, R., New, J., & Davis, C. (2013). Equine-assisted therapy and its impact on cortisol levels of children and horses: A pilot study and meta-analysis. *Early Child Development and Care, 183*(7), 874–894. https://doi.org/10.1080/03004430.2012.693486

# Part V
# Population Health Research and Applications

**Marcus K. Taylor, Section Editor**

# Chapter 24
# Salivary Bioscience in Military, Space, and Operational Research

**Peter G. Roma, Meaghan E. Beckner, Satish K. Mehta, Bradley C. Nindl, and Brian E. Crucian**

**Abstract** In this chapter, we discuss salivary bioscience research in military, space, and other operational contexts characterized by isolated, confined, and extreme environments. We consider historical developments and ongoing applications of salivary bioscience for assessing risk and providing insight into physical and behavioral health in individuals and teams in extreme settings. Future considerations and opportunities emphasize the importance of temporal dynamics and longitudinal studies, integrated multidisciplinary approaches, and technical innovations to enable operationally feasible assessment in the field and lifetime surveillance of physical and behavioral health for those who live, work, serve, and explore in extreme environments.

**Keywords** Military · Space · Operational research · Extreme environment · Behavioral health · Performance

---

P. G. Roma (✉)
Behavioral Health & Performance Laboratory, Biomedical Research and Environmental Sciences Division, Human Health and Performance Directorate, KBR/NASA Johnson Space Center, Houston, TX, USA
e-mail: pete.roma@nasa.gov

M. E. Beckner · B. C. Nindl
Neuromuscular Research Laboratory/Warrior Human Performance Research Center, School of Health and Rehabilitation Sciences, University of Pittsburgh, Pittsburgh, PA, USA

S. K. Mehta
Immunology Laboratory, Biomedical Research and Environmental Sciences Division, Human Health and Performance Directorate, JES Tech/NASA Johnson Space Center, Houston, TX, USA

B. E. Crucian
Immunology Laboratory, Biomedical Research and Environmental Sciences Division, Human Health and Performance Directorate, NASA Johnson Space Center, Houston, TX, USA



D. A. Granger, M. K. Taylor (eds.), *Salivary Bioscience*,
https://doi.org/10.1007/978-3-030-35784-9_24

## 24.1 Introduction

Working and living in operational environments such as military deployments, space exploration, and remote work sites pose a wide variety of risks to physical health, behavioral health, and performance, especially over long-duration missions. As such, these settings make a challenging but potentially fruitful research platform for understanding the limits of human performance and adaptability. A phrase used in operational communities is "isolated, confined, and extreme" (ICE) environments, which is the applied focus of this chapter. Space is the prototypical ICE operational environment, but we also include military and security deployments, remote work sites, and other long-duration team-based work settings.

Isolation in operational settings is primarily the physical separation from civilization and community, such as working and living in a remote location, military base, outpost, or vessel (e.g., submarine, spacecraft) at distances that are impractically or impossibly far from the crews' homes for anything resembling a standard workday "commute" allowing physical and social separation of professional and personal lives. However, isolation is not limited to physical proximity, as operational environments often have inherent or required communications constraints. For example, military environments (e.g., submarines) often operate on secure networks with limited accommodations for civilian communications. For future space exploration crews, all communications between Earth and Mars must accommodate an approximately 20 min delay each way due to the extreme distance the signals must travel between the planets. Taken together, physical separation combined with limited or lack of real-time communication with outside professional and social support systems both define and can intensify feelings of isolation.

The "confined" component of ICE environments refers to both the physical habitat and the social environment contained therein. ICE operations rely on constructed habitats, and as such are designed with a focus on basic life support and enabling execution of mission objectives, but are still subject to budgetary and logistical constraints in their design that can compound physical and psychosocial stress. Common physical confinement issues relevant to social stress in operational settings include limited privacy, and inadequate space for work, social, and recreational activities, with limited configurability and control (Kearney, 2016). Even with the most spacious and thoughtfully designed habitats, teams that work and live in isolation surrounded by hostile physical environments (e.g., polar, sea-bound, and space) are mostly confined to the habitat itself, or at least subject to extensive procedures for utilizing already limited resources outside scheduled work activities. In colloquial terms, one cannot spontaneously step out to take a break in most ICE environments. Similarly, confinement is not only limited to the physical environment, but crew members are socially limited to each other as the primary and often only source of social support, collegiality, and friendship.

As described above, the "extreme" portion of ICE typically refers to a dangerous external geophysical environment incompatible with human physiology, health, and well-being, including the lack of or toxic atmosphere, extreme heat or cold (or rapid shifts between them), extreme altitude (above or below sea level), non-24 h light–dark cycles and sleep deprivation, wildlife threats (e.g., predatory animals, micro-organisms, toxins), reduced gravity, and risk of exposure to radiation and other low-frequency/high-impact phenomena (e.g., solar flares, rough seas, dust storms, blizzards, high winds, and volcanism). In military operations, the hostile environment by definition includes the life-threatening presence of hostile enemy forces. However, even without imminent threat of physical danger, the isolation and confinement of operational settings can themselves make an environment extreme. This is especially true considering perhaps the most critical overarching contributor to the extremity of ICE environments: Time itself. For long-duration missions of several months to several years, biological and psychosocial stress effects are expected to accumulate and interact in ways the research and operations communities are only just beginning to recognize and understand (Goswami et al., 2013; Jonas et al., 2010; Landon et al., 2019).

## 24.2 Stress at the Core of Physical Health, Behavioral Health, and Performance in ICE Environments

Whether considering physical health, behavioral health, or mission performance, an essential framework for research in ICE environments biopsychosocial "stress" (Lloyd & Nemeroff, 2011; McEwen, Gray, & Nasca, 2015). Stressors can be physical challenges (e.g., combat, exercise, and noise), subjectively perceived psychological experiences (e.g., time pressure, social conflict, heavy workload, and threats to safety), and in operational environments, often a combined physiological and psychological experience (Dietz et al., 2010). The brain orchestrates responses to stress in that it interprets external and physical stimuli as threatening or potentially threatening, mediates biological and behavioral responses, and serves as the nexus of cyclical learning and adaptation processes (McEwen & Gianaros, 2010). The neurobiology of the stress system is accordingly complex, involving multiple brain circuits, neurotransmitters, inflammatory molecules, and peripheral organs, but is often associated with the hypothalamic–pituitary–adrenal (HPA) axis, the release of cortisol and epinephrine, and interplay between the sympathetic and parasympathetic nervous systems (Radley, Morilak, Viau, & Campeau, 2015; Thayer, Åhs, Fredrikson, Sollers, & Wager, 2012).

Stress and stress responses are not inherently negative (Lazarus & Folkman, 1984; McEwen & Karatsoreos, 2012)—indeed, success in response to the challenges of working and living in ICE environments provide opportunities for a sense of achievement, mastery, and resilience (Infurna & Luthar, 2018; Suedfeld, 2001; Troy & Mauss, 2011). However, chronic stress and individual differences in adaptation over time can alter brain architecture that yield dysregulated behavioral and cognitive systems associated with myriad physical and behavioral health dysfunctions (Byers & Yaffe, 2011; Gianaros et al., 2017; Radley et al., 2015).

Outside the HPA axis, but interacting with it, is the immune system, which plays a major role in the response to acute and chronic stress (Dantzer & Kelley, 1989). The immune system itself is complex, consisting of myriad cell types (neutrophils, monocytes, and lymphocytes) trafficking throughout the entire body with particular foci in immune organs such as lymph nodes, spleen, and thymus. These cells manifest various inflammatory responses depending on the pathogen or stressor type, body location, and individual endogenous predispositions. A key product of the immune system is cytokines, which are soluble proteins produced by immune and other cell types. They are pleiotropic in nature and may serve multiple functions, including recruiting and upregulating the function of immune cells, acting as growth factors, or conversely "shutting off" and preventing widespread effects of inflammation (Rothwell & Hopkins, 1995). Elevated inflammatory cytokines such as interleukin (IL)-1β, IL-6, tumor necrosis factor (TNF), and IL-8 have been associated with a variety of physical health conditions, including kidney disease, squamous cell carcinoma, and oral cancer (Katakura et al., 2007; SahebJamee, Eslami, AtarbashiMoghadam, & Sarafnejad, 2008; Thorman, Lundahl, Yucel-Lindberg, & Hylander, 2010), as well as behavioral health conditions such as bereavement, major depression, and cognitive dysfunction (Dowlati et al., 2010; O'Connor, Irwin, & Wellisch, 2009; Yirmiya & Goshen, 2011).

In addition to the HPA axis stress hormones, immune molecules, and inflammatory markers, dynamic levels of psychoactive estrogen (e.g., progesterone, estradiol) and androgen (e.g., testosterone) hormones interact with the stress and immune systems and also vary by gender, time of day (circadian rhythms), time of year (circannual rhythms), and as a function of physical and psychosocial factors, thus potentially conferring differential vulnerability to physical and behavioral health risks in ICE mission settings (Gatti & De Palo, 2011; Liening, Stanton, Saini, & Schultheiss, 2010; Smith, Coward, Kovac, & Lipshultz, 2013; Zhou, Liu, van Heerikhuize, Hofman, & Swaab, 2003). Although these and related gonadal hormones are classically associated with reproductive behaviors and mood, recent evidence suggests a moderating role in social processes, including group stability maintenance, social cognition, and team effectiveness (Derntl et al., 2013; Edwards et al., 2006; Eisenegger et al., 2010, 2011; Schultheiss, Dargel, & Rohde, 2003; Zyphur et al., 2009), all of which could be critical contributors to success in the mixed work/social settings of ICE mission environments. Clearly, there are many rich targets for modern salivary bioscience in ICE environments.

## 24.3 History of Salivary Bioscience in Military, Space, and Operational Research

The relative ease, noninvasiveness, and cost-effectiveness with which saliva may be sampled and processed makes it an appealing medium for research in operational settings. Although there is a growing appreciation for saliva as a source of biomarker data for assessing physical and behavioral health, the use of saliva applied to operational populations and extreme environments is not new. Saliva is quite naturally relevant to oral health, and indeed oral health was a significant concern for operational readiness during early war efforts in England, as an excess of missing or decaying teeth can not only lead to infection, but also potentially limit nutrient intake (e.g., protein and iron from meat) and subsequent performance risks when physically incapable of chewing and processing food with saliva (Westcott, 1907). Early direct application of salivary bioscience to operational health includes Wollstein's (1918) use of saliva from the parotid glands of US soldiers infected with mumps (parotitis) to develop an experimental laboratory animal model of parotitis, which had been burdening the service.

Beyond dental and oral health, Libbin (1943) recognized the broad potential value of salivary bioscience for physical and behavioral health assessment in ICE settings in his development of a "saliva ejector" apparatus for sample collection in the field. This device was not widely adopted, and indeed the limited subsequent attempts to use saliva for behavioral health biomarkers were of mixed value. For example, an early attempt to assess adrenal function and stress physiology in infantry soldiers 0–3 weeks post-combat included saliva samples for assessment of sodium and potassium (Davis & Taylor, 1954). Members of three different companies were tested, including a group exposed to 18 h of intense combat (60% causalities), a group exposed to 5 days of less intense "defensive" engagement (17% casualties), and a non-combat comparison company. Several blood and urine markers (e.g., lymphocytes, uric acid) showed adrenal responsiveness was higher in the high intensity, short duration combat group compared to the lower intensity, longer duration group. However, and unfortunately, contamination and insufficient sample quality rendered unusable what little salivary sodium and potassium samples they recovered. More successful efforts followed, including an investigation of adaptation to extreme cold and a high protein/high fat diet by Rodahl (1961), who used iodine-labeled pemmican to examine protein absorption and metabolism (cf. Lavik et al., 1952) in saliva of native Alaskan Eskimo populations and US soldiers, as iodine is selectively concentrated and secreted in saliva (Bruger & Member, 1943; Myant, 1960). Although there were no differences in salivary iodine excretion between the groups, this study was significant for its successful collection of analyzable saliva in an isolated and extreme environment. Work in extreme environments continued in the 1960s, with increasing emphasis on temporal dynamics and individual adaptation and recovery from missions in operational settings. For example, Hawkins and Zipkin (1964) examined the constituents of saliva in 29 US Navy personnel stationed at McMurdo base in Antarctica over the course of a year, with samples obtained just prior to deployment, then in-mission in the fall, winter, deep winter, and spring. No seasonal effects were observed; however, compared to pre-mission baseline, salivary protein, tyrosine, and tryptophan increased 25%, 17%, and 32%, respectively. Although not an aim of the study, the latter is particularly interesting from a behavioral health perspective, as tryptophan is a precursor molecule for serotonin and melatonin which are involved in mood regulation, sleep, and circadian rhythms (Campbell, Wilson, Walker, & Griffiths, 1985; Claustrat, Brun, & Chazot, 2005; Jenkins, Nguyen, Polglaze, & Bertrand, 2016), and a precursor for niacin (vitamin B) which may aid in reducing neuroinflammation (Wakade & Chong, 2014).

In what are likely the first applications of salivary bioscience to space exploration, Alekseyeva (1966) assessed innate immunity in the first Soyuz cosmonauts, with parallel efforts in endocrinology by C. L. Fischer as part of the US Gemini-Apollo program (see Johnson, 1986). In both cases, saliva sampling was limited to preflight and postflight with no in situ sampling in the spaceflight environment itself, and results (if any) are unknown. Similar and more successful assessments were later carried out by Taylor et al. (1977) as part of a joint US–Russian effort examining a variety of health-related outcomes of 9-day spaceflight missions in the Apollo and Soyuz spacecraft. For this, saliva samples were obtained from 10 astronauts at 45, 30, 15, 7, and 0 days preflight and in five astronauts at 0, 7, 15, and 30 days inflight, but revealed no effects of spaceflight on salivary flow rate, secretory immunoglobulin A (SIgA), or lysozyme content. Work in spaceflight and other extreme environments continued in the 1970s, with sustained focus on immune function and longer duration missions. For example, innate immune regulation was assessed in 102 construction workers adapting to living in the Arctic Circle for durations ranging from a few days to 2 years. Salivary lysozyme exhibited pronounced circadian rhythms peaking in the early morning, but over time the daily rhythms oscillated between steadily decreasing and u-shaped functions, with a decreasing rhythm at early mission (few days), a u-shaped rhythm at 6 months, decreasing rhythm at 18 months, and u-shaped rhythm at 2 years (Krasnozhenov, Vasil'ev, Smol'yaninov, & Fomina, 1974). A series of studies by Brown and colleagues (1974, 1975, 1977) examined the saliva of Skylab astronauts before and after their missions, as well as all phases of a 56-day isolation chamber analog mission. In the analog mission, they observed significant decreases in salivary flow rate and significantly higher IgA and lysozyme during the isolation and confinement period (Brown, Wheatcroft, Frome, & Rider, 1974). However, in the 28-, 59-, and 84-day Skylab spaceflight missions, they also observed significantly higher SIgA but lower lysozyme postflight compared to preflight (Brown, 1975; Brown et al., 1977).

Salivary bioscience in operational environments continued throughout the 1980s, although operational feasibility remained a challenge in ICE settings (e.g., Cotton, Little, & Johnston, 1998). However, Vickers, Hervig, Wallick, Poland, and Rubin (1987) found significant correlations between salivary cortisol and self-reported mood states including anger, depression, fear/anxiety, and happiness (negative correlation) in 39 Marine Corps recruits during 12-week basic training. Work in even more extreme environments came from Campbell et al. (1985), who successfully captured the circadian rhythms of cortisol in a team of four Arctic expeditioners traversing the Greenland ice cap on foot, and found an overall increase in cortisol compared to pre-mission baseline, with the peak of the rhythms tracking with changes in daily activity patterns. Conversely, a six-man team of climbers exhibited significant reductions in salivary glucocorticoids (cortisol + cortisone) and the sodium regulating hormone aldosterone after a 3-day ascent and 2 week stay at 4450 m (14,600 ft) altitude atop Mount Kenya (McLean, Booth, Tattersall, & Few, 1989).

The first known saliva collection in space was in 1984 as part of a series of studies by Cintrón and colleagues (Cintrón, 1987, 1991, Cintrón, Putcha, Chen, and Vanderploeg 1987; Cintrón, Putcha, and Parise, 1991; Cintrón, Putcha, and Vanderploeg, 1987), who investigated diurnal salivary cortisol rhythms and drug pharmacokinetics in astronauts during 7-day Space Shuttle missions (Johnson, 1986). In a precursor method to modern Salivette kits (Sarstedt, Inc.), samples were collected by chewing on a Teflon square to stimulate salivation, placing a cotton ball or dental roll between the jaws by the parotid gland to absorb saliva, then sealing the soaked cotton in a storage vial. Profound interindividual differences were observed in absolute levels of cortisol (ranging from ~0.10 to ~1.10 μg/dL), circadian rhythmicity, and changes in rhythms over the course of the missions, but overall cortisol levels were markedly higher in space than on Earth. The pharmacokinetic assessments following a 650 mg oral dose of acetaminophen revealed higher peak concentrations and longer time to reach peak in the space environment relative to preflight, supporting astronaut debrief reports of perceived longer latencies for medications to take effect in space. This was followed shortly by Leach, Inners, and Charles, who in 1985 used saliva samples to reveal a ~3% decrease in total body water in space relative to preflight (also see Leach, Inners, & Charles, 1991). Although the generalizability of these early studies was limited by modest sample size, this landmark work provided foundational spaceflight physiology data, confirmed that salivary bioscience was an operationally feasible and operationally acceptable approach in space, and set the stage for subsequent, ongoing, and future research in the final frontier.

## 24.4 Current Status of Knowledge in Military, Space, and Operational Research

The 1990s and early 2000s saw a dramatic surge in salivary bioscience research, particularly in multidisciplinary studies of stress, psychoneuroimmunology, circadian rhythms, and physical and behavioral health across laboratory, clinical, and field settings (Cacioppo et al., 2000; Kirschbaum & Hellhammer, 1994; Lawrence, 2002; Schwartz, Granger, Susman, Gunnar, & Laird, 1998). Application to extreme environments and populations was still relatively modest, but increasingly sophisticated and informative. In 1991–1992, Bernton, Hoover, Galloway, and Popp (1995) examined cortisol and DHEA responses to the extreme physical exertion, sleep restriction (~3.6 h/day), net caloric/energy loss, and hunger over 8-week US Army Ranger School classes, with saliva samples taken pre-mission and at the 4-week midpoint (Since approximately 50% of the incoming class does not complete the course). Under this extreme stress, salivary cortisol increased (0.30–0.71 μg/dL), with little to no increase in DHEA (22.3–24.5 ng/dL) relative to baseline, yielding a net significant increase in cortisol: DHEA ratio. Similar effects were observed by Hellhammer, Buchtal, Gutberlet, and Kirschbaum (1997) in 63 German Army recruits throughout their 6-week boot camp training, with evening salivary cortisol levels increasing throughout training (baseline to peak ~0.12 to ~0.24 μg/dL). Notably, responses to standardized social and physical stressors in the first week and post-training varied significantly by social dominance. Specifically, all subjects' responses were blunted post-training relative to early, with dominant group members exhibiting much higher cortisol reactivity than their subordinate counterparts in response to physical stress, but only early in training (peak ~1.18 vs. ~0.40 μg/d), with higher responses to social stress at both time points (peak ~0.76 vs. ~0.25 and ~0.45 vs. ~0.16, respectively).

In less physically extreme but more isolated and confined settings, Kelly et al. (1999) examined circadian rhythms in 12 US Navy submariners working alternating 18-h duty schedules in shifts of 6-h on/12-h off over a 6-week patrol mission. Salivary melatonin secretion phase-shifted by 13.5 h relative to clock time over the course of the mission, with an average endogenous circadian rhythm of 24.35 h that never entrained to 24 h despite social cues and knowledge of clock time. Interestingly, more abrupt changes in day to night shift schedule may impact melatonin secretion differently among individuals. Goh, Tong, Lim, Low, and Lee (2000) demonstrated a majority of naval personnel maintained a normal melatonin profile after a sudden change in work schedule, indicating high nocturnal levels of melatonin during working hours, which can contribute to sleepiness and decreased performance. A subgroup of individuals experienced abnormal disruptions in the melatonin rhythm during and after nightshift duties, suggesting these individuals are not suitable for night shift work. However, a small group of individuals had salivary melatonin profiles that indicated the beginnings of resynchronization and readjustment of the circadian rhythm to the new work schedule, making them potentially better candidates for night shift work. The various melatonin responses observed in this study demonstrate the variation in individual profiles that may be indicative of endogenous performance capacity and adaptability (Van Dongen, Baynard, Maislin, & Dinges, 2004). From 1995–1997, NASA conducted a series of 60- and 90-day isolation studies at Johnson Space Center, which included salivary assessments of physical and behavioral health parameters and further supported the utility of salivary bioscience in operational settings. For these eight crew members across both missions, the salivary melatonin peak in-mission was shifted by 2.7 h and correlated with reduced sleep duration (Putcha, Nimmagudda, & Rivera, 2002). Consistent with concurrent and future spaceflight studies, 35% of crew members' saliva samples revealed shedding of Epstein-Barr virus in-mission compared to 4% in healthy age-matched non-isolated control subjects, supporting the relevance of isolation and confinement stress per se on immune function in nonclinical populations independent of microgravity and other physiological effects of space (Pierson & Mehta, 2002). Focusing more on behavioral health, Sandal, Endresen, Vaernes, and Ursin (1999) examined the relationships between stress, coping styles, and salivary cortisone in submariners during 10- and 40-day missions. Compared with baseline, the crew showed higher levels of cortisone and self-reported stress during the missions; however, by the end of the 40-day mission, cortisone in crew with a positive coping profile was ~39% lower than that of their colleagues with a negative coping profile.

A vast majority of the current work in salivary biomarkers among military populations is concerned with biological signatures of behavioral health throughout the different stages of the military career. A recent study by Hellewell and Cernak (2018) investigated changes in salivary biomarkers across pre-deployment training, during deployment in Afghanistan, and readjustment 6–9 months after returning from deployment to determine markers of resilience and risk factors for maladaptation to stress. In conjunction with assessments of global health status, physical and mental functioning, resilience, and posttraumatic stress disorder, the investigators measured salivary cortisol, testosterone, cortisol/testosterone ratio, putative cancer biomarker chromogranin A (CgA), IgA, inflammatory marker C-reactive protein (CRP), and sympathetic nervous system activation marker alpha-amylase across the three-time points. Soldiers were categorized into one of three tertiles based on the number of salivary markers outside normal concentration ranges at pre-deployment. Individuals with few measures out-of-range and minimal changes across deployment were categorized as low stress-maladaptation risk. In contrast, individuals demonstrating several salivary markers outside of normal ranges, with significant alterations in salivary markers across deployment, were considered high risk for stress-maladaptation.

Developing behavioral health profiles using salivary biomarker technology can help identify soldiers who may be at an increased risk for adverse behavioral health based on occupational exposures. High levels of basal cortisol and testosterone concentrations have been shown to be potential protective factors against depression during deployment. Cobb, Josephs, Lancaster, Lee, and Telch (2018) measured pre-deployment levels of cortisol and testosterone in 120 US Army soldiers before and 30 min after a single inhalation of 35% $CO_2$/65% O2 gas, a technique used to elicit a physiological stress response, to examine the association of basal concentrations and reactivity to the 35% $CO_2$ challenge with reactions to war-zone stressors during deployment. Prior to deployment, soldiers were assessed for depressive symptoms and then asked to report occurrence of stressors and severity of depressive symptoms on a weekly basis during deployment. Higher basal testosterone and cortisol levels prior to deployment emerged as protective factors against depression during deployment, whereas elevated stress-induced testosterone levels predicted monthly depressive reaction increases amidst war-zone stressors. However, elevated stress-induced cortisol levels were only associated with depressive reactions during deployment in soldiers with elevated pre-deployment depression.

Cortisol concentrations in active duty military members of the Navy Sea, Air, and Land (SEAL) community were shown to increase on average by approximately 60% within the first 30 min of awakening across two consecutive weekdays (Taylor et al., 2016). Known as the cortisol awakening response (CAR), this increase in cortisol upon awakening to approximately 30 min thereafter has been associated with acute and chronic stress (Rodriguez, Bryce, Granger, & O'Haire, 2018; Taylor, Hernández, et al., 2016). Despite a nearly 60% increase in cortisol within the first half hour of awakening, concentrations at 60 min post awakening were similar to concentrations immediately upon awakening, indicating a quick recovery. In contrast, the anabolic hormones DHEA and testosterone appear to reach highest concentrations upon awakening and decrease across the day to evening nadir in this elite population (Taylor et al., 2016). CAR and other salivary biomarkers can also provide insight into military behavioral health post-deployment. In post-9/11 military veterans, having a PTSD service dog was associated with a significantly larger CAR compared to veterans on a service dog waitlist, even after controlling for factors that may influence cortisol output, suggesting that a higher CAR may be an indicator of better health and well-being (Rodriguez et al., 2018). Although assessing CAR may provide an important marker of health status and stress, more research is warranted to overcome sensitivity to confound and other methodological inconsistencies (Taylor, Hernández, et al., 2016). Another emerging application of salivary bioscience among the US veteran population is biological aging, which is known to be impacted by lifestyle factors (Blackburn, Epel, & Lin, 2015). Watkins et al. (2016) analyzed self-reported data and salivary samples from US veterans who participated in the National Health and Resilience in Veterans Study, particularly assessments of psychological distress and hostility in relation to telomere length, an indicator of biological aging. After controlling for sociodemographic, military, and clinical variables, greater scores of hostility were associated with telomere shortening. More specifically, each standard deviation in hostility score was associated with a 58% increase in likelihood of shorter telomere length. The investigators also discovered that hostility primarily manifested through difficulty controlling anger, highlighting the importance of behavioral health screening and early intervention to potentially mitigate long-term biological adverse effects (Maes, Galecki, Chang, & Berk, 2011).

The high physical demands of military operations are known to induce hormonal alterations (Oliver et al., 2015). In a 12-week block-periodized training program for Naval Special Warfare (NSW) Operators, salivary free testosterone and cortisol increased by approximately 20.3% and 20.8%, respectively, as intensity and training volume were increased during the second tertile of the training program. During the third block, free testosterone and cortisol concentrations decreased to similar levels observed at baseline, coinciding with a decrease in intensity and volume. This study demonstrates the synchrony between changes in training intensities and volumes and the hormonal response. One of the most extreme tests of operational performance, and the impact of physical and psychological stress, is military Survival, Evasion, Resistance, and Escape (SERE) school. This 2–3 week training prepares military personnel for the potential threat of being captured and how to survive in captivity and interrogation. Similar to Morgan et al. (2000) who demonstrated significant increases in salivary cortisol and decreases in testosterone during SERE training relative to baseline, Lieberman et al. (2016) showed significant increases in salivary cortisol, dehydroepiandrosterone sulfate (DHEA-s), and neuropeptide-Y (NPY) with a concurrent decrease in salivary testosterone during the capture and interrogation phases relative to pre-SERE baseline at the same time of day. A net effect of this was an inverted u-shaped function in cortisol: DHEA ratio, with increases during the capture and interrogation phases relative to baseline and post-mission. Although many of these biomarkers recovered upon cessation of the training relative to peak, a complete return to baseline was not observed and could potentially lead to long-term health consequences (Lieberman et al., 2016).

In spaceflight, Smith et al. (1999) used saliva samples to assess calcium metabolism and other markers of bone health following oral and intravenous doses of calcium preflight, inflight (at 110 days), and postflight in the three-man crew of the 115-day Mir 18 mission, and found significant reductions in calcium absorption from preflight to inflight. Although not a focus of the study, this work on bone health is also intriguing from a behavioral health perspective in light of recent evidence suggesting that the hormone osteocalcin, which is produced exclusively in bones (and detectable in saliva; McGehee & Johnson, 2004), can act on the brain and may mitigate anxiety and cognitive deficits (Obri, Khrimian, Karsenty, & Oury, 2018; Shan et al., 2019). This is particularly relevant to space exploration, where exposure to reduced gravity environments can decrease osteocalcin levels without sufficient exercise to produce and maintain it (Caillot-Augusseau et al., 2000; Garrett-Bakelman et al., 2019; Smith et al., 1999). Beginning with the Mir 18 mission, Payne, Mehta and colleagues assessed latent virus activation in spaceflight. In addition to preflight and postflight sampling, they used cotton roll Salivette kits for inflight collection of saliva in 11 astronauts over four 10–16 day shuttle missions and found significantly higher rates of Epstein-Barr virus shedding preflight (18% of samples) compared to the inflight (9%) and postflight (6%) mission phases (Payne, Mehta, Tyring, Stowe, & Pierson, 1999). Finally, starting in 1996, other projects by Sams, Shakleford, Stein, and colleagues that were part of the Shuttle-Mir program also used inflight saliva collection in pilot studies of acquired immunity, bone health, and protein metabolism in space (NASA, 1997; Uri & Lebedev, 2001).

Recent applications of salivary bioscience in spaceflight and other ICE environments follow from the foundational work described above, focusing largely on immunology, virology, and stress physiology, while expanding into other domains (e.g., Bilancio et al., 2019; Voorhies et al., 2017). Perhaps the biggest difference between all historical spaceflight missions and current and future operations is duration. Unlike early missions on the order of days or weeks up to 3 months, a standard mission length for the International Space Station (ISS) is 6 months. Initiatives to explore and potentially inhabit the Moon, Mars, and beyond only underscore the importance of mission duration on the order of years and the potential cumulative and interactive effects of space operations in the largely closed systems of spaceflight environments (Landon et al., 2019). This makes understanding the physical and behavioral health impacts of ICE environments that much more challenging, but all the more important.

Scientists at NASA Johnson Space Center developed a molecular-based, noninvasive assay that detects viral DNA in saliva. These studies were done in ground-based space analogs including Antarctica (Mehta, Pierson, Cooley, Dubow, & Lugg, 2000), NASA's underwater NEEMO habitat, and bed rest studies (analogs of spaceflight microgravity and fluid shift effects; Meck, Dreyer, & Warren, 2009). This work demonstrated that reactivation of Epstein-Barr virus—one of the most common latent herpes viruses—increased coincident with decreasing immunity (Payne et al., 1999; Pierson, Stowe, Phillips, Lugg, & Mehta, 2005). Subsequently, this noninvasive, sensitive, and specific assay was utilized with astronauts in spaceflight (Mehta et al., 2014, 2017). Epstein-Barr reactivated preflight and postflight, with the greatest amount of virus appearing inflight during shuttle missions, which correlated with decreased immunity. These studies were expanded to include cytomegalovirus (Mehta, Pierson, Cooley, Dubow, & Lugg, 2000; Mehta, Stowe, Feiveson, Tyring, & Pierson, 2000) and herpes simplex virus 1. Subsequent work shifted to Varicella-Zoster virus, the underlying cause of chickenpox and shingles, since this virus also represents an infectious disease risk among astronauts. Asymptomatic Varicella reactivation and shedding in saliva were demonstrated in astronauts (Mehta et al., 2004), and in a subsequent study was shown to be live and infectious (Cohrs, Mehta, Schmid, Gilden, & Pierson, 2008). Percent distributions of Epstein-Barr and Varicella viruses during shuttle and ISS spaceflight missions are presented in Fig. 24.1.

The most obvious sign of Varicella reactivation is vesicular rash and pain; however, in the absence of rash, the virus can also spread to the retina causing blindness, to the spinal cord causing paralysis and incontinence, and to cerebral arteries resulting in stroke (Kleinschmidt-DeMasters & Gilden, 2001; Orme et al., 2007). Although associating Varicella with disease is straightforward when rash and vesicles are apparent, correlating it with disease is difficult when the outward signs are absent. For example, when stroke occurs in the elderly, especially many months following zoster, the association with Varicella reactivation requires cerebral spinal fluid sampling for analysis of antibodies (Nagel et al., 2007). Likewise, detection of asymptomatic Varicella reactivation is often only seen as an increase in antibody titer against Varicella, and as such is difficult to detect in patients with acute herpes zoster, as the yield of cell DNA is greater in saliva collected by passive drool or synthetic swab than by cotton swab. The time to process saliva from collection to obtaining DNA is 1 h; Varicella DNA is present exclusively in the pelleted fraction of saliva and is found in 100% of patients before antiviral treatment. This rapid and sensitive method can be readily applied to subjects or patients with neurologic and other diseases that may be caused by Varicella in the absence of rash. For example, saliva assisted in a quick diagnosis of Varicella in a 41-year-old woman who unexpectedly developed skin lesions on her upper back and arm (Mehta et al., 2012). Immediate antiviral treatment led to a rapid recovery without the prolonged fatigue typically seen in adults with Varicella. This application of salivary bioscience

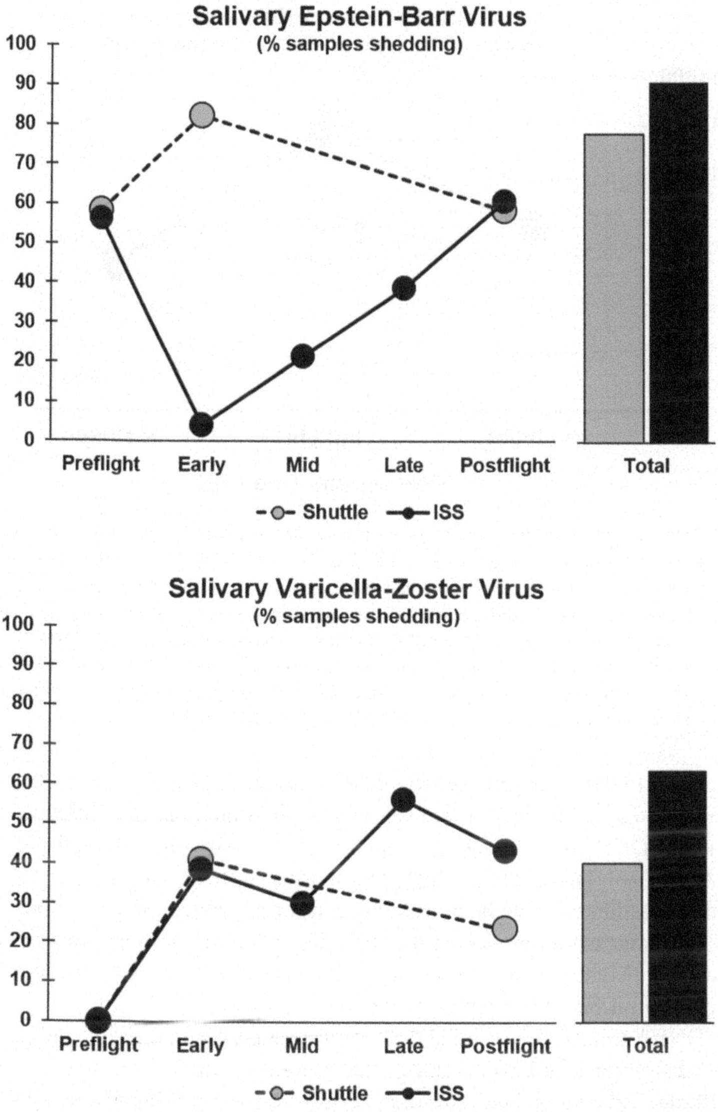


**Fig. 24.1** Prevalence rates (% of saliva samples) shedding Epstein-Barr virus (top) and Varicella-Zoster virus (bottom) preflight, inflight, postflight, and in total during spaceflight missions on the Space Shuttle (~14 days; $N = 89$) and the International Space Station (~6 months; $N = 23$). Shuttle missions included daily sampling inflight, whereas the longer ISS missions included inflight samplings at early, mid, and late mission time points. Figure adapted from Rooney, Crucian, Pierson, Laudenslager, and Mehta (2019)

could have considerable applied value as both a monitoring methodology and clinical application in isolated and confined operational environments.

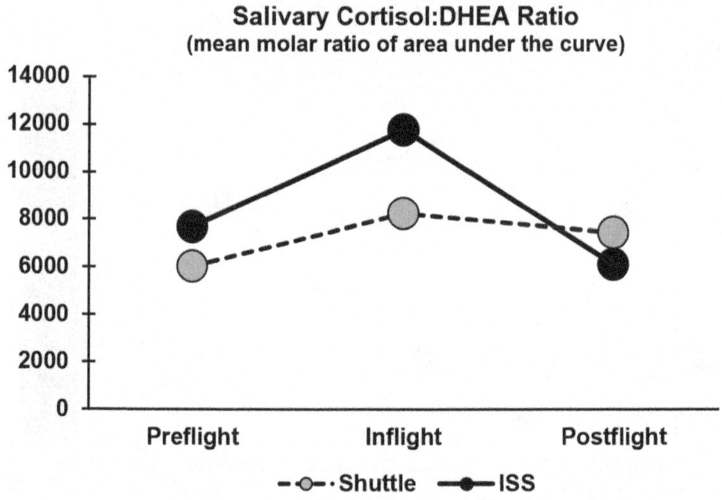


**Fig. 24.2** Salivary Cortisol:DHEA ratio in Astronauts before, during, and after spaceflight missions on the Space Shuttle (~14 days; $N = 17$) and the International Space Station (~6 months; $N = 10$). Each data point represents the mean molar ratio of area under the curve for Cortisol to DHEA. Both mission profiles produced significant increases from Preflight to Inflight, with a more pronounced increase in the longer-duration ISS missions. The increase in Cortisol:DHEA ratio may be associated with lower cellular immunity and innate immunity, potentially contributing to greater inflammatory cytokines that can negatively affect bone remodeling and growth and behavioral health. Figure adapted from data presented in Mehta et al. (2014, 2017)

Cortisol and DHEA are glucocorticoid steroid hormones released by the adrenal glands in response to stress. Cortisol is both anti-inflammatory and immunosuppressive, whereas DHEA is an antagonist to cortisol. Consequently, the ratio of cortisol to DHEA (cortisol: DHEA) is a valuable biomarker of stress and immune regulation. Recent flight studies tracked the diurnal rhythms of these hormones in saliva samples to evaluate changes as a function of flight phase, including preflight, launch, inflight, and postflight (Mehta et al., 2014). Salivary cortisol was present in significantly higher concentrations in samples taken inflight versus preflight and postflight. Salivary DHEA followed its normal daily decline at preflight, inflight, and postflight, but showed significantly lower waking concentrations inflight compared to preflight and postflight. Taken together, diurnal patterns of salivary cortisol were significantly higher inflight while DHEA was significantly lower. The cortisol area under the curve relative to ground (AUCg) did not significantly change inflight versus preflight, whereas the DHEA AUCg significantly declined. Ultimately, as seen in Fig. 24.2, this results in an increased cortisol: DHEA ratio during spaceflight, not unlike that observed in Army Ranger school, SERE training, and extended submarine operations training (Berntson et al., 1997; Lieberman et al., 2016; Lin et al., 2014). Increased cortisol: DHEA ratio has been linked to immune modulation (Christeff et al., 1997), including the increased inflammatory cytokine response and the T-helper cell shift observed in earlier spaceflight studies (Crucian et al., 2014; Mehta et al., 2013).

## 24.5 Methodological Issues, Challenges, and Considerations

As a research and clinical diagnostic tool, saliva has many advantages over blood and other biological samples. It is relatively easy to collect, store, and ship and can be obtained at low cost and in sufficient quantities for analysis. Nonetheless, salivary bioscience in ICE environments presents unique logistical challenges, especially for sample collection and analysis in the field. For both dimensions, critical constraints are limitations in power, mass, and volume requirements. This is especially relevant to spaceflight; in addition to inherent limits on the weight to fuel ratio necessary for sufficient thrust to exit Earth's gravity, the current cost of getting a payload to the ISS is approximately $10,000 USD per pound.

For saliva sample collection, NASA scientists developed a tool for the collection of human saliva samples for stress hormones assessments in spaceflight—the Saliva Procurement and Integrated Testing (SPIT) booklets (Fig. 24.3). SPIT booklets can be used in virtually any location, including space, and does not require any special conditions for storage or shipping. Saliva samples are collected on filter paper, five per booklet, and dried at room temperature. After drying, the booklets are placed in a

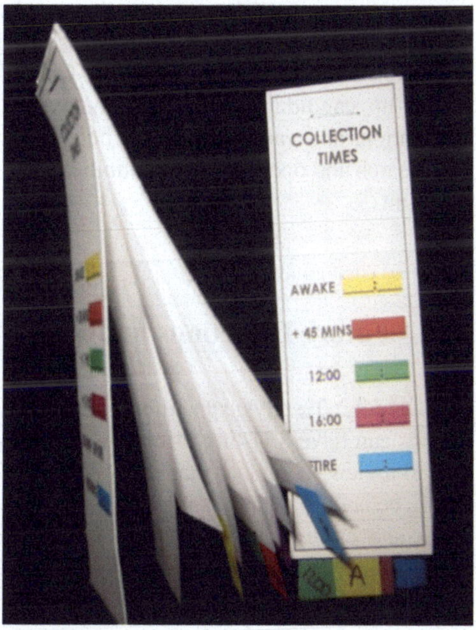


**Fig. 24.3** The NASA Saliva Procurement and Integrated Testing (SPIT) booklets for collection of saliva samples in field and operational settings

plastic zip bag and returned via mail. The samples are stable at room temperature for up to 6 months for cortisol and DHEA assay, thus eliminating the need for freezing. Filter extractions have been described previously (Neu, Goldstein, Gao, & Laudenslager, 2007), and steroid hormones are then measured using commercially available assay kits from Salimetrics, LLC. This novel method of collection of saliva samples is noninvasive, inexpensive, does not require specialized training, and robust to operational constraints and risks to sample integrity. The results are reliable, consistent, and correlate well with blood measurements.

For sample analysis, the continued miniaturization of sample analysis methodologies such as mass spectrometry, liquid chromatography, and genetic sequencing is all promising for field application (Chen et al., 2015; Mikheyev & Tin, 2014; Morrison, Watts, Hobbs, & Dawnay, 2018). However, ideally, sample collection and analysis would be one in the same process, with the ability to analyze samples in real time or near real time in a process akin to finger-stick tests for blood glucose or hemoglobin, requiring only a drop of saliva instead of a drop of blood. Aponte et al. (2006; Table 4) elegantly outlined the requirements for an ideal bioanalysis system in ICE environments, including: near real time, multiple analytes, minimally invasive, sensitive ($\leq$0.1 pg/mL), small sample volume (pL–μL), reduced or no bubble formation, gravity independent functionality, small size, low power consumption, operate at room temperature or have efficient temperature control, easily repaired or replaced, minimal waste, minimal sample preparation, driven by capillary action or micropump use, and minimal human interaction (automation desired). Not included in this list, but critically important for operational acceptability in ICE teams, is automating data analysis, optimizing data display/visualization, and ensuring safeguards and controls over data dissemination within and outside the crew. Advances in materials science, photonics, and smartphone technology as assay devices may lead to such a system (Kong et al., 2019; Krainak et al., 2018; Phare, Lee, Cardenas, & Lipson, 2015; Rey, Jain, Abdullah, Choudhury, & Erickson, 2018), but unfortunately, no candidate tools currently exist. Considerable research, development, verification, and validation would be required before deploying in operations, but the research and operations communities support this development and eagerly await the results.

## 24.6 Future Directions and Opportunities

Undoubtedly, the physical and psychological demands of working and living in ICE settings can have a lasting effect on physical and behavioral health. Even with our limited review, it is clear that multiple factors impact and are impacted by physical health, behavioral health, and performance factors in isolated, confined, and extreme teams and environments. Salivary bioscience has made invaluable contributions toward understanding the nature and effects of physical, environmental, and psychosocial stress in ICE mission settings.

Although the individual and combined effects of isolation, confinement, and various operationally relevant stressors on individual and team behavioral health and performance can be explored in laboratory experiments, analog environments, and field studies, the fact remains that one cannot model the passage of time. Scientists working with animal models have the luxury of examining the entire lifespan to better understand the development of basic biopsychosocial processes (Callaghan, Sullivan, Howell, & Tottenham, 2014), but compressing a long-duration operational team's lifecycle into a day, a week, a month, or a year is either not possible for those studying human performance, or carries with it considerable risk in extrapolation versus interpolation when the passage of time can serve as a significant transition event altering the trajectory of individual and team adaptation (Bliese, Adler, & Flynn, 2017). As Salas, Grossman, Hughes, and Coultas (2015) and Salas et al. (2015) note, research on mission-oriented and organizational team lifecycles is rarely longitudinal (e.g., Bartone & Adler, 1999), and mostly limited to cross-sectional designs. But if mission duration and the cumulative stress it naturally involves are as important as theory and the limited empirical evidence base suggest, then the need for continued long-duration studies and multidisciplinary operational behavioral health and performance monitoring is all the more critical (Kozlowski, 2015). To this end, we encourage continued work on team temporal dynamics and salivary biomarkers with high-fidelity space analog missions (Keeton et al., 2011) in NASA's Human Exploration Research Analog (HERA) facility, the Russian Ground-Based Experimental Complex (NEK/SIRIUS) chamber, the Hawaii Space Exploration Analog and Simulation (HI-SEAS) project, various polar outposts (e.g., Concordia and Neumayer stations), and military, first responders, elite athletes, medical units, and other high-performance teams in long-duration operational settings. In preparation for long-duration space exploration missions, we especially encourage the continued use of the International Space Station itself as a high-fidelity long-duration mission analog (Keeton, Slack, Schmidt, Ploutz-Snyder, & Baskin, 2012; Leveton et al., 2012), and developing permanent outposts on the Moon and near-Earth objects such as asteroids as ultrahigh fidelity biomedical and behavioral research analogs of long-duration deep space missions (Goswami et al., 2012).

In addition to continued work on temporal dynamics, an additional opportunity, if not imperative, for operational research is an integrated and multidisciplinary approach. At their most extreme, ICE operational settings are fully closed systems with highly interconnected components (Perrow, 1984), i.e., fully autonomous microsocieties within isolated ecosystems involving far more than just the physiological or psychological processes of the inhabitants (Anker, 2005; Brady, 1990, 2005, Brady & Emurian, 1982; Checinska et al., 2015; Emurian et al., 2009; Gitelson, Lisovsky, & MacElroy, 2003). A fully closed system with no outside input has even less flexibility than tightly coupled systems and potentially greater ripple effects of a disruption throughout the system. Insofar as ICE mission environments are closed systems, they are inherently "multidisciplinary" in that all components of the system can interact and potentially influence team dynamics through direct or indirect pathways. As such, integrating multidisciplinary data both horizontally across disciplines (i.e., physical health, behavioral health, performance, and environmental/engineering) as well as vertically through levels of analysis (molecular, physiological, behavioral, cognitive, and social) at high resolution over time can help support the identification of the most efficient, yet valid, methods of measuring constructs of interest, reducing measurement burden, and identifying targets for countermeasure development and implementation. Evidence-based, integrated monitoring and analysis, which should include noninvasive and objective biomarkers such as saliva, may help ICE teams and mission support personnel obtain comprehensive and more accurate assessments of team performance and functioning, individual health and well-being, and identify changing effects on the individuals within the team over time. Given the highly interactive nature of the largely closed systems that ICE environments are small interventions across multiple domains may be both more operationally acceptable and effective. For example, if the team collectively is fatigued due to an unexpected emergency waking them in the middle of the night, a multidisciplinary countermeasure package may include salivary biomarker measures to determine individual health and circadian status, adjust work and exercise schedules to allow recovery sleep based on need, adjust the habitat configuration for adequate privacy and lighting to support sleep, prioritize consumption of foods to enable sleep and provide sustained energy upon waking, etc., without any one countermeasure imposing an unacceptable or disruptive burden. Salivary bioscience can be a major contributor to understanding each individual crew member's unique biobehavioral systems and needs within a proactively individualized medicine approach, thereby allowing more effective countermeasures to be developed and customized at both the individual and team levels (Bradburne et al., 2015; Evans & Relling, 2004; Schmidt & Goodwin, 2013; Topol, 2014). Ultimately, the complexity in addressing the multiple pathways that increase risks to individual and team behavioral health and performance is challenging for researchers and practitioners alike. However, multiple pathways that increase risk also provide multiple pathways to reduce risk and maintain physical and behavioral health and performance for those who work, live, serve, and explore in extreme environments.

**Acknowledgments** PGR and SKM are supported by KBR's Human Health & Performance Contract NNJ15HK11B through the National Aeronautics and Space Administration. MEB, BCN, and PGR are supported in part by United States Army grant W81XWH-17-2-0070 (B. C. Nindl, PI) and NASA Human Research Program grant NNX15AK76A (D. F. Dinges, PI/PD). BEC is supported by the National Aeronautics and Space Administration, Human Health and Performance Directorate. The authors of this report are entirely responsible for its content and the decision to submit the work for publication. Any opinions, findings, and conclusions, or recommendations expressed in this material are those of the authors and do not necessarily reflect the views of the US Government, the National Aeronautics and Space Administration, the Department of Defense, KBR, JES Tech, or the University of Pittsburgh. The authors have no interests that may be perceived as conflicting with the work described or proposed here.

Author contributions: PGR conceived the project and organized the review. PGR, MEB, SKM, BCN, and BEC wrote the paper. All authors made substantial intellectual contributions and reviewed and approved the completed manuscript.

## References


Alekseyeva, O. G. (1966). *Some natural immunity factors and cosmonaut autoflora during the training period and following the flights of "Vostok," "Vostok 2," "Vostok 3," and "Vostok 4"*. NASA Technical Report 19660010006. https://ntrs.nasa.gov/search.jsp?R=19660010006

Anker, P. (2005). The ecological colonization of space. *Environmental History, 10*(2), 239–268.

Aponte, V. M., Finch, D. S., & Klaus, D. M. (2006). Considerations for non-invasive in-flight monitoring of astronaut immune status with potential use of MEMS and NEMS devices. *Life Sciences, 79*(14), 1317–1333.

Bartone, P. T., & Adler, A. B. (1999). Cohesion over time in a peacekeeping medical task force. *Military Psychology, 11*(1), 85–107.

Bernton, E., Hoover, D., Galloway, R., & Popp, K. (1995). Adaptation to chronic stress in military trainees: Adrenal androgens, testosterone, glucocorticoids, IGF-1, and immune function. *Annals of the New York Academy of Sciences, 774*(1), 217–231.

Berntson, G. G., Bigger, J. T., Eckberg, D. L., Grossman, P., Kaufmann, P. G., Malik, M., ... Van Der Molen, M. W. (1997). Heart rate variability: origins, methods, and interpretive caveats. *Psychophysiology, 34*(6), 623–648.

Bilancio, G., Cavallo, P., Lombardi, C., Guarino, E., Cozza, V., Giordano, F., & Cirillo, M. (2019). Urea and minerals monitoring in space missions by spot samples of saliva and urine. *Aerospace Medicine and Human Performance, 90*(1), 43–47.

Blackburn, E. H., Epel, E. S., & Lin, J. (2015). Human telomere biology: A contributory and interactive factor in aging, disease risks, and protection. *Science, 350*(6265), 1193–1198.

Bliese, P. D., Adler, A. B., & Flynn, P. J. (2017). Transition processes: A review and synthesis integrating methods and theory. *Annual Review of Organizational Psychology and Organizational Behavior, 4*, 263–286.

Bradburne, C., Graham, D., Kingston, H. M., Brenner, R., Pamuku, M., & Carruth, L. (2015). Overview of 'omics technologies for military occupational health surveillance and medicine. *Military Medicine, 180*(suppl_10), 34–48.

Brady, J. V. (1990). Toward applied behavior analysis of life aloft. *Behavioral Science, 35*(1), 11–23.

Brady, J. V. (2005). Behavioral health: The propaedeutic requirement. *Aviation, Space, and Environmental Medicine, 76*(6), B13–B24.

Brady, J. V., & Emurian, H. H. (1982). *Behavioral and biological interactions with small groups in confined microsocieties*. National Aeronautics and Space Administration technical report NASA-CR-169445. Houston, TX: NASA Johnson Space Center. https://ntrs.nasa.gov/archive/nasa/casi.ntrs.nasa.gov/19830002507.pdf

Brown, L. R. (1975). *Study to define and verify the personal oral hygiene requirements for extended manned space flight: Oral physiology and microbiology in Skylab manned space missions*. NASA Technical Report 19750021667. https://ntrs.nasa.gov/archive/nasa/casi.ntrs.nasa.gov/19750021667.pdf

Brown, L. R., Frome, W. J., Wheatcroft, M. G., Riggan, L. J., Bussell, N. E., & Johnston, D. A. (1977). The effect of Skylab on the chemical composition of saliva. *Journal of Dental Research, 56*(10), 1137–1143.

Brown, L. R., Wheatcroft, M. G., Frome, W. J., & Rider, L. J. (1974). Effects of a simulated skylab mission on the oral health of astronauts. *Journal of Dental Research, 53*(5), 1268–1275.

Bruger, M., & Member, S. (1943). On the excretion of iodine in the saliva. *American Journal of Physiology-Legacy Content, 139*(2), 212–216.

Byers, A. L., & Yaffe, K. (2011). Depression and risk of developing dementia. *Nature Reviews Neurology, 7*(6), 323–331.

Cacioppo, J. T., Ernst, J. M., Burleson, M. H., McClintock, M. K., Malarkey, W. B., Hawkley, L. C., ... Spiegel, D. (2000). Lonely traits and concomitant physiological processes: The MacArthur social neuroscience studies. *International Journal of Psychophysiology, 35*(2–3), 143–154.

Caillot-Augusseau, A., Vico, L., Heer, M., Voroviev, D., Souberbielle, J. C., Zitterman, A., … Lafage-Proust, M. H. (2000). Space flight is associated with rapid decreases of undercarboxylated osteocalcin and increases of markers of bone resorption without changes in their circadian variation: Observations in two cosmonauts. *Clinical Chemistry, 46*(8), 1136–1143.
Callaghan, B. L., Sullivan, R. M., Howell, B., & Tottenham, N. (2014). The international society for developmental psychobiology Sackler symposium: Early adversity and the maturation of emotion circuits—A cross-species analysis. *Developmental Psychobiology, 56*(8), 1635–1650.
Campbell, I. T., Wilson, D. W., Walker, R. F., & Griffiths, K. (1985). The use of salivary steroids to monitor circadian rhythmicity on expeditions in the arctic. *Chronobiology International, 2*(1), 55–59.
Checinska, A., Probst, A. J., Vaishampayan, P., White, J. R., Kumar, D., Stepanov, V. G., … Venkateswaran, K. (2015). Microbiomes of the dust particles collected from the International Space Station and Spacecraft Assembly Facilities. *Microbiome, 3*(1), 50.
Chen, C. H., Lin, Z., Tian, R., Shi, R., Cooks, R. G., & Ouyang, Z. (2015). Real-time sample analysis using a sampling probe and miniature mass spectrometer. *Analytical Chemistry, 87*(17), 8867–8873.
Christeff, N., Gherbi, N., Mammes, O., Dalle, M. T., Gharakhanian, S., Lortholary, O., … Nunez, E. A. (1997). Serum cortisol and DHEA concentrations during HIV infection. *Psychoneuroendocrinology, 22*, S11–S18.
Cintrón, N. M. (1987). Salivary cortisol levels during the acute phases of spaceflight. In M. W. Bungo, T. M. Bagian, M. A. Bowman, & B. M. Levitan (Eds.), *Results of the life sciences DSOs conducted aboard the space shuttle 1981–1986* (pp. 31–34). Houston, TX: National Aeronautics and Space Administration.
Cintrón, N. M. (1991). Salivary cortisol levels during the acute phases of spaceflight. In M. W. Bungo, T. M. Bagian, M. A. Bowman, & B. M. Levitan (Eds.), *Results of life sciences DSOs conducted aboard the space shuttle 1988–1990* (pp. 51–54). Houston, TX: National Aeronautics and Space Administration.
Cintrón, N. M., Putcha, L., Chen, Y.-M., & Vanderploeg, J. M. (1987). Inflight salivary pharmacokinetics of scopolamine and dextroamphetamine. In M. W. Bungo, T. M. Bagian, M. A. Bowman, & B. M. Levitan (Eds.), *Results of the life sciences DSOs conducted aboard the space shuttle 1981–1986* (pp. 25–29). Houston, TX: National Aeronautics and Space Administration.
Cintrón, N. M., Putcha, L., & Parise, C. (1991). In-flight pharmacokinetics of acetaminophen in saliva. In M. W. Bungo, T. M. Bagian, M. A. Bowman, & B. M. Levitan (Eds.), *Results of life sciences DSOs conducted aboard the space shuttle 1988–1990* (pp. 55–58). Houston, TX: National Aeronautics and Space Administration.
Cintrón, N. M., Putcha, L., & Vanderploeg, J. M. (1987). Inflight pharmacokinetics of acetaminophen in saliva. In M. W. Bungo, T. M. Bagian, M. A. Bowman, & B. M. Levitan (Eds.), *Results of the life sciences DSOs conducted aboard the space shuttle 1981–1986* (pp. 19–23). Houston, TX: National Aeronautics and Space Administration.
Claustrat, B., Brun, J., & Chazot, G. (2005). The basic physiology and pathophysiology of melatonin. *Sleep Medicine Reviews, 9*(1), 11–24.
Cobb, A. R., Josephs, R. A., Lancaster, C. L., Lee, H. J., & Telch, M. J. (2018). Cortisol, testosterone, and prospective risk for war-zone stress-evoked depression. *Military Medicine, 183*(11–12), e535–e545.
Cohrs, R. J., Mehta, S. K., Schmid, D. S., Gilden, D. H., & Pierson, D. L. (2008). Asymptomatic reactivation and shed of infectious varicella zoster virus in astronauts. *Journal of Medical Virology, 80*(6), 1116–1122.
Cotton, A. J., Little, G. P., & Johnston, I. D. (1998). *The measurement of combat stress in the field: A product development study* (No. AB-33-PAPER). Navy Advancement Center.
Crucian, B., Simpson, R. J., Mehta, S., Stowe, R., Chouker, A., Hwang, S. A., … Sams, C. (2014). Terrestrial stress analogs for spaceflight associated immune system dysregulation. *Brain, Behavior, and Immunity, 39*, 23–32.

Dantzer, R., & Kelley, K. W. (1989). Stress and immunity: An integrated view of relationships between the brain and the immune system. *Life Sciences, 44*(26), 1995–2008.

Davis, S. W., & Taylor, J. G. (1954). *Stress in infantry combat*. Technical memorandum ORO-T-295. Chevy Chase, MD: Johns Hopkins University Operations Research Office. http://www.dtic.mil/dtic/tr/fulltext/u2/069455.pdf

Derntl, B., Hack, R. L., Kryspin-Exner, I., & Habel, U. (2013). Association of menstrual cycle phase with the core components of empathy. *Hormones and Behavior, 63*(1), 97–104.

Dietz, A. S., Weaver, S. J., Sierra, M. J., Bedwell, W. L., Salas, E., Fiore, S. M., . . . Driskell, J. (2010). Unpacking the temporal and interactive effects of stress on individual and team performance. *Human Factors & Ergonomics Society Annual Proceedings, 54*, 1017–1021.

Dowlati, Y., Herrmann, N., Swardfager, W., Liu, H., Sham, L., Reim, E. K., & Lanctôt, K. L. (2010). A meta-analysis of cytokines in major depression. *Biological Psychiatry, 67*(5), 446–457.

Edwards, D. A., Wetzel, K., & Wyner, D. R. (2006). Intercollegiate soccer: Saliva cortisol and testosterone are elevated during competition, and testosterone is related to status and social connectedness with teammates. *Physiology & Behavior, 87*(1), 135–143.

Eisenegger, C., Naef, M., Snozzi, R., Heinrichs, M., & Fehr, E. (2010). Prejudice and truth about the effect of testosterone on human bargaining behaviour. *Nature, 463*(7279), 356–359.

Eisenegger, C., Haushofer, J., & Fehr, E. (2011). The role of testosterone in social interaction. *Trends in Cognitive Science, 15*(6), 263–271.

Emurian, H. H., Canfield, K., Roma, P. G., Gasior, E. D., Brinson, Z. S., Hienz, R. D., Hursh, S. R., & Brady, J. V. (2009). Behavioral systems management of confined microsocieties: An agenda for research and applications. In *Proceedings of the 39th international conference on environmental systems* (paper number 2009-01-2423). Warrendale, PA: SAE International.

Evans, W. E., & Relling, M. V. (2004). Moving towards individualized medicine with pharmacogenomics. *Nature, 429*(6990), 464–468.

Garrett-Bakelman, F. E., Darshi, M., Green, S. J., Gur, R. C., Lin, L., Macias, B. R., & Piening, B. D. (2019). The NASA twins study: A multidimensional analysis of a year-long human spaceflight. *Science, 364*(6436), eaau8650.

Gatti, R., & De Palo, E. F. (2011). An update: Salivary hormones and physical exercise. *Scandinavian Journal of Medicine and Science in Sports, 21*(2), 157–169.

Gianaros, P. J., Kuan, D. C. H., Marsland, A. L., Sheu, L. K., Hackman, D. A., Miller, K. G., & Manuck, S. B. (2017). Community socioeconomic disadvantage in midlife relates to cortical morphology via neuroendocrine and cardiometabolic pathways. *Cerebral Cortex, 27*(1), 460–473.

Gitelson, J. I., Lisovsky, G. M., & MacElroy, R. D. (2003). *Manmade closed ecological systems*. New York: CRC Press.

Goh, V. H. H., Tong, T. Y. Y., Lim, C. L., Low, E. C. T., & Lee, L. K. H. (2000). Circadian disturbances after night-shift work onboard a naval ship. *Military Medicine, 165*(2), 101–105.

Goswami, N., Batzel, J. J., Clément, G., Stein, T. P., Sharp, M. K., Blaber, A. P., . . . Hinghofer-Szalkay, H. G. (2013). Maximizing information from space data resources: A case for expanding integration between research disciplines. *European Journal of Applied Physiology, 113*(7), 1645–1654.

Goswami, N., Roma, P. G., De Boever, P., Clément, G., Hargens, A. R., Loeppky, J. A., . . . Hinghofer-Szalkay, H. G. (2012). Using the Moon as a high-fidelity analogue environment to study biological and behavioural mechanisms of long duration space exploration. *Planetary and Space Science, 74*(1), 111–120.

Hawkins, G. R., & Zipkin, I. (1964). Effects of prolonged residence in the Antarctica upon some organic constituents of human parotid saliva. *Proceedings of the Society for Experimental Biology and Medicine, 117*(3), 888–891.

Hellewell, S. C., & Cernak, I. (2018). Measuring resilience to operational stress in Canadian armed forces personnel. *Journal of Traumatic Stress, 31*(1), 89–101.

Hellhammer, D. H., Buchtal, J., Gutberlet, I., & Kirschbaum, C. (1997). Social hierarchy and adrenocortical stress reactivity in men. *Psychoneuroendocrinology, 22*(8), 643–650.

Infurna, F. J., & Luthar, S. S. (2018). Re-evaluating the notion that resilience is commonplace: A review and distillation of directions for future research, practice, and policy. *Clinical Psychology Review, 65*, 43–56.

Jenkins, T., Nguyen, J., Polglaze, K., & Bertrand, P. (2016). Influence of tryptophan and serotonin on mood and cognition with a possible role of the gut-brain axis. *Nutrients, 8*(1), 56. https://doi.org/10.3390/nu8010056

Johnson, P. (1986). Saliva: A way to obtain specimens in operational settings. *Aviation, Space, and Environmental Medicine, 57*(4), 397.

Jonas, W. B., O'Connor, F. G., Deuster, P., Peck, J., Shake, C., & Frost, S. S. (2010). Why total force fitness? *Military Medicine, 175*(suppl_8), 6–13.

Katakura, A., Kamiyama, I., Takano, N., Shibahara, T., Muramatsu, T., Ishihara, K., . . . Shouno, T. (2007). Comparison of salivary cytokine levels in oral cancer patients and healthy subjects. *The Bulletin of Tokyo Dental College, 48*(4), 199–203.

Kearney, A. R. (2016). *Team health and performance in spaceflight habitats*. NASA/TM-2016-219274. Houston, TX: NASA Johnson Space Center.

Keeton, K. E., Slack, K., Schmidt, L. L., Ploutz-Snyder, R., & Baskin, P. (2012). An examination of autonomy utilizing ISS as a testbed for analog research: Collaborations between research and operations. *Aviation, Space, and Environmental Medicine, 83*(3).

Keeton, K. E., Whitmire, A., Feiveson, A. H., Ploutz-Snyder, R., Leveton, L. B., & Shea, C. (2011). *Analog assessment tool*. NASA technical report NASA/TP2011-216146. Houston, TX: NASA Johnson Space Center.

Kelly, T. L., Neri, D. F., Grill, J. T., Ryman, D., Hunt, P. D., Dijk, D. J., . . . Czeisler, C. A. (1999). Nonentrained circadian rhythms of melatonin in submariners scheduled to an 18-hour day. *Journal of Biological Rhythms, 14*(3), 190–196.

Kirschbaum, C., & Hellhammer, D. H. (1994). Salivary cortisol in psychoneuroendocrine research: Recent developments and applications. *Psychoneuroendocrinology, 19*(4), 313–333.

Kleinschmidt-DeMasters, B. K., & Gilden, D. H. (2001). Varicella-zoster virus infections of the nervous system: Clinical and pathologic correlates. *Archives of Pathology and Laboratory Medicine, 125*(6), 770–780.

Kong, T., You, J. B., Zhang, B., Nguyen, B., Tarlan, F., Jarvi, K., & Sinton, D. (2019). Accessory-free quantitative smartphone imaging of colorimetric paper-based assays. *Lab on a Chip, 19*(11), 1991–1999.

Kozlowski, S. W. (2015). Advancing research on team process dynamics: Theoretical, methodological, and measurement considerations. *Organizational Psychology Review, 5*(4), 270–299.

Krainak, M., Stephen, M., Klamkin, J., Bergman, K., Lipson, M., Mookherjea, S., . . . & Matsko, A. (2018). NASA integrated photonics. In *2018 IEEE Photonics Conference (IPC)* (pp. 1–2). https://doi.org/10.1109/IPCon.2018.8527187

Krasnozhenov, E. P., Vasil'ev, N. V., Smol'yaninov, E. S., & Fomina, T. I. (1974). Dynamics of bactericidal properties of human saliva and blood serum during adaptation to arctic conditions. *Bulletin of Experimental Biology and Medicine, 78*(4), 1165–1167.

Landon, L. B., Douglas, G. L., Downs, M. E., Greene, M. R., Whitmire, A. M., Zwart, S. R., & Roma, P. G. (2019). The behavioral biology of teams: Multidisciplinary contributions to social dynamics in isolated, confined, and extreme environments. *Frontiers in Psychology*. https://doi.org/10.3389/fpsyg.2019.02571

Lavik, P. S., Matthews, L. W., Buckaloo, G. W., Lemm, F. J., Spector, S., & Friedell, H. L. (1952). Use of I131-labeled protein in the study of protein digestion and absorption in children with and without cystic fibrosis of the pancreas. *Pediatrics, 10*(6), 667–676.

Lawrence, H. P. (2002). Salivary markers of systemic disease: Noninvasive diagnosis of disease and monitoring of general health. *Journal of the Canadian Dental Association, 68*(3), 170–175.

Lazarus, R. S., & Folkman, S. (1984). Coping and adaptation. In W. D. Gentry (Ed.), *The handbook of behavioral medicine* (pp. 282–325). New York: Guilford.

Leach, C. S., Inners, L. D., & Charles, J. B. (1991). Changes in total body water during spaceflight. *The Journal of Clinical Pharmacology, 31*(10), 1001–1006.

Leveton, L. B., Keeton, K. E., Schmidt, L. L., Slack, K. J., Baskin, P., Ploutz-Snyder, R., & Palinkas, L. (2012). *Examination of communication delays on team performance: Utilizing the ISS as a Testbed for analog research*. Houston, TX: NASA Johnson Space Center. https://ntrs.nasa.gov/archive/nasa/casi.ntrs.nasa.gov/20120007707.pdf

Libbin, S. S. (1943). Construction of a simple saliva ejector apparatus in the field. *The Military Surgeon (United States), 92*(4), 435–436.

Lieberman, H. R., Farina, E. K., Caldwell, J., Williams, K. W., Thompson, L. A., Niro, P. J., … McClung, J. P. (2016). Cognitive function, stress hormones, heart rate and nutritional status during simulated captivity in military survival training. *Physiology and Behavior, 165*, 86–97.

Liening, S. H., Stanton, S. J., Saini, E. K., & Schultheiss, O. C. (2010). Salivary testosterone, cortisol, and progesterone: Two-week stability, interhormone correlations, and effects of time of day, menstrual cycle, and oral contraceptive use on steroid hormone levels. *Physiology and Behavior, 99*(1), 8–16.

Lin, S., Cheng, X., Zhao, M., Wei, N. I., Chen, G., Yang, Z., & Xiaodan, Y. U. (2014). Evaluation of performance stress among military submarine soldiers by using salivary cortisol, DHEA-S and cortisol/DHEA-S ratio assay. *Military Medical Sciences, 7*, 495–498.

Lloyd, R. B., & Nemeroff, C. B. (2011). The role of corticotropin-releasing hormone in the pathophysiology of depression: Therapeutic implications. *Current Topics in Medicinal Chemistry, 11*(6), 609–617.

Maes, M., Galecki, P., Chang, Y. S., & Berk, M. (2011). A review on the oxidative and nitrosative stress (O&NS) pathways in major depression and their possible contribution to the (neuro) degenerative processes in that illness. *Progress in Neuro-Psychopharmacology and Biological Psychiatry, 35*(3), 676–692.

McEwen, B. S., & Gianaros, P. J. (2010). Central role of the brain in stress and adaptation: Links to socioeconomic status, health, and disease. *Annals of the New York Academy of Sciences, 1186* (1), 190–222.

McEwen, B. S., Gray, J. D., & Nasca, C. (2015). Recognizing resilience: Learning from the effects of stress on the brain. *Neurobiology of Stress, 1*, 1–11.

McEwen, B. S., & Karatsoreos, I. N. (2012). What is stress? In *Stress challenges and immunity in space* (pp. 11–29). Berlin: Springer.

McGehee, J. W., Jr., & Johnson, R. B. (2004). Biomarkers of bone turnover can be assayed from human saliva. *The Journals of Gerontology Series A: Biological Sciences and Medical Sciences, 59*(3), B196–B200.

McLean, C. J., Booth, C. W., Tattersall, T., & Few, J. D. (1989). The effect of high altitude on saliva aldosterone and glucocorticoid concentrations. *European Journal of Applied Physiology and Occupational Physiology, 58*(4), 341–347.

Meck, J. V., Dreyer, S. A., & Warren, L. E. (2009). Long-duration head-down bed rest: Project overview, vital signs, and fluid balance. *Aviation, Space, and Environmental Medicine, 80*(5), A01–A08.

Mehta, S. K., Cohrs, R. J., Forghani, B., Zerbe, G., Gilden, D. H., & Pierson, D. L. (2004). Stress-induced subclinical reactivation of varicella zoster virus in astronauts. *Journal of Medical Virology, 72*(1), 174–179.

Mehta, S. K., Crucian, B. E., Stowe, R. P., Simpson, R. J., Ott, C. M., Sams, C. F., & Pierson, D. L. (2013). Reactivation of latent viruses is associated with increased plasma cytokines in astronauts. *Cytokine, 61*(1), 205–209.

Mehta, S. K., Gilden, D., Crucian, B. E., Sams, C. F., Cohrs, R. J., & Pierson, D. L. (2012). A case report: PCR-assisted diagnosis of varicella in an adult. *Open Journal of Medical Microbiology, 2*(3), 131–132. https://doi.org/10.4236/ojmm.2012.23019

Mehta, S. K., Laudenslager, M. L., Stowe, R. P., Crucian, B. E., Feiveson, A. H., Sams, C. F., & Pierson, D. L. (2017). Latent virus reactivation in astronauts on the international space station. *NPJ Microgravity, 3*(1), 11.

Mehta, S. K., Laudenslager, M. L., Stowe, R. P., Crucian, B. E., Sams, C. F., & Pierson, D. L. (2014). Multiple latent viruses reactivate in astronauts during space shuttle missions. *Brain, Behavior, and Immunity, 41*, 210–217.

Mehta, S. K., Pierson, D. L., Cooley, H., Dubow, R., & Lugg, D. (2000). Epstein-Barr virus reactivation associated with diminished cell-mediated immunity in Antarctic expeditioners. *Journal of Medical Virology, 61*(2), 235–240.

Mehta, S. K., Stowe, R. P., Feiveson, A. H., Tyring, S. K., & Pierson, D. L. (2000). Reactivation and shedding of cytomegalovirus in astronauts during spaceflight. *The Journal of Infectious Diseases, 182*(6), 1761–1764.

Mikheyev, A. S., & Tin, M. M. (2014). A first look at the Oxford Nanopore MinION sequencer. *Molecular Ecology Resources, 14*(6), 1097–1102.

Morgan, C. A., III, Wang, S., Mason, J., Southwick, S. M., Fox, P., Hazlett, G., ... Greenfield, G. (2000). Hormone profiles in humans experiencing military survival training. *Biological Psychiatry, 47*(10), 891–901.

Morrison, J., Watts, G., Hobbs, G., & Dawnay, N. (2018). Field-based detection of biological samples for forensic analysis: Established techniques, novel tools, and future innovations. *Forensic Science International, 285*, 147–160.

Myant, N. B. (1960). Iodine metabolism of salivary glands. *Annals of the New York Academy of Sciences, 85*(1), 208–214.

Nagel, M. A., Forghani, B., Mahalingam, R., Wellish, M. C., Cohrs, R. J., Russman, A. N., ... Gilden, D. H. (2007). The value of detecting anti-VZV IgG antibody in CSF to diagnose VZV vasculopathy. *Neurology, 68*(13), 1069–1073.

National Aeronautics and Space Administration. (1997). *Phase 1A research program Postflight science report*. Houston, TX: NASA Johnson Space Center. https://history.nasa.gov/SP-4225/science/issphase1sci.pdf

Neu, M., Goldstein, M., Gao, D., & Laudenslager, M. L. (2007). Salivary cortisol in preterm infants: Validation of a simple method for collecting saliva for cortisol determination. *Early Human Development, 83*(1), 47–54.

O'Connor, M. F., Irwin, M. R., & Wellisch, D. K. (2009). When grief heats up: Pro-inflammatory cytokines predict regional brain activation. *NeuroImage, 47*(3), 891–896.

Obri, A., Khrimian, L., Karsenty, G., & Oury, F. (2018). Osteocalcin in the brain: From embryonic development to age-related decline in cognition. *Nature Reviews Endocrinology, 14*(3), 174–182.

Oliver, J. M., Abt, J. P., Sell, T. C., Beals, K., Wood, D. E., & Lephart, S. M. (2015). Salivary hormone response to 12-week block-periodized training in naval special warfare operators. *The Journal of Strength and Conditioning Research, 29*(1), 66–73.

Orme, H. T., Smith, A. G., Nagel, M. A., Bert, R. J., Mickelson, T. S., & Gilden, D. H. (2007). VZV spinal cord infarction identified by diffusion-weighted MRI (DWI). *Neurology, 69*(4), 398–400.

Payne, D. A., Mehta, S. K., Tyring, S. K., Stowe, R. P., & Pierson, D. L. (1999). Incidence of Epstein-Barr virus in astronaut saliva during spaceflight. *Aviation, Space, and Environmental Medicine, 70*(12), 1211.

Perrow, C. (1984). *Normal accidents: Living with high risk systems*. New York: Basic Books.

Phare, C. T., Lee, Y. H. D., Cardenas, J., & Lipson, M. (2015). Graphene electro-optic modulator with 30 GHz bandwidth. *Nature Photonics, 9*(8), 511.

Pierson, D. L., & Mehta, S. K. (2002). Reactivation of latent viruses. In H. W. Lane, R. L. Sauer, & D. L. Feeback (Eds.), *Isolation: NASA experiments in closed-environment living—Advanced human life support enclosed system* (pp. 343–355). San Diego, CA: American Astronautical Society.

Pierson, D. L., Stowe, R. P., Phillips, T. M., Lugg, D. J., & Mehta, S. K. (2005). Epstein–Barr virus shedding by astronauts during space flight. *Brain, Behavior, and Immunity, 19*(3), 235–242.

Putcha, L., Nimmagudda, R., & Rivera, C. (2002). Assessment of sleep dynamics in a simulated space station environment. In H. W. Lane, R. L. Sauer, & D. L. Feeback (Eds.), *Isolation: NASA experiments in closed-environment living—Advanced human life support enclosed system* (pp. 131–139). San Diego, CA: American Astronautical Society.

Radley, J., Morilak, D., Viau, V., & Campeau, S. (2015). Chronic stress and brain plasticity: Mechanisms underlying adaptive and maladaptive changes and implications for stress-related CNS disorders. *Neuroscience and Biobehavioral Reviews, 58*, 79–91.

Rey, E., Jain, A., Abdullah, S., Choudhury, T., & Erickson, D. (2018). Personalized stress monitoring: A smartphone-enabled system for quantification of salivary cortisol. *Personal and Ubiquitous Computing, 22*(4), 867–877.

Rodahl, K. (1961). Human performance in the cold. In H. Spector, J. Brožek, & M. S. Peterson (Eds.), *Performance capacity—A symposium* (pp. 189–192). Washington, DC: Department of the Army, Research and Development Command.

Rodriguez, K. E., Bryce, C. I., Granger, D. A., & O'Haire, M. E. (2018). The effect of a service dog on salivary cortisol awakening response in a military population with posttraumatic stress disorder (PTSD). *Psychoneuroendocrinology, 98*, 202–210.

Rooney, B. V., Crucian, B. E., Pierson, D. L., Laudenslager, M. L., & Mehta, S. K. (2019). Herpes virus reactivation in astronauts during spaceflight and its application on earth. *Frontiers in Microbiology, 10*, 16. https://doi.org/10.3389/fmicb.2019.00016

Rothwell, N. J., & Hopkins, S. J. (1995). Cytokines and the nervous system II: Actions and mechanisms of action. *Trends in Neurosciences, 18*(3), 130–136.

SahebJamee, M., Eslami, M., AtarbashiMoghadam, F., & Sarafnejad, A. (2008). Salivary concentration of TNF alpha, IL1 alpha, IL6, and IL8 in oral squamous cell carcinoma. *Medicina Oral Patologia Oral y Cirugia Bucal, 13*(5), 292–295.

Salas, E., Grossman, R., Hughes, A. M., & Coultas, C. W. (2015). Measuring team cohesion: Observations from the science. *Human Factors, 57*(3), 365–374.

Salas, E., Tannenbaum, S. I., Kozlowski, S. W. J., Miller, C., Mathieu, J. E., & Vessey, W. B. (2015). Teams in space exploration: A new frontier for the science of team effectiveness. *Current Directions in Psychological Science, 24*(3), 200207.

Sandal, G. M., Endresen, I. M., Vaernes, R., & Ursin, H. (1999). Personality and coping strategies during submarine missions. *Military Psychology, 11*(4), 381–404.

Schmidt, M. A., & Goodwin, T. J. (2013). Personalized medicine in human space flight: Using Omics based analyses to develop individualized countermeasures that enhance astronaut safety and performance. *Metabolomics, 9*(6), 1134–1156.

Schultheiss, O. C., Dargel, A., & Rohde, W. (2003). Implicit motives and gonadal steroid hormones: Effects of menstrual cycle phase, oral contraceptive use, and relationship status. *Hormones and Behavior, 43*(2), 293–301.

Schwartz, E. B., Granger, D. A., Susman, E. J., Gunnar, M. R., & Laird, B. (1998). Assessing salivary cortisol in studies of child development. *Child Development, 69*(6), 1503–1513.

Shan, C., Ghosh, A., Guo, X. Z., Wang, S. M., Hou, Y. F., Li, S. T., & Liu, J. M. (2019). Roles for osteocalcin in brain signalling: Implications in cognition-and motor-related disorders. *Molecular Brain, 12*(1), 23. https://doi.org/10.1186/s13041-019-0444-5

Smith, R. P., Coward, R. M., Kovac, J. R., & Lipshultz, L. I. (2013). The evidence for seasonal variations of testosterone in men. *Maturitas, 74*(3), 208–212.

Smith, S. M., Wastney, M. E., Morukov, B. V., Larina, I. M., Nyquist, L. E., Abrams, S. A., … Rice, B. L. (1999). Calcium metabolism before, during, and after a 3-mo spaceflight: Kinetic and biochemical changes. *American Journal of Physiology-Regulatory, Integrative and Comparative Physiology, 277*(1), R1–R10.

Suedfeld, P. (2001). Applying positive psychology in the study of extreme environments. *Journal of Human Performance in Extreme Environments, 6*(1), 21–25.

Taylor, M. K., Hernández, L. M., Fuller, S. A., Sargent, P., Padilla, G. A., & Harris, E. (2016). Cortisol awakening response in elite military men: Summary parameters, stability measurement, and effect of compliance. *Military Medicine, 181*(11–12), e1600–e1607.

Taylor, G. R., Kropp, K. D., Henney, M. R., Ekblad, S. S., Baky, A. A., Groves, T. O., … Funderburk, N. R. (1977). Microbial exchange (experiment AR-002). In *Apollo-Soyuz test project summary science report, volume I: Astronomy, earth atmosphere and gravity field, life sciences, and materials processing* (pp. 237–256). Houston, TX: NASA Johnson Space Center.

Taylor, M. K., Kviatkovsky, S. A., Hernández, L. M., Sargent, P., Segal, S., & Granger, D. A. (2016). Anabolic hormone profiles in elite military men. *Steroids, 110*, 41–48.

Thayer, J. F., Åhs, F., Fredrikson, M., Sollers, J. J., III, & Wager, T. D. (2012). A meta-analysis of heart rate variability and neuroimaging studies: Implications for heart rate variability as a marker of stress and health. *Neuroscience & Biobehavioral Reviews, 36*(2), 747–756.

Thorman, R., Lundahl, J., Yucel-Lindberg, T., & Hylander, B. (2010). Inflammatory cytokines in saliva: Early signs of metabolic disorders in chronic kidney disease. A controlled cross-sectional study. *Oral Surgery, Oral Medicine, Oral Pathology, Oral Radiology, and Endodontology, 110* (5), 597–604.

Topol, E. J. (2014). Individualized medicine from prewomb to tomb. *Cell, 157*(1), 241–253.

Troy, A. S., & Mauss, I. B. (2011). Resilience in the face of stress: Emotion regulation as a protective factor. In S. M. Southwick, B. T. Litz, D. Charney, & M. J. Friedman (Eds.), *Resilience and mental health: challenges across the lifespan* (pp. 30–44). Cambridge: Cambridge University Press.

Uri, J., & Lebedev, O. (2001). Phase 1 research program overview. *Acta Astronautica, 48*(5–12), 845–851.

Van Dongen, P. A., Baynard, M. D., Maislin, G., & Dinges, D. F. (2004). Systematic interindividual differences in neurobehavioral impairment from sleep loss: Evidence of trait-like differential vulnerability. *Sleep, 27*(3), 423–433.

Vickers, R. R., Jr, Hervig, L. K., Wallick, M. T., Poland, R. E., & Rubin, R. T. (1987). *Psychological correlates of cortisol excretion in normal individuals under stress* (No. NHRC-87-1). San Diego, CA: Naval Health Research Center.

Voorhies, A., Torralba, M., Moncera, K., Lorenzi, H., Ott, C. M., Mehta, S., ... & Pierson, D. L. (2017). *Study of the Impact of Long-Term Space Travel on the Astronauts' Microbiome* (Microbiome). https://www.nasa.gov/mission_pages/station/research/experiments/explorer/Investigation.html?#id=982

Wakade, C., & Chong, R. (2014). A novel treatment target for Parkinson's disease. *Journal of the Neurological Sciences, 347*(1–2), 34–38.

Watkins, L. E., Harpaz-Rotem, I., Sippel, L. M., Krystal, J. H., Southwick, S. M., & Pietrzak, R. H. (2016). Hostility and telomere shortening among US military veterans: Results from the National Health and Resilience in Veterans Study. *Psychoneuroendocrinology, 74*, 251–257.

Westcott, S. (1907). The teeth of the soldier. *Journal of the Royal Army Medical Corps, 8*(2), 141–150.

Wollstein, M. (1918). A further study of experimental parotitis. *Journal of Experimental Medicine, 28*(4), 377–385.

Yirmiya, R., & Goshen, I. (2011). Immune modulation of learning, memory, neural plasticity and neurogenesis. *Brain, Behavior, and Immunity, 25*(2), 181–213.

Zhou, J. N., Liu, R. Y., van Heerikhuize, J., Hofman, M. A., & Swaab, D. F. (2003). Alterations in the circadian rhythm of salivary melatonin begin during middle-age. *Journal of Pineal Research, 34*(1), 11–16.

Zyphur, M. J., Narayanan, J., Koh, G., & Koh, D. (2009). Testosterone-status mismatch lowers collective efficacy in groups: Evidence from a slope-as-predictor multilevel structural equation model. *Organizational Behavior and Human Decision Processes, 110*, 70–79.

# Chapter 25
# Salivary Bioscience Research Related to Prenatal Adversity

**Shannon Shisler, Stephanie Godleski, Danielle S. Molnar, and Rina D. Eiden**

**Abstract** The prenatal period is a crucial point in development setting the stage for later health and well-being. Exciting new discoveries emerging from salivary bioscience have been particularly helpful in providing key insights into how prenatal adversity contours later health outcomes. Consequently, the aim of this chapter is to provide a critical review of the role of prenatal adversity in later stress and immunological functioning using salivary bioscience. We primarily focused on prenatal substance exposure and maternal psychosocial distress aspects of prenatal adversity, given that the majority of research on prenatal adversity using salivary bioscience has focused on these facets of adversity. We begin this chapter with a theoretical and historical overview of the prenatal adversity literature as it relates to salivary biosciences using the Fetal or Developmental Origins of Disease Model and Adaptive Calibration Model as our primary theoretical lenses. We then review the current status of knowledge regarding prenatal adversity effects on child stress (measured via salivary cortisol) and immune/inflammation (measured via salivary cytokines). Next, we discuss methodological issues that are particularly relevant with respect to the use of salivary bioscience to better understand the role of prenatal adversity in salient health outcomes. Finally, we discuss current gaps in the literature and provide insights for future directions in this exciting groundbreaking new area of research.

S. Shisler (✉)
Research Institute on Addictions, State University of New York at Buffalo, Buffalo, NY, USA
e-mail: smcasey@buffalo.edu

S. Godleski
Department of Psychology, Rochester Institute of Technology, Rochester, NY, USA

D. S. Molnar
Department of Child and Youth Studies, Brock University, St. Catharines, ON, Canada

R. D. Eiden
Department of Psychology, Pennsylvania State University, University Park, PA, USA



D. A. Granger, M. K. Taylor (eds.), *Salivary Bioscience*,
https://doi.org/10.1007/978-3-030-35784-9_25

## 25.1 Introduction

The prenatal period is a time of enhanced vulnerability during which adverse experiences may have long-lasting effects on child health and well-being. The study of prenatal adversity in the field of salivary biosciences has included a broad range of adversities. These include poverty (Appleton et al., 2013), malnutrition (Marques, O'Connor, Roth, Susser, & Bjørke-Monsen, 2013), maternal psychosocial distress (Beijers, Jansen, Riksen-Walraven, & De Weerth, 2010; Monk, Spicer, & Champagne, 2012), exposure to violence (Martinez-Torteya, Bogat, Levendosky, & Von Eye, 2016), natural disasters (Cao-Lei et al., 2014), maternal exposure to war stress (Mulligan, D'Errico, Stees, & Hughes, 2012), and maternal substance use during pregnancy (Molnar, Granger, Shisler, & Eiden, 2018). Many of these adverse experiences often co-occur (e.g., stress, depression, substance use, poverty), and contribute to a cascade of developmental processes that increase risk for maladaptive outcomes relevant for salivary bioscience throughout the life span. Thus, in line with transactional models of development (Rutter, 1987; Sameroff, Seifer, Baldwin, & Baldwin, 1993) that highlight the nested structure of risk factors, studies of prenatal adversity and developmental outcomes using salivary bioscience sometimes incorporate multiple measures of prenatal adversity. This approach utilizes a cumulative prenatal distress or risk score to address the often additive nature of prenatal adversity. This method has the advantage of providing greater external validity such that exposures to various forms of prenatal adversity tend to be highly intercorrelated. However, the majority of the literature on prenatal adversity of relevance to salivary bioscience has focused on maternal psychosocial distress (stress, depression, and anxiety) and maternal substance use during pregnancy. Thus, we have predominantly focused our discussion on these two broad areas of prenatal adversity.

The goal of our chapter is to present a review of the current state of the literature with respect to prenatal adversity and salivary bioscience. We will focus particularly on prenatal substance exposure and maternal psychosocial risks as these are the two most common areas of research using salivary biosciences. In addition to using salivary bioscience to measure maternal stress (e.g., measures of maternal salivary cortisol) and some aspects of maternal substance use during pregnancy (e.g., salivary cotinine, the predominant metabolite of nicotine, as a measure of tobacco use), salivary bioscience has been primarily utilized in studies of children's stress response systems and children's immune/inflammation systems. To achieve our objective, the chapter is organized as follows: (1) presentation of the historical/theoretical connection between prenatal adversity and salivary bioscience; (2) review of current status of knowledge regarding prenatal adversity effects on child stress and immune/inflammation; (3) review of methodological issues, particularly in the conceptualization and measurement of prenatal adversity and use of salivary bioscience in this area; and (4) a discussion of gaps in the literature with suggestions for future directions.

## 25.2 Historical and Theoretical Framework of Salivary Bioscience in the Prenatal Adversity Literature

Two major theoretical frameworks have relevance for understanding the links between prenatal adversity, stress, and immune functioning with respect to salivary bioscience that also provide a historical context. One framework, originally proposed by Barker (1999), and increasingly referenced in recent years is the fetal or developmental origins of disease (Barker et al., 1993; Barker, Eriksson, Forsén, & Osmond, 2002; Barker, Osmond, Kajantie, & Eriksson, 2009; Wadhwa, Buss, Entringer, & Swanson, 2009). The second, more recent framework is the Adaptive Calibration Model (Del Giudice, Ellis, & Shirtcliff, 2011). Two additional theories also have relevance to the literature on prenatal adversity effects in salivary bioscience and are described briefly. These include the Biological Embedding Model and the Allostatic Load Model. We discuss these frameworks in more detail below and connect them to a historical overview of this area.

**Fetal/Developmental Origins of Disease** This theory stemmed from Barker's observations that low birth weight is not only a significant birth outcome reflecting the fetal environment but also a significant marker for adult health (Barker, 1999). The theory proposes that there are specific sensitive periods in development that are marked by high sensitivity to the environment with the goal of providing the best fit between phenotype and environment. The prenatal period is one such sensitive period. Under conditions of high fetal stress (such as that posed by undernutrition, hypoxia–ischemia, or high glucocorticoid exposure), there will be alterations in the structure and function of various organs to promote survival. Thus, this fetal "programming" under conditions of high fetal stress results in lifelong changes in biology and behavior, including effects on neuroendocrine and metabolic dysfunctions (see Calkins & Devaskar, 2011; O'Donnell & Meaney, 2017; Buss, Entringer, & Wadhwa, 2012; reviews). Whereas these alterations may be adaptive in the short term by preparing the fetus for additional postpartum stressors, it may be dysfunctional in the long term since there is loss of plasticity in responses to environmental challenges (Cottrell & Seckl, 2009). It becomes especially critical if there is a mismatch between the fetal and long-term postnatal environments. In addition to direct effects of fetal malnutrition on fetal development, maternal glucocorticoid changes in response to stress in pregnancy and effects on pro-inflammatory cytokines have been implicated as significant mechanisms linking prenatal adversity to fetal and child outcomes (Buss et al., 2012; Cottrell & Seckl, 2009; O'Donnell et al., 2013).

**Adaptive Calibration Model** More recently, Del Guidice and colleagues (2011) proposed the Adaptive Calibration Model that is both evolutionary and developmental. The adaptive calibration model suggests that there are developmental changes in stress response systems; that periods of chronic elevation in the Hypothalamic–Pituitary–Adrenal (HPA) axis due to chronic stress exposure may often be followed by periods of blunted HPA responses; that these alterations may be long lasting; that males and females respond differently to various stressors; and that these responses have contrasting behavioral correlates. The theory suggests that while moderate stress may be associated with more regulated stress response systems, chronic stress or extreme stress may result in patterns of stress response that are adaptive and necessary for immediate survival, but may be associated with adverse outcomes in the long term. The model specifically references the prenatal period as a sensitive period. Indeed, the model suggests that infants are born with stress and immune systems that are predisposed to respond in particular ways based on prenatal experiences and genetic associations (see also Doom & Gunnar, 2013). Finally, the model suggests that males and females tend to respond differently to similar environmental cues because they use different life history strategies. Males are more likely to have a higher stress response to achievement-related tasks perhaps due to implications for social status. Females have higher stress responses to social rejection situations and these differences are associated with different behavioral correlates. *Consequently, this framework not only has relevance to our understanding of the literature on prenatal adversity and stress/immune functioning based on salivary bioscience, but is also relevant to our understanding of developmental changes in stress response systems.*

**Other Relevant Frameworks** In addition to these two major frameworks with direct implications for potential lifelong effects of prenatal adversity, the general model of allostatic load (McEwen & Stellar, 1993) also has some application in the context of prenatal adversity. This model suggests that the constant need to respond to ongoing stressors produces an allostatic load that wears on stress response systems over time resulting in dysregulation of autonomic, immune, and endocrine systems. This model has been applied to studies of chronic stress among children (Danese & McEwen, 2012), but chronicity of stressful environments may begin even earlier, in the prenatal period. Indeed, both the prenatal period and infancy have been suggested as critical periods for rapid changes and organization of the HPA axis and this has relevance for both stress and immune functioning (Doom & Gunnar, 2013).

Finally, the Biological Embedding Model is also of relevance here, although it is not specific to the prenatal period. The Biological Embedding Model (Miller, Chen, & Parker, 2011) posits that early childhood exposure to major psychological stressors becomes biologically embedded at the molecular level, increasing risk for morbidity and mortality. Authors of this framework cite evidence from children experiencing poverty, maltreated children, and from animal models indicating lifelong effects of early adversity on the development of chronic diseases and early aging. Similar to the fetal origins theory, this model suggests that early adversity establishes a proinflammatory phenotype at the cellular level in the immune system resulting in dysregulation of this system. This may be expressed as heightened immune response under threat and lower sensitivity to inhibitory signals. While this model focuses on early childhood stress effects such as poverty and child maltreatment, poverty effects would also occur prenatally, increasing risk for fetal malnourishment and high fetal stress. Similarly, risk factors associated with child maltreatment such as parental substance abuse and violence are also likely to exist prenatally with similar effects on increased risk for fetal stress, both directly due to effects of maternal substance use on fetal functioning (Day et al., 1992; Knopik, Neiderhiser, De Geus, & Boomsma, 2016; Schuetze, Eiden, Colder, Huestis, & Leonard, 2018; Shisler et al., 2017; USDHHS, 2014) and indirectly via effects on maternal stress and psychological functioning (Eiden, Foote, & Schuetze, 2007; Eiden, Schuetze, & Coles, 2011; Mark, Desai, & Terplan, 2016; Massey et al., 2016; Schuetze, Eiden, et al., 2018).

Taken together, these theoretical frameworks provide strong support for potential disruptions in stress response and immune systems beginning prenatally and for the critical role of postnatal experiences as moderators of this association. They differ in their discussion of immune system functioning, the degree to which they consider developmental changes, and in their consideration of sex-related differences. They also differ in the specificity of environmental experiences they reference, with the fetal/developmental origins model mostly focused on cardiovascular/metabolic risk; the adaptive calibration model mostly focused on linkages between stress and behavioral outcomes; and the biological embedding model mostly discussing immune system functioning and inflammatory processes. The study of prenatal adversity effects on stress regulation has predated these models; however, with one of the earliest studies on prenatal substance exposure effects on infant salivary cortisol (Magnano, Gardner, & Karmel, 1992) influenced by Gunnar's review of infant endocrine responses to challenge (Gunnar, 1986) and Selye's book on physiology and pathology of stress exposure (Selye, 1950).

Indeed, salivary techniques have been well-established for the measurement of HPA axis functioning using cortisol for over three decades, including the measurement of cortisol in children (Woolston, Gianfredi, Gertner, Paugus, & Mason, 1983). In addition to being cost-effective and noninvasive, salivary measures are particularly beneficial when measuring stress responses because it allows for samples to be taken without contaminating the results by adding the stress of a blood draw. Salivary bioscience also provides the opportunity for easy home sampling, allowing for research on daily rhythms outside of the laboratory. Utilizing saliva samples (as opposed to blood draws) may allow for increased participation, particularly for children whose parents might be reluctant to consent to a blood draw. Chapter 10 provides further details on contributions of salivary bioscience to developmental experiences from birth to adolescence.

While the study of prenatal effects on salivary cortisol has a long history, recent advances in salivary bioscience regarding measurement of salivary cytokines has opened new avenues of research on prenatal adversity. A plethora of literature linking prenatal adversity to immunological functioning has tended to rely on plasma biomarkers of cytokine production collected from cord blood. Whereas blood is generally regarded as the best body fluid for evaluation of systemic processes, collection of blood involves potential risks to research participants, including transient discomfort, bruising, and infection at the venipuncture site. Blood collection also tends to be less favored in research involving children for whom venous access is difficult. Additionally, the hesitation of parents and children with regard to giving consent to providing the blood samples needed to estimate immunologic parameters in young children may also hamper efforts to examine immunological functioning through such blood samples. Thus, the minimally invasive nature of saliva sampling and the assessment of cytokines in oral fluids provides a breakthrough in the field allowing researchers to study the critical health effects of prenatal adversity. Although these salivary cytokines reflect inflammation in the oral mucosal and respiratory systems only, they may be particularly useful in providing initial data and in some instances, critical data for some prenatal adversities such as tobacco exposure. In addition, maternal stress, and use of nicotine, tobacco, alcohol, cannabis, and illicit drugs can all be measured via maternal salivary assays during pregnancy (see Chap. 16 for an excellent review of the utility of salivary biosciences in detecting drugs of abuse). Evidence from all of these sources helps researchers to elucidate the ways in which prenatal adversity may have long-term consequences on these systems.

## 25.3 Current Status of Knowledge Regarding the Impact of Prenatal Adversity on Stress and Inflammation

This section will introduce the reader to previous work examining the relationship between prenatal adversity and stress and inflammation using salivary bioscience. The literature on prenatal adversity (substance exposure and maternal psychosocial distress) relating to both stress and inflammation will be discussed. Each of these sections is structured developmentally, whereby studies of neonates are discussed first, followed by infants, toddlers, children, and adolescents.

### *25.3.1 Prenatal Adversity and Salivary Cortisol*

As noted above, there is evidence from both human and animal studies that the prenatal environment can impact the developmental programming of the stress response system. While the fast acting autonomic nervous system regulates the rapid fight-or-flight response to stress, the HPA system activates a slower cascade of signals resulting in release of secretions from the adrenal gland such as cortisol (Granger, Johnson, Szanton, Out, & Schumann, 2012; Kirschbaum & Hellhammer, 1989); that can be reliably measured in saliva. As the regions associated with the HPA axis (e.g., hippocampus, amygdala, and frontal cortex) develop they may be particularly sensitive to prenatal adversity, such as maternal substance use and psychosocial distress (Lupien, McEwen, Gunnar, & Heim, 2009). This is particularly problematic as our ability to respond in a physiologically appropriate way to stress is quite crucial. Indeed, dysregulation of the stress response system has been implicated in the deterioration of other body organs when organisms are exposed to chronic adversity, including disruptions in brain circuitry, pulmonary and cardiovascular disease, liver cancer, and autoimmune diseases (Shonkoff et al., 2012).

The normal cycle of stress-induced cortisol secretion is to observe a peak level of cortisol 15–20 min post stressor, with a return to the pre-stressor baseline by 30–40 min post stressor (Gunnar, Brodersen, Krueger, & Rigatuso, 1996). This cycle can be quantified for researchers through cortisol levels in saliva, typically taken immediately before stressor (baseline), immediately after stressor, and at 20 and 40 min post stressor. In addition to the cortisol response to stress, there is a normal diurnal cortisol rhythm, marked by high levels of cortisol within 30 min of awakening (typically called the cortisol awakening response, or CAR), with a gradual decrease throughout the day, reaching the lowest point around midnight (King & Hegadoren, 2002; Schmidt, 1997). This typical diurnal pattern is important to our overall health because it supports brain development and helps us to sustain our normal levels of functioning. It too can be reliably measured using salivary cortisol, though the measurement of the diurnal rhythm is not as consistent among studies. There is evidence that as stress response patterns are disrupted by exposure to chronic adversity, the system becomes desensitized, which may result in blunted responses to stressors as well as decreased awakening cortisol (Gunnar & Vazquez, 2001). This atypical blunted pattern of cortisol responsivity has been associated with changes in motivation, sleep, attention, immune system functioning, and stress-related illnesses such as heart disease (Miller, Chen, & Zhou, 2007). However, there is also evidence that prenatal adversity is linked to an increase in cortisol reactivity (Haley, Handmaker, & Lowe, 2006; Jacobson, Bihun, & Chiodo, 1999; Schuetze, Lopez, Granger, & Eiden, 2008). Recent work has advanced the hypothesis that both highly reactive and highly blunted responses are indicators of a dysregulated system (Lovallo, 2011). Lovallo (2011) notes that one might imagine a normal curve that represents the typical response of a regulated system, in this case the stress response. Whereas the center of the distribution represents a "normal" stress response, the tails represent nonnormative responses that may be indicative of maladaptive dysregulation and may be predictive of poorer health outcomes.

### *25.3.2 Prenatal Substance Exposure and Child Salivary Cortisol*

One of the most significant exposures to prenatal adversity that a fetus might endure is exposure to maternal substance use in utero. While pregnant women report tobacco as the most used substance during pregnancy, with rates from 18 to 27% (USDHHS, 2014), alcohol, cannabis, and other illicit substance use is also common. Additionally, among substance-using women, polysubstance use during pregnancy is more the norm than the exception. For example, rates of tobacco and cannabis co-use have been reported to be as high as 45% among pregnant cannabis users (Chabarria et al., 2016). To compound the potential stress to the fetus, substance use during pregnancy is often confounded by comorbidity with maternal psychosocial distress as well as environmental and socioeconomic stressors, all of which can exacerbate the potential for negative maternal and fetal outcomes.

***Prenatal Tobacco Exposure and Child Salivary Cortisol*** As early as the first month of life, children prenatally exposed to cigarettes displayed both an attenuated baseline salivary cortisol as well as a blunted cortisol reactivity response compared with nonexposed children (Stroud et al., 2014). Similarly, at 2 months of age, infants prenatally exposed to cigarettes demonstrated less reactivity to a stressor than their nonexposed peers (Ramsay, Bendersky, & Lewis, 1996). However, in this particular study, this dampened response really reflected a higher level of pre-stressor salivary cortisol, making the change in cortisol from pre- to post stressor lower in magnitude for exposed infants. Yet other studies of infants have found quite the opposite. At 7 months of age, infants prenatally exposed to cigarettes showed an increased reactivity response to a stress-inducing lab procedure, with exposed boys showing the highest levels of reactivity as compared to both exposed girls, and nonexposed infants (Schuetze et al., 2008). In contrast, at 9 months of infant age, cigarette-exposed infants displayed a blunted response to acute laboratory stressor, with significantly lower overall cortisol compared to demographically similar nonexposed infants (Eiden et al., 2015). However, this association was moderated by child sex, such that cigarette-exposed boys had lower cortisol compared to nonexposed boys, but there were no differences among girls. At kindergarten age, these children continued to exhibit a blunted response to an acute laboratory stressor, but this was moderated by co-occurring cannabis exposure (Schuetze, Shisler, Eiden, & Granger, 2018). Demographically similar nonexposed children exhibited a normative cortisol response to a stressor with an initial increase post stress followed by stable values. Children exposed to a low number of cigarettes per day, but high number of joints per day had no cortisol response, exhibiting the blunted response pattern we have discussed previously. Finally, children exposed to high levels of daily cigarette use had a decreasing cortisol response to stress, regardless of level of cannabis exposure (Schuetze, Shisler, et al., 2018). These results, in particular, highlight the importance of examining joint, dose-response effects of tobacco and cannabis, especially given that these two substances co-occur most frequently during pregnancy. Finally, in one of the few studies on older children (Huijbregts, Van Berkel, Swaab-Barneveld, & Van Goozen, 2011) reported no associations between cigarette exposure and cortisol reactivity at 10.6 years of age, although the sample size was quite small ($n = 14$).

It has been posited that the HPA axis is continually developing through the third year of life, and thus results in the infant and toddler periods may be particularly inconsistent (Watamura, Donzella, Kertes, & Gunnar, 2004), with exposed children exhibiting responses that are either much higher or much lower than the normative response. However, both exaggerated and blunted responses are implicated in negative health outcomes (Carroll, Lovallo, & Phillips, 2009; Lovallo, 2011). The literature on prenatal tobacco exposure effects on children's salivary cortisol may be mixed for a number of additional reasons. First, there are significant differences in how tobacco exposure was measured and the relative sample sizes and concerns about power across different studies. For example, in the study by Ramsay et al. (1996), women who used any cigarettes or alcohol during pregnancy were grouped in the exposure group. In the study by Schuetze et al. (2008), prenatal tobacco exposure was measured by maternal self-reports alone. Finally, in the only study of older children, the sample size was limited and the prenatal tobacco exposure was measured using retrospective reports 8–13 years after delivery. This is especially concerning given changes in smoking across pregnancy and problems with retrospective recall (Eiden et al., 2013). Another potential issue is the nature of the comparison group. Without biochemical verification or detailed, reliable, prospective self-report measures (see Shisler et al., 2017), it is not clear that the comparison group were not exposed to tobacco or other substances. In addition, mothers who continue to smoke during pregnancy are more likely to be low income, single, and young (USDHHS, 2014). Consequently, the use of comparison groups that are not demographically similar are likely to lead to erroneous conclusions given that these demographics pose significant risk for chronic stress across the prenatal and postnatal periods.

***Prenatal Alcohol Exposure and Child Salivary Cortisol*** Similar results were found for prenatal alcohol exposure such that alcohol use from conception through pregnancy recognition was related to increased change in salivary cortisol during the still face procedure for 5–7 month old infants after controlling for child sex, maternal depression, and annual income (Haley et al., 2006), and to increased cortisol pre-stressor, and at recovery (Eiden, Veira, & Granger, 2009). Haley et al. (2006) also found sex differences such that boys of high frequency drinkers had higher salivary cortisol reactivity and did not exhibit the typical recovery response compared to boys of low frequency drinkers, but that this pattern was not found for girls. In 13-month old infants, alcohol-exposed children had higher levels of pre-stress cortisol (Jacobson et al., 1999). Moreover, children with heavy alcohol exposure also had elevated post-stress cortisol levels. In 19-month old toddlers, there was no direct relation between prenatal alcohol exposure and salivary cortisol reactivity to unfamiliar situations, but there were sex-related differences such that exposed boys had greater cortisol reactivity and had lower levels of baseline pretest cortisol than girls (Ouellet Morin et al., 2011). Fisher, Kim, Bruce, and Pears (2012) studied 10-year-old children who had been placed in foster care during preschool using the Trier Social Stress Test. Prenatal substance exposure was assessed using FAS Facial Photographic Analysis Software (Astley & Clarren, 2001) along with any mention of alcohol or drug use during pregnancy in the welfare system records. Children exposed to substances prenatally showed a decrease in salivary cortisol levels over time compared to nonexposed children (main effect; 68% showed decreased cortisol response after stressor), but there was also an interaction with experiencing physical abuse in preschool such that children who experienced both had a decreased cortisol response over time. In this sample, only 15% of participants exhibited a typical cortisol response, suggesting that prenatal substance exposure and early adversity create an allostatic load that disrupts HPA axis regulation.

Though the diurnal response of infants and toddlers prenatally exposed to alcohol is understudied, there is evidence of a disrupted diurnal cortisol rhythm in alcohol-exposed children aged 5–18 years. Afternoon and bedtime salivary cortisol was elevated in children with Fetal Alcohol Syndrome Disorders (FASD) compared to controls, and there was a trend for female FASD children to have a lower awakening cortisol than female control children (Keiver, Bertram, Orr, & Clarren, 2015). Afternoon cortisol was similarly higher in a sample of children with both prenatal alcohol exposure and early life adversity compared to control children (McLachlan et al., 2016). This same group also showed a flatter diurnal slope, again suggesting alteration to the normative trajectory of the daily cortisol rhythm.

***Prenatal Drug Exposure and Child Salivary Cortisol*** The impact of illicit drug use has also been studied through salivary bioscience in the context of altered stress responses. One of the first such studies involved newborns prenatally exposed to cocaine. They found similar basal cortisol levels between exposed and nonexposed infants, however, exposed infants had significantly lower cortisol levels 30 min after both noninvasive and invasive stressor (Magnano et al., 1992). However, these results must be viewed with caution as there was no information on the degree or duration of prenatal cocaine exposure, and the entire sample were infants that were NICU patients at the time of assessment. There is also some evidence of sex differences such that cocaine-exposed boys had greater cortisol reactivity than nonexposed boys (Eiden et al., 2009). In addition, higher caregiver instability combined with prenatal substance exposure was associated with higher levels of cortisol reactivity in comparison to nonexposed infants or infants with low caregiving instability. In 13-month-old infants, cocaine-exposed children had higher levels of pre-stress cortisol (Jacobson et al., 1999).

In early adolescence, Lester et al. (2010) found prenatal cocaine exposure (PCE) was related to a blunted cortisol response, and this blunted response was more likely when PCE children also experienced exposure to violence, and when the PCE was heavy (3 or more times per week during the first trimester). A study of adolescents similarly found that drug-exposed adolescents were less likely to have a reactivity response to a mild stressor than nondrug-exposed adolescents (Buckingham-Howes, Mazza, Wang, Granger, & Black, 2016). High levels of prenatal cocaine exposure were also related to smaller increases in cortisol overnight in 11 year olds compared to controls. However, there were no group differences on AM or PM cortisol levels (Bauer et al., 2011). Results support the presence of a blunted cortisol response among drug-exposed adolescents, and highlight the importance of considering the level of prenatal drug exposure. In one of the few studies examining sex differences, Chaplin et al. (2015) also noted a blunted response pattern among cocaine-exposed 14–17-year-old adolescents but this effect was not different for boys and girls. Animal models have helped to elucidate the possible mechanisms through which drug exposure, and more specifically prenatal cocaine exposure, may impact the stress response. First, prenatal cocaine exposure acts as a dopamine inhibitor, which attenuates catecholaminergic functioning (Spear et al., 1989), and exposure-related intrauterine arterial vasoconstriction may lead to an increased release of fetal catecholamine in response to hypoxia. Fetal hypoxia, or lack of oxygenated blood to the fetus, results in cell and tissue damage that is implicated in restricted intrauterine growth, infant respiratory distress, cognitive dysfunction, increased likelihood of perinatal mortality, and cardiac and circulatory birth defects (Hutter & Jaeggi, 2010; Rueda-Clausen, Morton, & Davidge, 2009). This prolonged exposure to toxic stress may result in the dysregulation of the stress response system (Woods, Plessinger, & Clark, 1987).

Taken together, evidence from studies of prenatal substance exposure suggest that there is an altered stress response pattern among exposed children. Both blunted and exaggerated responses have been predicted by prenatal substance exposure, both of which are maladaptive and indicative of pathophysiology. Both the dose of substance exposure and exposure to additional postnatal adversity seem to be particularly critical in predicting dysregulation of the stress response system in children, and thus should be taken into consideration in future studies whenever possible.

***Maternal Prenatal Anxiety/Stress and Child Salivary Cortisol*** Maternal stress during pregnancy may also predispose children to altered stress responses and disorders due to early interference with the development of the HPA axis, as hypothesized by the fetal origins of disease (Barker, 1999). Indeed, maternal anxiety during pregnancy has been found to be associated with cortisol reactivity in infants as young as 2 days (Leung et al., 2010) with evidence of continued impact on salivary cortisol levels extending through adolescence (O'Donnell et al., 2013).

In infancy, Leung et al. (2010) examined prenatal maternal stress (measured using maternal self-report of symptoms) and infant stress reactivity at 2 days and 10 months with a heel stick blood draw at 2 days and toy removal at 10 months. Infant salivary cortisol reactivity was fairly stable from 2 days to 10 months of infant age and greater maternal stress in pregnancy was associated with greater neonatal reactivity. Similarly, higher maternal plasma cortisol levels in pregnancy were associated with larger increases in infant salivary cortisol response to a heel stick procedure (Davis, Glynn, Waffarn, & Sandman, 2011). In particular, high maternal cortisol between 20 and 27 weeks of gestation was associated with elevated cortisol, with the strongest relationship at 25 weeks gestation (Davis et al., 2011). In toddlerhood, associations between prenatal cortisol at 17.2 weeks gestation measured using amniocentesis and toddler salivary cortisol response at 17 months before and after a separation paradigm were examined (O'Connor, Bergman, Sarkar, & Glover, 2013). Toddlers exposed to higher levels of cortisol prenatally exhibited high levels of pre-task cortisol and had only a slight decrease over time, whereas toddlers exposed to low levels of cortisol exhibited increases in cortisol over time. In one of the few studies examining sex moderation, Ping et al. (2015) reported a prenatal stress by child sex interaction at toddler age, such that girls exposed to prenatal maternal stress had significantly higher absolute change in cortisol from baseline to 45 min post stressor than nonexposed girls. However, for boys there was no significant difference in cortisol for those who were exposed versus those unexposed to maternal prenatal stress.

The literature on prenatal stress and anxiety exposure and adolescent outcomes is quite limited. In one of the few studies of adolescents, O'Donnell et al. (2013) examined diurnal cortisol among 15-year olds who had been prenatally exposed to maternal anxiety (as measured by maternal self-reports). Repeated measurements of maternal anxiety at 18 and 32 weeks of gestation and several times in the postnatal period were used. Adolescents with mothers who reported higher levels of prenatal anxiety reported diminished CAR and flatter diurnal rhythm (controlling for other variables including maternal postnatal anxiety). Importantly, this work also points to fetal origins of disease (Barker, 1999) as there was no effect of paternal prenatal anxiety on cortisol values and the results controlled for maternal postnatal anxiety. Thus, the authors discounted the possibility of a simple genetic transmission from parent anxiety to child stress regulation.

A similar pattern of results were reported with maternal anxiety. For instance, infants of mothers with prenatal anxiety disorder had significantly higher cortisol response to the Still Face Paradigm at 7 months of infant age compared to infants of non-anxious mothers (Grant et al., 2009). Similarly, in the toddler period, 14-month-old children of mothers with high levels of anxiety at 32 weeks gestation had significantly higher CAR than children of mothers with low anxiety (O'Connor et al., 2005). In contrast, infants of mothers with high levels of distress during pregnancy (e.g., Post-Traumatic Stress Disorder from being present at or near to the World Trade Center Collapse) had lower CAR and bedtime cortisol levels compared to infants of mothers without PTSD who were also present at or near the World Trade Center on 9/11 (Yehuda et al., 2005). It should be noted that the effect was strongest with third trimester exposure, which lends support to the possibility of fetal programming and emphasizes the importance of measuring timing of exposure. Similarly, high self-reported maternal anxiety in the first half of pregnancy was associated with lower than normal CAR among 14–15 year-old adolescents, but higher than expected levels in the evening (Van Den Bergh, Van Calster, Smits, Van Huffel, & Lagae, 2008). However, interestingly maternal anxiety in later pregnancy (21–40 weeks) was not associated with adolescent cortisol, again highlighting the importance of timing of exposure. Further, the kind of stress or distress during pregnancy is an important consideration, as measurement of stressors has ranged from daily stress and economic concerns (e.g., Luecken et al., 2013) to PTSD and other anxiety disorders during the prenatal period (e.g., Yehuda et al., 2005).

***Maternal Prenatal Depression and Child Salivary Cortisol*** In addition, several studies have examined the impact of prenatal depression and other psychopathology. Past research has suggested both unique and additive effects of maternal prenatal anxiety and depression (e.g., O'Connor, Heron, Glover, & Team, 2002; Yehuda et al., 2005). Some studies (e.g., O'Donnell et al., 2013) have found comparable results with anxiety and depression, while others have found different effects (e.g., Yehuda et al., 2005). In an investigation specifically examining maternal depression, Brennan et al. (2008) found that maternal lifetime history of depression predicted higher baseline and mean cortisol in 6-month-old infants compared to women with no lifetime history of mood or anxiety disorders, but was not related to cortisol reactivity to a stressor. However, maternal peripartum depression was related to increased cortisol reactivity. Indeed, there is a robust literature linking postnatal depression to other developmental outcomes as well (e.g., Pearson et al., 2013). Genetic or epigenetic influences may also play a role in the impact of prenatal psychopathology. Among mothers with high prenatal maternal psychological symptoms including anxiety and depressive symptoms, toddler genotype moderated the impact of prenatal maternal psychopathology symptoms (measured via maternal reports of global symptoms) on cortisol reactivity (Velders et al., 2012). Specifically, 14-month-old carriers of the minor allele at rs41423247 (CC) exhibited blunted cortisol reactivity in response to stress only if their mothers reported high psychological symptoms prenatally, but not among those with low levels of reported symptoms. Similarly, Stroud et al. (2016) examined prenatal maternal major depressive disorder (MDD) and infant cortisol responses (in response to neurobehavioral assessment) and investigated potential moderation by gene methylation and expression. The gene *HSD11B2* encodes HSD2, an enzyme that converts cortisol to inactive cortisone, which protects the fetus from increasing maternal glucocorticoids during pregnancy. The *SLC6A4* gene is targeted by psychomotor stimulants (such as cocaine), and implicated in susceptibility to depression. Results indicated that placental *HSD11B2* methylation (repression) and *SLC6A4* gene expression moderated the association between MDD and infant cortisol responses (see Chap. 6 on Genetics and Epigenetics (Nemoda et al.) for recent advances and technical issues in using saliva samples for such research)

***Prenatal Adversity and Salivary Alpha Amylase (sAA)*** In addition to measurement of the slow moving HPA responses to stress as reflected in salivary cortisol, prenatal adversity effects on the fast moving autonomic nervous system has also been the focus of a number of studies. While the majority of these studies have used cardiac measures of autonomic functioning, a few have used sAA as a marker of sympathetic nervous system (SNS) activity. sAA can be measured in saliva and secretion of sAA by the salivary glands is controlled by sympathetic innervation (Nater & Rohleder, 2009). Thus, increase in sAA in response to stressors is thought to be a marker for increased sympathetic activity. The literature on prenatal adversity effects on child sAA is small. However, including sAA in studies where collection of cortisol is already planned is labor and cost effective, and provides additional data regarding the stress response system. Studies of prenatal adversity that have included sAA are reviewed below.

In a study hypothesizing that maternal prenatal depression would induce changes in maternal sympathetic nervous system functioning, Braithwaite, Murphy, and Ramchandani (2016) examined diurnal measures of sAA across 2 days. Maternal reports of depressive symptoms were used.

Mothers who reported higher depressive symptoms in later pregnancy had higher awakening sAA levels compared to nondepressed mothers, and continued to exhibit higher levels across the day. Authors concluded that changes in sympathetic activity in prenatally depressed women may result in vasoconstriction and lower fetal blood flow, and partially explain the association between maternal prenatal depression and negative child outcomes. In addition to diurnal patterns, synchrony of sAA–cortisol reactivity to acute laboratory stressors has also been examined. For instance, maternal trauma experience due to intimate partner violence during pregnancy was examined as a predictor of infant sAA and cortisol responses to a laboratory stressor (Martinez-Torteya, Bogat, Lonstein, Granger, & Levendosky, 2017). These responses were construed as mediators of the association between prenatal trauma and child internalizing and externalizing problems. The association between maternal experience of intimate partner violence during pregnancy and infant internalizing problems was the strongest among infants who exhibited a pattern of high cortisol and low sAA and the least pronounced among infants who exhibited a synchronous pattern of high–high or low–low cortisol–sAA values. The results point to the value of assessing both the HPA and the sympathetic nervous system responses to stress and conceptualizing systemic dysregulation as discordance between the two systems.

In addition to maternal psychosocial distress, a few studies have examined sAA responses to maternal substance use during pregnancy. Measures of cortisol and sAA were collected at five time points before and after stressful neuropsychological tests in a small sample of tobacco-exposed and nonexposed children at 10.6 years of age (Huijbregts et al., 2011). Cortisol results have been described earlier and there were no differences between groups. However, nonexposed children had significantly higher sAA across all time points compared to tobacco-exposed children. Indeed, the authors noted that the trajectory of sAA responses for tobacco-exposed children was similar to those of children with disruptive behavior disorders. In contrast to sAA responses, tobacco-exposed children exhibited higher behavioral stress reactivity, again pointing to the mismatch between systems—in this study, the mismatch is between sympathetic response to stress and behavioral responses. In a study of prenatal cocaine exposure described earlier (Chaplin et al., 2015), measures of sAA were included in addition to salivary cortisol as mediators of the association between cocaine exposure and substance use in adolescence (14–17 years of age). Results indicated sex by stress response interaction effects on adolescent substance use such that blunted sAA responses predicted substance use for boys but not girls.

***Moderation*** The impact of prenatal adversity on child salivary outcomes may be mediated or moderated by other critical factors. Importantly, maternal substance use during pregnancy may co-occur (e.g., Grant et al., 2009; O'Connor et al., 2013) with higher levels of stress and psychopathology. Maternal parenting (e.g., sensitivity) and the parent–child relationship (e.g., attachment) have been examined as potential protective factors (Grant et al., 2009) as has maternal partner support (Luecken et al., 2013). For example, higher maternal economic stress during pregnancy was associated with higher infant salivary cortisol reactivity at 6 weeks of infant age to a frustration task among low-income Mexican American women (Luecken et al., 2013). However, this association was moderated by partner support such that maternal stress was only associated with higher infant cortisol when partner support was low. Thus, prenatal partner support may serve as a buffer against the negative influence of prenatal stress on infant cortisol reactivity. Similarly, using a twin design, Ouellet-Morin et al. (2011) examined the association between family adversity (seven factors including smoking, low birth weight, perinatal/postnatal development, low family income, low maternal education, single parenthood, young motherhood, and maternal hostility) and cortisol reactivity at 19 months of infant age. High family adversity was associated with higher cortisol reactivity to stress. However, among families with low adversity (exposure to 4 or few adversity factors), cortisol reactivity was primarily accounted for by genetic and unique environmental factors. In other words, it was the shared genes among twin pairs that explained the similarities in their cortisol reactivity. However, within the context of high family adversity cortisol reactivity was primarily accounted for by both shared and unique environmental factors, and not by genetic factors, suggesting that fetal programming of the HPA axis under conditions of toxic stress may play a significant role even after accounting for genetic factors in the development of the stress response system. As such it is critical to consider both potential genetic and environmental factors in developmental programming

***Summary*** Taken together, there is a significant body of literature using salivary bioscience that highlights prenatal adversity effects on children's stress responses along the HPA axis and a smaller literature on sympathetic functioning measured by sAA. Results seem to suggest developmental changes in these associations, highlight the importance of timing of exposure, point to potential sex differences, and reflect the diversity of maternal psychosocial distress variables that have been examined in relation to children's stress responses. A few studies have also considered synchrony or asynchrony in cortisol and sAA as indicators of regulated versus dysregulated stress response systems—an area that needs further study. Consistent with theory, results of the reviewed literature demonstrate disruptions in stress response and immune systems beginning with prenatal adversity and suggest the critical role of postnatal experiences as moderators of this association. Prenatal adversity, such as substance exposure during pregnancy, was found to be associated with both dysregulated stress response systems for children (i.e., highly reactive and highly blunted responses; Lovallo, 2011). The biopsychosocial perspective emphasizes the reciprocal interactions between biological influences, such as genetic and psychophysiological factors, and environmental influences, such as parenting (Calkins, Propper, & Mills-Koonce, 2013). As highlighted throughout, it is important to consider the moderating or mediating influences that may impact the developmental trajectories from prenatal adversity to alterations in stress and immune function. For example, the presence of other maternal or environmental risk factors that may co-occur with prenatal adversity could lead to potential additive or moderating effects.

### 25.3.3 Prenatal Adversity and Immune System Functioning

Blalock (2005) has referred to the immune system as individuals' "six sense" to illustrate the immune system's capacity to both perceive and respond to the environment. The immune system is organized so that it operates in stages. The innate response is the body's initial immune response to disease or trauma. The innate immune response is activated very quickly in reaction to disease or injury and a key component of the innate immune response is the inflammatory response, in which the immune system produces pro-inflammatory substances (Johnson, Riley, Granger, & Riis, 2013). If necessary, the second stage of the immune response, acquired immunity, is initiated. Acquired immunity entails the triggering of T and B lymphocytes that are "specific to the infecting agent and together they result in the production of antibodies that bind to and neutralize or kill the antigen (Johnson et al., 2013, p. 321)". Cytokines have a primary role in regulating both innate and acquired immunity (Blalock & Smith, 2007). Generally speaking, cytokines are cell signaling molecules that aid cell-to-cell communication in immune responses and stimulate the movement of cells toward sites of inflammation, infection, and trauma. In essence, cytokines regulate the duration and the intensity of the immune response. Type 1 cytokines (IFN-γ and IL-12), for example, facilitate the elimination of malignant cells and cells infected with viruses, while Type 2 cytokines (IL-4, IL-5, and IL-13) facilitate the elimination of soluble bacterial antigen (Duramad, Tager, & Holland, 2007). Cytokines have long been recognized as an early marker of health, with cytokines implicated in a variety of health conditions, such as autoimmune and inflammatory diseases, cardiovascular disease, cancers, atopic disorders, and susceptibility to viral and bacterial infections (Asadullah, Sterry, & Volk, 2003; Kieran, Kalluri, & Cho, 2012; Nater & Rohleder, 2009; Pirhonen, Sareneva, Kurimoto, Julkunen, & Matikainen, 1999; Schroder, Hertzog, Ravasi, & Hume, 2004; Sesso et al., 2003).

Two of the most commonly studied markers of inflammation as a consequence of prenatal adversity are IL-6 and C-Reactive Protein (CRP). IL-6 is an interleukin that is both pro-inflammatory and anti-inflammatory, is secreted by innate and adaptive immune cells during infection to activate an immune response, with chronic elevated levels associated with higher cardiovascular risk, cancer, and mortality (Berg et al., 2017; Volpato et al., 2001). Chronic inflammation increasing cytokines such as IL-6 also leads to the release of CRP by the liver, with chronic high levels of CRP an indicator of higher metabolic risk and coronary heart disease (see Berg et al., 2017). The American Heart Association and the Center for Disease Control and Prevention have recommended serum CRP as a primary inflammatory marker for use in clinical practice and research (Pearson, 2003).

To date, the small number of studies of prenatal adversity effects on cytokines have focused on prenatal tobacco exposure and the majority measure serum cytokines. Most of these studies report significant associations between prenatal and postnatal tobacco exposure and higher serum cytokines, although the specific cytokines vary by study. For instance, mothers who continued to smoke during pregnancy had newborns with higher levels of serum IL-8 compared to nonsmokers and women who quit during pregnancy (Chahal et al., 2017). At toddler age, significant associations between hair nicotine levels indicating secondhand exposure and high serum CRP have been reported (Groner et al., 2017). These associations continue to be significant at preschool age, even when secondhand or postnatal exposure was assessed via two parent report items (Kang et al., 2017), and among school age children (Merghani, Saeed, & Alawad, 2012). However, others have reported an inverse association between higher cotinine levels and interleukins such as IL-6 and IL-4 among adolescents, suggesting immunosuppressive effects of secondhand smoke exposure perhaps with increasing age (Matsunaga et al., 2014). It is worth noting that many children who experience postnatal exposure to secondhand smoke also experience prenatal exposure to tobacco. In addition to tobacco, a small number of studies have examined other prenatal adversities. For instance, in a large sample of Danish men and women, higher prenatal stressors such as single parent status and having an unwanted pregnancy and a cumulative stress index were associated with higher CRP and IL-6 among middle-aged adults, but there were no associations with IL-10 and TNF-α (Pedersen et al., 2018). Maternal depression during pregnancy has also been associated with higher serum CRP in adulthood—at 25 years of age (Plant, Pawlby, Sharp, Zunszain, & Pariante, 2016).

One reason why the literature on prenatal adversities and immune/inflammatory markers is small may be because blood draws are invasive, especially in children, thus presenting a significant barrier in research. However, there is growing evidence that salivary cytokines are measurable, valid, and reliable measures of inflammation (Riis, Granger, Dipietro, Bandeen-Roche, & Johnson, 2015), although more evidence is needed regarding associations with health risks, sex differences, and potential developmental changes (see Riis et al., 2014, 2015). For instance, salivary CRP is detectable as early as the neonatal period and there is some suggestion that they are reflective of serum CRP, including prediction of abnormal serum CRP levels (Iyengar, Paulus, Gerlanc, & Maron, 2014). While interleukins are also detectable in saliva, high levels of interleukins such as IL-1β and IL-6 may be more indicative of periodontal disease and mucosal inflammation rather than systemic inflammation (Äyräväinen et al., 2018; Riis et al., 2014, 2015). There are also sex differences in salivary cytokines with boys showing decreases in cytokines from pre- to post-stressor and girls having more stable levels across time (Riis et al., 2015). Taken together, the detection of IL-6 and CRP in saliva mark a significant advance in our ability to examine early warning signs of elevated health risk among prenatally exposed children, in addition to its clinical utility (Iyengar et al., 2014).

The mucosal immune system measured using salivary bioscience consists of a system of interrelated components that afford protection to mucosal surfaces. Although, the mucosal immune system matures throughout the first year of life and can be modified by a number of environmental factors (Gleeson & Cripps, 2004), research indicates that the structures comprising mucosal immunity (i.e., lymphoid-related structures of the nasal, bronchial, gastrointestinal, and urogenital tracts, along with the lactating mammary glands, salivary, lachrymal, and the synovium of joints) as well as the primary class of antibodies [i.e., immunoglobulin

A (IgA)] are completely developed in utero by 28 weeks gestation (Gleeson & Cripps, 2004). Thus, given that the components of the mucosal immune system develop in utero, it is not surprising that the prenatal environment captures a crucial period of potential vulnerability. Indeed, the placenta is a primary source of cytokines and other immune system regulators found in the fetus (Marzi et al., 1996; Saito et al., 1999). "As the immunologically active interface between the fetus and the mother, the placenta is vulnerable to these and other maternal influences, and has a significant capacity to modulate the effects of these on the fetus. Placental cells are also likely to be sensitive to adverse environmental exposures that may affect pathways underpinning the dramatic increase in immune disease in very early life (Prescott, 2008, p. 5)." Coe and colleagues have further postulated that the environment during the prenatal and perinatal periods of development not only refine and adjust immune system functioning, but that maturational processes strengthen the influence of early disturbances on the development of the immune system such as to change its lifelong trajectory (Coe & Lubach, 2003, 2005). Indeed, there is growing evidence that prenatal adversity can influence early immune system functioning by altering cytokine production (e.g., Almanzar et al., 2013; Noakes et al., 2006; Noakes, Holt, & Prescott, 2003; Prescott, 2008).

In one of the few studies examining the role of prenatal and postnatal cigarette and cannabis exposure on early immune system functioning, Molnar et al. (2018) reported higher Secretory Immunoglobulin A (SIgA) levels among early school age children who were exposed to cigarettes only and to both cigarettes and cannabis relative to the nonexposed children. Children who were exposed postnatally to cigarettes and cannabis also had higher levels of SIgA even after accounting for prenatal exposure and other potential confounds that are related to immune system functioning. Collectively, these results suggest that prenatal and postnatal adversity, operationally defined in this study as cigarette and cannabis exposure, may result in a hyperactive mucosal immune system in young children.

## 25.4 Methodological Issues, Challenges, and Considerations

There are several methodological issues to consider when using salivary assays to measure stress and immunity in children. Compliance may be a serious methodological issue, particularly with in-home sampling that is often employed for studies of diurnal cortisol rhythms. There is evidence that parental noncompliance is particularly related to findings of a blunted CAR, which is most impacted by noncompliance with the timing of the waking sample (Smith & Dougherty, 2014). Depending upon the sampling age, collection may also potentially be impacted by differences in diet, sleep, and school routines. Whether parental supervision of sampling is required (e.g., in early childhood), or adolescents are controlling their own samples, it may be beneficial to utilize color-coded collection tubes to denote specific collection times, provide reminders via text, and/or provide participants with additional incentives for completion of saliva collection. Researchers have also suggested electronic monitors to reduce noncompliance (Smith & Dougherty, 2014). For readers interested in specific recommendations related to saliva collection in children, there are several articles that provide thorough discussions of such issues (e.g., Hanrahan, McCarthy, Kleiber, Lutgendorf, & Tsalikian, 2006; Schwartz, Granger, Susman, Gunnar, & Laird, 1998).

Salivary bioscience has been extremely useful in measurement of prenatal adversity and the noninvasive nature of collecting saliva samples allows for repeated measurement during pregnancy. This not only enables researchers to examine timing effects, but also allows for exploration of changes in maternal stress or substance use across pregnancy as it relates to child outcomes. Indeed, studies of maternal prenatal stress are increasingly using saliva to measure maternal cortisol and changes in cortisol across pregnancy in relation to child outcomes (e.g., Davis et al., 2011; O'Connor et al., 2013). Similarly, salivary assays for tobacco and cannabis metabolites in particular allow for biochemical verification of maternal substance use during pregnancy and changes in use across time (e.g., Shisler et al., 2017). This is critically important given the high social desirability of reporting no substance use during pregnancy and fears of legal repercussions in some states when mothers are actively using. Collection of saliva samples for this purpose is particularly problematic in states where maternal substance use during pregnancy is considered to be child abuse (see Guttmacher Institute, 2018), but is vital from a scientific perspective. Biochemical verification in addition to intensive and reliable self-report measures (see Shisler et al., 2017) is also critical to ascertain that the comparison groups are appropriate. Recruitment of mothers who are in substance use treatment may partially mitigate these legal issues, but also has implications for generalizability.

Finally, while recent advances in salivary biosciences for measurement of salivary cytokines opens exciting possibilities for noninvasive measurement of risk markers for health and disease, further work is needed to understand reliability, validity, and interpretation of these markers. Studies linking prenatal adversity to salivary cytokines that also include measures of health and utilize repeated measures of salivary cytokines across development may help to shed more light in this area.

## 25.5 Future Directions and Opportunities

**Bidirectional Interactions Between Stress and Immune Function** In most of this chapter, we have reviewed prenatal adversity to child stress and immune/inflammation as it relates to salivary bioscience as two separate domains of functioning. This may be misleading since cortisol has a wide-ranging impact on regulating metabolic and immune functions in addition to responding to increased stress (Hostinar,

Nusslock, & Miller, 2018). Indeed, the Biological Embedding Model specifically proposes that major psychological stressors become biologically embedded at the molecular level increasing risk for morbidity and mortality via alterations in immune/inflammatory pathways. Recently, a neuroimmune network hypothesis has been proposed (Hostinar et al., 2018) that integrates research findings and argues for the existence of an integrated neuroimmune network that includes the brain, immune system, and behavior. This network is influenced by early life adversity that we suggest should include prenatal adversity, is self-perpetuating, and disease promoting. Few studies of prenatal adversity have examined the linkages between stress–immune/inflammatory mechanisms across development. This is now possible with recent advances in salivary bioscience that allow for examination of salivary cytokines. While high levels of salivary cytokines can only be interpreted in the context of the oral mucosal system, it does facilitate examination of linkages and changes in cytokines across development, across days, and across time when examining responses to acute stressors.

An additional direction for future studies may be integrating studies of stress and immune function with examination of epigenetic mechanisms. The chapter on Genetics and Epigenetics (Chap. 6, Nemoda et al.) has a systematic review of measurement issues and applications of using salivary samples in genetic analysis. Saliva is a reliable source for DNA and can be used for genetic testing (Nemoda, this volume), thus allowing examination of the complex interplay between stress–immune/inflammatory associations via epigenetic mechanisms.

**Potential Multiplicative Effects of Later Adversity or Protective Effects of Environmental Supports** A second major area for future research is examination of interactive effects of prenatal and postnatal adversity, or buffering effects of positive postnatal environments. In research conducted on prairie voles, Hartman, Freeman, Bales, & Belsky (2018) demonstrated that prenatal stress increased sensitivity to postnatal rearing and resulted in the highest behavioral and physiological reactivity under conditions of low quality rearing but the lowest behavioral and physiological reactivity under conditions of high quality rearing. Thus, prenatal stress may be conceptualized as a risk or an opportunity factor. Further, a number of human studies indicate stress buffering effects of secure infant attachment to parent with some suggestion that biologically, this may be mediated by increases in oxytocin during interactions with trusted caregivers (Doom & Gunnar, 2013 review). Theories that support stress buffering effects such as diathesis stress, differential susceptibility (Belsky & Pluess, 2009), and biological sensitivity to context (Boyce & Ellis, 2005), all support interactive effects of stressful experiences and positive or negative caregiving contexts. However, only a few studies of prenatal adversity effects on salivary outcomes have examined interactions with postnatal variables.

**Sex Differences** One issue that emerged continuously across different domains of prenatal adversity was the importance of examining sex differences. There are sex differences noted in changes in salivary cytokines (Riis et al., 2015) as well as salivary cortisol reactivity to stress (e.g., Eiden et al., 2015). However, few studies consider sex as a moderator of the association between prenatal adversity and salivary outcomes. In order to examine sex as a moderator, samples need to be large enough for sufficient statistical power to examine interactions of prenatal adversity and child sex.

**Timing of Exposure** A second issue was the importance of considering timing of exposure to adversity (Davis et al., 2011; O'Donnell et al., 2013; Van den Bergh et al., 2008; Yehuda et al., 2005). Although results were not consistent across studies regarding first half of pregnancy versus later half versus specific third trimester effects, consensus may only emerge if more studies examine timing of exposure carefully. This is only possible if study designs are prospective.

**Problems with Laboratory Measures and Importance of Diurnal Patterns** Finally, much of the literature on prenatal adversity effects on salivary outcomes have examined acute responses to stress. Only a small number of studies have examined diurnal patterns of stress response. This is critical given the issues surrounding accurate measurement of baseline cortisol in the laboratory (see Zmyj, Schneider, & Seehagen, 2017). It has been suggested that one reason many studies of stress induction in the laboratory do not see increases in cortisol may be because infants are not given sufficient time to acclimate to the laboratory environment and the stresses associated with arriving at the laboratory. Thus, the pretask cortisol is higher than normal baseline and this results in ceiling effects (Zmyj et al., 2017). Indeed, given sufficient time in the laboratory (e.g., 30 min), infants exhibit decreases in their cortisol levels. Thus, consideration of time spent in the laboratory before the first saliva collection is critical for examination of cortisol reactivity, and may well be important for other salivary markers as well.

**Parent–Child Synchrony and Synchrony Across Systems** Studies of prenatal adversity have not routinely examined mother–child or father–child synchrony in cortisol levels. While prenatal adversity alone clearly poses risk for children, most children who experience prenatal adversity continue to experience postnatal adversity (Eiden, Schuetze, Shisler, & Huestis, 2018). This postnatal adversity in the form of continued environmental stressors may also impact maternal stress and immune/inflammatory pathways, and these may mediate some of the association between prenatal adversity and child outcomes. This is an important area for study, with implications for treatment and intervention. Measures of cumulative risk (Evans, Li, & Whipple, 2013) may be particularly useful in capturing chronic exposure to postnatal stressors. Cumulative risk scores can capture meaningful differences in children's related experiences of risk on a broad range of environmental stressors by way of a single score. These scores can represent the aggregate exposure to multiple risks from multiple stressors, either at a single point in time, or across a specific period of time, and high levels of cumulative risk have been predictive of myriad types of developmental maladjustment. Finally, only a few studies have examined synchrony between stress response systems such as the HPA and the autonomic nervous system (e.g., Martinez-Torteya et al., 2017). Advances in salivary bioscience that allow for measurement of sAA and cortisol simultaneously across the same day or across the same tasks allow for analyses of factors related to synchrony and asynchrony of these systems. Results from a few studies indicate that asynchrony may be a marker for dysregulation of the stress response system, which has implications for other developmental outcomes. Advances in salivary bioscience have created opportunities for innovative and noninvasive assessment of stress response and immune/inflammation systems, which are critical components to understanding, and preventing, maladaptive health outcomes. As discussed, salivary bioscience has opened up exciting new avenues for future empirical and theoretical work on prenatal adversity and fetal origins of disease.

## References

Almanzar, G., Eberle, G., Lassacher, A., Specht, C., Koppelstaetter, C., Heinz-Erian, P., … Prelog, M. (2013). Maternal cigarette smoking and its effect on neonatal lymphocyte subpopulations and replication. *BMC Pediatrics, 13*, 57.

Appleton, A. A., Armstrong, D. A., Lesseur, C., Lee, J., Padbury, J. F., Lester, B. M., & Marsit, C. J. (2013). Patterning in placental 11-B hydroxysteroid dehydrogenase methylation according to prenatal socioeconomic adversity. *PLoS One, 8*, e74691.

Asadullah, K., Sterry, W., & Volk, H. (2003). Interleukin-10 therapy—Review of a new approach. *Pharmacological Reviews, 55*, 241–269.

Astley, S. J., & Clarren, S. K. (2001). Measuring the facial phenotype of individuals with prenatal alcohol exposure: Correlations with brain dysfunction. *Alcohol and Alcoholism, 36*, 147–159.

Äyräväinen, L., Heikkinen, A. M., Kuuliala, A., Ahola, K., Koivuniemi, R., Laasonen, L., … Meurman, J. H. (2018). Inflammatory biomarkers in saliva and serum of patients with rheumatoid arthritis with respect to periodontal status. *Annals of Medicine, 50*, 333–344.

Barker, D. J. (1999). Fetal origins of cardiovascular disease. *Annals of Medicine, 31*, 3–6.

Barker, D. J., Eriksson, J. G., Forsén, T., & Osmond, C. (2002). Fetal origins of adult disease: Strength of effects and biological basis. *International Journal of Epidemiology, 31*, 1235–1239.

Barker, D. J., Godfrey, K. M., Gluckman, P. D., Harding, J. E., Owens, J. A., & Robinson, J. S. (1993). Fetal nutrition and cardiovascular disease in adult life. *The Lancet, 341*, 938–941.

Barker, D. J., Osmond, C., Kajantie, E., & Eriksson, J. G. (2009). Growth and chronic disease: Findings in the Helsinki Birth Cohort. *Annals of Human Biology, 36*, 445–458.

Bauer, C. R., Lambert, B. L., Bann, C. M., Lester, B. M., Shankaran, S., Bada, H. S., … Higgins, R. D. (2011). Long-term impact of maternal substance use during pregnancy and extrauterine environmental adversity: Stress hormone levels of preadolescent children. *Pediatric Research, 70*, 213–219.

Beijers, R., Jansen, J., Riksen-Walraven, M., & De Weerth, C. (2010). Maternal prenatal anxiety and stress predict infant illnesses and health complaints. *Pediatrics, 126*(2), e401–e409.

Belsky, J., & Pluess, M. (2009). Beyond diathesis stress: Differential susceptibility to environmental influences. *Psychological Bulletin, 135*, 885.

Berg, C. J., Haardorfer, R., McBride, C. M., Kilaru, V., Ressler, K. J., Wingo, A. P., … Smith, A. (2017). Resilience and biomarkers of health risk in Black smokers and nonsmokers. *Health Psychology, 36*, 1047–1058.

Blalock, J. (2005). The immune system as the sixth sense. *Journal of Internal Medicine, 257*, 126–138.

Blalock, J. E., & Smith, E. M. (2007). Conceptual development of the immune system as a sixth sense. *Brain, Behavior, and Immunity, 21*, 23–33.

Boyce, W. T., & Ellis, B. J. (2005). Biological sensitivity to context: I. An evolutionary–developmental theory of the origins and functions of stress reactivity. *Development and Psychopathology, 17*, 271–301.

Braithwaite, E. C., Murphy, S. E., & Ramchandani, P. G. (2016). Effects of prenatal depressive symptoms on maternal and infant cortisol reactivity. *Archives of Women's Mental Health, 19*, 581–590.

Brennan, P. A., Pargas, R., Walker, E. F., Green, P., Newport, D. J., & Stowe, Z. (2008). Maternal depression and infant cortisol: Influences of timing, comorbidity and treatment. *Journal of Child Psychology and Psychiatry, 49*, 1099–1107.

Buckingham-Howes, S., Mazza, D., Wang, Y., Granger, D. A., & Black, M. M. (2016). Prenatal drug exposure and adolescent cortisol reactivity: Association with behavioral concerns. *Journal of Developmental and Behavioral Pediatrics: JDBP, 37*, 565.

Buss, C., Entringer, S., & Wadhwa, P. D. (2012). Fetal programming of brain development: Intrauterine stress and susceptibility to psychopathology. *Science Signaling, 5*, pt7.

Calkins, K., & Devaskar, S. U. (2011). Fetal origins of adult disease. *Current Problems in Pediatric and Adolescent Health Care, 41*, 158–176.

Calkins, S. D., Propper, C., & Mills-Koonce, W. R. (2013). A biopsychosocial perspective on parenting and developmental psychopathology. *Development and Psychopathology, 25*, 1399–1414.

Cao-Lei, L., Massart, R., Suderman, M. J., Machnes, Z., Elgbeili, G., Laplante, D. P., … King, S. (2014). DNA methylation signatures triggered by prenatal maternal stress exposure to a natural disaster: Project Ice Storm. *PLoS One, 9*, e107653.

Carroll, D., Lovallo, W. R., & Phillips, A. C. (2009). Are large physiological reactions to acute psychological stress always bad for health? *Social and Personality Psychology Compass, 3*, 725–743.

Chabarria, K. C., Racusin, D. A., Antony, K. M., Kahr, M., Suter, M. A., Mastrobattista, J. M., & Aagaard, K. M. (2016). Marijuana use and its effects in pregnancy. *American Journal of Obstetrics and Gynecology, 215*, 506.e1–506.e7.

Chahal, N., McLain, A. C., Ghassabian, A., Michels, K. A., Bell, E. M., Lawrence, D. A., & Yeung, E. H. (2017). Maternal smoking and newborn cytokine and immunoglobulin levels. *Nicotine and Tobacco Research, 19*, 789–796.

Chaplin, T. M., Visconti, K. J., Molfese, P. J., Susman, E. J., Klein, L. C., Sinha, R., & Mayes, L. C. (2015). Prenatal cocaine exposure differentially affects stress responses in girls and boys: Associations with future substance use. *Development and Psychopathology, 27*, 163–180.

Coe, C. L., & Lubach, G. R. (2003). Critical periods of special health relevance for psychoneuroimmunology. *Brain, Behavior, and Immunity, 17*, 3–12.

Coe, C. L., & Lubach, G. R. (2005). Prenatal origins of individual variation in behavior and immunity. *Neuroscience and Biobehavioral Reviews, 29*, 39–49.

Cottrell, E. C., & Seckl, J. R. (2009). Prenatal stress, glucocorticoids and the programming of adult disease. *Frontiers in Behavioral Neuroscience, 3*, 19.

Danese, A., & McEwen, B. S. (2012). Adverse childhood experiences, allostasis, allostatic load, and age-related disease. *Physiology and Behavior, 106*, 29–39.

Davis, E. P., Glynn, L. M., Waffarn, F., & Sandman, C. A. (2011). Prenatal maternal stress programs infant stress regulation. *Journal of Child Psychology and Psychiatry, 52*, 119–129.

Day, N., Cornelius, M., Goldschmidt, L., Richardson, G., Robles, N., & Taylor, P. (1992). The effects of prenatal tobacco and marijuana use on offspring growth from birth through 3 years of age. *Neurotoxicology and Teratology, 14*, 407–414.

Del Giudice, M., Ellis, B. J., & Shirtcliff, E. A. (2011). The Adaptive Calibration Model of stress responsivity. *Neuroscience and Biobehavioral Reviews, 35*, 1562–1592.

Doom, J. R., & Gunnar, M. R. (2013). Stress physiology and developmental psychopathology: Past, present, and future. *Development and Psychopathology, 25*, 1359–1373.

Duramad, P., Tager, I. B., & Holland, N. T. (2007). Cytokines and other immunological biomarkers in children's environmental health studies. *Toxicology Letters, 172*, 48–59.

Eiden, R. D., Foote, A., & Schuetze, P. (2007). Maternal cocaine use and caregiving status: Group differences in caregiver and infant risk variables. *Addictive Behaviors, 32*, 465–476.
Eiden, R. D., Homish, G. G., Colder, C. R., Schuetze, P., Gray, T. R., & Huestis, M. A. (2013). Changes in smoking patterns during pregnancy. *Substance Use and Misuse, 48*, 513–522.
Eiden, R. D., Molnar, D. S., Granger, D. A., Colder, C. R., Schuetze, P., & Huestis, M. A. (2015). Prenatal tobacco exposure and infant stress reactivity: Role of child sex and maternal behavior. *Developmental Psychobiology, 57*, 212–225.
Eiden, R. D., Schuetze, P., & Coles, C. D. (2011). Maternal cocaine use and mother–infant interactions: Direct and moderated associations. *Neurotoxicology and Teratology, 33*, 120–128.
Eiden, R. D., Schuetze, P., Shisler, S., & Huestis, M. A. (2018). Prenatal exposure to tobacco and cannabis: Effects on autonomic and emotion regulation. *Neurotoxicology and Teratology, 68*, 47–56.
Eiden, R. D., Veira, Y., & Granger, D. A. (2009). Prenatal cocaine exposure and infant cortisol reactivity. *Child Development, 80*, 528–543.
Evans, G. W., Li, D., & Whipple, S. S. (2013). Cumulative risk and child development. *Psychological Bulletin, 139*(6), 1342–1396. https://doi.org/10.1037/a0031808
Fisher, P. A., Kim, H. K., Bruce, J., & Pears, K. C. (2012). Cumulative effects of prenatal substance exposure and early adversity on Foster Children's HPA Axis reactivity during a psychosocial stressor. *International Journal of Behavioral Development, 36*, 29–35.
Gleeson, M., & Cripps, A. W. (2004). Development of mucosal immunity in the first year of life and relationship to sudden infant death syndrome. *FEMS Immunology and Medical Microbiology, 42*, 21–33.
Granger, D. A., Johnson, S. B., Szanton, S. L., Out, D., & Schumann, L. L. (2012). Incorporating salivary biomarkers into nursing research: An overview and review of best practices. *Biological Research for Nursing, 14*, 347–356.
Grant, K. A., McMahon, C., Austin, M. P., Reilly, N., Leader, L., & Ali, S. (2009). Maternal prenatal anxiety, postnatal caregiving and infants' cortisol responses to the still-face procedure. *Developmental Psychobiology, 51*, 625–637.
Groner, J. A., Huang, H., Joshi, M. S., Eastman, N., Nicholson, L., & Bauer, J. A. (2017). Secondhand smoke exposure and preclinical markers of cardiovascular risk in toddlers. *The Journal of Pediatrics, 189*, 155–161.
Gunnar, M. R. (1986). Human developmental psychoneuroendocrinology: A review of research on neuroendocrine responses to challenge and threat in infancy and childhood. *Advances in Developmental Psychology, 4*, 51–103.
Gunnar, M. R., Brodersen, L., Krueger, K., & Rigatuso, J. (1996). Dampening of adrenocortical responses during infancy: Normative changes and individual differences. *Child Development, 67*, 877–889.
Gunnar, M. R., & Vazquez, D. M. (2001). Low cortisol and a flattening of expected daytime rhythm: Potential indices of risk in human development. *Development and Psychopathology, 13*, 515–538.
Guttmacher Institute. (2018). *Substance use during pregnancy* [Online]. Retrieved from https://www.guttmacher.org/print/state-policy/explore/substance-use-during-pregnancy
Haley, D. W., Handmaker, N. S., & Lowe, J. (2006). Infant stress reactivity and prenatal alcohol exposure. *Alcoholism, Clinical and Experimental Research, 30*, 2055–2064.
Hanrahan, K., McCarthy, A. M., Kleiber, C., Lutgendorf, S., & Tsalikian, E. (2006). Strategies for salivary cortisol collection and analysis in research with children. *Applied Nursing Research, 19*, 95–101.
Hartman, S., Freeman, S. M., Bales, K. L., & Belsky, J. (2018). Prenatal stress as a risk—And an opportunity—Factor. *Psychological Science, 29*, 572–580.
Hostinar, C. E., Nusslock, R., & Miller, G. E. (2018). Future directions in the study of early-life stress and physical and emotional health: Implications of the Neuroimmune network hypothesis. *Journal of Clinical Child and Adolescent Psychology, 47*, 142–156.

Huijbregts, S. C., Van Berkel, S. R., Swaab-Barneveld, H., & Van Goozen, S. H. (2011). Neurobiological and behavioral stress reactivity in children prenatally exposed to tobacco. *Psychoneuroendocrinology, 36*, 913–918.

Hutter, D., & Jaeggi, E. (2010). Causes and mechanisms of intrauterine hypoxia and its impact on the fetal cardiovascular system; a review. *International Journal Of Pediatrics, 2010*, 401323.

Iyengar, A., Paulus, J. K., Gerlanc, D. J., & Maron, J. L. (2014). Detection and potential utility of C-reactive protein in saliva of neonates. *Frontiers in Pediatrics, 2*, 131.

Jacobson, S. W., Bihun, J. T., & Chiodo, L. M. (1999). Effects of prenatal alcohol and cocaine exposure on infant cortisol levels. *Development and Psychopathology, 11*, 195–208.

Johnson, S. B., Riley, A. W., Granger, D. A., & Riis, J. (2013). The science of early life toxic stress for pediatric practice and advocacy. *Pediatrics, 131*(2), 319–327.

Kang, E., Kim, S. Y., Chang, S. S., Lim, S., Kim, H. C., Lee, C. G., … Ha, M. (2017). Environmental tobacco smoke exposure at home and high-sensitivity C-reactive protein levels in three-to-five-year-old children. *International Journal of Environmental Research and Public Health, 14*, 1105.

Keiver, K., Bertram, C. P., Orr, A. P., & Clarren, S. (2015). Salivary cortisol levels are elevated in the afternoon and at bedtime in children with prenatal alcohol exposure. *Alcohol, 49*, 79–87.

Kieran, M. W., Kalluri, R., & Cho, Y.-J. (2012). The VEGF pathway in cancer and disease: Responses, resistance, and the path forward. *Cold Spring Harbor Perspectives in Medicine, 2* (12), a006593.

King, S. L., & Hegadoren, K. M. (2002). Stress hormones: How do they measure up? *Biological Research for Nursing, 4*, 92–103.

Kirschbaum, C., & Hellhammer, D. H. (1989). Salivary cortisol in psychobiological research: An overview. *Neuropsychobiology, 22*, 150–169.

Knopik, V. S., Neiderhiser, J. M., De Geus, E., & Boomsma, D. (2016). *The importance of the prenatal environment in behavioral genetics: Introduction to special issue*. New York: Springer.

Lester, B. M., Lagasse, L. L., Shankaran, S., Bada, H. S., Bauer, C. R., Lin, R., … Higgins, R. (2010). Prenatal cocaine exposure related to cortisol stress reactivity in 11-year-old children. *The Journal of Pediatrics, 157*(288–295), e1.

Leung, E., Tasker, S. L., Atkinson, L., Vaillancourt, T., Schulkin, J., & Schmidt, L. A. (2010). Perceived maternal stress during pregnancy and its relation to infant stress reactivity at 2 days and 10 months of postnatal life. *Clinical Pediatrics (Phila), 49*, 158–165.

Lovallo, W. R. (2011). Do low levels of stress reactivity signal poor states of health? *Biological Psychology, 86*, 121–128.

Luecken, L. J., Lin, B., Coburn, S. S., MacKinnon, D. P., Gonzales, N. A., & Crnic, K. A. (2013). Prenatal stress, partner support, and infant cortisol reactivity in low-income Mexican American families. *Psychoneuroendocrinology, 38*, 3092–3101.

Lupien, S. J., McEwen, B. S., Gunnar, M. R., & Heim, C. (2009). Effects of stress throughout the lifespan on the brain, behaviour and cognition. *Nature Reviews Neuroscience, 10*, 434.

Magnano, C. L., Gardner, J. M., & Karmel, B. Z. (1992). Differences in salivary cortisol levels in cocaine-exposed and noncocaine-exposed NICU infants. *Developmental Psychobiology: The Journal of the International Society for Developmental Psychobiology, 25*, 93–103.

Mark, K., Desai, A., & Terplan, M. (2016). Marijuana use and pregnancy: Prevalence, associated characteristics, and birth outcomes. *Archives of Women's Mental Health, 19*, 105–111.

Marques, A. H., O'Connor, T. G., Roth, C., Susser, E., & Bjørke-Monsen, A.-L. (2013). The influence of maternal prenatal and early childhood nutrition and maternal prenatal stress on offspring immune system development and neurodevelopmental disorders. *Frontiers in Neuroscience, 7*, 120.

Martinez-Torteya, C., Bogat, G. A., Levendosky, A. A., & Von Eye, A. (2016). The influence of prenatal intimate partner violence exposure on hypothalamic–pituitary–adrenal axis reactivity and childhood internalizing and externalizing symptoms. *Development and Psychopathology, 28*, 55–72.

Martinez-Torteya, C., Bogat, G. A., Lonstein, J. S., Granger, D. A., & Levendosky, A. A. (2017). Exposure to intimate partner violence in utero and infant internalizing behaviors: Moderation by salivary cortisol-alpha amylase asymmetry. *Early Human Development, 113*, 40–48.

Marzi, M., Vigano, A., Trabattoni, D., Villa, M., Salvaggio, A., Clerici, E., & Clerici, M. (1996). Characterization of type 1 and type 2 cytokine production profile in physiologic and pathologic human pregnancy. *Clinical and Experimental Immunology, 106*, 127–133.

Massey, S. H., Reiss, D., Neiderhiser, J. M., Leve, L. D., Shaw, D. S., & Ganiban, J. M. (2016). Maternal personality traits associated with patterns of prenatal smoking and exposure: Implications for etiologic and prevention research. *Neurotoxicology and Teratology, 53*, 48–54.

Matsunaga, Y., Vardavas, C. I., Plada, M., Warnberg, J., Gomez-Martinez, S., Tzatzarakis, M. N., … Kafatos, A. G. (2014). The relationship between cotinine concentrations and inflammatory markers among highly secondhand smoke exposed non-smoking adolescents. *Cytokine, 66*, 17–22.

McEwen, B. S., & Stellar, E. (1993). Stress and the individual: Mechanisms leading to disease. *Archives of Internal Medicine, 153*, 2093–2101.

McLachlan, K., Rasmussen, C., Oberlander, T. F., Loock, C., Pei, J., Andrew, G., … Weinberg, J. (2016). Dysregulation of the cortisol diurnal rhythm following prenatal alcohol exposure and early life adversity. *Alcohol, 53*, 9–18.

Merghani, T. H., Saeed, A., & Alawad, A. (2012). Changes in plasma IL4, TNFa and CRP in response to regular passive smoking at home among healthy school children in Khartoum, Sudan. *African Health Sciences, 12*, 41–47.

Miller, G. E., Chen, E., & Parker, K. J. (2011). Psychological stress in childhood and susceptibility to the chronic diseases of aging: Moving toward a model of behavioral and biological mechanisms. *Psychological Bulletin, 137*, 959.

Miller, G. E., Chen, E., & Zhou, E. S. (2007). If it goes up, must it come down? Chronic stress and the hypothalamic-pituitary-adrenocortical axis in humans. *Psychological Bulletin, 133*, 25.

Molnar, D. S., Granger, D. A., Shisler, S., & Eiden, R. D. (2018). Prenatal and postnatal cigarette and cannabis exposure: Effects on Secretory Immunoglobulin A in early childhood. *Neurotoxicology and Teratology, 67*, 31–36.

Monk, C., Spicer, J., & Champagne, F. A. (2012). Linking prenatal maternal adversity to developmental outcomes in infants: The role of epigenetic pathways. *Development and Psychopathology, 24*, 1361–1376.

Mulligan, C., D'Errico, N., Stees, J., & Hughes, D. (2012). Methylation changes at NR3C1 in newborns associate with maternal prenatal stress exposure and newborn birth weight. *Epigenetics, 7*, 853–857.

Nater, U. M., & Rohleder, N. (2009). Salivary alpha-amylase as a non-invasive biomarker for the sympathetic nervous system: Current state of research. *Psychoneuroendocrinology, 34*, 486–496.

Noakes, P. S., Hale, J., Thomas, R., Lane, C., Devadason, S. G., & Prescott, S. L. (2006). Maternal smoking is associated with impaired neonatal toll-like receptor (TLR) mediated immune responses. *European Respiratory Journal, 28*, 721.

Noakes, P., Holt, P., & Prescott, S. (2003). Maternal smoking in pregnancy alters neonatal cytokine responses. *Allergy, 58*, 1053–1058.

O'Connor, T. G., Ben-Shlomo, Y., Heron, J., Golding, J., Adams, D., & Glover, V. (2005). Prenatal anxiety predicts individual differences in cortisol in pre-adolescent children. *Biological Psychiatry, 58*, 211–217.

O'Connor, T. G., Bergman, K., Sarkar, P., & Glover, V. (2013). Prenatal cortisol exposure predicts infant cortisol response to acute stress. *Developmental Psychobiology, 55*, 145–155.

O'Connor, T. G., Heron, J., Glover, V., & Team, A. S. (2002). Antenatal anxiety predicts child behavioral/emotional problems independently of postnatal depression. *Journal of the American Academy of Child and Adolescent Psychiatry, 41*, 1470–1477.

O'Donnell, K. J., Glover, V., Jenkins, J., Browne, D., Ben-Shlomo, Y., Golding, J., & O'Connor, T. G. (2013). Prenatal maternal mood is associated with altered diurnal cortisol in adolescence. *Psychoneuroendocrinology, 38*, 1630–1638.

O'Donnell, K. J., & Meaney, M. J. (2017). Fetal origins of mental health: The developmental origins of health and disease hypothesis. *The American Journal of Psychiatry, 174*, 319–328.

Ouellet-Morin, I., Dionne, G., Lupien, S. J., Muckle, G., Cote, S., Perusse, D., ... Boivin, M. (2011). Prenatal alcohol exposure and cortisol activity in 19-month-old toddlers: An investigation of the moderating effects of sex and testosterone. *Psychopharmacology, 214*, 297–307.

Pearson, T. A. (2003). Markers of inflammation and cardiovascular disease: Application to clinical and public health practice: A statement for healthcare professionals from the Centers for Disease Control and Prevention and the American Heart Association. *Circulation, 107*, 499–511.

Pearson, R. M., Evans, J., Kounali, D., Lewis, G., Heron, J., Ramchandani, P. G., ... Stein, A. (2013). Maternal depression during pregnancy and the postnatal period: Risks and possible mechanisms for offspring depression at age 18 years. *JAMA Psychiatry, 70*, 1312–1319.

Pedersen, J. M., Mortensen, E. L., Christensen, D. S., Rozing, M., Brunsgaard, H., Meincke, R. H., ... Lund, R. (2018). Prenatal and early postnatal stress and later life inflammation. *Psychoneuroendocrinology, 88*, 158–166.

Ping, E. Y., Laplante, D. P., Elgbeili, G., Hillerer, K. M., Brunet, A., O'Hara, M. W., & King, S. (2015). Prenatal maternal stress predicts stress reactivity at 2½ years of age: The Iowa flood study. *Psychoneuroendocrinology, 56*, 62–78.

Pirhonen, J., Sareneva, T., Kurimoto, M., Julkunen, I., & Matikainen, S. (1999). Virus infection activates IL-1β and IL-18 production in human macrophages by a caspase-1-dependent pathway. *The Journal of Immunology, 162*, 7322–7329.

Plant, D. T., Pawlby, S., Sharp, D., Zunszain, P. A., & Pariante, C. M. (2016). Prenatal maternal depression is associated with offspring inflammation at 25 years: A prospective longitudinal cohort study. *Translational Psychiatry, 6*, e936.

Prescott, S. L. (2008). Effects of early cigarette smoke exposure on early immune development and respiratory disease. *Paediatric Respiratory Reviews, 9*, 3–10.

Ramsay, D. S., Bendersky, M. I., & Lewis, M. (1996). Effect of prenatal alcohol and cigarette exposure on two-and six-month-old infants' adrenocortical reactivity to stress. *Journal of Pediatric Psychology, 21*, 833–840.

Riis, J. L., Granger, D. A., Dipietro, J. A., Bandeen-Roche, K., & Johnson, S. B. (2015). Salivary cytokines as a minimally-invasive measure of immune functioning in young children: Correlates of individual differences and sensitivity to laboratory stress. *Developmental Psychobiology, 57*, 153–167.

Riis, J. L., Out, D., Dorn, L. D., Beal, S. J., Denson, L. A., Pabst, S., ... Granger, D. A. (2014). Salivary cytokines in healthy adolescent girls: Intercorrelations, stability, and associations with serum cytokines, age, and pubertal stage. *Developmental Psychobiology, 56*, 797–811.

Rueda-Clausen, C. F., Morton, J. S., & Davidge, S. T. (2009). Effects of hypoxia-induced intrauterine growth restriction on cardiopulmonary structure and function in adulthood. *Cardiovascular Research, 81*(4), 713–722.

Rutter, M. (1987). Psychosocial resilience and protective mechanisms. *American Journal of Orthopsychiatry, 57*, 316–331.

Saito, S., Sakai, M., Sasaki, Y., Tanebe, K., Tsuda, H., & Michimata, T. (1999). Quantitative analysis of peripheral blood Th0, Th1, Th2 and the Th1: Th2 cell ratio during normal human pregnancy and preeclampsia. *Clinical and Experimental Immunology, 117*, 550.

Sameroff, A. J., Seifer, R., Baldwin, A., & Baldwin, C. (1993). Stability of intelligence from preschool to adolescence: The influence of social and family risk factors. *Child Development, 64*, 80–97.

Schmidt, N. A. (1997). Salivary cortisol testing in children. *Issues in Comprehensive Pediatric Nursing, 20*, 183–190.

Schroder, K., Hertzog, P. J., Ravasi, T., & Hume, D. A. (2004). Interferon-γ: An overview of signals, mechanisms and functions. *Journal of Leukocyte Biology, 75*, 163–189.

Schuetze, P., Eiden, R. D., Colder, C. R., Huestis, M. A., & Leonard, K. E. (2018). Prenatal risk and infant regulation: Indirect pathways via fetal growth and maternal prenatal stress and anger. *Child Development, 89*, e123–e137.

Schuetze, P., Lopez, F. A., Granger, D. A., & Eiden, R. D. (2008). The association between prenatal exposure to cigarettes and cortisol reactivity and regulation in 7-month-old infants. *Developmental Psychobiology: The Journal of the International Society for Developmental Psychobiology, 50*, 819–834.

Schuetze, P., Shisler, S., Eiden, R. D., & Granger, D. A. (2018). *Effects of co-occurring prenatal tobacco and cannabis exposure on cortisol reactivity at early school age*. San Diego, CA: College for Problems in Drug Dependence.

Schwartz, E. B., Granger, D. A., Susman, E. J., Gunnar, M. R., & Laird, B. (1998). Assessing salivary cortisol in studies of child development. *Child Development, 69*, 1503–1513.

Selye, H. (1950). *The physiology and pathology of exposure to stress*. Montreal: ACTA.

Sesso, H. D., Buring, J. E., Rifai, N., Blake, G. J., Gaziano, J. M., & Ridker, P. M. (2003). C-reactive protein and the risk of developing hypertension. *JAMA, 290*, 2945–2951.

Shisler, S., Eiden, R. D., Molnar, D. S., Schuetze, P., Huestis, M., & Homish, G. (2017). Smoking in pregnancy and fetal growth: The case for more intensive assessment. *Nicotine and Tobacco Research, 19*, 525–531.

Shonkoff, J. P., Garner, A. S., Siegel, B. S., Dobbins, M. I., Earls, M. F., McGuinn, L., … Committee on Early Childhood, Adoption, and Dependent Care. (2012). The lifelong effects of early childhood adversity and toxic stress. *Pediatrics, 129*, e232–e246.

Smith, V. C., & Dougherty, L. R. (2014). Noisy spit: Parental noncompliance with child salivary cortisol sampling. *Developmental Psychobiology, 56*, 647–656.

Spear, L. P., Kirstein, C. L., Bell, J., Yoottanasumpun, V., Greenbaum, R., O'Shea, J., … Spear, N. (1989). Effects of prenatal cocaine exposure on behavior during the early postnatal period. *Neurotoxicology and Teratology, 11*, 57–63.

Stroud, L. R., Papandonatos, G. D., Parade, S. H., Salisbury, A. L., Phipps, M. G., Lester, B. M., … Marsit, C. J. (2016). Prenatal major depressive disorder, placenta glucocorticoid and serotonergic signaling, and infant cortisol response. *Psychosomatic Medicine, 78*, 979–990.

Stroud, L. R., Papandonatos, G. D., Rodriguez, D., McCallum, M., Salisbury, A. L., Phipps, M. G., … Marsit, C. J. (2014). Maternal smoking during pregnancy and infant stress response: Test of a prenatal programming hypothesis. *Psychoneuroendocrinology, 48*, 29–40.

US Department of Health and Human Services. (2014). *The health consequences of smoking—50 years of progress: A report of the Surgeon General* (p. 17). Atlanta, GA: US Department of Health and Human Services, Centers for Disease Control and Prevention, National Center for Chronic Disease Prevention and Health Promotion, Office on Smoking and Health.

Van Den Bergh, B. R., Van Calster, B., Smits, T., Van Huffel, S., & Lagae, L. (2008). Antenatal maternal anxiety is related to HPA-axis dysregulation and self-reported depressive symptoms in adolescence: A prospective study on the fetal origins of depressed mood. *Neuropsychopharmacology, 33*, 536–545.

Velders, F. P., Dieleman, G., Cents, R. A., Bakermans-Kranenburg, M. J., Jaddoe, V. W., Hofman, A., … Tiemeier, H. (2012). Variation in the glucocorticoid receptor gene at rs41423247 moderates the effect of prenatal maternal psychological symptoms on child cortisol reactivity and behavior. *Neuropsychopharmacology, 37*, 2541–2549.

Volpato, S., Guralnik, J. M., Ferrucci, L., Balfour, J., Chaves, P., Fried, L. P., & Harris, T. B. (2001). Cardiovascular disease, interleukin-6, and risk of mortality in older women: The women's health and aging study. *Circulation, 103*, 947–953.

Wadhwa, P. D., Buss, C., Entringer, S., & Swanson, J. M. (2009). Developmental origins of health and disease: Brief history of the approach and current focus on epigenetic mechanisms. *Seminars in Reproductive Medicine, 27*(5), 358–368.

Watamura, S. E., Donzella, B., Kertes, D. A., & Gunnar, M. R. (2004). Developmental changes in baseline cortisol activity in early childhood: Relations with napping and effortful control. *Developmental Psychobiology: The Journal of the International Society for Developmental Psychobiology, 45*, 125–133.

Woods, J. R., Plessinger, M. A., & Clark, K. E. (1987). Effect of cocaine on uterine blood flow and fetal oxygenation. *JAMA, 257*, 957–961.

Woolston, J. L., Gianfredi, S., Gertner, J. M., Paugus, J. A., & Mason, J. W. (1983). Salivary cortisol: A nontraumatic sampling technique for assaying cortisol dynamics. *Journal of the American Academy of Child Psychiatry, 22*, 474–476.

Yehuda, R., Engel, S. M., Brand, S. R., Seckl, J., Marcus, S. M., & Berkowitz, G. S. (2005). Transgenerational effects of posttraumatic stress disorder in babies of mothers exposed to the World Trade Center attacks during pregnancy. *The Journal of Clinical Endocrinology and Metabolism, 90*, 4115–4118.

Zmyj, N., Schneider, S., & Seehagen, S. (2017). Fifteen-month-old infants' cortisol levels decrease after a 30-min-warm-up procedure. *Psychoneuroendocrinology, 76*, 11–13.

# Chapter 26
# Salivary Bioscience and Pediatrics

**Sara B. Johnson and Kristin M. Voegtline**

**Abstract** Salivary biomarkers have particular utility in clinical and research applications in pediatrics given higher acceptability and fewer risks compared to blood collection and meet the special ethical considerations afforded child participants in research. The ease of oral fluid collection allows for repeated assessments that provide insight into dynamic interactions between individuals and their environments during development and hold enormous promise for future diagnostic platforms. Nonetheless, there are important considerations to ensure the reliability and validity of saliva-based assays. In this chapter, we highlight methodological challenges in salivary bioscience in pediatrics, including typical maturation of key biological systems indexed by common salivary assays and specific considerations for the use of a salivary assay with children at various ages. We discuss the growth of salivary diagnostic platforms in pediatrics using the examples of endocrine dysfunction, traumatic brain injury, and cardiometabolic disease. Finally, we highlight current barriers to broader adaption of salivary biomarkers in clinical settings as well as future directions and opportunities to use salivary bioscience to expand current understandings of human health and development in context, with a focus on the growing role of salivary "omics."

## 26.1 Introduction

Salivary biomarkers are useful in pediatric research and clinical applications particularly because saliva collection is more acceptable than blood collection and there are fewer associated risks. These advantages may also help to increase the scope and generalizability of studies with children that require biomonitoring. This expansion is consistent with the National Institutes of Health "Inclusion Across the Lifespan" policy and others like it that mandate the inclusion of children in clinical and behavioral research whenever scientifically appropriate (National Institutes of

S. B. Johnson (✉) · K. M. Voegtline
Department of Pediatrics, Johns Hopkins School of Medicine, Baltimore, MD, USA
e-mail: sjohnson@jhu.edu; kvoegtl1@jhu.edu



D. A. Granger, M. K. Taylor (eds.), *Salivary Bioscience*,
https://doi.org/10.1007/978-3-030-35784-9_26

Health, 2018). Blood collection is among the most common reasons for parents' declining participation in pediatric clinical research studies (Gattuso, Hinds, Tong, & Srivastava, 2006). Moreover, the feasibility of saliva collection can help to facilitate repeated assessments, which provides insight into stability and change in biomarkers of interest over time and with development. Repeated measurements also facilitate ecologically driven sampling approaches to capture dynamic interactions between individuals and their environments.

Salivary biomarkers are commonly used in two distinct but related areas of pediatric research and practice. First, they are used by developmental and behavioral scientists to describe the development and functioning of regulatory systems (e.g., the hypothalamic–pituitary–adrenal axis (HPA), gonadal system, and immune system). These studies often focus on how a given regulatory process changes in response to various stimuli (e.g., acute and chronic stress exposure, variations in parenting, and pubertal development) (Gunnar, Wewerka, Frenn, Long, & Griggs, 2009), and how changes in these regulatory systems are associated with health, developmental, or behavioral outcomes [e.g., how salivary cortisol reactivity to challenge is associated with child externalizing behavior (Kao, Doan, St John, Meyer, & Tarullo, 2018)]. Second, in a rapidly growing field of clinical research, salivary biomarkers are increasingly being used to diagnose and monitor pediatric health conditions and risk factors.

Since the 1970s, salivary diagnostic platforms for adults have burgeoned. Assays for infants, children, and adolescents, however, have lagged substantially (Hassaneen & Maron, 2017). A search of salivary biomarker studies between 2010 and 2016 in PubMed identified 1327 publications involving adults, but only 328 involving children and just 45 involving infants and toddlers (Hassaneen & Maron, 2017). The relative paucity of child- and infant-focused studies is surprising given the clear benefits of noninvasive collection in this vulnerable population (Hassaneen & Maron, 2017). The dearth of studies in children is explained, in part, by the fact that components of oral fluid change relatively rapidly with age and developmental stage, precluding a single set of normative values and reference ranges (Ben-Aryeh, Fisher, Szargel, & Laufer, 1990; Hassaneen & Maron, 2017). Researchers and clinicians must be aware of these expected changes in saliva composition, which, unaccounted for, can threaten the reliability and utility of assay results (Hassaneen & Maron, 2017).

In this chapter, we begin by discussing the growth of salivary bioscience in clinical pediatrics and the current state of the field. We use adrenal disorders, traumatic brain injury, and cardiometabolic disease to illustrate the promise of salivary bioscience for pediatric research and practice. Next, we discuss current barriers to wider adoption of salivary biomarkers in clinical settings, including briefly highlighting how typical maturation of select biological systems indexed by common salivary assays, as well as important considerations for collecting oral fluid from children at various ages. (For a detailed review of anticipated developmental changes in salivary analytes and their underlying physiological systems, see Chap. 10) Finally, we conclude by presenting a view on the future of the field of salivary bioscience, focusing on salivary "omics" for pediatric research and clinical care.

## 26.2 Using Salivary Biomarkers to Evaluate Child Health Outcomes

Salivary biomarkers have been developed for pediatric conditions including adrenal insufficiency and Cushing's disease, periodontal disease and caries, obesity, metabolic syndrome, type 1 and type 2 diabetes, cystic fibrosis, and viruses including cytomegalovirus, human herpes, and Epstein-Barr, measles, and rubella. However, large-scale cohort studies are needed to evaluate and validate potential biomarkers in children (Pappa, Kousvelari, & Vastardis, 2018). New opportunities in this area are emerging to address prevalent pediatric health concerns. As Hassaneen and Maron point out, the fact that children are now mandated to be included in many clinical trials is likely to drive clinical demand for noninvasive tools to monitor disease and recovery (Hassaneen & Maron, 2017). Here, we present three applications: endocrine dysfunction, pediatric traumatic brain injury, and cardiometabolic disease; the latter two examples are particularly fertile areas of current use of salivary biomarkers to evaluate child health in pediatric settings.

### *26.2.1 Salivary Biomarkers of Endocrine Dysfunction*

Salivary cortisol is increasingly used in studies of endocrine diseases such as adrenal insufficiency and Cushing's Syndrome. Cushing's Syndrome is characterized by disrupted cortisol production and failure of cortisol diurnal rhythm to reach its low point between 11 pm and 1 am. Historically, cortisol levels for a diagnosis of Cushing's Syndrome have been measured using either a 24-h dexamethasone suppression test, 24-h urine collection, or late-night serum cortisol collection, all of which are burdensome and impractical particularly for the pediatric patient (Gafni, Papanicolaou, & Nieman, 2000; Raff, 2009). Moreover, the stress of venipuncture, which predominates in children, may actually artificially inflate plasma cortisol levels (Gafni et al., 2000). Salivary cortisol is increasingly being used as an alternative. In a study of 67 children and adolescents, Gafni et al. (2000) found that the accuracy of midnight salivary and urine cortisol yielded 93% overlap of sensitivity for the identification of Cushing's syndrome. Moreover, salivary cortisol collected at the child's bedtime, while somewhat less accurate (83% accurate), offered an alternative approach (Gafni et al., 2000).

Increasingly, salivary cortisol is being used to evaluate adrenal function in a variety of pediatric settings (Cetinkaya, Ozon, & Yordam, 2007), including with critically ill children (Balbao, Costa, Castro, & Carlotti, 2014; Gunnala et al., 2015) and infants with low birth weight (Hochwald, Holsti, & Osiovich, 2016).

### 26.2.2 Salivary Biomarkers of Pediatric Traumatic Brain Injury and Recovery

Traumatic brain injury (TBI) morbidity and mortality are higher during childhood and adolescence than at any other life stage. Previous research shows that levels of specific biomarkers including the calcium-binding protein S100B, glial fibrillary acidic protein (GFAP), and neuron-specific enolase (NSE) increase in serum and cerebrospinal fluid following a TBI. Increases in these biomarkers are correlated with other indicators of brain injury based on neuroimaging and standardized clinical assessments (Yeung, Bhatia, Bhattarai, Ayutyanont, & Sinha, 2018). In a recent pilot study, Yeung et al. (2018) evaluated levels of S100B, GFAP, and NSE in saliva in the first 24 h following injury. They compared TBI-affected youth ($n = 24$) to those with nontraumatic injuries ($n = 25$) and those with musculoskeletal injuries ($n = 25$). Results of the analysis showed that salivary S100B levels, but not GFAP or NSE, were higher in children with TBI. The authors are currently conducting a larger study to evaluate this relationship in hopes of developing a salivary biomarker for TBI.

Similarly, salivary biomarkers are also being studied in hopes of predicting the subset of children and adolescents who will experience protracted symptoms following a concussion. Historically, prolonged concussion symptoms have been monitored clinically using questionnaires; however, this approach is time-consuming and cumbersome in the context of clinical practice. In a small prospective cohort study ($n = 52$), Johnson et al. (2018) evaluated the utility of salivary microRNAs measured in the first 2 weeks following a concussion for identifying youth (mean age 14, SD 3 years) at risk for prolonged symptoms. MircoRNAs are short, noncoding RNAs that are involved in protein translation after gene transcription (Cannell, Kong, & Bushell, 2008). Because they are transported around the body in extracellular space, they can be readily measured in saliva (Johnson et al., 2018). Concussion symptoms and salivary microRNAs were assessed at 4- and 8-weeks post-injury based on a validated self- or parent-reported symptom screener. Levels of five microRNAs (miR-320c-1, miR-133a-5p, miR-769-5p, let-7a-3p, and miR-1307-3p) were associated with prolonged concussion symptoms. A group of these microRNAs are known to be related to neurotrophic TRK signaling, which is involved in brain repair and neurotransmission. Further, microRNA levels were substantially better at identifying children with prolonged symptoms than child- and parent-reported symptoms (Johnson et al., 2018). The opportunity to monitor concussion symptoms and severity with noninvasive biomarkers could help pediatric providers identify youth who are most likely to need additional supportive therapies in the post-concussion period.

### *26.2.3 Salivary Biomarkers of Pediatric Cardiometabolic Disease*

Cardiometabolic disease, including obesity, hypertension, impaired glucose tolerance, type 2 diabetes, and high cholesterol, once seen as a disease of aging, is increasingly seen in childhood and adolescence (Sovio et al., 2013). Therefore, identifying noninvasive biomarkers of disease onset and progression in childhood is a pediatric public health priority (Sovio et al., 2013). To date, most studies of potential salivary cardiometabolic disease biomarkers have focused on comparing inflammatory and metabolic profiles in children and adolescents with and without overweight/obesity (Desai & Mathews, 2014; Goodson et al., 2014).

In a series of studies involving data from seven hundred and forty-four 11-year-olds, Goodson et al. investigated differences in 20 salivary biomarkers of inflammation and metabolism between children with underweight, normal weight, overweight, and obesity. Four of the 20 biomarkers probed varied by child weight status: insulin, C-reactive protein (CRP; a nonspecific inflammatory biomarker that is part of the acute phase response), adiponectin, and leptin (hormones involved in glucose regulation and appetite). CRP, insulin, and leptin levels were significantly higher in children with obesity compared to children with normal weight; whereas salivary adiponectin levels were lower (Goodson et al., 2014). Further, three distinct profiles of obesity emerged based on salivary biomarker profiles. Seventy-six percent of obese children had elevated inflammatory markers, 13% had elevated salivary insulin without elevated inflammatory markers, and 11% had elevated insulin with low levels of adiponectin. Intriguingly, 40% of children without obesity were also in one of these three groups, which the authors suggest might be a marker of future obesity risk (Goodson et al., 2014). These results hint at the promise of noninvasive biomarkers to help elucidate and predict risk of obesity, even before it is manifested clinically.

Using the same sample of seven hundred and forty-four 11-year-olds, Hartman et al. (2015) investigated saliva as a screening diagnostic for high fasting blood glucose. Significant serum and saliva glucose correlations were found ($r = 0.5216$, $p < 0.05$) in youth with obesity as well as those with normal weight. Interestingly, plasma glucose levels below 85 mg/dL were undetectable in saliva, but above this threshold, salivary and plasma levels of glucose exhibited high correlation (Hartman et al., 2015). These findings suggest that if glucose is measurable in saliva above a threshold, the individual is likely to be hyperglycemic. Thus, salivary glucose may be a good measure of hyperglycemia, but is not useful for monitoring lower glucose levels. Building on this study, Hartman et al. (2016) then evaluated the relationship between obesity and concentrations of 20 salivary metabolic biomarkers. They found that youth with obesity were six times more likely to have elevated salivary insulin concentrations suggestive of insulin resistance and hyperinsulinemia. However, obesity was not related to salivary glucose concentrations, and many of the biomarkers tested were poorly correlated.

Outside of salivary glucose and insulin, several studies have found elevated levels of salivary inflammatory biomarkers such as C-reactive protein (CRP), a nonspecific marker of the acute phase response, among children and adolescents with obesity compared to those with normal weight (Cook et al., 2000; Goodson et al., 2014; Naidoo, Konkol, Biccard, Dudose, & McKune, 2012; Pappa et al., 2018). Adiposity has been shown to be the primary contributor to CRP levels among children 9–11 years of age, with a smaller independent contribution of cardiorespiratory fitness (Cook et al., 2000; Naidoo et al., 2012). Importantly, salivary CRP levels are also significantly associated with other cardiometabolic disease markers including fibrinogen, HDL-cholesterol, and systolic blood pressure (Cook et al., 2000). Together, these studies suggest that salivary biomarkers of inflammation are related to obesity and fitness level in childhood and show convergent validity with other accepted markers of cardiometabolic risk.

In summary, evidence is accumulating that select salivary biomarkers may be sensitive to metabolic changes associated with cardiometabolic disease, particularly obesity and diabetes (Hartman et al., 2016; Pappa et al., 2018). Additional studies are needed to support the use of these biomarkers for diagnostics, but evidence to date is both promising and instructive (Hartman et al., 2016; Pappa et al., 2018)

## 26.3 Barriers to the Adoption of Salivary Diagnostic Application in a Pediatric Population

Saliva collection from children is highly acceptable and inexpensive, and there is a growing appreciation of the insight into human health than can be gained from saliva. It is perhaps surprising that there has not been more widespread integration of salivary bioscience into pediatric clinical and research applications. Segal and Wong point out that, saliva diagnostics is a "late bloomer," as only recently has there been a "growing appreciation that saliva can reflect virtually the entire spectrum of normal and disease states" (Segal & Wong, 2008, p. 24).

Efforts are only beginning to compare the diagnostic utility of saliva against other biospecimens such as blood or urine (Nunes, Mussavira, & Bindhu, 2015; Segal & Wong, 2008), and most, to date, have focused on adult samples. Multicenter studies must be undertaken to establish normative values (Hassaneen & Maron, 2017) and standardized collection protocols. Validating saliva against other biomarkers across a range of disease outcomes and establishing reference ranges will require time and resources. Moreover, given the expected, and sometimes marked, developmental patterns observed in salivary biomarker levels in pediatrics, the work involved to validate saliva as a biomarker across a wide range of ages and developmental stages is also substantial. This work must be completed before saliva can be routinely integrated into pediatric care (Hassaneen & Maron, 2017).

### 26.3.1 Anticipating Potential Sources of Variation in Salivary Diagnostic Results

Before salivary bioscience can be integrated more fully into pediatric practice, other sources of variation must also be evaluated. Understanding normative changes in key biomarkers is essential for measuring and interpreting salivary analytes in pediatric research and practice. Salivary cortisol indicators (basal levels, reactivity to a stressor, and circadian patterns) all demonstrate developmental changes that are driven by a constellation of biological and behavioral factors (e.g., sleep, feeding frequency) (Kiess et al., 1995; Tollenaar, Jansen, Beijers, Riksen-Walraven, & de Weerth, 2010; Watamura, Donzella, Kertes, & Gunnar, 2004). Similarly, gonadal hormones such as testosterone change markedly not only during the transition to adolescence (Harden, Kretsch, Tackett, & Tucker-Drob, 2014), but also even in the first 3 months of life, a phenomenon referred to as "mini-puberty." During this period, male infants exhibit higher levels of salivary testosterone than females (Contreras et al., 2017; Hines et al., 2016; Quigley, 2002). Indicators of sympathetic nervous system activity (e.g., salivary alpha-amylase) also change with age as a function of the development of the autonomic nervous system as well as factors such as changing diet (Nishizato, Fujisawa, Kosaka, & Tomoda, 2017). Emerging evidence suggests that immune biomarkers (e.g., salivary cytokines (e.g., TNF-a, IL-10 < IL-2, IL-6, and IL-8) may also exhibit sex-specific developmental patterns (Riis et al., 2014). These established and emerging developmental trends (see Hibel chapter for further detail) require careful attention to child age and sex in the evaluation of salivary biomarkers.

Furthermore, given developmental shifts in pediatric populations, the developmental periods of (1) infancy and toddlerhood, (2) childhood, and (3) puberty each require additional specific considerations. In early infancy, collecting sufficient volume of saliva can be challenging. In addition, in infancy and toddlerhood, naps and inconsistent sleep schedules are normative; thus, analytes with a strong diurnal pattern (e.g., cortisol) should be interpreted with this in mind. High within-person variability has been documented in this age group, with sleep and feeding frequency implicated as potential explanations (Tollenaar et al., 2010). Very young infants eat frequently; therefore, timing sample collection to align with common recommendations to avoid eating 30 min before sample collection may not be feasible (Granger, Johnson, Szanton, Out, & Schumann, 2012). Moreover, the impact of mode of infant feeding on most salivary analytes has not been well characterized; hormones including cortisol are measurable at moderate levels in animal and human milk (Magnano, Diamond, & Gardner, 1989). Thus, wake time, daytime napping, mode of feeding, and the time of last feeding should be considered when collecting saliva from infants and toddlers.

Some studies have found that the setting in which saliva is collected is associated with analyte levels and patterns. For example, children in childcare centers, particularly toddlers, show increases in cortisol from morning to afternoon while in childcare, but the same children do not show the same increases at home. Thus, attention to standardizing the setting in which saliva is collected is important for reducing bias (Watamura, Bonny, Jan, & Gunnar, 2003).

Finally, given rapid changes in some analyte levels and the behavioral routines that influence them over relatively short periods of time, relatively narrow age-matching should be strongly considered in comparative salivary bioscience studies with infants and toddlers.

In contrast to infancy and toddlerhood, during childhood, insufficient specimen volume and irregular sleep–wake cycles pose less of a threat to reliability (Granger et al., 2007). At this stage, researchers still need to be cognizant of child eating and drinking prior to sampling, which should be documented and accounted for in analyses. Contamination due to food or liquid residue can be reduced by asking children to complete a mouth rinse and waiting at least 10 min after the rinse to collect saliva to avoid sample dilution (Granger et al., 2007).

In childhood, salivary analyte levels may also be impacted by dental carries or blood contamination associated both with poor oral health and normative shedding of teeth (Granger et al., 2012). Saliva collection with children should include documentation of loose or recently shed teeth, as well as signs of tooth caps or fillings that may indicate a history of dental caries.

Some absorbent devices, including less widely used cotton-based collection methods, may underestimate cortisol levels (Harmon, Hibel, Rumyantseva, & Granger, 2007; Shirtcliff, Granger, Schwartz, & Curran, 2001). Therefore, researchers are advised to instead opt for polymer oral swabs, specifically designed for children's use (i.e., only a portion of the long swab is placed in the oral cavity, the remainder is external for grasping and to prevent choking hazard) or alternatively, passive drool which is intended for children 6 years and older (Granger et al., 2007).

Later changes in sleep coincident with the pubertal transition, particularly delayed sleep phase and disrupted sleep, are well documented (Sadeh, Dahl, Shahar, & Rosenblat-Stein, 2009). Changes in the "chronotype" may temporarily alter biological timekeeping in this period (Hagenauer & Lee, 2012). Thus, the collection of salivary biomarkers with known diurnal variation should include careful assessment of sleep and wake times as well as attention to standardizing collection times. Efforts to assess the stage of pubertal maturation are important during this period. The gold standard is a medical examination; however, such exams can be anxiety provoking for youth, expensive, and time-consuming. Self-reported pubertal maturation has been shown to be adequate to differentiate pre-and post-pubertal youth but is not a reliable indicator of pubertal stage at a more granular level (Rasmussen et al., 2015).

Finally, irrespective of the availability of normative values in children for the salivary assay under study and consideration of the child's developmental period, some of the challenges in integrating salivary bioscience into pediatrics have more to do with philosophical than technical challenges. Blood and urine have a long history and are well-established biomarkers in pediatric practice. Such entrenched patterns of practice change slowly and are codified in everything from billing to supplies routinely available to providers. Pediatric applications for salivary bioscience, however, may ultimately drive demand, given clear benefits to saliva collection over more invasive approaches for vulnerable pediatric patients.

## 26.4 Future Directions for Salivary Bioscience Integration in Pediatrics

Despite current barriers, there are several new technologies and applications for salivary bioscience that are likely to drive demand in the coming years. The "new frontier" of salivary bioscience research is likely to be facilitated by salivary "omics" research. Saliva, like blood, allows researchers a window into the genome and transcription factors (genomics and transcriptomics), proteins (proteomics), and metabolites (metabolomics).

### *26.4.1 Salivary Genomics*

Salivary DNA harvested from buccal cells offers a noninvasive way to collect cellular DNA, as well as cell-free and exosomal DNA (Hassaneen & Maron, 2017). A preliminary salivary assay has recently been developed for a genetic mutation in adults with lung cancer that shows perfect agreement with tissue genotyping approaches (Pu et al., 2016); while this technology has not yet been widely investigated in pediatrics, this work demonstrates the extraordinary promise of saliva to serve as a "liquid biopsy."

### *26.4.2 Salivary Transcriptomics*

Work by Maron et al. (2010) was among the first to apply salivary transcriptomics to characterize RNA and/or gene expression in the pediatric population. The investigators found that transcripts reflecting all of the major organ systems (e.g., gastrointestinal, nervous, and hematological) were detectable in the mouth of the newborn. Since then, saliva from newborns has been used to characterize a set of genes whose expression profile predicts neonatal oral feeding skills, a key developmental milestone critical to neonatal health and growth, particularly in preterm infants (Maron et al., 2015; Maron, Johnson, Dietz, Chen, & Bianchi, 2012).

### *26.4.3 Salivary Proteomics*

Proteomics is the large-scale investigation of proteins. Advances in mass spectrometry technology have facilitated high-throughput proteomic studies. Hundreds of proteins can now be examined from a single saliva sample (Hassaneen & Maron, 2017). Whereas DNA and RNA require extraction, proteins can be quantified immediately after collection, which simplifies processing. Moreover, proteins are more stable than DNA and RNA (Hassaneen & Maron, 2017). Proteomics platforms have now identified approximately 3000 proteins and peptides in human saliva and have helped to characterize diseases in adults including oral squamous cell carcinoma, schizophrenia, and bipolar disorder, and Down syndrome (Castagnola et al., 2017). These advances demonstrate that the salivary proteome can be harnessed to develop new diagnostics and to increase our understanding of basic physiological processes. To date, however, few studies have focused on children because the salivary proteome changes as children grow (Hassaneen & Maron, 2017). One notable exception is juvenile systemic lupus erythematosus (SLE), a debilitating autoimmune inflammatory condition that is often more aggressive than its adult analog. Proteins associated with juvenile SLE have been recently described in saliva and may ultimately contribute to a noninvasive way of monitoring the disease in children (Abrão et al., 2016).

### *26.4.4 Salivary Metabolomics*

Metabolomics is the study of the chemical fingerprints of metabolites or metabolite profiles (Zhang, Sun, & Wang, 2012). Saliva contains a large and diverse set of metabolites that can be queried to understand disease risk or progression. For example, in a study of young children with type 1 diabetes, de Oliveira et al. (2016) found that salivary metabolite profiles discriminated controlled from uncontrolled diabetics. In the future, diabetic glucose monitoring might be noninvasive, vastly improving the quality of life for children with the condition.

## 26.5 Summary and Conclusions

The promise of salivary bioscience in pediatrics is clear. Developmental differences in common salivary biomarkers are increasingly recognized and characterized, which will allow researchers and clinicians to devise reference ranges and best practices for collection and interpretation of salivary biomarker data based on age or stage of development. Advances in assay technology, coupled with the growth of "omics" across biomedical research, are likely to transform the diagnosis, monitoring, and treatment of a number of common pediatric conditions, which could lead to better medical care and better quality of life for children and their families.

## References

Abrão, A. L. P., Falcao, D. P., de Amorim, R. F. B., Bezerra, A. C. B., Pombeiro, G. A. N. M., Guimarães, L. J., … da Mota, L. M. H. (2016). Salivary proteomics: A new adjuvant approach to the early diagnosis of familial juvenile systemic lupus erythematosus. *Medical Hypotheses, 89*, 97–100. https://doi.org/10.1016/j.mehy.2016.02.010

Balbao, V. M., Costa, M. M., Castro, M., & Carlotti, A. P. (2014). Evaluation of adrenal function in critically ill children. *Clinical Endocrinology, 81*(4), 559–565. https://doi.org/10.1111/cen.12444

Ben-Aryeh, H., Fisher, M., Szargel, R., & Laufer, D. (1990). Composition of whole unstimulated saliva of healthy children: Changes with age. *Archives of Oral Biology, 35*(11), 929–931. https://doi.org/10.1016/0003-9969(90)90075-L

Cannell, I. G., Kong, Y. W., & Bushell, M. (2008). How do microRNAs regulate gene expression? *Biochemical Society Transactions, 36*(Pt 6), 1224–1231. https://doi.org/10.1042/bst0361224

Castagnola, M., Scarano, E., Passali, G. C., Messana, I., Cabras, T., Iavarone, F., … Paludetti, G. (2017). Salivary biomarkers and proteomics: Future diagnostic and clinical utilities. *Acta Otorhinolaryngologica Italica, 37*(2), 94–101. https://doi.org/10.14639/0392-100X-1598

Cetinkaya, S., Ozon, A., & Yordam, N. (2007). Diagnostic value of salivary cortisol in children with abnormal adrenal cortex functions. *Hormone Research, 67*(6), 301–306. https://doi.org/10.1159/000100363

Contreras, M., Raisingani, M., Chandler, D. W., Curtin, W. D., Barillas, J., Brar, P. C., … David, R. (2017). Salivary testosterone during the minipuberty of infancy. *Hormone Research in Pædiatrics, 87*(2), 111–115.

Cook, D. G., Mendall, M. A., Whincup, P. H., Carey, I. M., Ballam, L., Morris, J. E., … Strachan, D. P. (2000). C-reactive protein concentration in children: Relationship to adiposity and other cardiovascular risk factors. *Atherosclerosis, 149*(1), 139–150.

de Oliveira, L. R., Martins, C., Fidalgo, T. K., Freitas-Fernandes, L. B., de Oliveira Torres, R., Soares, A. L., … de Souza, I. P. (2016). Salivary metabolite fingerprint of Type 1 Diabetes in young children. *Journal of Proteome Research, 15*(8), 2491–2499. https://doi.org/10.1021/acs.jproteome.6b00007

Desai, G. S., & Mathews, S. T. (2014). Saliva as a non-invasive diagnostic tool for inflammation and insulin-resistance. *World Journal of Diabetes, 5*(6), 730–738. https://doi.org/10.4239/wjd.v5.i6.730

Gafni, R. I., Papanicolaou, D. A., & Nieman, L. K. (2000). Nighttime salivary cortisol measurement as a simple, noninvasive, outpatient screening test for Cushing's syndrome in children and adolescents. *Journal of Pediatrics, 137*(1), 30–35. https://doi.org/10.1067/mpd.2000.106226

Gattuso, J., Hinds, P., Tong, X., & Srivastava, K. (2006). Monitoring child and parent refusals to enrol in clinical research protocols. *Journal of Advanced Nursing, 53*(3), 319–326. https://doi.org/10.1111/j.1365-2648.2006.03724.x

Goodson, J. M., Kantarci, A., Hartman, M.-L., Denis, G. V., Stephens, D., Hasturk, H., … Welty, F. (2014). Metabolic disease risk in children by salivary biomarker analysis. *PLoS One, 9*(6), e98799. https://doi.org/10.1371/journal.pone.0098799

Granger, D. A., Johnson, S. B., Szanton, S. L., Out, D., & Schumann, L. L. (2012). Incorporating salivary biomarkers into nursing research: An overview and review of best practices. *Biological Research for Nursing, 14*(4), 347–356. https://doi.org/10.1177/1099800412443892

Granger, D. A., Kivlighan, K. T., Fortunato, C., Harmon, A. G., Hibel, L. C., Schwartz, E. B., & Whembolua, G.-L. (2007). Integration of salivary biomarkers into developmental and behaviorally-oriented research: Problems and solutions for collecting specimens. *Physiology and Behavior, 92*(4), 583–590.

Gunnala, V., Guo, R., Minutti, C., Durazo-Arvizu, R., Laporte, C., Mathews, H., … Bhatia, R. (2015). Measurement of salivary cortisol level for the diagnosis of critical illness-related corticosteroid insufficiency in children. *Pediatric Critical Care Medicine, 16*(4), e101–e106. https://doi.org/10.1097/pcc.0000000000000361

Gunnar, M. R., Wewerka, S., Frenn, K., Long, J. D., & Griggs, C. (2009). Developmental changes in hypothalamus-pituitary-adrenal activity over the transition to adolescence: Normative changes and associations with puberty. *Development and Psychopathology, 21*(1), 69–85.

Hagenauer, M. H., & Lee, T. M. (2012). The neuroendocrine control of the circadian system: Adolescent Chronotype. *Frontiers in Neuroendocrinology, 33*(3), 211–229. https://doi.org/10.1016/j.yfrne.2012.04.003

Harden, K. P., Kretsch, N., Tackett, J. L., & Tucker-Drob, E. M. (2014). Genetic and environmental influences on testosterone in adolescents: Evidence for sex differences. *Developmental Psychobiology, 56*(6), 1278–1289. https://doi.org/10.1002/dev.21207

Harmon, A. G., Hibel, L. C., Rumyantseva, O., & Granger, D. A. (2007). Measuring salivary cortisol in studies of child development: Watch out--what goes in may not come out of saliva collection devices. *Developmental Psychobiology, 49*(5), 495–500. https://doi.org/10.1002/dev.20231

Hartman, M. L., Goodson, J. M., Barake, R., Alsmadi, O., Al-Mutawa, S., Ariga, J., . . . Welty, F. (2015). Salivary glucose concentration exhibits threshold kinetics in normal-weight, overweight, and obese children. *Diabetes, Metabolic Syndrome and Obesity, 8*, 9–15. https://doi.org/10.2147/dmso.S72744

Hartman, M. L., Goodson, J. M., Shi, P., Vargas, J., Yaskell, T., Stephens, D., . . . Welty, F. (2016). Unhealthy phenotype as indicated by salivary biomarkers: Glucose, insulin, VEGF-A, and IL-12p70 in obese Kuwaiti adolescents. *Journal of Obesity, 2016*, 6860240. https://doi.org/10.1155/2016/6860240

Hassaneen, M., & Maron, J. L. (2017). Salivary diagnostics in pediatrics: Applicability, translatability, and limitations. *Frontiers in Public Health, 5*, 83. https://doi.org/10.3389/fpubh.2017.00083

Hines, M., Spencer, D., Kung, K. T., Browne, W. V., Constantinescu, M., & Noorderhaven, R. M. (2016). The early postnatal period, mini-puberty, provides a window on the role of testosterone in human neurobehavioural development. *Current Opinion in Neurobiology, 38*, 69–73. https://doi.org/10.1016/j.conb.2016.02.008

Hochwald, O., Holsti, L., & Osiovich, H. (2016). Salivary cortisol based early ACTH test for identifying adrenal insufficiency in low birth weight infants. *Indian Journal of Pediatrics, 83*(12–13), 1491–1492. https://doi.org/10.1007/s12098-016-2177-0

Johnson, J. J., Loeffert, A. C., Stokes, J., Olympia, R. P., Bramley, H., & Hicks, S. D. (2018). Association of salivary microRNA changes with prolonged concussion symptoms. *JAMA Pediatrics, 172*(1), 65–73. https://doi.org/10.1001/jamapediatrics.2017.3884

Kao, K., Doan, S. N., St John, A. M., Meyer, J. S., & Tarullo, A. R. (2018). Salivary cortisol reactivity in preschoolers is associated with hair cortisol and behavioral problems. *Stress, 21*(1), 28–35. https://doi.org/10.1080/10253890.2017.1391210

Kiess, W., Meidert, A., Dressendorfer, R. A., Schriever, K., Kessler, U., Konig, A., . . . Strasburger, C. J. (1995). Salivary cortisol levels throughout childhood and adolescence: Relation with age, pubertal stage, and weight. *Pediatric Research, 37*(4 Pt 1), 502–506. https://doi.org/10.1203/00006450-199504000-00020

Magnano, C. L., Diamond, E. J., & Gardner, J. M. (1989). Use of salivary cortisol measurements in young infants: A note of caution. *Child Development, 60*(5), 1099–1101.

Maron, J. L., Hwang, J. S., Pathak, S., Ruthazer, R., Russell, R. L., & Alterovitz, G. (2015). Computational gene expression modeling identifies salivary biomarker analysis that predict oral feeding readiness in the newborn. *Journal of Pediatrics, 166*(2), 282–288.e285. https://doi.org/10.1016/j.jpeds.2014.10.065

Maron, J. L., Johnson, K. L., Dietz, J. A., Chen, M. L., & Bianchi, D. W. (2012). Neuropeptide Y2 receptor (NPY2R) expression in saliva predicts feeding immaturity in the premature neonate. *PLoS One, 7*(5), e37870. https://doi.org/10.1371/journal.pone.0037870

Maron, J. L., Johnson, K. L., Rocke, D. M., Cohen, M. G., Liley, A. J., & Bianchi, D. W. (2010). Neonatal salivary analysis reveals global developmental gene expression changes in the premature infant. *Clinical Chemistry, 56*(3), 409–416.

Naidoo, T., Konkol, K., Biccard, B., Dudose, K., & McKune, A. J. (2012). Elevated salivary C-reactive protein predicted by low cardio-respiratory fitness and being overweight in African children. *Cardiovascular Journal of Africa, 23*(9), 501–506. https://doi.org/10.5830/cvja-2012-058

National Institutes of Health. (2018). *Inclusion across the lifespan - policy implementation NIH inclusion across the lifespan in research involving human subjects*. Retrieved from https://grants.nih.gov/grants/funding/lifespan/lifespan.htm

Nishizato, M., Fujisawa, T. X., Kosaka, H., & Tomoda, A. (2017). Developmental changes in social attention and oxytocin levels in infants and children. *Scientific Reports, 7*, 2540. https://doi.org/10.1038/s41598-017-02368-x

Nunes, L. A. S., Mussavira, S., & Bindhu, O. S. (2015). Clinical and diagnostic utility of saliva as a non-invasive diagnostic fluid: A systematic review. *Biochemia Medica, 25*(2), 177–192. https://doi.org/10.11613/BM.2015.018

Pappa, E., Kousvelari, E., & Vastardis, H. (2018). Saliva in the "omics" era: A promising tool in paediatrics. *Oral Diseases, 00*, 1–10. https://doi.org/10.1111/odi.12886

Pu, D., Liang, H., Wei, F., Akin, D., Feng, Z., Yan, Q., … Zhou, Q. (2016). Evaluation of a novel saliva-based epidermal growth factor receptor mutation detection for lung cancer: A pilot study. *Thorac Cancer, 7*(4), 428–436. https://doi.org/10.1111/1759-7714.12350

Quigley, C. A. (2002). Editorial: The postnatal gonadotropin and sex steroid surge-insights from the androgen insensitivity syndrome. *Journal of Clinical Endocrinology and Metabolism, 87*(1), 24–28. https://doi.org/10.1210/jcem.87.1.8265

Raff, H. (2009). Utility of salivary cortisol measurements in Cushing's syndrome and adrenal insufficiency. *Journal of Clinical Endocrinology and Metabolism, 94*(10), 3647–3655. https://doi.org/10.1210/jc.2009-1166

Rasmussen, A. R., Wohlfahrt-Veje, C., Tefre de Renzy-Martin, K., Hagen, C. P., Tinggaard, J., Mouritsen, A., … Main, K. M. (2015). Validity of self-assessment of pubertal maturation. *Pediatrics, 135*(1), 86.

Riis, J. L., Dorothee, O., Dorn, L. D., Beal, S. J., Denson, L. A., Stephanie, P., … Granger, D. A. (2014). Salivary cytokines in healthy adolescent girls: Intercorrelations, stability, and associations with serum cytokines, age, and pubertal stage. *Developmental Psychobiology, 56*(4), 797–811. https://doi.org/10.1002/dev.21149

Sadeh, A., Dahl, R. E., Shahar, G., & Rosenblat-Stein, S. (2009). Sleep and the transition to adolescence: A longitudinal study. *Sleep, 32*(12), 1602–1609. https://doi.org/10.1093/sleep/32.12.1602

Segal, A., & Wong, D. T. (2008). Salivary diagnostics: Enhancing disease detection and making medicine better. *Eur J Dent Educ, 12*(Suppl 1), 22–29. https://doi.org/10.1111/j.1600-0579.2007.00477.x

Shirtcliff, E. A., Granger, D. A., Schwartz, E., & Curran, M. J. (2001). Use of salivary biomarkers in biobehavioral research: Cotton-based sample collection methods can interfere with salivary immunoassay results. *Psychoneuroendocrinology, 26*(2), 165–173.

Sovio, U., Skow, A., Falconer, C., Park, M. H., Viner, R. M., & Kinra, S. (2013). Improving prediction algorithms for cardiometabolic risk in children and adolescents. *Journal of Obesity, 2013*, 684782. https://doi.org/10.1155/2013/684782

Tollenaar, M. S., Jansen, J., Beijers, R., Riksen-Walraven, J., & de Weerth, C. (2010). Cortisol in the first year of life: Normative values and intra-individual variability. *Early Human Development, 86*(1), 13–16. https://doi.org/10.1016/j.earlhumdev.2009.12.003

Watamura, S., Bonny, D., Jan, A., & Gunnar, M. R. (2003). Morning-to-afternoon increases in cortisol concentrations for infants and toddlers at child care: Age differences and behavioral correlates. *Child Development, 74*(4), 1006–1020. https://doi.org/10.1111/1467-8624.00583

Watamura, S., Donzella, B., Kertes, D., & Gunnar, M. (2004). Developmental changes in baseline cortisol activity in early childhood: Relations with napping and effortful control. *Developmental Psychobiology, 45*(3), 125 133.

Yeung, C., Bhatia, R., Bhattarai, B., Ayutyanont, N., & Sinha, M. (2018). Role of salivary biomarkers in predicting significant traumatic brain injury∗. *Pediatrics, 141*(1 MeetingAbstract), 357.

Zhang, A., Sun, H., & Wang, X. (2012). Saliva metabolomics opens door to biomarker discovery, disease diagnosis, and treatment. *Applied Biochemistry and Biotechnology, 168*(6), 1718–1727. https://doi.org/10.1007/s12010-012-9891-5

# Chapter 27
# Use of Saliva to Better Understand the Daily Experience of Adulthood and Aging

**David M. Almeida, Jennifer Piazza, Yin Liu, and Steven H. Zarit**

**Abstract** This chapter reviews the use of saliva to assess age-related changes in important biological systems, describes how saliva can be used to assess naturally occurring fluctuations of biomarkers in adults' daily lives, and offers cutting-edge statistical approaches that can help answer research questions that involve these multivariate and dynamic phenomena. This chapter highlights the use of saliva to assess day-to-day variability in biological markers across adulthood. Salivary biomarkers offer a unique and innovative window into investigating the daily experiences of midlife and older adults. Using findings from multiple daily diary studies where participants provide multiple saliva samples each day, we describe within-and across-day patterns of cortisol, dehydroepiandrosterone sulfate (DHEA-S), and salivary alpha-amylase (sAA). Using multilevel and latent state-trait modeling, we show differentiated patterns in each of these biomarkers across the day. Specific attention is paid to age differences in the daily patterning of these salivary biomarkers as well as their links to stressful events. The chapter also reviews recent research that links daily salivary biomarkers to long-term health and well-being. Recommendations for the design, collection, and statistical modeling of daily assessments of salivary biomarkers are also provided.

**Keywords** Cortisol · Dehydroepiandrosterone sulfate (DHEA-S) · Salivary alpha-amylase (sAA) · Daily stress · Multilevel modeling · Latent state-trait modeling

---

D. M. Almeida (✉) · S. H. Zarit
Department of Human Development and Family Studies, The Pennsylvania State University, State College, PA, USA
e-mail: dalmeida@psu.edu

J. Piazza
Department of Health Science, California State University, Fullerton, Fullerton, CA, USA

Y. Liu
Department of Human Development and Family Studies, Utah State University, Logan, UT, USA



D. A. Granger, M. K. Taylor (eds.), *Salivary Bioscience*,
https://doi.org/10.1007/978-3-030-35784-9_27

## 27.1 History of Salivary Bioscience in Aging

Saliva plays a critical role in maintaining health and can provide insight into underlying physiological processes. Composed primarily of water, saliva also contains several other compounds, including electrolytes and proteins, that help to ensure proper oral health and initiate digestion (Dodds, Johnson, & Yeh, 2005). Healthy individuals produce approximately 1–1.5 L of saliva per day. Alterations in saliva, whether by volume loss or compositional changes, have the potential to negatively impact oral health, which, in turn, has implications for overall systemic health (Varga, 2012). Aging is one factor that has been linked with both decreased salivary flow (Smith et al., 2013) and alterations in its constituent parts (Nagler & Hershkovich, 2005), even when not accompanied by other risk factors, such as medication use (Wolff et al., 2017) and chronic health conditions (Mortazavi, Baharvand, Movahhedian, Mohammadi, & Khodadoustan, 2014; Zhang et al., 2016).

Importantly, the relationship between saliva and aging is not static: research indicates that saliva flow rate (Affoo, Foley, Garrick, Siqueira, & Martin, 2015) and the composition of biological markers found in saliva (for review, see Piazza, Almeida, Dmitrieva, & Klein, 2010) fluctuate across daily life. Thus, for research scientists interested in aging, saliva provides a critical window into processes that until relatively recently could only be examined using invasive means. The goal of this chapter is threefold. First, we review the use of saliva to assess age-related changes in important biological systems, including the hypothalamic–pituitary–adrenal (HPA) axis and the sympathomedullary (SAM) pathway. Second, we describe how saliva can be used to assess naturally occurring fluctuations of biomarkers in adults' daily lives by summarizing findings from two research projects that collected and assessed multiple samples of saliva across a series of days. Third, we describe cutting-edge statistical approaches that can help answer research questions that involve these multivariate and dynamic phenomena.

**Age, Saliva, and Underlying Biological Processes** For years, salivary flow has been examined as a marker of aging (e.g., Affoo et al., 2015; Smith et al., 2013). Scientific developments, however, have enabled researchers to expand their inquiry into identifying components of saliva that are indicative of underlying physiological processes. Two systems that have been widely examined in conjunction with saliva and aging are the HPA axis and the SAM pathway. Both systems are critical for engaging the body's stress response, with the SAM pathway providing an immediate response to a stressor and the HPA axis providing a longer term hormonal response. Both systems also show several age-related changes, many of which mimic those that occur with increased stressor exposure (for review, see Piazza et al., 2010). Salivary biomarkers of the HPA axis commonly used in research include the hormones cortisol and dehydroepiandrosterone sulfate (DHEA-S), and for the SAM pathway, the digestive enzyme salivary alpha-amylase (sAA).

***Cortisol and Age-Associated Changes*** Cortisol is a product of the HPA axis that is essential for regulating metabolism, moderating inflammation, and mobilizing energy. It is released throughout the day in a pulsatile fashion, with increased pulses during times of acute stress (Tsigos, Kyrou, Kassi, & Chrousos, 2000). Release of cortisol in response to stress is a self-regulated, multistep process. When a stressor is perceived, the paraventricular nucleus of the hypothalamus triggers the release of corticotrophin-releasing hormone (CRH). Upon its release, CRH triggers the release of arginine vasopressin (AVP) and adrenocorticotropine hormone (ACTH). ACTH, which circulates to the adrenal glands, then stimulates the release of corticosteroids (e.g., cortisol) and other glucocorticoids (for review, see Klein & Corwin, 2007). After an optimal amount of corticosteroids have been released, the HPA axis restores itself through a negative feedback loop, whereby it dampens that release of CRH and ACTH, and, in turn, cortisol.

Cortisol exhibits a robust diurnal pattern: upon awakening, levels increase, reaching a peak between 30 and 45 min later, and then gradually decrease throughout the day until reaching a nadir in the evening hours (Kirschbaum & Hellhammer, 1989). Increasing age is associated with alterations in this pattern, including higher levels across the day, particularly in the evening. Older age is also associated with an attenuated awakening response, and an overall flatter diurnal pattern (for review, see Epel, Burke, & Wolkowitz, 2009). Research indicates that additional factors (e.g., stress) may exacerbate these age-associated changes (e.g., Piazza, Dmitrieva, Charles, Almeida, & Orona, 2018).

***DHEA-S and Age-Associated Changes*** DHEA-S is another commonly assessed salivary biomarker of the HPA axis. Unlike cortisol, which is a catabolic hormone, DHEA-S is an anabolic hormone primarily secreted by the adrenal cortex, as well as ovaries and testes (for review, see Maninger, Wolkowitz, Reus, Epel, & Mellon, 2009). DHEA-S is protective against a number of conditions experienced more often in later life, including decline in cognitive functioning, atherosclerosis, cancer, and diabetes (Krug, Ziegler, & Bornstein, 2008). Levels of DHEA-S reach their peak in the morning and steadily decline across the remainder of the day (Klein et al., 2008). DHEA-S significantly declines across the lifespan. After the age of 40 years, plasma concentrations decline by 2% each year; for men, the decline ranges between 1% and 4% per year. DHEA-S reaches its nadir between the ages of 65 and 70, which is also the time at which health typically declines (e.g., Tannenbaum, Barrett-Connor, Laughlin, & Platt, 2003).

***The SAM Pathway and sAA*** The SAM pathway, commonly referred to as the "fight or flight response," refers to the release of the catecholamines, epinephrine (EPI), and norepinephrine (NE) from the adrenal medulla upon activation of the sympathetic branch of the autonomic nervous system (for review, see Klein & Corwin, 2007). Although EPI and NE are found in saliva, direct assessment of the catecholamines is not recommended in field studies, due to the length of time it takes for them to transfer from blood to saliva and their stringent processing requirements (e.g., Rohleder, Nater, Wolf, Ehlert, & Kirschbaum, 2004). Given these constraints, researchers have turned to the digestive enzyme salivary α-amylase (sAA) as a proxy of SNS activation (Nater & Rohleder, 2009). The justification for using sAA is that when catecholamines are released in response to SNS activation, they trigger changes in salivary gland receptors, ultimately altering their activity (Nederfors & Dahlof, 1992). Although there is some conflicting evidence in the literature, several studies indicate that there is an association between sAA and the release of NE and EPI (e.g., Ditzen, Ehlert, & Nater, 2014; Thoma, Kirschbaum, Wolf, & Rohleder, 2012). As such, sAA has thus been used as a marker of SNS activity across a number of studies. sAA exhibits a diurnal pattern with levels moderately low upon awakening, dropping briefly at 30 min post awakening, and gradually increasing throughout the day (Nater, Rohleder, Schlotz, Ehlert, & Kirschbaum, 2007). With age, there is greater sAA output, but an attenuation of slope (e.g., attenuation of the diurnal increase), although basal levels are higher (Nater, Hoppmann, & Scott, 2013).

**Collection of Salivary Biomarkers** Assessment of biomarkers found in saliva has several benefits over self-reported measures, as well as other objective measures of biomarker collection, such as blood draw. Compared to more invasive assessments, salivary biomarkers are easy to collect, less burdensome to participants and researchers, and can be used in field studies, where frequent assessment of biomarkers is warranted. They also provide an objective indicator of physical health, which is difficult to attain through self-reported measures alone. Moreover, self-reported measures and salivary biomarkers in conjunction have the potential to provide a more comprehensive picture of an individual's state of health and the factors that influence it. To this aim, Sect. 27.2 of the chapter shows how the saliva is used to gather information on within- day and across-day assessment of biological functioning during adulthood.

## 27.2 State of the Knowledge

The collection of saliva has played a critical role in understanding how the daily lives of adults change as they age and how daily experiences shape health and well-being. These have been the overarching objectives of the National Study of Daily Experiences (NSDE) and the Daily Stress and Health study (DaSH). Both of these projects combine self-reports of daily experiences such as mood, physical symptoms, and stressors with multiple assessments of saliva to better understand the daily experiences of adulthood. The use of saliva in both projects is described below.

### *27.2.1 NSDE: Saliva as a Window into Daily Stress of Adulthood*

The NSDE is one of the projects of the Midlife in the United States study (MIDUS). The protocol involves two measurement bursts collected 10 years apart. Each burst consists of eight consecutive days of telephone interviews of daily stressors and well-being combined with multiple assessments of daily salivary cortisol

(4 occasions × 4 days). Consisting of over 25,000 daily interviews from 2022 adults ranging in age from 24 to 85, the NSDE is the largest longitudinal diary study of daily health and well-being in the USA. Early findings from the documented age and sociodemographic patterns of self-reported daily experiences include stressful experiences, mood, and physical health (Almeida, 2005; Almeida, Neupert, Banks, & Serido, 2005). The inclusion of saliva collection multiple times a day on multiple days has offered a multitude of opportunities to enhance this line of research. We first provide a description of how we collected saliva and then review only a few of these opportunities as they relate to aging research using salivary cortisol as an example.

**Saliva Collection Protocol and Compliance** All of our saliva was self-collected by participants in their homes. As part of recruitment, respondents received a Home Saliva Collection Kit one week prior to their initial phone call, containing 16 numbered and color-coded salivettes, as well as a detailed instruction sheet and a compact disk with video instructions. In addition to written and video instructions, telephone interviewers review the collection procedures and answer any questions the evening prior to the initial collection day. Respondents provide four saliva samples per day on days 2–5 of the eight-day period to be assayed for cortisol and sAA. To maximize compliance, our collection procedures are designed to be as convenient as possible. On saliva collection days, respondents will produce four saliva samples throughout the day: upon awakening, 30 min after getting out of bed, before lunch, and at bedtime. Data on the exact time respondents provide each saliva sample will be obtained from the nightly telephone interviews and from a form sent with the collection kit.

Compliance to the saliva collection protocol has been validated in a series of papers (Almeida, McGonagle, & King, 2009; Almeida, Piazza, & Stawski, 2009; Stawski, Cichy, Piazza, & Almeida, 2013). The correlations of self-reported times across collection occasions were all above 0.9. The correlations between self-reported times and times obtained from the "smart box" (a box provided to 25% of the sample that recorded when they placed their salivettes into the box) ranged from 0.75 for the evening occasion to 0.95 for the morning occasion. Of the 2022 respondents from the second wave of MIDUS Diary Project, 1736 participated in saliva collection (86%). Of the 27,776 possible saliva samples (1736 participants × 16 samples), there were 874 missed or unreliable samples, samples that could not be linked to a specific day, or samples with insufficient volume to detect cortisol, which resulted in 97% usable samples (26,902).

**Using Saliva to Assess Daily Stress in Adulthood** Most studies of psychosocial stress and physiology demonstrate elevated cortisol levels in response to laboratory-controlled acute psychological stressors (Dickerson & Kemeny, 2004). Less is known about the relationship between naturally occurring stressors and cortisol responses (Dettenborn et al., 2005; Polk et al., 2005). NSDE is the largest study to investigate the relationship between naturally occurring stressors and cortisol. Examining patterns of cortisol throughout the day (collected through saliva) has enabled us to determine HPA axis responses to external environments. For example, failure to activate the HPA axis in the morning and deactivate it in the evening may indicate difficulty disengaging from external demands, leading to inhibition of restoration and recovery processes (Sapolsky, Krey, & McEwen, 1986).

An initial set of analyses examined age and gender differences in diurnal cortisol. The entire daytime cortisol trajectory (including waking, peak, and nadir values and AUC) was higher in older than in younger participants and in men than in women (Almeida, Piazza, Stawski, & Klein, 2011; Karlamangla et al., 2018). There was a clear dose response with age, with the oldest group having significantly higher mean levels and total concentrations assessed by area under the curve across the day (AUC) than the middle group (ages 50–64 years), and the latter having significantly higher mean levels (and AUC) than the youngest group. The male–female differences in mean nadir and AUC were comparable to corresponding differences between the mid-age and youngest age groups (Karlamangla et al., 2018).

**Integrating Psychosocial and Biological Stress** Our dynamic assessment of diurnal cortisol in relation to both daily and chronic stress processes has resulted in many important findings (Almeida, McGonagle, et al., 2009; Almeida, Piazza, et al., 2009). Thus far, our work has linked elevated diurnal cortisol rhythms and cortisol levels to increasing age (Almeida, Piazza, et al., 2009), daily stressors (Stawski et al., 2013), negative affect (Piazza, Charles, Stawski, & Almeida, 2013), social strain (Friedman, Karlamangla, Almeida, & Seeman, 2012), poor cognitive performance (Stawski et al., 2011), early life adversity (Taylor, Karlamangla, Friedman, & Seeman, 2010), and cancer survivorship (Costanzo, Stawski, Ryff, Coe, & Almeida, 2012). We have also examined diurnal cortisol patterns to understand the day-to-day physiological effects of social role transitions, including widowhood (Ong, Fuller-Rowell, Bonanno, & Almeida, 2011), nonnormative parenting (Barker, Greenberg, Mailick Seltzer, & Almeida, 2012; Seltzer et al., 2010), and early and late retirement (Almeida & Wong, 2009). In contrast, positive psychosocial resources, such as perceived partner responsiveness among married couples (Slatcher, Selcuk, & Ong, 2015) and daily positive events (Sin, Graham, & Almeida, 2015), are associated with adaptive diurnal cortisol profiles.

Current analyses have focused on a specific diurnal rhythm characterized by both a low peak following waking and a failure to lessen cortisol output throughout the day, which we have identified as compressed dynamic range (CDR; Karlamangla et al., 2018). CDR is an innovative potential marker for chronic stress and is linked to lower education, minority status, and chronic health conditions (Dmitrieva, Almeida, Dmitrieva, Loken, & Pieper, 2013; Karlamangla et al., 2018).

Other recent work also reveals exciting linkages of daily experiences with psychophysiology, neuroscience, and biomarker assessments. New findings from the MIDUS Neuroscience Project show that sustained activity in the striatum and dorsolateral prefrontal cortex to positive stimuli is linked to better cortisol regulation and higher well-being (Heller et al., 2013).

Researchers have linked high levels of cortisol to a number of health problems, yet it remains unclear how or why daily patterns of cortisol change over time, how diurnal cortisol rhythms are influenced by changes in psychosocial factors, or the mechanisms whereby cortisol patterns influence long-term health outcomes. Our team and others are currently addressing these issues. For example, early life adversity appears to leave a long-term imprint on cortisol secretion dynamics, reducing diurnal dynamic range without increasing total secretion. This points to the importance of examining the adaptation capacity of physiological systems when studying the impact of early life and chronic stresses on adult health (Karlamangla et al., 2018). Another recent paper documented that older participants with dysregulated profiles across all interview days (i.e., all days elevated, flattened, or a combination of elevated and flattened) showed greater concurrent inflammation risk burden indexed by Il6 and CRP as well as more functional limitations 10 years later (Piazza et al., 2018).

In addition to cortisol, we have been assaying salivary alpha-amylase (sAA), which is a minimally invasive and easily obtainable surrogate marker of individuals' chronobiology and sympathetic nervous system (Granger, Kivlighan, El-Sheikh, Gordis, & Stroud, 2007; Keenan, Licinio, & Veldhuis, 2001). Studies suggest that sAA levels increase in response to stressors such as extreme exercise, heat and cold stress, written examinations (Chatterton, Vogelsong, Lu, Ellman, & Hudgens, 1996; Chatterton, Vogelsong, Lu, & Hudgens, 1997; Skosnik, Chatterton, Swisher, & Park, 2000), and laboratory-based stressors (Gordis, Granger, Susman, & Trickett, 2008; Nater et al., 2005, 2006). We have found that perceptions of and affective reactions to daily stressors were associated with vagal withdrawal and increased sympathetic predominance, as indexed by heart rate variability (Sin et al., 2015). The inclusion of salivary cortisol and sAA provides an integrated perspective on biological stress functioning by examining both HPA axis and SNS response (Granger et al., 2007; Lovallo & Thomas, 2000; McEwen, 2000). These systems interact with one another, leading to dynamic, synergistic effects on the body. Both sAA and cortisol follow diurnal patterns that adjust in response to stressful events (Granger et al., 2003; Nater et al., 2005) and may change from young adulthood to older ages (Nater et al., 2007).

### 27.2.2 DaSH Findings: Saliva as a Window into Caregiving

Biomarkers obtained from saliva can also be used in the study of a variety of problems in adulthood that involve exposure to stressors and in interventions designed to mitigate the effects of stress. One area where there is a growing interest in the use of biomarkers is family caregiving (Lovell & Wetherell, 2011; Von Känel et al., 2012). Family caregiving to individuals with dementia (IWDs), which usually is sustained for years, is both physically and emotionally challenging (Aneshensel, Pearlin, Mullan, Zarit, & Whitlatch, 1995). Caregiving to IWDs has been found to be more challenging compared to other types of chronic conditions, as consequence of the degenerative nature of the disease and the frequency that caregivers must contend with behavioral, cognitive, and emotional problems (Pearlin, Mullan, Semple, & Skaff, 1990). Thus, it is not surprising that dementia family caregivers are at the highest risks for adverse health changes compared to caregivers of persons with other types of disabilities (Vitaliano, Zhang, & Scanlan, 2003). Although various types of interventions to promote caregiver health have been tested, their effects have typically been modest (Sörensen, Pinquart, & Duberstein, 2002).

One promising approach for caregivers of IWDs is the use of adult day services (ADS). ADS and other types of respite care provide family caregivers time away from the IWD, thereby reducing their exposure to care-related stressors and giving them time to engage in other activities (Zarit et al., 2011). Building on this finding, the Daily Stress and Health (DaSH) study was designed to explore the effects of high and low stress days on caregivers' salivary stress biomarkers (Klein et al., 2016; Zarit, Kim, Femia, Almeida, & Klein, 2014). Participants were primary caregivers of IWDs and were using at least 2 days of ADS a week. Following an initial in-home interview, caregivers provided five saliva samples each day (before getting out of bed, 30 min after getting out of bed, before lunch, before dinner, and before bed) for eight consecutive days and completed brief telephone interviews at the end of each day that obtained information about daily stressors and affect and also addressed any problems associated with saliva collection. The 8-day period was selected in order to obtain data on both some high stress days when caregivers provided all the care and on low stress days when the IWD attended ADS. The comparison across multiple days functioned in effect as a classic treatment reversal design, where treatment benefits would be expected in the presence of an intervention (ADS use days) and not when the intervention was withdrawn. Saliva collection and assay followed standard and validated procedures for salivary cortisol, dehydroepiandrosterone sulfate (DHEA-S), and salivary alpha-amylase (sAA) as biomarkers of the hypothalamic–pituitary–adrenal (HPA) axis and autonomic nervous system (ANS), respectively. Follow-up interviews with participants were conducted at 6 and 12 months, and obtained information about affect and health, as well as transitions regarding caregiving (institutionalization and/or death of the IWD).

Initial analyses confirmed that the intervention had its intended effect in lowering caregivers' exposure to care-related stressors on days the IWD attended ADS, compared to days when caregivers provided all the care. Turning to HPA markers, caregivers had improved regulation of cortisol on ADS days and of DHEA-S on days following ADS use (Klein et al., 2016; Liu, Almeida, Rovine, & Zarit, 2016; Zarit et al., 2014). Typical of other chronic stress situations, many of the caregivers had blunted cortisol awakening responses (CAR) on days they provided all the care and increased CAR when the IWD attended ADS (Liu, Kong, Bangerter, Zarit, & Almeida, 2018). Likewise, caregivers had relatively low daily levels of DHEA-S, which is typical of other chronic stress situations (e.g., Lennartsson, Theorell, Kushnir, Bergquist, & Jonsdottir, 2013), but output increased on days following ADS use.

Additionally, ADS use impacted ANS regulation as indicated by diurnal trajectory of sAA. Previous laboratory studies had found that ANS dysregulation and poor sAA recovery from pre- to post-stress conditions relate to health problems such as fatigue and frequency of illness among children (Granger et al., 2007). Findings from the DaSH study showed that, controlling for daily ADS use, greater ADS use across 8 days was associated with a more prominent rise between 30 min after wake-up and before lunch, and a more prominent decline between before lunch and late afternoon, whereas fewer ADS days were associated with a more flattened and blunt sAA diurnal rhythm (Liu et al., 2017).

The DaSH study also explored the potential effects of the stress biomarkers on health over a 1-year period (Liu, Almeida, Rovine, & Zarit, 2017). During that time, some caregivers experienced major transitions, including placing the IWD into a residential care setting or turning caregiving over to another family member. Since those transitions alter stressor exposure, they were taken into account in the analyses. Among caregivers who transitioned out of the role, and who used fewer than average ADS days per week at baseline, lower daily cortisol total output and lower daily sAA total output were associated with increasing functional limitations over 12 months. Caregivers who experienced a transition but had more than average ADS days per week did not show such patterns of association. Among caregivers not experiencing a transition, functional health was stable and showed no association with baseline levels of any of the biomarkers. These findings suggest that ADS use may play a protective role in health outcomes (Liu, Almeida, et al., 2017).

## 27.3 Issues and Challenges: Assessing Daily Salivary Biomarkers in the Field

Conducting daily diary studies that incorporate saliva collection presents unique challenges due to the respondent burden and lack of investigator control in collecting salivary biomarkers. The next section provides evidence of the feasibility, reliability, and validity of the use of saliva in daily studies.

### *27.3.1 Challenge 1: Feasibility of Collecting Saliva in Field Studies*

Given the requirements of the daily designs that involve multiple interviews and multiple saliva collections throughout the day, a key issue is the feasibility of implementing this approach in large social surveys. Findings from the telephone-based NSDE, DaSH, and other studies suggest that overall, respondents are willing to participate and complete the protocol. Participants overwhelmingly participated in the saliva collection protocol. Of the 2022 respondents who completed the second wave of the NSDE, 1736 provided saliva samples (86%). Perhaps more remarkable was the rarity of missed saliva collections—approximately 3%. Rates of retention and successful saliva collection were similar in the DaSH study, despite the levels of stress experienced by the sample (Klein et al., 2016). The NSDE protocol has been adapted by other social surveys that include samples of elderly couples, mothers of children with autism, and hotel workers, their spouses, and children (Yorgason, Almeida, Neupert, Spiro, & Hoffman, 2006; O'Neill et al., 2009; Seltzer et al., 2010). Although the sample and protocol are somewhat different, the recruitment and retention rates are similar to the NSDE.

### 27.3.2 Challenge 2: Reliability

If respondents agree to participate in daily saliva studies, will they be compliant? Adherence to the saliva collection protocol is critical in obtaining reliable assessments of diurnal cortisol. We gauged compliance by contaminated samples and timing of collection. Of the 27,776 possible saliva samples (1736 $\times$ 16 samples), there were 874 missed or unreliable samples, samples that could not be linked to a specific day, or samples with insufficient volume to detect cortisol ($\sim$3%). These data resulted in final cortisol analyses based on 97% usable samples ($N = 26{,}902$).

Data on the exact time respondents provide each saliva sample were obtained from the nightly telephone interviews and on a paper-and-pencil log sent with the collection kit. In addition, approximately 25% of the respondents ($N = 430$) received a "smart box" to store their salivettes. These boxes contained a computer chip that recorded the time respondents opened and closed the box. The correlations of self-reported times across collection occasions were all above 0.9. The correlations between self-reported times and times obtained from the "smart box" ranged from 0.75 for the evening occasion to 0.95 for the morning occasion. Assessing diurnal rhythm also requires careful timing of collection. The biggest challenge we faced was collection of the second sample of the day (30 min after waking). Missing this time window could alter the assessment of the cortisol awakening response (CAR) parameter of the diurnal rhythm. On approximately 10% of our collection days, respondents either provided the sample too early or too late to capture the CAR. Additional protocols could be implemented to increase adherence to this critical time window, including alarm clocks, electronic time stampers, and additional instructions. Indeed, our team at Penn State recently produced an instructional video in collaboration with our local public television affiliate on how and when to collect saliva (a copy is available upon request to the first author). Future data collections will include test instructions on a DVD in the saliva collection kits.

### 27.3.3 Challenge 3: Validity

Lack of control over saliva collection in social surveys poses risks to the interpretation of cortisol. We used NSDE data to assess the validity of field assessments of diurnal cortisol. We compared components of the diurnal rhythm of cortisol in the NSDE with smaller samples in more controlled research settings and with more within-day assessments (Almeida, McGonagle, et al., 2009). We compared our cortisol values for the CAR to the findings of four studies combined and presented by Wüst et al. (2000).

Mean cortisol levels in these published studies were very similar to the NSDE for both awakening cortisol and for cortisol measured 30 min after awakening. We also compared the daily decline slopes from the NSDE with four studies reviewed in Stone et al. (2001). It is important to note that these previous studies had more control over the study protocol such as face-to-face instruction and telephone reminders for collection. Despite the differences between these four studies and NSDE in the number of participants, the number of saliva collections throughout the day, and the number of days assessed, values for the slopes are remarkably parallel (Almeida, McGonagle, et al., 2009; Almeida, Piazza, et al., 2009).

### 27.3.4 *Challenge 4: Costs*

The average cost of the NSDE and DaSH protocol was approximately $350 per respondent. The collection kits including salivettes, packaging materials, boxes, and postage cost $44. The cortisol assaying was conducted in the Biological Psychology Laboratory at the Technical University of Dresden at a cost of approximately $96 ($6 per sample × 16 samples). The interviewing cost via Penn State Survey Research Center was approximately $160 per person ($20 per interview × 8 interviews). Finally, the participants were given $50 as incentive to finish the protocol.

## 27.4 Future Directions in Analyses

Salivary stress biomarkers that are collected intensively within a day and repeatedly across multiple days have mostly been modeled using univariate approaches. However, the field has also seen some innovative models in recent years taking multivariate approaches, where more than one type of biomarkers either from the same person or spouses are modeled simultaneously. We will next contrast these varying modeling approaches to daily biomarkers.

The univariate approach entails modeling one and only one type of biomarker as the outcome, and there are three general approaches. The first approach utilizes a daily summary or composite score, which congregates one type of repeatedly measured biomarkers within a day as one daily measurement. Some methods include the average score within the day, the daily total outputs commonly calculated as the diurnal area under the curve with respect to ground (AUCg) (Pruessner, Kirschbaum, Meinlschmid, & Hellhammer, 2003), and the difference score between specific samples within a day such as the cortisol awakening response (CAR). The CAR can be calculated as the increase (CARi) in cortisol levels using the first two morning samples (Chida & Steptoe, 2009; Pruessner, Hellhammer, Pruessner, & Lupien, 2003), or more recently, as the rate of change in cortisol concentration that occurs during the first 30 min after waking up, which we will explain next (Leggett, Liu, Klein, & Zarit, 2016).

The second univariate approach aims to define the diurnal slope as the linear rate of change in biomarker levels across the day. This approach is most useful for modeling diurnal trajectories of salivary biomarkers that demonstrate circadian rhythms such as diurnal cortisol and salivary alpha-amylase. It typically utilizes each single, repeatedly measured biomarker sample within- and across days as the outcomes. Although this approach entails modeling more than one measurement within the day as the outcome, it is still considered as the univariate approach because only one type of biomarker is considered in the model. Multilevel growth curve modeling based on the linear mixed models, and specifically spline growth curve models, can be used to model these intensively measured biomarkers within a day and across days. Various diurnal slopes have been defined, including the cortisol awakening response (CAR) slope, the cortisol diurnal decline slope, and the salivary alpha-amylase morning decline and diurnal rise slopes. Both of the aforementioned univariate approaches are useful tools for understanding the impact of daily experiences and other personal traits on the daily physiology of one type of biomarker. Alternatively, the composite scores and diurnal slopes can also serve as daily physiological predictors of long-term physical health as detailed in some of the DaSH studies we mentioned earlier (Leggett et al., 2016; Liu, Granger, et al., 2017).

However, studies also show a major issue for these daily metrics of physiology. Specifically, they may fail to capture any stable individual differences in cortisol functioning over any period greater than 1 month. For example, the stability estimates were the lowest for CAR, moderate for diurnal slopes, and inconsistent for the AUCg across 8 months to 6 years (Doane, Chen, Sladek, Van Lenten, & Granger, 2015). Thus, the third univariate approach applies the latent state-trait modeling to diurnal cortisol levels to specifically capture the underlying individual differences by the latent trait cortisol (LTC) factor, as well as any state-specific situational influences and random errors by the latent state cortisol factor (Steyer, Mayer, Geiser, & Cole, 2015). In the first LTC study, Kirschbaum et al. (1990) showed that the variability in morning cortisol levels was largely accounted for by a latent trait factor, whereas the afternoon levels were largely accounted for by the latent state factor. Doane et al. (2015) further examined the reliability, validity, and stability of the LTC based on a study where salivary cortisol was repeatedly measured five times a day, over 3 days, and across three waves over 9 months. They used the first two morning samples to derive the LTC factor within and across waves. The study showed that the LTC was distinct from the CAR, differentially predictive of varying components of the diurnal slopes, and stable across the assessment waves (Doane et al., 2015).

Progressing forward, one innovative and potentially important approach for diurnal biomarkers, however, is to use multivariate models to characterize the neuroendocrine and autonomic impact of daily experiences. Guided by the Allostatic Load theory (Juster, McEwen, & Lupien, 2010; Marin et al., 2011; McEwen, 2003, 2004; Miller, Chen, & Zhou, 2007; Miller, Cohen, & Ritchey, 2002; Rohleder,

Marin, Ma, & Miller, 2009; Zarit et al., 2011), optimum hypothalamic–pituitary–adrenal (HPA) axis and autonomic nervous system (ANS) functioning in response to stress entail coordination across these systems, which is regulated by the central nervous system (Bauer, Quas, & Boyce, 2002). Although the HPA axis and ANS are interconnected physiologically, their responses to stressors may become asymmetrical under aging and/or chronic stressor exposure, (Gordis et al., 2008) such as dementia caregiving. Thus, the HPA–ANS dissociation may reflect inefficient central coordination, which can become a form of allostatic load (Bauer et al., 2002; McEwen, 2004), and further impact individuals' affective well-being (McEwen, 2003).

The HPA–ANS synchrony is the simultaneous reactivity of these two systems under chronic stressor exposure. The HPA axis and ANS activities can be measured by cortisol and salivary alpha-amylase (sAA) levels, respectively. The HPA–ANS synchrony can be measured by the dynamic covariations between daily total outputs of these biomarkers collected intensively over time. Synchrony presents when cortisol and sAA levels covary positively in the process of stress regulation, whereas asynchrony or asymmetry is indicated by a relatively strong response in cortisol or sAA levels over the other (Gordis et al., 2008).

Another type of multivariate approach serves the dyadic research perspective and is best for modeling the co-regulation or synchrony of the physiological processes between various forms of dyads as dynamic systems. In such models, one type of biomarker is modeled simultaneously with measurements typically collected from both parties of one dyad such as husbands and wives. Generally, daily summary or composite scores will be calculated first from each of the dyads respectively, and multilevel cross-lagged models can be utilized to show how the physiology of one person in the dyads impact that of the other person, on the same day and/or on the next day; covariates of the synchrony can also be modeled. For example, one study showed that on days when one experienced faster or slower cortisol decline, the spouse also showed similar cortisol regulation. For CAR, such positive synchrony was only found in couples having high levels of marital strains. Further, couples reporting more spousal support had stronger stability in CAR within each individual (Liu, Rovine, Klein, & Almeida, 2013).

## 27.5 Conclusion

Use of saliva-based biomarkers has shown considerable promise for examining the immediate physiological consequences of daily stressors. The approach is more effective when observations can be made over multiple days, during which there are likely to be sufficient occurrences of stressors. Reactivity of these biomarkers to stressors within- and across days may be a more valuable measure of biological effects than average daily scores. Use with special populations experiencing high levels of stressors is also feasible.

Saliva-based biomarkers may also be used as outcome measures for treatment studies. Although they do not replace self-reports of affect and subjective stress, biomarkers provide a complementary perspective on the stress response, and may be useful for identifying pathways from daily experiences to health outcomes. Future research could combine daily saliva-based biomarkers with more traditional immune system and cardiovascular measures to explore how stress reactivity under varying circumstances might lead to subsequent illness.

In summary, we hope that this chapter encourages other scientists to incorporate saliva into their research. Saliva can help uncover important age-related changes in important biological systems, including the hypothalamic–pituitary–adrenal (HPA) axis and the sympathomedullary (SAM) pathway. Saliva can also be used to assess naturally occurring fluctuations of biomarkers in daily life. Indeed, this chapter provides some initial evidence of the scientific benefits of collecting multiple samples of saliva across a series of days. Finally, we acknowledge the practical and analytic challenges of incorporating saliva bioscience and offer some cutting-edge statistical approaches that can help address these issues. In our view, the future of salivary bioscience is multivariate and dynamic and we will need to continue to develop statistical approaches to keep pace with our ability to assess such naturally occurring biopsychosocial phenomena.

**Acknowledgments** Since 1995 the MIDUS study has been funded by the John D. and Catherine T. MacArthur Foundation Research Network, National Institute on Aging (P01-AG020166), and National Institute on Aging (U19-AG051426). Biomarker data collection was further supported by the NIH National Center for Advancing Translational Sciences' (NCATS) Clinical and Translational Science Award (CTSA) program as "1UL1RR025011". The Daily Stress and Health Study (DaSH) was funded by National Institute on Aging (R01AG031758).

## References

Affoo, R. H., Foley, N., Garrick, R., Siqueira, W. L., & Martin, R. E. (2015). Meta-analysis of salivary flow rates in young and older adults. *Journal of the American Geriatrics Society, 63* (10), 2142–2151. https://doi.org/10.1111/jgs.13652

Almeida, D. M. (2005). Resilience and vulnerability to daily stressors assessed via diary methods. *Current Directions in Psychological Science, 14*, 64–68. https://doi.org/10.1111/j.0963-7214.2005.00336.x

Almeida, D. M., McGonagle, K., & King, H. (2009). Assessing daily stress processes in social surveys by combining stressor exposure and salivary cortisol. *Biodemography and Social Biology, 55*(2), 220–238. https://doi.org/10.1080/19485560903382338

Almeida, D. M., Neupert, S. D., Banks, S. R., & Serido, J. (2005). Do daily stress processes account for socioeconomic health disparities? *Journals of Gerontology: Social Science, 60*, 34–39. https://doi.org/10.1093/geronb/60.Special_Issue_2.S34

Almeida, D. M., Piazza, J. R., & Stawski, R. S. (2009). Interindividual differences and intraindividual variability in the cortisol awakening response: An examination of age and gender. *Psychology and Aging, 24*(4), 819–827. https://doi.org/10.1037/a0017910

Almeida, D. M., Piazza, J. R., Stawski, R. S., & Klein, L. C. (2011). The speedometer of life: Stress, health and aging. In K. W. Schaie & S. L. Willis (Eds.), *Handbook of the psychology of aging* (7th ed., pp. 191–206). San Diego, CA: Academic Press.

Almeida, D. M., & Wong, J. D. (2009). Life transition and stress: A life course perspective on daily stress processes. In G. H. Elder & J. Z. Giele (Eds.), *The craft of life course research*. New York: Guilford Press.

Aneshensel, C. S., Pearlin, L. I., Mullan, J. T., Zarit, S. H., & Whitlatch, C. J. (1995). *Profiles in caregiving: The unexpected career*. San Diego, CA: Academic Press.

Barker, E. T., Greenberg, J. S., Mailick Seltzer, M., & Almeida, D. M. (2012). Daily stress and cortisol patterns in parents of adult children with a serious mental illness. *Health Psychology, 31* (1), 130–134. https://doi.org/10.1037/a0025325

Bauer, A. M., Quas, J. A., & Boyce, W. T. (2002). Associations between physiological reactivity and children's behavior: Advantages of a multisystem approach. *Journal of Developmental & Behavioral Pediatrics, 23*(2), 102–113.

Chatterton, R. T., Vogelsong, K. M., Lu, Y., Ellman, A. B., & Hudgens, G. A. (1996). Salivary α-amylase as a measure of endogenous adrenergic activity. *Clinical Physiology and Functional Imaging, 16*, 433–448. https://doi.org/10.1111/j.1475-097X.1996.tb00731.x

Chatterton, R. T., Vogelsong, K. M., Lu, Y., & Hudgens, G. A. (1997). Hormonal responses to psychological stress in men preparing for skydiving. *The Journal of Clinical Endocrinology & Metabolism, 82*, 2503–2509. https://doi.org/10.1210/jcem.82.8.4133

Chida, Y., & Steptoe, A. (2009). Cortisol awakening response and psychosocial factors: A systematic review and meta-analysis. *Biological Psychology, 80*(3), 265–278. https://doi.org/10.1016/j.biopsycho.2008.10.004

Costanzo, E. S., Stawski, R. S., Ryff, C. D., Coe, C. L., & Almeida, D. M. (2012). Cancer survivors' responses to daily stressors: Implications for quality of life. *Health Psychology, 31*(3), 360–370. https://doi.org/10.1037/a0027018

Dettenborn, L., James, G. D., van Berge-Landry, H., Valdimarsdottir, H. B., Montgomery, G. H., & Bovbjerg, D. H. (2005). Heightened cortisol responses to daily stress in working women at familial risk for breast cancer. *Biological Psychology, 69*, 167–179. https://doi.org/10.1016/j.biopsycho.2004.07.004

Dickerson, S. S., & Kemeny, M. E. (2004). Acute stressors and cortisol responses: A theoretical integration and synthesis of laboratory research. *Psychological Bulletin, 130*, 355–391.

Ditzen, B., Ehlert, U., & Nater, U. M. (2014). Associations between salivary alpha-amylase and catecholamines—A multilevel modeling approach. *Biological Psychology, 103*, 15–18. https://doi.org/10.1016/j.biopsycho.2014.08.001

Dmitrieva, N. O., Almeida, D. M., Dmitrieva, J., Loken, E., & Pieper, C. F. (2013). A day-centered approach to modeling cortisol: Diurnal cortisol profiles and their associations among U.S. adults. *Psychoneuroendocrinology, 38*(10), 2354–2365. https://doi.org/10.1016/j.psyneuen.2013.05.003

Doane, L. D., Chen, F. R., Sladek, M. R., Van Lenten, S. A., & Granger, D. A. (2015). Latent trait cortisol (LTC) levels: Reliability, validity, and stability. *Psychoneuroendocrinology, 55*, 21–35. https://doi.org/10.1016/j.psyneuen.2015.01.017

Dodds, M. W., Johnson, D. A., & Yeh, C. K. (2005). Health benefits of saliva: A review. *Journal of Dentistry, 33*(3), 223–233. https://doi.org/10.1016/j.jdent.2004.10.009

Epel, E. S., Burke, H. M., & Wolkowitz, O. M. (2009). The psychoneuroendocrinology of aging: Anabolic and catabolic hormones. In C. M. Aldwin, C. L. Park, & A. Spiro (Eds.), *Handbook of health psychology and aging* (pp. 119–141). New York, NY: Guilford Press.

Friedman, E. M., Karlamangla, A. S., Almeida, D. M., & Seeman, T. E. (2012). Social strain and cortisol regulation in midlife in the US. *Social Science and Medicine, 74*, 607–615. https://doi.org/10.1016/j.socscimed.2011.11.003

Gordis, E. B., Granger, D. A., Susman, E. J., & Trickett, P. K. (2008). Salivary alpha amylase–cortisol asymmetry in maltreated youth. *Hormones and Behavior, 53*(1), 96–103. https://doi.org/10.1016/j.yhbeh.2007.09.002

Granger, D. A., Kivlighan, K. T., El-Sheikh, M., Gordis, E. B., & Stroud, L. R. (2007). Salivary α-amylase in biobehavioral research: Recent developments and applications. *Annals of the New York Academy of Sciences, 1098*(1), 122–144. https://doi.org/10.1196/annals.1384.008

Granger, D. A., Shirtcliff, E. A., Zahn-Waxler, C., Usher, B., Klimes-Dougan, B., & Hastings, P. (2003). Salivary testosterone diurnal variation and psychopathology in adolescent males and females: Individual differences and developmental effects. *Development and Psychopathology, 15*, 431–449. https://doi.org/10.1017/S0954579403000233

Heller, A. S., van Reekum, C. M., Schaefer, S. M., Lapate, R. C., Radler, B. T., Ryff, C. D., & Davidson, R. J. (2013). Sustained ventral striatal activity predicts eudaimonic well-being and cortisol output. *Psychological Science, 24*, 2191–2200. https://doi.org/10.1177/0956797613490744. PMCID: PMC386696.

Juster, R.-P., McEwen, B. S., & Lupien, S. J. (2010). Allostatic load biomarkers of chronic stress and impact on health and cognition. *Neuroscience & Biobehavioral Reviews, 35*(1), 2–16. https://doi.org/10.1016/j.neubiorev.2009.10.002

Karlamangla, A. S., Merkin, S. S., Almeida, D. M., Friedman, E. M., Mogle, J. A., & Seeman, T. E. (2018). Early-life adversity and dysregulation of adult diurnal cortisol rhythm. *The Journals of Gerontology: Series B, 74*(1), 160–169.

Keenan, D. M., Licinio, J., & Veldhuis, J. D. (2001). A feedback-controlled ensemble model of the stress-responsive hypothalamo-pituitary-adrenal axis. *PNAS, 98*, 4028–4033.

Kirschbaum, C., & Hellhammer, D. H. (1989). Salivary cortisol in psychobiological research: An overview. *Neuropsychobiology, 22*(3), 150–169. https://doi.org/10.1159/000118611

Kirschbaum, C., Steyer, R., Eid, M., Patalla, U., Schwenkmezger, P., & Hellhammer, D. H. (1990). Cortisol and behavior: 2. Application of a latent state-trait model to salivary cortisol. *Psychoneuroendocrinology, 15*(4), 297–307. https://doi.org/10.1016/0306-4530(90)90080-S

Klein, L. C., & Corwin, E. J. (2007). Homeostasis and the stress response. In E. J. Corwin (Ed.), *Handbook of pathophysiology* (3rd ed., pp. 159–172). Philadelphia, PA: Lippincott Williams & Wilkins.

Klein, L., Kim, K., Almeida, D. M., Femia, E. E., Rovine, M., & Zarit, S. H. (2016). Anticipating an easier day: Effects of adult day services on daily cortisol and stress. *The Gerontologist, 56*, 303–312. https://doi.org/10.1083/geront/gne060

Klein, L. C., Whetzel, C. A., Almeida, D. M., Bennett, J. M., Stawski, R. S., Banks, S. R., & Crouter, A. C. (2008). Salivary DHEA-S levels across the day: Evidence for a daily rhythm in a healthy adult population. *Psychosomatic Medicine, 70*(3), A-46.

Krug, A. W., Ziegler, C. G., & Bornstein, S. R. (2008). DHEA and DHEA-S and their functions in the brain and adrenal medulla. In M. S. Ritsner & A. Weizman (Eds.), *Neuroactive steroids in brain function, behavior and neuropsychiatric disorders* (pp. 227–239). Dordrecht: Springer. https://doi.org/10.1007/978-1-4020-6854-6_12

Leggett, A. N., Liu, Y., Klein, L. C., & Zarit, S. H. (2016). Sleep duration and the cortisol awakening response in dementia caregivers utilizing adult day services. *Health Psychology, 35*(5), 465–473. https://doi.org/10.1037/hea0000276

Lennartsson, A.-K., Theorell, T., Kushnir, M. M., Bergquist, J., & Jonsdottir, I. H. (2013). Perceived stress at work is associated with attenuated DHEA-S response during acute psychosocial stress. *Psychoneuroendocrinology, 38*(9), 1650–1657. https://doi.org/10.1016/j.psyneuen.2013.01.010

Liu, Y., Almeida, D. M., Rovine, M. J., & Zarit, S. H. (2016). Modeling cortisol daily rhythms of family caregivers of individuals with dementia: Daily stressors and adult day services use. *The Journals of Gerontology: Series B: Psychological Sciences and Social Sciences, 73*(3), 457–467. https://doi.org/10.1093/geronb/gbw140

Liu, Y., Almeida, D. M., Rovine, M. J., & Zarit, S. H. (2017). Care transitions and adult day services moderate the longitudinal links between stress biomarkers and family caregivers' functional health. *Gerontology, 63*(6), 538–549. https://doi.org/10.1159/000475557

Liu, Y., Kong, J., Bangerter, L. R., Zarit, S. H., & Almeida, D. M. (2018). Early parental abuse and daily assistance to aging parents with disability: Associations with the middle-aged adults' daily well-being. *The Journals of Gerontology: Series B, 73*(5), e59–e68. https://doi.org/10.1093/geronb/gbx173

Liu, Y., Granger, D. A., Kim, K., Klein, L. C., Almeida, D. M., & Zarit, S. H. (2017). Diurnal salivary alpha-amylase dynamics among dementia family caregivers. *Health Psychology, 36*(2), 160–168. https://doi.org/10.1037/hea0000430

Liu, S., Rovine, M. J., Klein, L. C., & Almeida, D. M. (2013). Synchrony of diurnal cortisol pattern in couples. *Journal of Family Psychology, 27*(4), 579–588. https://doi.org/10.1037/a0033735

Lovallo, W. R., & Thomas, T. L. (2000). Stress hormones in psychophysiological research: Emotional, behavioral, and cognitive implications. In J. T. Cacioppo, L. G. Tassinary, & G. G. Berntson (Eds.), *Handbook of psychophysiology* (pp. 342–367). New York, NY: Cambridge University Press.

Lovell, B., & Wetherell, M. A. (2011). The cost of caregiving: Endocrine and immune implications in elderly and non-elderly caregivers. *Neuroscience and Biobehavioral Reviews, 35*(6), 1342–1352. https://doi.org/10.1016/j.neubiorev.2011.02.007

Maninger, N., Wolkowitz, O. M., Reus, V. I., Epel, E. S., & Mellon, S. H. (2009). Neurobiological and neuropsychiatric effects of dehydroepiandrosterone (DHEA) and DHEA sulfate (DHEAS). *Frontiers in Neuroendocrinology, 30*(1), 65–91. https://doi.org/10.1016/j.yfrne.2008.11.002

Marin, M. F., Lord, C., Andrews, J., Juster, R. P., Sindi, S., Arsenault-Lapierre, G., . . . Lupien, S. J. (2011). Chronic stress, cognitive functioning and mental health. *Neurobiology of Learning and Memory, 96*(4), 583–595. https://doi.org/10.1016/j.nlm.2011.02.016

McEwen, B. S. (2000). The neurobiology of stress: From serendipity to clinical relevance. *Brain Research, 886*, 172–189. https://doi.org/10.1016/S0006-8993(00)02950-4

McEwen, B. S. (2003). Mood disorders and allostatic load. *Biological Psychiatry, 54*(3), 200–207. https://doi.org/10.1016/S0006-3223(03)00177-X

McEwen, B. S. (2004). Protection and damage from acute and chronic stress: Allostasis and allostatic overload and relevance to the pathophysiology of psychiatric disorders. *Annals of the New York Academy of Sciences, 1032*(1), 1–7. https://doi.org/10.1196/annals.1314.001

Miller, G. E., Chen, E., & Zhou, E. S. (2007). If it goes up, must it come down? Chronic stress and the hypothalamic-pituitary-adrenocortical axis in humans. *Psychological Bulletin, 133*(1), 25–45. https://doi.org/10.1037/0033-2909.133.1.25

Miller, G. E., Cohen, S., & Ritchey, A. K. (2002). Chronic psychological stress and the regulation of pro-inflammatory cytokines: A glucocorticoid-resistance model. *Health Psychology, 21*(6), 531–541. https://doi.org/10.1037/0278-6133.21.6.531

Mortazavi, H., Baharvand, M., Movahhedian, A., Mohammadi, M., & Khodadoustan, A. (2014). Xerostomia due to systemic disease: A review of 20 conditions and mechanisms. *Annals of Medical and Health Science Research, 4*(4), 503–510. https://doi.org/10.4103/2141-9248.139284

Nagler, R. M., & Hershkovich, O. (2005). Relationships between age, drugs, oral sensorial complaints and salivary profile. *Archives of Oral Biology, 50*(1), 7–16. https://doi.org/10.1016/j.archoralbio.2004.07.012

Nater, U. M., Hoppmann, C. A., & Scott, S. B. (2013). Diurnal profiles of salivary cortisol and alpha-amylase change across the adult lifespan: Evidence from repeated daily life assessments. *Psychoneuroendocrinology, 38*(12), 3167–3171. https://doi.org/10.1016/j.psyneuen.2013.09.008

Nater, U. M., La Marca, R., Florin, L., Moses, A., Langhans, W., Koller, M. M., & Ehlert, U. (2006). Stress induced changes in human salivary alpha-amylase activity. Associations with adrenergic activity. *Psychoneuroendocrinology, 31*, 49–58. https://doi.org/10.1016/j.psyneuen.2005.05.010

Nater, U. M., & Rohleder, N. (2009). Salivary alpha-amylase as a non-invasive biomarker for the sympathetic nervous system: Current state of research. *Psychoneuroendocrinology, 34*(4), 486–496. https://doi.org/10.1016/j.psyneuen.2009.01.014

Nater, U. M., Rohleder, N., Gaab, J., Berger, S., Jud, A., Kirschbaum, C., & Ehlert, U. (2005). Human salivary alpha-amylase reactivity in a psychosocial stress paradigm. *International Journal of Psychophysiology, 55*, 333–342. https://doi.org/10.1016/j.ijpsycho.2004.09.009

Nater, U. M., Rohleder, N., Schlotz, W., Ehlert, U., & Kirschbaum, C. (2007). Determinants of the diurnal course of salivary alpha-amylase. *Psychoneuroendocrinology, 32*(4), 392–401. https://doi.org/10.1016/j.psyneuen.2007.02.007

Nederfors, T., & Dahlof, C. (1992). Effects of the β-adrenoceptor antagonists atenolol and propranolol on human whole saliva flow rate and composition. *Archives of Oral Biology, 37*(7), 579–584. https://doi.org/10.1016/0003-9969(92)90141-T

O'Neill, J. W., Harrison, M. M., Cleveland, J. N., Almeida, D. M., Stawski, R. S., & Crouter, A. C. (2009). Work-family climate, organizational commitment, and turnover: Multilevel contagion effects of leaders. *Journal of Vocational Behavior, 74*, 18–29. https://doi.org/10.1016/j.jvb.2008.10.004

Ong, A. D., Fuller-Rowell, T. E., Bonanno, G. A., & Almeida, D. M. (2011). Spousal loss predicts alterations in diurnal cortisol activity through prospective changes in positive emotion. *Health Psychology, 30*(2), 220–227. https://doi.org/10.1037/a0022262

Pearlin, L. I., Mullan, J. T., Semple, S. J., & Skaff, M. M. (1990). Caregiving and the stress process: An overview of concepts and their measures. *The Gerontologist, 30*(5), 583–594. https://doi.org/10.1093/geront/30.5.583

Piazza, J. R., Almeida, D. M., Dmitrieva, N. O., & Klein, L. C. (2010). Frontiers in the use of biomarkers of health in research on stress and aging. *Journals of Gerontology Series B: Psychological Sciences and Social Sciences, 65*(5), 513–525. https://doi.org/10.1093/geronb/gbq049

Piazza, J. R., Charles, S. T., Stawski, R. S., & Almeida, D. M. (2013). Age and the association between negative affective states and diurnal cortisol. *Psychology and Aging, 28*(1), 47–56. https://doi.org/10.1037/a0029983

Piazza, J. R., Dmitrieva, N. D., Charles, S. T., Almeida, D. M., & Orona, G. O. (2018). Diurnal cortisol profiles, inflammation and functional limitations in aging: Findings from the MIDUS study. *Health Psychology, 37*(9), 839–849. https://doi.org/10.1037/hea0000629

Polk, D. E., Cohen, S., Doyle, W. J., Skoner, D. P., & Kirschbaum, C. (2005). State and trait affect as predictors of salivary cortisol in healthy adults. *Psychoneuroendocrinology, 30*, 261–272.

Pruessner, M., Hellhammer, D. H., Pruessner, J. C., & Lupien, S. J. (2003). Self-reported depressive symptoms and stress levels in healthy young men: Associations with the cortisol response to awakening. *Psychosomatic Medicine, 65*(1), 92–99.

Pruessner, J. C., Kirschbaum, C., Meinlschmid, G., & Hellhammer, D. H. (2003). Two formulas for computation of the area under the curve represent measures of total hormone concentration versus time-dependent change. *Psychoneuroendocrinology, 28*(7), 916–931. https://doi.org/10.1016/S0306-4530(02)00108-7

Rohleder, N., Marin, T. J., Ma, R., & Miller, G. E. (2009). Biologic cost of caring for a cancer patient: Dysregulation of pro- and anti-inflammatory signaling pathways. *Journal of Clinical Oncology, 27*(18), 2909–2915. https://doi.org/10.1200/jco.2008.18.7435

Rohleder, N., Nater, U. M., Wolf, J. M., Ehlert, U., & Kirschbaum, C. (2004). Psychosocial stress-induced activation of salivary alpha-amylase: An indicator of sympathetic activity? *Annals of the New York Academy of Sciences, 1032*(1), 258–263. https://doi.org/10.1196/annals.1314.033

Sapolsky, R. M., Krey, L. C., & McEwen, B. S. (1986). The neuroendocrinology of stress and aging: The glucocorticoid cascade hypothesis. *Endocrinology Review, 7*(3), 284–301. https://doi.org/10.1210/edrv-7-3-284

Seltzer, M. M., Greenberg, J. S., Hong, J., Smith, L. E., Almeida, D. M., Coe, C., & Stawski, R. S. (2010). Maternal cortisol levels and behavior problems in adolescents and adults with ASD. *Journal of Autism and Developmental Disorders, 40*(4), 457–469.

Sin, N. L., Graham, J. E., & Almeida, D. M. (2015). Daily positive events and inflammation: Findings from the National Study of Daily Experiences. *Brain, Behavior, and Immunity, 43*, 130–138. https://doi.org/10.1016/j.bbi.2014.07.015

Skosnik, P. D., Chatterton, R. T., Swisher, T., & Park, S. (2000). Modulation of attentional inhibition by norepinephrine and cortisol after psychological stress. *International Journal of Psychophysiology, 36*, 59–68. https://doi.org/10.1016/S0167-8760(99)00100-2

Slatcher, R. B., Selcuk, E., & Ong, A. D. (2015). Perceived partner responsiveness predicts diurnal cortisol profiles 10 years later. *Psychological Science, 26*(7), 972–982.

Smith, C. H., Boland, B., Daurreeawoo, Y., Donaldson, E., Small, K., & Tuomainen, J. (2013). Effect of aging on stimulated salivary flow in adults. *Journal of the American Geriatric Society, 61*(5), 805–808. https://doi.org/10.1111/jgs.12219

Sörensen, S., Pinquart, M., & Duberstein, P. (2002). How effective are interventions with caregivers? An updated meta-analysis. *The Gerontologist, 42*(3), 356–372. https://doi.org/10.1093/geront/42.3.356

Stawski, R. S., Almeida, D. M., Lachman, M. E., Tun, P. A., Rosnick, C. B., & Seeman, T. (2011). Associations between cognitive function and naturally occurring daily cortisol during middle adulthood: Timing is everything. *Journals of Gerontology. Series B, Psychological Sciences and Social Sciences, 66B*(suppl 1), i71–i81. https://doi.org/10.1093/geronb/gbq094

Stawski, R. S., Cichy, K. E., Piazza, J. R., & Almeida, D. M. (2013). Associations among daily stressors and salivary cortisol: Findings from the National Study of Daily Experiences. *Psychoneuroendocrinology, 38*(11), 2654–2665. https://doi.org/10.1016/j.psyneuen.2013.06.023

Steyer, R., Mayer, A., Geiser, C., & Cole, D. A. (2015). A theory of states and traits – revised. *Annual Review of Clinical Psychology, 11*(1), 71–98. https://doi.org/10.1146/annurev-clinpsy-032813-153719

Stone, A. A., Schwartz, J. E., Smyth, J., Kirschbaum, C., Cohen, S., Hellhammer, D., & Grossman, S. (2001). Individual differences in the diurnal cycle of salivary free cortisol: A replication of flattened cycles for some individuals. *Psychoneuroendocrinology, 26*, 295–306. https://doi.org/10.1016/S0306-4530(00)00057-3

Tannenbaum, C., Barrett-Connor, E., Laughlin, G. A., & Platt, R. W. (2003). A longitudinal study of dehydroepiandrosterone sulphate (DHEAS) change in older men and women: The Rancho Bernardo Study. *European Journal of Endocrinology, 151*(6), 717–725.

Taylor, S. E., Karlamangla, A. S., Friedman, E. M., & Seeman, T. E. (2010). Early environment affects neuroendocrine regulation in adulthood. *Social Cognitive and Affective Neuroscience, 6*(2), 244–251. https://doi.org/10.1093/scan/nsq037

Thoma, M. V., Kirschbaum, C., Wolf, J. M., & Rohleder, N. (2012). Acute responses in salivary alpha-amylase predict increases of plasma norepinehphrine. *Biological Psychology, 91*(3), 342–348. https://doi.org/10.1016/j.biopsycho.2012.07.008

Tsigos, C., Kyrou, I., Kassi, E., & Chrousos, G. P. (2000). Stress, endocrine physiology and pathophysiology. In L. J. De Groot, G. Chrousos, K. Dungan, K. R. Feingold, A. Grossman, J. M. Hershman, C. Koch, M. Korbonits, R. McLachlan, M. New, J. Purnell, R. Rebar, F. Singer, & A. Vinik (Eds.), *Endotext*. South Dartmouth, MA: MDText.com.

Varga, G. (2012). Physiology of the salivary glands. *Surgery (Oxford), 30*(11), 578–583. https://doi.org/10.1016/j.mpsur.2012.09.010

Vitaliano, P. P., Zhang, J., & Scanlan, J. M. (2003). Is caregiving hazardous to one's physical health? A meta-analysis. *Psychological Bulletin, 129*(6), 946–972. https://doi.org/10.1037/0033-2909.129.6.946

Von Känel, R., Mills, P. J., Mausbach, B. T., Dimsdale, J. E., Patterson, T. L., Ziegler, M. G., … Grant, I. (2012). Effect of Alzheimer caregiving on circulating levels of C-reactive protein and other biomarkers relevant to cardiovascular disease risk: A longitudinal study. *Gerontology, 58*(4), 354–365. https://doi.org/10.1159/000334219

Wolff, A., Joshi, R. K., Ekstrom, J., Aframian, D., Pedersen, A. M., Proctor, G., … Dawes, C. (2017). A guide to medications inducing salivary gland dysfunction, xerostomia, and subjective sialorrhea: A systematic review sponsored by the World Workshop on Oral Medicine VI. *Drugs in R&D, 17*(1), 1–28. https://doi.org/10.1007/s40268-016-0153-9

Wust, S., Wolf, J., Hellhammer, D. H., Federenko, I., Schommer, N., & Kirschbaum, C. (2000). The cortisol awakening response - normal values and confounds. *Noise Health, 2*, 79–88.

Yorgason, J. B., Almeida, D. M., Neupert, S. D., Spiro, A., & Hoffman, L. (2006). A dyadic examination of daily health symptoms and emotional well-being in later life couples. *Family Relations, 55*, 613–624. https://doi.org/10.1111/j.1741-3729.2006.00430.x

Zarit, S. H., Kim, K., Femia, E. E., Almeida, D. M., & Klein, L. C. (2014). The effects of adult day services on family caregivers' daily stress, affect, and health: Outcomes from the daily stress and health (DaSH) study. *The Gerontologist, 54*(4), 570–579. https://doi.org/10.1093/geront/gnt045

Zarit, S. H., Kim, K., Femia, E. E., Almeida, D. M., Savla, J., & Molenaar, P. C. M. (2011). Effects of adult day care on daily stress of caregivers: A within-person approach. *The Journals of Gerontology, Series B: Psychological Sciences and Social Sciences, 66*(5), 538–546. https://doi.org/10.1093/geronb/gbr030

Zarit, S. H., Whetzel, C. A., Kim, K., Femia, E. E., Almeida, D. M., Rovine, M. J., & Klein, L. C. (2014). Daily stressors and adult day service use by family caregivers: Effects on depressive symptoms, positive mood, and dehydroepiandrosterone-sulfate. *The American Journal of Geriatric Psychiatry, 22*(12), 1592–1602. https://doi.org/10.1016/j.jagp.2014.01.013

Zhang, C. Z., Cheng, X. Q., Li, J. Y., Zhang, P., Yi, P., Xu, X., & Zhou, X. D. (2016). Saliva in the diagnosis of diseases. *International Journal of Oral Science, 8*(3), 133–137. https://doi.org/10.1038/ijos.2016.38

# Chapter 28
# Salivary Bioscience and Research on Animal Welfare and Conservation Science

**Molly Staley and Lance J. Miller**

**Abstract** Animal welfare research strives to empirically assess how care and management practices impact the health and well-being of animals across diverse settings. This includes agricultural, biomedical, and companion animals as well as animals in professionally managed populations in zoos and aquariums. Tools developed for use with animals in professionally managed care may also have applications to conservation science, which is the interdisciplinary study and protection of biodiversity. Historically, interest in saliva arose out of a desire to find less invasive alternatives to blood that allow for near real-time monitoring of physiology and repeated sampling on shorter timescales. In the time since, applications for saliva have grown to include evaluating the response of animals to management practices and novel stimuli, reproductive profiling, and health monitoring. In this chapter, we emphasize the current state and challenges of implementing salivary research in animal welfare settings. Furthermore, we discuss how implementation of salivary research in zoos and aquariums for welfare purposes is facilitating novel research applications related to species conservation. Overall, our aim is to provide a critical examination of both the applications and limitations of salivary research to the fields of animal welfare and conservation science.

**Keywords** Animal welfare · Conservation science

M. Staley (✉)
Chicago Zoological Society – Brookfield Zoo, Brookfield, IL, USA

Department of Biology, Loyola University Chicago, Chicago, IL, USA
e-mail: mstaley1@luc.edu

L. J. Miller
Chicago Zoological Society – Brookfield Zoo, Brookfield, IL, USA
e-mail: Lance.Miller@czs.org



D. A. Granger, M. K. Taylor (eds.), *Salivary Bioscience*,
https://doi.org/10.1007/978-3-030-35784-9_28

## 28.1 History of Salivary Biosciences in Animal Welfare and Conservation Science

The field of salivary biosciences has witnessed tremendous growth in recent decades (Granger et al., 2012) and, with this, conservation scientists and animal welfare professionals have begun to realize the value of oral fluid (saliva). Despite lagging behind human health and medicine, in large part due to a historic focus on other noninvasive sample types, there is a growing awareness and appreciation for how saliva can aid in developing practices that promote species sustainability, mitigate the impacts of anthropogenic activities, and ensure animals are thriving under professionally managed care (Lacy, 1993; Ralls et al., 2018; Whitham & Wielebnowski, 2013). With salivary biosciences increasingly in the scientific spotlight, there is a clear need for both a synthesis and discussion regarding how saliva can aid in meeting conservation and welfare objectives. Here, we provide a historical perspective followed by an in-depth assessment of the current applications for salivary biosciences within the fields of animal welfare and conservation science.

To ensure animals are thriving in professionally managed care and natural settings, there must be a shift from simply reducing chronic stress toward understanding the conditions that facilitate positive affective states and expression of species-specific behaviors. On this animal welfare biologists and conservation scientists agree: A multifaceted approach that includes both positive and negative welfare indicators is essential, especially given the numerous studies demonstrating poor welfare can impair reproduction (Broom, 1991; Clubb & Mason, 2003; Meagher et al., 2014; Miller, Ivy, Vicino, & Schork, 2018). This includes using behavioral and physiological measures to empirically assess whether existing management practices are effective or, alternatively, if adjustments should be made to obtain the desired outcome (Barber, 2009; Brambell, 1965). However, this was not always the case and most early research focused on negative welfare indicators including stereotypic behavior and chronically elevated glucocorticoid concentrations (Mendl, Burman, & Paul, 2010). Even historically accepted welfare guidelines such as the five freedoms model focused on preventing negative situations (Farm Animal Welfare Council, 1993). This model, for example, stated that animals should experience:

1. "Freedom from hunger or thirst by ready access to fresh water and a diet to maintain full health and vigour.
2. Freedom from discomfort by providing an appropriate environment including shelter and a comfortable resting area.
3. Freedom from pain, injury or disease by prevention or rapid diagnosis and treatment.
4. Freedom to express (most) normal behaviour by providing sufficient space, proper facilities and company of the animal's own kind.
5. Freedom from fear and distress by ensuring conditions and treatment which avoid mental suffering." (Farm Animal Welfare Council, 1993).

Related to this third freedom, the freedom from pain, injury, or disease, there was recognition well before the model's publication that there was a need to move away from traditionally used, and putatively painful, blood sampling toward less invasive alternatives for regular physiological monitoring such as urine or feces (Palme, 2012). For example, urine samples were first used in the 1970s to examine ovarian cycles and diagnose pregnancies in laboratory-housed nonhuman primates, with zoos later adapting similar methods as well as developing protocols for using fecal steroid hormone metabolites to monitor reproduction (see Monfort, Holt, Pickard, Rodger, & Wildt, 2003 for review). The goal of this early research was to track reproductive cycles in order to properly time breeding introductions, monitor pregnancies, and predict approximate birth dates (Saunders, Harris, Traylor-Holzer, & Beck, 2014; Seal et al., 1985). Not until the 1990s did zoos and aquariums begin investigating saliva as a tool to improve reproductive success of threatened and endangered species (Czekala & Callison, 1996; Pietraszek & Atkinson, 1994; Theodorou & Atkinson, 1998). Work performed on Hawaiian monk seals (*Monachus schauinslandi*), for instance, was aimed at assessing whether saliva could replace blood sampling and vaginal cytology for estrous cycle monitoring (Pietraszek & Atkinson, 1994) and clarifying seasonal breeding patterns (Theodorou & Atkinson, 1998). On the other hand, for African black rhinos (*Diceros bicornis*), initial goals centered on measuring salivary bioactive progesterone metabolites to diagnose pregnancies (Czekala & Callison, 1996). As the need to improve sustainability of rhinos grew, scientists began investigating ways to optimize the success of mating introductions including using salivary androgen concentrations to predict behavioral estrous in female Indian rhinos (*Rhinoceros unicornis*) (Gomez, Jewell, Walker, & Brown, 2004). While research on noninvasive reproductive monitoring continues to hold value for promoting the sustainability of many threatened and endangered species (Amaral, Rosas, Graham, da Silva, & Oliveira, 2014; Illera et al., 2014), most saliva-based research is biased toward agricultural animals (Cook, 2012), albeit with tie-ins for how good versus poor welfare differentially influence reproductive success (Whitham & Wielebnowski, 2013) (Fig. 28.1).

Even before zoos took up the metaphorical saliva torch, agricultural researchers were implementing saliva-based methodology for use with pigs (*Sus scrofa*). The lack of large, superficial blood vessels in pigs made regular or repeated blood sampling for welfare monitoring difficult without prior surgical insertion of a venous catheter (Parrott, Misson, & Baldwin, 1989). These initial studies quickly expanded from asking whether cortisol concentrations in saliva and blood were correlated (Cook, Schaefer, Lepage, & Jones, 1996; Parrott et al., 1989) and performing adrenocorticotropic hormone (ACTH) challenges (Bushong, Friend, & Knabe, 2000, Cook et al., 1996; Parrott et al., 1989) to full-fledged investigations on welfare and reproduction. Indeed, researchers studying pig welfare were among the first in their field to use saliva to examine how welfare impacts reproduction, more specifically, how housing conditions affected gestating sows (Mendl, Zanella, & Broom, 1992). For example, middle-ranking sows (i.e., those sows displaying intermediate levels of aggression and that received the most aggression from high-ranking individuals) were found to have consistently higher basal salivary cortisol and the

**Fig. 28.1** There is great interest among zoological and other institutions in which animals are under professionally managed care to understand how animal care practices influence welfare and reproductive success

highest peak in salivary cortisol concentrations following an ACTH challenge compared to either high- or low-ranking individuals. These middle-ranking individuals also produced offspring with the lowest live birth weights, suggesting that being the recipient of high levels of aggression may negatively impact both welfare and reproduction (Mendl et al., 1992). As saliva grew into a tool of choice for pig producers aiming to improve welfare, researchers also began investigating novel salivary biomarkers for welfare monitoring and expanding saliva-based monitoring to new species, particularly working animals (i.e., dogs and horses), biomedical research animals (i.e., mice and rats), and nonhuman primates (e.g., Beerda, Schilder, Janssen, & Mol, 1996; Guhad & Hau, 1996; Pearson, Judge, & Reeder, 2008).

One limitation of these earlier studies was that use of glucocorticoids (i.e., cortisol) alone could not provide a complete picture of welfare. Rather, glucocorticoid concentrations were found to reflect only general arousal, highlighting the need for additional behavioral or physiological measures to infer whether hypothalamic pituitary adrenal (HPA) axis activation was associated with a stimulus of implicit positive or negative valence (Mendl et al., 2010). Consequently, in the mid-1990s researchers began exploring salivary biomarkers that had the potential to provide complementary information on welfare such as the antibody secretory Immunoglobulin A, which functions in both frontline disease prevention at mucosal surfaces and specific immune responses to mucosal-associated pathogens (Guhad & Hau, 1996; Skandakumar, Stodulski, & Hau, 1995; for review, see Staley, Conners, Hall, & Miller, 2018). Inspired by advances in human research, a number of biomarkers have since been investigated for their applicability to animal welfare monitoring. Most notably, these include biomarkers associated with sympathetic nervous system activation (chromogranin-A, salivary alpha-amylase) (Behringer et al., 2013; Ott et al., 2014), inflammation (C-reactive protein, haptoglobin) (Gómez-Laguna et al., 2010; Parra, Tecles, Subiela, & Cerón, 2005), and social behavior (oxytocin) (Maclean et al., 2017, 2018). However, understanding what these physiological biomarkers mean in the context of regular welfare monitoring will require careful evaluation with respect to each target species and other biologically relevant variables.

Despite saliva use in animal welfare dating back to the late 1980s and early 1990s, only recently has there been serious growth within conservation science toward implementing saliva for wildlife population health and management assessments. While urine or fecal samples are often preferred in both welfare and conservation research (Dantzer, Fletcher, Boonstra, & Sheriff, 2014; Palme, 2012), saliva collection may be desired if, for instance, the logistics of collecting other types of samples impedes regular monitoring in professionally managed care or when repeated sampling over a short period of time is necessary (Granger et al., 2007; Menargues, Urios, & Mauri, 2008). Conservation scientists have also begun employing salivary environmental DNA (eDNA) methods to aid in predator and herbivore identification as well as population health assessments (Barnes & Turner, 2016). Ultimately, whether oral fluid can be used to meet objectives in welfare and conservation research will depend on the research question and feasibility of sampling. Once decided upon, there are a number of factors researchers should consider to appropriately implement and validate use of saliva in the desired context (for more information, see Sect. 28.3). Assuming these conditions are met, saliva can become a powerful tool for aiding in species conservation, meeting welfare objectives, and enhancing reproductive success. In the next section, we highlight how salivary biosciences have been used to enhance the welfare and sustainability of species in professionally managed care. We then discuss emerging research on salivary biosciences within conservation science and novel directions within salivary biosciences that may have widespread applicability to both welfare and conservation research.

## 28.2 Current Status of Knowledge in Animal Welfare and Conservation Science

### *28.2.1 Saliva and Animal Welfare*

The animal welfare community has long agreed that poor welfare negatively impacts health, reproduction, and overall well-being of animals in professionally managed care (Broom, 1991). However, there is still much work to be done to identify best practices for promoting good welfare and positive outcomes such as improved fertility and live birth rates, lower incidence of disease, and increased longevity. Although the historical focus has been identifying conditions that lead to chronic stress (Dantzer et al., 2014; Von Borell, Dobson, & Prunier, 2007), there is presently a movement to provide more holistic welfare assessments by incorporating both positive and negative indicators of welfare. Here, we discuss how saliva has been used to help evaluate management practices, relate physiology to behavior and personality, and determine whether human interactions, particularly in regard to working animals, can play a role in promoting good welfare.

#### 28.2.1.1 Management Practices

Agricultural research, and more specifically research on pigs, in many cases has laid the foundation for understanding how salivary biosciences can be applied in welfare studies (Cook et al., 1996; Parrott et al., 1989). Over the course of their lives, for example, pigs may experience intensive husbandry practices that alter social groupings, environmental conditions, and a number of other parameters that can affect welfare (Couret, Otten, Puppe, Prunier, & Merlot, 2009; Coutellier et al., 2007; Stevens et al., 2015). Following repeated changes in social partners, young pigs were found to exhibit increased fighting, reduced daily weight gain, and increased salivary cortisol concentrations (Coutellier et al., 2007), an outcome comparable to what has been seen with gestating sows (Stevens et al., 2015). Such negative effects of recurrent psychosocial stress on animal welfare may also be additive with those of other stressors; for instance, piglets that simultaneously experienced both social and environmental changes displayed more pronounced negative behavioral shifts than piglets that experienced a change in a single condition. These behavioral changes included increased time lying awake inactive and fighting as well as reduced play and exploratory behavior (Colson, Martin, Orgeur, & Prunier, 2012). To facilitate resilience to stress under such circumstances, enrichment is increasingly being investigated for its positive effects on welfare. Although there was no difference between groups in salivary cortisol concentrations, enriched pigs (i.e., pigs given increased space, straw, a rooting area filled with peat, and varied enrichment toys) exhibited better memory performance in a spatial hole board discrimination task (i.e., knowing which holes contained the bait reward) (Grimberg-Henrici, Vermaak, Bolhuis, Nordquist, & van der Staay, 2016). Agricultural welfare research also leads in investigating salivary biomarkers beyond cortisol (i.e., salivary SIgA, chromogranin-A, C-reactive protein, haptoglobin, etc.), albeit much of this is still in pilot stages and, with the exception of SIgA, salivary biomarkers aside from cortisol have been limited in their applications to species other than pigs (Escribano et al., 2014; Escribano, Gutierrez, Tecles, & Ceron, 2015; Ott et al., 2014). As a result, expanded research in both agricultural and other animal welfare contexts is necessary to realize the full potential of salivary biosciences for improving management practices.

A major area of active research for salivary biosciences outside of, but complementary to, agricultural welfare studies is examining how social interactions with conspecifics influence individual welfare. For example, among professionally managed hamadryas baboons (*Papio hamadryas hamadryas*), both removal of outdoor habitat access (i.e., resulting in increased social density) and social isolation led to increased salivary cortisol concentrations (Pearson, Reeder, & Judge, 2015). Similarly, salivary cortisol was found to sharply increase upon removal of individual common marmosets (*Callithrix jacchus*) from their social group (Cross, Pines, & Rogers, 2004). Some primate species such as ring-tailed lemurs (*Lemur catta*) also exhibit seasonal changes in conflict rates, which in this species is associated with increased salivary and fecal androgens (von Engelhard, Kappeler, & Heistermann, 2000). Understanding how social dynamics impact welfare ultimately will aid in promoting good health and improving well-being of animals. For instance, group-housed male rats showed steady declines in salivary SIgA over nine days, indicative of chronic stress, whereas males maintained in isolation from social partners or with females exhibited steady or increased salivary SIgA, respectively. Overall, this suggests access to a mate provided the best welfare and positive health outcomes for male rats (Guhad & Hau, 1996). However, best practices for good welfare are going to be highly dependent on the species and individual (Fig. 28.2).

Thus, understanding how different aspects of management impact welfare will require welfare scientists to interpret physiological and behavioral reactions to diverse stimuli in a species-specific manner (Beerda et al., 1996; Colussi, Stefanon, Adorini, & Sandri, 2018).

As knowledge of salivary biosciences has grown, a number of new applications have emerged within agricultural, biomedical, and zoological research, although most studies are still predominantly focused on salivary cortisol. For zoological species, this includes evaluating the impact of zoo visitors on rhinos and elephants (Menargues et al., 2008), the effects of rides on dromedary camels (*Camelus dromedaries*) (Majchrzak, Mastromonaco, Korver, & Burness, 2015), and periods of unusually high human activity and noise levels on common marmosets (Cross et al., 2004). Salivary cortisol has also been investigated for diverse species to evaluate the impacts of diet (Nemeth et al., 2016), exercise training (Colussi et al., 2018; Pastore et al., 2011), weaning (Wulf, Beythien, Ille, Aurich, & Aurich, 2018; Yang et al., 2018), and transport (Behringer et al., 2014; Schmidt, Aurich, Möstl, Müller, & Aurich, 2010; Schmidt et al., 2010). More recently, the applications of salivary biosciences have been extended to include, for instance, evaluations of how common procedures in biomedical research affect the welfare of nonhuman primates

**Fig. 28.2** In order to understand how factors such as social interactions affect an animal's well-being, welfare specialists must take into account species-specific differences in social systems

(Pfefferle, Plümer, Burchardt, Treue, & Gail, 2018). While such strides in salivary biosciences are highly encouraging, continuing efforts are needed to expand beyond measuring salivary cortisol alone to include novel biomarkers as well as understand how biomarker concentrations may be influenced by intrinsic variables such as personality (Fig. 28.3).

#### 28.2.1.2 Personality and Behavior

It is well known that, within a species, interindividual variation in personality traits is often associated with consistent differences in behavioral responses to stimuli (Fig. 28.3) (Tetley & O'hara, 2012). However, whether personality is predictive of physiological responses remains debatable. In male chimpanzees, intensive vigilance toward speakers playing vocalizations of unfamiliar conspecifics was associated with increased salivary cortisol (Kutsukake et al., 2009) whereas African elephants' morning salivary cortisol concentrations were found to be positively correlated with the personality trait fearful (scored on a 5-point scale) (Grand, Kuhar, Leighty, Bettinger, & Laudenslager, 2012). In contrast, neither exploration tendency nor neophobia predicted physiological stress responses of pigs to social isolation (Adcock, Martin, & Walsh, 2015). In part, such contradictory findings may be due to the chosen stimulus; for example, of six stimuli trialed with dogs, only those that could not be anticipated (i.e., not visible to the dogs) induced a cortisol stress response and very low posture (Beerda, Schilder, Van Hooff, de Vries, & Mol, 1998). Similarly, even though orangutans showed behavioral signs of frustration in

**Fig. 28.3** Of great interest to welfare biologists is using salivary biosciences to improve understanding of how personality influences behavioral responses to stimuli

response to extended 20-s delays between computerized tasks, frustration was not associated with an increase in salivary cortisol (Elder & Menzel, 2001). Therefore, in such trials, researchers should identify stimuli appropriate for the species and to which animals have not been habituated nor formed a positive association with through positive reinforcement training.

#### 28.2.1.3 Human–Animal Interactions

Among working animals such as dogs as well as in zoos and aquariums, there is an interest in understanding how trainers' verbal and tactile interactions with animals can influence behavioral and physiological parameters (Csoltova, Martineau, Boissy, & Gilbert, 2017). Behavioral interventions in the form of regular human interactions and positive reinforcement training can aid in improving sociability and temperament; in dogs, such improvements have also been associated with a reduction in salivary cortisol (Bergamasco et al., 2010). This effect on salivary cortisol concentrations has been shown to depend on the nature of handler interactions, with authoritarian, or disciplinary, interactions being associated with increased cortisol and affiliative interactions leading to reduced cortisol (Horváth, Dóka, & Miklósi, 2008). The ability of animals to choose whether or not to engage in human interactions is also expected to affect welfare; for example, therapy dogs that were allowed to move freely about and decide whether to approach or avoid human contact (i.e., off-lead dogs) displayed reduced salivary cortisol compared to dogs worked with on lead and provided limited choice (Glenk et al., 2013). While there is heightened interest in using saliva to understand how human interactions, especially with respect to positive reinforcement training, impacts the welfare of zoological and aquarium species (Behringer et al., 2014; Vasconcellos et al., 2016), most physiological research in this area is still predominantly focused on cortisol. Promisingly though, recent research on dogs showed that increases in salivary oxytocin, a hormone associated with social bonding, were predicted by the extent of affiliative behavior between dogs and humans during a 10-min play session (Maclean et al., 2017). Furthermore, human interventions with cats displaying behaviors indicative of poor welfare led to increased fecal SIgA and lower incidence of upper respiratory disease (Gourkow, Hamon, & Phillips, 2014), suggesting either of these biomarkers in saliva may hold potential for improving understanding of the extent of human impact on welfare in diverse professionally managed settings.

### 28.2.2 *Saliva and Conservation Science*

Comparable to how animals under professionally managed care can benefit from positive interactions with humans, so too can humans benefit from experiencing nature and biodiversity. Poor ecosystem health can translate to measurable impacts on psychological well-being, physical health, and disease transmission (Sandifer, Sutton-Grier, & Ward, 2015). There is a heightened awareness among conservationists that the maintenance of functional biodiversity is necessary to sustain the ecosystem services that are the foundation upon which human civilization is built and animal life depends (Sandifer et al., 2015). To meet conservation goals, conservation initiatives are increasingly moving toward noninvasive sampling in order to reduce costs, minimize errors, and improve species management (Waits & Paetkau, 2005). While most studies have focused on sample types such as hair or feces (Palme, 2012), there is heightened recognition that specific conservation questions may benefit from salivary biosciences. The main areas of ongoing investigation include conservation physiology, environmental DNA (eDNA), and targeted population health assessments as discussed in greater detail below.

#### 28.2.2.1 Conservation Physiology

The burgeoning field of conservation physiology aims to use physiological measures to assess how species respond to environmental perturbations, inform on population viability, and aid in evaluating management plans. Conservation physiology emphasizes identifying conditions, especially human practices, that lead to chronic stress and negatively impact species' reproduction and/or survival (Dantzer et al., 2014). Although conservation physiologists have employed saliva sampling in controlled experimental tests, for example, of how amphibians respond to different habitat types (Janin, Léna, Deblois, & Joly, 2012) and noise pollution (Troïanowski, Mondy, Dumet, Arcanjo, & Lengagne, 2017), overall the applicability of saliva for assessing chronic stress in wild populations is limited. This is largely because saliva represents an instantaneous sampling of biomarker concentrations, comparable to blood, rather than a measure integrated over a period of time as with feces (Dantzer et al., 2014). The act of trapping or handling animals for saliva collection may therefore impact salivary biomarker concentrations (Dantzer et al., 2014; Dreschel & Granger, 2009) and recent research suggests that instantaneous sample types are inappropriate for identifying conditions that create chronic stress in wildlife; for example, a meta-analytical study found anthropogenic disturbances were consistently associated with increased fecal, but not necessarily plasma, glucocorticoid concentrations (Dantzer et al., 2014). Consequently, saliva may be used in controlled experiments aimed at evaluating short-term responses to stimuli (e.g., see Troïanowski et al., 2017), but its applicability on a broader scale within conservation physiology is likely limited (Dantzer et al., 2014).

#### 28.2.2.2 Environmental DNA (eDNA)

Environmental DNA, or eDNA, refers to free or trace DNA in the environment that is derived from diverse biological sample types, including saliva (Barnes & Turner, 2016; Bohmann et al., 2014). While eDNA methods are typically employed for the detection of rare or cryptic species, saliva-specific methods have been evaluated as a tool for mitigating human–wildlife conflicts, understanding predator–prey interactions, and developing novel strategies for species detection (Bohmann et al., 2014; Lobo, Godinho, Álvares, López-Bao, & Rodríguez, 2015; Waits & Paetkau, 2005). Typically, salivary eDNA samples are collected by swabbing the edges of wounds on prey carcasses with sterile cotton swabs (Caniglia, Fabbri, Mastrogiuseppe, & Randi, 2013; Mumma, Soulliere, Mahoney, & Waits, 2014; Van Bleijswijk et al., 2014) or collection of browsed or chewed plant material (Nichols, Königsson, Danell, & Spong, 2012) (Fig. 28.4). Investigators are also working to develop methods for monitoring carnivore populations; this includes placing bait on porous materials such as wood or cork from which salivary eDNA can be recovered (Lobo et al., 2015). Salivary eDNA has been used, for instance, to evaluate whether the increased stranding of harbor porpoises (*Phocoena phocoena*) with sharp-edged injuries could be due to active hunting by gray seals (*Halichoerus grypus*) rather than increases in ship strikes or fisheries bycatch (Leopold et al., 2015; Van Bleijswijk et al., 2014). Such data may be key to species conservation efforts; for example, eDNA analysis of swabs taken from wounds on the carcasses of 98 radio-collared woylies (*Bettongi penicillate,* a small Australian marsupial) revealed feral cats (*Felis catus*) caused 65% of mortalities, or three times those of foxes (*Vulpes vulpes*). Based on these findings, researchers concluded that feral cat control measures, in addition to ongoing fox control efforts, may be essential for the successful recovery of woylies and other endangered Australian fauna (Marlow et al., 2015).

**Fig. 28.4** Salivary environmental DNA (eDNA) can be recovered by swabbing browsed or chewed plant material

The major advantage of using eDNA to identify predators is that eDNA, when combined with existing management practices, can improve accuracy of predator identifications and help resolve human–wildlife conflicts. Traditionally, predator identification has hinged on kill site observations such as wound patterns, predator-associated samples (i.e., hair, feces) found near kill sites, and/or radio collar data (Caniglia et al., 2013; Mumma et al., 2014; Van Bleijswijk et al., 2014). However, conclusions based on observations alone may be unreliable due to human error associated with experience level, ambiguous or insufficient evidence, and/or scavenging by other species (Caniglia et al., 2013; Mumma et al., 2014; Sundqvist, Ellegren, & Vilà, 2008). In a study of caribou calf kills, for instance, observations of kill sites assigned 40% of kills to coyotes (*Canis latrans*), 45% to black bears (*Ursus americanus*), and 15% to other species. In contrast, eDNA evidence attributed 66.7% of calf kills to coyotes and 33.3% to black bears (Mumma et al., 2014). Environmental DNA can also help resolve disputes between conservation managers and livestock owners as well as change public perception of the impacts of wild carnivores. Livestock predation in Europe is generally attributed to wolves (*Canis lupus*) although there is increasing evidence that feral or free-ranging dogs also play a role. Furthermore, it is thought that compensation programs for livestock owners are leading to false predation reports, exacerbating the negative public perception of wolves (Caniglia et al., 2013; Sundqvist et al., 2008). Indeed, salivary eDNA testing of sheep carcasses revealed distinct incidences of killings by both wolves and dogs (Caniglia et al., 2013, Sundqvist et al., 2008) as well as cases of false reporting (Caniglia et al., 2013). Given human–wildlife conflicts pose a significant threat to many species (Dickman, 2010), such DNA-based technology is likely to be a valuable tool for ongoing species conservation efforts.

As eDNA research has gained momentum over the past decade, so too have studies specifically emphasizing saliva. These include using salivary eDNA to identify predators of ground-nesting birds and their eggs (Hopken, Orning, Young, & Piaggio, 2016; Steffens, Sanders, Gleenson, Pullen, & Stowe, 2012), recovering mountain gorilla (*Gorilla gorilla beringei*) eDNA from discarded pieces of wild celery (*Peucedanum linderi*) (Smiley et al., 2010), and experimentally testing the overwinter foraging preferences of four species of ungulates found in temperate forests (Nichols, Cromsigt, & Spong, 2015). Predators of endangered small carnivores such as the fisher (*Marte pennant*) can even be identified through salivary eDNA thanks to the development of methods that allow for specific amplification of felid or canid predator DNA without amplifying fisher DNA (Wengert, Gabriel, Foley, Kun, & Sacks, 2013). Yet, salivary eDNA is presently still considered a tool to augment rather than replace existing species management practices due to the potential for predator detectability biases. For example, the nests of ground-nesting bird species are especially susceptible to predation and identifying predators of those nests can help inform conservation strategies. Saliva-based studies on this topic are currently biased towards detecting mammalian predators due to both available molecular tools and other predators (i.e., avian, reptilian) being less likely to leave egg shell remains that can be sampled (Hopken et al., 2016; Steffens et al., 2012). And although researchers had greater success identifying individual brown bears (*Ursus arctos*) using salivary eDNA recovered from salmon (*Onchorhynchus* spp.) carcasses than feces, each approach detected some individuals that the other did not (Wheat, Allen, Miller, Wilmers, & Levi, 2016). Consequently, salivary eDNA may be insufficient by itself for accurately estimating population densities (Wheat et al., 2016). Despite this, as researchers gain a deeper understanding of the methodological considerations for using salivary eDNA (see Sect. 28.3), it still remains a promising tool for studying species ecology and interactions, especially as it relates to resolving predator–prey dynamics and human–wildlife conflicts.

#### 28.2.2.3 Wildlife Population Health Assessments

The long-term viability of wildlife populations will depend not only on management strategies that promote genetic diversity and account for species ecology, but also research on diseases and other factors that impact organismal health (Leendertz et al., 2006). With human population expansions and anthropogenic activities resulting in greater contact between humans and wildlife, some species, especially great apes, are at high risk for potentially devastating disease outbreaks of human or livestock origin (Gillespie & Chapman, 2008). Indeed, antibodies against a number of human pathogens have been detected in primates (Jones-Engel, Engel, Schillaci, Babo, & Froehlich, 2001) and viral strains isolated from chimpanzees that died during respiratory disease outbreaks were found to be closely related to contemporary strains circulating in humans (Köndgen et al., 2008). Of these, respiratory pathogens in particular can be difficult to detect using noninvasive sample types such as feces or urine, creating interest in developing PCR- or ELISA-based methods to screen for pathogens using saliva (Seeber, Soilemetzidou, East, Walzer, & Greenwood, 2017; Smiley et al., 2010). Initial steps toward this include testing saliva from cervid carcasses for infectious prions associated with chronic wasting disease (Henderson et al., 2013), developing salivary eDNA-based methods for use with discarded food (Smiley et al., 2010), and recovering salivary DNA from enrichment items given to Grevy's zebras (*Equus grevyi*) to screen for equine herpesvirus infection (Seeber et al., 2017). Developing this novel methodology will not only aid in understanding how human activities alter zoonotic disease transmission risks but also help inform on strategies to mitigate those risks with both wild and professionally managed populations (Leendertz et al., 2006; Smiley et al., 2010).

### 28.2.3 Bridging the Gap Between Animal Welfare Research and Conservation Science

With the heightened awareness surrounding use of oral fluid, a number of new lines of investigation have recently emerged that have applicability to both welfare and conservation research. These include testing hypotheses regarding the hormonal basis of different reproductive strategies and social systems in primates (Hohmann, Mundry, & Deschner, 2009; von Engelhard et al., 2000; Wobber, Hare, Lipson, Wrangham, & Ellison, 2013) and experimental studies of how maternal stress during the prenatal period impacts HPA axis development, health, and survival of offspring (Ison et al., 2010; Kapoor & Matthews, 2008, 2011; Rutherford et al., 2009; Sandercock et al., 2011). Among zoological institutions, there is also a push to validate salivary biomarkers (e.g., oxytocin) to study bonding between mothers and offspring (Carter et al., 2007; Rilling & Young, 2014) and, in highly social species, the responses of conspecifics to births (Behringer et al., 2009). Many institutions have initiated pilot studies to assess feasibility of using salivary biomarkers to evaluate the role of management practices in welfare generally (Box 28.1), an important step given welfare is known to affect reproductive success (Whitham & Wielebnowski, 2013). Overall, as research on conservation science and animal welfare expands, the potential for saliva-based physiological monitoring to augment existing practices is worth considering. To be successful in improving species sustainability and achieving long-term management goals, however, researchers must also recognize there is still much work to be done in terms of developing and validating salivary methodology.

**Box 28.1 A Case Study Incorporating Salivary Biosciences for Assessing Polar Bear (*Ursus maritimus*) Welfare**

The aim of this case study was to conduct a preliminary assessment of how salivary biomarker concentrations related to three behaviors of interest in a 7-year-old male polar bear at Brookfield Zoo in Brookfield, IL, USA: engagement in enrichment,

**Fig. 28.5** A 7-year-old male polar bear at Brookfield Zoo voluntarily engages in saliva sample collection

resting (non-alert), and pacing. Paired behavioral observations and salivary samples were collected from this individual across 12 days between Oct 1, 2015 and Nov 15, 2015. Prior to sample collection, the polar bear was trained to voluntarily participate in saliva collection using positive reinforcement training (for more details on positive reinforcement techniques, see Sect. 28.3). This training included approximating an open mouth behavior and allowing for a collection device to be inserted for approximately 10–15 s to secure a sample of sufficient volume for analysis (Fig. 28.5). Throughout each day, when the subject was engaged in one of the three focus behaviors for five continuous minutes, animal care staff were notified to collect a saliva sample, with sample collection occurring exactly 15 min after the observed behavior. Samples were not collected if the bear had been eating or drinking within the previous 15 min. Samples were then stored, processed, and analyzed for salivary alpha-amylase and cortisol using commercially available assay kits from Salimetrics and following protocols established by Granger et al. (2007). Although salivary alpha-amylase was below the detectable limit in these samples, salivary cortisol was successfully measured in these samples, with preliminary data supporting further investigation of salivary biomarkers in welfare assessments (Fig. 28.6) (Miller, unpublished data).

## 28.3 Methodological Issues, Challenges, and Considerations

Across the animal kingdom, there is widespread evidence of unique salivary adaptations such that salivary characteristics are not always comparable across species (Tucker, 2010). Saliva is known to serve diverse roles in vertebrates beyond the

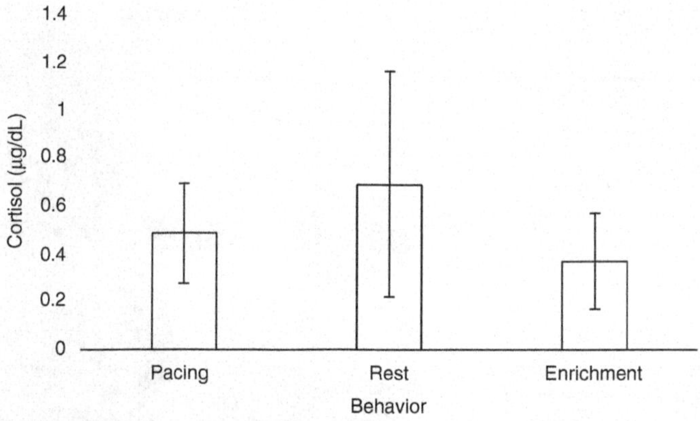


**Fig. 28.6** Preliminary data on salivary cortisol concentrations from a 7-year-old male polar bear at Brookfield Zoo. Mean salivary cortisol concentrations were lower after pacing or engagement in enrichment was observed than after resting

well-described functions of food lubrication, digestion, and mucosal immune function (Tucker, 2010). Kangaroos and wallabies, for instance, use saliva for evaporative cooling (Needham, Dawson, & Hales, 1974), frogs possess a non-Newtonian saliva that aids in capturing insects (Noel, Guo, Mandica, & Hu, 2017), and swifts construct and secure nests using their mucous-rich saliva that has glue-like properties (Johnston, 1958). Saliva plays a role in species and individual identification (reviewed in Brennan & Kendrick, 2006) and has properties that may be of biomedical interest such as the anticoagulant proteins found in vampire bat (*Desmodus rotundus*) saliva (Ware & Luck, 2017). While comparing salivary characteristics among species can ultimately help inform on evolutionary adaptations related to species ecology (Phillips & Baker, 2015; Ware & Luck, 2017), it is also equally important to recognize how these differences, rooted in species' evolutionary histories, can create challenges for applying methodology initially developed in humans to other animals. In this section, we discuss how species characteristics may influence the ability to use saliva, best practices for validating novel physiological biomarkers and implementing saliva collection with animals, and finally considerations for salivary eDNA-based studies.

### 28.3.1 Influence of Species Characteristics

As saliva sampling expands to novel species, researchers will be faced with a unique suite of challenges contingent on their study organism. In relation to both salivary gland anatomy and salivary biomarkers, most research to date has focused on the evolutionary influences of foraging strategies and dietary preferences (Behringer et al., 2013; Boehlke, Zierau, & Hannig, 2015; Hofmann, Streich, Fickel, Hummel, & Clauss, 2008). For example, ruminant species for which grasses comprise a greater percentage of the diet (i.e., grazers) have smaller salivary glands than species that primarily consume browse after accounting for metabolic body size (Hofmann et al., 2008). Similarly, evolutionary divergence in salivary gland characteristics between wallaby species occupying distinct habitats is hypothesized to be related to feeding ecology (i.e., total daily food intake and browse consumption) as well as the greater need for evaporative cooling in arid-adapted species (Lentle et al., 2002). In terms of the evolutionary influence of diet on salivary biomarkers specifically, the enzyme salivary alpha-amylase has received the most attention (Boehlke et al., 2015). In the oral cavity, salivary alpha-amylase functions in starch digestion and is also known to exhibit antimicrobial activity. Furthermore, salivary alpha-amylase is considered a biomarker for sympathetic nervous system activation, increasing in response to both psychological and physical stress (Nater & Rohleder, 2009). Cross-species comparisons of nonhuman primates have revealed variability in basal salivary alpha-amylase activity that corresponds with gene copy number and general dietary features (Behringer et al., 2013). Furthermore, this enzyme has predominantly been detected in omnivores as well as some herbivores, but not strict carnivores (for review, see Boehlke et al., 2015). Consequently, salivary biomarkers established for use of human health and medicine may not always translate well to animal welfare research.

Part of the difficulty in taking a physiological biomarker that has been validated for one species and applying it to a novel species is that inherent physiological differences must be taken into account. These include the well-described differences in the major glucocorticoid (i.e., cortisol vs. corticosterone) produced by different animal taxa (for review, see Sopinka et al., 2015) and differences with respect to reproductive hormones; for example, in both African elephants (*Loxodonata Africana*) and Asian elephants (*Elephas maximus*) the major progestin in circulation is not progesterone but rather 5a-reduced progestins (Hodges, 1998; Hodges, Van Aarde, Heistermann, & Hoppen, 1994). Similarly, both free progesterone and estrogen were found to be poor salivary biomarkers for rhino pregnancies, whereas progesterone metabolite concentrations displayed pregnancy-associated changes (Czekala & Callison, 1996). Much like salivary alpha-amylase, other salivary enzymes are not necessarily detectable across all species. For instance, both salivary alpha-amylase and salivary lysozyme activity could be detected in Asian and African elephants, but in contrast to findings on bovine species and nonhuman primates there was no detectable salivary peroxidase activity (Boehlke, Pötschke, Behringer, Hannig, & Zierau, 2017). For cetaceans in particular, the differences may be even more extreme given their unique salivary gland anatomy compared to terrestrial mammals (Leatherwood & Reeves, 2012). Monitoring cetacean reproduction through oral fluid remains questionable (Atkinson et al., 1999; Atkinson, Crocker, Houser, & Mashburn, 2015; de Mello & de Oliveira, 2016) and, while some studies have reported measurable concentrations of cortisol in bottlenose dolphins' (*Tursiops truncatus*) oral fluid (Monreal-Pawlowsky et al., 2017; Pedernera-Romano et al., 2006; Ugaz, Valdez, Romano, & Galindo, 2013), others have found cortisol concentrations to be generally at or below the detectable limit of the assay employed (M. Conners, *pers. comm.*). Study design should therefore account for species physiology and incorporate appropriate physiological and/or biological validations to ensure that measured salivary biomarker concentrations are biologically meaningful.

### *28.3.2 Validation of Salivary Physiological Biomarkers*

Before saliva can be implemented as part of regular physiological welfare monitoring protocols, the associated methodology must be validated for each biomarker and species. This process starts with laboratory validations to ensure that the assay performs appropriately with the samples of interest and is followed by physiological and/or biological validations to demonstrate that measured biomarker concentrations are biologically meaningful. These procedures should be performed not only when moving to a novel species, but also when working with a new sample type. Typically, laboratory validations involve testing for assay accuracy, specificity, and sensitivity to ensure reliable measurement of the biomarker of interest. This includes checking for cross-reactivity of the assay with other biomarkers, determining the assay detection limit, and spiking a sample with a known amount of biomarker to assess whether the assay performance is comparable between the sample matrix and standard diluent. To complement this information, a parallelism test is used to check that there is a linear relationship between sample and measured concentrations (i.e., the sample is serially diluted) (for review, see Buchanan & Goldsmith, 2004). Once completed, it is then imperative to follow up laboratory validations with either a physiological or biological validation.

Physiological validations typically involve challenging an animal with an exogenous hormone to stimulate activity of the physiological pathway of interest. For instance, when an animal is injected with adrenocorticotropic hormone (ACTH), a hormone normally produced by the anterior pituitary gland of the brain as part of HPA axis activation, this is expected to lead to increased release of glucocorticoids by the adrenal glands (Bushong et al., 2000; Cook et al., 1996; Parrott et al., 1989). If researchers were instead interested in validating reproductive hormone measures, then an injection of gonadotropin-releasing hormone (GnRH), also normally produced by the anterior pituitary gland, would be used to induce the secretion of reproductive hormones by the gonads (Amaral, Rosas, da Silva, Nichi, & Oliveira, 2013; de Souza Amaral et al., 2009). A post-challenge time series of saliva samples is then collected and assayed to determine whether salivary hormones of interest display the anticipated increase in concentrations. Often times, especially with glucocorticoids, researchers may opt to concurrently examine whether changes in salivary biomarker concentrations are correlated with those in blood. However, depending on the biomarker this should not necessarily be expected. While hormones such as cortisol are passively transferred into saliva, other biomarkers may be actively transported into saliva or, as in the case of salivary alpha-amylase and secretory Immunoglobulin A, independently expressed and released by the salivary glands (for review, see Gröschl, 2009). Furthermore, any blood collection may impact both an animal's physiological response to the challenge and welfare, and therefore, care should be taken to consider whether blood collection is necessary given the study's aims.

Biological validations offer both an alternative to and a way to bolster findings from physiological validations to ensure that measured biomarker concentrations have biological relevance. If the goal of a study is to validate physiological stress responses, for example, saliva samples can be collected before, during, and after putatively acutely stressful events. This can include scheduled transport events (Behringer et al., 2014) or changes to husbandry practices or social partners (Pearson et al., 2008). These events can be expected to induce acute physiological stress responses, or in other words, activation of the HPA axis as evidenced by increased glucocorticoid concentrations. However, such responses are not considered inherently bad as short-term elevations in stress hormones help organisms respond to environmental changes and energetically demanding situations (Palme, 2012). For researchers wanting to monitor reproductive hormones in saliva, particularly ovarian cycles, salivary hormone profiles can also be matched to other sample types (i.e., blood, vaginal cytology, urine) (Illera et al., 2014; Pietraszek & Atkinson, 1994) or periods of behavioral estrous (Gomez et al., 2004). Overall, performing laboratory validations followed by physiological and/or biological validations will be a necessary step to ensure appropriate use and interpretation of salivary biomarkers and any reported data.

### *28.3.3 Saliva Collection in Animals*

Positive reinforcement training is the preferred method used by zoological and other professionals to obtain the most reliable results while accounting for the welfare of the animals involved. This technique emphasizes connecting a cue to a specific behavioral response, such that when an animal performs the desired behavior they receive a reward, often in the form of food or play. This ultimately encourages voluntary participation in common husbandry and medical procedures while minimizing stress on animals (Fig. 28.7) (Laule, Bloomsmith, & Schapiro, 2003; Schapiro, Bloomsmith, & Laule, 2003). For example, neither bonobos nor orangutans exhibited a change in salivary cortisol concentrations following medical positive reinforcement training sessions (Behringer et al., 2014) while both human-socialized wolves (*Canis lupus*) and domestic dogs (*Canis lupus familiaris*) exhibited lower salivary cortisol concentrations post-training sessions (Vasconcellos et al., 2016). However, because this approach emphasizes voluntary participation, it must also be recognized that success can vary greatly and may require flexibility in approaches (Cross et al., 2004).

Nonhuman primates serve as an excellent model for examining potential challenges that may be encountered in training saliva collection due to the highly detailed recommendations published on both zoo-housed and biomedical research animals (Behringer & Deschner, 2017; Cross et al., 2004; Heintz, Santymire, Parr, &

**Fig. 28.7** Positive reinforcement training encourages animals to voluntarily participate in training and husbandry behaviors by connecting a cue (i.e., a hand signal) to a specific behavioral response, such that when the animal performs the requested behavior the connection is reinforced through a reward

Lonsdorf, 2011). Because these species often have complex social hierarchies, psychosocial variables are known to exert significant influence over whether animals will engage in training and/or perform the desired behavior. For instance, in a socially housed baboon colony, adult females were generally unwilling to participate in saliva sampling, a finding attributed to baboons' harem-based social system and the harem male being disinterested in participating (Pearson et al., 2008). To facilitate sample collection outside of training sessions, researchers have often turned to apparatuses upon which gauze or dental rope covered in flavoring can be mounted. The flavoring encourages animals to lick or chew on the material, allowing saliva to be recovered (Lutz, Tiefenbacher, Jorgensen, Meyer, & Novak, 2000). Yet, even under these circumstances animals have been shown to quickly lose interest in certain flavorings (Cross et al., 2004) or a specific method may be inappropriate for some individuals, for example, due to their tendency to bite off attached ropes (Lutz et al., 2000). Furthermore, any food particles or flavorings can influence biomarker concentrations and consequently must be tested to ensure validity of the results (Dreschel & Granger, 2009; Lutz et al., 2000). To circumvent this issue with positive reinforcement training, animals should first be familiarized to and encouraged to interact with the saliva collection devices. With primates, this can include initially covering the collection device in the food reward (i.e., banana). Once a positive association with the device is established, this can facilitate learning of the desired, more complex behavior of licking and chewing on the device sans banana, and then subsequently receiving the reward (Ash, Smith, Knight, & Buchanan-Smith, 2018). While, as evidenced by the above-described primate studies, a one-size-fits-all approach does not exist when working with animals, by considering both the species and individual as well as being willing to modify approaches researchers can achieve reasonable success in sample collection.

Although training saliva collection is preferred, it is not always feasible depending on the study organism and experimental conditions (Cook, 2012; Dreschel & Granger, 2009). Under such circumstances, researchers must account for any potential effects of handling, individual temperament, and other experimental conditions (Dreschel & Granger, 2009) as well as be aware of how the chosen sampling technique may affect biomarker concentrations (i.e., localized or pooled saliva collection) and flow rate calculations (Salimetrics, 2011). In dogs, for example, it takes approximately 4 min for handling to result in a measurable change in salivary cortisol concentrations (Kobelt, Hemsworth, Barnett, & Butler, 2003) and sampling is easily biased toward individuals whose temperament allows for easy handling (Dreschel & Granger, 2009). While this may be less of an issue for species such as horses (*Equus caballus*) that are accustomed to their mouths being regularly handled, for other agricultural species such as pigs and many biomedical species, restraint or handling to facilitate saliva collection can negatively impact welfare (Cook, 2012; Jarillo-Luna et al., 2007; Jarillo-Luna et al., 2015; Muneta et al., 2010). For pigs specifically, agricultural researchers have found success by fixing cotton swabs secured between rubber stoppers to one-meter-long aluminum poles that can then be inserted into a pig's mouth (for illustration, see Cook, 2012). This allows for quick sample collection ($< 2$ min) and, with the exception of uncooperative animals, repeated sampling was shown to have no measurable effect on welfare (Cook, Hayne, Rioja-Lang, Schaefer, & Gonyou, 2013). Furthermore, when oral fluid collection is desired for general purposes such as pathogen surveillance, collection can be simplified to minimize human impacts; for example, respiratory pathogen surveillance in pigs involves hanging cotton ropes in a clean area of the pen where pigs can chew on them (Prickett et al., 2008) whereas salivary eDNA methods for wildlife health assessments emphasize recovery of salivary DNA from inanimate objects or discarded food (Seeber et al., 2017; Smiley et al., 2010). Thus, even when training for saliva collection is not feasible, steps can be taken to minimize the impact the collection procedure has on both the welfare of the animal involved and the measured biomarker concentrations.

### 28.3.4 Salivary Environmental DNA (eDNA) Methodological Considerations

Environmental DNA (eDNA) techniques can enhance our understanding of human–wildlife interactions and provide a more complete picture of the challenges faced by threatened and endangered species. However, the quantity and quality of DNA

**Fig. 28.8** Zoological institutions can facilitate salivary environment DNA (eDNA) studies by providing high-quality reference and feeding samples for methodological development prior to researchers undertaking larger scale field studies

recovered can impact project success in a manner that is likely to depend on whether the goal is species, individual, and/or genetic sex identification. Species identification (i.e., via amplification of mitochondrial DNA) typically has higher success rates than individual or sex determination based on nuclear DNA (i.e., individual genotyping via microsatellites, amplification of loci located on the sex chromosomes). This is due to higher mitochondrial DNA copy numbers as well as structural features that influence degradation rates and amplification success (Foran, 2006; Waits & Paetkau, 2005). For instance, although predator species identification from feeding and killing wounds on freshly collected caribou calf carcasses was high (94% and 100%, respectively), individuals could be identified in only 70% of the cases (Mumma et al., 2014). The likelihood of success can be improved by collecting and testing multiple samples (i.e., multiple swabs from around a wound), conducting pilot studies to assess how sample degradation times and sample materials impact results, and using other recorded data and reports to clarify findings (Caniglia et al., 2013; Mumma et al., 2014; Van Bleijswijk et al., 2014).

Animals in professionally managed care can facilitate pilot studies on degradation time and sampling material influence by providing both high-quality reference samples and feeding samples (Fig. 28.8). In an experimental test, wolves (*Canis lupus*) and lynxes (*Lynx lynx*) were allowed to feed on freshly killed roe deer (*Capreolus capreolus*) pieces for 1 min, after which pieces were removed and stored for 1, 24, or 48 h before sampling for saliva traces. Individuals were accurately identified in 83% of the samples collected at 1 or 24 h using microsatellite loci. In contrast, at 48 h success dropped below 50%, with 7% of amplified samples having incorrect genotypes relative to the reference samples (Harms, Nowak, Carl, &

Muñoz-Fuentes, 2015). Similarly, known-browser twig samples collected from zoos were used to assess mitochondrial DNA degradation over a 26-week period. In this study, species identification success was shown to substantially decline after 12 weeks (Nichols et al., 2012). Pilot studies are also recommended for evaluating how the physical material from which saliva is collected or the type of wounds sampled may influence the quantity and quality of salivary DNA recovered. For instance, recovery of dog salivary DNA in bait apparatus trials was significantly greater when the bait was placed on Styrofoam than organic substrates. Furthermore, the specific combination of a wood substrate with sardine bait yielded the least amount of DNA (Lobo et al., 2015). Such considerations should be applied to field settings as well; for example, grey seal DNA could only be amplified from small, deep puncture wounds on harbor porpoise carcasses and not large, open wounds on the same carcass. This was attributed to a higher probability of DNA deposition with small wounds and also preservation given larger wounds often result in severe bleeding and may be subject to greater rinsing by sea water (Van Bleijswijk et al., 2014). By refining methods on a smaller scale first as did the above-described studies, researchers can improve the likelihood of success when undertaking more costly and larger scale fieldwork.

In the transition from controlled testing to field applications, there are likely to be additional challenges to accurately describing species interactions. For instance, proper understanding of predator–prey interactions may depend on the ability to distinguish predator kill wounds (i.e., hemorrhaged wounds) from scavenger feeding wounds (i.e., non-hemorrhaged wounds) (Heers et al., 2017; Mumma et al., 2014). Additionally, individual identification may be complicated when predators kill in pairs or groups, leaving behind mixed DNA samples for which only the species, but not the specific individuals nor sex of those individuals, can be determined (Blejwas, Williams, Shin, Mccullough, & Jaeger, 2010). The potential confounding effects of mixed individual or species DNA can be minimized by sampling only locations with a single, distinct bite mark or developing methods that amplify only the DNA of the species of interest (Heers et al., 2017; Van Bleijswijk et al., 2014; Wengert et al., 2013). For browsed twig samples, a similar approach can be taken, although it also recommended that, to avoid detection biases, samples be systematically collected from different heights and the number of samples collected be proportional to browsing intensity (Nichols et al., 2015). Overall, once these different methodological considerations are accounted for, salivary eDNA can be a powerful tool for understanding species interactions and resolving human–wildlife conflicts.

## 28.4 Future Directions

Although there are many promising applications for salivary biosciences within the fields of animal welfare and conservation science, ultimately whether saliva can or should be applied in these contexts will depend on the question at hand and the feasibility of collecting samples. More specifically, because saliva represents an instantaneous sampling of biomarker concentrations, much like blood, it can be used to measure near real-time responses of animals to stimuli on short time scales (Dantzer et al., 2014; Palme, 2012). Through positive reinforcement training or the development of alternative devices for minimally invasive sampling, daily saliva collection can be made feasible to allow for more regular, long-term welfare monitoring (Behringer & Deschner, 2017; Cross et al., 2004; Heintz et al., 2011). From a conservation perspective, saliva is also of great interest for eDNA-based studies aiming to describe species ecology, mitigate human–wildlife conflicts, and perform population health monitoring (Mumma et al., 2014; Nichols et al., 2012; Smiley et al., 2010). Physiological monitoring through saliva, on the other hand, is rare due to the need to trap animals, which may itself impact biomarker concentrations, and concerns that alternative methods for sample collection may result in habituation of animals to humans (Dantzer et al., 2014). However, saliva may still be of interest for experimental studies examining short-term responses to stimuli (Janin et al., 2012; Troïanowski et al., 2017) or for improving reproduction of threatened and endangered species under professionally managed care (Czekala & Callison, 1996; Pietraszek & Atkinson, 1994). Overall, and despite saliva's limitations, as the field of salivary biosciences continues to expand, saliva shows great promise for aiding in welfare assessments and species conservation.

Moving forward, animal welfare professionals and conservation scientists agree that a multifaceted approach incorporating both positive and negative indicators of health and well-being is necessary to ensure that animals are thriving. Currently, most salivary research is focused on reproductive hormones (Amaral et al., 2015; Illera et al., 2014) or the adrenal hormone cortisol, which, although indicative of arousal, is not necessarily informative about whether an animal perceives a stimulus positively or negatively (Mendl et al., 2010). Researchers have begun exploring novel physiological biomarkers in agricultural and companion animals that can provide information on general health, sympathetic nervous system activity, and social behaviors (Behringer et al., 2013; Gómez-Laguna et al., 2010; Maclean et al., 2018; Ott et al., 2014; Parra et al., 2005). Success in these contexts will hinge on taking appropriate steps to ensure validity of the study results, including accounting for how species characteristics may affect the ability to detect a given biomarker and how the employed methodology may influence results. Lastly, physiological or biological validations should be performed for each novel biomarker with each new species to ensure measured biomarker concentrations provide biologically relevant information.

New avenues for salivary research are frequently emerging, including understanding the role enrichment and human interactions play in welfare (Fig. 28.9), how maternal stress during gestation programs stress reactivity in offspring, and investigating questions related to species ecology and evolution. Logical next steps include improving inferences surrounding animals' emotional states through combined physiological and behavioral measures. For example, humans experiencing negative emotional states are more likely to make negative judgments about ambiguous stimuli than their happier counterparts (Harding, Paul, & Mendl, 2004). Using such an approach in animals, termed cognitive bias testing, alongside physiological

**Fig. 28.9** Future directions for salivary biosciences and animal welfare research include examining whether salivary biosciences can be applied to assessments of how management practices such as enrichment impact welfare

monitoring could aid in understanding how positive versus negative affective states impact measured biomarker concentrations and overall health. Ultimately, such measures will help shift the focus away from reducing instances of chronic stress toward promoting practices that enhance species sustainability and welfare.

**Acknowledgments** The images used in this chapter were provided courtesy of the Chicago Zoological Society—Brookfield Zoo.

## References

Adcock, S. J. J., Martin, G. M., & Walsh, C. J. (2015). The stress response and exploratory behaviour in Yucatan minipigs (*Sus scrofa*): Relations to sex and social rank. *Physiology & Behavior, 152*, 194–202.

Amaral, R. S., Rosas, F. C., da Silva, V. M., Graham, L. H., Viau, P., Nichi, M., & Oliveira, C. A. (2015). Seasonal variation in urinary and salivary reproductive hormone levels in Amazonian manatees (*Trichechus inunguis*). *Reproduction, Fertility and Development, 27*, 1065–1071.

Amaral, R. S., Rosas, F. C., da Silva, V. M., Nichi, M., & Oliveira, C. A. (2013). Endocrine monitoring of the ovarian cycle in captive female Amazonian manatees (*Trichechus inunguis*). *Animal Reproduction Science, 142*, 84–88.

Amaral, R. S., Rosas, F. C., Graham, L. H., da Silva, V. M., & Oliveira, C. A. (2014). First attempt to monitor luteinizing hormone and reproductive steroids in urine samples of the Amazonian manatee (*Trichechus inunguis*). *Journal of Zoo and Wildlife Medicine, 45*, 843–851.

Ash, H., Smith, T. E., Knight, S., & Buchanan-Smith, H. M. (2018). Measuring physiological stress in the common marmoset (*Callithrix jacchus*): Validation of a salivary cortisol collection and assay technique. *Physiology & Behavior, 185*, 14–22.

Atkinson, S., Combelles, C., Vincent, D., Nachtigall, P., Pawloski, J., & Breese, M. (1999). Monitoring of progesterone in captive female false killer whales, *Pseudorca crassidens. General and Comparative Endocrinology, 115*, 323–332.

Atkinson, S., Crocker, D., Houser, D., & Mashburn, K. (2015). Stress physiology in marine mammals: How well do they fit the terrestrial model? *Journal of Comparative Physiology B, 185*, 463–486.

Barber, J. C. (2009). Programmatic approaches to assessing and improving animal welfare in zoos and aquariums. *Zoo Biology: Published in affiliation with the American Zoo and Aquarium Association, 28*, 519–530.

Barnes, M. A., & Turner, C. R. (2016). The ecology of environmental DNA and implications for conservation genetics. *Conservation Genetics, 17*, 1–17.

Beerda, B., Schilder, M. B., Janssen, N. S., & Mol, J. A. (1996). The use of saliva cortisol, urinary cortisol, and catecholamine measurements for a noninvasive assessment of stress responses in dogs. *Hormones and Behavior, 30*, 272–279.

Beerda, B., Schilder, M. B., Van Hooff, J. A., de Vries, H. W., & Mol, J. A. (1998). Behavioural, saliva cortisol and heart rate responses to different types of stimuli in dogs. *Applied Animal Behaviour Science, 58*, 365–381.

Behringer, V., Borchers, C., Deschner, T., Mostl, E., Selzer, D., & Hohmann, G. (2013). Measurements of salivary alpha amylase and salivary cortisol in hominoid primates reveal within-species consistency and between-species differences. *PLoS One, 8*, e60773.

Behringer, V., Claus, W., Hachenburger, K., Kuchar, A., Möstl, E., & Selzer, D. (2009). Effect of giving birth on the cortisol level in a bonobo groups'(*Pan paniscus*) saliva. *Primates, 50*, 190.

Behringer, V., & Deschner, T. (2017). Non-invasive monitoring of physiological markers in primates. *Hormones and Behavior, 91*, 3–18.

Behringer, V., Stevens, J. M., Hohmann, G., Mostl, E., Selzer, D., & Deschner, T. (2014). Testing the effect of medical positive reinforcement training on salivary cortisol levels in bonobos and orangutans. *PLoS One, 9*, e108664.

Bergamasco, L., Osella, M. C., Savarino, P., Larosa, G., Ozella, L., Manassero, M., … Re, G. (2010). Heart rate variability and saliva cortisol assessment in shelter dog: Human–animal interaction effects. *Applied Animal Behaviour Science, 125*, 56–68.

Blejwas, K. M., Williams, C. L., Shin, G. T., Mccullough, D. R., & Jaeger, M. M. (2010). Salivary DNA evidence convicts breeding male coyotes of killing sheep. *The Journal of Wildlife Management, 70*, 1087–1093.

Boehlke, C., Pötschke, S., Behringer, V., Hannig, C., & Zierau, O. (2017). Does diet influence salivary enzyme activities in elephant species? *Journal of Comparative Physiology B, 187*, 213–226.

Boehlke, C., Zierau, O., & Hannig, C. (2015). Salivary amylase–the enzyme of unspecialized euryphagous animals. *Archives of Oral Biology, 60*, 1162–1176.

Bohmann, K., Evans, A., Gilbert, M. T., Carvalho, G. R., Creer, S., Knapp, M., … de Bruyn, M. (2014). Environmental DNA for wildlife biology and biodiversity monitoring. *Trends in Ecology & Evolution, 29*, 358–367.

Brambell, F. W. R. (1965). *Report on the technical committee to enquire into the welfare of livestock kept under intensive conditions*. London: HMSO.

Brennan, P. A., & Kendrick, K. M. (2006). Mammalian social odours: Attraction and individual recognition. *Philosophical Transactions of the Royal Society of London B: Biological Sciences, 361*, 2061–2078.

Broom, D. M. (1991). Animal welfare: Concepts and measurement. *Journal of Animal Science, 69*, 4167–4175.

Buchanan, K. L., & Goldsmith, A. R. (2004). Noninvasive endocrine data for behavioural studies: The importance of validation. *Animal Behaviour, 67*, 183–185.

Bushong, D., Friend, T., & Knabe, D. (2000). Salivary and plasma cortisol response to adrenocorticotropin administration in pigs. *Laboratory Animals, 34*, 171–181.

Caniglia, R., Fabbri, E., Mastrogiuseppe, L., & Randi, E. (2013). Who is who? Identification of livestock predators using forensic genetic approaches. *Forensic Science International: Genetics, 7*, 397–404.

Carter, C., Pournajafi-Nazarloo, H., Kramer, K. M., Ziegler, T. E., White-Traut, R., Bello, D., & Schwertz, D. (2007). Behavioral associations and potential as a salivary biomarker. *Annals of the New York Academy of Sciences, 1098*, 312–322.

Clubb, R., & Mason, G. (2003). Animal welfare: Captivity effects on wide-ranging carnivores. *Nature, 425*, 473–474.

Colson, V., Martin, E., Orgeur, P., & Prunier, A. (2012). Influence of housing and social changes on growth, behaviour and cortisol in piglets at weaning. *Physiology & Behavior, 107*, 59–64.

Colussi, A., Stefanon, B., Adorini, C., & Sandri, M. (2018). Variations of salivary cortisol in dogs exposed to different cognitive and physical activities. *Italian Journal of Animal Science, 17*, 1–8.

Cook, N. J. (2012). Minimally invasive sampling media and the measurement of corticosteroids as biomarkers of stress in animals. *Canadian Journal of Animal Science, 92*, 227–259.

Cook, N. J., Hayne, S. M., Rioja-Lang, F. C., Schaefer, A. L., & Gonyou, H. W. (2013). The collection of multiple saliva samples from pigs and the effect on adrenocortical activity. *Canadian Journal of Animal Science, 93*, 329–333.

Cook, N., Schaefer, A., Lepage, P., & Jones, S. M. (1996). Salivary vs. serum cortisol for the assessment of adrenal activity in swine. *Canadian Journal of Animal Science, 76*, 329–335.

Couret, D., Otten, W., Puppe, B., Prunier, A., & Merlot, E. (2009). Behavioural, endocrine and immune responses to repeated social stress in pregnant gilts. *Animal, 3*, 118–127.

Coutellier, L., Arnould, C., Boissy, A., Orgeur, P., Prunier, A., Veissier, I., & Meunier-Salaün, M.-C. (2007). Pig's responses to repeated social regrouping and relocation during the growing-finishing period. *Applied Animal Behaviour Science, 105*, 102–114.

Cross, N., Pines, M. K., & Rogers, L. J. (2004). Saliva sampling to assess cortisol levels in unrestrained common marmosets and the effect of behavioral stress. *American Journal of Primatology, 62*, 107–114.

Csoltova, E., Martineau, M., Boissy, A., & Gilbert, C. (2017). Behavioral and physiological reactions in dogs to a veterinary examination: Owner-dog interactions improve canine well-being. *Physiology & Behavior, 177*, 270–281.

Czekala, N., & Callison, L. (1996). Pregnancy diagnosis in the black rhinoceros (*Diceros bicornis*) by salivary hormone analysis. *Zoo Biology: Published in affiliation with the American Zoo and Aquarium Association, 15*, 37–44.

Dantzer, B., Fletcher, Q. E., Boonstra, R., & Sheriff, M. J. (2014). Measures of physiological stress: A transparent or opaque window into the status, management and conservation of species? *Conservation Physiology, 2*, cou023.

de Mello, D. M. D., & de Oliveira, C. A. (2016). Biological matrices for sampling free-ranging cetaceans and the implications of their use for reproductive endocrine monitoring. *Mammal Review, 46*, 77–91.

de Souza Amaral, R., Rosas, F. C. W., Viau, P., D'Affonsêca Neto, J. A., da Silva, V. M. F., & de Oliveira, C. A. (2009). Noninvasive monitoring of androgens in male Amazonian manatee (*Trichechus inunguis*): Biologic validation. *Journal of Zoo and Wildlife Medicine, 40*, 458–465.

Dickman, A. J. (2010). Complexities of conflict: The importance of considering social factors for effectively resolving human wildlife conflict. *Animal Conservation, 13*, 458–466.

Dreschel, N. A., & Granger, D. A. (2009). Methods of collection for salivary cortisol measurement in dogs. *Hormones and Behavior, 55*, 163–168.

Elder, C. M., & Menzel, C. R. (2001). Dissociation of cortisol and behavioral indicators of stress in an orangutan (*Pongo pygmaeus*) during a computerized task. *Primates, 42*, 345–357.

Escribano, D., Campos, P. H., Gutierrez, A. M., le Floc'h, N., Ceron, J. J., & Merlot, E. (2014). Effect of repeated administration of lipopolysaccharide on inflammatory and stress markers in saliva of growing pigs. *The Veterinary Journal, 200*, 393–397.

Escribano, D., Gutierrez, A. M., Tecles, F., & Ceron, J. J. (2015). Changes in saliva biomarkers of stress and immunity in domestic pigs exposed to a psychosocial stressor. *Research in Veterinary Science, 102*, 38–44.

Farm Animal Welfare Council. (1993). *Second report on priorities for research and development in farm animal welfare*. Tolworth: FAWC, Ministry of Agriculture, Fisheries and Food (Now DEFRA, London).

Foran, D. R. (2006). Relative degradation of nuclear and mitochondrial DNA: An experimental approach. *Journal of Forensic Sciences, 51*, 766–770.

Gillespie, T. R., & Chapman, C. A. (2008). Forest fragmentation, the decline of an endangered primate, and changes in host–parasite interactions relative to an unfragmented forest. *American Journal of Primatology, 70*, 222–230.

Glenk, L. M., Kothgassner, O. D., Stetina, B. U., Palme, R., Kepplinger, B., & Baran, H. (2013). Therapy dogs' salivary cortisol levels vary during animal-assisted interventions. *Animal Welfare, 22*, 369–378.

Gomez, A., Jewell, E., Walker, S. L., & Brown, J. L. (2004). Use of salivary steroid analyses to assess ovarian cycles in an Indian rhinoceros at the National Zoological Park. *Zoo Biology: Published in affiliation with the American Zoo and Aquarium Association, 23*, 501–512.

Gómez-Laguna, J., Gutiérrez, A., Pallarés, F. J., Salguero, F. J., Cerón, J. J., & Carrasco, L. (2010). Haptoglobin and C-reactive protein as biomarkers in the serum, saliva and meat juice of pigs experimentally infected with porcine reproductive and respiratory syndrome virus. *The Veterinary Journal, 185*, 83–87.

Gourkow, N., Hamon, S. C., & Phillips, C. J. (2014). Effect of gentle stroking and vocalization on behaviour, mucosal immunity and upper respiratory disease in anxious shelter cats. *Preventive Veterinary Medicine, 117*, 266–275.

Grand, A. P., Kuhar, C. W., Leighty, K. A., Bettinger, T. L., & Laudenslager, M. L. (2012). Using personality ratings and cortisol to characterize individual differences in African elephants (*Loxodonta africana*). *Applied Animal Behaviour Science, 142*, 69–75.

Granger, D. A., Fortunato, C. K., Beltzer, E. K., Virag, M., Bright, M. A., & Out, D. (2012). Focus on methodology: Salivary bioscience and research on adolescence: An integrated perspective. *Journal of Adolescence, 35*, 1081–1095.

Granger, D. A., Kivlighan, K. T., Fortunato, C., Harmon, A. G., Hibel, L. C., Schwartz, E. B., & Whembolua, G. L. (2007). Integration of salivary biomarkers into developmental and behaviorally-oriented research: Problems and solutions for collecting specimens. *Physiology & Behavior, 92*, 583–590.

Grimberg-Henrici, C. G., Vermaak, P., Bolhuis, J. E., Nordquist, R. E., & van der Staay, F. J. (2016). Effects of environmental enrichment on cognitive performance of pigs in a spatial holeboard discrimination task. *Animal Cognition, 19*, 271–283.

Gröschl, M. (2009). The physiological role of hormones in saliva. *BioEssays, 31*, 843–852.

Guhad, F. A., & Hau, J. (1996). Salivary IgA as a marker of social stress in rats. *Neuroscience Letters, 216*, 137–140.

Harding, E. J., Paul, E. S., & Mendl, M. (2004). Animal behaviour: Cognitive bias and affective state. *Nature, 427*, 312.

Harms, V., Nowak, C., Carl, S., & Muñoz-Fuentes, V. (2015). Experimental evaluation of genetic predator identification from saliva traces on wildlife kills. *Journal of Mammalogy, 96*, 138–143.

Heers, T., van Neer, A., Becker, A., Grilo, M. L., Siebert, U., & Abdulmawjood, A. (2017). Loop-mediated isothermal amplification (LAMP) assay—A rapid detection tool for identifying red fox (*Vulpes vulpes*) DNA in the carcasses of harbour porpoises (*Phocoena phocoena*). *PLoS One, 12*, e0184349.

Heintz, M. R., Santymire, R. M., Parr, L. A., & Lonsdorf, E. V. (2011). Validation of a cortisol enzyme immunoassay and characterization of salivary cortisol circadian rhythm in chimpanzees (*Pan troglodytes*). *American Journal of Primatology, 73*, 903–908.

Henderson, D. M., Manca, M., Haley, N. J., Denkers, N. D., Nalls, A. V., Mathiason, C. K., . . . Hoover, E. A. (2013). Rapid antemortem detection of CWD prions in deer saliva. *PLoS One, 8*, e74377.

Hodges, J. (1998). Endocrinology of the ovarian cycle and pregnancy in the Asian (*Elephas maximus*) and African (*Loxodonta africana*) elephant. *Animal Reproduction Science, 53*, 3–18.

Hodges, J., Van Aarde, R., Heistermann, M., & Hoppen, H. (1994). Progestin content and biosynthetic potential of the corpus luteum of the African elephant (*Loxodonta africana*). *Journal of Reproduction and Fertility, 102*, 163–168.

Hofmann, R. R., Streich, W. J., Fickel, J., Hummel, J., & Clauss, M. (2008). Convergent evolution in feeding types: Salivary gland mass differences in wild ruminant species. *Journal of Morphology, 269*, 240–257.

Hohmann, G., Mundry, R., & Deschner, T. (2009). The relationship between socio-sexual behavior and salivary cortisol in bonobos: Tests of the tension regulation hypothesis. *American Journal of Primatology: Official Journal of the American Society of Primatologists, 71*, 223–232.

Hopken, M. W., Orning, E. K., Young, J. K., & Piaggio, A. J. (2016). Molecular forensics in avian conservation: A DNA-based approach for identifying mammalian predators of ground-nesting birds and eggs. *BMC Research Notes, 9*, 14.

Horváth, Z., Dóka, A., & Miklósi, Á. (2008). Affiliative and disciplinary behavior of human handlers during play with their dog affects cortisol concentrations in opposite directions. *Hormones and Behavior, 54*, 107–114.

Illera, J. C., Silván, G., Cáceres, S., Carbonell, M. D., Gerique, C., Martínez-Fernández, L., . . . Casares, M. (2014). Assessment of ovarian cycles in the African elephant (*Loxodonta africana*) by measurement of salivary progesterone metabolites. *Zoo Biology, 33*, 245–249.

Ison, S. H., D'eath, R. B., Robson, S. K., Baxter, E. M., Ormandy, E., Douglas, A. J., . . . Jarvis, S. (2010). 'Subordination style'in pigs? The response of pregnant sows to mixing stress affects their offspring's behaviour and stress reactivity. *Applied Animal Behaviour Science, 124*, 16–27.

Janin, A., Léna, J. P., Deblois, S., & Joly, P. (2012). Use of stress-hormone levels and habitat selection to assess functional connectivity of a landscape for an amphibian. *Conservation Biology, 26*, 923–931.

Jarillo-Luna, A., Rivera-Aguilar, V., Garfias, H. R., Lara-Padilla, E., Kormanovsky, A., & Campos-Rodriguez, R. (2007). Effect of repeated restraint stress on the levels of intestinal IgA in mice. *Psychoneuroendocrinology, 32*, 681–692.

Jarillo-Luna, R. A., Rivera-Aguilar, V., Pacheco-Yepez, J., Godinez-Victoria, M., Oros-Pantoja, R., Miliar-Garcia, A., & Campos-Rodriguez, R. (2015). Nasal IgA secretion in a murine model of acute stress. The possible role of catecholamines. *Journal of Neuroimmunology, 278*, 223–231.

Johnston, D. W. (1958). Sex and age characters and salivary glands of the chimney swift. *The Condor, 60*, 73–84.

Jones-Engel, L., Engel, G. A., Schillaci, M. A., Babo, R., & Froehlich, J. (2001). Detection of antibodies to selected human pathogens among wild and pet macaques (*Macaca tonkeana*) in Sulawesi, Indonesia. *American Journal of Primatology, 54*, 171–178.

Kapoor, A., & Matthews, S. G. (2008). Prenatal stress modifies behavior and hypothalamic-pituitary-adrenal function in female Guinea pig offspring: Effects of timing of prenatal stress and stage of reproductive cycle. *Endocrinology, 149*, 6406–6415.

Kapoor, A., & Matthews, S. G. (2011). Testosterone is involved in mediating the effects of prenatal stress in male Guinea pig offspring. *The Journal of Physiology, 589*, 755–766.

Kobelt, A., Hemsworth, P., Barnett, J., & Butler, K. (2003). Sources of sampling variation in saliva cortisol in dogs. *Research in Veterinary Science, 75*, 157–161.

Köndgen, S., Kühl, H., N'goran, P. K., Walsh, P. D., Schenk, S., Ernst, N., ... Leendertz, F. H. (2008). Pandemic human viruses cause decline of endangered great apes. *Current Biology, 18*, 260–264.

Kutsukake, N., Ikeda, K., Honma, S., Teramoto, M., Mori, Y., Hayasaka, I., ... Hasegawa, T. (2009). Validation of salivary cortisol and testosterone assays in chimpanzees by liquid chromatography-tandem mass spectrometry. *American Journal of Primatology, 71*, 696–706.

Lacy, R. C. (1993). Impacts of inbreeding in natural and captive populations of vertebrates: Implications for conservation. *Perspectives in Biology and Medicine, 36*, 480–496.

Laule, G. E., Bloomsmith, M. A., & Schapiro, S. J. (2003). The use of positive reinforcement training techniques to enhance the care, management, and welfare of primates in the laboratory. *Journal of Applied Animal Welfare Science, 6*, 163–173.

Leatherwood, S., & Reeves, R. R. (2012). *The bottlenose dolphin*. Burlington: Elsevier.

Leendertz, F. H., Pauli, G., Maetz-Rensing, K., Boardman, W., Nunn, C., Ellerbrok, H., ... Christophe, B. (2006). Pathogens as drivers of population declines: The importance of systematic monitoring in great apes and other threatened mammals. *Biological Conservation, 131*, 325–337.

Lentle, R., Hume, I., Kennedy, M., Stafford, K., Potter, M., Springett, B., & Haslett, S. (2002). The histology and morphometrics of the major salivary glands of four species of wallabies (Marsupialia: Macropodiae) from Kawau Island, New Zealand. *Journal of Zoology, 257*, 403–410.

Leopold, M. F., Begeman, L., Van Bleijswijk, J. D. L., Ijsseldijk, L. L., Witte, H. J., & Gröne, A. (2015). Exposing the grey seal as a major predator of harbour porpoises. *Proceedings of the Royal Society B: Biological Sciences, 282*. https://doi.org/10.1098/rspb.2014.2429

Lobo, D., Godinho, R., Álvares, F., López-Bao, J. V., & Rodríguez, A. (2015). A new method for noninvasive genetic sampling of saliva in ecological research. *PLoS One, 10*, e0139765.

Lutz, C., Tiefenbacher, S., Jorgensen, M., Meyer, J., & Novak, M. (2000). Techniques for collecting saliva from awake, unrestrained, adult monkeys for cortisol assay. *American Journal of Primatology: Official Journal of the American Society of Primatologists, 52*, 93–99.

Maclean, E. L., Gesquiere, L. R., Gee, N. R., Levy, K., Martin, W. L., & Carter, C. S. (2017). Effects of affiliative human–animal interaction on dog salivary and plasma oxytocin and vasopressin. *Frontiers in Psychology, 8*, 1606.

Maclean, E. L., Gesquiere, L. R., Gee, N., Levy, K., Martin, W. L., & Carter, C. S. (2018). Validation of salivary oxytocin and vasopressin as biomarkers in domestic dogs. *Journal of Neuroscience Methods, 293*, 67–76.

Majchrzak, Y. N., Mastromonaco, G. F., Korver, W., & Burness, G. (2015). Use of salivary cortisol to evaluate the influence of rides in dromedary camels. *General and Comparative Endocrinology, 211*, 123–130.

Marlow, N. J., Thomas, N. D., Williams, A. A. E., Macmahon, B., Lawson, J., Hitchen, Y., ... Berry, O. (2015). Cats (*Felis catus*) are more abundant and are the dominant predator of woylies (*Bettongia penicillata*) after sustained fox (*Vulpes vulpes*) control. *Australian Journal of Zoology, 63*, 18–27.

Meagher, R. K., Ahloy Dallaire, J., Campbell, D. L., Ross, M., Moller, S. H., Hansen, S. W., ... Mason, G. J. (2014). Benefits of a ball and chain: Simple environmental enrichments improve welfare and reproductive success in farmed American mink (*Neovison vison*). *PLoS One, 9*, e110589.

Menargues, A., Urios, V., & Mauri, M. (2008). Welfare assessment of captive Asian elephants (*Elephas maximus*) and Indian rhinoceros (*Rhinoceros unicornis*) using salivary cortisol measurement. *Animal Welfare, 17*, 305–312.

Mendl, M., Burman, O. H., & Paul, E. S. (2010). An integrative and functional framework for the study of animal emotion and mood. *Proceedings of the Royal Society B: Biological Sciences, 277*, 2895–2904.

Mendl, M., Zanella, A. J., & Broom, D. M. (1992). Physiological and reproductive correlates of behavioural strategies in female domestic pigs. *Animal Behaviour, 44*, 1107–1121.

Miller, L. J., Ivy, J. A., Vicino, G. A., & Schork, I. G. (2018). Impacts of natural history and exhibit factors on carnivore welfare. *Journal of Applied Animal Welfare Science, 22*, 1–9.
Monfort, S. L., Holt, W., Pickard, A., Rodger, J., & Wildt, D. (2003). Non-invasive endocrine measures of reproduction and stress in wild populations. In W. Holt, A. Pickard, J. Rodger, & D. Wildt (Eds.), *Reproductive science and integrated conservation (Conservation biology series)* (pp. 147–165). Cambridge: Cambridge University Press.
Monreal-Pawlowsky, T., Carbajal, A., Tallo-Parra, O., Sabés-Alsina, M., Monclús, L., Almunia, J., . . . Lopez-Bejar, M. (2017). Daily salivary cortisol levels in response to stress factors in captive common bottlenose dolphins (*Tursiops truncatus*): A potential welfare indicator. *Veterinary Record, 180*(24), 593.
Mumma, M. A., Soulliere, C. E., Mahoney, S. P., & Waits, L. P. (2014). Enhanced understanding of predator-prey relationships using molecular methods to identify predator species, individual and sex. *Molecular Ecology Resources, 14*, 100–108.
Muneta, Y., Yoshikawa, T., Minagawa, Y., Shibahara, T., Maeda, R., & Omata, Y. (2010). Salivary IgA as a useful non-invasive marker for restraint stress in pigs. *Journal of Veterinary Medical Science, 72*, 1295–1300.
Nater, U. M., & Rohleder, N. (2009). Salivary alpha-amylase as a non-invasive biomarker for the sympathetic nervous system: Current state of research. *Psychoneuroendocrinology, 34*, 486–496.
Needham, A., Dawson, T., & Hales, J. (1974). Forelimb blood flow and saliva spreading in the thermoregulation of the red kangaroo, Megaleia rufa. *Comparative Biochemistry and Physiology Part A: Physiology, 49*, 555–565.
Nemeth, M., Millesi, E., Puehringer-Sturmayr, V., Kaplan, A., Wagner, K. H., Quint, R., & Wallner, B. (2016). Sex-specific effects of dietary fatty acids on saliva cortisol and social behavior in Guinea pigs under different social environmental conditions. *Biology of Sex Differences, 7*, 51.
Nichols, R. V., Cromsigt, J. P., & Spong, G. (2015). DNA left on browsed twigs uncovers bite-scale resource use patterns in European ungulates. *Oecologia, 178*, 275–284.
Nichols, R. V., Königsson, H., Danell, K., & Spong, G. (2012). Browsed twig environmental DNA: Diagnostic PCR to identify ungulate species. *Molecular Ecology Resources, 12*, 983–989.
Noel, A. C., Guo, H.-Y., Mandica, M., & Hu, D. L. (2017). Frogs use a viscoelastic tongue and non-Newtonian saliva to catch prey. *Journal of the Royal Society Interface, 14*, 20160764.
Ott, S., Soler, L., Moons, C., Kashiha, M., Bahr, C., Vandermeulen, J., . . . Cerón, J. (2014). Different stressors elicit different responses in the salivary biomarkers cortisol, haptoglobin, and chromogranin a in pigs. *Research in Veterinary Science, 97*, 124–128.
Palme, R. (2012). Monitoring stress hormone metabolites as a useful, non-invasive tool for welfare assessment in farm animals. *Animal Welfare-The UFAW Journal, 21*, 331.
Parra, M. D., Tecles, F., Subiela, S. M., & Cerón, J. J. (2005). C-reactive protein measurement in canine saliva. *Journal of Veterinary Diagnostic Investigation, 17*, 139–144.
Parrott, R. F., Misson, B. H., & Baldwin, B. A. (1989). Salivary cortisol in pigs following adrenocorticotrophic hormone stimulation: Comparison with plasma levels. *British Veterinary Journal, 145*, 362–366.
Pastore, C., Pirrone, F., Balzarotti, F., Faustini, M., Pierantoni, L., & Albertini, M. (2011). Evaluation of physiological and behavioral stress-dependent parameters in agility dogs. *Journal of Veterinary Behavior: Clinical Applications and Research, 6*, 188–194.
Pearson, B. L., Judge, P. G., & Reeder, D. M. (2008). Effectiveness of saliva collection and enzyme-immunoassay for the quantification of cortisol in socially housed baboons. *American Journal of Primatology: Official Journal of the American Society of Primatologists, 70*, 1145–1151.
Pearson, B. L., Reeder, D. M., & Judge, P. G. (2015). Crowding increases salivary cortisol but not self-directed behavior in captive baboons. *American Journal of Primatology, 77*, 462–467.

Pedernera-Romano, C., Valdez, R., Singh, S., Chiappa, X., Romano, M., & Galindo, F. (2006). Salivary cortisol in captive dolphins (*Tursiops truncatus*): A non-invasive technique. *Animal Welfare, 15*, 359–362.

Pfefferle, D., Plümer, S., Burchardt, L., Treue, S., & Gail, A. (2018). Assessment of stress responses in rhesus macaques (*Macaca mulatta*) to daily routine procedures in system neuroscience based on salivary cortisol concentrations. *PLoS One, 13*, e0190190.

Phillips, C. D., & Baker, R. J. (2015). Secretory gene recruitments in vampire bat salivary adaptation and potential convergences with sanguivorous leeches. *Frontiers in Ecology and Evolution, 3*, 122.

Pietraszek, J., & Atkinson, S. (1994). Concentrations of estrone sulfate and progesterone in plasma and saliva, vaginal cytology, and bioelectric impedance during the estrous cycle of the Hawaiian monk seal (Monachus schaunslandi). *Marine Mammal Science, 10*, 430–441.

Prickett, J., Simer, R., Christopher-Hennings, J., Yoon, K.-J., Evans, R. B., & Zimmerman, J. J. (2008). Detection of porcine reproductive and respiratory syndrome virus infection in porcine oral fluid samples: A longitudinal study under experimental conditions. *Journal of Veterinary Diagnostic Investigation, 20*, 156–163.

Ralls, K., Ballou, J. D., Dudash, M. R., Eldridge, M. D., Fenster, C. B., Lacy, R. C., . . . Frankham, R. (2018). Call for a paradigm shift in the genetic management of fragmented populations. *Conservation Letters, 11*, e12412.

Rilling, J. K., & Young, L. J. (2014). The biology of mammalian parenting and its effect on offspring social development. *Science, 345*, 771–776.

Rutherford, K. M., Robson, S. K., Donald, R. D., Jarvis, S., Sandercock, D. A., Scott, E. M., . . . Lawrence, A. B. (2009). Pre-natal stress amplifies the immediate behavioural responses to acute pain in piglets. *Biology Letters, 5*(4), 452–454.

Salimetrics, L. L. C. (2011). *Saliva collection and handling advice*. https://www.salimetrics.com/

Sandercock, D. A., Gibson, I. F., Rutherford, K. M., Donald, R. D., Lawrence, A. B., Brash, H. M., . . . Nolan, A. M. (2011). The impact of prenatal stress on basal nociception and evoked responses to tail-docking and inflammatory challenge in juvenile pigs. *Physiology & Behavior, 104*, 728–737.

Sandifer, P. A., Sutton-Grier, A. E., & Ward, B. P. (2015). Exploring connections among nature, biodiversity, ecosystem services, and human health and Well-being: Opportunities to enhance health and biodiversity conservation. *Ecosystem Services, 12*, 1–15.

Saunders, S. P., Harris, T., Traylor-Holzer, K., & Beck, K. G. (2014). Factors influencing breeding success, ovarian cyclicity, and cub survival in zoo-managed tigers (*Panthera tigris*). *Animal Reproduction Science, 144*, 38–47.

Schapiro, S. J., Bloomsmith, M. A., & Laule, G. E. (2003). Positive reinforcement training as a technique to alter nonhuman primate behavior: Quantitative assessments of effectiveness. *Journal of Applied Animal Welfare Science, 6*, 175–187.

Schmidt, A., Aurich, J., Möstl, E., Müller, J., & Aurich, C. (2010). Changes in cortisol release and heart rate and heart rate variability during the initial training of 3-year-old sport horses. *Hormones and Behavior, 58*, 628–636.

Schmidt, A., Möstl, E., Wehnert, C., Aurich, J., Müller, J., & Aurich, C. (2010). Cortisol release and heart rate variability in horses during road transport. *Hormones and Behavior, 57*, 209–215.

Seal, U. S., Plotka, E. D., Smith, J. D., Wright, F. H., Reindl, N. J., Taylor, R. S., & Seal, M. F. (1985). Immunoreactive luteinizing hormone, estradiol, progesterone, testosterone, and androstenedione levels during the breeding season and anestrus in Siberian tigers. *Biology of Reproduction, 32*, 361–368.

Seeber, P. A., Soilemetzidou, S. E., East, M. L., Walzer, C., & Greenwood, A. D. (2017). Equine behavioral enrichment toys as tools for non-invasive recovery of viral and host DNA. *Zoo Biology, 36*, 341–344.

Skandakumar, S., Stodulski, G., & Hau, J. (1995). Salivary IgA: A possible stress marker in dogs. *Animal Welfare, 4*, 339–350.

Smiley, T., Spelman, L., Lukasik-Braum, M., Mukherjee, J., Kaufman, G., Akiyoshi, D. E., & Cranfield, M. (2010). Noninvasive saliva collection techniques for free-ranging Mountain Gorillas and captive Eastern Gorillas. *Journal of Zoo and Wildlife Medicine, 41*, 201–209.

Sopinka, N. M., Patterson, L. D., Redfern, J. C., Pleizier, N. K., Belanger, C. B., Midwood, J. D., ... Cooke, S. J. (2015). Manipulating glucocorticoids in wild animals: Basic and applied perspectives. *Conservation Physiology, 3*, 1.

Staley, M., Conners, M. G., Hall, K., & Miller, L. J. (2018). Linking stress and immunity: Immunoglobulin a as a non-invasive physiological biomarker in animal welfare studies. *Hormones and Behavior, 102*, 55–68.

Steffens, K. E., Sanders, M. D., Gleenson, D. M., Pullen, K. M., & Stowe, C. J. (2012). Identification of predators at black-fronted tern *Chlidonias albostriatus* nests, using mtDNA analysis and digital video recorders. *New Zealand Journal of Ecology, 36*, 48–55.

Stevens, B., Karlen, G. M., Morrison, R., Gonyou, H. W., Butler, K. L., Kerswell, K. J., & Hemsworth, P. H. (2015). Effects of stage of gestation at mixing on aggression, injuries and stress in sows. *Applied Animal Behaviour Science, 165*, 40–46.

Sundqvist, A.-K., Ellegren, H., & Vilà, C. (2008). Wolf or dog? Genetic identification of predators from saliva collected around bite wounds on prey. *Conservation Genetics, 9*, 1275–1279.

Tetley, C. L., & O'hara, S. J. (2012). Ratings of animal personality as a tool for improving the breeding, management, and welfare of zoo mammals. *Animal Welfare, 21*, 463–476.

Theodorou, J., & Atkinson, S. (1998). Monitoring total androgen concentrations in saliva from captive Hawaiian monk seals (*Monachus schauinslandi*). *Marine Mammal Science, 14*, 304–310.

Troïanowski, M., Mondy, N., Dumet, A., Arcanjo, C., & Lengagne, T. (2017). Effects of traffic noise on tree frog stress levels, immunity, and color signaling. *Conservation Biology, 31*, 1132–1140.

Tucker, A. S. (2010). Salivary gland adaptations: Modification of the glands for novel uses. *Frontiers Oral Biology, 14*, 21–31.

Ugaz, C., Valdez, R. A., Romano, M. C., & Galindo, F. (2013). Behavior and salivary cortisol of captive dolphins (*Tursiops truncatus*) kept in open and closed facilities. *Journal of Veterinary Behavior: Clinical Applications and Research, 8*, 285–290.

Van Bleijswijk, J. D. L., Begeman, L., Witte, H. J., Ll, I. J., Brasseur, S., Gröne, A., & Leopold, M. F. (2014). Detection of grey seal *Halichoerus grypus* DNA in attack wounds on stranded harbour porpoises *Phocoena phocoena. Marine Ecology Progress Series, 513*, 277–281.

Vasconcellos, A. D. S., Virányi, Z., Range, F., Ades, C., Scheidegger, J. K., Möstl, E., & Kotrschal, K. (2016). Training reduces stress in human-socialised wolves to the same degree as in dogs. *PLoS One, 11*, e0162389.

Von Borell, E., Dobson, H., & Prunier, A. (2007). Stress, behaviour and reproductive performance in female cattle and pigs. *Hormones and Behavior, 52*, 130–138.

von Engelhard, N., Kappeler, P. M., & Heistermann, M. (2000). Androgen levels and female social dominance in *Lemur catta. Proceedings of the Royal Society of London. Series B: Biological Sciences, 267*, 1533–1539.

Waits, L. P., & Paetkau, D. (2005). Noninvasive genetic sampling tools for wildlife biologists: A review of applications and recommendations for accurate data collection. *The Journal of Wildlife Management, 69*, 1419–1433.

Ware, F. L., & Luck, M. R. (2017). Evolution of salivary secretions in haematophagous animals. *Bioscience Horizons: The International Journal of Student Research, 10*. https://doi.org/10.1093/biohorizons/hzw015

Wengert, G. M., Gabriel, M. W., Foley, J. E., Kun, T., & Sacks, B. N. (2013). Molecular techniques for identifying intraguild predators of fishers and other north American small carnivores. *Wildlife Society Bulletin, 37*, 659–663.

Wheat, R. E., Allen, J. M., Miller, S. D., Wilmers, C. C., & Levi, T. (2016). Environmental DNA from residual saliva for efficient noninvasive genetic monitoring of brown bears (*Ursus arctos*). *PLoS One, 11*, e0165259.

Whitham, J. C., & Wielebnowski, N. (2013). New directions for zoo animal welfare science. *Applied Animal Behaviour Science, 147*, 247–260.

Wobber, V., Hare, B., Lipson, S., Wrangham, R., & Ellison, P. (2013). Different ontogenetic patterns of testosterone production reflect divergent male reproductive strategies in chimpanzees and bonobos. *Physiology & Behavior, 116*, 44–53.

Wulf, M., Beythien, E., Ille, N., Aurich, J., & Aurich, C. (2018). The stress response of 6-month-old horses to abrupt weaning is influenced by their sex. *Journal of Veterinary Behavior, 23*, 19–24.

Yang, C.-H., Ko, H.-L., Salazar, L. C., Llonch, L., Manteca, X., Camerlink, I., & Llonch, P. (2018). Pre-weaning environmental enrichment increases piglets' object play behaviour on a large scale commercial pig farm. *Applied Animal Behaviour Science, 202*, 7–12.

# Part VI
# The Way Forward: Future Directions for the Science and Practice of Salivary Bioscience

Douglas A. Granger and Marcus K. Taylor

# Chapter 29
# Applications of Salivary Bioscience to Precision Medicine

**Kate Ryan Kuhlman and Zahra Mousavi**

**Abstract** Precision medicine is an approach to treating mental and physical illness that aims to optimize treatment effectiveness by using individual genetic, environmental, and lifestyle factors to determine which approaches will be effective for which patients. Salivary biomarkers enable noninvasive measurement of both state and trait individual differences in genetic, environmental, and lifestyle factors. In this chapter, the history and future directions of precision medicine will be summarized. Within this summary, we will review the role salivary bioscience has played in identifying moderators of treatment effects and explaining treatment efficacy across several prevalent and costly physical and mental conditions. Finally, this chapter will discuss the ways in which salivary bioscience could be used to advance the larger goals of precision medicine. We will highlight how salivary bioscience has the potential to overcome the feasibility barriers that have prevented many of the latest innovations in evidence-based precision medicine to be adopted by health care providers in primary and community settings.

**Keywords** Precision medicine · Risk factors · Treatment effectiveness · Treatment efficacy · Implementation science

Biomarkers are intended to be used for three broad purposes: to diagnose diseases, to determine the prognosis of a disease independent of treatment, and to predict treatment responses. Precision medicine focuses on the latter of these three purposes

K. R. Kuhlman (✉)
Department of Psychological Science, School of Social Ecology, University of California Irvine, Irvine, CA, USA

Institute for Interdisciplinary Salivary Bioscience Research, University of California, Irvine, CA, USA
e-mail: krkuhl@uci.edu

Z. Mousavi
Department of Psychological Science, School of Social Ecology, University of California Irvine, Irvine, CA, USA

 
D. A. Granger, M. K. Taylor (eds.), *Salivary Bioscience*,
https://doi.org/10.1007/978-3-030-35784-9_29

by taking into account individual variability in genes, environment, and lifestyle for each person (Advancing the Precision Medicine Initiative, 2015; Collins & Varmus, 2015; Ginsburg & Phillips, 2018).

Precision medicine has long been part of daily health care practices. Interest in developing more widely studied and utilized tools for precision medicine surged when the President of the United States of America, Barack Obama, announced a national $215 million dollar Precision Medicine initiative in his 2015 State of the Union Address (Obama, 2015). The basic premise of precision medicine is that even though two individuals may have the same diagnosis, they may not benefit equally from the same treatment. For example, corrective lenses are prescribed to individuals based upon their individual prescription which recognizes that impaired eyesight may be one construct while the prescription for near- and far-sightedness should differ. This simple idea is now being extended to individual differences in genotype, lifestyle, metabolic rates, and many other factors.

While personalized, individualized, and precision medicine have been used somewhat interchangeably over time, it is important to make distinctions between them. Many, if not most, health care providers have been providing personalized or individualized medicine to their patients for decades, even centuries; in the sense that their treatment of each patient's presenting symptoms are treated in a personal way based on their unique history and characteristics that include other preexisting conditions, age, personality, and clinical judgments about risks and benefits. However, precision medicine has a more empirically derived goal and definition. The established definition of precision medicine is: "treatments targeted to the needs of individual patients on the basis of genetic, biomarker, phenotypic, or psychosocial characteristics that distinguish a given patient from other patients with similar clinical presentations. Inherent in this definition is the goal of improving clinical outcomes for individual patients and minimizing unnecessary side effects for those less likely to have a response to a particular treatment" (Agusti et al., 2016; Jameson & Longo, 2015). Thus, the goal of precision medicine is not to tailor treatments to individual patients, but rather to tailor treatment decisions to established subcategories of a disease (Ashley, 2015), such as those identified by the presence of a specific symptom cluster, molecular atypicality, rate of drug metabolism, or genetic predisposition.

Some of the most impactful advances in precision medicine have emerged in the treatment of cancer. For example, breast cancer mortality dropped 24% between 1989 and 2007 from 33 to 25 women per 100,000 (Jatoi, Chen, Anderson, & Rosenberg, 2007). Much of this improvement in treatment and outcomes emerged out of a combination of rigorous prevention and innovative treatment through such means as mammography and systemic adjuvant treatments. Treatment outcomes further improved through application of precision medicine. From a treatment perspective, determining whether a breast tumor overexpresses the estrogen receptor (HER2) differentiated what type of treatments will delay disease progression and metastases (Slamon et al., 2001). From the preventative perspective, identification of mutations in BRCA1 and BRCA2 genes led to better monitoring of at-risk patients and in some cases reduced risk of disease via elective preventative mastectomies and oophorectomies (Nelson et al., 2014). Yet, these types of precision medicine-driven advances have not reached all aspects of health care due to costs and limited access to the technologies necessary to implement them. Even the summarized advances in breast cancer prevention and treatment have unevenly benefited society (Desai & Jena, 2016; Metcalfe et al., 2008).

## 29.1 History and Use of Salivary Bioscience Within Precision Medicine

Salivary biomarkers have the potential to be particularly valuable in precision medicine due to their ability to capture genetic and multisystem molecular markers, while also providing insight into lifestyle factors and hygiene (e.g., diet and oral health) (Zarco, Vess, & Ginsburg, 2012). Yet, salivary biomarkers have, so far, been used sparingly in research aimed at predicting treatment responses and outcomes. This research has been mostly limited to research on cancer and common psychiatric disorders such as anxiety and depression.

In keeping with the history of precision medicine, cancer researchers have developed many cancer biomarkers that have shown varying levels of utility in diagnosis, prognosis, and treatment response (Mehta et al., 2010). For example, saliva can be used to generate a transcriptome profile that distinguishes between mice with melanoma and lung cancer (Gao et al., 2009). Indeed, saliva has been termed a "liquid biopsy" that can detect the presence of cancers well outside the oral cavity in humans, including prostate, colorectal, breast, and more (See Rapado-González, Majem, Muinelo-Romay, López-López, & Suarez-Cunqueiro, 2016 for review). Chapter 8 details the use of saliva for liquid biopsies in cancer research and treatment.

In psychiatric research, the focus of the existing studies has been on functioning of the body's stress response system via salivary concentrations of cortisol. For example, in a study of older individuals with an anxiety disorder and cognitive dysfunction, higher salivary cortisol concentrations at baseline were associated with greater improvements in memory, executive function, and worry severity for up to 8 weeks after mifepristone treatment (Lenze et al., 2014). In veterans with PTSD, an increasing pattern of cortisol reactivity during prolonged exposure therapy sessions predicts better treatment response (Rauch, King, Liberzon, & Sripada, 2017). However, it is important to note that the few studies that have looked at salivary cortisol as predictors of treatment response for anxiety disorders have returned inconsistent results (Fischer & Cleare, 2017). Finally, adolescents with greater salivary cortisol reactivity to a laboratory conflict task showed greater improvements in depressive symptoms during interpersonal therapy (IPT-A) (Gunlicks-Stoessel, Mufson, Cullen, & Klimes Dougan, 2013). Thus, it appears that individual differences in the functioning of the HPA axis may be reliable and sensitive predictors of psychosocial treatment responses. Within the current definition of precision medicine, the next step in this research would be to determine for whom functioning of the HPA axis predicts response to treatment, which treatments, and why.

Salivary measures of cortisol can also be used to predict which patients may be at risk for psychiatric disorders while undergoing treatment for other diseases. For example, having a larger cortisol awakening response predicted the development of depression during the course of Interferon-α therapy for patients with chronic Hepatitis C (Eccles et al., 2012), and having a larger cortisol awakening response before the onset of adjuvant cancer treatment also predicted greater increases in depressive symptoms following treatment in women with breast cancer (Kuhlman et al., 2017). Indeed the importance of the HPA axis and glucocorticoids in precision medicine is growing such that salivary biomarkers can be used to differentiate neuronal endophenotypes (e.g., CRF-hyperactivity) during acute stress that indicate predisposition to subtypes of stress-related disorders as well as inform treatment (Hellhammer, Meinlschmidt, & Pruessner, 2018).

Despite the limited research integrating salivary bioscience and precision medicine, the public has enthusiastically embraced the use of commercially available saliva-based tests, such as 23andme® for ancestry and genetic health risk information and Nutriscan® which tests your dog or cat's saliva for food sensitivities. Indeed, 23andme® alone is estimated to have revenues of approximately $65 million dollars each year. The ease, cost-effectiveness, and noninvasiveness of saliva collection has likely contributed to the influx of consumer products aimed at improving health care one individual at a time. The growing consumer market for saliva-based biomarkers for home use is likely to drive opportunities for more research into the utility of salivary biomarkers in precision medicine.

## 29.2 Precision Medicine: A Salivary Bioscience Frontier

Despite the long-standing interest in integrating salivary bioscience into precision medicine, the field remains in the early stages of adoption, validation, and implementation. There are several important next steps in further developing salivary bioscience as a contributor to precision medicine.

First and foremost, more clinical trials can incorporate salivary measures into their baseline assessments in order to build a comprehensive body of knowledge about how different biomarkers may inform treatment responses. Saliva samples are minimally invasive for patients, can be collected in almost any setting, and are inexpensive to collect and store. Once the overall results of the clinical trials are discovered, additional funding can be secured to support collaborative efforts between experts in salivary bioscience and the clinical trial's principal investigators to develop biologically valid predictors of treatment response.

Second, there are a number of biomarkers that have already been identified as predictors of treatment response that are conventionally measured in blood or other tissues, but could be more widely assessed using salivary markers. One promising future direction for salivary bioscience and precision medicine is the use of genetic polymorphisms measured in saliva to predict treatment responses. Depressed patients with rs1360780 and possibly rs3800373 single nucleotide polymorphisms of the FKBP5 gene showed better responses to antidepressant treatment and lower recurrence rates (Binder et al., 2004). Similar observations have been made when looking at polymorphisms in the drug transporter gene ABCB1 (Uhr et al., 2008), while higher activity COMT 158 val/val genotype (Baune et al., 2008) and the G allele carriers of rs2270007 of CRHR2 gene (Papiol et al., 2007) have been linked to poorer response to antidepressant treatment. Some studies have even combined genetic indicators with individual differences in HPA axis functioning to show that treatment responses to antidepressants are worse for individuals carrying the Bcll polymorphism who also have high ACTH responses to the DEX/CHR test (Brouwer et al., 2006). Indeed, genetic polymorphisms are most commonly measured in blood, but can be reliably measured in saliva. See Chap. 6 for more information on the state of genetic research within salivary bioscience. However, implementing blood draws into standard psychiatric care is not as feasible or cost-effective as saliva collection. For this reason, salivary measures are being increasingly considered for therapeutic drug monitoring which is detailed in Chap. 17. The convenience and cost of saliva collection makes the widespread adoption of saliva-based genetic measurement a feasible option for improving our knowledge of genetic predictors of treatment response and the adoption of this knowledge in clinical settings.

Genetic predictors, however, are not the only existing biomarkers that can be extended to studies using saliva. For example, Raison et al. (2013) found that baseline elevations in the inflammatory marker, C-reactive protein (CRP), predicted better responses to an anti-inflammatory treatment for depressive symptoms. Currently, our ability to assess circulating inflammatory cytokines in saliva is fairly limited (see Chap. 9 for more on immune measurement in saliva); however, CRP is among the few inflammatory markers that has been validated in saliva (Ouellet-Morin, Danese, Williams, & Arseneault, 2011). Thus, more studies interested in treating the inflammatory mechanism involved in some depressive symptoms (see Dooley et al., 2018 for review) may be able to more widely assess the effectiveness of these treatments using salivary measures of CRP. Further, the presence of eosinophils in oral fluids has been identified as a criterion for specific treatments in chronic obstructive pulmonary disease and asthma (Agusti et al., 2016), despite salivary measures not being collected as part of standard practice by the providers who commonly treat these disorders. Finally, in areas of medicine where precision medicine is already well-established, there are existing guidelines of actionable traits that could systematically be tested for their efficacy if assessed using salivary biomarkers (e.g., see Table 1 in Robinson, 2012).

Measurement of microRNAs (miRNAs), proteomics, and metabolomics are also promising new directions for salivary bioscience. Unfortunately, there are no studies to our knowledge that have tested the predictive utility of saliva-based proteomics or metabolomics to treatment responses. Yet, miRNAs have already emerged as important cancer biomarkers, which can be effectively measured in saliva (Rapado-González et al., 2018). Circulating miRNAs are abundant gene regulating molecules. Cancer directly affects miRNAs which contribute to disease progression (Zhang et al., 2006), meaning they have the potential to measure early and pleiotropic disease processes. Indeed, miRNAs can be used to determine tumor subtypes as well as predict treatment response (Blenkiron & Miska, 2007). miRNAs are also particularly amenable to measurement in saliva due to their stability. Furthermore, metabolomics are an important new tool in precision medicine (Wishart, 2016) that can be implemented within saliva (Zhang, Sun, & Wang, 2012). Saliva can also be used for proteomics which has so far been useful in diagnosis of multiple forms of cancer, but may have growing utility as a predictive indicator of treatment response (Castagnola et al., 2017).

While there are several promising areas for salivary bioscience to contribute to precision medicine, there are some limitations to the current utility of salivary biomarkers that must be addressed before salivary bioscience can truly advance the field of precision medicine. For example, salivary cortisol is probably the most widely used salivary biomarker published in the literature today. Yet, biologically meaningful indices using salivary cortisol are quite unstable within individuals over time (Ross, Murphy, Adam, Chen, & Miller, 2014). That being said, there is currently a very active discourse within psychoneuroendocrinology to better understand these sources of instability so that salivary cortisol indices can more effectively and reliably be used in applied settings (Doane, Chen, Sladek, Van Lenten, & Granger, 2015; Kuhlman, Robles, Dickenson, Reynolds, & Repetti, 2019). Further, most salivary biomarkers to date have not been clinically validated for use in placing individuals in meaningful diagnostic or risk categories. Instead, many saliva-based biomarkers exist solely as continuous variables, meaning that their ability to predict treatment outcomes can only be interpreted as relatively higher risk based upon the mean and variability in that particular sample. This issue could be addressed by creating a stronger imperative within the field of salivary bioscience to develop clinically or biologically meaningful thresholds for salivary biomarkers.

Likewise, some of the future directions at the intersection of salivary bioscience and precision medicine pertain to the development of precision medicine as a broader goal within clinical research. Despite the push for precision medicine research and innovation, there are several barriers to progress even in the most promising areas of medicine (e.g., see Ashley (2015) for a summary of precision medicine advances within cystic fibrosis and oncology and the barriers to their progress and dissemination of findings so far). First, the aims of precision medicine require large sample sizes and in-depth data collection for each participant. Predicting treatment responses requires examination of multiple competing predictors and large sample sizes offer the highest likelihood that the results will be applicable to the general population. Ideally, these large-scale clinical studies will shed light on the effectiveness of different treatments for specific subgroups of individuals by testing the effectiveness of more than one treatment. Further, identification of biologically valid predictors of treatment responses, such as the deep and conceptual phenotypes, described earlier in this chapter, requires large datasets not only of a high number of participants, but with multidimensional data within each individual. This is articulated beautifully in Hellhammer et al.'s (2018) explanation of how social, behavioral, and biological data can be used in combination to identify conceptual phenotypes that inform treatment decisions.

One important consideration to be made within these large-scale studies will be the degree to which exploratory, rather than a priori hypothesis-driven, analyses will be tolerated by the field. Certainly, there is a wealth of evidence to support some theoretically defined endophenotypes; however, there are likely applications of computer science to Big Data from multidimensional and large-scale clinical trials that would more quickly identify factors that predict treatment response (e.g., see Robinson (2012) for justification of computational modeling to achieve deep phenotyping). The challenge with these empirically derived predictors will be whether they can be translated back to inform basic research on underlying mechanisms involved in the development of a disease, or are merely the best predictors of treatment response independent of their causal role in the disease and its symptoms. Currently, there is an active debate on the basic approach to precision medicine research. Some scholars believe that precision medicine should be based on biologically valid predictors of treatment responses that reflect disease subtypes, while other scholars remain agnostic to whether the best predictors of treatment responses need to be causally related to the development and course of an illness. As the field continues to develop and publish findings from both perspectives, this debate will undoubtedly become more rich and the merits of each ideological approach will become more apparent. Of course, the most impactful biomarkers will be those reflecting both underlying biological mechanisms of disease while also informing clinical care (Insel, 2014). To this end, Hellhammer et al. (2018) proposed a model for translating complex biological information into improvements in health care, specifically within psychiatry, using conceptual endophenotypes comprised of multiple levels of analysis (e.g., genetic, molecular, behavioral, social, and developmental). As more research of this nature is conducted, salivary bioscience will be particularly well-situated to better tailor interventions and therapies to individuals as a function of behavioral, social, and developmental levels of analysis. For example, salivary measures of testosterone can be used to estimate an adolescent's progress through puberty (Dorn & Biro, 2011) which is likely to be informative for treatment across a number of physical and mental conditions. Further, salivary measures of dyadic synchrony between mothers and their infants (Williams et al., 2013), or within married couples (Liu, Rovine, Klein, & Almeida, 2013), may be useful in tailoring behavioral interventions aimed at fostering better attachment and attunement.

Another barrier to the dissemination of breakthroughs in precision medicine today is its incorporation into the current electronic health systems used by most providers today. For example, researchers disseminate findings to academic journals, to their peers at scientific meetings, and to their trainees in educational settings. In parallel, the tools health care providers use in their daily practices are developed to organize and protect patient health records and provide a system for billing. If a known genetic polymorphism is identified to predict better responses to treatment A over treatment B for their diagnosis, how can the results of their genetic test be safely and ethically used to inform the health care providers' decisions and recommendations? Currently, it is up to the patient to provide the results of their genetic testing to their health care provider, and up to that provider and their training to decide whether and how to use that genetic information to guide their clinical practice. Precision medicine is unlikely to benefit as many patients as it could if researchers do not cooperate with industry to develop usable tools for health care providers. A good example of this type of tool is PracticeWise®, which is a flexible tool for health care providers within child mental health services that integrates the results of published trials into a user-friendly tool that guides treatment decisions. To our knowledge, this tool does not incorporate biological predictors of treatment responses into its tool; however, as more clinical trials include biological measures as predictors of treatment outcomes, that may change. Of course, innovation to this end is not limited to connecting cutting-edge research with industry, but will also inevitably involve federal and state legislation. Indeed, there are important legal and ethical considerations to be made whenever genetic information is used to guide health care decisions.

Finally, precision medicine is not limited to providing guidance on what works for whom, but could also be leveraged for prevention (Khoury, Iademarco, & Riley, 2016). More and more research results in the identification of risk factors and biological vulnerabilities for disorders. For example, genetic predictors of risk have been identified for mood disorders (Prendes-Alvarez & Nemeroff, 2018), Alzheimer's disease (Karch & Goate, 2015), breast cancer (Nelson et al., 2014), and many more. Certainly, public health overall will benefit from efforts to understand how to deliver effective preventions to the highest risk populations, and to make empirically-driven - decisions about which preventive strategies are best delivered universally and which to targeted groups.

## 29.3 Conclusions

Very little research to date has been published using salivary biomarkers to inform precision medicine. Yet, the research conducted to date, particularly with respect to cancer and psychiatric disorders, has been promising. First and foremost, more research is needed from large-scale treatment studies that test effectiveness of salivary biomarkers in predicting treatment responses. Advances in this field would benefit from systematic efforts to determine which established predictors of treatment responsiveness that are currently being measured using other tissues can be effectively measured in saliva. Additionally, there are several areas in which the fields of salivary bioscience and precision medicine could develop independently, such as the establishment of clinically and biologically meaningful values within important salivary biomarkers and the creation of more usable tools that translate findings of precision medicine into clinical practice. Nonetheless, precision medicine is an important new frontier in health care and public health to which salivary bioscience can make groundbreaking contributions.

| Summary of main points |
| --- |
| • Precision medicine aims to identify what treatments will be most effective for which patients. |
| • Saliva can be used to identify genetic, proteomic, molecular, and metabolomic markers that can be used to predict treatment response. |
| • Most salivary research within precision medicine has been limited to research on patients with cancer, depression, and anxiety. |
| • More research is needed to determine the utility of salivary biomarkers to precision medicine. |

## References


Advancing the Precision Medicine Initiative. (2015). *Cancer Discovery, 5*(12), 1230. https://doi.org/10.1158/2159-8290.CD-NB2015-146

Agusti, A., Bel, E., Thomas, M., Vogelmeier, C., Brusselle, G., Holgate, S., … Pavord, I. D. (2016). Treatable traits: Toward precision medicine of chronic airway diseases. *European Respiratory Journal, 47*(2), 410–419. https://doi.org/10.1183/13993003.01359-2015

Ashley, E. A. (2015). The precision medicine initiative: A new national effort. *JAMA, 313*(21), 2119–2120. https://doi.org/10.1001/jama.2015.3595

Baune, B. T., Hohoff, C., Berger, K., Neumann, A., Mortensen, S., Roehrs, T., … Domschke, K. (2008). Association of the COMT val158met variant with antidepressant treatment response in major depression. *Neuropsychopharmacology: Official Publication of the American College of Neuropsychopharmacology, 33*(4), 924–932. https://doi.org/10.1038/sj.npp.1301462

Binder, E. B., Salyakina, D., Lichtner, P., Wochnik, G. M., Ising, M., Pütz, B., … Muller-Myhsok, B. (2004). Polymorphisms in *FKBP5* are associated with increased recurrence of depressive episodes and rapid response to antidepressant treatment. *Nature Genetics, 36*(12), 1319–1325. https://doi.org/10.1038/ng1479

Blenkiron, C., & Miska, E. A. (2007). miRNAs in cancer: Approaches, aetiology, diagnostics and therapy. *Human Molecular Genetics, 16*(1), R106–R113. https://doi.org/10.1093/hmg/ddm056

Brouwer, J. P., Appelhof, B. C., van Rossum, E. F. C., Koper, J. W., Fliers, E., Huyser, J., … Hoogendijk, W. J. G. (2006). Prediction of treatment response by HPA-axis and glucocorticoid receptor polymorphisms in major depression. Psychoneuroendocrinology, 31(10), 1154–1163. doi:https://doi.org/10.1016/j.psyneuen.2006.08.001.

Castagnola, M., Scarano, E., Passali, G. C., Messana, I., Cabras, T., Iavarone, F., … Paludetti, G. (2017). Salivary biomarkers and proteomics: Future diagnostic and clinical utilities. *Acta Otorhinolaryngologica Italica, 37*(2), 94–101. https://doi.org/10.14639/0392-100X-1598

Collins, F. S., & Varmus, H. (2015). A new initiative on precision medicine. *The New England Journal of Medicine, 372*(9), 793–795. https://doi.org/10.1056/NEJMp1500523

Desai, S., & Jena, A. B. (2016). Christmas 2016: Famous figures: Do celebrity endorsements matter? Observational study of BRCA gene testing and mastectomy rates after Angelina Jolie's New York Times editorial. *The BMJ, 355*, i6357.

Doane, L. D., Chen, F. R., Sladek, M. R., Van Lenten, S. A., & Granger, D. A. (2015). Latent trait cortisol (LTC) levels: Reliability, validity, and stability. *Psychoneuroendocrinology, 55*, 21–35. https://doi.org/10.1016/j.psyneuen.2015.01.017

Dooley, L. N., Kuhlman, K. R., Robles, T. F., Eisenberger, N. I., Craske, M. G., & Bower, J. E. (2018). The role of inflammation in core features of depression: Insights from paradigms using exogenously-induced inflammation. *Neuroscience & Biobehavioral Reviews, 94*, 219–237. https://doi.org/10.1016/j.neubiorev.2018.09.006

Dorn, L. D., & Biro, F. M. (2011). Puberty and its measurement: A decade in review. *Journal of Research on Adolescence, 21*(1), 180–195. https://doi.org/10.1111/j.1532-7795.2010.00722.x

Eccles, J., Lallemant, C., Mushtaq, F., Greenwood, M., Keller, M., Golding, B., … Whale, R. (2012). Pre-treatment waking cortisol response and vulnerability to interferon α induced depression. *European Neuropsychopharmacology: The Journal of the European College of Neuropsychopharmacology, 22*(12), 892–896. https://doi.org/10.1016/j.euroneuro.2012.03.009

Fischer, S., & Cleare, A. J. (2017). Cortisol as a predictor of psychological therapy response in anxiety disorders-systematic review and meta-analysis. *Journal of Anxiety Disorders, 47*, 60–68. https://doi.org/10.1016/j.janxdis.2017.02.007

Gao, K., Zhou, H., Zhang, L., Lee, J. W., Zhou, Q., Hu, S., … Wong, D. T. (2009). Systemic disease-induced salivary biomarker profiles in mouse models of melanoma and non-small cell lung cancer. PLoS One, 4(6), e5875. doi:https://doi.org/10.1371/journal.pone.0005875.

Ginsburg, G. S., & Phillips, K. A. (2018). Precision medicine: From science to value. *Health Affairs (Project Hope), 37*(5), 694–701. https://doi.org/10.1377/hlthaff.2017.1624

Gunlicks-Stoessel, M., Mufson, L., Cullen, K. R., & Klimes-Dougan, B. (2013). A pilot study of depressed adolescents' cortisol patterns during parent-adolescent conflict and response to interpersonal psychotherapy (IPT-A). *Journal of Affective Disorders, 150*(3), 1125–1128. https://doi.org/10.1016/j.jad.2013.05.037

Hellhammer, D., Meinlschmidt, G., & Pruessner, J. C. (2018). Conceptual endophentoypes: A strategy to advance the impact of psychoneuroendocrinology in precision medicine. *Psychoneuroendocrinology, 89*, 147.

Insel, T. R. (2014). The NIMH Research Domain Criteria (RDoC) project: Precision medicine for psychiatry. *American Journal of Psychiatry, 171*(4), 395–397. https://doi.org/10.1176/appi.ajp.2014.14020138

Jameson, J. L., & Longo, D. L. (2015). Precision medicine–personalized, problematic, and promising. *The New England Journal of Medicine, 372*(23), 2229.

Jatoi, I., Chen, B. E., Anderson, W. F., & Rosenberg, P. S. (2007). Breast cancer mortality trends in the United States according to estrogen receptor status and age at diagnosis. *Journal of Clinical Oncology, 25*(13), 1683–1690. https://doi.org/10.1200/JCO.2006.09.2106

Karch, C. M., & Goate, A. M. (2015). Alzheimer's disease risk genes and mechanisms of disease pathogenesis. *Biological Psychiatry, 77*(1), 43–51. https://doi.org/10.1016/j.biopsych.2014.05.006

Khoury, M. J., Iademarco, M. F., & Riley, W. T. (2016). Precision public health for the era of precision medicine. *American Journal of Preventive Medicine, 50*(3), 398–401. https://doi.org/10.1016/j.amepre.2015.08.031

Kuhlman, K. R., Irwin, M. R., Ganz, P. A., Crespi, C. M., Petersen, L., Asher, A., & Bower, J. E. (2017). Cortisol awakening response as a prospective risk factor for depressive symptoms in women after treatment for breast cancer. *Psychosomatic Medicine, 79*(7), 763–769. https://doi.org/10.1097/PSY.0000000000000499

Kuhlman, K. R., Robles, T. F., Dickenson, L., Reynolds, B., & Repetti, R. L. (2019). Stability of diurnal cortisol measures across days, weeks, and years during middle childhood and early adolescence: Exploring the role of age, pubertal development, and sex. *Psychoneuroendocrinology, 100*, 67–74. https://doi.org/10.1016/j.psyneuen.2018.09.033

Lenze, E. J., Hershey, T., Newcomer, J. W., Karp, J. F., Blumberger, D., Anger, J., … Dixon, D. (2014). Antiglucocorticoid therapy for older adults with anxiety and co-occurring cognitive dysfunction: Results from a pilot study with mifepristone. *International Journal of Geriatric Psychiatry, 29*(9), 962–969. https://doi.org/10.1002/gps.4085

Liu, S., Rovine, M. J., Klein, L. C., & Almeida, D. M. (2013). Synchrony of diurnal cortisol pattern in couples. *Journal of Family Psychology: JFP: Journal of the Division of Family Psychology of the American Psychological Association (Division 43), 27*(4), 579.

Mehta, S., Shelling, A., Muthukaruppan, A., Lasham, A., Blenkiron, C., Laking, G., & Print, C. (2010). Predictive and prognostic molecular markers for cancer medicine. *Therapeutic Advances in Medical Oncology, 2*(2), 125–148. https://doi.org/10.1177/1758834009360519

Metcalfe, K. A., Birenbaum-Carmeli, D., Lubinski, J., Gronwald, J., Lynch, H., Moller, P., … Friedman, E. (2008). International variation in rates of uptake of preventive options in BRCA1 and BRCA2 mutation carriers. *International Journal of Cancer, 122*(9), 2017.

Nelson, H. D., Pappas, M., Zakher, B., Mitchell, J. P., Okinaka-Hu, L., & Fu, R. (2014). Risk assessment, genetic counseling, and genetic testing for BRCA-related cancer in women: A systematic review to update the U.S. preventive services task force recommendation. *Annals of Internal Medicine, 160*(4), 255–266. https://doi.org/10.7326/M13-1684

Obama, B. (2015). State of the Union 2015: Full Transcript. *CNN.*

Ouellet-Morin, I., Danese, A., Williams, B., & Arseneault, L. (2011). Validation of a high-sensitivity assay for C-reactive protein in human saliva. *Brain, Behavior, and Immunity, 25* (4), 640.

Papiol, S., Arias, B., Gastó, C., Gutiérrez, B., Catalán, R., & Fañanás, L. (2007). Genetic variability at HPA axis in major depression and clinical response to antidepressant treatment. *Journal of Affective Disorders, 104*(1–3), 83–90. https://doi.org/10.1016/j.jad.2007.02.017

Prendes-Alvarez, S., & Nemeroff, C. B. (2018). Personalized medicine: Prediction of disease vulnerability in mood disorders. *Neuroscience Letters, 669*, 10–13. https://doi.org/10.1016/j.neulet.2016.09.049

Raison, C. L., Rutherford, R. E., Woolwine, B. J., Shuo, C., Schettler, P., Drake, D. F., ... Miller, A. H. (2013). A randomized controlled trial of the tumor necrosis factor-alpha antagonist infliximab in treatment resistant depression: Role of baseline inflammatory biomarkers. *JAMA Psychiatry, 70*(1), 31–41. https://doi.org/10.1001/2013.jamapsychiatry.4

Rapado-González, Ó., Majem, B., Muinelo-Romay, L., Álvarez-Castro, A., Santamaría, A., Gil-Moreno, A., ... Suárez-Cunqueiro, M. M. (2018). Human salivary microRNAs in cancer. *Journal of Cancer, 9*(4), 638–649. https://doi.org/10.7150/jca.21180

Rapado-González, Ó., Majem, B., Muinelo-Romay, L., López-López, R., & Suarez-Cunqueiro, M. M. (2016). Cancer salivary biomarkers for tumours distant to the oral cavity. *International Journal of Molecular Sciences, 17*(9). https://doi.org/10.3390/ijms17091531

Rauch, S. A. M., King, A. P., Liberzon, I., & Sripada, R. K. (2017). Changes in salivary cortisol during psychotherapy for posttraumatic stress disorder: A pilot study in 30 veterans. *The Journal of Clinical Psychiatry, 78*(5), 599–603. https://doi.org/10.4088/JCP.15m10596

Robinson, P. N. (2012). Deep phenotyping for precision medicine. *Human Mutation, 33*(5), 777–780. https://doi.org/10.1002/humu.22080

Ross, K. M., Murphy, M. L., Adam, E. K., Chen, E., & Miller, G. E. (2014). How stable are diurnal cortisol activity indices in healthy individuals? Evidence from three multi-wave studies. *Psychoneuroendocrinology, 39*, 184–193. https://doi.org/10.1016/j.psyneuen.2013.09.016

Slamon, D. J., Leyland-Jones, B., Shak, S., Fuchs, H., Paton, V., Bajamonde, A., ... Norton, L. (2001). Use of chemotherapy plus a monoclonal antibody against HER2 for metastatic breast cancer that overexpresses HER2. *New England Journal of Medicine, 344*(11), 783–792. https://doi.org/10.1056/NEJM200103153441101

Uhr, M., Tontsch, A., Namendorf, C., Ripke, S., Lucae, S., Ising, M., ... Holsboer, F. (2008). Polymorphisms in the drug transporter gene ABCB1 predict antidepressant treatment response in depression. *Neuron, 57*(2), 203–209. https://doi.org/10.1016/j.neuron.2007.11.017

Williams, S. R., Cash, E., Daup, M., Geronimi, E. M., Sephton, S. E., & Woodruff-Borden, J. (2013). Exploring patterns in cortisol synchrony among anxious and nonanxious mother and child dyads: A preliminary study. *Biological Psychology, 93*(2), 287.

Wishart, D. S. (2016). Emerging applications of metabolomics in drug discovery and precision medicine. *Nature Reviews Drug Discovery, 15*(7), 473–484. https://doi.org/10.1038/nrd.2016.32

Zarco, M. F., Vess, T. J., & Ginsburg, G. S. (2012). The oral microbiome in health and disease and the potential impact on personalized dental medicine. *Oral Diseases, 18*(2), 109–120. https://doi.org/10.1111/j.1601-0825.2011.01851.x

Zhang, L., Huang, J., Yang, N., Greshock, J., Megraw, M. S., Giannakakis, A., ... Coukos, G. (2006). microRNAs exhibit high frequency genomic alterations in human cancer. *Proceedings of the National Academy of Sciences of the United States of America, 103*(24), 9136–9141. https://doi.org/10.1073/pnas.0508889103

Zhang, A., Sun, H., & Wang, X. (2012). Saliva metabolomics opens door to biomarker discovery, disease diagnosis, and treatment. *Applied Biochemistry and Biotechnology, 168*(6), 1718–1727. https://doi.org/10.1007/s12010-012-9891-5

# Chapter 30
# Public Health and Industry Applications of Salivary Bioscience

**Jenna L. Riis, Janeta Nikolovski, Kathryn Dean Luedtke, and Diana H. Fishbein**

**Abstract** Building on decades of basic science, clinical, and laboratory-based salivary bioscience research, salivary biomeasures are now being used in real-world settings to study population health, evaluate programs and policies, and develop and test the effectiveness of consumer products. In this chapter, we discuss the opportunities and challenges of taking salivary bioscience research outside the academic and clinical laboratory and into applied and translational research. The current state of salivary bioscience in large-scale, field-based, program evaluation and implementation, and industry research will be presented. The methodological, ethical, and communication challenges associated with salivary bioscience in public health and industry research and practice will be discussed, along with strategies for mitigating these obstacles. Finally, we highlight the unique potential of salivary bioscience to bring together interdisciplinary research teams, advance our understanding of population health, and inform products, policies, and programs that support health and wellness across the lifespan.

J. L. Riis (✉)
Department of Psychological Science, School of Social Ecology, University of California, Irvine, Irvine, CA, USA

Institute for Interdisciplinary Salivary Bioscience Research, University of California, Irvine, Irvine, CA, USA
e-mail: jriis@uci.edu

J. Nikolovski
Janssen Pharmaceutical Companies of Johnson and Johnson, New Brunswick, NJ, USA

K. D. Luedtke
Johnson & Johnson Consumer Companies, Inc., Skillman, NJ, USA

D. H. Fishbein
Department of Human Development and Family Studies, Program for Translational Research on Adversity and Neurodevelopment, The Pennsylvania State University, State College, PA, USA

National Prevention Science Coalition to Improve Lives, University Park, PA, USA



D. A. Granger, M. K. Taylor (eds.), *Salivary Bioscience*,
https://doi.org/10.1007/978-3-030-35784-9_30

## 30.1 Introduction

Decades of salivary bioscience research have enriched the social, developmental, behavioral, and health sciences by enabling the measurement of physiologic function in clinical and laboratory settings, as well as in the context of patients' and participants' everyday social worlds. Recent advancements in assay technology and in our understanding of salivary biomeasures have now helped push the field beyond the study of individual health in a laboratory or clinical setting to studying communities with implications for population-level health. The application of salivary bioscience is also rapidly expanding in industry and commodities research with the objective of providing consumers with ready access to health information. In response to these developments, the US National Institutes of Health is increasingly prioritizing the integration of biomedical techniques, such as those extending from salivary bioscience, into a wide range of clinical, social, and intervention sciences in both intramural research and extramural funding streams. At the same time, federal agencies and ethical review committees are working to address the rapidly evolving landscape of privacy and consumer protection issues arising from the scaling-up and translation of salivary bioscience research.

This chapter provides an overview of the field of salivary bioscience outside the academic research lab. We discuss ways in which salivary bioscience can: (a) meet a range of research objectives in a variety of real-world settings (e.g., community and industry settings); (b) be employed in corporate product research and personal use; (c) inform the decision-making of policymakers and agency leaders; and (d) achieve the potential for large-scale applications that support health on a population level. Distinctive opportunities and challenges facing researchers using salivary biomeasures in public health, policy, and industry settings will be considered. Through the discussion of case studies and findings from researchers working in these fields, we highlight key barriers and explore potential solutions and strategies for the successful integration of salivary biomeasures into real-world contexts, as well as the responsible dissemination of findings from these studies.

## 30.2 The Current State of Salivary Bioscience

### *30.2.1 Large-Scale Studies and Surveys: The Utility of Salivary Bioscience in a Public Health Arena*

Several large-scale, public health studies include assessments of salivary biomeasures and are currently in the field in the USA and abroad. As a minimally invasive, easy-to-collect, and low-cost method for obtaining biologic data, salivary biomeasures are well suited for studies of this scale. In addition, saliva collection is generally more accepted by participants across age and sociocultural groups and can therefore help increase participation in biologic data collection protocols on a large-scale. These studies generally pair salivary biomeasure data, such as viral infection status (e.g., HIV); hormonal activity levels (e.g., sex hormone levels, cortisol diurnal slopes); genetic markers; and/or indices of environmental exposures (e.g., cotinine, a nicotine metabolite, to assess tobacco smoke exposure), with other biologic, self-report, and/or clinical measures to assess the health of their participants. The integration of salivary biomeasures as indicators of exposure to environmental or pathogenic antigens, rather than as indicators of complex physiologic processes, may be easier to implement and provide more interpretable study findings. Assessing exposure on a large-scale with saliva is more cost-effective and feasible compared to serum assessments. Salivary exposure data is also particularly useful when community-level information is of interest, since saliva samples can be collected from large groups of people at the same time and these samples can be assayed for a wide range of analytes.

Many of the public health studies assessing salivary analytes have been in the field for several years. Readers are encouraged to review the research procedures and results from studies such as the National Longitudinal Study of Adolescent Health, the Los Angeles Family and Neighborhood Survey, the Chinese Longitudinal Healthy Longevity Survey, and the Adolescent Brain Cognitive Development study (Duke Center for the Study of Aging and Human Development, n.d.; RAND, 2019; Services, 2019), which have successfully integrated salivary biomeasures into their protocols. Researchers have also examined best practices for salivary bioscience methods in large-scale, public health studies, and readers are encouraged to review the paper by Adams and colleagues (Adam & Kumari, 2009) for methodological guidelines for conducting these studies.

Despite multiple long-running, large-scale studies that assess salivary biomeasures and disseminate these findings to the scientific community, we still lack population-level data for the majority of salivary biomeasures. Pooling information from across existing studies may allow us to establish population norms and reference ranges by age, sex, and race/ethnicity. This is an exciting area of opportunity for large-scale salivary bioscience researchers. In addition to advancing salivary bioscience as a field, reference ranges would allow laboratory and clinic-based researchers, as well as health professionals, to interpret individual-level salivary bioscience findings and provide individualized treatment. Salivary analyte norms would also allow intervention researchers to accurately assess program/policy impact at a physiologic level and establish valid, clinically significant programmatic goals.

### *30.2.2 Evaluation and Implementation Science: Individual- and Group-Level Information in Saliva*

Salivary bioscience methods are well suited for integration with evaluation and implementation studies examining environmental impacts on health and how changes in the environment (e.g., via intervention) may improve health. In the evaluation and implementation research context, salivary biomeasures are particularly useful in their ability to help investigators: (1) track exposures and design effective interventions to reduce exposure risk; (2) evaluate the process of intervention implementation and fidelity; and (3) elucidate the mechanisms by which interventions are associated with improved outcomes. These studies may focus on the individual- or group-level, and may be longitudinal, quasi-experimental, or tightly controlled randomized control trials. Each use of salivary biomeasures has direct implications for understanding how interventions work, for whom (moderating conditions), under what circumstances (contextual and experiential conditions), and why (mechanistic effects).

First, salivary biomeasures have clear applications as indices of environmental exposures such as toxic heavy metals and tobacco smoke. With relatively simple salivary collection, storage, and assay protocols for many environmental exposure analytes (see Chap. 15), salivary biomeasures can be used to assess levels of exposure before, during, and after intervention efforts, and exposures can be examined at the individual, family, and community-level. Importantly, salivary biomeasures of exposure are relatively inexpensive and easy to collect, store, and assay; this means that multiple samples can be assessed from individuals over time. The regular, repeated, and frequent monitoring afforded by salivary testing is particularly useful for assessing toxic exposures of uncertain origins and those with short half-lives that may be missed if examined with a single measurement using another more invasive, biospecimen.

Just as exogenous salivary analytes are used to index environmental exposures, several studies have worked to identify patterns of endogenous salivary analytes that are associated with psychosocial exposures, such as stress and adversity. The ability of researchers and clinicians to track change in context-relevant analytes to determine whether psychosocial exposure has impacted an individual at a biologic level is an interesting and important area of research. For example, salivary analytes may be reflective of the extent to which psychosocial trauma has altered an individual's hypothalamic–pituitary–adrenal (HPA) axis functioning. Differential patterns of change in salivary analytes across individuals in response to psychosocial exposures may help us to characterize heterogeneous responses to experiences. Not all individuals develop pathology in response to adversity, and resilience is not well understood. Thus, it is critical that we understand the biological basis for differential responsivity if we are to more effectively and efficiently identify individuals at risk of poor outcomes. This information is important for designing effective interventions as well as tracking exposures and health outcomes before, during, and after intervention.

Second, as early indicators of biologic change in response to experiences and exposures—including therapeutic and experimental exposures—salivary biomeasures may also provide important information regarding program implementation and fidelity. In large-scale randomized trials, when individual differences in responsivity to treatment/intervention are assumed to be random across intervention groups, salivary biomeasures may provide insight into the effects of implementation differences across groups or study sites. These data would complement traditional fidelity and process measures and allow for a more nuanced understanding of the impact deviation from implementation protocols has on the mechanisms underlying the desired health outcome. Salivary biomarkers can also help us validate self-report data collected during the implementation and evaluation process and assess the role of potential unintended biases in our data. For example, an evaluation of the truth® anti-smoking campaign used salivary cotinine to examine potential under-reporting of smoking behaviors due to campaign-related increases in social desirability bias (Messeri et al., 2007). The investigators found no evidence of under-reporting in the self-report data, supporting the continuation of their campaign and evaluation efforts (Messeri et al., 2007). Stagnant or unexpected change in salivary biomeasure concentrations in response to an intervention may either support the adaption of implementation protocols to fit the local culture and target population or provide early warning signs to program administrators that intervention efforts are not penetrating at the desired level or intensity. In both cases, assessing salivary biomeasures would allow for early and potentially ongoing feedback regarding program effectiveness and provide opportunities to improve the program. Even the most efficacious interventions do not benefit many recipients. Using salivary biomeasures to test whether an intervention has led to improvements in biological processes linked to favorable outcomes would help us more effectively and efficiently target at-risk individuals and disrupt pathways to negative outcomes. Salivary analytes have the potential to be biomarkers of diseases or disorders, as well as states of well-being and health. As such, salivary biomeasures can be quite valuable in informing intervention strategies that are likely to exert the greatest effect on the largest proportion of the target population.

Finally, the application of salivary biomeasures in evaluation research can help us to elucidate the physiologic mechanisms underlying changes associated with specific intervention approaches. One example of successful integration of salivary biomeasures in this field comes from studies of child maltreatment. Based on extensive research that enabled the characterization of healthy diurnal patterns of cortisol, investigators examined dysregulation in diurnal salivary cortisol patterns among children in foster care to assess the health consequences of early life instability and maltreatment (Van Andel, Jansen, Grietens, Knorth, & Van Der Gaag, 2014). Alterations in child salivary cortisol diurnal patterns were also used to assess the impact of caregiver-based interventions in supporting healthy child development during the foster care process (Van Andel et al., 2014). A normalization of cortisol's diurnal patterns correlated with behavioral improvements (Van Andel et al., 2014), further supporting the clinical significance of intervention-associated changes in salivary cortisol. While this example focuses on salivary cortisol, the number and type of measurable salivary analytes has increased in recent years and will continue to grow. This growth will greatly expand our ability to assess and track community health and changes in health from a multi- and cross-systems perspective.

### 30.2.3 Role of Salivary Bioscience in Advancing Industry Research

As minimally invasive indices of physiologic responses to therapeutic intervention, salivary biomeasures are also of great utility in corporate product development and effectiveness research. Consumer products companies often look for methods to evaluate the benefits of their products when used by consumers. For products that offer a more psychological benefit, one would typically use a questionnaire or survey tool to understand a consumer's perception of the usage experience. Survey and self-perception data are certainly useful, but oftentimes a more objective understanding of the customer's product experience is desired. Biometric tests of emotion and psychological state are possible using serum and urine. However, these methods have limited to no utility for consumer purposes as the needle stick required for blood sampling and the hassle of urine collection are inherently stressful, thus negating or invalidating the emotional benefit offered by the product experience. For industry researchers, salivary bioscience may represent an ideal approach for examining the biologic processes accompanying a consumer's product experience (e.g., reduced stress response and increased relaxation). For example, researchers in the Medical Device Division at Johnson & Johnson Consumer, Inc. have used salivary biomeasures like cortisol to help understand how surgical tool features influence the surgeon's stress level (Sutton et al., 2018).

Beyond Johnson & Johnson, many other companies have benefited by using salivary analytes to assess the efficacy, effectiveness, and impact of their products. For example, fragrance houses such as Takasago have used salivary cortisol and alpha-amylase to discover fragrance ingredients or combinations of ingredients that offer relaxing benefits (Kang & Cilia, 2017). Other personal care companies, like Shiseido, have also used salivary analytes for their research on the effects of stress on beauty ("Shiseido Conducts Joint Research on Stress in Closed-off Environment Simulating the International Space Station (ISS)", 2018). Nutritional supplement companies and sleep researchers have found markers such as salivary melatonin useful in investigating the effectiveness of supplements and other sleep products (Van Der Lely et al., 2015). Salivary biomeasures are also often used to evaluate the local oral health effects of consumer products, such as testing the impact of oral care products on salivary markers of inflammation like salivary secretory IgA. Further, industry researchers may perform chemical analysis of saliva to evaluate the presence and/or levels of product ingredients within saliva, such as triclosan, from certain personal care products.

In addition to product development and testing research, salivary biomeasures are increasingly being integrated into products that help consumers track their own health and behavior (i.e., the "quantified self" movement). These direct-to-consumer (DTC) product applications of salivary bioscience have flourished over the recent years. Dominated by consumer genetic testing kits that use a saliva-based sample, "quantified self" products have moved from niche to mainstream use. According to a report from Global Market Insights, Inc., the consumer market for genetic testing is projected to exceed USD 2.5 billion in the next 5 years (Global Market Insights, 2018). Movements like the Quantified Self (Quantified Self, 2019), exemplify the motivation for collecting data on oneself and the interest in taking control of one's health.

Saliva testing is amenable to a variety of consumer applications because of its minimally invasive nature, easy sample collection protocols, relative sample stability, and the opportunities it affords to assess a robust number of biomeasures in a single sample. As mentioned above, the popularity of DTC genetic testing kits using a saliva-based sample has increased considerably over recent years. Companies such as 23andMe (23andMe INC, 2019) and ORIG3N (Orig3n DNA Tests, 2019) offer individuals an analysis of their DNA and appeal to consumers' desire to discover aspects about their health, well-being, and ancestry. These products have also extended into the personalized medicine movement. In 2018, the US Food and Drug Administration (FDA) approved the reporting of genetic variants related to drug metabolism in 23andMe's consumer DNA reports (US Food and Drug Administration, 2018a). In their statement, the FDA acknowledges the medical potential of such reports and recommends "Any medical decisions should be made only after discussing the results with a licensed health care provider and results have been confirmed using clinical pharmacogenetic testing." (US Food and Drug Administration, 2018a).

Point-of-Care testing (POCT), also referred to as a liquid biopsy (Khan, Khurshid, & Yahya Ibrahim Asiri, 2017), is another emerging area at the intersection of salivary bioscience and medicine. Offering an alternative to traditional central laboratory testing, POCT using saliva could be a substitute for blood in the diagnosis of oral and systemic diseases (Khan et al., 2017). In the 1990s, Epitope, Inc. and Saliva Diagnostic Systems, Inc. were early leaders in oral diagnostics (Goswami, Mishra, Agrawal, & Agrawal, 2015). The US FDA approved Epitope's Orasure HIV test in 1996, which detected HIV antibodies in saliva using a mouth swab (Goswami et al., 2015). Over the years, more commercially available saliva tests have become available. In addition to DNA, POCT companies also offer tests for bacteria, enzymes, antigens, and specific metabolites. Several commercial salivary tests for drugs of abuse are available from companies like Cozart Biosciences, Securetec, and Mavand (Goswami et al., 2015). POCT of salivary biomarkers for cancer is also an active area of research (see Chaps. 8 and 19).

## 30.3 Methodological Challenges and Solutions

While integrating salivary bioscience into public health, program, and industry research provides opportunities to track and advance health and well-being on a large-scale, it also presents unique difficulties. Public health and industry researchers may face common challenges, including the need to adapt laboratory and clinic-based research protocols to real-world settings, manage a variety of stakeholders with potentially competing interests, and convey research findings to non-scientific

**Table 30.1** Considerations for incorporating salivary biomeasures in public health, program, and industry studies

| | |
|---|---|
| Conceptual | Which biomeasure(s) best reflect(s) the biologic processes of interest? |
| | Is there an objective measure of health? |
| | How are biomeasure(s) conceptualized (e.g., surrogate disease endpoint; exposure level or status)? |
| Logistical | What sample size is needed to power the salivary biomeasure research question(s)?<br>• Are subgroup analyses needed/planned (e.g., by sex)?<br>• What difference or change in biomeasure concentration is considered significant (both statistically and clinically/biologically)?<br>• What is the expected missingness and noncompliance rate?<br>• Will the scale or setting of the study affect compliance with salivary protocols? |
| | How many biomeasures will be assessed?<br>• How to maximize opportunities for future salivary bioscience research questions while balancing participant burden, costs, and ethical considerations? |
| | What additional information is needed to interpret salivary biomeasure data (e.g., time of day; physical/oral health of participants; presence and/or concentrations of other analytes)? |
| | What cold chain protocols are needed for the analyte(s) being studied, and can these be achieved in a non-laboratory setting? |
| | How does the analyte being studied vary by saliva collection protocols (e.g., collection technique; storage; time of day)? |
| | How to train participants to collect, store, and transport saliva appropriately? |
| Financial | Added cost of saliva collection, shipping, and assay materials |
| | Added cost and time for training research staff and participants |
| | Added cost of survey questions specific to saliva protocol and analytes |
| Ethical | How to ensure informed consent in large-scale studies with biospecimens? |
| | Will saliva be stored for future research? |
| | What types of analytes will be assessed in current and future studies (e.g., drug tests, genes, and hormones)? |
| | How will confidentiality and privacy of biologic data be ensured? |
| | Will the results of saliva analyses be shared with participants?<br>• For which analytes?<br>• Under what circumstances? |
| Communication | How to clearly present salivary bioscience findings to non-scientific audiences? |
| | How to responsibly communicate study findings and implications to the public/consumers, and policy and industry leaders? |

audiences. In addition to methodological and logistical hurdles, there are also ethical considerations that emerge as salivary bioscience research is translated from the laboratory to real-world research environments. Many of these issues are presented in Table 30.1, and a selection of important considerations is discussed in more detail in this section. We recommend reviewing Chap. 3 for in-depth discussions of salivary biomeasure protocols, as well as the relevant scientific literature for analyte-specific procedures (e.g., Adam & Kumari, 2009).

## 30.3.1 Conceptual Considerations

### 30.3.1.1 Biomeasure Selection and Conceptualization

Program implementation and evaluation studies, as well as consumer product development and testing research, may conceptualize salivary biomeasures as proxy endpoints for intervention outcomes (e.g., using salivary cortisol to index changes in HPA function in response to a trauma intervention or a stress-relieving product). These studies employ salivary markers as surrogate endpoints for health and diseased states and require a high level of understanding of the biologic processes underlying both the target outcome as well as the mechanisms linking the intervention to the desired outcome. While attractive to researchers, policymakers, and consumers alike, the role of biomeasures as surrogate endpoints of health and disease relies on the strength and consistency of the biomeasure findings across diverse populations and settings. For many analytes and health conditions, we currently do not have this level of understanding.

Additional research, including investigations of analyte validity and reliability and studies conducted in healthy and clinical populations, is needed for the responsible integration of salivary biomeasures into large-scale, public-oriented studies and products that conceptualize salivary analytes as surrogate endpoints for health and disease. Validation studies and population-level research with salivary biomeasures can run parallel to cutting-edge applications of salivary biomeasures in public health, intervention, and consumer research; however, the interpretation of these findings should reflect the current state of knowledge regarding the specific analyte being studied and its assessment in the population examined. For example, the number of research studies using salivary immune markers has increased rapidly in recent years. Many of these markers, however, have been shown to index oral, rather than systemic, immune processes (see Chap. 9). While validation and reliability studies are still needed, current studies of salivary immune markers should restrict the interpretation of their findings to local oral immune function as additional research is needed to develop a salivary biomeasure of systemic immune system functioning.

If salivary biomeasures are used as indices of exposure (e.g., exposure to environmental tobacco smoke or to specific components of a consumer product), however, a lower level of understanding of the analyte may be required. For these studies, information about analyte stability in saliva is crucial in order to design appropriate sample collection, storage, and transportation protocols (discussed below). However, a complete understanding of analyte serum-saliva associations may not be needed if simple exposure status and/or level is the key area of interest. Investigators should, however, consider all possible sources of exposure when designing their research studies and selecting their measures. For example, researchers using salivary cotinine to examine the effectiveness of a smoking cessation program found a 36% discordance rate between self-report and biologic indices of smoking behavior (Cha, Ganz, Cohn, Ehlke, & Graham, 2017). This discordance was largely related to the use of nicotine replacement products and e-cigarettes among study participants (Cha et al., 2017), suggesting that salivary cotinine lacked the specificity needed to assess program success in this study.

These conceptual issues should be addressed during the research question and study design process. A thorough understanding of the biologic processes and salivary analytes of interest will help guide appropriate study protocols, sampling schemas, and analyses. Detailed information about analyte measurement validity and reliability, stability in saliva, and associations with systemic physiology can be challenging to find in the published literature. Researchers are encouraged to supplement information from the scientific literature with discussions with their assay manufacturers and laboratory staff. Addressing these questions early in the research process will generate studies that yield the highest quality data.

## *30.3.2 Logistical Challenges*

### 30.3.2.1 Sample Collection, Handling, Transport, and Storage

Systematic variation in biomeasure determinations can result from variation in collection, handling, transport, and storage techniques. Unlike in precision medicine (Chap. 29), where there can be strict control over the collection and storage of saliva specimens, integrating salivary biomeasures into large-scale and field-based research requires additional considerations. These methodologic challenges include addressing questions regarding the sensitivity of biomeasures to temperature, and the feasibility of maintaining the cold chain during sample collection, handling, and transport. In addition, researchers should consider what type of sampling is needed to accurately assess physiologic functioning, and if the required sampling procedures are feasible in large-scale and field-based studies. Questions regarding analyte sensitivity to sample collection method (e.g., swab versus passive drool); stability across time (e.g., day, week, and season); and reliability and interpretability (e.g., number of assessments needed, and interactions with other analytes) should be considered when designing research questions and protocols.

For example, salivary cortisol, the most studied salivary marker, is highly variable acutely, diurnally, and over time. In most cases, a single measurement of salivary cortisol will not provide valid information regarding individual HPA functioning nor overall health. In addition to sensitivity to diurnal rhythms, salivary analytes may be impacted by the biochemical or immune environment of the oral cavity. Adiponectin, for example, may be highly sensitive to oral inflammation (Riis et al., 2017). Therefore, isolated assessments of adiponectin in saliva are likely confounded by oral health and hygiene practices. Assessing and statistically controlling for oral inflammation may improve the interpretability of salivary adiponectin, and many other analyte, concentrations (Riis et al., 2017).

## Consumer and Product Research Case Studies

Here, we offer two case studies involving the assessment of salivary cortisol that illustrate the challenges of collecting and handling saliva from special populations and across long periods of time. Researchers at Johnson & Johnson have used salivary cortisol as a biomarker of acute and chronic stress in many of their consumer and clinical studies. For studies assessing salivary cortisol, there are many saliva collection techniques that can be used in the laboratory and home setting. Johnson & Johnson researchers have experimented with absorbant swabs, passive drool, and a swab with string attached for saliva collection from babies. Overall, participants in these research studies have been open to all these methods. For research with adults and older children, Johnson & Johnson researchers, and others, tend to prefer passive drool collection in a plastic vial as it is the simplest collection method to explain to consumers and helps ensure that sufficient sample volume is collected. For babies, long absorbant rolls (like those used in dentistry) provide the safest and easiest collection process. One end is put gently in a baby's mouth while a caregiver holds the other end. Ideally, the caregiver holds the baby during sample collection to avoid added separation anxiety. Once the sample collection is complete, the researcher cuts off the end of the swab containing saliva and places it in a plastic vial for centrifugation and processing.

### *Short-Term Changes in Salivary Cortisol*

Much of the product benefit research conducted at Johnson & Johnson has focused on understanding the immediate, cortisol-reducing effects of relaxing product experiences—such as deep breathing exercises paired with fragrances, or warm baths using relaxing bath products. Figure 30.1 illustrates the findings from one of these studies, and below we provide a few lessons shared by members of the Johnson & Johnson research team for conducting studies of short-term changes in salivary cortisol.

- Conduct short-term salivary cortisol studies in the afternoon when the diurnal slope of cortisol change is the flattest.
- Include a placebo and/or baseline assessment to compare the effect of your intervention on cortisol levels to the natural decline in cortisol levels during a similar time period.
- Consider a stress challenge study design such as the Trier Social Stress Test or Takasago's variation on this approach (Kang & Cilia, 2017; Kirschbaum, Pirke, & Hellhammer, 2008). Understanding the effectiveness of your intervention in dampening the typical increase in cortisol during these types of standardized stress-inducing activities is as, if not more, beneficial than examining the relaxation effect of your intervention or product.

- Understand the time course of the physiologic effect(s) you aim to study. For example, to examine acute stress-related changes in salivary cortisol, a typical saliva sample collection protocol includes, at least, a sample collected immediately before an intervention, a sample collected immediately after an intervention, and a sample collected approximately 30 min later. These time points can be synced with psychological questionnaires to assess self-reported mood.

*Chronic Stress and Salivary Cortisol*

Salivary cortisol has also been useful in Johnson & Johnson's consumer research focused on chronic stress management. When assessing chronic stress, rather than focus on cortisol changes across the day, researchers have investigated changes in overall HPA axis activity across longer periods of time. For example, at Johnson & Johnson, researchers interested in chronic stress and HPA activity have examined salivary cortisol's diurnal rhythm and overall concentrations across the day, as well as changes in daily HPA activity patterns over time. Conducting longer term chronic

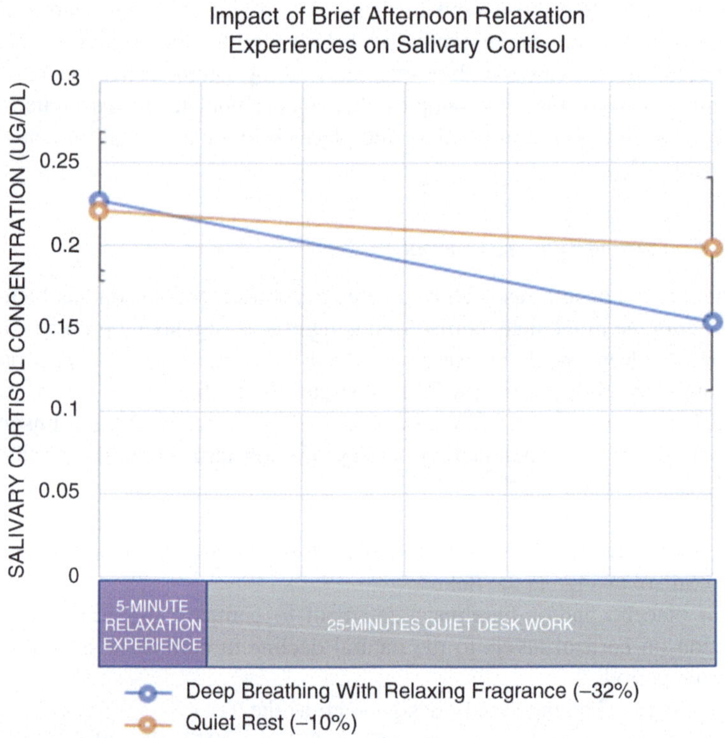


The participants assigned to Deep Breathing with a Relaxing Fragrance showed a significant decline in salivary cortisol compared to participants assigned to a Quiet Rest control group ($p < .05$ 30 min post-task)

**Fig. 30.1** Response of salivary cortisol to a five-minute relaxation task

stress research is significantly more challenging than acute stress studies because it requires asking participants to provide multiple saliva samples per day, for multiple days spread throughout the course of the study. To facilitate this, Johnson & Johnson researchers provide participants with detailed instruction cards, digital timers for sample collections (particularly for those collected 30 min after wakening and after lunch), and lunch boxes with freezer packs for temporary sample storage. They also use detailed labeling schemes to promote participant compliance with study protocols. In several of their studies, participants were required to freeze their own samples and return them to a central study center at different points throughout the study. While this complexity can be a cause of worry given the lack of researcher control over sample collection, handling, and transport, the level of concern varies by analyte. In studies where researchers are interested in relatively stable analytes (e.g., cortisol), this lack of control is less of a concern than when the focus is on highly temperature-sensitive analytes, such as oxytocin.

Logistical considerations regarding the collection, handling, transport, and storage of salivary biospecimens must be addressed for each analyte included in the study plan and be broadly considered if saliva samples are to be archived and used for future testing of planned, as well as new, yet-to-be discovered, biomeasures. For samples that will be stored for future studies, sample collection, handling, and storage procedures that maximize the ability to assay multiple salivary analytes need to be balanced with participant burden and ethical considerations. For example, while passive drool generally provides the most flexibility in our ability to accurately assay salivary biomeasures, asking participants to donate large volumes of saliva via this collection method may require too much time and be overly burdensome. Also, studies planning to archive saliva samples for future tests need to include this in study consent forms, and researchers should consider how this language may impact participation rates.

#### 30.3.2.2 Special Considerations for Environmental Exposures Research

In addition to the collection and storage considerations common to all analytes, there are special considerations for analytes indexing environmental exposures. First, the stability of the analyte in saliva (i.e., how long after exposure is it detectable in saliva) is important for designing appropriate sampling protocols. It is also important to understand factors that may impact analyte stability and metabolism, such as sample temperature and bacterial load, as these should be controlled and/or assessed during sample collection and handling procedures to maximize the validity of analyte measurements. Finally, understanding the analyte's mode of transmission between saliva and the blood and/or brain is needed to determine biologically significant levels. If exposure status is the only area of interest, this may not be needed. However, not understanding these processes may limit the clinical significance of the study findings.

#### 30.3.2.3 Special Considerations for Large-Scale and Longitudinal Population Studies

Public health surveillance and survey research that incorporates salivary biomeasures faces additional challenges due to the sheer scale of the studies, the thousands of data points that are often captured, the diversity within the populations studied, and the uncontrollable variability in the phenomena under study. Large-scale studies with dispersed and interdisciplinary research teams may be more vulnerable to issues of noncompliance than smaller studies with high levels of oversight of the investigators, data collectors, and study participants. Moreover, the response rate for returning saliva samples in large-scale studies range widely, from about 25% to greater than 90%, with an average of about 80% (Adam & Kumari, 2009). These sources of missingness should be anticipated and considered in the initial study design.

To minimize missingness researchers should invest in understanding and addressing participant concerns regarding saliva testing, developing easy-to-follow saliva collection protocols, and creating the infrastructure for follow-up appointments or testing in cases where rapid retesting is possible. Subgroup comparisons should also be planned and incorporated into the sampling design, particularly for analytes with known differences across sociodemographic characteristics or geographic region (e.g., sex, age, socioeconomic status, and urban versus rural). The analyses can then be segmented to determine whether background variables play a role in missingness (e.g., missingness related to cultural or ethnic differences in amenability to biologic testing), as well as "real" subgroup differences that may help to explain heterogeneity in health outcomes. Finally, for large-scale longitudinal studies, careful documentation of sample and assay protocols is important as advancements in assay technology and changes to or errors in sample handling can complicate cross-year and within-person comparisons of study findings.

### *30.3.3 Ethical Challenges*

Regulations are loosening to allow companies and researchers to both disseminate personal health data, as well as interpret these data for the public, such as informing individuals whether they are at risk for certain diseases. Balancing the individual's right to know about his/her own health with the state of the science and the analytical and clinical validity of the salivary tests has been a challenge for regulators and industry leaders. For example, there has been recent scrutiny of the terms of use and privacy practices of consumer genetic testing companies. The US Federal Trade Commission (FTC) posted a warning to consumers about DNA test kit companies' data sharing practices in 2017 (Fair, 2017). Unlike data collected in a medical setting, the information from DTC genetic tests is not subject to the Health Insurance Portability and Accountability Act (HIPAA) in the USA which protects patients' private medical and health information. Without such regulation, the DTC companies have wide-reaching freedoms in deciding their data privacy practices and agreements. For example, AncestryDNA's 1.5 million new users agreed to "grant Ancestry a sublicensable, worldwide, royalty-free license to host, store, copy, publish, distribute, provide access to, create derivative works of, and otherwise use such User Provided Content to the extent and in the form or context [AncestryDNA] deem[s] appropriate on or through any media or medium and with any technology or devices now known or hereafter developed or discovered." (Ancestry, 2019). Family Tree DNA's privacy policy informs users that if the company is sold, merged, or acquired, consumer "information will likely be among the assets transferred" (FamilyTreeDNA, 2019). As concerns regarding data sharing grow, there are opportunities for new companies to learn from previous successes and controversies. For example, Nebula Genomics is a new salivary DNA sequencing company that allows customers to anonymously share their data with research partners (Nebula Genomics, 2019). While shifting more control to the consumer, Nebula Genomics' privacy policy also states that consumers' personal/encrypted information will be included in the sale or transfer of the company (Nebula Genomics, 2019). With advances in DNA sequencing and encrypting technologies and the establishment of new companies and regulations, data security and privacy are critical considerations in the bioethics of salivary testing.

In addition to ethical considerations regarding the sharing of potentially sensitive and private information, there are also ethical concerns over the sharing of too little information with the public and/or consumer. With increasing knowledge of salivary biomeasures and their associations with health and disease risk, the integration of these measures into large-scale, public health and industry research will also likely increase. This presents the question of how much information about individual salivary biomeasure findings should be shared with participants/consumers and how to responsibly share this information with the public and/or consumer. In the USA, DTC tests that provide information for "moderate to high risk medical purposes" are reviewed by the FDA (US Food and Drug Administration, 2018b) to assess analytical and clinical validity as well as communication and marketing claims (US Food and Drug Administration, 2018b). The responsible and ethical sharing of important health-related information with study participants, consumers, and the public is an important area of evolving regulations and continued study and consideration that will only grow with advancements in the field of salivary bioscience.

### 30.3.4 Communication Challenges

The issue of information sharing is not only an ethical consideration, it also raises challenges regarding how best to communicate research findings with non-scientific audiences. Public health and industry researchers must discuss their research studies and findings with various stakeholders, including participants, the public and/or consumer, and industry and policy leaders. While this presents a unique challenge for researchers who may typically interact with academic colleagues, it is a necessary step to advance knowledge and health on a large-scale.

#### 30.3.4.1 Sharing Study Findings with the Public: Case Study from Johnson & Johnson—The Stress Thermometer

Over the course of conducting chronic stress and emotional benefit research using salivary cortisol, Johnson & Johnson researchers heard from their participants and consumers that they would like to understand their personal cortisol levels as a means of better understanding their current stress level and the effectiveness of their efforts to manage it. Johnson & Johnson developed a prototype chronic stress "thermometer" (Wiegand & McCulloch, 2002) which included four daily saliva collections as well as completion of the 10-item Perceived Stress Scale (Cohen, 1988; Cohen, Kamarck, & Mermelstein, 1983). Using data collected from hundreds of previous consumer study participants as well as the published literature, Johnson & Johnson researchers established high, medium, and low normative ranges for cortisol during key times in the diurnal rhythm (concentrations at waking, 30 min post-waking, after lunch, and before bedtime). Participants in the prototype research study were given a report that indicated how their cortisol rhythm compared to the general population (Fig. 30.2). The participant feedback was quite positive though it was noted that the concept of assessing a biomeasure's diurnal rhythm was too complex for many to understand.

Consumers craved a simple too high/too low measure as they are familiar with common well-being measures such as cholesterol and glucose. When asked about the value of the salivary biomeasure versus just a self-report questionnaire, consumers placed high value on seeing the biometric information, despite the complexity. As evident by the recent popularity of devices such as the Fitbit and the overall quantified self-movement, it is apparent that consumer interest in monitoring and understanding one's own biology is accelerating. Perhaps we will someday see an instant salivary biomarker self-monitoring device on the market that features consumer-relevant biomarkers such as salivary cortisol. We suspect that the key challenge will not be technological limitations, as sensor technology capabilities are rapidly expanding, but rather cracking the challenge of simplifying and synthesizing the meaning of concentrations of biomeasures with complex daily patterns and nuanced associations with health and well-being.

There are also communication challenges in the marketing of salivary testing products. US federal agencies have recently tightened controls on claims made by genetic testing companies regarding individual responses to specific drugs (CDRH's Division of Industry Communication and Education (DICE), 2019). Laboratory and DTC testing companies were found to be advertising medication recommendation information from their genetic tests that had not been approved by the US FDA (CDRH's Division of Industry Communication and Education (DICE), 2019). This

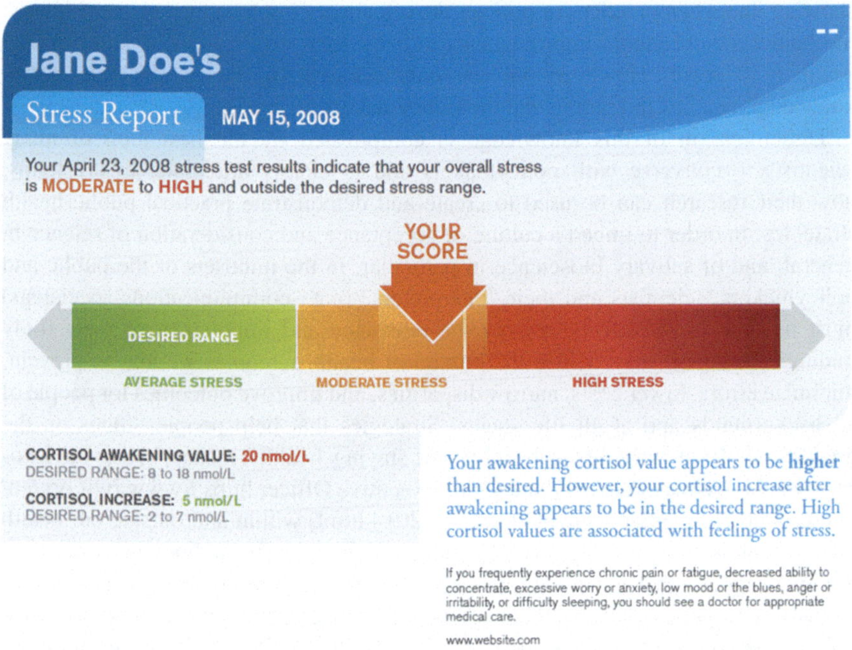


**Fig. 30.2** Example prototype cortisol report for consumers that was tested with research participants for feedback

is just one example of the challenges inherent in regulating the safety and marketing of saliva-based genetic tests; regulation processes must be dynamic, evolve with technological and scientific advancements, and be responsive to industry and consumer movements.

#### 30.3.4.2 "Making the Case for Salivary Biomarkers" to Policymakers

Although the utility of salivary bioscience is being increasingly demonstrated both in research and real-world settings, the study findings, and their public health implications, are virtually unknown to the public and private sectors of society. Salivary bioscience can help us understand how our homes, schools, workplaces, and communities get "under the skin" to influence health outcomes, and how we might remedy current public policy strategies to have positive impacts on a population level. As such, applications discussed in this chapter have the potential to inform the more effective and efficient design of the built environment and health care systems, as well as ensure that these designs target malleable biologic mechanisms of phenomenon we seek to either prevent (e.g., substance addiction) or strengthen (e.g., resilience to adversity). Continuing advances in salivary bioscience research and technology may further increase uptake of these programs and strategies by showing that when practices or policies are effectively implemented to alter biologic mechanisms, subsequent improvements in behavioral, psychological, and physical health may occur. However, we can only achieve this broad objective if this knowledge reaches the public and the policymakers.

Dissemination of this knowledge is complicated by the hesitation of many scientists to converse with policymakers and to clearly articulate, in lay terms, how their research can be used to create and demonstrate practical public health strategies. In order to imbed a culture of acceptance and consideration of science in general, and of salivary bioscience in particular, in the mindsets of the public and policymakers, scientists and their "interpreters" (e.g., communications specialists) must be able to effectively convey the relevance and importance of their study findings. Doing so may reduce the burden of health outcomes we aim to prevent, minimize errors, lower costs, narrow disparities, and improve outcomes for people of all backgrounds and at all life stages. Strategies that help people "focus on the grander whole of what being healthy and staying healthy means,"(Risa Lavizzo-Mourey, MD, MBA President and Chief Executive Officer https://www.rwjf.org/en/library/annual-reports/presidents-message-2014.html) within and outside our health care system, is an overriding goal. The burgeoning research and body of evidence in this area suggests that achieving this goal will require an empirically tested research-to-practice-to-policy protocol that experts, practitioners, and advocates across fields and sectors can use to increase public understanding of research, and in our case, salivary bioscience research in particular, and incentivize a public health system that supports it.

Constructing a research-to-practice-to-policy protocol to convey salivary bioscience findings to policymakers may involve several activities. First, scientists can work with communication and multimedia specialists to synthesize key messages from the science. This may include a series of statements that are written expressly and with specific legislative offices in mind; i.e., by first identifying current or proposed legislation, relevant caucuses and committees, and legislative office agendas, and then specifically targeting messages to provide evidence-based information to inform their decision-making. Imbedded in the messaging would be biosalivary evidence translated to relevant policy issues. Second, national framing summits for legislative staffers (who are responsible for providing their offices with background information on any given issue) and advocates can be held featuring health care experts, practitioners, and public health experts. These forums would enable attendees to learn about the science, its implications for policy and practice, and its limitations. A residual effect may be motivating additional funding for research that would further advance the field and its public health impacts. Additionally, conference leaders could export and promote new strategies across issue groups. Finally, a research-to-practice-to-policy protocol could include identifying and participating in ongoing conferences (e.g., national governor's associations), publications, and social media outlets that reach policy-making audiences. Additional resource materials could also be assembled and made widely available as a guide to understanding potential applications of the science and increasing its salience with policymakers who work on public health issues. These research-to-practice-to-policy efforts should be conducted in conjunction with individual research programs as bidirectional relations between research and practice promote more effective policies and programs as well as more informed and relevant research.

## 30.4 Future Directions and Opportunities

As salivary bioscience research and its application and translation to real-world issues grow, so too do the opportunities afforded by this unique field of study. Integrating a range of salivary biomeasures into large-scale, real-world, and industrial research will provide valuable population-level health data. These data will not only further the field of salivary bioscience, but they will also provide unique information that can be used to promote community and population well-being.

The integration of salivary bioscience research into epidemiologic studies provides the opportunity to track health on a large-scale. When salivary biomarkers of health and disease risk are clearly identified, and their collection, storage, and handling protocols empirically determined, salivary bioscience can provide a wealth of information about population health. Salivary bioscience has unique value in population health research because of the wide range of analytes that can be easily and simultaneously studied within a single, minimally invasive biologic sample. Compared to blood and urine sampling, saliva also offers a generally more accepted method of biospecimen collection across various cultures and age groups. One promising area of population health applications for salivary bioscience is the potential utility of salivary biomeasures to index individual infection and vaccination status (e.g., Heaney, Phillips, Carroll, & Drayson, 2018; Pisanic et al., 2017, 2018; see Chap. 13). Once fully scrutinized and confirmed, salivary bioscience could provide valuable, real-world, near-instant vaccination and infection risk information that could be used to strategically plan and implement targeted community health programs. Other biomarkers of exposure (e.g., bacterial, viral, environmental toxins) and health (e.g., immune function) at the population level would also help researchers and public health officials identify and target population-level health concerns and patterns in health and risk factors across geographic and sociodemographic boundaries.

Salivary biomeasure data collected in population surveys would add to traditional public health data derived from hospital admissions reports, health record information, and pen–paper, online, and telephone surveys. For example, population-level salivary data could provide local health departments and community health researchers with nuanced information about disease risk within a given population, including among those who do not have or utilize care from traditional health care professionals. By providing objective information about physiologic processes that may be related to a wide range of health conditions, these salivary data may also help researchers and health officials detect common pathophysiological mechanisms underlying health and disease within a community. These data could then be used to tailor and track the success of interventions that address the specific health needs of the community.

While still evolving, the integration of salivary biomeasures into public health, policy, and population-based research is not new, and researchers can adapt best practices from other fields to successfully incorporate salivary biomarkers into their studies. What is emerging, and is here today, is the public's individual interest in their own biomeasure levels, changes, and trends. We see this expanding beyond genetic testing to include assessments of other biomarkers of health and disease and with increasing integration of health information across various analytes and modalities (e.g., wearable sensors). Alongside these developments are the advancement of sophisticated analytics and machine learning technologies that guide the interpretation of the biometric results and its translation into meaningful conclusions or recommendations for the end user. Salivary bioscientists can be on the cutting edge of exciting research opportunities that capitalize on the growing public interest in biopsychosocial processes and health. By participating in this movement, salivary researchers can help ensure consumer products are borne from rigorous scientific research, developed to most effectively and ethically serve the public, and support individual- and population-level health.

Industrial advances in POCT for salivary biomeasures are also an area of rapid development and opportunity. POCT for drugs of abuse (e.g., cannabis, cocaine) is already developed and being used in the field. With cannabis legalization in the USA spreading, some states, such as California, are integrating roadside salivary "drugalysers" to their drugged driving enforcement efforts (Adams, 2017; Davis, 2017) (see Chap. 16 for a discussion of salivary bioscience and drugs of abuse). The use of salivary drug tests introduces several critical policy-related issues, such as the training of law enforcement officers in conducting the test and interpreting the results, privacy concerns over biospecimen collection, and drugged driving limits and penalties for each substance. As drug-testing technology advances and is more broadly adopted, these policy issues will likely gain more attention. Salivary bioscientists will be able to provide specialized knowledge regarding these tests and the interpretation of their results.

Other salivary POCT platforms suitable for the clinical or home setting, such as tests for cancer, and viral and environmental exposures, are also potential areas of rapid development (see Chaps. 8, 13, 15, and 19). A considerable amount of research is dedicated to discovering and validating salivary biomarkers for cancer. As this field advances, results will be translated into clinical and public health interventions, such as large-scale cancer screening programs that use salivary biomarkers to identify individuals in need of clinical and diagnostic testing (see Chaps. 8 and 19). Salivary POCT could also be developed to help address oral health problems (e.g., periodontal disease and oral cancers; see Chaps. 8 and 18 for a discussion of salivary biomarkers related to oral health). Poor oral health is a considerable public health problem in the USA. More than 40% of US adults have periodontitis, and there are vast disparities in the prevalence of oral health problems in the USA by race/ethnicity and socioeconomic status (Eke et al., 2018; U.S. Department of Health and Human Services, 2000). With access and availably barriers limiting oral and primary health care utilization in the USA (U.S. Department of Health and Human Services, n.d.-a, n.d.-b; Bersell, 2017), salivary POCT presents opportunities to expand disease screening for both physical and oral health problems in new settings (e.g., doctors' and dentists' offices, as well as in the community) and among individuals who might not otherwise be screened. From a health systems perspective, salivary POCT also provides opportunities to better integrate the primary and oral health care systems and move toward more patient-centered and comprehensive care (Atchison, Rozier, & Weintraub, 2018).

While some of the salivary point-of-care tests, such as Epitope's Orasure HIV test, have been widely adopted, the development of additional tests requires considerable investment with the prospect for future returns (e.g., from health insurance claims). Although rapid POCT has the potential to greatly expand public health efforts, upfront technological costs paired with concerns about the acceptance and adoption of these tests by the medical and public health systems are hurtles in their development. Salivary bioscientists can help advance POCT development by conducting research that aids in the identification of valid, reliable biomeasures of health, and sharing these findings with health policy, program, and medical professionals.

## 30.5 Concluding Thoughts

As salivary bioscience continues to advance as a field, the application of salivary biomeasures in interdisciplinary research, industry, and public health programs and policies will also grow. It is our responsibility to ensure that this growth and expansion of the field is done on a solid foundation of high-quality, ethically conducted research. Technological advances in assay technology have already outpaced our research efforts to fully validate all the analytes measurable in saliva. Basic methodologic studies that interrogate the validity, reliability, and utility of these analytes are needed to allow us to fully realize the great potential of salivary bioscience in advancing population health. While salivary bioscientists may come from a wide range of disciplines and sectors, they face common challenges and opportunities. As a research community linked together by salivary bioscience, we are poised to conduct the type of collaborative, innovative research needed to advance the field and support meaningful changes in health and wellness at the population level.

## References

23andMe INC. (2019). *23andMe* [WWW Document]. https://www.23andme.com/

Adam, E. K., & Kumari, M. (2009). Assessing salivary cortisol in large-scale, epidemiological research. *Psychoneuroendocrinology, 34*, 1423–1436. https://doi.org/10.1016/j.psyneuen.2009.06.011

Adams, L. (2017). Drug-driving limits and roadside testing to be introduced [WWW Document]. *BBC News*. https://www.bbc.com/news/uk-scotland-39655137

Ancestry. (2019). *Ancestry terms and conditions* [WWW Document]. https://www.ancestry.com/cs/legal/termsandconditions

Atchison, K., Rozier R, & Weintraub J (2018). *Integration of oral health and primary care: Communication, coordination, and referral*. NAM Perspect.

Bersell, C. (2017). Access to oral health care: A national crisis and call for reform. *Journal of Dental Hygiene, 91*, 2017.

CDRH's Division of Industry Communication and Education (DICE). (2019). *The FDA warns against the use of many genetic tests with unapproved claims to predict patient response to specific medications: FDA safety communication* [WWW Document]. US Food Drug Adm. https://www.fda.gov/MedicalDevices/Safety/AlertsandNotices/ucm624725.htm#actions

Cha, S., Ganz, O., Cohn, A. M., Ehlke, S. J., & Graham, A. L. (2017). Feasibility of biochemical verification in a web-based smoking cessation study. *Addictive Behaviors, 73*, 204–208. https://doi.org/10.1016/j.addbeh.2017.05.020

Cohen, S. (1988). Perceived stress in a probability sample of the United States. In S. Spacapan & S. Oskamp (Eds.), *The claremont symposium on applied social psychology*. Thousand Oaks, CA: Sage.

Cohen, S., Kamarck, T., & Mermelstein, R. (1983). A global measure of perceived stress. *Journal of Health and Social Behavior, 24*, 385–396. https://doi.org/10.1016/j.socscimed.2016.08.021

Davis, K. (2017). Police are using new mouth-swab tests to nab drivers under the influence of marijuana and other drugs [WWW Document]. *Los Angeles Times*. https://www.latimes.com/local/lanow/la-me-mouth-swab-drugs-test-police-pot-20170317-story.html

Duke Center for the Study of Aging and Human Development. (n.d.) *Chinese longitudinal healthy longevity survey (CLHLS)* [WWW Document]. Duke WordPress Sites. https://sites.duke.edu/centerforaging/programs/chinese-longitudinal-healthy-longevity-survey-clhls/

Eke, P. I., Thornton-Evans, G. O., Wei, L., Borgnakke, W. S., Dye, B. A., & Genco, R. J. (2018). Periodontitis in US adults: National health and nutrition examination survey 2009–2014. *Journal of the American Dental Association (1939), 149*, 576–588.e6. https://doi.org/10.1016/j.adaj.2018.04.023

Fair, L. (2017). *DNA test kits: Consider the privacy implications* [WWW Document]. https://www.consumer.ftc.gov/blog/2017/12/dna-test-kits-consider-privacy-implications

FamilyTreeDNA. (2019). *FamilyTreeDNA privacy statement* [WWW Document]. https://www.familytreedna.com/legal/privacy-statement

Global Market Insights. (2018). *Direct-to-consumer genetic testing market to hit $2.5 Bn by 2024*. Global Market Insights, Inc. [WWW Document]. https://www.prnewswire.com/news-releases/direct-to-consumer-genetic-testing-market-to-hit-2-5-bn-by-2024-global-market-insights-inc%2D%2D830436085.html

Goswami, Y., Mishra, R., Agrawal, A., & Agrawal, L. (2015). Salivary biomarkers: A review of powerful diagnostic tool. *IOSR Journal of Dental and Medical Sciences, 14*, 80–87. https://doi.org/10.9790/0853-14378087

Heaney, J. L. J., Phillips, A. C., Carroll, D., & Drayson, M. T. (2018). The utility of saliva for the assessment of anti-pneumococcal antibodies: Investigation of saliva as a marker of antibody status in serum. *Biomarkers, 23*, 115–122. https://doi.org/10.1080/1354750X.2016.1265009

Kang, R. R., & Cilia, A. (2017). *Musk compositions and methods of use thereof* (PCT/US2016/040223).

Khan, R., Khurshid, Z., & Yahya Ibrahim Asiri, F. (2017). Advancing Point-of-Care (PoC) testing using human saliva as liquid biopsy. *Diagnostics, 7.* https://doi.org/10.3390/diagnostics7030039

Kirschbaum, C., Pirke, K.-M., & Hellhammer, D. H. (2008). The 'trier social stress test' – A tool for investigating psychobiological stress responses in a laboratory setting. *Neuropsychobiology, 28* (1–2), 76–81. https://doi.org/10.1159/000119004

Messeri, P. A., Allen, J. A., Mowery, P. D., Healton, C. G., Haviland, M. L., Gable, J. M., & Pedrazzani, S. D. (2007). Do tobacco countermarketing campaigns increase adolescent underreporting of smoking? *Addictive Behaviors, 32*, 1532–1536. https://doi.org/10.1016/j.addbeh.2006.11.010

Nebula Genomics. (2019). *Nebula genomics* [WWW Document]. https://www.nebula.org/

Orig3n DNA Tests. (2019). *Orig3n* [WWW Document]. https://orig3n.com/

Pisanic, N., Ballard, S.-B., Colquechagua, F. D., François, R., Exum, N., Yori, P. P., . . . Kosek, M. N. (2018). Minimally invasive saliva testing to monitor norovirus infection in community settings. *The Journal of Infectious Diseases, 219*(8), 1234–1242. https://doi.org/10.1093/infdis/jiy638

Pisanic, N., Rahman, A., Saha, S. K., Labrique, A. B., Nelson, K. E., Granger, D. A., . . . Heaney, C. D. (2017). Development of an oral fluid immunoassay to assess past and recent hepatitis E virus (HEV) infection. *Journal of Immunological Methods, 448*, 1–8. https://doi.org/10.1016/j.jim.2017.04.012

Quantified Self. (2019). *Quantified self: Self knowledge through numbers* [WWW Document]. https://quantifiedself.com/

RAND. (2019) *L.A. FANS: The Los Angeles family and neighborhood survey* [WWW Document]. RAND Corp.https://lasurvey.rand.org/

Riis, J., Bryce, C., Ha, T., Hand, T., Stebbins, J., Matin, M., . . . Granger, D. A. (2017). Adiponectin: Serum-saliva associations and relations with oral and systemic markers of inflammation. *Peptides, 91*, 58–64. https://doi.org/10.1016/j.peptides.2017.03.006

Services, U.D. of H. and H. (2019). *Adolescent brain cognitive development study* [WWW Document]. https://abcdstudy.org/

Shiseido Conducts Joint Research on Stress in Closed-off Environment Simulating the International Space Station (ISS) [WWW Document]. (2018). *Shiseido*. https://www.shiseidogroup.com/news/detail.html?n=00000000002383

Sutton, N., Macdonald, M., Lombard, J., Ilie, B., Hinoul, P., & Granger, D. A. (2018). Preclinical evaluation of the effect of the combined use of the ethicon securestrap® open absorbable strap fixation device and ethicon physiomesh™ open flexible composite mesh device on surgeon stress during ventral hernia repair. *Medical Devices: Evidence and Research, 11*, 1–9. https://doi.org/10.2147/MDER.S146761

U.S. Department of Health and Human Services. (2000). Oral health in America: A report of the surgeon general. *Oral health in America: A report of the Surgeon General, 44*, 16–28. https://doi.org/10.1089/pop.2013.0038

U.S. Department of Health and Human Services. (n.d.-a) *Healthy people 2020: Access to primary care* [WWW Document]. https://www.healthypeople.gov/2020/topics-objectives/topic/social-determinants-health/interventions-resources/access-to-primary

U.S. Department of Health and Human Services. (n.d.-b) *Healthy people 2020: Access to health services* [WWW Document]. https://www.healthypeople.gov/2020/topics-objectives/topic/social-determinants-health/interventions-resources/access-to-health

U.S. Food and Drug Administration. (2018a). FDA authorizes first direct-to-consumer test for detecting genetic variants that may be associated with medication metabolism [WWW Document]. *FDA News Release*. https://www.fda.gov/NewsEvents/Newsroom/PressAnnouncements/ucm624753.htm

U.S. Food and Drug Administration. (2018b). Direct-to-consumer tests [WWW Document]. *Vitro Diagnositics*. https://www.fda.gov/MedicalDevices/ProductsandMedicalProcedures/InVitroDiagnostics/ucm624726.htm

Van Andel, H., Jansen, L., Grietens, H., Knorth, E., & Van Der Gaag, R. (2014). Salivary cortisol: A possible biomarker in evaluating stress and effects of interventions in young foster children? *European Child & Adolescent Psychiatry, 23*, 3–12. https://doi.org/10.1007/s00787-013-0439-1

Van Der Lely, S., Frey, S., Garbazza, C., Wirz-Justice, A., Jenni, O., Steiner, R., … Schmidt, C. (2015). Blue blocker glasses as a countermeasure for alerting effects of evening light-emitting diode screen exposure in male teenagers. *Journal of Adolescent Health, 56*, 113–119. https://doi.org/10.1016/j.jadohealth.2014.08.002

Wiegand, B., & McCulloch, L. (2002). Methods for measuring stress in mammals. US 2002/0090664 A1.

# Chapter 31
# Envisioning the Future of Salivary Bioscience

**Douglas A. Granger, Alyson K. Zalta, and Marcus K. Taylor**

There has been a remarkable evolution of salivary bioscience research from a narrow focus on basic oral biology and clinical dental research to applications across the spectrum of health. As summarized in this edited volume, the number of analytes available to be assayed in oral fluid specimens has expanded to the degree that we list them by category rather than individually and they include proteomics, metabolomics, microbiome, therapeutic drugs and drugs of abuse, DNA, epigenetics, markers of infectious disease and exposures, as well as exposure to environmental chemicals and elements. Undeniably, the range of measurement will continue to expand in the coming years. In addition to the expanding applications of salivary bioscience, significant disruptions have occurred during this time, most notably with the use of salivary bioscience as a commercial product to conduct DNA testing from one's own home. Yet, despite this progress, the promise of using salivary bioscience to revolutionize diagnostic testing and advance public health has not come to fruition as expected. As we look to the future of this field, it is important to reflect on the lessons we have learned, identify existing barriers and fruitful pathways for the

In the interest of full disclosure, Douglas A. Granger is the founder and chief scientific and strategy advisor of Salimetrics LLC and Salivabio LLC (Carlsbad, CA) and the nature of these relationships is managed by the policies of the committees on conflict of interest at Johns Hopkins University School of Medicine and the University of California at Irvine.

D. A. Granger (✉)
Institute for Interdisciplinary Salivary Bioscience Research, University of California at Irvine, Irvine, CA, USA

Johns Hopkins University, Baltimore, MD, USA
e-mail: dagrange@uci.edu

A. K. Zalta · M. K. Taylor
Institute for Interdisciplinary Salivary Bioscience Research, University of California at Irvine, Irvine, CA, USA



D. A. Granger, M. K. Taylor (eds.), *Salivary Bioscience*,
https://doi.org/10.1007/978-3-030-35784-9_31

translation of research, and consider what the next generation of scientists will need to make meaningful advances in the field.

## 31.1 Lessons Learned

As noted in many of the chapters, the minimally invasive nature of saliva collection is clearly among its main advantages as a biological specimen. It is apparent that early adopters of salivary bioscience working in behavioral and social sciences may have been so focused on *ease of use* they charged ahead without fully incorporating the basics of oral biology in the approach. As a consequence, there were many misuses related to sample collection, handling, and assay in the first waves of publications. These oversights were largely self-corrected in the literature that followed, but even today, some of those early publications are cited to justify methods and procedures that are no longer acceptable. A resulting challenge to the field is the perception of inconsistenting findings and ambiguities regarding the state of the science. To this end, we encourage reviewers to use the information presented in this edited volume to evaluate the quality of manuscripts submitted for publication and proposals they are asked to consider for funding. Research investigators might also use this information to be more skeptical and qualify conclusions drawn relative to the salivary bioscience methods employed in the existing literature. Efforts to this end would most certainly enable the field to clarify our collective view of the forest from the trees.

Another important lesson is that trends in the literature strongly suggest that simple main effects or simple direct effects (or even simple reciprocal effects) between biological and behavioral variables are infrequent. Instead, volumes of literature support the conclusion that these effects are moderated, mediated, or both. This means that *bigger data* will be needed to solve these big questions. Efforts such as the *All of Us* Research Program, which aims to gather data from one million individuals from across the USA, are underway to try and address these questions. However, individual investigators can also contribute to this effort through coordinated intra- and inter-institute efforts to create, establish, and maintain biorepositories that include oral fluid specimens. Moreover, by archiving, rather than discarding samples, investigators position themselves to take advantage of new opportunities and discoveries that are bound to come.

## 31.2 Roadblocks and Avenues for Future Research

One critical roadblock for the advancement of salivary bioscience research is that technological advances in salivary bioscience have, in this moment, outpaced theory. Nobel Laureate Edward Prescott considered science a race between theory and measurement (Prescott, 1986). That is, at any particular moment in the development of a scientific endeavor the balance between *theory* and *measurement* impacts scientific progress. Consider the case when a technical advance makes possible measurement at a new level of analysis or precision, without theory to guide in the interpretation of those measurements. Without a frame of reference, the value added to the literature of making these new measurements is negligible. Consider the opposite case when theory has advanced to a point beyond the scope of measurement. Here, research questions derived from the theory cannot be tested without the capability of the observations enabled by the appropriate measurement. In both situations, scientific progress slows down. These hypothetical scenarios have parallels in this moment in the evolution of salivary bioscience. The ability to measure multiple analytes in saliva simultaneously, in trillionth of a gram determinations, repeatedly from the same individual, within minutes at the point of care, and in the context of everyday life, has progressed beyond theory. Our capacity to clearly interpret the meaning of intra- and individual differences in these salivary metrics lags behind the technical capabilities of our measurements. This means that the future of salivary bioscience will depend to a large degree on observations that enable us to refine and fine-tune our theories such that future iterations of theory enable us to determine and interpret variation in more meaningful ways.

A second key challenge to the field is the translation of research findings in ways that will impact public health. As pointed out in several chapters in the volume, moving salivary research to saliva diagnostics has historically been a major hurdle. Our colleagues among the community of medical researchers often raise this challenge—most salivary bioscience literature is based on analysis of means derived from large groups of research participants. Their charge, and interest, is in metrics that can be derived from saliva that can enable them to make decisions about an individual. They desire a measurement they can make at this moment for this specific individual, or based on repeated measurements from that individual, that can inform their treatment decisions and observations. This moment, with few exceptions, remains elusive in the context of salivary bioscience. Most salivary determinations that have been FDA cleared for diagnostic/screening application of this type often require confirmation using a traditional biospecimen. On the other side of the spectrum, public health researchers, community officials, and policy makers often comment that they need a metric in saliva that can be obtained in a single sampling with a result returned within minutes, that costs pennies per determination, because they are working in communities (often third world or economically disadvantaged) and are unlikely to encounter the chance to sample repeatedly from the same person. In short, these anecdotal observations suggest that medical and public health researchers, clinicians, and policy makers await the next generation of advances in our understanding before salivary bioscience can serve their needs.

Where we think salivary bioscience holds the most promise is in leveraging the advantages of saliva to address important scientific questions that cannot feasibly be addressed using other tools. For example, the *context contingency effect* (Chen, Dariotis, & Granger, 2018) refers to a consistent pattern of findings across studies showing that the expression of associations between biological and behavioral variables largely depends on social context. In so much of the health research that has been conducted to date, the role of social context has largely been ignored despite the known impact of social determinants of health (Secretary's Advisory Committee, 2020). A key advantage of saliva as a research and diagnostic specimen is its portability, enabling measurements of biological parameters in ecologically valid settings, and thus the capability to explore the effects of context contingency. This means that salivary bioscience has the unique potential to address questions related to social context that will ultimately be critical for precision medicine.

## 31.3 Training the Next Generation

The utility of salivary bioscience is, and will continue, pulling future investigators from different disciplines together in teams to apply these tools in an effort to solve "wicked" problems (Stokols, 2018; Sapolky, 2018). Thus, the future of salivary bioscience will depend on training the next generation of scientists to approach with an interdisciplinary perspective and use sophisticated analytical strategies and tactics to model complex relationships. These skills are rarely part of the graduate curriculum in the biological sciences. They are, however, often part of the elective choices in graduate training programs in the social, behavioral, and public health-oriented disciplines. In the future, analytical strategy specialists and tacticians need to be trained and recruited as key contributors to the teams engaged in the science of salivary bioscience research. Moreover, effective strategies for promoting team science and interdisciplinary research need to be baked into our training programs to enable the next generation of researchers to tackle these complex problems.

## 31.4 Concluding Comment

The field of salivary bioscience has emerged to fill gaps in our understanding of biobehavioral health, to challenge and advance our theories, and to provide information about intra- and individual differences in biobehavioral processes that can be used by clinicians and policy makers to make a difference in the lives of individuals, communities, and populations. More than four decades of basic research has laid the groundwork of knowledge and described the basic phenomena. In the coming decades, we can anticipate that the investments made to develop that knowledge base will begin to be translated and pay dividends. The information summarized by this volume's writing teams serve in some respects as a "health and wellness check up" assessment for this emerging field. From these assessments, benchmarks can be set and used to evaluate future progress. No doubt the next phases in the evolution of salivary bioscience research will be marked by discovery, innovation, and translation. It will be fascinating to witness the uncovering of the details related to the "for whom, what, when, where and why" salivary bioscience research can be harnessed to make a difference.

## References

Chen, F. R., Dariotis, J. K., & Granger, D. A. (2018). Linking testosterone and antisocial behavior in at-risk transitional aged youth: Contextual effects of parentification. *Psychoneuroendocrinology, 91*, 1–10.

Prescott, E. C. (1986). *Theory ahead of business cycle measurement.* Staff Report 102, Federal Reserve Bank of Minneapolis.

Sapolky, R. M. (2018). Double edged swords in the biology of conflict. *Frontiers in Psychology, 9*, 2625.

Secretary's Advisory Committee in Health Promotion and Disease Prevention Objectives for 2020: Healthy People 2020. *An opportunity to address the social determinants of health in the United States*. Available July 26, 2010, from http://healthypeople.gov/2010/hp2020/advisory/Societal.DeterminantsHealth.htm

Stokols, D. (2018). *Social ecology in the digital age; solving complex problems in a globalized world*. London: Academic Press.

The manufacturer's authorised representative in the EU is Springer
Nature Customer Service Centre GmbH, Europaplatz 3, 69115 Heidelberg,
Germany. If you have any concerns regarding our products, please
contact ProductSafety@springernature.com

Printed and bound by CPI Group (UK) Ltd, Croydon, CR0 4YY
05/10/2026
02228123-0020